Ebook Set up and Use Ins

Use of the ebook is subject to the single user licence at the back of this book.

Electronic access to your price book is now provided as an ebook on the VitalSource® Bookshelf platform. You can access it online or offline on your PC/Mac, smartphone or tablet. You can browse and search the content across all the books you've got, make notes and highlights and share these notes with other users.

Setting up

1. **Create a VitalSource Bookshelf account** at https://online.vitalsource.com/user/new (or log into your existing account if you already have one).

2. **Retrieve the code** by scratching off the security-protected label inside the front cover of this book. Log in to Bookshelf and click the **Redeem Codes link in the Tools** menu at the top right of the screen. Enter the code in the **Redeem code** box and press **Redeem**. Once the code has been redeemed your Spon's Price Book will download and appear in your **library**. N.B. the code in the scratch-off panel can only be used once.

When you have created a Bookshelf account and redeemed the code you will be able to access the ebook online or offline on your smartphone, tablet or PC/Mac. Your notes and highlights will automatically stay in sync no matter where you make them.

Use ONLINE

1. Log in to your Bookshelf account at https://online.vitalsource.com).
2. Double-click on the title in your **library** to open the ebook.

Use OFFLINE

Download BookShelf to your PC, Mac, iOS device, Android device or Kindle Fire, and log in to your Bookshelf account to access your ebook, as follows:

On your PC/Mac
Go to https://support.vitalsource.com/hc/en-us/categories/360001056774-Bookshelf#360002383594 and follow the instructions to download the free VitalSource Bookshelf app to your PC or Mac. Double-click the VitalSource Bookshelf icon that appears on your desktop and log into your Bookshelf account. Select **All Titles** from the menu on the left – you should see your price book on your Bookshelf. If your Price Book does not appear, select **Update Booklist** from the **Account** menu. Double-click the price book to open it.

On your iPhone/iPod Touch/iPad
Download the free VitalSource Bookshelf App available via the iTunes App Store. Open the Bookshelf app and log into your Bookshelf account. Select **All Titles** – you should see your price book on your Bookshelf. Select the price book to open it. You can find more information at https://support.vitalsource.com/hc/en-us/articles/360014105113-iOS

On your Android™ smartphone or tablet
Download the free VitalSource Bookshelf App available via Google Play. Open the Bookshelf app and log into your Bookshelf account. You should see your price book on your Bookshelf. Select the price book to open it. You can find more information at https://support.vitalsource.com/hc/en-us/articles/360014007294-Android

On your Kindle Fire
Download the free VitalSource Bookshelf App from Amazon. Open the Bookshelf app and log into your Bookshelf account. Select All Titles – you should see your price book on your Bookshelf. Select the price book to open it. You can find more information at https://support.vitalsource.com/hc/en-us/articles/360014107953-Kindle-Fire

Support

If you have any questions about downloading Bookshelf, creating your account, or accessing and using your ebook edition, please visit https://support.vitalsource.com/

For questions or comments on content, please contact us on sponsonline@tandf.co.uk

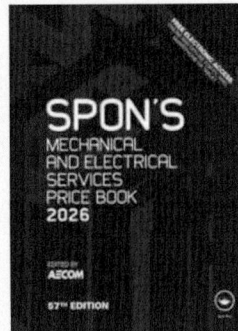

Spon's
Mechanical and
Electrical Services
Price Book

2026

Spon's Mechanical and Electrical Services Price Book

Edited by

A≡COM

2026

Fifty-seventh edition

CRC Press

Taylor & Francis Group

First edition 1968
Fifty-seventh edition published 2026
by CRC Press
2 Park Square, Milton Park, Abingdon, Oxon, OX14 4RN

and by CRC Press
Taylor & Francis, 2385 NW Executive Center Drive, Suite 320, Boca Raton FL 33431

CRC Press is an imprint of the Taylor & Francis Group, an informa business

British Library Cataloguing in Publication Data
A catalogue record for this book is available from the British Library

ISBN: 978-1-041-07940-8
Ebook: 978-1-003-64294-7
ISSN: 0305-4543

DOI: 10.1201/9781003642947

Typeset in Arial by Taylor & Francis Books
Printed and bound by CPI Group (UK) Ltd, Croydon CR0 4YY

Contents

Preface

The Fifty-Seventh Edition of *Spon's Mechanical and Electrical Services Price Book* continues to cover the widest range and depth of engineering services, reflecting the many alternative systems and products that are commonly used in the industry as well as current industry trends.

AECOM forecasts tender prices to remain at 3.4% during 2025, with inflation increasing to 4.6% in 2026.

Key Influences

- Inflation impacts continue to be driven by labour costs, including skills shortages, National Living Wage uplifts, and the change in National Insurance contributions.
- Decline in new work orders across most sectors. Private housing output has recovered.
- US tariffs may present opportunities for the UK, despite possible market restrictions for UK exporters.
- The June 2025 Spending Review may impact future growth, and the prospects for projects which remain dependent on investment.
- M&E tender prices are rising faster than building work, with prices increasing due to regulation, capacity constraints and high demand in other markets such as data centres, infrastructure and decarbonization.

Before referring to prices or other information in the book, readers are advised to study the 'Directions' which precede each section of the Materials Costs/Measured Work Prices. As before, no allowance has been made in any of the sections for Value Added Tax.

The order of the book reflects the order of the estimating process, from broad outline costs through to detailed unit rate items.

The approximate estimating section has been revised to provide up-to-date key data in terms of square metre rates, all-in rates for key elements and selected specialist activities and elemental analyses on a comprehensive range of building types.

The prime purpose of the Materials Costs/Measured Work Prices part is to provide industry average prices for mechanical and electrical services, giving a reasonably accurate indication of their likely cost. Supplementary information is included which will enable readers to make adjustments to suit their own requirements. It cannot be emphasized too strongly that it is not intended that these prices are used in the preparation of an actual tender without adjustment for the circumstances of the particular project in terms of productivity, locality, project size and current market conditions. Adjustments should be made to standard rates for time, location, local conditions, site constraints and any other factor likely to affect the costs of a specific scheme. Readers are referred to the build up of the gang rates, where allowances are included for supervision, labour-related insurances, and where the percentage allowances for overhead, profit and preliminaries are defined.

As with previous editions the Editors invite the views of readers, critical or otherwise, which might usefully be considered when preparing future editions of this work.

Whilst every effort is made to ensure the accuracy of the information given in this publication, neither the Editors nor Publishers in any way accept liability for loss of any kind resulting from the use made by any person of such information.

In conclusion, the Editors record their appreciation of the indispensable assistance received from the many individuals and organizations in compiling this book.

AECOM Ltd
Aldgate Tower
2 Leman Street
London E1 8FA

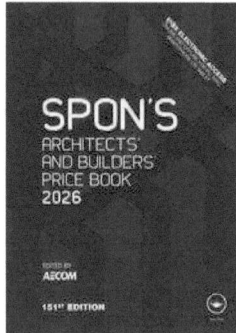

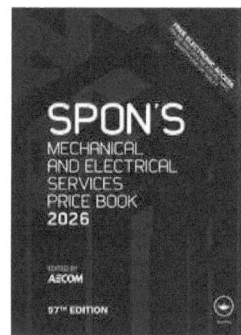

Special Acknowledgements

The Editors wish to record their appreciation of the special assistance given by the following organizations in the compilation of this edition.

HOTCHKISS AIR SUPPLY

Hampden Park Industrial Estate
Eastbourne
East Sussex
BN22 9 AX
Tel: 01323 501 234
Email: info@Hotchkiss.co.uk
www.Hotchkiss.co.uk

Abbey

Unit 14, The Capricorn Centre
Cranes Farm Road
Basildon, Essex
SS14 3JJ
Tel: 01268 572 116
Email: info@abbeythermal.com
www.abbeythermalinsulation.com

T.Clarke
BUILDING SERVICES GROUP

30 St Mary Axe
London
EC3A 8BF
Tel: 020 7997 7400
www.tclarke.co.uk

DORNAN

Dornan Engineering Ltd
114a Cromwell Road
Kensington
London
SW7 4ES
Tel: 020 7340 1030
Email: reception@dornangroup.com
www.dornan.ie

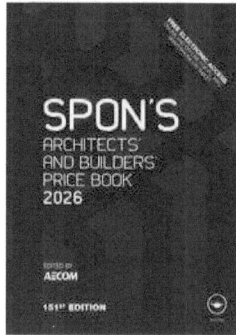

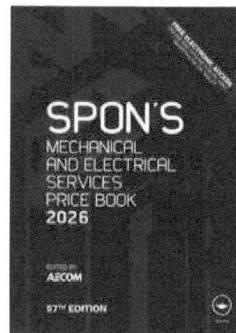

Acknowledgements

The editors wish to record their appreciation of the assistance given by many individuals and organizations in the compilation of this edition.

Manufacturers, Distributors and Subcontractors who have contributed this year include:

A1 Flue Systems
Systems House
Maun Way
Boughton Industrial Estate
New Ollerton, NR. Newark
Nottingham
NG22 9ZD
Tel: 01623 860 578
Email: info@a1flues.co.uk
www.a1flues.co.uk
Flue Systems

Amiblu
Amiblu Holding GmbH
Sterneckstraße 19
9020 Klagenfurt
Austria
Tel: 43 46348 2424
Email: info@amiblu.com
www.amiblu.com
Sewer System Pipework

Aquilar Ltd
Unit 30, Lawson Hunt Industrial Park
Broadbridge Heath
Horsham
West Sussex
RH12 3JR
Tel: 01403 216 100
Email: info@aquilar.co.uk
www.aquilar.co.uk
Leak Detection

Babcock Wanson
Elstree Way 7
Borehamwood
WD6 1SA
Tel: 020 8953 7111
www.babcock-wanson.com
Packaged Steam Generators

Balmoral Tanks
Rathbone Square
24 Tanfield Road
Croydon
CR0 1 AL
Tel: 020 8665 4100
Email: tanks-website@balmoral.co.uk
www.balmoral-group.com
Sprinkler Tanks

Biddle
St Mary's Road
Nuneaton
Warwickshire
CV11 5 AU
Tel: 024 7638 4233
Email: sales@biddle-air.co.uk
www.biddle-air.co.uk
Air Curtains

Bodet Ltd
6, Marchmont Gate
Boundary Way
Hemel Hempstead
HP2 7BF
Tel: 01442 418 800
www.bodet-time.co.uk
Clocks

Caice Acoustic Air Movement Ltd
Riverside House
Unit 3, Winnersh Fields
Wokingham
Berkshire
RG41 5QS
Tel: 01189 186 470
Email: enquiries@caice.co.uk
www.caice.co.uk
Silencers & Acoustic Treatment

DMS Flow Measurement and Controls Ltd
X-Cel House
Chrysalis Way
Langley Bridge
Eastwood
Nottinghamshire
NG16 3RY
Tel: 01773 534 555
Email: sales@dmsltd.com
www.dmsltd.com
Chilled Water Plant and Energy Meters

Frenger Systems Ltd
Riverside Road
Pride Park
Derby
DE24 8HY
Tel: 01332 295 678
Email: sales@frenger.co.uk
www.frenger.co.uk
Chilled Beams

Harlequin
Harlequin Manufacturing Ltd
21 Clarehill Road
Moira, County Armagh
Northern Ireland
BT67 0PB
Tel: 028 9261 1077
Email: info@harlequin-mfg.com
www.harlequinplastics.co.uk/
Fuel Oil Storage Tanks

Hoval Ltd
Northgate
Newark
Notts NG24 1JN
Tel: 01636 672 711
Email: sales.uk@hoval.com
www.hoval.co.uk
Storage Cylinders/Calorifiers, Commercial Oil Boilers, HIUs, MVHRs, CHP

Hudevad
Ambolten 37
DK−6000 Kolding
Denmark
Tel: 045 7542 0255
Email: contact@hudevad.com
www.hudevad.com
Heat Emitters, Radiators & Trench Heating

Hydrotec (UK) Ltd
Hydrotec House
5 Manor Courtyard
Hughenden Avenue
High Wycombe
HP13 5RE
Tel: 01494 796 040
Email: sales@hydrotec.co.uk
www.hydrotec.co.uk
Cleaning and Chemical Treatment

Integrated BMS
Brunel Drive
Newark
Nottinghamshire
NG24 2DE
Tel: 016 3667 4875
Email: controls@integratedbms.co.uk
BMS

JCB Broadcrown
Airfield Industrial Estate
Stafford
ST18 0PF
Tel: 080 0083 8015
www.jcb.com
Standby Generators

Lighting Controls Ltd
Unit 1 Bourne Mill Business Park
Farnham
Surrey
GU9 9PS
Tel: 01252 470027
Email: quotations@lightingcontrols.ltd.uk
Lighting Controls

Mitsubishi Electric
Travellers Lane
Hatfield
Hertfordshire
AL10 8XB
Tel: 01707 282 880
www.mitsubishielectric.com
Local Cooling Units

Ormandy Rycroft (formally known as HRS Hevac Ltd)
Hrs Group
Unit 3 A Abloy House
Watford
WD18 8 AJ
Tel: 01274 490 911
Email: sales@ormandygroup.com
www.ormandygroup.com
Heat Exchangers/Electric Water Heaters

Schneider Electric Ltd
80 Victoria Street
London
SW1E 5JL
Tel: 037 0608 8608
Email: gb-customerservices@gb.schneider-electric.com
www.se.com/uk/en
LV Switch & Distribution Boards/Breakers & Fusers

Socomec Ltd
Knowl Piece
Wilbury Way
Hitchin
Hertfordshire SG4 0TY
Tel: 033 3015 3002
Email: info.uk@socomec.com
www.socomec.co.uk/en-gb
Automatic Transfer Switches & UPS

Total Security Protection (TSP)
4 Century Court
Tolpits Lane
Watford,
Herts
WD18 9RS
Tel: 07976 455522
Email: chris.worrall@tsp.co.uk
Access Control Equipment

Utile Engineering Company Ltd
New Street
Irthlingborough
Northants
NN9 5UG
Tel: 01933 650 216
Email: sales@utileengineering.com
www.utileengineering.com
Gas Boosters

Vent Axia
Fleming Way
Crawley
West Sussex
RH10 9YX
Tel: 034 4856 0590
Email: info@vent-axia.com
www.vent-axia.com
Fans

Reimagining Chelsea Waterfront: Residential and commercial transformation

The Chelsea Waterfront project reimagines the River Thames' industrial past for a vibrant future, comprising a 711-apartment residential complex, including 277 affordable homes.

The major project involves the transformation of the century-old Lots Road Power Station — the largest steel-framed building in Britain which once powered the London Underground — into a new super prime residential complex.

AECOM has played a central role in this new neighbourhood, serving as cost manager and employers agent, while providing a range of additional, comprehensive services including MEP, resident engineers, vertical transportation, fire consultancy, acoustics, façade access and maintenance, and sustainability consultancy services.

Once completed, the project will provide abundant retail and commercial spaces while preserving an iconic architectural and industrial site.

AECOM

Chelsea Waterfront, London
Client: Hutchison Property Group
Image credit: ©Andy Stagg

Delivering a better world

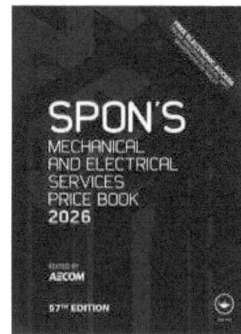

PART 1

Engineering Features

This section on Engineering Features, deals with current issues and/or technical advancements within the industry. These shall be complimented by cost models and/or itemized prices for items that form part of such.

The intention is that the book provides more than just a schedule of prices to assist the user in the preparation and evaluation of costs but can inform the reader on current and future trends in the industry.

A Practitioner's Guide to the JCT Design and Build Contract - 2024 Edition

Seán Mac Labhraí

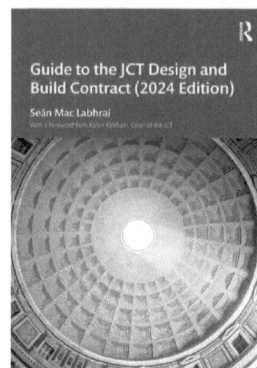

This book provides a comprehensive guide to the 2024 edition of the JCT Design and Build Contract. It offers a straightforward reference to help practitioners understand the rights and obligations of the parties, and to administrate the Contract in an accurate and timely manner.

Part 1 provides commentary on all the key provisions of the Contract, in plain English. It adopts the same structure as the Contract and includes the relevant extracts to reduce the need for cross-referencing. There is also guidance on current market practice, and recommendations for best practice.

Part 2 contains over 300 corresponding resources for the administration of the Contract, by both the Contractor and the Employer. It includes notice and other correspondence templates, checklists, and trackers to assist record keeping.

Written to meet the needs of busy construction practitioners, this is an accessible, practical reference for those acting on behalf of the Contractor and the Employer (and Employer's Agent), including project teams, consultants, and advisors. It is also ideally suited for students of construction law.

May 2025: 696pp
ISBN: 9781032744032

To Order
Tel: +44 (0) 1235 400524
Email: tandf@hachette.co.uk

For a complete listing of all our titles
visit: www.tandf.co.uk

Taylor & Francis Group
an **informa** business

Market Update

UK economic activity ended 2024 with a whimper and prospects for 2025 remain subdued. GDP is forecast to grow by one per cent in 2025, meaning yet another year of subdued growth is likely. All levels of contractor and client confidence have been trending lower recently, with an uncertain construction outlook following geopolitical tensions, global trade developments and restrictions, and closer to home, a lack of investment in infrastructure, have all given the sector mixed signals.

Output and new orders

The UK construction sector faced some significant challenges in 2024. After the industry emerged from technical recession in the second quarter, Office for National Statistics (ONS) data suggest activity remained largely flat with output increasing by just 0.4%, in real terms, during the year. Strong growth in repair and maintenance activity once again offset a contraction in new work activity. Repair and maintenance output rose by a further 8.5% in 2024, suggesting the volume of activity has risen by more than a third since 2019. New work output, in contrast, contracted by 5.3% in 2024 and output volumes remained down by 10.6% on 2019.

Public non-housing, boosted by the urgent need to increase prisons capacity, was the only new work sector to see output increase during 2024 but the fourth quarter delivered a long-awaited recovery in private house building volumes. Year-on-year, private housing output increased by 3.8% in Q4, alongside 6.8% growth in public non-housing activity and 2.3% growth in the industrial sector. Infrastructure output reduced by 5.6% compared with Q4 2023, with the cancellation of several high-profile road projects, a one-year delay to the publication of the new strategic road investment plan and the impact of local authorities constrained financial position on investment in local road networks. Commercial output reduced by 1.8% overall as weak consumer confidence weighed on investment in new leisure and entertainment facilities. The commercial offices subsector consolidated its recovery and output recorded double-digit growth in Q4.

Construction new work output (£m)

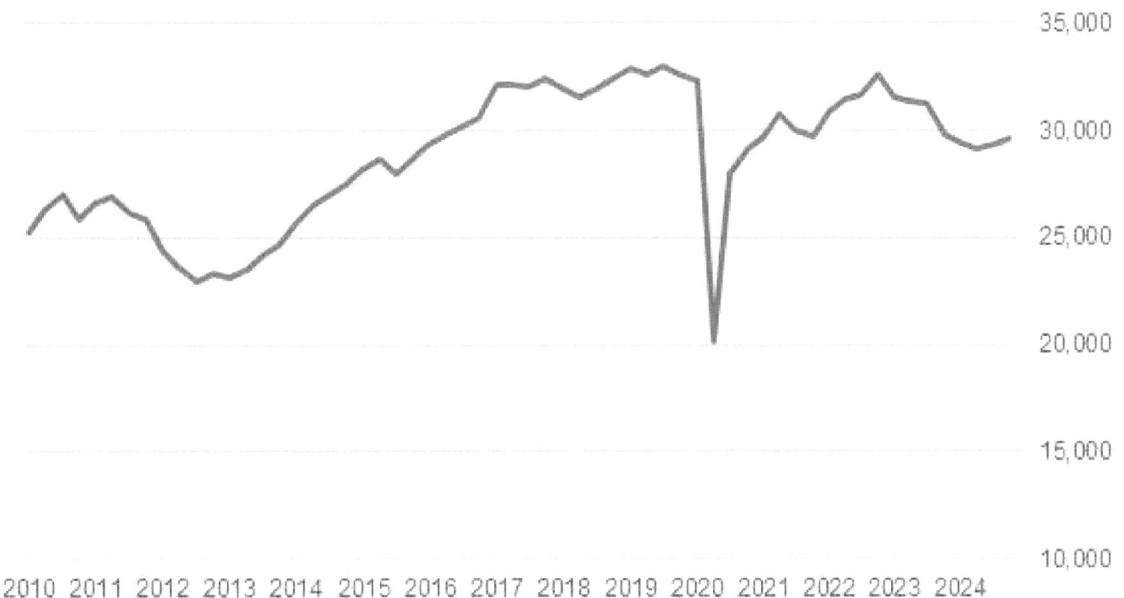

For new work orders, 2024 was a game of two halves. Optimism in the first six months, as the annual rate of consumer price inflation hit 2.0% for the first time since July 2021 boosting speculation about further base rate reductions, quickly receded as business tax increases announced in the Autumn Budget dented business confidence and delayed decision making. Overall, the 11.8% gain in order volumes in the first six months of 2024 was eroded by a 5.2% contraction in the second half, and order volumes rose by 3.5% during the year.

The commercial sector made the biggest gains in 2024, with orders rising by close to 16% in real terms, boosted by robust growth in commercial office projects amid soaring occupier demand for high-quality, well-located office space. Infrastructure orders rose by 13.7% and some progress on delayed plans to deliver new hospitals and prisons helped boost orders for public non-housing projects.

Construction new work orders (£m)

Activity indicators

Heightened geopolitical uncertainty, a marked escalation in global trade tensions and strengthening upside inflationary risks had a swift impact on business and consumer sentiment in the first quarter, as the UK economy struggled to escape from a prolonged stagnation.

Recent survey data reflect a marked deterioration in confidence but sentiment can be volatile, especially with the impact of exceptional events such as COVID–19 and the inflationary spike triggered by the Russian invasion of Ukraine, still fresh in our memories. Events over the next three months will determine the medium-term outlook for the UK economy and the construction sector.

The UK government faces exceptional financial challenges. Net debt, estimated at 95.3% of GDP, remains at levels not seen since the early 1960s, and government borrowing in the financial year to January 2025 was £11.6 billion higher than in the comparable part of the previous financial year. Already the financial assumptions underpinning last year's Autumn Budget seem overly optimistic and the urgent need to significantly increase spending on defence and security will shift investment priorities. Now at risk of missing at least one of the fiscal rules established in the Autumn Budget, the Chancellor will face difficult decisions in the 2025 Spending Review – a date for which has yet to be confirmed.

In the private sector, business confidence remains fragile. The decision to increase employers' National Insurance contributions took a heavy toll and broader global uncertainty is encouraging greater caution. Recent survey data from the Confederation of British Industry (CBI) suggests private sector activity is falling and with activity expected to continue to decline over the next three months the near-term outlook is weak. BDO tracked a fall in overall business output for the second month in a row in February and business optimism fell to its lowest level since January 2021. The employment index fell to levels not seen since after the financial crisis.

The Bank of England's decision to cut the base rate in February boosted consumer confidence – the NIQ GfK consumer confidence index rose by 2 points to −20 in February as respondents' views on the outlook for their personal financial situation improved. However, inflationary pressure is strengthening, making the Monetary Policy Committee's job this year even more challenging. Consumer price inflation (CPI) hit 2.8% in February, an increase on top of the 3.0% in January and slightly higher than in September 2024, when the inflation rate reached 1.7%. Upside risks have strengthened and the independent economic forecasters currently expect CPI to be 3.0% in 2025, above the Bank's 2.0% target.

Development project viability remains tight across most sectors and interest rate and inflation expectations are a vital consideration. Against an uncertain global backdrop, sentiment in the construction sector has weakened markedly. February's S&P Global Construction PMI fell to 44.6, its lowest level in nearly five years. The residential sector experienced the sharpest decline and the house building index fell to 39.3 – the fastest rate of decline since early 2009 – if the global pandemic is discounted. Delayed decision-making among clients triggered a sharp fall in new orders as budgetary concerns remained at the fore. Around 38% of the survey panel indicated a rise in their input prices, while only 3% noted a reduction.

Building costs

AECOM's Building Cost Index – a composite measure of labour and materials prices – increased by 3.3% in 2025 Q1 as persistently high wage inflation pushed construction costs higher. Wage growth increased to a month-on-year rate of 5.7% in January 2025 from a 4.8% rate recorded in December last year, but the rate of appreciation is still high in relatively subdued market conditions as skills shortages persist.

Materials prices are currently stable but recent gas price inflation, plus the US-led shift towards trade protectionism, potentially increase risks to cost inflation over the medium term.

Wholesale gas prices spiked in January as an agreement allowing Russia to pump gas to the EU through pipelines crossing Ukraine came to an end. European natural gas prices are now over 50% higher than in early 2024 and Russian gas is still an important source of EU and UK supply. Analysis by POLITICO suggesting that the EU-27 imported 837,300 metric tonnes of liquefied natural gas (LNG) from Russia in the first 15 days of 2025 and low levels of gas storage capacity makes the UK especially vulnerable to supply disruption.

While US tariffs disrupt global supply chains and reduce market access for UK exporters by discouraging the use of imported materials, they might present some opportunities for the UK and Europe.

The US accounts for a relatively small portion of UK steel and aluminium exports, representing only 7% of steel and 10% of aluminium exports by volume. The UK produces 5.6 million tonnes (Mt) of crude steel annually, meeting 70% of its domestic needs, with 40% of this being domestically produced. Therefore, the tariffs imposed on UK exports could allow British producers to increase their share of the local market. Furthermore, global excess steel capacity is substantial, estimated at 543 million tonnes in 2023 and projected to reach 630 million tonnes by 2026, over 100 times the UK's production. Most of this excess capacity originates from Asia, particularly China, potentially leading to a cheaper steel trade. Fundamentally, near-term prospects for global economic growth remain subdued suggesting relatively stable demand for base metals, barring further external shocks, but in a fractious global trading environment localized pricing volatility will occur.

Tender prices

Tender price inflation accelerated to 3.2% in the year to 2025 Q1, in response to rising delivery costs and relatively strong pipeline visibility. Main contractors' financial health has generally improved after recent inflationary shocks and insolvency rates have stabilized. Whilst the public and Infrastructure sectors await the publication of the Infrastructure strategy contractors are reporting reasonable pipeline visibility over the next 12 to 18 months, risk aversion remains elevated and tier one contractors seek to protect margins. At trade contractor level, pockets of capacity are emerging, due in part to the ongoing weakness of activity in the residential sector, to some extent due to persistent delays in the development of higher-risk buildings. Competitive pressure is building in these areas but demand for trades including cladding specialists, joinery, roofing and mechanical and electrical engineers remains strong and capacity is highly constrained.

Overall, AECOM forecasts tender prices will remain at 3.4% for 2025, with inflation accelerating to 4.6% in 2026 as firms throughout the supply chain seek to recover increases in building costs. This central view assumes no fundamental shift in the outlook for demand – that any pause proves temporary and that projects currently in the pipeline generally progress as planned – but risks are elevated and industry dynamics could quickly change.

Outlook

As economic forecasters digest recent developments, it's clear that their consensus view in early 2025, suggesting that economic growth will strengthen to 1% this year and 1.3% in 2026, now looks optimistic. Business and consumer confidence has dropped, and trade barriers constrain global growth – downward revisions are likely in the coming months but the tide can quickly turn if confidence returns.

How the Bank of England's Monetary Policy Committee balances future inflationary risks against the need to stimulate the ailing UK economy will be pivotal for construction demand. Further interest rate cuts would provide a much-needed boost to project viability. The recent gas price spike is baked into the Bank's central model, and the inflationary impact is expected to be temporary, however, further disruption to global supply chains from simmering regional tensions presents a real upside risk to inflation. Events in recent years demonstrated just how significant the commercial consequences of disruption to global supply chains can be. Conversely, an end to the conflict in Ukraine, if current negotiations prove successful, would strengthen energy security across Europe and restore full access to Russian metals and minerals.

The Construction Products Association's latest view on UK industry demand, published in early January, predicts 2.1% growth this year, accelerating to 4.0% in 2026, as house building recovers after two years of decline and momentum builds in the commercial sector. The elevated level of uncertainty, however, is likely to impact near-term decision-making by private sector clients, plus it is not yet clear if the government's highly constrained financial position, and shift in investment priorities, will further impact publicly financed capital programmes. A return of a role for private finance in the delivery of public sector assets is under consideration.

Downside risks are prevalent but there are positives. Amid much global uncertainty, the relative stability of UK property markets may prove increasingly attractive to US investors over the medium term. Recently published research from the Investment Property Forum (IPF) estimated that overseas investors accounted for 40% of the UK commercial real estate investment market in 2023, up from an estimated 24% in 2013. In the 12 months to January 2025, CBRE estimates that overseas investment in the Central London offices accounted for 51% of the total, with 15% of investment flowing from North America and 17% from Europe.

Government policy and the regulatory backdrop – from procurement and planning policy to building safety and building regulations and decarbonization – continues to evolve at pace, further complicating on the development viability equation. The Procurement Act 2023 came into force in February, setting out replacement rules for the procurement of public sector goods and services following the UK's departure from the EU and the government has accepted all findings of the 2024 Grenfell Inquiry and has committed to taking forward all recommendations. Alongside this, the Planning and Infrastructure Bill was recently presented to Parliament and final specifications and an implementation plan for the Future Homes and Buildings Standard are expected in 2025.

Further change to building safety legislation looms as changes from the 2022 Building Safety Act are still being embedded and lack of regulatory capacity is causing significant delays for high-rise projects seeking Gateway 2 and Gateway 3 approval. New, wide-ranging, commitments include system-wide reform of the construction products sector, the appointment of a chief construction advisor, creation of a single regulator across the construction sector with responsibility for all activity except product testing and certification and certifying building compliance. How higher-risk buildings are defined will also be reconsidered and government will set out plans for its ongoing review this summer. It also accepted the recommendations to professionalize fire engineers and assessors, to license principal contractors, and to review the role of building control. Phased delivery of regulatory change will help but the proposed timescale is tight. The first phase, running from 2025 to 2026 will focus on the delivery of the current programme of regulatory reform, with delivery of the Inquiry's recommendations and wider regulatory reform in a second phase spanning 2026 to 2028. From 2028, phase three will focus on implementation.

Heightened global political and economic uncertainty has already had a dramatic impact on UK business and consumer confidence. Global instability has clear implications for investment decisions at a time when development project viability is still challenging to achieve. From the perspective of demand, the UK construction sector is at a crossroads. If the global backdrop stabilizes and volatility subsides, demand could swifty recover and supply-side constraints will become important considerations. If volatility persists, risk aversion is high both among clients and contractors and the medium-term outlook could deteriorate markedly. Developments globally over the coming months will fundamentally shape the outlook for the UK construction sector.

While core materials price inflation remains low, upside risks exist. Experience over the past five years provides a stark reminder of the impacts when global supply chains break down. Building awareness of potential vulnerabilities is important and deepening understanding of supply chain resilience can help to minimize project risk in an increasingly protectionist and geopolitically unstable world.

Cost Model: ASHP Commercial Retrofit

The cost and considerations of replacing existing heat and cooling equipment with an air source heat pump using natural refrigerants. The cost model considers the retrofit of a 16-storey office in a city such as Birmingham.

Decarbonizing an office or commercial portfolio presents numerous obstacles, ranging from planning financial and legislative challenges to competition within the marketplace for green office space.

Continuous improvement of building performance in terms of Part L and reducing carbon emissions has resulted in building owners looking to replace non-renewable technology with electric plant and equipment. Tenants, too, are seeking green buildings that support environmental sustainability while offering a comfortable workspace that enhances employee wellbeing.

Large-scale capital investment in decarbonization poses significant financial challenges for landlords, including lost income when tenants vacate buildings during refurbishment.

The Market

The growing number of air source heat pump (ASHP) manufacturers in the marketplace is reducing the cost per kilowatt of large-scale heat pumps, making them a more attractive solution.

The latest models are available in smaller units that can be installed in modules. These operate more efficiently, matching actual building demand in real time. This approach does come with a cost premium, however.

Technical innovation and regulatory pressure to phase out refrigerants with high global warming potential (GWP) are driving the uptake of natural, low-GWP refrigerants such as propane and CO_2. The availability of these in ASHPs is limited at present and attract a cost premium. There are also risks around flammability and system complexity.

Considerations

Effective office decarbonization planning requires careful consideration of key factors to produce a robust variability study that supports investment decisions that enhance asset value. For ASHPs, these include:

1. **Structure**. Implications of increasing potential roof loads, as a result of placing additional plant on the roof, that requires additional steel work; interfaces with existing building elements, such as insulation; potential noise and vibration transfer.
2. **Architecture**. Requirement for additional plant screens or noise-mitigation measures around plant. Formation of new plantrooms at roof level, with structural and service implications. Major architectural interventions, such as recladding the building, as well as associated planning approval implications.
3. **MEP**. Requirements for wholesale replacement of existing equipment to accommodate ASHP installations, and replacement of obsolete/end-of-life systems. Upgrading of incoming power to support all-electric system.
4. **Programme of works**. Retrofitting ASHPs can involve a lengthy programme, meaning lost rental income. Consider: design period; planning applications; notice and decant of existing tenants; tender and construction period; testing and commissioning, marketing and reletting of the building.

Cost Model

The cost model looks at replacing existing gas-fired boilers and roof-mounted air-cooled chillers with roof-mounted ASHPs using natural refrigerants. The building is a 16-storey office block (one basement level, 15 storeys above ground) with circa 400,00 ft^2 GIA and 270,000 ft^2 NIA. Located in a city centre outside of London, the office reaches practical completion in 2025. The model assumes the building is vacant during works and that the incoming electrical supply does not need upgrading.

Description	Total Cost of Item (£)	Cost per GIA (ft²)
Validation and survey of existing installations by contractor	£29,000	£0.07
Drain down, strip out existing plant and the like, including cranage and disposal of refrigerant within chillers	£137,000	£0.34
Structural interventions to facilitate removal of plant and making good	£73,000	£0.18
Additional roof steel work and associated extraneous metal work	£87,000	£0.22
Minor modifications and adaptations to existing drainage and Cat 5 supplies	£33,000	£0.08
Formation of new plantrooms at roof level (prefabricated units), including cranage	£157,000	£0.39
Adaptions and modifications to existing LTHW and CHW installations	£387,000	£0.97
ASHP, including pumps, pressurizations, side-stream filtration, buffer vessels and the like	£1,800,000	£4.50
Adaptions and modifications to existing electrical installation	£157,000	£0.39
	£3,072,500	**£7.68**
Testing and commissioning	£130,889	£0.33
Subcontractor preliminaries	£400,808	£1.00
BWIC	£113,532	£0.28
	£3,717,729	**£9.29**

Building Regulations

England's Building Regulations provide an insight into how building design is evolving over time – the key challenges and preoccupations at a given moment. As a template for what a safe, healthy building should be, they are also an important indicator of the direction of travel government wants building design and the wider construction industry to move in.

These changes are reflective of a time where cutting carbon and creating well-ventilated spaces are paramount. Introduced by the Department for Levelling-Up, Housing and Communities (DLUHC), the new regulations are primarily driven by the pursuit of the UK's legally binding net zero 2050 goal and reflects the global need to decarbonize.

While 2050 may seem a long-distance target, buildings increasingly will need to meet stringent carbon emission standards if the UK is to meet its goal.

The updated rules have also been influenced by the Coronavirus pandemic. This has made clear the need for effective ventilation in buildings, particularly in high density shared spaces such as offices. Furthermore, office tenants are more aware than ever of a building's health elements.

Perhaps one of the most important things to recognize is that these updates are setting the stage for larger changes in two years' time. These uplifts are an interim – and preparatory – measure ahead of the much wider-reaching Future Homes Standard and Future Buildings Standard, which are due in 2025 and will apply to new homes and new non-domestic buildings respectively. The building regulations are dynamic – the Approved Documents that make up the Building Regulations often evolve as housing policy does, and in June 2022 changes were introduced which were intended to help make the transition to the more stringent new standards easier for the industry.

Current situation in the sector

The affected regulations are Part L and Part F. Broadly speaking, Part L addresses carbon emissions and energy efficiency, while Part F is concerned with delivering good indoor air quality by delivering sufficient ventilation and minimizing the ingress of external pollutants. The change also introduces a new regulation, Part O. This regulation is about reducing the risk of overheating, primarily in domestic buildings.

The previous major revision to UK building regulations was in 2013. Under the June 2022 updates, a key focus is on improving the building fabric and building services to make them more energy efficient. The headline element is the requirement for a 27% average cut in operational carbon for new non-domestic buildings (with a 31% average cut for new domestic buildings).

Two new Approved Documents were also introduced to promote advancements in the reduction of carbon emissions: Part O and Part S. Part O ensures that buildings do not overheat, and Part S details how buildings can provide infrastructure for electric vehicles. Starting from June 2022, all new residences must include provisions for charging for electric vehicles.

Thanks to increasing recognition of the need to reduce carbon emissions, some parts of the industry are already well prepared for the changes. Some developers are already designing buildings in ways that the new regulations will encourage. This is particularly true in London, as the Greater London Authority has been requiring higher standards for several years. The London Plan, for example, means that similar requirements to many of the changes introduced by the 2022 regulations are already in force. In the capital, it may well feel like business as usual for developers.

Design implications

Setting an effective regulatory environment, which incentivizes positive changes while minimizing the risk of unintended consequences, is extremely challenging given the technical complexity of the built environment and wide range of externalities faced on individual projects. A flexible approach is essential to ensure the most effective solutions are found based on individual project constraints.

In England, the notional building sits at the heart of the standard-setting process. It is used to set a feasible target across different building types. The CO_2 emissions and primary energy of the proposed building should be equal to or lower than that of the notional building, which has the same size and shape as the proposed building, but built to a standard recipe of U-values and building service efficiencies.

The specifications used to define the notional building are tougher this time, typically demonstrating a 27% improvement in operational carbon performance compared with its 2013 counterpart. This includes introducing the use of renewable technologies into the notional building.

The developer has the option to vary the specifications used in the notional building. This means it can relax some parameters relative to the notional specifications (but remaining at least as good as specified backstop values), as long as other parameters are improved beyond the notional specification, so that ultimately the CO_2 emissions and primary energy are as good or as better than the notional building. This is intended to allow for flexibility and choice in design – a 'mix and match' approach, enabling certain design choices to be offset by other elements of the building.

Costs will be impacted – but not to a major degree

Much has evolved in the time between 2013 when the regulations were last updated, and 2022.

Technological advancements and increases in supply chain capacity for more energy efficient products have eased pressure on the cost of meeting the new regulations. For example, the premium for triple-glazed windows has dropped in recent years as uptake has increased. The difference in cost versus double-glazed units is now approximately £20–£30 per m^2 – assuming that the same frame can accommodate both double and triple-glazed units. Frames are key to pricing. For some projects there is an uplift on the frame cost, which means that the cost differential per m^2 can rise to £50–£100 per m^2.

LED lighting was not widely undertaken as an option nine years ago. Technology has improved, and prices have dropped to make LED lighting a mainstream and accessible choice – in fact, the default option. Similarly, lighting and daylight control systems are now generally a standard feature in commercial office specifications throughout England unlike back in 2013.

Under the new regulations, there is a requirement to monitor indoor air quality in new build offices, and higher risk areas in other new non-domestic buildings, to reduce the transmission of airborne illnesses. Carbon dioxide monitors are expected to be installed to flag if CO_2 has risen beyond acceptable levels and trigger an increase in ventilation. Typically, this can be expected to add around £4 per m^2 to construction costs. This requirement applies (with a few caveats) to several types of space, including offices, gyms, theatres, nightclubs, pubs and chilled food processing areas.

The attractiveness of electricity as a power and heat source has also changed. The carbon intensity of electricity dropped significantly over the past decade, in large part due to the rapid increase of renewable electricity feeding into the UK energy system. Fossil fuels now account for less than 50% of total UK electricity generation, making electricity a more attractive source of power from a carbon reduction perspective.

Building regulations previously favoured the use of gas as an alternative to electricity – a situation which the rule change reverses. Heat pumps powered by electricity or even direct sources of electricity for heating water are now favoured over gas-fired boilers or combined heat and power (CHP).

Switching to electric power has clear environmental benefits but there can be cost implications. Electric heat pump technology in lieu of traditional gas fired boilers and chillers can add up to £50 per m^2 to construction costs. In addition, increasing the electricity supply increases the maximum demand level which can put in pressure on the grid, especially in densely populated urban areas and increase cost. Often local substations are unable to meet demand and connections to larger substations are required which increases complexity and can impact programme length and cost. Early engagement with the distribution network operator (DNO) is the key to minimizing the risk of potential impacts – addressing how loads can be secured and what is the best procurement approach.

Location and building type will strongly influence compliance choices

Because developers can choose the means by which their building design meets the requirement to match the operational carbon emission standards of a 'notional building' of the same shape and size, there are myriad ways to comply with the 27% average carbon reduction in the new Part L.

One part of a design solution can be to cover a building's roof in photovoltaic panels. This option, however, is only likely to be suitable for projects with large, simple footprints. In a London office project with a tight footprint, in contrast, it is far more likely that the roof has a terrace event space, or other area, to support occupant wellbeing or increase commercial potential, over installing solar panels. That said, solar panels might be the best choice for a large, out-of-town distribution centre building with swathes of available roof space.

As the public becomes increasingly aware of the shift towards net zero, clear indicators of a building's efforts to reduce its carbon output can be a selling point for potential tenants. Lighting controls to prevent unnecessary energy use are a good example. Installing solar PV makes a very clear statement of a building's green credentials – but it's important to note that as the UK's electricity system decarbonizes, the carbon saved by photovoltaics is reduced.

Less obvious or public-facing carbon cutting choices may derive a better carbon reduction, such as installing a more efficient heat pump. Another solution for an office building for which photovoltaics are not suitable could be to improve the U-value of a building's walls. The requirements may need to be met by multiple combined improvements. Hence it may need both improved U-value and low carbon heating. In particular, improving the U-value on its own is unlikely to comply with the demands of the new regulations.

There is likely to be a higher cost impact to meet requirements for out-of-London projects compared with those in London. This is because developers are less likely to have already established a policy of incorporating major carbon reduction measures, such as a shift from gas boilers to heat pumps, into their projects. This difference is primarily because planning requirements outside London are generally more relaxed on energy and carbon. One of the many reasons that developments are built in out-of-town locations is because the land is cheaper, and planning is quicker and simpler.

Ensuring the timing is right

There is a one-year transition period from June 2022. This is to provide developers with assurance about the standards to which they must build. They should not have to make material amendments to work which is already underway when new regulations come into force.

Effectively, developers have one year from when the regulations come into effect to commence work on each individual building on site if they still wish to adopt the current (2013) regulations. To do so, developers would also have needed to submit a building or initial notice or deposit plans.

Looking to the future

We have looked at three standard building types with basic specifications. In particular for larger, more complex projects we advise that developers should go through the regulations in close detail and commission indicative, early-stage modelling to identify appropriate solutions in order to build a picture of relative costs and benefits.

The most important takeaway is that the new regulations are part of a wider shift towards highly efficient buildings where low levels of carbon – both from an emissions and an embodied carbon perspective – is the rule, rather than the exception. In adapting to the new regulations, our key advice is:

- Timing is important. The rules do not change overnight – this year, developers will need to assess whether they will fall under the old regulations or must comply with the new.
- If developing a project that falls under the new Part L then it is vital that the impact of this is considered at the earliest opportunity.
- Our analysis suggests that Greater London Authority-compliant buildings may not need to change very much to meet the new regulations, but this is not a blanket rule.
- The regulations are likely to drive a national shift away from fossil fuelled heating towards heat pumps and other low carbon sources.

In the coming years, standards for carbon reduction and building efficiencies will only become higher. With the far more ambitious Future Homes and Future Buildings Standards set to be introduced in 2025, the industry needs to aim as high as possible when designing out carbon, and to drive up efficiencies and invest in innovation in order to remain compliant. As ever, it is best to be ahead of the curve rather than behind it.

The amendments to Approved Document B (ADB) were released on 29 March 2024 by the UK Government and are due to come into effect in phases from 2 March 2025 through to 2 September 2029. These updates form part of the Government's wider fire safety reform and seek to enhance fire resilience, particularly in higher-risk residential settings, care facilities, and tall buildings. The amendments primarily apply to Volume 1 of the document (residential buildings), with some specific changes impacting health care and the broader classification of fire performance.

Key amendments include the following:

- **Second escape stair requirement**: All new residential buildings with a floor height exceeding 18 m will be required to provide a second protected escape stair. This reflects a national rollout of a measure already mandated under The London Plan and addresses concerns raised in the aftermath of major fire incidents involving single-stair buildings.
- **Evacuation lifts and firefighting facilities**: In buildings above 18 m, evacuation lifts and associated refuge lobbies must be provided to support the safe evacuation of mobility-impaired occupants. This is accompanied by enhanced provisions for fire service access and firefighting infrastructure.
- **Evacuation Shafts**: Where evacuation lifts are provided, these should be located within an evacuation shaft containing a protected stairway, evacuation lift and evacuation lift lobby. An evacuation lift lobby should provide a refuge area for those waiting for the evacuation lift, have direct access to a protected stairway and not be directly accessible from any flat, maisonette, storage room or electrical equipment room. Evacuation shafts should be afforded the same level of minimum protection as the stairway. Any smoke control system designed to protect the staircase should extend the same level of protection to the evacuation lift and evacuation lift lobby.
- **Additional mechanical ventilation**: Areas critical to the means of escape – including common lobbies, stairwells, and corridors – must be equipped with mechanical smoke ventilation systems to support protected egress and firefighting operations.
- **Evacuation Alert System**: All new residential buildings with a top storey in excess of 18 m above ground level should be provided with an evacuation alert system which can enable the fire service to initiate simultaneous evacuation by floor, or zone or the entire building.
- **Secure Information Boxes (SIBs)**: All new residential buildings with a floor height over 11 m must include secure fire safety information boxes for use by the fire and rescue service, in line with Regulation 6 of the Fire Safety (England) Regulations 2022.
- **Ban on combustible materials**: The existing restrictions on combustible materials used in external walls have been expanded to ban the use of metal composite materials (MCMs) with a polyethylene core, reflecting expert findings on cladding combustibility and facade fire spread risks.
- **Mandatory sprinklers in care homes**: From 2 March 2025, all new care homes, regardless of height, will be required to incorporate an automatic fire suppression system (AFSS), such as a sprinkler system. Additional design requirements include a maximum compartment size of 10 beds and the provision of self-closing devices on all bedroom and communal area doors.
- **Regulation 38 enhancements**: Regulation 38 has been revised to require more comprehensive fire safety information to be handed over to the responsible person upon completion. This includes fire strategies, evacuation methodology, and system maintenance procedures, supporting the 'Golden Thread' principles introduced under the Building Safety Act 2022.
- **Transition from BS 476 to BS EN 13501**: The fire performance classification system in England is being transitioned from the national BS 476 series to the harmonized European standard BS EN 13501:
    - From 2 March 2025, BS 476 will no longer be accepted for reaction to fire and roof coverings.
    - From 2 September 2029, BS 476 will be phased out entirely for fire resistance classifications, impacting elements such as fire doors, structural frames, walls, and floors.

Regional alignment:

- In **Greater London**, many of the above requirements – including second staircases, evacuation lifts, and enhanced facade fire performance – are already standard through The London Plan.
- In **Scotland**, a separate regulatory framework (Scottish Technical Handbooks) already mandates AFSS in new flats, social housing, and care buildings. While the legislative mechanisms differ, Scotland remains aligned in spirit with the direction of travel in England.

These amendments are part of the broader national effort to address systemic fire safety risks, improve resilience in new developments, and implement key learnings from recent public inquiries and safety reviews. The phased transition allows developers and duty holders time to prepare, but early engagement with fire safety specialists, building control, and the supply chain is strongly recommended to mitigate programme and compliance risks.

New challenges and adaptability

The COVID pandemic started whilst work was already underway to revise the building regulations Parts L and F. As the ways the virus was spread became better understood, steps were taken to incorporate some of this learning into the regulations so that the impacts of future outbreaks might be managed more easily.

Part F is particularly relevant to this as it is concerned with the ventilation of buildings. Several changes have been made to this part of the regulations; some of these changes seek to future-proof some building types where the spread of airborne viruses might be reduced through improved ventilation standards.

For example there is a new requirement relating to recirculating ventilation systems in offices:

Recirculation of air within ventilation systems in offices 1.37 Ventilation systems that, under normal operation, recirculate air between more than one space, room or zone should also be able to operate in a mode that reduces the risk of the transmission of airborne infection. This can be achieved by one or more of the following. a. Systems capable of providing 100 per cent outdoor air to the levels specified in paragraphs 1.32 to 1.34 to all occupiable rooms and common spaces, without recirculating air. b. Systems incorporating a UV-C germicidal irradiation system that is able to disinfect the air that is being recirculated. This type of system is commonly located within the heating, ventilation and air conditioning (HVAC) system or ductwork. c. Systems designed so that they can incorporate HEPA filters, if required, which are able to provide filtration of the recirculated air. Note: For some system types some recirculation is necessary or desirable in normal operation. Use of any full outdoor air mode, UV-C germicidal irradiation of HEPA filtration may not be necessary under normal conditions of operation.

Source: Statutory guidance. Ventilation: Approved Document F. Department for Levelling Up, Housing and Communities

Recirculating ventilation systems can be an effective means of reducing energy demands compared with systems that provide 100% fresh air. This new clause in Part F seeks to strike a balance by allowing this system type to be used but requiring the system to have the functionality to mitigate spreading infections. This might be done by switching to a mode with no recirculation or incorporating a means of treating or filtering air before it is recirculated. Some of these features may require a system to be recalibrated or recommissioned when switching to a mode with an increased resistance to the ventilation flow.

The Building Safety Act

The Building Safety Act is the foundation of the new building safety regime in England, and is one of the biggest fundamental changes to our industry over the last 30 years. It introduces a clear, proportionate framework for the design, construction and management of safer and higher quality buildings for the UK in the years to come, and, as of 1 October 2023, these requirements are now materially active in England through the updating of existing and introduction of new Building Regulations. Wales is due to follow suit with its own regulations in 2026.

The BSA is not changing the technical requirements of building regulations. What is changing is the building control application process and the level of information required before undertaking building work, when making changes to an approved application, and on completion of building work.

Overview of the new regime

The BSA applies to all buildings covered by the Building Act 1984 (as amended) and includes additional requirements for Higher-Risk Buildings (HRBs). The Act applies to new-build projects and works to existing buildings including refurbishment, extensions, and alterations. When the material elements of the BSA came into force on 1 October 2023 there were transitional arrangements that applied where an initial notice or full plans had been submitted to a local authority before 1 October 2023 and work was sufficiently progressed by 6 April 2024.

New bodies

The Act creates three new bodies to oversee the new building safety regime:

- The Building Safety Regulator (BSR) oversees the safety and standards of all buildings. For non-high-risk buildings, the BSR will oversee safety standards, enforce building safety regulations, and maintain a register of Building Control Bodies and their Inspectors. For HRBs, the BSR will also be the Building Control authority.
- The National Regulator for Construction Products (NRCP) will oversee and enforce a more effective regulatory regime for construction products. The NRCP is in the process of being set up and in the interim the Office for Product Safety and Standards (OPSS) and Local Authority Trading Standards continue to act as the national regulator for consumer products including building materials.
- The New Homes Ombudsman will enable owners of new build homes to raise complaints.

New responsibilities

The BSA defines new responsibilities:

- Duty holders, known as Clients, Principal Designers, Designers, Principal Contractors, and Contractors are required to manage building safety risks during the design and construction of all building works, both new build and alteration/ refurbishment.
- Accountable Persons have a range of duties in relation to relevant occupied HRBs to ensure they are registered, building safety risks are managed and the concerns of residents are addressed.
- Building Owners, Landlords and Developers are required to pay for remediation of historical defects and may need to pay the Building Safety Levy on new residential projects.

New systems

The BSA introduces the following new systems:

- Building Control has become a new regulated profession.
- Competence is required for all individuals appointed to work on all buildings, and organizations must demonstrate they have the right organizational capability.
- For HRBs, new Gateways are decision points at three key stages of an HRB: before planning permission is granted, before building work can begin, and before a building can be occupied.
- For HRBs the Golden Thread of Information must be created, stored digitally and maintained and updated throughout the lifecycle of an HRB.
- For HRBs there must be Mandatory Occurrence Reporting to the BSR of occurrences which could cause a significant risk to life safety. This applies to both existing operational buildings and the design and construction of new buildings/ refurbishments.

Capital Allowances

Introduction

Capital Allowances provide tax relief by prescribing a statutory rate of depreciation for tax purposes in place of that used for accounting purposes. They are utilized by governments to provide an incentive to invest in capital equipment, including assets within commercial property, by allowing taxpayers a deduction from taxable profits for certain types of capital expenditure, thereby reducing, or deferring, tax liabilities.

The capital allowances most commonly applicable to real estate are those given for plant and machinery in all buildings, other than residential dwellings, except for common areas and Structures and Buildings Allowances (SBA) for the residual construction spend on non-residential property.

Enterprise Zone Allowances are also available for capital expenditure within designated areas only where there is a focus on high value manufacturing.

Freeports, introduced in 2020, are specifically designed to incentivize investment and job creation by including lower customs duties, simplified planning rules and tax breaks, such as an enhanced level of capital allowances, suspension of SDLT, business rates, customs and NIC obligations which was extended from a five-year to a ten-year period until September 2031, as announced within the 2023 Autumn Statement.

The Act

The primary legislation is contained in the Capital Allowances Act 2001. Major changes to the system were introduced in 2008, 2014, 2018 and 2020 affecting the treatment of tax relief to include plant and machinery allowances, SBA and the incentivized tax relief for the introduced Freeports.

Plant and Machinery

Qualifying expenditure on plant and machinery is allocated to either the main pool, or special rate pool, which provide for different rates of recovery and these are set out later.

Various legislative changes and case law precedents in recent years have introduced major changes to the availability of Capital Allowances on property expenditure. The Capital Allowances Act 2001 excludes expenditure on the provision of a building from qualifying for plant and machinery, with prescribed exceptions.

List A in Section 21 of the 2001 Act sets out those assets treated as parts of buildings:

- *Walls, floors, ceilings, doors, gates, shutters, windows and stairs.*
- *Mains services, and systems, for water, electricity and gas.*
- *Waste disposal systems.*
- *Sewerage and drainage systems.*
- *Shafts, or other structures, in which lifts, hoists, escalators and moving walkways are installed.*
- *Fire safety systems.*

Similarly, List B in Section 22 identifies excluded structures and other assets.

Both sections are, however, subject to Section 23. This section sets out expenditure, which, although being part of a building, may still be expenditure on the provision of Plant and Machinery.

List C in Section 23 is reproduced below:

Sections 21 and 22 do not affect the question whether expenditure on any item in List C is expenditure on the provision of Plant or Machinery

1.   Machinery (including devices for providing motive power) not within any other item in this list.
2.   Gas and sewerage systems provided mainly –
       a.   to meet the particular requirements of the qualifying activity, or
       b.   to serve particular plant or machinery used for the purposes of the qualifying activity.
3.   Omitted.
4.   Manufacturing or processing equipment; storage equipment (including cold rooms); display equipment; and counters, checkouts and similar equipment.
5.   Cookers, washing machines, dishwashers, refrigerators and similar equipment; washbasins, sinks, baths, showers, sanitary ware and similar equipment; and furniture and furnishings.
6.   Hoists.
7.   Sound insulation provided mainly to meet the particular requirements of the qualifying activity.
8.   Computer, telecommunication and surveillance systems (including their wiring or other links).
9.   Refrigeration or cooling equipment.
10.   Fire alarm systems; sprinkler and other equipment for extinguishing or containing fires.
11.   Burglar alarm systems.
12.   Strong rooms in bank or building society premises; safes.
13.   Partition walls, where moveable and intended to be moved in the course of the qualifying activity.
14.   Decorative assets provided for the enjoyment of the public in hotel, restaurant or similar trades.
15.   Advertising hoardings; signs, displays and similar assets.
16.   Swimming pools (including diving boards, slides & structures on which such boards or slides are mounted).
17.   Any glasshouse constructed so that the required environment (namely, air, heat, light, irrigation and temperature) for the growing of plants is provided automatically by means of devices forming an integral part of its structure.
18.   Cold stores.
19.   Caravans provided mainly for holiday lettings.
20.   Buildings provided for testing aircraft engines run within the buildings.
21.   Moveable buildings intended to be moved in the course of the qualifying activity.
22.   The alteration of land for the purpose only of installing Plant or Machinery.
23.   The provision of dry docks.
24.   The provision of any jetty or similar structure provided mainly to carry Plant or Machinery.
25.   The provision of pipelines or underground ducts or tunnels with a primary purpose of carrying utility conduits.
26.   The provision of towers to support floodlights.
27.   The provision of –
       a.   any reservoir incorporated into a water treatment works, or
       b.   any service reservoir of treated water for supply within any housing estate or other particular locality.
28.   The provision of –
       a.   silos provided for temporary storage, or
       b.   storage tanks.
29.   The provision of slurry pits or silage clamps.
30.   The provision of fish tanks or fish ponds.
31.   The provision of rails, sleepers and ballast for a railway or tramway.
32.   The provision of structures and other assets for providing the setting for any ride at an amusement park or exhibition.
33.   The provision of fixed zoo cages.

Case Law

The fact that an item appears in List C does not automatically mean that it will qualify for capital allowances. It only means that it may potentially qualify.

Guidance about what can qualify as plant is found in case law dating back to 1887 (*Yarmouth v France*). The case of *Wimpy International Ltd and Associated Restaurants Ltd v Warland* in the late 1980s is one of the most important case law references for determining what can qualify as plant.

The Judge in that case applied three tests when considering whether, or not, an item is plant.
1.   Is the item stock in trade? If the answer is yes, then the item is not plant.
2.   Is the item used for carrying on the business? In order to pass the business use test the item must be employed in carrying on the business; it is not enough for the asset to be simply used in the business. For example, product display lighting in a retail store may be plant but general lighting in a warehouse would fail the test. *(Please note, this case law relates to the pre-Integral Feature Legislation, which introduced lighting as an eligible asset.)*
3.   Is the item the business premises, or part of the business premises? An item cannot be plant if it fails the premises test, i.e. if the business use is the premises itself, or part of the premises, or a place in which the business is conducted. The meaning of 'part of the premises' in this context should not be confused with real property law. HMRC's internal manuals suggest there are four general factors to be considered, each of which is a question of fact and degree:

- Does the item appear visually to retain a separate identity?
- With what degree of permanence has it been attached to the building?
- To what extent is the structure complete without it?
- To what extent is it intended to be permanent, or alternatively, is it likely to be replaced within a short period?

Certain assets will qualify as plant in most cases. However, many others need to be considered on a case-by-case basis. For example, decorative assets in a hotel restaurant may be plant, but similar assets in an office reception area may be ineligible.

Main Pool Plant and Machinery

Capital Allowances on main pool plant and machinery are currently given in the form of writing down allowances at the current rate of 18% per annum on a reducing balance basis. For every £100 of qualifying expenditure £18 is claimable in year 1, £14.76 in year 2 and so on until either all of the allowances have been claimed, or the asset is sold.

Special Rate Pool Plant and Machinery

Capital Allowances on special rate pool plant and machinery are currently given in the form of writing down allowances at the current rate of 6% per annum on a reducing balance basis. For every £100 of qualifying expenditure £6 is claimable in year 1, £5.64 in year 2 and so on until either all of the allowances have been claimed, or the asset is sold.

The special rate pool specifically covers expenditure on integral features, thermal insulation, solar panels and long-life assets.

Integral Features

The category of qualifying expenditure on 'integral features' was introduced with effect from April 2008. The following items are integral features:
- An electrical system (including a lighting system)
- A cold water system
- A space or water heating system, a powered system of ventilation, air cooling or air purification, and any floor or ceiling comprised in such a system
- A lift, an escalator or a moving walkway
- External solar shading

Thermal Insulation

The installation of thermal insulation to buildings is allowable where it is provided to an existing building owned or occupied by the taxpayer. The allowance is not available for expenditure incurred on dwellings.

Long-Life Assets

A long-life asset is defined as plant and machinery that has an expected useful economic life of at least 25 years. The useful economic life is taken as the period from of first use until it is likely to cease to be used as a fixed asset of any business. It is important to note that this is likely to be a shorter period than an asset's physical life.

Plant and machinery provided for use in a building used wholly, or mainly, as a dwelling house, showroom, hotel, office, retail shop, or similar premises, or for purposes ancillary to such use, cannot be classified as a long-life asset.

In contrast certain plant and machinery assets in buildings such as factories, cinemas, hospitals etc. could potentially be treated as long-life assets.

Full Expensing and Enhanced First Year Allowance for Plant and Machinery Assets

The 2023 Autumn Statement made permanent the temporary measures introduced to stimulate investment in new and unused plant and machinery. These measures, 'full expensing' and a 50% first year allowance, replaced previous temporary enhanced reliefs, including the 'super deduction' that were introduced under Finance Act 2021.

These reliefs apply to capital expenditure incurred by companies within the charge to UK corporation tax from 1 April 2023 onwards.

These allowances apply as follows:

- 100% 'full expensing' for main pool plant and machinery (currently 18% per annum on a reducing balance basis). This replaces the 130% 'super deduction' for expenditure incurred between 1 April 2021 and 31 March 2023, where contracted after 3 March 2021; and
- 50% first year allowance for special rate plant and machinery. This is an extension of the 50% 'special rate allowance' introduced under Finance Act 2021.

In addition, unincorporated businesses will not benefit because qualifying expenditure must be incurred by a company within the charge to corporation tax.

Refurbishment Schemes

Building refurbishment projects will typically be a mixture of capital costs and revenue expenses, unless the works are so extensive that they are more appropriately classified as a redevelopment. A straightforward repair or a 'like for like' replacement of part of an asset would be a revenue expense, meaning that the entire amount can be deducted from taxable profits in the same year.

Where capital expenditure is incurred which is incidental to the installation of plant or machinery then Section 25 of the Capital Allowances Act 2001 allows it to be treated as part of the expenditure on the qualifying item. Incidental expenditure will often include parts of the building that would be otherwise disallowed, as shown in the Lists reproduced above. For example, the cost of forming a lift shaft inside an existing building would be deemed to be part of the expenditure on the provision of the new lift.

The extent of the application of section 25 was reviewed for the first time by the Special Commissioners in the case of JD Wetherspoon. The key areas of expenditure considered were overheads and preliminaries where it was held that such costs could be allocated on a pro-rata basis; decorative timber panelling which was found to be part of the premises and so ineligible for allowances; toilet lighting which was considered to provide an attractive ambience and qualified for allowances; and incidental building alterations of which enclosing walls to toilets and kitchens and floor finishes did not qualify except for tiled splashbacks, bespoke toilet cubicles and drainage did qualify, along with the related sanitary fittings and kitchen equipment.

Structures and Building Allowances

The 2018 Autumn Budget introduced the Structures and Buildings Allowance (SBA) for expenditure incurred on non-residential structures and buildings.

Initially, the SBA was available at a flat rate of 2% per annum over a 50-year period, but this increased to 3% per annum over a 33⅓ year period from April 2020. This relates to expenditure incurred on new commercial structures and buildings, including costs for new conversions and renovations, applicable where contracts, for the main construction works, were entered into on, or after 29 October 2018. A temporary 10% rate (over 10 years) applies for freeports between 1 April 2021 and 30 September 2026, as noted earlier.

The SBA extends to qualifying expenditure incurred under capital contributions – for example, a landlord contributing towards a tenant's fitting-out costs.

The relief will be available when the building, or structure, first comes into qualifying use and is available for both UK and overseas assets, provided the business is within the charge to UK tax. This relief is not applicable for expenditure on assets used for residential purposes but does include nursing and care homes.

The SBA expenditure will not qualify for the Annual Investment Allowance (AIA).

The relief will cease to be available if the building, or structure, is brought into residential use, or if it is demolished. However, SBA continues to be available for periods of temporary disuse.

If the building, or structure, is sold, the new owner would benefit from the unutilized residue of the SBA for the remaining period, via an Allowances Statement in accordance with s170IA CAA2001.

The costs of construction will include only the net direct costs related to the construction expenditure of the building, or structure, after any discounts, refunds, or other adjustments. This will include demolition costs, or any land alterations necessary for construction and direct costs required to bring the building, or structure, into existence. Excluded from the above expenditure are indirect costs not associated with the physical construction, such as finance, planning, legal and marketing costs, landscaping etc.

SBA qualifying expenditure will also be applicable to purchases of new and second-hand buildings, or structures. The basis of claim will be dependent upon the vendor's holding structure and whether the property has been used, or unused, at the time of acquisition, as well as being entitled to claim this tax relief.

Where a building, or structure, is acquired and subsequently altered, or renovated, this will trigger a new SBA qualification period. Similar additional expenditure on existing properties will also trigger new SBA streams of qualifying expenditure.

It is important to note that expenditure on plant and machinery does not qualify for SBA. A purchaser will be unable to reclassify expenditure on plant and machinery previously claimed as SBA to plant and machinery at a later date.

Expenditure on qualifying land remediation will also not qualify for SBA.

Annual Investment Allowance

AIA is available to all businesses of any size and allows a deduction for the whole AIA of qualifying expenditure on plant and machinery, including integral features and long-life assets. AIA rates have fluctuated over the year and is currently £1,000,000 per annum since 1 January 2019.

For accounting periods less, or greater, than 12 months, or if claiming in periods where the rates have changed, time apportionment rules will apply to calculate hybrid rates applicable to the period of claim.

Enterprise Zones

Enterprise zones benefit from a number of reliefs, including a 100% first year allowance for new and unused non-leased plant and machinery assets, where there is a focus on high-value manufacturing.

Freeports

Following a government bid process launched late in 2020, the first eight freeports in England were confirmed as follows:

- East Midlands Airport
- Felixstowe and Harwich
- Humber
- Liverpool City
- Plymouth and South Devon
- Solent
- Teesside
- Thames

In January 2023 the Scottish government announced two Green Freeports:

- Inverness and Cromarty Firth
- Firth of Forth

In March 2023 two freeports were announced for Wales:

- Anglesey
- Port Talbot and Milford Haven

In October 2024 one freeport was announced for Wales:

- Celtic

Further sites are expected to follow including Northern Ireland.

Dedicated 'tax sites' in freeports will benefit from generous tax allowances to 30 September 2026 including:

- An enhanced 10% rate of SBA, meaning that the relief can be claimed over 10 years, rather than the current 33⅓ years;
- An enhanced capital allowance of 100% for plant and machinery allowances, providing a full deduction, rather than the current rates of 18% (MP) and 6% (SRP) on a reducing balance basis. The allowance is only available to occupiers and not property owners or developers leasing premises;
- Full relief from Stamp Duty Land Tax (SDLT) on land and property purchased for a qualifying commercial purpose;
- Full Business Rates relief for new businesses and existing businesses that expand operations;
- Relief against employer National Insurance contributions, subject to Parliamentary process and approval.

Research and Development Allowances

Research and Development Allowance (RDA) gives relief for capital expenditure incurred on R&D by a business.

RDA is only due if the research and development expenditure is related to the trade being carried on or about to be carried on. Qualifying RDA expenditure is capital expenditure that a business incurs for providing facilities for carrying out R&D, for example, a laboratory in which to carry on R&D. RDA is only available to traders and is not available to a person carrying on a profession or vocation.

The rate of RDA is 100% of the qualifying expenditure. However, if a disposal value is brought into account on expenditure in a chargeable period in which an RDA is made, the RDA is 100% of the expenditure less the disposal value. RDA is made in the chargeable period in which the qualifying expenditure was incurred unless it was incurred before the trade began, in which case RDA is made in the chargeable period in which trade began.

Other Capital Allowances

Other types of allowances include those available for capital expenditure on Mineral Extraction, Know-How, Patents, Dredging and Assured Tenancy.

International Tax Depreciation

The UK is not the only tax regime offering investors, owners and occupiers valuable incentives to invest in plant and machinery and environmentally friendly equipment. Ireland, Australia, Malaysia, Hong Kong and Singapore also have capital allowances regimes which are broadly similar to the UK and provide comparable levels of tax relief to businesses.

Many other overseas countries have tax depreciation regimes based on accounting treatment, instead of capital allowances. Some use a systematic basis over the useful life of the asset and others have prescribed methods spreading the cost over a statutory period, not always equating to the asset's useful life. Some regimes have prescribed statutory rates, whilst others have rates which have become acceptable to the tax authorities through practice.

Contact

For further information or assistance, please contact one of the Fiscal Incentives team:

help.capitalallowances@aecom.com

Quantity Surveyor's Pocket Book, 4th edition

Duncan Cartlidge

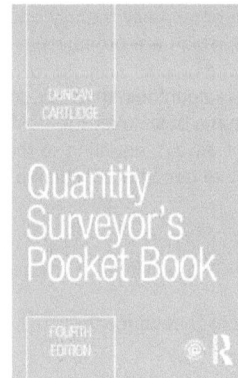

The fourth edition of Quantity Surveyor's Pocket Book remains a must-have guide for students and qualified practitioners. Its focused coverage of the data, techniques and skills essential to the quantity surveying role makes it an invaluable companion for everything from initial cost advice to the final account stage.

Key features and updates included in this new edition are as follows:

- The impact of Brexit on construction and public procurement;
- New developments in digital construction;
- Renewed focus on ethics in the industry;
- Up-to-date analysis of NRM1, 2 and 3;
- Latest practice guidance notes and government publications;
- Post-contract administration;
- A companion website with alternative approaches to taking off quantities using spreadsheets.

This text includes recommended formats for cost plans, developer's budgets, financial reports, financial statements and final accounts. This is the ideal concise reference for quantity surveyors, project and commercial managers, and students of any of the above.

March 2022: 518 pp
ISBN: 9781032061443

To Order
Tel:+44 (0) 1235 400524
Email: tandf@hachette.co.uk

For a complete listing of all our titles visit:
www.tandf.co.uk

Taylor & Francis
Taylor & Francis Group

LED Lighting

Background

LED lighting technology is now sufficiently developed to be widely accepted as the light source of choice for almost all lighting applications. The technology offers benefits including energy efficacy, high-quality light appearance, reduced operational temperatures, light colour adjustability, device connectivity and controllability. This makes LED luminaires suitable for all applications ranging from street lighting, public realm, museum and gallery, office, and residential lighting.

Currently available types of LED luminaires

Light emitting diodes (LEDs) are small, high efficacy light sources that can be packaged into a range of products suitable to both new build and retrofit applications. In new installations, LED chips are often integrated into the luminaire, which can offer technical benefits in terms of thermal properties, fixture size and efficacy.

Whereas in retrofit applications, LEDs are packaged into conventional lamp base formats that can easily be installed into existing light fixtures to replace traditional light sources. There are a wide range of retrofit LED lamps that are an alternative to halogen and compact fluorescent through to linear LED tubes that are a direct replacement to linear fluorescent lamps.

LED efficacy

There are LEDs commercially available with a light source efficacy over 200 lumens per watt and delivered light output from luminaires of over 140 lumens per watt. This makes them far more efficient than traditional halogen or incandescent lamps which are now banned in the UK. These were desirable in terms of light colour and colour rendering but had a typical efficacy range of 5–25 lumens per watt. LEDs are also more efficient than other commonly used light sources including fluorescent and discharge technology lamps with typical efficacy ranges of 40–100 lumens per watt.

LED light colour, appearance, and colour rendering

LEDs are available in a wide range of light colour appearances or 'temperatures'. This impacts ambience, mood, and how colours appear. It also has cultural context and relevance, so it is an essential feature to be considered when selecting light sources. Colour temperature, or Correlated Colour Temperature (CCT) is how we measure the hue of light, represented in Kelvins (K), ranging from warm (yellow/orange, low K) to cool (blue/white, high K).

The Kelvin scale in terms of light sources is derived as a black body temperature which ranges from 1,000 K to 10,000 K with very warm white typically having a colour temperature of 1,800 K to very cool white with a colour temperature in excess of 6,500 K. In addition, many saturated colour options are available that can be blended to create an infinite range of colour options.

Colour rendering is the ability of a light source to render colours faithfully is assessed by the calculation of the extent to which colours illuminated by the source have their colour appearance changed compared to those under a reference source. One of the most used metrics is the CIE colour rendering index (CRI), which was developed in the 1960s and measures how accurately a light source renders colour compared to natural daylight on a scale from 0–100.

CRI was largely designed to deal with the then new improvements in fluorescent lamps. However, the more recent emergence of LEDs into mainstream lighting applications has shown up some of the weaknesses in the CRI system and an alternative system known as the colour quality scale (CQS) has been developed. Whilst at present CQS does not have the same worldwide acceptance as CRI it does provide a useful metric that works well with most light sources.

Typical LED colour rendering characteristics vary from CRI >70 to >97 while specialized LED products are available with precise spectral properties tuned to specific applications which could include fashion, retail, film and television through to full spectrum LEDs.

LED binning

LED binning is the process of sorting LEDs based on various quality criteria such as colour, voltage, and brightness to ensure uniformity and consistency within a batch. The colour consistency (detectable differences of chromaticity across manufactured batches of LED) is quantified by the MacAdam's ellipse scale that defines Standard Deviation Colour Matching (SDCM). SCDM can have a significant bearing on LED costings with 1–2 SDCM being excellent quality with no discernible difference in batches and therefore more expensive, 3–4 SDCM being good and 4–7 SDCM acceptable in some applications where colour matching is less of a priority.

LED control

In both new build and retrofit applications, a wide range of LED control options are available, both wired and wireless. This makes it possible to dim LEDs and create programmable colour change effects, and in some situations support monitoring their performance or detecting issues within the luminaire.

Fluorescent lamp ban

In the EU, the production of CFL, T8 and T5 lamps was banned from 25 August 2023. Six months later, on 1 February 2024, the UK saw the same ban come into effect. Only High Pressure Discharge (HPD) lamps and special purpose lamps were exempt from the ban and will be produced for another 2–4 years.

Cost implications

Manufacturers and suppliers are keen to highlight potential energy saving from LED lighting technology. Although the principle of realizable savings is completely valid, a detailed study of the whole lighting system from procurement through operation, including drivers, must be undertaken. Review potential energy savings is essential to ensure the most beneficial and appropriate LED technologies and design approaches can be implemented.

Caution should be taken when considering manufacturer's efficacy figures and lumen output levels, and it should be noted that that there is a marked difference between lamp lumens (the lumen output of the source) and delivered lumens (the actual lumens delivered by the fixture). Conversely, the same caution should be applied to quoted watts figures as sometimes these vary between manufacturers and do not consider gains or losses from internal control gear and other luminaire features in all cases.

Sustainability (BREEAM / WELL / TM66 / TM65)

When selecting an LED source, in addition to its efficacy there are a myriad of other factors which should be considered, not least of all is its whole life in service and ability to be reused and repaired, recycling should be a last option. There are a number of sustainability accreditation credits that are useful to consider when costing lighting systems – luminaires, drivers and lighting control can all have a bearing – and building efficacy assessments can also reward the use of sustainable and efficient lighting systems, these include, but are not limited to:

- BREEAM – Although there is no specific lighting requirement under BREEAM assessment there are bonus points awarded for the use of efficient sustainable lighting.
- TM65 – Embodied carbon in building services: A calculation methodology (2021).
- TM66 – Creating a circular economy in the lighting industry (2021).
- WELL L03 – the use of circadian lighting in areas with no natural light.

Payback

For a new build project or new installation then LED fixtures are the sensible and only viable solution in terms of cost and longevity. While initially they may seem more expensive, the long-term energy savings, product longevity and reduced maintenance requirements soon balance out, therefore, the return on investment (ROI) is realised in a very short time period. If viable, retrofitting or reworking existing luminaires with LED sources is clearly a very sustainable solution and is usually more cost-effective than full replacement.

The following examples provide a return on investment (ROI) overview for a sample commercial scheme utilizing new LED Luminaire typologies and excludes the cost of installation labour which is assumed to be consistent.

Power cost (£)/kWh	0.28
Days in use/year	260
Hours in use/day	12

Note: Range of costs £/kWh varies greatly across suppliers/contracts, current in 2025 average ranges are 0.24–0.60 (£)/kWh, with some commercial bulk-buy deals going as low as 0.20 (£)/kWh (estimate).

	Light source type			
	Typical LED Downlight	Typical LED Retrofit Lamp	Compact Fluorescent Downlight**	LV Halogen (IRC) Downlight**
Input Power (watts)	6 W	7.5 W	12 W	35W
Lifetime (hours)	50,000	35,000	12,000	4,000
Replacement Lamp Cost including labour (£)	–	45.00	15.00	15.00
Annual Energy Cost per lamp (£)	5.24	6.55	10.48	30.58
Total Energy Cost for 25 luminaires per year (£)	131.04	163.80	262.08	764.40
Luminaire Unit Cost (£)	90.00	40.00* (0)	60.00	40.00
Supply only cost of 25 luminaires (£)	2,250.00	1,000.00* (0)	1,500.00	1,000.00
Allowance for 4 nr emergency battery packs (£)	480.00	480.00	480.00	480.00
Lamps (£)	Included	625.00	375.00	375.00
Total cost of luminaires/emergencies/lamps (£)	2,730.00	2105.00* (1105.00)	2,355.00	1,855.00
Extra over cost of LED luminaires and emergency luminaires compared to traditional luminaires is (£)	0	625* (1625.00)	375	875
Yearly cost of energy and yearly re-lamping allowance (£): • Compact fluorescent every 3.84 years*** • Halogen every 1.28 years*** • LED Retrofit every 11.54 years*** • (LED Luminaire replacement every 16 years)***	131.04	217.96	359.74	1057.37
The calculated yearly energy and re-lamping saving of using Dedicated LED luminaires when compared to retrofitting or traditional luminaires is (£)	NA	86.92	228.70	926.33
Therefore, time taken in years to 'pay' for the additional cost of the LED luminaires based on the energy and re-lamping costs of 'traditional' luminaires in **years** is	**NA**	**NA**	**1.64**	**0.94**

* The retrofit or aftermarket replacement LED would be installed into the existing halogen fixture so the cost here would be negated.

**We have included these sources for comparison but it should be noted these are now banned in the UK & EU and are no longer widely used.

***Figures based on 3120 hours use per year (12 hours a day 260 days per year).

Power cost (£)/kWh	0.28
Days in use/year	260
Hours in use/day	12

	Light source type		
	Recessed LED Office Luminaire 600 × 600 mm	Typical LED Retrofit For existing 5 ft Linear T8 fixtures	Typical Recessed 600 × 600 Fluorescent (CAT2) Office Fixture **
Input Power (watts)	26W	32 W (4 × 8W)	42W (4 × 18W)
Typical Lifetime (hours)	50,000	36,000	12,000
Replacement Lamp Cost including labour (£)	Included	58.00 (x4)	50.00 (x4)
Annual Energy Cost per lamp (£)	22.71	27.96 (4 × 8W)	62.90 (4 × 18W)
Total Energy Cost for 25 luminaires per year (£)	567.84	698.88	1572.48
Luminaire Unit Cost (£)	100.00	155.00***	155.00
Supply only cost of 25 lamps (base on 4 nr per fitting) (£)	Included	1450.00***	1250.00
Supply only cost of 25 luminaires including lamps (£)	4,625.00	4325.00***	4,375.00
Allowance for 4# emergency battery packs (£)	480.00	480.00	480.00
Total cost of luminaires/emergencies/lamps (£)	5,105.00	7,255.00	6,855.00
Extra over cost of LED luminaires and emergency luminaires compared to traditional luminaires is (£)	NA	4275.00***	3875.00
Yearly cost of energy and yearly re-lamping allowance (£) • compact fluorescent every 4.8 years**** • LED Retrofit every 14.4 years**** • (LED Luminaire replacement every 20 years)****	567.84	799.57	1832.90
The calculated yearly energy and re-lamping saving of using LED luminaires when compared to traditional luminaires is (£)	NA	231.73	1265.06
Therefore, time taken in years to 'pay' for the additional cost of the LED luminaires based on the energy and re-lamping costs of 'traditional' luminaires in **years** is	NA	NA	−3.06

**We have included these sources for comparison, but it should be noted these are now banned in the UK & EU and are no longer widely used.

*** Luminaire 'fixture' cost included for fair comparison, however existing fitting would typically be used and retrofitted with the LED source.

**** Office hours occupation based on 10 hours per day, 5 days per week, 50 weeks a year (2500 hours).

Notes and conclusions

1. LED luminaires are far more efficient at turning their energy into useful light and do save energy and are therefore the light source of choice for most applications.
2. Some LED luminaires may not be converted to emergency so 'additional' luminaires may be required.
3. Caution should be taken when considering manufacturer efficacy figures and lumen output levels; it should be noted that that there is a marked difference between lamp lumens (the lumen level of the source) and delivered lumens (the actual lumens delivered by the fixture).

    Conversely the same caution should be applied to quoted watt figures as sometime these do not consider gains or losses from internal control gear and other luminaire features.

    In addition, some less reputable manufacturers will, 'over run' their fittings meaning that they operate at a higher current, essentially working the LED chips harder than they ideally should which means although they may have a higher light output and apparent greater efficacy their life in service may not be as long.
4. LV Halogen (IRC) downlight utilized for comparison purposes as a desirable lamp type now banned in the UK.
5. In the UK, the production of CFL, T8 and T5 lamps were banned from 1 February 2024, their usage banned from September 2023.
6. Sustainability should not just look at figures and metrics for efficacy and the ability to be reused, reworked or recycled. Further consideration when designing and specifying is in the procurement, where and how the fittings are manufactured and the proximity of the manufacturing plant to site and how they will be shipped.

    Furthermore, it is important to understand the supplier's manufacturing processes, procedures, and the sustainability of their operation. Review of their full supply chain, including any third-party suppliers, to ensure ethical employment criteria and conditions, and that the use of slave labour or conflict minerals during production is outlawed.

Research and Development Tax Reliefs and Tax Credits

Introduction

Research and Development (R&D) tax reliefs and tax credits are aimed at encouraging expenditure by companies on innovative projects that advance the overall knowledge or capability in the fields of science and technology. R&D tax relief and tax credits are given against corporation tax and are therefore only available for companies chargeable to UK corporation tax.

The legislation covering R&D tax relief and tax credits is mainly found in Part 13 of the Corporation Tax Act 2009 (CTA 2009). The following legislative references are to CTA 2009 unless otherwise stated.

There are currently two schemes set out in the legislation, which applies for expenditure incurred before 1 April 2024, as follows:

- R&D expenditure credits (section 1040 A), commonly referred to as RDEC.
- Additional tax relief for small or medium-sized enterprises (section 1044), commonly referred to as the SME scheme.

The two schemes were effectively merged for expenditure incurred from 1 April 2024 by legislation set out in the Finance Act 2024. However, parts of the SME scheme are retained for loss making SMEs that incur substantial R&D expenditure – this is referred to as 'enhanced R&D intensive support' (ERIS).

Definition of R&D

In accordance with Section 1041, R&D has the meaning given by section 1138 of CTA 2010 which states that R&D means 'activities that fall to be treated as research and development in accordance with generally accepted accounting practice'. The activities must also fall within special definitions set out in the Department for Science, Innovation and Technology (DSIT) Guidelines for activities to be treated either as 'directly contributing' to seeking the advance in science or technology, or to be treated as a 'qualifying indirect activity'.

The company seeking to claim the tax relief must be able to explain the intended advance in science or technology and how the specific project seeks to achieve the desired outcome through the resolution of uncertainty. The project must seek an advance in the overall knowledge or capability in the specific field of science or technology and not simply an advance for the sole benefit of the company. Advance sought within arts, humanities, and social sciences including economics, will not qualify.

From 1 April 2023, mathematical advances can be treated as science for R&D tax purposes, whether or not they are advances in representing the nature and behaviour of the physical and material universe.

The RDEC before 1 April 2024

The RDEC is designed for companies that have 500 or more employees, and either a balance sheet that exceeds 86 million euros, or a turnover of more than 100 million euros. The RDEC is an 'above the line' expenditure credit, in addition to the normal 100% deduction available as a trading expense under section 87. It can also be claimed by SMEs who have been subcontracted to do R&D work by a large company or where SMEs have received certain types of funding.

Under RDEC, the company can obtain a tax credit of 20% (13% before 1 April 2023) of its R&D expenditure. The credit is taxable at the normal corporation tax rate, which effectively means that the tax benefit of the credit is 15% of the qualifying R&D expenditure for a company subject to the 25% tax rate. The tax credit is set off against the company's corporation tax liability. Any unutilized expenditure credit, after following steps 1 to 6 set out under section 104N(2), is paid to the company under step 7 of section 104N(2).

The payable amount under step 7 ensures that loss-making companies receive the same net benefit as profit makers where the credit is taxable. This is achieved by retaining a 'notional' tax of 25% so that the total cash benefit for all claimants is equal to the expenditure credit net of tax at the main rate of corporation tax. For example, £300,000 of qualifying R&D expenditure would give rise to an RDEC of £60,000 (i.e. 20% × £300,000), which would be taxed at 25% providing a net benefit of £45,000 or 15% of the R&D expenditure. If the company is loss-making, the payable credit (subject to cap below) would be calculated as 15% of the R&D expenditure.

The cap on the payable tax credit under RDEC is the total amount of expenditure on the R&D workers' Pay As You Earn (PAYE) and National Insurance contributions (NIC). The companies PAYE and NIC staffing costs consist of employees of the company engaged in R&D activity and any externally provided workers (EPWs) from a connected group company. No account is taken of the employees non-R&D activity but, for the EPWs, the cost of PAYE and NIC would be apportioned between the time spent on R&D activity and non-R&D activity.

The SME scheme before 1 April 2024

The SME scheme is only available to companies with less than 500 staff and a turnover of under 100 million euros or a balance sheet total under 86 million euros. Linked companies and partnerships must be included when deciding whether a company is an SME under this scheme, but certain relaxations apply from 1 April 2023.

The additional tax deduction available under the SME scheme is 86% for R&D expenditure incurred on or after 1 April 2023 (130% before 1 April 2023). This means that under the SME scheme a company can deduct 186% of its R&D expenditure (previously it was 230%) as 100% deduction is already available as a trading expense under section 87.

The payable tax credit under the SME scheme (which unlike RDEC is not taxed as income) is 10% of the surrenderable loss from 1 April 2023 (14.5% before 1 April 2023). The surrenderable loss being the lower of actual losses and 186% of the R&D expenditure. This means that the payable tax credit is worth a maximum of 18.6% of the company's R&D expenditure based on the 25% main rate of corporation tax.

Under the SME scheme the company can also obtain tax relief for pre-trading expenditure on R&D (which would not otherwise be available) by making an election under section 1045 to treat the pre-trading R&D expenditure as a trading loss equal to 186% of the qualifying R&D expenditure. The company can then surrender the loss for a payable tax credit under section 1058.

The amount of the payable tax credit that an SME company can receive in any one year is restricted to £20,000 plus three times the company's total PAYE and NIC liability. A company is exempt from the cap if its employees are creating, preparing to create or managing intellectual property and it does not spend more than 15% of its qualifying R&D expenditure on subcontracting R&D to, or the provision of EPWs by, connected persons.

For R&D intensive SMEs that incur at least 30% of their total annual expenditure on qualifying R&D, the payable tax credit is 14.5% from 1 April 2023 instead of the 10% rate that applies for other SMEs.

However, the additional deduction rate of 86% will still apply to R&D intensive SMEs from 1 April 2023. This means that the payable tax credit is worth 26.97% for loss-making intensive SMEs.

Qualifying expenditure

The revenue expenditure that qualifies for R&D tax relief and tax credits under either the RDEC scheme or the SME scheme can be grouped under the following categories:

- Staffing costs
- EPWs
- Subcontractors
- Consumables
- Software licence fees
- Data licence and cloud computing costs – for accounting periods beginning on or after 1 April 2023
- Payments to the subjects of clinical trials

Staffing costs include salaries, wages, overtime pay and cash bonuses, employer secondary Class 1 NIC, employer pension contributions and certain reimbursed business expenses such as travel and subsistence expenses provided that they are incurred and then reclaimed by the employee. Any benefits in kind, such as private medical cover and company cars, are specifically excluded from the staff costs category. An appropriate apportionment should be applied to staff costs if the staff concerned are carrying out R&D for only some of their time.

EPWs are individuals provided to the company through a staff provider. EPWs must operate through the staff provider instead of contracting directly with the company. These individuals would include agency staff, contractors and freelancers. EPWs must work under the supervision, direction and control of the company when carrying out R&D work. Again, an apportionment will be necessary between time spent on R&D and time spent on other activities. The costs for 'unconnected' EPWs, where the EPWs are provided by an agency or other party that is unconnected to the company, must be restricted to 65% to allow for the profit charged by the agency. The cost of connected EPWs does not have to be similarly restricted, but the cost is restricted to the lower of the cost incurred by the company and the cost incurred by the connected party that provides the EPWs.

Subcontractor costs incurred on R&D projects will normally only be allowable in the case of a claim under the SME scheme. As in the case of unconnected EPWs, unconnected subcontractor costs are restricted to 65%. For connected subcontractors, as for connected EPWs, the subcontractor costs that can be included in the claim are the lower of the cost incurred by the company and the costs incurred by the connected subcontractor.

In the case of RDEC claims, companies can only claim for expenditure on subcontracted R&D if the subcontractor is one of the following:

- An individual
- A partnership made up solely of individuals
- A qualifying body, including charities, universities and scientific research organizations.

The subcontracted costs included in an RDEC claim are not restricted in the same way as they are for claims under the SME scheme.

Consumable costs include the materials that are consumed or transformed in the R&D process, such as water, power, fuel and any materials used in the construction of prototypes. Any cost incurred after the R&D project has been completed, for example the cost of materials for fine-tuning or marketing, is not qualifying expenditure. Under Section 1126 A(1) CTA 2009, where a company sells or otherwise transfers ownership of items produced in the course of its R&D activity as part of its ordinary business, the cost of consumable items that form part of those products is excluded from expenditure qualifying for relief.

Software licence costs include software used for R&D and a reasonable share of the costs for software partly used in R&D activities.

For accounting periods beginning on or after 1 April 2023, qualifying expenditure includes data licence and cloud computing costs. A data licence is a licence to access and use a collection of digital data. Cloud computing includes the following:

- Data storage
- Hardware facilities
- Operating systems
- Software platforms

Data and cloud computing costs for qualifying indirect activities are not qualifying expenditure.

Expenditure on R&D carried out by the company that is met by any other party is subsidized expenditure under section 1138 and must be excluded from a claim under the SME scheme. However, the company can still claim under the RDEC scheme for subsidized expenditure.

Claim notification

For accounting periods beginning on or after 1 April 2023, if a company is making its first claim for R&D tax relief or tax credit, or it has not made a claim in any of the previous three calendar years, the company must submit a Claim Notification form otherwise its claim will not be valid.

If the Claim Notification is required, the form must be submitted not later than six months after the end of the period of account that includes the relevant accounting period. If the company's accounting period changes, or they decide not to claim until a later accounting period, the company may need to submit a new Claim Notification form.

The Claim Notification form is accessible on GOV.UK and should contain the following information:

- Unique Tax Reference (UTR) number of the company, as shown on its CT600 tax return
- Contact details of main internal R&D contact at the company, such as a company director
- Contact details of any agent involved in the R&D claim, including the agent reference number if possible

- The accounting period start and end dates for which the company is claiming the tax relief or expenditure credit, which must correlate with the company's tax return
- The period of account start and end dates

Notwithstanding the above, the company is not required to submit a Claim Notification form if it makes its R&D claim before the deadline for submitting a Claim Notification. Therefore, the purpose of the Claim Notification is to give HMRC early notice that the company will be making a claim, in advance of providing the information in the company's tax return.

Additional information

For R&D tax relief claims made on or after 8 August 2023, additional information must be provided in accordance with regulations made by HMRC. The guidance published by HMRC setting out what additional information is required became law under Schedule 1 of the Finance (No 2) Act 2023. This requirement applies to all R&D claims, including a claim for a payable tax credit.

HMRC guidance states that the additional information should be provided on the Additional Information form accessible on GOV.UK. Whilst additional information such as supporting reports can still be provided with the company's tax return, the guidance states that providing an R&D report and then inserting language into the Additional Information form such as 'see R&D report for more detail' does not meet the additional information requirements and will likely lead to the form being flagged for further investigation.

The requirement to provide an Additional Information form applies to each company making an R&D claim. If there are several companies within the same group making similar R&D claims, each company in the group will be required to submit a separate Additional Information form. Furthermore, for projects spanning several years, a new Additional Information form will need to be submitted for each year that a company claims R&D tax reliefs. Again, referring to previous forms would not meet the additional information requirements, although the information contained on earlier forms could be duplicated on later ones.

Whilst HMRC have stated that there is no pre-determined expectation of the amount of information required to be provided for any individual project when completing the Additional Information form, the information required must be sufficient to explain why the project qualifies for R&D tax relief and state under which scheme the claim is made.

The Additional Information form must be provided by the date that the claim is made or amended by the company in accordance with the Finance Act 1998 (FA 1998), Schedule 18, paragraph 83E.

The information required is substantial and includes the company's UTR, employer PAYE reference number, VAT registration number, business type such as the company's Standard Industrial Classification code, as well as contact details for the main senior internal R&D contact in the company, contact details of all agents involved in the R&D claim and the accounting period start and end dates for which the company is claiming the tax relief.

The company must also state the number of projects included within the R&D claim for the relevant accounting period. The level of information required for the projects will depend on the number of projects as follows:

- For 1 to 3 projects, a description is required for all the projects that cover 100% of the qualifying expenditure.
- For 4 to 10 projects, a description is required for those projects that account for at least 50% of the total expenditure, with a minimum of 3 projects described.
- For 11 to 100 (or more) projects, a description is required for those projects that account for at least 50% of the total expenditure, with a minimum of 3 projects described. If the qualifying expenditure is split across multiple smaller projects, a description is required for the 10 largest projects. If the company would have to select more than 10 projects to account for at least 50% of the qualifying expenditure, then it should select the 10 projects with the highest qualifying expenditure.

Some of the above information is the same as that required on the Claim Notification form. However, the project information required on the Additional Information form is far more onerous.

Furthermore, since 1 August 2023, all claims for R&D tax relief and tax credits must be made digitally through HMRC's tax return portal.

HMRC will remove claims if companies fail to comply with the pre-notification and/or additional information requirements. In such instances, companies will have 90 days to appeal by sending written representations to HMRC for consideration.

The merged scheme RDEC from 1 April 2024

The existing RDEC for large companies and the SME scheme are effectively merged into a single scheme for expenditure incurred in accounting periods starting on or after 1 April 2024. From that date, all claims will be made in accordance with the rules of the merged scheme, except for claims by SMEs that qualify for ERIS. The rules for the merged scheme are set out in Chapter 1 A of Part 13 of CTA 2009. Section 1042H confirms that the RDEC is still an 'above the line' credit which is subject to corporation tax and section 1042I sets out the 7 steps for claiming the credit.

For expenditure under the merged scheme, the rate of RDEC is 20%. This is the same as the rate under the old RDEC scheme for expenditure incurred on or after 1 April 2023. However, there are some notable differences between the existing RDEC scheme and the merged scheme, where aspects of the current SME scheme have been incorporated into the merged scheme. The most significant difference is the rules for contracted out R&D.

Generally, companies will not be able to make a claim for RDEC if the R&D has been contracted to them. However, where a company with a valid R&D project contracts a third-party subcontractor to undertake R&D activities on its behalf, the company (but not the subcontractor) may claim for the qualifying expenditure. The general rule is that the party who takes the decision to undertake R&D will be able to claim. Section 1133 of CTA 2009 defines when R&D is contracted out and ensures that the costs of contracting out R&D by a customer to a contractor can attract relief for the customer. Where the customer is not contracting out R&D, the contractor can claim relief, if it carries out relevant R&D.

There are some scenarios where a subcontractor can claim R&D tax relief. These include when a company carries out R&D for a contractor which is not a tax-paying entity, such as charities, universities, scientific research organizations, as well as overseas entities.

Costs incurred outside of the UK on subcontracted R&D, and on EPWs outside of the UK, will only be allowable for R&D tax relief where the work must be undertaken outside of the UK for reasons such as geography, environment or regulation. Both cost considerations and workforce availability are explicitly ruled out as reasons to qualify for overseas expenditure.

There are no restrictions on subsidized expenditure within the merged scheme.

If the company is loss-making, it can utilize the 'surrenderable loss' as calculated under section 1042I using a notional tax deduction rate of 19% (previously 25% under the old RDEC scheme) which means that the loss can be surrendered for a cash credit of up to 16.2% of the qualifying expenditure.

The cap on the amount of the payable tax credit that a company can receive in any one year under the merged scheme is the same as the cap that applied under the existing pre-April 2024 SME scheme, being £20,000 plus three times the company's total PAYE and NIC liability. This is more generous than the cap that applied under the pre-April 2024 RDEC scheme. Any excess over the cap is carried forward and treated as an amount of RDEC for the next accounting period.

ERIS from 1 April 2024

The rules for ERIS are set out in Chapter 2 of Part 13 of CTA 2009. Under ERIS, loss-making R&D intensive SMEs can deduct an extra 86% of their qualifying costs, just as SMEs could under the previous SME scheme, instead of claiming an RDEC under the merged scheme. However, if the company qualifies under ERIS, it can still opt to claim under the merged scheme but it cannot claim under both schemes.

The payable tax credit under ERIS (which unlike the merged RDEC is not taxed as income) is 14.5% of the surrenderable loss. The surrenderable loss being a maximum of 186% of the R&D expenditure. This means that the payable tax credit for loss-making and R&D intensive SMEs is still worth a maximum of 26.97% of the company's R&D expenditure under ERIS.

In addition, under ERIS, a loss-making R&D intensive SME can still obtain tax relief for pre-trading expenditure on R&D under section 1045, which is not available to other SMEs.

The threshold for companies to be considered 'R&D intensive' is 30% of their total annual expenditure on qualifying R&D in accordance with section 1045ZA of CTA 2009. Companies that incur less on qualifying R&D than the 30% threshold must claim tax relief under the merged scheme.

A one-year grace period will apply for companies who dip under the 30% threshold for ERIS but have met this intensity condition in their last 12-month accounting period and have made a valid claim to SME tax relief or ERIS in that period on expenditure incurred on or after 1 April 2023. This is intended to protect loss-making SMEs from moving in and out of ERIS retrospectively thereby avoiding uncertainty around financial planning and investment decisions.

The restrictions on contracted out R&D and on overseas expenditure, which apply under the merged scheme, also apply to ERIS. Also, in common with the merged scheme, under ERIS there will be no restrictions around subsidized R&D.

The cap on the amount of the payable tax credit that an SME company can receive in any one year is £20,000 plus three times the company's total PAYE and NIC liability. Any claim for a payable tax credit in excess of the cap is invalid. However, a company is exempt from the cap if its employees are creating, preparing to create or managing intellectual property and it does not spend more than 15% of its qualifying R&D expenditure on subcontracting R&D to, or the provision of EPWs by, connected persons.

Research and development allowances

Capital costs associated with the construction of dedicated R&D facilities may benefit from specific capital allowances – refer to section 05.

Contact

For further information or assistance, please contact AECOM's Fiscal Incentives team:

help.r&d@aecom.com

Value Added Tax

Introduction

Value Added Tax (VAT) is a tax on the consumption of goods and services. The UK introduced a domestic VAT regime when it joined the European Community in 1973. The principal source of European law in relation to VAT is Council Directive 2006/112/EC, a recast of Directive 77/388/EEC, which is currently restated and consolidated in the UK through the VAT Act 1994 and various Statutory Instruments, as amended by subsequent Finance Acts.

VAT Notice 708: Buildings and construction (January 2024) provides HMRC's interpretation of the VAT law in connection with construction works, however, the UK VAT legislation should always be referred to in conjunction with the publication. Recent VAT tribunals and court decisions since the date of this publication will affect the application of the VAT law in certain instances. The Notice is available on HM Revenue & Customs website at www.gov.uk/guidance/buildings-and-construction-vat-notice-708.

The Scope of VAT

VAT is payable on:

- Supplies of goods and services made in the UK;
- By a taxable person;
- In the course or furtherance of business; and
- Which are not specifically exempted or zero-rated.

Rates of VAT

There are three rates of VAT:

- A standard rate, currently 20%;
- A reduced rate, currently 5%; and
- A zero rate of 0%.

Additionally some supplies are exempt from VAT and others are considered outside the scope of VAT.

Recovery of VAT

When a taxpayer makes taxable supplies he must account for VAT, known as output VAT, at the appropriate rate of either 20%, 5% or 0%. Any VAT due then has to be declared and submitted on a VAT submission to HM Revenue & Customs and will normally be charged to the taxpayer's customers.

As a VAT registered person, the taxpayer is entitled to reclaim from HM Revenue & Customs, commonly referred to as input VAT, the VAT incurred on their purchases and expenses directly related to its business activities in respect of standard-rated, reduced-rated and zero-rated supplies. A taxable person cannot, however, reclaim VAT that relates to any non-business activities (but see below) or, depending on the amount of exempt supplies they made, input VAT may be restricted or not recoverable.

At predetermined intervals the taxpayer will pay to HM Revenue & Customs the excess of VAT collected over the VAT they can reclaim. However, if the VAT reclaimed is more than the VAT collected, the taxpayer, who will be in a net repayment position, can reclaim the difference from HM Revenue & Customs.

Example

X Ltd constructs a block of flats. It sells long leases to buyers for a premium. X Ltd has constructed a new building designed as a dwelling and will have granted a long lease. This first sale of a long lease is a VAT zero-rated supply. This means any VAT incurred in connection with the development, which X Ltd will have properly paid (e.g. payments for consultants and certain preliminary services) will be recoverable. For reasons detailed below, the contractor employed by X Ltd will have charged VAT on his construction services at the zero rate of VAT.

Use for Business and Non-Business Activities

Where a supply relates partly to business use and partly to non-business use, then the basic rule is that it must be apportioned on a fair and reasonable basis so that only the business element is potentially recoverable. In some cases, VAT on land, buildings and certain construction services, purchased for both business and non-business use, could be recovered in full by applying what is known as 'Lennartz' accounting, to reclaim VAT relating to the non-business use and account for VAT on the non-business use over a maximum period of 10 years. Following an ECJ case restricting the scope of this approach, its application to immovable property was removed completely in January 2011 by HMRC (business brief 53/10), when UK VAT law was amended to comply with EU Directive 2009/162/EU.

Taxable Persons

A taxable person is an individual, firm, company, etc., who is required to be registered for VAT. A person who makes taxable supplies above certain turnover limits is compulsorily required to be VAT registered. The current registration limit (since 1 April 2024), known as the VAT threshold, is £90,000. If the threshold is exceeded in any 12 month rolling period, or there is an expectation that the value of the taxable supplies in a single 30 day period, or goods are received into the UK from the EU worth more than the £90,000, then registration for UK VAT is compulsory.

A person who makes taxable supplies below the limit is still entitled to be registered on a voluntary basis if they wish, for example, in order to recover input VAT incurred in relation to those taxable supplies, however output VAT will then become due on the sales and must be accounted for.

VAT Exempt Supplies

Where a supply is exempt from VAT this means that no output VAT is payable – but equally the person making the exempt supply cannot normally recover any of the input VAT on their own costs relating to that exempt supply.

Generally commercial property transactions such as leasing of land and buildings are exempt unless a landlord chooses to standard-rate its interest in the property by a applying for an option to tax. This means that VAT is added to rental income and also that VAT incurred, on say, an expensive refurbishment, is recoverable.

Supplies outside the scope of VAT

Supplies are outside the scope of VAT if they are:

- Made by someone who is not a taxable person;
- Made outside the UK; or
- Not made in the course or furtherance of business.

In course or furtherance of business

VAT must be accounted for on all taxable supplies made in the course or furtherance of business, with the corresponding recovery of VAT on expenditure incurred.

If a taxpayer also carries out non-business activities, then VAT incurred in relation to such supplies is generally not recoverable.

In VAT terms, business means any activity continuously performed which is mainly concerned with making supplies for a consideration. This includes:

- Anyone carrying on a trade, vocation or profession;

- The provision of membership benefits by clubs, associations and similar bodies in return for a subscription or other consideration; and
- Admission to premises for a charge.

It may also include the activities of other bodies including charities and non-profit making organizations.

Examples of non-business activities are:

- Providing free services or information;
- Maintaining some museums or particular historic sites;
- Publishing religious or political views.

Construction Services

In general the provision of construction services by a contractor will be VAT standard rated at 20%, however, there are a number of exceptions for construction services provided in relation to certain relevant residential properties and charitable buildings.

The supply of building materials is VAT standard rated at 20%, however, where these materials are supplied and installed as part of the construction services, the VAT liability of those materials follows that of the construction services supplied.

Zero-rated construction services

The following construction services are VAT zero-rated, including the supply of related building materials.

The construction of new dwellings

The supply of services in the course of the construction of a new building designed for use as a dwelling or number of dwellings is zero-rated, other than the services of an architect, surveyor or any other person acting as a consultant or in a supervisory capacity.

The following basic conditions must ALL be satisfied in order for the works to qualify for zero-rating:

1. A qualifying building has been, is being, or will be constructed;
2. Services are made 'in the course of the construction' of that building;
3. Where necessary, you hold a valid certificate; and
4. Your services are not specifically excluded from zero-rating.

The construction of a new building for 'relevant residential or charitable' use

The supply of services in the course of the construction of a building designed for use as a relevant residential Purpose (RRP), or relevant charitable purpose (RCP), is zero-rated, other than the services of an architect, surveyor or any other person acting as a consultant or in a supervisory capacity.

A 'relevant residential' use building means:

1. A home or other institution providing residential accommodation for children;
2. A home or other institution providing residential accommodation with personal care for persons in need of personal care by reason of old age, disablement, past or present dependence on alcohol or drugs or past or present mental disorder;
3. A hospice;
4. Residential accommodation for students or school pupils;
5. Residential accommodation for members of any of the armed forces;
6. A monastery, nunnery, or similar establishment; or
7. An institution which is the sole or main residence of at least 90% of its residents.

A 'relevant residential' purpose building does not include use as a hospital, a prison, or similar institution, or as a hotel, inn, or similar establishment.

A 'relevant charitable' purpose means use by a charity in either, or both of the following ways:

1.   Otherwise than in the course or furtherance of a business; or
2.   As a village hall, or similarly in providing social or recreational facilities for a local community.

Non-qualifying use, which is not expected to exceed 10% of the time the building is normally available for use, can be ignored. The calculation of business use can be based on time, floor area, or head count, subject to approval being acquired from HM Revenue & Customs.

The construction services can only be zero-rated if a certificate is given by the end user to the contractor carrying out the works, confirming that the building is to be used for a qualifying purpose, i.e. for a 'relevant residential or charitable' purpose. It follows that such services can only be zero-rated when supplied to the end user and, unlike supplies relating to dwellings, supplies by subcontractors cannot be zero-rated.

The construction of an annex used for a 'relevant charitable' purpose

Construction services provided in the course of construction of an annexe for use entirely, or partly for a 'relevant charitable' purpose, can be zero-rated.

In order to qualify, the annexe must:

1.   Be capable of functioning independently from the existing building;
2.   Have its own main entrance; and
3.   Be covered by a qualifying use certificate.

The conversion of a non-residential building into dwellings, or the conversion of a building from non-residential use to 'relevant residential' use, where the supply is to a 'relevant' housing association

The supply to a 'relevant' housing association in the course of conversion of a non-residential building, or non-residential part of a building, into:

1.   A new eligible dwelling designed as a dwelling, or number of dwellings; or
2.   A building, or part of a building, for use solely for a relevant residential purpose.

Any services related to the conversion, other than the services of an architect, surveyor or any person acting as a consultant or in a supervisory capacity, are zero-rated.

A 'relevant' housing association is defined as:

1.   A private registered provider of social housing;
2.   A registered social landlord within the meaning of Part I of the Housing Act 1996 (Welsh registered social landlords);
3.   A registered social landlord within the meaning of the Housing (Scotland) Act 2001 (Scottish registered social land-lords); or
4.   A registered housing association within the meaning of Part II of the Housing (Northern Ireland) Order 1992 (Northern Irish registered housing associations).

If the building is to be used for a 'relevant residential' purpose, the housing association should issue a qualifying use certifi-cate to the contractor completing the works. Subcontractors' services that are not made directly to a relevant housing asso-ciation are standard-rated.

The development of a residential caravan park

The supply in the course of the construction of any civil engineering work 'necessary for' the development of a permanent park for residential caravans, and any services related to the construction, other than the services of an architect, surveyor or any person acting as a consultant or in a supervisory capacity, are zero-rated when a new permanent park is being developed, the civil engineering works are necessary for the development of the park and the services are not specifically excluded from zero-rating. This includes access roads, paths, drainage, sewerage and the installation of mains water, power and gas supplies.

Certain building alterations for 'disabled' persons

Certain goods and services supplied to a 'disabled' person, or a charity making these items and services available to 'disabled' persons, can be zero-rated. The recipient of these goods or services needs to give the supplier an appropriate written declaration that they are entitled to benefit from zero rating.

The following services (amongst others) are zero-rated:

1.   The installation of specialist lifts and hoists and their repair and maintenance;
2.   The construction of ramps, widening doorways or passageways including any preparatory work and making good work;
3.   The provision, extension and adaptation of a bathroom, washroom or lavatory; and
4.   Emergency alarm call systems.

Sale of reconstructed buildings

A protected building is not to be regarded as substantially reconstructed unless, when the reconstruction is completed, the reconstructed building incorporates no more of the original building than the external walls, together with other external features of architectural or historical interest.

DIY builders and converters

Private individuals who decide to construct their own home are able to reclaim VAT they pay on goods they use to construct their home by use of a special refund mechanism made by way of an application to HM Revenue & Customs. This also applies to goods provided in the conversion of an existing non-residential building to form a new dwelling.

The scheme is meant to ensure that private individuals do not suffer the burden of VAT if they decide to construct their own home.

Charities may also qualify for a refund on the purchase of materials incorporated into a building used for non-business purposes where they provide their own free labour for the construction of a 'relevant charitable' use building.

Installation of energy saving materials (ESM) to domestic properties

A temporary zero rate applies for the supply and installation of certain energy saving materials including insulation, draught stripping, central heating, hot water controls and solar panels in a residential building. The temporary zero rate runs from 1 April 2022 until 31 March 2027. From 1 April 2027, ESM will revert to the reduced rate.

The scope of ESM also includes wind and water turbines, which had previously been excluded.

Reduced-rated construction services

The following construction services are subject to the reduced rate of VAT of 5%, including the supply of related building materials.

Conversion – changing the number of dwellings

In order to qualify for the 5% rate, there must be a different number of 'single household dwellings' within a building than there were before commencement of the conversion works. A 'single household dwelling' is defined as a dwelling that is designed for occupation by a single household.

These conversions can be from 'relevant residential' purpose buildings, non-residential buildings and houses in multiple occupation.

A house in multiple occupation conversion

This relates to construction services provided in the course of converting a 'single household dwelling', a number of 'single household dwellings', a non-residential building or a 'relevant residential' purpose building into a house for multiple occupation, such as bedsit accommodation.

A special residential conversion

A special residential conversion involves the conversion of a 'single household dwelling', a house in multiple occupation, or a non-residential building into a 'relevant residential' purpose building. such as student accommodation or a care home.

Renovation of derelict dwellings

The provision of renovation services in connection with a dwelling or 'relevant residential' purpose building that has been empty for two or more years prior to the date of commencement of construction works can be carried out at a reduced rate of VAT of 5%.

Installation of energy saving materials

As stated above, ESM installed in domestic properties are subject to a zero rate until 31 March 2027, after which they will revert to the reduced rate of VAT.

Buildings that are used by charities for non-business purposes, and/or as village halls, are not eligible for the reduced rate for the supply of energy saving materials.

Grant-funded installation of heating equipment or connection of a gas supply

The grant-funded supply and installation of heating appliances, connection of a mains gas supply, supply, installation, maintenance and repair of central heating systems, and supply and installation of renewable source heating systems, to qualifying persons. A qualifying person is someone aged 60 or over or is in receipt of various specified benefits.

Grant-funded installation of security goods

The grant-funded supply and installation of security goods to a qualifying person.

Housing alterations for the elderly

Certain home adaptations that support the needs of elderly people are reduced rated.

Building Contracts

Design and build contracts

If a contractor provides a design and build service relating to works to which the reduced or zero rate of VAT is applicable, then any design costs incurred by the contractor will follow the VAT liability of the principal supply of construction services.

Management contracts

A management contractor acts as a main contractor for VAT purposes and the VAT liability of his services will follow that of the construction services provided. If the management contractor only provides advice without engaging trade contractors, his services will be VAT standard rated.

Construction Management and Project Management

The project manager or construction manager is appointed by the client to plan, manage and co-ordinate a construction project. This will involve establishing competitive bids for all the elements of the work and the appointment of trade contractors. The trade contractors are engaged directly by the client for their services. The VAT liability of the trade contractors will be determined by the nature of the construction services they provide and the building being constructed.

The fees of the construction manager or project manager will be VAT standard rated. If the construction manager also provides some construction services, these works may be zero or reduced rated if the works qualify.

Liquidated and Ascertained Damages

Liquidated damages are outside of the scope of VAT as compensation. The employer should not reduce the VAT amount due on a payment under a building contract on account of a deduction of damages. In contrast, an agreed reduction in the contract price will reduce the VAT amount.

Similarly, in certain circumstances, HM Revenue & Customs may agree that a claim by a contractor under a JCT, or other form of contract, is also compensation payment and outside the scope of VAT.

Reverse Charge for Building and Construction Services

A VAT reverse charge was introduced on 1 March 2021 for certain building and construction services. This measure has been introduced to combat missing trader fraud in the construction industry, removing the opportunity for fraudsters to charge VAT and then go missing, before paying it over to the HMRC.

For certain supplies of construction services, the customer will now be liable to account to HMRC for the VAT in respect of those purchases rather than the supplier. This reverse charge will apply through the supply chain where payments are required to be reported through the Construction Industry Scheme (CIS) up to the point where the customer receiving the supply is no longer a business that makes supplies of specified services (end users).

The introduction of a reverse charge does not change the liability of the supply of the specified services: it is the way in which the VAT on those supplies is accounted for. Rather than the supplier charging and accounting for the VAT, the recipient of those supplies accounts for the VAT its return instead of paying the VAT amount to its supplier. It will be able to reclaim that VAT amount as input tax, subject to the normal rules. The supplier will need to issue a VAT invoice that indicates the supplies are subject to the reverse charge.

The types of construction services covered by the reverse charge are based on the definition of 'construction operations' used in CIS under section 74 of the Finance Act 2004.

As well as excluding supplies of specified services to end users, the reverse charge does not capture supplies of specified services where the supplier and customer are connected in a particular way, and for supplies between landlords and tenants.

Contact

For further information or assistance, please contact one of the Fiscal Incentives team:

help.vat@aecom.com

Building Materials, Health and Indoor Air Quality: Volume 2

By Tom Woolley

In *Building Materials, Health and Indoor Air Quality: Volume 2* Tom Woolley uses new research to continue to advocate for limiting the use of hazardous materials in construction and raise awareness of the links between pollutants found in building materials, poor indoor air quality and health problems. Chapters in this volume reinforce previous arguments and present new ones covering:

- Further evidence of the health impacts of hazardous emissions from materials
- Hazardous materials to be avoided and why
- Fire and smoke toxicity – the Lakanal House and Grenfell Tower legacy
- Sub-standard retrofits leading to damp and mould in previously sound houses
- A critical review of recent reports from UK Government and others on air quality and health problems including policy changes on flame retardants
- Growing evidence of cancer risks and the failure of cancer research organisations to address these issues
- A critical review of recent climate change and zero carbon policies and a discussion on whether extreme energy efficiency is a good thing

This book asks some important and, for some, uncomfortable questions, but in doing so it brings to light important areas for research and provides much needed guidance for architects, engineers, construction professionals, students and researchers on hazardous materials and how to reduce their use and design and build healthier buildings for all occupants.

May 2024: 234 pp
ISBN: 9780367646691

To Order
Tel:+44 (0) 1235 400524
Email: tandf@hachette.co.uk

For a complete listing of all our titles visit:
www.routledge.com

Taylor & Francis
Taylor & Francis Group

Office Decarb

Decarbonizing the fit-out sector: creating net zero-enabled office spaces

When it comes to cutting whole-life embodied carbon, commercial office fit-outs lag behind other construction sectors. New solutions, attitudes and guidance is emerging to revolutionize and decarbonize the sector, as Lauren Lemcke, Danielle Rowley and Dave Cheshire report.

Current situation in the sector

Fit-outs have shorter lives than the shell and core of a building. As they are typically tied to specific tenants, carbon-intensive office elements from flooring to lighting, ceilings to air conditioning systems are often discarded and replaced far earlier than their natural operational lifespans.

These rapid replacement cycles mean that the embodied carbon emissions, raw material demand and waste arising from fit-outs quickly surpass that of the shell and core over a building's life.

The embodied carbon footprint of fit-outs is also difficult to fully quantify. Incoming tenants rarely comprehensively track what they have stripped out, and what they have replaced existing materials, products and building materials with. As a result, there is a lack of knowledge and evidence of the true embodied carbon cost of the sector.

Incoming legislation to drive change

The forthcoming release of the complete Net Zero Carbon Building Standard (currently in pilot form) is expected to set ambitious embodied carbon targets for fit-outs, which will be challenging for some aspects of the industry to adapt to. Under the Standard, acceptable embodied carbon targets reduce year-on-year. This calls for substantial and consistent decarbonization over time: the standard's upfront embodied carbon limit for a whole office (including shell and core) completing in 2050 is just 60 kg of CO_2/m^2, compared to 735 kg of CO_2/m^2 in 2025.

As a result of this clear standard emerging, it is likely that developers will take a stronger view of what their tenants can put in their buildings to try and minimize strip-outs for the next tenant. While ownership of the embodied carbon involved in a new fit-out sits with the tenant, we foresee greater numbers of landlords start to decrease the carbon footprint of their building to meet investor and environmental, social and governance demands.

Designing out embodied carbon

Factoring in cost and carbon from the outset is key to an efficient, net zero-enabled fit-out solution. Often, driving down carbon can lead to cost savings, but it is vital that this is calculated and communicated early. The key questions to ask are: 'what can be reused, re-purposed, or de-selected in favour of a lower-carbon material or process?'

Decarbonization of a fit-out can be approached at three levels. Strategic decisions on retention and reuse is the **primary method**.

The second level is systems selection, opting for systems that use considerably less embodied carbon.

The final level is component selection. At all levels, cost cutting and carbon cutting typically go hand in hand, as reductions in materials mean reductions in both project cost and embodied carbon content.

Typical elements of a fit-out should be considered in turn and in interplay with other elements. Raised access floors, suspended ceilings and partitions can be retained. Existing room layouts can be kept, as can meeting room pods. Every element can be interrogated: is a metal ceiling necessary? Is the existing kitchen functional? Is a new glass meeting room wall needed, or can the existing one be rebranded to fit the client's requirements?

Designing for disassembly is an essential element of a whole-life approach, allowing components and materials to be reclaimed for reuse at end of first life. Wet trades and chemical bonded joints should be designed out and replaced with reversible, mechanical connections.

This approach also enables ease of reconfiguration of interior spaces: designing in flexibility can also help increase the longevity of a space. Dry-lay floor tiles, for example, do not require VOC-heavy grouts, sealants or adhesives and provide fast, clean installation and de-installation.

Making the most of existing MEP

Retrofitting existing office buildings often involves ripping out and replacing mechanical, electrical and plant (MEP) systems, which has a significant embodied carbon impact. To decarbonize the MEP aspect of a fit-out, a staged approach is recommended:

- The first step is to retain the existing system and optimize the system performance to reduce operational energy use before considering replacing any MEP systems. This could include adjusting controls, upgrading older equipment, and ensuring the building is operating as efficiently as possible.
- Only after operational improvements have been made should the focus shift to replacing MEP systems. The aim is to minimize the amount of demolition and replacement required, reusing or refurbishing equipment where possible.
- Where new products are used, opting for more sustainable materials such as cardboard and fabric duct work can help reduce embodied carbon. These products are emerging on to the market as viable alternatives to plastics and metals.

Changing values

The fit-out sector is asking existential questions of itself. How do office strip-outs work in an industry which must carefully conserve resources if it is to decarbonize?

Change needs to be systemic. Everyone must be along for the ride – the decarbonization of the fit-out sector will not be achieved in isolation, by a handful of clients, manufacturers, or service providers. Buy-in is required from all parties in a project and across the industry.

We are moving away from an emphasis on the shiny and the brand-new. Instead, increased value is being placed on intelligent, resourceful, imaginative uses of what we already have – valuing existing products and fit-out elements as tried and tested, instead of used. The result is good-looking, fit-for-purpose, net zero-enabled office spaces.

Regenerative Design

Seven regenerative design principles to change our cities

Imagine a building that restores and regenerates natural systems and even metabolizes like a living organism. No longer a wild daydream, the concept of regenerative design is starting to flourish in the face of the biodiversity crisis and climate change. Its application, however, requires a major shift in thinking.

In this article, sustainability expert David Cheshire presents seven revolutionary regenerative design principles to guide the way.

In our cities, isolated from nature and surrounded by steel, concrete, glass and tarmac, it is easy to forget that we are utterly dependent on natural systems for every breath of fresh air, sip of clean water and mouthful of food. The ecosystem is already in crisis and every building we construct further undermines the very mechanisms that allow us to survive.

Buildings and cities are currently a huge burden on the planet – as well as requiring land to build on, they cast a long shadow on our ecosystems, from the mining of resources, through to capturing and treating water, emptying sewage into our rivers and polluting pristine environments with 'forever chemicals'.

We have been trying to reduce this harm by designing sustainable buildings which use less energy, water and materials, while generating less waste and pollutants. And this has been invaluable in raising awareness and starting the journey towards improving performance. However, this has only made the built environment less harmful and the climate and biodiversity crises are a stark reminder that this is no longer enough.

Instead, what if we could create buildings that have a positive impact on the planet? Ones that restore and regenerate natural systems and even metabolize like living organisms.

Imagine buildings that contribute energy and food, capture water, sequester carbon, clean the air, treat pollutants and reclaim nutrients from waste. All made from locally available resources. Imagine local infrastructure that defends against flooding, is a haven for pollinators, reduces overheating and reconnects humans with nature.

It sounds like a wild daydream, but there are already examples of regenerative buildings that have become an active part of the ecosystem and there are even design approaches, such as the Living Building Challenge, that are encouraging and certifying schemes that are achieving this aim.

Regenerative design as a concept has been widely discussed, but it is much harder to tie it down to a set of design principles for buildings. Based on extensive research and interviews with designers that are already creating regenerative buildings, my new book – Regenerative by Design – proposes seven regenerative design principles and maps them onto the hierarchy of scale from the site through to the ecosystem, all bounded by the planetary limits (as shown in Figure 1).

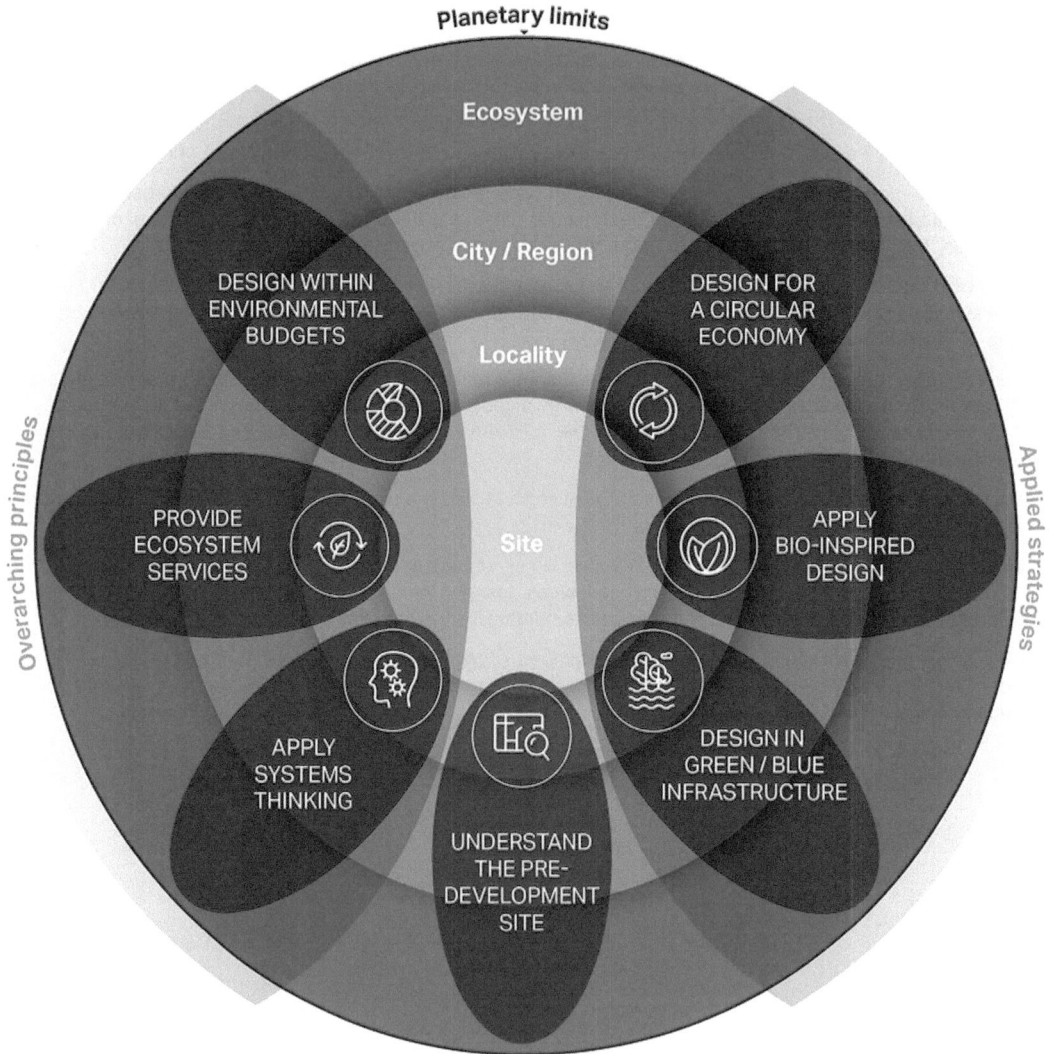

Figure 1: Synthesis of regenerative design principles (Regenerative by Design, David Cheshire, RIBA Publishing, 2024)

The design principles (the seven dark ellipses) straddle across all four of the concentric circles to demonstrate that the building and the site is intimately linked to the ecosystem, and that the wider context of each building has to be considered. The left-hand side of the diagram show the three overarching principles, and the right-hand sides shows the three applied strategies for a project.

Regenerative design requires three major shifts in thinking (shown as the overarching principles in the diagram).

Firstly, recognizing that there are absolute planetary limits in terms of the resources that can be used and the pollutants and waste that can be released. For design, this translates into creating buildings that live within their means, using only the rainwater that falls from the sky and the energy from the sun. It means using locally available materials and eliminating waste.

Secondly, it is the realization that the ecosystem provides free services that we are entirely dependent on for our survival, such as clear air, potable water, edible food, treatment of organic waste and even climate regulation.

These 'ecosystem services' are not recognized by our economic systems which makes it easy for them to be taken for granted. However, if we are to start to restore our ecosystem and reverse the climate crisis, we will have to begin to mimic and recreate these services in the built environment.

Thirdly, we need to think beyond the site boundary and apply systems thinking. This more holistic view looks beyond the scope and boundaries of a project and considers the impacts on natural cycles, the global supply and disposal of resources, and the impact on wider social systems.

For example, we grow salad, berries and beans in areas of water-scarcity, transport them around the world and then let them rot in our fridges when we could be growing fresh food locally, using building-integrated greenhouses that use the warm, wet, CO_2-laden air extracted from our buildings. Regenerative design requires three major shifts in thinking (shown as the overarching principles in the diagram).

Figure 2: What a regenerative approach to design could look like in central London. In this AI-generated image of the area around King's Cross, the River Fleet has been reintroduced, adding to biodiversity and supporting the cooling of the overheated city. Other bio-inspired features include algae sculptures that store carbon, urban forests, building-integrated rooftop greenhouses and floating self-sufficient communities, inspired by the Schoon Schip community in Amsterdam.

Once we have made this shift in thinking, then we can use three design strategies to create regenerative buildings: circular economy approaches that take advantage of locally available and biogenic materials; applying blue and green infrastructure to bring nature back into our towns and cities; using bio-inspired design that mimics natural systems and even uses nature directly. For example, microbial fuel cells treat wastewater and generate energy as a useful bioproduct.

This holistic approach that recognizes nature as a partner and something to be supported and even mimicked, will provide multiple benefits including increasing resilience to climate change, mitigating urban heating, less dependence on utilities infrastructure, and the wellbeing benefits of being surrounded by nature.

The technologies and design techniques also solve global environmental problems. Microbial fuel cells treat blackwater, which avoid polluting waterways, and generate renewable energy. Growing urban food reduces the huge impact of polytunnel farms, creates community spaces and provides fresh, healthy food for local people.

Regenerative design requires a new way of thinking. It needs new specialists – people who understand how ecosystems function and how we can enhance them. And it needs changes to the way that we calculate cost and value. It is a huge challenge, but one that I believe we have to rise to. The alternative is the decline and destruction of our life support system and there's no way we can build our way out of that.

UK's Building Safety Act

What does the UK's Building Safety Act mean for the industry?

The UK's Building Safety Act was given Royal Assent in April 2022, with secondary legislation following throughout 2022 and 2023.

Since then, the UK construction industry has been grappling with its implications – with early movers attempting real-world applications of the new rules.

The UK's Building Safety Act 2022 represents a significant shift in the design, construction, and management of buildings.

The Act was in part a response to the 2017 Grenfell Tower tragedy, in which 72 people died in the fire in the high-rise tower block in West London. The Act seeks to ensure that the safety of future building occupants is placed front and centre of the building control process.

The Act's requirements are wide-ranging and impose new obligations on all construction projects and duty holders. In this article, we will focus primarily on cost and programme issues for 'higher-risk buildings'.

Higher-risk buildings are broadly defined as buildings containing at least two units of accommodation in England, or one unit of accommodation in Wales, that are either 18 metres or higher or contain seven or more storeys.

An overview of the new regime for higher-risk buildings

One of the key changes under the Act was the creation of a Building Safety Regulator (BSR), part of the Health & Safety Executive. The BSR's key functions include overseeing safety and standards in all buildings and acting as the building control authority for higher-risk buildings.

To obtain building control approval, higher-risk buildings must now pass through three BSR 'gateways'.

Gateway 1 is at the planning approval stage; the second, design-focused ('Gateway 2') prior to starting on site; and Gateway 3 prior to occupation, to ensure that compliance with the building regulations is being achieved.

The gateways act as hard stops at key stages in the lifecycle of a construction project, requiring specific information to be submitted and approved by the regulator before the project can move forward or obtain its completion certificate.

This has particularly significant implications at the Gateway 2 design stage – with designs needing to be approved prior to work starting on site.

Change control in higher-risk buildings

Change control is therefore a key issue to consider from a cost and programme perspective when implementing the new rules.

Following Gateway 2 approval, most changes which alter the approved design must be notified to the regulator.

Changes requiring BSR involvement are split into two groups: major changes, and notifiable changes. For a major change, the BSR will need to approve the change prior to it being implemented. The period for reviewing and approving the change is six weeks.

Given these additional interactions with the BSR, it is highly likely that in the short term, clients will want to minimize any changes that occur after the submission of Gateway 2. This is likely to factor into the procurement route decisions taken at the outset.

How will the Building Safety Act affect construction programmes?

One of the Building Safety Act's biggest impacts is likely to be that typical project programmes will extend.

In addition to the Gateway 2 approval period – and the loss of overlap between early on-site works packages and the end of the design stage – there are further programme considerations to consider.

For example, the level of control the employer has over contractor start on site may have an impact on programme. The employer may not be able to give the instruction to commence until they have obtained gateway approval (depending on the length of any demolition or enabling works at the beginning of the programme).

This could potentially add several weeks to the programme – and possibly longer-term delays, depending on the contractor's ability and availability to mobilize.

Completion and handover may also be impacted. In larger schemes, groups of apartments or floors are typically completed and handed over early, prior to completion of the overall building.

The available guidance suggests that such handover strategies will no longer be suitable. Accordingly, clients will need to factor in higher hold costs at the end of the project, as earlier completed units will not be able to be handed over until the entirety of the building is signed off.

Addressing the long-term benefits

While there are undoubted cost implications for developing projects under the new regime, the less-discussed upside is increased value.

Assets which can demonstrate full compliance with building safety standards will be more attractive not only to eventual occupants, but also to investors in search of high-quality assets.

We may see a shift in the building ownership market over time because of the duties now in place under the Act. It is possible 'light touch' ownership may disappear, as the implications of the various duty holder requirements become more apparent.

Similarly, the Act's 'golden thread' requirement should also improve the quality of handover and as-built documentation. The payback – both from a safety and economic perspective – should be significant in the long term.

Typical Engineering Details

In addition to the Engineering Features, Typical Engineering Details are included. These are indicative schematics to assist in the compilation of costing exercises. The user should note that these are only examples and cannot be construed to reflect the design for each and every situation. They are merely provided to assist the user with gaining an understanding of the Engineering concepts and elements making up such.

Electrical

- Urban Network Mainly Underground
- Urban Network Mainly Underground with Reinforcement
- Urban Network Mainly Underground with Substation Reinforcement
- Typical Simple 11 kV Network Connection for LV Intakes up to 1000 kVA
- Typical 11 kV Network Connections for MV Intakes 1000 kVA to 6000 kVA
- Static UPS System – Simplified Single Line Schematic for a Single Module
- Typical Data Transmission (Structured Cabling)
- Typical Networked Lighting Control System
- Typical Standby Power System, Single Line Schematic
- Typical Fire Detection and Alarm Schematic
- Typical Block Diagram – Access Control System (ACS)
- Typical Block Diagram – Intruder Detection System (IDS)
- Typical Block Diagram – Digital CCTV

Mechanical

- Fan Coil Unit System
- Displacement Ventilation System
- Chilled Ceiling System (Passive System)
- Chilled Beam System (Passive or Active System)
- Variable Air Volume (VAV)
- Variable Refrigerant Volume System (VRV)
- Reverse Cycle Heat Pump

Urban Network Mainly Underground

Details: Connection to small housing development 10 houses, 60 m of LV cable from local 11 kV / 400 V substation route in footpath and verge, 10 m of service cable to each plot in verge

Supply Capacity: 200 kVA

Connection Voltage: LV

3 phase supply

Breakdown of Detailed Cost Information

	Labour	Plant	Materials	Overheads	Total
Mains Cable	£371	£124	£1,485	£394	£2,374
Service Cable	£371	£124	£619	£225	£1,339
Jointing	£1,485	£495	£1,856	£765	£4.601
Termination	Incl.	Incl.	Incl.	Incl.	Incl.
Trench/Reinstate	£2,228	£743	£1,856	£968	£5,794
Substation	–	–	–	–	–
HV Trench & Joint	–	–	–	–	–
HV Cable Install	–	–	–	–	–
Total Calculated Price	£4,455	£1,485	£5,816	£2,351	£14,108
Total Non-Contestable Elements and Associated Charges					£3,713
Grand Total Calculated Price excl. VAT					£17,820

Urban Network Mainly Underground with Reinforcement

Details: Connection to small housing development 10 houses, 60 m of LV cable from local 11 kV / 400 V substation route in footpath and verge, 10 m of service cable to each plot in unmade ground. Scheme includes reinforcement of LV distribution board at substation.

Supply Capacity: 200 kVA

Connection Voltage: LV

Single phase supply

Breakdown of Detailed Cost Information

	Labour	Plant	Materials	Overheads	Total
Mains Cable	£371	£124	£1,485	£394	£2,374
Service Cable	£371	£124	£619	£225	£1,339
Jointing	£1,485	£495	£1,865	£765	£4,601
Termination	Incl.	Incl.	Incl.	Incl.	Incl.
Trench/Reinstate	£2,228	£743	£1,856	£968	£5,794
Substation	–	–	–	–	–
HV Trench & Joint	–	–	–	–	–
HV Cable Install	–	–	–	–	–
Total Calculated Price	£4,455	£1,485	£5,816	£2,351	£14,108
Total Non-Contestable Elements and Associated Charges					£16,088
Grand Total Calculated Price excl. VAT					£30,195

Urban Network Mainly Underground with Substation Reinforcement

Details: Connection to small housing development 10 houses, 60 m of LV cable from local 11 kV / 400 V substation route in footpath and verge, 10 m of service cable to each plot in verge. Scheme includes reinforcement of LV distribution board and new substation and 20 m of HV cable.

Supply Capacity: 200 kVA

Connection Voltage: LV

Single phase

Breakdown of Detailed Cost Information

	Labour	Plant	Materials	Overheads	Total
Mains Cable	£371	£124	£1,856	£473	£2,824
Service Cable	£371	£124	£619	£225	£1,339
Jointing	£1,485	£495	£1,856	£765	£4,601
Termination	Incl.	Incl.	Incl.	Incl.	Incl.
Trench/Reinstate	£2,228	£743	£1,856	£968	£5,794
Substation	£12,375	£3,713	£99,000	£23,018	£138,105
HV Trench & Joint	£2,228	£743	£1,238	£844	£5,051
HV Cable Install	£371	£124	£1,395	£383	£2,273
Total Calculated Price	£19,429	£6,064	£107,820	£26,674	£159,986
Total Non-Contestable Elements and Associated Charges					£18,563
Grand Total Calculated Price excl. VAT					£178,549

Client LV Intake Switchboard (400 V)

DNO metering air circuit breaker (ACB)

DNO Transformer (typically iol filled, 500 to 1000 kVA, 11 kV/400 V)

DNO Ring main (RMU)

DNO 11 kV network

Client Demise

DNO* Demise

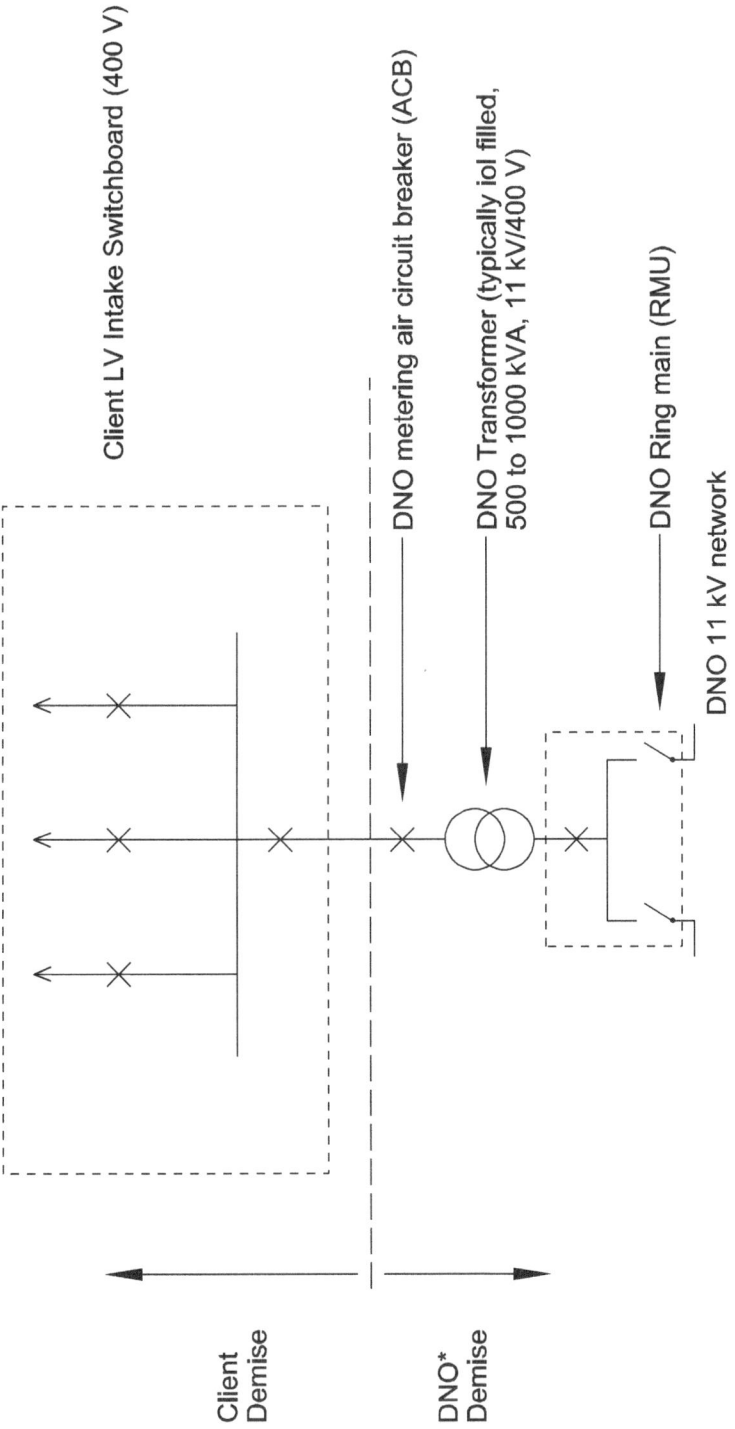

Note: *DNO – Distribution Network Operator

Typical Simple 11 kV Network Connection for LV Intakes up to 1000 kVA

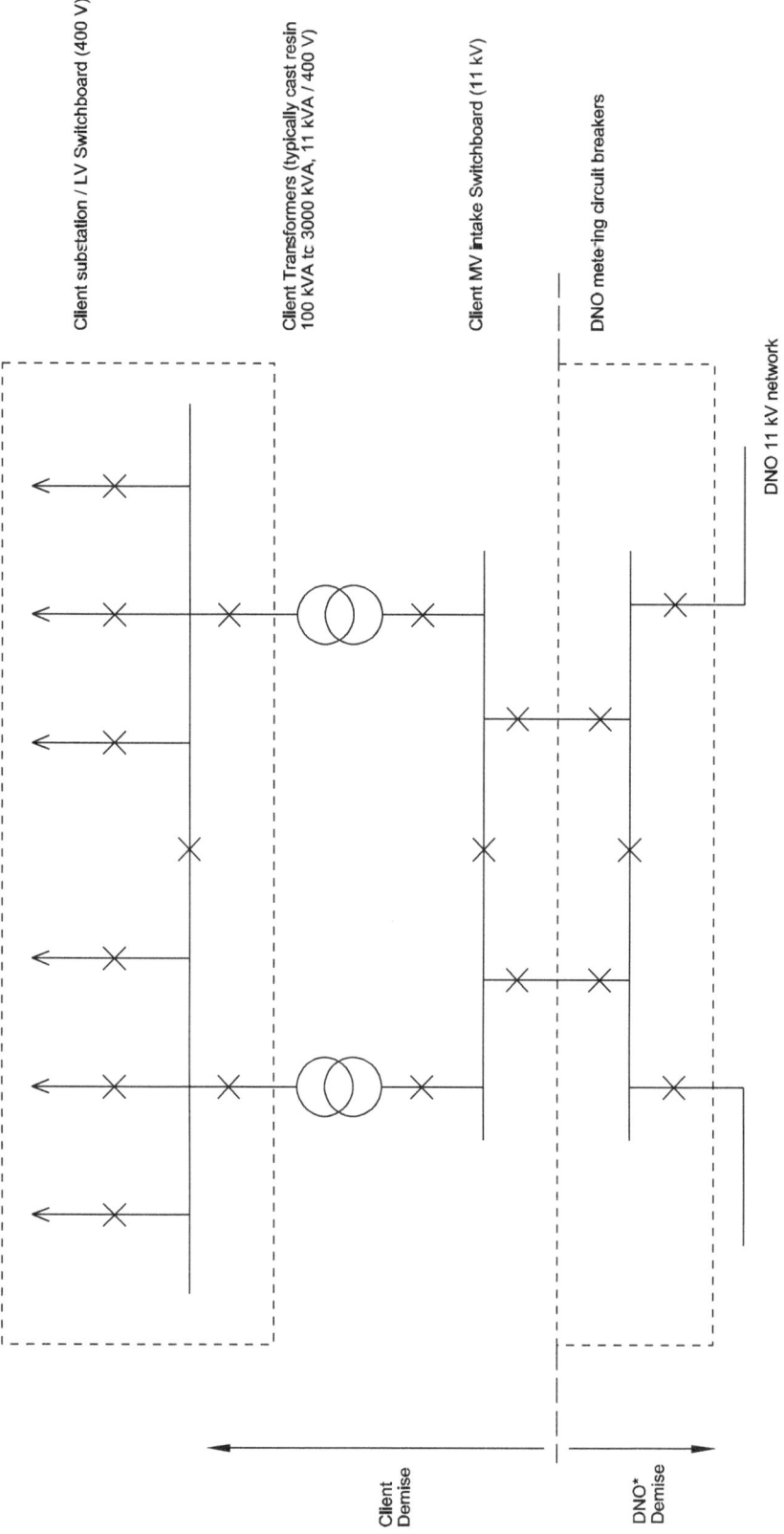

Client substation / LV Switchboard (400 V)

Client Transformers (typically cast resin 100 kVA to 3000 kVA, 11 kVA / 400 V)

Client MV Intake Switchboard (11 kV)

DNO metering circuit breakers

DNO 11 kV network

Client Demise

DNO* Demise

Note: *DNO - Distribution Network Operator

Typical 11 kV Network Connection for MV Intakes 1000 kVA up to 6000 kVA

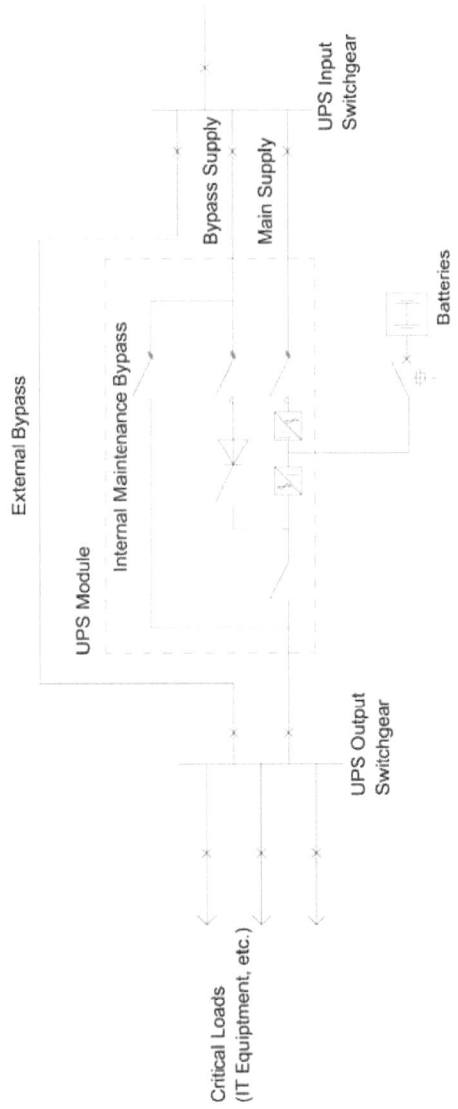

External Bypass

UPS Module

Internal Maintenance Bypass

Bypass Supply

Main Supply

UPS Input
Switchgear

Batteries

Critical Loads
(IT Equiptment, etc.)

UPS Output
Switchgear

Static UPS System – Simplified Schematic For Single Module

Sub-Equipment Room

Other Floors

Horizontal Flood Wiring (Cat 6)

First Floor

Raised Modular Floor

Vertical Backbone (Fibre)

Ground Floor

Floor Boxes or grommets

Main Equipment Room

IT Cabinet c/w Frame & Patch Panels

Basement

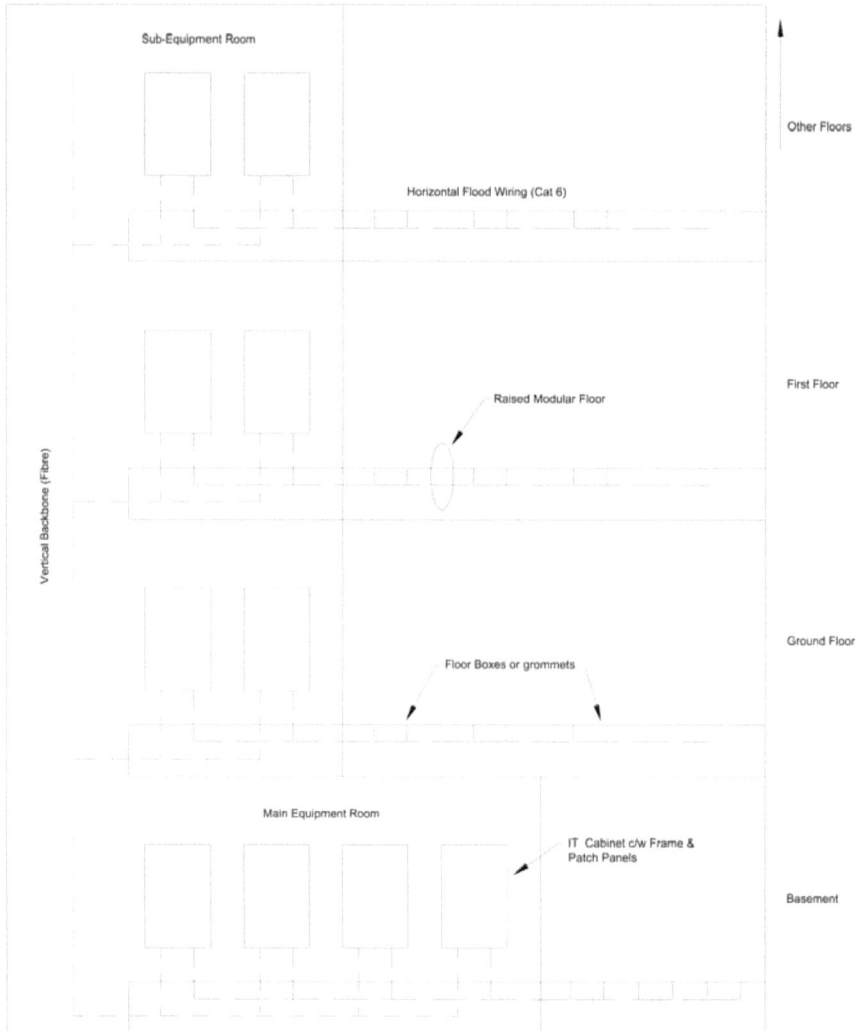

Typical Data Transmission (Structured Cabling)

58

Luminaires (or Luminaire Groups) Connected to Addressable Ports

Switch Plate

Screen Keyboard & Mouse

UPS

Optional Lighting Control Head End (Typically Located Within BMS Room)

ATS Touch Screen

Main Lighting Controls Processor

ATS Processor

BMS Interface

To Further Local Area Controllers.

Bus

Local Area Controller

Tenant DB

L.N.E.

Bus

230 V

MDB

Pre-Engineered Wiring Main Distribution Box (MDB)

Multi-Sensor

P A D

LCM

1 2 3 4 5 6 7 8 9

L.N.E.

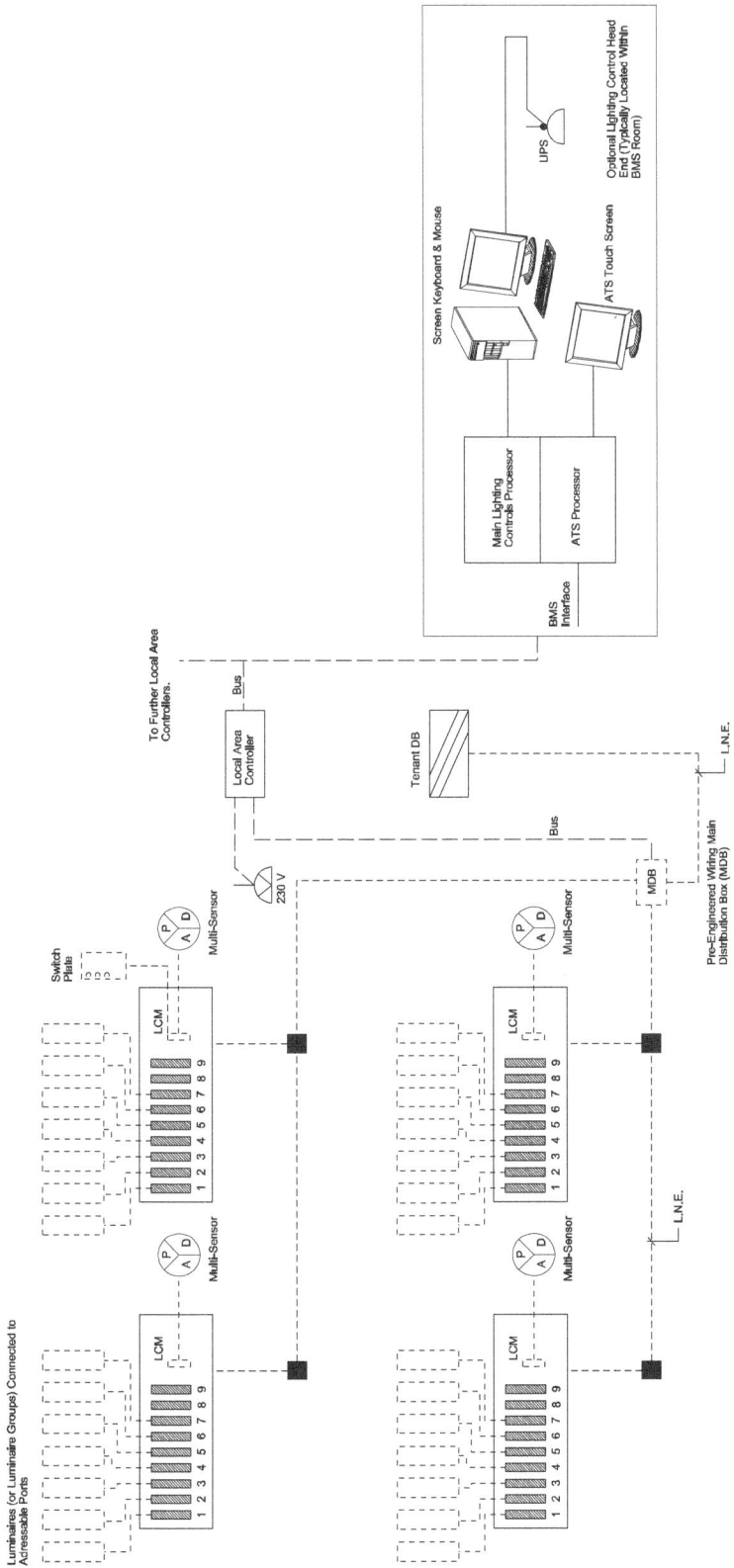

Typical Networked Lighting Control System

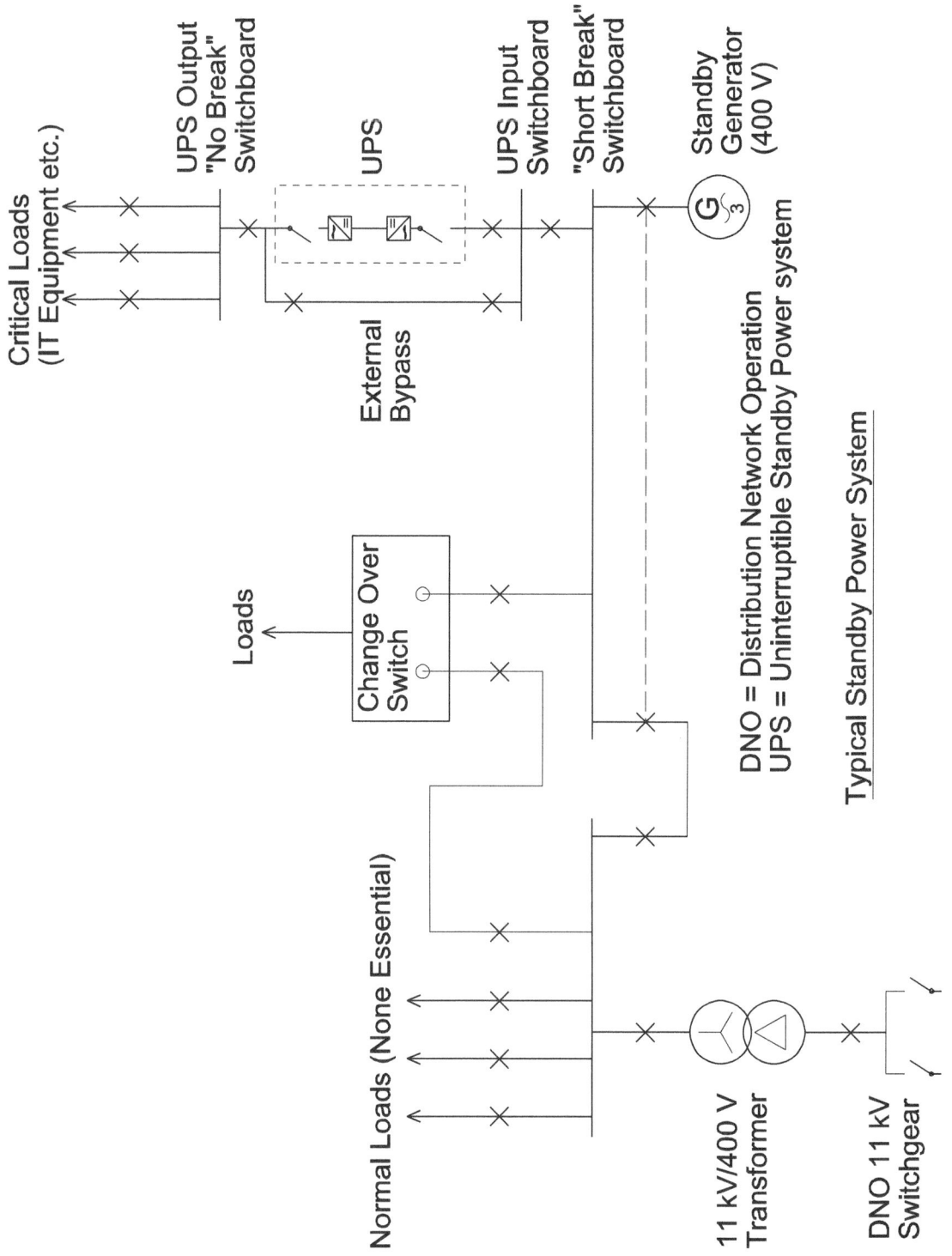

Critical Loads (IT Equipment etc.)

UPS Output "No Break" Switchboard

UPS

External Bypass

UPS Input Switchboard

"Short Break" Switchboard

Standby Generator (400 V)

Loads

Change Over Switch

Normal Loads (None Essential)

11 kV/400 V Transformer

DNO 11 kV Switchgear

DNO = Distribution Network Operation
UPS = Uninterruptible Standby Power system

Typical Standby Power System

60

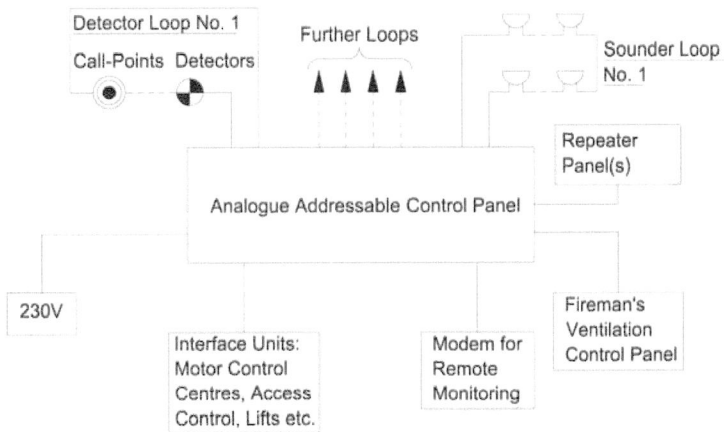

Detector Loop No. 1

Call-Points Detectors

Further Loops

Sounder Loop No. 1

Repeater Panel(s)

Analogue Addressable Control Panel

230V

Interface Units: Motor Control Centres, Access Control, Lifts etc.

Modem for Remote Monitoring

Fireman's Ventilation Control Panel

Typical Fire Detection and Alarm Schematic

Ethernet

230 V

RJ45

Access Controller

Fire Shutdown loop
input/output unit (IOU)

Reader Field Network

230 V

Reader
Module

230 V

Reader
Module

230 V

Reader
Module

Reader
Module

230 V

Controlled Door,
Magnetic Lock, BGU
and Proximity Reader

Controlled Barrier and Vehicle
Induction loop, Proximity
Reader, Intercom

RJ45

Ethernet
Network

RJ45

230 V

Intercom
Unit

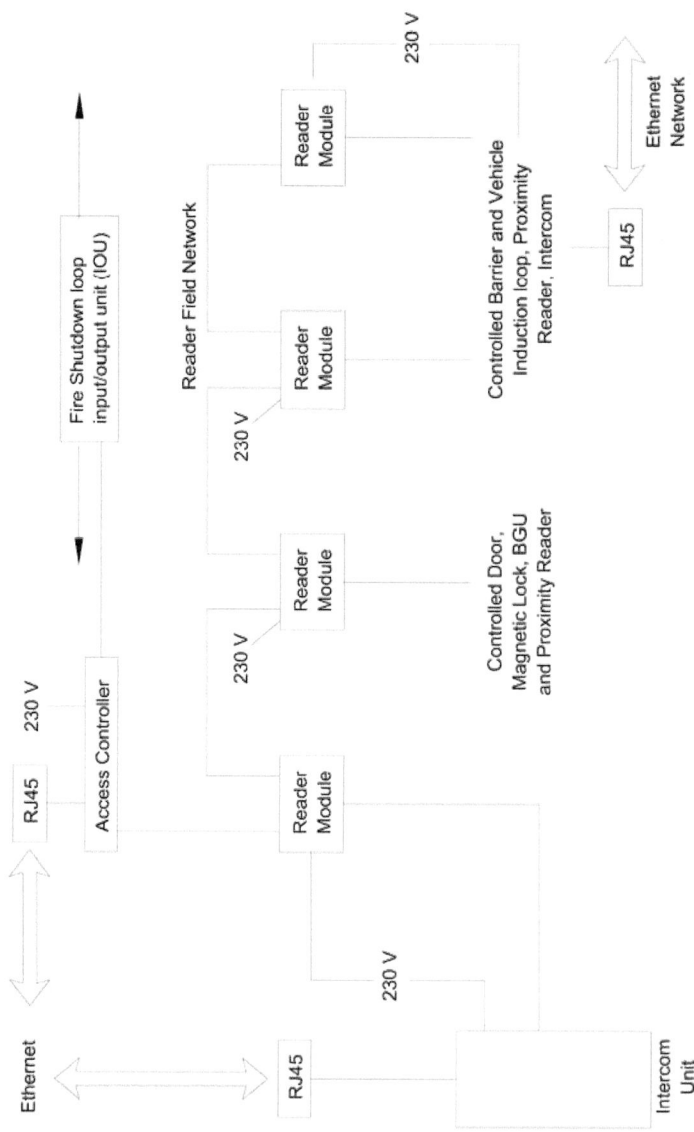

Typical Block Diagram – Access Control System

Passive Infared
(PIR) and
volumetric
detectors

Personal
attack
buttons

Contact loops
for doors and
windows

Multi-zone Control Panel
including keypad and
battery back-up

External Sounder

Internal Loudspeaker

230 V

Remote Keypad

Vibration Detectors

Final exit set button

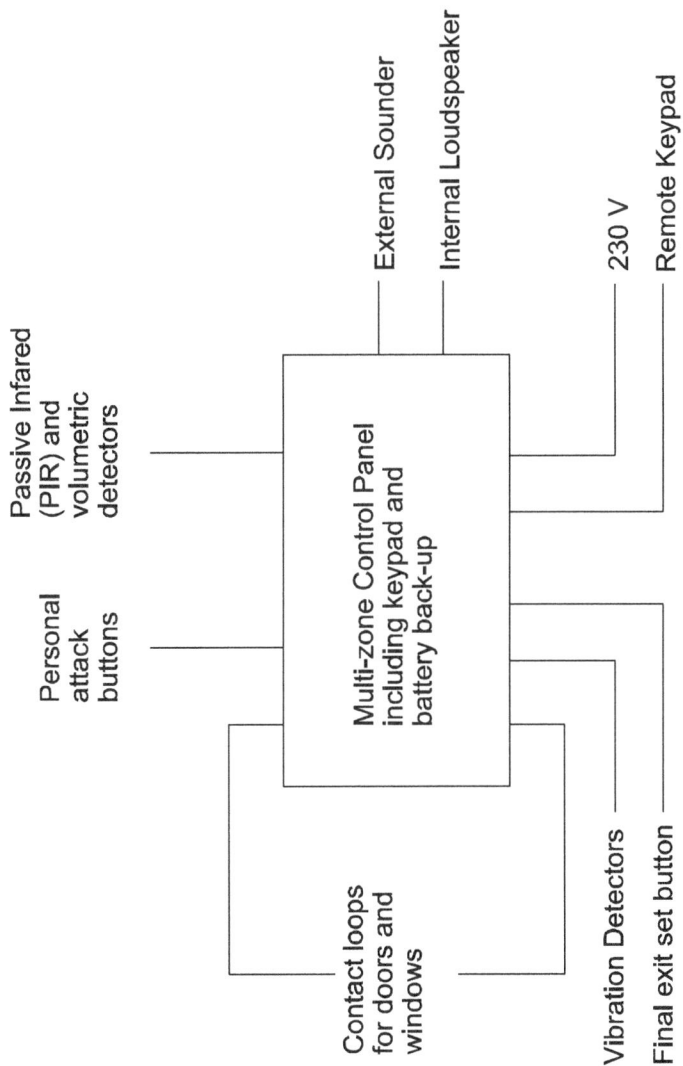

Typical Block Diagram – Intruder Detection System (IDS)

External Cameras — Internal Cameras — Sounders/speakers — Detectors (PIRs)

Automatic Number Plate Recognition

Multi Channel Digital Video Recorder

Microphone
Monitor
Mouse
Keyboard

Network Switch

WAN

Remote workstation

PC 1
Server
Keyboard/mouse switch
UPS

Microphone

keyboard

Keyboard/mouse Extender

Mouse

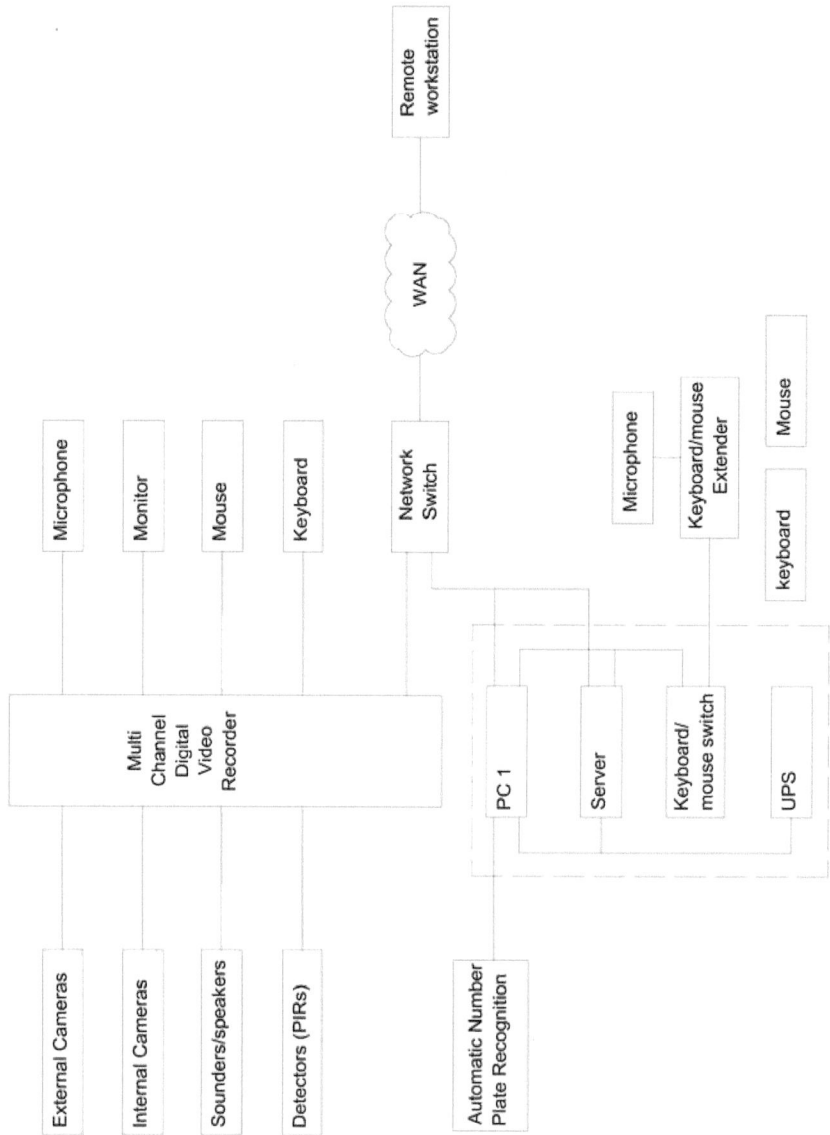

Typical Block Diagram – Digital CCTV

64

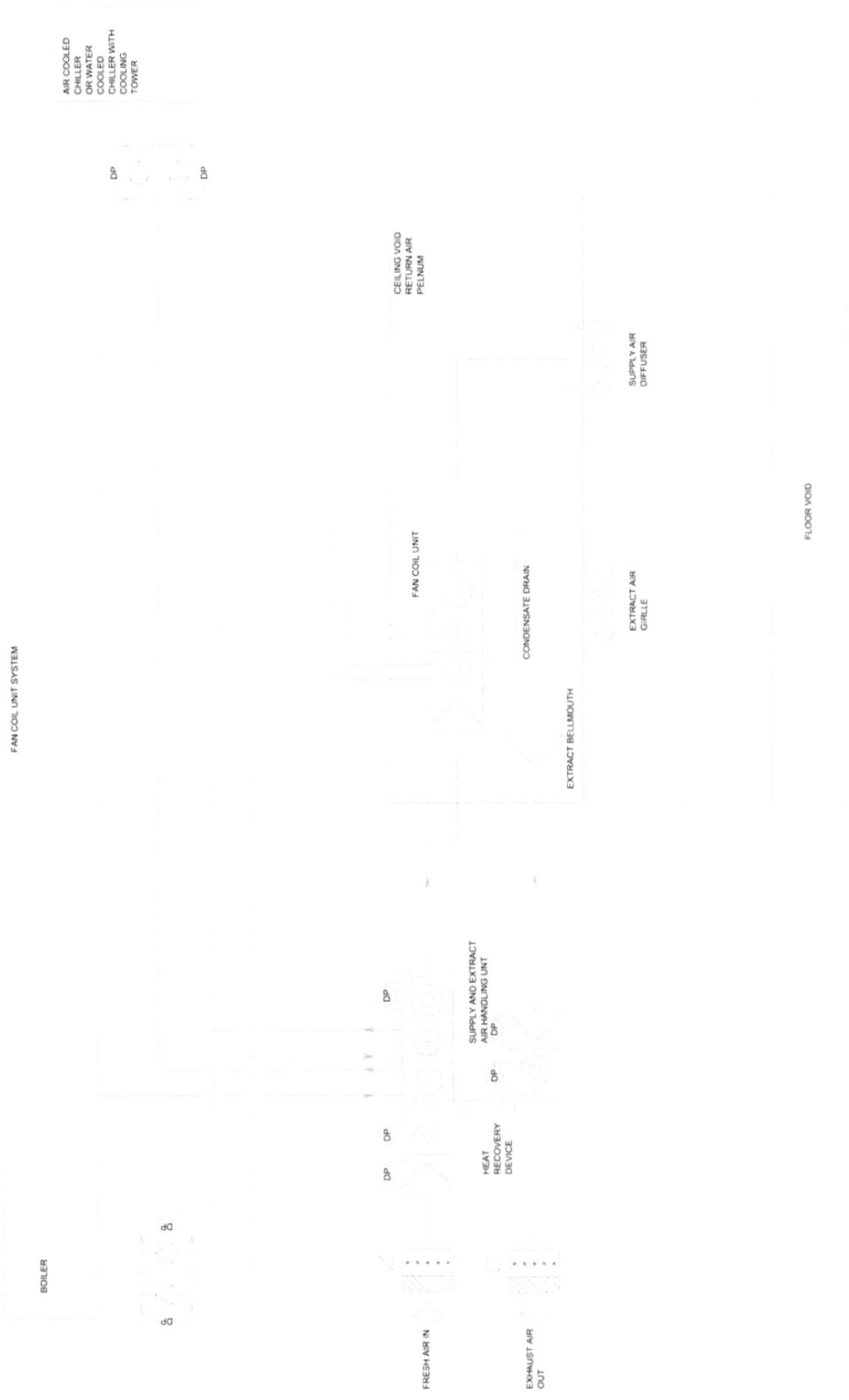

FAN COIL UNIT SYSTEM

BOILER

FRESH AIR IN

EXHAUST AIR OUT

HEAT RECOVERY DEVICE

DP

DP

DP

DP

SUPPLY AND EXTRACT AIR HANDLING UNIT

DP

DP

DP

DP

AIR COOLED CHILLER OR WATER COOLED CHILLER WITH COOLING TOWER

FAN COIL UNIT

CEILING VOID RETURN AIR PELNUM

SUPPLY AIR DIFFUSER

CONDENSATE DRAIN

EXTRACT BELLMOUTH

EXTRACT AIR GIRLLE

FLOOR VOID

DISPLACEMENT VENTILATION SYSTEM

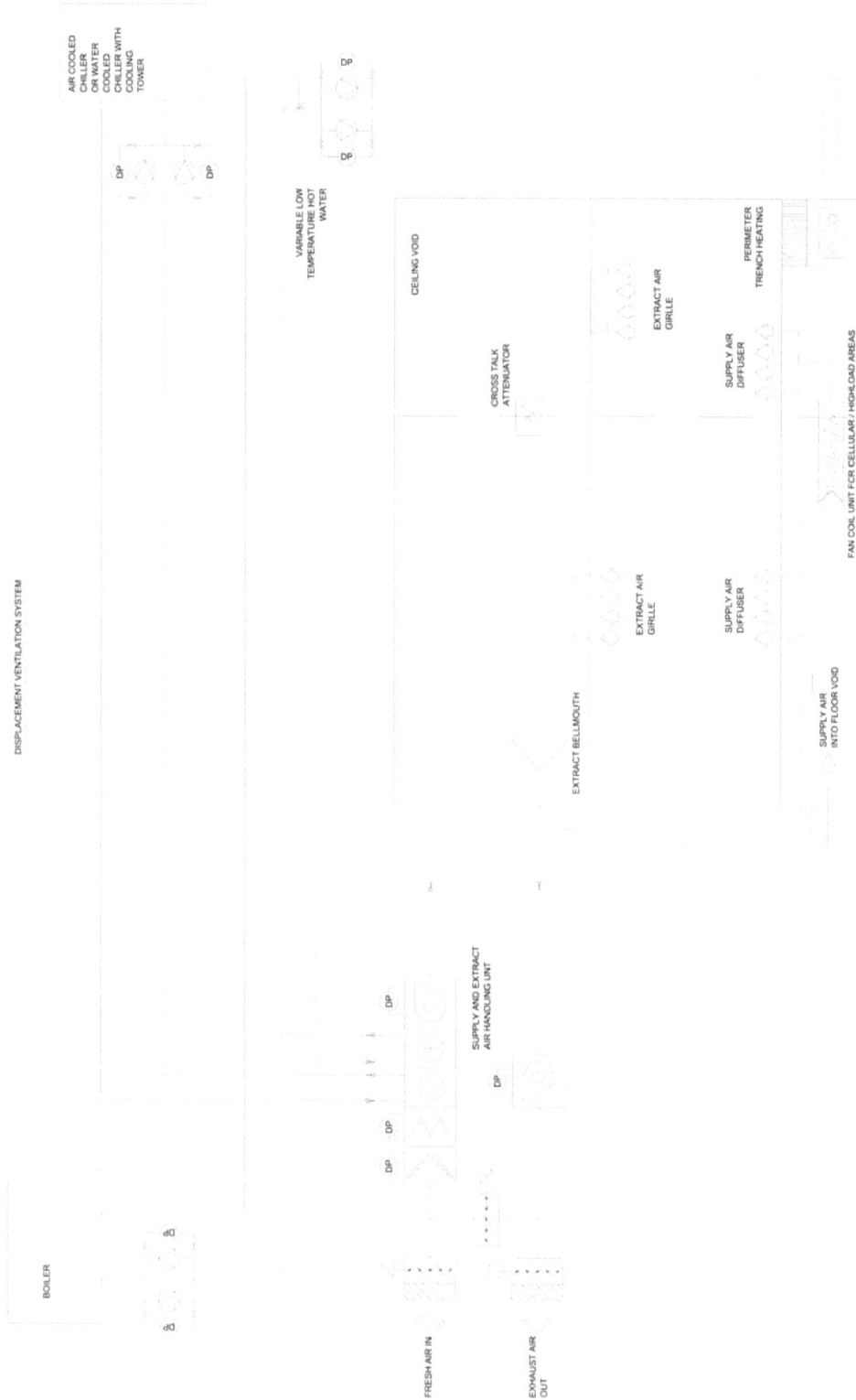

AIR COOLED
CHILLER
OR WATER
COOLED
CHILLER WITH
COOLING
TOWER

BOILER

VARIABLE LOW
TEMPERATURE HOT
WATER

CEILING VOID

CROSS TALK
ATTENUATOR

EXTRACT AIR
GIRLLE

EXTRACT AIR
GIRLLE

SUPPLY AIR
DIFFUSER

SUPPLY AIR
DIFFUSER

PERIMETER
TRENCH HEATING

EXTRACT BELLMOUTH

SUPPLY AND EXTRACT
AIR HANDLING UNIT

FRESH AIR IN

EXHAUST AIR
OUT

SUPPLY AIR
INTO FLOOR VOID

FAN COIL UNIT FOR CELLULAR / HIGH LOAD AREAS

DP

DP

DP

DP

DP

DP

DP

DP

DP

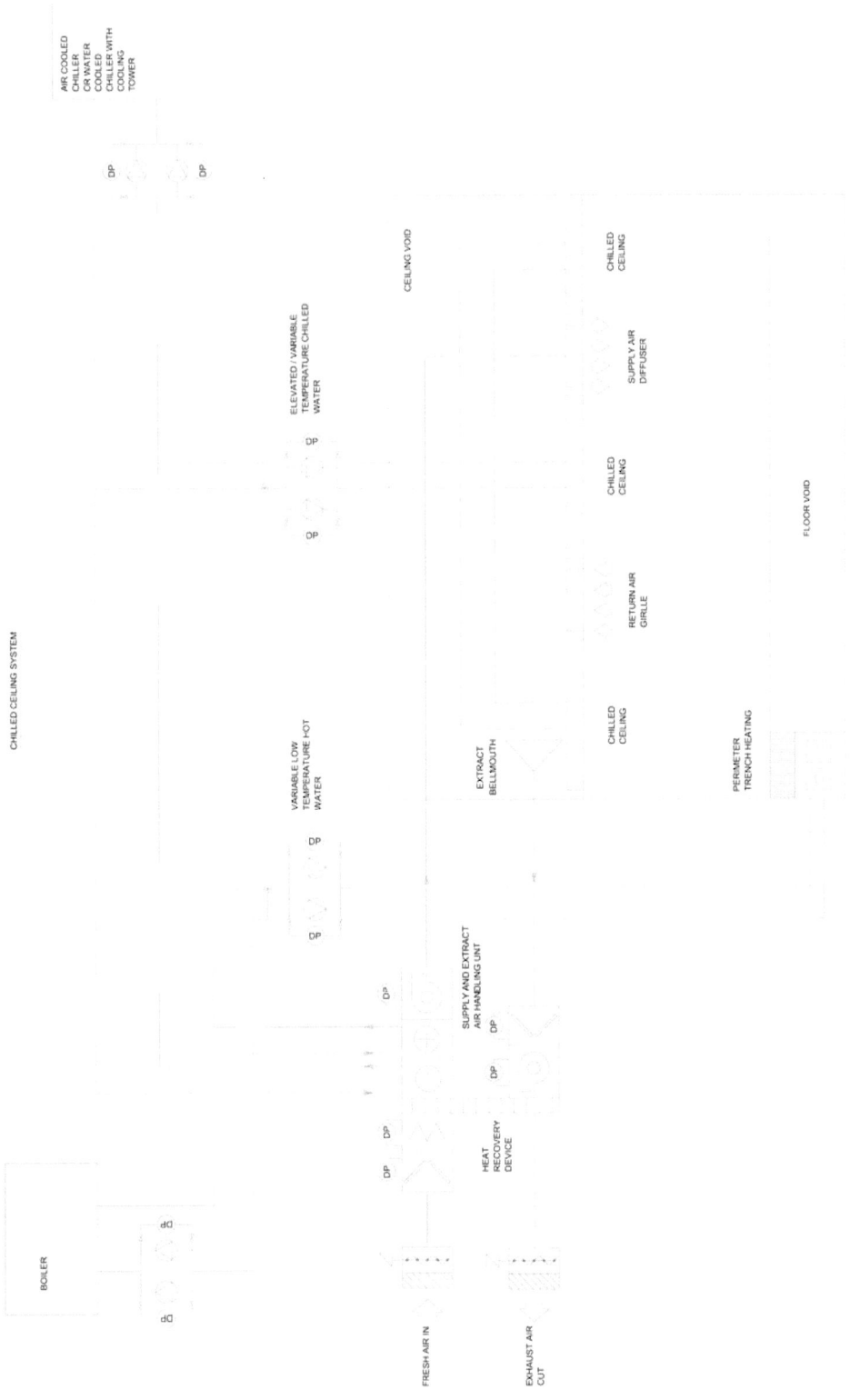

CHILLED CEILING SYSTEM

AIR COOLED CHILLER OR WATER COOLED CHILLER WITH COOLING TOWER

CEILING VOID

CHILLED CEILING

ELEVATED / VARIABLE TEMPERATURE CHILLED WATER

VARIABLE LOW TEMPERATURE HOT WATER

EXTRACT BELLMOUTH

CHILLED CEILING

RETURN AIR GRILLE

SUPPLY AIR DIFFUSER

CHILLED CEILING

PERIMETER TRENCH HEATING

FLOOR VOID

SUPPLY AND EXTRACT AIR HANDLING UNIT

HEAT RECOVERY DEVICE

BOILER

FRESH AIR IN

EXHAUST AIR OUT

DP

PASSIVE OR ACTIVE CHILLED BEAM SYSTEM

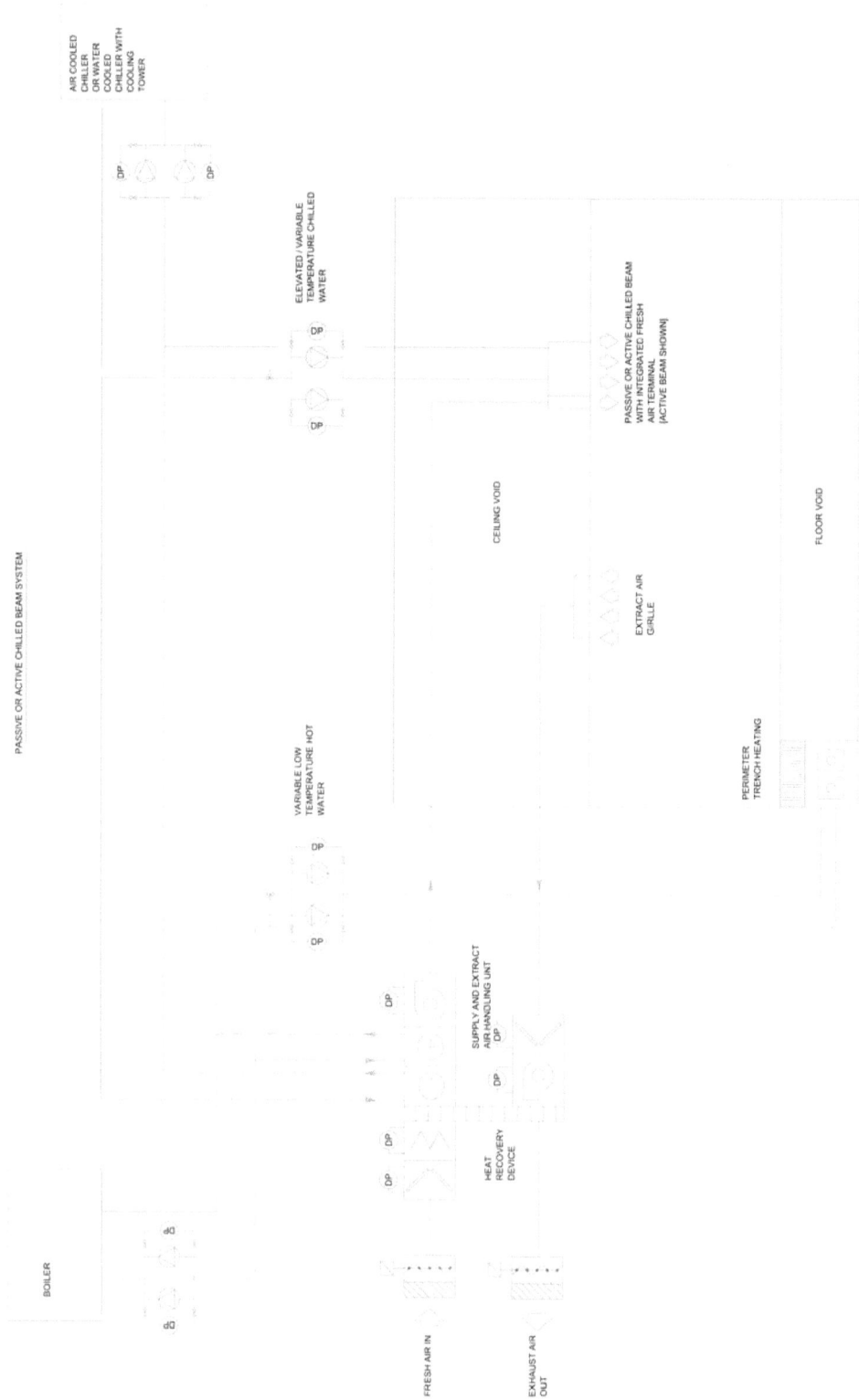

AIR COOLED
CHILLER
OR WATER
COOLED
CHILLER WITH
COOLING
TOWER

ELEVATED / VARIABLE
TEMPERATURE CHILLED
WATER

PASSIVE OR ACTIVE CHILLED BEAM
WITH INTEGRATED FRESH
AIR TERMINAL
(ACTIVE BEAM SHOWN)

CEILING VOID

EXTRACT AIR
GRILLE

VARIABLE LOW
TEMPERATURE HOT
WATER

FLOOR VOID

PERIMETER
TRENCH HEATING

BOILER

SUPPLY AND EXTRACT
AIR HANDLING UNIT

HEAT
RECOVERY
DEVICE

DP

FRESH AIR IN

EXHAUST AIR
OUT

VARIABLE AIR VOLUME SYSTEM

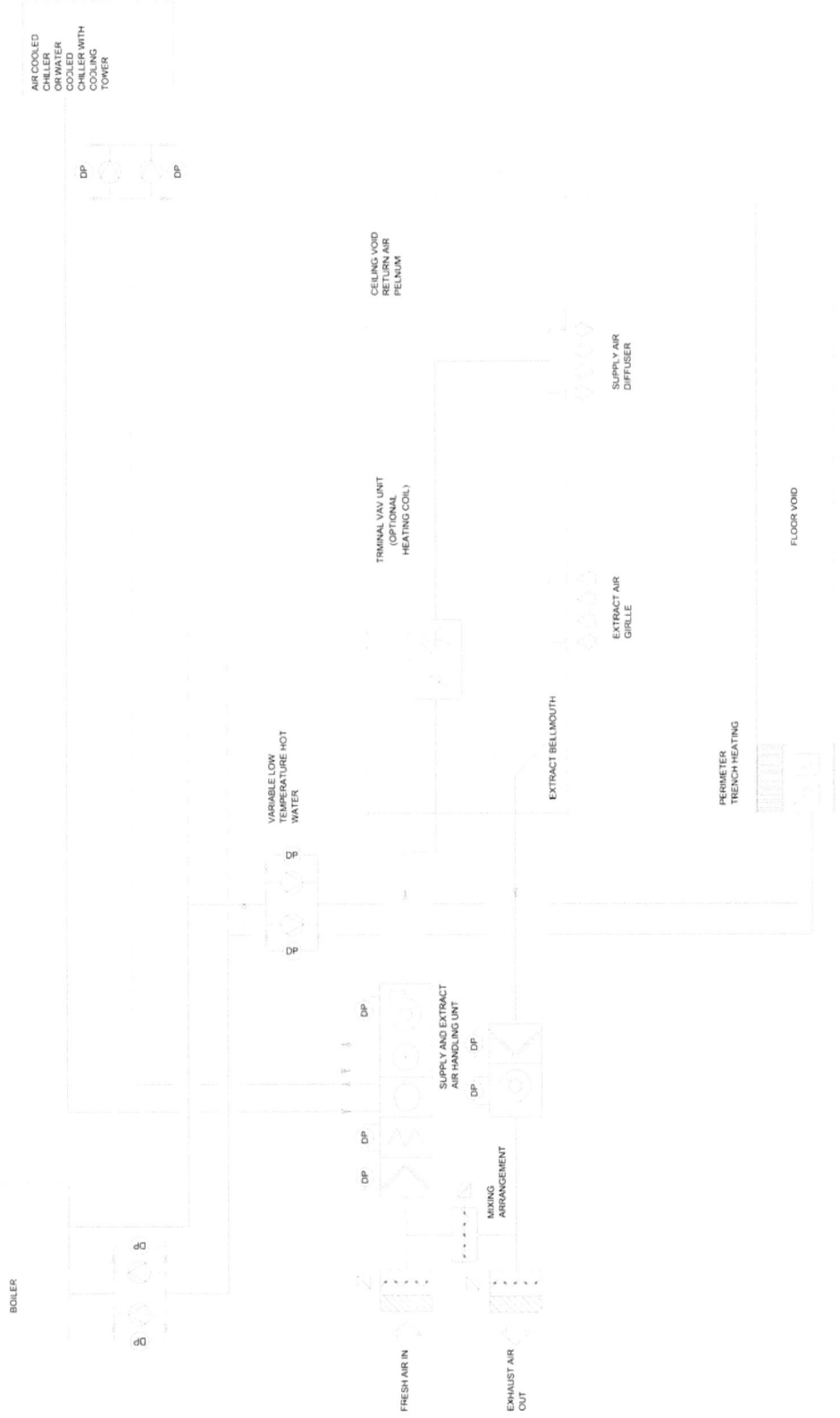

BOILER

AIR COOLED
CHILLER
OR WATER
COOLED
CHILLER WITH
COOLING
TOWER

CEILING VOID
RETURN AIR
PELNUM

TRMINAL VAV UNIT
(OPTIONAL
HEATING COIL)

SUPPLY AIR
DIFFUSER

FLOOR VOID

VARIABLE LOW
TEMPERATURE HOT
WATER

EXTRACT BELLMOUTH

EXTRACT AIR
GIRLLE

PERIMETER
TRENCH HEATING

SUPPLY AND EXTRACT
AIR HANDLING UNT

MIXING
ARRANGEMENT

FRESH AIR IN

EXHAUST AIR
OUT

DP

DP

DP

DP

DP

DP

DP

DP

DP

DP

DP

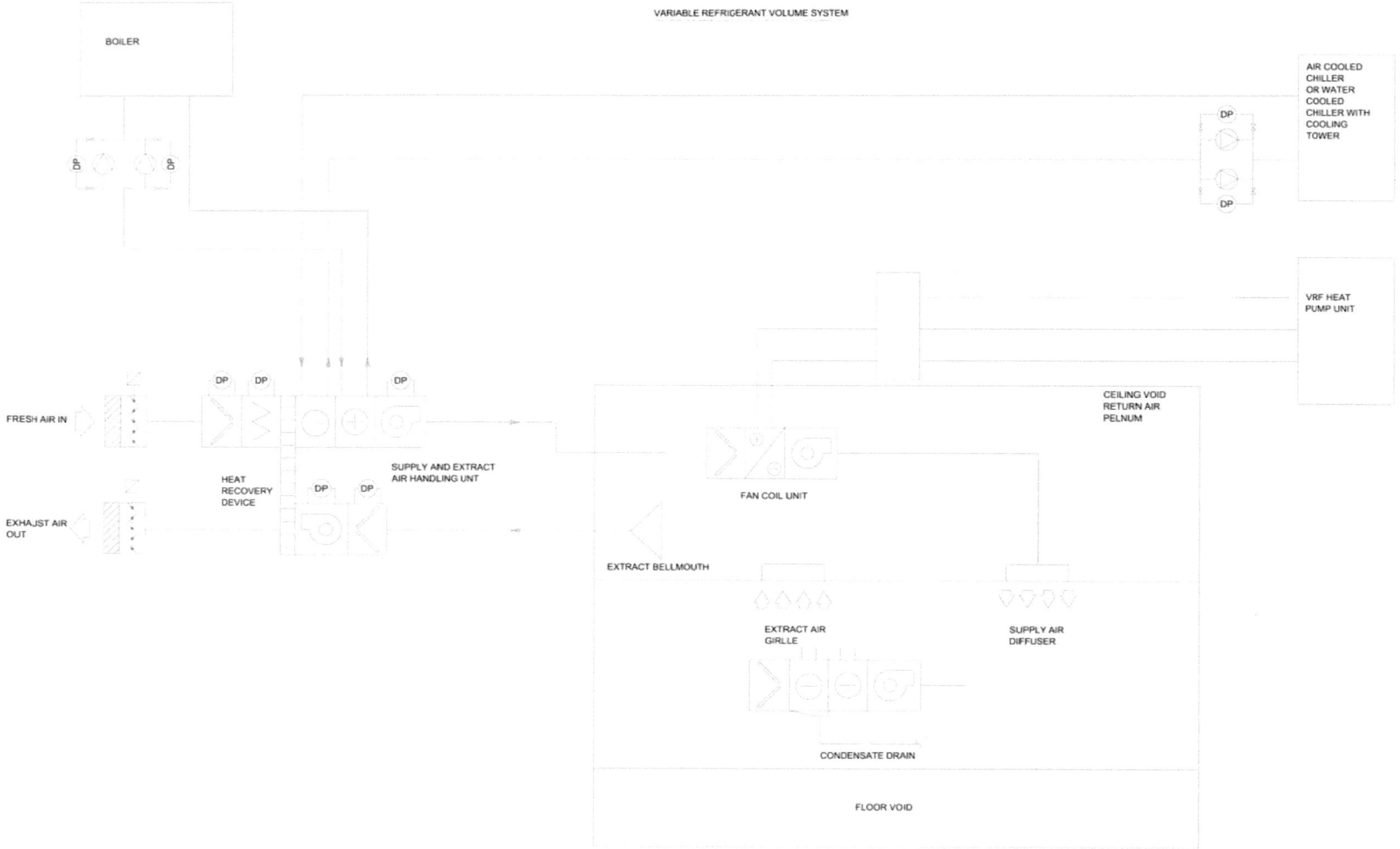

VARIABLE REFRIGERANT VOLUME SYSTEM

AIR COOLED
CHILLER
OR WATER
COOLED
CHILLER WITH
COOLING
TOWER

DP   DP

VRF HEAT
PUMP UNIT

CEILING VOID
RETURN AIR
PELNUM

SUPPLY AIR
DIFFUSER

FAN COIL UNIT

EXTRACT AIR
GRILLE

CONDENSATE DRAIN

FLOOR VOID

EXTRACT BELLMOUTH

BOILER

DP

DP

DP

DP

DP

DP

DP

DP

SUPPLY AND EXTRACT
AIR HANDLING UNIT

HEAT
RECOVERY
DEVICE

FRESH AIR IN

EXHAUST AIR
OUT

70

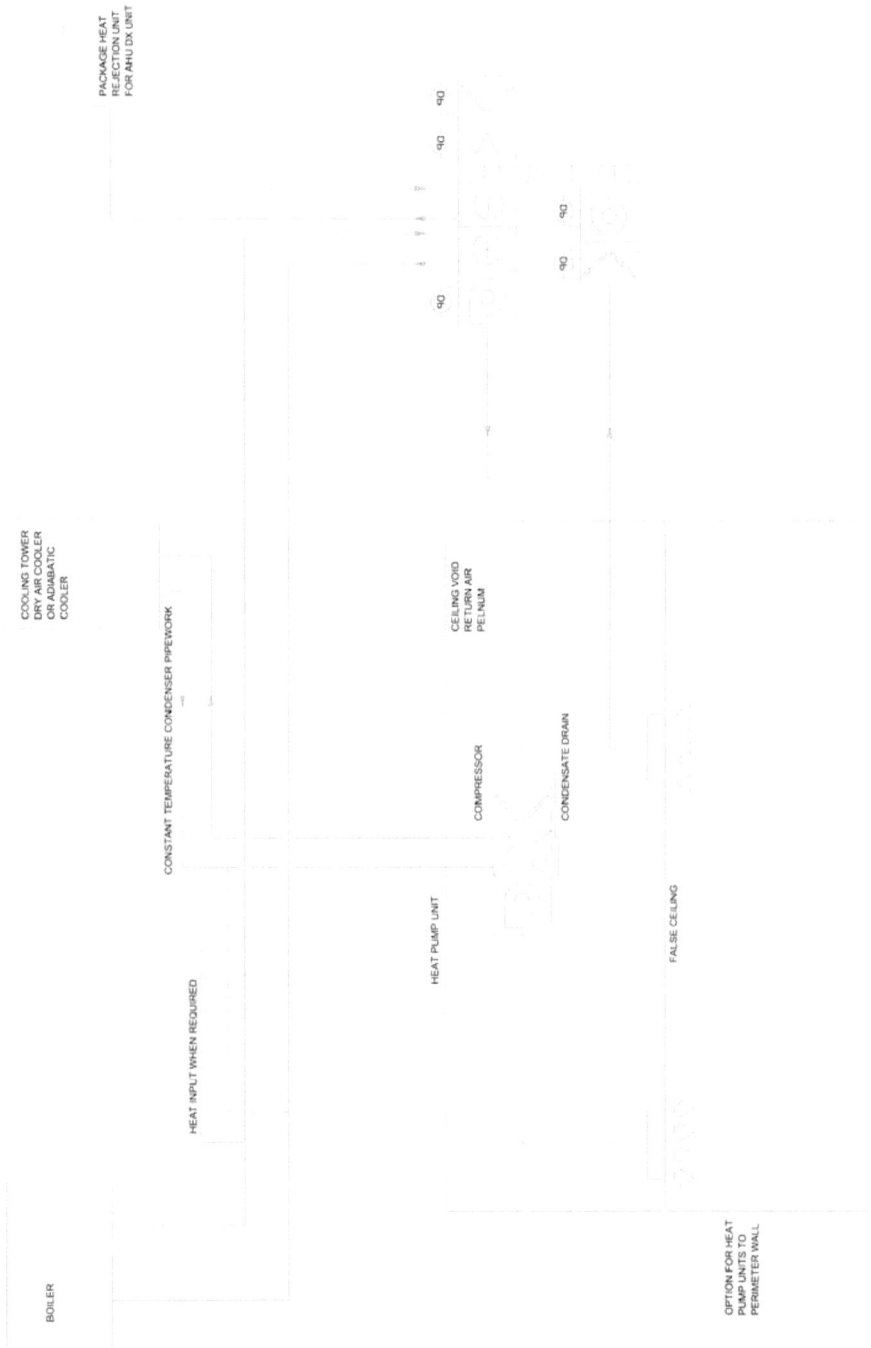

REVERSE CYCLE HEAT PUMP

BOILER

HEAT INPUT WHEN REQUIRED

CONSTANT TEMPERATURE CONDENSER PIPEWORK

COOLING TOWER
DRY AIR COOLER
OR ADIABATIC
COOLER

PACKAGE HEAT
REJECTION UNIT
FOR AHU DX UNIT

HEAT PUMP UNIT

COMPRESSOR

CEILING VOID
RETURN AIR
PLENUM

CONDENSATE DRAIN

FALSE CEILING

OPTION FOR HEAT
PUMP UNITS TO
PERIMETER WALL

Building Regulations in Brief, 10th edition

By Ray Tricker and Samantha Alford

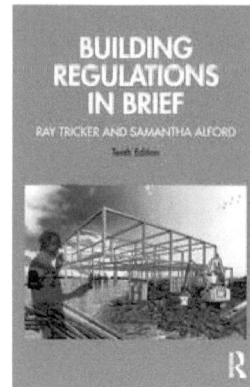

This tenth edition of the most popular and trusted guide reflects all the latest amendments to the Building Regulations, planning permission and the Approved Documents in England and Wales. This includes coverage of the recent changes to use classes, updated sections on planning permission, permitted development and application fees. We have included the revisions to Approved Document B (as a result of the Hackitt Review), as well as the latest changes to Approved Documents F and L, and the new documents O (overheating) and S (electric vehicle charging points), which come into effect in June 2022.

Giving practical information throughout on how to work with (and within) the Regulations, this book enables compliance in the simplest and most cost-effective manner possible. The no-nonsense approach of *Building Regulations in Brief* cuts through any confusion and explains the meaning of the Regulations. Consequently, it has become a favourite for anyone working in or studying the building industry, as well as those planning to have work carried out on their home. It is essential reading for all building contractors and subcontractors, site engineers, building engineers, building control officers, building surveyors, architects, construction site managers and DIYers.

May 2022: 1146 pp
ISBN: 9780367774233

To Order
Tel:+44 (0) 1235 400524
Email: tandf@hachette.co.uk

For a complete listing of all our titles visit:
www.tandf.co.uk

Taylor & Francis
Taylor & Francis Group

PART 2

Approximate Estimating

DIRECTIONS

The prices shown in this section of the book are average prices on a fixed price basis for typical buildings tendered during the third quarter of 2025. Unless otherwise noted, they exclude external services and professional fees.

The information in this section has been arranged to follow more closely the order in which estimates may be developed, in accordance with RIBA work stages.

a) Cost Indices and Regional Variations – These provide information regarding the adjustments to be made to estimates taking into account current pricing levels for different locations in the UK.

b) Feasibility Costs – These provide a range of data (based on a rate per square metre) for all-in engineering costs, excluding lifts, associated with a wide variety of building types. These would typically be used at work stage 0/1 (feasibility) of a project.

c) Elemental Rates – The outline costs for offices have been developed further to provide rates for the alternative solutions for each of the services elements. These would typically be used at work stage 2, outline proposal. Where applicable, costs have been identified as Shell and Core and Fit-Out to reflect projects where the choice of procurement has dictated that the project is divided into two distinctive contractual parts.
Such detail would typically be required at work stage 3, detailed proposals.

d) All-in Rates – These are provided for a number of items and complete parts of a system, i.e. boiler plant, duct-work, pipework, electrical switchgear and small power distribution, together with lifts and escalators. Refer to the relevant section for further guidance notes.

e) Elemental Costs – These are provided for a diverse range of building types; offices, laboratory, shopping mall, airport terminal building, supermarket, performing arts centre, sports hall, luxury hotel, hospital and secondary school. Also included is a separate analysis of a building management system for an office block. In each case, a full analysis of engineering services costs is given to show the division between all elements and their relative costs to the total building area. A regional variation factor has been applied to bring these analyses to a common London base.

Prices should be applied to the total floor area of all storeys of the building under consideration. The area should be measured between the external walls without deduction for internal walls and staircases/lift shafts, i.e. GIA (Gross Internal Area).

Although prices are reviewed in the light of recent tenders it has only been possible to provide a range of prices for each building type. This should serve to emphasize that these can only be average prices for typical requirements and that such prices can vary widely depending on variations in size, location, phasing, specification, site conditions, procurement route, programme, market conditions and net to gross area efficiencies. Rates per square metre should not therefore be used indiscriminately and each case needs to be assessed on its own merits.

The prices do not include for incidental builder's work nor for profit and attendance by a Main Contractor where the work is executed as a subcontract: they do however include for preliminaries, profit and overheads for the services contractor. Capital contributions to statutory authorities and public undertakings and the cost of work carried out by them have been excluded.

Where services works are procured indirectly, i.e. ductwork via a mechanical subcontractor, the reader should make due allowance for the addition of a further level of profit, etc.

COST INDICES

The following tables reflect the major changes in cost to contractors but do not necessarily reflect changes in tender levels. In addition to changes in labour and materials costs, tenders are affected by other factors such as the degree of competition in the particular industry, the area where the work is to be carried out, the availability of labour and the prevailing economic conditions. This has meant in years when there has been an abundance of work, tender levels have tended to increase at a greater rate than can be accounted for solely by increases in basic labour and material costs and, conversely, when there is a shortage of work this has tended to result in keener tenders. Allowances for these factors are impossible to assess on a general basis and can only be based on experience and knowledge of the particular circumstances.

In compiling the tables, the cost of labour has been calculated on the basis of a notional gang as set out elsewhere in the book. The proportion of labour to materials has been assumed as follows:

Mechanical Services – 30:70, Electrical Services – 50:50, (2015 = 100)

Mechanical Services

Year	First Quarter	Second Quarter	Third Quarter	Fourth Quarter	Annual Average	Annual Change
2023	139.3	138.8	138.5	140.7	139.1	4.3%
2024	141.3	143.3	142.1	143.4 (P)	142.5	2.4%
2025	144.7 (F)	146.1	147.5	148.9	146.3 (F)	2.7%
2026	150.4	151.9	153.4	155.0	152.2	4.0%

Electrical Services

Year	First Quarter	Second Quarter	Third Quarter	Fourth Quarter	Annual Average	Annual Change
2023	133.1	133.0	132.7	133.0	132.8	2.3%
2024	134.9	135.9	136.6	136.5 (P)	135.9	2.3%
2025	139.3 (F)	140.7	142.0	143.4	140.9 (F)	3.7%
2026	144.8	146.3	147.8	149.2	146.5	4.0%

(P = Provisional)

(F = Forecast)

Approximate Estimating

COST INDICES

Regional Variations

Prices throughout this Book apply to work in the London area (see Directions at the beginning of the Mechanical Installations and Electrical Installations sections). However, prices for mechanical and electrical services installations will of course vary from region to region, largely as a result of differing labour costs but also depending on the degree of accessibility, urbanization and local market conditions.

The following table of regional factors is intended to provide readers with indicative adjustments that may be made to the prices in the Book for locations outside of London. The figures are of necessity averages for regions and further adjustments should be considered for city centre or very isolated locations, or other known local factors.

Inner London	1.04	North East	0.84
Outer London	1.00	North West	0.90
South East	0.95 (Excl. GL)	Yorkshire and Humberside	0.87
South West	0.90	Scotland	0.88
East of England	0.93	Wales	0.88
East Midlands	0.89	Northern Ireland	0.77
West Midlands	0.89		

RIBA STAGE 0/1 PREPARATION

Item	Unit	Range £	
Typical Square Metre Rates for Engineering Services			
The following examples indicate the range of rates within each building type for engineering services, excluding lifts etc., utilities services and professional fees. Based on Gross Internal Area (GIA).			
Industrial Buildings			
Factories			
Owner occupation: Includes for rainwater, soil/waste, sprinklers, LTHW heating via HL Electric heaters, local air conditioning, BMS, HV/LV installations, lighting, fire alarms, security, earthing	m²	245.00 to	295.00
Warehouses & Distribution Centres			
High bay for owner occupation: Includes for rainwater, soil/waste, sprinklers, LTHW heating via HL radiant heaters, local air conditioning, BMS, HV/LV installations, lighting, fire alarms, security	m²	335.00 to	365.00
Offices see Stage 2/3			
Health and Welfare Facilities			
District General Hospitals			
Natural ventilation: Includes for rainwater, soil/waste, cold water, hot water, dry risers, sprinklers, medical gases, LTHW heating, WC/kitchen extract, BMS, LV installations, standby generation, lighting, small power, fire alarms, earthing/lightning protection, nurse call systems, security, IT wireways	m²	1575.00 to	1675.00
Natural ventilation: Includes for rainwater, soil/waste, cold water, hot water, dry risers, sprinklers, medical gases, LTHW heating, localized VAV air conditioning and mechanical ventilation, WC/kitchen extract, BMS, LV installations, standby generation, lighting, small power, fire alarms, earthing/lightning protection, nurse call systems, security, IT wireways	m²	1725.00 to	1800.00
Private Hospitals			
Natural ventilation: Includes for rainwater, soil/waste, cold water, hot water, dry risers, sprinklers, medical gases, LTHW heating, localized VAV air conditioning, WC/kitchen extract, BMS, LV installations, standby generation, lighting, small power, fire alarms, earthing/lightning protection, nurse call systems, security, IT wireways	m²	1750.00 to	1850.00
Air-conditioned: Includes for rainwater, soil/waste, cold water, hot water, dry risers, sprinklers, medical gases, LTHW heating, supply and extract ventilation, 4 pipe air conditioning, WC/kitchen extract, BMS, LV installations, standby generation, lighting, small power, fire alarms, earthing/lightning protection, nurse call systems, security, IT wireways	m²	2000.00 to	2075.00
GP Surgery			
Natural ventilation: Includes for rainwater, soil/waste, cold water, hot water, LTHW heating, WC/kitchen extract, BMS, LV installations, lighting, small power, fire alarms, earthing/lightning protection, nurse call systems, security, IT wireways	m²	1100.00 to	1225.00
Air-conditioned: Includes for rainwater, soil/waste, cold water, hot water, LTHW heating, supply and extract, BMS, LV installations, lighting, small power, fire alarms, earthing/lightning protection, nurse call systems, security, IT wireways	m²	1200.00 to	1500.00

RIBA STAGE 0/1 PREPARATION

Item	Unit	Range £
Typical Square Metre Rates for Engineering Services – cont		
Entertainment and Recreation Buildings		
Performing Arts (with Theatre)		
Air-Conditioned: Includes for rainwater, soil/waste, cold water, central hot water, sprinklers/dry risers, LTHW heating, air conditioning, kitchen/toilet extract, BMS, LV installations, lighting including enhanced dimming/scene setting, small power, fire alarms, earthing, security, IT wireways including production, audio and video recording	m²	1175.00 to 1500.00
Sports Halls		
Comfort cooled: Includes for rainwater, soil/waste, cold water, hot water via LTHW heat exchangers, LTHW heating, air conditioning via AHUs to limited areas, toilet extract, BMS, LV installations, lighting, small power, fire alarms, earthing, security	m²	495.00 to 610.00
Multi-Purpose Leisure Centre		
Comfort cooled: Includes for rainwater, soil/waste, cold water, hot water LTHW heat exchangers, LTHW heating, air conditioning via AHU to limited areas, pool hall supply/extract, kitchen/toilet extract, BMS, LV installations, lighting, small power, fire alarms, earthing, security, IT wireways	m²	870.00 to 1050.00
Retail Buildings		
Open Arcade		
Natural ventilation: Includes for rainwater, soil/waste, cold water, hot water, sprinklers/dry risers, LTHW heating, toilet extract, smoke extract, BMS, LV installations, life safety standby generators, lighting, small power, fire alarms, public address, earthing/lightning protection, security, IT wireways	m²	480.00 to 600.00
Department Stores		
Air-conditioned: Includes for sanitaryware, soil/waste, cold water, hot water, sprinklers/dry risers, LTHW heating, air conditioning via AHUs, toilet extract, smoke extract, BMS, LV installations, life safety standby generators, lighting, small power, fire alarms, public address, earthing, lightning protection, CCTV/security, IT installation wireways	m²	560.00 to 700.00
Food Stores		
Air-conditioned: Includes for rainwater, soil/waste, cold water, hot water, sprinklers, LTHW heating, air conditioning via AHUs, toilet extract, BMS, LV installations, lighting, small power, fire alarms, earthing, lightning protection, security, IT wireways, refrigeration	m²	800.00 to 950.00
Educational Buildings		
Secondary Schools (Academy)		
Natural ventilation: Includes for rainwater, soil/waste, cold water, central hot water, LTHW heating, toilet extract, BMS, LV installations, lighting, small power, fire alarms, earthing, security, IT wireways	m²	600.00 to 670.00
Natural vent with comfort cooling to selected areas (BB93 compliant): Includes for rainwater, soil/waste, cold water, central hot water, LTHW heating, DX air conditioning, general supply/extract, toilet extract, BMS, LV installations, lighting, small power, fire alarms, earthing, security, IT wireways	m²	670.00 to 770.00

RIBA STAGE 0/1 PREPARATION

Item	Unit	Range £
Scientific Buildings		
Educational Research		
Comfort cooled: Includes for rainwater, soil/waste, cold water, central hot water, dry risers, compressed air, medical gasses, LTHW heating, 4 pipe air conditioning, toilet extract, fume, BMS, LV installations, lighting, small power, fire alarms, earthing, lightning protection, security, IT wireways	m²	1125.00 to 1475.00
Air-conditioned: Includes for rainwater, soil/waste, laboratory waste, cold water, central hot water, specialist water, dry risers, compressed air, medical gasses, steam, LTHW heating, VAV air conditioning, Comms room cooling, toilet extract, fume extract, BMS, LV installations, UPS, standby generators, lighting, small power, fire alarms, earthing, lightning protection, security, IT wireways	m²	2125.00 to 2350.00
Commercial Research		
Air-conditioned; Includes for rainwater, soil/waste, laboratory waste, cold water, central hot water, specialist water, dry risers, compressed air, medical gasses, steam, LTHW heating, VAV air conditioning, Comm room cooling, toilet extract, fume extract, BMS, LV installations, UPS, standby generators, lighting, small power, fire alarms, earthing, lightning protection, security, IT wireways	m²	2025.00 to 2350.00
Laboratory Buildings		
Lab: Includes for sanitaryware, rainwater, soil/waste, water, sprinklers, ASHP, 4 pipe system, fume extract, HV/LV installations, lighting, fire alarms, security, access control and BMS, lab gases Excludes: Lifts and lab equipment	m²	1725.00 to 2750.00
Residential Facilities see Stage 2/3		

RIBA STAGE 2/3 DESIGN

Item	Unit	Range £	
Elemental Rates for Alternative Engineering Services Solutions – Offices			
The following examples of building types indicate the range of rates for alternative design solutions for each of the engineering services elements based on Gross Internal Area for the Shell and Core and Net Internal Area for the Fit-Out. Fit-Out is assumed to be to Cat A Standard.			
Consideration should be made for the size of the building height, sales point (£/ft^2) and location, which may affect the economies of scale for rates, i.e. the larger the building the lower the rates.			
5 Services			
Shell & Core			
5.1 Sanitary Installations	m² GIA	19.00 to	36.00
5.3 Disposal Installations	m² GIA	29.00 to	47.00
5.4 Water Installations	m² GIA	58.00 to	79.00
5.6 Space Heating and Air Conditioning			
Comfort cooling			
2 pipe variable refrigerant volume (VRV)	m² GIA	105.00 to	130.00
Full air conditioning, all electric ASHPs	m² GIA	210.00 to	250.00
5.7 Ventilation Systems	m² GIA	– to	–
5.8 Electrical Installations			
LV distribution	m² GIA	61.00 to	77.00
Lighting installations (including lighting controls and luminaires)	m² GIA	34.00 to	55.00
Small power	m² GIA	13.00 to	24.00
Electrical installations for mechanical plan	m² GIA	6.00 to	12.00
Containment	m² GIA	16.00 to	23.00
5.11 Fire and Lightning Protection			
Sprinkler installation	m² GIA	46.00 to	55.00
Wet risers	m² GIA	12.00 to	19.00
Earthing	m² GIA	2.00 to	4.00
Lightning protection	m² GIA	2.00 to	4.00
5.12 Communication, Security and Control Systems			
Fire alarms	m² GIA	16.00 to	23.00
Security	m² GIA	11.00 to	16.00
BMS controls; including MCC panels and control cabling	m² GIA	34.00 to	40.00
Fit-Out			
5.6 Space Heating and Air Conditioning			
Comfort cooling			
2 pipe variable refrigerant volume (VRV)	m² NIA	105.00 to	155.00
Full air conditioning			
4 pipe fan coil – concealed	m² NIA	230.00 to	375.00
4 pipe fan coil – exposed	m² NIA	260.00 to	410.00
4 pipe variable refrigerant volume	m² NIA	210.00 to	250.00
Mixed mode	m² NIA	375.00 to	450.00
Chilled ceiling and active chilled beam	m² NIA	295.00 to	450.00
Active chilled beam	m² NIA	320.00 to	470.00
Displacement	m² NIA	210.00 to	230.00

RIBA STAGE 2/3 DESIGN

Item	Unit	Range £	
5.8 Electrical Installations			
Tenant distribution boards	m² NIA	8.00 to	14.00
Lighting installations (including lighting controls and luminaires)	m² NIA	140.00 to	190.00
Electrical installations for mechanical plant	m² NIA	7.00 to	17.00
5.11 Fire and Lightning Protection			
Sprinkler installation	m² NIA	39.00 to	44.00
Earthing	m² NIA	2.00 to	4.00
5.12 Communication, Security and Control Systems			
Fire alarms	m² NIA	31.00 to	40.00
BMS controls; including MCC panels and control cabling	m² NIA	33.00 to	46.00

RIBA STAGE 2/3 DESIGN

Item	Unit	Range £	
Elemental Rates for Alternative Engineering Services Solutions – Hotels			
The following examples of building types indicate the range of rates for alternative design solutions for each of the engineering services elements based on Gross Internal Area for the Shell and Core and Net Internal Area for the Fit-Out. Fit-Out is assumed to be to Cat A Standard.			
Consideration should be made for the size of the building, which may affect the economies of scale for rates, i.e. the larger the building the lower the rates.			
5 Services			
Shell & Core			
5.1 Sanitary Installations			
2 to 3 star	m² GIA	39.00 to	49.00
4 to 5 star	m² GIA	58.00 to	81.00
5.3 Disposal Installations			
2 to 3 star	m² GIA	39.00 to	49.00
4 to 5 star	m² GIA	43.00 to	61.00
5.4 Water Installations			
2 to 3 star	m² GIA	56.00 to	71.00
4 to 5 star	m² GIA	80.00 to	110.00
5.6 Space Heating and Air Conditioning			
LPHW heating installation; including gas installations			
2 to 3 star	m² GIA	83.00 to	97.00
4 to 5 star	m² GIA	83.00 to	97.00
Air conditioning; including ventilation			
2 to 3 star – 4 pipe fan coil	m² GIA	320.00 to	435.00
4 to 5 star – 4 pipe fan coil	m² GIA	335.00 to	440.00
2 to 3 star – 3 pipe variable refrigerant volume	m² GIA	210.00 to	265.00
4 to 5 star – 3 pipe variable refrigerant volume	m² GIA	210.00 to	265.00
5.8 Electrical Installations			
LV installations			
Standby generators (life safety only)			
2 to 3 star	m² GIA	22.00 to	28.00
4 to 5 star	m² GIA	22.00 to	32.00
LV distribution			
2 to 3 star	m² GIA	45.00 to	61.00
4 to 5 star	m² GIA	62.00 to	84.00
Lighting installations			
2 to 3 star	m² GIA	33.00 to	42.00
4 to 5 star	m² GIA	50.00 to	67.00
Small power			
2 to 3 star	m² GIA	14.00 to	19.00
4 to 5 star	m² GIA	20.00 to	29.00
Electrical installations for mechanical plant			
2 to 3 star	m² GIA	11.00 to	17.00
4 to 5 star	m² GIA	11.00 to	17.00

RIBA STAGE 2/3 DESIGN

Item	Unit	Range £	
5.11 Fire and Lightning Protection			
Fire protection			
2 to 3 star – Dry risers	m² GIA	15.00 to	19.00
4 to 5 star – Dry risers	m² GIA	15.00 to	19.00
2 to 3 star – Sprinkler installation	m² GIA	47.00 to	60.00
4 to 5 star – Sprinkler installation	m² GIA	53.00 to	70.00
Earthing			
2 to 3 star	m² GIA	3.00 to	4.00
4 to 5 star	m² GIA	3.00 to	4.00
Lightning protection			
2 to 3 star	m² GIA	3.00 to	5.00
4 to 5 star	m² GIA	3.00 to	5.00
5.12 Communication, Security and Control Systems			
Fire alarms			
2 to 3 star	m² GIA	27.00 to	38.00
4 to 5 star	m² GIA	27.00 to	42.00
IT / AV / Containment/TV systems			
2 to 3 star	m² GIA	26.00 to	33.00
4 to 5 star	m² GIA	86.00 to	110.00
Security			
2 to 3 star	m² GIA	27.00 to	39.00
4 to 5 star	m² GIA	31.00 to	40.00
BMS controls; including MCC panels and control cabling			
2 to 3 star	m² GIA	19.00 to	25.00
4 to 5 star	m² GIA	48.00 to	65.00

RIBA STAGE 2/3 DESIGN

Item	Unit	Range £	
Elemental Rates for Alternative Engineering Services Solutions – Apartments			
The following examples of building types indicate the range of rates for alternative design solutions for each of the engineering services elements based on Gross Internal Area for the Shell and Core and Net Internal Area for the Fit-Out. Fit-Out is assumed to be to Cat A Standard.			
Consideration should be made for the size of the building, which may affect the economies of scale for rates, i.e. the larger the building the lower the rates.			
5 Services			
Shell & Core			
5.1 Sanitary Installations			
Affordable	m² GIA	1.00 to	1.00
Private	m² GIA	4.00 to	4.00
5.3 Disposal Installations			
Affordable	m² GIA	25.00 to	32.00
Private	m² GIA	32.00 to	41.00
5.4 Water Installations			
Affordable	m² GIA	27.00 to	34.00
Private	m² GIA	43.00 to	57.00
5.5 Heat Source (ASHP)			
Affordable	m² GIA	39.00 to	47.00
Private	m² GIA	39.00 to	48.00
5.6 Space Heating & Air Conditioning			
Affordable	m² GIA	39.00 to	44.00
Private	m² GIA	110.00 to	145.00
5.7 Ventilation Systems			
Affordable	m² GIA	27.00 to	29.00
Private	m² GIA	32.00 to	39.00
5.8 Electrical Installations			
Affordable	m² GIA	58.00 to	76.00
Private	m² GIA	87.00 to	120.00
5.11 Fire and Lightning Protection			
Affordable	m² GIA	43.00 to	50.00
Private	m² GIA	44.00 to	68.00
5.12 Communication, Security and Control Systems			
Affordable	m² GIA	50.00 to	62.00
Private	m² GIA	70.00 to	96.00
5.13 Special Installations			
Affordable	m² GIA	18.00 to	21.00
Private	m² GIA	55.00 to	77.00
Fit-Out			
5.1 Sanitary Installations			
Affordable	m² NIA	47.00 to	52.00
Private	m² NIA	150.00 to	175.00

RIBA STAGE 2/3 DESIGN

Item	Unit	Range £	
5.3 Disposal Installations			
Affordable	m² NIA	18.00 to	23.00
Private	m² NIA	36.00 to	46.00
5.4 Water Installations			
Affordable	m² NIA	41.00 to	48.00
Private	m² NIA	70.00 to	93.00
5.6 Space Heating and Air Conditioning			
Affordable	m² NIA	105.00 to	135.00
Private	m² NIA	290.00 to	355.00
5.7 Ventilating Systems			
Affordable (whole house vent)	m² NIA	58.00 to	70.00
Private (whole house vent)	m² NIA	70.00 to	84.00
5.8 Electrical Installations			
Affordable	m² NIA	81.00 to	100.00
Private	m² NIA	180.00 to	215.00
5.11 Fire and Lightning Protection			
Affordable	m² NIA	37.00 to	44.00
Private	m² NIA	44.00 to	54.00
5.12 Communication, Security and Control Systems			
Affordable	m² NIA	31.00 to	37.00
Private	m² NIA	150.00 to	180.00
5.13 Special Installations			
Affordable	m² NIA	18.00 to	21.00
Private	m² NIA	53.00 to	60.00

Note: The range in cost differs due to the vast diversity in services strategies available. The lower end of the scale reflects no comfort cooling, radiator heating in lieu of underfloor, etc. The high end of the scale is based on central plant installations with good quality apartments, which includes comfort cooling, sprinklers, home network installations, video entry, higher quality of sanitaryware and good quality LED lighting.

APPROXIMATE ESTIMATING RATES – 5 SERVICES

Item	Unit	Range £	
5.3 Disposal Installations			
ABOVE GROUND DRAINAGE			
Soil and waste	Point	630.00 to	800.00
5.4 Water Installations			
WATER INSTALLATIONS			
Cold water	Point	590.00 to	750.00
Hot water	Point	750.00 to	900.00
PIPEWORK			
HOT AND COLD WATER			
Excludes insulation, valves and ancillaries, etc.			
Light gauge copper tube to EN1057 R250 (TX) formerly BS 2871 Part 1 Table X with joints as described including allowance for waste, fittings and supports assuming average runs with capillary joints up to 54 mm and bronze welded thereafter			
Horizontal High Level Distribution			
15 mm	m	48.00 to	58.00
22 mm	m	49.00 to	59.00
28 mm	m	67.00 to	84.00
35 mm	m	81.00 to	97.00
42 mm	m	96.00 to	120.00
54 mm	m	125.00 to	150.00
67 mm	m	150.00 to	180.00
76 mm	m	170.00 to	210.00
108 mm	m	255.00 to	325.00
Risers			
15 mm	m	29.00 to	36.00
22 mm	m	34.00 to	39.00
28 mm	m	49.00 to	59.00
35 mm	m	58.00 to	70.00
42 mm	m	72.00 to	86.00
54 mm	m	79.00 to	105.00
67 mm	m	145.00 to	165.00
76 mm	m	160.00 to	195.00
108 mm	m	200.00 to	240.00
Toilet Areas, etc., at Low Level			
15 mm	m	80.00 to	94.00
22 mm	m	96.00 to	115.00
28 mm	m	135.00 to	165.00

APPROXIMATE ESTIMATING RATES – 5 SERVICES

Item	Unit	Range £	
5.5 Heat Source			
HEAT SOURCE			
Gas fired boilers including gas train and controls	kW	46.00 to	57.00
Gas fired boilers including gas train, controls, flue, plantroom pipework, valves and insulation, pumps and pressurization unit	kW	140.00 to	175.00
5.6 Space Heating and Air Conditioning			
CHILLED WATER			
Air cooled R134a refrigerant chiller including control panel, anti-vibration mountings	kW	245.00 to	325.00
Air cooled R134a refrigerant chiller including control panel, anti-vibration mountings, plantroom pipework, valves, insulation, pumps and pressurization units	kW	355.00 to	435.00
Water cooled R134a refrigerant chiller including control panel, anti-vibration mountings	kW	135.00 to	205.00
Water cooled R134a refrigerant chiller including control panel, anti-vibration mountings, plantroom pipework, valves, insulation, pumps and pressurization units	kW	270.00 to	315.00
Absorption steam medium chiller including control panel, anti-vibration mountings, plantroom pipework, valves, insulation, pumps and pressurization units	kW	460.00 to	560.00
HEAT REJECTION			
Open circuit, forced draft cooling tower	kW	165.00 to	200.00
Closed circuit, forced draft cooling tower	kW	150.00 to	180.00
Dry air cooler	kW	93.00 to	98.00
PUMPS			
Pumps including flexible connections, anti-vibration mountings	kPa	59.00 to	76.00
Pumps including flexible connections, anti-vibration mountings, plantroom pipework, valves, insulation and accessories	kPa	135.00 to	165.00
DUCTWORK			
The rates below allow for ductwork and for all other labour and material in fabrication, fittings, supports and jointing to equipment, stop and capped ends, elbows, bends, diminishing and transition pieces, regular and reducing couplings, volume control dampers, branch diffuser and 'snap on' grille connections, ties, 'Ys', crossover spigots, etc., turning vanes, regulating dampers, access doors and openings, hand-holes, test holes and covers, blanking plates, flanges, stiffeners, tie rods and all supports and brackets fixed to structure.			
Rectangular galvanized mild steel ductwork as HVCA DW 144 up to 1000 mm longest side	m^2	84.00 to	99.00
Rectangular galvanized mild steel ductwork as HVCA DW 144 up to 2500 mm longest side	m^2	91.00 to	110.00
Rectangular galvanized mild steel ductwork as HVCA DW 144 up to 3000 mm longest side and above	m^2	125.00 to	145.00
Circular galvanized mild steel ductwork as HVCA DW 144	m^2	110.00 to	130.00
Flat oval galvanized mild steel ductwork as HVCA DW 144 up to 545 mm wide	m^2	81.00 to	98.00
Flat oval galvanized mild steel ductwork as HVCA DW 144 up to 880 mm wide	m^2	88.00 to	110.00
Flat oval galvanized mild steel ductwork as HVCA DW 144 up to 1785 mm wide	m^2	100.00 to	130.00
PACKAGED AIR HANDLING UNITS			
Air handling unit including LPHW preheater coil, pre-filter panel, LPHW heater coils, chilled water coil, filter panels, inverter drive, motorized volume control dampers, sound attenuation, flexible connections to ductwork and all anti-vibration mountings	m^3/s	11500.00 to	14500.00

APPROXIMATE ESTIMATING RATES – 5 SERVICES

Item	Unit	Range £	
5.6 Space Heating and Air Conditioning – cont			
LTHW AND CHILLED WATER			
Excludes insulation, valves and ancillaries, etc.			
Black heavy weight mild steel tube to BS 1387 with joints in the running length, allowance for waste, fittings and supports assuming average runs			
Horizontal Distribution – Basements, etc.			
15 mm	m	79.00 to	95.00
20 mm	m	83.00 to	100.00
25 mm	m	96.00 to	115.00
32 mm	m	110.00 to	140.00
40 mm	m	125.00 to	150.00
50 mm	m	160.00 to	205.00
65 mm	m	160.00 to	210.00
80 mm	m	210.00 to	255.00
100 mm	m	260.00 to	310.00
125 mm	m	345.00 to	420.00
150 mm	m	410.00 to	480.00
200 mm	m	610.00 to	740.00
250 mm	m	750.00 to	940.00
300 mm	m	1100.00 to	1350.00
Risers			
15 mm	m	55.00 to	67.00
20 mm	m	57.00 to	67.00
25 mm	m	64.00 to	78.00
32 mm	m	74.00 to	90.00
40 mm	m	86.00 to	99.00
50 mm	m	110.00 to	130.00
65 mm	m	125.00 to	150.00
80 mm	m	140.00 to	175.00
100 mm	m	185.00 to	225.00
125 mm	m	235.00 to	270.00
150 mm	m	290.00 to	355.00
200 mm	m	445.00 to	560.00
250 mm	m	550.00 to	670.00
300 mm	m	860.00 to	1050.00
On Floor Distribution			
15 mm	m	87.00 to	105.00
20 mm	m	95.00 to	110.00
25 mm	m	105.00 to	130.00
32 mm	m	120.00 to	150.00
40 mm	m	135.00 to	165.00
50 mm	m	180.00 to	215.00
Plantroom Areas, etc.			
15 mm	m	85.00 to	100.00
20 mm	m	88.00 to	105.00
25 mm	m	98.00 to	115.00
32 mm	m	110.00 to	135.00
40 mm	m	125.00 to	150.00
50 mm	m	170.00 to	195.00
65 mm	m	185.00 to	210.00

APPROXIMATE ESTIMATING RATES – 5 SERVICES

Item	Unit	Range £	
80 mm	m	235.00 to	295.00
100 mm	m	305.00 to	365.00
125 mm	m	400.00 to	485.00
150 mm	m	465.00 to	560.00
200 mm	m	690.00 to	840.00
250 mm	m	860.00 to	1000.00
300 mm	m	1225.00 to	1500.00

5.7 Ventilating Systems
EXTRACT FANS

Extract fan including inverter drive, sound attenuation, flexible connections to ductwork and all anti-vibration mountings	m³/s	4150.00 to	4775.00

5.8 Electrical Installations

HV/LV INSTALLATIONS
The cost of HV/LV equipment will vary according to the electricity supplier's requirements, the duty required and the actual location of the site. For estimating purposes, the items indicated below are typical of the equipment required in an HV substation incorporated into a building.

Item	Unit	Range £	
RING MAIN UNIT			
Ring main unit, 11 kV including electrical terminations	Point	23000.00 to	28000.00
TRANSFORMERS			
Oil filled transformers, 11 kV to 415 V including electrical terminations	kVA	30.00 to	36.00
Cast resin transformers, 11 kV to 415 V including electrical terminations	kVA	33.00 to	37.00
Midal filled transformers, 11 kV to 415 V including electrical terminations	kVA	35.00 to	41.00
HV SWITCHGEAR			
Cubicle section HV switchpanel, Form 4 type 6 including air circuit breakers, meters and electrical terminations	Section	29500.00 to	43500.00
LV SWITCHGEAR			
LV switchpanel, Form 3 including all isolators, fuses, meters and electrical terminations	Isolator	4700.00 to	5700.00
LV switchpanel, Form 4 type 5 including all isolators, fuses, meters and electrical terminations	Isolator	7100.00 to	8900.00
EXTERNAL PACKAGED SUBSTATION			
Extra over cost for prefabricated packaged substation housing, excludes base and protective security fencing	Each	48500.00 to	58000.00

APPROXIMATE ESTIMATING RATES – 5 SERVICES

Item	Unit	Range £	
5.8 Electrical Installations – cont			
STANDBY GENERATING SETS			
Diesel powered including control panel, flue, oil day tank and attenuation			
Approximate installed cost, LV	kVA	400.00 to	485.00
Approximate installed cost, HV	kVA	465.00 to	570.00
UNINTERRUPTIBLE POWER SUPPLY			
Rotary UPS including control panel and choke transformer (excludes distribution)			
Approximate installed cost (range 1000 kVA to 2500 kVA)	kVA	620.00 to	760.00
Static UPS including control panel, automatic bypass, DC isolator and batteries for 30 minutes standby (excludes distribution)			
Approximate installed cost (range 500 kVA to 1000 kVA)	kVA	290.00 to	400.00
SMALL POWER			
Approximate prices for wiring of power points of length not exceeding 20 m, including accessories, wireways but excluding distribution boards			
13 amp Accessories			
Wired in PVC insulated twin and earth cable in ring main circuit			
Domestic properties	Point	86.00 to	105.00
Commercial properties	Point	115.00 to	135.00
Industrial properties	Point	115.00 to	135.00
Wired in PVC insulated twin and earth cable in radial circuit			
Domestic properties	Point	115.00 to	135.00
Commercial properties	Point	135.00 to	175.00
Industrial properties	Point	135.00 to	175.00
Wired in LSF insulated single cable in ring main circuit			
Commercial properties	Point	140.00 to	180.00
Industrial properties	Point	140.00 to	180.00
Wired in LSF insulated single cable in radial circuit			
Commercial properties	Point	170.00 to	205.00
Industrial properties	Point	170.00 to	205.00
45 amp wired in PVC insulated twin and earth cable			
Domestic properties	Point	155.00 to	195.00
Low Voltage Power Circuits			
Three phase four wire radial circuit feeding an individual load, wired in LSF insulated single cable including all wireways, isolator, not exceeding 10 metres; in commercial properties			
1.5 mm	Point	280.00 to	365.00
2.5 mm	Point	320.00 to	385.00
4 mm	Point	340.00 to	380.00
6 mm	Point	365.00 to	460.00
10 mm	Point	430.00 to	540.00
16 mm	Point	480.00 to	570.00
Three phase four core radial circuit feeding an individual load item, wired in LSF/SWA/XLPE insulated cable including terminations, isolator, clipped to surface, not exceeding 10 metres; in commercial properties			
1.5 mm	Point	210.00 to	255.00
2.5 mm	Point	230.00 to	275.00
4 mm	Point	250.00 to	315.00

APPROXIMATE ESTIMATING RATES – 5 SERVICES

Item	Unit	Range £	
6 mm	Point	275.00 to	345.00
10 mm	Point	420.00 to	520.00
16 mm	Point	520.00 to	640.00

LIGHTING
Approximate prices for wiring of lighting points including rose, wireways but excluding distribution boards, luminaires and switches

Final Circuits			
Wired in PVC insulated twin and earth cable			
Domestic properties	Point	64.00 to	77.00
Commercial properties	Point	79.00 to	99.00
Industrial properties	Point	82.00 to	100.00
Wired in LSF insulated single cable			
Commercial properties	Point	120.00 to	140.00
Industrial properties	Point	120.00 to	150.00

ELECTRICAL WORKS IN CONNECTION WITH MECHANICAL SERVICES
The cost of electrical connections to mechanical services equipment will vary depending on the type of building and complexity of the equipment.

Typical rate for power wiring, isolators and associated wireways	m^2	11.00 to	16.00

FIRE ALARMS
Cost per point for two core FP200 wired system including all terminations, supports and wireways

Call point	Point	370.00 to	435.00
Smoke detector	Point	305.00 to	365.00
Smoke/heat detector	Point	345.00 to	425.00
Heat detector	Point	330.00 to	400.00
Heat detector and sounder	Point	310.00 to	345.00
Input/output/relay units	Point	420.00 to	495.00
Alarm sounder	Point	330.00 to	400.00
Alarm sounder/beacon	Point	395.00 to	480.00
Speakers/voice sounders	Point	395.00 to	480.00
Speakers/voice sounders (weatherproof)	Point	420.00 to	495.00
Beacon/strobe	Point	305.00 to	360.00
Beacon/strobe (weatherproof)	Point	430.00 to	530.00
Door release units	Point	430.00 to	530.00
Beam detector	Point	1325.00 to	1450.00

For costs for zone control panel, battery chargers and batteries, see 'Prices for Measured Work' section.

EXTERNAL LIGHTING

Estate Road Lighting
Post type road lighting lantern 70 watt CDM-T 3000k complete with 5 m high column with hinged lockable door, control gear and cut-out including 2.5 mm two core butyl cable internal wiring, interconnections and earthing fed by 16 mm^2 four core XLPE/SWA/LSF cable and terminations. Approximate installed price per metre road length (based on 300 metres run) including time switch but excluding builder's work in connection

Columns erected at 30 m intervals along road (cost per m of road)	m	110.00 to	135.00

APPROXIMATE ESTIMATING RATES – 5 SERVICES

Item	Unit	Range £	
5.8 Electrical Installations – cont			
Bollard Lighting			
Bollard lighting fitting 26 watt TC-D 3500k including control gear, all internal wiring, interconnections, earthing and 25 metres of 2.5 mm² three core XLPE/SWA/LSF cable			
Approximate installed price excluding builder's work in connection	Each	1600.00 to	2000.00
Outdoor Flood Lighting			
Wall mounted outdoor floodlight fitting complete with tungsten halogen lamp, mounting bracket, wire guard and all internal wiring and containment, fixed to brickwork or concrete and connected			
Installed price 500 watt	Point	350.00 to	425.00
Installed price 1000 watt	Point	415.00 to	500.00
Pedestal mounted outdoor flood light fitting complete with 1000 watt MBF/U lamp, control gear, contained in weatherproof steel box, all internal wiring and containment, interconnections and earthing, fixed to brickwork or concrete and connected			
Approximate installed price excluding builder's work in connection	Each	1575.00 to	2025.00
BUILDING MANAGEMENT INSTALLATIONS			
Category A Fit-Out			
Option 1 – 185 nr four pipe fan coil – 740 points			
1.0 Field Equipment			
Network devices; Valves/actuators; Sensing devices	Point	105.00 to	110.00
2.0 Cabling			
Power – from local isolator to DDC controller; Control – from DDC controller to field equipment	Point	57.00 to	60.00
3.0 Programming			
Software – central facility; Software – network devices; Graphics	Point	29.00 to	32.00
4.0 On-site testing and commissioning			
Equipment; Programming/graphics; Power and control cabling	Point	29.00 to	32.00
Total Option 1 – Four pipe fan coil	Point	220.00 to	235.00
Cost/FCU	Each	890.00 to	950.00
Option 2 – 185 nr two pipe fan coil system with electric heating – 740 points			
1.0 Field Equipment			
Network devices; Valves/actuators/thyristors; Sensing devices	Point	140.00 to	150.00
2.0 Cabling			
Power – from local isolator to DDC controller; Control – from DDC controller to field equipment	Point	57.00 to	60.00
3.0 Programming			
Software – central facility; Software – network devices; Graphics	Point	30.00 to	34.00
4.0 On-site testing and commissioning			
Equipment; Programming/graphics; Power and control cabling	Point	31.00 to	35.00
Total Option 2 – Two pipe fan coil with electric heating	Point	255.00 to	280.00
Cost/FCU	Each	1050.00 to	1125.00

APPROXIMATE ESTIMATING RATES – 5 SERVICES

Item	Unit	Range £	
Option 3 – 180 nr chilled beams with perimeter heating – 567 points			
1.0 Field Equipment			
Network devices; Valves/actuators; Sensing devices	Point	135.00 to	135.00
2.0 Cabling			
Power – from local isolator to DDC controller; Control – from DDC controller to field equipment	Point	80.00 to	82.00
3.0 Programming			
Cost/Point	Point	44.00 to	46.00
4.0 On-site testing and commissioning			
Equipment; Programming/Graphics; Power and control cabling	Point	45.00 to	47.00
Total Option 3 – Chilled beams with perimeter heating	Point	300.00 to	310.00
Cost/Chilled Beam	Each	950.00 to	960.00
Shell & Core Only			
Main Plant – Cost/BMS Points	Each	950.00 to	960.00
Landlord FCUs – Cost/Terminal Units	Each	950.00 to	960.00
Trade Contract Preliminaries	%	21.00 to	32.00
Notes: The following are included in points, rates			
- DDC Controllers/Control Enclosures/Control Panels			
- Motor Control Centre (MCC)			
- Field Devices			
- Control and Power Cabling from DDC Controllers/MCC			
- Programming			
- On-Site Testing and Commissioning			

APPROXIMATE ESTIMATING RATES – 5 SERVICES

Item	Unit	Range £
5.10 Lift and Conveyor Systems		
LIFT INSTALLATIONS		
The cost of lift installations will vary depending upon a variety of circumstances. The following prices assume a car height of 2.2 metres, manufacturer's standard car finish, brushed stainless steel 2 panel centre opening doors to BSEN81–20 and Lift Regulations 2016.		
Passenger Lifts, machine room above		
Electrically operated AC drive serving 2 levels with directional collective controls and a speed of 1.0 m/s		
8 Person	Item	90000.00 to 115000.00
10 Person	Item	98000.00 to 120000.00
13 Person	Item	105000.00 to 130000.00
17 Person	Item	120000.00 to 140000.00
21 Person	Item	130000.00 to 160000.00
26 Person	Item	145000.00 to 180000.00
Electrically operated AC drive serving 4 levels and a speed of 1.0 m/s		
8 Person	Item	105000.00 to 130000.00
10 Person	Item	115000.00 to 135000.00
13 Person	Item	120000.00 to 145000.00
17 Person	Item	135000.00 to 155000.00
21 Person	Item	145000.00 to 180000.00
26 Person	Item	170000.00 to 210000.00
Electrically operated AC drive serving 6 levels and a speed of 1.0 m/s		
8 Person	Item	120000.00 to 145000.00
10 Person	Item	125000.00 to 150000.00
13 Person	Item	135000.00 to 155000.00
17 Person	Item	145000.00 to 180000.00
21 Person	Item	170000.00 to 210000.00
26 Person	Item	185000.00 to 225000.00
Electrically operated AC drive serving 8 levels and a speed of 1.0 m/s		
8 Person	Item	135000.00 to 165000.00
10 Person	Item	140000.00 to 165000.00
13 Person	Item	145000.00 to 180000.00
17 Person	Item	170000.00 to 210000.00
21 Person	Item	185000.00 to 220000.00
26 Person	Item	210000.00 to 260000.00
Electrically operated AC drive serving 10 levels and a speed of 1.0 m/s		
8 Person	Item	145000.00 to 180000.00
10 Person	Item	150000.00 to 190000.00
13 Person	Item	150000.00 to 190000.00
17 Person	Item	185000.00 to 220000.00
21 Person	Item	205000.00 to 240000.00
26 Person	Item	235000.00 to 285000.00
Electrically operated AC drive serving 12 levels and a speed of 1.0 m/s		
8 Person	Item	150000.00 to 190000.00
10 Person	Item	170000.00 to 210000.00
13 Person	Item	170000.00 to 210000.00
17 Person	Item	195000.00 to 240000.00
21 Person	Item	220000.00 to 275000.00
26 Person	Item	250000.00 to 300000.00

APPROXIMATE ESTIMATING RATES – 5 SERVICES

Item	Unit	Range £	
Electrically operated AC drive serving 14 levels and a speed of 1.0 m/s			
8 Person	Item	185000.00 to	220000.00
10 Person	Item	185000.00 to	220000.00
13 Person	Item	195000.00 to	240000.00
17 Person	Item	220000.00 to	275000.00
21 Person	Item	235000.00 to	285000.00
26 Person	Item	280000.00 to	330000.00
Add to above for			
Increase speed from 1.0 m/s to 1.6 m/s			
8 Person	Item	6000.00 to	7500.00
10 Person	Item	6400.00 to	7900.00
13 Person	Item	6600.00 to	7900.00
17 Person	Item	6600.00 to	7900.00
21 Person	Item	6600.00 to	7900.00
26 Person	Item	6600.00 to	7900.00
Increase speed from 1.6 m/s to 2.0 m/s			
8 Person	Item	3750.00 to	4300.00
10 Person	Item	3750.00 to	4300.00
13 Person	Item	4625.00 to	5500.00
17 Person	Item	4625.00 to	5500.00
21 Person	Item	4625.00 to	5800.00
26 Person	Item	4625.00 to	5800.00
Increase speed from 2.0 m/s to 2.5 m/s			
8 Person	Item	3250.00 to	3825.00
10 Person	Item	3250.00 to	3825.00
13 Person	Item	3800.00 to	4625.00
17 Person	Item	3800.00 to	4625.00
21 Person	Item	4625.00 to	5500.00
26 Person	Item	4625.00 to	5500.00
Enhanced finish to car – Centre mirror, flat ceiling, carpet			
8 Person	Item	4525.00 to	5300.00
10 Person	Item	4775.00 to	5800.00
13 Person	Item	4575.00 to	6000.00
17 Person	Item	5300.00 to	6600.00
21 Person	Item	6200.00 to	7600.00
26 Person	Item	7200.00 to	8800.00
Bottom motor room			
8 Person	Item	11000.00 to	13000.00
10 Person	Item	11000.00 to	13000.00
13 Person	Item	11000.00 to	13000.00
17 Person	Item	13500.00 to	15500.00
21 Person	Item	13500.00 to	15500.00
26 Person	Item	13500.00 to	16500.00
Fire fighting control			
8 Person	Item	8700.00 to	10500.00
10 Person	Item	8700.00 to	10500.00
13 Person	Item	8700.00 to	10500.00
17 Person	Item	8700.00 to	10500.00
21 Person	Item	8700.00 to	10500.00
26 Person	Item	8700.00 to	10500.00

APPROXIMATE ESTIMATING RATES – 5 SERVICES

Item	Unit	Range £
5.10 Lift and Conveyor Systems – cont		
Add to above for – cont		
Glass back		
8 Person	Item	3950.00 to 4725.00
10 Person	Item	4525.00 to 5300.00
13 Person	Item	5200.00 to 6400.00
17 Person	Item	6200.00 to 7600.00
21 Person	Item	6200.00 to 7600.00
26 Person	Item	6200.00 to 7600.00
Glass doors		
8 Person	Item	30000.00 to 36000.00
10 Person	Item	30000.00 to 36000.00
13 Person	Item	32500.00 to 39000.00
17 Person	Item	34500.00 to 43000.00
21 Person	Item	32500.00 to 39000.00
26 Person	Item	32500.00 to 39000.00
Painting to entire pit		
8 Person	Item	3075.00 to 3850.00
10 Person	Item	3075.00 to 3850.00
13 Person	Item	3075.00 to 3850.00
17 Person	Item	3075.00 to 3850.00
21 Person	Item	3075.00 to 3850.00
26 Person	Item	3075.00 to 3850.00
Dual seal shaft		
8 Person	Item	6200.00 to 7600.00
10 Person	Item	6200.00 to 7600.00
13 Person	Item	6200.00 to 7600.00
17 Person	Item	7400.00 to 9200.00
21 Person	Item	7400.00 to 9200.00
26 Person	Item	7400.00 to 9200.00
Dust sealing machine room		
8 Person	Item	1225.00 to 1500.00
10 Person	Item	1225.00 to 1500.00
13 Person	Item	2050.00 to 2450.00
17 Person	Item	2050.00 to 2450.00
21 Person	Item	2050.00 to 2450.00
26 Person	Item	2050.00 to 2450.00
Intercom to reception desk and security room		
8 Person	Item	600.00 to 750.00
10 Person	Item	600.00 to 750.00
13 Person	Item	600.00 to 750.00
17 Person	Item	600.00 to 750.00
21 Person	Item	600.00 to 750.00
26 Person	Item	600.00 to 750.00
Heating, cooling and ventilation to machine room		
8 Person	Item	1575.00 to 1925.00
10 Person	Item	1575.00 to 1925.00
13 Person	Item	1575.00 to 1925.00
17 Person	Item	1575.00 to 1925.00
21 Person	Item	1575.00 to 1925.00
26 Person	Item	1575.00 to 1925.00

APPROXIMATE ESTIMATING RATES – 5 SERVICES

Item	Unit	Range £	
Shaft lighting/small power			
8 Person	Item	5300.00 to	6600.00
10 Person	Item	5300.00 to	6600.00
13 Person	Item	5300.00 to	6600.00
17 Person	Item	5300.00 to	6600.00
21 Person	Item	5300.00 to	6600.00
26 Person	Item	5300.00 to	6600.00
Motor room lighting/small power			
8 Person	Item	2000.00 to	2350.00
10 Person	Item	2000.00 to	2350.00
13 Person	Item	2450.00 to	3000.00
17 Person	Item	2450.00 to	3000.00
21 Person	Item	2450.00 to	3000.00
26 Person	Item	2450.00 to	3000.00
Lifting beams			
8 Person	Item	2300.00 to	2875.00
10 Person	Item	2300.00 to	2875.00
13 Person	Item	2300.00 to	2875.00
17 Person	Item	2625.00 to	3200.00
21 Person	Item	2625.00 to	3200.00
26 Person	Item	2625.00 to	3200.00
10 mm equipotential bonding of all entrance metalwork			
8 Person	Item	1225.00 to	1500.00
10 Person	Item	1225.00 to	1500.00
13 Person	Item	1225.00 to	1500.00
17 Person	Item	1225.00 to	1500.00
21 Person	Item	1225.00 to	1500.00
26 Person	Item	1225.00 to	1500.00
Shaft secondary steelwork			
8 Person	Item	9000.00 to	11500.00
10 Person	Item	9200.00 to	11500.00
13 Person	Item	9400.00 to	12000.00
17 Person	Item	9800.00 to	12000.00
21 Person	Item	9800.00 to	12000.00
26 Person	Item	9800.00 to	12000.00
Independent insurance inspection			
8 Person	Item	2925.00 to	3575.00
10 Person	Item	2925.00 to	3575.00
13 Person	Item	2925.00 to	3575.00
17 Person	Item	2925.00 to	3575.00
21 Person	Item	2925.00 to	3575.00
26 Person	Item	2925.00 to	3575.00
12 month warranty service			
8 Person	Item	1575.00 to	1925.00
10 Person	Item	1575.00 to	1925.00
13 Person	Item	1575.00 to	1925.00
17 Person	Item	1575.00 to	1925.00
21 Person	Item	1575.00 to	1925.00
26 Person	Item	1575.00 to	1925.00

APPROXIMATE ESTIMATING RATES – 5 SERVICES

Item	Unit	Range £
5.10 Lift and Conveyor Systems – cont		
Passenger Lifts, machine room less		
Electrically operated AC drive serving 2 levels with directional collective controls and a speed of 1.0 m/s		
8 Person	Item	83000.00 to 99000.00
10 Person	Item	90000.00 to 115000.00
13 Person	Item	96000.00 to 120000.00
17 Person	Item	115000.00 to 140000.00
21 Person	Item	125000.00 to 150000.00
26 Person	Item	140000.00 to 165000.00
Electrically operated AC drive serving 4 levels and a speed of 1.0 m/s		
8 Person	Item	95000.00 to 115000.00
10 Person	Item	105000.00 to 130000.00
13 Person	Item	110000.00 to 135000.00
17 Person	Item	130000.00 to 160000.00
21 Person	Item	140000.00 to 165000.00
26 Person	Item	150000.00 to 190000.00
Electrically operated AC drive serving 6 levels and a speed of 1.0 m/s		
8 Person	Item	110000.00 to 135000.00
10 Person	Item	120000.00 to 140000.00
13 Person	Item	125000.00 to 150000.00
17 Person	Item	140000.00 to 165000.00
21 Person	Item	150000.00 to 190000.00
26 Person	Item	165000.00 to 200000.00
Electrically operated AC drive serving 8 levels and a speed of 1.0 m/s		
8 Person	Item	125000.00 to 150000.00
10 Person	Item	130000.00 to 150000.00
13 Person	Item	135000.00 to 165000.00
17 Person	Item	150000.00 to 190000.00
21 Person	Item	165000.00 to 200000.00
26 Person	Item	185000.00 to 225000.00
Electrically operated AC drive serving 10 levels and a speed of 1.0 m/s		
8 Person	Item	135000.00 to 165000.00
10 Person	Item	140000.00 to 165000.00
13 Person	Item	150000.00 to 180000.00
17 Person	Item	165000.00 to 200000.00
21 Person	Item	185000.00 to 225000.00
26 Person	Item	210000.00 to 260000.00
Electrically operated AC drive serving 12 levels and a speed of 1.0 m/s		
8 Person	Item	150000.00 to 180000.00
10 Person	Item	150000.00 to 180000.00
13 Person	Item	165000.00 to 200000.00
17 Person	Item	185000.00 to 225000.00
21 Person	Item	195000.00 to 240000.00
26 Person	Item	220000.00 to 275000.00

APPROXIMATE ESTIMATING RATES – 5 SERVICES

Item	Unit	Range £	
Electrically operated AC drive serving 14 levels and a speed of 1.0 m/s			
8 Person	Item	165000.00 to	200000.00
10 Person	Item	165000.00 to	200000.00
13 Person	Item	170000.00 to	210000.00
17 Person	Item	210000.00 to	260000.00
21 Person	Item	220000.00 to	275000.00
26 Person	Item	250000.00 to	300000.00
Add to above for			
Increase speed from 1.0 m/s to 1.6 m/s			
8 Person	Item	4000.00 to	4875.00
10 Person	Item	4000.00 to	4875.00
13 Person	Item	4200.00 to	5100.00
17 Person	Item	6500.00 to	7800.00
21 Person	Item	7500.00 to	9000.00
26 Person	Item	9400.00 to	12000.00
Enhanced finish to car – Centre mirror, flat ceiling, carpet			
8 Person	Item	4350.00 to	5300.00
10 Person	Item	4600.00 to	5700.00
13 Person	Item	4600.00 to	5700.00
17 Person	Item	5300.00 to	6500.00
21 Person	Item	6000.00 to	7500.00
26 Person	Item	7300.00 to	8800.00
Fire fighting control			
8 Person	Item	8300.00 to	10500.00
10 Person	Item	8300.00 to	10500.00
13 Person	Item	8300.00 to	10500.00
17 Person	Item	8300.00 to	10500.00
21 Person	Item	8300.00 to	10500.00
26 Person	Item	8300.00 to	10500.00
Painting to entire pit			
8 Person	Item	3050.00 to	3725.00
10 Person	Item	3050.00 to	3725.00
13 Person	Item	3050.00 to	3725.00
17 Person	Item	3050.00 to	3725.00
21 Person	Item	3050.00 to	3725.00
26 Person	Item	3050.00 to	3725.00
Dual seal shaft			
8 Person	Item	6300.00 to	7500.00
10 Person	Item	6300.00 to	7500.00
13 Person	Item	6300.00 to	7500.00
17 Person	Item	7500.00 to	9000.00
21 Person	Item	7500.00 to	9000.00
26 Person	Item	7500.00 to	9000.00
Shaft lighting/small power			
8 Person	Item	5300.00 to	6600.00
10 Person	Item	5300.00 to	6600.00
13 Person	Item	5300.00 to	6600.00
17 Person	Item	5300.00 to	6600.00
21 Person	Item	5300.00 to	6600.00
26 Person	Item	5300.00 to	6600.00

APPROXIMATE ESTIMATING RATES – 5 SERVICES

Item	Unit	Range £	
5.10 Lift and Conveyor Systems – cont			
Add to above for – cont			
Intercom to reception desk and security room			
8 Person	Item	600.00 to	750.00
10 Person	Item	600.00 to	750.00
13 Person	Item	600.00 to	750.00
17 Person	Item	600.00 to	750.00
21 Person	Item	600.00 to	750.00
26 Person	Item	600.00 to	750.00
Lifting beams			
8 Person	Item	2300.00 to	2800.00
10 Person	Item	2300.00 to	2800.00
13 Person	Item	2300.00 to	2800.00
17 Person	Item	2625.00 to	3200.00
21 Person	Item	2625.00 to	3200.00
26 Person	Item	2625.00 to	3200.00
10 mm equipotential bonding of all entrance metalwork			
8 Person	Item	1225.00 to	1500.00
10 Person	Item	1225.00 to	1500.00
13 Person	Item	1225.00 to	1500.00
17 Person	Item	1225.00 to	1500.00
21 Person	Item	1225.00 to	1500.00
26 Person	Item	1225.00 to	1500.00
Shaft secondary steelwork			
8 Person	Item	8900.00 to	10500.00
10 Person	Item	8800.00 to	11500.00
13 Person	Item	9200.00 to	11500.00
17 Person	Item	9400.00 to	12000.00
21 Person	Item	9900.00 to	12500.00
26 Person	Item	9900.00 to	12500.00
Independent insurance inspection			
8 Person	Item	2800.00 to	3425.00
10 Person	Item	2800.00 to	3425.00
13 Person	Item	2800.00 to	3425.00
17 Person	Item	2800.00 to	3425.00
21 Person	Item	2800.00 to	3425.00
26 Person	Item	2800.00 to	3425.00
12 month warranty service			
8 Person	Item	1525.00 to	1825.00
10 Person	Item	1525.00 to	1825.00
13 Person	Item	1525.00 to	1825.00
17 Person	Item	1525.00 to	1825.00
21 Person	Item	1525.00 to	1825.00
26 Person	Item	1525.00 to	1825.00

APPROXIMATE ESTIMATING RATES – 5 SERVICES

Item	Unit	Range £
Goods Lifts, machine room above		
Electrically operated two speed serving 2 levels to take 1000 kg load, prime coated internal finish and a speed of 1.0 m/s		
2000 kg	Item	150000.00 to 190000.00
2250 kg	Item	165000.00 to 200000.00
2500 kg	Item	165000.00 to 200000.00
3000 kg	Item	185000.00 to 230000.00
Electrically operated two speed serving 4 levels and a speed of 1.0 m/s		
2000 kg	Item	165000.00 to 200000.00
2250 kg	Item	185000.00 to 230000.00
2500 kg	Item	190000.00 to 230000.00
3000 kg	Item	220000.00 to 270000.00
Electrically operated two speed serving 6 levels and a speed of 1.0 m/s		
2000 kg	Item	190000.00 to 230000.00
2250 kg	Item	210000.00 to 260000.00
2500 kg	Item	210000.00 to 260000.00
3000 kg	Item	235000.00 to 285000.00
Electrically operated two speed serving 8 levels and a speed of 1.0 m/s		
2000 kg	Item	210000.00 to 260000.00
2250 kg	Item	220000.00 to 280000.00
2500 kg	Item	240000.00 to 285000.00
3000 kg	Item	265000.00 to 325000.00
Electrically operated two speed serving 10 levels and a speed of 1.0 m/s		
2000 kg	Item	240000.00 to 285000.00
2250 kg	Item	255000.00 to 300000.00
2500 kg	Item	255000.00 to 300000.00
3000 kg	Item	285000.00 to 350000.00
Electrically operated two speed serving 12 levels and a speed of 1.0 m/s		
2000 kg	Item	255000.00 to 300000.00
2250 kg	Item	265000.00 to 325000.00
2500 kgl	Item	265000.00 to 325000.00
3000 kg	Item	315000.00 to 385000.00
Electrically operated two speed serving 14 levels and a speed of 1.0 m/s		
2000 kg	Item	280000.00 to 335000.00
2250 kg	Item	295000.00 to 360000.00
2500 kg	Item	295000.00 to 360000.00
3000 kg	Item	335000.00 to 410000.00
Add to above for		
Increase speed of travel from 1.0 m/s to 1.6 m/s		
2000 kg	Item	2025.00 to 2375.00
Enhanced finish to car – Centre mirror, flat ceiling, carpet		
2000 kg	Item	5400.00 to 6600.00
2250 kg	Item	5400.00 to 6600.00
2500 kg	Item	5400.00 to 6600.00
Bottom motor room		
2000 kg	Item	13500.00 to 15500.00

APPROXIMATE ESTIMATING RATES – 5 SERVICES

Item	Unit	Range £
5.10 Lift and Conveyor Systems – cont		
Add to above for – cont		
Painting to entire pit		
2000 kg	Item	3075.00 to 3750.00
2250 kg	Item	3075.00 to 3750.00
2500 kg	Item	3075.00 to 3750.00
3000 kg	Item	3275.00 to 4000.00
Dual seal shaft		
2000 kg	Item	6200.00 to 7600.00
2250 kg	Item	6200.00 to 7600.00
2500 kg	Item	6200.00 to 7600.00
3000 kg	Item	6200.00 to 7600.00
Intercom to reception desk and security room		
2000 kg	Item	620.00 to 760.00
2250 kg	Item	620.00 to 760.00
2500 kg	Item	620.00 to 760.00
3000 kg	Item	620.00 to 760.00
Heating, cooling and ventilation to machine room		
2000 kg	Item	1700.00 to 2075.00
2250 kg	Item	1700.00 to 2075.00
2500 kg	Item	1700.00 to 2075.00
3000 kg	Item	1700.00 to 2075.00
Lifting beams		
2000 kg	Item	2275.00 to 2825.00
2250 kg	Item	2275.00 to 2825.00
2500 kg	Item	2275.00 to 2825.00
3000 kg	Item	2275.00 to 2825.00
10 mm equipotential bonding of all entrance metalwork		
2000 kg	Item	1250.00 to 1500.00
2250 kg	Item	1250.00 to 1500.00
2500 kg	Item	1250.00 to 1500.00
3000 kg	Item	1250.00 to 1500.00
Independent insurance inspection		
2000 kg	Item	2925.00 to 3575.00
2250 kg	Item	2925.00 to 3575.00
2500 kg	Item	2925.00 to 3575.00
3000 kg	Item	2925.00 to 3575.00
12 month warranty service		
2000 kg	Item	1575.00 to 1925.00
2250 kg	Item	1875.00 to 2250.00
2500 kg	Item	1875.00 to 2250.00
3000 kg	Item	1875.00 to 2250.00
Goods Lift, machine room less		
Electrically operated two speed serving 2 levels to take 1000 kg load, prime coated internal finish and a speed of 1.0 m/s		
2000 kg	Item	135000.00 to 165000.00
2250 kg	Item	140000.00 to 165000.00
2500 kg	Item	150000.00 to 190000.00

APPROXIMATE ESTIMATING RATES – 5 SERVICES

Item	Unit	Range £
Electrically operated two speed serving 4 levels and a speed of 1.0 m/s		
2000 kg	Item	150000.00 to 190000.00
2250 kg	Item	155000.00 to 200000.00
2500 kg	Item	170000.00 to 210000.00
Electrically operated two speed serving 6 levels and a speed of 1.0 m/s		
2000 kg	Item	170000.00 to 210000.00
2250 kg	Item	170000.00 to 210000.00
2500 kg	Item	195000.00 to 245000.00
Electrically operated two speed serving 8 levels and a speed of 1.0 m/s		
2000 kg	Item	185000.00 to 230000.00
2250 kg	Item	195000.00 to 245000.00
2500 kg	Item	210000.00 to 260000.00
Electrically operated two speed serving 10 levels and a speed of 1.0 m/s		
2000 kg	Item	205000.00 to 250000.00
2250 kg	Item	210000.00 to 260000.00
2500 kg	Item	220000.00 to 280000.00
Electrically operated two speed serving 12 levels and a speed of 1.0 m/s		
2000 kg	Item	185000.00 to 230000.00
2250 kg	Item	240000.00 to 285000.00
2500 kg	Item	255000.00 to 300000.00
Electrically operated two speed serving 14 levels and a speed of 1.0 m/s		
2000 kg	Item	240000.00 to 285000.00
2250 kg	Item	255000.00 to 300000.00
2500 kg	Item	265000.00 to 325000.00
Add to above for		
Increase speed of travel from 1.0 m/s to 1.6 m/s		
2000 kg	Item	7400.00 to 9300.00
2250 kg	Item	12000.00 to 14500.00
Enhanced finish to car – Centre mirror, flat ceiling, carpet		
2000 kg	Item	6300.00 to 7700.00
2250 kg	Item	8300.00 to 10500.00
2500 kg	Item	12500.00 to 15000.00
Painting to entire pit		
2000 kg	Item	3125.00 to 3825.00
2250 kg	Item	3125.00 to 3825.00
2500 kg	Item	3125.00 to 3825.00
Dual seal shaft		
2000 kg	Item	6200.00 to 7600.00
2250 kg	Item	6200.00 to 7600.00
2500 kg	Item	6200.00 to 7600.00
Intercom to reception desk and security room		
2000 kg	Item	620.00 to 760.00
2250 kg	Item	620.00 to 760.00
2500 kg	Item	620.00 to 760.00
Lifting beams		
2000 kg	Item	2350.00 to 2875.00
2250 kg	Item	2350.00 to 2875.00
2500 kg	Item	2350.00 to 2875.00

APPROXIMATE ESTIMATING RATES – 5 SERVICES

Item	Unit	Range £	
5.10 Lift and Conveyor Systems – cont			
Add to above for – cont			
10 mm equipotential bonding of all entrance metalwork			
2000 kg	Item	1250.00 to	1500.00
2250 kg	Item	1250.00 to	1500.00
2500 kg	Item	1250.00 to	1500.00
Independent insurance inspection			
2000 kg	Item	2900.00 to	3625.00
2250 kg	Item	2900.00 to	3625.00
2500 kg	Item	2900.00 to	3625.00
12 month warranty service			
2000 kg	Item	1575.00 to	1925.00
2250 kg	Item	1575.00 to	1925.00
2500 kg	Item	1575.00 to	1925.00
ESCALATOR INSTALLATIONS			
30Ø Pitch escalator with a rise of 3 to 6 metres with standard balustrades			
1000 mm step width	Item	135000.00 to	160000.00
Add to above for			
Balustrade lighting	Item	4175.00 to	5200.00
Skirting lighting	Item	16500.00 to	20500.00
Emergency stop button pedestals	Item	7400.00 to	9300.00
Truss cladding – stainless steel	Item	42000.00 to	52000.00
Truss cladding – spray painted steel	Item	37000.00 to	45500.00

APPROXIMATE ESTIMATING RATES – 5 SERVICES

Item	Unit	Range £
5.11 Fire and Lightning Protection		
SPRINKLER INSTALLATION		
Recommended maximum area coverage per sprinkler head:		
Extra light hazard, 21 m² of floor area		
Ordinary hazard, 12 m² of floor area		
Extra high hazard, 9 m² of floor area		
Equipment		
Sprinkler equipment installation, pipework, valve sets, booster pumps and water storage	Item	185000.00 to 215000.00
Price per sprinkler head; including pipework, valves and supports	Point	570.00 to 630.00
HOSE REELS AND DRY RISERS		
Wall mounted concealed hose reel with 36 metre hose including approximately 15 metres of pipework and isolating valve		
Price per hose reel	Point	2450.00 to 3050.00
100 mm dry riser main including 2 way breeching valve and box, 65 mm landing valve, complete with padlock and leather strap and automatic air vent and drain valve		
Price per landing	Point	2975.00 to 3550.00
5.12 Communication, Security and Control Systems		
ACCESS CONTROL SYSTEMS		
Door mounted access control unit inclusive of door furniture, lock plus software; including up to 50 metres of cable and termination; including documentation testing and commissioning		
Internal single leaf door	Point	1900.00 to 2325.00
Internal double door	Point	2075.00 to 2550.00
External single leaf door	Point	2250.00 to 2750.00
External double leaf door	Point	2575.00 to 3200.00
Management control PC with printer software and commissioning up to 1000 users	Point	21500.00 to 25500.00
CCTV INSTALLATIONS		
CCTV equipment inclusive of 50 m of cable including testing and commissioning		
Internal camera with bracket	Point	1325.00 to 1550.00
Internal camera with housing	Point	1600.00 to 1950.00
Internal PTZ camera with bracket	Point	2425.00 to 3025.00
External fixed camera with housing	Point	1750.00 to 2150.00
External PTZ camera dome	Point	3475.00 to 4150.00
External PTZ camera dome with power	Point	4250.00 to 5100.00

APPROXIMATE ESTIMATING RATES – 5 SERVICES

Item	Unit	Range £
5.12 Communication, Security and Control Systems – cont		
TURNSTILES		
Physical Access Control Barrier system – standard security level comprising unit caseworks in stainless steel finish, standard lane width – 650 mm, restricting panels standard 900 mm high, standard level detection sensor system, including provision to integrate Access Control Card Readers (issued by others), including LED Pictogram – Green Arrow/Red Cross, closed base for ease of installation and cable management		
Including delivery, installation and commissioning to a site in London; budget cost land configurations		
single lane	Item	13500.00 to 14500.00
double lane	Item	21500.00 to 22500.00
triple lane	Item	29500.00 to 31500.00
Physical Access Control Barrier system – medium security level comprising unit caseworks in stainless steel finish, standard lane width – 650 mm, restricting panels standard 900 mm high, medium level detection sensor system, including provision to integrate Access Control Card Readers (issued by others), including LED Pictogram – Green Arrow/Red Cross, closed base for ease of installation and cable management		
Including delivery, installation and commissioning to a site in London; budget cost land configurations		
single lane	Item	15500.00 to 17500.00
double lane	Item	23000.00 to 25000.00
triple lane	Item	33000.00 to 34500.00
Physical Access Control Barrier system – high security level comprising unit caseworks in stainless steel finish, standard lane width – 650 mm, restricting panels standard 1600 mm or 1800 mm high (same cost), higher level detection sensor system, including provision to integrate Access Control Card Readers (issued by others), including LED Pictogram – Green Arrow/Red Cross, closed base for ease of installation and cable management		
Including delivery, installation and commissioning to a site in London; budget cost land configurations		
single lane	Item	15500.00 to 17500.00
double lane	Item	26500.00 to 29500.00
triple lane	Item	38000.00 to 40500.00
High security full height security revolving door – 2300 mm high, four wing T25 sections, no centre column, positioning drive with horizontal and vertical safety strips at door leaves, controlled via electronic Access Control Card Readers (supplied by others), sensor system in ceiling monitors door segments, secure simultaneous bi-directional use possible, finish – standard finish – PPC to RAL colour, card reader mounting boxes, 4 LED lights in ceiling, rubber matting		
Including delivery, installation and commissioning to a site in London		
buget cost – 1800 mm dia., per door	Item	34500.00 to 36000.00
buget cost – 2000 mm dia., per door	Item	34500.00 to 36000.00
additional cost – stainless steel finish	Item	2450.00 to 2450.00

APPROXIMATE ESTIMATING RATES – 5 SERVICES

Item	Unit	Range £	
IT INSTALLATIONS			
DATA CABLING			
Complete channel link including patch leads, cable, panels, testing and documentation (excludes cabinets and/or frames, patch cords, backbone/harness connectivity as well as containment)			
Low Level			
Cat 5e (up to 5,000 outlets)	Point	76.00 to	87.00
Cat 5e (5,000 to 15,000 outlets)	Point	50.00 to	62.00
Cat 6 (up to 5,000 outlets)	Point	87.00 to	100.00
Cat 6 (5,000 to 15,000 outlets)	Point	76.00 to	87.00
Cat 6a (up to 5,000 outlets)	Point	100.00 to	120.00
Cat 6a (5,000 to 15,000 outlets)	Point	100.00 to	110.00
Cat 7 (up to 5,000 outlets)	Point	120.00 to	155.00
Cat 7 (5,000 to 15,000 outlets)	Point	110.00 to	145.00

Note: LSZH cable based on average of 50 metres false floor low level installation assuming 1 workstation in 2.5 m × 2.5 m (to 3.2 m × 3.2 m) density, with 4 data points per workstation. Not applicable for installations with less than 250 No. outlets.

Item	Unit	Range £	
High Level			
Cat 5e (up to 500 outlets)	Point	87.00 to	100.00
Cat 5e (over 500 outlets)	Point	76.00 to	87.00
Cat 6 (up to 500 outlets)	Point	100.00 to	110.00
Cat 6 (over 500 outlets)	Point	89.00 to	110.00
Cat 6a (up to 500 outlets)	Point	110.00 to	145.00
Cat 6a (over 500 outlets)	Point	110.00 to	140.00
Cat 7 (up to 500 outlets)	Point	135.00 to	165.00
Cat 7 (over 500 outlets)	Point	135.00 to	165.00

Note: High level at 10 × 10 m grid.

BUILDING MODELS – ELEMENTAL COST SUMMARIES

Item	Unit	Range £	
AIRPORT TERMINAL BUILDING			
New build airport terminal building, premium quality, located in the South East of England, handling both domestic and international flights with a gross internal floor area (GIFA) of 25,000 m². These costs exclude baggage handling, check-in systems, pre-check-in and boarding security systems, vertical transportation and travelators, pre-conditioned air systems to aircraft, services to stands and visual docking systems. Costs assume that the works are undertaken under landside access/logistics environment.			
5 Services			
5.1 Sanitary Installations	m²	8.00 to	14.00
5.3 Disposal Installations			
rainwater	m²	8.00 to	13.00
soil and waste	m²	12.00 to	19.00
condensate	m²	2.00 to	3.00
5.4 Water Installations			
domestic hot and cold water services	m²	26.00 to	29.00
5.5 Heat Source	m²	26.00 to	32.00
5.6 Space Heating and Air Conditioning			
LTHW heating system	m²	110.00 to	130.00
chilled water system	m²	130.00 to	145.00
supply and extract air conditioning system	m²	150.00 to	185.00
local cooling; DX systems to IT rooms	m²	7.00 to	13.00
5.7 Ventilation Systems			
mechanical ventilation to baggage handling and plantrooms	m²	39.00 to	45.00
toilet extract ventilation	m²	13.00 to	16.00
smoke extract installation	m²	16.00 to	24.00
kitchen extract system	m²	3.00 to	6.00
5.8 Electrical Installations			
main HV/MV Installations including switchgear, transformers (Cast Resin)	m²	90.00 to	115.00
low voltage (LV) – Incoming LV switchgear, distribution boards and distribution systems	m²	63.00 to	77.00
generators, life safety – Containerized standby diesel generators and 24 hour capacity			
belly tanks	m²	48.00 to	63.00
small power installation	m²	63.00 to	77.00
lighting installations; warehouse, office, ancillary and plant spaces	m²	170.00 to	205.00
emergency lighting installation	m²	16.00 to	26.00
power to mechanical services	m²	10.00 to	21.00
5.9 Fuel Installations (gas)	m²	2.00 to	3.00
5.11 Fire and Lightning Protection			
lightning protection	m²	1.00 to	2.00
earthing and bonding	m²	1.00 to	2.00
sprinkler installations	m²	81.00 to	100.00
dry riser and hosereels installation	m²	8.00 to	12.00
fire suppression to IT rooms	m²	6.00 to	7.00

BUILDING MODELS – ELEMENTAL COST SUMMARIES

Item	Unit	Range £	
5.12 Communication, Security and Control Systems			
fire alarm systems; DDI	m^2	48.00 to	62.00
voice/public address systems	m^2	29.00 to	37.00
other alarm systems, e.g. disabled refuge	m^2	2.00 to	4.00
security installations; CCTV, access control and intruder detection	m^2	51.00 to	64.00
wireways for IT, comms and FA systems	m^2	20.00 to	25.00
structured IT cabling; fibre optic backbone and copper Cat 6 to office and warehouse	m^2	16.00 to	21.00
BMS systems	m^2	79.00 to	91.00
flight information display system	m^2	40.00 to	49.00
Total Cost/m² (based on GIFA of 25,000 m²)	m^2	1400.00 to	1750.00

BUILDING MODELS – ELEMENTAL COST SUMMARIES

Item	Unit	Range £	
SHOPPING MALL (TENANT'S FIT-OUT EXCLUDED)			
Natural ventilation shopping mall with approximately 33,000 m² two storey retail area and a 13,000 m² above ground, mechanically ventilated, covered car park, situated in a town centre in South East England.			
RETAIL BUILDING			
5 Services			
5.1 Sanitary Installations	m²	1.00 to	2.00
5.3 Disposal Installations			
rainwater	m²	8.00 to	9.00
soil, waste and vent	m²	8.00 to	9.00
5.4 Water Installations			
cold water installation	m²	8.00 to	9.00
hot water installation	m²	7.00 to	8.00
5.6 Space Heating and Air Conditioning			
condenser water system	m²	46.00 to	48.00
LTHW installation	m²	6.00 to	7.00
air conditioning system	m²	63.00 to	83.00
over door heaters at entrances	m²	1.00 to	2.00
5.7 Ventilation Systems			
public toilet ventilation	m²	1.00 to	2.00
plantroom ventilation	m²	6.00 to	7.00
supply and extract systems to shop units	m²	21.00 to	21.00
toilet extract systems to shop units	m²	2.00 to	3.00
smoke ventilation to mall area	m²	15.00 to	16.00
service corridor ventilation	m²	2.00 to	3.00
other miscellaneous ventilation	m²	27.00 to	28.00
5.8 Electrical Installations			
LV distribution	m²	41.00 to	51.00
standby power	m²	8.00 to	10.00
general lighting	m²	94.00 to	98.00
external lighting	m²	7.00 to	8.00
emergency lighting	m²	20.00 to	20.00
small power	m²	15.00 to	16.00
mechanical services power supplies	m²	4.00 to	6.00
general earthing	m²	1.00 to	2.00
UPS for security and CCTV equipment	m²	1.00 to	2.00
5.9 Fuel Installations (gas)			
gas supply and boilers	m²	2.00 to	3.00
gas supplies to anchor (major) stores	m²	1.00 to	2.00
5.11 Fire and Lightning Protection			
lightning protection	m²	1.00 to	2.00
sprinkler installations	m²	61.00 to	80.00
5.12 Communication, Security and Control Systems			
fire alarm installation	m²	13.00 to	13.00
public address/voice alarm	m²	7.00 to	8.00
security installation	m²	16.00 to	18.00
general containment	m²	18.00 to	20.00

BUILDING MODELS – ELEMENTAL COST SUMMARIES

Item	Unit	Range £	
5.13 Specialist Installations			
BMS/Controls	m²	30.00 to	51.00
Total Cost/m² (based on GIFA of 33,000 m²)	m²	560.00 to	670.00
CAR PARK – 13,000 m²			
5 Services			
5.3 Disposal Intallations			
car park drainage	m²	6.00 to	7.00
5.7 Ventilation Systems			
car park ventilation (impulse fans)	m²	46.00 to	51.00
5.8 Electrical Installations			
LV distribution	m²	20.00 to	25.00
general lighting	m²	32.00 to	34.00
emergency lighting	m²	7.00 to	8.00
small power	m²	4.00 to	4.00
mechanical services power supplies	m²	7.00 to	8.00
general earthing	m²	1.00 to	2.00
ramp frost protection	m²	2.00 to	3.00
5.11 Fire and Lightning Protection			
sprinkler installation	m²	41.00 to	51.00
fire alarm installations	m²	30.00 to	33.00
5.12 Communication, Security and Control Systems			
security installations	m²	14.00 to	14.00
BMS/Controls	m²	8.00 to	9.00
5.13 Specialist Installations			
entry/exit barriers, pay stations	m²	8.00 to	9.00
Total Cost/m² (based on GIFA of 13,000 m²)	m²	225.00 to	260.00

BUILDING MODELS – ELEMENTAL COST SUMMARIES

Item	Unit	Range £	
SUPERMARKET			
Supermarket located in the South East with a total gross floor area of 4,000 m², including a sales area of 2,350 m². The building is on one level and incorporates a main sales, coffee shop, bakery, offices and amenities areas and warehouse.			
5 Services			
5.1 Sanitary Installations	m²	2.00 to	3.00
5.3 Disposal Installations	m²	4.00 to	6.00
5.4 Water Installations			
hot and cold water services	m²	39.00 to	40.00
5.6 Space Heating and Air Conditioning			
heating & ventilation with cooling via DX units	m²	25.00 to	40.00
5.7 Ventilation Systems			
supply and extract sytems	m²	10.00 to	15.00
5.8 Electrical Installations			
panels/boards	m²	51.00 to	61.00
containment	m²	5.00 to	10.00
general lighting	m²	30.00 to	51.00
small power	m²	15.00 to	16.00
mechanical services wiring	m²	3.00 to	4.00
5.11 Fire and Lightning Protection			
lightning protection	m²	1.00 to	2.00
5.12 Communication, Security and Control Systems			
fire alarms, detection and public address	m²	5.00 to	10.00
CCTV	m²	5.00 to	10.00
intruder alarm, detection and store security	m²	3.00 to	5.00
telecom and structured cabling	m²	3.00 to	5.00
BMS	m²	15.00 to	20.00
data cabinet	m²	5.00 to	10.00
controls wiring	m²	5.00 to	6.00
5.13 Specialist Installations			
refrigeration			
installation	m²	42.00 to	45.00
plant	m²	42.00 to	45.00
cold store	m²	15.00 to	16.00
cabinets	m²	115.00 to	125.00
Total Cost/m² (based on GIFA of 4,000 m²)	m²	440.00 to	550.00

BUILDING MODELS – ELEMENTAL COST SUMMARIES

Item	Unit	Range £	
OFFICE BUILDING			
Speculative 15 storey office in Central London for multiple tenant occupancy with gross internal area of 20,000m^2. 1 level of basement of approx. 1,600m^2, total NIA of 13,000m^2. An allowance has been made for building smart technology, wired enabled and BREEAM Excellent.			
SHELL & CORE – 20,000 m^2			
5 Services			
5.1 Sanitary Installations	m^2	20.00 to	35.00
5.3 Disposal Installations			
rainwater/soil and waste	m^2	32.00 to	36.00
condensate	m^2	2.00 to	3.00
5.4 Water Installations			
hot and cold water services	m^2	65.00 to	70.00
5.5 Heat Source – (included in Section 5.6)	N/A		
5.6 Space Heating and Air Conditioning	m^2	165.00 to	175.00
5.7 Ventilation Systems	m^2	100.00 to	120.00
5.8 Electrical Installations	m^2	215.00 to	240.00
5.11 Fire and Lighting Protection	m^2	38.00 to	43.00
5.12 Communication, Security and Control Systems	m^2	150.00 to	170.00
Total Cost/m^2 (based on GIFA of 20,000 m^2)	m^2	790.00 to	900.00
CATEGORY 'A' FIT-OUT – 13,000 m^2 NIA			
5 Services			
5.6 Space Heating and Air Conditioning	m^2	250.00 to	260.00
5.8 Electrical Installations	m^2	185.00 to	195.00
5.11 Fire and Lightning Protection	m^2	33.00 to	41.00
5.12 Communication, Security and Control Systems	m^2	52.00 to	64.00
Total Cost/m^2 (based on NIFA of 13,000 m^2)	m^2	520.00 to	570.00

BUILDING MODELS – ELEMENTAL COST SUMMARIES

Item	Unit	Range £	
BUSINESS PARK			
New build office in South East within M25 part of a speculative business park with a gross floor area of 10,000m². A full air displacement system with roof mounted air source heat pumps and air handling plant. Total M&E services value approximately £3,700,000 (shell & core) and £1,500,000 (Cat A fit-out). Excluding lifts.			
SHELL & CORE – 10,000 m² GIA			
5 Services			
5.1 Sanitary Installations	m²	10.00 to	14.00
5.3 Disposal Installations			
condensate	m²	3.00 to	4.00
rainwater, soil and waste	m²	15.00 to	20.00
5.4 Water Installations			
hot and cold water services	m²	16.00 to	21.00
5.5 Heat Source	m²	10.00 to	14.00
5.6 Space Heating and Air Conditioning			
LTHW heating; plantroom and risers	m²	16.00 to	21.00
chilled water; plantroom and risers	m²	43.00 to	49.00
ductwork; plantroom and risers	m²	81.00 to	94.00
5.7 Ventilation Systems			
toilet and miscellaneous ventilation	m²	16.00 to	21.00
5.8 Electrical Installations			
LV supply/distribution	m²	35.00 to	43.00
general lighting	m²	28.00 to	35.00
general power	m²	6.00 to	9.00
electrical services in connection with mechanical services	m²	4.00 to	6.00
security (wireways)	m²	2.00 to	4.00
voice and data (wireways)	m²	2.00 to	4.00
5.9 Fuel Installations (gas)	m²	2.00 to	4.00
5.11 Fire and Lightning Protection			
earthing and bonding	m²	3.00 to	6.00
lightning protection	m²	3.00 to	6.00
dry risers	m²	2.00 to	4.00
5.12 Communication, Security and Control Systems			
fire alarms	m²	14.00 to	18.00
BMS	m²	32.00 to	36.00
Total Cost/m² (based on GIFA of 10,000 m²)	m²	345.00 to	430.00

BUILDING MODELS – ELEMENTAL COST SUMMARIES

Item	Unit	Range £	
CATEGORY 'A' FIT-OUT – 8,000 m² NIA			
5 Services			
5.6 Space Heating and Air Conditioning			
LTHW heating and perimeter heaters	m²	49.00 to	60.00
floor swirl diffusers and supply ductwork	m²	30.00 to	36.00
5.8 Electrical Installations			
distribution boards	m²	2.00 to	4.00
general lightning, recessed including lighting controls	m²	66.00 to	80.00
5.11 Fire and Lightning Protection			
earthing and bonding	m²	2.00 to	4.00
5.12 Communication, Security and Control Systems			
fire alarms	m²	10.00 to	12.00
BMS	m²	7.00 to	9.00
Total Cost/m² (based on NIFA of 8,000 m²)	m²	180.00 to	220.00

BUILDING MODELS – ELEMENTAL COST SUMMARIES

Item	Unit	Range £	
PERFORMING ARTS CENTRE (MEDIUM SPECIFICATION)			
Performing Arts centre with a Gross Internal Area (GIA) of approximately 8,000 m², based on a medium specification with cooling to the Auditorium.			
The development comprises dance studios and a theatre auditorium in the outer London area. The theatre would require all the necessary stage lighting, machinery and equipment installed in a modern professional theatre (these are excluded from the model, as assumed to be FF&E, but the containment and power wiring is included).			
5 Services			
5.1 Sanitary Installations	m²	13.00 to	16.00
5.3 Disposal Installations			
soil, waste and rainwater	m²	21.00 to	27.00
5.4 Water Installations			
cold water services	m²	20.00 to	26.00
hot water services	m²	16.00 to	21.00
5.5 Heat Source	m²	15.00 to	20.00
5.6 Space Heating and Air Conditioning			
heating	m²	98.00 to	105.00
chilled water system	m²	85.00 to	115.00
supply and extract air systems	m²	155.00 to	180.00
5.7 Ventilation Systems			
ventilation and extract systems to toilets, kitchen and workshop	m²	28.00 to	38.00
5.8 Electrical Installations			
LV supply/distribution	m²	58.00 to	63.00
general lighting	m²	140.00 to	170.00
small power	m²	43.00 to	49.00
power to mechanical plant	m²	12.00 to	13.00
5.9 Fuel Installations (gas)	m²	4.00 to	7.00
5.11 Fire and Lightning Protection			
lightning protection	m²	4.00 to	6.00
5.12 Communication, Security and Control Systems			
fire alarms and detection	m²	38.00 to	43.00
voice and data complete installation (excluding active equipment)	m²	43.00 to	56.00
security; access; control; disabled alarms; staff paging	m²	41.00 to	56.00
BMS	m²	63.00 to	77.00
5.13 Specialist Installations			
theatre systems includes for containment and power wiring	m²	38.00 to	43.00
Total Cost/m² (based on GIFA of 8,000 m²)	m²	1025.00 to	1250.00

BUILDING MODELS – ELEMENTAL COST SUMMARIES

Item	Unit	Range £	
SPORTS HALL Single storey sports hall, located in the South East, with a gross internal area of 1,200 m² (40 m × 30 m).			
5 Services			
5.1 Sanitary Installations	m²	15.00 to	16.00
5.3 Disposal Installations			
rainwater	m²	4.00 to	6.00
soil and waste	m²	9.00 to	13.00
5.4 Water Installations			
hot and cold water services	m²	21.00 to	22.00
5.5 Heat Source			
boilers, flues, pumps and controls	m²	32.00 to	42.00
5.6 Space Heating and Air Conditioning			
warm air heating to sports hall area	m²	27.00 to	28.00
radiator heating to ancillary areas	m²	30.00 to	31.00
5.7 Ventilation Systems			
ventilation to changing, fitness and sports hall areas	m²	29.00 to	30.00
5.8 Electrical Installations			
main switchgear and sub-mains	m²	21.00 to	31.00
small power	m²	15.00 to	16.00
lighting and luminaires to sports areas	m²	36.00 to	38.00
lighting and luminaires to ancillary areas	m²	42.00 to	43.00
5.11 Fire and Lightning Protection			
lightning protection	m²	6.00 to	7.00
5.12 Communication, Security and Control Systems			
fire, smoke detection and alarm system, intruder detection	m²	18.00 to	19.00
CCTV installation	m²	20.00 to	21.00
public address and music systems	m²	9.00 to	13.00
wireways for voice and data	m²	4.00 to	6.00
Total Cost/m² (based on GIFA of 1,200 m²)	m²	340.00 to	380.00

BUILDING MODELS – ELEMENTAL COST SUMMARIES

Item	Unit	Range £	
STADIUM – NEW			
A three storey stadium, located in Greater London with a gross internal area of 85,000 m² and incorporating 60,000 spectator seats.			
5 Services			
5.1 Sanitary Installations	m²	20.00 to	25.00
5.3 Disposal Installations			
rainwater	m²	10.00 to	15.00
above ground drainage	m²	20.00 to	30.00
5.4 Water Installations			
hot and cold water services	m²	30.00 to	32.00
5.5 Heat Source	m²	31.00 to	41.00
5.6 Space Heating and Air Conditioning			
heating	m²	8.00 to	9.00
cooling	m²	21.00 to	23.00
5.7 Ventilation Systems	m²	86.00 to	89.00
5.8 Electrical Installations			
HV/LV supply	m²	15.00 to	16.00
LV distribution	m²	51.00 to	61.00
general lighting	m²	89.00 to	93.00
small power	m²	29.00 to	30.00
earthing and bonding	m²	2.00 to	3.00
power supply to mechanical equipment	m²	5.00 to	10.00
pitch lighting	m²	15.00 to	20.00
5.9 Fuel Installations (gas)	m²	2.00 to	3.00
5.11 Fire and Lightning Protection			
lightning protection	m²	2.00 to	3.00
hydrants	m²	5.00 to	10.00
5.12 Communication, Security and Control Systems			
wireways for data, TV, telecom and PA	m²	15.00 to	20.00
public address	m²	24.00 to	25.00
security	m²	21.00 to	23.00
data voice installations	m²	46.00 to	48.00
fire alarms	m²	13.00 to	15.00
disabled/refuse alarm/call systems	m²	4.00 to	5.00
BMS	m²	40.00 to	50.00
Total Cost/m² (based on GIFA of 85,000 m²)	m²	610.00 to	700.00
Total Cost/seat (based on 60,000 seats)	Each	860.00 to	990.00

BUILDING MODELS – ELEMENTAL COST SUMMARIES

Item	Unit	Range £	
HOTELS			
200 Bedroom, four star hotel, situated in Central London, with a gross internal floor area of 16,500 m².			
The development comprises a 10 storey building with large suites on each guest floor, together with banqueting, meeting rooms and leisure facilities.			
5 Services			
5.1 Sanitary Installations	m²	61.00 to	75.00
5.3 Disposal Installations			
rainwater	m²	7.00 to	9.00
soil and waste	m²	38.00 to	48.00
5.4 Water Installations			
hot and cold water services	m²	80.00 to	100.00
5.5 Heat Source			
air source heat pumps	m²	80.00 to	97.00
5.6 Space Heating and Air Conditioning			
air conditioning and space heating system; chillers, pumps, CHW and LTHW pipework, insulation, 4 pipe FCU, ductwork, grilles and diffusers, to guest rooms, public areas, meeting and banquet rooms	m²	230.00 to	320.00
5.7 Ventilation Systems			
general bathroom extract from guest suites, ventilation to kitchens and bathrooms, etc.	m²	65.00 to	84.00
smoke extract	m²	20.00 to	26.00
5.8 Electrical Installations			
HV/LV installation, standby power, lighting, emergency lighting and small power to guest rooms and public areas, including earthing and bonding	m²	175.00 to	225.00
5.11 Fire and Lightning Protection			
dry risers and sprinkler installation	m²	59.00 to	73.00
lightning protection	m²	3.00 to	4.00
5.12 Communication, Security and Control Systems			
fire/smoke detection and fire alarm system	m²	23.00 to	35.00
security/access control	m²	30.00 to	40.00
integrated sound and AV system	m²	29.00 to	35.00
telephone and data and TV installation	m²	24.00 to	27.00
containment	m²	21.00 to	26.00
wire ways for voice and data (no hotel management and head end equipment)	m²	14.00 to	17.00
BMS	m²	50.00 to	62.00
Total Cost/m² (based on GIFA of 16,500 m²)	m²	1000.00 to	1300.00

BUILDING MODELS – ELEMENTAL COST SUMMARIES

Item	Unit	Range £	
PRIVATE HOSPITAL			
New build project building. The works consist of a new 80 bed hospital of approximately 15,000 m², eight storeys with a plantroom.			
All heat is provided from existing steam boiler plant, medical gases are also served from existing plant. The project includes the provision of additional standby electrical generation to serve the wider site requirements.			
This hospital has six operating theatres, ITU/HDU department, pathology facilities, diagnostic imaging, outpatient facilities and physiotherapy.			
5 Services			
5.1 Sanitary Installations	m²	21.00 to	25.00
5.3 Disposal Installations			
rainwater	m²	3.00 to	11.00
soil and waste	m²	63.00 to	64.00
5.4 Water Installations			
hot and cold water services	m²	145.00 to	150.00
5.5 Heat Source (included in 5.6)	N/A		
5.6 Space Heating and Air Conditioning			
LPHW heating	m²	135.00 to	180.00
chilled water	m²	– to	–
steam and condensate	m²	330.00 to	465.00
ventilation, comfort cooling and air conditioning	m²	335.00 to	470.00
5.7 Ventilation Systems (included in 5.6)	N/A		
5.8 Electrical Installations			
HV distribution	m²	75.00 to	81.00
LV supply/distribution	m²	140.00 to	370.00
standby power	m²	66.00 to	125.00
UPS	m²	43.00 to	50.00
general lighting	m²	140.00 to	155.00
general power	m²	68.00 to	81.00
emergency lighting	m²	42.00 to	45.00
theatre lighting	m²	32.00 to	36.00
specialist lighting	m²	32.00 to	37.00
external lighting	m²	4.00 to	11.00
electrical supplies for mechanical services	m²	26.00 to	34.00
earthing and bonding	m²	8.00 to	12.00
5.9 Fuel Installations			
gas installations	m²	8.00 to	10.00
oil installations	m²	11.00 to	13.00
5.11 Fire and Lightning Protection			
dry risers	m²	18.00 to	24.00
sprinkler systems/gaseous fire suppression	m²	125.00 to	150.00
lightning protection	m²	2.00 to	5.00

BUILDING MODELS – ELEMENTAL COST SUMMARIES

Item	Unit	Range £	
5.12 Communication, Security and Control Systems			
fire alarms and detection	m^2	59.00 to	72.00
voice and data	m^2	90.00 to	100.00
data containment	m^2	21.00 to	24.00
security and CCTV	m^2	110.00 to	115.00
nurse call and cardiac alarm system	m^2	70.00 to	83.00
TV systems	m^2	58.00 to	92.00
disabled WC alarm	m^2	12.00 to	14.00
BMS	m^2	120.00 to	140.00
5.13 Specialist Installations			
pneumatic tube conveying system	m^2	9.00 to	13.00
medical gases	m^2	130.00 to	155.00
Total Cost/m² (based on GIFA of 15,000 m²)	m^2	2400.00 to	3225.00

BUILDING MODELS – ELEMENTAL COST SUMMARIES

Item	Unit	Range £	
SCHOOL			
New build secondary school (Academy) located in Southern England, with a gross internal floor area of 10,000 m². Total M&E services value approximately £5,000,000 and extensive use of off-site manufacturing.			
The building comprises a three storey teaching block, including provision for music, drama, catering, sports hall, science laboratories, food technology, workshops and reception/admin (BB93 compliant). Excludes IT active systems, renewable energy and sprinkler protection.			
5 Services			
5.1 Sanitary Installations			
toilet cores and changing facilities and lab sinks	m²	19.00 to	22.00
5.3 Disposal Installations			
rainwater installations	m²	6.00 to	8.00
soil and waste	m²	21.00 to	23.00
5.4 Water Installations			
potable hot and cold water services	m²	43.00 to	49.00
non-potable hot and cold water services to labs and art rooms	m²	14.00 to	21.00
5.5 Heat Source			
gas fired boiler installation	m²	21.00 to	22.00
5.6 Space Heating and Air Conditioning			
LTHW heating system (primary)	m²	52.00 to	57.00
LTHW heating system (secondary)	m²	13.00 to	17.00
DX cooling system to ICT server rooms	m²	12.00 to	15.00
mechanical supply and extract ventilation including DX type cooling to Music, Drama, Kitchen/Dining and Sports Hall	m²	79.00 to	100.00
5.7 Ventilation Systems			
toilet extract systems	m²	13.00 to	18.00
changing area extract systems	m²	8.00 to	11.00
extract ventilation from design/food technology and science labs	m²	16.00 to	23.00
5.8 Electrical Installations			
mains and sub-mains distribution	m²	53.00 to	65.00
lighting and luminaires including emergency fittings	m²	99.00 to	120.00
small power installation	m²	51.00 to	58.00
earthing and bonding	m²	3.00 to	4.00
5.9 Fuel Installations (gas)	m²	2.00 to	3.00
5.11 Fire and Lightning Protection			
lightning protection	m²	6.00 to	6.00
5.12 Communication, Security and Control Systems			
containment for telephone, IT data, AV and security systems	m²	8.00 to	15.00
fire, smoke detection and alarm system	m²	28.00 to	38.00
security installations including CCTV, access control and intruder alarm	m²	30.00 to	34.00
disabled toilet, refuge and induction loop systems	m²	4.00 to	7.00
BMS – to plant	m²	34.00 to	41.00
BMS – to opening vents/windows	m²	14.00 to	16.00
Total Cost/m² (based on GIFA of 10,000 m²)	m²	640.00 to	790.00

BUILDING MODELS – ELEMENTAL COST SUMMARIES

Item	Unit	Range £	
AFFORDABLE RESIDENTIAL DEVELOPMENT			
An 8 storey, 117 affordable residential development with a gross internal area of 11,400 m² and a net internal area of 8,400 m², situated within the Central London area. The development has no car park or communal facilities and achieves a net to gross efficiency of 74%.			
Based upon radiator LTHW heating within each apartment, with plate heat exchanger, whole house ventilation, plastic cold and hot water services pipework, sprinkler protection to apartments. Excludes remote metering. Based upon 2 bed units with 1 bathroom. Heating provided by air source heat pumps.			
SHELL & CORE			
5 Services			
5.1 Sanitary Installations	m²	1.00 to	1.00
5.3 Disposal Installations	m²	27.00 to	35.00
5.4 Water Installations	m²	26.00 to	33.00
5.5 Heat Source	m²	39.00 to	47.00
5.6 Space Heating and Air Conditioning	m²	35.00 to	41.00
5.7 Ventilation Systems	m²	27.00 to	29.00
5.8 Electrical Installations	m²	58.00 to	75.00
5.9 Fuel Installations – Gas	N/A		
5.11 Fire and Lightning Protection	m²	43.00 to	51.00
5.12 Communication, Security and Control Systems	m²	51.00 to	62.00
5.13 Specialist Installations	m²	22.00 to	22.00
Total Cost/m² (based on GIFA of 11,400 m²)	m²	330.00 to	400.00
FITTING OUT			
5 Services			
5.1 Sanitary Installations	m²	47.00 to	52.00
5.3 Disposal Installations	m²	18.00 to	23.00
5.4 Water Installations	m²	41.00 to	48.00
5.5 Heat Source	m²	47.00 to	61.00
5.6 Space Heating and Air Conditioning	m²	57.00 to	74.00
5.7 Ventilation Systems	m²	59.00 to	70.00
5.8 Electrical Installations	m²	82.00 to	100.00
5.11 Fire and Lightning Protection	m²	37.00 to	44.00
5.12 Communication, Security and Control Systems	m²	31.00 to	37.00
5.13 Specialist Installations	m²	22.00 to	22.00
Total Cost/m² (based on an NIA of 8,400 m²)	m²	445.00 to	540.00
Total cost per apartment – Shell & Core and Fitting Out	Each	64000.00 to	77000.00

Approximate Estimating

BUILDING MODELS – ELEMENTAL COST SUMMARIES

Item	Unit	Range £
PRIVATE RESIDENTIAL DEVELOPMENT		
A 24 storey, 203 apartment private residential development with a gross internal area of 27,000 m² and a net internal area of 18,000 m², situated within London's prime residential area. The development has limited car park facilities, residents' communal areas such as gym, cinema room, self-stimulated, etc.		
Based upon LTHW underfloor heating, 4 pipe fan coils to principal rooms, with plate heat and cooling exchangers, whole house ventilation with summer house facility, copper hot and cold water services pipework, sprinkler protection to apartments, lighting control, TV/data outlets. Flood wiring for apartment, home automation and sound system only. Based upon a combination of 2/3 bedroom apartments with family bathroom and ensuite.		
SHELL & CORE		
5 Services		
5.1 Sanitary Installations	m²	4.00 to 4.00
5.3 Disposal Installations	m²	34.00 to 44.00
5.4 Water Installations	m²	43.00 to 57.00
5.5 Heat Source	m²	38.00 to 48.00
5.6 Space Heating and Air Conditioning	m²	110.00 to 145.00
5.7 Ventilation Systems	m²	32.00 to 39.00
5.8 Electrical Installations	m²	87.00 to 120.00
5.9 Fuel Installations – Gas	N/A	
5.11 Fire and Lightning Protection	m²	45.00 to 68.00
5.12 Communication, Security and Control Systems	m²	69.00 to 94.00
5.13 Specialist Installations	m²	57.00 to 81.00
Total Cost/m² (based on GIFA of 27,000 m²)	m²	520.00 to 700.00
FITTING OUT		
5 Services		
5.1 Sanitary Installations	m²	150.00 to 175.00
5.3 Disposal Installations	m²	36.00 to 46.00
5.4 Water Installations	m²	70.00 to 93.00
5.5 Heat Source	m²	82.00 to 105.00
5.6 Space Heating and Air Conditioning	m²	210.00 to 250.00
5.7 Ventilation Systems	m²	70.00 to 84.00
5.8 Electrical Installations	m²	180.00 to 215.00
5.11 Fire and Lightning Protection	m²	44.00 to 54.00
5.12 Communication, Security and Control Systems	m²	150.00 to 180.00
5.13 Specialist Installations	m²	57.00 to 62.00
Total Cost/m² (based on NIA of 18,000 m²)	m²	1050.00 to 1250.00
Total cost per apartment – Shell& Core and Fitting Out	Each	295000.00 to 380000.00

BUILDING MODELS – ELEMENTAL COST SUMMARIES

Item	Unit	Range £	
DISTRIBUTION CENTRE			
New build distribution centre located in the South East of England with a total gross floor area of 75,000 m², including a refrigerated cold box of 17,500 m².			
The building is on one level (no mezzanine decks) and incorporates a small office area at ground level, vehicle recovery unit, electric forklift truck docking bay, gatehouse and associated plantrooms.			
5 Services			
5.3 Disposal Installations			
soil and waste	m²	8.00 to	9.00
rainwater	m²	3.00 to	4.00
5.4 Water Installations			
domestic hot and cold water services	m²	3.00 to	4.00
emergency drench showers & Cat 5 vehicle wash downs	m²	1.00 to	2.00
5.6 Space Heating and Air Conditioning			
heating with ventilation to offices, displacement system to main warehouse	m²	47.00 to	63.00
local cooling; DX systems to refrigerated cold box	m²	50.00 to	61.00
5.7 Ventilation Systems			
smoke extract system	m²	12.00 to	13.00
specialist extract to battery charging and maintenance areas	m²	2.00 to	3.00
5.8 Electrical Installations			
generator, life safety – containerized standby diesel generator and 24 hour capacity belly tanks	m²	6.00 to	7.00
main HV/MV installations including switchgear, transformers (Cast Resin)	m²	30.00 to	31.00
low voltage (LV) – incoming LV switchgear, distribution boards and distribution systems	m²	24.00 to	25.00
lighting installations; warehouse, office, ancillary and plant spaces	m²	20.00 to	25.00
small power installations incl. mech/HVAC plant power	m²	21.00 to	31.00
5.9 Fuel Installations (gas)			
gas mains services to plantroom	m²	1.00 to	2.00
5.11 Fire and Lightning Protection			
sprinklers including rack protection	m²	78.00 to	90.00
lightning protection	m²	1.00 to	2.00
earthing and bonding	m²	1.00 to	2.00
5.12 Communication, Security and Control Systems			
fire alarm systems; DDI	m²	16.00 to	20.00
security installations; CCTV, access control and intruder detection	m²	9.00 to	12.00
BMS systems	m²	7.00 to	8.00
other alarm systems, e.g. disabled refuge	m²	1.00 to	2.00
structured IT cabling; fibre-optic backbone ad copper Cat 6 to office and warehouse	m²	16.00 to	21.00
5.13 Specialist Installations			
testing and commissioning	m²	3.00 to	4.00
Total Cost/m² (based on GIFA of 75,000 m²)	m²	365.00 to	445.00

Approximate Estimating

BUILDING MODELS – ELEMENTAL COST SUMMARIES

Item	Unit	Range £	
DATA CENTRE			
New build data centre in a single storey 'warehouse' type construction in the Greater London area/home counties proximity to the M25. Total Net Technical Area (NTA) of 2,000 m² (2 nr data halls of 1,000 m² NTA each) with typically other areas of 300 m² for NOC/Security Room, office space, 350 m² ancillary space and 1,000 m² of internal plant areas. Total GIA of 3,650 m².			
Power and cooling to Technical Spaces designed to 1,500 w/m² of NTA = 3,000 kW IT load. Critical infrastructure built to UTI Tier III resilience standard HVAC generally N+1 and Electrical on a dual 'A' and 'B' string supplies from MV/LV switchgear to IT cab (Critical UPS's N+1 on each string) with all heat rejection and standby generators (N+1) located in a secure, external plant farm.			
Total MEP services value of approximately £24,000,000 to £30,000,000.			
All costs expressed as £/m² against Net Technical Area – NTA.			
5 Services			
5.3 Disposal Installations			
soil and waste	m²	21.00 to	24.00
rainwater	m²	14.00 to	16.00
condensate	m²	14.00 to	20.00
5.4 Water Installations			
domestic hot and cold water services	m²	29.00 to	41.00
5.5 Heat Source			
for office and ancillary spaces only	m²	23.00 to	36.00
5.6 Space Heating and Air Conditioning			
chilled water plant with redundancy of N+1 to technical space; packaged air-cooled Turbocor chillers with free cooling capability	m²	690.00 to	900.00
chilled water distribution systems; plant, header and primary distribution to technical space CRAC units, office and ancillary space cooling units; single distribution ring with redundancy N+1 on pumps, pressurization, dosing and buffer vessels	m²	1375.00 to	1600.00
chilled water final connections to technical space CRAC office/ancillary space cooling units	m²	180.00 to	225.00
computer room air conditioning (CRAC) units; 100 kW single coil units with 30% of units with humidification; N+2 redundancy	m²	335.00 to	420.00
CRAC units to critical ancillary space, e.g. UPS module and battery rooms	m²	60.00 to	85.00
cooling units to office, support and other spaces; single coil FCUs	m²	47.00 to	73.00
5.7 Ventilation Systems			
supply and extract ventilation systems to data halls, switchrooms and ancillary spaces including dedicated gas extract to gaseous fire suppression protected spaces	m²	130.00 to	190.00
computer room floor grilles; heavy duty 600 × 600 mm metal grilles with dampers	m²	180.00 to	215.00
general supply and extract to office, support and ancillary spaces	m²	71.00 to	115.00
5.8 Electrical Installations			
main HV/MV installations including switchgear, transformers (Cast Resin); dual MV supply from local authority (REC supply cost excluded)	m²	260.00 to	325.00
low voltage (LV) switchgear – Incoming LV switchgear, UPS Input and Output LV boards, mechanical services and support/ancillary area panels and distribution boards	m²	1350.00 to	1475.00
LV distribution including busbars, cabling and containment systems to provide full dual 'A' and 'B' supplies to data halls and supplies to mechanical services, office and ancillary areas	m²	1025.00 to	1200.00

BUILDING MODELS – ELEMENTAL COST SUMMARIES

Item	Unit	Range £	
generators – containerized standby diesel generators, DC rated with acoustic treatment and exhaust systems; On an N+1 redundancy for critical IT electrical and cooling loads including synchronization panel and 48 hour capacity belly tanks	m²	1250.00 to	1375.00
uninterruptible power supplies (UPS) – Static UPS systems to provide 2 × (N+1) redundancy; critical IT loads – 600 kVA modules with 10 minute battery autonomy, 0.9 PFC, harmonic filtration and static bypass; 500 kVA Static UPS with 10 minute battery autonomy for mechanical loads on N load basis	m²	1050.00 to	1250.00
critical Power Distribution Units (PDUs) to data halls; Dual 'A' and 'B' units to provide diverse supplies to each IT cabinet	m²	660.00 to	1050.00
final critical IT power supplies. 2 nr supplies to each cabinet from 'A' and 'B' PDUs	m²	560.00 to	760.00
lighting installations; data halls, office, support, ancillary and plant supplies	m²	180.00 to	225.00
5.9 Fuel Installations			
manual fill point and distribution pipework to generator belly tanks	m²	56.00 to	84.00
5.10 Lift and Conveyor Systems			
dock levellers and DDA access platform lifts only	m²	53.00 to	73.00
5.11 Fire and Lightning Protection			
gaseous fire suppression to technical, UPS and switchgear rooms	m²	415.00 to	465.00
lightning protection	m²	14.00 to	28.00
earthing and clean earth	m²	28.00 to	43.00
leak detection; to chilled water circuits within data hall service corridor	m²	73.00 to	100.00
5.12 Communication, Security and Control Systems			
fire alarm systems; DDI	m²	73.00 to	100.00
very early smoke detection alarm (VESDA) to technical areas	m²	170.00 to	215.00
other alarm systems, e.g. disabled refuge	m²	13.00 to	26.00
structured IT cabling; fibre-optic backbone and copper Cat 6 to office and support spaces only. IT cabling to data halls EXCLUDED (Clientfit-out item)	m²	160.00 to	235.00
security installations; CCTV, Access Control and Intruder Detection	m²	700.00 to	970.00
BMS/EMS systems	m²	830.00 to	1200.00
5.13 Specialist Installations			
factory acceptance testing, site acceptance testing, integrated systems testing and 'black building' testing @ 2% of MEP value	m²	240.00 to	300.00
Total Cost/m² (based on NTA of 2,000 m²)	m²	12500.00 to	15500.00
Total Cost/kW IT Load (based om 3,000 kW IT Load)	m²	8200.00 to	10500.00

Notes: The above includes: MEP preliminaries and OH&P allowances.
The above EXCLUDES: Main contractor's preliminaries and OH&P; Utility connections; Electric, water, drainage and Fibre Connections; Contingency/Risk Allowances; Foreign Exchange fluctuations; VAT at prevailing rates

BUILDING MODELS – ELEMENTAL COST SUMMARIES

Item	Unit	Range £	

GYM

Gym located in Central London on the ground floor of a mixed use development. This model is based on a total of 2,000 m², with costs based on connection to central plant within plant rooms, water source heat pump. The model includes for an AHU dedicated to providing ventilation to the gym. Services provided also include for 4 pipe fan coil units providing heating and cooling, additional underfloor heating to changing rooms. Excluded from the gym fit-out costs are FFE, active IT equipment and gym equipment.

5 Services

Item	Unit	Range £	
5.1 Sanitary Installations			
sanitary appliances including WCs, wash hand basins, cleaner's sinks, water fountains, urinals, showers and the provision of disabled toilets and accessible showers	m²	51.00 to	60.00
5.3 Disposal Installations			
soil, waste and vent installation to all sanitaryware points, provision of stub stacks, condensate drainage for fan coil units including insulation, rainwater disposal excluded as part of base build installations	m²	24.00 to	33.00
5.4 Water Installations			
installation of mains cold water services including meter, storage tanks, pumps, electromagnetic water conditioner and connections to sanitaryware, hot water including connection to base built plate heat exchanger for hot water generation, bulk storage, distribution, pump sets and connections to sanitaryware, miscellaneous water points to drinking fountains and plant supplies	m²	35.00 to	42.00
5.5 Heat Source			
connection to base build plate heat exchanger, energy meter, buffer vessels, associated pump sets and primary distribution pipework and insulation and water source heat pump	m²	71.00 to	84.00
5.6 Space Heating and Air Conditioning			
CHW distribution pipework to AHU and FCUs, LTHW distribution to underfloor heating, AHU and FCUs, intake and exhaust air ductwork to AHU. Air handling unit and associated supply and extract ductwork distribution	m²	325.00 to	400.00
5.7 Ventilation Systems			
intake and exhaust air ductwork to fans, toilet and shower supply and extract system. Extract to stores and cleaner's cupboards. MVHR unit to office/staff rest room.	m²	47.00 to	61.00
5.8 Electrical Installations			
LV distribution system including switchgear, containment and cabling, small power and lighting including emergency lighting and controls, floor boxes to gym equipment, enhanced lighting to lobbies, reception areas and toilet/changing areas, earthing and bonding	m²	270.00 to	315.00
5.11 Fire and Lightning Protection			
connection to base build installations, zone valve, concealed sprinkler heads throughout	m²	41.00 to	51.00
5.12 Communication, Security and Control Systems			
fire alarm system, PA/VA installation, access control including card reader turnstiles at reception controlling entrance to the gym, security system comprising of security cameras, access control and intruder alarms to back of houses, TV and data installations to gym, Wi-Fi installation, complete sound system including speakers, cabling and central music generation and smart device docks	m²	305.00 to	355.00
Total Cost/m² (based on GIFA of 2,000 m²)	m²	1150.00 to	1425.00

BUILDING MODELS – ELEMENTAL COST SUMMARIES

Item	Unit	Range £	
BAR Bar located in Central London on the 15th floor of a mixed use development. This model is based on a total GIA of 500 m², with costs based on a complete stand-alone plant. The model includes for a displacement ventilation installation, heating and cooling provided by a VRF installation via terminal units and an AHU. Excluded from the cost model is the bar and back bar unit, lifts with connections only provided for electrical, comms, water and drainage, FFE and active IT equipment.			
5 Services			

5.1 Sanitary Installations
    sanitary appliances including WCs, wash hand basins, urinals and the provision of
    disabled toilets
	m²	83.00 to	97.00

5.3 Disposal Installations
    soil, waste and vent installation to all sanitaryware points, provision of stub stacks
    capped connections to bar areas, condensate drainage for fan coil units including
    insulation, rainwater gullies to terrace areas
	m²	36.00 to	41.00

5.4 Water Installations
    installation of mains cold water services including meter, storage tanks, pumps,
    electromagnetic water conditioner and connections to sanitaryware, hot water
    generation, plant bulk storage, distribution, pump sets and connections to sanitaryware,
    capped connections to bar areas, miscellaneous water points and plant supplies
	m²	190.00 to	210.00

5.6 Space Heating and Air Conditioning
    VRF installation serving the AHU and terminal units provide heating and cooling to bar
    areas stand-alone cooling only VRF units to bar cellar and stores, specialist temperature
    control and humidity control units to wine cellar, Intake and exhaust air ductwork to AHU.
    Air handling unit and associated supply and extract ductwork distribution
	m²	350.00 to	405.00

5.7 Ventilation Systems
    supply and extract installations to WC areas and back of house areas. Extract to
    cleaner's cupboard, stores and the like
	m²	79.00 to	98.00

5.8 Electrical Installations
    LV distribution system including switchgear, containment and cabling small power and
    lighting including emergency lighting and controls, provision of enhanced lighting to
    lobbies, bar areas and toilet areas. Allowance for statement lighting, full lighting control
    and scene setting, capped electrical supply to bar area, earthing and bonding
	m²	620.00 to	820.00

5.11 Fire and Lightning Protection connection to base build installations, zone valve, concealed sprinkler heads throughout	m²	46.00 to	54.00

5.12 Communication, Security and Control Systems
    fire alarm system, PA/VA installation, security system comprising of security cameras,
    access control and intruder alarms to back of houses, TV and data installations to gym,
    Wi-Fi installation, complete sound system including speakers, cabling and central music
    generation and smart device docks
	m²	500.00 to	550.00

Total Cost/m² (based on GIFA of 500 m²)	m²	1900.00 to	2275.00

BUILDING MODELS – ELEMENTAL COST SUMMARIES

Item	Unit	Range £	

SPA

High end luxury spa located in Central London within the basement level of a mixed use development. This model is based on a total GIA of 1,050 m², with costs based on connection to central plant within plant rooms. The model includes for a water source heat pump, an AHU which is only providing ventilation to the spa area and swimming pool hall. Services provided also include for 4 pipe fan coil units providing heating and cooling, displacement ventilation to the pool hall, additional underfloor heating to changing rooms. Included within the cost plan are costs for swimming pool plant, steam rooms and a sauna. Excluded from the spa fit-out costs are FFE and active IT equipment. Locating elements such as spas within the basements of developments will attract a cost premium due to the difficulty associated with bringing services into these areas. A holistic approach to costing must be considered for spa areas, as basement digs to accommodate swimming pools and the associated structural and waterproofing required.

5 Services

5.1 Sanitary Installations

sanitary appliances including WCs, wash hand basins, cleaner's sinks, water fountains, urinals, showers, ice fountain and the provision of disabled toilets and accessible showers	m²	135.00 to	165.00

5.3 Disposal Installations

soil, waste and vent installation to all sanitaryware points, provision of stub stacks, condensate drainage for fan coil units including insulation, back wash drainage to pool plant, rainwater gullies to terrace areas	m²	39.00 to	49.00

5.4 Water Installations

installation of mains cold water services including meter, storage tanks, pumps, electromagnetic water conditioner and connections to sanitaryware, hot water including connection to base built plate heat exchanger for hot water generation, bulk storage, distribution, pump sets and connections to sanitaryware, softened water to experience shower, saunas and steam rooms, miscellaneous water points to drinking fountains and plant supplies	m²	81.00 to	100.00

5.5 Heat Source

connection to base build plate heat exchanger, energy meter, buffer vessels, associated pump sets and primary distribution pipework and insulation	m²	130.00 to	155.00

5.6 Space Heating and Air Conditioning

CHW distribution pipework to AHU and FCUs, VRF installation to comms/media rooms, LTHW distribution to underfloor heating, AHU and FCUs, supply and extract installation to pool hall, intake and exhaust air ductwork to AHU. Air handling unit and associated supply and extract ductwork distribution	m²	870.00 to	980.00

5.7 Ventilation Systems

intake and exhaust air ductwork to fans, toilet and shower supply and extract system. Extract to stores and cleaner's cupboards. MVHR unit to office/staff rest room. Extra over for smoke extract	m²	175.00 to	205.00

5.8 Electrical Installations

LV distribution system including switchgear, containment and cabling, small power and lighting including emergency lighting and controls, floor boxes to gym equipment, enhanced lighting to lobbies, reception areas and toilet/changing areas, allowance for statement lighting, full lighting control, fibre optic lighting with pool and scene setting, earthing and bonding	m²	455.00 to	560.00

BUILDING MODELS – ELEMENTAL COST SUMMARIES

Item	Unit	Range £	
5.11 Fire and Lightning Protection			
connection to base build installations, zone valve, concealed sprinkler heads throughout	m²	43.00 to	50.00
5.12 Communication, Security and Control Systems			
fire alarm system, PA/VA installation, access control including card reader turnstiles at reception controlling entrance to the gym, security system comprising of security cameras, access control and intruder alarms to back of houses, TV and data installations, Wi-Fi installation, complete sound system including speakers, cabling and central music generation and smart device docks, installation of central building management system including central control panels and BMS to plant and equipment	m²	490.00 to	570.00
5.13 Specialist Installations			
saunas and steam rooms. Pool plant, including all pumps, water treatment, filters and controls	m²	790.00 to	870.00
Total Cost/m² (based on GIFA of 1,050 m²)	m²	3200.00 to	3700.00

Building Regulations Pocket Book, 2nd edition

Ray Tricker and Samantha Alford

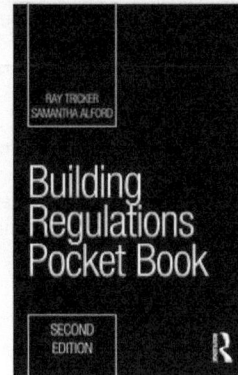

The new edition of the *Building Regulations Pocket Book* has been fully updated with recent changes to the UK Building Regulations and Planning Law. This handy guide provides you with all the information you need to comply with the UK Building Regulations and Approved Documents. On site, in the van, in the office – wherever you are – this is the book you'll refer to time and time again to check the regulations on your current job.

- Part 1 provides an overview of the Building Act.
- Part 2 offers a handy guide to the dos and don'ts of gaining the Local Council's approval for Planning Permission and Building Regulations Approval.
- Part 3 presents an overview of the requirements of the Approved Documents associated with the Building Regulations.
- Part 4 is an easy-to-read explanation of the essential requirements of the Building Regulations that any architect, builder or DIYer needs to know to keep their work safe and compliant on both domestic and non-domestic jobs.

Key new updates to this second edition include, but are not limited to: changes to the fire regulations as a result of the Hackitt Review, updates to Approved Document F and L, new Approved Documents covering *Overheating* (AD-O) and *Infrastructure for the charging of electric vehicles* (AD-S), amendments to and the reinstatement of the *Manual to the Building Regulations*. This book is essential reading for all building contractors and sub-contractors, site engineers, building engineers, building control officers, building surveyors, architects, construction site managers as well as DIYers and those who are supervising work in their own home.

September 2022: 686 pp
ISBN: 9780367774172

To Order
Tel:+44 (0) 1235 400524
Email: tandf@hachette.co.uk

For a complete listing of all our titles visit:
www.tandf.co.uk

Taylor & Francis
Taylor & Francis Group

PART 3

Material Costs/Measured Work Prices – Mechanical Installations

Building Services Design for Energy Efficient Buildings, 2nd edition

Paul Tymkow *et al.*

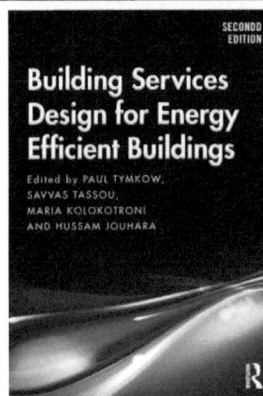

The role and influence of building services engineers are undergoing rapid change and are pivotal to achieving low-carbon buildings.

The essential conceptual design issues for planning the principal building services systems that influence energy efficiency are examined in detail, namely HVAC and electrical systems. In addition, the following issues are addressed:
- background issues on climate change, whole-life performance and design collaboration
- generic strategies for energy efficient, low-carbon design
- health and wellbeing and post occupancy evaluation
- building ventilation
- air conditioning and HVAC system selection
- thermal energy generation and distribution systems
- low-energy approaches for thermal control
- electrical systems, data collection, controls and monitoring
- building thermal load assessment
- building electric power load assessment
- space planning and design integration with other disciplines.

In order to deliver buildings that help mitigate climate change impacts, a new perspective is required for building services engineers, from the initial conceptual design and throughout the design collaboration with other disciplines. This book provides a contemporary introduction and guide to this new approach, for students and practitioners alike.

July 2020: 376 pp
ISBN: 9780815365617

To Order
Tel: +44 (0) 1235 400524
Email: tandf@hachette.co.uk

For a complete listing of all our titles
visit: www.tandf.co.uk

Taylor & Francis
Taylor & Francis Group

Material Costs/Measured Work Prices

DIRECTIONS

The following explanations are given for each of the column headings and letter codes.

Unit	Prices for each unit are given as singular (i.e. 1 metre, 1 nr) unless stated otherwise.
Net price	Industry tender prices, plus nominal allowance for fixings (unless measured separately), waste and applicable trade discounts.
Material cost	Net price plus percentage allowance for overheads (7%), profit (5%) and preliminaries (13%).
Labour norms	In man-hours for each operation.
Labour cost	Labour constant multiplied by the appropriate all-in man-hour cost based on gang rate (See also relevant Rates of Wages Section) plus percentage allowance for overheads, profit and preliminaries.
Measured work Price (total rate)	Material cost plus Labour cost.

MATERIAL COSTS

The Material Costs given are based at Third Quarter 2025 but exclude any charges in respect of VAT. The average rate of copper during this quarter is US $9,093 / UK £6,864 per tonne.

MEASURED WORK PRICES

These prices are intended to apply to new work in the London area. The prices are for reasonable quantities of work and the user should make suitable adjustments if the quantities are especially small or especially large. Adjustments may also be required for locality (e.g. outside London – refer to cost indices in the approximate estimating section for details of adjustment factors) and for the market conditions, e.g. volume of work secured or being tendered) at the time of use.

MECHANICAL INSTALLATIONS

The labour rate has been based on average gang rates per man hour effective from 7 October 2019. To this rate has been added 13% and 7% to cover preliminary items, site and head office overheads together with 5% for profit, resulting in an inclusive rate of £38.65 per man hour. The rate has been calculated on a working year of 2,007 hours; a detailed build-up of the rate is given at the end of these directions.

The rates and allowances will be subject to increases as covered by the National Agreements.

DIRECTIONS

DUCTWORK INSTALLATIONS

The labour rate basis is as per Mechanical above and to this rate has been added 13% plus 7% to cover site and head office overheads only (factory overhead is included in the material rate) and preliminary items together with 5% for profit, resulting in an inclusive rate of £36.92 per man hour. The rate has been calculated on a working year of 2,007 hours; a detailed build-up of the rate is given at the end of these directions.

In calculating the 'Measured Work Prices' the following assumptions have been made:

(a)   That the work is carried out as a subcontract under the Standard Form of Building Contract.
(b)   That, unless otherwise stated, the work is being carried out in open areas at a height which would not require more than simple scaffolding.
(c)   That the building in which the work is being carried out is no more than six storey's high.

Where these assumptions are not valid, as for example where work is carried out in ducts and similar confined spaces or in multi-storey structures when additional time is needed to get to and from upper floors, then an appropriate adjustment must be made to the prices. Such adjustment will normally be to the labour element only.

Note: The rates do not include for any uplift applied if the ductwork package is procured via the Mechanical Subcontractor.

DIRECTIONS

LABOUR RATE – MECHANICAL & PUBLIC HEALTH

The annual cost of a notional 12 man gang

	FOREMAN	SENIOR CRAFTSMAN (+2 Welding skill)	SENIOR CRAFTSMAN	CRAFTSMAN	INSTALLER	MATE (Over 18)	SUB TOTALS
	1 NR	1 NR	2 NR	4 NR	2 NR	2 NR	
Hourly Rate from 7 October 2024	20.29	18.10	16.78	15.45	13.95	11.76	
Working hours per annum per man	1,672.50	1,672.50	1,672.50	1,672.50	1,672.50	1,672.50	
× Hourly rate × nr of men = £ per annum	33,935.03	30,272.25	56,129.10	103,360.50	46,662.75	39,337.20	309,696.83
Overtime Rate	28.59	25.55	24.57	21.76	19.67	16.57	
Overtime hours per annum per man	334.50	334.50	334.50	334.50	334.50	334.50	
× Hourly rate × nr of men = £ per annum	9,563.36	8,546.48	16,437.33	29,114.88	13,159.23	11,085.33	87,906.60
Total	43,498.38	38,818.73	72,566.43	132,475.38	59,821.98	50,422.53	397,603.43
Incentive schemes 0.00%	0.00	0.00	0.00	0.00	0.00	0.00	0.00
Daily Travel Time Allowance (15–20 miles each way)	12.05	12.05	12.05	12.05	12.05	12.05	
Days per annum per man	223.00	223.00	223.00	223.00	223.00	223.00	
× nr of men = £ per annum	2,687.15	2,687.15	5,374.30	10,748.60	5,374.30	5,374.30	32,245.80
Daily Travel Fare (15–20 miles each way)	19.80	19.80	19.80	19.80	19.80	19.80	
Days per annum per man	223.00	223.00	223.00	223.00	223.00	223.00	
× nr of men = £ per annum	4,415.40	4,415.40	8,830.80	17,661.60	8,830.80	8,830.80	52,984.80
Employer's Pension contributions at 7 October 2024:							
£ Contributions/annum	2,373.80	2,117.96	3,927.04	7,230.08	3,264.56	2,751.84	21,665.28
National Insurance Contributions:							
Gross pay – subject to NI	56,842.85	51,491.68	97,184.65	179,895.50	82,611.40	71,863.15	
% of NI Contributions	15	15	15	15	15	15	
£ Contributions/annum	6,269.21	5,466.53	10,063.25	17,955.44	7,877.27	6,265.03	53,896.71
Holiday Credit and Welfare contributions:							
Number of weeks	52	52	52	52	52	52	
Total weekly £ contribution each	105.84	102.60	89.57	83.38	76.43	66.32	
× nr of men = £ Contributions/annum	5,503.64	5,335.38	9,315.52	17,343.59	7,949.20	6,897.31	52,344.64
Holiday Top-up Funding including overtime	14.20	4.52	10.55	8.01	6.11	3.25	
Cost	738.28	235.02	1,097.60	1,666.33	635.12	338.21	4,710.56

SUBTOTAL		615,451.22
TRAINING (INCLUDING ANY TRADE REGISTRATIONS) – SAY	1.00%	6,154.51
SEVERANCE PAY AND SUNDRY COSTS – SAY	1.50%	9,324.09
EMPLOYER'S LIABILITY AND THIRD PARTY INSURANCE – SAY	2.00%	12,618.60
ANNUAL COST OF NOTIONAL GANG		643,548.41
THEREFORE ANNUAL COST PER PRODUCTIVE MAN		61,290.33
AVERAGE NR OF HOURS WORKED PER MAN = 2007		
THEREFORE ALL-IN MAN-HOURS		30.54
PRELIMINARY ITEMS – SAY	13%	3.97
SITE AND HEAD OFFICE OVERHEADS AND PROFIT (7% & 5% RESPECTIVELY) – SAY	12%	4.14
THEREFORE INCLUSIVE MAN-HOUR RATE		38.65

MEN ACTUALLY WORKING = 10.5

DIRECTIONS

Notes:
1)   The following assumptions have been made in the above calculations:
       a)   Increase in Hourly Rate from 7 October 2024, in accordance with the latest BESA Wage Agreement.
       b)   The working week of 37.5 hours, i.e. the normal working week as defined by the National Agreement.
       c)   The actual hours worked are five days of 7.5 hours each.
       d)   A working year of 2,007 hours (including overtime).
       e)   Five days in the year are lost through sickness or similar reason.
       f)   Annual holiday entitlement calculated at 24 days.
2)   National insurance contributions are those effective from 6 April 2025.
3)   Weekly Holiday Credit/Welfare Stamp values are those effective from 7 October 2019.
4)   Rates are based from 7 October 2024.
5)   Overtime rates are based on Premium Rate 1.
6)   Fares with Oyster Card (New Malden to Waterloo (underground) + Zone 1 – Anytime fare) current at 2 March 2025 (TFL 0343 222 1234).

DIRECTIONS

LABOUR RATE – DUCTWORK

The annual cost of a notional eight man gang

		FOREMAN	SENIOR CRAFTSMAN	CRAFTSMAN	INSTALLER	SUB TOTALS
		1 NR	1 NR	4 NR	2 NR	
Hourly Rate from 7 October 2024		**20.29**	**16.78**	**15.45**	**13.95**	
Working hours per annum per man		1,672.50	1,672.50	1,672.50	1,672.50	
× Hourly rate × nr of men = £ per annum		**33,935.03**	**28,064.55**	**103,360.50**	**46,662.75**	**212,022.83**
Overtime Rate		28.59	24.57	21.76	19.67	
Overtime hours per annum per man		334.50	334.50	334.50	334.50	
× hourly rate × nr of men = £ per annum		**9,563.36**	**8,218.67**	**29,114.88**	**13,159.23**	**60,056.13**
Total		**43,498.38**	**36,283.22**	**132,475.38**	**59,821.98**	**272,078.96**
Incentive schemes	0.00%	**0.00**	**0.00**	**0.00**	**0.00**	**0.00**
Daily Travel Time Allowance (15–20 miles each way)		12.05	12.05	12.05	12.05	
Days per annum per man		223.00	223.00	223.00	223.00	
× nr of men = £ per annum		**2,687.15**	**2,687.15**	**10,748.60**	**5,374.30**	**21,497.20**
Daily Travel Fare (15–20 miles each way)		19.80	19.80	19.80	19.80	
Days per annum per man		223.00	223.00	223.00	223.00	
× nr of men = £ per annum		**4,415.00**	**4,415.00**	**17,661.60**	**8,830.80**	**35,323.20**
Employer's Pension Contributions from 7 October 2019						
£ Contributions/annum		**2,373.80**	**1,963.52**	**7,230.08**	**3,264.56**	**14,831.96**
National Insurance Contributions:						
Gross pay – subject to NI		56,842.85	48,592.33	179,895.50	82,611.40	
% of NI Contributions		15	15	15	15	
£ Contributions/annum		**5,543.52**	**4,437.15**	**15,812.44**	**6,920.20**	**32,713.31**
Holiday Credit and Welfare contributions:						
Number of weeks		52	52	52	52	
Total weekly £ contribution each		105.84	89.57	83.38	76.43	
× nr of men = £ Contributions/annum		**5,503.64**	**4,657.76**	**17,343.59**	**7,949.20**	**35,454.19**
Holiday Top-up Funding including overtime		**14.20**	**10.55**	**8.01**	**6.11**	
Cost		**738.28**	**548.80**	**1,666.33**	**635.12**	**3,588.53**

SUBTOTAL		419,907.57
TRAINING (INCLUDING ANY TRADE REGISTRATIONS) – SAY	1.00%	4,199.08
SEVERANCE PAY AND SUNDRY COSTS – SAY	1.50%	6,361.60
EMPLOYER'S LIABILITY AND THIRD PARTY INSURANCE – SAY	2.00%	8,609.36
ANNUAL COST OF NOTIONAL GANG		439,077.61
THEREFORE ANNUAL COST PER PRODUCTIVE MAN		58,543.68

MEN ACTUALLY WORKING = 7.5

AVERAGE NR OF HOURS WORKED PER MAN = 2007		
THEREFORE ALL-IN MAN-HOURS		29.17
PRELIMINARY ITEMS – SAY	13%	3.79
SITE AND HEAD OFFICE OVERHEADS AND PROFIT (7% & 5% RESPECTIVELY) – SAY	12%	3.96
THEREFORE INCLUSIVE MAN-HOUR RATE		36.92

DIRECTIONS

Notes:

1)   The following assumptions have been made in the above calculations:
     a)   Increase in Hourly Rate from 7 October 2024, in accordance with the latest BESA Wage Agreement.
     b)   The working week of 37.5 hours, i.e. the normal working week as defined by the National Agreement.
     c)   The actual hours worked are five days of 7.5 hours each.
     d)   A working year of 2,007 hours (including overtime).
     e)   Five days in the year are lost through sickness or similar reason.
     f)   Annual holiday entitlement calculated at 24 days.
2)   National insurance contributions are those effective from 6 April 2025.
3)   Weekly Holiday Credit/Welfare Stamp values are those effective from 7 October 2019.
4)   Rates are based from 7 October 2024.
5)   Overtime rates are based on Premium Rate 1.
6)   Fares with Oyster Card (New Malden to Waterloo (underground) + Zone 1 – Anytime fare) current at 2 March 2025 (TFL 0343 222 1234).

33 ABOVE GROUND DRAINAGE

Item	Net Price £	Material £	Labour hours	Labour £	Unit	Total rate £
RAINWATER PIPEWORK/GUTTERS						
PVC-u gutters: push fit joints; fixed with brackets to backgrounds; BS 4576 BS EN 607						
Half round gutter, with brackets measured separately						
75 mm	6.52	8.25	0.69	27.13	m	**35.39**
100 mm	13.12	16.60	0.64	25.17	m	**41.77**
150 mm	15.12	19.14	0.82	32.25	m	**51.38**
Brackets: including fixing to backgrounds. For minimum fixing distances, refer to the Tables and Memoranda at the rear of the book						
75 mm; Fascia	2.41	3.05	0.15	5.90	nr	**8.95**
100 mm; Jointing	3.84	4.87	0.16	6.29	nr	**11.16**
100 mm; Support	1.55	1.96	0.16	6.29	nr	**8.25**
150 mm; Fascia	4.66	5.90	0.16	6.29	nr	**12.19**
Bracket supports: including fixing to backgrounds. For minimum fixing distances, refer to the Tables and Memoranda at the rear of the book						
Side rafter	6.11	7.73	0.16	6.29	nr	**14.03**
Top rafter	6.11	7.73	0.16	6.29	nr	**14.03**
Rise and fall	6.61	8.36	0.16	6.29	nr	**14.65**
Extra over fittings half round PVC-u gutter						
Union						
75 mm	3.57	4.51	0.19	7.47	nr	**11.99**
100 mm	4.08	5.17	0.24	9.44	nr	**14.61**
150 mm	15.30	19.36	0.28	11.01	nr	**30.38**
Rainwater pipe outlets						
Running: 75 × 53 mm dia.	8.79	11.12	0.12	4.72	nr	**15.84**
Running: 100 × 68 mm dia.	6.00	7.59	0.12	4.72	nr	**12.31**
Running: 150 × 110 mm dia.	28.07	35.52	0.12	4.72	nr	**40.24**
Stop end: 100 × 68 mm dia.	6.73	8.52	0.12	4.72	nr	**13.24**
Internal stop ends: short						
75 mm	1.83	2.31	0.09	3.54	nr	**5.85**
100 mm	3.57	4.51	0.09	3.54	nr	**8.05**
150 mm	10.13	12.82	0.09	3.54	nr	**16.36**
External stop ends: short						
75 mm	7.25	9.18	0.09	3.54	nr	**12.72**
100 mm	9.81	12.41	0.09	3.54	nr	**15.95**
150 mm	10.13	12.82	0.09	3.54	nr	**16.36**
Angles						
75 mm; 45°	9.74	12.32	0.20	7.87	nr	**20.19**
75 mm; 90°	9.74	12.32	0.20	7.87	nr	**20.19**
100 mm; 90°	6.72	8.51	0.20	7.87	nr	**16.37**
100 mm; 120°	7.43	9.40	0.20	7.87	nr	**17.26**
100 mm; 135°	7.43	9.40	0.20	7.87	nr	**17.26**
100 mm; Prefabricated to special angle	41.94	53.08	0.23	9.04	nr	**62.12**
100 mm; Prefabricated to raked angle	48.38	61.22	0.23	9.04	nr	**70.27**
150 mm; 90°	24.09	30.49	0.20	7.87	nr	**38.35**

33 ABOVE GROUND DRAINAGE

Item	Net Price £	Material £	Labour hours	Labour £	Unit	Total rate £
RAINWATER PIPEWORK/GUTTERS – cont						
Extra over fittings half round PVC-u gutter – cont						
Gutter adaptors						
100 mm; Stainless steel clip	3.68	4.66	0.16	6.29	nr	**10.95**
100 mm; Cast iron spigot	10.73	13.57	0.23	9.04	nr	**22.62**
100 mm; Cast iron socket	10.73	13.57	0.23	9.04	nr	**22.62**
100 mm; Cast iron 'ogee' spigot	10.89	13.78	0.23	9.04	nr	**22.83**
100 mm; Cast iron 'ogee' socket	10.89	13.78	0.23	9.04	nr	**22.83**
100 mm; Half round to square PVC-u	18.33	23.20	0.23	9.04	nr	**32.24**
100 mm; Gutter overshoot guard	22.66	28.68	0.58	22.81	nr	**51.49**
Square gutter, with brackets measured separately						
120 mm square	6.20	7.85	0.82	32.25	m	**40.10**
Brackets: including fixing to backgrounds. For minimum fixing distances, refer to the Tables and Memoranda at the rear of the book						
Jointing	4.01	5.07	0.16	6.29	nr	**11.36**
Support	1.74	2.20	0.16	6.29	nr	**8.49**
Bracket support: including fixing to backgrounds. For minimum fixing distances, refer to the Tables and Memoranda at the rear of the book						
Side rafter	6.11	7.73	0.16	6.29	nr	**14.03**
Top rafter	6.11	7.73	0.16	6.29	nr	**14.03**
Rise and fall	6.61	8.36	0.16	6.29	nr	**14.65**
Extra over fittings square PVC-u gutter						
Rainwater pipe outlets						
Running: 62 mm square	6.02	7.61	0.12	4.72	nr	**12.33**
Stop end: 62 mm square	6.83	8.64	0.12	4.72	nr	**13.36**
Stop ends: short						
External	2.92	3.70	0.09	3.54	nr	**7.24**
Angles						
90°	7.28	9.21	0.20	7.87	nr	**17.07**
120°	18.19	23.02	0.20	7.87	nr	**30.89**
135°	8.50	10.76	0.20	7.87	nr	**18.63**
Prefabricated to special angle	43.02	54.45	0.23	9.04	nr	**63.49**
Prefabricated to raked angle	53.02	67.10	0.23	9.04	nr	**76.15**
Gutter adaptors						
Cast iron	31.23	39.53	0.23	9.04	nr	**48.57**
High capacity square gutter, with brackets measured separately						
137 mm	13.26	16.78	0.82	32.25	m	**49.03**
Brackets: including fixing to backgrounds. For minimum fixing distances, refer to the Tables and Memoranda at the rear of the book						
Jointing	12.60	15.95	0.16	6.29	nr	**22.24**
Support	5.26	6.66	0.16	6.29	nr	**12.96**
Overslung	4.88	6.18	0.16	6.29	nr	**12.47**

33 ABOVE GROUND DRAINAGE

Item	Net Price £	Material £	Labour hours	Labour £	Unit	Total rate £
Bracket supports: including fixing to backgrounds. For minimum fixing distances, refer to the Tables and Memoranda at the rear of the book						
Side rafter	7.50	9.49	0.16	6.29	nr	**15.78**
Top rafter	7.50	9.49	0.16	6.29	nr	**15.78**
Rise and fall	12.05	15.24	0.16	6.29	nr	**21.54**
Extra over fittings high capacity square UPV-C						
Rainwater pipe outlets						
Running: 75 mm square	20.66	26.14	0.12	4.72	nr	**30.86**
Running: 82 mm dia.	20.66	26.14	0.12	4.72	nr	**30.86**
Running: 110 mm dia.	20.65	26.14	0.12	4.72	nr	**30.86**
Screwed outlet adaptor						
75 mm square pipe	13.14	16.63	0.23	9.04	nr	**25.68**
Stop ends: short						
External	7.25	9.18	0.09	3.54	nr	**12.72**
Angles						
90°	18.07	22.86	0.20	7.87	nr	**30.73**
135°	21.26	26.91	0.20	7.87	nr	**34.77**
Prefabricated to special angle	61.41	77.72	0.23	9.04	nr	**86.77**
Prefabricated to raked internal angle	106.93	135.33	0.23	9.04	nr	**144.38**
Prefabricated to raked external angle	106.93	135.33	0.23	9.04	nr	**144.38**
Deep elliptical gutter, with brackets measured separately						
137 mm	7.18	9.08	0.82	32.25	m	**41.33**
Brackets: including fixing to backgrounds. For minimum fixing distances, refer to the Tables and Memoranda at the rear of the book						
Jointing	5.69	7.21	0.16	6.29	nr	**13.50**
Support	2.44	3.09	0.16	6.29	nr	**9.38**
Bracket support: including fixing to backgrounds. For minimum fixing distances, refer to the Tables and Memoranda at the rear of the book						
Side rafter	7.87	9.97	0.16	6.29	nr	**16.26**
Top rafter	7.87	9.97	0.16	6.29	nr	**16.26**
Rise and fall	12.65	16.01	0.16	6.29	nr	**22.30**
Extra over fittings deep elliptical PVC-u gutter						
Rainwater pipe outlets						
Running: 68 mm dia.	9.01	11.40	0.12	4.72	nr	**16.12**
Running: 82 mm dia.	9.17	11.61	0.12	4.72	nr	**16.33**
Stop end: 68 mm dia.	9.18	11.62	0.12	4.72	nr	**16.34**
Stop ends: short						
External	4.43	5.61	0.09	3.54	nr	**9.15**
Angles						
90°	9.12	11.54	0.20	7.87	nr	**19.40**
135°	10.84	13.71	0.20	7.87	nr	**21.58**
Prefabricated to special angle	31.71	40.13	0.23	9.04	nr	**49.18**

33 ABOVE GROUND DRAINAGE

Item	Net Price £	Material £	Labour hours	Labour £	Unit	Total rate £
RAINWATER PIPEWORK/GUTTERS – cont						
Extra over fittings deep elliptical PVC-u gutter – cont						
Gutter adaptors						
Stainless steel clip	7.79	9.86	0.16	6.29	nr	**16.15**
Marley deepflow	6.30	7.97	0.23	9.04	nr	**17.02**
Ogee profile PVC-u gutter, with brackets measured separately						
122 mm	7.56	9.57	0.82	32.25	m	**41.81**
Brackets: including fixing to backgrounds. For minimum fixing distances, refer to the Tables and Memoranda at the rear of the book						
Jointing	6.01	7.60	0.16	6.29	nr	**13.90**
Support	2.32	2.94	0.16	6.29	nr	**9.23**
Overslung	2.32	2.94	0.16	6.29	nr	**9.23**
Extra over fittings Ogee profile PVC-u gutter						
Rainwater pipe outlets						
Running: 68 mm dia.	8.58	10.86	0.12	4.72	nr	**15.58**
Stop ends: short						
Internal/External: left or right hand	4.36	5.51	0.09	3.54	nr	**9.05**
Angles						
90°: internal or external	8.68	10.99	0.20	7.87	nr	**18.85**
135°: internal or external	8.68	10.99	0.20	7.87	nr	**18.85**
PVC-u rainwater pipe: dry push fit joints; fixed with brackets to backgrounds; BS 4576/ BS EN 607						
Pipe: circular, with brackets measured separately						
53 mm	9.77	12.36	0.61	23.99	m	**36.35**
68 mm	10.17	12.88	0.61	23.99	m	**36.87**
Pipe clip: including fixing to backgrounds. For minimum fixing distances, refer to the Tables and Memoranda at the rear of the book						
68 mm	2.28	2.89	0.16	6.29	nr	**9.18**
Pipe clip adjustable: including fixing to backgrounds. For minimum fixing distances, refer to the Tables and Memoranda at the rear of the book						
53 mm	3.48	4.40	0.16	6.29	nr	**10.70**
68 mm	4.87	6.16	0.16	6.29	nr	**12.45**
Pipe clip drive in: including fixing to backgrounds. For minimum fixing distances, refer to the Tables and Memoranda at the rear of the book						
68 mm	5.47	6.92	0.16	6.29	nr	**13.21**

33 ABOVE GROUND DRAINAGE

Item	Net Price £	Material £	Labour hours	Labour £	Unit	Total rate £
Extra over fittings circular pipework PVC-u						
Pipe coupler: PVC-u to PVC-u						
68 mm	2.84	3.59	0.12	4.72	nr	**8.31**
Pipe coupler: PVC-u to cast iron						
68 mm: to 3" cast iron	12.60	15.94	0.17	6.69	nr	**22.63**
68 mm: to 3 ¾" cast iron	43.31	54.82	0.17	6.69	nr	**61.50**
Access pipe: single socket						
68 mm	17.76	22.48	0.15	5.90	nr	**28.38**
Bend: short radius						
53 mm: 67.5°	4.14	5.24	0.20	7.87	nr	**13.10**
68 mm: 92.5°	6.08	7.70	0.20	7.87	nr	**15.56**
68 mm: 112.5°	6.08	7.70	0.20	7.87	nr	**15.56**
Bend: long radius						
68 mm: 112°	5.76	7.29	0.20	7.87	nr	**15.15**
Branch						
68 mm: 92°	36.05	45.62	0.23	9.04	nr	**54.67**
68 mm: 112°	36.17	45.77	0.23	9.04	nr	**54.82**
Double branch						
68 mm: 112°	73.95	93.60	0.24	9.44	nr	**103.03**
Shoe						
53 mm	6.13	7.75	0.12	4.72	nr	**12.47**
68 mm	9.05	11.45	0.12	4.72	nr	**16.17**
Rainwater head: including fixing to backgrounds						
68 mm	17.76	22.48	0.29	11.40	nr	**33.88**
Pipe: square, with brackets measured separately						
62 mm	6.23	7.88	0.45	17.70	m	**25.58**
75 mm	11.84	14.99	0.45	17.70	m	**32.68**
Pipe clip: including fixing to backgrounds. For minimum fixing distances, refer to the Tables and Memoranda at the rear of the book						
62 mm	2.08	2.63	0.16	6.29	nr	**8.92**
75 mm	4.62	5.85	0.16	6.29	nr	**12.14**
Pipe clip adjustable: including fixing to backgrounds. For minimum fixing distances, refer to the Tables and Memoranda at the rear of the book						
62 mm	6.45	8.16	0.16	6.29	nr	**14.45**
Extra over fittings square pipework PVC-u						
Pipe coupler: PVC-u to PVC-u						
62 mm	3.06	3.87	0.20	7.87	nr	**11.73**
75 mm	4.74	5.99	0.20	7.87	nr	**13.86**
Square to circular adaptor: single socket						
62 mm to 68 mm	5.95	7.53	0.20	7.87	nr	**15.40**
Square to circular adaptor: single socket						
75 mm to 62 mm	7.69	9.73	0.20	7.87	nr	**17.60**
Access pipe						
62 mm	35.57	45.01	0.16	6.29	nr	**51.31**
75 mm	37.54	47.51	0.16	6.29	nr	**53.80**

33 ABOVE GROUND DRAINAGE

Item	Net Price £	Material £	Labour hours	Labour £	Unit	Total rate £
RAINWATER PIPEWORK/GUTTERS – cont						
Extra over fittings square pipework PVC-u – cont						
Bends						
62 mm: 92.5°	5.50	6.96	0.20	7.87	nr	**14.82**
62 mm: 112.5°	4.01	5.07	0.20	7.87	nr	**12.93**
75 mm: 112.5°	10.03	12.69	0.20	7.87	nr	**20.56**
Bends: prefabricated special angle						
62 mm	35.74	45.23	0.23	9.04	nr	**54.28**
75 mm	49.66	62.85	0.23	9.04	nr	**71.89**
Offset						
62 mm	8.61	10.90	0.20	7.87	nr	**18.76**
75 mm	23.71	30.01	0.20	7.87	nr	**37.87**
Offset: prefabricated special angle						
62 mm	35.74	45.23	0.23	9.04	nr	**54.28**
Shoe						
62 mm	5.06	6.40	0.12	4.72	nr	**11.12**
75 mm	8.65	10.95	0.12	4.72	nr	**15.67**
Branch						
62 mm	12.50	15.82	0.23	9.04	nr	**24.86**
75 mm	60.93	77.11	0.23	9.04	nr	**86.16**
Double branch						
62 mm	67.04	84.85	0.24	9.44	nr	**94.28**
Rainwater head						
62 mm	56.22	71.15	0.29	11.40	nr	**82.56**
75 mm	59.15	74.86	3.45	135.61	nr	**210.46**
PVC-u rainwater pipe: solvent welded joints; fixed with brackets to backgrounds; BS 4576/ BS EN 607						
Pipe: circular, with brackets measured separately						
82 mm	19.18	24.27	0.35	13.76	m	**38.04**
Pipe clip: galvanized; including fixing to backgrounds. For minimum fixing distances, refer to the Tables and Memoranda at the rear of the book						
82 mm	7.81	9.88	0.58	22.81	nr	**32.69**
Pipe clip: galvanized plastic coated; including fixing to backgrounds. For minimum fixing distances, refer to the Tables and Memoranda at the rear of the book						
82 mm	10.80	13.67	0.58	22.81	nr	**36.48**
Pipe clip: PVC-u including fixing to backgrounds. For minimum fixing distances, refer to the Tables and Memoranda at the rear of the book						
82 mm	5.80	7.35	0.58	22.81	nr	**30.15**
Pipe clip: PVC-u adjustable: including fixing to backgrounds. For minimum fixing distances, refer to the Tables and Memoranda at the rear of the book						
82 mm	10.61	13.43	0.58	22.81	nr	**36.24**

33 ABOVE GROUND DRAINAGE

Item	Net Price £	Material £	Labour hours	Labour £	Unit	Total rate £
Extra over fittings circular pipework PVC-u						
Pipe coupler: PVC-u to PVC-u						
82 mm	9.72	12.30	0.21	8.26	nr	**20.56**
Access pipe						
82 mm	59.88	75.78	0.23	9.04	nr	**84.83**
Bend						
82 mm: 92, 112.5 and 135°	24.44	30.93	0.29	11.40	nr	**42.34**
Shoe						
82 mm	15.09	19.09	0.29	11.40	nr	**30.50**
110 mm	18.93	23.96	0.32	12.58	nr	**36.54**
Branch						
82 mm: 92, 112.5 and 135°	36.43	46.10	0.35	13.76	nr	**59.87**
Rainwater head						
82 mm	29.11	36.84	0.58	22.81	nr	**59.65**
110 mm	31.98	40.47	0.58	22.81	nr	**63.28**
Roof outlets: 178 dia.; Flat						
50 mm	31.96	40.44	1.15	45.22	nr	**85.67**
82 mm	31.96	40.44	1.15	45.22	nr	**85.67**
Roof outlets: 178 mm dia.; Domed						
50 mm	31.96	40.44	1.15	45.22	nr	**85.67**
82 mm	31.96	40.44	1.15	45.22	nr	**85.67**
Roof outlets: 406 mm dia.; Flat						
82 mm	62.41	78.98	1.15	45.22	nr	**124.21**
110 mm	62.41	78.98	1.15	45.22	nr	**124.21**
Roof outlets: 406 mm dia.; Domed						
82 mm	62.41	78.98	1.15	45.22	nr	**124.21**
110 mm	62.41	78.98	1.15	45.22	nr	**124.21**
Roof outlets: 406 mm dia.; Inverted						
82 mm	137.43	173.93	1.15	45.22	nr	**219.15**
110 mm	137.43	173.93	1.15	45.22	nr	**219.15**
Roof outlets: 406 mm dia.; Vent Pipe						
82 mm	90.27	114.24	1.15	45.22	nr	**159.47**
110 mm	90.27	114.24	1.15	45.22	nr	**159.47**
Balcony outlets: screed						
82 mm	53.42	67.60	1.15	45.22	nr	**112.83**
Balcony outlets: asphalt						
82 mm	53.14	67.25	1.15	45.22	nr	**112.48**
Adaptors						
82 mm × 62 mm square pipe	7.15	9.05	0.21	8.26	nr	**17.31**
82 mm × 68 mm circular pipe	7.15	9.05	0.21	8.26	nr	**17.31**
For 110 mm dia. pipework and fittings refer to 33 Drainage Above Ground						
Cast iron gutters: mastic and bolted joints; BS 460; fixed with brackets to backgrounds						
Half round gutter, with brackets measured separately						
100 mm	29.64	37.51	0.85	33.43	m	**70.94**
115 mm	30.91	39.11	0.97	38.15	m	**77.26**
125 mm	36.16	45.76	0.97	38.15	m	**83.91**
150 mm	61.81	78.23	1.12	44.04	m	**122.27**

33 ABOVE GROUND DRAINAGE

Item	Net Price £	Material £	Labour hours	Labour £	Unit	Total rate £
RAINWATER PIPEWORK/GUTTERS – cont						
Half round gutter, with brackets measured separately – cont						
Brackets; fixed to backgrounds. For minimum fixing distances, refer to the Tables and Memoranda at the rear of the book						
Fascia						
100 mm	6.96	8.81	0.16	6.29	nr	**15.10**
115 mm	6.96	8.81	0.16	6.29	nr	**15.10**
125 mm	6.96	8.81	0.16	6.29	nr	**15.10**
150 mm	8.80	11.14	0.16	6.29	nr	**17.43**
Rise and fall						
100 mm	13.81	17.48	0.39	15.34	nr	**32.82**
115 mm	13.81	17.48	0.39	15.34	nr	**32.82**
125 mm	14.20	17.97	0.39	15.34	nr	**33.30**
150 mm	14.44	18.28	0.39	15.34	nr	**33.62**
Top rafter						
100 mm	8.50	10.76	0.16	6.29	nr	**17.05**
115 mm	8.50	10.76	0.16	6.29	nr	**17.05**
125 mm	11.51	14.57	0.16	6.29	nr	**20.86**
150 mm	15.64	19.79	0.16	6.29	nr	**26.08**
Side rafter						
100 mm	8.50	10.76	0.16	6.29	nr	**17.05**
115 mm	8.50	10.76	0.16	6.29	nr	**17.05**
125 mm	11.51	14.57	0.16	6.29	nr	**20.86**
150 mm	15.64	19.79	0.16	6.29	nr	**26.08**
Extra over fittings half round gutter cast iron BS 460						
Union						
100 mm	16.31	20.64	0.39	15.34	nr	**35.97**
115 mm	20.35	25.75	0.48	18.88	nr	**44.63**
125 mm	22.98	29.09	0.48	18.88	nr	**47.96**
150 mm	25.79	32.64	0.55	21.63	nr	**54.27**
Stop end; internal						
100 mm	8.33	10.54	0.12	4.72	nr	**15.26**
115 mm	10.77	13.63	0.15	5.90	nr	**19.53**
125 mm	10.77	13.63	0.15	5.90	nr	**19.53**
150 mm	14.95	18.92	0.20	7.87	nr	**26.79**
Stop end; external						
100 mm	8.33	10.54	0.12	4.72	nr	**15.26**
115 mm	10.52	13.32	0.15	5.90	nr	**19.22**
125 mm	10.77	13.63	0.15	5.90	nr	**19.53**
150 mm	14.95	18.92	0.20	7.87	nr	**26.79**
90° angle; single socket						
100 mm	24.72	31.29	0.39	15.34	nr	**46.62**
115 mm	25.42	32.18	0.43	16.91	nr	**49.09**
125 mm	29.99	37.96	0.43	16.91	nr	**54.87**
150 mm	54.80	69.36	0.50	19.66	nr	**89.02**

33 ABOVE GROUND DRAINAGE

Item	Net Price £	Material £	Labour hours	Labour £	Unit	Total rate £
90° angle; double socket						
100 mm	29.97	37.94	0.39	15.34	nr	**53.27**
115 mm	31.80	40.25	0.43	16.91	nr	**57.16**
125 mm	41.13	52.06	0.43	16.91	nr	**68.97**
135° angle; single socket						
100 mm	25.25	31.95	0.39	15.34	nr	**47.29**
115 mm	25.48	32.24	0.43	16.91	nr	**49.15**
125 mm	37.65	47.65	0.43	16.91	nr	**64.56**
150 mm	55.87	70.71	0.50	19.66	nr	**90.38**
Running outlet						
65 mm outlet						
100 mm	24.12	30.53	0.39	15.34	nr	**45.87**
115 mm	26.25	33.22	0.43	16.91	nr	**50.13**
125 mm	29.99	37.96	0.43	16.91	nr	**54.87**
75 mm outlet						
100 mm	24.12	30.53	0.39	15.34	nr	**45.87**
115 mm	26.25	33.22	0.43	16.91	nr	**50.13**
125 mm	29.99	37.96	0.43	16.91	nr	**54.87**
150 mm	51.94	65.73	0.50	19.66	nr	**85.39**
100 mm outlet						
150 mm	51.94	65.73	0.50	19.66	nr	**85.39**
Stop end outlet; socket						
65 mm outlet						
100 mm	28.27	35.78	0.39	15.34	nr	**51.12**
115 mm	31.70	40.12	0.43	16.91	nr	**57.03**
75 mm outlet						
125 mm	28.27	35.78	0.43	16.91	nr	**52.69**
150 mm	59.44	140.96	0.50	19.66	nr	**160.62**
100 mm outlet						
150 mm	51.94	65.73	0.50	19.66	nr	**85.39**
Stop end outlet; spigot						
65 mm outlet						
100 mm	28.27	35.78	0.39	15.34	nr	**51.12**
115 mm	31.70	40.12	0.43	16.91	nr	**57.03**
75 mm outlet						
125 mm	28.27	35.78	0.43	16.91	nr	**52.69**
150 mm	59.44	75.23	0.50	19.66	nr	**94.89**
100 mm outlet						
150 mm	59.44	75.23	0.50	19.66	nr	**94.89**
Half round; 3 mm thick double beaded gutter, with brackets measured separately						
100 mm	30.19	38.20	0.85	33.43	m	**71.63**
115 mm	31.50	39.87	0.85	33.43	m	**73.30**
125 mm	36.14	45.74	0.97	38.15	m	**83.89**
Brackets; fixed to backgrounds. For minimum fixing distances, refer to the Tables and Memoranda at the rear of the book						

33 ABOVE GROUND DRAINAGE

Item	Net Price £	Material £	Labour hours	Labour £	Unit	Total rate £
RAINWATER PIPEWORK/GUTTERS – cont						
Half round; 3 mm thick double beaded gutter – cont						
Fascia						
100 mm	6.96	8.81	0.16	6.29	nr	**15.10**
115 mm	6.96	8.81	0.16	6.29	nr	**15.10**
125 mm	6.96	8.81	0.16	6.29	nr	**15.10**
Extra over fittings half round 3 mm thick gutter BS 460						
Union						
100 mm	16.31	20.64	0.38	14.94	nr	**35.58**
115 mm	19.87	25.15	0.38	14.94	nr	**40.09**
125 mm	22.98	29.09	0.43	16.91	nr	**46.00**
Stop end; internal						
100 mm	8.33	10.54	0.12	4.72	nr	**15.26**
115 mm	10.77	13.63	0.12	4.72	nr	**18.35**
125 mm	10.81	13.68	0.15	5.90	nr	**19.57**
Stop end; external						
100 mm	8.33	10.54	0.12	4.72	nr	**15.26**
115 mm	10.77	13.63	0.12	4.72	nr	**18.35**
125 mm	10.81	13.68	0.15	5.90	nr	**19.57**
90° angle; single socket						
100 mm	25.71	32.53	0.38	14.94	nr	**47.48**
115 mm	26.02	32.93	0.38	14.94	nr	**47.88**
125 mm	29.99	37.96	0.43	16.91	nr	**54.87**
135° angle; single socket						
100 mm	25.25	31.95	0.38	14.94	nr	**46.90**
115 mm	25.42	32.18	0.38	14.94	nr	**47.12**
125 mm	31.71	40.14	0.43	16.91	nr	**57.05**
Running outlet						
65 mm outlet						
100 mm	25.90	32.78	0.38	14.94	nr	**47.72**
115 mm	26.76	33.87	0.38	14.94	nr	**48.81**
125 mm	35.44	44.85	0.43	16.91	nr	**61.76**
75 mm outlet						
115 mm	26.76	33.87	0.38	14.94	nr	**48.81**
125 mm	31.05	39.29	0.43	16.91	nr	**56.20**
Stop end outlet; socket						
65 mm outlet						
100 mm	28.27	35.78	0.38	14.94	nr	**50.72**
115 mm	31.70	40.12	0.38	14.94	nr	**55.06**
125 mm	35.44	44.85	0.43	16.91	nr	**61.76**
75 mm outlet						
125 mm	36.09	45.67	0.43	16.91	nr	**62.58**
Stop end outlet; spigot						

33 ABOVE GROUND DRAINAGE

Item	Net Price £	Material £	Labour hours	Labour £	Unit	Total rate £
65 mm outlet						
100 mm	28.27	35.78	0.38	14.94	nr	**50.72**
115 mm	31.70	40.12	0.38	14.94	nr	**55.06**
125 mm	35.44	44.85	0.43	16.91	nr	**61.76**
Deep half round gutter, with brackets measured separately						
100 × 75 mm	49.65	62.84	0.85	33.43	m	**96.27**
125 × 75 mm	64.15	81.19	0.97	38.15	m	**119.33**
Brackets; fixed to backgrounds. For minimum fixing distances, refer to the Tables and Memoranda at the rear of the book						
Fascia						
100 × 75 mm	23.46	29.69	0.16	6.29	nr	**35.98**
125 × 75 mm	28.89	36.56	0.16	6.29	nr	**42.85**
Extra over fittings deep half round gutter BS 460						
Union						
100 × 75 mm	27.27	34.51	0.38	14.94	nr	**49.46**
125 × 75 mm	28.89	36.56	0.43	16.91	nr	**53.47**
Stop end; internal						
100 × 75 mm	23.88	30.22	0.12	4.72	nr	**34.94**
125 × 75 mm	29.45	37.27	0.15	5.90	nr	**43.17**
Stop end; external						
100 × 75 mm	23.88	30.22	0.12	4.72	nr	**34.94**
125 × 75 mm	29.45	37.27	0.15	5.90	nr	**43.17**
90° angle; single socket						
100 × 75 mm	68.05	86.12	0.38	14.94	nr	**101.07**
125 × 75 mm	86.41	109.36	0.43	16.91	nr	**126.27**
135° angle; single socket						
100 × 75 mm	68.05	86.12	0.38	14.94	nr	**101.07**
125 × 75 mm	86.41	109.36	0.43	16.91	nr	**126.27**
Running outlet						
65 mm outlet						
100 × 75 mm	68.05	86.12	0.38	14.94	nr	**101.07**
125 × 75 mm	86.41	109.36	0.43	16.91	nr	**126.27**
75 mm outlet						
100 × 75 mm	62.71	79.36	0.38	14.94	nr	**94.31**
125 × 75 mm	68.05	86.12	0.43	16.91	nr	**103.03**
Stop end outlet; socket						
65 mm outlet						
100 × 75 mm	91.93	116.34	0.38	14.94	nr	**131.29**
75 mm outlet						
100 × 75 mm	91.93	116.34	0.38	14.94	nr	**131.29**
125 × 75 mm	91.93	116.34	0.43	16.91	nr	**133.25**
Stop end outlet; spigot						
65 mm outlet						
100 × 75 mm	91.93	116.34	0.38	14.94	nr	**131.29**

33 ABOVE GROUND DRAINAGE

Item	Net Price £	Material £	Labour hours	Labour £	Unit	Total rate £
RAINWATER PIPEWORK/GUTTERS – cont						
Extra over fittings deep half round gutter BS 460 – cont						
75 mm outlet						
100 × 75 mm	91.93	116.34	0.38	14.94	nr	**131.29**
125 × 75 mm	91.93	116.34	0.43	16.91	nr	**133.25**
Ogee gutter, with brackets measured separately						
100 mm	33.07	41.85	0.85	33.43	m	**75.28**
115 mm	36.35	46.01	0.97	38.15	m	**84.15**
125 mm	38.14	48.28	0.97	38.15	m	**86.42**
Brackets; fixed to backgrounds. For minimum fixing distances, refer to the Tables and Memoranda at the rear of the book						
Fascia						
100 mm	7.57	9.58	0.16	6.29	nr	**15.88**
115 mm	7.57	9.58	0.16	6.29	nr	**15.88**
125 mm	8.52	10.78	0.16	6.29	nr	**17.08**
Extra over fittings ogee cast iron gutter BS 460						
Union						
100 mm	16.32	20.66	0.38	14.94	nr	**35.60**
115 mm	19.87	25.15	0.43	16.91	nr	**42.06**
125 mm	22.98	29.09	0.43	16.91	nr	**46.00**
Stop end; internal						
100 mm	8.50	10.76	0.12	4.72	nr	**15.48**
115 mm	10.98	13.90	0.15	5.90	nr	**19.80**
125 mm	10.98	13.90	0.15	5.90	nr	**19.80**
Stop end; external						
100 mm	8.50	10.76	0.12	4.72	nr	**15.48**
115 mm	10.98	13.90	0.15	5.90	nr	**19.80**
125 mm	10.98	13.90	0.15	5.90	nr	**19.80**
90° angle; internal						
100 mm	25.79	32.64	0.38	14.94	nr	**47.59**
115 mm	27.95	35.38	0.43	16.91	nr	**52.29**
125 mm	30.52	38.63	0.43	16.91	nr	**55.54**
90° angle; external						
100 mm	25.79	32.64	0.38	14.94	nr	**47.59**
115 mm	27.95	35.38	0.43	16.91	nr	**52.29**
125 mm	30.52	38.63	0.43	16.91	nr	**55.54**
135° angle; internal						
100 mm	26.78	33.89	0.38	14.94	nr	**48.83**
115 mm	28.53	36.11	0.43	16.91	nr	**53.02**
125 mm	37.58	47.56	0.43	16.91	nr	**64.47**
135° angle; external						
100 mm	26.78	33.89	0.38	14.94	nr	**48.83**
115 mm	28.53	36.11	0.43	16.91	nr	**53.02**
125 mm	37.58	47.56	0.43	16.91	nr	**64.47**

33 ABOVE GROUND DRAINAGE

Item	Net Price £	Material £	Labour hours	Labour £	Unit	Total rate £
Running outlet						
65 mm outlet						
100 mm	26.27	33.24	0.38	14.94	nr	**48.19**
115 mm	27.97	35.40	0.43	16.91	nr	**52.31**
125 mm	30.52	38.63	0.43	16.91	nr	**55.54**
75 mm outlet						
125 mm	30.52	38.63	0.43	16.91	nr	**55.54**
Stop end outlet; socket						
65 mm outlet						
100 mm	41.48	52.50	0.38	14.94	nr	**67.44**
115 mm	41.48	52.50	0.43	16.91	nr	**69.41**
125 mm	41.48	52.50	0.43	16.91	nr	**69.41**
75 mm outlet						
125 mm	41.48	52.50	0.43	16.91	nr	**69.41**
Stop end outlet; spigot						
65 mm outlet						
100 mm	41.48	52.50	0.38	14.94	nr	**67.44**
115 mm	41.48	52.50	0.43	16.91	nr	**69.41**
125 mm	41.48	52.50	0.43	16.91	nr	**69.41**
75 mm outlet						
125 mm	41.48	52.50	0.43	16.91	nr	**69.41**
Notts ogee gutter, with brackets measured separately						
115 mm	58.74	74.34	0.85	33.43	m	**107.76**
Brackets; fixed to backgrounds. For minimum fixing distances, refer to the Tables and Memoranda at the rear of the book						
Fascia						
115 mm	22.91	29.00	0.16	6.29	nr	**35.29**
Extra over fittings Notts ogee cast iron gutter BS 460						
Union						
115 mm	27.81	35.20	0.38	14.94	nr	**50.14**
Stop end; internal						
115 mm	22.91	29.00	0.16	6.29	nr	**35.29**
Stop end; external						
115 mm	22.91	29.00	0.16	6.29	nr	**35.29**
90° angle; internal						
115 mm	66.45	84.10	0.43	16.91	nr	**101.01**
90° angle; external						
115 mm	66.45	84.10	0.43	16.91	nr	**101.01**
135° angle; internal						
115 mm	67.72	85.70	0.43	16.91	nr	**102.61**
135° angle; external						
115 mm	67.72	85.70	0.43	16.91	nr	**102.61**
Running outlet						
65 mm outlet						
115 mm	79.66	100.82	0.43	16.91	nr	**117.73**

33 ABOVE GROUND DRAINAGE

Item	Net Price £	Material £	Labour hours	Labour £	Unit	Total rate £
RAINWATER PIPEWORK/GUTTERS – cont						
Extra over fittings Notts ogee cast iron gutter BS 460 – cont						
75 mm outlet						
115 mm	79.66	100.82	0.43	16.91	nr	**117.73**
Stop end outlet; socket						
65 mm outlet						
115 mm	102.56	129.80	0.43	16.91	nr	**146.71**
Stop end outlet; spigot						
65 mm outlet						
115 mm	102.56	129.80	0.43	16.91	nr	**146.71**
No 46 moulded gutter, with brackets measured separately						
100 × 75 mm	56.80	71.89	0.85	33.43	m	**105.32**
125 × 100 mm	83.28	105.40	0.97	38.15	m	**143.55**
Brackets; fixed to backgrounds. For minimum fixing distances, refer to the Tables and Memoranda at the rear of the book						
Fascia						
100 × 75 mm	12.67	16.03	0.16	6.29	nr	**22.33**
125 × 100 mm	12.67	16.03	0.16	6.29	nr	**22.33**
Extra over fittings						
Union						
100 × 75 mm	26.76	33.87	0.38	14.94	nr	**48.81**
125 × 100 mm	31.01	39.25	0.43	16.91	nr	**56.16**
Stop end; internal						
100 × 75 mm	23.97	30.33	0.12	4.72	nr	**35.05**
125 × 100 mm	31.01	39.25	0.15	5.90	nr	**45.15**
Stop end; external						
100 × 75 mm	23.97	30.33	0.12	4.72	nr	**35.05**
125 × 100 mm	31.01	39.25	0.15	5.90	nr	**45.15**
90° angle; internal						
100 × 75 mm	62.74	79.41	0.38	14.94	nr	**94.35**
125 × 100 mm	90.21	114.16	0.43	16.91	nr	**131.07**
90° angle; external						
100 × 75 mm	62.74	79.41	0.38	14.94	nr	**94.35**
125 × 100 mm	90.21	114.16	0.43	16.91	nr	**131.07**
135° angle; internal						
100 × 75 mm	63.96	80.94	0.38	14.94	nr	**95.89**
125 × 100 mm	90.21	114.16	0.43	16.91	nr	**131.07**
135° angle; external						
100 × 75 mm	63.96	80.94	0.38	14.94	nr	**95.89**
125 × 100 mm	90.21	114.16	0.43	16.91	nr	**131.07**
Running outlet						
65 mm outlet						
100 × 75 mm	63.96	80.94	0.38	14.94	nr	**95.89**
125 × 100 mm	90.21	114.16	0.43	16.91	nr	**131.07**

33 ABOVE GROUND DRAINAGE

Item	Net Price £	Material £	Labour hours	Labour £	Unit	Total rate £
75 mm outlet						
100 × 75 mm	63.96	80.94	0.38	14.94	nr	**95.89**
125 × 100 mm	90.21	114.16	0.43	16.91	nr	**131.07**
100 mm outlet						
100 × 75 mm	63.96	80.94	0.38	14.94	nr	**95.89**
125 × 100 mm	90.21	114.16	0.43	16.91	nr	**131.07**
100 × 75 mm outlet						
125 × 100 mm	63.96	80.94	0.43	16.91	nr	**97.85**
Stop end outlet; socket						
65 mm outlet						
100 × 75 mm	121.22	153.41	0.38	14.94	nr	**168.36**
75 mm outlet						
125 × 100 mm	121.22	153.41	0.43	16.91	nr	**170.32**
Stop end outlet; spigot						
65 mm outlet						
100 × 75 mm	121.22	153.41	0.38	14.94	nr	**168.36**
75 mm outlet						
125 × 100 mm	121.22	153.41	0.43	16.91	nr	**170.32**
Box gutter, with brackets measured separately						
100 × 75 mm	92.12	116.59	0.85	33.43	m	**150.01**
Brackets; fixed to backgrounds. For minimum fixing distances, refer to the Tables and Memoranda at the rear of the book						
Fascia						
100 × 75 mm	14.60	18.48	0.16	6.29	nr	**24.77**
Extra over fittings box cast iron gutter BS 460						
Union						
100 × 75 mm	19.36	24.50	0.38	14.94	nr	**39.45**
Stop end; external						
100 × 75 mm	14.60	18.48	0.12	4.72	nr	**23.20**
90° angle						
100 × 75 mm	73.62	93.17	0.38	14.94	nr	**108.12**
135° angle						
100 × 75 mm	73.62	93.17	0.38	14.94	nr	**108.12**
Running outlet						
65 mm outlet						
100 × 75 mm	73.62	93.17	0.38	14.94	nr	**108.12**
75 mm outlet						
100 × 75 mm	73.62	93.17	0.38	14.94	nr	**108.12**
100 × 75 mm outlet						
100 × 75 mm	73.62	93.17	0.38	14.94	nr	**108.12**
Cast iron rainwater pipe; dry joints; BS 460; fixed to backgrounds: Circular						
Plain socket pipe, with brackets measured separately						
65 mm	53.31	67.47	0.69	27.13	m	**94.60**
75 mm	53.31	67.47	0.69	27.13	m	**94.60**
100 mm	72.76	92.08	0.69	27.13	m	**119.22**

33 ABOVE GROUND DRAINAGE

Item	Net Price £	Material £	Labour hours	Labour £	Unit	Total rate £
RAINWATER PIPEWORK/GUTTERS – cont						
Cast iron rainwater pipe – cont						
Bracket; fixed to backgrounds. For minimum fixing distances, refer to the Tables and Memoranda at the rear of the book						
65 mm	19.47	24.64	0.29	11.40	nr	**36.04**
75 mm	19.56	24.75	0.29	11.40	nr	**36.15**
100 mm	19.80	25.06	0.29	11.40	nr	**36.47**
Eared socket pipe, with wall spacers measured separately						
65 mm	56.96	72.09	0.62	24.38	m	**96.47**
75 mm	56.96	72.09	0.62	24.38	m	**96.47**
100 mm	76.45	96.75	0.62	24.38	m	**121.13**
Wall spacer plate; eared pipework						
65 mm	12.58	15.92	0.16	6.29	nr	**22.21**
75 mm	12.79	16.19	0.16	6.29	nr	**22.48**
100 mm	19.80	25.06	0.16	6.29	nr	**31.35**
Extra over fittings circular cast iron pipework BS 460						
Loose sockets						
Plain socket						
65 mm	15.41	19.50	0.23	9.04	nr	**28.55**
75 mm	15.41	19.50	0.23	9.04	nr	**28.55**
100 mm	23.81	30.13	0.23	9.04	nr	**39.18**
Eared socket						
65 mm	22.26	28.17	0.29	11.40	nr	**39.58**
75 mm	22.26	28.17	0.29	11.40	nr	**39.58**
100 mm	30.08	38.07	0.29	11.40	nr	**49.47**
Shoe; front projection						
Plain socket						
65 mm	47.91	60.64	0.23	9.04	nr	**69.68**
75 mm	47.91	60.64	0.23	9.04	nr	**69.68**
100 mm	64.61	81.76	0.23	9.04	nr	**90.81**
Eared socket						
65 mm	55.54	70.29	0.29	11.40	nr	**81.70**
75 mm	55.54	70.29	0.29	11.40	nr	**81.70**
100 mm	73.72	93.31	0.29	11.40	nr	**104.71**
Access pipe						
65 mm	86.52	109.49	0.23	9.04	nr	**118.54**
75 mm	90.82	114.94	0.23	9.04	nr	**123.99**
100 mm	161.10	203.89	0.23	9.04	nr	**212.93**
100 mm; eared	181.90	230.22	0.29	11.40	nr	**241.62**
Bends; any degree						
65 mm	34.61	43.81	0.23	9.04	nr	**52.85**
75 mm	41.24	52.19	0.23	9.04	nr	**61.23**
100 mm	58.28	73.76	0.23	9.04	nr	**82.80**

33 ABOVE GROUND DRAINAGE

Item	Net Price £	Material £	Labour hours	Labour £	Unit	Total rate £
Branch						
92.5°						
65 mm	66.78	84.52	0.29	11.40	nr	**95.93**
75 mm	73.65	93.22	0.29	11.40	nr	**104.62**
100 mm	85.83	108.63	0.29	11.40	nr	**120.03**
112.5°						
65 mm	66.78	84.52	0.29	11.40	nr	**95.93**
75 mm	73.65	93.22	0.29	11.40	nr	**104.62**
135°						
65 mm	66.78	84.52	0.29	11.40	nr	**95.93**
75 mm	73.65	93.22	0.29	11.40	nr	**104.62**
Offsets						
75 to 150 mm projection						
65 mm	52.01	65.82	0.25	9.83	nr	**75.65**
75 mm	52.01	65.82	0.25	9.83	nr	**75.65**
100 mm	98.13	124.19	0.25	9.83	nr	**134.02**
225 mm projection						
65 mm	53.01	67.09	0.25	9.83	nr	**76.92**
75 mm	53.01	67.09	0.25	9.83	nr	**76.92**
100 mm	98.13	124.19	0.25	9.83	nr	**134.02**
305 mm projection						
65 mm	70.93	89.77	0.25	9.83	nr	**99.60**
75 mm	74.46	94.24	0.25	9.83	nr	**104.07**
100 mm	121.18	153.37	0.25	9.83	nr	**163.20**
380 mm projection						
65 mm	141.56	179.16	0.25	9.83	nr	**188.99**
75 mm	141.56	179.16	0.25	9.83	nr	**188.99**
100 mm	193.20	244.52	0.25	9.83	nr	**254.35**
455 mm projection						
65 mm	165.70	209.72	0.25	9.83	nr	**219.55**
75 mm	165.70	209.72	0.25	9.83	nr	**219.55**
100 mm	240.27	304.09	0.25	9.83	nr	**313.92**
Cast iron rainwater pipe; dry joints; BS 460; fixed to backgrounds: Rectangular						
Plain socket						
100 × 75 mm	150.61	190.61	1.04	40.90	m	**231.51**
Bracket; fixed to backgrounds. For minimum fixing distances, refer to the Tables and Memoranda at the rear of the book						
100 × 75 mm; build in holderbat	85.71	108.47	0.35	13.76	nr	**122.24**
100 × 75 mm; trefoil earband	68.30	86.43	0.29	11.40	nr	**97.84**
100 × 75 mm; plain earband	66.05	83.59	0.29	11.40	nr	**94.99**
Eared socket, with wall spacers measured separately						
100 × 75 mm	153.11	193.77	1.16	45.62	m	**239.39**
Wall spacer plate; eared pipework						
100 × 75	19.80	25.06	0.16	6.29	nr	**31.35**

33 ABOVE GROUND DRAINAGE

Item	Net Price £	Material £	Labour hours	Labour £	Unit	Total rate £
RAINWATER PIPEWORK/GUTTERS – cont						
Extra over fittings rectangular cast iron pipework BS 460						
Loose socket						
100 × 75 mm; plain	63.89	80.85	0.23	9.04	nr	**89.90**
100 × 75 mm; eared	105.83	133.93	0.29	11.40	nr	**145.34**
Shoe; front						
100 × 75 mm; plain	169.45	214.45	0.23	9.04	nr	**223.50**
100 × 75 mm; eared	207.05	262.04	0.29	11.40	nr	**273.44**
Shoe; side						
100 × 75 mm; plain	205.59	260.19	0.23	9.04	nr	**269.24**
100 × 75 mm; eared	257.02	325.28	0.29	11.40	nr	**336.68**
Bends; side; any degree						
100 × 75 mm; plain	155.90	197.31	0.25	9.83	nr	**207.14**
100 × 75 mm; 135°; plain	159.24	201.53	0.25	9.83	nr	**211.36**
Bends; side; any degree						
100 × 75 mm; eared	197.14	249.50	0.25	9.83	nr	**259.33**
Bends; front; any degree						
100 × 75 mm; plain	147.68	186.90	0.25	9.83	nr	**196.73**
100 × 75 mm; eared	167.32	211.76	0.25	9.83	nr	**221.59**
Offset; side						
Plain socket						
75 mm projection	205.43	259.99	0.25	9.83	nr	**269.82**
115 mm projection	213.65	270.40	0.25	9.83	nr	**280.23**
225 mm projection	266.17	336.87	0.25	9.83	nr	**346.70**
305 mm projection	306.79	388.28	0.25	9.83	nr	**398.11**
Offset; front						
Plain socket						
75 mm projection	156.29	197.80	0.25	9.83	nr	**207.63**
150 mm projection	172.71	218.59	0.25	9.83	nr	**228.42**
225 mm projection	213.81	270.60	0.25	9.83	nr	**280.43**
305 mm projection	254.40	321.97	0.25	9.83	nr	**331.80**
Eared socket						
75 mm projection	199.91	253.01	0.25	9.83	nr	**262.84**
150 mm projection	215.59	272.85	0.25	9.83	nr	**282.68**
225 mm projection	255.03	322.77	0.25	9.83	nr	**332.60**
305 mm projection	296.73	375.54	0.25	9.83	nr	**385.37**
Offset; plinth						
115 mm projection; plain	161.31	204.16	0.25	9.83	nr	**213.99**
115 mm projection; eared	207.75	262.93	0.25	9.83	nr	**272.76**
Rainwater heads						
Flat hopper						
210 × 160 × 185 mm; 65 mm outlet	122.73	155.32	0.40	15.73	nr	**171.05**
210 × 160 × 185 mm; 75 mm outlet	122.73	155.32	0.40	15.73	nr	**171.05**
250 × 215 × 215 mm; 100 mm outlet	145.46	184.10	0.40	15.73	nr	**199.83**
Flat rectangular						
225 × 125 × 125 mm; 65 mm outlet	170.64	215.96	0.40	15.73	nr	**231.69**
225 × 125 × 125 mm; 75 mm outlet	170.64	215.96	0.40	15.73	nr	**231.69**
280 × 150 × 130 mm; 100 mm outlet	235.60	298.17	0.40	15.73	nr	**313.90**

33 ABOVE GROUND DRAINAGE

Item	Net Price £	Material £	Labour hours	Labour £	Unit	Total rate £
Rectangular						
250 × 180 × 175 mm; 75 mm outlet	159.10	201.35	0.40	15.73	nr	**217.08**
250 × 180 × 175 mm; 100 mm outlet	159.10	201.35	0.40	15.73	nr	**217.08**
300 × 250 × 200 mm; 65 mm outlet	221.59	280.45	0.40	15.73	nr	**296.18**
300 × 250 × 200 mm; 75 mm outlet	221.59	280.45	0.40	15.73	nr	**296.18**
300 × 250 × 200 mm; 100 mm outlet	221.59	280.45	0.40	15.73	nr	**296.18**
300 × 250 × 200 mm; 100 × 75 mm outlet	221.59	280.45	0.40	15.73	nr	**296.18**
Castellated rectangular						
250 × 180 × 175 mm; 65 mm outlet	159.10	201.35	0.40	15.73	nr	**217.08**

33 ABOVE GROUND DRAINAGE

Item	Net Price £	Material £	Labour hours	Labour £	Unit	Total rate £
DISPOSAL SYSTEMS						
Pricing note: degree angles are only indicated where material prices differ						
PVC-u overflow pipe; solvent welded joints; fixed with clips to backgrounds						
Pipe, with brackets measured separately						
19 mm	2.20	2.78	0.21	8.26	m	**11.04**
FIXINGS						
Pipe clip: including fixing to backgrounds. For minimum fixing distances, refer to the Tables and Memoranda at the rear of the book						
19 mm	0.89	1.12	0.18	7.08	nr	**8.20**
Extra over fittings overflow pipework PVC-u						
Straight coupler						
19 mm	2.35	2.97	0.17	6.69	nr	**9.65**
Bend						
19 mm: 91.25°	2.75	3.48	0.17	6.69	nr	**10.17**
19 mm: 135°	2.77	3.50	0.17	6.69	nr	**10.19**
Tee						
19 mm	2.99	3.79	0.18	7.08	nr	**10.87**
Reverse nut connector						
19 mm	1.10	1.39	0.15	5.90	nr	**7.29**
BSP adaptor: solvent welded socket to threaded socket						
19 mm × ¾"	4.03	5.10	0.14	5.51	nr	**10.61**
Straight tank connector						
19 mm	1.84	2.33	0.28	11.01	nr	**13.34**
32 mm	1.84	2.33	0.28	11.01	nr	**13.34**
40 mm	1.84	2.33	0.30	11.80	nr	**14.13**
Bent tank connector						
19 mm	4.30	5.44	0.21	8.26	nr	**13.70**
Tundish						
19 mm	59.24	74.97	0.38	14.94	nr	**89.92**
MuPVC waste pipe; solvent welded joints; fixed with clips to backgrounds; BS 5255						
Pipe, with brackets measured separately						
32 mm	3.96	5.01	0.23	9.04	m	**14.06**
40 mm	4.88	6.18	0.23	9.04	m	**15.22**
50 mm	7.41	9.38	0.26	10.22	m	**19.60**
FIXINGS						
Pipe clip: including fixing to backgrounds. For minimum fixing distances, refer to the Tables and Memoranda at the rear of the book						
32 mm	0.80	1.01	0.13	5.11	nr	**6.12**
40 mm	0.90	1.13	0.13	5.11	nr	**6.25**
50 mm	1.21	1.53	0.13	5.11	nr	**6.65**

33 ABOVE GROUND DRAINAGE

Item	Net Price £	Material £	Labour hours	Labour £	Unit	Total rate £
Pipe clip: expansion: including fixing to backgrounds. For minimum fixing distances, refer to the Tables and Memoranda at the rear of the book						
32 mm	0.95	1.20	0.13	5.11	nr	**6.31**
40 mm	0.98	1.24	0.13	5.11	nr	**6.35**
50 mm	2.31	2.92	0.13	5.11	nr	**8.03**
Pipe clip: metal; including fixing to backgrounds. For minimum fixing distances, refer to the Tables and Memoranda at the rear of the book						
32 mm	3.45	4.37	0.13	5.11	nr	**9.48**
40 mm	4.10	5.19	0.13	5.11	nr	**10.30**
50 mm	5.23	6.62	0.13	5.11	nr	**11.73**
Extra over fittings waste pipework MuPVC						
Screwed access plug						
32 mm	2.24	2.84	0.18	7.08	nr	**9.91**
40 mm	2.24	2.84	0.18	7.08	nr	**9.91**
50 mm	3.22	4.07	0.25	9.83	nr	**13.91**
Straight coupling						
32 mm	2.39	3.02	0.27	10.62	nr	**13.64**
40 mm	2.39	3.02	0.27	10.62	nr	**13.64**
50 mm	4.40	5.57	0.27	10.62	nr	**16.18**
Expansion coupling						
32 mm	4.25	5.38	0.27	10.62	nr	**15.99**
40 mm	5.09	6.45	0.27	10.62	nr	**17.07**
50 mm	6.92	8.76	0.27	10.62	nr	**19.38**
MuPVC to copper coupling						
32 mm	4.25	5.38	0.27	10.62	nr	**15.99**
40 mm	5.09	6.45	0.27	10.62	nr	**17.07**
50 mm	6.92	8.76	0.27	10.62	nr	**19.38**
Spigot and socket coupling						
32 mm	4.25	5.38	0.27	10.62	nr	**15.99**
40 mm	5.09	6.45	0.27	10.62	nr	**17.07**
50 mm	6.92	8.76	0.27	10.62	nr	**19.38**
Union						
32 mm	10.07	12.75	0.28	11.01	nr	**23.76**
40 mm	13.19	16.70	0.28	11.01	nr	**27.71**
50 mm	15.02	19.01	0.28	11.01	nr	**30.02**
Reducer: socket						
32 × 19 mm	1.52	1.93	0.27	10.62	nr	**12.54**
40 × 32 mm	2.39	3.02	0.27	10.62	nr	**13.64**
50 × 32 mm	3.47	4.39	0.27	10.62	nr	**15.01**
50 × 40 mm	4.25	5.38	0.27	10.62	nr	**15.99**
Reducer: level invert						
40 × 32 mm	2.99	3.78	0.27	10.62	nr	**14.40**
50 × 32 mm	3.68	4.66	0.27	10.62	nr	**15.28**
50 × 40 mm	3.68	4.66	0.27	10.62	nr	**15.28**
Swept bend						
32 mm	2.46	3.11	0.27	10.62	nr	**13.73**
32 mm: 165°	2.54	3.21	0.27	10.62	nr	**13.83**
40 mm	2.74	3.47	0.27	10.62	nr	**14.08**

33 ABOVE GROUND DRAINAGE

Item	Net Price £	Material £	Labour hours	Labour £	Unit	Total rate £
DISPOSAL SYSTEMS – cont						
FIXINGS – cont						
Extra over fittings waste pipework MuPVC – cont						
Swept bend – cont						
40 mm: 165°	4.75	6.01	0.27	10.62	nr	**16.62**
50 mm	4.80	6.07	0.30	11.80	nr	**17.87**
50 mm: 165°	6.34	8.02	0.30	11.80	nr	**19.82**
Knuckle bend						
32 mm	2.24	2.84	0.27	10.62	nr	**13.45**
40 mm	2.47	3.13	0.27	10.62	nr	**13.75**
Spigot and socket bend						
32 mm	4.00	5.06	0.27	10.62	nr	**15.68**
32 mm: 150°	4.15	5.25	0.27	10.62	nr	**15.87**
40 mm	4.58	5.80	0.27	10.62	nr	**16.41**
50 mm	6.52	8.25	0.30	11.80	nr	**20.05**
Swept tee						
32 mm: 91.25°	3.27	4.14	0.31	12.19	nr	**16.33**
32 mm: 135°	3.93	4.98	0.31	12.19	nr	**17.17**
40 mm: 91.25°	4.17	5.27	0.31	12.19	nr	**17.46**
40 mm: 135°	5.19	6.57	0.31	12.19	nr	**18.76**
50 mm	8.11	10.27	0.31	12.19	nr	**22.46**
Swept cross						
40 mm: 91.25°	10.12	12.81	0.31	12.19	nr	**25.00**
50 mm: 91.25°	10.59	13.40	0.31	12.19	nr	**25.59**
50 mm: 135°	13.36	16.91	0.43	16.91	nr	**33.82**
Male iron adaptor						
32 mm	3.65	4.62	0.28	11.01	nr	**15.63**
40 mm	4.30	5.44	0.28	11.01	nr	**16.45**
Female iron adaptor						
32 mm	4.30	5.44	0.28	11.01	nr	**16.45**
40 mm	4.30	5.44	0.28	11.01	nr	**16.45**
50 mm	6.17	7.81	0.31	12.19	nr	**20.00**
Reverse nut adaptor						
32 mm	5.46	6.91	0.20	7.87	nr	**14.77**
40 mm	5.46	6.91	0.20	7.87	nr	**14.77**
Automatic air admittance valve						
32 mm	30.68	38.83	0.27	10.62	nr	**49.45**
40 mm	30.68	38.83	0.28	11.01	nr	**49.84**
50 mm	30.68	38.83	0.31	12.19	nr	**51.02**
MuPVC to metal adpator: including heat shrunk joint to metal						
50 mm	13.23	16.74	0.38	14.94	nr	**31.68**
Caulking bush: including joint to metal						
32 mm	5.76	7.29	0.31	12.19	nr	**19.48**
40 mm	5.76	7.29	0.31	12.19	nr	**19.48**
50 mm	5.76	7.29	0.32	12.58	nr	**19.87**
Weathering apron						
50 mm	5.06	6.41	0.65	25.56	nr	**31.97**
Vent cowl						
50 mm	5.77	7.31	0.19	7.47	nr	**14.78**

33 ABOVE GROUND DRAINAGE

Item	Net Price £	Material £	Labour hours	Labour £	Unit	Total rate £
ABS waste pipe; solvent welded joints; fixed with clips to backgrounds; BS 5255						
Pipe, with brackets measured separately						
32 mm	3.17	4.01	0.23	9.04	m	**13.05**
40 mm	3.96	5.01	0.23	9.04	m	**14.05**
50 mm	4.97	6.30	0.26	10.22	m	**16.52**
FIXINGS						
Pipe clip: including fixing to backgrounds. For minimum fixing distances, refer to the Tables and Memoranda at the rear of the book						
32 mm	0.51	0.64	0.17	6.69	nr	**7.33**
40 mm	0.63	0.80	0.17	6.69	nr	**7.49**
50 mm	1.90	2.40	0.17	6.69	nr	**9.09**
Pipe clip: expansion: including fixing to backgrounds. For minimum fixing distances, refer to the Tables and Memoranda at the rear of the book						
32 mm	0.51	0.64	0.17	6.69	nr	**7.33**
40 mm	0.63	0.80	0.17	6.69	nr	**7.49**
50 mm	1.90	2.40	0.17	6.69	nr	**9.09**
Pipe clip: metal; including fixing to backgrounds. For minimum fixing distances, refer to the Tables and Memoranda at the rear of the book						
32 mm	2.88	3.65	0.17	6.69	nr	**10.33**
40 mm	3.42	4.33	0.17	6.69	nr	**11.01**
50 mm	4.36	5.52	0.17	6.69	nr	**12.20**
Extra over fittings waste pipework ABS						
Screwed access plug						
32 mm	1.97	2.49	0.18	7.08	nr	**9.57**
40 mm	1.97	2.49	0.18	7.08	nr	**9.57**
50 mm	4.07	5.14	0.25	9.83	nr	**14.98**
Straight coupling						
32 mm	1.97	2.49	0.27	10.62	nr	**13.11**
40 mm	1.97	2.49	0.27	10.62	nr	**13.11**
50 mm	4.07	5.14	0.27	10.62	nr	**15.76**
Expansion coupling						
32 mm	4.07	5.14	0.27	10.62	nr	**15.76**
40 mm	4.07	5.14	0.27	10.62	nr	**15.76**
50 mm	6.23	7.89	0.27	10.62	nr	**18.50**
ABS to copper coupling						
32 mm	3.85	4.88	0.27	10.62	nr	**15.50**
40 mm	3.85	4.88	0.27	10.62	nr	**15.50**
50 mm	5.89	7.46	0.27	10.62	nr	**18.08**
Reducer: socket						
40 × 32 mm	1.97	2.49	0.27	10.62	nr	**13.11**
50 × 32 mm	4.47	5.66	0.27	10.62	nr	**16.28**
50 × 40 mm	4.47	5.66	0.27	10.62	nr	**16.28**

33 ABOVE GROUND DRAINAGE

Item	Net Price £	Material £	Labour hours	Labour £	Unit	Total rate £
DISPOSAL SYSTEMS – cont						
Extra over fittings waste pipework ABS – cont						
Swept bend						
32 mm	1.97	2.49	0.27	10.62	nr	**13.11**
40 mm	1.97	2.49	0.27	10.62	nr	**13.11**
50 mm	4.07	5.14	0.30	11.80	nr	**16.94**
Knuckle bend						
32 mm	1.97	2.49	0.27	10.62	nr	**13.11**
40 mm	1.97	2.49	0.27	10.62	nr	**13.11**
Swept tee						
32 mm	2.81	3.56	0.31	12.19	nr	**15.75**
40 mm	2.81	3.56	0.31	12.19	nr	**15.75**
50 mm	7.44	9.42	0.31	12.19	nr	**21.61**
Swept cross						
40 mm	9.48	12.00	0.23	9.04	nr	**21.04**
50 mm	10.83	13.71	0.43	16.91	nr	**30.62**
Male iron adaptor						
32 mm	4.07	5.14	0.28	11.01	nr	**16.16**
40 mm	4.07	5.14	0.28	11.01	nr	**16.16**
Female iron adapator						
32 mm	4.07	5.14	0.28	11.01	nr	**16.16**
40 mm	4.07	5.14	0.28	11.01	nr	**16.16**
50 mm	6.09	7.71	0.31	12.19	nr	**19.90**
Tank connectors						
32 mm	2.80	3.54	0.29	11.40	nr	**14.95**
40 mm	3.09	3.92	0.29	11.40	nr	**15.32**
Caulking bush: including joint to pipework						
50 mm	4.81	6.09	0.50	19.66	nr	**25.75**
Polypropylene waste pipe; push fit joints; fixed with clips to backgrounds; BS 5254. Pipe, with brackets measured separately						
32 mm	2.07	2.63	0.21	8.26	m	**10.88**
40 mm	2.58	3.26	0.21	8.26	m	**11.52**
50 mm	4.30	5.44	0.38	14.94	m	**20.39**
FIXINGS						
Pipe clip: saddle; including fixing to backgrounds. For minimum fixing distances, refer to the Tables and Memoranda at the rear of the book						
32 mm	0.70	0.88	0.17	6.69	nr	**7.57**
40 mm	0.70	0.88	0.17	6.69	nr	**7.57**
Pipe clip: including fixing to backgrounds. For minimum fixing distances, refer to the Tables and Memoranda at the rear of the book						
50 mm	1.64	2.08	0.17	6.69	nr	**8.76**

33 ABOVE GROUND DRAINAGE

Item	Net Price £	Material £	Labour hours	Labour £	Unit	Total rate £
Extra over fittings waste pipework polypropylene						
Screwed access plug						
32 mm	1.94	2.46	0.16	6.29	nr	**8.75**
40 mm	1.94	2.46	0.16	6.29	nr	**8.75**
50 mm	3.32	4.20	0.20	7.87	nr	**12.07**
Straight coupling						
32 mm	1.94	2.46	0.19	7.47	nr	**9.93**
40 mm	1.94	2.46	0.19	7.47	nr	**9.93**
50 mm	3.32	4.20	0.20	7.87	nr	**12.07**
Universal waste pipe coupler						
32 mm dia.	3.29	4.16	0.20	7.87	nr	**12.02**
40 mm dia.	3.73	4.73	0.20	7.87	nr	**12.59**
Reducer						
40 × 32 mm	3.32	4.20	0.19	7.47	nr	**11.67**
50 × 32 mm	3.50	4.43	0.19	7.47	nr	**11.90**
50 × 40 mm	3.60	4.56	0.20	7.87	nr	**12.42**
Swept bend						
32 mm	1.94	2.46	0.19	7.47	nr	**9.93**
40 mm	1.94	2.46	0.19	7.47	nr	**9.93**
50 mm	3.32	4.20	0.20	7.87	nr	**12.07**
Knuckle bend						
32 mm	1.94	2.46	0.19	7.47	nr	**9.93**
40 mm	1.94	2.46	0.19	7.47	nr	**9.93**
50 mm	3.32	4.20	0.20	7.87	nr	**12.07**
Spigot and socket bend						
32 mm	1.94	2.46	0.19	7.47	nr	**9.93**
40 mm	1.94	2.46	0.19	7.47	nr	**9.93**
Swept tee						
32 mm	2.14	2.71	0.22	8.65	nr	**11.36**
40 mm	2.14	2.71	0.22	8.65	nr	**11.36**
50 mm	3.60	4.56	0.23	9.04	nr	**13.60**
Male iron adaptor						
32 mm	1.94	2.46	0.13	5.11	nr	**7.57**
40 mm	1.94	2.46	0.15	5.90	nr	**8.36**
50 mm	3.32	4.20	0.19	7.47	nr	**11.67**
Tank connector						
32 mm	1.94	2.46	0.24	9.44	nr	**11.90**
40 mm	1.94	2.46	0.24	9.44	nr	**11.90**
50 mm	3.32	4.20	0.35	13.76	nr	**17.96**
Polypropylene traps; including fixing to appliance and connection to pipework; BS 3943						
Tubular P trap; 75 mm seal						
32 mm dia.	7.92	10.02	0.20	7.87	nr	**17.88**
40 mm dia.	9.13	11.55	0.20	7.87	nr	**19.42**
Tubular S trap; 75 mm seal						
32 mm dia.	9.94	12.58	0.20	7.87	nr	**20.45**
40 mm dia.	11.75	14.87	0.20	7.87	nr	**22.73**

33 ABOVE GROUND DRAINAGE

Item	Net Price £	Material £	Labour hours	Labour £	Unit	Total rate £
DISPOSAL SYSTEMS – cont						
Polypropylene traps – cont						
Running tubular P trap; 75 mm seal						
32 mm dia.	12.11	15.33	0.20	7.87	nr	**23.20**
40 mm dia.	13.24	16.76	0.20	7.87	nr	**24.62**
Running tubular S trap; 75 mm seal						
32 mm dia.	14.60	18.48	0.20	7.87	nr	**26.35**
40 mm dia.	15.71	19.89	0.20	7.87	nr	**27.75**
Spigot and socket bend; converter from P to S Trap						
32 mm	3.10	3.93	0.20	7.87	nr	**11.79**
40 mm	3.30	4.18	0.21	8.26	nr	**12.44**
Bottle P trap; 75 mm seal						
32 mm dia.	8.81	11.15	0.20	7.87	nr	**19.02**
40 mm dia.	10.55	13.36	0.20	7.87	nr	**21.22**
Bottle S trap; 75 mm seal						
32 mm dia.	10.64	13.46	0.20	7.87	nr	**21.33**
40 mm dia.	12.91	16.34	0.25	9.83	nr	**26.17**
Bottle P trap; resealing; 75 mm seal						
32 mm dia.	10.97	13.88	0.20	7.87	nr	**21.75**
40 mm dia.	12.81	16.21	0.25	9.83	nr	**26.04**
Bottle S trap; resealing; 75 mm seal						
32 mm dia.	12.55	15.88	0.20	7.87	nr	**23.74**
40 mm dia.	14.57	18.44	0.25	9.83	nr	**28.27**
Bath trap, low level; 38 mm seal						
40 mm dia.	11.00	13.92	0.25	9.83	nr	**23.76**
Bath trap, low level; 38 mm seal complete with overflow hose						
40 mm dia.	16.98	21.48	0.25	9.83	nr	**31.32**
Bath trap; 75 mm seal complete with overlow hose						
40 mm dia.	16.93	21.42	0.25	9.83	nr	**31.25**
Bath trap; 75 mm seal complete with overflow hose and overflow outlet						
40 mm dia.	28.89	36.56	0.20	7.87	nr	**44.43**
Bath trap; 75 mm seal complete with overflow hose, overflow outlet and ABS chrome waste						
40 mm dia.	39.91	50.51	0.20	7.87	nr	**58.37**
Washing machine trap; 75 mm seal including stand pipe						
40 mm dia.	22.97	29.07	0.25	9.83	nr	**38.90**
Washing machine standpipe						
40 mm dia.	11.98	15.16	0.25	9.83	nr	**24.99**
Plastic unslotted chrome plated basin/sink waste including plug						
32 mm	14.42	18.25	0.34	13.37	nr	**31.62**
40 mm	19.30	24.42	0.34	13.37	nr	**37.80**
Plastic slotted chrome plated basin/sink waste including plug						
32 mm	11.45	14.49	0.34	13.37	nr	**27.86**
40 mm	19.18	24.28	0.34	13.37	nr	**37.65**

33 ABOVE GROUND DRAINAGE

Item	Net Price £	Material £	Labour hours	Labour £	Unit	Total rate £
Bath overflow outlet; plastic; white						
42 mm	10.07	12.75	0.37	14.55	nr	**27.30**
Bath overlow outlet; plastic; chrome plated						
42 mm	13.28	16.80	0.37	14.55	nr	**31.35**
Combined cistern and bath overflow outlet; plastic; white						
42 mm	20.63	26.10	0.39	15.34	nr	**41.44**
Combined cistern and bath overlow outlet; plastic; chrome plated						
42 mm	20.63	26.10	0.39	15.34	nr	**41.44**
Cistern overflow outlet; plastic; white						
42 mm	16.63	21.04	0.15	5.90	nr	**26.94**
Cistern overlow outlet; plastic; chrome plated						
42 mm	14.98	18.96	0.15	5.90	nr	**24.86**
PVC-u soil and waste pipe; solvent welded joints; fixed with clips to backgrounds; BS 4514/ BS EN 607						
Pipe, with brackets measured separately						
82 mm	14.80	18.73	0.35	13.76	m	**32.49**
110 mm	15.09	19.10	0.41	16.12	m	**35.22**
160 mm	39.10	49.49	0.51	20.06	m	**69.54**
FIXINGS						
Galvanized steel pipe clip: including fixing to backgrounds. For minimum fixing distances, refer to the Tables and Memoranda at the rear of the book						
82 mm	5.93	7.51	0.18	7.08	nr	**14.58**
110 mm	6.15	7.78	0.18	7.08	nr	**14.86**
160 mm	14.82	18.76	0.18	7.08	nr	**25.84**
Plastic coated steel pipe clip: including fixing to backgrounds. For minimum fixing distances, refer to the Tables and Memoranda at the rear of the book						
82 mm	8.61	10.89	0.18	7.08	nr	**17.97**
110 mm	11.72	14.83	0.18	7.08	nr	**21.91**
160 mm	14.78	18.71	0.18	7.08	nr	**25.79**
Plastic pipe clip: including fixing to backgrounds. For minimum fixing distances, refer to the Tables and Memoranda at the rear of the book						
82 mm	4.54	5.75	0.18	7.08	nr	**12.83**
110 mm	8.75	11.07	0.18	7.08	nr	**18.15**
Plastic coated steel pipe clip: adjustable; including fixing to backgrounds. For minimum fixing distances, refer to the Tables and Memoranda at the rear of the book						
82 mm	6.93	8.78	0.20	7.87	nr	**16.64**
110 mm	8.68	10.98	0.20	7.87	nr	**18.85**
Galvanized steel pipe clip: drive in; including fixing to backgrounds. For minimum fixing distances, refer to the Tables and Memoranda at the rear of the book						
110 mm	12.78	16.18	0.22	8.65	nr	**24.83**

33 ABOVE GROUND DRAINAGE

Item	Net Price £	Material £	Labour hours	Labour £	Unit	Total rate £
DISPOSAL SYSTEMS – cont						
Extra over fittings solvent welded pipework PVC-u						
Straight coupling						
82 mm	7.85	9.93	0.21	8.26	nr	**18.19**
110 mm	9.82	12.43	0.22	8.65	nr	**21.08**
160 mm	28.30	35.82	0.24	9.44	nr	**45.26**
Expansion coupling						
82 mm	11.75	14.86	0.21	8.26	nr	**23.12**
110 mm	12.03	15.22	0.22	8.65	nr	**23.87**
160 mm	36.15	45.75	0.24	9.44	nr	**55.19**
Slip coupling; double ring socket						
82 mm	24.42	30.90	0.21	8.26	nr	**39.16**
110 mm	30.57	38.68	0.22	8.65	nr	**47.33**
160 mm	48.54	61.43	0.24	9.44	nr	**70.87**
Puddle flanges						
110 mm	238.50	301.84	0.45	17.70	nr	**319.54**
160 mm	385.56	487.96	0.55	21.63	nr	**509.59**
Socket reducer						
82 to 50 mm	11.96	15.13	0.18	7.08	nr	**22.21**
110 to 50 mm	14.90	18.85	0.18	7.08	nr	**25.93**
110 to 82 mm	15.36	19.44	0.22	8.65	nr	**28.09**
160 to 110 mm	31.09	39.34	0.26	10.22	nr	**49.57**
Socket plugs						
82 mm	9.26	11.71	0.15	5.90	nr	**17.61**
110 mm	13.53	17.13	0.20	7.87	nr	**24.99**
160 mm	24.91	31.53	0.27	10.62	nr	**42.14**
Access door; including cutting into pipe						
82 mm	26.25	33.22	0.28	11.01	nr	**44.23**
110 mm	26.25	33.22	0.34	13.37	nr	**46.59**
160 mm	46.88	59.33	0.46	18.09	nr	**77.42**
Screwed access cap						
82 mm	18.57	23.50	0.15	5.90	nr	**29.40**
110 mm	21.89	27.70	0.20	7.87	nr	**35.56**
160 mm	41.21	52.16	0.27	10.62	nr	**62.78**
Access pipe: spigot and socket						
110 mm	32.37	40.96	0.22	8.65	nr	**49.61**
Access pipe: double socket						
110 mm	32.37	40.96	0.22	8.65	nr	**49.61**
Swept bend						
82 mm	19.69	24.92	0.29	11.40	nr	**36.33**
110 mm	23.07	29.19	0.32	12.58	nr	**41.78**
160 mm	57.47	72.74	0.49	19.27	nr	**92.01**
Bend; special angle						
82 mm	37.96	48.05	0.29	11.40	nr	**59.45**
110 mm	45.26	57.29	0.32	12.58	nr	**69.87**
160 mm	76.43	96.73	0.49	19.27	nr	**116.00**
Spigot and socket bend						
82 mm	19.06	24.12	0.26	10.22	nr	**34.35**
110 mm	22.34	28.27	0.32	12.58	nr	**40.85**
110 mm: 135°	25.11	31.78	0.32	12.58	nr	**44.36**
160 mm: 135°	55.59	70.35	0.44	17.30	nr	**87.66**

33 ABOVE GROUND DRAINAGE

Item	Net Price £	Material £	Labour hours	Labour £	Unit	Total rate £
Variable bend: single socket						
110 mm	43.27	54.76	0.33	12.98	nr	**67.74**
Variable bend: double socket						
110 mm	43.17	54.63	0.33	12.98	nr	**67.61**
Access bend						
110 mm	64.04	81.05	0.33	12.98	nr	**94.03**
Single branch: two bosses						
82 mm	27.54	34.86	0.35	13.76	nr	**48.62**
82 mm: 104°	29.37	37.17	0.35	13.76	nr	**50.93**
110 mm	30.52	38.63	0.42	16.52	nr	**55.15**
110 mm: 135°	31.84	40.30	0.42	16.52	nr	**56.82**
160 mm	64.82	82.03	0.50	19.66	nr	**101.69**
160 mm: 135°	66.28	83.88	0.50	19.66	nr	**103.54**
Single branch; four bosses						
110 mm	37.16	47.03	0.42	16.52	nr	**63.55**
Single access branch						
82 mm	52.25	66.13	0.35	13.76	nr	**79.90**
110 mm	89.30	113.02	0.42	16.52	nr	**129.54**
Unequal single branch						
160 × 160 × 110 mm	73.17	92.60	0.50	19.66	nr	**112.27**
160 × 160 × 110 mm: 135°	78.64	99.53	0.50	19.66	nr	**119.19**
Double branch						
110 mm	78.85	99.80	0.42	16.52	nr	**116.31**
110 mm: 135°	75.42	95.45	0.42	16.52	nr	**111.97**
Corner branch						
110 mm	133.25	168.63	0.42	16.52	nr	**185.15**
Unequal double branch						
160 × 160 × 110 mm	136.28	172.48	0.50	19.66	nr	**192.14**
Single boss pipe; single socket						
110 × 110 × 32 mm	8.24	10.43	0.24	9.44	nr	**19.87**
110 × 110 × 40 mm	8.24	10.43	0.24	9.44	nr	**19.87**
110 × 110 × 50 mm	8.71	11.02	0.24	9.44	nr	**20.46**
Single boss pipe; triple socket						
110 × 110 × 40 mm	13.39	16.95	0.24	9.44	nr	**26.39**
Waste boss; including cutting into pipe						
82 to 32 mm	10.79	13.65	0.29	11.40	nr	**25.06**
82 to 40 mm	10.79	13.65	0.29	11.40	nr	**25.06**
110 to 32 mm	10.79	13.65	0.29	11.40	nr	**25.06**
110 to 40 mm	10.79	13.65	0.29	11.40	nr	**25.06**
110 to 50 mm	11.18	14.15	0.29	11.40	nr	**25.56**
160 to 32 mm	15.25	19.30	0.30	11.80	unit	**31.09**
160 to 40 mm	15.25	19.30	0.35	13.76	nr	**33.06**
160 to 50 mm	15.25	19.30	0.40	15.73	nr	**35.03**
Self-locking waste boss; including cutting into pipe						
110 to 32 mm	14.40	18.23	0.30	11.80	nr	**30.03**
110 to 40 mm	15.05	19.05	0.30	11.80	nr	**30.85**
110 to 50 mm	19.45	24.62	0.30	11.80	nr	**36.42**
Adaptor saddle; including cutting to pipe						
82 to 32 mm	7.17	9.08	0.29	11.40	nr	**20.48**
110 to 40 mm	8.86	11.21	0.29	11.40	nr	**22.62**
160 to 50 mm	16.05	20.31	0.29	11.40	nr	**31.72**

33 ABOVE GROUND DRAINAGE

Item	Net Price £	Material £	Labour hours	Labour £	Unit	Total rate £
DISPOSAL SYSTEMS – cont						
Extra over fittings solvent welded pipework PVC-u – cont						
Branch boss adaptor						
32 mm	4.29	5.43	0.26	10.22	nr	**15.65**
40 mm	4.29	5.43	0.26	10.22	nr	**15.65**
50 mm	6.13	7.76	0.26	10.22	nr	**17.99**
Branch boss adaptor bend						
32 mm	5.87	7.42	0.26	10.22	nr	**17.65**
40 mm	6.39	8.08	0.26	10.22	nr	**18.31**
50 mm	7.65	9.68	0.26	10.22	nr	**19.91**
Automatic air admittance valve						
82 to 110 mm	67.28	85.15	0.19	7.47	nr	**92.62**
PVC-u to metal adpator: including heat shrunk joint to metal						
110 mm	18.74	23.71	0.57	22.42	nr	**46.13**
Caulking bush: including joint to pipework						
82 mm	18.44	23.34	0.46	18.09	nr	**41.43**
110 mm	18.44	23.34	0.46	18.09	nr	**41.43**
Vent cowl						
82 mm	5.57	7.05	0.13	5.11	nr	**12.16**
110 mm	5.61	7.10	0.13	5.11	nr	**12.22**
160 mm	14.68	18.58	0.13	5.11	nr	**23.70**
Weathering apron; to lead slates						
82 mm	5.57	7.05	1.15	45.22	nr	**52.27**
110 mm	6.37	8.06	1.15	45.22	nr	**53.29**
160 mm	19.17	24.26	1.15	45.22	nr	**69.49**
Weathering apron; to asphalt						
82 mm	23.41	29.62	1.10	43.26	nr	**72.88**
110 mm	23.41	29.62	1.10	43.26	nr	**72.88**
Weathering slate; flat; 406 × 406 mm						
82 mm	66.04	83.58	1.04	40.90	nr	**124.48**
110 mm	66.04	83.58	1.04	40.90	nr	**124.48**
Weathering slate; flat; 457 × 457 mm						
82 mm	67.75	85.75	1.04	40.90	nr	**126.65**
110 mm	67.75	85.75	1.04	40.90	nr	**126.65**
Weathering slate; angled; 610 × 610 mm						
82 mm	91.55	115.87	1.04	40.90	nr	**156.77**
110 mm	91.55	115.87	1.04	40.90	nr	**156.77**
PVC-u soil and waste pipe; ring seal joints; fixed with clips to backgrounds; BS 4514/BS EN 607						
Pipe, with brackets measured separately						
82 mm dia.	13.05	16.52	0.35	13.76	m	**30.28**
110 mm dia.	13.15	16.64	0.41	16.12	m	**32.77**
160 mm dia.	47.16	59.69	0.51	20.06	m	**79.75**

33 ABOVE GROUND DRAINAGE

Item	Net Price £	Material £	Labour hours	Labour £	Unit	Total rate £
FIXINGS						
Galvanized steel pipe clip: including fixing to backgrounds. For minimum fixing distances, refer to the Tables and Memoranda at the rear of the book						
82 mm	5.93	7.51	0.18	7.08	nr	**14.58**
110 mm	6.15	7.78	0.18	7.08	nr	**14.86**
160 mm	14.82	18.76	0.18	7.08	nr	**25.84**
Plastic coated steel pipe clip: including fixing to backgrounds. For minimum fixing distances, refer to the Tables and Memoranda at the rear of the book						
82 mm	8.61	10.89	0.18	7.08	nr	**17.97**
110 mm	11.72	14.83	0.18	7.08	nr	**21.91**
160 mm	14.78	18.71	0.18	7.08	nr	**25.79**
Plastic pipe clip: including fixing to backgrounds. For minimum fixing distances, refer to the Tables and Memoranda at the rear of the book						
82 mm	4.54	5.75	0.18	7.08	nr	**12.83**
110 mm	8.75	11.07	0.18	7.08	nr	**18.15**
Plastic coated steel pipe clip: adjustable; including fixing to backgrounds. For minimum fixing distances, refer to the Tables and Memoranda at the rear of the book						
82 mm	6.93	8.78	0.20	7.87	nr	**16.64**
110 mm	8.68	10.98	0.20	7.87	nr	**18.85**
Galvanized steel pipe clip: drive in; including fixing to backgrounds. For minimum fixing distances, refer to the Tables and Memoranda at the rear of the book						
110 mm	12.78	16.18	0.22	8.65	nr	**24.83**
Extra over fittings ring seal pipework PVC-u						
Straight coupling						
82 mm	8.55	10.82	0.21	8.26	nr	**19.08**
110 mm	9.58	12.12	0.22	8.65	nr	**20.77**
160 mm	32.37	40.96	0.24	9.44	nr	**50.40**
Straight coupling; double socket						
82 mm	15.09	19.10	0.21	8.26	nr	**27.36**
110 mm	21.29	26.94	0.22	8.65	nr	**35.60**
160 mm	32.37	40.96	0.24	9.44	nr	**50.40**
Reducer; socket						
82 to 50 mm	14.07	17.80	0.15	5.90	nr	**23.70**
110 to 50 mm	19.62	24.83	0.15	5.90	nr	**30.73**
110 to 82 mm	19.62	24.83	0.19	7.47	nr	**32.31**
160 to 110 mm	28.27	35.78	0.31	12.19	nr	**47.97**
Access cap						
82 mm	15.32	19.39	0.15	5.90	nr	**25.28**
110 mm	18.05	22.84	0.17	6.69	nr	**29.52**
Access cap; pressure plug						
160 mm	42.44	53.71	0.33	12.98	nr	**66.69**
Access pipe						
82 mm	37.60	47.58	0.22	8.65	nr	**56.24**
110 mm	64.00	81.00	0.22	8.65	nr	**89.65**
160 mm	78.78	99.71	0.24	9.44	nr	**109.15**

33 ABOVE GROUND DRAINAGE

Item	Net Price £	Material £	Labour hours	Labour £	Unit	Total rate £
DISPOSAL SYSTEMS – cont						
Extra over fittings ring seal pipework PVC-u – cont						
Bend						
82 mm	19.12	24.19	0.29	11.40	nr	**35.60**
82 mm; adjustable radius	38.37	48.56	0.29	11.40	nr	**59.97**
110 mm	20.58	26.04	0.32	12.58	nr	**38.63**
110 mm; adjustable radius	43.11	54.56	0.32	12.58	nr	**67.15**
160 mm	61.76	78.17	0.49	19.27	nr	**97.44**
160 mm; adjustable radius	81.02	102.54	0.49	19.27	nr	**121.81**
Bend; spigot and socket						
110 mm	20.76	26.28	0.32	12.58	nr	**38.86**
Bend; offset						
82 mm	16.84	21.31	0.21	8.26	nr	**29.57**
110 mm	27.01	34.18	0.32	12.58	nr	**46.76**
160 mm	59.57	75.39	0.32	12.58	nr	**87.97**
Bend; access						
110 mm	56.66	71.70	0.33	12.98	nr	**84.68**
Single branch						
82 mm	30.02	37.99	0.35	13.76	nr	**51.75**
110 mm	56.65	71.69	0.42	16.52	nr	**88.21**
110 mm; 45°	28.55	36.14	0.31	12.19	nr	**48.33**
160 mm	71.57	90.57	0.50	19.66	nr	**110.24**
Single branch; access						
82 mm	44.98	56.93	0.35	13.76	nr	**70.69**
110 mm	65.12	82.42	0.42	16.52	nr	**98.94**
Unequal single branch						
160 × 160 × 110 mm	73.13	92.55	0.50	19.66	nr	**112.21**
160 × 160 × 110 mm; 45°	78.64	99.53	0.50	19.66	nr	**119.19**
Double branch; 4 bosses						
110 mm	72.65	91.95	0.49	19.27	nr	**111.22**
Corner branch; 2 bosses						
110 mm	133.22	168.60	0.49	19.27	nr	**187.87**
Multi-branch; 4 bosses						
110 mm	39.78	50.34	0.52	20.45	nr	**70.79**
Boss branch						
110 × 32 mm	8.23	10.41	0.34	13.37	nr	**23.78**
110 × 40 mm	8.23	10.41	0.34	13.37	nr	**23.78**
Strap on boss						
110 × 32 mm	9.17	11.61	0.30	11.80	nr	**23.40**
110 × 40 mm	9.17	11.61	0.30	11.80	nr	**23.40**
110 × 50 mm	9.24	11.70	0.30	11.80	nr	**23.49**
Patch boss						
82 × 32 mm	10.79	13.65	0.31	12.19	nr	**25.84**
82 × 40 mm	10.79	13.65	0.31	12.19	nr	**25.84**
82 × 50 mm	10.79	13.65	0.31	12.19	nr	**25.84**
Boss pipe; collar 4 boss						
110 mm	13.39	16.95	0.35	13.76	nr	**30.71**

33 ABOVE GROUND DRAINAGE

Item	Net Price £	Material £	Labour hours	Labour £	Unit	Total rate £
Boss adaptor; rubber; push fit						
32 mm	5.01	6.34	0.26	10.22	nr	**16.56**
40 mm	5.01	6.34	0.26	10.22	nr	**16.56**
50 mm	5.67	7.17	0.26	10.22	nr	**17.40**
WC connector; cap and seal; solvent socket						
110 mm	13.95	17.66	0.23	9.04	nr	**26.70**
110 mm; 90°	22.42	28.38	0.27	10.62	nr	**38.99**
Vent terminal						
82 mm	5.57	7.05	0.13	5.11	nr	**12.16**
110 mm	5.61	7.10	0.13	5.11	nr	**12.22**
160 mm	14.71	18.62	0.13	5.11	nr	**23.73**
Weathering slate; inclined; 610 × 610 mm						
82 mm	91.55	115.87	1.04	40.90	nr	**156.77**
110 mm	91.55	115.87	1.04	40.90	nr	**156.77**
Weathering slate; inclined; 450 × 450 mm						
82 mm	67.75	85.75	1.04	40.90	nr	**126.65**
110 mm	67.75	85.75	1.04	40.90	nr	**126.65**
Weathering slate; flat; 400 × 400 mm						
82 mm	66.07	83.61	1.04	40.90	nr	**124.51**
110 mm	66.07	83.61	1.04	40.90	nr	**124.51**
Air admittance valve						
82 mm	67.28	85.15	0.19	7.47	nr	**92.62**
110 mm	67.28	85.15	0.19	7.47	nr	**92.62**
Cast iron pipe; nitrile rubber gasket joint with continuity clip BSEN 877; fixed vertically to backgrounds						
Pipe, with brackets and jointing couplings measured separately						
50 mm	41.61	52.66	0.31	12.19	m	**64.85**
75 mm	48.48	61.35	0.34	13.37	m	**74.72**
100 mm	53.85	68.16	0.37	14.55	m	**82.71**
150 mm	112.48	142.36	0.60	23.60	m	**165.96**
FIXINGS						
Brackets; fixed to backgrounds. For minimum fixing distances, refer to the Tables and Memoranda at the rear of the book						
50 mm	15.51	19.64	0.15	5.90	nr	**25.53**
75 mm	16.46	20.84	0.18	7.08	nr	**27.91**
100 mm	16.87	21.35	0.18	7.08	nr	**28.43**
150 mm	31.26	39.56	0.20	7.87	nr	**47.42**
Extra over fittings nitrile gasket cast iron pipework BS 416/6087, with jointing couplings measured separately						
Standard coupling						
50 mm	20.12	25.46	0.17	6.69	nr	**32.15**
75 mm	22.03	27.88	0.17	6.69	nr	**34.57**
100 mm	28.76	36.40	0.17	6.69	nr	**43.09**
150 mm	57.40	72.65	0.17	6.69	nr	**79.33**

33 ABOVE GROUND DRAINAGE

Item	Net Price £	Material £	Labour hours	Labour £	Unit	Total rate £
DISPOSAL SYSTEMS – cont						
Extra over fittings nitrile gasket cast iron pipework BS 416/6087, with jointing couplings measured separately – cont						
Conversion coupling						
65 × 75 mm	23.35	29.55	0.60	23.60	nr	**53.16**
70 × 75 mm	23.35	29.55	0.60	23.60	nr	**53.16**
90 × 100 mm	30.06	38.05	0.67	26.36	nr	**64.40**
Access pipe; round door						
50 mm	58.83	74.45	0.41	16.12	nr	**90.57**
75 mm	84.57	107.03	0.46	18.09	nr	**125.12**
100 mm	88.89	112.50	0.67	26.39	nr	**138.89**
150 mm	147.85	187.12	0.83	32.66	nr	**219.79**
Access pipe; square door						
100 mm	175.16	221.68	0.67	26.39	nr	**248.07**
150 mm	268.12	339.33	0.83	32.66	nr	**372.00**
Taper reducer						
75 mm	48.34	61.17	0.60	23.60	nr	**84.78**
100 mm	61.36	77.65	0.67	26.36	nr	**104.01**
150 mm	119.48	151.21	0.83	32.66	nr	**183.87**
Blank cap						
50 mm	12.21	15.45	0.24	9.44	nr	**24.89**
75 mm	14.79	18.72	0.26	10.22	nr	**28.95**
100 mm	15.22	19.26	0.32	12.58	nr	**31.84**
150 mm	21.37	27.04	0.40	15.73	nr	**42.77**
Blank cap; 50 mm screwed tapping						
75 mm	30.05	38.03	0.26	10.22	nr	**48.25**
100 mm	34.86	44.12	0.32	12.58	nr	**56.70**
150 mm	38.71	48.99	0.40	15.73	nr	**64.72**
Universal connector						
50 × 56/48/40 mm	18.15	22.97	0.33	12.98	nr	**35.95**
Change piece; BS416						
100 mm	38.71	48.99	0.47	18.48	nr	**67.47**
WC connector						
100 mm	62.27	78.81	0.49	19.27	nr	**98.08**
Boss pipe; 2″ BSPT socket						
50 mm	49.72	62.93	0.58	22.81	nr	**85.74**
75 mm	72.78	92.10	0.65	25.56	nr	**117.67**
100 mm	86.94	110.03	0.79	31.07	nr	**141.10**
150 mm	135.45	171.42	0.86	33.82	nr	**205.24**
Boss pipe; 2″ BSPT socket; 135°						
100 mm	104.75	132.58	0.79	31.07	nr	**163.64**
Boss pipe; 2 × 2″ BSPT socket; opposed						
75 mm	96.43	122.04	0.65	25.56	nr	**147.60**
100 mm	106.77	135.13	0.79	31.07	nr	**166.20**
Boss pipe; 2 × 2″ BSPT socket; in line						
100 mm	113.52	143.67	0.79	31.07	nr	**174.74**

33 ABOVE GROUND DRAINAGE

Item	Net Price £	Material £	Labour hours	Labour £	Unit	Total rate £
Boss pipe; 2 × 2" BSPT socket; 90°						
100 mm	112.41	142.27	0.79	31.07	nr	**173.34**
Bend; short radius						
50 mm	35.18	44.52	0.50	19.66	nr	**64.18**
75 mm	39.80	50.37	0.60	23.60	nr	**73.97**
100 mm	48.65	61.57	0.67	26.36	nr	**87.93**
100 mm; 11°	41.94	53.08	0.67	26.36	nr	**79.44**
100 mm; 67°	48.65	61.57	0.67	26.36	nr	**87.93**
150 mm	86.94	110.03	0.83	32.66	nr	**142.69**
Access bend; short radius						
50 mm	86.64	109.65	0.50	19.66	nr	**129.31**
75 mm	93.98	118.94	0.60	23.60	nr	**142.55**
100 mm	102.96	130.31	0.67	26.36	nr	**156.66**
100 mm; 45°	102.96	130.31	0.67	26.36	nr	**156.66**
150 mm	146.17	184.99	0.83	32.66	nr	**217.65**
150 mm; 45°	146.17	184.99	0.83	32.66	nr	**217.65**
Long radius bend						
75 mm	66.38	84.01	0.60	23.60	nr	**107.61**
100 mm	78.84	99.78	0.67	26.36	nr	**126.13**
100 mm; 5°	48.65	61.57	0.67	26.36	nr	**87.93**
150 mm	171.80	217.43	0.83	32.66	nr	**250.10**
150 mm; 22.5°	180.11	227.95	0.83	32.66	nr	**260.61**
Access bend; long radius						
75 mm	117.79	149.08	0.60	23.60	nr	**172.68**
100 mm	133.15	168.51	0.67	26.36	nr	**194.87**
150 mm	234.47	296.75	0.83	32.66	nr	**329.41**
Long tail bend						
100 × 250 mm long	62.87	79.56	0.70	27.54	nr	**107.10**
100 × 815 mm long	199.84	252.92	0.70	27.54	nr	**280.46**
Offset						
75 mm projection						
75 mm	51.20	64.80	0.53	20.84	nr	**85.64**
100 mm	51.88	65.67	0.66	25.96	nr	**91.62**
115 mm projection						
75 mm	57.02	72.16	0.53	20.84	nr	**93.00**
100 mm	63.94	80.92	0.66	25.96	nr	**106.87**
150 mm projection						
75 mm	67.00	84.79	0.53	20.84	nr	**105.63**
100 mm	67.00	84.79	0.66	25.96	nr	**110.74**
225 mm projection						
100 mm	73.25	92.71	0.66	25.96	nr	**118.66**
300 mm projection						
100 mm	78.84	99.78	0.66	25.96	nr	**125.73**
Branch; equal and unequal						
50 mm	52.89	66.93	0.78	30.67	nr	**97.61**
75 mm	59.86	75.76	0.85	33.44	nr	**109.20**
100 mm	75.22	95.20	1.00	39.33	nr	**134.52**
150 mm	186.45	235.98	1.20	47.21	nr	**283.19**
150 × 100 mm; 87.5°	142.69	180.59	1.21	47.39	nr	**227.97**
150 × 100 mm; 45°	209.30	264.88	1.21	47.39	nr	**312.27**

33 ABOVE GROUND DRAINAGE

Item	Net Price £	Material £	Labour hours	Labour £	Unit	Total rate £
DISPOSAL SYSTEMS – cont						
Extra over fittings nitrile gasket cast iron pipework BS 416/6087, with jointing couplings measured separately – cont						
Branch; 2" BSPT screwed socket						
100 mm	100.77	127.53	1.00	39.33	nr	**166.85**
Branch; long tail						
100 × 915 mm long	208.35	263.68	1.00	39.33	nr	**303.01**
Access branch; equal and unequal						
50 mm	115.21	145.81	0.78	30.67	nr	**176.48**
75 mm	115.21	145.81	0.85	33.44	nr	**179.25**
100 mm	129.53	163.93	1.02	40.13	nr	**204.06**
150 mm	275.69	348.92	1.20	47.21	nr	**396.13**
150 × 100 mm; 87.5°	231.70	293.24	1.20	47.21	nr	**340.45**
150 × 100 mm; 45°	251.17	317.88	1.20	47.21	nr	**365.09**
Parallel branch						
100 mm	78.84	99.78	1.00	39.33	nr	**139.10**
Double branch						
75 mm	88.89	112.50	0.95	37.36	nr	**149.86**
100 mm	93.03	117.74	1.30	51.14	nr	**168.88**
150 × 100 mm	261.99	331.57	1.56	61.35	nr	**392.92**
Double access branch						
100 mm	147.33	186.46	1.43	56.26	nr	**242.72**
Corner branch						
100 mm	132.27	167.40	1.30	51.14	nr	**218.54**
Puddle flange; grey epoxy coated						
100 mm	88.45	111.94	1.00	39.33	nr	**151.27**
Roof vent connector; asphalt						
75 mm	84.06	106.38	0.90	35.39	nr	**141.77**
100 mm	107.63	136.22	0.97	38.15	nr	**174.37**
P trap						
100 mm	77.99	98.71	1.00	39.33	nr	**138.03**
P trap with access						
50 mm	119.37	151.08	0.77	30.28	nr	**181.36**
75 mm	119.37	151.08	0.90	35.39	nr	**186.47**
100 mm	132.30	167.44	1.16	45.62	nr	**213.07**
150 mm	231.01	292.37	1.77	69.61	nr	**361.98**
Bellmouth gully inlet						
100 mm	112.89	142.87	1.08	42.47	nr	**185.34**
Balcony gully inlet						
100 mm	340.84	431.37	1.08	42.47	nr	**473.84**
Roof outlet						
Flat grate						
75 mm	210.40	266.29	0.83	32.64	nr	**298.93**
100 mm	296.00	374.62	1.08	42.47	nr	**417.10**
Dome grate						
75 mm	210.40	266.29	0.83	32.64	nr	**298.93**
100 mm	332.02	420.21	1.08	42.47	nr	**462.68**
Top hat						
100 mm	449.50	568.88	1.08	42.47	nr	**611.36**

33 ABOVE GROUND DRAINAGE

Item	Net Price £	Material £	Labour hours	Labour £	Unit	Total rate £
Cast iron pipe; EPDM rubber gasket joint with continuity clip; BS EN877; fixed to backgrounds						
Pipe, with brackets and jointing couplings measured separately						
50 mm	42.68	54.01	0.31	12.19	m	**66.20**
70 mm	49.88	63.13	0.34	13.37	m	**76.50**
100 mm	59.37	75.14	0.37	14.55	m	**89.69**
125 mm	95.27	120.57	0.65	25.56	m	**146.13**
150 mm	117.65	148.90	0.70	27.53	m	**176.43**
200 mm	196.61	248.83	1.14	44.83	m	**293.66**
250 mm	274.71	347.67	1.25	49.16	m	**396.83**
300 mm	342.09	432.95	1.53	60.17	m	**493.12**
FIXINGS						
Brackets; fixed to backgrounds. For minimum fixing distances, refer to the Tables and Memoranda at the rear of the book						
Ductile iron						
50 mm	18.13	22.95	0.10	3.93	nr	**26.88**
70 mm	18.13	22.95	0.10	3.93	nr	**26.88**
100 mm	20.96	26.53	0.15	5.90	nr	**32.43**
150 mm	38.80	49.10	0.20	7.87	nr	**56.96**
200 mm	143.93	182.16	0.25	9.83	nr	**192.00**
Mild steel; vertical						
125 mm	35.33	44.72	0.15	5.90	nr	**50.62**
Mild steel; stand off						
250 mm	76.90	97.33	0.25	9.83	nr	**107.16**
300 mm	84.72	107.23	0.25	9.83	nr	**117.06**
Stack support; rubber seal						
70 mm	71.93	91.04	0.55	21.63	nr	**112.67**
100 mm	80.03	101.29	0.65	25.56	nr	**126.85**
125 mm	88.76	112.34	0.74	29.10	nr	**141.44**
150 mm	126.66	160.31	0.86	33.82	nr	**194.13**
Wall spacer plate; cast iron (eared sockets)						
50 mm	15.44	19.55	0.10	3.93	nr	**23.48**
70 mm	15.44	19.55	0.10	3.93	nr	**23.48**
100 mm	15.44	19.55	0.10	3.93	nr	**23.48**
Extra over fittings EPDM rubber jointed cast iron pipework BS EN 877, with jointing couplings measured separately						
Coupling						
50 mm	19.38	24.53	0.10	3.93	nr	**28.46**
70 mm	21.33	27.00	0.10	3.93	nr	**30.93**
100 mm	27.78	35.16	0.10	3.93	nr	**39.09**
125 mm	34.51	43.67	0.15	5.90	nr	**49.57**
150 mm	55.66	70.45	0.30	11.80	nr	**82.24**
200 mm	124.52	157.59	0.35	13.76	nr	**171.36**
250 mm	178.37	225.75	0.40	15.73	nr	**241.48**
300 mm	206.40	261.22	0.50	19.66	nr	**280.88**

33 ABOVE GROUND DRAINAGE

Item	Net Price £	Material £	Labour hours	Labour £	Unit	Total rate £
DISPOSAL SYSTEMS – cont						
Extra over fittings EPDM rubber jointed cast iron pipework BS EN 877, with jointing couplings measured separately – cont						
Plain socket						
50 mm	48.72	61.66	0.10	3.93	nr	**65.60**
70 mm	48.72	61.66	0.10	3.93	nr	**65.60**
100 mm	55.93	70.78	0.10	3.93	nr	**74.71**
150 mm	94.32	119.37	0.10	3.93	nr	**123.30**
Eared socket						
50 mm	50.16	63.49	0.25	9.83	nr	**73.32**
70 mm	50.16	63.49	0.25	9.83	nr	**73.32**
100 mm	60.78	76.92	0.25	9.83	nr	**86.75**
150 mm	99.66	126.13	0.25	9.83	nr	**135.96**
Slip socket						
50 mm	63.09	79.85	0.25	9.83	nr	**89.68**
70 mm	63.09	79.85	0.25	9.83	nr	**89.68**
100 mm	73.69	93.26	0.25	9.83	nr	**103.09**
150 mm	110.90	140.36	0.25	9.83	nr	**150.19**
Stack support pipe						
70 mm	46.54	58.91	0.74	29.10	nr	**88.01**
100 mm	63.94	80.92	0.88	34.61	nr	**115.53**
125 mm	68.38	86.55	1.00	39.33	nr	**125.87**
150 mm	92.70	117.32	1.19	46.80	nr	**164.12**
Access pipe; round door						
50 mm	89.40	113.14	0.27	10.62	nr	**123.76**
70 mm	94.58	119.70	0.30	11.80	nr	**131.50**
100 mm	103.96	131.57	0.32	12.58	nr	**144.16**
150 mm	188.18	238.16	0.71	27.92	nr	**266.08**
Access pipe; square door						
100 mm	201.02	254.41	0.32	12.58	nr	**267.00**
125 mm	209.21	264.77	0.67	26.35	nr	**291.12**
150 mm	314.75	398.35	0.71	27.92	nr	**426.27**
200 mm	625.23	791.30	1.21	47.58	nr	**838.88**
250 mm	983.77	1245.06	1.31	51.52	nr	**1296.58**
300 mm	1227.17	1553.11	1.41	55.45	nr	**1608.56**
Taper reducer						
70 mm	51.62	65.33	0.30	11.80	nr	**77.13**
100 mm	60.69	76.81	0.32	12.58	nr	**89.39**
125 mm	61.02	77.23	0.64	25.17	nr	**102.40**
150 mm	116.51	147.45	0.67	26.35	nr	**173.80**
200 mm	189.20	239.45	1.15	45.22	nr	**284.67**
250 mm	390.80	494.59	1.25	49.16	nr	**543.75**
300 mm	537.23	679.91	1.37	53.88	nr	**733.79**
Blank cap						
50 mm	13.51	17.10	0.24	9.44	nr	**26.54**
70 mm	14.27	18.06	0.26	10.22	nr	**28.28**
100 mm	16.57	20.97	0.32	12.58	nr	**33.55**

33 ABOVE GROUND DRAINAGE

Item	Net Price £	Material £	Labour hours	Labour £	Unit	Total rate £
125 mm	23.46	29.69	0.35	13.76	nr	**43.45**
150 mm	23.98	30.35	0.40	15.73	nr	**46.08**
200 mm	105.88	134.00	0.60	23.60	nr	**157.59**
250 mm	216.36	273.82	0.65	25.56	nr	**299.39**
300 mm	308.34	390.23	0.72	28.31	nr	**418.55**
Blank cap; 50 mm screwed tapping						
70 mm	31.80	40.25	0.26	10.22	nr	**50.47**
100 mm	34.35	43.47	0.32	12.58	nr	**56.06**
150 mm	41.25	52.21	0.40	15.73	nr	**67.94**
Universal connector; EPDM rubber						
50 × 56/48/40 mm	27.59	34.91	0.30	11.80	nr	**46.71**
Blank end; push fit						
100 × 38/32 mm	30.78	38.96	0.32	12.58	nr	**51.54**
Boss pipe; 2" BSPT socket						
50 mm	75.59	95.66	0.27	10.62	nr	**106.28**
75 mm	75.59	95.66	0.30	11.80	nr	**107.46**
100 mm	92.37	116.90	0.32	12.58	nr	**129.48**
150 mm	150.61	190.61	0.71	27.92	nr	**218.54**
Boss pipe; 2 × 2" BSPT socket; opposed						
100 mm	119.35	151.05	0.32	12.58	nr	**163.64**
Boss pipe; 2 × 2" BSPT socket; 90°						
100 mm	119.35	151.05	0.32	12.58	nr	**163.64**
Manifold connector						
100 mm	184.26	233.20	0.64	25.17	nr	**258.37**
150 mm	256.75	324.95	1.00	39.33	nr	**364.27**
Bend; short radius						
50 mm	33.56	42.47	0.27	10.62	nr	**53.09**
70 mm	37.78	47.81	0.30	11.80	nr	**59.61**
100 mm	44.70	56.57	0.32	12.58	nr	**69.15**
125 mm	79.31	100.38	0.62	24.38	nr	**124.76**
150 mm	80.30	101.62	0.67	26.35	nr	**127.97**
200 mm; 45°	239.04	302.53	1.21	47.58	nr	**350.12**
250 mm; 45°	466.44	590.32	1.31	51.52	nr	**641.84**
300 mm; 45°	656.14	830.41	1.43	56.24	nr	**886.65**
Access bend; short radius						
70 mm	73.20	92.64	0.30	11.80	nr	**104.44**
100 mm	106.91	135.31	0.32	12.58	nr	**147.90**
150 mm	188.16	238.13	0.67	26.35	nr	**264.48**
Bend; long radius bend						
100 mm; 88°	113.61	143.78	0.32	12.58	nr	**156.37**
100 mm; 22°	83.79	106.05	0.32	12.58	nr	**118.63**
150 mm; 88°	325.72	412.23	0.67	26.35	nr	**438.57**
Access bend; long radius						
100 mm	138.31	175.05	0.32	12.58	nr	**187.63**
150 mm	336.01	425.26	0.32	12.58	nr	**437.84**
Bend; long tail						
100 mm	78.36	99.18	0.32	12.58	nr	**111.76**
Bend; long tail double						
70 mm	121.41	153.66	0.30	11.80	nr	**165.45**
100 mm	133.94	169.51	0.32	12.58	nr	**182.10**

33 ABOVE GROUND DRAINAGE

Item	Net Price £	Material £	Labour hours	Labour £	Unit	Total rate £
DISPOSAL SYSTEMS – cont						
Extra over fittings EPDM rubber jointed cast iron pipework BS EN 877, with jointing couplings measured separately – cont						
Offset; 75 mm projection						
100 mm	68.72	86.97	0.32	12.58	nr	**99.55**
Offset; 130 mm projection						
50 mm	56.84	71.94	0.27	10.62	nr	**82.55**
70 mm	86.25	109.16	0.30	11.80	nr	**120.96**
100 mm	113.29	143.38	0.32	12.58	nr	**155.97**
125 mm	143.57	181.70	0.67	26.35	nr	**208.05**
Branch; equal and unequal						
50 mm	53.80	68.09	0.37	14.55	nr	**82.64**
70 mm	56.82	71.91	0.40	15.73	nr	**87.64**
100 mm	77.96	98.66	0.42	16.52	nr	**115.18**
125 mm	156.25	197.75	0.76	29.89	nr	**227.64**
150 mm	170.01	215.16	0.97	38.15	nr	**253.31**
200 mm	436.11	551.94	1.51	59.38	nr	**611.32**
250 mm	557.33	705.35	1.63	64.10	nr	**769.45**
300 mm	919.61	1163.85	1.77	69.61	nr	**1233.46**
Branch; radius; equal and unequal						
70 mm	69.23	87.61	0.40	15.73	nr	**103.34**
100 mm	91.14	115.34	0.42	16.52	nr	**131.86**
150 mm	197.10	249.45	1.37	53.88	nr	**303.33**
200 mm	554.64	701.95	1.51	59.38	nr	**761.33**
Branch; long tail						
100 mm	254.22	321.74	0.52	20.45	nr	**342.19**
Access branch; radius; equal and unequal						
70 mm	101.52	128.48	0.40	15.73	nr	**144.21**
100 mm	137.61	174.16	0.42	16.52	nr	**190.68**
150 mm	328.54	415.81	0.97	38.15	nr	**453.95**
Double branch; equal and unequal						
100 mm	76.75	97.13	0.52	20.45	nr	**117.58**
100 mm; 70°	114.49	144.90	0.52	20.45	nr	**165.34**
150 mm	328.54	415.81	1.37	53.88	nr	**469.68**
200 mm	561.89	711.13	1.51	59.38	nr	**770.52**
Double branch; radius; equal and unequal						
100 mm	98.99	125.28	0.52	20.45	nr	**145.73**
150 mm	405.27	512.91	1.37	53.88	nr	**566.79**
Corner branch						
100 mm	200.25	253.43	0.52	20.45	nr	**273.88**
150 mm	221.40	280.21	0.52	20.45	nr	**300.66**
Corner branch; long tail						
100 mm	298.75	378.09	0.52	20.45	nr	**398.54**
Roof vent connector; asphalt						
100 mm	125.98	159.44	0.32	12.58	nr	**172.02**
Movement connector						
100 mm	160.01	202.51	0.32	12.58	nr	**215.09**
150 mm	296.22	374.89	0.67	26.35	nr	**401.24**

33 ABOVE GROUND DRAINAGE

Item	Net Price £	Material £	Labour hours	Labour £	Unit	Total rate £
Expansion plugs						
70 mm	35.60	45.05	0.32	12.58	nr	**57.64**
100 mm	44.36	56.15	0.39	15.34	nr	**71.48**
150 mm	79.63	100.78	0.55	21.63	nr	**122.41**
P trap						
100 mm dia.	82.98	105.02	0.32	12.58	nr	**117.61**
P trap with access						
50 mm	126.73	160.39	0.27	10.62	nr	**171.01**
70 mm	126.73	160.39	0.30	11.80	nr	**172.19**
100 mm	137.28	173.74	0.32	12.58	nr	**186.32**
150 mm	245.44	310.63	0.67	26.35	nr	**336.97**
Branch trap						
100 mm	300.78	380.67	0.42	16.52	nr	**397.19**
Stench trap						
100 mm	584.93	740.29	0.42	16.52	nr	**756.80**
Balcony gully inlet						
100 mm	332.01	420.19	1.00	39.33	nr	**459.51**
Roof outlet						
Flat grate						
70 mm	204.96	259.39	1.00	39.33	nr	**298.72**
100 mm	288.31	364.88	1.00	39.33	nr	**404.21**
Dome grate						
70 mm	204.96	259.39	1.00	39.33	nr	**298.72**
100 mm	323.36	409.25	1.00	39.33	nr	**448.57**
Top hat						
100 mm	437.81	554.10	1.00	39.33	nr	**593.42**
Floor drains; for cast iron pipework BS 416 and BS EN877						
Adjustable clamp plate body						
100 mm; 165 mm nickel bronze grate and frame	195.17	247.01	0.50	19.66	nr	**266.67**
100 mm; 165 mm nickel bronze rodding eye	227.11	287.43	0.50	19.66	nr	**307.10**
100 mm; 150 × 150 mm nickel bronze grate and frame	218.98	277.14	0.50	19.66	nr	**296.80**
100 mm; 150 × 150 mm nickel bronze rodding eye	227.11	287.43	0.50	19.66	nr	**307.10**
Deck plate body						
100 mm; 165 mm nickel bronze grate and frame	195.17	247.01	0.50	19.66	nr	**266.67**
100 mm; 165 mm nickel bronze rodding eye	227.11	287.43	0.50	19.66	nr	**307.10**
100 mm; 150 × 150 mm nickel bronze grate and frame	218.98	277.14	0.50	19.66	nr	**296.80**
100 mm; 150 × 150 mm nickel bronze rodding eye	227.11	287.43	0.50	19.66	nr	**307.10**
Extra for						
100 mm; screwed extension piece	76.11	96.33	0.30	11.80	nr	**108.13**
100 mm; grating extension piece; screwed or spigot	57.74	73.07	0.30	11.80	nr	**84.87**
100 mm; brewery trap	2100.44	2658.32	2.00	78.65	nr	**2736.97**
High density polyethylene (HDPE) pipes and fittings						
Pipe, with brackets and jointing couplings measured separately						
40 mm dia.	5.74	7.26	0.47	18.49	m	**25.75**
50 mm dia.	6.81	8.61	0.53	20.85	m	**29.47**
56 mm dia.	8.17	10.34	0.60	23.60	m	**33.95**

33 ABOVE GROUND DRAINAGE

Item	Net Price £	Material £	Labour hours	Labour £	Unit	Total rate £
DISPOSAL SYSTEMS – cont						
High density polyethylene (HDPE) pipes and fittings – cont						
Pipe, with brackets and jointing couplings measured separately – cont						
63 mm dia.	9.38	11.87	0.60	23.60	m	**35.48**
75 mm dia.	10.80	13.67	0.90	35.40	m	**49.06**
90 mm dia.	16.05	20.31	0.90	35.40	m	**55.71**
110 mm dia.	20.32	25.71	1.10	43.26	m	**68.97**
125 mm dia.	28.83	36.49	1.20	47.21	m	**83.70**
160 mm dia.	46.54	58.90	1.48	58.20	m	**117.11**
200 mm dia.	60.31	76.33	1.75	68.87	m	**145.20**
250 mm dia.	96.68	122.36	1.77	69.61	m	**191.96**
315 mm dia.	149.39	189.07	1.90	74.72	m	**263.78**
Extra over fittings						
Coupler						
40 mm dia.	8.36	10.58	0.48	18.88	nr	**29.46**
50 mm dia.	8.61	10.90	0.52	20.45	nr	**31.35**
56 mm dia.	8.66	10.96	0.58	22.81	nr	**33.77**
63 mm dia.	11.55	14.62	0.58	22.81	nr	**37.43**
90 mm dia.	12.09	15.29	0.67	26.35	nr	**41.64**
110 mm dia.	12.92	16.35	0.74	29.10	nr	**45.45**
125 mm dia.	22.24	28.15	0.83	32.64	nr	**60.79**
160 mm dia.	30.94	39.15	1.00	39.33	nr	**78.48**
200 mm dia.	215.40	272.61	1.35	53.09	nr	**325.70**
250 mm dia.	356.10	450.68	1.50	58.99	nr	**509.67**
315 mm dia.	403.91	511.19	1.80	70.79	nr	**581.98**
Cap						
40 mm dia.	4.79	6.06	0.24	9.44	nr	**15.50**
50 mm dia.	4.09	5.18	0.26	10.22	nr	**15.40**
56 mm dia.	4.09	5.18	0.30	11.80	nr	**16.97**
63 mm dia.	4.09	5.18	0.32	12.58	nr	**17.76**
75 mm dia.	4.09	5.18	0.35	13.76	nr	**18.94**
90 mm dia.	5.17	6.54	0.37	14.55	nr	**21.09**
110 mm dia.	9.67	12.24	0.40	15.73	nr	**27.97**
125 mm dia.	12.15	15.38	0.46	18.09	nr	**33.47**
160 mm dia.	38.85	49.17	0.50	19.66	nr	**68.83**
200 mm dia.	55.04	69.66	0.68	26.74	nr	**96.40**
250 mm dia.	71.26	90.19	0.75	29.49	nr	**119.69**
315 mm dia.	80.99	102.50	0.90	35.39	nr	**137.89**
Reducer						
50 mm dia.	4.27	5.41	0.54	21.24	nr	**26.64**
56 mm dia.	4.67	5.91	0.60	23.60	nr	**29.51**
63 mm dia.	4.67	5.91	0.67	26.35	nr	**32.26**
75 mm dia.	5.47	6.92	0.67	26.35	nr	**33.27**
90 mm dia.	7.00	8.86	0.67	26.35	nr	**35.21**
110 mm dia.	8.58	10.86	0.74	29.10	nr	**39.96**
125 mm dia.	13.61	17.23	0.83	32.64	nr	**49.87**
160 mm dia.	15.24	19.29	1.10	43.26	nr	**62.55**
200 mm dia.	97.00	122.76	1.40	55.06	nr	**177.81**

33 ABOVE GROUND DRAINAGE

Item	Net Price £	Material £	Labour hours	Labour £	Unit	Total rate £
Bend; 45°						
40 mm dia.	3.82	4.84	0.45	17.70	nr	**22.54**
50 mm dia.	3.84	4.86	0.50	19.66	nr	**24.52**
56 mm dia.	4.50	5.70	0.55	21.63	nr	**27.33**
63 mm dia.	5.78	7.32	0.58	22.81	nr	**30.13**
75 mm dia.	7.41	9.38	0.67	26.35	nr	**35.73**
90 mm dia.	10.62	13.44	0.67	26.35	nr	**39.79**
110 mm dia.	10.62	13.44	0.74	29.10	nr	**42.54**
125 mm dia.	17.12	21.67	0.83	32.64	nr	**54.31**
160 mm dia.	44.97	56.91	1.00	39.33	nr	**96.23**
200 mm dia.	124.01	156.95	1.40	55.04	nr	**211.99**
250 mm dia.	201.74	255.32	1.80	70.79	nr	**326.11**
315 mm dia.	279.78	354.10	2.40	94.38	nr	**448.48**
Bend; 90°						
40 mm dia.	3.82	4.84	0.40	15.73	nr	**20.57**
50 mm dia.	4.50	5.70	0.45	17.70	nr	**23.40**
56 mm dia.	5.47	6.92	0.50	19.66	nr	**26.58**
63 mm dia.	6.43	8.14	0.58	22.81	nr	**30.95**
75 mm dia.	9.67	12.24	0.64	25.17	nr	**37.41**
90 mm dia.	12.95	16.39	0.67	26.35	nr	**42.74**
110 mm dia.	12.95	16.39	0.74	29.10	nr	**45.49**
125 mm dia.	20.03	25.35	0.83	32.64	nr	**57.99**
160 mm dia.	43.67	55.27	1.00	39.33	nr	**94.59**
200 mm dia.	213.39	270.07	1.40	55.06	nr	**325.12**
250 mm dia.	356.27	450.89	1.80	70.79	nr	**521.68**
315 mm dia.	518.21	655.85	2.40	94.38	nr	**750.23**
Equal tee						
40 mm dia.	7.63	9.66	0.60	23.60	nr	**33.25**
50 mm dia.	9.67	12.24	0.70	27.53	nr	**39.77**
56 mm dia.	9.67	12.24	0.70	27.53	nr	**39.77**
63 mm dia.	10.06	12.73	0.75	29.49	nr	**42.22**
75 mm dia.	11.90	15.06	0.83	32.64	nr	**47.70**
90 mm dia.	15.16	19.19	0.87	34.21	nr	**53.40**
110 mm dia.	17.44	22.07	1.00	39.33	nr	**61.40**
125 mm dia.	25.83	32.69	1.08	42.47	nr	**75.17**
160 mm dia.	97.13	122.93	1.35	53.09	nr	**176.02**
200 mm dia.	192.65	243.81	1.90	74.72	nr	**318.53**
250 mm dia.	275.30	348.42	2.70	106.18	nr	**454.59**
315 mm dia.	388.65	491.88	3.60	141.57	nr	**633.45**
Reducing tee						
50 mm dia.	9.67	12.24	0.70	27.53	nr	**39.77**
56 mm dia.	9.67	12.24	0.70	27.53	nr	**39.77**
63 mm dia.	10.06	12.73	0.75	29.49	nr	**42.22**
75 mm dia.	11.90	15.06	0.83	32.64	nr	**47.70**
90 mm dia.	15.16	19.19	0.87	34.21	nr	**53.40**
110 mm dia.	17.44	22.07	1.00	39.33	nr	**61.40**
125 mm dia.	25.83	32.69	1.08	42.47	nr	**75.17**
160 mm dia.	97.13	122.93	1.35	53.09	nr	**176.02**
200 mm dia.	192.65	243.81	1.90	74.72	nr	**318.53**
250 mm dia.	275.30	348.42	2.70	106.18	nr	**454.59**
315 mm dia.	388.65	491.88	3.60	141.57	nr	**633.45**

33 ABOVE GROUND DRAINAGE

Item	Net Price £	Material £	Labour hours	Labour £	Unit	Total rate £
DISPOSAL SYSTEMS – cont						
Extra over fittings – cont						
Y-branch 45°						
40 mm dia.	9.33	11.80	0.65	25.56	nr	**37.36**
50 mm dia.	9.67	12.24	0.70	27.53	nr	**39.77**
56 mm dia.	9.67	12.24	0.70	27.53	nr	**39.77**
63 mm dia.	10.06	12.73	0.75	29.49	nr	**42.22**
75 mm dia.	11.29	14.29	0.83	32.64	nr	**46.93**
90 mm dia.	16.12	20.41	0.87	34.21	nr	**54.62**
110 mm dia.	19.37	24.51	1.00	39.33	nr	**63.84**
125 mm dia.	25.83	32.69	1.08	42.47	nr	**75.17**
160 mm dia.	89.37	113.10	1.35	53.09	nr	**166.19**
200 mm dia.	288.23	364.78	1.90	74.72	nr	**439.50**
250 mm dia.	485.83	614.87	2.70	106.18	nr	**721.05**
315 mm dia.	712.55	901.81	3.60	141.57	nr	**1043.38**
Reducing Y-branch 45°						
50 mm dia.	9.67	12.24	0.70	27.53	nr	**39.77**
56 mm dia.	9.67	12.24	0.70	27.53	nr	**39.77**
63 mm dia.	10.06	12.73	0.75	29.49	nr	**42.22**
75 mm dia.	11.29	14.29	0.83	32.64	nr	**46.93**
90 mm dia.	16.81	21.27	0.87	34.21	nr	**55.48**
110 mm dia.	19.37	24.51	1.00	39.33	nr	**63.84**
125 mm dia.	25.83	32.69	1.08	42.47	nr	**75.17**
160 mm dia.	120.45	152.44	1.35	53.09	nr	**205.53**
200 mm dia.	192.65	243.81	1.90	74.72	nr	**318.53**
250 mm dia.	356.27	450.89	2.70	106.18	nr	**557.07**
315 mm dia.	599.17	758.30	3.60	141.57	nr	**899.88**

33 BELOW GROUND DRAINAGE

Item	Net Price £	Material £	Labour hours	Labour £	Unit	Total rate £
SEWER SYSTEM PIPEWORK						
GRP PN1 gravity sewer pipe; laid in trenches (trenches not included), stiffness class SN10,000, manufactured to EN 14364, push fit coupling; confirmation is required for use in public foul and surface water drainage system						
Labour cost excluded; correct installation of pipes requires individual calculations and comprehensive planning by certified engineers. In addition to the applicable standards and guidelines, the requirements for each installation and the operating conditions for each project shall be evaluated by qualified engineers						
Pricing note: excludes builder's work (excavation, backfill, compacting, disposal, etc.)						
Pipe, 3 m lengths						
600 mm nominal diameter	–	228.00	–	–	m	**228.00**
800 mm nominal diameter	–	331.79	–	–	m	**331.79**
1000 mm nominal diameter	–	451.45	–	–	m	**451.45**
1200 mm nominal diameter	–	551.36	–	–	m	**551.37**
1400 mm nominal diameter	–	765.35	–	–	m	**765.35**
1600 mm nominal diameter	–	1106.53	–	–	m	**1106.53**
1800 mm nominal diameter	–	1265.09	–	–	m	**1265.09**
2000 mm nominal diameter	–	1430.05	–	–	m	**1430.05**
Pipe, 6 m lengths						
600 mm nominal diameter	–	196.92	–	–	m	**196.92**
800 mm nominal diameter	–	294.96	–	–	m	**294.96**
1000 mm nominal diameter	–	408.43	–	–	m	**408.43**
1200 mm nominal diameter	–	500.67	–	–	m	**500.67**
1400 mm nominal diameter	–	704.45	–	–	m	**704.45**
1600 mm nominal diameter	–	1012.48	–	–	m	**1012.48**
1800 mm nominal diameter	–	1159.64	–	–	m	**1159.64**
2000 mm nominal diameter	–	1311.92	–	–	m	**1311.92**
Extra over for fittings						
11.25°, 15°, 22.5°, 30° bend						
600 mm nominal diameter	–	664.35	–	–	nr	**664.35**
800 mm nominal diameter	–	848.70	–	–	nr	**848.70**
1000 mm nominal diameter	–	1086.93	–	–	nr	**1086.93**
1200 mm nominal diameter	–	1477.93	–	–	nr	**1477.93**
1400 mm nominal diameter	–	1906.03	–	–	nr	**1906.03**
1600 mm nominal diameter	–	2516.75	–	–	nr	**2516.75**
1800 mm nominal diameter	–	3177.96	–	–	nr	**3177.96**
2000 mm nominal diameter	–	3653.38	–	–	nr	**3653.38**

33 BELOW GROUND DRAINAGE

Item	Net Price £	Material £	Labour hours	Labour £	Unit	Total rate £
SEWER SYSTEM PIPEWORK – cont						
Extra over for fittings – cont						
Segmental elbow – PN1 , 45° & 60° bend						
600 mm nominal diameter	–	895.43	–	–	nr	895.43
800 mm nominal diameter	–	1234.56	–	–	nr	1234.56
1000 mm nominal diameter	–	1655.18	–	–	nr	1655.18
1200 mm nominal diameter	–	2387.50	–	–	nr	2387.50
1400 mm nominal diameter	–	3153.55	–	–	nr	3153.55
1600 mm nominal diameter	–	4199.41	–	–	nr	4199.41
1800 mm nominal diameter	–	5473.44	–	–	nr	5473.44
2000 mm nominal diameter	–	6442.01	–	–	nr	6442.01
Segmental elbow – PN1, 90° bend						
600 mm nominal diameter, 90° bend	–	1111.28	–	–	nr	1111.28
800 mm nominal diameter, 90° bend	–	1602.12	–	–	nr	1602.12
1000 mm nominal diameter, 90° bend	–	2241.20	–	–	nr	2241.20
1200 mm nominal diameter, 90° bend	–	3255.67	–	–	nr	3255.67
1400 mm nominal diameter, 90° bend	–	4413.29	–	–	nr	4413.29
1600 mm nominal diameter, 90° bend	–	5918.60	–	–	nr	5918.60
1800 mm nominal diameter, 90° bend	–	7759.53	–	–	nr	7759.53
2000 mm nominal diameter, 90° bend	–	9159.06	–	–	nr	9159.06
T-piece concentric – PN1						
600 mm to 400 mm nominal diameter	–	895.70	–	–	nr	895.70
800 mm to 400 mm nominal diameter	–	1063.10	–	–	nr	1063.10
1000 mm to 400 mm nominal diameter	–	1105.62	–	–	nr	1105.62
1200 mm to 400 mm nominal diameter	–	1231.94	–	–	nr	1231.94
1400 mm to 400 mm nominal diameter	–	1395.54	–	–	nr	1395.54
1600 mm to 400 mm nominal diameter	–	1687.01	–	–	nr	1687.01

38 PIPED SUPPLY SYSTEMS

Item	Net Price £	Material £	Labour hours	Labour £	Unit	Total rate £
COLD WATER PIPELINES: COPPER PIPEWORK						
Copper pipe; capillary or compression joints in the running length; EN1057 R250 (TX) formerly BS 2871 Table X						
Fixed vertically or at low level, with brackets measured separately						
12 mm dia.	2.62	3.31	0.39	15.34	m	**18.65**
15 mm dia.	2.94	3.72	0.40	15.73	m	**19.45**
22 mm dia.	5.91	7.48	0.47	18.48	m	**25.97**
28 mm dia.	7.49	9.48	0.51	20.06	m	**29.54**
35 mm dia.	17.77	22.48	0.58	22.81	m	**45.29**
42 mm dia.	21.64	27.39	0.66	25.96	m	**53.34**
54 mm dia.	27.82	35.20	0.72	28.31	m	**63.52**
67 mm dia.	36.39	46.05	0.75	29.49	m	**75.55**
76 mm dia.	51.43	65.10	0.76	29.89	m	**94.98**
108 mm dia.	74.02	93.68	0.78	30.67	m	**124.35**
133 mm dia.	96.03	121.53	1.05	41.29	m	**162.82**
159 mm dia.	152.21	192.63	1.15	45.22	m	**237.86**
Fixed horizontally at high level or suspended, with brackets measured separately						
12 mm dia.	2.62	3.31	0.45	17.70	m	**21.01**
15 mm dia.	2.94	3.72	0.46	18.09	m	**21.81**
22 mm dia.	5.91	7.48	0.54	21.24	m	**28.72**
28 mm dia.	7.49	9.48	0.59	23.20	m	**32.69**
35 mm dia.	17.77	22.48	0.67	26.35	m	**48.83**
42 mm dia.	21.64	27.39	0.76	29.89	m	**57.27**
54 mm dia.	27.82	35.20	0.83	32.64	m	**67.84**
67 mm dia.	36.39	46.05	0.86	33.82	m	**79.87**
76 mm dia.	51.43	65.10	0.87	34.21	m	**99.31**
108 mm dia.	74.02	93.68	0.90	35.39	m	**129.07**
133 mm dia.	96.03	121.53	1.21	47.58	m	**169.12**
159 mm dia.	152.21	192.63	1.32	51.91	m	**244.54**
Copper pipe; capillary or compression joints in the running length; EN1057 R250 (TY) formerly BS 2871 Table Y						
Fixed vertically or at low level with brackets measured separately (Refer to Copper Pipe Table X Section)						
12 mm dia.	3.38	4.28	0.41	16.12	m	**20.41**
15 mm dia.	4.83	6.12	0.43	16.91	m	**23.03**
22 mm dia.	8.51	10.77	0.50	19.66	m	**30.44**
28 mm dia.	11.00	13.92	0.54	21.24	m	**35.15**
35 mm dia.	16.05	20.31	0.60	23.60	m	**43.91**
42 mm dia.	19.45	24.62	0.62	24.38	m	**49.00**
54 mm dia.	33.17	41.98	0.71	27.92	m	**69.90**
67 mm dia.	44.06	55.76	0.78	30.67	m	**86.44**
76 mm dia.	64.38	81.48	0.82	32.25	m	**113.73**
108 mm dia.	89.39	113.13	0.88	34.61	m	**147.74**

38 PIPED SUPPLY SYSTEMS

Item	Net Price £	Material £	Labour hours	Labour £	Unit	Total rate £
COLD WATER PIPELINES: COPPER PIPEWORK – cont						
Copper pipe; capillary or compression joints in the running length; EN1057 R250 (TX) formerly BS 2871 Table X						
Plastic coated gas and cold water service pipe for corrosive environments, fixed vertically or at low level with brackets measured separtely						
15 mm dia. (white)	6.75	8.55	0.59	23.20	m	**31.75**
22 mm dia. (white)	11.99	15.17	0.68	26.75	m	**41.92**
28 mm dia. (white)	12.37	15.66	0.74	29.11	m	**44.77**
FIXINGS						
Saddle band						
6 mm dia.	0.08	0.10	0.05	2.01	nr	**2.11**
8 mm dia.	0.08	0.10	0.07	2.70	nr	**2.80**
10 mm dia.	0.12	0.15	0.09	3.36	nr	**3.52**
12 mm dia.	0.12	0.15	0.11	4.21	nr	**4.36**
15 mm dia.	0.12	0.15	0.13	5.11	nr	**5.27**
22 mm dia.	0.12	0.15	0.13	5.11	nr	**5.27**
28 mm dia.	0.26	0.32	0.16	6.29	nr	**6.62**
35 mm dia.	0.46	0.58	0.18	7.08	nr	**7.66**
42 mm dia.	0.96	1.21	0.21	8.26	nr	**9.47**
54 mm dia.	1.28	1.62	0.21	8.26	nr	**9.88**
Single spacing clip						
15 mm dia.	0.27	0.34	0.14	5.51	nr	**5.85**
22 mm dia.	0.27	0.34	0.15	5.90	nr	**6.24**
28 mm dia.	0.63	0.80	0.17	6.69	nr	**7.49**
Two piece spacing clip						
8 mm dia. Bottom	0.22	0.27	0.11	4.33	nr	**4.60**
8 mm dia. Top	0.22	0.27	0.11	4.33	nr	**4.60**
12 mm dia. Bottom	0.22	0.27	0.13	5.11	nr	**5.39**
12 mm dia. Top	0.22	0.27	0.13	5.11	nr	**5.39**
15 mm dia. Bottom	0.26	0.32	0.13	5.11	nr	**5.44**
15 mm dia. Top	0.26	0.32	0.13	5.11	nr	**5.44**
22 mm dia. Bottom	0.26	0.32	0.14	5.51	nr	**5.83**
22 mm dia. Top	0.27	0.34	0.14	5.51	nr	**5.85**
28 mm dia. Bottom	0.26	0.32	0.16	6.29	nr	**6.62**
28 mm dia. Top	0.45	0.56	0.16	6.29	nr	**6.86**
35 mm dia. Bottom	0.27	0.34	0.21	8.26	nr	**8.60**
35 mm dia. Top	0.62	0.79	0.21	8.26	nr	**9.04**
Single pipe bracket						
15 mm dia.	2.24	2.83	0.14	5.51	nr	**8.34**
22 mm dia.	2.27	2.87	0.14	5.51	nr	**8.37**
28 mm dia.	3.12	3.94	0.17	6.69	nr	**10.63**
Single pipe ring						
15 mm dia.	0.47	0.60	0.26	10.22	nr	**10.82**
22 mm dia.	0.57	0.72	0.26	10.22	nr	**10.94**
28 mm dia.	0.62	0.79	0.31	12.19	nr	**12.98**

38 PIPED SUPPLY SYSTEMS

Item	Net Price £	Material £	Labour hours	Labour £	Unit	Total rate £
35 mm dia.	0.84	1.06	0.32	12.58	nr	**13.64**
42 mm dia.	1.25	1.59	0.32	12.58	nr	**14.17**
54 mm dia.	1.93	2.44	0.34	13.37	nr	**15.81**
67 mm dia.	2.21	2.80	0.35	13.76	nr	**16.56**
76 mm dia.	2.59	3.28	0.42	16.52	nr	**19.79**
108 mm dia.	5.23	6.62	0.42	16.52	nr	**23.14**
Double pipe ring						
15 mm dia.	0.50	0.63	0.26	10.22	nr	**10.86**
22 mm dia.	0.81	1.02	0.26	10.22	nr	**11.25**
28 mm dia.	0.86	1.09	0.31	12.19	nr	**13.28**
35 mm dia.	0.90	1.14	0.32	12.58	nr	**13.73**
42 mm dia.	1.89	2.39	0.32	12.58	nr	**14.97**
54 mm dia.	2.02	2.56	0.34	13.37	nr	**15.93**
67 mm dia.	2.83	3.58	0.35	13.76	nr	**17.35**
76 mm dia.	3.72	4.71	0.42	16.52	nr	**21.23**
108 mm dia.	6.54	8.28	0.42	16.52	nr	**24.80**
Wall bracket						
15 mm dia.	4.42	5.60	0.05	1.97	nr	**7.57**
22 mm dia.	5.81	7.36	0.05	1.97	nr	**9.32**
28 mm dia.	6.93	8.77	0.05	1.97	nr	**10.74**
35 mm dia.	10.37	13.13	0.05	1.97	nr	**15.09**
42 mm dia.	13.74	17.40	0.05	1.97	nr	**19.36**
54 mm dia.	21.41	27.09	0.05	1.97	nr	**29.06**
Hospital bracket						
15 mm dia.	7.31	9.25	0.26	10.22	nr	**19.48**
22 mm dia.	8.66	10.96	0.26	10.22	nr	**21.18**
28 mm dia.	10.16	12.85	0.31	12.19	nr	**25.05**
35 mm dia.	10.75	13.61	0.32	12.58	nr	**26.19**
42 mm dia.	15.36	19.44	0.32	12.58	nr	**32.03**
54 mm dia.	20.56	26.02	0.34	13.37	nr	**39.39**
Screw on backplate, female						
All sizes 15 mm to 54 mm × 10 mm	2.33	2.95	0.10	3.93	nr	**6.89**
Screw on backplate, male						
All sizes 15 mm to 54 mm × 10 mm	3.13	3.96	0.10	3.93	nr	**7.89**
Pipe joist clips, single						
15 mm dia.	1.48	1.88	0.08	3.15	nr	**5.02**
22 mm dia.	1.48	1.88	0.08	3.15	nr	**5.02**
Pipe joist clips, double						
15 mm dia.	1.12	1.42	0.08	3.15	nr	**4.56**
22 mm dia.	2.06	2.61	0.08	3.15	nr	**5.76**
Extra over channel sections for fabricated hangers and brackets						
Galvanized steel; including inserts, bolts, nuts, washers; fixed to backgrounds						
41 × 21 mm	9.61	12.16	0.29	11.40	m	**23.57**
41 × 41 mm	11.52	14.58	0.29	11.40	m	**25.99**
Threaded rods; metric thread; including nuts, washers, etc.						
10 mm dia. × 600 mm long for ring clips up to 54 mm	3.10	3.92	0.18	7.08	nr	**11.00**
12 mm dia. × 600 mm long for ring clips from 54 mm	4.85	6.14	0.18	7.08	nr	**13.22**

38 PIPED SUPPLY SYSTEMS

Item	Net Price £	Material £	Labour hours	Labour £	Unit	Total rate £
COLD WATER PIPELINES: COPPER PIPEWORK – cont						
Extra over copper pipes; capillary fittings; BS 864						
Stop end						
15 mm dia.	2.29	2.89	0.13	5.11	nr	**8.01**
22 mm dia.	4.22	5.34	0.14	5.51	nr	**10.84**
28 mm dia.	7.57	9.58	0.17	6.69	nr	**16.26**
35 mm dia.	16.68	21.11	0.19	7.47	nr	**28.59**
42 mm dia.	28.75	36.39	0.22	8.65	nr	**45.04**
54 mm dia.	40.14	50.80	0.23	9.04	nr	**59.85**
Straight coupling; copper to copper						
6 mm dia.	0.27	0.34	0.23	9.04	nr	**9.39**
8 mm dia.	1.25	1.58	0.23	9.04	nr	**10.63**
10 mm dia.	2.26	2.86	0.23	9.04	nr	**11.90**
15 mm dia.	2.31	2.92	0.23	9.04	nr	**11.97**
22 mm dia.	2.39	3.02	0.26	10.22	nr	**13.24**
28 mm dia.	2.43	3.07	0.30	11.80	nr	**14.87**
35 mm dia.	6.83	8.64	0.34	13.37	nr	**22.01**
42 mm dia.	11.40	14.43	0.38	14.94	nr	**29.37**
54 mm dia.	21.04	26.63	0.42	16.52	nr	**43.15**
67 mm dia.	62.43	79.01	0.53	20.84	nr	**99.86**
Adaptor coupling; imperial to metric						
½" × 15 mm dia.	4.34	5.50	0.27	10.62	nr	**16.12**
¾" × 22 mm dia.	4.96	6.27	0.31	12.19	nr	**18.46**
1" × 28 mm dia.	8.50	10.76	0.36	14.16	nr	**24.92**
1 ¼" × 35 mm dia.	14.21	17.99	0.41	16.12	nr	**34.11**
1 ½" × 42 mm dia.	18.10	22.91	0.46	18.09	nr	**41.00**
Reducing coupling						
15 × 10 mm dia.	5.20	6.58	0.23	9.04	nr	**15.62**
22 × 10 mm dia.	7.62	9.65	0.26	10.22	nr	**19.87**
22 × 15 mm dia.	8.15	10.31	0.27	10.62	nr	**20.93**
28 × 15 mm dia.	9.37	11.86	0.28	11.01	nr	**22.87**
28 × 22 mm dia.	9.44	11.95	0.30	11.80	nr	**23.75**
35 × 28 mm dia.	13.55	17.14	0.34	13.37	nr	**30.51**
42 × 35 mm dia.	19.88	25.16	0.38	14.94	nr	**40.10**
54 × 35 mm dia.	34.83	44.08	0.42	16.52	nr	**60.60**
54 × 42 mm dia.	38.02	48.12	0.42	16.52	nr	**64.64**
Straight female connector						
15 mm × ½" dia.	5.17	6.54	0.27	10.62	nr	**17.16**
22 mm × ¾" dia.	7.47	9.45	0.31	12.19	nr	**21.64**
28 mm × 1" dia.	14.13	17.88	0.36	14.16	nr	**32.04**
35 mm × 1¼" dia.	24.44	30.93	0.41	16.12	nr	**47.05**
42 mm × 1½" dia.	31.69	40.11	0.46	18.09	nr	**58.20**
54 mm × 2" dia.	50.29	63.65	0.52	20.45	nr	**84.10**
Straight male connector						
15 mm × ½" dia.	4.40	5.57	0.27	10.62	nr	**16.19**
22 mm × ¾" dia.	7.87	9.96	0.31	12.19	nr	**22.15**
28 mm × 1" dia.	12.67	16.03	0.36	14.16	nr	**30.19**

38 PIPED SUPPLY SYSTEMS

Item	Net Price £	Material £	Labour hours	Labour £	Unit	Total rate £
35 mm × 1¼" dia.	22.26	28.18	0.41	16.12	nr	**44.30**
42 mm × 1½" dia.	28.68	36.30	0.46	18.09	nr	**54.39**
54 mm × 2" dia.	43.53	55.10	0.52	20.45	nr	**75.54**
67 mm × 2½" dia.	69.52	87.98	0.63	24.78	nr	**112.76**
Female reducing connector						
15 mm × ¾" dia.	12.82	16.23	0.27	10.62	nr	**26.84**
Male reducing connector						
15 mm × ¾" dia.	11.47	14.52	0.27	10.62	nr	**25.14**
22 mm × 1" dia.	17.45	22.08	0.31	12.19	nr	**34.28**
Flanged connector						
28 mm dia.	81.97	103.74	0.36	14.16	nr	**117.90**
35 mm dia.	103.78	131.34	0.41	16.12	nr	**147.46**
42 mm dia.	124.02	156.97	0.46	18.09	nr	**175.06**
54 mm dia.	187.51	237.31	0.52	20.45	nr	**257.76**
67 mm dia.	220.49	279.05	0.61	23.99	nr	**303.04**
Tank connector						
15 mm × ½" dia.	10.99	13.91	0.25	9.83	nr	**23.74**
22 mm × ¾" dia.	16.73	21.17	0.28	11.01	nr	**32.18**
28 mm × 1" dia.	21.99	27.84	0.32	12.58	nr	**40.42**
35 mm × 1¼" dia.	28.21	35.71	0.37	14.55	nr	**50.26**
42 mm × 1½" dia.	36.99	46.81	0.43	16.91	nr	**63.72**
54 mm × 2" dia.	56.54	71.56	0.46	18.09	nr	**89.65**
Tank connector with long thread						
15 mm × ½" dia.	14.21	17.99	0.30	11.80	nr	**29.79**
22 mm × ¾" dia.	20.26	25.64	0.33	12.98	nr	**38.62**
28 mm × 1" dia.	25.06	31.72	0.39	15.34	nr	**47.05**
Reducer						
15 × 10 mm dia.	1.24	1.57	0.23	9.04	nr	**10.61**
22 × 15 mm dia.	1.75	2.21	0.26	10.22	nr	**12.44**
28 × 15 mm dia.	4.61	5.84	0.28	11.01	nr	**16.85**
28 × 22 mm dia.	9.22	11.67	0.30	11.80	nr	**23.47**
35 × 22 mm dia.	13.13	16.62	0.34	13.37	nr	**29.99**
42 × 22 mm dia.	23.73	30.03	0.36	14.16	nr	**44.18**
42 × 35 mm dia.	32.04	40.55	0.38	14.94	nr	**55.49**
54 × 35 mm dia.	38.53	48.77	0.40	15.73	nr	**64.50**
54 × 42 mm dia.	41.87	52.99	0.42	16.52	nr	**69.50**
67 × 54 mm dia.	45.18	57.18	0.53	20.84	nr	**78.02**
Adaptor; copper to female iron						
15 mm × ½" dia.	8.90	11.27	0.27	10.62	nr	**21.89**
22 mm × ¾" dia.	13.56	17.16	0.31	12.19	nr	**29.35**
28 mm × 1" dia.	19.10	24.17	0.36	14.16	nr	**38.33**
35 mm × 1¼" dia.	34.57	43.76	0.41	16.12	nr	**59.88**
42 mm × 1½" dia.	43.53	55.10	0.46	18.09	nr	**73.19**
54 mm × 2" dia.	52.36	66.27	0.52	20.45	nr	**86.72**
Adaptor; copper to male iron						
15 mm × ½" dia.	9.07	11.48	0.27	10.62	nr	**22.10**
22 mm × ¾" dia.	11.61	14.70	0.31	12.19	nr	**26.89**
28 mm × 1" dia.	19.41	24.56	0.36	14.16	nr	**38.72**
35 mm × 1¼" dia.	28.23	35.72	0.41	16.12	nr	**51.85**
42 mm × 1½" dia.	39.00	49.36	0.46	18.09	nr	**67.45**
54 mm × 2" dia.	52.36	66.27	0.52	20.45	nr	**86.72**

38 PIPED SUPPLY SYSTEMS

Item	Net Price £	Material £	Labour hours	Labour £	Unit	Total rate £
COLD WATER PIPELINES: COPPER PIPEWORK – cont						
Extra over copper pipes – cont						
Union coupling						
15 mm dia.	12.27	15.53	0.41	16.12	nr	**31.65**
22 mm dia.	19.65	24.87	0.45	17.70	nr	**42.57**
28 mm dia.	28.68	36.30	0.51	20.06	nr	**56.36**
35 mm dia.	37.64	47.64	0.64	25.17	nr	**72.81**
42 mm dia.	54.98	69.58	0.68	26.74	nr	**96.32**
54 mm dia.	104.63	132.42	0.78	30.67	nr	**163.09**
67 mm dia.	177.17	224.23	0.96	37.75	nr	**261.98**
Elbow						
15 mm dia.	0.48	0.61	0.23	9.04	nr	**9.66**
22 mm dia.	1.28	1.62	0.26	10.22	nr	**11.84**
28 mm dia.	3.37	4.26	0.31	12.19	nr	**16.45**
35 mm dia.	14.62	18.51	0.35	13.76	nr	**32.27**
42 mm dia.	24.15	30.57	0.41	16.12	nr	**46.69**
54 mm dia.	49.89	63.15	0.44	17.30	nr	**80.45**
67 mm dia.	129.51	163.90	0.54	21.24	nr	**185.14**
Backplate elbow						
15 mm dia.	9.22	11.66	0.51	20.06	nr	**31.72**
22 mm dia.	19.82	25.09	0.54	21.24	nr	**46.32**
Overflow bend						
22 mm dia.	27.91	35.33	0.26	10.22	nr	**45.55**
Return bend						
15 mm dia.	13.82	17.48	0.23	9.04	nr	**26.53**
22 mm dia.	27.15	34.36	0.26	10.22	nr	**44.58**
28 mm dia.	34.69	43.90	0.31	12.19	nr	**56.09**
Obtuse elbow						
15 mm dia.	1.79	2.26	0.23	9.04	nr	**11.31**
22 mm dia.	3.69	4.67	0.26	10.22	nr	**14.90**
28 mm dia.	7.09	8.97	0.31	12.19	nr	**21.16**
35 mm dia.	22.08	27.94	0.36	14.16	nr	**42.10**
42 mm dia.	39.30	49.74	0.41	16.12	nr	**65.86**
54 mm dia.	71.04	89.90	0.44	17.30	nr	**107.21**
67 mm dia.	128.89	163.13	0.54	21.24	nr	**184.37**
Straight tap connector						
15 mm × ½" dia.	2.37	3.00	0.13	5.11	nr	**8.11**
22 mm × ¾" dia.	3.52	4.46	0.14	5.51	nr	**9.96**
Bent tap connector						
15 mm × ½" dia.	3.51	4.44	0.13	5.11	nr	**9.55**
22 mm × ¾" dia.	10.76	13.62	0.14	5.51	nr	**19.13**
Bent male union connector						
15 mm × ½" dia.	17.99	22.77	0.41	16.12	nr	**38.89**
22 mm × ¾" dia.	23.34	29.54	0.45	17.70	nr	**47.24**
28 mm × 1" dia.	33.44	42.32	0.51	20.06	nr	**62.37**
35 mm × 1¼" dia.	54.54	69.02	0.64	25.17	nr	**94.19**
42 mm × 1½" dia.	88.64	112.19	0.68	26.74	nr	**138.93**
54 mm × 2" dia.	140.01	177.20	0.78	30.67	nr	**207.87**

38 PIPED SUPPLY SYSTEMS

Item	Net Price £	Material £	Labour hours	Labour £	Unit	Total rate £
Bent female union connector						
15 mm × ¾" dia.	17.99	22.77	0.41	16.12	nr	**38.89**
22 mm × 1" dia.	23.34	29.54	0.45	17.70	nr	**47.24**
28 mm × 1¼" dia.	33.44	42.32	0.51	20.06	nr	**62.37**
35 mm × 1½" dia.	54.54	69.02	0.64	25.17	nr	**94.19**
42 mm × 2" dia.	88.64	112.19	0.68	26.74	nr	**138.93**
54 mm × 2½" dia.	140.01	177.20	0.78	30.67	nr	**207.87**
Straight union adaptor						
15 mm × ¾" dia.	7.68	9.72	0.41	16.12	nr	**25.85**
22 mm × 1" dia.	10.90	13.80	0.45	17.70	nr	**31.50**
28 mm × 1¼" dia.	17.63	22.32	0.51	20.06	nr	**42.37**
35 mm × 1½" dia.	27.15	34.36	0.64	25.17	nr	**59.53**
42 mm × 2" dia.	34.28	43.38	0.68	26.74	nr	**70.12**
54 mm × 2½" dia.	52.92	66.97	0.78	30.67	nr	**97.65**
Straight male union connector						
15 mm × ½" dia.	15.32	19.39	0.41	16.12	nr	**35.51**
22 mm × ¾" dia.	19.88	25.16	0.45	17.70	nr	**42.85**
28 mm × 1" dia.	29.59	37.45	0.51	20.06	nr	**57.51**
35 mm × 1¼" dia.	42.62	53.95	0.64	25.17	nr	**79.11**
42 mm × 1½" dia.	66.97	84.76	0.68	26.74	nr	**111.50**
54 mm × 2" dia.	96.24	121.80	0.78	30.67	nr	**152.47**
Straight female union connector						
15 mm × ½" dia.	15.32	19.39	0.41	16.12	nr	**35.51**
22 mm × ¾" dia.	19.88	25.16	0.45	17.70	nr	**42.85**
28 mm × 1" dia.	29.59	37.45	0.51	20.06	nr	**57.51**
35 mm × 1¼" dia.	42.62	53.95	0.64	25.17	nr	**79.11**
42 mm × 1½" dia.	66.97	84.76	0.68	26.74	nr	**111.50**
54 mm × 2" dia.	96.24	121.80	0.78	30.67	nr	**152.47**
Male nipple						
¾ × ½" dia.	3.61	4.56	0.24	9.33	nr	**13.89**
1 × ¾" dia.	4.34	5.50	0.32	12.58	nr	**18.08**
1¼ × 1" dia.	4.76	6.02	0.37	14.55	nr	**20.57**
1 ½ × 1¼" dia.	17.69	22.39	0.42	16.52	nr	**38.91**
2 × 1 ½" dia.	36.21	45.82	0.46	18.09	nr	**63.91**
2 ½ × 2" dia.	47.17	59.70	0.56	22.02	nr	**81.72**
Female nipple						
¾ × ½" dia.	5.59	7.08	0.19	7.47	nr	**14.55**
1 × ¾" dia.	8.75	11.07	0.32	12.58	nr	**23.65**
1¼× 1" dia.	11.96	15.13	0.37	14.55	nr	**29.68**
1½ × 1¼" dia.	17.69	22.39	0.42	16.52	nr	**38.91**
2 × 1½" dia.	36.21	45.82	0.46	18.09	nr	**63.91**
2½ × 2" dia.	47.17	59.70	0.56	22.02	nr	**81.72**
Equal tee						
10 mm dia.	0.49	0.61	0.25	9.83	nr	**10.45**
15 mm dia.	1.27	1.61	0.36	14.16	nr	**15.77**
22 mm dia.	4.79	6.07	0.39	15.34	nr	**21.40**
28 mm dia.	9.34	11.82	0.43	16.91	nr	**28.73**
35 mm dia.	23.81	30.14	0.57	22.42	nr	**52.55**
42 mm dia.	38.19	48.34	0.60	23.60	nr	**71.93**
54 mm dia.	77.01	97.47	0.65	25.56	nr	**123.03**
67 mm dia.	104.79	132.62	0.78	30.67	nr	**163.29**

38 PIPED SUPPLY SYSTEMS

Item	Net Price £	Material £	Labour hours	Labour £	Unit	Total rate £
COLD WATER PIPELINES: COPPER PIPEWORK – cont						
Extra over copper pipes – cont						
Female tee, reducing branch FI						
15 × 15 mm × ¼" dia.	11.53	14.59	0.36	14.16	nr	**28.75**
22 × 22 mm × ½" dia.	14.10	17.84	0.39	15.36	nr	**33.20**
28 × 28 mm × ¾" dia.	27.62	34.95	0.43	16.91	nr	**51.86**
35 × 35 mm × ¾" dia.	39.87	50.46	0.47	18.48	nr	**68.94**
42 × 42 mm × ½" dia.	47.89	60.61	0.60	23.60	nr	**84.21**
Backplate tee						
15 × 15 mm × ½" dia.	21.79	27.58	0.62	24.38	nr	**51.97**
Heater tee						
½ × ½ × 15 mm dia.	19.59	24.80	0.36	14.16	nr	**38.96**
Union heater tee						
½ × ½ × 15 mm dia.	19.59	24.80	0.36	14.16	nr	**38.96**
Sweep tee – equal						
15 mm dia.	15.58	19.71	0.36	14.16	nr	**33.87**
22 mm dia.	20.05	25.37	0.39	15.34	nr	**40.71**
28 mm dia.	33.75	42.71	0.43	16.91	nr	**59.62**
35 mm dia.	47.86	60.58	0.57	22.42	nr	**82.99**
42 mm dia.	70.98	89.83	0.60	23.60	nr	**113.43**
54 mm dia.	78.60	99.48	0.65	25.56	nr	**125.04**
67 mm dia.	107.16	135.62	0.78	30.67	nr	**166.29**
Sweep tee – reducing						
22 × 22 × 15 mm dia.	16.80	21.26	0.39	15.34	nr	**36.60**
28 × 28 × 22 mm dia.	28.57	36.16	0.43	16.91	nr	**53.07**
35 × 35 × 22 mm dia.	47.86	60.58	0.57	22.42	nr	**82.99**
Sweep tee – double						
15 mm dia.	17.63	22.32	0.36	14.16	nr	**36.48**
22 mm dia.	24.00	30.37	0.39	15.34	nr	**45.71**
28 mm dia.	36.48	46.16	0.43	16.91	nr	**63.07**
Cross						
15 mm dia.	23.33	29.52	0.48	18.88	nr	**48.40**
22 mm dia.	26.04	32.96	0.53	20.84	nr	**53.80**
28 mm dia.	37.40	47.33	0.61	23.99	nr	**71.32**
Extra over copper pipes; high duty capillary fittings; BS 864						
Stop end						
15 mm dia.	10.37	13.13	0.16	6.29	nr	**19.42**
Straight coupling; copper to copper						
15 mm dia.	4.73	5.99	0.27	10.62	nr	**16.61**
22 mm dia.	7.61	9.63	0.32	12.58	nr	**22.21**
28 mm dia.	10.17	12.87	0.37	14.55	nr	**27.42**
35 mm dia.	18.99	24.04	0.43	16.91	nr	**40.95**
42 mm dia.	20.77	26.29	0.50	19.66	nr	**45.95**
54 mm dia.	30.55	38.67	0.54	21.24	nr	**59.90**

38 PIPED SUPPLY SYSTEMS

Item	Net Price £	Material £	Labour hours	Labour £	Unit	Total rate £
Reducing coupling						
15 × 12 mm dia.	8.94	11.32	0.27	10.62	nr	**21.94**
22 × 15 mm dia.	10.37	13.13	0.32	12.58	nr	**25.71**
28 × 22 mm dia.	14.28	18.08	0.37	14.55	nr	**32.63**
Straight female connector						
15 mm × ½" dia.	11.65	14.75	0.32	12.58	nr	**27.33**
22 mm × ¾" dia.	13.16	16.66	0.36	14.16	nr	**30.82**
28 mm × 1" dia.	19.42	24.58	0.42	16.52	nr	**41.10**
Straight male connector						
15 mm × ½" dia.	11.36	14.37	0.32	12.58	nr	**26.96**
22 mm × ¾" dia.	13.16	16.66	0.36	14.16	nr	**30.82**
28 mm × 1" dia.	19.42	24.58	0.42	16.52	nr	**41.10**
42 mm × 1½" dia.	37.89	47.95	0.53	20.84	nr	**68.80**
54 mm × 2" dia.	61.59	77.95	0.62	24.38	nr	**102.33**
Reducer						
15 × 12 mm dia.	5.73	7.26	0.27	10.62	nr	**17.87**
22 × 15 mm dia.	5.88	7.44	0.32	12.58	nr	**20.03**
28 × 22 mm dia.	10.37	13.13	0.37	14.55	nr	**27.68**
35 × 28 mm dia.	13.16	16.66	0.39	15.34	nr	**32.00**
42 × 35 mm dia.	16.97	21.48	0.43	16.91	nr	**38.39**
54 × 42 mm dia.	27.34	34.60	0.50	19.66	nr	**54.27**
Straight union adaptor						
15 mm × ¾" dia.	9.48	12.00	0.27	10.62	nr	**22.62**
22 mm × 1" dia.	12.85	16.27	0.32	12.58	nr	**28.85**
28 mm × 1¼" dia.	16.97	21.48	0.37	14.55	nr	**36.03**
35 mm × 1½" dia.	30.75	38.92	0.43	16.91	nr	**55.83**
42 mm × 2" dia.	38.97	49.32	0.50	19.66	nr	**68.98**
Bent union adaptor						
15 mm × ¾" dia.	24.66	31.21	0.27	10.62	nr	**41.82**
22 mm × 1" dia.	33.29	42.13	0.32	12.58	nr	**54.72**
28 mm × 1¼" dia.	44.81	56.71	0.37	14.55	nr	**71.26**
Adaptor; male copper to FI						
15 mm × ½" dia.	11.10	14.05	0.27	10.62	nr	**24.67**
22 mm × ¾" dia.	19.05	24.10	0.32	12.58	nr	**36.69**
Union coupling						
15 mm dia.	21.38	27.06	0.54	21.24	nr	**48.29**
22 mm dia.	27.34	34.60	0.60	23.60	nr	**58.20**
28 mm dia.	38.00	48.09	0.68	26.74	nr	**74.83**
35 mm dia.	66.31	83.92	0.83	32.64	nr	**116.56**
42 mm dia.	78.10	98.84	0.89	35.00	nr	**133.84**
Elbow						
15 mm dia.	13.76	17.41	0.27	10.62	nr	**28.03**
22 mm dia.	14.72	18.62	0.32	12.58	nr	**31.21**
28 mm dia.	21.85	27.66	0.37	14.55	nr	**42.21**
35 mm dia.	34.17	43.24	0.43	16.91	nr	**60.15**
42 mm dia.	42.56	53.86	0.50	19.66	nr	**73.52**
54 mm dia.	74.01	93.67	0.52	20.45	nr	**114.12**
Return bend						
28 mm dia.	32.95	41.70	0.37	14.55	nr	**56.26**
35 mm dia.	40.61	51.40	0.43	16.91	nr	**68.31**

38 PIPED SUPPLY SYSTEMS

Item	Net Price £	Material £	Labour hours	Labour £	Unit	Total rate £
COLD WATER PIPELINES: COPPER PIPEWORK – cont						
Extra over copper pipes – cont						
Bent male union connector						
15 mm × ½" dia.	31.91	40.39	0.54	21.24	nr	**61.63**
22 mm × ¾" dia.	42.99	54.41	0.60	23.60	nr	**78.00**
28 mm × 1" dia.	78.10	98.84	0.68	26.74	nr	**125.58**
Composite flange						
35 mm dia.	6.43	8.14	0.38	14.95	nr	**23.09**
42 mm dia.	8.36	10.58	0.41	16.12	nr	**26.71**
54 mm dia.	10.67	13.50	0.43	16.91	nr	**30.42**
Equal tee						
15 mm dia.	15.81	20.01	0.44	17.30	nr	**37.31**
22 mm dia.	19.90	25.18	0.47	18.48	nr	**43.66**
28 mm dia.	26.22	33.19	0.53	20.84	nr	**54.03**
35 mm dia.	44.81	56.71	0.70	27.53	nr	**84.24**
42 mm dia.	57.06	72.21	0.84	33.03	nr	**105.24**
54 mm dia.	89.83	113.69	0.79	31.07	nr	**144.76**
Reducing tee						
15 × 12 mm dia.	21.68	27.43	0.44	17.30	nr	**44.74**
22 × 15 mm dia.	25.60	32.40	0.47	18.48	nr	**50.88**
28 × 22 mm dia.	36.58	46.30	0.53	20.84	nr	**67.14**
35 × 28 mm dia.	57.99	73.39	0.73	28.71	nr	**102.10**
42 × 28 mm dia.	74.23	93.94	0.84	33.03	nr	**126.98**
54 × 28 mm dia.	117.23	148.36	1.01	39.72	nr	**188.08**
Extra over copper pipes; compression fittings; BS 864						
Stop end						
15 mm dia.	2.34	2.96	0.10	3.93	nr	**6.89**
22 mm dia.	2.78	3.52	0.12	4.72	nr	**8.24**
28 mm dia.	3.19	4.03	0.15	5.90	nr	**9.93**
Straight connector; copper to copper						
15 mm dia.	4.85	6.14	0.18	7.08	nr	**13.22**
22 mm dia.	6.42	8.13	0.21	8.26	nr	**16.39**
28 mm dia.	8.54	10.81	0.24	9.44	nr	**20.25**
Straight connector; copper to imperial copper						
22 mm dia.	8.00	10.12	0.21	8.26	nr	**18.38**
Male coupling; copper to MI (BSP)						
15 mm dia.	1.38	1.74	0.19	7.47	nr	**9.22**
22 mm dia.	2.17	2.75	0.23	9.04	nr	**11.79**
28 mm dia.	5.04	6.38	0.26	10.22	nr	**16.61**
Male coupling with long thread and backnut						
15 mm dia.	8.88	11.24	0.19	7.47	nr	**18.71**
22 mm dia.	11.27	14.27	0.23	9.04	nr	**23.31**
Female coupling; copper to FI (BSP)						
15 mm dia.	1.70	2.16	0.19	7.47	nr	**9.63**
22 mm dia.	2.43	3.08	0.23	9.04	nr	**12.12**
28 mm dia.	7.05	8.92	0.27	10.62	nr	**19.54**

38 PIPED SUPPLY SYSTEMS

Item	Net Price £	Material £	Labour hours	Labour £	Unit	Total rate £
Elbow						
15 mm dia.	1.81	2.29	0.18	7.08	nr	**9.37**
22 mm dia.	3.09	3.92	0.21	8.26	nr	**12.17**
28 mm dia.	10.43	13.20	0.24	9.44	nr	**22.64**
Male elbow; copper to FI (BSP)						
15 mm × ½" dia.	3.48	4.41	0.19	7.47	nr	**11.88**
22 mm × ¾" dia.	4.52	5.73	0.23	9.04	nr	**14.77**
28 mm × 1" dia.	10.88	13.77	0.27	10.62	nr	**24.39**
Female elbow; copper to FI (BSP)						
15 mm × ½" dia.	5.33	6.75	0.19	7.47	nr	**14.22**
22 mm × ¾" dia.	7.71	9.76	0.23	9.04	nr	**18.80**
28 mm × 1" dia.	13.59	17.20	0.27	10.62	nr	**27.81**
Backplate elbow						
15 mm × ½" dia.	7.71	9.76	0.50	19.66	nr	**29.42**
Tank coupling; long thread						
22 mm dia.	11.27	14.27	0.46	18.09	nr	**32.36**
Tee equal						
15 mm dia.	2.57	3.26	0.28	11.01	nr	**14.27**
22 mm dia.	4.30	5.45	0.30	11.80	nr	**17.24**
28 mm dia.	19.63	24.85	0.34	13.37	nr	**38.22**
Tee reducing						
22 mm dia.	11.10	14.05	0.30	11.80	nr	**25.85**
Backplate tee						
15 mm dia.	27.12	34.33	0.62	24.38	nr	**58.71**
Extra over fittings; silver brazed welded joints						
Reducer						
76 × 67 mm dia.	43.50	55.05	1.40	55.06	nr	**110.10**
108 × 76 mm dia.	91.17	115.39	1.80	70.79	nr	**186.18**
133 × 108 mm dia.	181.80	230.09	2.20	86.52	nr	**316.61**
159 × 133 mm dia.	232.52	294.27	2.60	102.25	nr	**396.52**
90° elbow						
76 mm dia.	105.09	133.00	1.60	62.92	nr	**195.92**
108 mm dia.	193.47	244.85	2.00	78.65	nr	**323.50**
133 mm dia.	408.97	517.60	2.40	94.38	nr	**611.98**
159 mm dia.	507.73	642.58	2.80	110.11	nr	**752.70**
45° elbow						
76 mm dia.	98.48	124.63	1.60	62.92	nr	**187.56**
108 mm dia.	154.43	195.45	2.00	78.65	nr	**274.10**
133 mm dia.	394.90	499.78	2.40	94.38	nr	**594.16**
159 mm dia.	568.41	719.38	2.80	110.11	nr	**829.49**
Equal tee						
76 mm dia.	127.98	161.98	2.40	94.38	nr	**256.36**
108 mm dia.	198.81	251.62	3.00	117.98	nr	**369.60**
133 mm dia.	468.20	592.56	3.60	141.57	nr	**734.13**
159 mm dia.	558.40	706.72	4.20	165.17	nr	**871.88**

38 PIPED SUPPLY SYSTEMS

Item	Net Price £	Material £	Labour hours	Labour £	Unit	Total rate £
COLD WATER PIPELINES: COPPER PIPEWORK – cont						
Extra over copper pipes; dezincification resistant compression fittings; BS 864						
Stop end						
15 mm dia.	3.35	4.25	0.10	3.93	nr	8.18
22 mm dia.	4.85	6.14	0.13	5.11	nr	11.25
28 mm dia.	10.39	13.15	0.15	5.90	nr	19.05
35 mm dia.	16.27	20.59	0.18	7.08	nr	27.66
42 mm dia.	27.12	34.33	0.20	7.87	nr	42.19
Straight coupling; copper to copper						
15 mm dia.	2.68	3.39	0.18	7.08	nr	10.47
22 mm dia.	4.39	5.56	0.21	8.26	nr	13.82
28 mm dia.	9.93	12.57	0.24	9.44	nr	22.01
35 mm dia.	21.02	26.61	0.29	11.40	nr	38.01
42 mm dia.	27.64	34.98	0.33	12.98	nr	47.96
54 mm dia.	41.33	52.31	0.38	14.94	nr	67.25
Straight swivel connector; copper to imperial copper						
22 mm dia.	9.89	12.52	0.20	7.87	nr	20.39
Male coupling; copper to MI (BSP)						
15 mm × ½" dia.	2.39	3.03	0.19	7.47	nr	10.50
22 mm × ¾" dia.	3.63	4.59	0.23	9.04	nr	13.64
28 mm × 1" dia.	7.03	8.90	0.26	10.22	nr	19.13
35 mm × 1¼" dia.	15.97	20.21	0.32	12.58	nr	32.79
42 mm × 1½" dia.	23.98	30.34	0.37	14.55	nr	44.89
54 mm × 2" dia.	35.37	44.76	0.57	22.42	nr	67.17
Male coupling with long thread and backnuts						
22 mm dia.	13.59	17.20	0.23	9.04	nr	26.24
28 mm dia.	15.06	19.06	0.24	9.44	nr	28.49
Female coupling; copper to FI (BSP)						
15 mm × ½" dia.	2.86	3.62	0.19	7.47	nr	11.09
22 mm × ¾" dia.	4.19	5.30	0.23	9.04	nr	14.34
28 mm × 1" dia.	9.11	11.54	0.27	10.62	nr	22.15
35 mm × 1¼" dia.	19.20	24.30	0.32	12.58	nr	36.89
42 mm × 1½" dia.	25.78	32.63	0.37	14.55	nr	47.18
54 mm × 2" dia.	37.84	47.88	0.42	16.52	nr	64.40
Elbow						
15 mm dia.	3.22	4.08	0.18	7.08	nr	11.16
22 mm dia.	5.16	6.53	0.21	8.26	nr	14.79
28 mm dia.	12.82	16.22	0.24	9.44	nr	25.66
35 mm dia.	28.37	35.91	0.29	11.40	nr	47.31
42 mm dia.	38.42	48.63	0.33	12.98	nr	61.60
54 mm dia.	66.13	83.69	0.38	14.94	nr	98.63
Male elbow; copper to MI (BSP)						
15 mm × ½" dia.	5.60	7.09	0.19	7.47	nr	14.56
22 mm × ¾" dia.	6.27	7.93	0.23	9.04	nr	16.98
28 mm × 1" dia.	11.74	14.86	0.27	10.62	nr	25.48

38 PIPED SUPPLY SYSTEMS

Item	Net Price £	Material £	Labour hours	Labour £	Unit	Total rate £
Female elbow; copper to FI (BSP)						
15 mm × ½" dia.	5.98	7.57	0.19	7.47	nr	**15.04**
22 mm × ¾" dia.	8.63	10.93	0.23	9.04	nr	**19.97**
28 mm × 1" dia.	14.32	18.12	0.27	10.62	nr	**28.74**
Backplate elbow						
15 mm × ½" dia.	8.74	11.06	0.50	19.66	nr	**30.72**
Straight tap connector						
15 mm dia.	4.95	6.27	0.13	5.11	nr	**11.38**
22 mm dia.	10.74	13.59	0.15	5.90	nr	**19.49**
Tank coupling						
15 mm dia.	6.83	8.64	0.19	7.47	nr	**16.11**
22 mm dia.	7.54	9.54	0.23	9.04	nr	**18.59**
28 mm dia.	15.94	20.17	0.27	10.62	nr	**30.79**
35 mm dia.	28.19	35.67	0.31	12.19	nr	**47.87**
42 mm dia.	45.81	57.97	0.32	12.58	nr	**70.56**
54 mm dia.	59.00	74.67	0.37	14.55	nr	**89.22**
Tee equal						
15 mm dia.	4.54	5.74	0.28	11.01	nr	**16.75**
22 mm dia.	7.49	9.48	0.30	11.80	nr	**21.28**
28 mm dia.	20.44	25.87	0.34	13.37	nr	**39.24**
35 mm dia.	36.90	46.70	0.43	16.91	nr	**63.61**
42 mm dia.	58.03	73.44	0.46	18.09	nr	**91.53**
54 mm dia.	93.21	117.97	0.54	21.24	nr	**139.20**
Tee reducing						
22 mm dia.	11.97	15.16	0.30	11.80	nr	**26.95**
28 mm dia.	19.75	25.00	0.34	13.37	nr	**38.37**
35 mm dia.	36.05	45.63	0.43	16.91	nr	**62.54**
42 mm dia.	55.74	70.54	0.46	18.09	nr	**88.63**
54 mm dia.	93.21	117.97	0.54	21.24	nr	**139.20**
Extra over copper pipes; bronze one piece brazing flanges; metric, including jointing ring and bolts						
Bronze flange; PN6						
15 mm dia.	48.87	61.85	0.27	10.62	nr	**72.47**
22 mm dia.	58.23	73.70	0.32	12.58	nr	**86.28**
28 mm dia.	63.97	80.97	0.36	14.16	nr	**95.12**
35 mm dia.	76.42	96.72	0.47	18.48	nr	**115.21**
42 mm dia.	83.38	105.53	0.54	21.24	nr	**126.77**
54 mm dia.	96.72	122.41	0.63	24.78	nr	**147.18**
67 mm dia.	114.01	144.30	0.77	30.28	nr	**174.58**
76 mm dia.	137.41	173.91	0.93	36.57	nr	**210.48**
108 mm dia.	181.56	229.78	1.14	44.83	nr	**274.61**
133 mm dia.	248.77	314.85	1.41	55.45	nr	**370.30**
159 mm dia.	358.89	454.21	1.74	68.43	nr	**522.64**
Bronze flange; PN10						
15 mm dia.	55.95	70.81	0.27	10.62	nr	**81.43**
22 mm dia.	59.55	75.37	0.32	12.58	nr	**87.95**
28 mm dia.	65.49	82.88	0.38	14.94	nr	**97.83**
35 mm dia.	79.18	100.22	0.47	18.48	nr	**118.70**
42 mm dia.	88.75	112.33	0.54	21.24	nr	**133.56**
54 mm dia.	97.12	122.92	0.63	24.78	nr	**147.70**
67 mm dia.	114.12	144.43	0.77	30.28	nr	**174.71**

38 PIPED SUPPLY SYSTEMS

Item	Net Price £	Material £	Labour hours	Labour £	Unit	Total rate £
COLD WATER PIPELINES: COPPER PIPEWORK – cont						
Extra over copper pipes – cont						
Bronze flange; PN10 – cont						
76 mm dia.	137.41	173.91	0.93	36.57	nr	**210.48**
108 mm dia.	173.67	219.80	1.14	44.83	nr	**264.63**
133 mm dia.	231.88	293.47	1.41	55.45	nr	**348.92**
159 mm dia.	293.92	371.99	1.74	68.43	nr	**440.42**
Bronze flange; PN16						
15 mm dia.	61.56	77.91	0.27	10.62	nr	**88.53**
22 mm dia.	65.53	82.94	0.32	12.58	nr	**95.53**
28 mm dia.	72.04	91.18	0.38	14.94	nr	**106.12**
35 mm dia.	87.10	110.24	0.47	18.48	nr	**128.72**
42 mm dia.	97.65	123.59	0.54	21.24	nr	**144.82**
54 mm dia.	106.81	135.18	0.63	24.78	nr	**159.96**
67 mm dia.	125.52	158.86	0.77	30.28	nr	**189.14**
76 mm dia.	151.14	191.28	0.93	36.57	nr	**227.86**
108 mm dia.	191.04	241.78	1.14	44.83	nr	**286.61**
133 mm dia.	255.07	322.82	1.41	55.45	nr	**378.27**
159 mm dia.	323.29	409.16	1.74	68.43	nr	**477.59**
Extra over copper pipes; bronze blank flanges; metric, including jointing ring and bolts						
Gunmetal blank flange; PN6						
15 mm dia.	41.03	51.92	0.27	10.62	nr	**62.54**
22 mm dia.	52.33	66.23	0.27	10.62	nr	**76.85**
28 mm dia.	53.79	68.08	0.27	10.62	nr	**78.69**
35 mm dia.	88.03	111.42	0.32	12.58	nr	**124.00**
42 mm dia.	118.77	150.31	0.32	12.58	nr	**162.90**
54 mm dia.	130.95	165.73	0.34	13.37	nr	**179.10**
67 mm dia.	161.13	203.93	0.36	14.16	nr	**218.08**
76 mm dia.	207.57	262.70	0.37	14.55	nr	**277.25**
108 mm dia.	329.79	417.38	0.41	16.12	nr	**433.50**
133 mm dia.	389.04	492.37	0.58	22.81	nr	**515.18**
159 mm dia.	486.27	615.42	0.61	23.99	nr	**639.41**
Gunmetal blank flange; PN10						
15 mm dia.	49.60	62.78	0.27	10.62	nr	**73.40**
22 mm dia.	64.20	81.25	0.27	10.62	nr	**91.87**
28 mm dia.	71.07	89.95	0.27	10.62	nr	**100.56**
35 mm dia.	88.03	111.42	0.32	12.58	nr	**124.00**
42 mm dia.	165.00	208.82	0.32	12.58	nr	**221.41**
54 mm dia.	188.11	238.08	0.34	13.37	nr	**251.45**
67 mm dia.	201.79	255.39	0.46	18.09	nr	**273.48**
76 mm dia.	268.17	339.39	0.47	18.48	nr	**357.88**
108 mm dia.	314.71	398.30	0.51	20.06	nr	**418.36**
133 mm dia.	418.09	529.14	0.58	22.81	nr	**551.95**
159 mm dia.	569.40	720.63	0.71	27.92	nr	**748.55**
Gunmetal blank flange; PN16						
15 mm dia.	49.60	62.78	0.27	10.62	nr	**73.40**
22 mm dia.	65.17	82.49	0.27	10.62	nr	**93.10**
28 mm dia.	71.07	89.95	0.27	10.62	nr	**100.56**

38 PIPED SUPPLY SYSTEMS

Item	Net Price £	Material £	Labour hours	Labour £	Unit	Total rate £
35 mm dia.	88.03	111.42	0.32	12.58	nr	**124.00**
42 mm dia.	165.00	208.82	0.32	12.58	nr	**221.41**
54 mm dia.	188.11	238.08	0.34	13.37	nr	**251.45**
67 mm dia.	234.39	296.64	0.46	18.09	nr	**314.73**
76 mm dia.	268.17	339.39	0.47	18.48	nr	**357.88**
108 mm dia.	314.71	398.30	0.51	20.06	nr	**418.36**
133 mm dia.	722.85	914.84	0.58	22.81	nr	**937.64**
159 mm dia.	863.41	1092.73	0.71	27.92	nr	**1120.65**
Extra over copper pipes; bronze screwed flanges; metric, including jointing ring and bolts						
Gunmetal screwed flange; 6 BSP						
15 mm dia.	41.03	51.92	0.35	13.76	nr	**65.69**
22 mm dia.	47.41	60.01	0.47	18.48	nr	**78.49**
28 mm dia.	49.42	62.55	0.52	20.45	nr	**83.00**
35 mm dia.	66.63	84.33	0.62	24.38	nr	**108.71**
42 mm dia.	79.98	101.22	0.70	27.53	nr	**128.75**
54 mm dia.	108.70	137.58	0.84	33.03	nr	**170.61**
67 mm dia.	136.41	172.64	1.03	40.51	nr	**213.15**
76 mm dia.	164.65	208.39	1.22	47.98	nr	**256.36**
108 mm dia.	260.76	330.02	1.41	55.45	nr	**385.47**
133 mm dia.	309.12	391.22	1.75	68.82	nr	**460.04**
159 mm dia.	393.49	498.01	2.21	86.91	nr	**584.92**
Gunmetal screwed flange; 10 BSP						
15 mm dia.	49.68	62.87	0.35	13.76	nr	**76.64**
22 mm dia.	57.85	73.22	0.47	18.48	nr	**91.70**
28 mm dia.	63.88	80.85	0.52	20.45	nr	**101.30**
35 mm dia.	91.05	115.23	0.62	24.38	nr	**139.61**
42 mm dia.	111.18	140.71	0.70	27.53	nr	**168.24**
54 mm dia.	158.08	200.07	0.84	33.03	nr	**233.11**
67 mm dia.	185.07	234.22	1.03	40.51	nr	**274.73**
76 mm dia.	208.92	264.41	1.22	47.98	nr	**312.39**
108 mm dia.	277.32	350.97	1.41	55.45	nr	**406.42**
133 mm dia.	336.10	425.37	1.75	68.82	nr	**494.19**
159 mm dia.	596.43	754.84	2.21	86.91	nr	**841.75**
Gunmetal screwed flange; 16 BSP						
15 mm dia.	39.03	49.40	0.35	13.76	nr	**63.16**
22 mm dia.	49.35	62.46	0.47	18.48	nr	**80.94**
28 mm dia.	57.12	72.29	0.52	20.45	nr	**92.74**
35 mm dia.	87.30	110.49	0.62	24.38	nr	**134.87**
42 mm dia.	105.07	132.98	0.70	27.53	nr	**160.51**
54 mm dia.	135.13	171.03	0.84	33.03	nr	**204.06**
67 mm dia.	195.33	247.21	1.03	40.51	nr	**287.71**
76 mm dia.	223.99	283.49	1.22	47.98	nr	**331.46**
108 mm dia.	262.09	331.71	1.41	55.45	nr	**387.16**
133 mm dia.	424.69	537.49	1.75	68.82	nr	**606.31**
159 mm dia.	525.31	664.84	2.21	86.91	nr	**751.75**

38 PIPED SUPPLY SYSTEMS

Item	Net Price £	Material £	Labour hours	Labour £	Unit	Total rate £
COLD WATER PIPELINES: COPPER PIPEWORK – cont						
Extra over copper pipes; labour						
Male bend						
15 mm dia.	–	–	0.26	10.22	nr	**10.22**
22 mm dia.	–	–	0.28	11.01	nr	**11.01**
28 mm dia.	–	–	0.31	12.19	nr	**12.19**
35 mm dia.	–	–	0.42	16.52	nr	**16.52**
42 mm dia.	–	–	0.51	20.06	nr	**20.06**
54 mm dia.	–	–	0.58	22.81	nr	**22.81**
67 mm dia.	–	–	0.69	27.13	nr	**27.13**
76 mm dia.	–	–	0.80	31.46	nr	**31.46**
Bronze butt weld						
15 mm dia.	–	–	0.25	9.83	nr	**9.83**
22 mm dia.	–	–	0.31	12.19	nr	**12.19**
28 mm dia.	–	–	0.37	14.55	nr	**14.55**
35 mm dia.	–	–	0.49	19.27	nr	**19.27**
42 mm dia.	–	–	0.58	22.81	nr	**22.81**
54 mm dia.	–	–	0.72	28.31	nr	**28.31**
67 mm dia.	–	–	0.88	34.61	nr	**34.61**
76 mm dia.	–	–	1.08	42.47	nr	**42.47**
108 mm dia.	–	–	1.37	53.88	nr	**53.88**
133 mm dia.	–	–	1.73	68.03	nr	**68.03**
159 mm dia.	–	–	2.03	79.83	nr	**79.83**
Press fit (copper fittings); Mechanical press fit joints; butyl rubber O ring						
Coupler						
15 mm dia.	1.67	2.11	0.36	14.16	nr	**16.27**
22 mm dia.	2.65	3.35	0.36	14.16	nr	**17.51**
28 mm dia.	5.43	6.87	0.44	17.30	nr	**24.17**
35 mm dia.	6.85	8.67	0.44	17.30	nr	**25.97**
42 mm dia.	12.32	15.59	0.52	20.45	nr	**36.04**
54 mm dia.	15.72	19.89	0.60	23.60	nr	**43.49**
Stop end						
22 mm dia.	4.25	5.37	0.18	7.08	nr	**12.45**
28 mm dia.	6.72	8.51	0.22	8.65	nr	**17.16**
35 mm dia.	11.53	14.59	0.22	8.65	nr	**23.24**
42 mm dia.	17.31	21.91	0.26	10.22	nr	**32.14**
54 mm dia.	20.86	26.39	0.30	11.80	nr	**38.19**
Reducer						
22 × 15 mm dia.	1.96	2.48	0.36	14.16	nr	**16.63**
28 × 15 mm dia.	5.17	6.54	0.40	15.73	nr	**22.27**
28 × 22 mm dia.	5.37	6.79	0.40	15.73	nr	**22.52**
35 × 22 mm dia.	6.43	8.14	0.40	15.73	nr	**23.87**
35 × 28 mm dia.	7.10	8.98	0.44	17.30	nr	**26.28**
42 × 22 mm dia.	11.10	14.04	0.44	17.30	nr	**31.35**
42 × 28 mm dia.	10.58	13.39	0.48	18.88	nr	**32.26**

38 PIPED SUPPLY SYSTEMS

Item	Net Price £	Material £	Labour hours	Labour £	Unit	Total rate £
42 × 35 mm dia.	10.58	13.39	0.48	18.88	nr	**32.26**
54 × 35 mm dia.	14.32	18.12	0.52	20.45	nr	**38.57**
54 × 42 mm dia.	14.32	18.12	0.56	22.02	nr	**40.15**
90° elbow						
15 mm dia.	1.83	2.31	0.36	14.16	nr	**16.47**
22 mm dia.	3.05	3.86	0.36	14.16	nr	**18.02**
28 mm dia.	6.55	8.29	0.44	17.30	nr	**25.59**
35 mm dia.	13.53	17.12	0.44	17.30	nr	**34.43**
42 mm dia.	25.42	32.17	0.52	20.45	nr	**52.62**
54 mm dia.	35.29	44.66	0.60	23.60	nr	**68.26**
45° elbow						
15 mm dia.	2.25	2.84	0.36	14.16	nr	**17.00**
22 mm dia.	3.09	3.92	0.36	14.16	nr	**18.07**
28 mm dia.	9.30	11.77	0.44	17.30	nr	**29.07**
35 mm dia.	13.26	16.78	0.44	17.30	nr	**34.08**
42 mm dia.	22.06	27.92	0.52	20.45	nr	**48.37**
54 mm dia.	31.39	39.73	0.60	23.60	nr	**63.32**
Equal tee						
15 mm dia.	2.89	3.66	0.54	21.24	nr	**24.90**
22 mm dia.	5.27	6.67	0.54	21.24	nr	**27.90**
28 mm dia.	9.47	11.99	0.66	25.96	nr	**37.94**
35 mm dia.	16.38	20.73	0.66	25.96	nr	**46.68**
42 mm dia.	32.34	40.93	0.78	30.67	nr	**71.60**
54 mm dia.	40.34	51.06	0.90	35.39	nr	**86.45**
Reducing tee						
22 × 15 mm dia.	4.27	5.41	0.54	21.24	nr	**26.65**
28 × 15 mm dia.	8.33	10.55	0.62	24.38	nr	**34.93**
28 × 22 mm dia.	11.18	14.15	0.62	24.38	nr	**38.54**
35 × 22 mm dia.	14.59	18.47	0.62	24.38	nr	**42.85**
35 × 28 mm dia.	16.24	20.55	0.62	24.38	nr	**44.93**
42 × 28 mm dia.	29.36	37.16	0.70	27.53	nr	**64.69**
42 × 35 mm dia.	29.36	37.16	0.70	27.53	nr	**64.69**
54 × 35 mm dia.	49.60	62.77	0.82	32.25	nr	**95.02**
54 × 42 mm dia.	49.60	62.77	0.82	32.25	nr	**95.02**
Male iron connector; BSP thread						
15 mm dia.	6.12	7.74	0.18	7.08	nr	**14.82**
22 mm dia.	9.10	11.51	0.18	7.08	nr	**18.59**
28 mm dia.	12.15	15.37	0.22	8.65	nr	**24.03**
35 mm dia.	21.98	27.82	0.22	8.65	nr	**36.47**
42 mm dia.	29.48	37.31	0.26	10.22	nr	**47.53**
54 mm dia.	56.85	71.95	0.30	11.80	nr	**83.75**
90° elbow; male iron BSP thread						
15 mm dia.	9.90	12.53	0.36	14.16	nr	**26.69**
22 mm dia.	15.53	19.65	0.36	14.16	nr	**33.81**
28 mm dia.	23.78	30.09	0.44	17.30	nr	**47.40**
35 mm dia.	30.93	39.15	0.44	17.30	nr	**56.45**
42 mm dia.	40.31	51.02	0.52	20.45	nr	**71.47**
54 mm dia.	58.90	74.54	0.60	23.60	nr	**98.13**

38 PIPED SUPPLY SYSTEMS

Item	Net Price £	Material £	Labour hours	Labour £	Unit	Total rate £
COLD WATER PIPELINES: COPPER PIPEWORK – cont						
Press fit (copper fittings) – cont						
Female iron connector; BSP thread						
15 mm dia.	6.99	8.85	0.18	7.08	nr	**15.93**
22 mm dia.	9.24	11.69	0.18	7.08	nr	**18.77**
28 mm dia.	12.42	15.72	0.22	8.65	nr	**24.37**
35 mm dia.	24.32	30.78	0.22	8.65	nr	**39.44**
42 mm dia.	34.76	43.99	0.26	10.22	nr	**54.22**
54 mm dia.	59.63	75.47	0.30	11.80	nr	**87.26**
90° elbow; female iron BSP thread						
15 mm dia.	8.35	10.57	0.36	14.16	nr	**24.72**
22 mm dia.	12.32	15.59	0.36	14.16	nr	**29.75**
28 mm dia.	20.32	25.72	0.44	17.30	nr	**43.02**
35 mm dia.	26.28	33.26	0.44	17.30	nr	**50.56**
42 mm dia.	35.77	45.27	0.52	20.45	nr	**65.71**
54 mm dia.	52.59	66.56	0.60	23.60	nr	**90.15**

38 PIPED SUPPLY SYSTEMS

Item	Net Price £	Material £	Labour hours	Labour £	Unit	Total rate £
COLD WATER PIPELINES: STAINLESS STEEL PIPEWORK						
Stainless steel pipes; capillary or compression joints; BS 4127, vertical or at low level, with brackets measured separately						
Grade 304; satin finish						
15 mm dia.	7.51	9.50	0.41	16.12	m	**25.62**
22 mm dia.	10.53	13.33	0.51	20.06	m	**33.40**
28 mm dia.	14.34	18.15	0.58	22.81	m	**40.96**
35 mm dia.	21.67	27.43	0.65	25.57	m	**53.00**
42 mm dia.	27.48	34.78	0.71	27.93	m	**62.71**
54 mm dia.	38.31	48.49	0.80	31.46	m	**79.95**
Grade 316; satin finish						
15 mm dia.	9.63	12.19	0.61	23.99	m	**36.18**
22 mm dia.	18.04	22.83	0.76	29.91	m	**52.74**
28 mm dia.	21.39	27.07	0.87	34.23	m	**61.29**
35 mm dia.	38.81	49.12	0.98	38.55	m	**87.67**
42 mm dia.	50.25	63.59	1.06	41.70	m	**105.30**
54 mm dia.	58.55	74.10	1.16	45.62	m	**119.73**
FIXINGS						
Single pipe ring						
15 mm dia.	18.59	23.53	0.26	10.22	nr	**33.76**
22 mm dia.	21.66	27.41	0.26	10.22	nr	**37.64**
28 mm dia.	22.69	28.71	0.31	12.19	nr	**40.90**
35 mm dia.	25.77	32.61	0.32	12.58	nr	**45.20**
42 mm dia.	29.29	37.07	0.32	12.58	nr	**49.65**
54 mm dia.	33.43	42.30	0.34	13.37	nr	**55.68**
Screw on backplate, female						
All sizes 15 mm to 54 mm dia.	16.78	21.23	0.10	3.93	nr	**25.17**
Screw on backplate, male						
All sizes 15 mm to 54 mm dia.	19.10	24.17	0.10	3.93	nr	**28.10**
Stainless steel threaded rods; metric thread; including nuts, washers, etc.						
10 mm dia. × 600 mm long	20.20	25.56	0.18	7.08	nr	**32.64**
Extra over stainless steel pipes; capillary fittings						
Straight coupling						
15 mm dia.	2.44	3.09	0.25	9.83	nr	**12.92**
22 mm dia.	3.93	4.97	0.28	11.01	nr	**15.98**
28 mm dia.	5.19	6.57	0.33	12.98	nr	**19.55**
35 mm dia.	12.00	15.19	0.37	14.55	nr	**29.74**
42 mm dia.	13.85	17.52	0.42	16.52	nr	**34.04**
54 mm dia.	20.82	26.36	0.45	17.70	nr	**44.05**
45° bend						
15 mm dia.	13.10	16.57	0.25	9.83	nr	**26.41**
22 mm dia.	17.20	21.76	0.30	11.66	nr	**33.43**
28 mm dia.	21.11	26.71	0.33	12.98	nr	**39.69**
35 mm dia.	25.03	31.68	0.37	14.55	nr	**46.24**
42 mm dia.	32.22	40.77	0.42	16.52	nr	**57.29**
54 mm dia.	39.83	50.41	0.45	17.70	nr	**68.11**

38 PIPED SUPPLY SYSTEMS

Item	Net Price £	Material £	Labour hours	Labour £	Unit	Total rate £
COLD WATER PIPELINES: STAINLESS STEEL PIPEWORK – cont						
Extra over stainless steel pipes – cont						
90° bend						
15 mm dia.	6.79	8.59	0.28	11.01	nr	**19.61**
22 mm dia.	9.14	11.56	0.28	11.01	nr	**22.58**
28 mm dia.	12.88	16.30	0.33	12.98	nr	**29.27**
35 mm dia.	31.40	39.74	0.37	14.55	nr	**54.30**
42 mm dia.	43.21	54.69	0.42	16.52	nr	**71.21**
54 mm dia.	58.63	74.20	0.45	17.70	nr	**91.89**
Reducer						
22 × 15 mm dia.	15.60	19.74	0.28	11.01	nr	**30.75**
28 × 22 mm dia.	17.40	22.02	0.33	12.98	nr	**35.00**
35 × 28 mm dia.	21.33	26.99	0.37	14.55	nr	**41.54**
42 × 35 mm dia.	22.94	29.03	0.42	16.52	nr	**45.55**
54 × 42 mm dia.	68.08	86.16	0.48	18.91	nr	**105.06**
Tap connector						
15 mm dia.	32.94	41.68	0.13	5.11	nr	**46.79**
22 mm dia.	43.54	55.11	0.14	5.51	nr	**60.61**
28 mm dia.	60.44	76.49	0.17	6.69	nr	**83.18**
Tank connector						
15 mm dia.	42.53	53.82	0.13	5.11	nr	**58.93**
22 mm dia.	63.29	80.10	0.13	5.11	nr	**85.21**
28 mm dia.	77.93	98.63	0.15	5.90	nr	**104.53**
35 mm dia.	112.70	142.63	0.18	7.08	nr	**149.71**
42 mm dia.	148.83	188.35	0.21	8.26	nr	**196.61**
54 mm dia.	225.30	285.14	0.24	9.44	nr	**294.58**
Tee equal						
15 mm dia.	12.13	15.35	0.37	14.55	nr	**29.90**
22 mm dia.	15.08	19.09	0.40	15.73	nr	**34.82**
28 mm dia.	18.24	23.09	0.45	17.70	nr	**40.79**
35 mm dia.	43.79	55.42	0.59	23.21	nr	**78.64**
42 mm dia.	54.07	68.43	0.62	24.39	nr	**92.83**
54 mm dia.	109.16	138.16	0.67	26.36	nr	**164.51**
Unequal tee						
22 × 15 mm dia.	24.58	31.11	0.37	14.55	nr	**45.66**
28 × 15 mm dia.	27.69	35.05	0.45	17.70	nr	**52.75**
28 × 22 mm dia.	27.69	35.05	0.45	17.70	nr	**52.75**
35 × 22 mm dia.	48.39	61.25	0.59	23.21	nr	**84.46**
35 × 28 mm dia.	48.39	61.25	0.59	23.21	nr	**84.46**
42 × 28 mm dia.	59.50	75.31	0.62	24.39	nr	**99.70**
42 × 35 mm dia.	59.50	75.31	0.62	24.39	nr	**99.70**
54 × 35 mm dia.	123.17	155.88	0.67	26.36	nr	**182.23**
54 × 42 mm dia.	123.17	155.88	0.67	26.36	nr	**182.23**
Union, conical seat						
15 mm dia.	54.54	69.03	0.25	9.83	nr	**78.86**
22 mm dia.	85.88	108.69	0.28	11.01	nr	**119.70**
28 mm dia.	111.04	140.53	0.33	12.98	nr	**153.51**
35 mm dia.	145.74	184.45	0.37	14.55	nr	**199.01**
42 mm dia.	183.83	232.65	0.42	16.52	nr	**249.17**
54 mm dia.	243.22	307.82	0.45	17.70	nr	**325.51**

38 PIPED SUPPLY SYSTEMS

Item	Net Price £	Material £	Labour hours	Labour £	Unit	Total rate £
Union, flat seat						
15 mm dia.	56.95	72.08	0.25	9.83	nr	**81.91**
22 mm dia.	88.67	112.22	0.28	11.01	nr	**123.23**
28 mm dia.	114.62	145.07	0.33	12.98	nr	**158.04**
35 mm dia.	149.78	189.56	0.37	14.55	nr	**204.12**
42 mm dia.	188.71	238.83	0.42	16.52	nr	**255.34**
54 mm dia.	253.01	320.21	0.45	17.70	nr	**337.91**
Extra over stainless steel pipes; compression fittings						
Straight coupling						
15 mm dia.	47.41	60.00	0.18	7.08	nr	**67.08**
22 mm dia.	90.31	114.30	0.22	8.65	nr	**122.95**
28 mm dia.	121.59	153.88	0.25	9.83	nr	**163.71**
35 mm dia.	187.69	237.54	0.30	11.80	nr	**249.34**
42 mm dia.	219.08	277.26	0.40	15.73	nr	**292.99**
90° bend						
15 mm dia.	59.75	75.62	0.18	7.08	nr	**82.70**
22 mm dia.	118.72	150.26	0.22	8.65	nr	**158.91**
28 mm dia.	161.89	204.89	0.25	9.83	nr	**214.72**
35 mm dia.	327.83	414.90	0.33	12.98	nr	**427.88**
42 mm dia.	479.10	606.35	0.35	13.76	nr	**620.11**
Reducer						
22 × 15 mm dia.	86.01	108.85	0.28	11.01	nr	**119.86**
28 × 22 mm dia.	117.81	149.11	0.28	11.01	nr	**160.12**
35 × 28 mm dia.	172.14	217.86	0.30	11.80	nr	**229.66**
42 × 35 mm dia.	229.00	289.82	0.37	14.55	nr	**304.37**
Stud coupling						
15 mm dia.	49.27	62.36	0.42	16.52	nr	**78.87**
22 mm dia.	83.33	105.46	0.25	9.83	nr	**115.30**
28 mm dia.	115.70	146.43	0.25	9.83	nr	**156.26**
35 mm dia.	185.34	234.57	0.37	14.55	nr	**249.13**
42 mm dia.	219.08	277.26	0.42	16.52	nr	**293.78**
Equal tee						
15 mm dia.	84.14	106.49	0.37	14.55	nr	**121.05**
22 mm dia.	173.83	220.00	0.40	15.73	nr	**235.73**
28 mm dia.	237.77	300.93	0.45	17.70	nr	**318.62**
35 mm dia.	472.78	598.35	0.59	23.21	nr	**621.56**
42 mm dia.	655.57	829.69	0.62	24.39	nr	**854.08**
Running tee						
15 mm dia.	103.59	131.11	0.37	14.55	nr	**145.66**
22 mm dia.	186.61	236.17	0.40	15.73	nr	**251.90**
28 mm dia.	315.96	399.87	0.59	23.21	nr	**423.09**
Press fit (stainless steel); Press fit jointing system; butyl rubber O ring mechanical joint						
Pipework						
15 mm dia.	9.29	11.76	0.46	18.09	m	**29.85**
22 mm dia.	14.85	18.80	0.48	18.88	m	**37.68**
28 mm dia.	18.30	23.15	0.52	20.45	m	**43.60**
35 mm dia.	26.96	34.12	0.56	22.02	m	**56.15**
42 mm dia.	33.17	41.98	0.58	22.81	m	**64.79**
54 mm dia.	42.24	53.46	0.66	25.96	m	**79.42**

38 PIPED SUPPLY SYSTEMS

Item	Net Price £	Material £	Labour hours	Labour £	Unit	Total rate £
COLD WATER PIPELINES: STAINLESS STEEL PIPEWORK – cont						
Press fit (stainless steel) – cont						
FIXINGS						
For stainless steel pipes						
Refer to fixings for stainless steel pipes; capillary or compression joints; BS 4127						
Extra over stainless steel pipes; Press fit jointing system						
Coupling						
15 mm dia.	11.64	14.73	0.36	14.16	nr	**28.89**
22 mm dia.	14.66	18.56	0.36	14.16	nr	**32.71**
28 mm dia.	16.49	20.87	0.44	17.30	nr	**38.17**
35 mm dia.	20.48	25.92	0.44	17.30	nr	**43.23**
42 mm dia.	27.98	35.41	0.52	20.45	nr	**55.86**
54 mm dia.	33.66	42.60	0.60	23.60	nr	**66.20**
Stop end						
22 mm dia.	11.05	13.98	0.18	7.08	nr	**21.06**
28 mm dia.	12.88	16.29	0.22	8.65	nr	**24.95**
35 mm dia.	21.05	26.64	0.22	8.65	nr	**35.30**
42 mm dia.	29.59	37.45	0.26	10.22	nr	**47.68**
54 mm dia.	34.23	43.32	0.30	11.80	nr	**55.12**
Reducer						
22 × 15 mm dia.	13.86	17.55	0.36	14.16	nr	**31.70**
28 × 15 mm dia.	15.71	19.88	0.40	15.73	nr	**35.61**
28 × 22 mm dia.	16.26	20.58	0.40	15.73	nr	**36.31**
35 × 22 mm dia.	19.85	25.13	0.40	15.73	nr	**40.86**
35 × 28 mm dia.	24.59	31.12	0.44	17.30	nr	**48.42**
42 × 35 mm dia.	25.94	32.83	0.48	18.88	nr	**51.71**
54 × 42 mm dia.	29.65	37.52	0.56	22.02	nr	**59.55**
90° bend						
15 mm dia.	16.66	21.08	0.36	14.16	nr	**35.24**
22 mm dia.	23.28	29.46	0.36	14.16	nr	**43.62**
28 mm dia.	29.36	37.16	0.44	17.30	nr	**54.47**
35 mm dia.	46.21	58.49	0.44	17.30	nr	**75.79**
42 mm dia.	77.16	97.65	0.52	20.45	nr	**118.10**
54 mm dia.	106.56	134.86	0.60	23.60	nr	**158.46**
45° bend						
15 mm dia.	22.59	28.59	0.36	14.16	nr	**42.75**
22 mm dia.	28.09	35.55	0.36	14.16	nr	**49.71**
28 mm dia.	32.69	41.38	0.44	17.30	nr	**58.68**
35 mm dia.	38.36	48.55	0.44	17.30	nr	**65.85**
42 mm dia.	61.71	78.10	0.52	20.45	nr	**98.55**
54 mm dia.	80.18	101.48	0.60	23.60	nr	**125.07**
Equal tee						
15 mm dia.	27.31	34.56	0.54	21.24	nr	**55.80**
22 mm dia.	33.51	42.41	0.54	21.24	nr	**63.65**
28 mm dia.	39.12	49.51	0.66	25.96	nr	**75.47**

38 PIPED SUPPLY SYSTEMS

Item	Net Price £	Material £	Labour hours	Labour £	Unit	Total rate £
35 mm dia.	49.58	62.75	0.66	25.96	nr	**88.70**
42 mm dia.	70.37	89.06	0.78	30.67	nr	**119.73**
54 mm dia.	84.19	106.55	0.90	35.39	nr	**141.95**
Reducing tee						
22 × 15 mm dia.	28.64	36.25	0.54	21.24	nr	**57.48**
28 × 15 mm dia.	34.71	43.93	0.62	24.38	nr	**68.31**
28 × 22 mm dia.	37.58	47.56	0.62	24.38	nr	**71.94**
35 × 22 mm dia.	44.54	56.37	0.62	24.38	nr	**80.75**
35 × 28 mm dia.	46.48	58.83	0.62	24.38	nr	**83.21**
42 × 28 mm dia.	66.09	83.64	0.70	27.53	nr	**111.17**
42 × 35 mm dia.	68.08	86.17	0.70	27.53	nr	**113.70**
54 × 35 mm dia.	76.91	97.34	0.82	32.25	nr	**129.58**
54 × 42 mm dia.	79.08	100.08	0.82	32.25	nr	**132.33**

FIXINGS

For stainless steel pipes

Refer to fixings for stainless steel pipes; capillary or compression joints; BS 4127

38 PIPED SUPPLY SYSTEMS

Item	Net Price £	Material £	Labour hours	Labour £	Unit	Total rate £
COLD WATER PIPELINES: MEDIUM DENSITY POLYETHYLENE – BLUE PIPEWORK **Note:** MDPE is sized on Outside Dia. i.e. OD not ID						
Pipes for water distribution; laid underground; electrofusion joints in the running length; BS 6572						
Coiled service pipe						
20 mm dia.	1.44	1.82	0.37	14.55	m	**16.38**
25 mm dia.	1.87	2.37	0.41	16.12	m	**18.49**
32 mm dia.	3.14	3.98	0.47	18.49	m	**22.46**
50 mm dia.	7.56	9.56	0.53	20.85	m	**30.41**
63 mm dia.	11.84	14.99	0.60	23.60	m	**38.59**
Mains service pipe						
90 mm dia.	17.91	22.67	0.90	35.40	m	**58.07**
110 mm dia.	26.79	33.90	1.10	43.26	m	**77.16**
125 mm dia.	33.89	42.89	1.20	47.21	m	**90.11**
160 mm dia.	54.01	68.36	1.48	58.20	m	**126.56**
180 mm dia.	70.34	89.02	1.50	59.05	m	**148.07**
225 mm dia.	106.89	135.29	1.75	68.87	m	**204.16**
250 mm dia.	134.92	170.76	1.77	69.61	m	**240.36**
315 mm dia.	208.32	263.65	1.90	74.72	m	**338.36**
Extra over fittings; MDPE blue; electrofusion joints						
Coupler						
20 mm dia.	12.18	15.41	0.36	14.16	nr	**29.57**
25 mm dia.	12.18	15.41	0.40	15.73	nr	**31.14**
32 mm dia.	12.18	15.41	0.44	17.30	nr	**32.71**
40 mm dia.	17.24	21.82	0.48	18.88	nr	**40.70**
50 mm dia.	17.96	22.73	0.52	20.45	nr	**43.18**
63 mm dia.	22.55	28.54	0.58	22.81	nr	**51.35**
90 mm dia.	33.15	41.95	0.67	26.35	nr	**68.30**
110 mm dia.	53.32	67.48	0.74	29.10	nr	**96.58**
125 mm dia.	60.16	76.14	0.83	32.64	nr	**108.78**
160 mm dia.	95.68	121.09	1.00	39.33	nr	**160.42**
180 mm dia.	112.50	142.38	1.25	49.16	nr	**191.54**
225 mm dia.	179.79	227.54	1.35	53.09	nr	**280.63**
250 mm dia.	263.26	333.18	1.50	58.99	nr	**392.17**
315 mm dia.	434.15	549.46	1.80	70.79	nr	**620.24**
Extra over fittings; MDPE blue; butt fused joints						
Cap						
25 mm dia.	23.09	29.22	0.20	7.87	nr	**37.09**
32 mm dia.	23.09	29.22	0.22	8.65	nr	**37.87**
40 mm dia.	24.46	30.95	0.24	9.44	nr	**40.39**
50 mm dia.	35.67	45.15	0.26	10.22	nr	**55.37**
63 mm dia.	40.92	51.79	0.32	12.58	nr	**64.37**
90 mm dia.	67.51	85.44	0.37	14.55	nr	**99.99**
110 mm dia.	104.46	132.20	0.40	15.73	nr	**147.93**
125 mm dia.	123.72	156.58	0.46	18.09	nr	**174.67**
160 mm dia.	168.66	213.45	0.50	19.66	nr	**233.12**
180 mm dia.	207.54	262.66	0.60	23.60	nr	**286.25**
225 mm dia.	245.15	310.26	0.68	26.74	nr	**337.00**

38 PIPED SUPPLY SYSTEMS

Item	Net Price £	Material £	Labour hours	Labour £	Unit	Total rate £
250 mm dia.	359.83	455.40	0.75	29.49	nr	**484.89**
315 mm dia.	463.99	587.22	0.90	35.39	nr	**622.61**
Reducer						
63 × 32 mm dia.	31.19	39.48	0.54	21.24	nr	**60.71**
63 × 50 mm dia.	36.48	46.17	0.60	23.60	nr	**69.77**
90 × 63 mm dia.	46.16	58.43	0.67	26.35	nr	**84.77**
110 × 90 mm dia.	63.62	80.51	0.74	29.10	nr	**109.61**
125 × 90 mm dia.	92.52	117.10	0.83	32.64	nr	**149.74**
125 × 110 mm dia.	101.95	129.03	1.00	39.33	nr	**168.35**
160 × 110 mm dia.	156.61	198.20	1.10	43.26	nr	**241.46**
180 × 125 mm dia.	170.11	215.28	1.25	49.16	nr	**264.44**
225 × 160 mm dia.	213.11	269.71	1.40	55.06	nr	**324.77**
250 × 180 mm dia.	245.16	310.28	1.80	70.79	nr	**381.06**
315 × 250 mm dia.	277.38	351.05	2.40	94.38	nr	**445.43**
Bend; 45 °						
50 mm dia.	47.61	60.25	0.50	19.66	nr	**79.91**
63 mm dia.	57.59	72.88	0.58	22.81	nr	**95.69**
90 mm dia.	89.19	112.87	0.67	26.35	nr	**139.22**
110 mm dia.	130.77	165.50	0.74	29.10	nr	**194.60**
125 mm dia.	146.14	184.96	0.83	32.64	nr	**217.60**
160 mm dia.	270.84	342.77	1.00	39.33	nr	**382.10**
180 mm dia.	308.31	390.20	1.25	49.16	nr	**439.36**
225 mm dia.	392.99	497.37	1.40	55.04	nr	**552.41**
250 mm dia.	425.26	538.21	1.80	70.79	nr	**609.00**
315 mm dia.	529.83	670.55	2.40	94.38	nr	**764.93**
Bend; 90 °						
50 mm dia.	47.61	60.25	0.50	19.66	nr	**79.91**
63 mm dia.	57.59	72.88	0.58	22.81	nr	**95.69**
90 mm dia.	89.19	112.87	0.67	26.35	nr	**139.22**
110 mm dia.	130.77	165.50	0.74	29.10	nr	**194.60**
125 mm dia.	146.14	184.96	0.83	32.64	nr	**217.60**
160 mm dia.	270.84	342.77	1.00	39.33	nr	**382.10**
180 mm dia.	526.05	665.77	1.25	49.16	nr	**714.93**
225 mm dia.	665.13	841.79	1.40	55.06	nr	**896.85**
250 mm dia.	727.55	920.78	1.80	70.79	nr	**991.57**
315 mm dia.	910.79	1152.69	2.40	94.38	nr	**1247.08**
Equal tee						
50 mm dia.	52.06	65.88	0.70	27.53	nr	**93.41**
63 mm dia.	56.81	71.89	0.75	29.49	nr	**101.39**
90 mm dia.	103.75	131.31	0.87	34.21	nr	**165.52**
110 mm dia.	154.73	195.82	1.00	39.33	nr	**235.15**
125 mm dia.	199.75	252.80	1.08	42.47	nr	**295.27**
160 mm dia.	328.44	415.67	1.35	53.09	nr	**468.76**
180 mm dia.	336.09	425.36	1.63	64.10	nr	**489.46**
225 mm dia.	407.31	515.50	1.90	74.72	nr	**590.22**
250 mm dia.	562.03	711.30	2.70	106.18	nr	**817.48**
315 mm dia.	1401.11	1773.25	3.60	141.57	nr	**1914.82**

38 PIPED SUPPLY SYSTEMS

Item	Net Price £	Material £	Labour hours	Labour £	Unit	Total rate £
COLD WATER PIPELINES: MEDIUM DENSITY POLYETHYLENE – BLUE PIPEWORK – cont						
Extra over plastic fittings, compression joints						
Straight connector						
20 mm dia.	6.00	7.59	0.38	14.95	nr	**22.54**
25 mm dia.	6.28	7.95	0.45	17.70	nr	**25.65**
32 mm dia.	14.93	18.89	0.50	19.66	nr	**38.55**
50 mm dia.	34.38	43.51	0.68	26.75	nr	**70.26**
63 mm dia.	51.76	65.51	0.85	33.44	nr	**98.95**
Reducing connector						
25 mm dia.	12.33	15.60	0.38	14.95	nr	**30.55**
32 mm dia.	19.90	25.19	0.45	17.70	nr	**42.88**
50 mm dia.	55.13	69.77	0.50	19.66	nr	**89.44**
63 mm dia.	76.94	97.38	0.62	24.39	nr	**121.77**
Straight connector; polyethylene to MI						
20 mm dia.	5.43	6.87	0.31	12.19	nr	**19.06**
25 mm dia.	9.25	11.70	0.35	13.76	nr	**25.47**
32 mm dia.	10.02	12.68	0.40	15.73	nr	**28.41**
50 mm dia.	25.56	32.35	0.55	21.63	nr	**53.99**
63 mm dia.	36.05	45.63	0.65	25.57	nr	**71.20**
Straight connector; polyethylene to FI						
20 mm dia.	7.30	9.24	0.31	12.19	nr	**21.44**
25 mm dia.	7.91	10.01	0.35	13.76	nr	**23.77**
32 mm dia.	9.45	11.96	0.40	15.73	nr	**27.69**
50 mm dia.	30.04	38.01	0.55	21.63	nr	**59.65**
63 mm dia.	42.06	53.24	0.75	29.50	nr	**82.74**
Elbow						
20 mm dia.	7.99	10.11	0.38	14.95	nr	**25.06**
25 mm dia.	11.81	14.95	0.45	17.70	nr	**32.64**
32 mm dia.	17.22	21.80	0.50	19.66	nr	**41.46**
50 mm dia.	39.94	50.54	0.68	26.75	nr	**77.30**
63 mm dia.	54.29	68.71	0.80	31.46	nr	**100.17**
Elbow; polyethylene to MI						
25 mm dia.	10.17	12.87	0.35	13.76	nr	**26.63**
Elbow; polyethylene to FI						
20 mm dia.	7.27	9.20	0.31	12.19	nr	**21.40**
25 mm dia.	9.88	12.51	0.35	13.76	nr	**26.27**
32 mm dia.	14.81	18.74	0.42	16.52	nr	**35.26**
50 mm dia.	35.16	44.50	0.50	19.66	nr	**64.16**
63 mm dia.	46.09	58.33	0.55	21.63	nr	**79.96**
Tank coupling						
25 mm dia.	15.26	19.31	0.42	16.52	nr	**35.83**
Equal tee						
20 mm dia.	10.75	13.61	0.53	20.85	nr	**34.46**
25 mm dia.	16.82	21.29	0.55	21.63	nr	**42.92**
32 mm dia.	21.07	26.67	0.64	25.18	nr	**51.85**
50 mm dia.	49.13	62.18	0.75	29.50	nr	**91.69**
63 mm dia.	76.11	96.32	0.87	34.23	nr	**130.54**

38 PIPED SUPPLY SYSTEMS

Item	Net Price £	Material £	Labour hours	Labour £	Unit	Total rate £
Equal tee; FI branch						
20 mm dia.	10.42	13.19	0.45	17.70	nr	**30.88**
25 mm dia.	16.60	21.01	0.50	19.66	nr	**40.67**
32 mm dia.	20.20	25.57	0.60	23.60	nr	**49.17**
50 mm dia.	46.55	58.92	0.68	26.75	nr	**85.67**
63 mm dia.	65.37	82.73	0.81	31.87	nr	**114.60**
Equal tee; MI branch						
25 mm dia.	16.32	20.65	0.50	19.66	nr	**40.31**

38 PIPED SUPPLY SYSTEMS

Item	Net Price £	Material £	Labour hours	Labour £	Unit	Total rate £
COLD WATER PIPELINES: ABS PIPEWORK						
Pipes; solvent welded joints in the running length, brackets measured separately						
Class C (9 bar pressure)						
1" dia.	10.09	12.77	0.30	11.80	m	**24.57**
1 ¼" dia.	16.97	21.48	0.33	12.98	m	**34.46**
1 ½" dia.	21.53	27.25	0.36	14.16	m	**41.41**
2" dia.	29.01	36.71	0.39	15.34	m	**52.05**
3" dia.	59.77	75.65	0.46	18.09	m	**93.74**
4" dia.	98.52	124.68	0.53	20.84	m	**145.52**
6" dia.	194.76	246.49	0.76	29.89	m	**276.38**
8" dia.	334.49	423.34	0.97	38.15	m	**461.48**
Class E (15 bar pressure)						
½" dia.	7.70	9.74	0.24	9.44	m	**19.18**
¾" dia.	11.85	15.00	0.27	10.62	m	**25.61**
1" dia.	15.61	19.76	0.30	11.80	m	**31.55**
1 ¼" dia.	23.30	29.49	0.33	12.98	m	**42.47**
1 ½" dia.	30.72	38.88	0.36	14.16	m	**53.03**
2" dia.	38.46	48.67	0.39	15.34	m	**64.01**
3" dia.	77.36	97.91	0.49	19.27	m	**117.17**
4" dia.	124.44	157.49	0.57	22.42	m	**179.90**
FIXINGS						
Refer to steel pipes; galvanized iron. For minimum fixing dimensions, refer to the Tables and Memoranda at the rear of the book						
Extra over fittings; solvent welded joints						
Cap						
½" dia.	2.98	3.77	0.16	6.29	nr	**10.07**
¾" dia.	3.45	4.37	0.19	7.47	nr	**11.84**
1" dia.	3.96	5.01	0.22	8.65	nr	**13.67**
1 ¼" dia.	6.62	8.37	0.25	9.83	nr	**18.21**
1 ½" dia.	10.20	12.90	0.28	11.01	nr	**23.91**
2" dia.	12.91	16.34	0.31	12.19	nr	**28.53**
3" dia.	38.79	49.10	0.36	14.16	nr	**63.25**
4" dia.	59.33	75.08	0.44	17.30	nr	**92.39**
Elbow 90°						
½" dia.	4.16	5.27	0.29	11.40	nr	**16.67**
¾" dia.	4.98	6.31	0.34	13.37	nr	**19.68**
1" dia.	6.97	8.82	0.40	15.73	nr	**24.55**
1 ¼" dia.	11.77	14.90	0.45	17.70	nr	**32.59**
1 ½" dia.	15.29	19.35	0.51	20.06	nr	**39.41**
2" dia.	23.29	29.47	0.56	22.02	nr	**51.50**
3" dia.	66.90	84.66	0.65	25.56	nr	**110.22**
4" dia.	99.91	126.45	0.80	31.46	nr	**157.91**
6" dia.	402.19	509.02	1.21	47.58	nr	**556.60**
8" dia.	613.80	776.82	1.45	57.02	nr	**833.84**

38 PIPED SUPPLY SYSTEMS

Item	Net Price £	Material £	Labour hours	Labour £	Unit	Total rate £
Elbow 45°						
½" dia.	8.07	10.21	0.29	11.40	nr	**21.61**
¾" dia.	8.16	10.33	0.34	13.37	nr	**23.70**
1" dia.	10.20	12.90	0.40	15.73	nr	**28.63**
1 ¼" dia.	14.94	18.91	0.45	17.70	nr	**36.60**
1 ½" dia.	18.49	23.40	0.51	20.06	nr	**43.45**
2" dia.	25.69	32.51	0.56	22.02	nr	**54.53**
3" dia.	60.49	76.56	0.65	25.56	nr	**102.12**
4" dia.	125.43	158.74	0.80	31.46	nr	**190.20**
6" dia.	259.98	329.03	1.21	47.58	nr	**376.62**
8" dia.	559.44	708.03	1.45	57.02	nr	**765.05**
Reducing bush						
¾" × ½" dia.	3.07	3.88	0.42	16.52	nr	**20.40**
1" × ½" dia.	3.96	5.01	0.45	17.70	nr	**22.71**
1" × ¾" dia.	3.96	5.01	0.45	17.70	nr	**22.71**
1 ¼" × 1" dia.	5.33	6.74	0.48	18.88	nr	**25.62**
1 ½" × ¾" dia.	6.97	8.82	0.51	20.06	nr	**28.88**
1 ½" × 1" dia.	6.97	8.82	0.51	20.06	nr	**28.88**
1 ½" × 1 ¼" dia.	6.97	8.82	0.51	20.06	nr	**28.88**
2" × 1" dia.	9.14	11.57	0.56	22.02	nr	**33.60**
2" × 1 ¼" dia.	9.14	11.57	0.56	22.02	nr	**33.60**
2" × 1 ½" dia.	9.14	11.57	0.56	22.02	nr	**33.60**
3" × 1 ½" dia.	25.69	32.51	0.65	25.56	nr	**58.07**
3" × 2" dia.	25.69	32.51	0.65	25.56	nr	**58.07**
4" × 3" dia.	35.44	44.86	0.80	31.46	nr	**76.32**
6" × 4" dia.	109.11	138.09	1.21	47.58	nr	**185.67**
Union						
½" dia.	16.56	20.95	0.34	13.37	nr	**34.33**
¾" dia.	17.86	22.61	0.39	15.34	nr	**37.94**
1" dia.	24.05	30.44	0.43	16.91	nr	**47.35**
1 ¼" dia.	29.48	37.31	0.50	19.66	nr	**56.97**
1 ½" dia.	40.61	51.40	0.57	22.42	nr	**73.81**
2" dia.	52.97	67.03	0.62	24.38	nr	**91.41**
Sockets						
½" dia.	3.07	3.88	0.34	13.37	nr	**17.25**
¾" dia.	3.45	4.37	0.39	15.34	nr	**19.70**
1" dia.	3.96	5.01	0.43	16.91	nr	**21.92**
1 ¼" dia.	6.97	8.82	0.50	19.66	nr	**28.49**
1 ½" dia.	8.41	10.64	0.57	22.42	nr	**33.05**
2" dia.	11.77	14.90	0.62	24.38	nr	**39.28**
3" dia.	47.38	59.97	0.70	27.53	nr	**87.50**
4" dia.	67.24	85.09	0.70	27.53	nr	**112.62**
6" dia.	168.03	212.65	1.26	49.55	nr	**262.20**
8" dia.	335.58	424.71	1.55	60.95	nr	**485.67**
Barrel nipple						
½" dia.	5.79	7.33	0.34	13.37	nr	**20.70**
¾" dia.	7.53	9.52	0.39	15.34	nr	**24.86**
1" dia.	9.76	12.35	0.43	16.91	nr	**29.26**
1 ¼" dia.	13.53	17.13	0.50	19.66	nr	**36.79**
1 ½" dia.	15.96	20.20	0.57	22.42	nr	**42.62**
2" dia.	19.31	24.44	0.62	24.38	nr	**48.82**
3" dia.	51.35	64.98	0.70	27.53	nr	**92.51**

38 PIPED SUPPLY SYSTEMS

Item	Net Price £	Material £	Labour hours	Labour £	Unit	Total rate £
COLD WATER PIPELINES: ABS PIPEWORK – cont						
Extra over fittings – cont						
Tee, 90°						
½" dia.	4.74	6.00	0.41	16.12	nr	**22.13**
¾" dia.	6.62	8.37	0.47	18.48	nr	**26.86**
1" dia.	9.14	11.57	0.55	21.63	nr	**33.20**
1 ¼" dia.	13.15	16.64	0.64	25.17	nr	**41.81**
1 ½" dia.	19.31	24.44	0.71	27.92	nr	**52.36**
2" dia.	29.48	37.31	0.78	30.67	nr	**67.98**
3" dia.	85.99	108.83	0.91	35.79	nr	**144.62**
4" dia.	126.24	159.76	1.12	44.04	nr	**203.81**
6" dia.	441.20	558.38	1.69	66.46	nr	**624.84**
8" dia.	687.93	870.65	2.03	79.83	nr	**950.48**
Full face flange						
½" dia.	51.39	65.04	0.10	3.93	nr	**68.97**
¾" dia.	52.60	66.57	0.13	5.11	nr	**71.68**
1" dia.	56.97	72.10	0.15	5.90	nr	**78.00**
1 ¼" dia.	63.33	80.15	0.18	7.08	nr	**87.23**
1 ½" dia.	76.15	96.38	0.21	8.26	nr	**104.64**
2" dia.	103.08	130.45	0.29	11.40	nr	**141.86**
3" dia.	176.74	223.69	0.37	14.55	nr	**238.24**
4" dia.	231.64	293.16	0.41	16.12	nr	**309.29**

38 PIPED SUPPLY SYSTEMS

Item	Net Price £	Material £	Labour hours	Labour £	Unit	Total rate £
COLD WATER PIPELINES: PVC-u PIPEWORK						
Pipes; solvent welded joints in the running length,						
brackets measured separately						
Class C (9 bar pressure)						
2" dia.	23.92	30.27	0.41	16.12	m	**46.39**
3" dia.	45.84	58.02	0.47	18.48	m	**76.50**
4" dia.	81.32	102.92	0.50	19.66	m	**122.59**
6" dia.	175.98	222.73	1.76	69.21	m	**291.94**
Class D (12 bar pressure)						
1 ¼" dia.	13.93	17.63	0.41	16.12	m	**33.76**
1 ½" dia.	19.15	24.24	0.42	16.52	m	**40.76**
2" dia.	29.70	37.59	0.45	17.70	m	**55.28**
3" dia.	63.65	80.55	0.48	18.88	m	**99.43**
4" dia.	106.51	134.80	0.53	20.84	m	**155.64**
6" dia.	197.68	250.18	0.58	22.81	m	**272.99**
Class E (15 bar pressure)						
½" dia.	6.82	8.63	0.38	14.94	m	**23.57**
¾" dia.	9.75	12.34	0.40	15.73	m	**28.07**
1" dia.	11.33	14.34	0.41	16.12	m	**30.46**
1 ¼" dia.	16.63	21.04	0.41	16.12	m	**37.17**
1 ½" dia.	21.64	27.39	0.42	16.52	m	**43.91**
2" dia.	33.82	42.80	0.45	17.70	m	**60.50**
3" dia.	73.28	92.75	0.47	18.48	m	**111.23**
4" dia.	120.32	152.27	0.50	19.66	m	**171.94**
6" dia.	260.62	329.83	0.53	20.84	m	**350.68**
Class 7						
½" dia.	12.05	15.25	0.32	12.58	m	**27.84**
¾" dia.	16.93	21.42	0.33	12.98	m	**34.40**
1" dia.	25.80	32.65	0.40	15.73	m	**48.38**
1 ¼" dia.	35.44	44.85	0.40	15.73	m	**60.58**
1 ½" dia.	43.87	55.52	0.41	16.12	m	**71.64**
2" dia.	72.92	92.29	0.43	16.91	m	**109.20**
FIXINGS						
Refer to steel pipes; galvanized iron. For minimum fixing						
dimensions, refer to the Tables and Memoranda at the						
rear of the book						
Extra over fittings; solvent welded joints						
End cap						
½" dia.	2.10	2.66	0.17	6.69	nr	**9.35**
¾" dia.	2.46	3.11	0.19	7.47	nr	**10.58**
1" dia.	2.78	3.52	0.22	8.65	nr	**12.17**
1 ¼" dia.	4.35	5.50	0.25	9.83	nr	**15.33**
1 ½" dia.	7.26	9.18	0.28	11.01	nr	**20.19**
2" dia.	8.90	11.27	0.31	12.19	nr	**23.46**
3" dia.	27.25	34.49	0.36	14.16	nr	**48.64**
4" dia.	42.07	53.25	0.44	17.30	nr	**70.55**
6" dia.	101.70	128.71	0.67	26.35	nr	**155.06**

38 PIPED SUPPLY SYSTEMS

Item	Net Price £	Material £	Labour hours	Labour £	Unit	Total rate £
COLD WATER PIPELINES: PVC-u PIPEWORK – cont						
Extra over fittings – cont						
Socket						
½" dia.	2.23	2.82	0.31	12.19	nr	**15.01**
¾" dia.	2.46	3.11	0.35	13.76	nr	**16.87**
1" dia.	2.87	3.63	0.42	16.52	nr	**20.15**
1 ¼" dia.	5.17	6.54	0.45	17.70	nr	**24.24**
1 ½" dia.	6.06	7.67	0.51	20.06	nr	**27.73**
2" dia.	8.62	10.91	0.56	22.02	nr	**32.93**
3" dia.	32.94	41.69	0.65	25.56	nr	**67.25**
4" dia.	47.75	60.44	0.80	31.46	nr	**91.90**
6" dia.	119.79	151.60	1.21	47.58	nr	**199.19**
Reducing socket						
¾" × ½" dia.	2.60	3.29	0.31	12.19	nr	**15.48**
1" × ¾" dia.	3.28	4.15	0.35	13.76	nr	**17.92**
1 ¼" × 1" dia.	6.25	7.91	0.42	16.52	nr	**24.42**
1 ½" × 1¼" dia.	6.97	8.82	0.45	17.70	nr	**26.52**
2" × 1 ½" dia.	10.52	13.32	0.51	20.06	nr	**33.37**
3" × 2" dia.	32.05	40.56	0.56	22.02	nr	**62.58**
4" × 3" dia.	47.41	60.01	0.65	25.56	nr	**85.57**
6" × 4" dia.	172.87	218.78	0.80	31.46	nr	**250.24**
8" × 6" dia.	267.72	338.83	1.21	47.58	nr	**386.41**
Elbow, 90°						
½" dia.	2.94	3.72	0.31	12.19	nr	**15.91**
¾" dia.	3.54	4.47	0.35	13.76	nr	**18.24**
1" dia.	4.94	6.25	0.42	16.52	nr	**22.77**
1 ¼" dia.	8.62	10.91	0.45	17.70	nr	**28.61**
1 ½" dia.	11.09	14.04	0.45	17.70	nr	**31.73**
2" dia.	16.46	20.83	0.56	22.02	nr	**42.85**
3" dia.	47.41	60.01	0.65	25.56	nr	**85.57**
4" dia.	71.44	90.41	0.80	31.46	nr	**121.87**
6" dia.	282.84	357.97	1.21	47.58	nr	**405.55**
Elbow 45°						
½" dia.	5.62	7.12	0.31	12.19	nr	**19.31**
¾" dia.	5.98	7.57	0.35	13.76	nr	**21.33**
1" dia.	7.26	9.18	0.45	17.70	nr	**26.88**
1 ¼" dia.	10.39	13.15	0.45	17.70	nr	**30.85**
1 ½" dia.	13.05	16.52	0.51	20.06	nr	**36.57**
2" dia.	18.39	23.27	0.56	22.02	nr	**45.30**
3" dia.	43.28	54.78	0.65	25.56	nr	**80.34**
4" dia.	88.93	112.55	0.80	31.46	nr	**144.01**
6" dia.	183.49	232.22	1.21	47.58	nr	**279.81**
Bend 90° (long radius)						
3" dia.	132.23	167.35	0.65	25.56	nr	**192.91**
4" dia.	267.11	338.06	0.80	31.46	nr	**369.52**
6" dia.	586.99	742.89	1.21	47.58	nr	**790.48**
Bend 45° (long radius)						
1 ½" dia.	31.40	39.73	0.51	20.06	nr	**59.79**
2" dia.	51.30	64.93	0.56	22.02	nr	**86.95**
3" dia.	109.66	138.79	0.65	25.56	nr	**164.35**
4" dia.	213.44	270.12	0.80	31.46	nr	**301.59**

38 PIPED SUPPLY SYSTEMS

Item	Net Price £	Material £	Labour hours	Labour £	Unit	Total rate £
Socket union						
½" dia.	11.40	14.43	0.34	13.37	nr	**27.80**
¾" dia.	13.05	16.52	0.39	15.34	nr	**31.85**
1" dia.	16.94	21.44	0.45	17.70	nr	**39.14**
1 ¼" dia.	21.07	26.67	0.50	19.66	nr	**46.33**
1 ½" dia.	28.85	36.52	0.57	22.42	nr	**58.93**
2" dia.	37.35	47.26	0.62	24.38	nr	**71.65**
3" dia.	139.04	175.97	0.70	27.53	nr	**203.50**
4" dia.	188.26	238.26	0.89	35.00	nr	**273.26**
Saddle plain						
2" × 1 ¼" dia.	29.34	37.13	0.42	16.52	nr	**53.65**
3" × 1 ½" dia.	41.24	52.19	0.48	18.88	nr	**71.06**
4" × 2" dia.	46.43	58.77	0.68	26.74	nr	**85.51**
6" × 2" dia.	54.54	69.03	0.91	35.79	nr	**104.81**
Straight tank connector						
½" dia.	7.58	9.60	0.13	5.11	nr	**14.71**
¾" dia.	8.58	10.85	0.14	5.51	nr	**16.36**
1" dia.	18.20	23.04	0.14	5.51	nr	**28.54**
1 ¼" dia.	46.26	58.55	0.16	6.29	nr	**64.84**
1 ½" dia.	50.72	64.19	0.18	7.08	nr	**71.27**
2" dia.	60.75	76.88	0.24	9.44	nr	**86.32**
3" dia.	62.29	78.84	0.29	11.40	nr	**90.24**
Equal tee						
½" dia.	3.42	4.33	0.44	17.30	nr	**21.63**
¾" dia.	4.35	5.50	0.48	18.88	nr	**24.38**
1" dia.	6.50	8.23	0.54	21.24	nr	**29.47**
1 ¼" dia.	9.22	11.66	0.70	27.53	nr	**39.19**
1 ½" dia.	13.28	16.80	0.74	29.10	nr	**45.90**
2" dia.	21.07	26.67	0.80	31.46	nr	**58.13**
3" dia.	61.07	77.29	1.04	40.90	nr	**118.19**
4" dia.	89.53	113.31	1.28	50.34	nr	**163.65**
6" dia.	311.88	394.72	1.93	75.90	nr	**470.62**

38 PIPED SUPPLY SYSTEMS

Item	Net Price £	Material £	Labour hours	Labour £	Unit	Total rate £
COLD WATER PIPELINES: PVC-C PIPEWORK						
Pipes; solvent welded in the running length, brackets measured separately						
Pipe; 3 m long; PN25						
16 × 2.0 mm	7.32	9.26	0.20	7.87	m	**17.13**
20 × 2.3 mm	11.05	13.99	0.20	7.87	m	**21.86**
25 × 2.8 mm	14.34	18.14	0.20	7.87	m	**26.01**
32 × 3.6 mm	21.29	26.95	0.20	7.87	m	**34.81**
Pipe; 5 m long; PN25						
40 × 4.5 mm	25.96	32.86	0.20	7.87	m	**40.73**
50 × 5.6 mm	39.08	49.45	0.20	7.87	m	**57.32**
63 × 7.0 mm	60.34	76.36	0.20	7.87	m	**84.23**
FIXINGS						
Refer to steel pipes; galvanized iron. For minimum fixing dimensions, refer to the Tables and Memoranda at the rear of the book						
Extra over fittings; solvent welded joints						
Straight coupling; PN25						
16 mm	1.13	1.44	0.20	7.87	nr	**9.30**
20 mm	1.60	2.03	0.20	7.87	nr	**9.89**
25 mm	2.00	2.53	0.20	7.87	nr	**10.39**
32 mm	6.08	7.69	0.20	7.87	nr	**15.56**
40 mm	7.80	9.88	0.20	7.87	nr	**17.74**
50 mm	10.45	13.22	0.20	7.87	nr	**21.09**
63 mm	18.46	23.37	0.20	7.87	nr	**31.23**
Elbow; 90°; PN25						
16 mm	1.83	2.32	0.20	7.87	nr	**10.18**
20 mm	2.80	3.54	0.20	7.87	nr	**11.41**
25 mm	3.49	4.42	0.20	7.87	nr	**12.29**
32 mm	7.27	9.21	0.20	7.87	nr	**17.07**
40 mm	11.21	14.18	0.20	7.87	nr	**22.05**
50 mm	15.50	19.62	0.20	7.87	nr	**27.48**
63 mm	26.51	33.55	0.20	7.87	nr	**41.42**
Elbow; 45°; PN25						
20 mm	2.80	3.54	0.20	7.87	nr	**11.41**
25 mm	3.49	4.42	0.20	7.87	nr	**12.29**
32 mm	7.27	9.21	0.20	7.87	nr	**17.07**
40 mm	11.21	14.18	0.20	7.87	nr	**22.05**
50 mm	15.50	19.62	0.20	7.87	nr	**27.48**
63 mm	26.51	33.55	0.20	7.87	nr	**41.42**
Reducer fitting; single stage reduction						
20/16 mm	2.00	2.53	0.20	7.87	nr	**10.39**
25/20 mm	2.39	3.02	0.20	7.87	nr	**10.89**
32/25 mm	4.79	6.07	0.20	7.87	nr	**13.93**
40/32 mm	6.28	7.94	0.20	7.87	nr	**15.81**
50/40 mm	7.27	9.21	0.20	7.87	nr	**17.07**
63/50 mm	11.01	13.93	0.20	7.87	nr	**21.80**

38 PIPED SUPPLY SYSTEMS

Item	Net Price £	Material £	Labour hours	Labour £	Unit	Total rate £
Equal tee; 90°; PN25						
16 mm	3.04	3.85	0.20	7.87	nr	**11.71**
20 mm	4.19	5.30	0.20	7.87	nr	**13.17**
25 mm	5.31	6.72	0.20	7.87	nr	**14.58**
32 mm	8.63	10.93	0.20	7.87	nr	**18.79**
40 mm	14.93	18.89	0.20	7.87	nr	**26.76**
50 mm	22.35	28.29	0.20	7.87	nr	**36.15**
63 mm	37.67	47.67	0.20	7.87	nr	**55.54**
Cap; PN25						
20 mm	2.09	2.64	0.20	7.87	nr	**10.51**
25 mm	2.80	3.54	0.20	7.87	nr	**11.41**
32 mm	4.07	5.15	0.20	7.87	nr	**13.01**
40 mm	5.58	7.06	0.20	7.87	nr	**14.93**
50 mm	7.80	9.88	0.20	7.87	nr	**17.74**
63 mm	12.43	15.73	0.20	7.87	nr	**23.60**

38 PIPED SUPPLY SYSTEMS

Item	Net Price £	Material £	Labour hours	Labour £	Unit	Total rate £
COLD WATER PIPELINES: SCREWED STEEL PIPEWORK						
Galvanized steel pipes; screwed and socketed joints; BS 1387: 1985						
Galvanized; medium, fixed vertically, with brackets measured separately, screwed joints are within the running length, but any flanges are additional						
10 mm dia.	7.76	9.82	0.51	20.06	m	**29.87**
15 mm dia.	7.83	9.90	0.52	20.45	m	**30.35**
20 mm dia.	7.89	9.99	0.55	21.63	m	**31.62**
25 mm dia.	11.03	13.96	0.60	23.60	m	**37.56**
32 mm dia.	13.66	17.29	0.67	26.35	m	**43.64**
40 mm dia.	15.87	20.09	0.75	29.49	m	**49.58**
50 mm dia.	25.97	32.87	0.85	33.43	m	**66.30**
65 mm dia.	35.20	44.54	0.93	36.57	m	**81.12**
80 mm dia.	45.58	57.68	1.07	42.08	m	**99.76**
100 mm dia.	64.50	81.63	1.46	57.42	m	**139.05**
125 mm dia.	102.55	129.79	1.72	67.64	m	**197.43**
150 mm dia.	119.09	150.72	1.96	77.08	m	**227.80**
Galvanized; heavy, fixed vertically, with brackets measured separately, screwed joints are within the running length, but any flanges are additional						
15 mm dia.	8.30	10.50	0.52	20.45	m	**30.95**
20 mm dia.	9.36	11.85	0.55	21.63	m	**33.48**
25 mm dia.	13.38	16.93	0.60	23.60	m	**40.53**
32 mm dia.	16.59	21.00	0.67	26.35	m	**47.35**
40 mm dia.	19.38	24.52	0.75	29.49	m	**54.02**
50 mm dia.	31.34	39.66	0.85	33.43	m	**73.09**
65 mm dia.	42.56	53.87	0.93	36.57	m	**90.44**
80 mm dia.	54.05	68.40	1.07	42.08	m	**110.48**
100 mm dia.	75.35	95.37	1.46	57.42	m	**152.78**
125 mm dia.	109.18	138.18	1.72	67.64	m	**205.82**
150 mm dia.	127.70	161.62	1.96	77.08	m	**238.70**
Galvanized; medium, fixed horizontaly or suspended at high level, with brackets measured separately, screwed joints are within the running length, but any flanges are additional						
10 mm dia.	7.76	9.82	0.51	20.06	m	**29.87**
15 mm dia.	7.83	9.90	0.52	20.45	m	**30.35**
20 mm dia.	7.89	9.99	0.55	21.63	m	**31.62**
25 mm dia.	11.03	13.96	0.60	23.60	m	**37.56**
32 mm dia.	13.66	17.29	0.67	26.35	m	**43.64**
40 mm dia.	15.87	20.09	0.75	29.49	m	**49.58**
50 mm dia.	25.97	32.87	0.85	33.43	m	**66.30**
65 mm dia.	35.20	44.54	0.93	36.57	m	**81.12**
80 mm dia.	45.58	57.68	1.07	42.08	m	**99.76**
100 mm dia.	64.50	81.63	1.46	57.42	m	**139.05**
125 mm dia.	102.55	129.79	1.72	67.64	m	**197.43**
150 mm dia.	119.09	150.72	1.96	77.08	m	**227.80**

38 PIPED SUPPLY SYSTEMS

Item	Net Price £	Material £	Labour hours	Labour £	Unit	Total rate £
Galvanized; heavy, fixed horizontaly or suspended at high level, with brackets measured separately, screwed joints are within the running length, but any flanges are additional						
15 mm dia.	8.30	10.50	0.52	20.45	m	**30.95**
20 mm dia.	9.36	11.85	0.55	21.63	m	**33.48**
25 mm dia.	13.38	16.93	0.60	23.60	m	**40.53**
32 mm dia.	16.59	21.00	0.67	26.35	m	**47.35**
40 mm dia.	19.38	24.52	0.75	29.49	m	**54.02**
50 mm dia.	31.34	39.66	0.85	33.43	m	**73.09**
65 mm dia.	42.56	53.87	0.93	36.57	m	**90.44**
80 mm dia.	54.05	68.40	1.07	42.08	m	**110.48**
100 mm dia.	75.35	95.37	1.46	57.42	m	**152.78**
125 mm dia.	109.18	138.18	1.72	67.64	m	**205.82**
150 mm dia.	127.70	161.62	1.96	77.08	m	**238.70**

FIXINGS

For steel pipes; galvanized iron. For minimum fixing dimensions, refer to the Tables and Memoranda at the rear of the book

Item	Net Price £	Material £	Labour hours	Labour £	Unit	Total rate £
Single pipe bracket, screw on, galvanized iron; screwed to wood						
15 mm dia.	2.15	2.72	0.14	5.51	nr	**8.23**
20 mm dia.	2.39	3.02	0.14	5.51	nr	**8.53**
25 mm dia.	2.79	3.53	0.17	6.69	nr	**10.21**
32 mm dia.	3.81	4.82	0.19	7.47	nr	**12.29**
40 mm dia.	5.65	7.16	0.22	8.65	nr	**15.81**
50 mm dia.	7.50	9.50	0.22	8.65	nr	**18.15**
65 mm dia.	8.86	11.21	0.28	11.01	nr	**22.22**
80 mm dia.	13.87	17.56	0.32	12.58	nr	**30.14**
100 mm dia.	20.05	25.38	0.35	13.76	nr	**39.14**
Single pipe bracket, screw on, galvanized iron; plugged and screwed						
15 mm dia.	2.15	2.72	0.25	9.83	nr	**12.55**
20 mm dia.	2.39	3.02	0.25	9.83	nr	**12.86**
25 mm dia.	2.79	3.53	0.30	11.80	nr	**15.33**
32 mm dia.	3.81	4.82	0.32	12.58	nr	**17.40**
40 mm dia.	5.65	7.16	0.32	12.58	nr	**19.74**
50 mm dia.	7.50	9.50	0.32	12.58	nr	**22.08**
65 mm dia.	8.86	11.21	0.35	13.76	nr	**24.97**
80 mm dia.	13.87	17.56	0.42	16.52	nr	**34.08**
100 mm dia.	20.05	25.38	0.42	16.52	nr	**41.90**
Single pipe bracket for building in, galvanized iron						
15 mm dia.	2.28	2.88	0.10	3.93	nr	**6.82**
20 mm dia.	2.56	3.25	0.11	4.33	nr	**7.57**
25 mm dia.	2.79	3.53	0.12	4.72	nr	**8.25**
32 mm dia.	2.95	3.73	0.14	5.51	nr	**9.24**
40 mm dia.	3.81	4.82	0.15	5.90	nr	**10.72**
50 mm dia.	4.59	5.81	0.16	6.29	nr	**12.10**

38 PIPED SUPPLY SYSTEMS

Item	Net Price £	Material £	Labour hours	Labour £	Unit	Total rate £
COLD WATER PIPELINES: SCREWED STEEL PIPEWORK – cont						
For steel pipes – cont						
Pipe ring, single socket, galvanized iron						
15 mm dia.	2.28	2.88	0.10	3.93	nr	6.82
20 mm dia.	2.56	3.25	0.11	4.33	nr	7.57
25 mm dia.	2.79	3.53	0.12	4.72	nr	8.25
32 mm dia.	2.80	3.55	0.15	5.90	nr	9.45
40 mm dia.	3.68	4.66	0.15	5.90	nr	10.56
50 mm dia.	4.81	6.09	0.16	6.29	nr	12.38
65 mm dia.	6.69	8.47	0.30	11.80	nr	20.26
80 mm dia.	7.87	9.96	0.35	13.76	nr	23.72
100 mm dia.	12.65	16.01	0.40	15.73	nr	31.74
125 mm dia.	25.58	32.38	0.60	23.60	nr	55.98
150 mm dia.	30.89	39.09	0.77	30.28	nr	69.37
Pipe ring, double socket, galvanized iron						
15 mm dia.	20.13	25.48	0.10	3.93	nr	29.41
20 mm dia.	23.16	29.31	0.11	4.33	nr	33.64
25 mm dia.	25.53	32.32	0.12	4.72	nr	37.03
32 mm dia.	28.15	35.62	0.14	5.51	nr	41.13
40 mm dia.	35.55	45.00	0.15	5.90	nr	50.89
50 mm dia.	40.46	51.20	0.16	6.29	nr	57.50
Screw on backplate (Male), galvanized iron; plugged and screwed						
All sizes 15 mm to 50 mm × M12	1.80	2.28	0.10	3.93	nr	6.21
Screw on backplate (Female), galvanized iron; plugged and screwed						
All sizes 15 mm to 50 mm × M12	1.80	2.28	0.10	3.93	nr	6.21
Extra over channel sections for fabricated hangers and brackets						
Galvanized steel; including inserts, bolts, nuts, washers; fixed to backgrounds						
41 × 21 mm	9.61	12.16	0.29	11.40	m	23.57
41 × 41 mm	11.52	14.58	0.29	11.40	m	25.99
Threaded rods; metric thread; including nuts, washers etc.						
10 mm dia. × 600 mm long	3.10	3.92	0.18	7.08	nr	11.00
12 mm dia. × 600 mm long	4.85	6.14	0.18	7.08	nr	13.22
Extra over steel flanges, screwed and drilled; metric; BS 4504						
Screwed flanges; PN6						
15 mm dia.	26.10	33.03	0.35	13.76	nr	46.79
20 mm dia.	26.10	33.03	0.47	18.48	nr	51.51
25 mm dia.	26.10	33.03	0.53	20.84	nr	53.87
32 mm dia.	26.10	33.03	0.62	24.38	nr	57.41
40 mm dia.	26.10	33.03	0.70	27.53	nr	60.55
50 mm dia.	27.83	35.22	0.84	33.03	nr	68.25
65 mm dia.	38.69	48.97	1.03	40.51	nr	89.48

38 PIPED SUPPLY SYSTEMS

Item	Net Price £	Material £	Labour hours	Labour £	Unit	Total rate £
80 mm dia.	54.65	69.17	1.23	48.37	nr	**117.54**
100 mm dia.	64.62	81.79	1.41	55.45	nr	**137.24**
125 mm dia.	118.53	150.01	1.77	69.61	nr	**219.62**
150 mm dia.	118.53	150.01	2.21	86.91	nr	**236.92**
Screwed flanges; PN16						
15 mm dia.	33.19	42.01	0.35	13.76	nr	**55.77**
20 mm dia.	33.19	42.01	0.47	18.48	nr	**60.49**
25 mm dia.	33.19	42.01	0.53	20.84	nr	**62.85**
32 mm dia.	35.44	44.86	0.62	24.38	nr	**69.24**
40 mm dia.	35.44	44.86	0.70	27.53	nr	**72.38**
50 mm dia.	43.24	54.72	0.84	33.03	nr	**87.76**
65 mm dia.	53.98	68.32	1.03	40.51	nr	**108.83**
80 mm dia.	65.75	83.21	1.23	48.37	nr	**131.58**
100 mm dia.	73.74	93.33	1.41	55.45	nr	**148.78**
125 mm dia.	127.00	160.73	1.77	69.61	nr	**230.33**
150 mm dia.	129.08	163.36	2.21	86.91	nr	**250.27**
Extra over steel flanges, screwed and drilled; imperial; BS 10						
Screwed flanges; Table E						
½" dia.	44.58	56.42	0.35	13.76	nr	**70.18**
¾" dia.	44.58	56.42	0.47	18.48	nr	**74.90**
1" dia.	44.58	56.42	0.53	20.84	nr	**77.26**
1 ¼" dia.	44.58	56.42	0.62	24.38	nr	**80.80**
1 ½" dia.	44.58	56.42	0.70	27.53	nr	**83.94**
2" dia.	44.58	56.42	0.84	33.03	nr	**89.45**
2 ½" dia.	53.07	67.17	1.03	40.51	nr	**107.67**
3" dia.	63.94	80.92	1.23	48.37	nr	**129.29**
4" dia.	81.34	102.95	1.41	55.45	nr	**158.39**
5" dia.	172.41	218.20	1.77	69.61	nr	**287.81**
Extra over steel flange connections						
Bolted connection between pair of flanges; including gasket, bolts, nuts and washers						
50 mm dia.	85.52	108.24	0.53	20.84	nr	**129.08**
65 mm dia.	107.75	136.37	0.53	20.84	nr	**157.22**
80 mm dia.	123.62	156.45	0.53	20.84	nr	**177.29**
100 mm dia.	147.52	186.70	0.53	20.84	nr	**207.55**
125 mm dia.	278.64	352.64	0.61	23.99	nr	**376.63**
150 mm dia.	286.59	362.71	0.90	35.39	nr	**398.11**
Extra over heavy steel tubular fittings; BS 1387						
Long screw connection with socket and backnut						
15 mm dia.	9.11	11.53	0.63	24.78	nr	**36.30**
20 mm dia.	11.44	14.47	0.84	33.03	nr	**47.51**
25 mm dia.	14.95	18.92	0.95	37.36	nr	**56.28**
32 mm dia.	19.68	24.91	1.11	43.65	nr	**68.56**
40 mm dia.	23.96	30.33	1.28	50.34	nr	**80.66**
50 mm dia.	35.36	44.75	1.53	60.17	nr	**104.92**
65 mm dia.	81.23	102.80	1.87	73.54	nr	**176.34**
80 mm dia.	105.04	132.94	2.21	86.91	nr	**219.85**
100 mm dia.	119.17	150.82	3.05	119.94	nr	**270.76**

38 PIPED SUPPLY SYSTEMS

Item	Net Price £	Material £	Labour hours	Labour £	Unit	Total rate £
COLD WATER PIPELINES: SCREWED STEEL PIPEWORK – cont						
Extra over heavy steel tubular fittings – cont						
Long screw connection with socket and backnut – cont						
Running nipple						
15 mm dia.	2.29	2.89	0.50	19.66	nr	22.56
20 mm dia.	2.84	3.59	0.68	26.74	nr	30.33
25 mm dia.	3.05	3.87	0.77	30.28	nr	34.15
32 mm dia.	4.92	6.23	0.90	35.39	nr	41.63
40 mm dia.	6.65	8.41	1.03	40.51	nr	48.92
50 mm dia.	10.10	12.79	1.23	48.37	nr	61.16
65 mm dia.	21.72	27.48	1.50	58.99	nr	86.47
80 mm dia.	33.87	42.86	1.78	70.00	nr	112.86
100 mm dia.	53.08	67.18	2.38	93.60	nr	160.77
Barrel nipple						
15 mm dia.	1.91	2.42	0.50	19.66	nr	22.08
20 mm dia.	2.88	3.64	0.68	26.74	nr	30.39
25 mm dia.	3.23	4.09	0.77	30.28	nr	34.37
32 mm dia.	5.35	6.78	0.90	35.39	nr	42.17
40 mm dia.	5.96	7.54	1.03	40.51	nr	48.05
50 mm dia.	8.52	10.78	1.23	48.37	nr	59.15
65 mm dia.	18.20	23.04	1.50	58.99	nr	82.03
80 mm dia.	25.42	32.17	1.78	70.00	nr	102.17
100 mm dia.	45.97	58.19	2.38	93.60	nr	151.78
125 mm dia.	85.40	108.08	2.87	112.87	nr	220.94
150 mm dia.	134.53	170.26	3.39	133.31	nr	303.58
Close taper nipple						
15 mm dia.	2.70	3.42	0.50	19.66	nr	23.09
20 mm dia.	3.51	4.44	0.68	26.74	nr	31.19
25 mm dia.	4.60	5.82	0.77	30.28	nr	36.10
32 mm dia.	6.86	8.68	0.90	35.39	nr	44.08
40 mm dia.	8.49	10.74	1.03	40.51	nr	51.25
50 mm dia.	13.05	16.52	1.23	48.37	nr	64.89
65 mm dia.	20.61	26.09	1.50	58.99	nr	85.08
80 mm dia.	33.78	42.76	1.78	70.00	nr	112.76
100 mm dia.	64.23	81.29	2.38	93.60	nr	174.89
90° bend with socket						
15 mm dia.	7.45	9.43	0.64	25.17	nr	34.60
20 mm dia.	10.04	12.70	0.85	33.43	nr	46.13
25 mm dia.	15.35	19.43	0.97	38.15	nr	57.58
32 mm dia.	22.03	27.87	1.12	44.04	nr	71.92
40 mm dia.	26.90	34.04	1.29	50.73	nr	84.77
50 mm dia.	41.83	52.94	1.55	60.95	nr	113.90
65 mm dia.	84.33	106.73	1.89	74.33	nr	181.06
80 mm dia.	125.17	158.41	2.24	88.09	nr	246.50
100 mm dia.	221.95	280.90	3.09	121.52	nr	402.41
125 mm dia.	543.65	688.04	3.92	154.16	nr	842.20
150 mm dia.	816.20	1032.98	4.74	186.40	nr	1219.38

38 PIPED SUPPLY SYSTEMS

Item	Net Price £	Material £	Labour hours	Labour £	Unit	Total rate £
Extra over heavy steel fittings; BS 1740						
Plug						
15 mm dia.	2.06	2.61	0.28	11.01	nr	**13.62**
20 mm dia.	3.22	4.07	0.38	14.94	nr	**19.01**
25 mm dia.	5.65	7.15	0.44	17.30	nr	**24.46**
32 mm dia.	8.77	11.10	0.51	20.06	nr	**31.16**
40 mm dia.	9.67	12.24	0.59	23.20	nr	**35.45**
50 mm dia.	13.83	17.51	0.70	27.53	nr	**45.03**
65 mm dia.	33.07	41.86	0.85	33.43	nr	**75.28**
80 mm dia.	61.88	78.31	1.00	39.33	nr	**117.64**
100 mm dia.	118.79	150.34	1.44	56.63	nr	**206.97**
Socket						
15 mm dia.	2.29	2.89	0.64	25.17	nr	**28.06**
20 mm dia.	2.60	3.29	0.85	33.43	nr	**36.71**
25 mm dia.	3.67	4.65	0.97	38.15	nr	**42.79**
32 mm dia.	5.29	6.69	1.12	44.04	nr	**50.74**
40 mm dia.	6.42	8.12	1.29	50.73	nr	**58.85**
50 mm dia.	9.90	12.53	1.55	60.95	nr	**73.49**
65 mm dia.	19.67	24.90	1.89	74.33	nr	**99.22**
80 mm dia.	25.44	32.20	2.24	88.09	nr	**120.29**
100 mm dia.	47.90	60.62	3.09	121.52	nr	**182.14**
150 mm dia.	114.32	144.69	4.74	186.40	nr	**331.09**
Elbow, female/female						
15 mm dia.	13.15	16.64	0.64	25.17	nr	**41.81**
20 mm dia.	17.14	21.69	0.85	33.43	nr	**55.12**
25 mm dia.	23.26	29.44	0.97	38.15	nr	**67.59**
32 mm dia.	43.31	54.81	1.12	44.04	nr	**98.86**
40 mm dia.	51.65	65.37	1.29	50.73	nr	**116.10**
50 mm dia.	84.67	107.16	1.55	60.95	nr	**168.11**
65 mm dia.	206.88	261.82	1.89	74.33	nr	**336.15**
80 mm dia.	246.68	312.19	2.24	88.09	nr	**400.28**
100 mm dia.	427.20	540.66	3.09	121.52	nr	**662.18**
Equal tee						
15 mm dia.	16.35	20.69	0.91	35.79	nr	**56.48**
20 mm dia.	18.98	24.03	1.22	47.98	nr	**72.00**
25 mm dia.	27.99	35.42	1.40	55.06	nr	**90.47**
32 mm dia.	57.80	73.15	1.62	63.71	nr	**136.86**
40 mm dia.	62.90	79.61	1.86	73.15	nr	**152.75**
50 mm dia.	102.28	129.45	2.21	86.91	nr	**216.36**
65 mm dia.	247.36	313.06	2.72	106.97	nr	**420.03**
80 mm dia.	265.49	336.00	3.21	126.24	nr	**462.23**
100 mm dia.	427.26	540.75	4.44	174.61	nr	**715.35**
Extra over malleable iron fittings; BS 143						
Cap						
15 mm dia.	6.37	8.06	0.32	12.58	nr	**20.64**
20 mm dia.	6.66	8.43	0.43	16.91	nr	**25.34**
25 mm dia.	12.05	15.26	0.49	19.27	nr	**34.53**
32 mm dia.	22.88	28.95	0.58	22.81	nr	**51.76**
40 mm dia.	24.55	31.07	0.66	25.96	nr	**57.02**
50 mm dia.	39.02	49.39	0.78	30.67	nr	**80.06**
65 mm dia.	67.07	84.88	0.96	37.75	nr	**122.63**

38 PIPED SUPPLY SYSTEMS

Item	Net Price £	Material £	Labour hours	Labour £	Unit	Total rate £
COLD WATER PIPELINES: SCREWED STEEL PIPEWORK – cont						
Extra over malleable iron fittings – cont						
Cap – cont						
80 mm dia.	110.64	140.02	1.13	44.44	nr	**184.46**
100 mm dia.	188.60	238.70	1.70	66.85	nr	**305.55**
Plain plug, hollow						
15 mm dia.	3.10	3.93	0.28	11.01	nr	**14.94**
20 mm dia.	4.81	6.09	0.38	14.94	nr	**21.03**
25 mm dia.	8.43	10.67	0.44	17.30	nr	**27.97**
32 mm dia.	13.06	16.53	0.51	20.06	nr	**36.59**
40 mm dia.	14.43	18.26	0.59	23.20	nr	**41.46**
50 mm dia.	20.67	26.15	0.70	27.53	nr	**53.68**
65 mm dia.	49.34	62.45	0.85	33.43	nr	**95.87**
80 mm dia.	92.36	116.89	1.00	39.33	nr	**156.22**
100 mm dia.	177.33	224.43	1.44	56.63	nr	**281.06**
Plain plug, solid						
15 mm dia.	3.85	4.87	0.29	11.40	nr	**16.28**
20 mm dia.	4.08	5.16	0.38	14.94	nr	**20.10**
25 mm dia.	5.43	6.87	0.44	17.30	nr	**24.17**
32 mm dia.	8.37	10.59	0.51	20.06	nr	**30.65**
40 mm dia.	11.26	14.25	0.59	23.20	nr	**37.45**
50 mm dia.	14.80	18.73	0.70	27.53	nr	**46.26**
Elbow, male/female						
15 mm dia.	2.00	2.53	0.64	25.17	nr	**27.70**
20 mm dia.	2.70	3.41	0.85	33.43	nr	**36.84**
25 mm dia.	4.47	5.65	0.97	38.15	nr	**43.80**
32 mm dia.	10.12	12.81	1.12	44.04	nr	**56.85**
40 mm dia.	13.99	17.70	1.29	50.73	nr	**68.43**
50 mm dia.	18.03	22.82	1.55	60.95	nr	**83.78**
65 mm dia.	40.19	50.87	1.89	74.33	nr	**125.19**
80 mm dia.	54.98	69.58	2.24	88.09	nr	**157.67**
100 mm dia.	96.10	121.62	3.09	121.52	nr	**243.14**
Elbow						
15 mm dia.	1.79	2.26	0.64	25.17	nr	**27.43**
20 mm dia.	2.45	3.10	0.85	33.43	nr	**36.53**
25 mm dia.	3.82	4.83	0.97	38.15	nr	**42.98**
32 mm dia.	7.94	10.05	1.12	44.04	nr	**54.10**
40 mm dia.	11.88	15.03	1.29	50.73	nr	**65.76**
50 mm dia.	13.92	17.62	1.55	60.95	nr	**78.58**
65 mm dia.	31.05	39.29	1.89	74.33	nr	**113.62**
80 mm dia.	45.59	57.70	2.24	88.09	nr	**145.79**
100 mm dia.	78.26	99.04	3.09	121.52	nr	**220.56**
125 mm dia.	187.82	237.71	4.44	174.61	nr	**412.32**
150 mm dia.	349.69	442.56	5.79	227.70	nr	**670.26**
45° elbow						
15 mm dia.	4.61	5.84	0.64	25.17	nr	**31.01**
20 mm dia.	5.72	7.24	0.85	33.43	nr	**40.66**
25 mm dia.	7.86	9.95	0.97	38.15	nr	**48.10**

38 PIPED SUPPLY SYSTEMS

Item	Net Price £	Material £	Labour hours	Labour £	Unit	Total rate £
32 mm dia.	18.29	23.15	1.12	44.04	nr	**67.20**
40 mm dia.	21.53	27.24	1.29	50.73	nr	**77.97**
50 mm dia.	29.49	37.32	1.55	60.95	nr	**98.27**
65 mm dia.	41.44	52.45	1.89	74.33	nr	**126.78**
80 mm dia.	62.37	78.94	2.24	88.09	nr	**167.03**
100 mm dia.	120.22	152.15	3.09	121.52	nr	**273.67**
150 mm dia.	365.87	463.04	5.79	227.70	nr	**690.74**
Bend, male/female						
15 mm dia.	3.53	4.46	0.64	25.17	nr	**29.63**
20 mm dia.	5.80	7.34	0.85	33.43	nr	**40.77**
25 mm dia.	8.17	10.34	0.97	38.15	nr	**48.49**
32 mm dia.	13.81	17.48	1.12	44.04	nr	**61.52**
40 mm dia.	20.24	25.62	1.29	50.73	nr	**76.35**
50 mm dia.	38.05	48.15	1.55	60.95	nr	**109.11**
65 mm dia.	58.23	73.69	1.89	74.33	nr	**148.02**
80 mm dia.	89.68	113.50	2.24	88.09	nr	**201.59**
100 mm dia.	222.07	281.05	3.09	121.52	nr	**402.57**
Bend, male						
15 mm dia.	8.11	10.26	0.64	25.17	nr	**35.43**
20 mm dia.	9.10	11.51	0.85	33.43	nr	**44.94**
25 mm dia.	13.35	16.90	0.97	38.15	nr	**55.05**
32 mm dia.	29.54	37.38	1.12	44.04	nr	**81.43**
40 mm dia.	41.43	52.43	1.29	50.73	nr	**103.16**
50 mm dia.	62.86	79.55	1.55	60.95	nr	**140.51**
Bend, female						
15 mm dia.	3.64	4.61	0.64	25.17	nr	**29.77**
20 mm dia.	5.18	6.56	0.85	33.43	nr	**39.99**
25 mm dia.	7.28	9.21	0.97	38.15	nr	**47.36**
32 mm dia.	14.18	17.95	1.12	44.04	nr	**62.00**
40 mm dia.	16.88	21.36	1.29	50.73	nr	**72.09**
50 mm dia.	26.60	33.66	1.55	60.95	nr	**94.61**
65 mm dia.	58.23	73.69	1.89	74.33	nr	**148.02**
80 mm dia.	86.35	109.28	2.24	88.09	nr	**197.37**
100 mm dia.	181.18	229.30	3.09	121.52	nr	**350.82**
125 mm dia.	365.53	462.61	4.44	174.61	nr	**637.22**
150 mm dia.	803.04	1016.33	5.79	227.70	nr	**1244.03**
Return bend						
15 mm dia.	18.54	23.46	0.64	25.17	nr	**48.63**
20 mm dia.	29.97	37.94	0.85	33.43	nr	**71.36**
25 mm dia.	37.40	47.33	0.97	38.15	nr	**85.48**
32 mm dia.	51.92	65.71	1.12	44.04	nr	**109.76**
40 mm dia.	70.33	89.01	1.29	50.73	nr	**139.74**
50 mm dia.	107.39	135.91	1.55	60.95	nr	**196.87**
Equal socket, parallel thread						
15 mm dia.	1.74	2.20	0.64	25.17	nr	**27.37**
20 mm dia.	2.27	2.88	0.85	33.43	nr	**36.31**
25 mm dia.	2.88	3.64	0.97	38.15	nr	**41.79**
32 mm dia.	6.40	8.10	1.12	44.04	nr	**52.15**
40 mm dia.	8.69	11.00	1.29	50.73	nr	**61.73**
50 mm dia.	12.57	15.91	1.55	60.95	nr	**76.87**
65 mm dia.	19.95	25.25	1.89	74.33	nr	**99.58**

38 PIPED SUPPLY SYSTEMS

Item	Net Price £	Material £	Labour hours	Labour £	Unit	Total rate £
COLD WATER PIPELINES: SCREWED STEEL PIPEWORK – cont						
Extra over malleable iron fittings – cont						
Equal socket, parallel thread – cont						
80 mm dia.	28.25	35.76	2.24	88.09	nr	**123.85**
100 mm dia.	46.64	59.03	3.09	121.52	nr	**180.55**
Concentric reducing socket						
20 × 15 mm dia.	2.78	3.52	0.76	29.89	nr	**33.40**
25 × 15 mm dia.	3.64	4.61	0.86	33.82	nr	**38.43**
25 × 20 mm dia.	4.50	5.70	0.86	33.82	nr	**39.52**
32 × 25 mm dia.	6.58	8.33	1.01	39.72	nr	**48.05**
40 × 25 mm dia.	8.69	11.00	1.16	45.62	nr	**56.62**
40 × 32 mm dia.	9.63	12.19	1.16	45.62	nr	**57.81**
50 × 25 mm dia.	16.67	21.10	1.38	54.27	nr	**75.37**
50 × 40 mm dia.	19.24	24.35	1.38	54.27	nr	**78.62**
65 × 50 mm dia.	23.54	29.79	1.69	66.46	nr	**96.25**
80 × 50 mm dia.	29.37	37.18	2.00	78.65	nr	**115.83**
100 × 50 mm dia.	58.54	74.08	2.75	108.15	nr	**182.23**
100 × 80 mm dia.	175.12	221.63	2.75	108.15	nr	**329.78**
150 × 100 mm dia.	162.72	205.94	4.10	161.24	nr	**367.18**
Eccentric reducing socket						
20 × 15 mm dia.	6.30	7.98	0.76	29.89	nr	**37.87**
25 × 15 mm dia.	17.87	22.62	0.86	33.82	nr	**56.44**
25 × 20 mm dia.	20.29	25.68	0.86	33.82	nr	**59.50**
32 × 25 mm dia.	26.24	33.21	1.01	39.72	nr	**72.93**
40 × 25 mm dia.	30.06	38.04	1.16	45.62	nr	**83.66**
40 × 32 mm dia.	32.74	41.43	1.16	45.62	nr	**87.05**
50 × 25 mm dia.	32.82	41.53	1.18	46.40	nr	**87.94**
50 × 40 mm dia.	32.90	41.64	1.28	50.34	nr	**91.97**
65 × 50 mm dia.	33.34	42.19	1.69	66.46	nr	**108.65**
80 × 50 mm dia.	35.51	44.95	2.00	78.65	nr	**123.60**
Hexagon bush						
20 × 15 mm dia.	1.56	1.97	0.37	14.55	nr	**16.52**
25 × 15 mm dia.	2.14	2.71	0.43	16.91	nr	**19.62**
25 × 20 mm dia.	2.31	2.93	0.43	16.91	nr	**19.84**
32 × 25 mm dia.	2.71	3.43	0.51	20.06	nr	**23.49**
40 × 25 mm dia.	4.05	5.12	0.58	22.81	nr	**27.93**
40 × 32 mm dia.	4.05	5.12	0.58	22.81	nr	**27.93**
50 × 25 mm dia.	8.56	10.84	0.71	27.92	nr	**38.76**
50 × 40 mm dia.	10.88	13.77	0.71	27.92	nr	**41.69**
65 × 50 mm dia.	14.74	18.65	0.84	33.03	nr	**51.68**
80 × 50 mm dia.	22.24	28.15	1.00	39.33	nr	**67.47**
100 × 50 mm dia.	51.52	65.20	1.52	59.78	nr	**124.98**
100 × 80 mm dia.	42.86	54.24	1.52	59.78	nr	**114.02**
150 × 100 mm dia.	153.98	194.88	2.48	97.53	nr	**292.41**
Hexagon nipple						
15 mm dia.	1.71	2.16	0.28	11.01	nr	**13.17**
20 mm dia.	1.90	2.41	0.38	14.94	nr	**17.35**
25 mm dia.	2.70	3.41	0.44	17.30	nr	**20.72**

38 PIPED SUPPLY SYSTEMS

Item	Net Price £	Material £	Labour hours	Labour £	Unit	Total rate £
32 mm dia.	5.74	7.26	0.51	20.06	nr	**27.31**
40 mm dia.	6.58	8.33	0.59	23.20	nr	**31.53**
50 mm dia.	12.01	15.19	0.70	27.53	nr	**42.72**
65 mm dia.	20.10	25.43	0.85	33.43	nr	**58.86**
80 mm dia.	27.78	35.16	1.00	39.33	nr	**74.49**
100 mm dia.	49.31	62.40	1.44	56.63	nr	**119.03**
150 mm dia.	136.50	172.76	2.32	91.24	nr	**263.99**
Union, male/female						
15 mm dia.	8.43	10.67	0.64	25.17	nr	**35.84**
20 mm dia.	10.35	13.10	0.85	33.43	nr	**46.52**
25 mm dia.	12.01	15.19	0.97	38.15	nr	**53.34**
32 mm dia.	21.22	26.85	1.12	44.04	nr	**70.90**
40 mm dia.	27.15	34.36	1.29	50.73	nr	**85.09**
50 mm dia.	42.68	54.02	1.55	60.95	nr	**114.97**
65 mm dia.	95.51	120.88	1.89	74.33	nr	**195.21**
Union, female						
15 mm dia.	7.98	10.10	0.64	25.17	nr	**35.26**
20 mm dia.	8.77	11.10	0.85	33.43	nr	**44.53**
25 mm dia.	10.28	13.02	0.97	38.15	nr	**51.16**
32 mm dia.	17.74	22.45	1.12	44.04	nr	**66.50**
40 mm dia.	20.03	25.35	1.29	50.73	nr	**76.08**
50 mm dia.	29.86	37.79	1.55	60.95	nr	**98.75**
65 mm dia.	76.47	96.78	1.89	74.33	nr	**171.11**
80 mm dia.	101.09	127.93	2.24	88.09	nr	**216.02**
100 mm dia.	192.49	243.61	3.09	121.52	nr	**365.13**
Union elbow, male/female						
15 mm dia.	12.72	16.10	0.64	25.17	nr	**41.27**
20 mm dia.	15.95	20.19	0.85	33.43	nr	**53.62**
25 mm dia.	22.40	28.35	0.97	38.15	nr	**66.50**
Twin elbow						
15 mm dia.	12.23	15.48	0.91	35.79	nr	**51.27**
20 mm dia.	13.53	17.13	1.22	47.98	nr	**65.11**
25 mm dia.	21.92	27.74	1.39	54.66	nr	**82.40**
32 mm dia.	43.91	55.58	1.62	63.71	nr	**119.29**
40 mm dia.	55.60	70.36	1.86	73.15	nr	**143.51**
50 mm dia.	71.45	90.43	2.21	86.91	nr	**177.34**
65 mm dia.	115.48	146.15	2.72	106.97	nr	**253.12**
80 mm dia.	131.06	165.87	3.21	126.24	nr	**292.11**
Equal tee						
15 mm dia.	2.45	3.10	0.91	35.79	nr	**38.89**
20 mm dia.	3.57	4.52	1.22	47.98	nr	**52.50**
25 mm dia.	5.15	6.52	1.39	54.66	nr	**61.18**
32 mm dia.	10.90	13.80	1.62	63.71	nr	**77.50**
40 mm dia.	14.88	18.83	1.86	73.15	nr	**91.98**
50 mm dia.	21.41	27.10	2.21	86.91	nr	**114.01**
65 mm dia.	50.23	63.58	2.72	106.97	nr	**170.54**
80 mm dia.	58.54	74.08	3.21	126.24	nr	**200.32**
100 mm dia.	106.14	134.33	4.44	174.61	nr	**308.93**
125 mm dia.	196.89	249.18	5.38	211.57	nr	**460.76**
150 mm dia.	470.84	595.89	6.31	248.15	nr	**844.04**

38 PIPED SUPPLY SYSTEMS

Item	Net Price £	Material £	Labour hours	Labour £	Unit	Total rate £
COLD WATER PIPELINES: SCREWED STEEL PIPEWORK – cont						
Extra over malleable iron fittings – cont						
Tee reducing on branch						
20 × 15 mm dia.	3.67	4.65	1.22	47.98	nr	**52.62**
25 × 15 mm dia.	5.02	6.35	1.39	54.66	nr	**61.02**
25 × 20 mm dia.	5.72	7.24	1.39	54.66	nr	**61.90**
32 × 25 mm dia.	11.03	13.96	1.62	63.71	nr	**77.67**
40 × 25 mm dia.	13.99	17.70	1.86	73.15	nr	**90.85**
40 × 32 mm dia.	20.54	25.99	1.86	73.15	nr	**99.14**
50 × 25 mm dia.	18.59	23.52	2.21	86.91	nr	**110.43**
50 × 40 mm dia.	28.90	36.58	2.21	86.91	nr	**123.49**
65 × 50 mm dia.	44.60	56.44	2.72	106.97	nr	**163.41**
80 × 50 mm dia.	60.29	76.30	3.21	126.24	nr	**202.54**
100 × 50 mm dia.	99.80	126.31	4.44	174.61	nr	**300.92**
100 × 80 mm dia.	153.98	194.88	4.44	174.61	nr	**369.49**
150 × 100 mm dia.	249.77	316.11	6.28	246.97	nr	**563.08**
Equal pitcher tee						
15 mm dia.	8.53	10.79	0.91	35.79	nr	**46.58**
20 mm dia.	10.51	13.30	1.22	47.98	nr	**61.28**
25 mm dia.	15.73	19.90	1.39	54.66	nr	**74.57**
32 mm dia.	24.39	30.86	1.62	63.71	nr	**94.57**
40 mm dia.	37.76	47.78	1.86	73.15	nr	**120.93**
50 mm dia.	52.96	67.03	2.21	86.91	nr	**153.94**
65 mm dia.	75.38	95.41	2.72	106.97	nr	**202.37**
80 mm dia.	117.62	148.86	3.21	126.24	nr	**275.10**
100 mm dia.	264.65	334.95	4.44	174.61	nr	**509.55**
Cross						
15 mm dia.	7.07	8.94	1.00	39.33	nr	**48.27**
20 mm dia.	11.08	14.02	1.33	52.30	nr	**66.33**
25 mm dia.	14.02	17.74	1.51	59.38	nr	**77.13**
32 mm dia.	20.86	26.40	1.77	69.61	nr	**96.01**
40 mm dia.	28.09	35.55	2.02	79.44	nr	**114.99**
50 mm dia.	43.67	55.27	2.42	95.17	nr	**150.44**
65 mm dia.	62.37	78.94	2.97	116.80	nr	**195.73**
80 mm dia.	82.91	104.93	3.50	137.64	nr	**242.57**
100 mm dia.	171.11	216.55	4.84	190.34	nr	**406.89**

38 PIPED SUPPLY SYSTEMS

Item	Net Price £	Material £	Labour hours	Labour £	Unit	Total rate £
COLD WATER PIPELINES: PIPEWORK ANCILLARIES – VALVES						
Regulators						
Gunmetal; self-acting two port thermostat; single seat; screwed; normally closed; with adjustable or fixed bleed device						
25 mm dia.	1249.70	1581.62	1.46	57.42	nr	**1639.03**
32 mm dia.	1285.75	1627.25	1.45	57.02	nr	**1684.27**
40 mm dia.	1374.31	1739.33	1.55	60.97	nr	**1800.30**
50 mm dia.	1654.40	2093.81	1.68	66.07	nr	**2159.88**
Self-acting temperature regulator for storage calorifier; integral sensing element and pocket; screwed ends						
15 mm dia.	1055.82	1336.24	1.32	51.91	nr	**1388.15**
25 mm dia.	1158.93	1466.74	1.52	59.78	nr	**1526.52**
32 mm dia.	1496.58	1894.07	1.79	70.39	nr	**1964.46**
40 mm dia.	1830.78	2317.04	1.99	78.26	nr	**2395.30**
50 mm dia.	2140.06	2708.46	2.26	88.88	nr	**2797.34**
Self-acting temperature regulator for storage calorifier; integral sensing element and pocket; flanged ends; bolted connection						
15 mm dia.	1549.93	1961.59	0.61	23.99	nr	**1985.58**
25 mm dia.	1773.89	2245.04	0.72	28.31	nr	**2273.35**
32 mm dia.	2236.06	2829.95	0.94	36.97	nr	**2866.92**
40 mm dia.	2648.43	3351.86	1.03	40.51	nr	**3392.36**
50 mm dia.	3075.03	3891.75	1.18	46.40	nr	**3938.16**
Chrome plated thermostatic mixing valves including non-return valves and inlet swivel connections with strainers; copper compression fittings						
15 mm dia.	171.08	216.51	0.69	27.13	nr	**243.65**
Chrome plated thermostatic mixing valves including non-return valves and inlet swivel connections with angle pattern combined isolating valves and strainers; copper compression fittings						
15 mm dia.	280.74	355.31	0.69	27.13	nr	**382.44**
Gunmetal thermostatic mixing valves including non-return valves and inlet swivel connections with strainers; copper compression fittings						
15 mm dia.	355.05	449.35	0.69	27.13	nr	**476.49**
Gunmetal thermostatic mixing valves including non-return valves and inlet swivel connections with angle pattern combined isolating valves and strainers; copper compression fittings						
15 mm dia.	373.28	472.43	0.69	27.13	nr	**499.56**
Ball float valves						
Bronze, equilibrium; copper float; working pressure cold services up to 16 bar; flanged ends; BS 4504 Table 16/21; bolted connections						
25 mm dia.	254.46	322.04	1.04	40.90	nr	**362.94**
32 mm dia.	353.00	446.75	1.22	47.98	nr	**494.73**
40 mm dia.	478.79	605.96	1.38	54.27	nr	**660.23**

38 PIPED SUPPLY SYSTEMS

Item	Net Price £	Material £	Labour hours	Labour £	Unit	Total rate £
COLD WATER PIPELINES: PIPEWORK ANCILLARIES – VALVES – cont						
Ball float valves – cont						
Bronze, equilibrium – cont						
50 mm dia.	752.35	952.17	1.66	65.28	nr	**1017.45**
65 mm dia.	806.80	1021.08	1.93	75.90	nr	**1096.98**
80 mm dia.	899.62	1138.56	2.16	84.94	nr	**1223.50**
Heavy, equilibrium; with long tail and backnut; copper float; screwed for iron						
25 mm dia.	183.61	232.37	1.58	62.13	nr	**294.51**
32 mm dia.	252.07	319.02	1.78	70.00	nr	**389.02**
40 mm dia.	355.37	449.76	1.90	74.72	nr	**524.48**
50 mm dia.	466.67	590.62	2.65	104.21	nr	**694.84**
Brass, ball valve; BS 1212; copper float; screwed						
15 mm dia.	11.91	15.07	0.25	9.83	nr	**24.90**
22 mm dia.	21.64	27.39	0.29	11.40	nr	**38.79**
28 mm dia.	94.27	119.31	0.35	13.76	nr	**133.07**
Gate valves						
DZR copper alloy wedge non-rising stem; capillary joint to copper						
15 mm dia.	25.31	32.03	0.84	33.03	nr	**65.07**
22 mm dia.	31.06	39.31	1.01	39.72	nr	**79.03**
28 mm dia.	42.00	53.15	1.19	46.80	nr	**99.95**
35 mm dia.	75.39	95.41	1.38	54.27	nr	**149.68**
42 mm dia.	128.08	162.09	1.62	63.71	nr	**225.80**
54 mm dia.	179.46	227.12	1.94	76.29	nr	**303.41**
Cocks; capillary joints to copper						
Stopcock; brass head with gunmetal body						
15 mm dia.	11.69	14.80	0.45	17.70	nr	**32.50**
22 mm dia.	21.81	27.60	0.46	18.09	nr	**45.69**
28 mm dia.	62.01	78.48	0.54	21.24	nr	**99.72**
Lockshield stopcocks; brass head with gunmetal body						
15 mm dia.	13.49	17.07	0.45	17.70	nr	**34.77**
22 mm dia.	13.49	17.07	0.46	18.09	nr	**35.16**
28 mm dia.	13.49	17.07	0.54	21.24	nr	**38.31**
DZR stopcock; brass head with gunmetal body						
15 mm dia.	32.00	40.50	0.45	17.70	nr	**58.20**
22 mm dia.	55.42	70.14	0.46	18.09	nr	**88.23**
28 mm dia.	92.37	116.91	0.54	21.24	nr	**138.15**
Gunmetal stopcock						
35 mm dia.	120.73	152.79	0.69	27.13	nr	**179.93**
42 mm dia.	160.34	202.93	0.71	27.92	nr	**230.85**
54 mm dia.	239.49	303.10	0.81	31.85	nr	**334.95**
Double union stopcock						
15 mm dia.	34.40	43.54	0.60	23.60	nr	**67.13**
22 mm dia.	48.37	61.22	0.60	23.60	nr	**84.81**
28 mm dia.	86.05	108.90	0.69	27.13	nr	**136.04**

38 PIPED SUPPLY SYSTEMS

Item	Net Price £	Material £	Labour hours	Labour £	Unit	Total rate £
Double union DZR stopcock						
15 mm dia.	56.15	71.06	0.60	23.60	nr	**94.66**
22 mm dia.	69.03	87.37	0.61	23.99	nr	**111.36**
28 mm dia.	127.58	161.47	0.69	27.13	nr	**188.60**
Double union gunmetal stopcock						
35 mm dia.	212.56	269.02	0.63	24.78	nr	**293.79**
42 mm dia.	291.73	369.21	0.67	26.35	nr	**395.56**
54 mm dia.	458.94	580.83	0.85	33.43	nr	**614.26**
Double union stopcock with easy clean cover						
15 mm dia.	59.60	75.43	0.60	23.60	nr	**99.02**
22 mm dia.	74.37	94.12	0.61	23.99	nr	**118.11**
28 mm dia.	138.22	174.93	0.69	27.13	nr	**202.07**
Combined stopcock and drain						
15 mm dia.	59.12	74.82	0.67	26.35	nr	**101.17**
22 mm dia.	72.13	91.28	0.68	26.74	nr	**118.03**
Combined DZR stopcock and drain						
15 mm dia.	78.13	98.88	0.67	26.35	nr	**125.23**
Gate valve						
DZR copper alloy wedge non-rising stem; compression joint to copper						
15 mm dia.	25.31	32.03	0.84	33.03	nr	**65.07**
22 mm dia.	31.06	39.31	1.01	39.72	nr	**79.03**
28 mm dia.	42.00	53.15	1.19	46.80	nr	**99.95**
35 mm dia.	75.39	95.41	1.38	54.27	nr	**149.68**
42 mm dia.	128.08	162.09	1.62	63.71	nr	**225.80**
54 mm dia.	179.46	227.12	1.94	76.29	nr	**303.41**
Cocks; compression joints to copper						
Stopcock; brass head gunmetal body						
15 mm dia.	10.63	13.45	0.42	16.52	nr	**29.97**
22 mm dia.	19.81	25.08	0.42	16.52	nr	**41.59**
28 mm dia.	56.37	71.34	0.45	17.70	nr	**89.03**
Lockshield stopcock; brass head gunmetal body						
15 mm dia.	11.25	14.23	0.42	16.52	nr	**30.75**
22 mm dia.	16.10	20.38	0.42	16.52	nr	**36.90**
28 mm dia.	28.65	36.26	0.45	17.70	nr	**53.95**
DZR stopcock						
15 mm dia.	26.67	33.75	0.38	14.94	nr	**48.69**
22 mm dia.	46.20	58.47	0.39	15.34	nr	**73.81**
28 mm dia.	76.95	97.39	0.40	15.73	nr	**113.12**
35 mm dia.	120.73	152.79	0.52	20.45	nr	**173.24**
42 mm dia.	160.34	202.93	0.54	21.24	nr	**224.16**
54 mm dia.	239.49	303.10	0.63	24.78	nr	**327.88**
DZR lockshield stopcock						
15 mm dia.	11.25	14.23	0.38	14.94	nr	**29.18**
22 mm dia.	17.74	22.45	0.39	15.34	nr	**37.79**
Combined stop/draincock						
15 mm dia.	59.12	74.82	0.22	8.65	nr	**83.47**
22 mm dia.	76.07	96.28	0.45	17.70	nr	**113.97**

38 PIPED SUPPLY SYSTEMS

Item	Net Price £	Material £	Labour hours	Labour £	Unit	Total rate £
COLD WATER PIPELINES: PIPEWORK ANCILLARIES – VALVES – cont						
Cocks – cont						
DZR combined stop/draincock						
15 mm dia.	78.13	98.88	0.41	16.12	nr	**115.00**
22 mm dia.	101.24	128.13	0.42	16.52	nr	**144.64**
Stopcock to polyethylene						
15 mm dia.	40.85	51.70	0.38	14.94	nr	**66.64**
20 mm dia.	51.45	65.11	0.39	15.34	nr	**80.45**
25 mm dia.	68.43	86.61	0.40	15.73	nr	**102.34**
Draw off coupling						
15 mm dia.	24.81	31.40	0.38	14.94	nr	**46.34**
DZR draw off coupling						
15 mm dia.	24.81	31.40	0.38	14.94	nr	**46.34**
22 mm dia.	28.71	36.33	0.39	15.34	nr	**51.67**
Draw off elbow						
15 mm dia.	27.22	34.45	0.38	14.94	nr	**49.40**
22 mm dia.	33.53	42.44	0.39	15.34	nr	**57.78**
Lockshield drain cock						
15 mm dia.	28.48	36.04	0.41	16.12	nr	**52.16**
Check valves						
DZR copper alloy and bronze, WRC approved cartridge double check valve; BS 6282; working pressure cold services up to 10 bar at 65°C; screwed ends						
32 mm dia.	158.61	200.73	1.38	54.27	nr	**255.00**
40 mm dia.	180.77	228.78	1.62	63.71	nr	**292.48**
50 mm dia.	296.38	375.09	1.94	76.29	nr	**451.39**

38 PIPED SUPPLY SYSTEMS

Item	Net Price £	Material £	Labour hours	Labour £	Unit	Total rate £
COLD WATER PIPELINES: PIPEWORK ANCILLARIES – PUMPS						
Packaged cold water pressure booster set; fully automatic; 3 phase supply; includes fixing in position; electrical work elsewhere						
Pressure booster set						
0.75 l/s @ 30 m head	7140.80	9037.39	9.38	368.88	nr	**9406.27**
1.5 l/s @ 30 m head	8479.72	10731.93	9.38	368.88	nr	**11100.80**
3 l/s @ 30 m head	10360.53	13112.29	10.38	408.20	nr	**13520.49**
6 l/s @ 30 m head	23398.90	29613.65	10.38	408.20	nr	**30021.85**
12 l/s @ 30 m head	29232.65	36996.84	12.38	486.85	nr	**37483.69**
0.75 l/s @ 50 m head	8129.06	10288.14	9.38	368.88	nr	**10657.02**
1.5 l/s @ 50 m head	10360.53	13112.29	9.38	368.88	nr	**13481.16**
3 l/s @ 50 m head	11444.42	14484.06	10.38	408.20	nr	**14892.27**
6 l/s @ 50 m head	26134.07	33075.28	10.38	408.20	nr	**33483.48**
12 l/s @ 50 m head	32101.72	40627.93	12.38	486.85	nr	**41114.78**
0.75 l/s @ 70 m head	8894.13	11256.41	9.38	368.88	nr	**11625.28**
1.5 l/s @ 70 m head	11438.07	14476.02	9.38	368.88	nr	**14844.90**
3 l/s @ 70 m head	12177.62	15412.00	10.38	408.20	nr	**15820.20**
6 l/s @ 70 m head	28531.34	36109.26	10.38	408.20	nr	**36517.47**
12 l/s @ 70 m head	34683.85	43895.89	12.38	486.85	nr	**44382.74**
Automatic sump pump for clear and drainage water; single stage centrifugal pump, presure tight electric motor; single phase supply; includes fixing in position; electrical work elsewhere						
Single pump						
1 l/s @ 2.68 m total head	389.93	493.49	3.50	137.64	nr	**631.13**
1 l/s @ 4.68 m total head	429.22	543.22	3.50	137.64	nr	**680.86**
1 l/s @ 6.68 m total head	566.65	717.15	3.50	137.64	nr	**854.79**
2 l/s @ 4.38 m total head	566.65	717.15	4.00	157.30	nr	**874.45**
2 l/s @ 6.38 m total head	566.65	717.15	4.00	157.30	nr	**874.45**
2 l/s @ 8.38 m total head	729.34	923.06	4.00	157.30	nr	**1080.36**
3 l/s @ 3.7 m total head	566.65	717.15	4.50	176.97	nr	**894.11**
3 l/s @ 5.7 m total head	729.34	923.06	4.50	176.97	nr	**1100.02**
4 l/s @ 2.9 m total head	566.65	717.15	5.00	196.63	nr	**913.78**
4 l/s @ 4.9 m total head	729.34	923.06	5.00	196.63	nr	**1119.69**
4 l/s @ 6.9 m total head	2022.55	2559.74	5.00	196.63	nr	**2756.37**
Extra for high level alarm box with single float switch, local alarm and volt free contacts for remote alarm.	593.49	751.12		–	nr	**751.12**
Duty/standby pump unit						
1 l/s @ 2.68 m total head	740.57	937.26	5.00	196.63	nr	**1133.89**
1 l/s @ 4.68 m total head	813.54	1029.61	5.00	196.63	nr	**1226.24**
1 l/s @ 6.68 m total head	1099.64	1391.71	5.00	196.63	nr	**1588.34**
2 l/s @ 4.38 m total head	1099.64	1391.71	5.50	216.29	nr	**1608.00**
2 l/s @ 6.38 m total head	1099.64	1391.71	5.50	216.29	nr	**1608.00**
2 l/s @ 8.38 m total head	1408.24	1782.26	5.50	216.29	nr	**1998.55**
3 l/s @ 3.7 m total head	1099.64	1391.71	6.00	235.95	nr	**1627.66**

38 PIPED SUPPLY SYSTEMS

Item	Net Price £	Material £	Labour hours	Labour £	Unit	Total rate £
COLD WATER PIPELINES: PIPEWORK ANCILLARIES – PUMPS – cont						
Automatic sump pump for clear and drainage water – cont						
3 l/s @ 5.7 m total head	1408.24	1782.26	6.00	235.95	nr	**2018.22**
4 l/s @ 2.9 m total head	1099.64	1391.71	6.50	255.62	nr	**1647.32**
4 l/s @ 4.9 m total head	1408.24	1782.26	6.50	255.62	nr	**2037.88**
4 l/s @ 6.9 m total head	3742.13	4736.04	7.00	275.28	nr	**5011.33**
Extra for 4 nr float switches to give pump on, off and high level alarm	620.09	784.78		–	nr	**784.78**
Extra for dual pump control panel, internal wall mounted IP54, including volt free contacts	2952.85	3737.13	4.00	157.30	nr	**3894.43**

38 PIPED SUPPLY SYSTEMS

Item	Net Price £	Material £	Labour hours	Labour £	Unit	Total rate £
COLD WATER PIPELINES: PIPEWORK ANCILLARIES – TANKS						
Cisterns; fibreglass; complete with ball valve, screened fittings, outlets, drain provisions, fixing plate and fitted covers						
Rectangular						
60 litre capacity	585.74	741.31	1.33	52.30	nr	**793.62**
100 litre capacity	667.41	844.67	1.40	55.06	nr	**899.73**
150 litre capacity	763.15	965.85	1.61	63.31	nr	**1029.16**
250 litre capacity	822.29	1040.69	1.61	63.31	nr	**1104.00**
420 litre capacity	1018.01	1288.39	1.99	78.26	nr	**1366.65**
730 litre capacity	1437.60	1819.42	3.31	130.17	nr	**1949.59**
800 litre capacity	1484.06	1878.23	3.60	141.57	nr	**2019.80**
1700 litre capacity	2100.78	2658.75	13.32	523.82	nr	**3182.57**
2250 litre capacity	2472.50	3129.20	20.18	793.59	nr	**3922.79**
3400 litre capacity	3345.48	4234.04	24.50	963.48	nr	**5197.52**
4500 litre capacity	3872.08	4900.51	29.91	1176.23	nr	**6076.74**
Cisterns; polypropylene; complete with ball valve, fixing plate and cover; includes placing in position						
Rectangular						
18 litre capacity	110.67	140.07	1.00	39.33	nr	**179.39**
68 litre capacity	149.14	188.75	1.00	39.33	nr	**228.08**
91 litre capacity	150.10	189.96	1.00	39.33	nr	**229.29**
114 litre capacity	167.01	211.36	1.00	39.33	nr	**250.69**
182 litre capacity	220.54	279.12	1.00	39.33	nr	**318.44**
227 litre capacity	280.90	355.51	1.00	39.33	nr	**394.84**
Circular						
114 litre capacity	156.21	197.70	1.00	39.33	nr	**237.02**
227 litre capacity	159.30	201.62	1.00	39.33	nr	**240.94**
318 litre capacity	264.96	335.34	1.00	39.33	nr	**374.66**
455 litre capacity	304.53	385.41	1.00	39.33	nr	**424.74**
Steel sectional water storage tank; hot pressed steel tank to BS 1564 TYPE 1; 5mm plate; pre-insulated and complete with all connections and fittings to comply with BSEN 13280; 2001 and WRAS water supply (water fittings) regulations 1999; externally flanged base and sides; cost of erection (on prepared base) is included within the net price, labour cost allows for offloading and positioning materials						
Note: Prices are based on the most economical tank size for each volume, and the cost will vary with differing tank dimensions, for the same volume						
Volume, size						
4,900 litres, 3.66 m × 1.22 m × 1.22 m (h)	9517.99	12045.97	6.00	235.95	nr	**12281.93**
20,300 litres, 3.66 m × 2.4 m × 2.4 m (h)	20278.27	25664.17	12.00	471.91	nr	**26136.08**
52,000 litres, 6.1 m × 3.6 m × 2.4 m (h)	36153.02	45755.27	19.00	747.19	nr	**46502.46**
94,000 litres, 7.3 m × 3.6 m × 3.6 m (h)	63653.32	80559.65	28.00	1101.12	nr	**81660.77**
140,000 litres, 9.7 m × 6.1 m × 2.44 m (h)	71021.01	89884.20	28.00	1101.12	nr	**90985.32**

38 PIPED SUPPLY SYSTEMS

Item	Net Price £	Material £	Labour hours	Labour £	Unit	Total rate £
COLD WATER PIPELINES: PIPEWORK ANCILLARIES – TANKS – cont						
GRP sectional water storage tank; pre-insulated and complete with twin compartment and basic connections and fittings to comply with BSEN 13280; 2001 and WRAS water supply (water fittings) regulations 1999; externally flanged base and sides; cost of erection (on prepared base) is included within the net price, labour cost allows for offloading and positioning materials						
Note: Prices are based on the most economical tank size for each volume, and the cost will vary with differing tank dimensions, for the same volume						
Volume, size						
4,500 litres, 3 m × 1 m × 1.5 m (h)	10393.98	13154.62	5.00	196.63	nr	**13351.24**
10,000 litres, 2.5 m × 2 m × 2 m (h)	14865.49	18813.77	7.00	275.28	nr	**19089.05**
20,000 litres, 4 m × 2.5 m × 2 m (h)	19242.06	24352.75	10.00	393.26	nr	**24746.01**
30,000 litres 5 m × 3 m × 2 m (h)	23968.75	30334.85	12.00	471.91	nr	**30806.76**
40,000 litres, 5 m × 4 m × 2 m (h)	27977.39	35408.18	12.00	471.91	nr	**35880.09**
50,000 litres, 5 m × 4 m × 2.5 m (h)	34218.82	43307.34	14.00	550.56	nr	**43857.90**
60,000 litres, 6 m × 4 m × 2.5 m (h)	36398.20	46065.56	16.00	629.21	nr	**46694.78**
70,000 litres, 7 m × 4 m × 2.5 m (h)	41136.76	52062.69	16.00	629.21	nr	**52691.90**
80,000 litres, 8 m × 4 m × 2.5 m (h)	43522.36	55081.90	16.00	629.21	nr	**55711.11**
90,000 litres, 6 m × 5 m × 3 m (h)	46894.54	59349.73	16.00	629.21	nr	**59978.95**
105,000 litres, 7 m × 5 m × 3 m (h)	52541.06	66495.96	24.00	943.82	nr	**67439.78**
120,000 litres, 8 m × 5 m × 3 m (h)	54913.30	69498.28	24.00	943.82	nr	**70442.10**
135,000 litres, 9 m × 6 m × 2.5 m (h)	60930.71	77113.91	24.00	943.82	nr	**78057.73**
144,000 litres, 8 m × 6 m × 3 m (h)	61013.79	77219.06	24.00	943.82	nr	**78162.88**
CLEANING AND CHEMICAL TREATMENT						
Electromagnetic water conditioner; WRAS approved; complete with control box; maximum inlet pressure 16 bar; electrical work elsewhere						
Connection size, nominal flow rate at 50 mbar						
20 mm dia., 0.3 l/s	1845.12	2335.18	1.25	49.16	nr	**2384.34**
25 mm dia., 0.6 l/s	2422.77	3066.26	1.45	57.02	nr	**3123.28**
32 mm dia., 1.2 l/s	3454.69	4372.26	1.55	60.95	nr	**4433.21**
40 mm dia., 1.7 l/s	4127.68	5224.00	1.65	64.89	nr	**5288.88**
50 mm dia., 3.4 l/s	5406.37	6842.30	1.75	68.82	nr	**6911.12**
65 mm dia., 5.2 l/s	5916.72	7488.20	1.90	74.72	nr	**7562.92**
100 mm dia., 30.5 l/s, 595 mbar	11048.28	13982.71	3.00	117.98	nr	**14100.68**
Ultraviolet water sterilizing unit; WRAS approved; complete with control unit; UV lamp housed in quartz tube; unit complete with UV intensity sensor, flushing and discharge valve and facilities for remote alarm; electrical work elsewhere						

38 PIPED SUPPLY SYSTEMS

Item	Net Price £	Material £	Labour hours	Labour £	Unit	Total rate £
Maximum flow rate (@ 400 J/m² exposure in accordance with BSEN 14897), connection size						
0.8 l/s, 25 mm dia.	3437.87	4350.96	1.98	77.87	nr	**4428.83**
1.25 l/s, 40 mm dia.	4346.41	5500.81	1.98	77.87	nr	**5578.68**
2.57 l/s, 50 mm dia.	4946.49	6260.28	1.98	77.87	nr	**6338.14**
4.46 l/s, 80 mm dia.	11295.05	14295.01	1.98	77.87	nr	**14372.88**
10.5 l/s, 100 mm dia.	14031.88	17758.75	1.98	77.87	nr	**17836.61**
20.1 l/s, 100 mm dia.	15781.66	19973.27	3.60	141.57	nr	**20114.84**
26.9 l/s 125 mm dia.	17722.12	22429.11	3.60	141.57	nr	**22570.69**
33.7 l/s 125 mm dia.	19999.07	25310.83	3.60	141.57	nr	**25452.40**
Base exchange water softener; WRAS approved; complete with resin tank, brine tank and consumption data monitoring facilities						
Capacities of softeners are based on 300 ppm hardness and quoted in m³ of softened water produced. Design flow rates are recommended for continuous use						
Simplex configuration						
Design flow rate, min–max softenend water produced						
0.97 l/s, 7.7 m³	3336.92	4223.20	10.00	393.26	nr	**4616.46**
1.25 l/s, 11.7 m³	3421.04	4329.67	12.00	471.91	nr	**4801.58**
1.8 l/s, 19.8 m³	5097.91	6451.92	12.00	471.91	nr	**6923.83**
2.23 l/s, 23.4 m³	5720.43	7239.78	12.00	471.91	nr	**7711.69**
2.92 l/s, 31.6 m³	5916.72	7488.20	15.00	589.89	nr	**8078.09**
3.75 l/s, 39.8 m³	7066.41	8943.25	15.00	589.89	nr	**9533.14**
4.72 l/s, 60.3 m³	8025.43	10156.98	18.00	707.86	nr	**10864.85**
6.66 l/s, 96.6 m³	9477.97	11995.32	20.00	786.52	nr	**12781.83**
Duplex configuration						
Design flow rate, min–max softenend water produced						
0.97 l/s, 7.7 m³	4924.06	6231.89	12.00	471.91	nr	**6703.80**
1.25 l/s, 11.7 m³	5125.95	6487.41	12.00	471.91	nr	**6959.32**
1.81 l/s, 19.8 m³	8597.47	10880.96	15.00	589.89	nr	**11470.85**
2.23 l/s, 23.4 m³	9853.72	12470.87	15.00	589.89	nr	**13060.76**
2.29 l/s, 31.6 m³	10156.57	12854.15	18.00	707.86	nr	**13562.02**
3.75 l/s, 39.8 m³	12629.81	15984.29	18.00	707.86	nr	**16692.16**
4.72 l/s, 60.3 m³	14413.24	18241.40	18.00	707.86	nr	**18949.26**
6.66 l/s, 96.6 m³	14413.24	18241.40	23.00	904.49	nr	**19145.89**
THERMAL INSULATION						
Flexible closed cell walled insulation; Class 1/Class O; adhesive joints; including around fittings						
6 mm wall thickness						
15 mm dia.	2.05	2.60	0.15	5.90	m	**8.49**
22 mm dia.	2.43	3.07	0.15	5.90	m	**8.97**
28 mm dia.	3.07	3.88	0.15	5.90	m	**9.78**

38 PIPED SUPPLY SYSTEMS

Item	Net Price £	Material £	Labour hours	Labour £	Unit	Total rate £
COLD WATER PIPELINES: PIPEWORK ANCILLARIES – TANKS – cont						
Flexible closed cell walled insulation – cont						
9 mm wall thickness						
15 mm dia.	2.36	2.99	0.15	5.90	m	8.88
22 mm dia.	2.95	3.73	0.15	5.90	m	9.63
28 mm dia.	3.16	4.00	0.15	5.90	m	9.90
35 mm dia.	3.58	4.53	0.15	5.90	m	10.43
42 mm dia.	4.09	5.17	0.15	5.90	m	11.07
54 mm dia.	4.24	5.36	0.15	5.90	m	11.26
13 mm wall thickness						
15 mm dia.	3.02	3.82	0.15	5.90	m	9.72
22 mm dia.	3.70	4.68	0.15	5.90	m	10.58
28 mm dia.	4.40	5.56	0.15	5.90	m	11.46
35 mm dia.	4.74	6.00	0.15	5.90	m	11.90
42 mm dia.	5.58	7.06	0.15	5.90	m	12.96
54 mm dia.	7.77	9.84	0.15	5.90	m	15.74
67 mm dia.	11.49	14.54	0.15	5.90	m	20.44
76 mm dia.	13.51	17.10	0.15	5.90	m	23.00
108 mm dia.	14.17	17.93	0.15	5.90	m	23.83
19 mm wall thickness						
15 mm dia.	4.89	6.19	0.15	5.90	m	12.09
22 mm dia.	5.96	7.55	0.15	5.90	m	13.45
28 mm dia.	8.10	10.25	0.15	5.90	m	16.14
35 mm dia.	9.37	11.86	0.15	5.90	m	17.76
42 mm dia.	11.06	13.99	0.15	5.90	m	19.89
54 mm dia.	13.99	17.71	0.15	5.90	m	23.61
67 mm dia.	16.77	21.22	0.15	5.90	m	27.12
76 mm dia.	19.38	24.53	0.22	8.65	m	33.18
108 mm dia.	26.43	33.45	0.22	8.65	m	42.10
25 mm wall thickness						
15 mm dia.	9.61	12.16	0.15	5.90	m	18.06
22 mm dia.	10.51	13.30	0.15	5.90	m	19.20
28 mm dia.	11.90	15.06	0.15	5.90	m	20.96
35 mm dia.	13.19	16.69	0.15	5.90	m	22.59
42 mm dia.	14.07	17.81	0.15	5.90	m	23.71
54 mm dia.	16.53	20.91	0.15	5.90	m	26.81
67 mm dia.	22.27	28.19	0.15	5.90	m	34.09
76 mm dia.	25.12	31.79	0.22	8.65	m	40.44
32 mm wall thickness						
15 mm dia.	12.02	15.21	0.15	5.90	m	21.11
22 mm dia.	13.19	16.69	0.15	5.90	m	22.59
28 mm dia.	15.29	19.35	0.15	5.90	m	25.25
35 mm dia.	15.79	19.98	0.15	5.90	m	25.88
42 mm dia.	18.39	23.27	0.15	5.90	m	29.17
54 mm dia.	23.13	29.28	0.15	5.90	m	35.18
76 mm dia.	34.59	43.78	0.22	8.65	m	52.43

38 PIPED SUPPLY SYSTEMS

Item	Net Price £	Material £	Labour hours	Labour £	Unit	Total rate £
Note: For mineral fibre sectional insulation; bright class O foil faced; bright class O foil taped joints; 19 mm aluminium bands rates, refer to section – Thermal Insulation						
Note: For mineral fibre sectional insulation; bright class O foil faced; bright class O foil taped joints; 22 SWG plain/ embossed aluminium cladding; pop riveted rates, refer to section – Thermal Insulation						
Note: For mineral fibre sectional insulation; bright class O foil faced; bright class O foil taped joints; 0.8 mm polyisobutylene sheeting; welded joints rates, refer to section – Thermal Insulation						

38 PIPED SUPPLY SYSTEMS

Item	Net Price £	Material £	Labour hours	Labour £	Unit	Total rate £
HOT WATER						
PIPELINES						
Note: For pipework prices refer to section – Cold Water						
PIPELINE ANCILLARIES						
Note: For prices for ancillaries refer to section – Cold Water						
STORAGE CYLINDERS/CALORIFIERS						
Insulated copper storage cylinders; BS 699; includes placing in position						
Grade 3 (maximum 10 m working head)						
BS size 6; 115 litre capacity; 400 mm dia.; 1050 mm high	269.28	340.80	1.50	59.05	nr	**399.85**
BS size 7; 120 litre capacity; 450 mm dia.; 900 mm high	318.38	402.94	2.00	78.65	nr	**481.60**
BS size 8; 144 litre capacity; 450 mm dia.; 1050 mm high	338.45	428.34	2.80	110.16	nr	**538.50**
Grade 4 (maximum 6 m working head)						
BS size 2; 96 litre capacity; 400 mm dia.; 900 mm high	208.50	263.88	1.50	59.05	nr	**322.93**
BS size 7; 120 litre capacity; 450 mm dia.; 900 mm high	219.12	277.31	1.50	59.05	nr	**336.36**
BS size 8; 144 litre capacity; 450 mm dia.; 1050 mm high	261.56	331.02	1.50	59.05	nr	**390.07**
BS size 9; 166 litre capacity; 450 mm dia.; 1200 mm high	382.44	484.02	1.50	59.05	nr	**543.07**
Storage cylinders; brazed copper construction; to BS 699; screwed bosses; includes placing in position						
Tested to 2.2 bar, 15 m maximum head						
144 litres	1185.32	1500.15	3.00	118.10	nr	**1618.24**
160 litres	1339.97	1695.87	3.00	118.10	nr	**1813.96**
200 litres	1381.16	1747.99	3.76	147.84	nr	**1895.83**
255 litres	1571.83	1989.30	3.76	147.84	nr	**2137.14**
290 litres	2123.26	2687.20	3.76	147.84	nr	**2835.04**
370 litres	2432.53	3078.61	4.50	177.14	nr	**3255.75**
450 litres	3313.59	4193.67	5.00	196.63	nr	**4390.30**
Tested to 2.55 bar, 17 m maximum head						
550 litres	3587.93	4540.89	5.00	196.63	nr	**4737.52**
700 litres	4194.66	5308.76	6.02	236.90	nr	**5545.66**
800 litres	4851.64	6140.23	6.54	257.03	nr	**6397.26**
900 litres	5249.97	6644.36	8.00	314.61	nr	**6958.96**
1000 litres	5540.19	7011.66	8.00	314.61	nr	**7326.27**
1250 litres	6067.78	7679.38	13.16	517.45	nr	**8196.83**
1500 litres	9293.91	11762.37	15.15	595.84	nr	**12358.22**
2000 litres	11153.52	14115.89	17.24	678.03	nr	**14793.93**
3000 litres	15666.86	19827.97	24.39	959.16	nr	**20787.14**

38 PIPED SUPPLY SYSTEMS

Item	Net Price £	Material £	Labour hours	Labour £	Unit	Total rate £
Indirect cylinders; copper; bolted top; up to 5 tappings for connections; BS 1586; includes placing in position						
Grade 3, tested to 1.45 bar, 10 m maximum head						
74 litre capacity	543.47	687.81	1.50	59.05	nr	**746.86**
96 litre capacity	553.29	700.24	1.50	59.05	nr	**759.29**
114 litre capacity	568.13	719.02	1.50	59.05	nr	**778.07**
117 litre capacity	589.66	746.28	2.00	78.65	nr	**824.93**
140 litre capacity	607.62	769.01	2.50	98.31	nr	**867.32**
162 litre capacity	849.72	1075.41	3.00	118.10	nr	**1193.50**
190 litre capacity	928.81	1175.50	3.51	137.99	nr	**1313.49**
245 litre capacity	1086.89	1375.57	3.80	149.53	nr	**1525.10**
280 litre capacity	1926.69	2438.42	4.00	157.30	nr	**2595.73**
360 litre capacity	2084.81	2638.54	4.50	177.14	nr	**2815.68**
440 litre capacity	2420.74	3063.69	4.50	177.14	nr	**3240.84**
Grade 2, tested to 2.2 bar, 15 m maximum head						
117 litre capacity	785.47	994.09	2.00	78.65	nr	**1072.74**
140 litre capacity	854.69	1081.70	2.50	98.31	nr	**1180.02**
162 litre capacity	978.17	1237.97	2.80	110.16	nr	**1348.13**
190 litre capacity	1136.33	1438.14	3.00	118.10	nr	**1556.24**
245 litre capacity	1373.43	1738.22	4.00	157.30	nr	**1895.52**
280 litre capacity	2193.48	2776.07	4.00	157.30	nr	**2933.37**
360 litre capacity	2420.74	3063.69	4.50	177.14	nr	**3240.84**
440 litre capacity	2865.34	3626.37	4.50	177.14	nr	**3803.52**
Grade 1, tested 3.65 bar, 25 m maximum head						
190 litre capacity	1689.62	2138.38	3.00	118.10	nr	**2256.48**
245 litre capacity	1921.71	2432.11	3.00	118.10	nr	**2550.21**
280 litre capacity	2717.18	3438.86	4.00	157.30	nr	**3596.16**
360 litre capacity	3438.50	4351.77	4.50	177.14	nr	**4528.91**
440 litre capacity	4174.52	5283.27	4.50	177.14	nr	**5460.42**
Indirect cylinders, including manhole; BS 853						
Grade 3, tested to 1.5 bar, 10 m maximum head						
550 litre capacity	3513.94	4447.24	5.21	204.82	nr	**4652.06**
700 litre capacity	3890.47	4923.78	6.02	236.90	nr	**5160.68**
800 litre capacity	4517.94	5717.91	6.54	257.03	nr	**5974.94**
1000 litre capacity	5647.48	7147.44	7.04	276.94	nr	**7424.39**
1500 litre capacity	6525.94	8259.23	10.00	393.26	nr	**8652.49**
2000 litre capacity	9035.97	11435.92	16.13	634.29	nr	**12070.20**
Grade 2, tested to 2.55 bar, 15 m maximum head						
550 litre capacity	3895.87	4930.61	5.21	204.82	nr	**5135.43**
700 litre capacity	4875.94	6170.99	6.02	236.90	nr	**6407.90**
800 litre capacity	5145.45	6512.09	6.54	257.03	nr	**6769.12**
1000 litre capacity	6370.59	8062.62	7.04	276.94	nr	**8339.56**
1500 litre capacity	7840.71	9923.20	10.00	393.26	nr	**10316.46**
2000 litre capacity	9800.92	12404.04	16.13	634.29	nr	**13038.33**
Grade 1, tested to 4 bar, 25 m maximum head						
550 litre capacity	4532.90	5736.84	5.21	204.82	nr	**5941.66**
700 litre capacity	5145.45	6512.09	6.02	236.90	nr	**6748.99**
800 litre capacity	5513.02	6977.28	6.54	257.03	nr	**7234.31**

38 PIPED SUPPLY SYSTEMS

Item	Net Price £	Material £	Labour hours	Labour £	Unit	Total rate £
HOT WATER – cont						
Indirect cylinders, including manhole – cont						
Grade 1, tested to 4 bar, 25 m maximum head – cont						
1000 litre capacity	7350.71	9303.05	7.04	276.94	nr	**9580.00**
1500 litre capacity	8820.76	11163.55	10.00	393.26	nr	**11556.81**
2000 litre capacity	10780.97	13644.39	16.13	634.29	nr	**14278.68**
Storage calorifiers; copper; heater battery capable of raising temperature of contents from 10°C to 65°C in one hour; static head not exceeding 1.35 bar; BS 853; includes fixing in position on cradles or legs						
Horizontal; primary LPHW at 82°C/71°C						
400 litre capacity	5420.65	6860.37	7.04	276.94	nr	**7137.32**
1000 litre capacity	8672.99	10976.53	8.00	314.61	nr	**11291.14**
2000 litre capacity	17346.09	21953.21	14.08	553.88	nr	**22507.09**
3000 litre capacity	21411.53	27098.44	25.00	983.15	nr	**28081.58**
4000 litre capacity	26019.04	32929.70	40.00	1573.03	nr	**34502.73**
4500 litre capacity	29322.34	37110.35	50.00	1966.29	nr	**39076.64**
Vertical; primary LPHW at 82°C/71°C						
400 litre capacity	6109.09	7731.66	7.04	276.94	nr	**8008.60**
1000 litre capacity	9818.21	12425.92	8.00	314.61	nr	**12740.53**
2000 litre capacity	18701.24	23668.29	14.08	553.88	nr	**24222.18**
3000 litre capacity	23376.56	29585.38	25.00	983.15	nr	**30568.52**
4000 litre capacity	28675.14	36291.26	40.00	1573.03	nr	**37864.29**
4500 litre capacity	32415.44	41024.98	50.00	1966.29	nr	**42991.27**
Storage calorifiers; galvanized mild steel; heater battery capable of raising temperature of contents from 10°C to 65°C in one hour; static head not exceeding 1.35 bar; BS 853; includes fixing in position on cradles or legs						
Horizontal; primary LPHW at 82°C/71°C						
400 litre capacity	5420.65	6860.37	7.04	276.94	nr	**7137.32**
1000 litre capacity	8672.99	10976.53	8.00	314.61	nr	**11291.14**
2000 litre capacity	17346.09	21953.21	14.08	553.88	nr	**22507.09**
3000 litre capacity	21411.53	27098.44	25.00	983.15	nr	**28081.58**
4000 litre capacity	26019.04	32929.70	40.00	1573.03	nr	**34502.73**
4500 litre capacity	29322.34	37110.35	50.00	1966.29	nr	**39076.64**
Vertical; primary LPHW at 82°C/71°C						
400 litre capacity	6109.09	7731.66	7.04	276.94	nr	**8008.60**
1000 litre capacity	9818.21	12425.92	8.00	314.61	nr	**12740.53**
2000 litre capacity	18701.24	23668.29	14.08	553.88	nr	**24222.18**
3000 litre capacity	23376.56	29585.38	25.00	983.15	nr	**30568.52**
4000 litre capacity	28675.14	36291.26	40.00	1573.03	nr	**37864.29**
4500 litre capacity	32415.44	41024.98	50.00	1966.29	nr	**42991.27**

38 PIPED SUPPLY SYSTEMS

Item	Net Price £	Material £	Labour hours	Labour £	Unit	Total rate £
Indirect cylinders; mild steel, welded throughout, galvanized, with connections. Tested to 4 bar, 95C. Includes sensors, with full insulation and cases, includes delivery						
222 litre nominal content	1465.95	1855.31	2.50	98.31	nr	**1953.62**
278 litre nominal content	1628.08	2060.50	2.80	110.16	nr	**2170.65**
474 litre nominal content	2035.44	2576.05	3.00	118.10	nr	**2694.15**
765 litre nominal content	2279.31	2884.70	3.00	118.06	nr	**3002.75**
956 litre nominal content	2686.67	3400.25	3.00	118.10	nr	**3518.35**
1365 litre nominal content	3483.68	4408.95	4.00	157.30	nr	**4566.25**
2039 litre nominal content	4444.18	5624.56	5.00	196.63	nr	**5821.19**
2361 litre nominal content	5746.65	7272.95	6.02	236.90	nr	**7509.86**
Indirect cylinders; mild steel welded throughout, galvanized, with connections. Tested to 4 bar, 95C. Includes sensors. Includes delivery						
4021 litre nominal content	7415.60	9385.18	1.50	58.99	nr	**9444.17**
5897 litre nominal content	8791.63	11126.69	1.50	58.99	nr	**11185.68**
8000 litre nominal content	13836.64	17511.65	1.50	58.99	nr	**17570.64**
10170 litre nominal content	16256.28	20573.94	2.00	78.65	nr	**20652.59**
Local electric hot water heaters						
Unvented multipoint water heater; providing hot water for one or more outlets; used with conventional taps or mixers; factory fitted temperature and pressure relief valve; externally adjustable thermostat; elemental 'on' indicator; fitted with 1 m of 3 core cable; electrical supply and connection excluded						
5 litre capacity; 2.2 kW rating	301.24	381.24	1.50	58.99	nr	**440.23**
10 litre capacity; 2.2 kW rating	373.30	472.45	1.50	58.99	nr	**531.44**
15 litre capacity; 2.2 kW rating	665.32	842.03	1.50	58.99	nr	**901.02**
30 litre capacity; 2.2 kW rating	665.32	842.03	2.00	78.65	nr	**920.68**
50 litre capacity; 2.2 kW rating	700.46	886.51	2.00	78.65	nr	**965.16**
80 litre capacity; 2.2 kW rating	1378.32	1744.40	2.00	78.65	nr	**1823.05**
100 litre capacity; 2.2 kW rating	1479.71	1872.72	2.00	78.65	nr	**1951.37**
Accessories						
Pressure reducing valve and expansion kit	240.57	304.46	2.00	78.65	nr	**383.11**
Thermostatic blending valve	128.02	162.03	1.00	39.33	nr	**201.35**
SOLAR THERMAL PACKAGES						
Packages include solar collectors, mild steel storage vessel, fresh water module (including plate heat exchanger), roof fixing kit, glycol, stratified pump station (including plate heat exchanger), solar explansion vessel, controls, anti-legionella protection; delivery and commissioning						
No. of panels; storage size						
6 solar collectors, 1000 l storage vessel	10285.91	13017.85	11.00	432.58	nr	**13450.43**
8 solar collectors, 1500 l storage vessel	11892.69	15051.39	13.00	511.24	nr	**15562.63**
10 solar collectors, 2000 l storage vessel	12484.34	15800.18	15.00	589.89	nr	**16390.07**

38 PIPED SUPPLY SYSTEMS

Item	Net Price £	Material £	Labour hours	Labour £	Unit	Total rate £
HOT WATER – cont						
SOLAR THERMAL PACKAGES – cont						
No. of panels; storage size – cont						
12 solar collectors, 2000 l storage vessel	13688.80	17324.55	17.00	668.54	nr	**17993.09**
14 solar collectors, 2000 l storage vessel	14901.99	18859.96	18.00	707.86	nr	**19567.82**
16 solar collectors, 2500 l storage vessel	16146.31	20434.77	20.00	786.52	nr	**21221.29**

38 PIPED SUPPLY SYSTEMS

Item	Net Price £	Material £	Labour hours	Labour £	Unit	Total rate £
NATURAL GAS: PIPELINES: MEDIUM DENSITY POLYETHYLENE – YELLOW						
Pipe; laid underground; electrofusion joints in the running length; BS 6572; BGT PL2 standards						
Coiled service pipe						
20 mm dia.	2.67	3.38	0.37	14.55	m	**17.94**
25 mm dia.	3.47	4.39	0.41	16.12	m	**20.51**
32 mm dia.	5.73	7.26	0.47	18.49	m	**25.74**
63 mm dia.	21.87	27.68	0.60	23.60	m	**51.28**
90 mm dia.	28.67	36.28	0.90	35.40	m	**71.68**
Mains service pipe						
63 mm dia.	21.24	26.88	0.60	23.60	m	**50.47**
90 mm dia.	27.83	35.23	0.90	35.39	m	**70.62**
125 mm dia.	53.76	68.03	1.20	47.19	m	**115.23**
180 mm dia.	111.02	140.50	1.50	58.99	m	**199.49**
250 mm dia.	204.33	258.60	1.75	68.82	m	**327.42**
Extra over fittings, electrofusion joints						
Straight connector						
32 mm dia.	16.64	21.06	0.47	18.48	nr	**39.54**
63 mm dia.	31.23	39.53	0.58	22.81	nr	**62.34**
90 mm dia.	46.07	58.30	0.67	26.35	nr	**84.65**
125 mm dia.	86.28	109.19	0.83	32.64	nr	**141.83**
180 mm dia.	155.38	196.65	1.25	49.16	nr	**245.81**
Reducing connector						
90 × 63 mm dia.	64.34	81.43	0.67	26.35	nr	**107.78**
125 × 90 mm dia.	129.04	163.31	0.83	32.64	nr	**195.95**
180 × 125 mm dia.	236.73	299.61	1.25	49.16	nr	**348.76**
Bend; 45°						
90 mm dia.	124.35	157.38	0.67	26.35	nr	**183.73**
125 mm dia.	203.42	257.45	0.83	32.64	nr	**290.09**
180 mm dia.	461.43	583.99	1.25	49.16	nr	**633.15**
Bend; 90°						
63 mm dia.	80.56	101.96	0.58	22.81	nr	**124.76**
90 mm dia.	124.35	157.38	0.67	26.35	nr	**183.73**
125 mm dia.	203.42	257.45	0.83	32.64	nr	**290.09**
180 mm dia.	461.43	583.99	1.25	49.16	nr	**633.15**
Extra over malleable iron fittings, compression joints						
Straight connector						
20 mm dia.	23.33	29.52	0.38	14.95	nr	**44.47**
25 mm dia.	25.41	32.16	0.45	17.70	nr	**49.86**
32 mm dia.	28.53	36.11	0.50	19.66	nr	**55.77**
63 mm dia.	57.28	72.50	0.85	33.44	nr	**105.94**
Straight connector; polyethylene to MI						
20 mm dia.	19.82	25.08	0.31	12.19	nr	**37.28**
25 mm dia.	21.63	27.37	0.35	13.76	nr	**41.13**
32 mm dia.	24.13	30.54	0.40	15.73	nr	**46.27**
63 mm dia.	40.45	51.19	0.65	25.57	nr	**76.76**

38 PIPED SUPPLY SYSTEMS

Item	Net Price £	Material £	Labour hours	Labour £	Unit	Total rate £
NATURAL GAS: PIPELINES: MEDIUM DENSITY POLYETHYLENE – YELLOW – cont						
Extra over malleable iron fittings, compression joints – cont						
Straight connector; polyethylene to FI						
20 mm dia.	19.82	25.08	0.31	12.19	nr	**37.28**
25 mm dia.	24.13	30.54	0.35	13.76	nr	**44.30**
32 mm dia.	24.13	30.54	0.40	15.73	nr	**46.27**
63 mm dia.	40.45	51.19	0.75	29.50	nr	**80.69**
Elbow						
20 mm dia.	30.33	38.39	0.38	14.95	nr	**53.34**
25 mm dia.	33.09	41.88	0.45	17.70	nr	**59.58**
32 mm dia.	37.06	46.90	0.50	19.66	nr	**66.57**
63 mm dia.	74.48	94.26	0.80	31.46	nr	**125.72**
Equal tee						
20 mm dia.	40.45	51.19	0.53	20.85	nr	**72.04**
25 mm dia.	47.26	59.81	0.55	21.63	nr	**81.45**
32 mm dia.	59.55	75.37	0.64	25.18	nr	**100.54**
GAS BOOSTERS						
Complete skid mounted gas booster set, including AV mounts, flexible connections, low pressure switch, control panel and NRV (for run/standby unit); 3 phase supply; in accordance with IGE/UP/2; includes delivery, offloading and positioning						
Single unit						
Flow, pressure range						
0–200 m³/hour, 0.1–2.6 kPa	5670.72	7176.86	10.00	393.26	nr	**7570.12**
0–200 m³/hour, 0.1–4.0 kPa	6470.45	8189.00	10.00	393.26	nr	**8582.25**
0–200 m³/hour, 0.1–7 kPa	7529.76	9529.67	10.00	393.26	nr	**9922.92**
0–200 m³/hour, 0.1–9.5 kPa	7841.38	9924.05	10.00	393.26	nr	**10317.31**
0–200 m³/hour 0.1–11.0 kPa	8760.55	11087.35	10.00	393.26	nr	**11480.61**
0–400 m³/hour, 0.1–4.0 kPa	9528.49	12059.26	10.00	393.26	nr	**12452.52**
0–1000 m³/hour, 0.1–7.4 kPa	9897.74	12526.58	10.00	393.26	nr	**12919.84**
50–1000 m³/hour, 0.1–16.0 kPa	18681.90	23643.82	20.00	786.52	nr	**24430.34**
50–1000 m³/hour, 0.1–24.5 kPa	20906.77	26459.61	20.00	786.52	nr	**27246.13**
50–1000 m³/hour, 0.1–31.0 kPa	23778.53	30094.11	20.00	786.52	nr	**30880.62**
50–1000 m³/hour, 0.1–41.0 kPa	26245.22	33215.95	20.00	786.52	nr	**34002.46**
50–1000 m³/hour, 0.1–51.0 kPa	27231.78	34464.55	20.00	786.52	nr	**35251.06**
100–1800 m³/hour, 3.5–23.5 kPa	27943.31	35365.05	20.00	786.52	nr	**36151.56**
100–1800 m³/hour, 4.5–27.0 kPa	30139.91	38145.07	20.00	786.52	nr	**38931.59**
100–1800 m³/hour, 6.0–32.5 kPa	33380.30	42246.10	20.00	786.52	nr	**43032.62**
100–1800 m³/hour, 7.2–39.0 kPa	38401.88	48601.42	20.00	786.52	nr	**49387.93**
100–1800 m³/hour, 9.0–42.0 kPa	40364.83	51085.73	20.00	786.52	nr	**51872.25**
Run/Standby unit						
Flow, pressure range						
0–200 m³/hour, 0.1–2.6 kPa	27910.76	35323.86	16.00	629.21	nr	**35953.07**
0–200 m³/hour, 0.1–4.0 kPa	28760.81	36399.69	16.00	629.21	nr	**37028.90**
0–200 m³/hour, 0.1–7 kPa	29426.95	37242.75	16.00	629.21	nr	**37871.96**

38 PIPED SUPPLY SYSTEMS

Item	Net Price £	Material £	Labour hours	Labour £	Unit	Total rate £
0–200 m³/hour, 0.1–9.5 kPa	30156.12	38165.58	16.00	629.21	nr	**38794.80**
0–200 m³/hour 0.1–11.0 kPa	30655.73	38797.89	16.00	629.21	nr	**39427.10**
0–400 m³/hour, 0.1–4.0 kPa	33806.36	42785.33	16.00	629.21	nr	**43414.55**
0–1000 m³/hour, 0.1–7.4 kPa	43091.71	54536.87	25.00	983.15	nr	**55520.01**
50–1000 m³/hour, 0.1–16.0 kPa	59060.94	74747.53	25.00	983.15	nr	**75730.68**
50–1000 m³/hour, 0.1–24.5 kPa	66771.03	84505.42	25.00	983.15	nr	**85488.56**
50–1000 m³/hour, 0.1–31.0 kPa	74463.04	94240.43	25.00	983.15	nr	**95223.57**
50–1000 m³/hour, 0.1–41.0 kPa	82177.63	104004.01	25.00	983.15	nr	**104987.15**
50–1000 m³/hour, 0.1–51.0 kPa	82546.67	104471.07	25.00	983.15	nr	**105454.22**
100–1800 m³/hour, 3.5–23.5 kPa	85883.71	108694.43	25.00	983.15	nr	**109677.57**
100–1800 m³/hour, 4.5–27.0 kPa	88238.34	111674.44	25.00	983.15	nr	**112657.59**
100–1800 m³/hour, 6.0–32.5 kPa	93744.89	118643.54	25.00	983.15	nr	**119626.68**
100–1800 m³/hour, 7.2–39.0 kPa	99641.12	126105.80	25.00	983.15	nr	**127088.94**
100–1800 m³/hour, 9.0–42.0 kPa	104736.18	132554.11	25.00	983.15	nr	**133537.25**

38 PIPED SUPPLY SYSTEMS

Item	Net Price £	Material £	Labour hours	Labour £	Unit	Total rate £
NATURAL GAS: PIPELINES: SCREWED STEEL						
For prices for steel pipework refer to section – Low Temperature Hot Water Heating						
PIPE IN PIPE						
Note: For pipe in pipe, a sleeve size two pipe sizes bigger than actual pipe size has been allowed. All rates refer to actual pipe size						
Black steel pipes – Screwed and socketed joints; BS 1387: 1985 up to 50 mm pipe size. Butt welded joints; BS 1387: 1985 65 mm pipe size and above						
Pipe dia.						
25 mm	25.33	32.06	1.73	68.03	m	**100.10**
32 mm	33.64	42.58	1.95	76.69	m	**119.26**
40 mm	43.27	54.76	2.16	84.94	m	**139.70**
50 mm	56.76	71.84	2.44	95.96	m	**167.80**
65 mm	77.86	98.54	2.95	116.01	m	**214.55**
80 mm	94.12	119.12	3.42	134.49	m	**253.61**
100 mm	117.54	148.75	4.00	157.30	m	**306.06**
Extra over black steel pipes – Screwed pipework; black malleable iron fittings; BS 143. Welded pipework; butt welded steel fittings; BS 1965						
Bend, 90°						
25 mm	14.62	18.50	2.91	114.44	m	**132.94**
32 mm	22.25	28.16	3.45	135.67	m	**163.83**
40 mm	31.17	39.45	5.34	210.00	m	**249.45**
50 mm	35.13	44.46	6.53	256.80	m	**301.26**
65 mm	47.44	60.05	8.84	347.64	m	**407.69**
80 mm	80.89	102.37	10.73	421.97	m	**524.34**
100 mm	107.42	135.95	12.76	501.80	m	**637.75**
Bend, 45°						
25 mm	18.47	23.38	2.91	114.44	m	**137.82**
32 mm	23.92	30.28	3.45	135.67	m	**165.95**
40 mm	26.75	33.85	5.34	210.00	m	**243.85**
50 mm	31.91	40.38	6.53	256.80	m	**297.18**
65 mm	41.67	52.74	8.84	347.64	m	**400.38**
80 mm	71.27	90.20	10.73	421.97	m	**512.17**
100 mm	91.30	115.54	12.76	501.80	m	**617.34**
Equal tee						
25 mm	122.97	155.63	4.18	164.38	m	**320.02**
32 mm	126.27	159.81	4.94	194.27	m	**354.08**
40 mm	155.34	196.60	7.28	286.29	m	**482.89**
50 mm	159.38	201.71	8.48	333.48	m	**535.19**
65 mm	212.20	268.55	11.47	451.07	m	**719.62**
80 mm	364.06	460.76	14.23	559.61	m	**1020.37**
100 mm	412.03	521.47	17.92	704.72	m	**1226.18**

38 PIPED SUPPLY SYSTEMS

Item	Net Price £	Material £	Labour hours	Labour £	Unit	Total rate £
Copper pipe; capillary or compression joints in the running length; EN1057 R250 (TX) formerly BS 2871 Table X						
Plastic coated gas service pipe for corrosive environments, fixed vertically or at low level with brackets measured separtely						
15 mm dia. (yellow)	13.97	17.68	0.85	33.44	m	**51.12**
22 mm dia. (yellow)	27.52	34.83	0.96	37.78	m	**72.61**
28 mm dia. (yellow)	34.96	44.24	1.06	41.69	m	**85.92**
Copper pipe; capillary or compression joints in the running length; EN1057 R250 (TY) formerly BS 2871 Table Y						
Plastic coated gas and cold water service pipe for corrosive environments, fixed vertically or at low level with brackets measured separately (Refer to Copper Pipe Table X Section)						
15 mm dia. (yellow)	16.14	20.43	0.61	23.99	m	**44.42**
22 mm dia. (yellow)	29.32	37.11	0.69	27.13	m	**64.25**

FIXINGS
Refer to Section – Cold Water

Extra over copper pipes; capillary fittings; BS 864
Refer to Section – Cold Water

38 PIPED SUPPLY SYSTEMS

Item	Net Price £	Material £	Labour hours	Labour £	Unit	Total rate £
FUEL OIL STORAGE/DISTRIBUTION						
PIPELINES						
For pipework prices refer to Section – Low Temperature Hot Water Heating						
TANKS						
Fuel storage tanks; mild steel; with all necessary screwed bosses; oil resistant joint rings; includes placing in position. Rectangular shape						
1360 litre (300 gallon) capacity; 2 mm plate	558.31	706.60	12.03	473.09	nr	**1179.69**
2730 litre (600 gallon) capacity; 2.5 mm plate	746.05	944.20	18.60	731.46	nr	**1675.66**
4550 litre (1000 gallon) capacity; 3 mm plate	1564.17	1979.61	25.00	983.15	nr	**2962.76**
Fuel storage tanks; 5 mm plate mild steel to BS 799 type J; complete with raised neck manhole with bolted cover, screwed connections, vent and fill connections, drain valve, gauge and overfill alarm; includes placing in position; excludes pumps and control panel. Nominal capacity size						
5,600 litres, 2.5 m × 1.5 m × 1.5 m high	3650.86	4620.53	20.00	786.52	nr	**5407.05**
Extra for bund unit (internal use)	2099.51	2657.14	30.00	1179.77	nr	**3836.92**
Extra for external use with bund (watertight)	1329.68	1682.84	2.00	78.65	nr	**1761.50**
10,200 litres, 3.05 m × 1.83 m × 1.83 m high	4537.31	5742.43	30.00	1179.77	nr	**6922.20**
Extra for bund unit (internal use)	3055.98	3867.64	40.00	1573.03	nr	**5440.68**
Extra for external use with bund (watertight)	1632.96	2066.67	2.00	78.65	nr	**2145.33**
15,000 litres, 3.75 m × 2 m × 2 m high	5738.71	7262.91	40.00	1573.03	nr	**8835.94**
Extra for bund unit (internal use)	4105.76	5196.25	55.00	2162.92	nr	**7359.17**
Extra for external use with bund (watertight)	1912.92	2420.99	2.00	78.65	nr	**2499.64**
20,000 litres, 4 m × 2.5 m × 2 m high	7569.98	9580.56	50.00	1966.29	nr	**11546.86**
Extra for bund unit (internal use)	4980.59	6303.44	65.00	2556.18	nr	**8859.61**
Extra for external use with bund (watertight)	2262.82	2863.83	2.00	78.65	nr	**2942.48**
Extra for BMS output (all tank sizes)	793.16	1003.82		–	nr	**1003.82**
Fuel storage tanks; plastic; with all necessary screwed bosses; oil resistant joint rings; includes placing in position						
Cylindrical; horizontal						
1350 litre (300 gallon) capacity	786.12	994.92	4.30	169.10	nr	**1164.02**
2500 litre (550 gallon) capacity	1221.79	1546.30	4.88	191.91	nr	**1738.21**
Cylindrical; vertical						
1365 litre (300 gallon) capacity	553.29	700.24	3.73	146.69	nr	**846.93**
2600 litre (570 gallon) capacity	601.57	761.34	4.88	191.91	nr	**953.25**
Bunded tanks						
1135 litre (250 gallon) capacity	1701.87	2153.89	4.30	169.10	nr	**2322.99**
1590 litre (350 gallon) capacity	1806.29	2286.04	4.88	191.91	nr	**2477.95**
2500 litre (550 gallon) capacity	2165.14	2740.20	5.95	233.99	nr	**2974.19**

38 PIPED SUPPLY SYSTEMS

Item	Net Price £	Material £	Labour hours	Labour £	Unit	Total rate £
FIRE HOSE REELS: PIPEWORK ANCILLARIES						
PIPELINES						
For pipework prices refer to section – Cold Water						
PIPELINE ANCILLARIES						
For prices for ancillaries refer to section – Cold Water						
Hose reels; automatic; connection to 25 mm screwed joint; reel with 30.5 m, 19 mm rubber hose; suitable for working pressure up to 7 bar						
Reels						
Non-swing pattern	205.81	260.47	3.75	147.47	nr	**407.95**
Swinging pattern	286.05	362.03	3.75	147.47	nr	**509.50**
Hose reels; manual; connection to 25 mm screwed joint; reel with 30.5 m, 19 mm rubber hose; suitable for working pressure up to 7 bar						
Reels						
Non-swing pattern	205.81	260.47	3.25	127.81	nr	**388.28**
Swinging pattern	286.05	362.03	3.25	127.81	nr	**489.84**

38 PIPED SUPPLY SYSTEMS

Item	Net Price £	Material £	Labour hours	Labour £	Unit	Total rate £
DRY RISERS: PIPEWORK ANCILLARIES						
PIPELINES						
For pipework prices refer to section – Cold Water						
VALVES (BS 5041, Parts 2 and 3)						
Bronze/gunmetal inlet breeching for pumping in with 65 mm dia. instantaneous male coupling; with cap, chain and 25 mm drain valve						
Double inlet with back pressure valve, flanged to steel	407.13	515.26	1.75	68.82	nr	**584.08**
Quadruple inlet with back pressure valve, flanged to steel	911.58	1153.69	1.75	68.82	nr	**1222.51**
Bronze/gunmetal gate type outlet valve with 65 mm dia. instantaneous female coupling; cap and chain; wheel head secured by padlock and leather strap						
Flanged to BS 4504 PN6 (bolted connection to counter flanges measured separately)	326.50	413.22	1.75	68.82	nr	**482.04**
Bronze/gunmetal landing type outlet valve, with 65 mm dia. instantaneous female coupling; cap and chain; wheelhead secured by padlock and leather strap; bolted connections to counter flanges measured separately						
Horizontal, flanged to BS 4504 PN6	355.59	450.04	1.50	58.99	nr	**509.03**
Oblique, flanged to BS 4504 PN6	355.59	450.04	1.50	58.99	nr	**509.03**
Air valve, screwed joints to steel						
25 mm dia.	70.67	89.44	0.55	21.63	nr	**111.07**
INLET BOXES (BS 5041, Part 5)						
Steel dry riser inlet box with hinged wire glazed door suitably lettered (fixing by others)						
610 × 460 × 325 mm; double inlet	431.64	546.29	3.00	117.98	nr	**664.26**
610 × 610 × 356 mm; quadruple inlet	805.54	1019.50	3.00	117.98	nr	**1137.47**
OUTLET BOXES (BS 5041, Part 5)						
Steel dry riser outlet box with hinged wire glazed door suitably lettered (fixing by others)						
610 × 460 × 325 mm; single outlet	421.48	533.43	3.00	117.98	nr	**651.41**

38 PIPED SUPPLY SYSTEMS

Item	Net Price £	Material £	Labour hours	Labour £	Unit	Total rate £
SPRINKLERS						
PIPELINES						
Prefabricated black steel pipework; screwed joints, including all coupliings, unions and the like to BS 1387:1985; includes fixing to backgrounds, with brackets measured separately						
Heavy weight						
25 mm dia. – pipe plus allowance for coupling every 6 m	9.88	12.50	0.47	18.48	m	**30.99**
32 mm dia. – pipe plus allowance for coupling every 6 m	12.27	15.53	0.53	20.84	m	**36.38**
40 mm dia. – pipe plus allowance for coupling every 6 m	14.28	18.07	0.58	22.81	m	**40.88**
50 mm dia. – pipe plus allowance for coupling every 6 m	19.83	25.10	0.63	24.78	m	**49.87**
FIXINGS						
For steel pipes; black malleable iron. For minimum fixing distances, refer to the Tables and Memoranda at the rear of the book						
Pipe ring, single socket, black malleable iron						
25 mm dia.	1.90	2.41	0.12	4.72	nr	**7.13**
32 mm dia.	2.03	2.57	0.14	5.51	nr	**8.07**
40 mm dia.	2.60	3.29	0.15	5.90	nr	**9.19**
50 mm dia.	3.28	4.15	0.16	6.29	nr	**10.44**
Extra over channel sections for fabricated hangers and brackets						
Galvanized steel; including inserts, bolts, nuts, washers; fixed to backgrounds						
41 × 21 mm	9.61	12.16	0.29	11.40	m	**23.57**
41 × 41 mm	11.52	14.58	0.29	11.40	m	**25.99**
Threaded rods; metric thread; including nuts, washers etc.						
12 mm dia. × 600 mm long	4.85	6.14	0.18	7.08	nr	**13.22**
Extra over for black malleable iron fittings; BS 143						
Plain plug, solid						
25 mm dia.	3.51	4.45	0.40	15.73	nr	**20.18**
32 mm dia.	5.06	6.41	0.44	17.30	nr	**23.71**
40 mm dia.	7.13	9.02	0.48	18.88	nr	**27.90**
50 mm dia.	10.92	13.82	0.56	22.02	nr	**35.85**
Concentric reducing socket						
32 mm dia.	4.17	5.28	0.48	18.88	nr	**24.16**
40 mm dia.	5.43	6.88	0.55	21.63	nr	**28.51**
50 mm dia.	8.92	11.28	0.60	23.60	nr	**34.88**

38 PIPED SUPPLY SYSTEMS

Item	Net Price £	Material £	Labour hours	Labour £	Unit	Total rate £
SPRINKLERS – cont						
Extra over for black malleable iron fittings – cont						
Elbow; 90° female/female						
25 mm dia.	2.44	3.09	0.44	17.30	nr	**20.39**
32 mm dia.	4.50	5.70	0.53	20.84	nr	**26.54**
40 mm dia.	7.53	9.53	0.60	23.60	nr	**33.12**
50 mm dia.	10.30	13.04	0.65	25.56	nr	**38.60**
Tee						
25 mm dia. equal	3.36	4.25	0.51	20.06	nr	**24.30**
32 mm dia. reducing to 25 mm dia.	6.09	7.71	0.54	21.24	nr	**28.95**
40 mm dia.	9.36	11.85	0.65	25.56	nr	**37.41**
50 mm dia.	15.71	19.88	0.78	30.67	nr	**50.55**
Cross tee						
25 mm dia. equal	9.12	11.54	1.16	45.62	nr	**57.16**
32 mm dia.	13.26	16.79	1.40	55.06	nr	**71.84**
40 mm dia.	17.03	21.56	1.60	62.92	nr	**84.48**
50 mm dia.	32.29	40.86	1.68	66.07	nr	**106.93**
Prefabricated black steel pipework; welded joints, including all coupliings, unions and the like to BS 1387:1985; fixing to backgrounds						
Heavy weight						
65 mm dia. – pipe length only, welded joints on runs assumed	30.26	38.30	0.65	25.56	m	**63.86**
80 mm dia. – pipe length only, welded joints on runs assumed	38.52	48.76	0.70	27.53	m	**76.29**
100 mm dia. – pipe length only, welded joints on runs assumed	53.75	68.02	0.85	33.43	m	**101.45**
150 mm dia. – pipe length only, welded joints on runs assumed	72.72	92.03	1.15	45.22	m	**137.26**
FIXINGS						
For steel pipes; black malleable iron. For minimum fixing distances, refer to the Tables and Memoranda at the rear of the book						
Pipe ring, single socket, black malleable iron						
65 mm dia.	4.75	6.01	0.30	11.80	nr	**17.81**
80 mm dia.	5.69	7.21	0.35	13.76	nr	**20.97**
100 mm dia.	8.62	10.91	0.40	15.73	nr	**26.64**
150 mm dia.	19.51	24.69	0.77	30.28	nr	**54.97**
Extra over fittings						
Reducer (one size down)						
65 mm dia.	18.49	23.40	2.70	106.18	nr	**129.58**
80 mm dia.	24.04	30.43	2.86	112.47	nr	**142.90**
100 mm dia.	47.94	60.68	3.22	126.63	nr	**187.31**
150 mm dia.	111.96	141.70	4.20	165.17	nr	**306.87**

38 PIPED SUPPLY SYSTEMS

Item	Net Price £	Material £	Labour hours	Labour £	Unit	Total rate £
Elbow; 90°						
65 mm dia.	21.35	27.01	3.06	120.34	nr	**147.35**
80 mm dia.	31.38	39.72	3.40	133.71	nr	**173.42**
100 mm dia.	39.24	49.67	3.70	145.51	nr	**195.17**
150 mm dia.	58.82	74.44	5.20	204.49	nr	**278.93**
Branch bend						
65 mm dia.	21.75	27.52	3.60	141.57	nr	**169.10**
80 mm dia.	25.46	32.22	3.80	149.44	nr	**181.66**
100 mm dia.	39.16	49.56	5.10	200.56	nr	**250.12**
150 mm dia.	101.34	128.25	7.50	294.94	nr	**423.20**
Prefabricated black steel pipe; Victaulic Firelock joints; including all couplings and the like to BS 1387: 1985; fixing to backgrounds						
Heavy weight						
65 mm dia. – pipe plus allowance for Victaulic coupling every 6 m	27.18	34.40	0.70	27.53	m	**61.93**
80 mm dia. – pipe plus allowance for Victaulic coupling every 6 m	35.34	44.73	0.78	30.67	m	**75.40**
100 mm dia. – pipe plus allowance for Victaulic coupling every 6 m	51.52	65.20	0.93	36.57	m	**101.78**
150 mm dia. – pipe plus allowance for Victaulic coupling every 6 m	102.57	129.81	1.25	49.16	m	**178.97**
FIXINGS						
For fixings refer to For steel pipes; black malleable iron						
Extra over fittings						
Coupling						
65 mm dia. – Victaulic coupling	22.73	28.76	0.26	10.22	nr	**38.99**
80 mm dia. – Victaulic coupling	24.36	30.83	0.26	10.22	nr	**41.05**
100 mm dia. – Victaulic coupling	31.92	40.40	0.32	12.58	nr	**52.98**
150 mm dia. – Victaulic coupling	56.18	71.10	0.35	13.76	nr	**84.87**
Reducer						
65 mm dia. – Victaulic excl. couplings	18.49	23.40	0.48	18.88	nr	**42.28**
80 mm dia. – Victaulic excl. couplings	24.04	30.43	0.43	16.91	nr	**47.34**
100 mm dia. – Victaulic excl. couplings	47.94	60.68	0.45	17.70	nr	**78.37**
150 mm dia. – Victaulic excl. couplings	71.92	91.03	0.46	18.09	nr	**109.12**
Elbow; any degree						
65 mm dia. – Victaulic excl. couplings	25.39	32.14	0.56	22.02	nr	**54.16**
80 mm dia. – Victaulic excl. couplings	37.34	47.25	0.63	24.78	nr	**72.03**
100 mm dia. – Victaulic excl. couplings	72.00	91.13	0.71	27.92	nr	**119.05**
150 mm dia. – Victaulic excl. couplings	108.01	136.69	0.80	31.46	nr	**168.15**
Equal tee						
65 mm dia. – Victaulic excl. couplings	39.39	49.85	0.74	29.10	nr	**78.95**
80 mm dia. – Victaulic excl. couplings	47.41	60.00	0.83	32.64	nr	**92.64**
100 mm dia. – Victaulic excl. couplings	86.89	109.97	0.94	36.97	nr	**146.93**
150 mm dia. – Victaulic excl. couplings	130.38	165.00	1.05	41.29	nr	**206.29**

38 PIPED SUPPLY SYSTEMS

Item	Net Price £	Material £	Labour hours	Labour £	Unit	Total rate £
SPRINKLERS – cont						
PIPELINE ANCILLARIES						
SPRINKLER HEADS						
Sprinkler heads; brass body; frangible glass bulb; manufactured to standard operating temperature of 57–141°C; quick response; RTI<50						
Conventional pattern; 15 mm dia.	5.95	7.54	0.15	5.90	nr	**13.44**
Sidewall pattern; 15 mm dia.	8.73	11.05	0.15	5.90	nr	**16.95**
Conventional pattern; 15 mm dia.; satin chrome plated	6.99	8.84	0.15	5.90	nr	**14.74**
Sidewall pattern; 15 mm dia.; satin chrome plated	9.07	11.48	0.15	5.90	nr	**17.38**
Fully concealed; fusible link; 15 mm dia.	21.81	27.60	0.15	5.90	nr	**33.50**
VALVES						
Wet system alarm valves; including internal non-return valve; working pressure up to 12.5 bar; BS4504 PN16 flanged ends; bolted connections						
100 mm dia.	2015.05	2550.25	25.00	983.15	nr	**3533.39**
150 mm dia.	2461.45	3115.21	25.00	983.15	nr	**4098.35**
Wet system bypass alarm valves; including internal non-return valve; working pressure up to 12.5 bar; BS4504 PN16 flanged ends; bolted connections						
100 mm dia.	3551.98	4495.38	25.00	983.15	nr	**5478.53**
150 mm dia.	4490.91	5683.70	25.00	983.15	nr	**6666.85**
Alternate system wet/dry alarm station; including butterfly valve, wet alarm valve, dry pipe differential pressure valve and pressure gauges; working pressure up to 10.5 bar; BS4505 PN16 flanged ends; bolted connections						
100 mm dia.	4402.16	5571.37	40.00	1573.03	nr	**7144.40**
150 mm dia.	5130.19	6492.77	40.00	1573.03	nr	**8065.81**
Alternate system wet/dry alarm station; including electrically operated butterfly valve, water supply accelerator set, wet alarm valve, dry pipe differential pressure valve and pressure gauges; working pressure up to 10.5 bar; BS4505 PN16 flanged ends; bolted connections						
100 mm dia.	5017.91	6350.67	45.00	1769.66	nr	**8120.33**
150 mm dia.	5752.89	7280.86	45.00	1769.66	nr	**9050.52**
ALARM/GONGS						
Water operated motor alarm and gong; stainless steel and aluminum body and gong; screwed connections						
Connection to sprinkler system and drain pipework	599.60	758.85	6.00	235.95	nr	**994.80**

38 PIPED SUPPLY SYSTEMS

Item	Net Price £	Material £	Labour hours	Labour £	Unit	Total rate £
WATER TANKS						
Note: Prices are based on the most economical tank size for each volume, and the cost will vary with differing tank dimensions, for the same volume						
GRP sectional sprinkler tank; ordinary hazard, life safety classification; two compartment tank, complete with all fittings and accessories to comply with LPCB type A requirements; cost of erection (on prepared supports)is included within net price, labour cost allows for offloading and positioning of materials						
Volume, size						
55 m³, 6 m × 4 m × 3 m (h)	55184.12	69841.02	14.00	550.56	nr	**70391.58**
70 m³, 6 m × 5 m × 3 m (h)	61699.86	78087.34	16.00	629.21	nr	**78716.56**
80 m³, 8 m × 4 m × 3 m (h)	63943.14	80926.44	16.00	629.21	nr	**81555.65**
105 m³, 10 m × 4 m × 3 m (h)	72356.63	91574.55	24.00	943.82	nr	**92518.37**
125 m³, 10 m × 5 m × 3 m (h)	81602.24	103275.80	24.00	943.82	nr	**104219.62**
140 m³, 8 m × 7 m × 3 m (h)	84981.80	107552.96	24.00	943.82	nr	**108496.78**
135 m³, 9 m × 6 m × 3 m (h)	87551.06	110804.62	24.00	943.82	nr	**111748.44**
160 m³, 13 m × 5 m × 3 m (h)	96122.72	121652.91	24.00	943.82	nr	**122596.73**
185 m³, 12 m × 6 m × 3 m (h)	101679.79	128685.95	24.00	943.82	nr	**129629.77**
GRP sectional sprinkler tank; ordinary hazard, property protection classification; single compartment tank, complete with all fittings and accessories to comply with LPCB type A requirements; cost of erection (on prepared supports) is within net price, labour cost allows for offloading and positioning of materials						
Volume, size						
55 m³, 6 m × 4 m × 3 m (h)	42454.40	53730.28	14.00	550.56	nr	**54280.85**
70 m³, 6 m × 5 m × 3 m (h)	47962.83	60701.76	16.00	629.21	nr	**61330.97**
80 m³, 8 m × 4 m × 3 m (h)	51193.91	64791.01	16.00	629.21	nr	**65420.22**
105 m³, 10 m × 4 m × 3 m (h)	59974.83	75904.14	24.00	943.82	nr	**76847.96**
125 m³, 10 m × 5 m × 3 m (h)	67818.99	85831.71	24.00	943.82	nr	**86775.53**
140 m³, 8 m × 7 m × 3 m (h)	70208.24	88855.55	24.00	943.82	nr	**89799.36**
135 m³, 9 m × 6 m × 3 m (h)	73765.73	93357.90	24.00	943.82	nr	**94301.72**
160 m³, 13 m × 5 m × 3 m (h)	82697.12	104661.48	24.00	943.82	nr	**105605.30**
185 m³, 12 m × 6 m × 3 m (h)	86874.68	109948.60	24.00	943.82	nr	**110892.42**

38 PIPED SUPPLY SYSTEMS

Item	Net Price £	Material £	Labour hours	Labour £	Unit	Total rate £
WATER MIST SYSTEM						
PIPELINES						
For pipework prices refer to section – Cold Water: Stainless Steel						
PIPELINE ANCILLARIES						
Nozzles						
Semco CJ nozzles	129.49	163.88	0.75	29.49	nr	**193.38**
Extra over for rosettes (1 per nozzle)	16.99	21.50	0.20	7.87	nr	**29.36**
Extra over for protection caps (1 per nozzle)	1.58	2.00	0.10	3.93	nr	**5.93**
PUMP SETS						
Automatic pump units (APU); includes fixing in position; electrical work elsewhere						
15 l/s pump unit	2368.69	2997.82	16.00	629.21	nr	**3627.03**
18 l/s pump unit	2763.48	3497.47	16.00	629.21	nr	**4126.68**
Extra over for control module to duty pump	394.15	498.84	2.00	78.65	nr	**577.49**

38 PIPED SUPPLY SYSTEMS

Item	Net Price £	Material £	Labour hours	Labour £	Unit	Total rate £
FIRE EXTINGUISHERS AND HYDRANTS						
EXTINGUISHERS						
Fire extinguishers; hand held; BS 5423; placed in position						
Water type; cartridge operated; for Class A fires						
Water type, 9 litre capacity; 55 gm CO_2 cartridge; Class A fires (fire-rating 13 A)	44.24	55.99	1.00	39.33	nr	**95.32**
Foam type, 9 litre capacity; 75 gm CO_2 cartridge; Class A & B fires (fire-rating 13 A:183B)	57.63	72.93	1.00	39.33	nr	**112.26**
Dry powder type; cartridge operated; for Class A, B & C fires and electrical equipment fires						
Dry powder type, 1 kg capacity; 12 gm CO_2 cartridge; Class A, B & C fires (fire-rating 5 A:34B)	45.00	56.95	1.00	39.33	nr	**96.28**
Dry powder type, 2 kg capacity; 28 gm CO_2 cartridge; Class A, B & C fires (fire-rating 13 A:55B)	60.60	76.69	1.00	39.33	nr	**116.02**
Dry powder type, 4 kg capacity; 90 gm CO_2 cartridge; Class A, B & C fires (fire-rating 21 A:183B)	107.95	136.62	1.00	39.33	nr	**175.94**
Dry powder type, 9 kg capacity; 190 gm CO_2 cartridge; Class A, B & C fires (fire-rating 43 A:233B)	144.64	183.06	1.00	39.33	nr	**222.39**
Dry powder type; stored pressure type; for Class A, B & C fires and electrical equipment fires						
Dry powder type, 1 kg capacity; Class A, B & C fires (fire-rating 5 A:34B)	19.57	24.77	1.00	39.33	nr	**64.09**
Dry powder type, 2 kg capacity; Class A, B & C fires (fire-rating 13 A:55B)	17.96	22.73	1.00	39.33	nr	**62.05**
Dry powder type, 4 kg capacity; Class A, B & C fires (fire-rating 21 A:183B)	37.18	47.05	1.00	39.33	nr	**86.38**
Dry powder type, 9 kg capacity; Class A, B & C fires (fire-rating 43 A:233B)	39.40	49.87	1.00	39.33	nr	**89.20**
Carbon dioxide type; for Class B fires and electrical equipment fires						
CO_2 type with hose and horn, 2 kg capacity, Class B fires (fire-rating 34B)	38.67	48.94	1.00	39.33	nr	**88.27**
CO_2 type with hose and horn, 5 kg capacity, Class B fires (fire-rating 55B)	77.46	98.03	1.00	39.33	nr	**137.36**
Glass fibre blanket, in GRP container						
1100 × 1100 mm	14.86	18.81	0.50	19.66	nr	**38.47**
1200 × 1200 mm	17.34	21.95	0.50	19.66	nr	**41.61**
1800 × 1200 mm	21.81	27.60	0.50	19.66	nr	**47.26**

38 PIPED SUPPLY SYSTEMS

Item	Net Price £	Material £	Labour hours	Labour £	Unit	Total rate £
FIRE EXTINGUISHERS AND HYDRANTS – cont						
HYDRANTS						
Fire hydrants; bolted connections						
Underground hydrants, complete with frost plug to BS 750						
Sluice valve pattern type 1	294.26	372.42	4.50	176.97	nr	**549.39**
Screw down pattern type 2	213.74	270.51	4.50	176.97	nr	**447.47**
Stand pipe for underground hydrant; screwed base; light alloy						
Single outlet	169.24	214.19	1.00	39.33	nr	**253.52**
Double outlet	247.64	313.41	1.00	39.33	nr	**352.73**
64 mm dia. bronze/gunmetal outlet valves						
Oblique flanged landing valve	185.25	234.45	1.00	39.33	nr	**273.78**
Oblique screwed landing valve	185.25	234.45	1.00	39.33	nr	**273.78**
Cast iron surface box; fixing by others						
400 × 200 × 100 mm	162.20	205.28	1.00	39.33	nr	**244.60**
500 × 200 × 150 mm	218.54	276.58	1.00	39.33	nr	**315.91**
Frost plug	39.36	49.81	0.25	9.83	nr	**59.64**

38 MECHANICAL/COOLING/HEATING SYSTEMS

Item	Net Price £	Material £	Labour hours	Labour £	Unit	Total rate £
GAS/OIL FIRED BOILERS – DOMESTIC						
Domestic water boilers; stove enamelled casing; electric controls; placing in position; assembling and connecting; electrical work elsewhere						
Gas fired; floor standing; connected to conventional flue						
9–12 kW	804.89	1018.66	8.59	337.81	nr	**1356.47**
12–15 kW	846.73	1071.62	8.59	337.81	nr	**1409.43**
15–18 kW	901.85	1141.38	8.88	349.21	nr	**1490.60**
18–21 kW	1047.76	1326.05	9.92	390.11	nr	**1716.16**
21–23 kW	1161.55	1470.06	10.66	419.21	nr	**1889.27**
23–29 kW	1506.51	1906.64	11.81	464.44	nr	**2371.08**
29–37 kW	1781.84	2255.09	11.81	464.44	nr	**2719.53**
37–41 kW	1853.04	2345.20	12.68	498.65	nr	**2843.85**
Gas fired; wall hung; connected to conventional flue						
9–12 kW	707.43	895.32	8.59	337.81	nr	**1233.13**
12–15 kW	913.31	1155.88	8.59	337.81	nr	**1493.69**
13–18 kW	1159.07	1466.91	8.59	337.81	nr	**1804.72**
Gas fired; floor standing; connected to balanced flue						
9–12 kW	1005.42	1272.47	9.16	360.22	nr	**1632.69**
12–15 kW	1046.11	1323.96	10.95	430.62	nr	**1754.58**
15–18 kW	1126.04	1425.12	11.98	471.12	nr	**1896.24**
18–21 kW	1326.94	1679.38	12.78	502.58	nr	**2181.96**
21–23 kW	1530.44	1936.92	12.78	502.58	nr	**2439.51**
23–29 kW	1950.73	2468.85	15.45	607.58	nr	**3076.43**
29–37 kW	3750.07	4746.09	17.65	694.10	nr	**5440.20**
Gas fired; wall hung; connected to balanced flue						
6–9 kW	759.14	960.77	9.16	360.22	nr	**1321.00**
9–12 kW	868.71	1099.43	9.16	360.22	nr	**1459.66**
12–15 kW	983.68	1244.95	9.45	371.63	nr	**1616.58**
15–18 kW	1179.49	1492.76	9.74	383.03	nr	**1875.79**
18–22 kW	1251.99	1584.52	9.74	383.03	nr	**1967.55**
Gas fired; wall hung; connected to fan flue (including flue kit)						
6–9 kW	868.23	1098.84	9.16	360.22	nr	**1459.06**
9–12 kW	967.56	1224.54	9.16	360.22	nr	**1584.77**
12–15 kW	1050.31	1329.27	10.95	430.62	nr	**1759.89**
15–18 kW	1131.36	1431.85	11.98	471.12	nr	**1902.97**
18–23 kW	1474.13	1865.66	12.78	502.58	nr	**2368.25**
23–29 kW	1908.98	2416.00	15.45	607.58	nr	**3023.59**
29–35 kW	2343.67	2966.15	17.65	694.10	nr	**3660.25**

38 MECHANICAL/COOLING/HEATING SYSTEMS

Item	Net Price £	Material £	Labour hours	Labour £	Unit	Total rate £
GAS/OIL FIRED BOILERS – DOMESTIC – cont						
Domestic water boilers – cont						
Oil fired; floor standing; connected to conventional flue						
12–15 kW	1507.50	1907.90	10.38	408.20	nr	**2316.10**
15–19 kW	1572.55	1990.22	12.20	479.77	nr	**2470.00**
21–25 kW	1794.73	2271.41	14.30	562.36	nr	**2833.77**
26–32 kW	1967.69	2490.31	15.80	621.35	nr	**3111.66**
35–50 kW	2221.59	2811.64	20.46	804.61	nr	**3616.25**
Fire place mounted natural gas fire and back boiler; cast iron water boiler; electric control box; fire output 3 kW with wood surround						
10.50 kW	419.21	530.56	8.88	349.21	nr	**879.77**

38 MECHANICAL/COOLING/HEATING SYSTEMS

Item	Net Price £	Material £	Labour hours	Labour £	Unit	Total rate £
GAS/OIL FIRED BOILERS – COMMERCIAL; FORCED DRAFT						
Commercial steel shell floor standing boilers; pressure jet burner; 8 bar max working pressure; including controls, enamelled jacket, insulation, placing in position and commissioning; electrical work elsewhere						
Natural gas; burner and boiler (high/low type), connected to conventional flue						
411–500 kW	18548.92	23475.51	12.00	471.91	nr	**23947.42**
601–750 kW	21896.22	27711.86	12.00	471.91	nr	**28183.77**
1151–1250 kW	29741.80	37641.22	14.00	550.56	nr	**38191.78**
1426–1500 kW	36415.77	46087.79	14.00	550.56	nr	**46638.36**
1625–1760 kW	42910.38	54307.38	14.00	550.56	nr	**54857.94**
Natural gas; burner and boiler (modulating type), connected to conventional flue						
411–500 kW	19880.66	25160.96	12.00	471.91	nr	**25632.87**
601–750 kW	22970.28	29071.19	12.00	471.91	nr	**29543.10**
1151–1250 kW	33613.57	42541.33	12.00	471.91	nr	**43013.24**
1426–1500 kW	40124.44	50781.49	14.00	550.56	nr	**51332.05**
1625–1760 kW	48257.47	61074.65	14.00	550.56	nr	**61625.21**
1900–2050 kW	50632.84	64080.92	14.00	550.56	nr	**64631.48**
2450–2650 kW	59428.69	75212.95	14.00	550.56	nr	**75763.51**
Natural gas; low Nox burner and boiler (modulating type), connected to conventional flue						
420 kW	20841.98	26377.61	12.00	471.91	nr	**26849.52**
530 kW	22466.08	28433.07	12.00	471.91	nr	**28904.98**
620 kW	25917.45	32801.12	12.00	471.91	nr	**33273.03**
750 kW	26925.85	34077.36	14.00	550.56	nr	**34627.92**
1000 kW	33615.50	42543.78	14.00	550.56	nr	**43094.34**
1250 kW	42113.84	53299.28	14.00	550.56	nr	**53849.84**
1500 kW	51234.07	64841.83	14.00	550.56	nr	**65392.39**
1800 kW	60349.34	76378.12	14.00	550.56	nr	**76928.68**
2200 kW	64768.22	81970.66	14.00	550.56	nr	**82521.22**
2700 kW	71653.61	90684.81	14.00	550.56	nr	**91235.37**
Oil; burner and boiler (high/low type), connected to conventional flue						
411–500 kW	17914.64	22672.77	12.00	471.91	nr	**23144.68**
601–750 kW	20221.33	25592.12	12.00	471.91	nr	**26064.03**
1151–1250 kW	27325.20	34582.77	12.00	471.91	nr	**35054.68**
1426–1500 kW	33273.87	42111.41	14.00	550.56	nr	**42661.97**
1625–1760 kW	39637.90	50165.73	14.00	550.56	nr	**50716.29**

38 MECHANICAL/COOLING/HEATING SYSTEMS

Item	Net Price £	Material £	Labour hours	Labour £	Unit	Total rate £
GAS/OIL FIRED BOILERS – COMMERCIAL; FORCED DRAFT – cont						
Commercial steel shell floor standing boilers – cont						
Oil; burner and boiler (modulating type), connected to conventional flue						
420 kW	19442.11	24605.94	12.00	471.91	nr	**25077.85**
530 kW	19585.82	24787.81	12.00	471.91	nr	**25259.72**
620 kW	21529.53	27247.77	12.00	471.91	nr	**27719.68**
750 kW	23395.20	29608.97	14.00	550.56	nr	**30159.53**
1000 kW	29065.30	36785.04	14.00	550.56	nr	**37335.61**
1250 kW	38259.86	48421.67	14.00	550.56	nr	**48972.24**
1500 kW	41811.62	52916.78	14.00	550.56	nr	**53467.34**
1800 kW	49570.32	62736.20	14.00	550.56	nr	**63286.76**
2200 kW	71748.88	90805.38	14.00	550.56	nr	**91355.94**
2700 kW	80448.57	101815.71	14.00	550.56	nr	**102366.27**
Dual fuel; burner and boiler (high/low oil and modulating gast type), connected to conventional flue						
411–500 kW	22970.28	29071.19	12.00	471.91	nr	**29543.10**
601–750 kW	25799.76	32652.18	12.00	471.91	nr	**33124.09**
1151–1250 kW	39095.32	49479.04	12.00	471.91	nr	**49950.95**
1426–1500 kW	47020.62	59509.30	14.00	550.56	nr	**60059.86**
1625–1760 kW	54428.99	68885.33	14.00	550.56	nr	**69435.90**
2450–2650-kW	70834.60	89648.27	14.00	550.56	nr	**90198.83**
Dual fuel; low Nox burner and boiler (high/low oil and modulating gast type), connected to conventional flue						
420 kW	22606.07	28610.24	12.00	471.91	nr	**29082.15**
530 kW	24883.03	31491.96	12.00	471.91	nr	**31963.87**
620 kW	25740.30	32576.92	12.00	471.91	nr	**33048.83**
750 kW	27963.99	35391.22	14.00	550.56	nr	**35941.78**
1000 kW	39672.12	50209.03	14.00	550.56	nr	**50759.59**
1250 kW	48004.45	60754.43	14.00	550.56	nr	**61304.99**
1500 kW	52166.90	66022.43	14.00	550.56	nr	**66572.99**
1800 kW	68039.96	86111.37	14.00	550.56	nr	**86661.93**
2200 kW	73261.60	92719.89	14.00	550.56	nr	**93270.45**
2700 kW	79544.92	100672.05	14.00	550.56	nr	**101222.61**

38 MECHANICAL/COOLING/HEATING SYSTEMS

Item	Net Price £	Material £	Labour hours	Labour £	Unit	Total rate £
GAS/OIL FIRED BOILERS – COMMERCIAL; CONDENSING						
Gas boiler; low Nox wall mounted condensing boiler, with high efficiency modulating premix burner; Aluminium heat exchanger; includes integral control panel; delivery included						
Maximum output						
35 kW	4138.66	5237.89	11.00	432.58	nr	5670.48
45 kW	4771.95	6039.38	11.00	432.58	nr	6471.97
60 kW	4851.24	6139.73	11.00	432.58	nr	6572.31
80 kW	5426.05	6867.21	11.00	432.58	nr	7299.79
100 kW	5598.25	7085.14	11.00	432.58	nr	7517.73
120 kW	6527.37	8261.03	11.00	432.58	nr	8693.62
Commissioning rate	309.71	391.96		–	nr	391.96
Gas boiler; low Nox floor standing condensing boiler with high efficiency modulating premix burner; Stainless steel heat exchanger; includes controls; delivery and commissioning						
Maximum output						
50 kW	5806.37	7348.54	8.00	314.61	nr	7663.15
70 kW	6414.63	8118.36	8.00	314.61	nr	8432.97
100 kW	6942.37	8786.27	8.00	314.61	nr	9100.87
125 kW	7169.08	9073.18	10.00	393.26	nr	9466.44
150 kW	7996.61	10120.51	10.00	393.26	nr	10513.77
200 kW	9204.47	11649.17	10.00	393.26	nr	12042.43
250 kW	10552.31	13355.00	10.00	393.26	nr	13748.26
300 kW	11767.59	14893.07	10.00	393.26	nr	15286.33
350 kW	13276.48	16802.72	10.00	393.26	nr	17195.97
400 kW	14937.75	18905.21	10.00	393.26	nr	19298.47
450 kW	16585.38	20990.46	10.00	393.26	nr	21383.72
500 kW	18101.70	22909.52	10.00	393.26	nr	23302.77
575 kW	20363.80	25772.42	10.00	393.26	nr	26165.68
650 kW	21726.50	27497.06	10.00	393.26	nr	27890.32
720 kW	25343.87	32075.20	10.00	393.26	nr	32468.46
850 kW	31305.10	39619.73	12.00	471.91	nr	40091.64
1000 kW	34885.30	44150.83	12.00	471.91	nr	44622.74
1150 kW	41560.08	52598.44	12.00	471.91	nr	53070.35
1300 kW	45070.91	57041.75	14.00	550.56	nr	57592.31
1440 kW	51280.73	64900.89	14.00	550.56	nr	65451.46
1550 kW	60609.49	76707.37	14.00	550.56	nr	77257.93
1700 kW	63707.59	80628.33	14.00	550.56	nr	81178.89
2000 kW	71468.16	90450.10	14.00	550.56	nr	91000.66
2300 kW	74289.37	94020.63	14.00	550.56	nr	94571.19
3100 kW	95202.42	120488.19	14.00	550.56	nr	121038.75

38 MECHANICAL/COOLING/HEATING SYSTEMS

Item	Net Price £	Material £	Labour hours	Labour £	Unit	Total rate £
GAS/OIL FIRED BOILERS – COMMERCIAL; CONDENSING – cont						
Gas boilers, wall hung cascade; low Nox frame mounted, hung condensing boiler with high efficiency modulating premix burner; Aluminium heat exchanger; includes integral control panel; pumps and headers included, delivery included						
Maximum cascade output						
160 kW (2 units)	14644.14	18533.63	10.00	393.26	nr	**18926.89**
200 kW (2 units)	17093.30	21633.28	10.00	393.26	nr	**22026.54**
240 kW (3 units)	21327.60	26992.21	10.00	393.26	nr	**27385.47**
300 kW (3 units)	25404.57	32152.03	12.00	471.91	nr	**32623.94**
360 kW (3 units)	25937.27	32826.21	12.00	471.91	nr	**33298.12**
400 kW (4 units)	29700.82	37589.35	12.00	471.91	nr	**38061.26**
480 kW (4 units)	33885.57	42885.57	12.00	471.91	nr	**43357.48**
600 kW (5 units)	43163.12	54627.25	14.00	550.56	nr	**55177.81**
720 kW (6 units)	51769.24	65519.15	14.00	550.56	nr	**66069.71**
Commissioning rate	235.38	297.89		–	nr	**297.89**
Oil boiler; low Nox floor standing condensing boiler with high efficiency modulating premix burner; Stainless steel heat exchanger; includes controls; delivery and commissioning						
50 kW	6829.64	8643.59	8.00	314.61	nr	**8958.20**
80 kW	9040.94	11442.22	8.00	314.61	nr	**11756.82**
110 kW	12251.97	15506.10	8.00	314.61	nr	**15820.71**
130 kW	12621.14	15973.32	10.00	393.26	nr	**16366.58**
160 kW	13296.30	16827.80	10.00	393.26	nr	**17221.06**
200 kW	14187.02	17955.09	10.00	393.26	nr	**18348.35**
250 kW	17848.98	22589.67	10.00	393.26	nr	**22982.93**
300 kW	20088.78	25424.36	10.00	393.26	nr	**25817.62**
320 kW	28230.33	35728.31	10.00	393.26	nr	**36121.57**
400 kW	31362.08	39691.85	12.00	471.91	nr	**40163.76**
500 kW	38829.72	49142.89	12.00	471.91	nr	**49614.80**
600 kW	43309.31	54812.26	12.00	471.91	nr	**55284.17**

38 MECHANICAL/COOLING/HEATING SYSTEMS

Item	Net Price £	Material £	Labour hours	Labour £	Unit	Total rate £
GAS/OIL FIRED BOILERS – FLUE SYSTEMS						
Flues; suitable for domestic, medium sized industrial and commercial oil and gas appliances; stainless steel, twin wall, insulated; for use internally or externally						
Straight length; 120 mm long; including one locking band						
127 mm dia.	60.16	76.14	0.49	19.27	nr	**95.41**
152 mm dia.	67.41	85.31	0.51	20.06	nr	**105.36**
178 mm dia.	78.24	99.02	0.54	21.24	nr	**120.26**
203 mm dia.	89.02	112.66	0.58	22.81	nr	**135.47**
254 mm dia.	105.28	133.24	0.70	27.53	nr	**160.77**
304 mm dia.	131.90	166.93	0.74	29.10	nr	**196.03**
355 mm dia.	188.88	239.05	0.80	31.46	nr	**270.51**
Straight length; 300 mm long; including one locking band						
127 mm dia.	92.80	117.45	0.52	20.45	nr	**137.90**
152 mm dia.	104.90	132.77	0.52	20.45	nr	**153.22**
178 mm dia.	120.55	152.56	0.55	21.63	nr	**174.19**
203 mm dia.	136.41	172.64	0.64	25.17	nr	**197.81**
254 mm dia.	151.28	191.46	0.79	31.07	nr	**222.53**
304 mm dia.	181.48	229.69	0.86	33.82	nr	**263.51**
355 mm dia.	199.15	252.04	0.94	36.97	nr	**289.01**
400 mm dia.	213.11	269.71	1.03	40.51	nr	**310.21**
450 mm dia.	243.77	308.52	1.03	40.51	nr	**349.03**
500 mm dia.	261.45	330.89	1.10	43.26	nr	**374.15**
550 mm dia.	288.53	365.16	1.10	43.26	nr	**408.42**
600 mm dia.	318.37	402.93	1.10	43.26	nr	**446.19**
Straight length; 500 mm long; including one locking band						
127 mm dia.	109.07	138.03	0.55	21.63	nr	**159.66**
152 mm dia.	121.59	153.88	0.55	21.63	nr	**175.51**
178 mm dia.	136.92	173.29	0.63	24.78	nr	**198.07**
203 mm dia.	160.05	202.56	0.63	24.78	nr	**227.34**
254 mm dia.	185.92	235.30	0.86	33.82	nr	**269.12**
304 mm dia.	222.72	281.87	0.95	37.36	nr	**319.23**
355 mm dia.	250.13	316.56	1.03	40.51	nr	**357.07**
400 mm dia.	273.10	345.64	1.12	44.04	nr	**389.68**
450 mm dia.	315.41	399.18	1.12	44.04	nr	**443.22**
500 mm dia.	339.85	430.11	1.19	46.80	nr	**476.91**
550 mm dia.	374.64	474.15	1.19	46.80	nr	**520.94**
600 mm dia.	388.08	491.16	1.19	46.80	nr	**537.95**
Straight length; 1000 mm long; including one locking band						
127 mm dia.	194.95	246.73	0.62	24.38	nr	**271.11**
152 mm dia.	217.29	275.01	0.68	26.74	nr	**301.75**
178 mm dia.	244.48	309.42	0.74	29.10	nr	**338.52**
203 mm dia.	287.97	364.46	0.80	31.46	nr	**395.92**
254 mm dia.	326.35	413.02	0.87	34.21	nr	**447.24**
304 mm dia.	376.70	476.76	1.06	41.69	nr	**518.44**
355 mm dia.	432.27	547.08	1.16	45.62	nr	**592.70**

38 MECHANICAL/COOLING/HEATING SYSTEMS

Item	Net Price £	Material £	Labour hours	Labour £	Unit	Total rate £
GAS/OIL FIRED BOILERS – FLUE SYSTEMS – cont						
Flues – cont						
Straight length; 1000 mm long – cont						
400 mm dia.	462.92	585.87	1.26	49.55	nr	**635.43**
450 mm dia.	488.46	618.20	1.26	49.55	nr	**667.75**
500 mm dia.	530.08	670.87	1.33	52.30	nr	**723.18**
550 mm dia.	582.98	737.82	1.33	52.30	nr	**790.12**
600 mm dia.	612.46	775.13	1.33	52.30	nr	**827.44**
Adjustable length; boiler removal; internal use only; including one locking band						
127 mm dia.	89.38	113.11	0.52	20.45	nr	**133.56**
152 mm dia.	100.97	127.79	0.55	21.63	nr	**149.42**
178 mm dia.	114.45	144.85	0.59	23.20	nr	**168.05**
203 mm dia.	131.25	166.12	0.64	25.17	nr	**191.28**
254 mm dia.	194.60	246.29	0.79	31.07	nr	**277.35**
304 mm dia.	233.09	295.00	0.86	33.82	nr	**328.82**
355 mm dia.	261.38	330.80	0.99	38.93	nr	**369.73**
400 mm dia.	457.38	578.86	0.91	35.79	nr	**614.65**
450 mm dia.	489.58	619.62	0.91	35.79	nr	**655.40**
500 mm dia.	533.98	675.80	0.99	38.93	nr	**714.73**
550 mm dia.	582.53	737.25	0.99	38.93	nr	**776.18**
600 mm dia.	609.82	771.79	0.99	38.93	nr	**810.72**
Inspection length; 500 mm long; including one locking band						
127 mm dia.	229.70	290.71	0.55	21.63	nr	**312.34**
152 mm dia.	237.86	301.03	0.55	21.63	nr	**322.66**
178 mm dia.	250.53	317.07	0.63	24.78	nr	**341.84**
203 mm dia.	265.29	335.75	0.63	24.78	nr	**360.53**
254 mm dia.	343.01	434.11	0.86	33.82	nr	**467.93**
304 mm dia.	371.74	470.48	0.95	37.36	nr	**507.84**
355 mm dia.	419.74	531.23	1.03	40.51	nr	**571.73**
400 mm dia.	673.40	852.25	1.12	44.04	nr	**896.30**
450 mm dia.	691.72	875.44	1.12	44.04	nr	**919.49**
500 mm dia.	759.04	960.64	1.19	46.80	nr	**1007.44**
550 mm dia.	793.71	1004.52	1.19	46.80	nr	**1051.32**
600 mm dia.	807.19	1021.58	1.19	46.80	nr	**1068.38**
Adapters						
127 mm dia.	19.21	24.32	0.49	19.27	nr	**43.59**
152 mm dia.	21.02	26.60	0.51	20.06	nr	**46.65**
178 mm dia.	22.49	28.46	0.54	21.24	nr	**49.69**
203 mm dia.	25.59	32.39	0.58	22.81	nr	**55.20**
254 mm dia.	28.16	35.63	0.70	27.53	nr	**63.16**
304 mm dia.	35.13	44.46	0.74	29.10	nr	**73.56**
355 mm dia.	42.65	53.98	0.80	31.46	nr	**85.44**
400 mm dia.	47.84	60.55	0.89	35.00	nr	**95.55**
450 mm dia.	51.11	64.68	0.89	35.00	nr	**99.68**
500 mm dia.	54.25	68.66	0.96	37.75	nr	**106.41**
550 mm dia.	63.06	79.81	0.96	37.75	nr	**117.57**
600 mm dia.	75.79	95.93	0.96	37.75	nr	**133.68**

38 MECHANICAL/COOLING/HEATING SYSTEMS

Item	Net Price £	Material £	Labour hours	Labour £	Unit	Total rate £
Fittings for flue system						
90° insulated tee; including two locking bands						
127 mm dia.	223.32	282.64	1.89	74.33	nr	**356.96**
152 mm dia.	257.75	326.21	2.04	80.22	nr	**406.44**
178 mm dia.	281.72	356.55	2.39	93.99	nr	**450.54**
203 mm dia.	329.86	417.48	2.56	100.67	nr	**518.15**
254 mm dia.	334.15	422.91	2.95	116.01	nr	**538.92**
304 mm dia.	415.54	525.91	3.41	134.10	nr	**660.01**
355 mm dia.	529.76	670.47	3.77	148.26	nr	**818.72**
400 mm dia.	698.08	883.48	4.25	167.13	nr	**1050.62**
450 mm dia.	727.60	920.85	4.76	187.19	nr	**1108.04**
500 mm dia.	823.33	1042.00	5.12	201.35	nr	**1243.35**
550 mm dia.	885.02	1120.09	5.61	220.62	nr	**1340.70**
600 mm dia.	920.87	1165.45	5.98	235.17	nr	**1400.62**
135° insulated tee; including two locking bands						
127 mm dia.	289.65	366.58	1.89	74.33	nr	**440.90**
152 mm dia.	313.18	396.36	2.04	80.22	nr	**476.58**
178 mm dia.	342.40	433.34	2.39	93.99	nr	**527.33**
203 mm dia.	434.55	549.96	2.56	100.67	nr	**650.64**
254 mm dia.	493.77	624.92	2.95	116.01	nr	**740.93**
304 mm dia.	614.09	777.19	3.41	134.10	nr	**911.29**
355 mm dia.	732.04	926.47	3.77	148.26	nr	**1074.73**
400 mm dia.	963.63	1219.57	4.25	167.13	nr	**1386.70**
450 mm dia.	1043.20	1320.27	4.76	187.19	nr	**1507.46**
500 mm dia.	1215.30	1538.08	5.12	201.35	nr	**1739.43**
550 mm dia.	1246.91	1578.09	5.61	220.62	nr	**1798.70**
600 mm dia.	1315.98	1665.50	5.98	235.17	nr	**1900.67**
Wall sleeve; for 135° tee through wall						
127 mm dia.	29.64	37.51	1.89	74.33	nr	**111.83**
152 mm dia.	39.78	50.34	2.04	80.22	nr	**130.57**
178 mm dia.	41.65	52.71	2.39	93.99	nr	**146.70**
203 mm dia.	46.87	59.31	2.56	100.67	nr	**159.99**
254 mm dia.	52.26	66.15	2.95	116.01	nr	**182.16**
304 mm dia.	60.91	77.09	3.41	134.10	nr	**211.19**
355 mm dia.	67.66	85.63	3.77	148.26	nr	**233.89**
15° insulated elbow; including two locking bands						
127 mm dia.	154.98	196.14	1.57	61.74	nr	**257.88**
152 mm dia.	172.24	217.99	1.79	70.39	nr	**288.39**
178 mm dia.	184.22	233.14	2.05	80.62	nr	**313.76**
203 mm dia.	195.30	247.17	2.33	91.63	nr	**338.80**
254 mm dia.	201.15	254.57	2.45	96.35	nr	**350.92**
304 mm dia.	254.29	321.83	3.43	134.89	nr	**456.72**
355 mm dia.	340.19	430.54	4.71	185.22	nr	**615.76**

38 MECHANICAL/COOLING/HEATING SYSTEMS

Item	Net Price £	Material £	Labour hours	Labour £	Unit	Total rate £
GAS/OIL FIRED BOILERS – FLUE SYSTEMS – cont						
Fittings for flue system – cont						
30° insulated elbow; including two locking bands						
127 mm dia.	154.98	196.14	1.44	56.63	nr	**252.77**
152 mm dia.	172.24	217.99	1.62	63.71	nr	**281.70**
178 mm dia.	184.22	233.14	1.89	74.33	nr	**307.47**
203 mm dia.	195.30	247.17	2.16	84.94	nr	**332.11**
254 mm dia.	201.15	254.57	2.17	85.34	nr	**339.91**
304 mm dia.	254.29	321.83	2.74	107.75	nr	**429.58**
355 mm dia.	339.54	429.72	3.17	124.66	nr	**554.39**
400 mm dia.	340.19	430.54	3.53	138.82	nr	**569.36**
450 mm dia.	394.27	498.98	3.88	152.58	nr	**651.57**
500 mm dia.	413.35	523.14	4.24	166.74	nr	**689.88**
550 mm dia.	443.61	561.43	4.61	181.29	nr	**742.72**
600 mm dia.	483.71	612.18	4.96	195.06	nr	**807.24**
45° insulated elbow; including two locking bands						
127 mm dia.	154.98	196.14	1.44	56.63	nr	**252.77**
152 mm dia.	172.24	217.99	1.51	59.38	nr	**277.37**
178 mm dia.	184.22	233.14	1.58	62.13	nr	**295.28**
203 mm dia.	195.30	247.17	1.66	65.28	nr	**312.45**
254 mm dia.	201.15	254.57	1.72	67.64	nr	**322.21**
304 mm dia.	254.29	321.83	1.80	70.79	nr	**392.61**
355 mm dia.	340.19	430.54	1.94	76.29	nr	**506.83**
400 mm dia.	433.83	549.05	2.01	79.04	nr	**628.10**
450 mm dia.	454.41	575.10	2.09	82.19	nr	**657.29**
500 mm dia.	487.43	616.89	2.16	84.94	nr	**701.84**
550 mm dia.	531.62	672.81	2.23	87.70	nr	**760.51**
600 mm dia.	545.08	689.86	2.30	90.45	nr	**780.31**
Flue supports						
Wall support, galvanized; including plate and brackets						
127 mm dia.	94.09	119.08	2.24	88.09	nr	**207.17**
152 mm dia.	103.45	130.92	2.44	95.96	nr	**226.88**
178 mm dia.	114.21	144.54	2.52	99.10	nr	**243.64**
203 mm dia.	118.92	150.51	2.77	108.93	nr	**259.44**
254 mm dia.	141.82	179.49	2.98	117.19	nr	**296.68**
304 mm dia.	160.35	202.94	3.46	136.07	nr	**339.01**
355 mm dia.	214.64	271.65	4.08	160.45	nr	**432.10**
400 mm dia.; with 300 mm support length and collar	551.33	697.77	4.80	188.76	nr	**886.53**
450 mm dia.; with 300 mm support length and collar	587.80	743.92	5.62	221.01	nr	**964.93**
500 mm dia.; with 300 mm support length and collar	642.16	812.72	6.24	245.39	nr	**1058.12**
550 mm dia.; with 300 mm support length and collar	702.57	889.18	6.97	274.10	nr	**1163.28**
600 mm dia.; with 300 mm support length and collar	741.66	938.64	7.49	294.55	nr	**1233.19**
Ceiling/floor support						
127 mm dia.	29.13	36.87	1.86	73.15	nr	**110.02**
152 mm dia.	32.38	40.98	1.93	75.90	nr	**116.88**
178 mm dia.	46.26	58.54	2.14	84.16	nr	**142.70**
203 mm dia.	58.37	73.88	2.74	107.75	nr	**181.63**
254 mm dia.	66.59	84.28	3.21	126.24	nr	**210.52**
304 mm dia.	76.10	96.32	3.68	144.72	nr	**241.04**
355 mm dia.	90.70	114.79	4.28	168.31	nr	**283.11**

38 MECHANICAL/COOLING/HEATING SYSTEMS

Item	Net Price £	Material £	Labour hours	Labour £	Unit	Total rate £
400 mm dia.	115.86	146.63	4.86	191.12	nr	**337.75**
450 mm dia.	125.82	159.23	5.46	214.72	nr	**373.95**
500 mm dia.	152.75	193.32	6.04	237.53	nr	**430.85**
550 mm dia.	186.87	236.50	6.65	261.52	nr	**498.02**
600 mm dia.	268.15	339.38	7.24	284.72	nr	**624.09**
Ceiling/floor firestop spacer						
127 mm dia.	5.67	7.18	0.66	25.96	nr	**33.13**
152 mm dia.	6.31	7.99	0.69	27.13	nr	**35.13**
178 mm dia.	7.13	9.02	0.70	27.53	nr	**36.55**
203 mm dia.	8.45	10.70	0.87	34.21	nr	**44.91**
254 mm dia.	8.74	11.06	0.91	35.79	nr	**46.84**
304 mm dia.	10.49	13.27	0.95	37.36	nr	**50.63**
355 mm dia.	19.10	24.17	0.99	38.93	nr	**63.10**
Wall band; internal or external use						
127 mm dia.	34.29	43.40	1.03	40.51	nr	**83.90**
152 mm dia.	35.81	45.32	1.07	42.08	nr	**87.40**
178 mm dia.	37.23	47.11	1.11	43.65	nr	**90.77**
203 mm dia.	39.30	49.74	1.18	46.40	nr	**96.14**
254 mm dia.	40.90	51.76	1.30	51.12	nr	**102.89**
304 mm dia.	44.20	55.94	1.45	57.02	nr	**112.96**
355 mm dia.	47.02	59.51	1.65	64.89	nr	**124.40**
400 mm dia.	58.32	73.81	1.85	72.75	nr	**146.56**
450 mm dia.	62.14	78.64	2.25	88.48	nr	**167.12**
500 mm dia.	74.70	94.54	2.39	93.99	nr	**188.53**
550 mm dia.	78.24	99.02	2.45	96.35	nr	**195.37**
600 mm dia.	82.71	104.68	2.66	104.61	nr	**209.29**
Flashings and terminals						
Insulated top stub; including one locking band						
127 mm dia.	84.49	106.93	1.49	58.60	nr	**165.53**
152 mm dia.	95.55	120.93	1.90	74.72	nr	**195.65**
178 mm dia.	102.87	130.19	1.92	75.51	nr	**205.69**
203 mm dia.	109.47	138.54	2.20	86.52	nr	**225.06**
254 mm dia.	116.23	147.10	2.49	97.92	nr	**245.02**
304 mm dia.	158.87	201.06	2.79	109.72	nr	**310.78**
355 mm dia.	209.74	265.45	3.19	125.45	nr	**390.90**
400 mm dia.	202.24	255.96	3.59	141.18	nr	**397.14**
450 mm dia.	219.72	278.07	3.97	156.12	nr	**434.20**
500 mm dia.	254.17	321.68	4.38	172.25	nr	**493.93**
550 mm dia.	265.96	336.60	4.78	187.98	nr	**524.58**
600 mm dia.	275.24	348.34	5.17	203.31	nr	**551.66**
Rain cap; including one locking band						
127 mm dia.	45.18	57.18	1.49	58.60	nr	**115.77**
152 mm dia.	47.21	59.75	1.54	60.56	nr	**120.32**
178 mm dia.	51.98	65.79	1.72	67.64	nr	**133.43**
203 mm dia.	62.16	78.67	2.00	78.65	nr	**157.32**
254 mm dia.	81.73	103.44	2.49	97.92	nr	**201.36**
304 mm dia.	115.02	145.56	2.80	110.11	nr	**255.68**
355 mm dia.	147.65	186.86	3.19	125.45	nr	**312.31**

38 MECHANICAL/COOLING/HEATING SYSTEMS

Item	Net Price £	Material £	Labour hours	Labour £	Unit	Total rate £
GAS/OIL FIRED BOILERS – FLUE SYSTEMS – cont						
Flashings and terminals – cont						
Rain cap; including one locking band – cont						
400 mm dia.	154.25	195.22	3.45	135.67	nr	330.89
450 mm dia.	160.20	202.74	3.97	156.12	nr	358.87
500 mm dia.	173.20	219.20	4.38	172.25	nr	391.45
550 mm dia.	185.84	235.20	4.78	187.98	nr	423.18
600 mm dia.	198.70	251.47	5.17	203.31	nr	454.79
Round top; including one locking band						
127 mm dia.	86.37	109.31	1.49	58.60	nr	167.91
152 mm dia.	94.17	119.18	1.65	64.89	nr	184.07
178 mm dia.	107.34	135.85	1.92	75.51	nr	211.35
203 mm dia.	126.33	159.89	2.20	86.52	nr	246.40
254 mm dia.	149.37	189.05	2.49	97.92	nr	286.97
304 mm dia.	196.19	248.29	2.80	110.11	nr	358.41
355 mm dia.	261.72	331.24	3.19	125.45	nr	456.69
Coping cap; including one locking band						
127 mm dia.	48.37	61.22	1.49	58.60	nr	119.82
152 mm dia.	50.67	64.12	1.65	64.89	nr	129.01
178 mm dia.	55.74	70.55	1.92	75.51	nr	146.06
203 mm dia.	66.90	84.67	2.20	86.52	nr	171.19
254 mm dia.	81.73	103.44	2.49	97.92	nr	201.36
304 mm dia.	110.20	139.47	2.79	109.72	nr	249.19
355 mm dia.	147.65	186.86	3.19	125.45	nr	312.31
Storm collar						
127 mm dia.	9.16	11.60	0.52	20.45	nr	32.04
152 mm dia.	9.81	12.41	0.55	21.63	nr	34.04
178 mm dia.	10.88	13.76	0.57	22.42	nr	36.18
203 mm dia.	11.39	14.42	0.66	25.96	nr	40.37
254 mm dia.	14.28	18.07	0.66	25.96	nr	44.02
304 mm dia.	14.86	18.80	0.72	28.31	nr	47.12
355 mm dia.	15.89	20.11	0.77	30.28	nr	50.39
400 mm dia.	42.14	53.33	0.82	32.25	nr	85.58
450 mm dia.	46.38	58.69	0.87	34.21	nr	92.91
500 mm dia.	50.56	63.99	0.92	36.18	nr	100.17
550 mm dia.	54.80	69.36	0.98	38.54	nr	107.90
600 mm dia.	59.00	74.68	1.03	40.51	nr	115.18
Flat flashing; including storm collar and sealant						
127 mm dia.	51.50	65.18	1.49	58.60	nr	123.78
152 mm dia.	53.22	67.35	1.65	64.89	nr	132.24
178 mm dia.	55.83	70.66	1.92	75.51	nr	146.17
203 mm dia.	61.16	77.40	2.20	86.52	nr	163.92
254 mm dia.	83.55	105.74	2.49	97.92	nr	203.66
304 mm dia.	99.83	126.34	2.80	110.11	nr	236.45
355 mm dia.	157.32	199.11	3.20	125.84	nr	324.95
400 mm dia.	219.46	277.75	3.59	141.18	nr	418.93
450 mm dia.	252.12	319.09	3.97	156.12	nr	475.21
500 mm dia.	273.11	345.65	4.38	172.25	nr	517.90
550 mm dia.	290.01	367.03	4.78	187.98	nr	555.01
600 mm dia.	300.48	380.29	5.17	203.31	nr	583.61

38 MECHANICAL/COOLING/HEATING SYSTEMS

Item	Net Price £	Material £	Labour hours	Labour £	Unit	Total rate £
5°–30° rigid adjustable flashing; including storm collar and sealant						
127 mm dia.	87.71	111.01	1.49	58.60	nr	**169.61**
152 mm dia.	92.42	116.96	1.65	64.89	nr	**181.85**
178 mm dia.	98.27	124.37	1.92	75.51	nr	**199.87**
203 mm dia.	103.33	130.78	2.20	86.52	nr	**217.29**
254 mm dia.	108.79	137.69	2.49	97.92	nr	**235.61**
304 mm dia.	134.62	170.37	2.80	110.11	nr	**280.48**
355 mm dia.	153.11	193.78	3.19	125.45	nr	**319.22**
400 mm dia.	493.59	624.69	3.59	141.18	nr	**765.87**
450 mm dia.	576.05	729.05	3.97	156.12	nr	**885.17**
500 mm dia.	614.98	778.31	4.38	172.25	nr	**950.56**
550 mm dia.	643.54	814.47	4.77	187.58	nr	**1002.05**
600 mm dia.	700.16	886.13	5.17	203.31	nr	**1089.44**
Domestic and small commercial; twin walled gas vent system suitable for gas fired appliances; domestic gas boilers; small commercial boilers with internal or external flues						
152 mm long						
100 mm dia.	9.38	11.87	0.52	20.45	nr	**32.32**
125 mm dia.	11.52	14.58	0.52	20.45	nr	**35.03**
150 mm dia.	12.49	15.80	0.52	20.45	nr	**36.25**
305 mm long						
100 mm dia.	14.25	18.04	0.52	20.45	nr	**38.49**
125 mm dia.	16.71	21.15	0.52	20.45	nr	**41.60**
150 mm dia.	19.82	25.08	0.52	20.45	nr	**45.53**
457 mm long						
100 mm dia.	15.75	19.93	0.55	21.63	nr	**41.56**
125 mm dia.	17.71	22.41	0.55	21.63	nr	**44.04**
150 mm dia.	21.91	27.72	0.55	21.63	nr	**49.35**
914 mm long						
100 mm dia.	28.14	35.62	0.62	24.38	nr	**60.00**
125 mm dia.	32.81	41.52	0.62	24.38	nr	**65.90**
150 mm dia.	37.61	47.60	0.62	24.38	nr	**71.99**
1524 mm long						
100 mm dia.	40.64	51.44	0.82	32.25	nr	**83.68**
125 mm dia.	50.02	63.31	0.84	33.03	nr	**96.34**
150 mm dia.	53.71	67.97	0.84	33.03	nr	**101.01**
Adjustable length 305 mm long						
100 mm dia.	18.03	22.82	0.56	22.02	nr	**44.84**
125 mm dia.	20.23	25.60	0.56	22.02	nr	**47.63**
150 mm dia.	25.50	32.27	0.56	22.02	nr	**54.30**
Adjustable length 457 mm long						
100 mm dia.	24.30	30.76	0.56	22.02	nr	**52.78**
125 mm dia.	29.47	37.30	0.56	22.02	nr	**59.32**
150 mm dia.	32.78	41.49	0.56	22.02	nr	**63.51**
Adjustable elbow 0°–90°						
100 mm dia.	20.55	26.01	0.48	18.88	nr	**44.89**
125 mm dia.	24.30	30.76	0.48	18.88	nr	**49.63**
150 mm dia.	30.42	38.50	0.48	18.88	nr	**57.38**

38 MECHANICAL/COOLING/HEATING SYSTEMS

Item	Net Price £	Material £	Labour hours	Labour £	Unit	Total rate £
GAS/OIL FIRED BOILERS – FLUE SYSTEMS – cont						
Domestic and small commercial – cont						
Draughthood connector						
100 mm dia.	6.34	8.02	0.48	18.88	nr	**26.90**
125 mm dia.	7.14	9.03	0.48	18.88	nr	**27.91**
150 mm dia.	7.76	9.82	0.48	18.88	nr	**28.69**
Adaptor						
100 mm dia.	15.39	19.47	0.48	18.88	nr	**38.35**
125 mm dia.	15.71	19.88	0.48	18.88	nr	**38.76**
150 mm dia.	16.04	20.30	0.48	18.88	nr	**39.18**
Support plate						
100 mm dia.	11.09	14.04	0.48	18.88	nr	**32.92**
125 mm dia.	11.80	14.94	0.48	18.88	nr	**33.81**
150 mm dia.	12.64	16.00	0.48	18.88	nr	**34.87**
Wall band						
100 mm dia.	10.04	12.70	0.48	18.88	nr	**31.58**
125 mm dia.	10.70	13.54	0.48	18.88	nr	**32.41**
150 mm dia.	13.56	17.16	0.48	18.88	nr	**36.03**
Firestop						
100 mm dia.	4.34	5.50	0.48	18.88	nr	**24.37**
125 mm dia.	4.34	5.50	0.48	18.88	nr	**24.37**
150 mm dia.	5.00	6.33	0.48	18.88	nr	**25.20**
Flat flashing						
125 mm dia.	29.57	37.43	0.55	21.63	nr	**59.06**
150 mm dia.	41.17	52.10	0.55	21.63	nr	**73.73**
Adjustable flashing 5°–30°						
100 mm dia.	77.10	97.57	0.55	21.63	nr	**119.20**
125 mm dia.	120.35	152.32	0.55	21.63	nr	**173.95**
Storm collar						
100 mm dia.	6.15	7.78	0.55	21.63	nr	**29.41**
125 mm dia.	6.29	7.96	0.55	21.63	nr	**29.59**
150 mm dia.	6.44	8.15	0.55	21.63	nr	**29.78**
Gas vent terminal						
100 mm dia.	23.47	29.70	0.55	21.63	nr	**51.33**
125 mm dia.	25.78	32.63	0.55	21.63	nr	**54.26**
150 mm dia.	33.08	41.86	0.55	21.63	nr	**63.49**
Twin wall galvanized steel flue box, 125 mm dia.; fitted for gas fire, where no chimney exists						
Free-standing	156.19	197.67	2.15	84.55	nr	**282.22**
Recess	156.19	197.67	2.15	84.55	nr	**282.22**
Back boiler	115.07	145.63	2.40	94.38	nr	**240.01**

38 MECHANICAL/COOLING/HEATING SYSTEMS

Item	Net Price £	Material £	Labour hours	Labour £	Unit	Total rate £
PACKAGED STEAM GENERATORS						
Packaged steam boilers; boiler mountings centrifugal water feed pump; insulation; and sheet steel wrap around casing; plastic coated						
Gas fired						
276 kW	20140.29	25489.55	86.00	3382.02	nr	**28871.57**
1384 kW	76916.71	97345.79	148.00	5820.22	nr	**103166.01**
2940 KW	86945.69	110038.47	207.00	8140.44	nr	**118178.91**
4843 kW	130083.79	164634.04	295.00	11601.11	nr	**176235.16**
5213 kW	150678.96	190699.29	310.00	12191.00	nr	**202890.29**
Gas oil fired						
276 kW	16039.61	20299.73	86.00	3382.02	nr	**23681.75**
1384 kW	80253.41	101568.71	148.00	5820.22	nr	**107388.93**
2940 kW	96630.47	122295.52	207.00	8140.44	nr	**130435.97**
4843 kW	139811.76	176945.76	295.00	11601.11	nr	**188546.88**
5213 kW	168706.82	213515.35	310.00	12191.00	nr	**225706.35**

38 MECHANICAL/COOLING/HEATING SYSTEMS

Item	Net Price £	Material £	Labour hours	Labour £	Unit	Total rate £
LOW TEMPERATURE HOT WATER HEATING; PIPELINE: SCREWED STEEL						
Black steel pipes; screwed and socketed joints; BS 1387: 1985. Fixed vertically, brackets measured separately. Screwed joints are within the running length, but any flanges are additional						
Medium weight						
10 mm dia.	3.78	4.78	0.37	14.55	m	19.33
15 mm dia.	4.60	5.82	0.37	14.55	m	20.37
20 mm dia.	5.42	6.85	0.37	14.55	m	21.41
25 mm dia.	7.79	9.86	0.41	16.12	m	25.99
32 mm dia.	9.64	12.20	0.48	18.88	m	31.07
40 mm dia.	11.20	14.17	0.52	20.45	m	34.62
50 mm dia.	15.74	19.92	0.62	24.38	m	44.31
65 mm dia.	21.37	27.05	0.65	25.56	m	52.61
80 mm dia.	27.78	35.16	1.10	43.26	m	78.42
100 mm dia.	39.33	49.78	1.31	51.52	m	101.30
125 mm dia.	50.07	63.37	1.66	65.28	m	128.65
150 mm dia.	58.13	73.57	1.88	73.93	m	147.50
Heavy weight						
15 mm dia.	5.47	6.93	0.37	14.55	m	21.48
20 mm dia.	6.49	8.21	0.37	14.55	m	22.76
25 mm dia.	9.51	12.04	0.41	16.12	m	28.16
32 mm dia.	11.80	14.94	0.48	18.88	m	33.81
40 mm dia.	13.77	17.43	0.52	20.45	m	37.88
50 mm dia.	19.09	24.16	0.62	24.38	m	48.54
65 mm dia.	25.97	32.87	0.64	25.17	m	58.04
80 mm dia.	33.06	41.84	1.10	43.26	m	85.10
100 mm dia.	46.14	58.40	1.31	51.52	m	109.92
125 mm dia.	53.40	67.59	1.66	65.28	m	132.87
150 mm dia.	62.41	78.98	1.88	73.93	m	152.91
200 mm dia.	98.38	124.51	2.99	117.58	m	242.09
250 mm dia.	123.33	156.09	3.49	137.25	m	293.34
300 mm dia.	146.86	185.86	3.91	153.76	m	339.63
Black steel pipes; screwed and socketed joints; BS 1387: 1985. Fixed at high level or suspended, brackets measured separately. Screwed joints are within the running length, but any flanges are additional						
Medium weight						
10 mm dia.	3.78	4.78	0.58	22.81	m	27.59
15 mm dia.	4.60	5.82	0.58	22.81	m	28.63
20 mm dia.	5.42	6.85	0.58	22.81	m	29.66
25 mm dia.	7.79	9.86	0.60	23.60	m	33.46
32 mm dia.	9.64	12.20	0.68	26.74	m	38.94
40 mm dia.	11.20	14.17	0.73	28.71	m	42.88
50 mm dia.	15.74	19.92	0.85	33.43	m	53.35
65 mm dia.	21.37	27.05	0.88	34.61	m	61.65
80 mm dia.	27.78	35.16	1.45	57.02	m	92.19
100 mm dia.	39.33	49.78	1.74	68.43	m	118.21
125 mm dia.	50.07	63.37	2.21	86.91	m	150.28
150 mm dia.	58.13	73.57	2.50	98.31	m	171.88

38 MECHANICAL/COOLING/HEATING SYSTEMS

Item	Net Price £	Material £	Labour hours	Labour £	Unit	Total rate £
Heavy weight						
15 mm dia.	5.47	6.93	0.58	22.81	m	**29.74**
20 mm dia.	6.49	8.21	0.58	22.81	m	**31.02**
25 mm dia.	9.51	12.04	0.60	23.60	m	**35.63**
32 mm dia.	11.80	14.94	0.68	26.74	m	**41.68**
40 mm dia.	13.77	17.43	0.73	28.71	m	**46.14**
50 mm dia.	19.09	24.16	0.85	33.43	m	**57.59**
65 mm dia.	25.97	32.87	0.88	34.61	m	**67.47**
80 mm dia.	33.06	41.84	1.45	57.02	m	**98.86**
100 mm dia.	46.14	58.40	1.74	68.43	m	**126.83**
125 mm dia.	53.40	67.59	2.21	86.91	m	**154.50**
150 mm dia.	62.41	78.98	2.50	98.31	m	**177.30**
200 mm dia.	98.38	124.51	2.99	117.58	m	**242.09**
250 mm dia.	123.33	156.09	3.49	137.25	m	**293.34**
300 mm dia.	146.86	185.86	3.91	153.76	m	**339.63**

FIXINGS

For steel pipes; black malleable iron. For minimum fixing distances, refer to the Tables and Memoranda to the rear of the book

Item	Net Price £	Material £	Labour hours	Labour £	Unit	Total rate £
Single pipe bracket, screw on, black malleable iron; screwed to wood						
15 mm dia.	1.32	1.66	0.14	5.51	nr	**7.17**
20 mm dia.	1.47	1.86	0.14	5.51	nr	**7.37**
25 mm dia.	1.72	2.17	0.17	6.69	nr	**8.86**
32 mm dia.	3.13	3.96	0.19	7.47	nr	**11.43**
40 mm dia.	3.71	4.69	0.22	8.65	nr	**13.35**
50 mm dia.	4.18	5.29	0.22	8.65	nr	**13.94**
65 mm dia.	5.48	6.93	0.28	11.01	nr	**17.94**
80 mm dia.	7.71	9.76	0.32	12.58	nr	**22.34**
100 mm dia.	10.99	13.90	0.35	13.76	nr	**27.67**
Single pipe bracket, screw on, black malleable iron; plugged and screwed						
15 mm dia.	1.32	1.66	0.25	9.83	nr	**11.50**
20 mm dia.	1.47	1.86	0.25	9.83	nr	**11.69**
25 mm dia.	1.72	2.17	0.30	11.80	nr	**13.97**
32 mm dia.	3.13	3.96	0.32	12.58	nr	**16.54**
40 mm dia.	3.71	4.69	0.32	12.58	nr	**17.28**
50 mm dia.	4.18	5.29	0.32	12.58	nr	**17.87**
65 mm dia.	5.48	6.93	0.35	13.76	nr	**20.70**
80 mm dia.	7.71	9.76	0.42	16.52	nr	**26.27**
100 mm dia.	10.99	13.90	0.42	16.52	nr	**30.42**
Single pipe bracket for building in, black malleable iron						
15 mm dia.	3.97	5.02	0.10	3.93	nr	**8.95**
20 mm dia.	3.85	4.88	0.11	4.33	nr	**9.20**
25 mm dia.	3.85	4.88	0.12	4.72	nr	**9.60**
32 mm dia.	4.38	5.54	0.14	5.51	nr	**11.05**
40 mm dia.	4.41	5.58	0.15	5.90	nr	**11.48**
50 mm dia.	4.41	5.58	0.16	6.29	nr	**11.87**

38 MECHANICAL/COOLING/HEATING SYSTEMS

Item	Net Price £	Material £	Labour hours	Labour £	Unit	Total rate £
LOW TEMPERATURE HOT WATER HEATING; PIPELINE: SCREWED STEEL – cont						
For steel pipes – cont						
Pipe ring, single socket, black malleable iron						
15 mm dia.	1.53	1.94	0.10	3.93	nr	**5.87**
20 mm dia.	1.70	2.15	0.11	4.33	nr	**6.48**
25 mm dia.	1.90	2.41	0.12	4.72	nr	**7.13**
32 mm dia.	2.03	2.57	0.14	5.51	nr	**8.07**
40 mm dia.	2.60	3.29	0.15	5.90	nr	**9.19**
50 mm dia.	3.28	4.15	0.16	6.29	nr	**10.44**
65 mm dia.	4.75	6.01	0.30	11.80	nr	**17.81**
80 mm dia.	5.69	7.21	0.35	13.76	nr	**20.97**
100 mm dia.	8.62	10.91	0.40	15.73	nr	**26.64**
125 mm dia.	17.08	21.62	0.60	23.60	nr	**45.22**
150 mm dia.	19.51	24.69	0.77	30.28	nr	**54.97**
200 mm dia.	57.33	72.55	0.90	35.39	nr	**107.94**
250 mm dia.	67.45	85.36	1.10	43.26	nr	**128.62**
300 mm dia.	86.19	109.09	1.25	49.16	nr	**158.24**
350 mm dia.	97.14	122.94	1.50	58.99	nr	**181.93**
400 mm dia.	118.74	150.28	1.75	68.82	nr	**219.10**
Pipe ring, double socket, black malleable iron						
15 mm dia.	1.76	2.23	0.10	3.93	nr	**6.17**
20 mm dia.	2.04	2.58	0.11	4.33	nr	**6.91**
25 mm dia.	2.27	2.88	0.12	4.72	nr	**7.60**
32 mm dia.	2.69	3.41	0.14	5.51	nr	**8.91**
40 mm dia.	2.94	3.72	0.15	5.90	nr	**9.62**
50 mm dia.	4.58	5.80	0.16	6.29	nr	**12.09**
Screw on backplate (Male), black malleable iron; plugged and screwed						
M12	0.96	1.22	0.10	3.93	nr	**5.15**
Screw on backplate (Female), black malleable iron; plugged and screwed						
M12	0.96	1.22	0.10	3.93	nr	**5.15**
Extra over channel sections for fabricated hangers and brackets						
Galvanized steel; including inserts, bolts, nuts, washers; fixed to backgrounds						
41 × 21 mm	9.61	12.16	0.29	11.40	m	**23.57**
41 × 41 mm	11.52	14.58	0.29	11.40	m	**25.99**
Threaded rods; metric thread; including nuts, washers etc.						
12 mm dia. × 600 mm long	4.85	6.14	0.18	7.08	nr	**13.22**
Pipe roller and chair						
Roller and chair; black malleable						
Up to 50 mm dia.	25.81	32.67	0.20	7.87	nr	**40.53**
65 mm dia.	26.59	33.65	0.20	7.87	nr	**41.51**
80 mm dia.	28.88	36.55	0.20	7.87	nr	**44.42**

38 MECHANICAL/COOLING/HEATING SYSTEMS

Item	Net Price £	Material £	Labour hours	Labour £	Unit	Total rate £
100 mm dia.	29.40	37.20	0.20	7.87	nr	**45.07**
125 mm dia.	41.18	52.11	0.20	7.87	nr	**59.98**
150 mm dia.	41.18	52.11	0.30	11.80	nr	**63.91**
175 mm dia.	69.47	87.92	0.30	11.80	nr	**99.72**
200 mm dia.	106.41	134.67	0.30	11.80	nr	**146.47**
250 mm dia.	138.80	175.66	0.30	11.80	nr	**187.46**
300 mm dia.	174.79	221.21	0.30	11.80	nr	**233.01**
Roller bracket; black malleable						
25 mm dia.	5.29	6.69	0.20	7.87	nr	**14.56**
32 mm dia.	5.53	7.00	0.20	7.87	nr	**14.87**
40 mm dia.	5.93	7.51	0.20	7.87	nr	**15.37**
50 mm dia.	6.25	7.91	0.20	7.87	nr	**15.77**
65 mm dia.	8.24	10.43	0.20	7.87	nr	**18.30**
80 mm dia.	11.85	15.00	0.20	7.87	nr	**22.87**
100 mm dia.	13.20	16.71	0.20	7.87	nr	**24.57**
125 mm dia.	21.77	27.55	0.20	7.87	nr	**35.42**
150 mm dia.	21.77	27.55	0.30	11.80	nr	**39.35**
175 mm dia.	48.52	61.40	0.30	11.80	nr	**73.20**
200 mm dia.	48.52	61.40	0.30	11.80	nr	**73.20**
250 mm dia.	64.51	81.65	0.30	11.80	nr	**93.44**
300 mm dia.	79.22	100.26	0.30	11.80	nr	**112.06**
350 mm dia.	127.39	161.22	0.30	11.80	nr	**173.02**
400 mm dia.	145.61	184.28	0.30	11.80	nr	**196.08**
Extra over black steel screwed pipes; black steel flanges, screwed and drilled; metric; BS 4504						
Screwed flanges; PN6						
15 mm dia.	12.23	15.48	0.35	13.76	nr	**29.24**
20 mm dia.	12.23	15.48	0.47	18.48	nr	**33.96**
25 mm dia.	12.23	15.48	0.53	20.84	nr	**36.32**
32 mm dia.	12.23	15.48	0.62	24.38	nr	**39.86**
40 mm dia.	12.23	15.48	0.70	27.53	nr	**43.01**
50 mm dia.	13.03	16.49	0.84	33.03	nr	**49.52**
65 mm dia.	18.13	22.94	1.03	40.51	nr	**63.45**
80 mm dia.	25.57	32.36	1.23	48.37	nr	**80.73**
100 mm dia.	30.23	38.26	1.41	55.45	nr	**93.71**
125 mm dia.	55.47	70.21	1.77	69.61	nr	**139.81**
150 mm dia.	55.47	70.21	2.21	86.91	nr	**157.12**
Screwed flanges; PN16						
15 mm dia.	15.52	19.65	0.35	13.76	nr	**33.41**
20 mm dia.	15.52	19.65	0.47	18.48	nr	**38.13**
25 mm dia.	15.52	19.65	0.53	20.84	nr	**40.49**
32 mm dia.	16.59	21.00	0.62	24.38	nr	**45.38**
40 mm dia.	16.59	21.00	0.70	27.53	nr	**48.53**
50 mm dia.	20.23	25.61	0.84	33.03	nr	**58.64**
65 mm dia.	25.28	32.00	1.03	40.51	nr	**72.50**
80 mm dia.	30.79	38.96	1.23	48.37	nr	**87.34**
100 mm dia.	34.51	43.67	1.41	55.45	nr	**99.12**
125 mm dia.	59.43	75.21	1.77	69.61	nr	**144.82**
150 mm dia.	60.39	76.43	2.21	86.91	nr	**163.34**

38 MECHANICAL/COOLING/HEATING SYSTEMS

Item	Net Price £	Material £	Labour hours	Labour £	Unit	Total rate £
LOW TEMPERATURE HOT WATER HEATING; PIPELINE: SCREWED STEEL – cont						
Extra over black steel screwed pipes; black steel flanges, screwed and drilled; imperial; BS 10						
Screwed flanges; Table E						
½" dia.	23.18	29.33	0.35	13.76	nr	**43.09**
¾" dia.	23.20	29.36	0.47	18.48	nr	**47.85**
1" dia.	23.20	29.36	0.53	20.84	nr	**50.21**
1 ¼" dia.	23.20	29.36	0.62	24.38	nr	**53.75**
1 ½" dia.	23.20	29.36	0.70	27.53	nr	**56.89**
2" dia.	23.20	29.36	0.84	33.03	nr	**62.40**
2 ½" dia.	27.60	34.93	1.03	40.51	nr	**75.44**
3" dia.	31.27	39.57	1.23	48.37	nr	**87.94**
4" dia.	42.32	53.56	1.41	55.45	nr	**109.00**
5" dia.	89.64	113.45	1.77	69.61	nr	**183.06**
6" dia.	89.64	113.45	2.21	86.91	nr	**200.36**
Extra over black steel screwed pipes; black steel flange connections						
Bolted connection between pair of flanges; including gasket, bolts, nuts and washers						
50 mm dia.	14.35	18.16	0.53	20.84	nr	**39.00**
65 mm dia.	17.01	21.53	0.53	20.84	nr	**42.37**
80 mm dia.	25.51	32.29	0.53	20.84	nr	**53.13**
100 mm dia.	27.71	35.07	0.61	23.99	nr	**59.05**
125 mm dia.	32.61	41.27	0.61	23.99	nr	**65.26**
150 mm dia.	46.59	58.96	0.90	35.39	nr	**94.36**
Extra over black steel screwed pipes; black heavy steel tubular fittings; BS 1387						
Long screw connection with socket and backnut						
15 mm dia.	10.12	12.81	0.63	24.78	nr	**37.59**
20 mm dia.	12.75	16.13	0.84	33.03	nr	**49.17**
25 mm dia.	16.73	21.18	0.95	37.36	nr	**58.54**
32 mm dia.	22.43	28.38	1.11	43.65	nr	**72.04**
40 mm dia.	26.97	34.13	1.28	50.34	nr	**84.47**
50 mm dia.	39.74	50.29	1.53	60.17	nr	**110.46**
65 mm dia.	55.57	70.33	1.87	73.54	nr	**143.86**
80 mm dia.	127.06	160.81	2.21	86.91	nr	**247.72**
100 mm dia.	205.85	260.52	3.05	119.94	nr	**380.46**
Running nipple						
15 mm dia.	2.62	3.32	0.50	19.66	nr	**22.99**
20 mm dia.	3.31	4.19	0.68	26.74	nr	**30.94**
25 mm dia.	4.05	5.13	0.77	30.28	nr	**35.41**
32 mm dia.	5.73	7.25	0.90	35.39	nr	**42.64**
40 mm dia.	7.71	9.76	1.03	40.51	nr	**50.26**
50 mm dia.	11.73	14.85	1.23	48.37	nr	**63.22**
65 mm dia.	25.20	31.89	1.50	58.99	nr	**90.88**
80 mm dia.	39.33	49.77	1.78	70.00	nr	**119.77**
100 mm dia.	61.57	77.92	2.38	93.60	nr	**171.52**

38 MECHANICAL/COOLING/HEATING SYSTEMS

Item	Net Price £	Material £	Labour hours	Labour £	Unit	Total rate £
Barrel nipple						
15 mm dia.	1.49	1.89	0.50	19.66	nr	**21.55**
20 mm dia.	2.38	3.01	0.68	26.74	nr	**29.75**
25 mm dia.	3.15	3.99	0.77	30.28	nr	**34.27**
32 mm dia.	4.68	5.92	0.90	35.39	nr	**41.31**
40 mm dia.	5.77	7.31	1.03	40.51	nr	**47.81**
50 mm dia.	8.25	10.44	1.23	48.37	nr	**58.81**
65 mm dia.	17.62	22.30	1.50	58.99	nr	**81.29**
80 mm dia.	24.58	31.10	1.78	70.00	nr	**101.10**
100 mm dia.	44.51	56.33	2.38	93.60	nr	**149.93**
125 mm dia.	82.59	104.52	2.87	112.87	nr	**217.39**
150 mm dia.	130.16	164.74	3.39	133.31	nr	**298.05**
Close taper nipple						
15 mm dia.	3.15	3.99	0.50	19.66	nr	**23.65**
20 mm dia.	4.05	5.13	0.68	26.74	nr	**31.87**
25 mm dia.	5.30	6.71	0.77	30.28	nr	**36.99**
32 mm dia.	7.94	10.05	0.90	35.39	nr	**45.44**
40 mm dia.	9.84	12.46	1.03	40.51	nr	**52.96**
50 mm dia.	15.19	19.23	1.23	48.37	nr	**67.60**
65 mm dia.	29.45	37.27	1.50	58.99	nr	**96.26**
80 mm dia.	36.29	45.93	1.78	70.00	nr	**115.93**
100 mm dia.	74.58	94.39	2.38	93.60	nr	**187.99**
Extra over black steel screwed pipes; black malleable iron fittings; BS 143						
Cap						
15 mm dia.	1.03	1.31	0.32	12.58	nr	**13.89**
20 mm dia.	1.19	1.51	0.43	16.91	nr	**18.42**
25 mm dia.	1.51	1.91	0.49	19.27	nr	**21.18**
32 mm dia.	2.63	3.33	0.58	22.81	nr	**26.14**
40 mm dia.	3.12	3.95	0.66	25.96	nr	**29.91**
50 mm dia.	6.49	8.21	0.78	30.67	nr	**38.89**
65 mm dia.	10.83	13.71	0.96	37.75	nr	**51.47**
80 mm dia.	12.26	15.52	1.13	44.44	nr	**59.95**
100 mm dia.	26.88	34.02	1.70	66.85	nr	**100.88**
Plain plug, hollow						
15 mm dia.	0.73	0.93	0.28	11.01	nr	**11.94**
20 mm dia.	0.94	1.19	0.38	14.94	nr	**16.13**
25 mm dia.	1.28	1.62	0.44	17.30	nr	**18.92**
32 mm dia.	1.77	2.23	0.51	20.06	nr	**22.29**
40 mm dia.	3.34	4.23	0.59	23.20	nr	**27.43**
50 mm dia.	4.68	5.93	0.70	27.53	nr	**33.46**
65 mm dia.	7.63	9.66	0.85	33.43	nr	**43.08**
80 mm dia.	11.92	15.09	1.00	39.33	nr	**54.41**
100 mm dia.	21.93	27.75	1.44	56.63	nr	**84.38**
Plain plug, solid						
15 mm dia.	2.20	2.78	0.28	11.01	nr	**13.79**
20 mm dia.	2.27	2.87	0.38	14.94	nr	**17.81**
25 mm dia.	3.27	4.14	0.44	17.30	nr	**21.44**
32 mm dia.	4.71	5.96	0.51	20.06	nr	**26.02**
40 mm dia.	6.61	8.37	0.59	23.20	nr	**31.57**
50 mm dia.	8.70	11.01	0.70	27.53	nr	**38.54**

38 MECHANICAL/COOLING/HEATING SYSTEMS

Item	Net Price £	Material £	Labour hours	Labour £	Unit	Total rate £
LOW TEMPERATURE HOT WATER HEATING; PIPELINE: SCREWED STEEL – cont						
Extra over black steel screwed pipes – cont						
90° elbow, male/female						
15 mm dia.	1.19	1.51	0.64	25.17	nr	**26.68**
20 mm dia.	1.62	2.04	0.85	33.43	nr	**35.47**
25 mm dia.	2.69	3.40	0.97	38.15	nr	**41.55**
32 mm dia.	4.93	6.24	1.12	44.04	nr	**50.28**
40 mm dia.	8.26	10.45	1.29	50.73	nr	**61.18**
50 mm dia.	10.62	13.44	1.55	60.95	nr	**74.39**
65 mm dia.	22.92	29.01	1.89	74.33	nr	**103.33**
80 mm dia.	31.32	39.64	2.24	88.09	nr	**127.73**
100 mm dia.	54.78	69.34	3.09	121.52	nr	**190.85**
90° elbow						
15 mm dia.	1.09	1.37	0.64	25.17	nr	**26.54**
20 mm dia.	1.48	1.87	0.85	33.43	nr	**35.30**
25 mm dia.	2.27	2.87	0.97	38.15	nr	**41.02**
32 mm dia.	4.18	5.29	1.12	44.04	nr	**49.34**
40 mm dia.	6.99	8.85	1.29	50.73	nr	**59.58**
50 mm dia.	8.21	10.40	1.55	60.95	nr	**71.35**
65 mm dia.	17.70	22.41	1.89	74.33	nr	**96.73**
80 mm dia.	25.99	32.89	2.24	88.09	nr	**120.98**
100 mm dia.	50.14	63.46	3.09	121.52	nr	**184.98**
125 mm dia.	121.97	154.36	4.44	174.61	nr	**328.97**
150 mm dia.	227.01	287.31	5.79	227.70	nr	**515.00**
45° elbow						
15 mm dia.	2.59	3.28	0.64	25.17	nr	**28.45**
20 mm dia.	3.16	4.00	0.85	33.43	nr	**37.43**
25 mm dia.	4.71	5.96	0.97	38.15	nr	**44.11**
32 mm dia.	9.63	12.18	1.12	44.04	nr	**56.23**
40 mm dia.	11.80	14.93	1.29	50.73	nr	**65.66**
50 mm dia.	16.17	20.47	1.55	60.95	nr	**81.42**
65 mm dia.	23.67	29.95	1.89	74.33	nr	**104.28**
80 mm dia.	35.55	44.99	2.24	88.09	nr	**133.08**
100 mm dia.	77.95	98.65	3.09	121.52	nr	**220.17**
150 mm dia.	217.69	275.50	5.79	227.70	nr	**503.20**
90° bend, male/female						
15 mm dia.	2.12	2.68	0.64	25.17	nr	**27.85**
20 mm dia.	3.10	3.92	0.85	33.43	nr	**37.34**
25 mm dia.	4.55	5.76	0.97	38.15	nr	**43.90**
32 mm dia.	7.24	9.16	1.12	44.04	nr	**53.20**
40 mm dia.	11.43	14.47	1.29	50.73	nr	**65.20**
50 mm dia.	19.97	25.28	1.55	60.95	nr	**86.23**
65 mm dia.	33.21	42.03	1.89	74.33	nr	**116.36**
80 mm dia.	51.11	64.68	2.24	88.09	nr	**152.77**
100 mm dia.	126.65	160.29	3.09	121.52	nr	**281.80**
90° bend, male						
15 mm dia.	4.87	6.17	0.64	25.17	nr	**31.34**
20 mm dia.	5.46	6.91	0.85	33.43	nr	**40.33**
25 mm dia.	8.00	10.12	0.97	38.15	nr	**48.27**

38 MECHANICAL/COOLING/HEATING SYSTEMS

Item	Net Price £	Material £	Labour hours	Labour £	Unit	Total rate £
32 mm dia.	17.37	21.98	1.12	44.04	nr	66.02
40 mm dia.	24.38	30.86	1.29	50.73	nr	81.59
50 mm dia.	37.03	46.86	1.55	60.95	nr	107.81
90° bend, female						
15 mm dia.	1.94	2.46	0.64	25.17	nr	27.63
20 mm dia.	2.77	3.51	0.85	33.43	nr	36.93
25 mm dia.	3.88	4.91	0.97	38.15	nr	43.06
32 mm dia.	7.39	9.35	1.12	44.04	nr	53.39
40 mm dia.	9.83	12.44	1.29	50.73	nr	63.17
50 mm dia.	13.86	17.54	1.55	60.95	nr	78.50
65 mm dia.	29.57	37.43	1.89	74.33	nr	111.75
80 mm dia.	47.13	59.64	2.24	88.09	nr	147.73
100 mm dia.	103.32	130.77	3.09	121.52	nr	252.28
125 mm dia.	243.73	308.46	4.44	174.61	nr	483.07
150 mm dia.	443.25	560.97	5.79	227.70	nr	788.67
Return bend						
15 mm dia.	11.12	14.07	0.64	25.17	nr	39.24
20 mm dia.	17.99	22.77	0.85	33.43	nr	56.19
25 mm dia.	22.43	28.39	0.97	38.15	nr	66.53
32 mm dia.	34.79	44.02	1.12	44.04	nr	88.07
40 mm dia.	41.41	52.41	1.29	50.73	nr	103.14
50 mm dia.	63.20	79.99	1.55	60.95	nr	140.94
Equal socket, parallel thread						
15 mm dia.	1.03	1.31	0.64	25.17	nr	26.47
20 mm dia.	1.28	1.62	0.85	33.43	nr	35.04
25 mm dia.	1.75	2.22	0.97	38.15	nr	40.36
32 mm dia.	3.52	4.45	1.12	44.04	nr	48.50
40 mm dia.	5.11	6.46	1.29	50.73	nr	57.19
50 mm dia.	7.40	9.37	1.55	60.95	nr	70.32
65 mm dia.	11.36	14.38	1.89	74.33	nr	88.71
80 mm dia.	16.13	20.41	2.24	88.09	nr	108.50
100 mm dia.	26.57	33.63	3.09	121.52	nr	155.14
Concentric reducing socket						
20 × 15 mm dia.	1.66	2.10	0.76	29.89	nr	31.98
25 × 15 mm dia.	1.94	2.46	0.85	33.43	nr	35.88
25 × 20 mm dia.	2.06	2.61	0.86	33.82	nr	36.43
32 × 25 mm dia.	3.87	4.90	1.01	39.72	nr	44.62
40 × 25 mm dia.	4.87	6.17	1.16	45.62	nr	51.79
40 × 32 mm dia.	5.05	6.39	1.16	45.62	nr	52.01
50 × 25 mm dia.	7.26	9.19	1.38	54.27	nr	63.46
50 × 40 mm dia.	9.36	11.85	1.38	54.27	nr	66.12
65 × 50 mm dia.	12.87	16.29	1.69	66.46	nr	82.75
80 × 50 mm dia.	16.74	21.19	2.00	78.65	nr	99.84
100 × 50 mm dia.	33.41	42.29	2.75	108.15	nr	150.43
100 × 80 mm dia.	46.07	58.30	2.75	108.15	nr	166.45
150 × 100 mm dia.	92.76	117.40	4.10	161.24	nr	278.63
Eccentric reducing socket						
20 × 15 mm dia.	2.93	3.71	0.73	28.71	nr	32.42
25 × 15 mm dia.	8.46	10.71	0.85	33.43	nr	44.13
25 × 20 mm dia.	9.57	12.11	0.85	33.43	nr	45.54

38 MECHANICAL/COOLING/HEATING SYSTEMS

Item	Net Price £	Material £	Labour hours	Labour £	Unit	Total rate £
LOW TEMPERATURE HOT WATER HEATING; PIPELINE: SCREWED STEEL – cont						
Extra over black steel screwed pipes – cont						
Eccentric reducing socket – cont						
32 × 25 mm dia.	13.05	16.51	1.01	39.72	nr	**56.23**
40 × 25 mm dia.	14.95	18.92	1.16	45.62	nr	**64.54**
40 × 32 mm dia.	15.47	19.58	1.16	45.62	nr	**65.20**
50 × 25 mm dia.	15.60	19.75	1.38	54.27	nr	**74.02**
50 × 40 mm dia.	18.68	23.64	1.38	54.27	nr	**77.91**
65 × 50 mm dia.	24.03	30.42	1.69	66.46	nr	**96.88**
80 × 50 mm dia.	27.21	34.44	2.00	78.65	nr	**113.09**
Hexagon bush						
20 × 15 mm dia.	0.94	1.19	0.37	14.55	nr	**15.74**
25 × 15 mm dia.	1.14	1.44	0.43	16.91	nr	**18.35**
25 × 20 mm dia.	1.19	1.51	0.43	16.91	nr	**18.42**
32 × 25 mm dia.	1.53	1.94	0.51	20.06	nr	**22.00**
40 × 25 mm dia.	2.44	3.09	0.58	22.81	nr	**25.90**
40 × 32 mm dia.	4.72	5.97	0.58	22.81	nr	**28.78**
50 × 25 mm dia.	5.05	6.39	0.71	27.92	nr	**34.31**
50 × 40 mm dia.	5.84	7.39	0.71	27.92	nr	**35.31**
65 × 50 mm dia.	7.82	9.90	0.85	33.43	nr	**43.32**
80 × 50 mm dia.	12.68	16.05	1.00	39.33	nr	**55.38**
100 × 50 mm dia.	29.34	37.13	1.52	59.78	nr	**96.91**
100 × 80 mm dia.	40.49	51.24	1.52	59.78	nr	**111.02**
150 × 100 mm dia.	87.79	111.11	2.57	101.07	nr	**212.18**
Hexagon nipple						
15 mm dia.	1.00	1.27	0.28	11.01	nr	**12.28**
20 mm dia.	1.14	1.44	0.38	14.94	nr	**16.39**
25 mm dia.	1.62	2.04	0.44	17.30	nr	**19.35**
32 mm dia.	3.34	4.23	0.51	20.06	nr	**24.28**
40 mm dia.	3.87	4.90	0.59	23.20	nr	**28.10**
50 mm dia.	7.03	8.90	0.70	27.53	nr	**36.43**
65 mm dia.	11.46	14.50	0.85	33.43	nr	**47.93**
80 mm dia.	16.56	20.96	1.00	39.33	nr	**60.29**
100 mm dia.	28.11	35.57	1.44	56.63	nr	**92.20**
150 mm dia.	62.74	79.40	2.32	91.24	nr	**170.64**
Union, male/female						
15 mm dia.	5.28	6.68	0.64	25.17	nr	**31.85**
20 mm dia.	6.46	8.18	0.85	33.43	nr	**41.61**
25 mm dia.	8.08	10.22	0.97	38.15	nr	**48.37**
32 mm dia.	12.45	15.76	1.12	44.04	nr	**59.80**
40 mm dia.	15.93	20.16	1.29	50.73	nr	**70.89**
50 mm dia.	25.10	31.77	1.55	60.95	nr	**92.73**
65 mm dia.	47.90	60.62	1.89	74.33	nr	**134.95**
80 mm dia.	65.86	83.36	2.24	88.09	nr	**171.45**
Union, female						
15 mm dia.	4.34	5.50	0.64	25.17	nr	**30.67**
20 mm dia.	4.71	5.96	0.85	33.43	nr	**39.39**
25 mm dia.	5.49	6.94	0.97	38.15	nr	**45.09**

38 MECHANICAL/COOLING/HEATING SYSTEMS

Item	Net Price £	Material £	Labour hours	Labour £	Unit	Total rate £
32 mm dia.	10.33	13.08	1.12	44.04	nr	**57.12**
40 mm dia.	11.69	14.79	1.29	50.73	nr	**65.53**
50 mm dia.	19.36	24.50	1.55	60.95	nr	**85.46**
65 mm dia.	41.75	52.84	1.89	74.33	nr	**127.17**
80 mm dia.	57.65	72.96	2.24	88.09	nr	**161.05**
100 mm dia.	109.75	138.89	3.09	121.52	nr	**260.41**
Union elbow, male/female						
15 mm dia.	7.64	9.67	0.55	21.63	nr	**31.30**
20 mm dia.	9.59	12.13	0.85	33.43	nr	**45.56**
25 mm dia.	13.44	17.01	0.97	38.15	nr	**55.16**
Twin elbow						
15 mm dia.	6.19	7.84	0.91	35.79	nr	**43.62**
20 mm dia.	6.87	8.69	1.22	47.98	nr	**56.67**
25 mm dia.	11.09	14.04	1.39	54.66	nr	**68.70**
32 mm dia.	22.77	28.82	1.62	63.71	nr	**92.52**
40 mm dia.	28.82	36.48	1.86	73.15	nr	**109.63**
50 mm dia.	37.07	46.91	2.21	86.91	nr	**133.82**
65 mm dia.	65.84	83.32	2.72	106.97	nr	**190.29**
80 mm dia.	112.20	142.00	3.21	126.24	nr	**268.24**
Equal tee						
15 mm dia.	1.48	1.87	0.91	35.79	nr	**37.66**
20 mm dia.	2.16	2.73	1.22	47.98	nr	**50.71**
25 mm dia.	3.11	3.94	1.39	54.66	nr	**58.60**
32 mm dia.	5.65	7.15	1.62	63.71	nr	**70.86**
40 mm dia.	8.70	11.01	1.86	73.15	nr	**84.16**
50 mm dia.	12.52	15.84	2.21	86.91	nr	**102.75**
65 mm dia.	27.41	34.69	2.72	106.97	nr	**141.66**
80 mm dia.	33.41	42.29	3.21	126.24	nr	**168.52**
100 mm dia.	60.54	76.62	4.44	174.61	nr	**251.23**
125 mm dia.	175.83	222.53	5.38	211.57	nr	**434.10**
150 mm dia.	280.14	354.55	6.31	248.15	nr	**602.69**
Tee reducing on branch						
20 × 15 mm dia.	1.94	2.46	1.22	47.98	nr	**50.43**
25 × 15 mm dia.	2.69	3.40	1.39	54.66	nr	**58.07**
25 × 20 mm dia.	3.19	4.04	1.39	54.66	nr	**58.70**
32 × 25 mm dia.	5.55	7.03	1.62	63.71	nr	**70.74**
40 × 25 mm dia.	7.35	9.30	1.86	73.15	nr	**82.44**
40 × 32 mm dia.	9.65	12.22	1.86	73.15	nr	**85.36**
50 × 25 mm dia.	10.92	13.82	2.21	86.91	nr	**100.73**
50 × 40 mm dia.	16.29	20.62	2.21	86.91	nr	**107.53**
65 × 50 mm dia.	25.44	32.20	2.72	106.97	nr	**139.17**
80 × 50 mm dia.	34.38	43.51	3.21	126.24	nr	**169.74**
100 × 50 mm dia.	56.93	72.05	4.44	174.61	nr	**246.66**
100 × 80 mm dia.	87.79	111.11	4.44	174.61	nr	**285.72**
150 × 100 mm dia.	206.36	261.17	6.31	248.15	nr	**509.32**
Equal pitcher tee						
15 mm dia.	4.52	5.72	0.91	35.79	nr	**41.51**
20 mm dia.	5.62	7.11	1.22	47.98	nr	**55.09**
25 mm dia.	8.46	10.71	1.39	54.66	nr	**65.37**

38 MECHANICAL/COOLING/HEATING SYSTEMS

Item	Net Price £	Material £	Labour hours	Labour £	Unit	Total rate £
LOW TEMPERATURE HOT WATER HEATING; PIPELINE: SCREWED STEEL – cont						
Extra over black steel screwed pipes – cont						
Equal pitcher tee – cont						
32 mm dia.	12.83	16.24	1.62	63.71	nr	**79.95**
40 mm dia.	19.78	25.04	1.86	73.15	nr	**98.18**
50 mm dia.	27.79	35.17	2.21	86.91	nr	**122.08**
65 mm dia.	42.97	54.39	2.72	106.97	nr	**161.35**
80 mm dia.	67.09	84.90	3.21	126.24	nr	**211.14**
100 mm dia.	150.90	190.98	4.44	174.61	nr	**365.58**
Cross						
15 mm dia.	4.41	5.58	1.00	39.33	nr	**44.91**
20 mm dia.	6.64	8.40	1.33	52.30	nr	**60.71**
25 mm dia.	8.46	10.71	1.51	59.38	nr	**70.09**
32 mm dia.	12.29	15.55	1.77	69.61	nr	**85.16**
40 mm dia.	15.82	20.02	2.02	79.44	nr	**99.46**
50 mm dia.	25.70	32.53	2.42	95.17	nr	**127.70**
65 mm dia.	35.55	44.99	2.97	116.80	nr	**161.78**
80 mm dia.	47.26	59.82	3.50	137.64	nr	**197.46**
100 mm dia.	97.55	123.46	4.84	190.34	nr	**313.80**

38 MECHANICAL/COOLING/HEATING SYSTEMS

Item	Net Price £	Material £	Labour hours	Labour £	Unit	Total rate £
LOW TEMPERATURE HOT WATER HEATING; PIPELINE: BLACK WELDED STEEL						
Black steel pipes; butt welded joints; BS 1387: 1985; including protective painting. Fixed vertically with brackets measured separately (Refer to Screwed Steel Section). Welded butt joints are within the running length, but any flanges are additional						
Medium weight						
10 mm dia.	4.32	5.47	0.37	14.55	m	**20.02**
15 mm dia.	4.60	5.82	0.37	14.55	m	**20.37**
20 mm dia.	5.42	6.85	0.37	14.55	m	**21.41**
25 mm dia.	7.79	9.86	0.41	16.12	m	**25.99**
32 mm dia.	9.64	12.20	0.48	18.88	m	**31.07**
40 mm dia.	11.20	14.17	0.52	20.45	m	**34.62**
50 mm dia.	15.74	19.92	0.62	24.38	m	**44.31**
65 mm dia.	21.37	27.05	0.64	25.17	m	**52.21**
80 mm dia.	27.78	35.16	1.10	43.26	m	**78.42**
100 mm dia.	39.33	49.78	1.31	51.52	m	**101.30**
125 mm dia.	50.07	63.37	1.66	65.28	m	**128.65**
150 mm dia.	58.13	73.57	1.88	73.93	m	**147.50**
Heavy weight						
15 mm dia.	5.47	6.93	0.37	14.55	m	**21.48**
20 mm dia.	6.49	8.21	0.37	14.55	m	**22.76**
25 mm dia.	9.51	12.04	0.41	16.12	m	**28.16**
32 mm dia.	11.80	14.94	0.48	18.88	m	**33.81**
40 mm dia.	13.77	17.43	0.52	20.45	m	**37.88**
50 mm dia.	19.09	24.16	0.62	24.38	m	**48.54**
65 mm dia.	25.97	32.87	0.64	25.17	m	**58.04**
80 mm dia.	33.06	41.84	1.10	43.26	m	**85.10**
100 mm dia.	46.14	58.40	1.31	51.52	m	**109.92**
125 mm dia.	53.40	67.59	1.66	65.28	m	**132.87**
150 mm dia.	62.41	78.98	1.88	73.93	m	**152.91**
Black steel pipes; butt welded joints; BS 1387: 1985; including protective painting. Fixed at high level or suspended with brackets measured separately (Refer to Screwed Steel Section). Welded butt joints are within the running length, but any flanges are additional						
Medium weight						
10 mm dia.	4.32	5.47	0.58	22.81	m	**28.28**
15 mm dia.	4.60	5.82	0.58	22.81	m	**28.63**
20 mm dia.	5.42	6.85	0.58	22.81	m	**29.66**
25 mm dia.	7.79	9.86	0.60	23.60	m	**33.46**
32 mm dia.	9.64	12.20	0.68	26.74	m	**38.94**
40 mm dia.	11.20	14.17	0.73	28.71	m	**42.88**
50 mm dia.	15.74	19.92	0.85	33.43	m	**53.35**

38 MECHANICAL/COOLING/HEATING SYSTEMS

Item	Net Price £	Material £	Labour hours	Labour £	Unit	Total rate £
LOW TEMPERATURE HOT WATER HEATING; PIPELINE: BLACK WELDED STEEL – cont						
Black steel pipes – cont						
Medium weight – cont						
65 mm dia.	21.37	27.05	0.88	34.61	m	**61.65**
80 mm dia.	27.78	35.16	1.45	57.02	m	**92.19**
100 mm dia.	39.33	49.78	1.74	68.43	m	**118.21**
125 mm dia.	50.07	63.37	2.21	86.91	m	**150.28**
150 mm dia.	58.13	73.57	2.50	98.31	m	**171.88**
Heavy weight						
15 mm dia.	5.47	6.93	0.58	22.81	m	**29.74**
20 mm dia.	6.49	8.21	0.58	22.81	m	**31.02**
25 mm dia.	9.51	12.04	0.60	23.60	m	**35.63**
32 mm dia.	11.80	14.94	0.68	26.74	m	**41.68**
40 mm dia.	13.77	17.43	0.73	28.71	m	**46.14**
50 mm dia.	19.09	24.16	0.85	33.43	m	**57.59**
65 mm dia.	25.97	32.87	0.88	34.61	m	**67.47**
80 mm dia.	33.06	41.84	1.45	57.02	m	**98.86**
100 mm dia.	46.14	58.40	1.74	68.43	m	**126.83**
125 mm dia.	53.40	67.59	2.21	86.91	m	**154.50**
150 mm dia.	62.41	78.98	2.50	98.31	m	**177.30**
FIXINGS						
Refer to steel pipes; black malleable iron. For minimum fixing distances, refer to the Tables and Memoranda to the rear of the book						
Extra over black steel butt welded pipes; black steel flanges, welded and drilled; metric; BS 4504						
Welded flanges; PN6						
15 mm dia.	6.23	7.89	0.59	23.20	nr	**31.09**
20 mm dia.	6.23	7.89	0.69	27.13	nr	**35.02**
25 mm dia.	6.23	7.89	0.84	33.03	nr	**40.92**
32 mm dia.	10.33	13.07	1.00	39.33	nr	**52.40**
40 mm dia.	11.04	13.97	1.11	43.65	nr	**57.62**
50 mm dia.	13.23	16.75	1.37	53.88	nr	**70.62**
65 mm dia.	19.45	24.61	1.54	60.56	nr	**85.17**
80 mm dia.	29.22	36.98	1.67	65.67	nr	**102.66**
100 mm dia.	31.91	40.38	2.22	87.30	nr	**127.69**
125 mm dia.	41.81	52.92	2.61	102.64	nr	**155.56**
150 mm dia.	42.26	53.48	2.99	117.58	nr	**171.07**

38 MECHANICAL/COOLING/HEATING SYSTEMS

Item	Net Price £	Material £	Labour hours	Labour £	Unit	Total rate £
Welded flanges; PN16						
15 mm dia.	12.48	15.79	0.59	23.20	nr	**38.99**
20 mm dia.	12.48	15.79	0.69	27.13	nr	**42.93**
25 mm dia.	12.48	15.79	0.84	33.03	nr	**48.83**
32 mm dia.	16.82	21.29	1.00	39.33	nr	**60.62**
40 mm dia.	16.82	21.29	1.11	43.65	nr	**64.94**
50 mm dia.	25.78	32.63	1.37	53.88	nr	**86.51**
65 mm dia.	29.76	37.66	1.54	60.56	nr	**98.23**
80 mm dia.	37.84	47.88	1.67	65.67	nr	**113.56**
100 mm dia.	38.62	48.88	2.22	87.30	nr	**136.18**
125 mm dia.	61.29	77.57	2.61	102.64	nr	**180.21**
150 mm dia.	73.18	92.62	2.99	117.58	nr	**210.20**
Blank flanges, slip on for welding; PN6						
15 mm dia.	3.90	4.93	0.48	18.88	nr	**23.81**
20 mm dia.	3.90	4.93	0.55	21.63	nr	**26.56**
25 mm dia.	3.90	4.93	0.64	25.17	nr	**30.10**
32 mm dia.	8.01	10.14	0.76	29.89	nr	**40.03**
40 mm dia.	8.01	10.14	0.84	33.03	nr	**43.17**
50 mm dia.	8.51	10.76	1.01	39.72	nr	**50.48**
65 mm dia.	16.85	21.32	1.30	51.12	nr	**72.45**
80 mm dia.	17.58	22.25	1.41	55.45	nr	**77.70**
100 mm dia.	18.07	22.87	1.78	70.00	nr	**92.87**
125 mm dia.	36.08	45.66	2.06	81.01	nr	**126.67**
150 mm dia.	34.15	43.22	2.35	92.42	nr	**135.63**
Blank flanges, slip on for welding; PN16						
15 mm dia.	3.55	4.49	0.48	18.88	nr	**23.36**
20 mm dia.	5.43	6.88	0.55	21.63	nr	**28.51**
25 mm dia.	6.62	8.37	0.64	25.17	nr	**33.54**
32 mm dia.	6.77	8.57	0.76	29.89	nr	**38.45**
40 mm dia.	8.76	11.09	0.84	33.03	nr	**44.13**
50 mm dia.	15.85	20.06	1.01	39.72	nr	**59.77**
65 mm dia.	19.45	24.61	1.30	51.12	nr	**75.73**
80 mm dia.	25.23	31.93	1.41	55.45	nr	**87.38**
100 mm dia.	31.26	39.57	1.78	70.00	nr	**109.57**
125 mm dia.	45.89	58.08	2.06	81.01	nr	**139.09**
150 mm dia.	59.22	74.94	2.35	92.42	nr	**167.36**

38 MECHANICAL/COOLING/HEATING SYSTEMS

Item	Net Price £	Material £	Labour hours	Labour £	Unit	Total rate £
LOW TEMPERATURE HOT WATER HEATING; **PIPELINE: BLACK WELDED STEEL – cont**						
Extra over black steel butt welded pipes; black steel **flanges, welding and drilled; imperial; BS 10**						
Welded flanges; Table E						
15 mm dia.	12.80	16.20	0.59	23.20	nr	**39.40**
20 mm dia.	12.80	16.20	0.69	27.13	nr	**43.33**
25 mm dia.	12.80	16.20	0.84	33.03	nr	**49.23**
32 mm dia.	12.80	16.20	1.00	39.33	nr	**55.53**
40 mm dia.	13.14	16.63	1.11	43.65	nr	**60.28**
50 mm dia.	16.81	21.27	1.37	53.88	nr	**75.15**
65 mm dia.	20.44	25.87	1.54	60.56	nr	**86.44**
80 mm dia.	26.87	34.01	1.67	65.67	nr	**99.69**
100 mm dia.	39.69	50.23	2.22	87.30	nr	**137.53**
125 mm dia.	77.23	97.74	2.61	102.64	nr	**200.38**
150 mm dia.	77.23	97.74	2.99	117.58	nr	**215.33**
Blank flanges, slip on for welding; Table E						
15 mm dia.	8.81	11.15	0.48	18.88	nr	**30.03**
20 mm dia.	8.81	11.15	0.55	21.63	nr	**32.78**
25 mm dia.	8.81	11.15	0.64	25.17	nr	**36.32**
32 mm dia.	11.07	14.01	0.76	29.89	nr	**43.89**
40 mm dia.	12.05	15.25	0.84	33.03	nr	**48.28**
50 mm dia.	13.29	16.82	1.01	39.72	nr	**56.54**
65 mm dia.	15.33	19.40	1.30	51.12	nr	**70.53**
80 mm dia.	33.44	42.33	1.41	55.45	nr	**97.78**
100 mm dia.	37.99	48.08	1.78	70.00	nr	**118.08**
125 mm dia.	62.48	79.08	2.06	81.01	nr	**160.09**
150 mm dia.	80.33	101.67	2.35	92.42	nr	**194.08**
Extra over black steel butt welded pipes; black steel **flange connections**						
Bolted connection between pair of flanges; including gasket, bolts, nuts and washers						
50 mm dia.	14.35	18.16	0.50	19.66	nr	**37.82**
65 mm dia.	17.01	21.53	0.50	19.66	nr	**41.19**
80 mm dia.	25.51	32.29	0.50	19.66	nr	**51.95**
100 mm dia.	27.71	35.07	0.50	19.66	nr	**54.73**
125 mm dia.	32.61	41.27	0.50	19.66	nr	**60.93**
150 mm dia.	46.59	58.96	0.50	19.66	nr	**78.63**
Extra over fittings; BS 1965; butt welded						
Cap						
25 mm dia.	36.21	45.83	0.47	18.48	nr	**64.31**
32 mm dia.	36.21	45.83	0.59	23.20	nr	**69.03**
40 mm dia.	36.43	46.11	0.70	27.53	nr	**73.63**
50 mm dia.	42.58	53.89	0.99	38.93	nr	**92.82**
65 mm dia.	50.04	63.33	1.35	53.09	nr	**116.42**
80 mm dia.	50.94	64.47	1.66	65.28	nr	**129.75**
100 mm dia.	66.58	84.26	2.23	87.70	nr	**171.96**
125 mm dia.	93.64	118.50	3.03	119.16	nr	**237.66**
150 mm dia.	108.31	137.08	3.79	149.04	nr	**286.13**

38 MECHANICAL/COOLING/HEATING SYSTEMS

Item	Net Price £	Material £	Labour hours	Labour £	Unit	Total rate £
Concentric reducer						
20 × 15 mm dia.	20.17	25.52	0.69	27.13	nr	**52.66**
25 × 15 mm dia.	25.18	31.86	0.87	34.21	nr	**66.08**
25 × 20 mm dia.	26.01	32.92	0.87	34.21	nr	**67.14**
32 × 25 mm dia.	26.82	33.94	1.08	42.47	nr	**76.41**
40 × 25 mm dia.	29.02	36.72	1.38	54.27	nr	**90.99**
40 × 32 mm dia.	30.71	38.86	1.38	54.27	nr	**93.13**
50 × 25 mm dia.	31.05	39.30	1.82	71.57	nr	**110.87**
50 × 40 mm dia.	35.04	44.35	1.82	71.57	nr	**115.93**
65 × 50 mm dia.	42.57	53.88	2.52	99.10	nr	**152.98**
80 × 50 mm dia.	47.10	59.61	3.24	127.42	nr	**187.03**
100 × 50 mm dia.	54.81	69.37	4.08	160.45	nr	**229.82**
100 × 80 mm dia.	71.20	90.11	4.08	160.45	nr	**250.56**
125 × 80 mm dia.	82.35	104.22	4.71	185.22	nr	**289.44**
150 × 100 mm dia.	89.07	112.72	5.33	209.61	nr	**322.33**
Eccentric reducer						
20 × 15 mm dia.	30.00	37.97	0.69	27.13	nr	**65.11**
25 × 15 mm dia.	31.14	39.41	0.87	34.21	nr	**73.63**
25 × 20 mm dia.	32.97	41.72	0.87	34.21	nr	**75.94**
32 × 25 mm dia.	36.43	46.11	1.08	42.47	nr	**88.58**
40 × 25 mm dia.	44.98	56.92	1.38	54.27	nr	**111.19**
40 × 32 mm dia.	51.25	64.86	1.38	54.27	nr	**119.13**
50 × 25 mm dia.	52.01	65.82	1.82	71.57	nr	**137.39**
50 × 40 mm dia.	57.72	73.05	1.82	71.57	nr	**144.63**
65 × 50 mm dia.	75.81	95.94	2.52	99.10	nr	**195.04**
80 × 50 mm dia.	86.66	109.68	3.24	127.42	nr	**237.09**
100 × 50 mm dia.	90.38	114.38	4.08	160.45	nr	**274.83**
100 × 80 mm dia.	144.53	182.91	4.08	160.45	nr	**343.36**
125 × 80 mm dia.	178.12	225.42	4.71	185.22	nr	**410.65**
150 × 100 mm dia.	235.12	297.57	5.33	209.61	nr	**507.18**

38 MECHANICAL/COOLING/HEATING SYSTEMS

Item	Net Price £	Material £	Labour hours	Labour £	Unit	Total rate £
LOW TEMPERATURE HOT WATER HEATING; PIPELINE: BLACK WELDED STEEL – cont						
Extra over fittings – cont						
45° elbow, long radius						
15 mm dia.	7.57	9.58	0.56	22.02	nr	**31.60**
20 mm dia.	7.66	9.70	0.75	29.49	nr	**39.19**
25 mm dia.	9.96	12.61	0.93	36.57	nr	**49.18**
32 mm dia.	12.10	15.31	1.17	46.01	nr	**61.33**
40 mm dia.	12.25	15.50	1.46	57.42	nr	**72.92**
50 mm dia.	16.74	21.19	1.97	77.47	nr	**98.66**
65 mm dia.	19.95	25.24	2.70	106.18	nr	**131.42**
80 mm dia.	21.57	27.30	3.32	130.56	nr	**157.86**
100 mm dia.	30.15	38.16	4.09	160.84	nr	**199.00**
125 mm dia.	64.11	81.14	4.94	194.27	nr	**275.41**
150 mm dia.	79.66	100.81	5.78	227.30	nr	**328.12**
90° elbow, long radius						
15 mm dia.	7.83	9.91	0.56	22.02	nr	**31.93**
20 mm dia.	7.64	9.67	0.75	29.49	nr	**39.17**
25 mm dia.	10.17	12.87	0.93	36.57	nr	**49.44**
32 mm dia.	12.14	15.36	1.17	46.01	nr	**61.37**
40 mm dia.	12.27	15.52	1.46	57.42	nr	**72.94**
50 mm dia.	16.74	21.19	1.97	77.47	nr	**98.66**
65 mm dia.	21.57	27.30	2.70	106.18	nr	**133.47**
80 mm dia.	23.43	29.65	3.32	130.56	nr	**160.21**
100 mm dia.	35.47	44.89	4.09	160.84	nr	**205.74**
125 mm dia.	73.82	93.42	4.94	194.27	nr	**287.69**
150 mm dia.	93.67	118.55	5.78	227.30	nr	**345.85**
Branch bend (based on branch and pipe sizes being the same)						
15 mm dia.	31.18	39.46	0.85	33.43	nr	**72.89**
20 mm dia.	31.20	39.48	0.85	33.43	nr	**72.91**
25 mm dia.	31.22	39.51	1.02	40.11	nr	**79.62**
32 mm dia.	33.07	41.85	1.11	43.65	nr	**85.50**
40 mm dia.	35.19	44.54	1.36	53.48	nr	**98.02**
50 mm dia.	37.84	47.89	1.70	66.85	nr	**114.74**
65 mm dia.	41.83	52.94	1.78	70.00	nr	**122.94**
80 mm dia.	65.35	82.70	1.82	71.57	nr	**154.27**
100 mm dia.	85.24	107.88	1.87	73.54	nr	**181.41**
125 mm dia.	150.20	190.09	2.21	86.91	nr	**277.00**
150 mm dia.	230.96	292.30	2.65	104.21	nr	**396.51**

38 MECHANICAL/COOLING/HEATING SYSTEMS

Item	Net Price £	Material £	Labour hours	Labour £	Unit	Total rate £
Equal tee						
15 mm dia.	73.91	93.54	0.82	32.25	nr	**125.79**
20 mm dia.	73.91	93.54	1.10	43.26	nr	**136.80**
25 mm dia.	73.91	93.54	1.35	53.09	nr	**146.63**
32 mm dia.	73.91	93.54	1.63	64.10	nr	**157.64**
40 mm dia.	73.91	93.54	2.14	84.16	nr	**177.70**
50 mm dia.	77.85	98.53	3.02	118.76	nr	**217.29**
65 mm dia.	113.58	143.75	3.61	141.97	nr	**285.71**
80 mm dia.	113.80	144.03	4.18	164.38	nr	**308.41**
100 mm dia.	141.54	179.13	5.24	206.07	nr	**385.20**
125 mm dia.	323.97	410.01	6.70	263.48	nr	**673.49**
150 mm dia.	353.91	447.91	8.45	332.30	nr	**780.21**
Extra over black steel butt welded pipes; labour						
Made bend						
15 mm dia.	–	–	0.42	16.52	nr	**16.52**
20 mm dia.	–	–	0.42	16.52	nr	**16.52**
25 mm dia.	–	–	0.50	19.66	nr	**19.66**
32 mm dia.	–	–	0.62	24.38	nr	**24.38**
40 mm dia.	–	–	0.74	29.10	nr	**29.10**
50 mm dia.	–	–	0.89	35.00	nr	**35.00**
65 mm dia.	–	–	1.05	41.29	nr	**41.29**
80 mm dia.	–	–	1.13	44.44	nr	**44.44**
100 mm dia.	–	–	2.90	114.04	nr	**114.04**
125 mm dia.	–	–	3.56	140.00	nr	**140.00**
150 mm dia.	–	–	4.18	164.38	nr	**164.38**
Splay cut end						
15 mm dia.	–	–	0.14	5.51	nr	**5.51**
20 mm dia.	–	–	0.16	6.29	nr	**6.29**
25 mm dia.	–	–	0.18	7.08	nr	**7.08**
32 mm dia.	–	–	0.25	9.83	nr	**9.83**
40 mm dia.	–	–	0.27	10.62	nr	**10.62**
50 mm dia.	–	–	0.31	12.19	nr	**12.19**
65 mm dia.	–	–	0.35	13.76	nr	**13.76**
80 mm dia.	–	–	0.40	15.73	nr	**15.73**
100 mm dia.	–	–	0.48	18.88	nr	**18.88**
125 mm dia.	–	–	0.56	22.02	nr	**22.02**
150 mm dia.	–	–	0.64	25.17	nr	**25.17**
Screwed joint to fitting						
15 mm dia.	–	–	0.30	11.80	nr	**11.80**
20 mm dia.	–	–	0.40	15.73	nr	**15.73**
25 mm dia.	–	–	0.46	18.09	nr	**18.09**
32 mm dia.	–	–	0.53	20.84	nr	**20.84**
40 mm dia.	–	–	0.61	23.99	nr	**23.99**
50 mm dia.	–	–	0.73	28.71	nr	**28.71**
65 mm dia.	–	–	0.89	35.00	nr	**35.00**
80 mm dia.	–	–	1.05	41.29	nr	**41.29**
100 mm dia.	–	–	1.46	57.42	nr	**57.42**
125 mm dia.	–	–	2.10	82.58	nr	**82.58**
150 mm dia.	–	–	2.73	107.36	nr	**107.36**

38 MECHANICAL/COOLING/HEATING SYSTEMS

Item	Net Price £	Material £	Labour hours	Labour £	Unit	Total rate £
LOW TEMPERATURE HOT WATER HEATING; PIPELINE: BLACK WELDED STEEL – cont						
Extra over black steel butt welded pipes – cont						
Straight butt weld						
15 mm dia.	–	–	0.31	12.19	nr	**12.19**
20 mm dia.	–	–	0.42	16.52	nr	**16.52**
25 mm dia.	–	–	0.52	20.45	nr	**20.45**
32 mm dia.	–	–	0.69	27.13	nr	**27.13**
40 mm dia.	–	–	0.83	32.64	nr	**32.64**
50 mm dia.	–	–	1.22	47.98	nr	**47.98**
65 mm dia.	–	–	1.57	61.74	nr	**61.74**
80 mm dia.	–	–	1.95	76.69	nr	**76.69**
100 mm dia.	–	–	2.38	93.60	nr	**93.60**
125 mm dia.	–	–	2.83	111.29	nr	**111.29**
150 mm dia.	–	–	3.27	128.60	nr	**128.60**

38 MECHANICAL/COOLING/HEATING SYSTEMS

Item	Net Price £	Material £	Labour hours	Labour £	Unit	Total rate £
LOW TEMPERATURE HOT WATER HEATING; PIPELINE: CARBON WELDED STEEL						
Hot finished seamless carbon steel pipe; BS 806 and BS 3601; wall thickness to BS 3600; butt welded joints; including protective painting, fixed vertically or at low level, brackets measured separately (Refer to Screwed Pipework Section). Welded butt joints are within the running length, but any flanges are additional						
Pipework						
200 mm dia.	180.85	228.89	2.04	80.22	m	**309.11**
250 mm dia.	281.82	356.67	2.59	101.85	m	**458.52**
300 mm dia.	382.78	484.45	2.99	117.58	m	**602.03**
350 mm dia.	403.98	511.28	3.52	138.43	m	**649.70**
400 mm dia.	704.34	891.41	4.08	160.45	m	**1051.86**
Hot finished seamless carbon steel pipe; BS 806 and BS 3601; wall thickness to BS 3600; butt welded joints; including protective painting, fixed at high level or suspended, brackets measured separately (Refer to Screwed Pipework Section). Welded butt joints are within the running length, but any flanges are additional						
Pipework						
200 mm dia.	177.83	225.06	3.70	145.51	m	**370.57**
250 mm dia.	281.82	356.67	4.73	186.01	m	**542.68**
300 mm dia.	382.78	484.45	5.65	222.19	m	**706.64**
350 mm dia.	403.98	511.28	6.68	262.70	m	**773.97**
400 mm dia.	704.34	891.41	7.70	302.81	m	**1194.22**
FIXINGS						
Refer to steel pipes; black malleable iron. For minimum fixing distances, refer to the Tables and Memoranda to the rear of the book						
Extra over fittings; BS 1965 part 1; butt welded						
Cap						
200 mm dia.	148.05	187.38	3.70	145.51	nr	**332.88**
250 mm dia.	284.13	359.60	4.73	186.01	nr	**545.61**
300 mm dia.	310.75	393.28	5.65	222.19	nr	**615.47**
350 mm dia.	478.76	605.92	6.68	262.70	nr	**868.62**
400 mm dia.	553.78	700.87	7.70	302.81	nr	**1003.68**
Concentric reducer						
200 mm × 150 mm dia.	161.75	204.70	7.27	285.90	nr	**490.60**
250 mm × 150 mm dia.	234.55	296.84	9.05	355.90	nr	**652.74**
250 mm × 200 mm dia.	252.73	319.85	9.10	357.86	nr	**677.72**
300 mm × 150 mm dia.	270.92	342.88	10.75	422.75	nr	**765.63**
300 mm × 200 mm dia.	280.57	355.09	10.75	422.75	nr	**777.85**
300 mm × 250 mm dia.	353.02	446.79	11.15	438.48	nr	**885.27**

38 MECHANICAL/COOLING/HEATING SYSTEMS

Item	Net Price £	Material £	Labour hours	Labour £	Unit	Total rate £
LOW TEMPERATURE HOT WATER HEATING; PIPELINE: CARBON WELDED STEEL – cont						
Extra over fittings – cont						
Concentric reducer – cont						
350 mm × 200 mm dia.	425.48	538.49	12.50	491.57	nr	1030.06
350 mm × 250 mm dia.	481.55	609.45	12.70	499.44	nr	1108.89
350 mm × 300 mm dia.	501.50	634.70	13.00	511.24	nr	1145.93
400 mm × 250 mm dia.	521.61	660.15	14.46	568.65	nr	1228.80
400 mm × 300 mm dia.	645.96	817.53	14.51	570.62	nr	1388.15
400 mm × 350 mm dia.	770.30	974.89	15.16	596.18	nr	1571.07
Eccentric reducer						
200 mm × 150 mm dia.	292.12	369.71	7.27	285.90	nr	655.61
250 mm × 150 mm dia.	389.25	492.63	9.05	355.90	nr	848.53
250 mm × 200 mm dia.	428.11	541.81	9.10	357.86	nr	899.68
300 mm × 150 mm dia.	466.97	590.99	10.75	422.75	nr	1013.75
300 mm × 200 mm dia.	538.59	681.64	10.75	422.75	nr	1104.39
300 mm × 250 mm dia.	570.51	722.04	11.15	438.48	nr	1160.53
350 mm × 200 mm dia.	602.44	762.45	12.50	491.57	nr	1254.03
350 mm × 250 mm dia.	634.37	802.86	12.70	499.44	nr	1302.30
350 mm × 300 mm dia.	666.30	843.27	13.00	511.24	nr	1354.51
400 mm × 250 mm dia.	698.24	883.70	14.46	568.65	nr	1452.35
400 mm × 300 mm dia.	730.17	924.10	14.51	570.62	nr	1494.72
400 mm × 350 mm dia.	762.11	964.53	15.16	596.18	nr	1560.70
45° elbow						
200 mm dia.	154.70	195.79	7.75	304.78	nr	500.57
250 mm dia.	290.53	367.69	10.05	395.22	nr	762.92
300 mm dia.	424.50	537.24	12.20	479.77	nr	1017.02
350 mm dia.	497.72	629.92	14.65	576.12	nr	1206.04
400 mm dia.	570.96	722.61	17.12	673.26	nr	1395.87
90° elbow						
200 mm dia.	182.01	230.36	7.75	304.78	nr	535.13
250 mm dia.	341.88	432.69	10.05	395.22	nr	827.91
300 mm dia.	499.45	632.10	12.20	479.77	nr	1111.87
350 mm dia.	686.59	868.95	14.65	576.12	nr	1445.07
400 mm dia.	876.99	1109.92	17.12	673.26	nr	1783.17
Equal tee						
200 mm dia.	480.80	608.50	11.25	442.42	nr	1050.92
250 mm dia.	824.45	1043.43	14.53	571.40	nr	1614.83
300 mm dia.	1023.18	1294.94	17.55	690.17	nr	1985.11
350 mm dia.	1372.75	1737.35	20.98	825.06	nr	2562.41
400 mm dia.	1568.85	1985.54	24.38	958.76	nr	2944.30
Extra over black steel butt welded pipes; labour						
Straight butt weld						
200 mm dia.	–	–	4.08	160.45	nr	160.45
250 mm dia.	–	–	5.20	204.49	nr	204.49
300 mm dia.	–	–	6.22	244.61	nr	244.61
350 mm dia.	–	–	7.33	288.26	nr	288.26
400 mm dia.	–	–	8.41	330.73	nr	330.73

38 MECHANICAL/COOLING/HEATING SYSTEMS

Item	Net Price £	Material £	Labour hours	Labour £	Unit	Total rate £
Branch weld						
100 mm dia.	–	–	3.46	136.07	nr	**136.07**
125 mm dia.	–	–	4.23	166.35	nr	**166.35**
150 mm dia.	–	–	5.00	196.63	nr	**196.63**
Extra over black steel butt welded pipes; black steel flanges, welding and drilled; metric; BS 4504						
Welded flanges; PN16						
200 mm dia.	101.46	128.41	4.10	161.24	nr	**289.65**
250 mm dia.	145.75	184.46	5.33	209.61	nr	**394.06**
300 mm dia.	204.17	258.40	6.40	251.69	nr	**510.08**
350 mm dia.	394.88	499.76	7.43	292.19	nr	**791.95**
400 mm dia.	521.37	659.84	8.45	332.30	nr	**992.15**
Welded flanges; PN25						
200 mm dia.	336.93	426.42	4.10	161.24	nr	**587.65**
250 mm dia.	403.87	511.14	5.33	209.61	nr	**720.75**
300 mm dia.	546.12	691.17	6.40	251.69	nr	**942.85**
Blank flanges, slip on for welding; PN16						
200 mm dia.	234.62	296.94	2.70	106.18	nr	**403.12**
250 mm dia.	356.59	451.30	3.48	136.85	nr	**588.15**
300 mm dia.	380.99	482.18	4.20	165.17	nr	**647.35**
350 mm dia.	596.25	754.61	4.78	187.98	nr	**942.59**
400 mm dia.	788.38	997.77	5.35	210.39	nr	**1208.17**
Blank flanges, slip on for welding; PN25						
200 mm dia.	234.62	296.94	2.70	106.18	nr	**403.12**
250 mm dia.	356.59	451.30	3.48	136.85	nr	**588.15**
300 mm dia.	380.99	482.18	4.20	165.17	nr	**647.35**
Extra over black steel butt welded pipes; black steel flange connections						
Bolted connection between pair of flanges; including gasket, bolts, nuts and washers						
200 mm dia.	196.90	249.19	3.83	150.62	nr	**399.81**
250 mm dia.	304.72	385.65	4.93	193.88	nr	**579.53**
300 mm dia.	415.49	525.84	5.90	232.02	nr	**757.86**

38 MECHANICAL/COOLING/HEATING SYSTEMS

Item	Net Price £	Material £	Labour hours	Labour £	Unit	Total rate £
LOW TEMPERATURE HOT WATER HEATING; PIPELINE: PRESS FIT						
Press fit jointing system; operating temperature −20°C to +120°C; operating pressure 16 bar; butyl rubber 'O' ring mechanical joint. With brackets measured separately (Refer to Screwed Steel section)						
Carbon steel						
Pipework						
15 mm dia.	3.07	3.88	0.46	18.09	m	**21.97**
20 mm dia.	4.94	6.25	0.48	18.88	m	**25.12**
25 mm dia.	6.98	8.83	0.52	20.45	m	**29.28**
32 mm dia.	9.01	11.40	0.56	22.02	m	**33.42**
40 mm dia.	12.27	15.53	0.58	22.81	m	**38.33**
50 mm dia.	15.93	20.17	0.66	25.96	m	**46.12**
Extra over for carbon steel press fit fittings						
Coupling						
15 mm dia.	1.69	2.14	0.36	14.16	nr	**16.30**
20 mm dia.	2.08	2.63	0.36	14.16	nr	**16.79**
25 mm dia.	2.62	3.32	0.44	17.30	nr	**20.62**
32 mm dia.	4.38	5.55	0.44	17.30	nr	**22.85**
40 mm dia.	5.86	7.41	0.52	20.45	nr	**27.86**
50 mm dia.	6.92	8.75	0.60	23.60	nr	**32.35**
Reducer						
20 × 15 mm dia.	1.58	1.99	0.36	14.16	nr	**16.15**
25 × 15 mm dia.	2.08	2.63	0.40	15.73	nr	**18.36**
25 × 20 mm dia.	2.16	2.74	0.40	15.73	nr	**18.47**
32 × 20 mm dia.	2.42	3.06	0.40	15.73	nr	**18.79**
32 × 25 mm dia.	2.58	3.26	0.44	17.30	nr	**20.57**
40 × 32 mm dia.	5.60	7.09	0.48	18.88	nr	**25.96**
50 × 20 mm dia.	16.20	20.50	0.48	18.88	nr	**39.37**
50 × 25 mm dia.	16.33	20.66	0.52	20.45	nr	**41.11**
50 × 40 mm dia.	17.17	21.73	0.56	22.02	nr	**43.75**
90° elbow						
15 mm dia.	2.45	3.10	0.36	14.16	nr	**17.26**
20 mm dia.	3.24	4.10	0.36	14.16	nr	**18.25**
25 mm dia.	4.44	5.62	0.44	17.30	nr	**22.92**
32 mm dia.	11.08	14.03	0.44	17.30	nr	**31.33**
40 mm dia.	17.71	22.42	0.52	20.45	nr	**42.87**
50 mm dia.	21.15	26.77	0.60	23.60	nr	**50.37**
45° elbow						
15 mm dia.	2.94	3.72	0.36	14.16	nr	**17.87**
20 mm dia.	3.28	4.15	0.36	14.16	nr	**18.31**
25 mm dia.	4.45	5.64	0.44	17.30	nr	**22.94**
32 mm dia.	8.79	11.13	0.44	17.30	nr	**28.43**
40 mm dia.	11.01	13.94	0.52	20.45	nr	**34.39**
50 mm dia.	12.46	15.77	0.60	23.60	nr	**39.36**

38 MECHANICAL/COOLING/HEATING SYSTEMS

Item	Net Price £	Material £	Labour hours	Labour £	Unit	Total rate £
Equal tee						
15 mm dia.	4.71	5.96	0.54	21.24	nr	**27.20**
20 mm dia.	5.46	6.91	0.54	21.24	nr	**28.14**
25 mm dia.	7.32	9.26	0.66	25.96	nr	**35.22**
32 mm dia.	11.38	14.41	0.66	25.96	nr	**40.36**
40 mm dia.	16.80	21.26	0.78	30.67	nr	**51.93**
50 mm dia.	20.15	25.50	0.90	35.39	nr	**60.89**
Reducing tee						
20 × 15 mm dia.	5.36	6.78	0.54	21.24	nr	**28.01**
25 × 15 mm dia.	7.25	9.17	0.62	24.38	nr	**33.55**
25 × 20 mm dia.	7.86	9.95	0.62	24.38	nr	**34.33**
32 × 15 mm dia.	10.61	13.43	0.62	24.38	nr	**37.81**
32 × 20 mm dia.	11.46	14.50	0.62	24.38	nr	**38.88**
32 × 25 mm dia.	11.64	14.73	0.62	24.38	nr	**39.12**
40 × 20 mm dia.	15.34	19.41	0.70	27.53	nr	**46.94**
40 × 25 mm dia.	15.55	19.68	0.70	27.53	nr	**47.21**
40 × 32 mm dia.	15.91	20.14	0.70	27.53	nr	**47.67**
50 × 20 mm dia.	18.29	23.14	0.82	32.25	nr	**55.39**
50 × 25 mm dia.	18.66	23.62	0.82	32.25	nr	**55.86**
50 × 32 mm dia.	19.22	24.32	0.82	32.25	nr	**56.57**
50 × 40 mm dia.	20.13	25.48	0.82	32.25	nr	**57.73**

38 MECHANICAL/COOLING/HEATING SYSTEMS

Item	Net Price £	Material £	Labour hours	Labour £	Unit	Total rate £
LOW TEMPERATURE HOT WATER HEATING; PIPELINE: MECHANICAL GROOVED						
MECHANICAL GROOVED						
Mechanical grooved jointing system; working temperature not exceeding 82°C BS 5750; pipework complete with grooved joints; painted finish. With brackets measured separately (Refer to Screwed Steel section)						
Grooved joints						
65 mm	24.00	30.37	0.58	22.81	m	53.18
80 mm	25.15	31.83	0.68	26.74	m	58.57
100 mm	32.62	41.29	0.79	31.07	m	72.35
125 mm	51.22	64.82	1.02	40.11	m	104.94
150 mm	53.47	67.67	1.15	45.22	m	112.90
Extra over mechanical grooved system fittings						
Couplings						
65 mm	24.00	30.37	0.41	16.12	nr	46.50
80 mm	25.15	31.83	0.41	16.12	nr	47.96
100 mm	32.62	41.29	0.66	25.96	nr	67.24
125 mm	51.22	64.82	0.68	26.74	nr	91.57
150 mm	53.47	67.67	0.80	31.46	nr	99.13
Concentric reducers (one size down)						
80 mm	38.52	48.75	0.59	23.20	nr	71.95
100 mm	38.61	48.86	0.71	27.92	nr	76.78
125 mm	66.53	84.20	0.85	33.43	nr	117.63
150 mm	86.27	109.19	0.98	38.54	nr	147.72
Short radius elbow; 90°						
65 mm	44.11	55.83	0.53	20.84	nr	76.67
80 mm	45.00	56.95	0.61	23.99	nr	80.94
100 mm	60.25	76.25	0.80	31.46	nr	107.71
125 mm	99.37	125.76	0.90	35.39	nr	161.15
150 mm	128.96	163.21	0.94	36.97	nr	200.18
Short radius elbow; 45°						
65 mm	37.75	47.78	0.53	20.84	nr	68.62
80 mm	42.42	53.68	0.61	23.99	nr	77.67
100 mm	52.72	66.72	0.80	31.46	nr	98.18
125 mm	89.20	112.89	0.90	35.39	nr	148.28
150 mm	98.21	124.30	0.94	36.97	nr	161.26
Equal tee						
65 mm	79.39	100.47	0.83	32.64	nr	133.11
80 mm	84.10	106.43	0.93	36.57	nr	143.01
100 mm	94.04	119.02	1.18	46.40	nr	165.42
125 mm	249.43	315.67	1.37	53.88	nr	369.55
150 mm	231.26	292.69	1.43	56.24	nr	348.92

38 MECHANICAL/COOLING/HEATING SYSTEMS

Item	Net Price £	Material £	Labour hours	Labour £	Unit	Total rate £
LOW TEMPERATURE HOT WATER HEATING; PIPELINE: PLASTIC PIPEWORK						
Polypropylene PP-R 80 pipe, mechanically stabilized by fibre compound mixture in middle layer; suitable for continuous working temperatures of 0–90°C; thermally fused joints in the running length						
Pipe; 4 m long; PN 20						
20 mm dia.	5.56	7.04	0.35	13.76	m	**20.81**
25 mm dia.	8.38	10.61	0.39	15.34	m	**25.94**
32 mm dia.	9.57	12.11	0.43	16.91	m	**29.02**
40 mm dia.	12.77	16.16	0.47	18.48	m	**34.65**
50 mm dia.	18.58	23.52	0.51	20.06	m	**43.57**
63 mm dia.	30.67	38.82	0.52	20.45	m	**59.27**
75 mm dia.	39.68	50.22	0.60	23.60	m	**73.82**
90 mm dia.	61.22	77.48	0.69	27.13	m	**104.61**
110 mm dia.	92.01	116.45	0.69	27.13	m	**143.59**
125 mm dia.	98.57	124.76	0.85	33.43	m	**158.18**
FIXINGS						
Refer to steel pipes; black malleable iron. For minimum fixing distances, refer to the Tables and Memoranda to the rear of the book						
Extra over fittings; thermally fused joints						
Overbridge bow						
20 mm dia.	3.62	4.58	0.51	20.06	nr	**24.64**
25 mm dia.	6.66	8.42	0.56	22.02	nr	**30.45**
32 mm dia.	13.33	16.87	0.65	25.56	nr	**42.43**
Elbow 90°						
20 mm dia.	1.22	1.54	0.44	17.30	nr	**18.84**
25 mm dia.	1.61	2.04	0.52	20.45	nr	**22.49**
32 mm dia.	2.32	2.93	0.59	23.20	nr	**26.13**
40 mm dia.	3.65	4.62	0.66	25.96	nr	**30.58**
50 mm dia.	7.89	9.99	0.73	28.71	nr	**38.70**
63 mm dia.	12.06	15.26	0.85	33.43	nr	**48.69**
75 mm dia.	26.71	33.81	0.85	33.43	nr	**67.23**
90 mm dia.	49.36	62.47	1.04	40.90	nr	**103.36**
110 mm dia.	70.15	88.78	1.04	40.90	nr	**129.68**
125 mm dia.	108.11	136.83	1.30	51.12	nr	**187.95**
Long bend 90°						
20 mm dia.	6.48	8.21	0.48	18.88	nr	**27.08**
25 mm dia.	6.81	8.62	0.57	22.42	nr	**31.04**
32 mm dia.	7.79	9.86	0.65	25.56	nr	**35.42**
40 mm dia.	14.57	18.43	0.73	28.71	nr	**47.14**
Elbow 90°, female/male						
20 mm dia.	1.25	1.58	0.44	17.30	nr	**18.89**
25 mm dia.	1.65	2.08	0.52	20.45	nr	**22.53**
32 mm dia.	2.35	2.97	0.59	23.20	nr	**26.18**

38 MECHANICAL/COOLING/HEATING SYSTEMS

Item	Net Price £	Material £	Labour hours	Labour £	Unit	Total rate £
LOW TEMPERATURE HOT WATER HEATING; PIPELINE: PLASTIC PIPEWORK – cont						
Extra over fittings – cont						
Elbow 45°						
20 mm dia.	1.25	1.58	0.44	17.30	nr	**18.89**
25 mm dia.	1.65	2.08	0.52	20.45	nr	**22.53**
32 mm dia.	2.35	2.97	0.59	23.20	nr	**26.18**
40 mm dia.	3.67	4.65	0.66	25.96	nr	**30.60**
50 mm dia.	7.89	9.99	0.73	28.71	nr	**38.70**
63 mm dia.	12.09	15.31	0.85	33.43	nr	**48.73**
75 mm dia.	26.73	33.83	0.85	33.43	nr	**67.25**
90 mm dia.	49.36	62.47	1.04	40.90	nr	**103.36**
110 mm dia.	70.18	88.82	1.04	40.90	nr	**129.72**
125 mm dia.	108.13	136.85	1.30	51.12	nr	**187.97**
Elbow 45°, female/male						
20 mm dia.	1.25	1.58	0.44	17.30	nr	**18.89**
25 mm dia.	1.65	2.08	0.52	20.45	nr	**22.53**
32 mm dia.	2.35	2.97	0.59	23.20	nr	**26.18**
T-Piece 90°						
20 mm dia.	1.70	2.15	0.61	23.99	nr	**26.14**
25 mm dia.	2.30	2.91	0.72	28.31	nr	**31.22**
32 mm dia.	2.99	3.78	0.83	32.64	nr	**36.42**
40 mm dia.	4.63	5.86	0.92	36.18	nr	**42.04**
50 mm dia.	13.14	16.63	1.01	39.72	nr	**56.35**
63 mm dia.	18.82	23.82	1.11	43.65	nr	**67.47**
75 mm dia.	31.38	39.71	1.18	46.40	nr	**86.12**
90 mm dia.	57.76	73.10	1.46	57.42	nr	**130.52**
110 mm dia.	90.12	114.05	1.46	57.42	nr	**171.47**
125 mm dia.	119.76	151.57	1.82	71.57	nr	**223.15**
T-Piece 90° reducing						
25 × 20 × 20 mm	2.35	2.97	0.72	28.31	nr	**31.29**
25 × 20 × 25 mm	2.35	2.97	0.72	28.31	nr	**31.29**
32 × 20 × 32 mm	2.99	3.78	0.83	32.64	nr	**36.42**
32 × 25 × 32 mm	2.99	3.78	0.83	32.64	nr	**36.42**
40 × 20 × 40 mm	4.63	5.86	0.92	36.18	nr	**42.04**
40 × 25 × 40 mm	4.63	5.86	0.92	36.18	nr	**42.04**
40 × 32 × 40 mm	4.63	5.86	0.92	36.18	nr	**42.04**
50 × 25 × 50 mm	13.14	16.63	1.01	39.72	nr	**56.35**
50 × 32 × 50 mm	13.14	16.63	1.01	39.72	nr	**56.35**
50 × 40 × 50 mm	13.14	16.63	1.01	39.72	Unit	**56.35**
63 × 20 × 63 mm	17.72	22.43	1.11	43.65	nr	**66.08**
63 × 25 × 63 mm	17.72	22.43	1.11	43.65	nr	**66.08**
63 × 32 × 63 mm	17.72	22.43	1.11	43.65	nr	**66.08**
63 × 40 × 63 mm	17.72	22.43	1.01	39.72	nr	**62.15**
63 × 50 × 63 mm	17.72	22.43	1.01	39.72	nr	**62.15**
75 × 20 × 75 mm	28.80	36.45	1.18	46.40	nr	**82.86**
75 × 25 × 75 mm	28.80	36.45	1.18	46.40	nr	**82.86**
75 × 32 × 75 mm	28.80	36.45	1.18	46.40	nr	**82.86**
75 × 40 × 75 mm	28.80	36.45	1.18	46.40	nr	**82.86**

38 MECHANICAL/COOLING/HEATING SYSTEMS

Item	Net Price £	Material £	Labour hours	Labour £	Unit	Total rate £
75 × 50 × 75 mm	28.80	36.45	1.18	46.40	nr	**82.86**
75 × 63 × 75 mm	28.80	36.45	1.18	46.40	nr	**82.86**
90 × 63 × 90 mm	57.76	73.10	1.46	57.42	nr	**130.52**
110 × 75 × 110 mm	90.12	114.05	1.46	57.42	nr	**171.47**
110 × 90 × 110 mm	90.12	114.05	1.46	57.42	nr	**171.47**
125 × 90 × 125 mm	107.21	135.68	1.82	71.57	nr	**207.25**
125 × 110 × 125 mm	109.40	138.46	1.82	71.57	nr	**210.03**
Reducer						
25 × 20 mm	1.36	1.72	0.59	23.20	nr	**24.92**
32 × 20 mm	1.75	2.21	0.62	24.38	nr	**26.60**
32 × 25 mm	1.75	2.21	0.62	24.38	nr	**26.60**
40 × 20 mm	2.76	3.50	0.66	25.96	nr	**29.45**
40 × 25 mm	2.76	3.50	0.66	25.96	nr	**29.45**
40 × 32 mm	2.76	3.50	0.66	25.96	nr	**29.45**
50 × 20 mm	4.43	5.60	0.73	28.71	nr	**34.31**
50 × 25 mm	4.43	5.60	0.73	28.71	nr	**34.31**
50 × 32 mm	4.43	5.60	0.73	28.71	nr	**34.31**
50 × 40 mm	4.43	5.60	0.73	28.71	nr	**34.31**
63 × 40 mm	8.94	11.31	0.78	30.67	nr	**41.99**
63 × 25 mm	8.94	11.31	0.78	30.67	nr	**41.99**
63 × 32 mm	8.94	11.31	0.78	30.67	nr	**41.99**
63 × 50 mm	8.94	11.31	0.78	30.67	nr	**41.99**
75 × 50 mm	9.98	12.64	0.85	33.43	nr	**46.06**
75 × 63 mm	9.98	12.64	0.85	33.43	nr	**46.06**
90 × 63 mm	22.30	28.23	1.04	40.90	nr	**69.12**
90 × 75 mm	22.30	28.23	1.04	40.90	nr	**69.12**
110 × 90 mm	36.03	45.60	1.17	46.01	nr	**91.61**
125 × 110 mm	56.29	71.24	1.43	56.24	nr	**127.47**
Socket						
20 mm dia.	1.20	1.52	0.51	20.06	nr	**21.58**
25 mm dia.	1.36	1.72	0.56	22.02	nr	**23.74**
32 mm dia.	1.73	2.19	0.65	25.56	nr	**27.75**
40 mm dia.	2.13	2.69	0.74	29.10	nr	**31.79**
50 mm dia.	4.39	5.56	0.81	31.85	nr	**37.41**
63 mm dia.	8.94	11.31	0.86	33.82	nr	**45.13**
75 mm dia.	9.98	12.64	0.91	35.79	nr	**48.42**
90 mm dia.	25.77	32.61	0.91	35.79	nr	**68.40**
110 mm dia.	43.78	55.41	0.91	35.79	nr	**91.20**
125 mm dia.	60.99	77.19	1.30	51.12	nr	**128.31**
End cap						
20 mm dia.	1.87	2.37	0.25	9.83	nr	**12.20**
25 mm dia.	2.30	2.91	0.29	11.40	nr	**14.31**
32 mm dia.	2.85	3.60	0.33	12.98	nr	**16.58**
40 mm dia.	4.56	5.78	0.36	14.16	nr	**19.93**
50 mm dia.	6.31	7.99	0.40	15.73	nr	**23.72**
63 mm dia.	10.62	13.44	0.44	17.30	nr	**30.74**
75 mm dia.	15.27	19.32	0.47	18.48	nr	**37.81**
90 mm dia.	34.60	43.79	0.57	22.42	nr	**66.21**
110 mm dia.	41.60	52.65	0.57	22.42	nr	**75.07**
125 mm dia.	63.37	80.20	0.85	33.43	nr	**113.63**

38 MECHANICAL/COOLING/HEATING SYSTEMS

Item	Net Price £	Material £	Labour hours	Labour £	Unit	Total rate £
LOW TEMPERATURE HOT WATER HEATING; PIPELINE: PLASTIC PIPEWORK – cont						
Extra over fittings – cont						
Stub flange with gasket						
32 mm dia.	41.83	52.94	0.23	9.04	nr	**61.99**
40 mm dia.	52.64	66.62	0.27	10.62	nr	**77.24**
50 mm dia.	63.67	80.58	0.38	14.94	nr	**95.53**
63 mm dia.	76.43	96.73	0.43	16.91	nr	**113.64**
75 mm dia.	89.69	113.52	0.48	18.88	nr	**132.39**
90 mm dia.	121.39	153.63	0.53	20.84	nr	**174.47**
110 mm dia.	170.00	215.15	0.53	20.84	nr	**235.99**
125 mm dia.	243.66	308.37	0.75	29.49	nr	**337.87**
Weld in saddle with female thread						
40 × ½"	3.02	3.82	0.36	14.16	nr	**17.98**
50 × ½"	3.02	3.82	0.36	14.16	nr	**17.98**
63 × ½"	3.02	3.82	0.40	15.73	nr	**19.55**
75 × ½"	3.02	3.82	0.40	15.73	nr	**19.55**
90 × ½"	3.02	3.82	0.46	18.09	nr	**21.91**
110 × ½"	3.02	3.82	0.46	18.09	nr	**21.91**
Weld in saddle with male thread						
50 × ½"	3.02	3.82	0.36	14.16	nr	**17.98**
63 × ½"	3.02	3.82	0.40	15.73	nr	**19.55**
75 × ½"	3.02	3.82	0.40	15.73	nr	**19.55**
90 × ½"	3.02	3.82	0.46	18.09	nr	**21.91**
110 × ½"	3.02	3.82	0.46	18.09	nr	**21.91**
Transition piece, round with female thread						
20 × ½"	7.07	8.95	0.29	11.40	nr	**20.35**
20 × ¾"	9.33	11.81	0.29	11.40	nr	**23.22**
25 × ½"	9.33	11.81	0.33	12.98	nr	**24.79**
25 × ¾"	9.33	11.81	0.33	12.98	nr	**24.79**
Transition piece, hexagon with female thread						
32 × 1"	26.30	33.28	0.36	14.16	nr	**47.44**
40 × 1¼"	41.64	52.70	0.36	14.16	nr	**66.85**
50 × 1½"	48.33	61.16	0.36	14.16	nr	**75.32**
63 × 2"	74.85	94.73	0.40	15.73	nr	**110.46**
75 × 2"	78.11	98.86	0.40	15.73	nr	**114.59**
125 × 5"	149.61	189.35	0.51	20.06	nr	**209.41**
Stop valve for surface assembly						
20 mm dia.	29.18	36.93	0.25	9.83	nr	**46.76**
25 mm dia.	29.18	36.93	0.29	11.40	nr	**48.34**
32 mm dia.	54.97	69.57	0.33	12.98	nr	**82.54**
Ball valve						
20 mm dia.	102.76	130.06	0.25	9.83	nr	**139.89**
25 mm dia.	110.12	139.37	0.29	11.40	nr	**150.78**
32 mm dia.	132.32	167.47	0.33	12.98	nr	**180.45**
40 mm dia.	168.88	213.74	0.36	14.16	nr	**227.89**
50 mm dia.	231.84	293.42	0.40	15.73	nr	**309.15**
63 mm dia.	261.42	330.86	0.44	17.30	nr	**348.16**

38 MECHANICAL/COOLING/HEATING SYSTEMS

Item	Net Price £	Material £	Labour hours	Labour £	Unit	Total rate £
Floor or ceiling cover plates						
Plastic						
15 mm dia.	0.67	0.84	0.16	6.29	nr	**7.14**
20 mm dia.	0.75	0.95	0.22	8.65	nr	**9.60**
25 mm dia.	0.80	1.01	0.22	8.65	nr	**9.67**
32 mm dia.	0.87	1.11	0.24	9.44	nr	**10.54**
40 mm dia.	1.95	2.47	0.26	10.22	nr	**12.70**
50 mm dia.	2.14	2.70	0.26	10.22	nr	**12.93**
Chromium plated						
15 mm dia.	4.53	5.73	0.16	6.29	nr	**12.02**
20 mm dia.	4.84	6.13	0.17	6.69	nr	**12.81**
25 mm dia.	5.00	6.33	0.21	8.26	nr	**14.59**
32 mm dia.	5.12	6.48	0.22	8.65	nr	**15.13**
40 mm dia.	5.79	7.33	0.26	10.22	nr	**17.55**
50 mm dia.	6.92	8.75	0.26	10.22	nr	**18.98**

38 MECHANICAL/COOLING/HEATING SYSTEMS

Item	Net Price £	Material £	Labour hours	Labour £	Unit	Total rate £
LOW TEMPERATURE HOT WATER HEATING; PIPELINE: PIPELINE ANCILLARIES						
EXPANSION JOINTS						
Axial movement bellows expansion joints; stainless steel						
Screwed ends for steel pipework; up to 6 bar G at 100°C						
15 mm dia.	124.51	157.58	0.68	26.74	nr	**184.32**
20 mm dia.	133.00	168.33	0.81	31.85	nr	**200.18**
25 mm dia.	138.16	174.85	0.93	36.57	nr	**211.43**
32 mm dia.	146.28	185.14	1.06	41.69	nr	**226.82**
40 mm dia.	154.94	196.09	1.16	45.62	nr	**241.71**
50 mm dia.	154.35	195.35	1.19	46.80	nr	**242.15**
Screwed ends for steel pipework; aluminium and steel outer sleeves; up to 16 bar G at 120°C						
20 mm dia.	135.64	171.67	1.32	51.91	nr	**223.58**
25 mm dia.	138.16	174.85	1.52	59.78	nr	**234.63**
32 mm dia.	146.28	185.14	1.80	70.79	nr	**255.92**
40 mm dia.	154.94	196.09	2.03	79.83	nr	**275.92**
50 mm dia.	157.43	199.24	2.26	88.88	nr	**288.12**
Flanged ends for steel pipework; aluminium and steel outer sleeves; up to 16 bar G at 120°C						
20 mm dia.	120.02	151.90	0.53	20.84	nr	**172.74**
25 mm dia.	128.58	162.73	0.64	25.17	nr	**187.90**
32 mm dia.	177.18	224.23	0.74	29.10	nr	**253.34**
40 mm dia.	191.45	242.30	0.82	32.25	nr	**274.55**
50 mm dia.	227.16	287.50	0.89	35.00	nr	**322.50**
Flanged ends for steel pipework; up to 16 bar G at 120°C						
65 mm dia.	210.02	265.80	1.10	43.26	nr	**309.05**
80 mm dia.	225.73	285.69	1.31	51.52	nr	**337.21**
100 mm dia.	258.57	327.25	1.78	70.00	nr	**397.25**
150 mm dia.	395.75	500.86	3.08	121.12	nr	**621.99**
Screwed ends for non-ferrous pipework; up to 6 bar G at 100°C						
20 mm dia.	142.30	180.10	0.72	28.31	nr	**208.41**
25 mm dia.	144.94	183.44	0.84	33.03	nr	**216.48**
32 mm dia.	153.47	194.23	1.02	40.11	nr	**234.34**
40 mm dia.	162.55	205.72	1.11	43.65	nr	**249.37**
50 mm dia.	165.16	209.03	1.18	46.40	nr	**255.43**
Flanged ends for steel, copper ornon-ferrous pipework; up to 16 bar G at 120°C						
65 mm dia.	283.29	358.53	0.87	34.21	nr	**392.75**
80 mm dia.	292.27	369.90	0.95	37.36	nr	**407.26**
100 mm dia.	337.26	426.84	1.15	45.22	nr	**472.06**
150 mm dia.	553.06	699.96	1.36	53.48	nr	**753.44**

38 MECHANICAL/COOLING/HEATING SYSTEMS

Item	Net Price £	Material £	Labour hours	Labour £	Unit	Total rate £
Angular movement bellows expansion joints; stainless steel						
Flanged ends for steel pipework; up to 16 bar G at 120°C						
50 mm dia.	311.27	393.94	0.71	27.92	nr	**421.86**
65 mm dia.	377.76	478.09	0.83	32.64	nr	**510.74**
80 mm dia.	441.45	558.70	0.91	35.79	nr	**594.49**
100 mm dia.	569.89	721.26	0.97	38.15	nr	**759.40**
125 mm dia.	752.77	952.71	1.16	45.62	nr	**998.33**
150 mm dia.	895.11	1132.85	1.18	46.40	nr	**1179.26**

38 MECHANICAL/COOLING/HEATING SYSTEMS

Item	Net Price £	Material £	Labour hours	Labour £	Unit	Total rate £
LOW TEMPERATURE HOT WATER HEATING; PIPELINE: VALVES						
Gate valves						
Bronze gate valve; non-rising stem; BS 5154, series B, PN 16; working pressure up to 16 from −10°C to 100°C; 7 bar for saturated steam; screwed ends to steel						
15 mm dia.	79.37	100.46	1.18	46.40	nr	**146.86**
20 mm dia.	108.42	137.21	1.24	48.76	nr	**185.98**
25 mm dia.	137.98	174.63	1.31	51.52	nr	**226.14**
32 mm dia.	187.95	237.87	1.43	56.24	nr	**294.11**
40 mm dia.	250.92	317.57	1.53	60.17	nr	**377.74**
50 mm dia.	346.21	438.16	1.63	64.10	nr	**502.26**
Bronze gate valve; non-rising stem; BS 5154, series B, PN 32; working pressure up to 14 bar for saturated steam, 32 bar from −10°C to 100°C; screwed ends to steel						
15 mm dia.	69.46	87.91	1.11	43.65	nr	**131.56**
20 mm dia.	90.10	114.03	1.28	50.34	nr	**164.37**
25 mm dia.	122.05	154.47	1.49	58.60	nr	**213.07**
32 mm dia.	168.58	213.35	1.88	73.93	nr	**287.29**
40 mm dia.	226.33	286.45	2.31	90.84	nr	**377.29**
50 mm dia.	318.03	402.50	2.80	110.11	nr	**512.61**
Bronze gate valve; non-rising stem; BS 5154, series B, PN 20; working pressure up to 9 bar for saturated steam, 20 bar from −10°C to 100°C; screwed ends to steel						
15 mm dia.	47.38	59.96	0.84	33.03	nr	**92.99**
20 mm dia.	67.31	85.19	1.01	39.72	nr	**124.91**
25 mm dia.	87.10	110.23	1.19	46.80	nr	**157.03**
32 mm dia.	124.29	157.30	1.38	54.27	nr	**211.57**
40 mm dia.	170.11	215.29	1.62	63.71	nr	**278.99**
50 mm dia.	243.08	307.64	1.94	76.29	nr	**383.94**
Bronze gate valve; non-rising stem; BS 5154, series B, PN 16; working pressure up to 7 bar for saturated steam, 16 bar from −10°C to 100°C; BS4504 flanged ends; bolted connections						
15 mm dia.	213.32	269.98	1.18	46.40	nr	**316.38**
20 mm dia.	235.23	297.70	1.24	48.76	nr	**346.47**
25 mm dia.	309.77	392.05	1.31	51.52	nr	**443.56**
32 mm dia.	388.64	491.87	1.43	56.24	nr	**548.10**
40 mm dia.	469.89	594.69	1.53	60.17	nr	**654.86**
50 mm dia.	649.61	822.15	1.63	64.10	nr	**886.25**
65 mm dia.	1008.91	1276.87	1.71	67.25	nr	**1344.12**
80 mm dia.	1427.90	1807.16	1.88	73.93	nr	**1881.09**
100 mm dia.	2594.94	3284.16	2.03	79.83	nr	**3363.99**

38 MECHANICAL/COOLING/HEATING SYSTEMS

Item	Net Price £	Material £	Labour hours	Labour £	Unit	Total rate £
Cast iron gate valve; bronze trim; non-rising stem; BS 5150, PN6; working pressure 6 bar from −10°C to 120°C; BS4504 flanged ends; bolted connections						
50 mm dia.	402.72	509.69	1.85	72.75	nr	**582.44**
65 mm dia.	420.93	532.72	2.00	78.65	nr	**611.38**
80 mm dia.	486.46	615.66	2.27	89.27	nr	**704.93**
100 mm dia.	615.37	778.81	2.76	108.54	nr	**887.35**
125 mm dia.	880.80	1114.74	6.05	237.92	nr	**1352.66**
150 mm dia.	1013.54	1282.74	8.03	315.79	nr	**1598.53**
200 mm dia.	1930.55	2443.30	9.17	360.62	nr	**2803.92**
250 mm dia.	2970.54	3759.52	10.72	421.57	nr	**4181.09**
300 mm dia.	3523.27	4459.06	11.75	462.08	nr	**4921.14**
Cast iron gate valve; bronze trim; non-rising stem; BS 5150, PN10; working pressure up to 8.4 bar for saturated steam, 10 bar from −10°C to 120°C; BS4504 flanged ends; bolted connections						
50 mm dia.	474.52	600.55	1.85	72.75	nr	**673.30**
65 mm dia.	589.69	746.32	2.00	78.65	nr	**824.97**
80 mm dia.	656.55	830.93	2.27	89.27	nr	**920.20**
100 mm dia.	873.97	1106.09	2.76	108.54	nr	**1214.63**
125 mm dia.	1221.90	1546.44	6.05	237.92	nr	**1784.36**
150 mm dia.	1441.65	1824.56	8.03	315.79	nr	**2140.34**
200 mm dia.	2648.74	3352.25	9.17	360.62	nr	**3712.86**
250 mm dia.	3983.49	5041.51	10.72	421.57	nr	**5463.08**
300 mm dia.	4482.59	5673.17	11.75	462.08	nr	**6135.25**
350 mm dia.	5213.76	6598.53	12.67	498.26	nr	**7096.79**
Cast iron gate valve; bronze trim; non-rising stem; BS 5163 series A, PN16; working pressure for cold water services up to 16 bar; BS4504 flanged ends; bolted connections						
50 mm dia.	474.52	600.55	1.85	72.75	nr	**673.30**
65 mm dia.	589.69	746.32	2.00	78.65	nr	**824.97**
80 mm dia.	656.55	830.93	2.27	89.27	nr	**920.20**
100 mm dia.	873.97	1106.09	2.76	108.54	nr	**1214.63**
125 mm dia.	1221.90	1546.44	6.05	237.92	nr	**1784.36**
150 mm dia.	1441.65	1824.56	8.03	315.79	nr	**2140.34**
Ball valves						
Malleable iron body; lever operated stainless steel ball and stem; working pressure up to 12 bar; flanged ends to BS 4504 16/11; bolted connections						
40 mm dia.	585.86	741.47	1.54	60.56	nr	**802.03**
50 mm dia.	738.24	934.32	1.64	64.49	nr	**998.81**
80 mm dia.	1194.18	1511.35	1.92	75.51	nr	**1586.85**
100 mm dia.	1234.65	1562.57	2.80	110.11	nr	**1672.68**
150 mm dia.	2282.22	2888.38	12.05	473.88	nr	**3362.25**
Malleable iron body; lever operated stainless steel ball and stem; working pressure up to 16 bar; screwed ends to steel						
15 mm dia.	52.59	66.55	1.34	52.72	nr	**119.27**
20 mm dia.	69.57	88.05	1.34	52.72	nr	**140.77**
25 mm dia.	71.57	90.57	1.40	55.08	nr	**145.65**

38 MECHANICAL/COOLING/HEATING SYSTEMS

Item	Net Price £	Material £	Labour hours	Labour £	Unit	Total rate £
LOW TEMPERATURE HOT WATER HEATING; PIPELINE: VALVES – cont						
Ball valves – cont						
Malleable iron body – cont						
32 mm dia.	98.59	124.78	1.46	57.49	nr	**182.27**
40 mm dia.	98.59	124.78	1.54	60.59	nr	**185.37**
50 mm dia.	117.90	149.22	1.64	64.57	nr	**213.79**
Carbon steel body; lever operated stainless steel ball and stem; Class 150; working pressure up to 19 bar; screwed ends to steel						
15 mm dia.	118.30	149.72	0.84	33.05	nr	**182.76**
20 mm dia.	132.97	168.29	1.14	44.84	nr	**213.13**
25 mm dia.	143.97	182.21	1.30	51.14	nr	**233.35**
32 mm dia.	185.66	234.97	1.42	55.84	nr	**290.82**
40 mm dia.	232.07	293.70	1.56	61.35	nr	**355.05**
50 mm dia.	290.10	367.15	1.68	66.07	nr	**433.22**
Globe valves						
Bronze; rising stem; renewable disc; BS 5154 series B, PN32; working pressure up to 14 bar for saturated steam, 32 bar from −10°C to 100°C; screwed ends to steel						
15 mm dia.	75.40	95.43	0.77	30.28	nr	**125.71**
20 mm dia.	103.12	130.51	1.03	40.51	nr	**171.01**
25 mm dia.	147.95	187.24	1.19	46.80	nr	**234.04**
32 mm dia.	208.29	263.61	1.38	54.27	nr	**317.88**
40 mm dia.	290.05	367.09	1.61	63.31	nr	**430.41**
50 mm dia.	445.22	563.46	1.62	63.71	nr	**627.17**
Bronze; needle valve; rising stem; BS 5154, series B, PN32; working pressure up to 14 bar for saturated steam, 32 bar from −10°C to 100°C; screwed ends to steel						
15 mm dia.	57.33	72.56	1.07	42.08	nr	**114.64**
20 mm dia.	97.06	122.84	1.18	46.40	nr	**169.24**
25 mm dia.	136.85	173.20	1.27	49.94	nr	**223.14**
32 mm dia.	284.84	360.50	1.35	53.09	nr	**413.59**
40 mm dia.	448.75	567.93	1.47	57.81	nr	**625.74**
50 mm dia.	568.39	719.35	1.61	63.31	nr	**782.67**
Bronze; rising stem; renewable disc; BS 5154, series B, PN16; working pressure up to 7 bar for saturated steam, 16 bar from −10°C to 100°C; BS4504 flanged ends; bolted connections						
15 mm dia.	152.37	192.84	1.16	45.62	nr	**238.46**
20 mm dia.	176.25	223.06	1.26	49.55	nr	**272.61**
25 mm dia.	309.06	391.15	1.38	54.27	nr	**445.42**
32 mm dia.	390.15	493.77	1.47	57.81	nr	**551.58**
40 mm dia.	443.47	561.26	1.56	61.35	nr	**622.61**
50 mm dia.	636.27	805.26	1.71	67.25	nr	**872.51**

38 MECHANICAL/COOLING/HEATING SYSTEMS

Item	Net Price £	Material £	Labour hours	Labour £	Unit	Total rate £
Bronze; rising stem; renewable disc; BS 2060, class 250; working pressure up to 24 bar for saturated steam, 38 bar from −10°C to 100°C; flanged ends (BS 10 table H); bolted connections						
15 mm dia.	388.04	491.10	1.16	45.62	nr	**536.72**
20 mm dia.	450.98	570.76	1.26	49.55	nr	**620.31**
25 mm dia.	619.73	784.32	1.38	54.27	nr	**838.59**
32 mm dia.	809.59	1024.62	1.47	57.81	nr	**1082.43**
40 mm dia.	969.47	1226.96	1.56	61.35	nr	**1288.31**
50 mm dia.	1507.49	1907.88	1.71	67.25	nr	**1975.13**
65 mm dia.	1802.81	2281.64	1.88	73.93	nr	**2355.57**
80 mm dia.	1899.79	2404.37	2.03	79.83	nr	**2484.20**
Check valves						
Bronze; swing pattern; BS 5154 series B, PN 25; working pressure up to 10.5 bar for saturated steam, 25 bar from −10°C to 100°C; screwed ends to steel						
15 mm dia.	31.72	40.14	0.77	30.28	nr	**70.42**
20 mm dia.	37.69	47.70	1.03	40.51	nr	**88.20**
25 mm dia.	52.23	66.10	1.19	46.80	nr	**112.90**
32 mm dia.	88.50	112.01	1.38	54.27	nr	**166.28**
40 mm dia.	110.09	139.32	1.62	63.71	nr	**203.03**
50 mm dia.	168.86	213.71	1.94	76.29	nr	**290.00**
65 mm dia.	319.02	403.75	2.45	96.35	nr	**500.10**
80 mm dia.	451.09	570.90	2.83	111.29	nr	**682.20**
Bronze; vertical lift pattern; BS 5154 series B, PN32; working pressure up to 14 bar for saturated steam, 32 bar from −10°C to 100°C; screwed ends to steel						
15 mm dia.	41.34	52.32	0.96	37.75	nr	**90.07**
20 mm dia.	58.79	74.40	1.07	42.08	nr	**116.48**
25 mm dia.	86.32	109.25	1.17	46.01	nr	**155.26**
32 mm dia.	131.50	166.42	1.33	52.30	nr	**218.73**
40 mm dia.	171.35	216.86	1.41	55.45	nr	**272.31**
50 mm dia.	256.71	324.90	1.55	60.95	nr	**385.85**
65 mm dia.	696.53	881.53	1.80	70.79	nr	**952.32**
80 mm dia.	1041.32	1317.89	1.99	78.26	nr	**1396.15**
Bronze; oblique swing pattern; BS 5154 series A, PN32; working pressure up to 14 bar for saturated steam, 32 bar from −10°C to 120°C; screwed connections to steel						
15 mm dia.	52.52	66.46	0.96	37.75	nr	**104.22**
20 mm dia.	62.38	78.95	1.07	42.08	nr	**121.02**
25 mm dia.	86.48	109.45	1.17	46.01	nr	**155.46**
32 mm dia.	129.32	163.67	1.33	52.30	nr	**215.97**
40 mm dia.	166.68	210.95	1.41	55.45	nr	**266.40**
50 mm dia.	276.14	349.48	1.55	60.95	nr	**410.43**
Cast iron; swing pattern; BS 5153 PN6; working pressure up to 6 bar from −10°C to 120°C; BS 4504 flanged ends; bolted connections						
50 mm dia.	1055.27	1335.55	1.86	73.15	nr	**1408.70**
65 mm dia.	1155.38	1462.25	2.00	78.65	nr	**1540.90**
80 mm dia.	1300.49	1645.91	2.56	100.67	nr	**1746.58**

38 MECHANICAL/COOLING/HEATING SYSTEMS

Item	Net Price £	Material £	Labour hours	Labour £	Unit	Total rate £
LOW TEMPERATURE HOT WATER HEATING; PIPELINE: VALVES – cont						
Check valves – cont						
Cast iron; swing pattern – cont						
100 mm dia.	1665.98	2108.47	2.76	108.54	nr	**2217.01**
125 mm dia.	2485.41	3145.53	6.05	237.92	nr	**3383.45**
150 mm dia.	2794.53	3536.75	8.11	318.93	nr	**3855.68**
200 mm dia.	6180.04	7821.46	9.26	364.16	nr	**8185.62**
250 mm dia.	7724.86	9776.58	10.72	421.57	nr	**10198.16**
300 mm dia.	9269.85	11731.92	11.75	462.08	nr	**12193.99**
Cast iron; horizontal lift pattern; BS 5153 PN16; working pressure up to 13 bar for saturated steam, 16 bar from −10°C to 120°C; BS 4504 flanged ends; bolted connections						
50 mm dia.	952.93	1206.03	1.86	73.15	nr	**1279.17**
65 mm dia.	952.93	1206.03	2.00	78.65	nr	**1284.68**
80 mm dia.	1179.81	1493.17	2.56	100.67	nr	**1593.85**
100 mm dia.	1373.93	1738.85	2.96	116.40	nr	**1855.26**
125 mm dia.	2049.89	2594.34	7.76	305.17	nr	**2899.51**
150 mm dia.	2304.68	2916.80	10.50	412.92	nr	**3329.72**
Cast iron; semi-lugged butterfly valve; BS5155 PN16; working pressure 16 bar from −10°C to 120°C; BS 4504 flanged ends; bolted connections						
50 mm dia.	291.68	369.15	2.20	86.52	nr	**455.67**
65 mm dia.	301.01	380.96	2.31	90.84	nr	**471.81**
80 mm dia.	354.58	448.75	2.88	113.26	nr	**562.01**
100 mm dia.	492.91	623.82	3.11	122.30	nr	**746.13**
125 mm dia.	719.42	910.49	5.02	197.42	nr	**1107.91**
150 mm dia.	825.77	1045.10	6.98	274.49	nr	**1319.59**
200 mm dia.	1318.21	1668.32	8.25	324.44	nr	**1992.76**
250 mm dia.	1945.63	2462.39	10.47	411.74	nr	**2874.13**
300 mm dia.	2918.52	3693.68	11.48	451.46	nr	**4145.14**
Cast iron; semi-lugged butterfly valve; BS5155 PN20; working pressure up to 20 bar from −10°C to 120°C; BS 4504 flanged ends; bolted connections						
50 mm dia.	2612.06	3305.83	2.20	86.52	nr	**3392.34**
65 mm dia.	2702.47	3420.25	2.31	90.84	nr	**3511.09**
80 mm dia.	2885.20	3651.51	2.88	113.26	nr	**3764.77**
100 mm dia.	3513.41	4446.57	3.11	122.30	nr	**4568.88**
125 mm dia.	4509.66	5707.43	5.02	197.42	nr	**5904.84**
150 mm dia.	5117.95	6477.28	5.02	197.42	nr	**6674.70**
200 mm dia.	5886.36	7449.78	8.25	324.44	nr	**7774.22**
250 mm dia.	9281.76	11747.00	10.47	411.74	nr	**12158.74**
300 mm dia.	11246.67	14233.78	11.48	451.46	nr	**14685.24**
Cast iron; semi-lugged butterfly valve; BS5155 PN20; working pressure up to 30 bar from −10°C to 120°C; BS 4504 flanged ends; bolted connections						
50 mm dia.	2612.06	3305.83	2.20	86.52	nr	**3392.34**
65 mm dia.	2702.47	3420.25	2.31	90.84	nr	**3511.09**
80 mm dia.	2885.20	3651.51	2.88	113.26	nr	**3764.77**
100 mm dia.	3513.41	4446.57	3.11	122.30	nr	**4568.88**

38 MECHANICAL/COOLING/HEATING SYSTEMS

Item	Net Price £	Material £	Labour hours	Labour £	Unit	Total rate £
125 mm dia.	4509.66	5707.43	5.02	197.42	nr	**5904.84**
150 mm dia.	5117.95	6477.28	5.02	197.42	nr	**6674.70**
200 mm dia.	5886.36	7449.78	8.25	324.44	nr	**7774.22**
250 mm dia.	9281.76	11747.00	10.47	411.74	nr	**12158.74**
300 mm dia.	11246.67	14233.78	11.48	451.46	nr	**14685.24**
Commissioning valves						
Bronze commissioning set; metering station; double regulating valve; BS5154 PN20 Series B; working pressure 20 bar from −10°C to 100°C; screwed ends to steel						
15 mm dia.	137.18	173.62	1.08	42.47	nr	**216.09**
20 mm dia.	220.43	278.97	1.46	57.42	nr	**336.39**
25 mm dia.	262.11	331.73	1.68	66.07	nr	**397.79**
32 mm dia.	351.43	444.76	1.95	76.69	nr	**521.45**
40 mm dia.	509.66	645.03	2.27	89.27	nr	**734.30**
50 mm dia.	769.18	973.47	2.73	107.36	nr	**1080.83**
Cast iron commissioning set; metering station; double regulating valve; BS5152 PN16; working pressure 16 bar from −10°C to 90°C; flanged ends (BS 4504, Part 1, Table 16); bolted connections						
65 mm dia.	1157.18	1464.52	1.80	70.79	nr	**1535.31**
80 mm dia.	1385.56	1753.57	2.56	100.67	nr	**1854.24**
100 mm dia.	1928.96	2441.30	2.30	90.45	nr	**2531.75**
125 mm dia.	2825.43	3575.87	2.44	95.96	nr	**3671.82**
150 mm dia.	3721.97	4710.53	2.90	114.04	nr	**4824.58**
200 mm dia.	9526.75	12057.05	8.26	324.83	nr	**12381.88**
250 mm dia.	11917.85	15083.23	10.49	412.53	nr	**15495.76**
300 mm dia.	12943.03	16380.70	11.49	451.85	nr	**16832.55**
Cast iron variable orifice double regulating valve; orifice valve; BS5152 PN16; working pressure 16 bar from −10° to 90°C; flanged ends (BS 4504, Part 1, Table 16); bolted connections						
65 mm dia.	847.23	1072.26	2.00	78.65	nr	**1150.91**
80 mm dia.	1037.00	1312.43	2.56	100.67	nr	**1413.10**
100 mm dia.	1423.09	1801.06	2.96	116.40	nr	**1917.46**
125 mm dia.	2131.07	2697.08	7.76	305.17	nr	**3002.25**
150 mm dia.	2737.29	3464.32	8.26	324.83	nr	**3789.15**
200 mm dia.	7174.20	9079.67	10.49	412.53	nr	**9492.20**
250 mm dia.	10908.31	13805.56	10.50	412.92	nr	**14218.48**
300 mm dia.	19567.33	24764.41	11.49	451.85	nr	**25216.27**
Cast iron globe valve with double regulating feature; BS5152 PN16; working pressure 16 bar from −10°C to 120°C; flanged ends (BS 4504, Part 1, Table 16); bolted connections						
65 mm dia.	1009.26	1277.31	2.00	78.65	nr	**1355.96**
80 mm dia.	1220.51	1544.68	2.56	100.67	nr	**1645.35**
100 mm dia.	1607.76	2034.79	2.96	116.40	nr	**2151.19**
125 mm dia.	2241.65	2837.03	7.76	305.17	nr	**3142.20**
150 mm dia.	2837.13	3590.67	10.50	412.92	nr	**4003.59**
200 mm dia.	7345.02	9295.86	8.26	324.83	nr	**9620.69**
250 mm dia.	11112.86	14064.44	10.49	412.53	nr	**14476.97**
300 mm dia.	17459.93	22097.29	11.49	451.85	nr	**22549.14**

38 MECHANICAL/COOLING/HEATING SYSTEMS

Item	Net Price £	Material £	Labour hours	Labour £	Unit	Total rate £
LOW TEMPERATURE HOT WATER HEATING; PIPELINE: VALVES – cont						
Commissioning valves – cont						
Bronze autoflow commissioning valve; PN25 ; working pressure 25 bar up to 100°C; screwed ends to steel						
15 mm dia.	155.56	196.88	0.82	32.25	nr	**229.12**
20 mm dia.	164.71	208.45	1.08	42.47	nr	**250.92**
25 mm dia.	205.08	259.55	1.27	49.94	nr	**309.50**
32 mm dia.	320.31	405.38	1.50	58.99	nr	**464.37**
40 mm dia.	350.72	443.88	1.76	69.21	nr	**513.09**
50 mm dia.	518.58	656.31	2.13	83.76	nr	**740.08**
Ductile iron autoflow commissioning valves; PN16; working pressure 16 bar from −10°C to 120°C; for ANSI 150 flanged ends						
65 mm dia.	1356.11	1716.29	2.31	90.84	nr	**1807.14**
80 mm dia.	1499.33	1897.55	2.88	113.26	nr	**2010.81**
100 mm dia.	1915.38	2424.10	3.11	122.30	nr	**2546.40**
150 mm dia.	3945.68	4993.66	6.98	274.49	nr	**5268.15**
200 mm dia.	5747.22	7273.68	8.26	324.83	nr	**7598.51**
250 mm dia.	7982.06	10102.10	10.49	412.53	nr	**10514.63**
300 mm dia.	10194.32	12901.93	11.49	451.85	nr	**13353.78**
Strainers						
Bronze strainer; Y type; PN32; working pressure 32 bar from −10°C to 100°C; screwed ends to steel						
15 mm dia.	64.01	81.01	0.82	32.25	nr	**113.26**
20 mm dia.	81.29	102.88	1.08	42.47	nr	**145.35**
25 mm dia.	107.86	136.50	1.27	49.94	nr	**186.45**
32 mm dia.	174.20	220.47	1.50	58.99	nr	**279.46**
40 mm dia.	225.31	285.15	1.76	69.21	nr	**354.37**
50 mm dia.	377.35	477.58	2.13	83.76	nr	**561.34**
Cast iron strainer; Y type; PN16; working pressure 16 bar from −10°C to 120°C; BS 4504 flanged ends						
65 mm dia.	417.65	528.58	2.31	90.84	nr	**619.42**
80 mm dia.	494.33	625.63	2.88	113.26	nr	**738.88**
100 mm dia.	723.32	915.44	3.11	122.30	nr	**1037.74**
125 mm dia.	1333.40	1687.55	5.02	197.42	nr	**1884.96**
150 mm dia.	1710.77	2165.14	6.98	274.49	nr	**2439.64**
200 mm dia.	2871.44	3634.09	8.26	324.83	nr	**3958.93**
250 mm dia.	4152.25	5255.09	10.49	412.53	nr	**5667.62**
300 mm dia.	6761.85	8557.80	11.49	451.85	nr	**9009.65**
Regulators						
Gunmetal; self-acting two port thermostatic regulator; single seat; screwed ends; complete with sensing element, 2 m long capillary tube						
15 mm dia.	1233.33	1560.90	1.37	53.88	nr	**1614.78**
20 mm dia.	1273.02	1611.13	1.24	48.76	nr	**1659.89**
25 mm dia.	1306.42	1653.41	1.34	52.70	nr	**1706.10**

38 MECHANICAL/COOLING/HEATING SYSTEMS

Item	Net Price £	Material £	Labour hours	Labour £	Unit	Total rate £
Gunmetal; self-acting two port thermostatic regulator; double seat; flanged ends (BS 4504 PN25); with sensing element, 2 m long capillary tube; steel body						
65 mm dia.	4573.91	5788.75	1.23	56.22	nr	**5844.96**
80 mm dia.	5398.48	6832.32	1.62	74.04	nr	**6906.36**
Control valves						
Electrically operated (electrical work elsewhere)						
Pressure independent control valve (PICV)						
Brass; pressure independent control valve; rotary type; PN 25; maximum working pressure 25 bar at from −10°C to 120°C; screwed ends to steel; electrical work elsewhere						
15 mm dia.	127.54	161.41	1.08	42.47	nr	**203.89**
20 mm dia.	152.70	193.25	1.18	46.40	nr	**239.66**
25 mm dia.	224.86	284.58	1.35	53.09	nr	**337.67**
32 mm dia.	327.23	414.14	1.46	57.42	nr	**471.55**
40 mm dia.	362.45	458.72	1.53	60.17	nr	**518.89**
50 mm dia.	632.59	800.61	1.61	63.31	nr	**863.92**
Brass; pressure independent control valve; rotary type with electronic actuator; 24 V motor; PN 25; maximum working pressure 25 bar at from −10°C to 120°C; screwed ends to steel; electrical work elsewhere						
15 mm dia.	266.81	337.68	2.06	81.01	nr	**418.69**
20 mm dia.	291.97	369.52	2.15	84.55	nr	**454.07**
25 mm dia.	364.13	460.84	2.27	89.27	nr	**550.11**
32 mm dia.	513.47	649.85	2.35	92.42	nr	**742.26**
40 mm dia.	548.70	694.43	2.47	97.13	nr	**791.57**
50 mm dia.	837.31	1059.70	2.55	100.28	nr	**1159.98**
Differential pressure control valves (DPCV)						
Brass; differential control valve; PN 25; maximum working pressure 25 bar at from −20°C to 120°C; screwed ends to steel						
15 mm dia.	350.72	443.87	1.99	78.26	nr	**522.13**
20 mm dia.	377.67	477.98	2.03	79.83	nr	**557.81**
25 mm dia.	419.67	531.13	2.11	82.98	nr	**614.11**
32 mm dia.	475.05	601.22	2.17	85.34	nr	**686.56**
40 m m dia.	818.82	1036.30	2.26	88.88	nr	**1125.18**
50 mm dia.	983.18	1244.31	2.31	90.84	nr	**1335.16**
Cast iron; butterfly type; two position electrically controlled 240 V motor and linkage mechanism; for low pressure hot water; maximum pressure 6 bar at 120°C; flanged ends; electrical work elsewhere						
25 mm dia.	1318.33	1668.48	1.47	57.81	nr	**1726.28**
32 mm dia.	1362.21	1724.01	1.52	59.78	nr	**1783.79**
40 mm dia.	1903.47	2409.04	1.61	63.31	nr	**2472.35**
50 mm dia.	1957.55	2477.47	1.71	67.25	nr	**2544.72**
65 mm dia.	2006.58	2539.53	2.51	98.71	nr	**2638.23**
80 mm dia.	2085.04	2638.83	2.69	105.79	nr	**2744.61**
100 mm dia.	2188.09	2769.25	2.81	110.51	nr	**2879.75**

38 MECHANICAL/COOLING/HEATING SYSTEMS

Item	Net Price £	Material £	Labour hours	Labour £	Unit	Total rate £
LOW TEMPERATURE HOT WATER HEATING; PIPELINE: VALVES – cont						
Differential pressure control valves (DPCV) – cont						
Cast iron; butterfly type – cont						
125 mm dia.	2423.58	3067.28	2.94	115.62	nr	**3182.90**
150 mm dia.	2626.33	3323.89	3.33	130.95	nr	**3454.84**
200 mm dia.	3257.54	4122.74	3.67	144.33	nr	**4267.07**
Cast iron; three way 240 V motorized; for low pressure hot water; maximum pressure 6 bar 120°C; flanged ends, drilled (BS 10, Table F)						
25 mm dia.	1473.46	1864.81	1.99	78.26	nr	**1943.07**
40 mm dia.	1533.04	1940.22	2.13	83.76	nr	**2023.98**
50 mm dia.	1552.75	1965.16	3.21	126.24	nr	**2091.40**
65 mm dia.	1704.80	2157.60	3.23	127.02	nr	**2284.62**
80 mm dia.	1910.25	2417.62	3.50	137.64	nr	**2555.26**
2 port control valves						
Cast iron; 2 port; motorized control valve; normally open for heating applications; PN 16; screwed ends to steel						
15 mm dia.	59.85	75.75	1.24	48.76	nr	**124.51**
22 mm dia.	91.39	115.67	1.31	51.52	nr	**167.18**
28 mm dia.	109.30	138.33	1.37	53.88	nr	**192.21**
32 mm dia.	127.71	161.63	1.45	57.02	nr	**218.65**
40 mm dia.	159.63	202.03	1.52	59.78	nr	**261.81**
50 mm dia.	199.53	252.53	1.60	62.92	nr	**315.45**
Two port normally closed motorized valve; electric actuator; spring return; domestic usage						
22 mm dia.	203.78	257.90	1.18	46.40	nr	**304.30**
28 mm dia.	287.55	363.93	1.35	53.09	nr	**417.02**
Two port on/off motorized valve; electric actuator; spring return; domestic usage						
22 mm dia.	203.78	257.90	1.18	46.40	nr	**304.30**
3 port control valves						
Cast iron; 3 port; motorized control valve with temperatyre control unit; PN 25; screwed ends to steel						
22 mm dia.	93.13	117.86	2.00	78.65	nr	**196.51**
25 mm dia.	105.57	133.61	2.13	83.76	nr	**217.38**
32 mm dia.	135.16	171.06	2.19	86.12	nr	**257.18**
40 mm dia.	168.94	213.81	2.26	88.88	nr	**302.68**
Three port motorized valve; electric actuator; spring return; domestic usage						
22 mm dia.	309.55	391.77	1.18	46.40	nr	**438.18**
4 port control valves						
Cast iron; 4 port; fixed orifice double regulating and control valve; PN 16; screwed ends to steel						
22 mm dia.	82.53	104.45	2.89	113.65	nr	**218.11**

38 MECHANICAL/COOLING/HEATING SYSTEMS

Item	Net Price £	Material £	Labour hours	Labour £	Unit	Total rate £
Safety and relief valves						
Bronze relief valve; spring type; side outlet; working pressure up to 20.7 bar at 120°C; screwed ends to steel						
15 mm dia.	257.29	325.63	0.26	10.22	nr	**335.86**
20 mm dia.	282.18	357.12	0.36	14.16	nr	**371.28**
Bronze relief valve; spring type; side outlet; working pressure up to 17.2 bar at 120°C; screwed ends to steel						
25 mm dia.	385.93	488.43	0.38	14.94	nr	**503.37**
32 mm dia.	522.88	661.76	0.48	18.88	nr	**680.64**
Bronze relief valve; spring type; side outlet; working pressure up to 13.8 bar at 120°C; screwed ends to steel						
40 mm dia.	682.64	863.95	0.64	25.17	nr	**889.12**
50 mm dia.	890.11	1126.52	0.76	29.89	nr	**1156.41**
65 mm dia.	1648.07	2085.80	0.94	36.97	nr	**2122.76**
80 mm dia.	2162.24	2736.53	1.10	43.26	nr	**2779.79**
Cocks; screwed joints to steel						
Bronze gland cock; complete with malleable iron lever; working pressure up to 10 bar at 100°C; screwed ends to steel						
15 mm dia.	89.03	112.68	0.77	30.28	nr	**142.96**
20 mm dia.	126.64	160.27	1.03	40.51	nr	**200.78**
25 mm dia.	170.64	215.96	1.19	46.80	nr	**262.76**
32 mm dia.	257.85	326.33	1.38	54.27	nr	**380.60**
40 mm dia.	360.05	455.68	1.62	63.71	nr	**519.39**
50 mm dia.	549.49	695.44	1.94	76.29	nr	**771.73**
Bronze three-way plug cock; complete with malleable iron lever; working pressure up to 10 bar at 100°C; screwed ends to steel						
15 mm dia.	180.46	228.39	0.77	30.28	nr	**258.67**
20 mm dia.	208.88	264.36	1.03	40.51	nr	**304.87**
25 mm dia.	291.98	369.53	1.19	46.80	nr	**416.33**
32 mm dia.	414.11	524.10	1.38	54.27	nr	**578.37**
40 mm dia.	499.30	631.92	1.62	63.71	nr	**695.62**
Air vents; including regulating, adjusting and testing						
Automatic air vent; maximum pressure up to 7 bar at 93°C; screwed ends to steel						
15 mm dia.	18.69	23.66	0.80	31.46	nr	**55.12**
Automatic air vent; maximum pressure up to 7 bar at 93°C; lockhead isolating valve; screwed ends to steel						
15 mm dia.	18.69	23.66	0.83	32.64	nr	**56.30**
Automatic air vent; maximum pressure up to 17 bar at 200°C; flanged ends (BS10, Table H); bolted connections to counter flange (measured separately)						
15 mm dia.	673.03	851.79	0.83	32.64	nr	**884.43**

38 MECHANICAL/COOLING/HEATING SYSTEMS

Item	Net Price £	Material £	Labour hours	Labour £	Unit	Total rate £
LOW TEMPERATURE HOT WATER HEATING; PIPELINE: VALVES – cont						
Radiator valves						
Bronze; wheelhead or lockshield; chromium plated finish; screwed joints to steel						
Straight						
15 mm dia.	24.56	31.09	0.59	23.20	nr	**54.29**
20 mm dia.	38.22	48.37	0.73	28.71	nr	**77.08**
25 mm dia.	47.97	60.71	0.85	33.43	nr	**94.14**
Angled						
15 mm dia.	83.11	105.18	0.59	23.20	nr	**128.38**
20 mm dia.	109.56	138.66	0.73	28.71	nr	**167.37**
25 mm dia.	140.88	178.30	0.85	33.43	nr	**211.72**
Bronze; wheelhead or lockshield; chromium plated finish; compression joints to copper						
Straight						
15 mm dia.	44.00	55.68	0.59	23.20	nr	**78.89**
20 mm dia.	55.44	70.17	0.73	28.71	nr	**98.88**
25 mm dia.	69.67	88.17	0.85	33.43	nr	**121.60**
Angled						
15 mm dia.	33.19	42.01	0.59	23.20	nr	**65.21**
20 mm dia.	34.84	44.09	0.73	28.71	nr	**72.80**
25 mm dia.	43.19	54.66	0.85	33.43	nr	**88.08**
Twin entry						
8 mm dia.	46.31	58.61	0.23	9.04	nr	**67.65**
10 mm dia.	52.55	66.50	0.23	9.04	nr	**75.55**
Bronze; thermostatic head; chromium plated finish; compression joints to copper						
Straight						
15 mm dia.	18.19	23.02	0.59	23.20	nr	**46.22**
20 mm dia.	26.67	33.76	0.73	28.71	nr	**62.47**
Angled						
15 mm dia.	16.91	21.40	0.59	23.20	nr	**44.61**
20 mm dia.	24.80	31.38	0.73	28.71	nr	**60.09**

38 MECHANICAL/COOLING/HEATING SYSTEMS

Item	Net Price £	Material £	Labour hours	Labour £	Unit	Total rate £
LOW TEMPERATURE HOT WATER HEATING; PIPELINE: EQUIPMENT						
GAUGES						
Thermometers and pressure gauges						
Dial thermometer; coated steel case and dial; glass window; brass pocket; BS 5235; pocket length 100 mm; screwed end						
Back/bottom entry						
100 mm dia. face	92.10	116.57	0.81	31.87	nr	**148.44**
150 mm dia. face	150.97	191.07	0.81	31.87	nr	**222.94**
Dial thermometer; coated steel case and dial; glass window; brass pocket; BS 5235; pocket length 100 mm; screwed end						
100 mm dia. face	136.55	172.81	0.81	31.87	nr	**204.68**
150 mm dia. face	152.91	193.53	0.81	31.87	nr	**225.40**
PRESSURIZATION UNITS						
LTHW pressurization unit complete with expansion vessel(s), interconnecting pipework and all necessary isolating and drain valves; includes placing in position; electrical work elsewhere. Selection based on a final working pressure of 4 bar, a 3 m static head and system operating temperatures of 82/71°C						
System volume						
2,400 litres	3000.87	3797.90	15.00	589.89	nr	**4387.79**
6,000–20,000 litres	5444.62	6890.71	22.00	865.17	nr	**7755.87**
25,000 litres	6203.23	7850.81	22.00	865.17	nr	**8715.98**
DIRT SEPARATORS						
Dirt separator; maximum operating temperature and pressure of 110°C and 10 bar; fitted with drain valve						
Bore size, flow rate (at 1.0 m/s velocity); threaded connections						
32 mm dia., 3.7 m³/h	133.52	168.99	2.29	90.06	nr	**259.04**
40 mm dia., 5.0 m³/h	158.95	201.17	2.45	96.35	nr	**297.52**
Bore size, flow rate (at 1.5 m/s velocity); flanged connections to PN16						
50 mm dia., 13.0 m³/h	898.61	1137.28	3.00	117.98	nr	**1255.26**
65 mm dia., 21.0 m³/h	936.78	1185.59	3.00	117.98	nr	**1303.57**
80 mm dia., 29.0 m³/h	1318.25	1668.38	3.84	151.01	nr	**1819.39**
100 mm dia., 49.0 m³/h	1394.55	1764.94	4.44	174.61	nr	**1939.54**
125 mm dia., 74.0 m³/h	2672.53	3382.36	11.64	457.75	nr	**3840.11**
150 mm dia., 109.0 m³/h	2786.97	3527.19	15.75	619.38	nr	**4146.57**

38 MECHANICAL/COOLING/HEATING SYSTEMS

Item	Net Price £	Material £	Labour hours	Labour £	Unit	Total rate £
LOW TEMPERATURE HOT WATER HEATING; PIPELINE: EQUIPMENT – cont						
Dirt separator – cont						
Bore size, flow rate – cont						
200 mm dia., 181.0 m³/h	4200.58	5316.25	15.75	619.38	nr	**5935.63**
250 mm dia., 288.0 m³/h	7445.35	9422.84	15.75	619.38	nr	**10042.22**
300 mm dia., 407.0 m³/h	10882.94	13773.44	17.24	677.98	nr	**14451.42**
MICROBUBBLE DEAERATORS						
Microbubble deaerator; maximum operating temperature and pressure of 110°C and 10 bar; fitted with drain valve						
Bore size, flow rate (at 1.0 m/s velocity); threaded connections						
32 mm dia., 3.7 m³/h	122.92	155.57	2.29	90.06	nr	**245.62**
40 mm dia., 5.0 m³/h	146.25	185.09	2.45	96.35	nr	**281.44**
Bore size, flow rate (at 1.5 m/s velocity); flanged connections to PN16						
50 mm dia., 13.0 m³/h	898.61	1137.28	3.00	117.98	nr	**1255.26**
65 mm dia., 21.0 m³/h	934.98	1183.31	3.00	117.98	nr	**1301.29**
80 mm dia., 29.0 m³/h	1318.25	1668.38	3.84	151.01	nr	**1819.39**
100 mm dia., 49.0 m³/h	1394.55	1764.94	4.44	174.61	nr	**1939.54**
125 mm dia., 74.0 m³/h	2672.53	3382.36	11.64	457.75	nr	**3840.11**
150 mm dia., 109.0 m³/h	2786.97	3527.19	15.75	619.38	nr	**4146.57**
200 mm dia., 181.0 m³/h	4200.58	5316.25	15.75	619.38	nr	**5935.63**
250 mm dia., 288.0 m³/h	7445.35	9422.84	15.75	619.38	nr	**10042.22**
300 mm dia., 407.0 m³/h	10921.09	13821.73	17.24	677.98	nr	**14499.71**
COMBINED MICROBUBBLE DEAERATORS AND DIRT SEPARATORS						
Combined deaerator and dirt separators; maximum operating temperature and pressure of 110°C and 10 bar; fitted with drain valve						
Bore size, flow rate (at 1.5 m/s velocity); threaded connections						
25 mm dia., 2.0 m³/h	169.55	214.59	2.75	108.15	nr	**322.73**
Bore size, flow rate (at 1.5 m/s velocity); flanged connections to PN16						
50 mm dia., 13.0 m³/h	1146.57	1451.10	3.60	141.57	nr	**1592.68**
65 mm dia., 21.0 m³/h	1184.73	1499.39	3.60	141.57	nr	**1640.96**
80 mm dia., 29.0 m³/h	1604.37	2030.49	4.61	181.29	nr	**2211.78**
100 mm dia., 49.0 m³/h	1681.11	2127.61	5.33	209.61	nr	**2337.22**
125 mm dia., 74.0 m³/h	3206.61	4058.28	13.97	549.38	nr	**4607.67**
150 mm dia., 109.0 m³/h	3323.17	4205.80	18.90	743.26	nr	**4949.06**
200 mm dia., 181.0 m³/h	5154.30	6523.29	18.90	743.26	nr	**7266.54**
250 mm dia., 288.0 m³/h	9087.86	11501.59	18.90	743.26	nr	**12244.85**
300 mm dia., 407.0 m³/h	13936.95	17638.60	20.69	813.65	nr	**18452.26**

38 MECHANICAL/COOLING/HEATING SYSTEMS

Item	Net Price £	Material £	Labour hours	Labour £	Unit	Total rate £
LOW TEMPERATURE HOT WATER HEATING; PIPELINE: PUMPS						
Centrifugal heating and chilled water pump; belt drive; 3 phase, 1450 rpm motor; max pressure 1000 kN/m²; max temperature 125°C; bed plate; coupling guard; bolted connections; supply only mating flanges; includes fixing on prepared concrete base; electrical work elsewhere						
40 mm pump size; 4.0 l/s at 70 kPa max head; 0.25 kW max motor rating	2430.34	3075.84	7.59	298.48	nr	**3374.32**
40 mm pump size; 4.0 l/s at 130 kPa max head; 1.5 kW max motor rating	3041.54	3849.37	8.09	318.15	nr	**4167.52**
50 mm pump size; 8.5 l/s at 90 kPa max head; 2.2 kW max motor rating	3136.67	3969.77	8.67	340.95	nr	**4310.73**
50 mm pump size; 8.5 l/s at 190 kPa max head; 3 kW max motor rating	5451.68	6899.65	11.20	440.45	nr	**7340.10**
50 mm pump size; 8.5 l/s at 215 kPa max head; 4 kW max motor rating	6030.43	7632.12	11.70	460.11	nr	**8092.23**
65 mm pump size; 14.0 l/s at 90 kPa max head; 3 kW max motor rating	3125.17	3955.21	11.70	460.11	nr	**4415.32**
65 mm pump size; 14.0 l/s at 160 kPa max head; 4 kW max motor rating	3465.33	4385.72	11.70	460.11	nr	**4845.83**
80 mm pump size; 14.5 l/s at 210 kPa max head; 5.5 kW max motor rating	5316.20	6728.19	11.70	460.11	nr	**7188.30**
80 mm pump size; 22.0 l/s at 130 kPa max head; 5.5 kW max motor rating	5316.20	6728.19	13.64	536.40	nr	**7264.59**
80 mm pump size; 22.0 l/s at 200 kPa max head; 7.5 kW max motor rating	5567.01	7045.61	13.64	536.40	nr	**7582.02**
100 mm pump size; 22.0 l/s at 250 kPa max head; 11 kW max motor rating	7812.82	9887.91	13.64	536.40	nr	**10424.31**
100 mm pump size; 30.0 l/s at 100 kPa max head; 4.0 kW max motor rating	4373.47	5535.06	19.15	753.09	nr	**6288.15**
100 mm pump size; 36.0 l/s at 250 kPa max head; 15.0 kW max motor rating	8274.16	10471.77	19.15	753.09	nr	**11224.86**
100 mm pump size; 36.0 l/s at 550 kPa max head; 30.0 kW max motor rating	11229.16	14211.63	19.15	753.09	nr	**14964.72**
Centrifugal heating and chilled water pump; twin head belt drive; 3 phase, 1450 rpm motor; max pressure 1000 kN/m²; max temperature 125°C; bed plate; coupling guard; bolted connections; supply only mating flanges; includes fixing on prepared concrete base; electrical work elsewhere						
40 mm pump size; 4.0 l/s at 70 kPa max head; 0.75 kW max motor rating	5206.65	6589.54	7.59	298.48	nr	**6888.02**
40 mm pump size; 4.0 l/s at 130 kPa max head; 1.5 kW max motor rating	6267.59	7932.27	8.09	318.15	nr	**8250.41**
50 mm pump size; 8.5 l/s at 90 kPa max head; 2.2 kW max motor rating	6403.10	8103.77	8.67	340.95	nr	**8444.72**

38 MECHANICAL/COOLING/HEATING SYSTEMS

Item	Net Price £	Material £	Labour hours	Labour £	Unit	Total rate £
LOW TEMPERATURE HOT WATER HEATING; PIPELINE: PUMPS – cont						
Centrifugal heating and chilled water pump – cont						
50 mm pump size; 8.5 l/s at 190 kPa max head; 4 kW max motor rating	7380.40	9340.64	11.20	440.45	nr	**9781.09**
65 mm pump size; 8.5 l/s at 215 kPa max head; 4 kW max motor rating	10142.22	12835.99	11.70	460.11	nr	**13296.11**
65 mm pump size; 14.0 l/s at 90 kPa max head; 3 kW max motor rating	10649.35	13477.81	11.70	460.11	nr	**13937.93**
65 mm pump size; 14.0 l/s at 160 kPa max head; 4 kW max motor rating	11181.83	14151.72	11.70	460.11	nr	**14611.84**
80 mm pump size; 14.5 l/s at 210 kPa max head; 7.5 kW max motor rating	11563.59	14634.88	13.64	536.40	nr	**15171.28**
Centrifugal heating and chilled water pump; close coupled; 3 phase, 1450 rpm motor; max pressure 1000 kN/m²; max temperature 110°C; bed plate; coupling guard; bolted connections; supply only mating flanges; includes fixing on prepared concrete base; electrical work elsewhere						
40 mm pump size; 4.0 l/s at 23 kPa max head; 0.55 kW max motor rating	1629.76	2062.62	7.31	287.47	nr	**2350.09**
50 mm pump size; 4.0 l/s at 75 kPa max head; 0.75 kW max motor rating	1806.54	2286.36	7.31	287.47	nr	**2573.83**
50 mm pump size; 7.0 l/s at 65 kPa max head; 0.75 kW max motor rating	1806.54	2286.36	8.01	315.00	nr	**2601.36**
65 mm pump size; 10.0 l/s at 33 kPa max head; 0.75 kW max motor rating	2047.68	2591.55	8.01	315.00	nr	**2906.55**
50 mm pump size; 4.0 l/s at 120 kPa max head; 1.5 kW max motor rating	2465.50	3120.34	8.01	315.00	nr	**3435.34**
80 mm pump size; 16.0 l/s at 80 kPa max head; 2.2 kW max motor rating	3153.49	3991.06	12.35	485.67	nr	**4476.73**
80 mm pump size; 16.0 l/s at 120 kPa max head; 4.0 kW max motor rating	3192.00	4039.80	12.35	485.67	nr	**4525.47**
100 mm pump size; 28.0 l/s at 40 kPa max head; 2.2 kW max motor rating	3198.50	4048.02	17.86	702.36	nr	**4750.38**
100 mm pump size; 28.0 l/s at 90 kPa max head; 4.0 kW max motor rating	3320.59	4202.54	17.86	702.36	nr	**4904.90**
125 mm pump size; 40.0 l/s at 50 kPa max head; 3.0 kW max motor rating	3481.29	4405.92	25.85	1016.57	nr	**5422.50**
125 mm pump size; 40.0 l/s at 120 kPa max head; 7.5 kW max motor rating	4098.51	5187.07	25.85	1016.57	nr	**6203.64**
150 mm pump size; 70.0 l/s at 75 kPa max head; 11 kW max motor rating	6023.99	7623.96	30.43	1196.68	nr	**8820.64**
150 mm pump size; 70.0 l/s at 120 kPa max head; 15.0 kW max motor rating	6438.68	8148.79	30.43	1196.68	nr	**9345.48**
150 mm pump size; 70.0 l/s at 150 kPa max head; 15.0 kW max motor rating	6438.68	8148.79	30.43	1196.68	nr	**9345.48**

38 MECHANICAL/COOLING/HEATING SYSTEMS

Item	Net Price £	Material £	Labour hours	Labour £	Unit	Total rate £
Centrifugal heating & chilled water pump; close coupled; 3 phase, variable speed motor; max system pressure 1000 kN/m²; max temperature 110°C; bed plate; coupling guard; bolted connections; supply only mating flanges; includes fixing on prepared concrete base; electrical work elsewhere						
40 mm pump size; 4.0 l/s at 23 kPa max head; 0.55 kW max motor rating	2941.29	3722.49	7.31	287.47	nr	**4009.96**
40 mm pump size; 4.0 l/s at 75 kPa max head; 0.75 kW max motor rating	3246.66	4108.98	7.31	287.47	nr	**4396.45**
50 mm pump size; 7.0 l/s at 65 kPa max head; 1.5 kW max motor rating	3982.78	5040.60	8.01	315.00	nr	**5355.60**
50 mm pump size; 10.0 l/s at 33 kPa max head; 1.5 kW max motor rating	3982.78	5040.60	8.01	315.00	nr	**5355.60**
50 mm pump size; 4.0 l/s at 120 kPa max head; 1.5 kW max motor rating	3982.78	5040.60	8.01	315.00	nr	**5355.60**
80 mm pump size; 16.0 l/s at 80 kPa max head; 2.2 kW max motor rating	5191.45	6570.30	12.35	485.67	nr	**7055.98**
80 mm pump size; 16.0 l/s at 120 kPa max head; 3.0 kW max motor rating	5599.68	7086.96	12.35	485.67	nr	**7572.63**
100 mm pump size; 28.0 l/s at 40 kPa max head; 2.2 kW max motor rating	5320.01	6733.00	17.86	702.36	nr	**7435.36**
100 mm pump size; 28.0 l/s at 90 kPa max head; 4.0. kW max motor rating	6345.45	8030.80	17.86	702.36	nr	**8733.16**
125 mm pump size; 40.0 l/s at 50 kPa max head; 3.0 kW max motor rating	6091.39	7709.26	25.85	1016.57	nr	**8725.84**
125 mm pump size; 40.0 l/s at 120 kPa max head; 7.5 kW max motor rating	7518.41	9515.31	25.85	1016.57	nr	**10531.88**
150 mm pump size; 70.0 l/s at 75 kPa max head; 7.5 kW max motor rating	9910.34	12542.53	30.43	1196.68	nr	**13739.22**
Glandless domestic heating pump; for low pressure domestic hot water heating systems; 240 volt; 50 Hz electric motor; max working pressure 1000 N/m² and max temperature of 130°C; includes fixing in position; electrical work elsewhere						
1" BSP unions – 2 speed	234.71	297.05	1.58	62.13	nr	**359.19**
1.25" BSP unions – 3 speed	345.86	437.73	1.58	62.13	nr	**499.86**
Glandless pumps; for hot water secondary supply; silent running; 3 phase; max pressure 1000 kN/m²; max temperature 130°C; bolted connections; supply only mating flanges; including fixing in position; electrical elsewhere						
1" BSP unions – 3 speed	481.94	609.94	1.58	62.13	nr	**672.08**

38 MECHANICAL/COOLING/HEATING SYSTEMS

Item	Net Price £	Material £	Labour hours	Labour £	Unit	Total rate £
LOW TEMPERATURE HOT WATER HEATING; PIPELINE: PUMPS – cont						
Pipeline mounted circulator; for heating and chilled water; silent running; 3 phase; 1450 rpm motor; max pressure 1000 kN/m²; max temperature 120°C; bolted connections; supply only mating flanges; includes fixing in position; electrical elsewhere						
32 mm pump size; 2.0 l/s at 17 kPa max head; 0.2 kW max motor rating	963.91	1219.92	6.44	253.26	nr	**1473.18**
50 mm pump size; 3.0 l/s at 20 kPa max head; 0.2 kW max motor rating	963.91	1219.92	6.86	269.78	nr	**1489.70**
65 mm pump size; 5.0 l/s at 30 kPa max head; 0.37 kW max motor rating	1524.80	1929.79	7.48	294.16	nr	**2223.95**
65 mm pump size; 8.0 l/s at 37 kPa max head; 0.75 kW max motor rating	1707.24	2160.69	7.48	294.16	nr	**2454.84**
80 mm pump size; 12.0 l/s at 42 kPa max head; 1.1 kW max motor rating	1985.00	2512.21	8.01	315.00	nr	**2827.21**
100 mm pump size; 25.0 l/s at 37 kPa max head; 2.2 kW max motor rating	2829.08	3580.49	9.11	358.26	nr	**3938.75**
Dual pipeline mounted circulator; for heating & chilled water; silent running; 3 phase; 1450 rpm motor; max pressure 1000 kN/m²; max temperature 120°C; bolted connections; supply only mating flanges; includes fixing in position; electrical work elsewhere						
40 mm pump size; 2.0 l/s at 17 kPa max head; 0.8 kW max motor rating	1778.06	2250.31	7.88	309.89	nr	**2560.20**
50 mm pump size; 3.0 l/s at 20 kPa max head; 0.2 kW max motor rating	1783.48	2257.18	8.01	315.00	nr	**2572.18**
65 mm pump size; 5.0 l/s at 30 kPa max head; 0.37 kW max motor rating	2894.42	3663.18	9.20	361.80	nr	**4024.98**
65 mm pump size; 8.0 l/s at 37 kPa max head; 0.75 kW max motor rating	3346.43	4235.25	9.20	361.80	nr	**4597.04**
100 mm pump size; 12.0 l/s at 42 kPa max head; 1.1 kW max motor rating	3784.79	4790.03	9.45	371.63	nr	**5161.66**
Glandless accelerator pumps; for low and medium pressure heating services; silent running; 3 phase; 1450 rpm motor; max pressure 1000 kN/m²; max temperature 130°C; bolted connections; supply only mating flanges; includes fixing in position; electrical work elsewhere						
40 mm pump size; 4.0 l/s at 15 kPa max head; 0.35 kW max motor rating	702.25	888.77	6.94	272.92	nr	**1161.69**
50 mm pump size; 6.0 l/s at 20 kPa max head; 0.45 kW max motor rating	748.39	947.16	7.35	289.04	nr	**1236.21**
80 mm pump size; 13.0 l/s at 28 kPa max head; 0.58 kW max motor rating	1474.72	1866.40	7.76	305.17	nr	**2171.57**

38 MECHANICAL/COOLING/HEATING SYSTEMS

Item	Net Price £	Material £	Labour hours	Labour £	Unit	Total rate £
Air source heat pumps						
Two stage heat pump (air to water) in monoblock design, cascadable for outdoor installation. Providing heating and cooling. Unit and controller only						
Heat output & cooling capacity						
17 kW heating, 18 kW cooling	15187.58	19221.41	12.00	471.91	nr	**19693.32**
24 kW heating, 24 kW cooling	17261.98	21846.77	12.00	471.91	nr	**22318.68**
32 kW heating, 31 kW cooling	19319.40	24450.64	12.00	471.91	nr	**24922.55**
64 kW heating, 62 kW cooling	28610.00	36208.82	12.00	471.91	nr	**36680.73**

38 MECHANICAL/COOLING/HEATING SYSTEMS

Item	Net Price £	Material £	Labour hours	Labour £	Unit	Total rate £
LOW TEMPERATURE HOT WATER HEATING; PIPELINE: HEAT EXCHANGERS						
Plate heat exchanger; for use in LTHW systems; painted carbon steel frame; stainless steel plates, nitrile rubber gaskets; design pressure of 10 bar and operating temperature of 110/135°C						
Primary side; 80°C in, 69°C out; secondary side; 82°C in, 71°C out						
107 kW, 2.38 l/s	3096.87	3919.40	10.00	393.26	nr	**4312.66**
245 kW, 5.46 l/s	4672.60	5913.64	10.00	393.26	nr	**6306.90**
287 kW, 6.38 l/s	5027.39	6362.66	10.00	393.26	nr	**6755.92**
328 kW, 7.31 l/s	5713.98	7231.62	10.00	393.26	nr	**7624.88**
364 kW, 8.11 l/s	5892.10	7457.04	10.00	393.26	nr	**7850.30**
403 kW, 8.96 l/s	6334.51	8016.95	10.00	393.26	nr	**8410.21**
453 kW, 10.09 l/s	6776.92	8576.87	10.00	393.26	nr	**8970.12**
490 kW, 10.89 l/s	7450.59	9429.46	10.00	393.26	nr	**9822.72**
1000 kW, 21.7 l/s	12477.97	15792.12	12.00	471.91	nr	**16264.03**
1500 kW, 32.6 l/s	16328.95	20665.92	12.00	471.91	nr	**21137.83**
2000 kW, 43.4 l/s	19553.66	24747.11	15.00	589.89	nr	**25337.00**
2500 kW, 54.3 l/s	24664.36	31215.21	15.00	589.89	nr	**31805.10**
Note: For temperature conditions different to those above, the cost of the units can vary significantly, and so manufacturer's advice should be sought.						
District heating HIUs						
District heating station; for conventional connection to a district or local heating network; offering primary F/R and secondary F/R connections, with stainless steel plate heat exchanger, integrated control panel, heat meter and combination valve to adjust output						
Heat output						
149 kW	7183.75	9091.75	12.00	471.91	nr	**9563.66**
194 kW	8021.83	10152.43	12.00	471.91	nr	**10624.34**
240 kW	8129.66	10288.90	12.00	471.91	nr	**10760.81**
400 kW	9663.87	12230.59	12.00	471.91	nr	**12702.50**
494 kW	11522.93	14583.43	12.00	471.91	nr	**15055.34**
572 kW	13271.44	16796.33	12.00	471.91	nr	**17268.24**
915 kW	16185.61	20484.51	14.00	550.56	nr	**21035.07**
1417 kW	20797.79	26321.68	14.00	550.56	nr	**26872.25**

38 MECHANICAL/COOLING/HEATING SYSTEMS

Item	Net Price £	Material £	Labour hours	Labour £	Unit	Total rate £
LOW TEMPERATURE HOT WATER HEATING; PIPELINE: CALORIFIERS						
Non-storage calorifiers; mild steel; heater battery duty 116°C/90°C to BS 853, maximum test on shell 11.55 bar, tubes 26.25 bar						
Horizontal or vertical; primary water at 116°C on, 90°C off						
40 kW capacity	1537.96	1946.45	3.00	117.98	nr	**2064.42**
88 kW capacity	1791.72	2267.60	5.00	196.63	nr	**2464.23**
176 kW capacity	2053.22	2598.56	7.04	276.94	nr	**2875.50**
293 kW capacity	2847.41	3603.68	9.01	354.29	nr	**3957.97**
586 kW capacity	4242.06	5368.75	22.22	873.91	nr	**6242.65**
879 kW capacity	5907.87	7477.00	28.57	1123.59	nr	**8600.59**
1465 kW capacity	9491.33	12012.23	50.00	1966.29	nr	**13978.52**
2000 kW capacity	17275.47	21863.83	60.00	2359.55	nr	**24223.38**

38 MECHANICAL/COOLING/HEATING SYSTEMS

Item	Net Price £	Material £	Labour hours	Labour £	Unit	Total rate £
LOW TEMPERATURE HOT WATER HEATING; PIPELINE: HEAT EMITTERS						
Perimeter convector heating; metal casing with standard finish; aluminium extruded grille; including backplates						
Top/sloping/flat front outlet						
60 × 200 mm	56.44	71.43	2.00	78.65	m	**150.08**
60 × 300 mm	61.40	77.71	2.00	78.65	m	**156.36**
60 × 450 mm	76.11	96.32	2.00	78.65	m	**174.97**
60 × 525 mm	81.02	102.54	2.00	78.65	m	**181.19**
60 × 600 mm	88.38	111.85	2.00	78.65	m	**190.50**
90 × 260 mm	61.40	77.71	2.00	78.65	m	**156.36**
90 × 300 mm	63.83	80.78	2.00	78.65	m	**159.43**
90 × 450 mm	78.60	99.47	2.00	78.65	m	**178.12**
90 × 525 mm	85.93	108.76	2.00	78.65	m	**187.41**
90 × 600 mm	90.86	115.00	2.00	78.65	m	**193.65**
Extra over for dampers						
Damper	26.38	33.38	0.25	9.83	nr	**43.22**
Extra over for fittings						
60 mm end caps	21.11	26.72	0.25	9.83	nr	**36.55**
90 mm end caps	33.22	42.04	0.25	9.83	nr	**51.88**
60 mm corners	44.88	56.80	0.25	9.83	nr	**66.63**
90 mm corners	66.01	83.54	0.25	9.83	nr	**93.37**

38 MECHANICAL/COOLING/HEATING SYSTEMS

Item	Net Price £	Material £	Labour hours	Labour £	Unit	Total rate £
LOW TEMPERATURE HOT WATER HEATING; PIPELINE: RADIATORS						
Radiant strip heaters						
Suitable for connection to hot water system; aluminium sheet panels with steel pipe clamped to upper surface; including insulation, sliding brackets, cover plates, end closures; weld or screwed BSP ends						
One pipe						
1500 mm long	103.59	131.11	3.11	122.30	nr	**253.41**
3000 mm long	163.83	207.35	3.11	122.30	nr	**329.65**
4500 mm long	222.26	281.29	3.11	122.30	nr	**403.59**
6000 mm long	306.38	387.75	3.11	122.30	nr	**510.05**
Two pipe						
1500 mm long	193.39	244.76	4.15	163.20	nr	**407.96**
3000 mm long	305.47	386.60	4.15	163.20	nr	**549.81**
4500 mm long	417.14	527.94	4.15	163.20	nr	**691.14**
6000 mm long	561.99	711.25	4.15	163.20	nr	**874.45**
Pressed steel panel type radiators; fixed with and including brackets; taking down once for decoration; refixing						
300 mm high; single panel						
500 mm length	28.94	36.62	2.03	79.83	nr	**116.46**
1000 mm length	57.82	73.18	2.03	79.83	nr	**153.01**
1500 mm length	75.77	95.89	2.03	79.83	nr	**175.72**
2000 mm length	85.00	107.58	2.47	97.13	nr	**204.72**
2500 mm length	94.21	119.24	2.97	116.80	nr	**236.03**
3000 mm length	112.74	142.69	3.22	126.63	nr	**269.32**
300 mm high; double panel; convector						
500 mm length	55.62	70.40	2.13	83.76	nr	**154.16**
1000 mm length	111.32	140.89	2.13	83.76	nr	**224.65**
1500 mm length	166.96	211.30	2.13	83.76	nr	**295.07**
2000 mm length	222.65	281.79	2.57	101.07	nr	**382.86**
2500 mm length	278.26	352.17	3.07	120.73	nr	**472.90**
3000 mm length	333.94	422.64	3.31	130.17	nr	**552.81**
450 mm high; single panel						
500 mm length	26.99	34.15	2.08	81.80	nr	**115.95**
1000 mm length	54.03	68.38	2.08	81.80	nr	**150.17**
1600 mm length	86.43	109.39	2.53	99.49	nr	**208.88**
2000 mm length	108.07	136.77	2.97	116.80	nr	**253.57**
2400 mm length	129.64	164.07	3.47	136.46	nr	**300.53**
3000 mm length	162.04	205.08	3.82	150.22	nr	**355.30**
450 mm high; double panel; convector						
500 mm length	49.47	62.60	2.18	85.73	nr	**148.33**
1000 mm length	98.92	125.19	2.18	85.73	nr	**210.92**
1600 mm length	181.17	229.29	2.63	103.43	nr	**332.72**
2000 mm length	304.56	385.46	3.06	120.34	nr	**505.79**
2400 mm length	365.50	462.58	3.37	132.53	nr	**595.11**
3000 mm length	456.88	578.23	3.92	154.16	nr	**732.39**

38 MECHANICAL/COOLING/HEATING SYSTEMS

Item	Net Price £	Material £	Labour hours	Labour £	Unit	Total rate £
LOW TEMPERATURE HOT WATER HEATING; PIPELINE: RADIATORS – cont						
Pressed steel panel type radiators – cont						
600 mm high; single panel						
500 mm length	36.18	45.79	2.18	85.73	nr	**131.52**
1000 mm length	72.35	91.56	2.43	95.56	nr	**187.12**
1600 mm length	115.74	146.48	3.13	123.09	nr	**269.57**
2000 mm length	144.68	183.10	3.77	148.26	nr	**331.36**
2400 mm length	173.63	219.74	4.07	160.06	nr	**379.80**
3000 mm length	217.05	274.70	5.11	200.95	nr	**475.65**
600 mm high; double panel; convector						
500 mm length	62.28	78.82	2.28	89.66	nr	**168.49**
1000 mm length	124.56	157.65	2.28	89.66	nr	**247.31**
1600 mm length	228.14	288.74	3.23	127.02	nr	**415.76**
2000 mm length	383.54	485.40	3.87	152.19	nr	**637.60**
2400 mm length	460.24	582.49	4.17	163.99	nr	**746.47**
3000 mm length	575.30	728.10	5.24	206.07	nr	**934.17**
700 mm high; single panel						
500 mm length	42.31	53.55	2.23	87.70	nr	**141.24**
1000 mm length	84.62	107.09	2.83	111.29	nr	**218.39**
1600 mm length	135.41	171.38	3.73	146.69	nr	**318.06**
2000 mm length	169.28	214.24	4.46	175.39	nr	**389.64**
2400 mm length	203.17	257.13	4.48	176.18	nr	**433.31**
3000 mm length	253.89	321.32	5.24	206.07	nr	**527.39**
700 mm high; double panel; convector						
500 mm length	80.94	102.44	2.33	91.63	nr	**194.07**
1000 mm length	217.79	275.64	3.08	121.12	nr	**396.76**
1600 mm length	348.43	440.97	3.83	150.62	nr	**591.59**
2000 mm length	435.54	551.22	4.17	163.99	nr	**715.21**
2400 mm length	522.67	661.49	4.37	171.85	nr	**833.34**
3000 mm length	653.33	826.86	4.82	189.55	nr	**1016.41**
Flat panel type steel radiators; fixed with and including brackets; taking down once for decoration; refixing						
300 mm high; single panel (44 mm deep)						
500 mm length	297.89	377.01	2.03	79.83	nr	**456.84**
1000 mm length	467.87	592.14	2.03	79.83	nr	**671.97**
1500 mm length	637.87	807.28	2.03	79.83	nr	**887.11**
2000 mm length	807.86	1022.43	2.47	97.13	nr	**1119.56**
2400 mm length	943.85	1194.54	2.97	116.80	nr	**1311.34**
3000 mm length	1357.97	1718.65	3.22	126.63	nr	**1845.27**
300 mm high; double panel (100 mm deep)						
500 mm length	590.91	747.85	2.03	79.83	nr	**827.69**
1000 mm length	926.02	1171.97	2.03	79.83	nr	**1251.80**
1500 mm length	1261.16	1596.13	2.03	79.83	nr	**1675.96**
2000 mm length	1596.27	2020.24	2.47	97.13	nr	**2117.38**
2400 mm length	1864.37	2359.55	2.97	116.80	nr	**2476.35**
3000 mm length	2476.65	3134.45	3.22	126.63	nr	**3261.08**

38 MECHANICAL/COOLING/HEATING SYSTEMS

Item	Net Price £	Material £	Labour hours	Labour £	Unit	Total rate £
500 mm high; single panel (44 mm deep)						
500 mm length	351.31	444.61	2.13	83.76	nr	**528.38**
1000 mm length	574.73	727.38	2.13	83.76	nr	**811.14**
1500 mm length	798.14	1010.12	2.13	83.76	nr	**1093.89**
2000 mm length	1021.55	1292.87	2.57	101.07	nr	**1393.93**
2400 mm length	1200.28	1519.07	3.07	120.73	nr	**1639.80**
3000 mm length	1678.53	2124.34	3.31	130.17	nr	**2254.51**
500 mm high; double panel (100 mm deep)						
500 mm length	688.04	870.79	2.08	81.80	nr	**952.59**
1000 mm length	1120.31	1417.86	2.08	81.80	nr	**1499.66**
1500 mm length	1552.52	1964.88	2.53	99.49	nr	**2064.37**
2000 mm length	1984.81	2511.98	2.97	116.80	nr	**2628.78**
2400 mm length	2330.62	2949.63	3.47	136.46	nr	**3086.09**
3000 mm length	3059.48	3872.07	3.82	150.22	nr	**4022.30**
600 mm high; single panel (44 mm deep)						
500 mm length	396.64	501.99	2.18	85.73	nr	**587.72**
1000 mm length	649.20	821.63	2.18	85.73	nr	**907.36**
1500 mm length	901.74	1141.25	2.63	103.43	nr	**1244.67**
2000 mm length	1154.30	1460.89	3.06	120.34	nr	**1581.22**
2400 mm length	1356.36	1716.61	3.37	132.53	nr	**1849.13**
3000 mm length	1869.55	2366.10	3.92	154.16	nr	**2520.26**
600 mm high; double panel (100 mm deep)						
500 mm length	783.24	991.27	2.18	85.73	nr	**1077.00**
1000 mm length	1278.64	1618.25	2.43	95.56	nr	**1713.81**
1500 mm length	1774.02	2245.20	3.13	123.09	nr	**2368.29**
2000 mm length	2269.44	2872.20	3.77	148.26	nr	**3020.46**
2400 mm length	2665.75	3373.77	4.07	160.06	nr	**3533.82**
3000 mm length	3470.35	4392.08	5.11	200.95	nr	**4593.03**
700 mm high; single panel (44 mm deep)						
500 mm length	430.64	545.01	2.28	89.66	nr	**634.68**
1000 mm length	717.20	907.68	2.28	89.66	nr	**997.35**
1500 mm length	1003.74	1270.33	3.23	127.02	nr	**1397.36**
2000 mm length	1290.29	1632.98	3.87	152.19	nr	**1785.18**
2400 mm length	1519.99	1923.70	4.17	163.99	nr	**2087.69**
3000 mm length	2073.54	2624.27	5.24	206.07	nr	**2830.34**
700 mm high; double panel (100 mm deep)						
500 mm length	846.39	1071.19	2.23	87.70	nr	**1158.89**
1000 mm length	1404.93	1778.07	2.83	111.29	nr	**1889.37**
1500 mm length	1963.46	2484.96	3.73	146.69	nr	**2631.64**
2000 mm length	2521.98	3191.82	4.46	175.39	nr	**3367.21**
2400 mm length	2968.81	3757.33	4.48	176.18	nr	**3933.51**
Fan convector; sheet metal casing with lockable access panel; centrifugal fan; air filter; LPHW heating coil; extruded aluminium grilles; 3 speed; includes fixing in position; electrical work elsewhere						
Free-standing flat top, 695 mm high, medium speed rating						
Entering air temperature, 18°C						
695 mm long, 1 row 1.94 kW, 75 l/s	961.75	1217.19	2.73	107.36	nr	**1324.55**
695 mm long, 2 row 2.64 kW, 75 l/s	961.75	1217.19	2.73	107.36	nr	**1324.55**
895 mm long, 1 row 4.02 kW, 150 l/s	1083.74	1371.58	2.73	107.36	nr	**1478.94**

38 MECHANICAL/COOLING/HEATING SYSTEMS

Item	Net Price £	Material £	Labour hours	Labour £	Unit	Total rate £
LOW TEMPERATURE HOT WATER HEATING; PIPELINE: RADIATORS – cont						
Fan convector – cont						
Free-standing flat top – cont						
895 mm long, 2 row 5.62 kW, 150 l/s	1083.74	1371.58	2.73	107.36	nr	**1478.94**
1195 mm long, 1 row 6.58 kW, 250 l/s	1233.85	1561.57	3.00	117.98	nr	**1679.54**
1195 mm long, 2 row 9.27 kW, 250 l/s	1233.85	1561.57	3.00	117.98	nr	**1679.54**
1495 mm long, 1 row 9.04 kW, 340 l/s	1376.96	1742.68	3.26	128.20	nr	**1870.88**
1495 mm long, 2 row 12.73 kW, 340 l/s	1376.96	1742.68	3.26	128.20	nr	**1870.88**
Free-standing flat top, 695 mm high, medium speed rating, c/w floor plinth						
695 mm long, 1 row 1.94 kW, 75 l/s	1009.82	1278.03	2.73	107.36	nr	**1385.39**
695 mm long, 2 row 2.64 kW, 75 l/s	1009.82	1278.03	2.73	107.36	nr	**1385.39**
895 mm long, 1 row 4.02 kW, 150 l/s	1137.92	1440.15	2.73	107.36	nr	**1547.51**
895 mm long, 2 row 5.62 kW, 150 l/s	1137.92	1440.15	2.73	107.36	nr	**1547.51**
1195 mm long, 1 row 6.58 kW, 250 l/s	1295.58	1639.68	3.00	117.98	nr	**1757.66**
1195 mm long, 2 row 9.27 kW, 250 l/s	1295.58	1639.68	3.00	117.98	nr	**1757.66**
1495 mm long, 1 row 9.04 kW, 340 l/s	1445.80	1829.81	3.26	128.20	nr	**1958.01**
1495 mm long, 2 row 12.73 kW, 340 l/s	1445.80	1829.81	3.26	128.20	nr	**1958.01**
Free-standing sloping top, 695 mm high, medium speed rating, c/w floor plinth						
695 mm long, 1 row 1.94 kW, 75 l/s	1045.03	1322.58	2.73	107.36	nr	**1429.94**
695 mm long, 2 row 2.64 kW, 75 l/s	1045.03	1322.58	2.73	107.36	nr	**1429.94**
895 mm long, 1 row 4.02 kW, 150 l/s	1173.11	1484.69	2.73	107.36	nr	**1592.05**
895 mm long, 2 row 5.62 kW, 150 l/s	1173.11	1484.69	2.73	107.36	nr	**1592.05**
1195 mm long, 1 row 6.58 kW, 250 l/s	1330.75	1684.20	3.00	117.98	nr	**1802.18**
1195 mm long, 2 row 9.27 kW, 250 l/s	1330.75	1684.20	3.00	117.98	nr	**1802.18**
1495 mm long, 1 row 9.04 kW, 340 l/s	1480.98	1874.32	3.26	128.20	nr	**2002.52**
1495 mm long, 2 row 12.73 kW, 340 l/s	1480.98	1874.32	3.26	128.20	nr	**2002.52**
Wall mounted high level sloping discharge						
695 mm long, 1 row 1.94 kW, 75 l/s	1055.58	1335.95	2.73	107.36	nr	**1443.31**
695 mm long, 2 row 2.64 kW, 75 l/s	1055.58	1335.95	2.73	107.36	nr	**1443.31**
895 mm long, 1 row 4.02 kW, 150 l/s	1086.11	1374.58	2.73	107.36	nr	**1481.94**
895 mm long, 2 row 5.62 kW, 150 l/s	1086.11	1374.58	2.73	107.36	nr	**1481.94**
1195 mm long, 1 row 6.58 kW, 250 l/s	1321.80	1672.87	3.00	117.98	nr	**1790.84**
1195 mm long, 2 row 9.27 kW, 250 l/s	1321.80	1672.87	3.00	117.98	nr	**1790.84**
1495 mm long, 1 row 9.04 kW, 340 l/s	1433.24	1813.91	3.26	128.20	nr	**1942.11**
1495 mm long, 2 row 12.73 kW, 340 l/s	1433.24	1813.91	3.26	128.20	nr	**1942.11**
Ceiling mounted sloping inlet/outlet 665 mm wide						
895 mm long, 1 row 4.02 kW, 150 l/s	1193.97	1511.09	4.15	163.20	nr	**1674.29**
895 mm long, 2 row 5.62 kW, 150 l/s	1193.97	1511.09	4.15	163.20	nr	**1674.29**
1195 mm long, 1 row 6.58 kW, 250 l/s	1341.79	1698.17	4.15	163.20	nr	**1861.37**
1195 mm long, 2 row 9.27 kW, 250 l/s	1341.79	1698.17	4.15	163.20	nr	**1861.37**
1495 mm long, 1 row 9.04 kW, 340 l/s	1473.12	1864.39	4.15	163.20	nr	**2027.59**
1495 mm long, 2 row 12.73 kW, 340 l/s	1473.12	1864.39	4.15	163.20	nr	**2027.59**
Free-standing unit, extended height 1700/1900/2100 mm						
895 mm long, 1 row 4.02 kW, 150 l/s	1382.12	1749.21	3.11	122.30	nr	**1871.51**
895 mm long, 2 row 5.62 kW, 150 l/s	1381.64	1748.61	3.11	122.30	nr	**1870.91**
1195 mm long, 1 row 6.58 kW, 250 l/s	1613.88	2042.52	3.11	122.30	nr	**2164.83**
1195 mm long, 2 row 9.27 kW, 250 l/s	1613.88	2042.52	3.11	122.30	nr	**2164.83**

38 MECHANICAL/COOLING/HEATING SYSTEMS

Item	Net Price £	Material £	Labour hours	Labour £	Unit	Total rate £
1495 mm long, 1 row 9.04 kW, 340 l/s	1775.75	2247.39	3.11	122.30	nr	**2369.69**
1495 mm long, 2 row 12.73 kW, 340 l/s	1775.75	2247.39	3.11	122.30	nr	**2369.69**
LTHW trench heating; water temperatures 90°C/70°C; room air temperature 20°C; convector with copper tubes and aluminium fins within steel duct; Includes fixing within floor screed; electrical work elsewhere						
Natural convection type						
Normal capacity, 182 mm width, complete with linear, natural anodized aluminium grille (grille also costed separately below)						
90 mm deep						
1300 mm long, 287 W output	427.14	540.59	2.00	78.65	nr	**619.24**
2300 mm long, 576 W output	697.95	883.33	4.00	157.30	nr	**1040.63**
3300 mm long, 864 W output	955.41	1209.16	5.00	196.63	nr	**1405.79**
4500 mm long, 1210 W output	1307.77	1655.11	7.00	275.28	nr	**1930.39**
4900 mm long, 1325 W output	1404.06	1776.98	8.00	314.61	nr	**2091.59**
110 mm deep						
1300 mm long, 331 W output	430.80	545.23	2.00	78.65	nr	**623.88**
2300 mm long, 662 W output	704.08	891.09	4.00	157.30	nr	**1048.39**
3300 mm long, 993 W output	964.02	1220.06	5.00	196.63	nr	**1416.69**
4500 mm long, 1389 W output	1319.77	1670.30	7.00	275.28	nr	**1945.58**
4900 mm long, 1522 W output	1416.50	1792.72	8.00	314.61	nr	**2107.33**
140 mm deep						
1300 mm long, 551 W output	507.22	641.94	2.00	78.65	nr	**720.59**
2300 mm long, 1100 W output	833.96	1055.46	4.00	157.30	nr	**1212.76**
3300 mm long, 1651 W output	1145.51	1449.75	5.00	196.63	nr	**1646.38**
4500 mm long, 2310 W output	1561.29	1975.97	7.00	275.28	nr	**2251.25**
4900 mm long, 2530 W output	1678.63	2124.47	8.00	314.61	nr	**2439.08**
190 mm deep						
1300 mm long, 625 W output	524.34	663.60	2.00	78.65	nr	**742.25**
2300 mm long, 1250 W output	864.81	1094.50	4.00	157.30	nr	**1251.80**
3300 mm long, 1874 W output	1190.08	1506.17	5.00	196.63	nr	**1702.80**
4500 mm long, 2624 W output	1623.44	2054.62	7.00	275.28	nr	**2329.90**
4900 mm long, 2875 W output	1745.98	2209.71	8.00	314.61	nr	**2524.32**
Fan assisted type (outputs assume fan at 50%)						
Normal capacity, 182 mm width, complete with Natural anodized aluminium grille						
140 mm deep						
1300 mm long, 1439 W output	833.14	1054.42	2.00	78.65	nr	**1133.07**
2300 mm long, 3331 W output	1377.79	1743.73	4.00	157.30	nr	**1901.04**
3300 mm long, 5225 W output	1901.85	2406.98	5.00	196.63	nr	**2603.61**
4300 mm long, 5753 W output	2232.26	2825.15	7.00	275.28	nr	**3100.43**
4900 mm long, 6070 W output	2407.52	3046.95	8.00	314.61	nr	**3361.56**
Linear grille anodized aluminium, 170 mm width (if supplied as a separate item)	226.78	287.02	–	–	m	**287.02**
Roll up grille, natural anodized aluminium	226.78	287.02	–	–	m	**287.02**
Thermostatic valve with remote regulator (c/w valve body)	93.64	118.50	4.00	157.30	nr	**275.81**
Fan speed controller	45.76	57.92	2.00	78.65	nr	**136.57**

38 MECHANICAL/COOLING/HEATING SYSTEMS

Item	Net Price £	Material £	Labour hours	Labour £	Unit	Total rate £
LOW TEMPERATURE HOT WATER HEATING; PIPELINE: RADIATORS – cont						
Fan assisted type (outputs assume fan at 50%) – cont						
Note: As an alternative to thermostatic control, the system can be controlled via two port valves. Refer to Valve section for valve prices.						
LTHW underfloor heating; water flow and return temperatures of 60°C and 70°C; pipework at 300 mm centres; pipe fixings; flow and return manifolds and zone actuators; wiring block; insulation; includes fixing in position; excludes secondary pump, mixing valve, zone thermostats and floor finishes; electrical work elsewhere						
Note: All rates are expressed on a m² basis, for the following example areas						
Screeded floor with 15–25 mm stone/marble finish (producing 80–100 W/m²)						
250 m² area (single zone)	38.04	48.15	0.10	3.93	m²	**52.08**
1000 m² area (single zone)	38.09	48.21	0.12	4.72	m²	**52.93**
5000 m² area (multi-zone)	38.39	48.58	0.14	5.51	m²	**54.09**
Screeded floor with 10 mm carpet tile (producing 80–100 W/m²)						
250 m² area (single zone)	38.04	48.15	0.10	3.93	m²	**52.08**
1000 m² area (single zone)	38.09	48.21	0.12	4.72	m²	**52.93**
5000 m² area (multi-zone)	38.39	48.58	0.14	5.51	m²	**54.09**
Floating timber floor with 20 mm timber finish (producing 70–80 Wm²)						
250 m² area (single zone)	49.77	62.99	0.10	3.93	m²	**66.93**
1000 m² area (single zone)	49.95	63.22	0.12	4.72	m²	**67.94**
5000 m² area (multi-zone)	50.97	64.51	0.14	5.51	m²	**70.02**
Floating timber floor with 10 mm carpet tile (producing 70–80 W/m²)						
250 m² area (single zone)	57.47	72.74	0.10	3.93	m²	**76.67**
1000 m² area (single zone)	60.07	76.03	0.12	4.72	m²	**80.75**
5000 m² area (multi-zone)	64.79	82.00	0.16	6.29	m²	**88.29**

38 MECHANICAL/COOLING/HEATING SYSTEMS

Item	Net Price £	Material £	Labour hours	Labour £	Unit	Total rate £
LOW TEMPERATURE HOT WATER HEATING; PIPELINE: PIPE FREEZING						
Freeze isolation of carbon steel or copper pipelines containing static water, either side of work location, freeze duration not exceeding 4 hours assuming that flow and return circuits are treated concurrently and activities undertaken during normal working hours						
Up to 4 freezes						
50 mm dia.	585.37	740.85	1.00	39.33	nr	**780.18**
65 mm dia.	585.37	740.85	1.00	39.33	nr	**780.18**
80 mm dia.	669.21	846.95	1.00	39.33	nr	**886.27**
100 mm dia.	753.07	953.09	1.00	39.33	nr	**992.42**
150 mm dia.	1253.07	1585.88	1.00	39.33	nr	**1625.21**
200 mm dia.	1920.71	2430.85	1.00	39.33	nr	**2470.17**

38 MECHANICAL/COOLING/HEATING SYSTEMS

Item	Net Price £	Material £	Labour hours	Labour £	Unit	Total rate £
LOW TEMPERATURE HOT WATER HEATING; PIPELINE: ENERGY METERS						
Ultrasonic						
Energy meter for measuring energy use in LTHW systems; includes ultrasonic flow meter (with sensor and signal converter), energy calculator, pair of temperature sensors with brass pockets, and 3 m of interconnecting cable; includes fixing in position; electrical work elsewhere						
Pipe size (flanged connections to PN16); maximum flow rate						
50 mm, 36 m³/hr	1715.13	2170.67	1.80	70.79	nr	**2241.46**
65 mm, 60 m³/hr	1890.57	2392.71	2.32	91.24	nr	**2483.95**
80 mm, 100 m³/hr	2119.20	2682.05	2.56	100.67	nr	**2782.73**
125 mm, 250 m³/hr	2455.29	3107.42	3.60	141.57	nr	**3248.99**
150 mm, 360 m³/hr	2665.37	3373.29	4.80	188.76	nr	**3562.06**
200 mm, 600 m³/hr	2977.99	3768.94	6.24	245.39	nr	**4014.33**
250 mm, 1000 m³/hr	3436.41	4349.13	9.60	377.53	nr	**4726.65**
300 mm, 1500 m³/hr	4039.43	5112.31	10.80	424.72	nr	**5537.03**
350 mm, 2000 m³/hr	4867.34	6160.10	13.20	519.10	nr	**6679.20**
400 mm, 2500 m³/hr	5571.68	7051.52	15.60	613.48	nr	**7665.00**
500 mm, 3000 m³/hr	6322.96	8002.34	24.00	943.82	nr	**8946.16**
600 mm, 3500 m³/hr	7103.91	8990.70	28.00	1101.12	nr	**10091.83**

38 MECHANICAL/COOLING/HEATING SYSTEMS

Item	Net Price £	Material £	Labour hours	Labour £	Unit	Total rate £
LOW TEMPERATURE HOT WATER HEATING: CONTROL COMPONENTS MECHANICAL						
Room thermostats; light and medium duty; installed and connected						
Range 3°C to 27°C; 240 volt						
1 amp; on/off type	46.20	58.47	0.30	11.80	nr	**70.26**
Range 0°C to +15°C; 240 volt						
6 amp; frost thermostat	31.15	39.42	0.30	11.80	nr	**51.22**
Range 3°C to 27°C; 250 volt						
2 amp; changeover type; dead zone	77.27	97.80	0.30	11.80	nr	**109.59**
2 amp; changeover type	35.87	45.39	0.30	11.80	nr	**57.19**
2 amp; changeover type; concealed setting	45.20	57.20	0.30	11.80	nr	**69.00**
6 amp; on/off type	27.90	35.31	0.30	11.80	nr	**47.10**
6 amp; temperature setback	57.73	73.07	0.30	11.80	nr	**84.87**
16 amp; on/off type	42.99	54.40	0.30	11.80	nr	**66.20**
16 amp; on/off type; concealed setting	46.84	59.28	0.30	11.80	nr	**71.08**
20 amp; on/off type; concealed setting	39.91	50.51	0.30	11.80	nr	**62.31**
20 amp; indicated 'off' position	50.32	63.68	0.30	11.80	nr	**75.48**
20 amp; manual; double pole on/off and neon indicator	91.60	115.93	0.30	11.80	nr	**127.73**
20 amp; indicated 'off' position	59.94	75.86	0.30	11.80	nr	**87.66**
Range 10°C to 40°C; 240 volt						
20 amp; changeover contacts	54.63	69.14	0.30	11.80	nr	**80.94**
2 amp; 'heating-cooling' switch	118.14	149.52	0.30	11.80	nr	**161.32**
Surface thermostats						
Cylinder thermostat						
6 amp; changeover type; with cable	24.79	31.38	0.25	9.83	nr	**41.21**
Electrical thermostats; installed and connected						
Range 5°C to 30°C; 230 volt standard port single time						
10 amp with sensor	39.48	49.96	0.30	11.80	nr	**61.76**
Range 5°C to 30°C; 230 volt standard port double time						
10 amp with sensor	44.72	56.60	0.30	11.80	nr	**68.40**
10 amp with sensor and on/off switch	68.23	86.36	0.30	11.80	nr	**98.16**
Radiator thermostats						
Angled valve body; thermostatic head; built in sensor						
15 mm; liquid filled	27.46	34.75	0.84	33.05	nr	**67.79**
15 mm; wax filled	27.46	34.75	0.84	33.05	nr	**67.79**
Immersion thermostats; stem type; domestic water boilers; fitted; electrical work elsewhere						
Temperature range 0°C to 40°C						
Non-standard; 280 mm stem	16.62	21.04	0.25	9.83	nr	**30.87**
Temperature range 18°C to 88°C						
13 amp; 178 mm stem	11.04	13.97	0.25	9.83	nr	**23.80**
20 amp; 178 mm stem	17.18	21.74	0.25	9.83	nr	**31.57**
Non-standard; pocket clip; 280 mm stem	15.93	20.16	0.25	9.83	nr	**29.99**

38 MECHANICAL/COOLING/HEATING SYSTEMS

Item	Net Price £	Material £	Labour hours	Labour £	Unit	Total rate £
LOW TEMPERATURE HOT WATER HEATING: CONTROL COMPONENTS MECHANICAL – cont						
Immersion thermostats – cont						
Temperature range 40°C to 80°C						
13 amp; 178 mm stem	6.08	7.70	0.25	9.83	nr	**17.53**
20 amp; 178 mm stem	11.47	14.52	0.25	9.83	nr	**24.35**
Non-standard; pocket clip; 280 mm stem	17.87	22.61	0.25	9.83	nr	**32.44**
13 amp; 457 mm stem	7.19	9.10	0.25	9.83	nr	**18.93**
20 amp; 457 mm stem	12.55	15.89	0.25	9.83	nr	**25.72**
Temperature range 50°C to 100°C						
Non-standard; 1780 mm stem	15.01	19.00	0.25	9.83	nr	**28.83**
Non-standard; 280 mm stem	15.66	19.82	0.25	9.83	nr	**29.65**
Pockets for thermostats						
For 178 mm stem	19.67	24.89	0.25	9.83	nr	**34.72**
For 280 mm stem	20.00	25.31	0.25	9.83	nr	**35.14**
Immersion thermostats; stem type; industrial installations; fitted; electrical work elsewhere						
Temperature range 5°C to 105°C						
For 305 mm stem	246.99	312.59	0.50	19.66	nr	**332.25**

38 MECHANICAL/COOLING/HEATING SYSTEMS

Item	Net Price £	Material £	Labour hours	Labour £	Unit	Total rate £
THERMAL INSULATION						
For flexible closed cell insulation see Section – Cold Water						
Mineral fibre sectional insulation; bright class O foil faced; bright class O foil taped joints; 19 mm aluminium bands						
Concealed pipework						
20 mm thick						
15 mm dia.	5.68	7.19	0.15	5.90	m	**13.09**
20 mm dia.	6.06	7.67	0.15	5.90	m	**13.57**
25 mm dia.	6.49	8.21	0.15	5.90	m	**14.11**
32 mm dia.	7.26	9.19	0.15	5.90	m	**15.09**
40 mm dia.	7.77	9.84	0.15	5.90	m	**15.74**
50 mm dia.	8.89	11.25	0.15	5.90	m	**17.14**
Extra over for fittings concealed insulation						
Flange/union						
15 mm dia.	2.84	3.60	0.13	5.11	nr	**8.71**
20 mm dia.	3.03	3.83	0.13	5.11	nr	**8.95**
25 mm dia.	3.27	4.14	0.13	5.11	nr	**9.25**
32 mm dia.	3.63	4.60	0.13	5.11	nr	**9.71**
40 mm dia.	3.86	4.88	0.13	5.11	nr	**10.00**
50 mm dia.	4.44	5.61	0.13	5.11	nr	**10.73**
Valves						
15 mm dia.	5.68	7.19	0.15	5.90	nr	**13.09**
20 mm dia.	6.49	8.21	0.15	5.90	nr	**14.11**
25 mm dia.	6.49	8.21	0.15	5.90	nr	**14.11**
32 mm dia.	7.26	9.19	0.15	5.90	nr	**15.09**
40 mm dia.	7.77	9.84	0.15	5.90	nr	**15.74**
50 mm dia.	8.89	11.25	0.15	5.90	nr	**17.14**
Expansion bellows						
15 mm dia.	11.35	14.37	0.22	8.65	nr	**23.02**
20 mm dia.	12.13	15.35	0.22	8.65	nr	**24.00**
25 mm dia.	13.04	16.50	0.22	8.65	nr	**25.16**
32 mm dia.	14.50	18.35	0.22	8.65	nr	**27.00**
40 mm dia.	15.53	19.66	0.22	8.65	nr	**28.31**
50 mm dia.	17.74	22.46	0.22	8.65	nr	**31.11**
25 mm thick						
15 mm dia.	6.25	7.90	0.15	5.90	m	**13.80**
20 mm dia.	6.71	8.50	0.15	5.90	m	**14.40**
25 mm dia.	7.55	9.55	0.15	5.90	m	**15.45**
32 mm dia.	8.22	10.40	0.15	5.90	m	**16.30**
40 mm dia.	8.78	11.11	0.15	5.90	m	**17.01**
50 mm dia.	10.05	12.72	0.15	5.90	m	**18.62**
65 mm dia.	11.50	14.55	0.15	5.90	m	**20.45**
80 mm dia.	12.60	15.94	0.22	8.65	m	**24.60**
100 mm dia.	16.59	21.00	0.22	8.65	m	**29.65**
125 mm dia.	19.19	24.29	0.22	8.65	m	**32.94**
150 mm dia.	22.86	28.94	0.22	8.65	m	**37.59**

38 MECHANICAL/COOLING/HEATING SYSTEMS

Item	Net Price £	Material £	Labour hours	Labour £	Unit	Total rate £
THERMAL INSULATION – cont						
25 mm thick – cont						
200 mm dia.	32.31	40.90	0.25	9.83	m	**50.73**
250 mm dia.	38.68	48.95	0.25	9.83	m	**58.78**
300 mm dia.	41.37	52.36	0.25	9.83	m	**62.19**
Extra over for fittings concealed insulation						
Flange/union						
15 mm dia.	3.12	3.95	0.13	5.11	nr	**9.06**
20 mm dia.	3.35	4.24	0.13	5.11	nr	**9.35**
25 mm dia.	3.77	4.77	0.13	5.11	nr	**9.88**
32 mm dia.	4.10	5.19	0.13	5.11	nr	**10.30**
40 mm dia.	4.37	5.53	0.13	5.11	nr	**10.64**
50 mm dia.	5.01	6.34	0.13	5.11	nr	**11.46**
65 mm dia.	5.74	7.26	0.13	5.11	nr	**12.37**
80 mm dia.	6.31	7.99	0.18	7.08	nr	**15.07**
100 mm dia.	8.31	10.52	0.18	7.08	nr	**17.60**
125 mm dia.	9.60	12.14	0.18	7.08	nr	**19.22**
150 mm dia.	11.45	14.49	0.18	7.08	nr	**21.56**
200 mm dia.	16.15	20.44	0.22	8.65	nr	**29.09**
250 mm dia.	19.35	24.49	0.22	8.65	nr	**33.14**
300 mm dia.	20.67	26.16	0.22	8.65	nr	**34.81**
Valves						
15 mm dia.	6.25	7.90	0.15	5.90	nr	**13.80**
20 mm dia.	6.71	8.50	0.15	5.90	nr	**14.40**
25 mm dia.	7.55	9.55	0.15	5.90	nr	**15.45**
32 mm dia.	8.22	10.40	0.15	5.90	nr	**16.30**
40 mm dia.	8.78	11.11	0.15	5.90	nr	**17.01**
50 mm dia.	10.05	12.72	0.15	5.90	nr	**18.62**
65 mm dia.	11.50	14.55	0.15	5.90	nr	**20.45**
80 mm dia.	12.60	15.94	0.20	7.87	nr	**23.81**
100 mm dia.	16.59	21.00	0.20	7.87	nr	**28.86**
125 mm dia.	19.19	24.29	0.20	7.87	nr	**32.15**
150 mm dia.	22.86	28.94	0.20	7.87	nr	**36.80**
200 mm dia.	32.31	40.90	0.25	9.83	nr	**50.73**
250 mm dia.	38.68	48.95	0.25	9.83	nr	**58.78**
300 mm dia.	41.37	52.36	0.25	9.83	nr	**62.19**
Expansion bellows						
15 mm dia.	12.49	15.81	0.22	8.65	nr	**24.46**
20 mm dia.	13.50	17.08	0.22	8.65	nr	**25.73**
25 mm dia.	15.05	19.05	0.22	8.65	nr	**27.70**
32 mm dia.	16.39	20.74	0.22	8.65	nr	**29.40**
40 mm dia.	17.57	22.24	0.22	8.65	nr	**30.89**
50 mm dia.	20.12	25.46	0.22	8.65	nr	**34.11**
65 mm dia.	22.96	29.06	0.22	8.65	nr	**37.71**
80 mm dia.	25.17	31.85	0.29	11.40	nr	**43.26**
100 mm dia.	33.18	42.00	0.29	11.40	nr	**53.40**
125 mm dia.	38.41	48.61	0.29	11.40	nr	**60.02**
150 mm dia.	45.77	57.93	0.29	11.40	nr	**69.33**

38 MECHANICAL/COOLING/HEATING SYSTEMS

Item	Net Price £	Material £	Labour hours	Labour £	Unit	Total rate £
200 mm dia.	64.64	81.81	0.36	14.16	nr	**95.96**
250 mm dia.	77.41	97.97	0.36	14.16	nr	**112.13**
300 mm dia.	82.75	104.72	0.36	14.16	nr	**118.88**
30 mm thick						
15 mm dia.	8.12	10.28	0.15	5.90	m	**16.18**
20 mm dia.	8.70	11.01	0.15	5.90	m	**16.91**
25 mm dia.	9.19	11.64	0.15	5.90	m	**17.53**
32 mm dia.	10.01	12.67	0.15	5.90	m	**18.57**
40 mm dia.	10.63	13.45	0.15	5.90	m	**19.35**
50 mm dia.	12.14	15.37	0.15	5.90	m	**21.27**
65 mm dia.	13.75	17.40	0.15	5.90	m	**23.30**
80 mm dia.	15.02	19.01	0.22	8.65	m	**27.67**
100 mm dia.	19.39	24.54	0.22	8.65	m	**33.20**
125 mm dia.	22.38	28.33	0.22	8.65	m	**36.98**
150 mm dia.	26.26	33.23	0.22	8.65	m	**41.88**
200 mm dia.	36.68	46.42	0.25	9.83	m	**56.26**
250 mm dia.	43.61	55.19	0.25	9.83	m	**65.03**
300 mm dia.	46.25	58.54	0.25	9.83	m	**68.37**
350 mm dia.	50.82	64.32	0.25	9.83	m	**74.15**
Extra over for fittings concealed insulation						
Flange/union						
15 mm dia.	4.05	5.12	0.13	5.11	nr	**10.23**
20 mm dia.	4.36	5.51	0.13	5.11	nr	**10.63**
25 mm dia.	4.57	5.78	0.13	5.11	nr	**10.90**
32 mm dia.	5.01	6.34	0.13	5.11	nr	**11.46**
40 mm dia.	5.29	6.70	0.13	5.11	nr	**11.81**
50 mm dia.	6.08	7.70	0.13	5.11	nr	**12.81**
65 mm dia.	6.89	8.72	0.13	5.11	nr	**13.83**
80 mm dia.	7.48	9.46	0.18	7.08	nr	**16.54**
100 mm dia.	9.73	12.31	0.18	7.08	nr	**19.39**
125 mm dia.	11.15	14.11	0.18	7.08	nr	**21.19**
150 mm dia.	13.13	16.62	0.18	7.08	nr	**23.70**
200 mm dia.	18.37	23.25	0.22	8.65	nr	**31.91**
250 mm dia.	21.83	27.63	0.22	8.65	nr	**36.28**
300 mm dia.	23.15	29.29	0.22	8.65	nr	**37.95**
350 mm dia.	25.41	32.16	0.22	8.65	nr	**40.81**
Valves						
15 mm dia.	8.70	11.01	0.15	5.90	nr	**16.91**
20 mm dia.	8.70	11.01	0.15	5.90	nr	**16.91**
25 mm dia.	9.19	11.64	0.15	5.90	nr	**17.53**
32 mm dia.	10.01	12.67	0.15	5.90	nr	**18.57**
40 mm dia.	10.63	13.45	0.15	5.90	nr	**19.35**
50 mm dia.	12.14	15.37	0.15	5.90	nr	**21.27**
65 mm dia.	13.75	17.40	0.15	5.90	nr	**23.30**
80 mm dia.	15.02	19.01	0.20	7.87	nr	**26.88**
100 mm dia.	19.39	24.54	0.20	7.87	nr	**32.41**
125 mm dia.	22.38	28.33	0.20	7.87	nr	**36.19**
150 mm dia.	26.26	33.23	0.20	7.87	nr	**41.09**

38 MECHANICAL/COOLING/HEATING SYSTEMS

Item	Net Price £	Material £	Labour hours	Labour £	Unit	Total rate £
THERMAL INSULATION – cont						
Extra over for fittings concealed insulation – cont						
Valves – cont						
200 mm dia.	36.68	46.42	0.25	9.83	nr	**56.26**
250 mm dia.	43.61	55.19	0.25	9.83	nr	**65.03**
300 mm dia.	46.25	58.54	0.25	9.83	nr	**68.37**
350 mm dia.	50.82	64.32	0.25	9.83	nr	**74.15**
Expansion bellows						
15 mm dia.	16.26	20.57	0.22	8.65	nr	**29.23**
20 mm dia.	17.37	21.98	0.22	8.65	nr	**30.63**
25 mm dia.	18.40	23.29	0.22	8.65	nr	**31.94**
32 mm dia.	20.10	25.44	0.22	8.65	nr	**34.09**
40 mm dia.	21.27	26.92	0.22	8.65	nr	**35.57**
50 mm dia.	24.31	30.77	0.22	8.65	nr	**39.42**
65 mm dia.	27.50	34.81	0.22	8.65	nr	**43.46**
80 mm dia.	29.99	37.96	0.29	11.40	nr	**49.37**
100 mm dia.	38.85	49.17	0.29	11.40	nr	**60.58**
125 mm dia.	44.68	56.55	0.29	11.40	nr	**67.96**
150 mm dia.	52.50	66.44	0.29	11.40	nr	**77.84**
200 mm dia.	73.36	92.85	0.36	14.16	nr	**107.01**
250 mm dia.	87.30	110.49	0.36	14.16	nr	**124.65**
300 mm dia.	92.56	117.14	0.36	14.16	nr	**131.30**
350 mm dia.	101.62	128.61	0.36	14.16	nr	**142.76**
40 mm thick						
15 mm dia.	10.39	13.15	0.15	5.90	m	**19.04**
20 mm dia.	10.72	13.57	0.15	5.90	m	**19.47**
25 mm dia.	11.54	14.60	0.15	5.90	m	**20.50**
32 mm dia.	12.29	15.55	0.15	5.90	m	**21.45**
40 mm dia.	12.95	16.39	0.15	5.90	m	**22.28**
50 mm dia.	14.64	18.52	0.15	5.90	m	**24.42**
65 mm dia.	16.40	20.76	0.15	5.90	m	**26.66**
80 mm dia.	17.83	22.56	0.22	8.65	m	**31.21**
100 mm dia.	23.20	29.36	0.22	8.65	m	**38.01**
125 mm dia.	26.20	33.16	0.22	8.65	m	**41.81**
150 mm dia.	30.56	38.67	0.22	8.65	m	**47.33**
200 mm dia.	42.23	53.45	0.25	9.83	m	**63.28**
250 mm dia.	49.48	62.62	0.25	9.83	m	**72.46**
300 mm dia.	52.90	66.95	0.25	9.83	m	**76.78**
350 mm dia.	58.51	74.06	0.25	9.83	m	**83.89**
400 mm dia.	65.19	82.50	0.25	9.83	m	**92.33**
Extra over for fittings concealed insulation						
Flange/union						
15 mm dia.	5.23	6.62	0.13	5.11	nr	**11.73**
20 mm dia.	5.35	6.77	0.13	5.11	nr	**11.88**
25 mm dia.	5.75	7.28	0.13	5.11	nr	**12.39**
32 mm dia.	6.14	7.77	0.13	5.11	nr	**12.88**
40 mm dia.	6.45	8.16	0.13	5.11	nr	**13.27**
50 mm dia.	7.32	9.26	0.13	5.11	nr	**14.37**
65 mm dia.	8.22	10.40	0.13	5.11	nr	**15.51**

38 MECHANICAL/COOLING/HEATING SYSTEMS

Item	Net Price £	Material £	Labour hours	Labour £	Unit	Total rate £
80 mm dia.	8.94	11.31	0.18	7.08	nr	**18.39**
100 mm dia.	11.58	14.66	0.18	7.08	nr	**21.73**
125 mm dia.	13.08	16.55	0.18	7.08	nr	**23.63**
150 mm dia.	15.29	19.35	0.18	7.08	nr	**26.43**
200 mm dia.	21.12	26.73	0.22	8.65	nr	**35.38**
250 mm dia.	24.73	31.29	0.22	8.65	nr	**39.95**
300 mm dia.	26.42	33.43	0.22	8.65	nr	**42.08**
350 mm dia.	29.26	37.03	0.22	8.65	nr	**45.68**
400 mm dia.	32.63	41.30	0.22	8.65	nr	**49.95**
Valves						
15 mm dia.	10.39	13.15	0.15	5.90	nr	**19.04**
20 mm dia.	10.72	13.57	0.15	5.90	nr	**19.47**
25 mm dia.	11.54	14.60	0.15	5.90	nr	**20.50**
32 mm dia.	12.29	15.55	0.15	5.90	nr	**21.45**
40 mm dia.	12.95	16.39	0.15	5.90	nr	**22.28**
50 mm dia.	14.64	18.52	0.15	5.90	nr	**24.42**
65 mm dia.	16.40	20.76	0.15	5.90	nr	**26.66**
80 mm dia.	17.83	22.56	0.20	7.87	nr	**30.42**
100 mm dia.	23.20	29.36	0.20	7.87	nr	**37.23**
125 mm dia.	26.20	33.16	0.20	7.87	nr	**41.03**
150 mm dia.	30.56	38.67	0.20	7.87	nr	**46.54**
200 mm dia.	42.23	53.45	0.25	9.83	nr	**63.28**
250 mm dia.	49.48	62.62	0.25	9.83	nr	**72.46**
300 mm dia.	52.90	66.95	0.25	9.83	nr	**76.78**
350 mm dia.	63.49	80.36	0.25	9.83	nr	**90.19**
400 mm dia.	65.19	82.50	0.25	9.83	nr	**92.33**
Expansion bellows						
15 mm dia.	20.85	26.39	0.22	8.65	nr	**35.04**
20 mm dia.	21.42	27.11	0.22	8.65	nr	**35.76**
25 mm dia.	23.09	29.23	0.22	8.65	nr	**37.88**
32 mm dia.	24.58	31.11	0.22	8.65	nr	**39.76**
40 mm dia.	25.85	32.72	0.22	8.65	nr	**41.37**
50 mm dia.	29.26	37.03	0.22	8.65	nr	**45.68**
65 mm dia.	32.78	41.49	0.22	8.65	nr	**50.14**
80 mm dia.	35.73	45.22	0.29	11.40	nr	**56.63**
100 mm dia.	46.32	58.62	0.29	11.40	nr	**70.03**
125 mm dia.	52.36	66.27	0.29	11.40	nr	**77.68**
150 mm dia.	58.51	74.06	0.29	11.40	nr	**85.46**
200 mm dia.	61.11	77.35	0.36	14.16	nr	**91.50**
250 mm dia.	84.46	106.89	0.36	14.16	nr	**121.05**
300 mm dia.	98.98	125.26	0.36	14.16	nr	**139.42**
350 mm dia.	63.49	80.36	0.36	14.16	nr	**94.52**
400 mm dia.	130.42	165.06	0.36	14.16	nr	**179.21**
50 mm thick						
15 mm dia.	14.47	18.32	0.15	5.90	m	**24.22**
20 mm dia.	15.21	19.25	0.15	5.90	m	**25.15**
25 mm dia.	16.18	20.47	0.15	5.90	m	**26.37**
32 mm dia.	16.90	21.39	0.15	5.90	m	**27.29**
40 mm dia.	17.76	22.47	0.15	5.90	m	**28.37**
50 mm dia.	19.94	25.24	0.15	5.90	m	**31.14**
65 mm dia.	21.79	27.58	0.15	5.90	m	**33.48**

38 MECHANICAL/COOLING/HEATING SYSTEMS

Item	Net Price £	Material £	Labour hours	Labour £	Unit	Total rate £
THERMAL INSULATION – cont						
50 mm thick – cont						
80 mm dia.	23.31	29.50	0.22	8.65	m	**38.15**
100 mm dia.	29.83	37.76	0.22	8.65	m	**46.41**
125 mm dia.	33.47	42.35	0.22	8.65	m	**51.01**
150 mm dia.	38.63	48.88	0.22	8.65	m	**57.54**
200 mm dia.	52.74	66.75	0.25	9.83	m	**76.58**
250 mm dia.	60.90	77.08	0.25	9.83	m	**86.91**
300 mm dia.	64.53	81.67	0.25	9.83	m	**91.50**
350 mm dia.	71.23	90.15	0.25	9.83	m	**99.98**
400 mm dia.	79.01	99.99	0.25	9.83	m	**109.82**
Extra over for fittings concealed insulation						
Flange/union						
15 mm dia.	7.25	9.18	0.13	5.11	nr	**14.29**
20 mm dia.	7.63	9.65	0.13	5.11	nr	**14.76**
25 mm dia.	8.11	10.26	0.13	5.11	nr	**15.37**
32 mm dia.	8.47	10.72	0.13	5.11	nr	**15.83**
40 mm dia.	8.90	11.26	0.13	5.11	nr	**16.38**
50 mm dia.	9.94	12.59	0.13	5.11	nr	**17.70**
65 mm dia.	10.90	13.79	0.13	5.11	nr	**18.90**
80 mm dia.	11.66	14.76	0.18	7.08	nr	**21.84**
100 mm dia.	14.93	18.90	0.18	7.08	nr	**25.97**
125 mm dia.	16.74	21.19	0.18	7.08	nr	**28.26**
150 mm dia.	19.34	24.48	0.18	7.08	nr	**31.55**
200 mm dia.	26.35	33.35	0.22	8.65	nr	**42.00**
250 mm dia.	30.45	38.54	0.22	8.65	nr	**47.19**
300 mm dia.	32.27	40.84	0.22	8.65	nr	**49.50**
350 mm dia.	35.65	45.12	0.22	8.65	nr	**53.77**
400 mm dia.	39.51	50.00	0.22	8.65	nr	**58.66**
Valves						
15 mm dia.	14.47	18.32	0.15	5.90	nr	**24.22**
20 mm dia.	15.21	19.25	0.15	5.90	nr	**25.15**
25 mm dia.	16.18	20.47	0.15	5.90	nr	**26.37**
32 mm dia.	16.90	21.39	0.15	5.90	nr	**27.29**
40 mm dia.	17.76	22.47	0.15	5.90	nr	**28.37**
50 mm dia.	19.94	25.24	0.15	5.90	nr	**31.14**
65 mm dia.	21.79	27.58	0.15	5.90	nr	**33.48**
80 mm dia.	23.31	29.50	0.20	7.87	nr	**37.36**
100 mm dia.	29.83	37.76	0.20	7.87	nr	**45.62**
125 mm dia.	33.47	42.35	0.20	7.87	nr	**50.22**
150 mm dia.	38.63	48.88	0.20	7.87	nr	**56.75**
200 mm dia.	52.74	66.75	0.25	9.83	nr	**76.58**
250 mm dia.	60.90	77.08	0.25	9.83	nr	**86.91**
300 mm dia.	64.53	81.67	0.25	9.83	nr	**91.50**
350 mm dia.	71.23	90.15	0.25	9.83	nr	**99.98**
400 mm dia.	79.01	99.99	0.25	9.83	nr	**109.82**

38 MECHANICAL/COOLING/HEATING SYSTEMS

Item	Net Price £	Material £	Labour hours	Labour £	Unit	Total rate £
Expansion bellows						
15 mm dia.	28.88	36.55	0.22	8.65	nr	**45.20**
20 mm dia.	30.41	38.49	0.22	8.65	nr	**47.14**
25 mm dia.	32.34	40.93	0.22	8.65	nr	**49.58**
32 mm dia.	33.81	42.79	0.22	8.65	nr	**51.45**
40 mm dia.	35.57	45.02	0.22	8.65	nr	**53.67**
50 mm dia.	39.86	50.45	0.22	8.65	nr	**59.10**
65 mm dia.	43.56	55.13	0.22	8.65	nr	**63.78**
80 mm dia.	46.67	59.06	0.29	11.40	nr	**70.47**
100 mm dia.	59.60	75.43	0.29	11.40	nr	**86.83**
125 mm dia.	66.92	84.69	0.29	11.40	nr	**96.10**
150 mm dia.	77.25	97.77	0.29	11.40	nr	**109.17**
200 mm dia.	105.50	133.52	0.36	14.16	nr	**147.68**
250 mm dia.	121.79	154.13	0.36	14.16	nr	**168.29**
300 mm dia.	129.10	163.39	0.36	14.16	nr	**177.55**
350 mm dia.	142.48	180.32	0.36	14.16	nr	**194.48**
400 mm dia.	157.99	199.95	0.36	14.16	nr	**214.11**
Mineral fibre sectional insulation; bright class O foil faced; bright class O foil taped joints; 22 SWG plain/ embossed aluminium cladding; pop riveted						
Plantroom pipework						
20 mm thick						
15 mm dia.	9.27	11.74	0.44	17.30	m	**29.04**
20 mm dia.	9.80	12.40	0.44	17.30	m	**29.70**
25 mm dia.	10.49	13.28	0.44	17.30	m	**30.58**
32 mm dia.	11.37	14.38	0.44	17.30	m	**31.69**
40 mm dia.	11.98	15.16	0.44	17.30	m	**32.47**
50 mm dia.	13.34	16.88	0.44	17.30	m	**34.18**
Extra over for fittings plantroom insulation						
Flange/union						
15 mm dia.	10.36	13.11	0.58	22.81	nr	**35.92**
20 mm dia.	11.00	13.93	0.58	22.81	nr	**36.73**
25 mm dia.	11.78	14.91	0.58	22.81	nr	**37.72**
32 mm dia.	12.87	16.28	0.58	22.81	nr	**39.09**
40 mm dia.	13.59	17.20	0.58	22.81	nr	**40.01**
50 mm dia.	15.25	19.30	0.58	22.81	nr	**42.11**
Bends						
15 mm dia.	5.07	6.41	0.44	17.30	nr	**23.72**
20 mm dia.	5.41	6.85	0.44	17.30	nr	**24.16**
25 mm dia.	5.78	7.31	0.44	17.30	nr	**24.61**
32 mm dia.	6.25	7.90	0.44	17.30	nr	**25.21**
40 mm dia.	6.55	8.29	0.44	17.30	nr	**25.60**
50 mm dia.	7.34	9.30	0.44	17.30	nr	**26.60**
Tees						
15 mm dia.	3.04	3.85	0.44	17.30	nr	**21.15**
20 mm dia.	3.26	4.12	0.44	17.30	nr	**21.43**
25 mm dia.	3.46	4.38	0.44	17.30	nr	**21.68**

38 MECHANICAL/COOLING/HEATING SYSTEMS

Item	Net Price £	Material £	Labour hours	Labour £	Unit	Total rate £
THERMAL INSULATION – cont						
Extra over for fittings plantroom insulation – cont						
Tees – cont						
32 mm dia.	3.75	4.75	0.44	17.30	nr	**22.05**
40 mm dia.	3.97	5.02	0.44	17.30	nr	**22.32**
50 mm dia.	4.41	5.58	0.44	17.30	nr	**22.88**
Valves						
15 mm dia.	5.68	7.19	0.78	30.67	nr	**37.87**
20 mm dia.	6.49	8.21	0.78	30.67	nr	**38.88**
25 mm dia.	6.49	8.21	0.78	30.67	nr	**38.88**
32 mm dia.	7.26	9.19	0.78	30.67	nr	**39.87**
40 mm dia.	7.77	9.84	0.78	30.67	nr	**40.51**
50 mm dia.	8.89	11.25	0.78	30.67	nr	**41.92**
Pumps						
15 mm dia.	30.45	38.54	2.34	92.02	nr	**130.56**
20 mm dia.	32.29	40.86	2.34	92.02	nr	**132.88**
25 mm dia.	34.67	43.88	2.34	92.02	nr	**135.90**
32 mm dia.	37.83	47.88	2.34	92.02	nr	**139.91**
40 mm dia.	39.97	50.58	2.34	92.02	nr	**142.60**
50 mm dia.	44.94	56.87	2.34	92.02	nr	**148.90**
Expansion bellows						
15 mm dia.	24.35	30.82	1.05	41.29	nr	**72.11**
20 mm dia.	25.84	32.70	1.05	41.29	nr	**73.99**
25 mm dia.	27.76	35.13	1.05	41.29	nr	**76.42**
32 mm dia.	30.26	38.30	1.05	41.29	nr	**79.59**
40 mm dia.	31.94	40.42	1.05	41.29	nr	**81.71**
50 mm dia.	35.95	45.49	1.05	41.29	nr	**86.78**
25 mm thick						
15 mm dia.	10.17	12.87	0.44	17.30	m	**30.18**
20 mm dia.	11.00	13.93	0.44	17.30	m	**31.23**
25 mm dia.	12.05	15.25	0.44	17.30	m	**32.55**
32 mm dia.	12.83	16.23	0.44	17.30	m	**33.54**
40 mm dia.	13.80	17.47	0.44	17.30	m	**34.77**
50 mm dia.	15.47	19.57	0.44	17.30	m	**36.88**
65 mm dia.	17.48	22.12	0.44	17.30	m	**39.42**
80 mm dia.	18.94	23.97	0.52	20.45	m	**44.42**
100 mm dia.	23.51	29.75	0.52	20.45	m	**50.20**
125 mm dia.	27.30	34.55	0.52	20.45	m	**55.00**
150 mm dia.	31.78	40.22	0.52	20.45	m	**60.67**
200 mm dia.	43.41	54.94	0.60	23.60	m	**78.54**
250 mm dia.	51.59	65.29	0.60	23.60	m	**88.88**
300 mm dia.	57.75	73.09	0.60	23.60	m	**96.68**
Extra over for fittings plantroom insulation						
Flange/union						
15 mm dia.	11.37	14.38	0.58	22.81	nr	**37.19**
20 mm dia.	12.29	15.55	0.58	22.81	nr	**38.36**
25 mm dia.	13.56	17.17	0.58	22.81	nr	**39.97**

38 MECHANICAL/COOLING/HEATING SYSTEMS

Item	Net Price £	Material £	Labour hours	Labour £	Unit	Total rate £
32 mm dia.	14.58	18.45	0.58	22.81	nr	**41.26**
40 mm dia.	15.63	19.78	0.58	22.81	nr	**42.59**
50 mm dia.	17.61	22.29	0.58	22.81	nr	**45.10**
65 mm dia.	19.98	25.29	0.58	22.81	nr	**48.10**
80 mm dia.	21.73	27.50	0.67	26.35	nr	**53.84**
100 mm dia.	27.50	34.81	0.67	26.35	nr	**61.15**
125 mm dia.	31.92	40.40	0.67	26.35	nr	**66.75**
150 mm dia.	37.42	47.36	0.67	26.35	nr	**73.71**
200 mm dia.	51.71	65.44	0.87	34.21	nr	**99.65**
250 mm dia.	61.58	77.94	0.87	34.21	nr	**112.15**
300 mm dia.	68.08	86.17	0.87	34.21	nr	**120.38**
Bends						
15 mm dia.	5.59	7.07	0.44	17.30	nr	**24.38**
20 mm dia.	6.06	7.67	0.44	17.30	nr	**24.97**
25 mm dia.	6.62	8.38	0.44	17.30	nr	**25.68**
32 mm dia.	7.06	8.94	0.44	17.30	nr	**26.24**
40 mm dia.	7.63	9.65	0.44	17.30	nr	**26.95**
50 mm dia.	8.51	10.77	0.44	17.30	nr	**28.07**
65 mm dia.	9.60	12.14	0.44	17.30	nr	**29.45**
80 mm dia.	10.40	13.16	0.52	20.45	nr	**33.61**
100 mm dia.	12.95	16.39	0.52	20.45	nr	**36.83**
125 mm dia.	15.05	19.05	0.52	20.45	nr	**39.50**
150 mm dia.	17.48	22.12	0.52	20.45	nr	**42.57**
200 mm dia.	23.87	30.21	0.60	23.60	nr	**53.80**
250 mm dia.	28.37	35.91	0.60	23.60	nr	**59.50**
300 mm dia.	31.76	40.20	0.60	23.60	nr	**63.80**
Tees						
15 mm dia.	3.35	4.24	0.44	17.30	nr	**21.54**
20 mm dia.	3.61	4.56	0.44	17.30	nr	**21.87**
25 mm dia.	3.99	5.05	0.44	17.30	nr	**22.36**
32 mm dia.	4.24	5.36	0.44	17.30	nr	**22.66**
40 mm dia.	4.54	5.75	0.44	17.30	nr	**23.05**
50 mm dia.	5.07	6.41	0.44	17.30	nr	**23.72**
65 mm dia.	5.75	7.28	0.44	17.30	nr	**24.58**
80 mm dia.	6.25	7.90	0.52	20.45	nr	**28.35**
100 mm dia.	7.76	9.82	0.52	20.45	nr	**30.27**
125 mm dia.	9.03	11.43	0.52	20.45	nr	**31.88**
150 mm dia.	10.49	13.28	0.52	20.45	nr	**33.73**
200 mm dia.	14.30	18.10	0.60	23.60	nr	**41.69**
250 mm dia.	17.05	21.58	0.60	23.60	nr	**45.17**
300 mm dia.	19.09	24.15	0.60	23.60	nr	**47.75**
Valves						
15 mm dia.	18.07	22.86	0.78	30.67	nr	**53.54**
20 mm dia.	19.53	24.71	0.78	30.67	nr	**55.39**
25 mm dia.	21.56	27.29	0.78	30.67	nr	**57.97**
32 mm dia.	23.13	29.28	0.78	30.67	nr	**59.95**
40 mm dia.	24.82	31.41	0.78	30.67	nr	**62.09**
50 mm dia.	27.93	35.35	0.78	30.67	nr	**66.02**
65 mm dia.	31.74	40.17	0.78	30.67	nr	**70.84**

38 MECHANICAL/COOLING/HEATING SYSTEMS

Item	Net Price £	Material £	Labour hours	Labour £	Unit	Total rate £
THERMAL INSULATION – cont						
Extra over for fittings plantroom insulation – cont						
Valves – cont						
80 mm dia.	34.52	43.69	0.92	36.18	nr	**79.87**
100 mm dia.	43.64	55.23	0.92	36.18	nr	**91.41**
125 mm dia.	50.74	64.22	0.92	36.18	nr	**100.40**
150 mm dia.	59.44	75.23	0.92	36.18	nr	**111.41**
200 mm dia.	82.13	103.94	1.12	44.04	nr	**147.99**
250 mm dia.	97.84	123.82	1.12	44.04	nr	**167.87**
300 mm dia.	108.10	136.82	1.12	44.04	nr	**180.86**
Pumps						
15 mm dia.	33.45	42.34	2.34	92.02	nr	**134.36**
20 mm dia.	36.15	45.75	2.34	92.02	nr	**137.77**
25 mm dia.	39.94	50.55	2.34	92.02	nr	**142.57**
32 mm dia.	42.79	54.16	2.34	92.02	nr	**146.18**
40 mm dia.	45.94	58.15	2.34	92.02	nr	**150.17**
50 mm dia.	51.79	65.54	2.34	92.02	nr	**157.56**
65 mm dia.	58.74	74.34	2.34	92.02	nr	**166.37**
80 mm dia.	63.92	80.89	2.76	108.54	nr	**189.43**
100 mm dia.	80.87	102.35	2.76	108.54	nr	**210.89**
125 mm dia.	93.92	118.87	2.76	108.54	nr	**227.41**
150 mm dia.	110.06	139.29	2.76	108.54	nr	**247.83**
200 mm dia.	152.09	192.48	3.36	132.13	nr	**324.62**
250 mm dia.	181.17	229.29	3.36	132.13	nr	**361.43**
300 mm dia.	200.15	253.31	3.36	132.13	nr	**385.45**
Expansion bellows						
15 mm dia.	26.75	33.86	1.05	41.29	nr	**75.15**
20 mm dia.	28.90	36.57	1.05	41.29	nr	**77.86**
25 mm dia.	31.92	40.40	1.05	41.29	nr	**81.70**
32 mm dia.	34.27	43.37	1.05	41.29	nr	**84.66**
40 mm dia.	36.78	46.54	1.05	41.29	nr	**87.84**
50 mm dia.	41.44	52.45	1.05	41.29	nr	**93.74**
65 mm dia.	47.02	59.50	1.05	41.29	nr	**100.79**
80 mm dia.	51.16	64.74	1.26	49.55	nr	**114.29**
100 mm dia.	64.68	81.86	1.26	49.55	nr	**131.41**
125 mm dia.	75.09	95.04	1.26	49.55	nr	**144.59**
150 mm dia.	88.03	111.41	1.26	49.55	nr	**160.96**
200 mm dia.	121.68	154.00	1.53	60.17	nr	**214.17**
250 mm dia.	144.96	183.46	1.53	60.17	nr	**243.63**
300 mm dia.	160.10	202.63	1.53	60.17	nr	**262.80**
30 mm thick						
15 mm dia.	12.54	15.88	0.44	17.30	m	**33.18**
20 mm dia.	13.25	16.78	0.44	17.30	m	**34.08**
25 mm dia.	14.05	17.78	0.44	17.30	m	**35.08**
32 mm dia.	15.40	19.49	0.44	17.30	m	**36.79**
40 mm dia.	16.02	20.27	0.44	17.30	m	**37.57**
50 mm dia.	17.76	22.47	0.44	17.30	m	**39.78**
65 mm dia.	20.06	25.39	0.44	17.30	m	**42.70**

38 MECHANICAL/COOLING/HEATING SYSTEMS

Item	Net Price £	Material £	Labour hours	Labour £	Unit	Total rate £
80 mm dia.	21.93	27.75	0.52	20.45	m	**48.20**
100 mm dia.	26.94	34.09	0.52	20.45	m	**54.54**
125 mm dia.	30.76	38.93	0.52	20.45	m	**59.38**
150 mm dia.	35.84	45.36	0.52	20.45	m	**65.81**
200 mm dia.	48.22	61.03	0.60	23.60	m	**84.62**
250 mm dia.	57.01	72.16	0.60	23.60	m	**95.75**
300 mm dia.	62.76	79.43	0.60	23.60	m	**103.03**
350 mm dia.	69.88	88.44	0.60	23.60	m	**112.04**
Extra over for fittings plantroom insulation						
Flange/union						
15 mm dia.	14.25	18.03	0.58	22.81	nr	**40.84**
20 mm dia.	15.13	19.15	0.58	22.81	nr	**41.96**
25 mm dia.	16.04	20.30	0.58	22.81	nr	**43.11**
32 mm dia.	17.57	22.24	0.58	22.81	nr	**45.05**
40 mm dia.	18.39	23.27	0.58	22.81	nr	**46.08**
50 mm dia.	20.57	26.04	0.58	22.81	nr	**48.85**
65 mm dia.	23.25	29.43	0.58	22.81	nr	**52.24**
80 mm dia.	25.41	32.16	0.67	26.35	nr	**58.51**
100 mm dia.	31.74	40.17	0.67	26.35	nr	**66.51**
125 mm dia.	36.32	45.97	0.67	26.35	nr	**72.32**
150 mm dia.	42.49	53.77	0.67	26.35	nr	**80.12**
200 mm dia.	57.86	73.22	0.87	34.21	nr	**107.44**
250 mm dia.	68.59	86.81	0.87	34.21	nr	**121.02**
300 mm dia.	74.62	94.44	0.87	34.21	nr	**128.66**
350 mm dia.	82.80	104.79	0.87	34.21	nr	**139.00**
Bends						
15 mm dia.	6.89	8.72	0.44	17.30	nr	**26.02**
20 mm dia.	7.28	9.21	0.44	17.30	nr	**26.51**
25 mm dia.	7.73	9.79	0.44	17.30	nr	**27.09**
32 mm dia.	8.44	10.69	0.44	17.30	nr	**27.99**
40 mm dia.	8.79	11.13	0.44	17.30	nr	**28.43**
50 mm dia.	9.80	12.40	0.44	17.30	nr	**29.70**
65 mm dia.	11.04	13.98	0.44	17.30	nr	**31.28**
80 mm dia.	12.04	15.23	0.52	20.45	nr	**35.68**
100 mm dia.	14.81	18.74	0.52	20.45	nr	**39.19**
125 mm dia.	16.91	21.41	0.52	20.45	nr	**41.86**
150 mm dia.	19.73	24.97	0.52	20.45	nr	**45.42**
200 mm dia.	26.51	33.55	0.60	23.60	nr	**57.15**
250 mm dia.	31.39	39.73	0.60	23.60	nr	**63.32**
300 mm dia.	34.52	43.69	0.60	23.60	nr	**67.29**
350 mm dia.	38.46	48.68	0.60	23.60	nr	**72.28**
Tees						
15 mm dia.	4.11	5.21	0.44	17.30	nr	**22.51**
20 mm dia.	4.37	5.53	0.44	17.30	nr	**22.83**
25 mm dia.	4.61	5.83	0.44	17.30	nr	**23.14**
32 mm dia.	5.05	6.39	0.44	17.30	nr	**23.70**
40 mm dia.	5.28	6.68	0.44	17.30	nr	**23.99**
50 mm dia.	5.86	7.41	0.44	17.30	nr	**24.72**
65 mm dia.	6.61	8.36	0.44	17.30	nr	**25.67**

38 MECHANICAL/COOLING/HEATING SYSTEMS

Item	Net Price £	Material £	Labour hours	Labour £	Unit	Total rate £
THERMAL INSULATION – cont						
Extra over for fittings plantroom insulation – cont						
Tees – cont						
80 mm dia.	7.25	9.18	0.52	20.45	nr	**29.63**
100 mm dia.	8.90	11.26	0.52	20.45	nr	**31.71**
125 mm dia.	10.17	12.87	0.52	20.45	nr	**33.32**
150 mm dia.	11.86	15.01	0.52	20.45	nr	**35.46**
200 mm dia.	15.91	20.13	0.60	23.60	nr	**43.73**
250 mm dia.	18.84	23.85	0.60	23.60	nr	**47.44**
300 mm dia.	20.71	26.21	0.60	23.60	nr	**49.80**
350 mm dia.	23.09	29.23	0.60	23.60	nr	**52.82**
Valves						
15 mm dia.	22.65	28.67	0.78	30.67	nr	**59.34**
20 mm dia.	24.02	30.40	0.78	30.67	nr	**61.07**
25 mm dia.	25.48	32.24	0.78	30.67	nr	**62.92**
32 mm dia.	27.92	35.33	0.78	30.67	nr	**66.01**
40 mm dia.	29.19	36.94	0.78	30.67	nr	**67.62**
50 mm dia.	32.69	41.37	0.78	30.67	nr	**72.04**
65 mm dia.	36.95	46.76	0.78	30.67	nr	**77.44**
80 mm dia.	40.34	51.06	0.92	36.18	nr	**87.24**
100 mm dia.	50.39	63.78	0.92	36.18	nr	**99.96**
125 mm dia.	57.71	73.04	0.92	36.18	nr	**109.22**
150 mm dia.	67.47	85.39	0.92	36.18	nr	**121.57**
200 mm dia.	91.87	116.27	1.12	44.04	nr	**160.32**
250 mm dia.	108.93	137.87	1.12	44.04	nr	**181.91**
300 mm dia.	118.53	150.01	1.12	44.04	nr	**194.06**
350 mm dia.	131.49	166.41	1.12	44.04	nr	**210.46**
Pumps						
15 mm dia.	41.96	53.11	2.34	92.02	nr	**145.13**
20 mm dia.	44.52	56.35	2.34	92.02	nr	**148.37**
25 mm dia.	47.16	59.69	2.34	92.02	nr	**151.71**
32 mm dia.	51.68	65.41	2.34	92.02	nr	**157.43**
40 mm dia.	54.07	68.42	2.34	92.02	nr	**160.45**
50 mm dia.	60.57	76.65	2.34	92.02	nr	**168.67**
65 mm dia.	68.39	86.56	2.34	92.02	nr	**178.58**
80 mm dia.	74.68	94.51	2.76	108.54	nr	**203.05**
100 mm dia.	93.29	118.07	2.76	108.54	nr	**226.61**
125 mm dia.	106.86	135.24	2.76	108.54	nr	**243.78**
150 mm dia.	124.96	158.15	2.76	108.54	nr	**266.69**
200 mm dia.	170.16	215.35	3.36	132.13	nr	**347.48**
250 mm dia.	201.77	255.36	3.36	132.13	nr	**387.50**
300 mm dia.	219.50	277.80	3.36	132.13	nr	**409.94**
350 mm dia.	243.49	308.17	3.36	132.13	nr	**440.30**
Expansion bellows						
15 mm dia.	33.57	42.49	1.05	41.29	nr	**83.78**
20 mm dia.	35.62	45.08	1.05	41.29	nr	**86.38**
25 mm dia.	37.71	47.73	1.05	41.29	nr	**89.02**
32 mm dia.	41.35	52.33	1.05	41.29	nr	**93.62**
40 mm dia.	43.25	54.74	1.05	41.29	nr	**96.03**
50 mm dia.	48.41	61.27	1.05	41.29	nr	**102.56**
65 mm dia.	54.72	69.26	1.05	41.29	nr	**110.55**

38 MECHANICAL/COOLING/HEATING SYSTEMS

Item	Net Price £	Material £	Labour hours	Labour £	Unit	Total rate £
80 mm dia.	59.80	75.68	1.26	49.55	nr	**125.23**
100 mm dia.	74.65	94.48	1.26	49.55	nr	**144.03**
125 mm dia.	85.52	108.23	1.26	49.55	nr	**157.79**
150 mm dia.	99.98	126.54	1.26	49.55	nr	**176.09**
200 mm dia.	136.14	172.30	1.53	60.17	nr	**232.47**
250 mm dia.	161.42	204.29	1.53	60.17	nr	**264.46**
300 mm dia.	175.60	222.24	1.53	60.17	nr	**282.40**
350 mm dia.	194.79	246.53	1.53	60.17	nr	**306.69**
40 mm thick						
15 mm dia.	15.33	19.40	0.44	17.30	m	**36.71**
20 mm dia.	16.10	20.37	0.44	17.30	m	**37.67**
25 mm dia.	17.14	21.69	0.44	17.30	m	**39.00**
32 mm dia.	17.97	22.75	0.44	17.30	m	**40.05**
40 mm dia.	19.10	24.17	0.44	17.30	m	**41.47**
50 mm dia.	21.15	26.77	0.44	17.30	m	**44.07**
65 mm dia.	23.55	29.80	0.44	17.30	m	**47.11**
80 mm dia.	25.34	32.08	0.52	20.45	m	**52.52**
100 mm dia.	35.12	44.45	0.52	20.45	m	**64.90**
125 mm dia.	37.12	46.98	0.52	20.45	m	**67.43**
150 mm dia.	40.56	51.33	0.52	20.45	m	**71.78**
200 mm dia.	53.65	67.90	0.60	23.60	m	**91.49**
250 mm dia.	62.83	79.52	0.60	23.60	m	**103.11**
300 mm dia.	69.45	87.90	0.60	23.60	m	**111.49**
350 mm dia.	77.63	98.24	0.60	23.60	m	**121.84**
400 mm dia.	87.32	110.51	0.60	23.60	m	**134.10**
Extra over for fittings plantroom insulation						
Flange/union						
15 mm dia.	17.70	22.41	0.58	22.81	nr	**45.22**
20 mm dia.	18.48	23.39	0.58	22.81	nr	**46.20**
25 mm dia.	19.74	24.98	0.58	22.81	nr	**47.79**
32 mm dia.	20.81	26.34	0.58	22.81	nr	**49.15**
40 mm dia.	22.07	27.94	0.58	22.81	nr	**50.75**
50 mm dia.	24.58	31.11	0.58	22.81	nr	**53.92**
65 mm dia.	27.47	34.77	0.58	22.81	nr	**57.58**
80 mm dia.	29.63	37.50	0.67	26.35	nr	**63.85**
100 mm dia.	35.12	44.45	0.67	26.35	nr	**70.80**
125 mm dia.	41.57	52.62	0.67	26.35	nr	**78.96**
150 mm dia.	48.49	61.37	0.67	26.35	nr	**87.72**
200 mm dia.	65.07	82.35	0.87	34.21	nr	**116.56**
250 mm dia.	76.27	96.53	0.87	34.21	nr	**130.74**
300 mm dia.	83.44	105.61	0.87	34.21	nr	**139.82**
350 mm dia.	93.04	117.75	0.87	34.21	nr	**151.96**
400 mm dia.	104.31	132.02	0.87	34.21	nr	**166.23**
Bends						
15 mm dia.	8.43	10.67	0.44	17.30	nr	**27.97**
20 mm dia.	8.83	11.18	0.44	17.30	nr	**28.48**
25 mm dia.	9.44	11.94	0.44	17.30	nr	**29.24**
32 mm dia.	9.86	12.48	0.44	17.30	nr	**29.79**
40 mm dia.	10.51	13.30	0.44	17.30	nr	**30.60**
50 mm dia.	11.62	14.71	0.44	17.30	nr	**32.01**
65 mm dia.	12.99	16.44	0.44	17.30	nr	**33.74**

38 MECHANICAL/COOLING/HEATING SYSTEMS

Item	Net Price £	Material £	Labour hours	Labour £	Unit	Total rate £
THERMAL INSULATION – cont						
Extra over for fittings plantroom insulation – cont						
Bends – cont						
80 mm dia.	13.93	17.62	0.52	20.45	nr	**38.07**
100 mm dia.	17.20	21.76	0.52	20.45	nr	**42.21**
125 mm dia.	19.13	24.20	0.52	20.45	nr	**44.65**
150 mm dia.	22.30	28.22	0.52	20.45	nr	**48.67**
200 mm dia.	29.49	37.32	0.60	23.60	nr	**60.91**
250 mm dia.	34.32	43.44	0.60	23.60	nr	**67.04**
300 mm dia.	38.21	48.36	0.60	23.60	nr	**71.95**
350 mm dia.	42.73	54.07	0.60	23.60	nr	**77.67**
400 mm dia.	48.02	60.77	0.60	23.60	nr	**84.37**
Tees						
15 mm dia.	5.05	6.39	0.44	17.30	nr	**23.70**
20 mm dia.	5.28	6.68	0.44	17.30	nr	**23.99**
25 mm dia.	5.66	7.16	0.44	17.30	nr	**24.46**
32 mm dia.	5.92	7.50	0.44	17.30	nr	**24.80**
40 mm dia.	6.31	7.99	0.44	17.30	nr	**25.29**
50 mm dia.	6.96	8.80	0.44	17.30	nr	**26.11**
65 mm dia.	7.77	9.84	0.44	17.30	nr	**27.14**
80 mm dia.	8.36	10.58	0.52	20.45	nr	**31.03**
100 mm dia.	10.32	13.06	0.52	20.45	nr	**33.51**
125 mm dia.	11.50	14.55	0.52	20.45	nr	**35.00**
150 mm dia.	13.40	16.96	0.52	20.45	nr	**37.41**
200 mm dia.	17.70	22.41	0.60	23.60	nr	**46.00**
250 mm dia.	20.77	26.29	0.60	23.60	nr	**49.89**
300 mm dia.	22.90	28.99	0.60	23.60	nr	**52.58**
350 mm dia.	25.64	32.45	0.60	23.60	nr	**56.04**
400 mm dia.	28.80	36.45	0.60	23.60	nr	**60.05**
Valves						
15 mm dia.	28.13	35.60	0.78	30.67	nr	**66.28**
20 mm dia.	29.35	37.15	0.78	30.67	nr	**67.82**
25 mm dia.	31.39	39.73	0.78	30.67	nr	**70.40**
32 mm dia.	33.05	41.83	0.78	30.67	nr	**72.50**
40 mm dia.	35.02	44.32	0.78	30.67	nr	**75.00**
50 mm dia.	39.04	49.41	0.78	30.67	nr	**80.08**
65 mm dia.	43.60	55.18	0.78	30.67	nr	**85.85**
80 mm dia.	47.08	59.59	0.92	36.18	nr	**95.77**
100 mm dia.	59.00	74.67	0.92	36.18	nr	**110.85**
125 mm dia.	65.98	83.50	0.92	36.18	nr	**119.68**
150 mm dia.	77.04	97.50	0.92	36.18	nr	**133.68**
200 mm dia.	103.33	130.78	1.12	44.04	nr	**174.82**
250 mm dia.	121.13	153.30	1.12	44.04	nr	**197.35**
300 mm dia.	132.56	167.77	1.12	44.04	nr	**211.82**
350 mm dia.	147.76	187.01	1.12	44.04	nr	**231.05**
400 mm dia.	165.69	209.70	1.12	44.04	nr	**253.75**
Pumps						
15 mm dia.	52.05	65.88	2.34	92.02	nr	**157.90**
20 mm dia.	54.32	68.75	2.34	92.02	nr	**160.77**
25 mm dia.	58.11	73.55	2.34	92.02	nr	**165.57**

38 MECHANICAL/COOLING/HEATING SYSTEMS

Item	Net Price £	Material £	Labour hours	Labour £	Unit	Total rate £
32 mm dia.	61.13	77.36	2.34	92.02	nr	**169.39**
40 mm dia.	64.88	82.11	2.34	92.02	nr	**174.14**
50 mm dia.	72.32	91.53	2.34	92.02	nr	**183.55**
65 mm dia.	80.78	102.23	2.34	92.02	nr	**194.25**
80 mm dia.	87.20	110.35	2.76	108.54	nr	**218.89**
100 mm dia.	109.20	138.21	2.76	108.54	nr	**246.75**
125 mm dia.	122.18	154.63	2.76	108.54	nr	**263.16**
150 mm dia.	142.64	180.53	2.76	108.54	nr	**289.07**
200 mm dia.	191.36	242.18	3.36	132.13	nr	**374.32**
250 mm dia.	224.33	283.91	3.36	132.13	nr	**416.04**
300 mm dia.	417.61	528.53	3.36	132.13	nr	**660.66**
350 mm dia.	514.25	650.84	3.36	132.13	nr	**782.97**
400 mm dia.	610.89	773.15	3.36	132.13	nr	**905.28**
Expansion bellows						
15 mm dia.	41.63	52.68	1.05	41.29	nr	**93.98**
20 mm dia.	43.50	55.06	1.05	41.29	nr	**96.35**
25 mm dia.	46.48	58.82	1.05	41.29	nr	**100.12**
32 mm dia.	48.93	61.93	1.05	41.29	nr	**103.22**
40 mm dia.	51.91	65.69	1.05	41.29	nr	**106.99**
50 mm dia.	57.86	73.22	1.05	41.29	nr	**114.52**
65 mm dia.	64.64	81.81	1.05	41.29	nr	**123.10**
80 mm dia.	69.75	88.27	1.26	49.55	nr	**137.82**
100 mm dia.	87.37	110.58	1.26	49.55	nr	**160.13**
125 mm dia.	97.72	123.67	1.26	49.55	nr	**173.22**
150 mm dia.	114.11	144.41	1.26	49.55	nr	**193.96**
200 mm dia.	153.05	193.71	1.53	60.17	nr	**253.87**
250 mm dia.	179.50	227.17	1.53	60.17	nr	**287.34**
300 mm dia.	196.36	248.51	1.53	60.17	nr	**308.68**
350 mm dia.	218.87	277.01	1.53	60.17	nr	**337.17**
400 mm dia.	245.45	310.64	1.53	60.17	nr	**370.81**
50 mm thick						
15 mm dia.	20.00	25.31	0.44	17.30	m	**42.61**
20 mm dia.	21.07	26.66	0.44	17.30	m	**43.97**
25 mm dia.	22.33	28.26	0.44	17.30	m	**45.56**
32 mm dia.	23.48	29.72	0.44	17.30	m	**47.02**
40 mm dia.	24.55	31.07	0.44	17.30	m	**48.38**
50 mm dia.	27.10	34.30	0.44	17.30	m	**51.60**
65 mm dia.	29.11	36.84	0.44	17.30	m	**54.14**
80 mm dia.	31.27	39.57	0.52	20.45	m	**60.02**
100 mm dia.	38.63	48.88	0.52	20.45	m	**69.33**
125 mm dia.	42.77	54.13	0.52	20.45	m	**74.58**
150 mm dia.	49.12	62.17	0.52	20.45	m	**82.62**
200 mm dia.	64.38	81.49	0.60	23.60	m	**105.08**
250 mm dia.	74.42	94.19	0.60	23.60	m	**117.79**
300 mm dia.	81.02	102.54	0.60	23.60	m	**126.13**
350 mm dia.	90.29	114.27	0.60	23.60	m	**137.87**
400 mm dia.	100.97	127.79	0.60	23.60	m	**151.39**

38 MECHANICAL/COOLING/HEATING SYSTEMS

Item	Net Price £	Material £	Labour hours	Labour £	Unit	Total rate £
THERMAL INSULATION – cont						
Extra over for fittings plantroom insulation						
Flange/union						
15 mm dia.	23.55	29.80	0.58	22.81	nr	**52.61**
20 mm dia.	24.86	31.46	0.58	22.81	nr	**54.27**
25 mm dia.	26.34	33.33	0.58	22.81	nr	**56.14**
32 mm dia.	27.64	34.98	0.58	22.81	nr	**57.78**
40 mm dia.	28.96	36.65	0.58	22.81	nr	**59.46**
50 mm dia.	32.15	40.69	0.58	22.81	nr	**63.50**
65 mm dia.	34.74	43.97	0.58	22.81	nr	**66.77**
80 mm dia.	37.22	47.10	0.67	26.35	nr	**73.45**
100 mm dia.	46.56	58.93	0.67	26.35	nr	**85.27**
125 mm dia.	51.77	65.52	0.67	26.35	nr	**91.87**
150 mm dia.	59.59	75.41	0.67	26.35	nr	**101.76**
200 mm dia.	79.18	100.21	0.87	34.21	nr	**134.42**
250 mm dia.	91.54	115.85	0.87	34.21	nr	**150.06**
300 mm dia.	98.74	124.96	0.87	34.21	nr	**159.17**
350 mm dia.	109.81	138.97	0.87	34.21	nr	**173.18**
400 mm dia.	122.48	155.02	0.87	34.21	nr	**189.23**
Bends						
15 mm dia.	11.03	13.96	0.44	17.30	nr	**31.26**
20 mm dia.	11.59	14.67	0.44	17.30	nr	**31.98**
25 mm dia.	12.29	15.55	0.44	17.30	nr	**32.86**
32 mm dia.	12.95	16.39	0.44	17.30	nr	**33.69**
40 mm dia.	13.50	17.08	0.44	17.30	nr	**34.38**
50 mm dia.	14.92	18.88	0.44	17.30	nr	**36.18**
65 mm dia.	16.02	20.27	0.44	17.30	nr	**37.57**
80 mm dia.	17.20	21.76	0.52	20.45	nr	**42.21**
100 mm dia.	21.24	26.88	0.52	20.45	nr	**47.33**
125 mm dia.	23.51	29.75	0.52	20.45	nr	**50.20**
150 mm dia.	27.03	34.21	0.52	20.45	nr	**54.66**
200 mm dia.	35.41	44.81	0.60	23.60	nr	**68.41**
250 mm dia.	40.93	51.80	0.60	23.60	nr	**75.40**
300 mm dia.	44.55	56.38	0.60	23.60	nr	**79.98**
350 mm dia.	49.66	62.84	0.60	23.60	nr	**86.44**
400 mm dia.	55.54	70.29	0.60	23.60	nr	**93.89**
Tees						
15 mm dia.	6.61	8.36	0.44	17.30	nr	**25.67**
20 mm dia.	6.94	8.79	0.44	17.30	nr	**26.09**
25 mm dia.	7.36	9.31	0.44	17.30	nr	**26.62**
32 mm dia.	7.76	9.82	0.44	17.30	nr	**27.12**
40 mm dia.	8.11	10.26	0.44	17.30	nr	**27.57**
50 mm dia.	8.94	11.31	0.44	17.30	nr	**28.62**
65 mm dia.	9.60	12.14	0.44	17.30	nr	**29.45**
80 mm dia.	10.32	13.06	0.52	20.45	nr	**33.51**
100 mm dia.	12.75	16.13	0.52	20.45	nr	**36.58**
125 mm dia.	14.11	17.86	0.52	20.45	nr	**38.31**
150 mm dia.	16.19	20.49	0.52	20.45	nr	**40.94**

38 MECHANICAL/COOLING/HEATING SYSTEMS

Item	Net Price £	Material £	Labour hours	Labour £	Unit	Total rate £
200 mm dia.	21.24	26.88	0.60	23.60	nr	**50.48**
250 mm dia.	24.57	31.09	0.60	23.60	nr	**54.69**
300 mm dia.	26.75	33.86	0.60	23.60	nr	**57.45**
350 mm dia.	29.81	37.72	0.60	23.60	nr	**61.32**
400 mm dia.	33.33	42.18	0.60	23.60	nr	**65.78**
Valves						
15 mm dia.	37.45	47.39	0.78	30.67	nr	**78.07**
20 mm dia.	39.46	49.94	0.78	30.67	nr	**80.61**
25 mm dia.	41.83	52.94	0.78	30.67	nr	**83.61**
32 mm dia.	43.91	55.57	0.78	30.67	nr	**86.24**
40 mm dia.	45.98	58.20	0.78	30.67	nr	**88.87**
50 mm dia.	51.04	64.59	0.78	30.67	nr	**95.27**
65 mm dia.	55.14	69.78	0.78	30.67	nr	**100.46**
80 mm dia.	59.14	74.85	0.92	36.18	nr	**111.03**
100 mm dia.	73.95	93.60	0.92	36.18	nr	**129.78**
125 mm dia.	82.22	104.06	0.92	36.18	nr	**140.24**
150 mm dia.	94.63	119.77	0.92	36.18	nr	**155.95**
200 mm dia.	125.82	159.24	1.12	44.04	nr	**203.28**
250 mm dia.	145.38	183.99	1.12	44.04	nr	**228.03**
300 mm dia.	157.48	199.30	1.12	44.04	nr	**243.35**
350 mm dia.	174.39	220.71	1.12	44.04	nr	**264.75**
400 mm dia.	194.56	246.24	1.12	44.04	nr	**290.28**
Pumps						
15 mm dia.	69.34	87.76	2.34	92.02	nr	**179.78**
20 mm dia.	73.10	92.51	2.34	92.02	nr	**184.53**
25 mm dia.	77.43	97.99	2.34	92.02	nr	**190.01**
32 mm dia.	81.31	102.91	2.34	92.02	nr	**194.93**
40 mm dia.	85.20	107.83	2.34	92.02	nr	**199.85**
50 mm dia.	94.47	119.57	2.34	92.02	nr	**211.59**
65 mm dia.	135.10	170.98	2.34	92.02	nr	**263.00**
80 mm dia.	136.14	172.29	2.76	108.54	nr	**280.83**
100 mm dia.	136.92	173.28	2.76	108.54	nr	**281.82**
125 mm dia.	152.28	192.72	2.76	108.54	nr	**301.26**
150 mm dia.	175.22	221.76	2.76	108.54	nr	**330.30**
200 mm dia.	232.97	294.85	3.36	132.13	nr	**426.98**
250 mm dia.	269.20	340.70	3.36	132.13	nr	**472.83**
300 mm dia.	290.48	367.63	3.36	132.13	nr	**499.77**
350 mm dia.	322.90	408.67	3.36	132.13	nr	**540.80**
400 mm dia.	360.31	456.01	3.36	132.13	nr	**588.14**
Expansion bellows						
15 mm dia.	55.45	70.17	1.05	41.29	nr	**111.46**
20 mm dia.	58.45	73.97	1.05	41.29	nr	**115.26**
25 mm dia.	61.93	78.38	1.05	41.29	nr	**119.67**
32 mm dia.	65.03	82.30	1.05	41.29	nr	**123.59**
40 mm dia.	68.14	86.23	1.05	41.29	nr	**127.53**
50 mm dia.	75.59	95.67	1.05	41.29	nr	**136.96**
65 mm dia.	81.67	103.37	1.05	41.29	nr	**144.66**

38 MECHANICAL/COOLING/HEATING SYSTEMS

Item	Net Price £	Material £	Labour hours	Labour £	Unit	Total rate £
THERMAL INSULATION – cont						
Extra over for fittings plantroom insulation – cont						
Expansion bellows – cont						
80 mm dia.	87.66	110.95	1.26	49.55	nr	**160.50**
100 mm dia.	109.55	138.65	1.26	49.55	nr	**188.20**
125 mm dia.	121.81	154.17	1.26	49.55	nr	**203.72**
150 mm dia.	140.20	177.44	1.26	49.55	nr	**226.99**
200 mm dia.	186.39	235.89	1.53	60.17	nr	**296.06**
250 mm dia.	215.36	272.56	1.53	60.17	nr	**332.73**
300 mm dia.	232.44	294.17	1.53	60.17	nr	**354.34**
350 mm dia.	258.33	326.94	1.53	60.17	nr	**387.11**
400 mm dia.	288.22	364.77	1.53	60.17	nr	**424.94**
Mineral fibre sectional insulation; bright class O foil faced; bright class O foil taped joints; 0.8 mm polyisobutylene sheeting; welded joints						
External pipework						
20 mm thick						
15 mm dia.	8.98	11.36	0.30	11.80	m	**23.16**
20 mm dia.	9.56	12.09	0.30	11.80	m	**23.89**
25 mm dia.	10.32	13.06	0.30	11.80	m	**24.86**
32 mm dia.	11.32	14.33	0.30	11.80	m	**26.13**
40 mm dia.	12.09	15.30	0.30	11.80	m	**27.10**
50 mm dia.	13.66	17.28	0.30	11.80	m	**29.08**
Extra over for fittings external insulation						
Flange/union						
15 mm dia.	13.75	17.40	0.75	29.49	nr	**46.90**
20 mm dia.	14.61	18.49	0.75	29.49	nr	**47.98**
25 mm dia.	15.68	19.85	0.75	29.49	nr	**49.34**
32 mm dia.	17.10	21.64	0.75	29.49	nr	**51.14**
40 mm dia.	18.05	22.85	0.75	29.49	nr	**52.34**
50 mm dia.	20.22	25.60	0.75	29.49	nr	**55.09**
Bends						
15 mm dia.	2.28	2.88	0.30	11.80	nr	**14.68**
20 mm dia.	2.40	3.04	0.30	11.80	nr	**14.83**
25 mm dia.	2.59	3.27	0.30	11.80	nr	**15.07**
32 mm dia.	2.84	3.60	0.30	11.80	nr	**15.39**
40 mm dia.	3.03	3.83	0.30	11.80	nr	**15.63**
50 mm dia.	3.40	4.31	0.30	11.80	nr	**16.11**
Tees						
15 mm dia.	2.28	2.88	0.30	11.80	nr	**14.68**
20 mm dia.	2.40	3.04	0.30	11.80	nr	**14.83**
25 mm dia.	2.59	3.27	0.30	11.80	nr	**15.07**
32 mm dia.	2.84	3.60	0.30	11.80	nr	**15.39**
40 mm dia.	3.03	3.83	0.30	11.80	nr	**15.63**
50 mm dia.	3.40	4.31	0.30	11.80	nr	**16.11**

38 MECHANICAL/COOLING/HEATING SYSTEMS

Item	Net Price £	Material £	Labour hours	Labour £	Unit	Total rate £
Valves						
15 mm dia.	21.86	27.66	1.03	40.51	nr	**68.17**
20 mm dia.	23.21	29.38	1.03	40.51	nr	**69.88**
25 mm dia.	24.92	31.53	1.03	40.51	nr	**72.04**
32 mm dia.	27.13	34.33	1.03	40.51	nr	**74.84**
40 mm dia.	28.64	36.25	1.03	40.51	nr	**76.75**
50 mm dia.	32.10	40.62	1.03	40.51	nr	**81.13**
Expansion bellows						
15 mm dia.	32.35	40.95	1.42	55.84	nr	**96.79**
20 mm dia.	34.40	43.54	1.42	55.84	nr	**99.38**
25 mm dia.	36.96	46.78	1.42	55.84	nr	**102.62**
32 mm dia.	40.21	50.89	1.42	55.84	nr	**106.73**
40 mm dia.	42.43	53.70	1.42	55.84	nr	**109.54**
50 mm dia.	47.57	60.20	1.42	55.84	nr	**116.04**
25 mm thick						
15 mm dia.	9.90	12.53	0.30	11.80	m	**24.33**
20 mm dia.	10.63	13.45	0.30	11.80	m	**25.25**
25 mm dia.	11.69	14.79	0.30	11.80	m	**26.59**
32 mm dia.	12.65	16.01	0.30	11.80	m	**27.81**
40 mm dia.	13.47	17.05	0.30	11.80	m	**28.84**
50 mm dia.	15.21	19.25	0.30	11.80	m	**31.05**
65 mm dia.	17.25	21.83	0.30	11.80	m	**33.63**
80 mm dia.	18.88	23.90	0.40	15.73	m	**39.63**
100 mm dia.	23.83	30.16	0.40	15.73	m	**45.89**
125 mm dia.	27.45	34.74	0.40	15.73	m	**50.47**
150 mm dia.	32.19	40.74	0.40	15.73	m	**56.47**
200 mm dia.	43.64	55.23	0.50	19.66	m	**74.89**
250 mm dia.	52.12	65.97	0.50	19.66	m	**85.63**
300 mm dia.	56.68	71.73	0.50	19.66	m	**91.40**
Extra over for fittings external insulation						
Flange/union						
15 mm dia.	15.13	19.15	0.75	29.49	nr	**48.64**
20 mm dia.	16.32	20.66	0.75	29.49	nr	**50.15**
25 mm dia.	17.92	22.68	0.75	29.49	nr	**52.17**
32 mm dia.	19.15	24.24	0.75	29.49	nr	**53.73**
40 mm dia.	20.51	25.95	0.75	29.49	nr	**55.45**
50 mm dia.	22.99	29.09	0.75	29.49	nr	**58.58**
65 mm dia.	25.99	32.89	0.75	29.49	nr	**62.38**
80 mm dia.	28.27	35.77	0.89	35.00	nr	**70.77**
100 mm dia.	35.13	44.46	0.89	35.00	nr	**79.46**
125 mm dia.	40.68	51.48	0.89	35.00	nr	**86.48**
150 mm dia.	47.35	59.93	0.89	35.00	nr	**94.93**
200 mm dia.	64.06	81.08	1.15	45.22	nr	**126.30**
250 mm dia.	76.27	96.53	1.15	45.22	nr	**141.76**
300 mm dia.	84.73	107.23	1.15	45.22	nr	**152.46**
Bends						
15 mm dia.	2.47	3.12	0.30	11.80	nr	**14.92**
20 mm dia.	2.67	3.38	0.30	11.80	nr	**15.17**
25 mm dia.	2.89	3.66	0.30	11.80	nr	**15.46**

38 MECHANICAL/COOLING/HEATING SYSTEMS

Item	Net Price £	Material £	Labour hours	Labour £	Unit	Total rate £
THERMAL INSULATION – cont						
Extra over for fittings external insulation – cont						
Bends – cont						
32 mm dia.	3.14	3.97	0.30	11.80	nr	**15.77**
40 mm dia.	3.36	4.26	0.30	11.80	nr	**16.06**
50 mm dia.	3.81	4.82	0.30	11.80	nr	**16.62**
65 mm dia.	4.33	5.48	0.30	11.80	nr	**17.28**
80 mm dia.	4.72	5.97	0.40	15.73	nr	**21.70**
100 mm dia.	5.95	7.53	0.40	15.73	nr	**23.26**
125 mm dia.	6.85	8.67	0.40	15.73	nr	**24.40**
150 mm dia.	8.07	10.21	0.40	15.73	nr	**25.94**
200 mm dia.	10.90	13.79	0.50	19.66	nr	**33.45**
250 mm dia.	13.04	16.50	0.50	19.66	nr	**36.17**
300 mm dia.	14.17	17.93	0.50	19.66	nr	**37.59**
Tees						
15 mm dia.	2.47	3.12	0.30	11.80	nr	**14.92**
20 mm dia.	2.67	3.38	0.30	11.80	nr	**15.17**
25 mm dia.	2.89	3.66	0.30	11.80	nr	**15.46**
32 mm dia.	3.14	3.97	0.30	11.80	nr	**15.77**
40 mm dia.	3.36	4.26	0.30	11.80	nr	**16.06**
50 mm dia.	3.81	4.82	0.30	11.80	nr	**16.62**
65 mm dia.	4.33	5.48	0.30	11.80	nr	**17.28**
80 mm dia.	4.72	5.97	0.40	15.73	nr	**21.70**
100 mm dia.	5.95	7.53	0.40	15.73	nr	**23.26**
125 mm dia.	6.85	8.67	0.40	15.73	nr	**24.40**
150 mm dia.	8.07	10.21	0.40	15.73	nr	**25.94**
200 mm dia.	10.90	13.79	0.50	19.66	nr	**33.45**
250 mm dia.	13.04	16.50	0.50	19.66	nr	**36.17**
300 mm dia.	14.17	17.93	0.50	19.66	nr	**37.59**
Valves						
15 mm dia.	24.06	30.45	1.03	40.51	nr	**70.95**
20 mm dia.	25.92	32.80	1.03	40.51	nr	**73.31**
25 mm dia.	28.43	35.98	1.03	40.51	nr	**76.48**
32 mm dia.	30.46	38.55	1.03	40.51	nr	**79.06**
40 mm dia.	32.54	41.18	1.03	40.51	nr	**81.69**
50 mm dia.	36.55	46.26	1.03	40.51	nr	**86.76**
65 mm dia.	41.33	52.31	1.03	40.51	nr	**92.82**
80 mm dia.	44.91	56.84	1.25	49.16	nr	**106.00**
100 mm dia.	55.83	70.66	1.25	49.16	nr	**119.82**
125 mm dia.	64.61	81.77	1.25	49.16	nr	**130.93**
150 mm dia.	75.23	95.21	1.25	49.16	nr	**144.36**
200 mm dia.	101.79	128.83	1.55	60.95	nr	**189.78**
250 mm dia.	121.13	153.30	1.55	60.95	nr	**214.26**
300 mm dia.	134.56	170.30	1.55	60.95	nr	**231.25**
Expansion bellows						
15 mm dia.	35.65	45.12	1.42	55.84	nr	**100.96**
20 mm dia.	38.41	48.61	1.42	55.84	nr	**104.46**
25 mm dia.	42.12	53.31	1.42	55.84	nr	**109.15**
32 mm dia.	45.18	57.18	1.42	55.84	nr	**113.02**
40 mm dia.	48.26	61.08	1.42	55.84	nr	**116.92**
50 mm dia.	54.09	68.46	1.42	55.84	nr	**124.30**

38 MECHANICAL/COOLING/HEATING SYSTEMS

Item	Net Price £	Material £	Labour hours	Labour £	Unit	Total rate £
65 mm dia.	61.20	77.45	1.42	55.84	nr	**133.29**
80 mm dia.	66.53	84.20	1.75	68.82	nr	**153.02**
100 mm dia.	82.71	104.67	1.75	68.82	nr	**173.49**
125 mm dia.	95.72	121.14	1.75	68.82	nr	**189.96**
150 mm dia.	111.43	141.02	1.75	68.82	nr	**209.84**
200 mm dia.	150.78	190.82	2.17	85.34	nr	**276.16**
250 mm dia.	179.44	227.10	2.17	85.34	nr	**312.44**
300 mm dia.	199.35	252.29	3.17	124.66	nr	**376.96**
30 mm thick						
15 mm dia.	12.23	15.48	0.30	11.80	m	**27.28**
20 mm dia.	12.99	16.44	0.30	11.80	m	**28.23**
25 mm dia.	13.75	17.40	0.30	11.80	m	**29.20**
32 mm dia.	14.93	18.90	0.30	11.80	m	**30.69**
40 mm dia.	15.72	19.90	0.30	11.80	m	**31.69**
50 mm dia.	17.70	22.41	0.30	11.80	m	**34.20**
65 mm dia.	19.90	25.19	0.30	11.80	m	**36.99**
80 mm dia.	21.63	27.38	0.40	15.73	m	**43.11**
100 mm dia.	27.06	34.25	0.40	15.73	m	**49.98**
125 mm dia.	30.99	39.22	0.40	15.73	m	**54.95**
150 mm dia.	35.99	45.54	0.40	15.73	m	**61.27**
200 mm dia.	48.46	61.33	0.50	19.66	m	**81.00**
250 mm dia.	57.51	72.78	0.50	19.66	m	**92.45**
300 mm dia.	62.03	78.50	0.50	19.66	m	**98.16**
350 mm dia.	67.82	85.83	0.50	19.66	m	**105.49**
Extra over for fittings external insulation						
Flange/union						
15 mm dia.	18.52	23.44	0.75	29.49	nr	**52.94**
20 mm dia.	19.61	24.82	0.75	29.49	nr	**54.31**
25 mm dia.	20.81	26.34	0.75	29.49	nr	**55.84**
32 mm dia.	22.70	28.73	0.75	29.49	nr	**58.23**
40 mm dia.	23.71	30.01	0.75	29.49	nr	**59.50**
50 mm dia.	26.46	33.48	0.75	29.49	nr	**62.98**
65 mm dia.	29.79	37.71	0.75	29.49	nr	**67.20**
80 mm dia.	32.46	41.08	0.89	35.00	nr	**76.08**
100 mm dia.	39.97	50.58	0.89	35.00	nr	**85.58**
125 mm dia.	45.68	57.81	0.89	35.00	nr	**92.81**
150 mm dia.	53.05	67.14	0.89	35.00	nr	**102.14**
200 mm dia.	70.86	89.68	1.15	45.22	nr	**134.90**
250 mm dia.	83.91	106.20	1.15	45.22	nr	**151.42**
300 mm dia.	91.97	116.39	1.15	45.22	nr	**161.62**
350 mm dia.	101.58	128.55	1.15	45.22	nr	**173.78**
Bends						
15 mm dia.	3.04	3.85	0.30	11.80	nr	**15.65**
20 mm dia.	3.24	4.10	0.30	11.80	nr	**15.90**
25 mm dia.	3.46	4.38	0.30	11.80	nr	**16.17**
32 mm dia.	3.73	4.72	0.30	11.80	nr	**16.51**
40 mm dia.	3.93	4.97	0.30	11.80	nr	**16.77**
50 mm dia.	4.44	5.61	0.30	11.80	nr	**17.41**
65 mm dia.	4.99	6.31	0.30	11.80	nr	**18.11**

38 MECHANICAL/COOLING/HEATING SYSTEMS

Item	Net Price £	Material £	Labour hours	Labour £	Unit	Total rate £
THERMAL INSULATION – cont						
Extra over for fittings external insulation – cont						
Bends – cont						
80 mm dia.	5.43	6.87	0.40	15.73	nr	**22.60**
100 mm dia.	6.79	8.60	0.40	15.73	nr	**24.33**
125 mm dia.	7.76	9.82	0.40	15.73	nr	**25.55**
150 mm dia.	8.99	11.38	0.40	15.73	nr	**27.11**
200 mm dia.	12.13	15.35	0.50	19.66	nr	**35.01**
250 mm dia.	14.41	18.23	0.50	19.66	nr	**37.90**
300 mm dia.	15.49	19.61	0.50	19.66	nr	**39.27**
350 mm dia.	16.95	21.46	0.50	19.66	nr	**41.12**
Tees						
15 mm dia.	3.04	3.85	0.30	11.80	nr	**15.65**
20 mm dia.	3.24	4.10	0.30	11.80	nr	**15.90**
25 mm dia.	3.46	4.38	0.30	11.80	nr	**16.17**
32 mm dia.	3.73	4.72	0.30	11.80	nr	**16.51**
40 mm dia.	3.93	4.97	0.30	11.80	nr	**16.77**
50 mm dia.	4.44	5.61	0.30	11.80	nr	**17.41**
65 mm dia.	4.99	6.31	0.30	11.80	nr	**18.11**
80 mm dia.	5.43	6.87	0.40	15.73	nr	**22.60**
100 mm dia.	6.79	8.60	0.40	15.73	nr	**24.33**
125 mm dia.	7.76	9.82	0.40	15.73	nr	**25.55**
150 mm dia.	8.99	11.38	0.40	15.73	nr	**27.11**
200 mm dia.	12.13	15.35	0.50	19.66	nr	**35.01**
250 mm dia.	14.41	18.23	0.50	19.66	nr	**37.90**
300 mm dia.	15.49	19.61	0.50	19.66	nr	**39.27**
350 mm dia.	16.95	21.46	0.50	19.66	nr	**41.12**
Valves						
15 mm dia.	29.42	37.23	1.03	40.51	nr	**77.74**
20 mm dia.	31.17	39.45	1.03	40.51	nr	**79.96**
25 mm dia.	33.06	41.85	1.03	40.51	nr	**82.35**
32 mm dia.	36.00	45.56	1.03	40.51	nr	**86.07**
40 mm dia.	37.67	47.68	1.03	40.51	nr	**88.19**
50 mm dia.	42.02	53.18	1.03	40.51	nr	**93.68**
65 mm dia.	47.31	59.88	1.03	40.51	nr	**100.38**
80 mm dia.	51.55	65.24	1.25	49.16	nr	**114.39**
100 mm dia.	63.42	80.26	1.25	49.16	nr	**129.42**
125 mm dia.	72.49	91.75	1.25	49.16	nr	**140.90**
150 mm dia.	84.22	106.59	1.25	49.16	nr	**155.75**
200 mm dia.	112.58	142.48	1.55	60.95	nr	**203.44**
250 mm dia.	133.30	168.70	1.55	60.95	nr	**229.66**
300 mm dia.	146.05	184.83	1.55	60.95	nr	**245.79**
350 mm dia.	161.34	204.19	1.55	60.95	nr	**265.14**
Expansion bellows						
15 mm dia.	43.56	55.13	1.42	55.84	nr	**110.97**
20 mm dia.	46.20	58.47	1.42	55.84	nr	**114.31**
25 mm dia.	48.95	61.95	1.42	55.84	nr	**117.79**
32 mm dia.	53.35	67.53	1.42	55.84	nr	**123.37**
40 mm dia.	55.83	70.66	1.42	55.84	nr	**126.51**
50 mm dia.	62.24	78.77	1.42	55.84	nr	**134.61**
65 mm dia.	70.09	88.71	1.42	55.84	nr	**144.55**

38 MECHANICAL/COOLING/HEATING SYSTEMS

Item	Net Price £	Material £	Labour hours	Labour £	Unit	Total rate £
80 mm dia.	76.41	96.70	1.75	68.82	nr	**165.52**
100 mm dia.	93.98	118.94	1.75	68.82	nr	**187.76**
125 mm dia.	107.41	135.93	1.75	68.82	nr	**204.75**
150 mm dia.	124.76	157.90	1.75	68.82	nr	**226.72**
200 mm dia.	166.77	211.06	2.17	85.34	nr	**296.40**
250 mm dia.	197.48	249.93	2.17	85.34	nr	**335.27**
300 mm dia.	216.37	273.83	2.17	85.34	nr	**359.17**
350 mm dia.	239.04	302.53	2.17	85.34	nr	**387.87**
40 mm thick						
15 mm dia.	15.25	19.30	0.30	11.80	m	**31.10**
20 mm dia.	15.79	19.98	0.30	11.80	m	**31.78**
25 mm dia.	16.86	21.34	0.30	11.80	m	**33.14**
32 mm dia.	17.96	22.73	0.30	11.80	m	**34.53**
40 mm dia.	18.76	23.75	0.30	11.80	m	**35.54**
50 mm dia.	20.92	26.48	0.30	11.80	m	**38.28**
65 mm dia.	23.36	29.56	0.30	11.80	m	**41.36**
80 mm dia.	25.32	32.04	0.40	15.73	m	**47.77**
100 mm dia.	31.59	39.98	0.40	15.73	m	**55.71**
125 mm dia.	35.60	45.05	0.40	15.73	m	**60.78**
150 mm dia.	41.11	52.02	0.40	15.73	m	**67.75**
200 mm dia.	54.83	69.39	0.50	19.66	m	**89.05**
250 mm dia.	64.21	81.27	0.50	19.66	m	**100.93**
300 mm dia.	69.50	87.96	0.50	19.66	m	**107.63**
350 mm dia.	76.41	96.70	0.50	19.66	m	**116.36**
400 mm dia.	85.02	107.61	0.50	19.66	m	**127.27**
Extra over for fittings external insulation						
Flange/union						
15 mm dia.	22.80	28.85	0.75	29.49	nr	**58.35**
20 mm dia.	23.82	30.14	0.75	29.49	nr	**59.64**
25 mm dia.	25.38	32.13	0.75	29.49	nr	**61.62**
32 mm dia.	26.75	33.86	0.75	29.49	nr	**63.35**
40 mm dia.	28.24	35.74	0.75	29.49	nr	**65.23**
50 mm dia.	31.28	39.59	0.75	29.49	nr	**69.08**
65 mm dia.	34.86	44.12	0.75	29.49	nr	**73.61**
80 mm dia.	37.58	47.56	0.89	35.00	nr	**82.56**
100 mm dia.	46.28	58.57	0.89	35.00	nr	**93.57**
125 mm dia.	51.77	65.52	0.89	35.00	nr	**100.52**
150 mm dia.	59.99	75.92	0.89	35.00	nr	**110.92**
200 mm dia.	79.10	100.11	1.15	45.22	nr	**145.33**
250 mm dia.	92.73	117.36	1.15	45.22	nr	**162.58**
300 mm dia.	101.88	128.94	1.15	45.22	nr	**174.17**
350 mm dia.	113.01	143.02	1.15	45.22	nr	**188.25**
400 mm dia.	126.54	160.15	1.15	45.22	nr	**205.38**
Bends						
15 mm dia.	3.81	4.82	0.30	11.80	nr	**16.62**
20 mm dia.	3.95	5.00	0.30	11.80	nr	**16.80**
25 mm dia.	4.22	5.34	0.30	11.80	nr	**17.14**
32 mm dia.	4.48	5.67	0.30	11.80	nr	**17.46**
40 mm dia.	4.72	5.97	0.30	11.80	nr	**17.77**
50 mm dia.	5.27	6.67	0.30	11.80	nr	**18.46**
65 mm dia.	5.86	7.41	0.30	11.80	nr	**19.21**

38 MECHANICAL/COOLING/HEATING SYSTEMS

Item	Net Price £	Material £	Labour hours	Labour £	Unit	Total rate £
THERMAL INSULATION – cont						
Extra over for fittings external insulation – cont						
Bends – cont						
80 mm dia.	6.35	8.04	0.40	15.73	nr	**23.77**
100 mm dia.	7.89	9.99	0.40	15.73	nr	**25.72**
125 mm dia.	8.90	11.26	0.40	15.73	nr	**26.99**
150 mm dia.	10.29	13.03	0.40	15.73	nr	**28.76**
200 mm dia.	13.71	17.35	0.50	19.66	nr	**37.01**
250 mm dia.	16.04	20.30	0.50	19.66	nr	**39.97**
300 mm dia.	17.37	21.98	0.50	19.66	nr	**41.65**
350 mm dia.	19.09	24.15	0.50	19.66	nr	**43.82**
400 mm dia.	21.27	26.92	0.50	19.66	nr	**46.58**
Tees						
15 mm dia.	3.81	4.82	0.30	11.80	nr	**16.62**
20 mm dia.	3.95	5.00	0.30	11.80	nr	**16.80**
25 mm dia.	4.22	5.34	0.30	11.80	nr	**17.14**
32 mm dia.	4.48	5.67	0.30	11.80	nr	**17.46**
40 mm dia.	4.72	5.97	0.30	11.80	nr	**17.77**
50 mm dia.	5.27	6.67	0.30	11.80	nr	**18.46**
65 mm dia.	5.86	7.41	0.30	11.80	nr	**19.21**
80 mm dia.	6.35	8.04	0.40	15.73	nr	**23.77**
100 mm dia.	7.89	9.99	0.40	15.73	nr	**25.72**
125 mm dia.	8.90	11.26	0.40	15.73	nr	**26.99**
150 mm dia.	10.29	13.03	0.40	15.73	nr	**28.76**
200 mm dia.	13.71	17.35	0.50	19.66	nr	**37.01**
250 mm dia.	16.04	20.30	0.50	19.66	nr	**39.97**
300 mm dia.	17.37	21.98	0.50	19.66	nr	**41.65**
350 mm dia.	19.09	24.15	0.50	19.66	nr	**43.82**
400 mm dia.	21.27	26.92	0.50	19.66	nr	**46.58**
Valves						
15 mm dia.	36.19	45.80	1.03	40.51	nr	**86.30**
20 mm dia.	37.82	47.87	1.03	40.51	nr	**88.37**
25 mm dia.	40.30	51.00	1.03	40.51	nr	**91.51**
32 mm dia.	42.45	53.72	1.03	40.51	nr	**94.22**
40 mm dia.	44.88	56.81	1.03	40.51	nr	**97.31**
50 mm dia.	49.74	62.95	1.03	40.51	nr	**103.45**
65 mm dia.	55.38	70.09	1.03	40.51	nr	**110.59**
80 mm dia.	59.68	75.53	1.25	49.16	nr	**124.69**
100 mm dia.	73.51	93.04	1.25	49.16	nr	**142.19**
125 mm dia.	82.22	104.06	1.25	49.16	nr	**153.22**
150 mm dia.	95.32	120.63	1.25	49.16	nr	**169.79**
200 mm dia.	125.65	159.02	1.55	60.95	nr	**219.97**
250 mm dia.	147.21	186.31	1.55	60.95	nr	**247.27**
300 mm dia.	161.85	204.83	1.55	60.95	nr	**265.79**
350 mm dia.	179.50	227.17	1.55	60.95	nr	**288.13**
400 mm dia.	200.97	254.34	1.55	60.95	nr	**315.30**
Expansion bellows						
15 mm dia.	53.60	67.83	1.42	55.84	nr	**123.67**
20 mm dia.	56.04	70.92	1.42	55.84	nr	**126.76**
25 mm dia.	59.69	75.55	1.42	55.84	nr	**131.39**

38 MECHANICAL/COOLING/HEATING SYSTEMS

Item	Net Price £	Material £	Labour hours	Labour £	Unit	Total rate £
32 mm dia.	62.90	79.60	1.42	55.84	nr	**135.45**
40 mm dia.	66.49	84.15	1.42	55.84	nr	**139.99**
50 mm dia.	73.70	93.27	1.42	55.84	nr	**149.12**
65 mm dia.	82.02	103.81	1.42	55.84	nr	**159.65**
80 mm dia.	88.42	111.90	1.75	68.82	nr	**180.72**
100 mm dia.	108.92	137.85	1.75	68.82	nr	**206.67**
125 mm dia.	121.81	154.17	1.75	68.82	nr	**222.99**
150 mm dia.	141.22	178.73	1.75	68.82	nr	**247.55**
200 mm dia.	186.15	235.59	2.17	85.34	nr	**320.92**
250 mm dia.	218.08	276.01	2.17	85.34	nr	**361.34**
300 mm dia.	239.77	303.45	2.17	85.34	nr	**388.79**
350 mm dia.	265.93	336.56	2.17	85.34	nr	**421.90**
400 mm dia.	297.76	376.84	2.17	85.34	nr	**462.18**
50 mm thick						
15 mm dia.	20.10	25.44	0.30	11.80	m	**37.24**
20 mm dia.	21.10	26.70	0.30	11.80	m	**38.50**
25 mm dia.	22.30	28.22	0.30	11.80	m	**40.02**
32 mm dia.	24.07	30.46	0.30	11.80	m	**42.26**
40 mm dia.	24.46	30.96	0.30	11.80	m	**42.75**
50 mm dia.	27.10	34.30	0.30	11.80	m	**46.09**
65 mm dia.	29.54	37.38	0.30	11.80	m	**49.18**
80 mm dia.	31.59	39.98	0.40	15.73	m	**55.71**
100 mm dia.	39.12	49.51	0.40	15.73	m	**65.24**
125 mm dia.	43.77	55.40	0.40	15.73	m	**71.13**
150 mm dia.	50.04	63.34	0.40	15.73	m	**79.07**
200 mm dia.	66.26	83.86	0.50	19.66	m	**103.52**
250 mm dia.	76.54	96.87	0.50	19.66	m	**116.53**
300 mm dia.	82.09	103.89	0.50	19.66	m	**123.56**
350 mm dia.	90.10	114.04	0.50	19.66	m	**133.70**
400 mm dia.	99.81	126.32	0.50	19.66	m	**145.98**
Extra over for fittings external insulation						
Flange/union						
15 mm dia.	29.63	37.50	0.75	29.49	nr	**67.00**
20 mm dia.	31.17	39.45	0.75	29.49	nr	**68.95**
25 mm dia.	32.94	41.69	0.75	29.49	nr	**71.19**
32 mm dia.	34.58	43.76	0.75	29.49	nr	**73.26**
40 mm dia.	36.19	45.80	0.75	29.49	nr	**75.29**
50 mm dia.	39.89	50.48	0.75	29.49	nr	**79.97**
65 mm dia.	43.13	54.58	0.75	29.49	nr	**84.08**
80 mm dia.	46.24	58.52	0.89	35.00	nr	**93.52**

38 MECHANICAL/COOLING/HEATING SYSTEMS

Item	Net Price £	Material £	Labour hours	Labour £	Unit	Total rate £
THERMAL INSULATION – cont						
Extra over for fittings external insulation – cont						
Flange/union – cont						
100 mm dia.	56.85	71.95	0.89	35.00	nr	**106.95**
125 mm dia.	63.19	79.98	0.89	35.00	nr	**114.98**
150 mm dia.	72.37	91.59	0.89	35.00	nr	**126.59**
200 mm dia.	94.66	119.80	1.15	45.22	nr	**165.03**
250 mm dia.	109.38	138.43	1.15	45.22	nr	**183.65**
300 mm dia.	118.68	150.20	1.15	45.22	nr	**195.42**
350 mm dia.	131.33	166.21	1.15	45.22	nr	**211.44**
400 mm dia.	146.34	185.21	1.15	45.22	nr	**230.43**
Bends						
15 mm dia.	5.03	6.36	0.30	11.80	nr	**18.16**
20 mm dia.	5.28	6.68	0.30	11.80	nr	**18.48**
25 mm dia.	5.58	7.06	0.30	11.80	nr	**18.85**
32 mm dia.	5.86	7.41	0.30	11.80	nr	**19.21**
40 mm dia.	6.14	7.77	0.30	11.80	nr	**19.57**
50 mm dia.	6.79	8.60	0.30	11.80	nr	**20.40**
65 mm dia.	7.37	9.33	0.30	11.80	nr	**21.13**
80 mm dia.	7.87	9.96	0.40	15.73	nr	**25.69**
100 mm dia.	9.78	12.38	0.40	15.73	nr	**28.11**
125 mm dia.	10.91	13.81	0.40	15.73	nr	**29.54**
150 mm dia.	12.53	15.86	0.40	15.73	nr	**31.59**
200 mm dia.	16.57	20.96	0.50	19.66	nr	**40.63**
250 mm dia.	19.14	24.22	0.50	19.66	nr	**43.88**
300 mm dia.	20.55	26.00	0.50	19.66	nr	**45.67**
350 mm dia.	22.53	28.51	0.50	19.66	nr	**48.18**
400 mm dia.	24.93	31.55	0.50	19.66	nr	**51.21**
Tees						
15 mm dia.	5.03	6.36	0.30	11.80	nr	**18.16**
20 mm dia.	5.28	6.68	0.30	11.80	nr	**18.48**
25 mm dia.	5.58	7.06	0.30	11.80	nr	**18.85**
32 mm dia.	5.86	7.41	0.30	11.80	nr	**19.21**
40 mm dia.	6.14	7.77	0.30	11.80	nr	**19.57**
50 mm dia.	6.79	8.60	0.30	11.80	nr	**20.40**
65 mm dia.	7.37	9.33	0.30	11.80	nr	**21.13**
80 mm dia.	7.87	9.96	0.40	15.73	nr	**25.69**
100 mm dia.	9.78	12.38	0.40	15.73	nr	**28.11**
125 mm dia.	10.91	13.81	0.40	15.73	nr	**29.54**
150 mm dia.	12.53	15.86	0.40	15.73	nr	**31.59**
200 mm dia.	16.57	20.96	0.50	19.66	nr	**40.63**
250 mm dia.	19.14	24.22	0.50	19.66	nr	**43.88**
300 mm dia.	20.55	26.00	0.50	19.66	nr	**45.67**
350 mm dia.	22.53	28.51	0.50	19.66	nr	**48.18**
400 mm dia.	24.93	31.55	0.50	19.66	nr	**51.21**
Valves						
15 mm dia.	47.06	59.55	1.03	40.51	nr	**100.06**
20 mm dia.	49.49	62.64	1.03	40.51	nr	**103.15**
25 mm dia.	52.32	66.22	1.03	40.51	nr	**106.73**

38 MECHANICAL/COOLING/HEATING SYSTEMS

Item	Net Price £	Material £	Labour hours	Labour £	Unit	Total rate £
32 mm dia.	54.94	69.53	1.03	40.51	nr	**110.03**
40 mm dia.	57.48	72.75	1.03	40.51	nr	**113.26**
50 mm dia.	63.39	80.23	1.03	40.51	nr	**120.74**
65 mm dia.	68.55	86.76	1.03	40.51	nr	**127.27**
80 mm dia.	73.44	92.95	1.25	49.16	nr	**142.11**
100 mm dia.	90.29	114.27	1.25	49.16	nr	**163.43**
125 mm dia.	100.36	127.01	1.25	49.16	nr	**176.17**
150 mm dia.	114.88	145.40	1.25	49.16	nr	**194.56**
200 mm dia.	150.32	190.25	1.55	60.95	nr	**251.20**
250 mm dia.	173.71	219.84	1.55	60.95	nr	**280.80**
300 mm dia.	188.48	238.54	1.55	60.95	nr	**299.49**
350 mm dia.	208.58	263.98	1.55	60.95	nr	**324.93**
400 mm dia.	232.38	294.10	1.55	60.95	nr	**355.06**
Expansion bellows						
15 mm dia.	69.71	88.22	1.42	55.84	nr	**144.06**
20 mm dia.	73.34	92.82	1.42	55.84	nr	**148.66**
25 mm dia.	77.52	98.11	1.42	55.84	nr	**153.95**
32 mm dia.	81.43	103.06	1.42	55.84	nr	**158.90**
40 mm dia.	85.15	107.76	1.42	55.84	nr	**163.60**
50 mm dia.	93.86	118.78	1.42	55.84	nr	**174.63**
65 mm dia.	101.54	128.50	1.42	55.84	nr	**184.35**
80 mm dia.	108.83	137.73	1.75	68.82	nr	**206.55**
100 mm dia.	133.77	169.30	1.75	68.82	nr	**238.12**
125 mm dia.	148.69	188.18	1.75	68.82	nr	**257.00**
150 mm dia.	170.22	215.43	1.75	68.82	nr	**284.25**
200 mm dia.	222.72	281.87	2.17	85.34	nr	**367.21**
250 mm dia.	257.36	325.72	2.17	85.34	nr	**411.06**
300 mm dia.	279.22	353.39	2.17	85.34	nr	**438.72**
350 mm dia.	309.00	391.08	2.17	85.34	nr	**476.41**
400 mm dia.	344.31	435.75	2.17	85.34	nr	**521.09**

38 MECHANICAL/COOLING/HEATING SYSTEMS

Item	Net Price £	Material £	Labour hours	Labour £	Unit	Total rate £
STEAM HEATING						
PIPELINES						
For pipework prices refer to Section – Low Temperature Hot Water Heating						
PIPELINE ANCILLARIES						
Steam traps and accessories						
Cast iron; inverted bucket type; steam trap pressure range up to 17 bar at 210°C; screwed ends						
½" dia.	172.27	218.02	0.85	33.43	nr	**251.45**
¾" dia.	172.27	218.02	1.13	44.44	nr	**262.46**
1" dia.	290.26	367.35	1.35	53.09	nr	**420.44**
1½" dia.	896.14	1134.15	1.80	70.79	nr	**1204.94**
2" dia.	1003.08	1269.49	2.18	85.73	nr	**1355.22**
Cast iron; inverted bucket type; steam trap pressure range up to 17 bar at 210°C; flanged ends to BS 4504 PN16; bolted connections						
15 mm dia.	280.12	354.53	1.15	45.22	nr	**399.75**
20 mm dia.	348.67	441.27	1.25	49.16	nr	**490.43**
25 mm dia.	432.85	547.81	1.33	52.30	nr	**600.11**
40 mm dia.	918.24	1162.12	1.46	57.42	nr	**1219.54**
50 mm dia.	1025.18	1297.46	1.60	62.92	nr	**1360.39**
Steam traps and strainers						
Stainless steel; thermodynamic trap with pressure range up to 42 bar; temperature range to 400°C; screwed ends to steel						
15 mm dia.	111.98	141.72	0.84	33.05	nr	**174.76**
20 mm dia.	126.74	160.40	1.14	44.84	nr	**205.25**
Stainless steel; thermodynamic trap with pressure range up to 24 bar; temperature range to 288°C; flanged ends to DIN 2456 PN64; bolted connections						
15 mm dia.	234.45	296.72	1.24	48.79	nr	**345.51**
20 mm dia.	256.99	325.24	1.34	52.72	nr	**377.96**
25 mm dia.	297.08	375.99	1.40	55.08	nr	**431.07**
Malleable iron pipeline strainer; max steam working pressure 14 bar and temperature range to 230°C; screwed ends to steel						
½" dia.	33.59	42.51	0.84	33.05	nr	**75.55**
¾" dia.	41.98	53.13	1.14	44.84	nr	**97.97**
1" dia.	65.51	82.90	1.30	51.14	nr	**134.04**
1½" dia.	134.53	170.26	1.50	59.05	nr	**229.31**
2" dia.	188.88	239.05	1.74	68.51	nr	**307.56**
Bronze pipeline strainer; max steam working pressure 25 bar; flanged ends to BS 4504 PN25; bolted connections						
15 mm dia.	288.54	365.18	1.24	48.79	nr	**413.97**
20 mm dia.	351.96	445.44	1.34	52.72	nr	**498.16**
25 mm dia.	403.53	510.71	1.40	55.08	nr	**565.78**

38 MECHANICAL/COOLING/HEATING SYSTEMS

Item	Net Price £	Material £	Labour hours	Labour £	Unit	Total rate £
32 mm dia.	626.43	792.81	1.46	57.49	nr	**850.31**
40 mm dia.	710.89	899.71	1.54	60.59	nr	**960.30**
50 mm dia.	1093.34	1383.74	1.64	64.57	nr	**1448.31**
65 mm dia.	1210.61	1532.15	2.50	98.31	nr	**1630.47**
80 mm dia.	1506.24	1906.30	2.91	114.32	nr	**2020.62**
100 mm dia.	2608.95	3301.89	3.51	137.99	nr	**3439.87**
Balanced pressure thermostatic steam trap and strainer; max working pressure up to 13 bar; screwed ends to steel						
½" dia.	232.09	293.73	1.26	49.59	nr	**343.32**
¾" dia.	232.69	294.49	1.71	67.34	nr	**361.83**
Bimetallic thermostatic steam trap and strainer; max working pressure up to 21 bar; flanged ends						
15 mm	301.91	382.10	1.24	48.79	nr	**430.89**
20 mm	331.69	419.79	1.34	52.72	nr	**472.51**
Sight glasses						
Pressed brass; straight; single window; screwed ends to steel						
15 mm dia.	67.96	86.02	0.84	33.05	nr	**119.06**
20 mm dia.	75.43	95.46	1.14	44.84	nr	**140.30**
25 mm dia.	94.27	119.31	1.30	51.14	nr	**170.45**
Gunmetal; straight; double window; screwed ends to steel						
15 mm dia.	109.60	138.71	0.84	33.05	nr	**171.75**
20 mm dia.	120.61	152.65	1.14	44.84	nr	**197.49**
25 mm dia.	149.08	188.68	1.30	51.14	nr	**239.82**
32 mm dia.	245.60	310.83	1.35	53.14	nr	**363.97**
40 mm dia.	245.60	310.83	1.74	68.51	nr	**379.34**
50 mm dia.	298.22	377.43	2.08	81.93	nr	**459.35**
SG iron flanged; BS 4504, PN 25						
15 mm dia.	223.65	283.05	1.00	39.33	nr	**322.38**
20 mm dia.	263.13	333.02	1.25	49.16	nr	**382.18**
25 mm dia.	335.51	424.62	1.50	59.05	nr	**483.67**
32 mm dia.	370.54	468.95	1.70	66.88	nr	**535.84**
40 mm dia.	484.58	613.28	2.00	78.65	nr	**691.93**
50 mm dia.	585.45	740.94	2.30	90.61	nr	**831.56**
Check valve and sight glass; gun metal; screwed						
15 mm dia.	110.69	140.09	0.84	33.05	nr	**173.14**
20 mm dia.	117.30	148.45	1.14	44.84	nr	**193.30**
25 mm dia.	197.33	249.74	1.30	51.14	nr	**300.88**
Pressure reducing valves						
Pressure reducing valve for steam; maximum range of 17 bar and 232°C; screwed ends to steel						
15 mm dia.	861.46	1090.26	0.87	34.21	nr	**1124.48**
20 mm dia.	932.08	1179.65	0.91	35.79	nr	**1215.43**
25 mm dia.	1005.02	1271.95	1.35	53.09	nr	**1325.04**
Pressure reducing valve for steam; maximum range of 17 bar and 232°C; flanged ends to BS 4504 PN 25						
25 mm dia.	1210.16	1531.58	1.70	66.85	nr	**1598.43**
32 mm dia.	1374.23	1739.23	1.87	73.54	nr	**1812.77**
40 mm dia.	1640.90	2076.73	2.12	83.37	nr	**2160.10**
50 mm dia.	1893.86	2396.87	2.57	101.07	nr	**2497.93**

38 MECHANICAL/COOLING/HEATING SYSTEMS

Item	Net Price £	Material £	Labour hours	Labour £	Unit	Total rate £
STEAM HEATING – cont						
Safety and relief valves						
Bronze safety valve; 'pop' type; side outlet; including easing lever; working pressure saturated steam up to 20.7 bar; screwed ends to steel						
15 mm dia.	308.91	390.95	0.32	12.58	nr	**403.54**
20 mm dia.	349.19	441.93	0.40	15.73	nr	**457.66**
Bronze safety valve; 'pop' type; side outlet; including easing lever; working pressure saturated steam up to 17.2 bar; screwed ends to steel						
25 mm dia.	496.92	628.90	0.47	18.48	nr	**647.39**
32 mm dia.	593.67	751.35	0.56	22.02	nr	**773.37**
Bronze safety valve; 'pop' type; side outlet; including easing lever; working pressure saturated steam up to 13.8 bar; screwed ends to steel						
40 mm dia.	768.24	972.29	0.64	25.17	nr	**997.46**
50 mm dia.	1069.12	1353.08	0.76	29.89	nr	**1382.96**
65 mm dia.	1520.45	1924.29	0.94	36.97	nr	**1961.25**
80 mm dia.	1971.73	2495.42	1.10	43.26	nr	**2538.68**
EQUIPMENT						
Calorifiers						
Non-storage calorifiers; mild steel shell construction with indirect steam heating for secondary LPHW at 82°C flow and 71°C return to BS 853; maximum test on shell 11 bar, tubes 26 bar						
Horizontal/vertical, for steam at 3 bar – 5.5 bar						
88 kW capacity	1122.12	1420.16	8.00	314.61	nr	**1734.76**
176 kW capacity	1633.51	2067.37	12.05	473.81	nr	**2541.18**
293 kW capacity	1779.65	2252.32	14.08	553.88	nr	**2806.21**
586 kW capacity	2598.31	3288.42	37.04	1456.51	nr	**4744.93**
879 kW capacity	3235.22	4094.49	40.00	1573.03	nr	**5667.52**
1465 kW capacity	3945.19	4993.03	45.45	1787.54	nr	**6780.57**

38 MECHANICAL/COOLING/HEATING SYSTEMS

Item	Net Price £	Material £	Labour hours	Labour £	Unit	Total rate £
LOCAL HEATING UNITS						
Warm air unit heater for connection to LTHW or steam supplies; suitable for heights up to 3 m; recirculating type; mild steel casing; heating coil; adjustable discharge louvre; axial fan; horizontal or vertical discharge; normal speed; entering air temperature 15°C; complete with enclosures; includes fixing in position; includes connections to primary heating supply; electrical work elsewhere						
Low pressure hot water						
7.5 kW, 265 l/s	974.40	1233.20	6.53	256.80	nr	**1490.00**
15.4 kW, 575 l/s	1179.52	1492.80	7.54	296.52	nr	**1789.32**
26.9 kW, 1040 l/s	1598.31	2022.82	8.65	340.17	nr	**2362.99**
48.0 kW, 1620 l/s	2106.85	2666.43	9.35	367.70	nr	**3034.13**
Steam, 2 bar						
9.2 kW, 265 l/s	1387.11	1755.52	6.53	256.80	nr	**2012.32**
18.8 kW, 575 l/s	1502.00	1900.93	6.82	268.20	nr	**2169.13**
34.8 kW, 1040 l/s	1727.55	2186.39	6.82	268.20	nr	**2454.59**
51.6 kW, 1625 l/s	2340.31	2961.89	7.10	279.21	nr	**3241.11**

38 MECHANICAL/COOLING/HEATING SYSTEMS

Item	Net Price £	Material £	Labour hours	Labour £	Unit	Total rate £
CENTRAL REFRIGERATION PLANT						
CHILLERS; Air cooled						
Selection of air cooled chillers based on chilled water flow and return temperatures 6°C and 12°C, and an outdoor temperature of 35°C						
Air cooled liquid chiller; refrigerant 410 A; scroll compressors; twin circuit; integral controls; includes placing in position; electrical work elsewhere						
Cooling load						
100 kW	28283.69	35795.84	8.00	314.61	nr	**36110.45**
150 kW	35644.56	45111.75	8.00	314.61	nr	**45426.36**
200 kW	44120.07	55838.36	8.00	314.61	nr	**56152.97**
Air cooled liquid chiller; refrigerant 410 A; reciprocating compressors; twin circuit; integral controls; includes placing in position; electrical work elsewhere						
Cooling load						
400 kW	78706.38	99610.79	8.00	314.61	nr	**99925.40**
550 kW	112232.51	142041.46	8.00	314.61	nr	**142356.07**
700 kW	130384.68	165014.85	9.00	353.93	nr	**165368.78**
Air cooled liquid chiller; refrigerant R134a; screw compressors; twin circuit; integral controls; includes placing in position; electrical work elsewhere						
Cooling load						
250 kW	79360.50	100438.65	8.00	314.61	nr	**100753.25**
400 kW	95074.70	120326.53	8.00	314.61	nr	**120641.14**
600 kW	127630.94	161529.72	8.00	314.61	nr	**161844.33**
800 kW	153401.47	194144.91	9.00	353.93	nr	**194498.84**
1000 kW	179867.52	227640.33	9.00	353.93	nr	**227994.26**
1200 kW	228472.52	289154.82	10.00	393.26	nr	**289548.08**
Air cooled liquid chiller; ductable for indoor installation; refrigerant 410 A; scroll compressors; integral controls; includes placing in position; electrical work elsewhere						
Cooling load						
40 kW	16821.35	21289.10	6.00	235.95	nr	**21525.05**
80 kW	26341.98	33338.41	6.00	235.95	nr	**33574.37**

38 MECHANICAL/COOLING/HEATING SYSTEMS

Item	Net Price £	Material £	Labour hours	Labour £	Unit	Total rate £
CHILLERS; Higher efficiency air cooled						
Selection of air cooled chillers based on chilled water flow and return temperatures of 6°C and 12°C and an outdoor temperature of 25°C						
These machines have significantly higher part load operating efficiencies than conventional air cooled machines						
Air cooled liquid chiller, refrigerant R410 A; scroll compressors; complete with free cooling facility; integral controls; including placing in position; electrical work elsewhere						
Cooling load						
250 kW	71824.80	90901.47	8.00	314.61	nr	**91216.07**
300 kW	79247.71	100295.91	8.00	314.61	nr	**100610.51**
350 kW	91576.62	115899.36	8.00	314.61	nr	**116213.97**
400 kW	103352.85	130803.37	8.00	314.61	nr	**131117.97**
450 kW	107392.31	135915.71	8.00	314.61	nr	**136230.31**
500 kW	116745.64	147753.28	8.00	314.61	nr	**148067.89**
600 kW	119805.79	151626.20	8.00	314.61	nr	**151940.81**
650 kW	139929.74	177095.09	9.00	353.93	nr	**177449.02**
700 kW	148204.14	187567.16	9.00	353.93	nr	**187921.09**
CHILLERS; Water cooled						
Selection of water cooled chillers based on chilled water flow and return temperatures of 6°C and 12°C, and condenser entering and leaving temperatures of 27°C and 33°C						
Water cooled liquid chiller; refrigerant 410 A; reciprocating compressors; twin circuit; integral controls; includes placing in position; electrical work elsewhere						
Cooling load						
200 kW	36356.96	46013.37	8.00	314.61	nr	**46327.98**
350 kW	64339.89	81428.57	8.00	314.61	nr	**81743.17**
500 kW	73003.38	92393.08	8.00	314.61	nr	**92707.68**
650 kW	84614.23	107087.77	9.00	353.93	nr	**107441.70**
750 kW	103065.25	130439.38	9.00	353.93	nr	**130793.31**
Water cooled condenserless liquid chiller; refrigerant 410 A; reciprocating compressors; twin circuit; integral controls; includes placing in position; electrical work elsewhere						
Cooling load						
200 kW	31378.53	39712.67	8.00	314.61	nr	**40027.27**
350 kW	55083.09	69713.15	8.00	314.61	nr	**70027.76**
500 kW	69617.99	88108.53	8.00	314.61	nr	**88423.14**
650 kW	85144.44	107758.81	9.00	353.93	nr	**108112.74**
750 kW	96885.41	122618.17	9.00	353.93	nr	**122972.11**

38 MECHANICAL/COOLING/HEATING SYSTEMS

Item	Net Price £	Material £	Labour hours	Labour £	Unit	Total rate £
CENTRAL REFRIGERATION PLANT – cont						
Selection of water cooled chillers based on chilled water flow and return temperatures of 6°C and 12°C – cont						
Water cooled liquid chiller; refrigerant R134a; screw compressors; twin circuit; integral controls; includes placing in position; electrical work elsewhere						
Cooling load						
300 kW	60625.62	76727.79	8.00	314.61	nr	77042.40
500 kW	77270.28	97793.27	8.00	314.61	nr	98107.88
700 kW	97437.47	123316.87	9.00	353.93	nr	123670.80
900 kW	136578.25	172853.44	9.00	353.93	nr	173207.37
1100 kW	150871.42	190942.87	10.00	393.26	nr	191336.13
1300 kW	174438.96	220769.94	10.00	393.26	nr	221163.20
Water cooled liquid chiller; refrigerant R134a; centrifugal compressors; twin circuit; integral controls; includes placing in position; electrical work elsewhere						
Cooling load						
1600 kW	246089.22	311450.51	11.00	432.58	nr	311883.09
1900 kW	299713.50	379317.40	11.00	432.58	nr	379749.99
2200 kW	357078.84	451918.98	13.00	511.24	nr	452430.21
2500 kW	368926.12	466912.90	13.00	511.24	nr	467424.13
3000 kW	448946.63	568186.86	15.00	589.89	nr	568776.75
3500 kW	611066.24	773365.43	15.00	589.89	nr	773955.32
4000 kW	649006.35	821382.43	20.00	786.52	nr	822168.95
4500 kW	692421.18	876328.24	20.00	786.52	nr	877114.76
5000 kW	844470.66	1068762.07	25.00	983.15	nr	1069745.22
CHILLERS; Absorption						
Absorption chiller, for operation using low pressure steam; selection based on chilled water flow and return temperatures of 6°C and 12°C, steam at 1 bar gauge and condenser entering and leaving temperatures of 27°C and 33°C; integral controls; includes placing in position; electrical work elsewhere						
Cooling load						
400 kW	106026.26	134186.83	8.00	314.61	nr	134501.44
700 kW	134627.89	170385.05	9.00	353.93	nr	170738.99
1000 kW	153591.72	194385.68	10.00	393.26	nr	194778.93
1300 kW	183294.97	231978.11	12.00	471.91	nr	232450.02
1600 kW	216928.57	274544.80	14.00	550.56	nr	275095.36
2000 kW	257101.01	325387.03	15.00	589.89	nr	325976.92
Absorption chiller, for operation using low pressure hot water; selection based on chilled water flow and return temperatures of 6°C and 12°C, cooling water temperatures of 27°C and 33°C and hot water at 90°C; integral controls; includes placing in position; electrical work elsewhere						

38 MECHANICAL/COOLING/HEATING SYSTEMS

Item	Net Price £	Material £	Labour hours	Labour £	Unit	Total rate £
Cooling load						
700 kW	153591.72	194385.68	9.00	353.93	nr	**194739.61**
1000 kW	191995.30	242989.25	10.00	393.26	nr	**243382.51**
1300 kW	223303.11	282612.41	12.00	471.91	nr	**283084.32**
1600 kW	257101.01	325387.03	14.00	550.56	nr	**325937.59**
HEAT REJECTION						
Dry air liquid coolers						
Flat coil configuration						
500 kW	47537.45	60163.40	15.00	589.89	nr	**60753.28**
Extra for inverter panels (factory wired and mounted on units)	3654.37	4624.97	15.00	589.89	nr	**5214.85**
800 kW	80095.37	101368.70	15.00	589.89	nr	**101958.59**
Extra for inverter panels (factory wired and mounted on units)	–	8196.62	15.00	589.89	nr	**8786.50**
1100 kW	104519.18	132279.48	15.00	589.89	nr	**132869.37**
Extra for inverter panels (factory wired and mounted on units)	7308.73	9249.93	15.00	589.89	nr	**9839.82**
1400 kW	157045.72	198757.06	15.00	589.89	nr	**199346.95**
Extra for inverter panels (factory wired and mounted on units)	10125.45	12814.77	15.00	589.89	nr	**13404.66**
1700 kW	157045.72	198757.06	15.00	589.89	nr	**199346.95**
Extra for inverter panels (factory wired and mounted on units)	10963.10	13874.90	15.00	589.89	nr	**14464.79**
2000 kW	191154.68	241925.36	15.00	589.89	nr	**242515.24**
Extra for inverter panels (factory wired and mounted on units)	12638.40	15995.16	15.00	589.89	nr	**16585.05**
Note: Heat rejection capacities above 500 kW require multiple units. Prices are therefore for total number of units						
'Vee' type coil configuration						
500 kW	50128.64	63442.81	15.00	589.89	nr	**64032.70**
Extra for inverter panels (factory wired and mounted on units)	3375.15	4271.59	15.00	589.89	nr	**4861.48**
800 kW	81391.74	103009.38	15.00	589.89	nr	**103599.27**
Extra for inverter panels (factory wired and mounted on units)	4492.02	5685.10	15.00	589.89	nr	**6274.98**
1100 kW	106987.65	135403.56	15.00	589.89	nr	**135993.45**
Extra for inverter panels (factory wired and mounted on units)	5050.45	6391.85	15.00	589.89	nr	**6981.74**
1400 kW	145561.00	184222.01	15.00	589.89	nr	**184811.89**
Extra for inverter panels (factory wired and mounted on units)	7867.17	9956.69	15.00	589.89	nr	**10546.57**
1700 kW	165004.94	208830.25	15.00	589.89	nr	**209420.13**
Extra for inverter panels (factory wired and mounted on units)	9542.47	12076.95	15.00	589.89	nr	**12666.84**
2000 kW	196366.21	248521.08	15.00	589.89	nr	**249110.97**
Extra for inverter panels (factory wired and mounted on units)	10100.90	12783.70	15.00	589.89	nr	**13373.59**

38 MECHANICAL/COOLING/HEATING SYSTEMS

Item	Net Price £	Material £	Labour hours	Labour £	Unit	Total rate £
CENTRAL REFRIGERATION PLANT – cont						
Dry air liquid coolers – cont						
'Vee' type coil configuration – cont						
Note: Heat rejection capacities above 1100 kW require multiple units. Prices are for total number of units						
Air cooled condensers						
Air cooled condenser; refrigerant 410 A; selection based on condensing temperature of 45°C at 32°C dry bulb ambient; includes placing in position; electrical work elsewhere						
Flat coil configuration						
500 kW	28831.98	36489.76	15.00	589.89	nr	**37079.65**
Extra for inverter panels (factory wired and mounted on units)	13192.71	16696.69	15.00	589.89	nr	**17286.58**
800 kW	47301.75	59865.10	15.00	589.89	nr	**60454.99**
Extra for inverter panels (factory wired and mounted on units)	20196.81	25561.08	15.00	589.89	nr	**26150.97**
1100 kW	63324.92	80144.02	15.00	589.89	nr	**80733.90**
Extra for inverter panels (factory wired and mounted on units)	28712.09	36338.02	15.00	589.89	nr	**36927.91**
1400 kW	79060.19	100058.58	15.00	589.89	nr	**100648.47**
Extra for inverter panels (factory wired and mounted on units)	35308.45	44686.37	15.00	589.89	nr	**45276.26**
1700 kW	102950.95	130294.72	15.00	589.89	nr	**130884.61**
Extra for inverter panels (factory wired and mounted on units)	52962.66	67029.54	15.00	589.89	nr	**67619.43**
2000 kW	118590.30	150087.89	15.00	589.89	nr	**150677.78**
Extra for inverter panels (factory wired and mounted on units)	52962.66	67029.54	15.00	589.89	nr	**67619.43**
Note: Heat rejection capacities above 500 kW require multiple units. Prices are for total number of units						
'Vee' type coil configuration						
500 kW	28304.30	35821.92	15.00	589.89	nr	**36411.81**
Extra for inverter panels (factory wired and mounted on units)	10098.43	12780.57	15.00	589.89	nr	**13370.46**
800 kW	42576.38	53884.67	15.00	589.89	nr	**54474.56**
Extra for inverter panels (factory wired and mounted on units)	20676.55	26168.24	15.00	589.89	nr	**26758.13**
1100 kW	62485.39	79081.52	15.00	589.89	nr	**79671.40**
Extra for inverter panels (factory wired and mounted on units)	35164.53	44504.22	15.00	589.89	nr	**45094.11**
1400 kW	72751.69	92074.54	15.00	589.89	nr	**92664.43**
Extra for inverter panels (factory wired and mounted on units)	41353.08	52336.46	15.00	589.89	nr	**52926.35**
1700 kW	93728.06	118622.23	15.00	589.89	nr	**119212.12**
Extra for inverter panels (factory wired and mounted on units)	52770.76	66786.67	15.00	589.89	nr	**67376.56**
2000 kW	109127.56	138111.84	15.00	589.89	nr	**138701.73**

38 MECHANICAL/COOLING/HEATING SYSTEMS

Item	Net Price £	Material £	Labour hours	Labour £	Unit	Total rate £
Extra for inverter panels (factory wired and mounted on units)	62029.63	78504.70	15.00	589.89	nr	**79094.59**

Note: Heat rejection capacities above 1100 kW require multiple units. Prices are for total number of units

COOLING TOWERS

Cooling towers; forced draught, centrifugal fan, conterflow design; based on water temperatures of 35°C on and 29°C off at 21°C wet bulb ambient temperature; includes placing in position; electrical work elsewhere

Item	Net Price £	Material £	Labour hours	Labour £	Unit	Total rate £
Cooling towers: Open circuit type						
900 kW	27232.38	34465.30	20.00	786.52	nr	**35251.82**
Extra for stainless steel construction	12799.07	16198.50	–	–	nr	**16198.50**
Extra for intake and discharge sound attenuation	12888.59	16311.80	–	–	nr	**16311.80**
1500 kW	34219.46	43308.15	20.00	786.52	nr	**44094.66**
Extra for stainless steel construction	16082.95	20354.58	–	–	nr	**20354.58**
Extra for intake and discharge sound attenuation	19890.04	25172.84	–	–	nr	**25172.84**
2100 kW	43980.79	55662.09	20.00	786.52	nr	**56448.61**
Extra for stainless steel construction	20670.96	26161.17	–	–	nr	**26161.17**
Extra for intake and discharge sound attenuation	27117.45	34319.85	–	–	nr	**34319.85**
2700 kW	56573.75	71599.74	20.00	786.52	nr	**72386.25**
Extra for stainless steel construction	26589.59	33651.78	–	–	nr	**33651.78**
Extra for intake and discharge sound attenuation	28751.54	36387.95	–	–	nr	**36387.95**
3300 kW	69009.38	87338.28	20.00	786.52	nr	**88124.79**
Extra for stainless steel construction	32433.98	41048.44	–	–	nr	**41048.44**
Extra for intake and discharge sound attenuation	32086.35	40608.48	–	–	nr	**40608.48**
3900 kW	84346.58	106749.03	20.00	786.52	nr	**107535.55**
Extra for stainless steel construction	39642.22	50171.19	–	–	nr	**50171.19**
Extra for intake and discharge sound attenuation	38109.02	48230.77	–	–	nr	**48230.77**
4500 kW	86868.35	109940.58	23.00	904.49	nr	**110845.08**
Extra for stainless steel construction	40827.81	51671.68	–	–	nr	**51671.68**
Extra for intake and discharge sound attenuation	36345.73	45999.16	–	–	nr	**45999.16**
5100 kW	110055.82	139286.64	23.00	904.49	nr	**140191.14**
Extra for stainless steel construction	51726.19	65464.67	–	–	nr	**65464.67**
Extra for intake and discharge sound attenuation	50405.66	63793.41	–	–	nr	**63793.41**
5700 kW	122761.16	155366.52	30.00	1179.77	nr	**156546.29**
Extra for stainless steel construction	57697.71	73022.22	–	–	nr	**73022.22**
Extra for intake and discharge sound attenuation	65247.49	82577.23	–	–	nr	**82577.23**
6300 kW	132677.36	167916.47	30.00	1179.77	nr	**169096.24**
Extra for stainless steel construction	62357.88	78920.14	–	–	nr	**78920.14**
Extra for intake and discharge sound attenuation	65247.95	82577.80	–	–	nr	**82577.80**

38 MECHANICAL/COOLING/HEATING SYSTEMS

Item	Net Price £	Material £	Labour hours	Labour £	Unit	Total rate £
CENTRAL REFRIGERATION PLANT – cont						
Cooling towers: Closed circuit type (includes 20% ethylene glycol)						
900 kW	67669.61	85642.66	20.00	786.52	nr	**86429.18**
Extra for stainless steel construction	42293.29	53526.39	–	–	nr	**53526.39**
Extra for intake and discharge sound attenuation	18968.45	24006.47	–	–	nr	**24006.47**
1500 kW	108839.26	137746.97	20.00	786.52	nr	**138533.49**
Extra for stainless steel construction	68024.09	86091.29	–	–	nr	**86091.29**
Extra for intake and discharge sound attenuation	36722.51	46476.01	–	–	nr	**46476.01**
2100 kW	127402.41	161240.49	20.00	786.52	nr	**162027.01**
Extra for stainless steel construction	79626.20	100774.91	–	–	nr	**100774.91**
Extra for intake and discharge sound attenuation	36722.98	46476.61	–	–	nr	**46476.61**
2700 kW	184895.68	234003.98	25.00	983.15	nr	**234987.12**
Extra for stainless steel construction	115558.93	146251.38	–	–	nr	**146251.38**
Extra for intake and discharge sound attenuation	48536.52	61427.82	–	–	nr	**61427.82**
3300 kW	231215.82	292626.74	25.00	983.15	nr	**293609.88**
Extra for stainless steel construction	144509.05	182890.66	–	–	nr	**182890.66**
Extra for intake and discharge sound attenuation	73169.06	92602.76	–	–	nr	**92602.76**
3900 kW	253024.88	320228.29	25.00	983.15	nr	**321211.43**
Extra for stainless steel construction	158139.78	200141.71	–	–	nr	**200141.71**
Extra for intake and discharge sound attenuation	74198.87	93906.08	–	–	nr	**93906.08**
4500 kW	301993.10	382202.47	40.00	1573.03	nr	**383775.50**
Extra for stainless steel construction	188744.91	238875.56	–	–	nr	**238875.56**
Extra for intake and discharge sound attenuation	109566.91	138667.88	–	–	nr	**138667.88**
5100 kW	342280.67	433190.42	40.00	1573.03	nr	**434763.45**
Extra for stainless steel construction	213925.07	270743.57	–	–	nr	**270743.57**
Extra for intake and discharge sound attenuation	121819.93	154175.31	–	–	nr	**154175.31**
5700 kW	413808.65	523716.22	40.00	1573.03	nr	**525289.26**
Extra for stainless steel construction	258628.89	327320.73	–	–	nr	**327320.73**
Extra for intake and discharge sound attenuation	121820.11	154175.53	–	–	nr	**154175.53**
6300 kW	462431.14	585252.85	40.00	1573.03	nr	**586825.88**
Extra for stainless steel construction	289019.03	365782.48	–	–	nr	**365782.48**
Extra for intake and discharge sound attenuation	146337.39	185204.59	–	–	nr	**185204.59**

38 MECHANICAL/COOLING/HEATING SYSTEMS

Item	Net Price £	Material £	Labour hours	Labour £	Unit	Total rate £
CHILLED WATER						
SCREWED STEEL						
PIPELINES						
For pipework prices refer to Section – Low Temperature Hot Water Heating, with the exception of chilled water blocks within brackets as detailed hereafter. For minimum fixing distances, refer to the Tables and Memoranda to the rear of the book						
FIXINGS						
For steel pipes; black malleable iron						
Oversized pipe clip, to contain 30 mm insulation block for vapour barrier						
15 mm dia.	7.22	9.13	0.10	3.93	nr	**13.07**
20 mm dia.	7.28	9.22	0.11	4.33	nr	**13.54**
25 mm dia.	7.19	9.11	0.12	4.72	nr	**13.83**
32 mm dia.	7.87	9.97	0.14	5.51	nr	**15.47**
40 mm dia.	8.30	10.51	0.15	5.90	nr	**16.41**
50 mm dia.	11.87	15.02	0.16	6.29	nr	**21.31**
65 mm dia.	12.81	16.21	0.30	11.80	nr	**28.01**
80 mm dia.	13.58	17.19	0.35	13.76	nr	**30.95**
100 mm dia.	21.54	27.26	0.40	15.73	nr	**42.99**
125 mm dia.	24.99	31.63	0.60	23.60	nr	**55.23**
150 mm dia.	39.09	49.48	0.77	30.28	nr	**79.76**
200 mm dia.	45.03	56.98	0.90	35.39	nr	**92.38**
250 mm dia.	58.48	74.01	1.10	43.26	nr	**117.27**
300 mm dia.	69.59	88.07	1.25	49.16	nr	**137.23**
350 mm dia.	80.75	102.20	1.50	58.99	nr	**161.18**
400 mm dia.	90.43	114.44	1.75	68.82	nr	**183.26**
Screw on backplate (Male), black malleable iron; plugged and screwed						
M12	0.96	1.22	0.10	3.93	nr	**5.15**
Screw on backplate (Female), black malleable iron; plugged and screwed						
M12	0.96	1.22	0.10	3.93	nr	**5.15**
Extra over channel sections for fabricated hangers and brackets						
Galvanized steel; including inserts, bolts, nuts, washers; fixed to backgrounds						
41 × 21 mm	9.61	12.16	0.29	11.40	m	**23.57**
41 × 41 mm	11.52	14.58	0.29	11.40	m	**25.99**
Threaded rods; metric thread; including nuts, washers etc.						
12 mm dia. × 600 mm long	4.85	6.14	0.18	7.08	nr	**13.22**

38 MECHANICAL/COOLING/HEATING SYSTEMS

Item	Net Price £	Material £	Labour hours	Labour £	Unit	Total rate £
CHILLED WATER – cont						
For plastic pipework suitable for chilled water systems, refer to ABS pipework details in Section – Cold Water, with the exception of chilled water blocks within brackets as detailed for the aforementioned steel pipe. For minimum fixing distances refer to the Tables and Memoranda to the rear of the book						
For copper pipework, refer to Section – Cold Water with the exception of chilled water blocks within brackets as detailed hereafter. For minimum fixing distances refer to the Tables and Memoranda to the rear of the book						
FIXINGS						
For copper pipework						
Oversized pipe clip, to contain 30 mm insulation block for vapour barrier						
15 mm dia.	7.22	9.13	0.10	3.93	nr	**13.07**
22 mm dia.	7.28	9.22	0.11	4.33	nr	**13.54**
28 mm dia.	7.19	9.11	0.12	4.72	nr	**13.83**
35 mm dia.	7.87	9.97	0.14	5.51	nr	**15.47**
42 mm dia.	8.30	10.51	0.15	5.90	nr	**16.41**
54 mm dia.	11.87	15.02	0.16	6.29	nr	**21.31**
67 mm dia.	12.81	16.21	0.30	11.80	nr	**28.01**
76 mm dia.	13.58	17.19	0.35	13.76	nr	**30.95**
108 mm dia.	15.45	19.56	0.40	15.73	nr	**35.29**
133 mm dia.	24.99	31.63	0.60	23.60	nr	**55.23**
159 mm dia.	39.09	49.48	0.77	30.28	nr	**79.76**
Screw on backplate, female						
All sizes 15 mm to 54 mm × 10 mm	2.33	2.95	0.10	3.93	nr	**6.89**
Screw on backplate, male						
All sizes 15 mm to 54 mm × 10 mm	3.13	3.96	0.10	3.93	nr	**7.89**
Extra over channel sections for fabricated hangers and brackets						
Galvanized steel; including inserts, bolts, nuts, washers; fixed to backgrounds						
41 × 21 mm	9.61	12.16	0.29	11.40	m	**23.57**
41 × 41 mm	11.52	14.58	0.29	11.40	m	**25.99**
Threaded rods; metric thread; including nuts, washers etc.						
10 mm dia. × 600 mm long for ring clips up to 54 mm	3.10	3.92	0.18	7.08	nr	**11.00**
12 mm dia. × 600 mm long for ring clips from 54 mm	4.85	6.14	0.18	7.08	nr	**13.22**

38 MECHANICAL/COOLING/HEATING SYSTEMS

Item	Net Price £	Material £	Labour hours	Labour £	Unit	Total rate £
PIPELINE ANCILLARIES						
For prices for ancillaries refer to Section: Low Temperature Hot Water Heating						
HEAT EXCHANGERS						
Plate heat exchanger; for use in CHW systems; painted carbon steel frame, stainless steel plates, nitrile rubber gaskets, design pressure of 10 bar and operating temperature of 110/135°C						
Primary side; 13°C in, 8°C out; secondary side; 6°C in, 11°C out						
264 kW, 12.60 l/s	5739.84	7264.34	10.00	393.26	nr	**7657.60**
290 kW, 13.85 l/s	6067.34	7678.82	12.00	471.91	nr	**8150.73**
316 kW, 15.11 l/s	6393.40	8091.49	12.00	471.91	nr	**8563.40**
350 kW, 16.69 l/s	8916.77	11285.07	16.00	629.21	nr	**11914.28**
395 kW, 18.88 l/s	8369.88	10592.92	16.00	629.21	nr	**11222.13**
454 kW, 21.69 l/s	9116.81	11538.23	16.00	629.21	nr	**12167.44**
475 kW, 22.66 l/s	10161.07	12859.85	16.00	629.21	nr	**13489.06**
527 kW, 25.17 l/s	11058.81	13996.03	16.00	629.21	nr	**14625.25**
554 kW, 26.43 l/s	11506.97	14563.22	16.00	629.21	nr	**15192.43**
580 kW, 27.68 l/s	11880.43	15035.88	16.00	629.21	nr	**15665.09**
633 kW, 30.19 l/s	13000.82	16453.84	16.00	629.21	nr	**17083.05**
661 kW, 31.52 l/s	13598.36	17210.09	16.00	629.21	nr	**17839.30**
713 kW, 34.04 l/s	15801.79	19998.75	18.00	707.86	nr	**20706.61**
740 kW, 35.28 l/s	16439.55	20805.90	18.00	707.86	nr	**21513.76**
804 kW, 38.33 l/s	17332.99	21936.63	18.00	707.86	nr	**22644.50**
1925 kW, 91.82 l/s	31126.70	39393.96	20.00	786.52	nr	**40180.47**
2710 kW, 129.26 l/s	49223.86	62297.72	20.00	786.52	nr	**63084.23**
3100 kW, 147.87 l/s	56269.38	71214.53	20.00	786.52	nr	**72001.05**
Note: For temperature conditions different to those above, the cost of the units can vary significantly, and therefore the manufacturers advice should be sought.						
TRACE HEATING						
Trace heating; for freeze protection or temperature maintenance of pipework; to BS 6351; including fixing to parent structures by plastic pull ties						
Straight laid F-S-C Wintergaurd						
15 mm	35.65	45.12	0.27	10.62	m	**55.74**
25 mm	35.65	45.12	0.27	10.62	m	**55.74**
28 mm	35.65	45.12	0.27	10.62	m	**55.74**
32 mm	35.65	45.12	0.30	11.80	m	**56.92**
35 mm	35.65	45.12	0.31	12.19	m	**57.31**
50 mm	35.65	45.12	0.34	13.37	m	**58.49**
100 mm	35.65	45.12	0.40	15.73	m	**60.85**
150 mm F-S-B Wintergaurd (to larger sizes)	35.65	45.12	0.40	15.73	m	**60.85**

38 MECHANICAL/COOLING/HEATING SYSTEMS

Item	Net Price £	Material £	Labour hours	Labour £	Unit	Total rate £
CHILLED WATER – cont						
Trace heating – cont						
Helically wound F-S-C Wintergaurd						
15 mm	45.45	57.52	1.00	39.33	m	**96.84**
25 mm	45.45	57.52	1.00	39.33	m	**96.84**
28 mm	45.45	57.52	1.00	39.33	m	**96.84**
32 mm	45.45	57.52	1.00	39.33	m	**96.84**
35 mm	45.45	57.52	1.00	39.33	m	**96.84**
50 mm	45.45	57.52	1.00	39.33	m	**96.84**
100 mm	45.45	57.52	1.00	39.33	m	**96.84**
150 mm F-S-B Wintergaurd (to larger sizes)	45.45	57.52	1.00	39.33	m	**96.84**
Accessories for trace heating; weatherproof; polycarbonate enclosure to IP standards; fully installed						
Connection junction box						
100 × 100 × 75 mm junction box incl. power connecting kit	106.79	135.15	1.40	55.06	nr	**190.21**
Single air thermostat						
150 × 150 × 75 mm AT-TS−13	233.94	296.07	1.42	55.84	nr	**351.91**
Single capillary thermostat						
150 × 150 × 75 mm AT-TS−13	336.58	425.98	1.46	57.42	nr	**483.39**
Twin capillary thermostat						
150 × 150 × 75 mm	603.83	764.21	1.46	57.42	nr	**821.63**
PRESSURIZATION UNITS						
Chilled water packaged pressurization unit complete with expansion vessel(s), interconnecting pipework and necessary isolating and drain valves; includes placing in position; electrical work elsewhere. Selection based on a final working pressure of 4 bar, a 3 m static head and system operating temperatures of 6°/12°C						
System volume						
1800 litres	1647.42	2084.98	8.00	314.61	nr	**2399.58**
4500 litres	1793.59	2269.96	8.00	314.61	nr	**2584.57**
7200 litres	1888.91	2390.60	10.00	393.26	nr	**2783.86**
9900 litres	2532.32	3204.90	10.00	393.26	nr	**3598.16**
15300 litres	2648.30	3351.69	13.00	511.24	nr	**3862.92**
22500 litres	2767.44	3502.48	20.00	786.52	nr	**4288.99**
27000 litres	2767.44	3502.48	20.00	786.52	nr	**4288.99**

38 MECHANICAL/COOLING/HEATING SYSTEMS

Item	Net Price £	Material £	Labour hours	Labour £	Unit	Total rate £
CHILLED BEAMS						
Static (passive) beams; cooling data based on 9 K waterside cooling dt (e.g. water 14°C flow and 16°C return, room temperature 24°C)						
Static passive beam in perforated metal casing for exposed mounting from solid ceiling or roof soffit, 2 pipe cooling only						
Length/Cooling output						
1200 mm/300 W	504.89	638.99	1.31	51.52	m	**690.51**
1800 mm/450 W	577.35	730.69	1.53	60.17	m	**790.86**
2400 mm/600 W	666.44	843.45	1.75	68.82	m	**912.27**
3000 mm/750 W	769.21	973.51	2.01	79.04	m	**1052.56**
3600 mm/900 W	941.61	1191.70	2.18	85.73	m	**1277.43**
Static passive beam, chassis type for mounting above open grid or perforated metal ceiling or roof soffit, 2 pipe cooling only						
Length/Cooling output						
1200 mm/300 W	323.80	409.80	1.31	51.52	m	**461.32**
1800 mm/450 W	386.92	489.69	1.53	60.17	m	**549.86**
2400 mm/600 W	456.21	577.38	1.75	68.82	m	**646.20**
3000 mm/750 W	536.89	679.49	2.01	79.04	m	**758.54**
3600 mm/900 W	622.82	788.24	2.18	85.73	m	**873.97**
Multi-service beams; for exposed mounting on solid ceiling slabs, with Integrated lighting and lighting control, provision of fascia openings for other services (electrical work and other services elsewhere). Performance criteria equivalent to standard beams as listed above						
2 way discharge exposed linear multi-service climate beam for exposed on solid ceiling; closed type with integrated secondary air circulation, 2 pipe cooling only, with primary air						
Length/Cooling output						
1200 mm/550 W	1053.42	1333.21	2.26	88.88	m	**1422.09**
1800 mm/900 W	1213.28	1535.53	2.49	97.92	m	**1633.45**
2400 mm/1200 W	1431.93	1812.25	2.67	105.00	m	**1917.25**
3000 mm/1550 W	1602.48	2028.09	2.84	111.69	m	**2139.78**
3600 mm/1850 W	1822.16	2306.13	3.08	121.12	m	**2427.25**
2 way discharge exposed linear multi-service climate beam for exposed on solid ceiling; closed type with integrated secondary air circulation, 4 pipe cooling and heating, with primary air						
Length/Cooling output/Heating output						
1200 mm/550 W/495 W	1070.75	1355.14	2.59	101.85	m	**1456.99**
1800 mm/900 W/810 W	1241.77	1571.59	2.82	110.90	m	**1682.49**
2400 mm/1200 W/1080 W	1447.19	1831.56	3.00	117.98	m	**1949.54**
3000 mm/1550 W/1395 W	1644.89	2081.78	3.18	125.06	m	**2206.83**
3600 mm/1850 W/1665 W	1828.51	2314.17	3.41	134.10	m	**2448.27**

38 MECHANICAL/COOLING/HEATING SYSTEMS

Item	Net Price £	Material £	Labour hours	Labour £	Unit	Total rate £
CHILLED WATER – cont						
Multi-service beams – cont						
Static passive multi-service beam for exposed mounting on solid ceiling, 2 pipe cooling only						
Length/Cooling output						
1200 mm/300 W	753.73	953.92	1.81	71.18	m	**1025.09**
1800 mm/450 W	1031.17	1305.05	2.03	79.83	m	**1384.88**
2400 mm/600 W	1308.62	1656.18	2.25	88.48	m	**1744.67**
3000 mm/750 W	1581.44	2001.47	2.51	98.71	m	**2100.17**
3600 mm/900 W	1905.12	2411.12	2.68	105.39	m	**2516.52**
Ventilated (active) beams; cooling data based on 9 K airside and 9 K waterside cooling dt (e.g. water 14°C flow and 16°C return, primary air 15°C, room temperature 24°C), heating data based on –4 K airside and 20 K waterside heating dt (e.g. water at 45°C flow and 35°C return, primary air 16°C, room temperature 20°C)						
1 way discharge recessed linear active climate beam mounted within a false ceiling; closed type with integrated secondary air circulation; 300 mm wide, 2 pipe cooling only, 5 l/s/m primary air						
Length/Cooling output						
1200 mm/340 W	454.73	575.50	1.76	69.21	m	**644.72**
1800 mm/530 W	554.99	702.39	1.99	78.26	m	**780.65**
2400 mm/730 W	677.14	856.99	2.17	85.34	m	**942.33**
3000 mm/900 W	766.91	970.60	2.34	92.02	m	**1062.62**
3600 mm/1100 W	887.42	1123.12	2.58	101.46	m	**1224.58**
1 way discharge recessed linear active climate beam mounted within a false ceiling; closed type with integrated secondary air circulation; 300 mm wide, 4-pipe cooling and heating, with 5 l/s/m primary air						
Length/Cooling output/Heating output						
1200 mm/340 W/180 W	444.17	562.15	2.09	82.19	m	**644.34**
1800 mm/530 W/310 W	535.56	677.80	2.32	91.24	m	**769.03**
2400 mm/730 W/450 W	628.76	795.76	2.50	98.31	m	**894.07**
3000 mm/900 W/600 W	708.67	896.89	2.68	105.39	m	**1002.29**
3600 mm/1100 W/750 W	759.62	961.37	2.91	114.44	m	**1075.81**
2 way discharge recessed linear active climate beam mounted within a false ceiling; closed type with integrated secondary air circulation, 600 mm wide, 2-pipe cooling only, with 10 l/s/m primary air						
Length/Cooling output						
1200 mm/550 W	424.77	537.59	1.76	69.21	m	**606.80**
1800 mm/900 W	476.34	602.85	1.99	78.26	m	**681.11**
2400 mm/1200 W	579.86	733.87	2.17	85.34	m	**819.21**
3000 mm/1550 W	678.33	858.49	2.34	92.02	m	**950.51**
3600 mm/1850 W	770.47	975.10	2.58	101.46	m	**1076.56**

38 MECHANICAL/COOLING/HEATING SYSTEMS

Item	Net Price £	Material £	Labour hours	Labour £	Unit	Total rate £
2 way discharge recessed linear active climate beam mounted within a false ceiling; closed type with integrated secondary air circulation; 600 mm wide, 4-pipe cooling and heating, with 10 l/s/m primary air						
Length/Cooling output/Heating output						
1200 mm/550 W/495 W	436.64	552.61	2.09	82.19	m	**634.80**
1800 mm/900 W/810 W	531.09	672.14	2.32	91.24	m	**763.38**
2400 mm/1200 W/1080 W	651.93	825.08	2.50	98.31	m	**923.39**
3000 mm/1550 W/1395 W	746.20	944.40	2.68	105.39	m	**1049.79**
3600 mm/1850 W/1665 W	844.59	1068.91	2.91	114.44	m	**1183.35**
4 way discharge recessed modular active climate beam mounted within a false ceiling; closed type with integrated secondary air circulation; 600 mm wide, 2-pipe cooling only, with primary air at 10 l/s/m of beam discharge (600 × 600 unit, 12 l/s, 1200 × 600 unit 18 l/s)						
Length/Cooling output						
600 × 600/500 W	323.82	409.83	1.56	61.35	m	**471.18**
1200 × 600/850 W	456.13	577.28	1.76	69.21	m	**646.49**
4 way discharge recessed modular active climate beam mounted within a false ceiling; closed type with integrated secondary air circulation; 600 mm wide, 4-pipe cooling and heating, with primary air at 10 l/s/m of beam discharge (600 × 600 unit 12 l/s, 1200 × 600 unit 18 l/s)						
Length/Cooling output/Heating output						
600 × 600/500 W/450 W	337.47	427.10	1.89	74.33	m	**501.43**
1200 × 600/850 W/765 W	489.58	619.62	2.09	82.19	m	**701.81**
1 way sidewall discharge active climate beam mounted within a ceiling bulkhead; closed type with integrated secondary air cisculation, 2-pipe cooling only, with 20 l/s/m primary air						
Length/Cooling output						
900 mm/700 W	588.93	745.35	1.89	74.33	m	**819.67**
1100 mm/900 W	611.60	774.04	1.89	74.33	m	**848.37**
1300 mm/1100 W	659.09	834.14	2.09	82.19	m	**916.33**
1500 mm/1200 W	688.80	871.75	2.09	82.19	m	**953.94**
1 way sidewall discharge active climate beam mounted within a ceiling bulkhead; closed type with integrated secondary air circulation, 4-pipe only, with 20 l/s/m primary air						
Length/Cooling output/Heating output						
900 mm/700 W/400 W	590.33	747.12	1.89	74.33	m	**821.45**
1100 mm/900 W/600 W	618.53	782.81	1.89	74.33	m	**857.14**
1300 mm/1100 W/800 W	659.89	835.16	2.09	82.19	m	**917.35**
1500 mm/1200 W/900 W	700.29	886.29	2.09	82.19	m	**968.48**
CHILLED CEILINGS						
Traditional (60% Radiant Absorption, 40% Convection) with bonded elements in perforated metal ceiling tiles (typically 30% free area and 2.5 mm perforated hole sizes) / tartan grid arrangement. Maximum waterside cooling effect approx. 65 W/m² at 9 dTK						
Radiant elements and metal ceilings combined	171.39	216.91	1.75	68.82	m	**285.73**

38 MECHANICAL/COOLING/HEATING SYSTEMS

Item	Net Price £	Material £	Labour hours	Labour £	Unit	Total rate £
CHILLED WATER – cont						
CHILLED CEILINGS – cont						
Hybrid (40% Radiant Absorption, 60% Convection) with radiant cooling chilled beam elements positioned above a perforated metal ceiling tiles (typically 30% free area and 2.5 mm perforated hole sizes) /tartan grid arrangement. Maximum waterside cooling effect approx. 100 W/m² at 9 dTK						
Radiant chilled beams	73.45	92.96	0.60	23.60	m	**116.56**
Metal ceilings	42.85	54.23	1.15	45.22	m	**99.45**
Convective beam (5% Radiant Absorption, 95% Convection) with convective cooling fin coil batteries positioned above a perforated metal ceiling tiles (typically 50% free area and 5.0 mm perforated hole sizes) / tartan grid arrangement. Maximum waterside cooling effect approx. 100 W/m² at 9 dTK						
Convective fin coil batteries	79.57	100.71	0.60	23.60	m	**124.30**
Metal ceilings	42.85	54.23	1.15	45.22	m	**99.45**
LEAK DETECTION						
Leak detection system consisting of a central control module connected by a leader cable to water sensing cables						
Control modules						
Alarm only (1 zone)	748.45	947.24	4.00	182.81	nr	**1130.06**
Alarm only (8 zone)	1465.78	1855.09	4.00	182.81	nr	**2037.90**
Alarm and location	4515.50	5714.81	8.00	365.63	nr	**6080.44**
Cables						
Sensing – 3 m length	181.51	229.72	4.00	182.81	nr	**412.53**
Sensing – 5 m length	223.04	282.28	4.00	182.81	nr	**465.10**
Sensing – 7.5 m length	316.87	401.04	4.00	182.81	nr	**583.85**
Sensing – 15 m length	443.01	560.67	8.00	365.63	nr	**926.30**
Sensing – 25 m length	615.29	778.71	12.00	471.91	nr	**1250.62**
Leader – 3.5 m length	72.96	92.34	2.00	91.41	nr	**183.75**
End terminal						
End terminal	39.59	50.11	0.05	2.29	nr	**52.40**
ENERGY METERS						
Ultrasonic						
Energy meter for measuring energy use in chilled water systems; includes ultrasonic flow meter (with sensor and signal converter), energy calculator, pair of temperature sensors with brass pockets, and 3 m of interconnecting cable; includes fixing in position; electrical work elsewhere						

38 MECHANICAL/COOLING/HEATING SYSTEMS

Item	Net Price £	Material £	Labour hours	Labour £	Unit	Total rate £
Pipe size (flanged connections to PN16); maximum flow rate						
50 mm, 36 m³/hr	1715.13	2170.67	1.80	70.79	nr	**2241.46**
65 mm, 60 m³/hr	1890.57	2392.71	2.32	91.24	nr	**2483.95**
80 mm, 100 m³/hr	2119.20	2682.05	2.56	100.67	nr	**2782.73**
125 mm, 250 m³/hr	2455.29	3107.42	3.60	141.57	nr	**3248.99**
150 mm, 360 m³/hr	2665.37	3373.29	4.80	188.76	nr	**3562.06**
200 mm, 600 m³/hr	2977.99	3768.94	6.24	245.39	nr	**4014.33**
250 mm, 1000 m³/hr	3436.41	4349.13	9.60	377.53	nr	**4726.65**
300 mm, 1500 m³/hr	4039.43	5112.31	10.80	424.72	nr	**5537.03**
350 mm, 2000 m³/hr	4867.34	6160.10	13.20	519.10	nr	**6679.20**
400 mm, 2500 m³/hr	5571.68	7051.52	15.60	613.48	nr	**7665.00**
500 mm, 3000 m³/hr	6322.96	8002.34	24.00	943.82	nr	**8946.16**
600 mm, 3500 m³/hr	7103.91	8990.70	28.00	1101.12	nr	**10091.83**

Electromagnetic

Energy meter for measuring energy use in chilled water systems; includes electromagnetic flow meter (with sensor and signal converter), energy calculator, pair of temperature sensors with brass pockets, and 3 m of interconnecting cable; includes fixing in position; electrical work elsewhere

Item	Net Price £	Material £	Labour hours	Labour £	Unit	Total rate £
Pipe size (flanged connections to PN40); maximum flow rate						
25 mm, 17.7 m³/hr	1229.62	1556.21	1.48	58.20	nr	**1614.41**
40 mm, 45 m³/hr	1241.37	1571.08	1.55	60.95	nr	**1632.04**
Pipe size (flanged connections to PN16); maximum flow rate						
50 mm, 70 m³/hr	1255.28	1588.69	1.80	70.79	nr	**1659.47**
65 mm, 120 m³/hr	1260.63	1595.45	2.32	91.24	nr	**1686.69**
80 mm, 180 m³/hr	1267.06	1603.59	2.56	100.67	nr	**1704.27**
125 mm, 450 m³/hr	1379.32	1745.67	3.60	141.57	nr	**1887.24**
150 mm, 625 m³/hr	1463.80	1852.58	4.80	188.76	nr	**2041.35**
200 mm, 1100 m³/hr	1575.00	1993.32	6.24	245.39	nr	**2238.71**
250 mm, 1750 m³/hr	1767.46	2236.90	9.60	377.53	nr	**2614.43**
300 mm, 2550 m³/hr	2252.90	2851.27	10.80	424.72	nr	**3275.99**
350 mm, 3450 m³/hr	2930.79	3709.21	13.20	519.10	nr	**4228.31**
400 mm, 4500 m³/hr	3343.51	4231.54	15.60	613.48	nr	**4845.03**

38 MECHANICAL/COOLING/HEATING SYSTEMS

Item	Net Price £	Material £	Labour hours	Labour £	Unit	Total rate £
LOCAL COOLING UNITS						
Split system with ceiling void evaporator unit and external condensing unit						
Ceiling mounted 4 way blow cassette heat pump unit with remote fan speed and load control; refrigerant 410 A; includes outdoor unit						
Cooling 3.6 kW, heating 4.1 kW	2127.30	2692.31	35.00	1376.40	nr	**4068.71**
Cooling 4.9 kW, heating 5.5 kW	2344.25	2966.88	35.00	1376.40	nr	**4343.29**
Cooling 7.1 kW, heating 8.2 kW	2837.50	3591.14	35.00	1376.40	nr	**4967.54**
Cooling 10 kW, heating 11.2 kW	3356.19	4247.60	35.00	1376.40	nr	**5624.00**
Cooling 12.20 kW, heating 14.60 kW	3671.47	4646.61	35.00	1376.40	nr	**6023.01**
Ceiling mounted 4 way blow cooling only unit with remote fan speed and load control; refrigerant 410 A; includes outdoor unit						
Cooling 3.80 kW	1925.57	2437.01	35.00	1376.40	nr	**3813.41**
Cooling 5.20 kW	2151.00	2722.31	35.00	1376.40	nr	**4098.71**
Cooling 7.10 kW	2652.75	3357.32	35.00	1376.40	nr	**4733.72**
Cooling 10 kW	4248.12	5376.41	35.00	1376.40	nr	**6752.82**
Cooling 12.2 kW	4897.45	6198.21	35.00	1376.40	nr	**7574.61**
In ceiling, ducted heat pump unit with remote fan speed and load control; refrigerant 410 A; includes outdoor unit						
Cooling 3.60 kW, heating 4.10 kW	1559.45	1973.64	35.00	1376.40	nr	**3350.04**
Cooling 4.90 kW, heating 5.50 kW	1793.36	2269.68	35.00	1376.40	nr	**3646.08**
Cooling 7.10 kW, heating 8.20 kW	1793.36	2269.68	35.00	1376.40	nr	**3646.08**
Cooling 10 kW, heating 11.20 kW	1793.36	2269.68	35.00	1376.40	nr	**3646.08**
Cooling 12.20 kW, heating 14.50 kW	3615.55	4575.84	35.00	1376.40	nr	**5952.24**
In ceiling, ducted cooling only unit with remote fan speed and load control; refrigerant 410 A; includes outdoor unit						
Cooling 3.70 kW	1406.89	1780.56	35.00	1376.40	nr	**3156.97**
Cooling 4.90 kW	1673.02	2117.38	35.00	1376.40	nr	**3493.78**
Cooling 7.10 kW	2525.63	3196.44	35.00	1376.40	nr	**4572.84**
Cooling 10 kW	2879.89	3644.79	35.00	1376.40	nr	**5021.20**
Cooling 12.3 kW	3210.41	4063.10	35.00	1376.40	nr	**5439.50**
Room units						
Ceiling mounted 4 way blow cassette heat pump unit with remote fan speed and load control; refrigerant 410 A; excludes outdoor unit						
Cooling 3.6 kW, heating 4.1 kW	1342.48	1699.04	17.00	668.54	nr	**2367.58**
Cooling 4.9 kW, heating 5.5 kW	1374.69	1739.80	17.00	668.54	nr	**2408.34**
Cooling 7.1 kW, heating 8.2 kW	1488.26	1883.54	17.00	668.54	nr	**2552.08**
Cooling 10 kW, heating 11.2 kW	1612.00	2040.15	17.00	668.54	nr	**2708.69**
Cooling 12.20 kW, heating 14.60 kW	1747.61	2211.78	17.00	668.54	nr	**2880.31**
Ceiling mounted 4 way blow cooling unit with remote fan speed and load control; refrigerant 410 A; excludes outdoor unit						
Cooling 3.80 kW	1267.88	1604.63	17.00	668.54	nr	**2273.17**
Cooling 5.20 kW	1278.06	1617.52	17.00	668.54	nr	**2286.05**
Cooling 7.10 kW	1488.26	1883.54	17.00	668.54	nr	**2552.08**
Cooling 10 kW	1612.00	2040.15	17.00	668.54	nr	**2708.69**
Cooling 12.2 kW	1747.61	2211.78	17.00	668.54	nr	**2880.31**

38 MECHANICAL/COOLING/HEATING SYSTEMS

Item	Net Price £	Material £	Labour hours	Labour £	Unit	Total rate £
In ceiling, ducted heat pump unit with remote fan speed and load control; refrigerant 410 A; excludes outdoor unit						
Cooling 3.60 kW, heating 4.10 kW	774.64	980.38	17.00	668.54	nr	**1648.92**
Cooling 4.90 kW, heating 5.50 kW	823.80	1042.60	17.00	668.54	nr	**1711.14**
Cooling 7.10 kW, heating 8.20 kW	1361.13	1722.65	17.00	668.54	nr	**2391.19**
Cooling 10 kW, heating 11.20 kW	1410.29	1784.86	17.00	668.54	nr	**2453.40**
Cooling 12.20 kW, heating 14.50 kW	1691.62	2140.91	17.00	668.54	nr	**2809.45**
In ceiling, ducted cooling unit only with remote fan speed and load control; refrigerant 410 A; excludes outdoor unit						
Cooling 3.70 kW	749.21	948.20	17.00	668.54	nr	**1616.74**
Cooling 4.90 kW	800.07	1012.57	17.00	668.54	nr	**1681.11**
Cooling 7.10 kW	1361.13	1722.65	17.00	668.54	nr	**2391.19**
Cooling 10 kW	1410.29	1784.86	17.00	668.54	nr	**2453.40**
Cooling 12.3 kW	1691.68	2140.98	17.00	668.54	nr	**2809.52**
External condensing units suitable for connection to multiple indoor units; inverter driven; refrigerant 410 A						
Cooling only						
9 kW	2930.74	3709.15	17.00	668.54	nr	**4377.68**
Heat pump						
Cooling 5.20 kW, heating 6.10 kW	2088.33	2642.99	17.00	668.54	nr	**3311.53**
Cooling 6.80 kW, heating 2.50 kW	2662.92	3370.19	17.00	668.54	nr	**4038.73**
Cooling 8 kW, heating 9.60 kW	3083.31	3902.24	17.00	668.54	nr	**4570.78**
Cooling 14.50 kW, heating 16.50 kW	5117.36	6476.53	21.00	825.84	nr	**7302.37**

38 VENTILATION/AIR CONDITIONING SYSTEMS

Item	Net Price £	Material £	Labour hours	Labour £	Unit	Total rate £
DUCTWORK: CIRCULAR						
AIR DUCTLINES						
Galvanized sheet metal DW144 class B spirally wound circular section ductwork; including all necessary stiffeners, joints, couplers in the running length and duct supports						
Straight duct						
80 mm dia.	8.73	11.05	0.87	32.68	m	**43.73**
100 mm dia.	8.98	11.37	0.87	32.68	m	**44.05**
160 mm dia.	12.66	16.02	0.87	32.68	m	**48.70**
200 mm dia.	16.47	20.85	0.87	32.68	m	**53.53**
250 mm dia.	19.54	24.73	1.21	45.45	m	**70.19**
315 mm dia.	23.89	30.24	1.21	45.45	m	**75.69**
355 mm dia.	35.63	45.09	1.21	45.45	m	**90.54**
400 mm dia.	36.24	45.87	1.21	45.45	m	**91.32**
450 mm dia.	43.22	54.69	1.21	45.45	m	**100.15**
500 mm dia.	46.95	59.41	1.21	45.45	m	**104.87**
630 mm dia.	97.38	123.25	1.39	52.21	m	**175.46**
710 mm dia.	108.64	137.50	1.39	52.21	m	**189.71**
800 mm dia.	115.60	146.31	1.44	54.09	m	**200.40**
900 mm dia.	137.54	174.07	1.46	54.84	m	**228.91**
1000 mm dia.	174.94	221.41	1.65	61.98	m	**283.39**
1120 mm dia.	208.72	264.16	2.43	91.28	m	**355.44**
1250 mm dia.	227.32	287.70	2.43	91.28	m	**378.98**
1400 mm dia.	255.73	323.65	2.77	104.05	m	**427.70**
1600 mm dia.	304.36	385.20	3.06	114.94	m	**500.14**
Extra over fittings; circular duct class B						
End cap						
80 mm dia.	2.35	2.97	0.15	5.63	nr	**8.61**
100 mm dia.	2.55	3.23	0.15	5.63	nr	**8.86**
160 mm dia.	3.81	4.83	0.15	5.63	nr	**10.46**
200 mm dia.	4.83	6.11	0.20	7.51	nr	**13.62**
250 mm dia.	7.63	9.65	0.29	10.89	nr	**20.54**
315 mm dia.	8.90	11.26	0.29	10.89	nr	**22.15**
355 mm dia.	13.46	17.04	0.44	16.53	nr	**33.56**
400 mm dia.	14.77	18.70	0.44	16.53	nr	**35.22**
450 mm dia.	16.62	21.04	0.44	16.53	nr	**37.57**
500 mm dia.	17.48	22.12	0.44	16.53	nr	**38.65**
630 mm dia.	90.61	114.67	0.58	21.79	nr	**136.46**
710 mm dia.	102.57	129.81	0.69	25.92	nr	**155.73**
800 mm dia.	107.80	136.43	0.81	30.43	nr	**166.85**
900 mm dia.	132.31	167.45	0.92	34.56	nr	**202.01**
1000 mm dia.	194.93	246.70	1.04	39.07	nr	**285.77**
1120 mm dia.	227.85	288.37	1.16	43.57	nr	**331.95**
1250 mm dia.	257.19	325.50	1.16	43.57	nr	**369.07**
1400 mm dia.	319.72	404.64	1.16	43.57	nr	**448.21**
1600 mm dia.	354.65	448.84	1.16	43.57	nr	**492.41**

38 VENTILATION/AIR CONDITIONING SYSTEMS

Item	Net Price £	Material £	Labour hours	Labour £	Unit	Total rate £
Reducer						
80 mm dia.	12.91	16.34	0.29	10.89	nr	**27.24**
100 mm dia.	13.29	16.82	0.29	10.89	nr	**27.71**
160 mm dia.	17.71	22.42	0.29	10.89	nr	**33.31**
200 mm dia.	21.54	27.26	0.44	16.53	nr	**43.79**
250 mm dia.	21.40	27.09	0.58	21.79	nr	**48.87**
315 mm dia.	24.53	31.05	0.58	21.79	nr	**52.84**
355 mm dia.	26.70	33.79	0.87	32.68	nr	**66.47**
400 mm dia.	31.21	39.49	0.87	32.68	nr	**72.17**
450 mm dia.	33.57	42.49	0.87	32.68	nr	**75.17**
500 mm dia.	35.88	45.40	0.87	32.68	nr	**78.09**
630 mm dia.	138.01	174.66	0.87	32.68	nr	**207.34**
710 mm dia.	162.13	205.19	0.96	36.06	nr	**241.26**
800 mm dia.	174.61	220.99	1.06	39.82	nr	**260.81**
900 mm dia.	199.99	253.10	1.16	43.57	nr	**296.68**
1000 mm dia.	290.18	367.25	1.25	46.95	nr	**414.20**
1120 mm dia.	291.80	369.31	3.47	130.35	nr	**499.65**
1250 mm dia.	362.97	459.37	3.47	130.35	nr	**589.72**
1400 mm dia.	389.21	492.58	4.05	152.13	nr	**644.72**
1600 mm dia.	470.59	595.57	4.62	173.54	nr	**769.12**
90° segmented radius bend						
80 mm dia.	5.42	6.86	0.29	10.89	nr	**17.76**
100 mm dia.	5.70	7.22	0.29	10.89	nr	**18.11**
160 mm dia.	9.65	12.22	0.29	10.89	nr	**23.11**
200 mm dia.	11.92	15.09	0.44	16.53	nr	**31.62**
250 mm dia.	16.52	20.90	0.58	21.79	nr	**42.69**
315 mm dia.	16.64	21.05	0.58	21.79	nr	**42.84**
355 mm dia.	15.36	19.43	0.87	32.68	nr	**52.11**
400 mm dia.	18.75	23.72	0.87	32.68	nr	**56.40**
450 mm dia.	21.14	26.76	0.87	32.68	nr	**59.44**
500 mm dia.	26.97	34.13	0.87	32.68	nr	**66.81**
630 mm dia.	98.36	124.48	0.87	32.68	nr	**157.16**
710 mm dia.	119.89	151.73	0.96	36.06	nr	**187.79**
800 mm dia.	120.01	151.89	1.06	39.82	nr	**191.71**
900 mm dia.	165.67	209.67	1.16	43.57	nr	**253.25**
1000 mm dia.	258.80	327.53	1.25	46.95	nr	**374.49**
1120 mm dia.	305.79	387.00	3.47	130.35	nr	**517.35**
1250 mm dia.	326.88	413.70	3.47	130.35	nr	**544.04**
1400 mm dia.	674.24	853.32	4.05	152.13	nr	**1005.46**
1600 mm dia.	1001.90	1268.00	4.62	173.54	nr	**1441.55**
45° radius bend						
80 mm dia.	4.61	5.83	0.29	10.89	nr	**16.72**
100 mm dia.	4.96	6.28	0.29	10.89	nr	**17.17**
160 mm dia.	8.32	10.53	0.29	10.89	nr	**21.42**
200 mm dia.	10.30	13.03	0.40	15.03	nr	**28.06**
250 mm dia.	14.32	18.13	0.58	21.79	nr	**39.91**
315 mm dia.	15.11	19.12	0.58	21.79	nr	**40.91**
355 mm dia.	17.40	22.02	0.87	32.68	nr	**54.70**
400 mm dia.	18.80	23.79	0.87	32.68	nr	**56.47**
450 mm dia.	19.97	25.28	0.87	32.68	nr	**57.96**
500 mm dia.	23.42	29.63	0.87	32.68	nr	**62.31**
630 mm dia.	114.47	144.87	0.87	32.68	nr	**177.55**

38 VENTILATION/AIR CONDITIONING SYSTEMS

Item	Net Price £	Material £	Labour hours	Labour £	Unit	Total rate £
DUCTWORK: CIRCULAR – cont						
Extra over fittings – cont						
45° radius bend – cont						
710 mm dia.	128.15	162.19	0.96	36.06	nr	**198.25**
800 mm dia.	139.98	177.16	1.06	39.82	nr	**216.98**
900 mm dia.	158.72	200.88	1.16	43.57	nr	**244.45**
1000 mm dia.	264.84	335.19	1.25	46.95	nr	**382.14**
1120 mm dia.	303.83	384.53	3.47	130.35	nr	**514.87**
1250 mm dia.	320.34	405.43	3.47	130.35	nr	**535.77**
1400 mm dia.	466.43	590.32	4.05	152.13	nr	**742.45**
1600 mm dia.	482.34	610.45	4.62	173.54	nr	**783.99**
90° equal twin bend						
80 mm dia.	29.72	37.62	0.58	21.79	nr	**59.41**
100 mm dia.	30.23	38.26	0.58	21.79	nr	**60.05**
160 mm dia.	39.78	50.35	0.58	21.79	nr	**72.13**
200 mm dia.	51.72	65.46	0.87	32.68	nr	**98.14**
250 mm dia.	57.50	72.77	1.16	43.57	nr	**116.35**
315 mm dia.	88.51	112.02	1.16	43.57	nr	**155.59**
355 mm dia.	104.62	132.41	1.73	64.99	nr	**197.39**
400 mm dia.	121.44	153.70	1.73	64.99	nr	**218.69**
450 mm dia.	142.35	180.15	1.73	64.99	nr	**245.14**
500 mm dia.	150.51	190.48	1.73	64.99	nr	**255.47**
630 mm dia.	283.17	358.38	1.73	64.99	nr	**423.36**
710 mm dia.	337.98	427.75	1.82	68.37	nr	**496.12**
800 mm dia.	457.25	578.69	1.93	72.50	nr	**651.19**
900 mm dia.	599.70	758.98	2.02	75.88	nr	**834.85**
1000 mm dia.	923.34	1168.58	2.11	79.26	nr	**1247.83**
1120 mm dia.	1336.35	1691.29	4.62	173.54	nr	**1864.83**
1250 mm dia.	1348.22	1706.31	4.62	173.54	nr	**1879.85**
1400 mm dia.	1955.10	2474.38	4.62	173.54	nr	**2647.92**
1600 mm dia.	2079.41	2631.70	4.62	173.54	nr	**2805.24**
Conical branch						
80 mm dia.	27.93	35.35	0.58	21.79	nr	**57.13**
100 mm dia.	28.16	35.64	0.58	21.79	nr	**57.42**
160 mm dia.	31.41	39.75	0.58	21.79	nr	**61.53**
200 mm dia.	36.74	46.50	0.87	32.68	nr	**79.18**
250 mm dia.	40.41	51.14	1.16	43.57	nr	**94.71**
315 mm dia.	63.87	80.83	1.16	43.57	nr	**124.40**
355 mm dia.	73.55	93.09	1.73	64.99	nr	**158.07**
400 mm dia.	86.92	110.01	1.73	64.99	nr	**175.00**
450 mm dia.	94.89	120.10	1.73	64.99	nr	**185.08**
500 mm dia.	104.62	132.41	1.73	64.99	nr	**197.40**
630 mm dia.	173.70	219.84	1.73	64.99	nr	**284.83**
710 mm dia.	244.07	308.89	1.82	68.37	nr	**377.26**
800 mm dia.	278.90	352.97	1.93	72.50	nr	**425.47**
900 mm dia.	299.49	379.04	2.02	75.88	nr	**454.92**
1000 mm dia.	356.59	451.29	2.11	79.26	nr	**530.55**
1120 mm dia.	418.28	529.37	4.62	173.54	nr	**702.91**
1250 mm dia.	449.65	569.08	5.20	195.33	nr	**764.41**
1400 mm dia. .	634.86	803.48	5.20	195.33	nr	**998.81**
1600 mm dia.	783.40	991.47	5.20	195.33	nr	**1186.80**

38 VENTILATION/AIR CONDITIONING SYSTEMS

Item	Net Price £	Material £	Labour hours	Labour £	Unit	Total rate £
45° branch						
80 mm dia.	25.19	31.88	0.58	21.79	nr	**53.67**
100 mm dia.	25.40	32.14	0.58	21.79	nr	**53.93**
160 mm dia.	26.94	34.10	0.58	21.79	nr	**55.89**
200 mm dia.	29.08	36.80	0.87	32.68	nr	**69.48**
250 mm dia.	32.52	41.16	1.16	43.57	nr	**84.73**
315 mm dia.	37.57	47.54	1.16	43.57	nr	**91.12**
355 mm dia.	40.96	51.83	1.73	64.99	nr	**116.82**
400 mm dia.	46.13	58.38	1.73	64.99	nr	**123.37**
450 mm dia.	51.96	65.76	1.73	64.99	nr	**130.75**
500 mm dia.	57.66	72.97	1.73	64.99	nr	**137.95**
630 mm dia.	122.61	155.17	1.73	64.99	nr	**220.16**
710 mm dia.	142.99	180.97	1.82	68.37	nr	**249.34**
800 mm dia.	153.45	194.21	2.13	80.01	nr	**274.22**
900 mm dia.	170.25	215.47	2.31	86.77	nr	**302.24**
1000 mm dia.	251.45	318.24	2.31	86.77	nr	**405.01**
1120 mm dia.	303.57	384.20	4.62	173.54	nr	**557.74**
1250 mm dia.	393.21	497.65	4.62	173.54	nr	**671.19**
1400 mm dia.	440.95	558.07	4.62	173.54	nr	**731.61**
1600 mm dia.	499.88	632.65	4.62	173.54	nr	**806.19**

For galvanized sheet metal DW144 class C rates, refer to galvanized sheet metal DW144 class B

38 VENTILATION/AIR CONDITIONING SYSTEMS

Item	Net Price £	Material £	Labour hours	Labour £	Unit	Total rate £
DUCTWORK: FLAT OVAL						
AIR DUCTLINES						
Galvanized sheet metal DW144 class B spirally wound flat oval section ductwork; including all necessary stiffeners, joints, couplers in the running length and duct supports						
Straight duct						
345 × 102 mm	84.12	106.46	2.71	101.80	m	208.26
427 × 102 mm	89.24	112.94	2.99	112.32	m	225.26
508 × 102 mm	94.60	119.72	3.14	117.95	m	237.67
559 × 152 mm	99.95	126.49	3.43	128.84	m	255.33
531 × 203 mm	99.95	126.49	3.43	128.84	m	255.33
851 × 203 mm	149.55	189.27	3.62	135.98	m	325.25
582 × 254 mm	105.88	134.01	3.93	147.63	m	281.63
823 × 254 mm	149.67	189.42	5.72	214.86	m	404.29
1303 × 254 mm	275.33	348.46	5.72	214.86	m	563.32
632 × 305 mm	111.05	140.54	5.80	217.87	m	358.41
1275 × 305 mm	268.53	339.85	8.12	305.02	m	644.86
765 × 356 mm	126.71	160.37	8.12	305.02	m	465.39
1247 × 356 mm	277.18	350.80	8.13	305.39	m	656.19
1727 × 356 mm	330.66	418.49	10.41	391.04	m	809.52
737 × 406 mm	121.69	154.01	5.72	214.86	m	368.87
818 × 406 mm	147.83	187.10	6.21	233.27	m	420.37
978 × 406 mm	194.29	245.89	6.92	259.94	m	505.83
1379 × 406 mm	290.84	368.08	8.75	328.68	m	696.77
1699 × 406 mm	332.98	421.42	10.41	391.04	m	812.46
709 × 457 mm	131.78	166.78	5.72	214.86	m	381.65
1189 × 457 mm	253.63	321.00	8.80	330.56	m	651.56
1671 × 457 mm	329.66	417.22	10.31	387.28	m	804.50
678 × 508 mm	131.49	166.42	5.72	214.86	m	381.28
919 × 508 mm	165.70	209.71	7.30	274.21	m	483.93
1321 × 508 mm	290.10	367.15	8.75	328.68	m	695.83
Extra over fittings; flat oval duct class B						
End cap						
345 × 102 mm	40.82	51.67	0.20	7.51	nr	59.18
427 × 102 mm	51.20	64.79	0.20	7.51	nr	72.31
508 × 102 mm	55.12	69.75	0.20	7.51	nr	77.27
559 × 152 mm	57.30	72.51	0.29	10.89	nr	83.41
531 × 203 mm	57.32	72.54	0.29	10.89	nr	83.43
851 × 203 mm	70.79	89.60	0.44	16.53	nr	106.12
582 × 254 mm	61.03	77.24	0.44	16.53	nr	93.77
823 × 254 mm	70.81	89.62	0.44	16.53	nr	106.15
1303 × 254 mm	218.80	276.91	0.69	25.92	nr	302.83
632 × 305 mm	63.71	80.63	0.69	25.92	nr	106.55
1275 × 305 mm	205.79	260.44	0.69	25.92	nr	286.36
765 × 356 mm	70.81	89.62	0.69	25.92	nr	115.54
1247 × 356 mm	218.82	276.94	0.69	25.92	nr	302.86
1727 × 356 mm	326.01	412.60	0.69	25.92	nr	438.52
737 × 406 mm	70.88	89.70	0.69	25.92	nr	115.62

38 VENTILATION/AIR CONDITIONING SYSTEMS

Item	Net Price £	Material £	Labour hours	Labour £	Unit	Total rate £
818 × 406 mm	77.36	97.91	0.69	25.92	nr	**123.83**
978 × 406 mm	105.05	132.95	1.04	39.07	nr	**172.01**
1379 × 406 mm	246.39	311.84	1.04	39.07	nr	**350.90**
1699 × 406 mm	326.35	413.02	1.04	39.07	nr	**452.09**
709 × 457 mm	70.83	89.64	1.04	39.07	nr	**128.70**
1189 × 457 mm	217.51	275.28	1.04	39.07	nr	**314.35**
1671 × 457 mm	320.25	405.31	1.04	39.07	nr	**444.37**
678 × 508 mm	70.81	89.62	1.04	39.07	nr	**128.69**
919 × 508 mm	105.03	132.93	1.04	39.07	nr	**172.00**
1321 × 508 mm	246.35	311.78	1.04	39.07	nr	**350.85**
Reducer						
345 × 102 mm	86.19	109.08	0.95	35.69	nr	**144.77**
427 × 102 mm	93.47	118.30	1.06	39.82	nr	**158.11**
508 × 102 mm	107.82	136.45	1.13	42.45	nr	**178.90**
559 × 152 mm	126.72	160.38	1.26	47.33	nr	**207.71**
531 × 203 mm	128.15	162.19	1.26	47.33	nr	**209.52**
851 × 203 mm	140.88	178.29	1.34	50.34	nr	**228.63**
582 × 254 mm	132.19	167.30	1.34	50.34	nr	**217.64**
823 × 254 mm	142.97	180.94	1.34	50.34	nr	**231.28**
1303 × 254 mm	380.52	481.58	1.34	50.34	nr	**531.92**
632 × 305 mm	134.98	170.83	0.70	26.29	nr	**197.13**
1275 × 305 mm	380.63	481.72	1.16	43.57	nr	**525.29**
765 × 356 mm	156.49	198.05	1.16	43.57	nr	**241.63**
1247 × 356 mm	379.52	480.32	1.16	43.57	nr	**523.89**
1727 × 356 mm	435.57	551.26	1.25	46.95	nr	**598.21**
737 × 406 mm	160.30	202.88	1.16	43.57	nr	**246.45**
818 × 406 mm	174.10	220.34	1.27	47.71	nr	**268.05**
978 × 406 mm	190.41	240.98	1.44	54.09	nr	**295.07**
1379 × 406 mm	380.25	481.24	1.44	54.09	nr	**535.34**
1699 × 406 mm	438.65	555.16	1.44	54.09	nr	**609.25**
709 × 457 mm	158.21	200.23	1.16	43.57	nr	**243.80**
1189 × 457 mm	394.87	499.75	1.34	50.34	nr	**550.09**
1671 × 457 mm	433.67	548.85	1.44	54.09	nr	**602.94**
678 × 508 mm	160.26	202.82	1.16	43.57	nr	**246.40**
919 × 508 mm	210.39	266.27	1.26	47.33	nr	**313.60**
1321 × 508 mm	388.31	491.45	1.44	54.09	nr	**545.54**
90° radius bend						
345 × 102 mm	221.10	279.82	0.29	10.89	nr	**290.71**
427 × 102 mm	245.33	310.49	0.58	21.79	nr	**332.28**
508 × 102 mm	302.55	382.91	0.58	21.79	nr	**404.70**
559 × 152 mm	351.44	444.78	0.58	21.79	nr	**466.57**
531 × 203 mm	355.88	450.40	0.87	32.68	nr	**483.08**
851 × 203 mm	361.92	458.05	0.87	32.68	nr	**490.73**
582 × 254 mm	365.40	462.46	0.87	32.68	nr	**495.14**
823 × 254 mm	368.27	466.09	0.87	32.68	nr	**498.77**
1303 × 254 mm	387.36	490.24	0.96	36.06	nr	**526.30**
632 × 305 mm	391.30	495.23	0.87	32.68	nr	**527.91**
1275 × 305 mm	405.56	513.27	0.96	36.06	nr	**549.34**

38 VENTILATION/AIR CONDITIONING SYSTEMS

Item	Net Price £	Material £	Labour hours	Labour £	Unit	Total rate £
DUCTWORK: FLAT OVAL – cont						
Extra over fittings – cont						
90° radius bend – cont						
765 × 356 mm	397.70	503.34	0.87	32.68	nr	**536.02**
1247 × 356 mm	400.85	507.32	0.96	36.06	nr	**543.38**
1727 × 356 mm	776.53	982.78	1.25	46.95	nr	**1029.73**
737 × 406 mm	400.12	506.40	0.87	32.68	nr	**539.08**
818 × 406 mm	403.71	510.93	0.96	36.06	nr	**547.00**
978 × 406 mm	333.80	422.46	0.96	36.06	nr	**458.52**
1379 × 406 mm	648.97	821.33	1.16	43.57	nr	**864.91**
1699 × 406 mm	784.72	993.15	1.25	46.95	nr	**1040.10**
709 × 457 mm	416.26	526.82	0.87	32.68	nr	**559.50**
1189 × 457 mm	465.28	588.86	0.96	36.06	nr	**624.92**
1671 × 457 mm	1129.84	1429.92	1.25	46.95	nr	**1476.88**
678 × 508 mm	399.99	506.23	0.87	32.68	nr	**538.91**
919 × 508 mm	408.18	516.60	0.96	36.06	nr	**552.66**
1321 × 508 mm	755.32	955.94	1.16	43.57	nr	**999.51**
45° radius bend						
345 × 102 mm	109.34	138.38	0.79	29.68	nr	**168.06**
427 × 102 mm	143.50	181.61	0.85	31.93	nr	**213.54**
508 × 102 mm	176.82	223.78	0.95	35.69	nr	**259.46**
559 × 152 mm	205.32	259.86	0.79	29.68	nr	**289.53**
531 × 203 mm	207.55	262.67	0.85	31.93	nr	**294.60**
851 × 203 mm	220.07	278.52	0.98	36.81	nr	**315.33**
582 × 254 mm	218.74	276.83	0.76	28.55	nr	**305.38**
823 × 254 mm	223.36	282.69	0.95	35.69	nr	**318.37**
1303 × 254 mm	417.22	528.03	1.16	43.57	nr	**571.61**
632 × 305 mm	229.99	291.07	0.58	21.79	nr	**312.86**
1275 × 305 mm	422.15	534.27	1.16	43.57	nr	**577.84**
765 × 356 mm	244.63	309.61	0.87	32.68	nr	**342.29**
1247 × 356 mm	419.98	531.52	1.16	43.57	nr	**575.10**
1727 × 356 mm	742.57	939.79	1.26	47.33	nr	**987.12**
737 × 406 mm	211.98	268.28	0.69	25.92	nr	**294.20**
818 × 406 mm	243.35	307.98	0.78	29.30	nr	**337.28**
978 × 406 mm	250.72	317.32	0.87	32.68	nr	**350.00**
1379 × 406 mm	601.58	761.36	1.16	43.57	nr	**804.94**
1699 × 406 mm	747.01	945.42	1.27	47.71	nr	**993.12**
709 × 457 mm	246.92	312.50	0.81	30.43	nr	**342.93**
1189 × 457 mm	448.53	567.66	0.95	35.69	nr	**603.34**
1671 × 457 mm	791.26	1001.42	1.26	47.33	nr	**1048.75**
678 × 508 mm	243.15	307.73	0.92	34.56	nr	**342.29**
919 × 508 mm	250.52	317.06	1.10	41.32	nr	**358.38**
1321 × 508 mm	614.92	778.24	1.25	46.95	nr	**825.20**
90° hard bend with turning vanes						
345 × 102 mm	93.79	118.70	0.55	20.66	nr	**139.36**
427 × 102 mm	161.43	204.30	1.16	43.57	nr	**247.88**
508 × 102 mm	219.10	277.29	1.16	43.57	nr	**320.87**
559 × 152 mm	246.32	311.74	1.16	43.57	nr	**355.32**
531 × 203 mm	246.79	312.34	1.73	64.99	nr	**377.33**
851 × 203 mm	360.58	456.35	1.73	64.99	nr	**521.33**
582 × 254 mm	264.35	334.56	1.73	64.99	nr	**399.54**

38 VENTILATION/AIR CONDITIONING SYSTEMS

Item	Net Price £	Material £	Labour hours	Labour £	Unit	Total rate £
823 × 254 mm	360.58	456.35	1.73	64.99	nr	**521.33**
1303 × 254 mm	639.50	809.35	1.82	68.37	nr	**877.71**
632 × 305 mm	281.81	356.66	1.73	64.99	nr	**421.64**
1275 × 305 mm	635.43	804.20	1.82	68.37	nr	**872.56**
765 × 356 mm	343.69	434.97	1.73	64.99	nr	**499.96**
1247 × 356 mm	635.81	804.69	1.82	68.37	nr	**873.05**
1727 × 356 mm	839.72	1062.75	1.82	68.37	nr	**1131.12**
737 × 406 mm	343.69	434.97	1.73	64.99	nr	**499.96**
818 × 406 mm	390.16	493.79	1.73	64.99	nr	**558.77**
978 × 406 mm	450.40	570.02	1.73	64.99	nr	**635.01**
1379 × 406 mm	700.78	886.90	1.82	68.37	nr	**955.27**
1699 × 406 mm	840.04	1063.16	2.11	79.26	nr	**1142.42**
709 × 457 mm	343.70	434.99	1.73	64.99	nr	**499.97**
1189 × 457 mm	632.01	799.88	1.82	68.37	nr	**868.24**
1671 × 457 mm	833.80	1055.26	2.11	79.26	nr	**1134.52**
678 × 508 mm	343.69	434.97	1.82	68.37	nr	**503.34**
919 × 508 mm	450.40	570.02	1.82	68.37	nr	**638.39**
1321 × 508 mm	700.19	886.16	2.11	79.26	nr	**965.41**
90° branch						
345 × 102 mm	176.85	223.82	0.58	21.79	nr	**245.61**
427 × 102 mm	179.34	226.97	0.58	21.79	nr	**248.76**
508 × 102 mm	189.23	239.49	1.16	43.57	nr	**283.06**
559 × 152 mm	220.69	279.30	1.16	43.57	nr	**322.88**
531 × 203 mm	223.51	282.88	1.16	43.57	nr	**326.45**
851 × 203 mm	327.40	414.36	1.73	64.99	nr	**479.34**
582 × 254 mm	234.79	297.15	1.73	64.99	nr	**362.13**
823 × 254 mm	331.55	419.62	1.73	64.99	nr	**484.60**
1303 × 254 mm	454.90	575.72	1.82	68.37	nr	**644.08**
632 × 305 mm	246.31	311.73	1.73	64.99	nr	**376.71**
1275 × 305 mm	464.14	587.42	1.82	68.37	nr	**655.78**
765 × 356 mm	361.02	456.91	1.73	64.99	nr	**521.89**
1247 × 356 mm	460.21	582.44	1.82	68.37	nr	**650.81**
1727 × 356 mm	690.06	873.34	2.11	79.26	nr	**952.60**
737 × 406 mm	363.43	459.95	1.73	64.99	nr	**524.94**
818 × 406 mm	369.16	467.21	1.73	64.99	nr	**532.19**
978 × 406 mm	412.31	521.82	1.82	68.37	nr	**590.19**
1379 × 406 mm	464.59	587.99	1.93	72.50	nr	**660.49**
1699 × 406 mm	695.27	879.93	2.11	79.26	nr	**959.19**
709 × 457 mm	363.86	460.51	1.73	64.99	nr	**525.49**
1189 × 457 mm	501.20	634.32	1.82	68.37	nr	**702.68**
1671 × 457 mm	967.32	1224.23	2.11	79.26	Unit	**1303.49**
678 × 508 mm	367.98	465.71	1.73	64.99	nr	**530.70**
919 × 508 mm	406.12	513.98	1.82	68.37	nr	**582.35**
1321 × 508 mm	588.04	744.23	2.11	79.26	nr	**823.49**
45° branch						
345 × 102 mm	111.86	141.58	0.58	21.79	nr	**163.36**
427 × 102 mm	192.87	244.10	0.58	21.79	nr	**265.89**
508 × 102 mm	261.94	331.52	1.16	43.57	nr	**375.09**
559 × 152 mm	295.04	373.41	1.73	64.99	nr	**438.39**
531 × 203 mm	295.06	373.42	1.73	64.99	nr	**438.41**
851 × 203 mm	410.72	519.80	1.73	64.99	nr	**584.79**
582 × 254 mm	315.95	399.87	1.73	64.99	nr	**464.85**

38 VENTILATION/AIR CONDITIONING SYSTEMS

Item	Net Price £	Material £	Labour hours	Labour £	Unit	Total rate £
DUCTWORK: FLAT OVAL – cont						
Extra over fittings – cont						
45° branch – cont						
823 × 254 mm	410.74	519.83	1.82	68.37	nr	**588.20**
1303 × 254 mm	716.48	906.78	1.92	72.12	nr	**978.90**
632 × 305 mm	336.78	426.22	1.73	64.99	nr	**491.21**
1275 × 305 mm	712.41	901.63	1.82	68.37	nr	**969.99**
765 × 356 mm	410.74	519.83	1.73	64.99	nr	**584.82**
1247 × 356 mm	712.80	902.13	1.82	68.37	nr	**970.49**
1727 × 356 mm	942.46	1192.78	1.82	68.37	nr	**1261.14**
737 × 406 mm	410.74	519.83	1.73	64.99	nr	**584.82**
818 × 406 mm	444.43	562.46	1.73	64.99	nr	**627.45**
978 × 406 mm	513.05	649.31	1.73	64.99	nr	**714.30**
1379 × 406 mm	786.35	995.21	1.93	72.50	nr	**1067.70**
1699 × 406 mm	942.81	1193.22	2.19	82.26	nr	**1275.49**
709 × 457 mm	410.75	519.85	1.73	64.99	nr	**584.83**
1189 × 457 mm	709.01	897.32	1.82	68.37	nr	**965.68**
1671 × 457 mm	936.56	1185.30	2.11	79.26	nr	**1264.56**
678 × 508 mm	410.74	519.83	1.73	64.99	nr	**584.82**
919 × 508 mm	513.04	649.30	1.82	68.37	nr	**717.66**
1321 × 508 mm	785.75	994.45	1.93	72.50	nr	**1066.95**
For rates for access doors refer to ancillaries in Ductwork Ancillaries: Access Hatches						

38 VENTILATION/AIR CONDITIONING SYSTEMS

Item	Net Price £	Material £	Labour hours	Labour £	Unit	Total rate £
DUCTWORK: FLEXIBLE						
AIR DUCTLINES						
Aluminium foil flexible ductwork, DW 144 class B; multiply aluminium polyester laminate fabric, with high tensile steel wire helix						
Duct						
102 mm dia.	2.14	2.71	0.33	12.40	m	**15.10**
152 mm dia.	2.20	2.78	0.33	12.40	m	**15.18**
203 mm dia.	2.91	3.68	0.33	12.40	m	**16.07**
254 mm dia.	4.02	5.09	0.33	12.40	m	**17.48**
304 mm dia.	4.02	5.09	0.33	12.40	m	**17.49**
355 mm dia.	6.62	8.38	0.33	12.40	m	**20.78**
406 mm dia.	6.64	8.40	0.33	12.40	m	**20.79**
Insulated aluminium foil flexible ductwork, DW144 class B; laminate construction of aluminium and polyester multiply inner core with 25 mm insulation; outer layer of multiply aluminium polyester laminate, with high tensile steel wire helix						
Duct						
102 mm dia.	2.14	2.71	0.50	18.78	m	**21.49**
152 mm dia.	2.20	2.79	0.50	18.78	m	**21.57**
203 mm dia.	2.91	3.68	0.50	18.78	m	**22.47**
254 mm dia.	4.00	5.06	0.50	18.78	m	**23.84**
304 mm dia.	4.01	5.08	0.50	18.78	m	**23.86**
355 mm dia.	6.62	8.38	0.50	18.78	m	**27.16**
406 mm dia.	6.65	8.41	0.50	18.78	m	**27.19**

38 VENTILATION/AIR CONDITIONING SYSTEMS

Item	Net Price £	Material £	Labour hours	Labour £	Unit	Total rate £
DUCTWORK: PLASTIC						
AIR DUCTLINES						
Rigid grey PVC DW 154 circular section ductwork; solvent welded or filler rod welded joints; excludes couplers and supports (these are detailed separately); ductwork to conform to curent HSE regulations						
Straight duct (standard length 6 m)						
110 mm	11.83	14.97	1.71	64.23	m	**79.20**
160 mm	18.08	22.89	1.74	65.36	m	**88.25**
200 mm	22.86	28.94	1.79	67.24	m	**96.18**
225 mm	27.68	35.03	1.96	73.62	m	**108.65**
250 mm	30.60	38.73	1.98	74.38	m	**113.11**
315 mm	39.23	49.66	2.23	83.77	m	**133.42**
355 mm	47.88	60.60	2.25	84.52	m	**145.12**
400 mm	56.14	71.05	2.39	89.78	m	**160.83**
450 mm	72.41	91.65	2.95	110.81	m	**202.46**
500 mm	96.43	122.04	2.98	111.94	m	**233.98**
600 mm	186.36	235.86	3.12	117.20	m	**353.06**
Extra for supports (BZP finish); Horizontal – Maximum 2.4 m centres; Vertical – Maximum 4.0 m centres						
Duct size						
110 mm	44.96	56.90	0.59	22.16	m	**79.06**
160 mm	45.12	57.10	0.59	22.16	m	**79.26**
200 mm	45.26	57.28	0.59	22.16	m	**79.44**
225 mm	46.31	58.61	0.63	23.67	m	**82.28**
250 mm	48.06	60.82	0.63	23.67	m	**84.49**
315 mm	48.07	60.84	0.63	23.67	m	**84.51**
355 mm	48.42	61.28	0.78	29.30	m	**90.58**
400 mm	62.87	79.57	0.80	30.05	m	**109.62**
450 mm	64.63	81.80	0.80	30.05	m	**111.85**
500 mm	64.93	82.17	0.81	30.43	m	**112.60**
600 mm	68.96	87.27	0.85	31.93	m	**119.20**
Note: These are maximum figures and may be reduced subject to local conditions (i.e. a high number of changes of direction)						
Extra over fittings; Rigid grey PVC						
90° bend						
110 mm	33.35	42.21	1.11	41.70	m	**83.91**
160 mm	36.11	45.69	1.34	50.34	m	**96.03**
200 mm	39.85	50.43	1.79	67.24	m	**117.67**
225 mm	51.19	64.78	2.00	75.13	m	**139.91**
250 mm	54.49	68.96	2.01	75.50	m	**144.46**
315 mm	94.08	119.07	2.56	96.16	m	**215.24**
355 mm	112.03	141.78	2.71	101.80	m	**243.58**

38 VENTILATION/AIR CONDITIONING SYSTEMS

Item	Net Price £	Material £	Labour hours	Labour £	Unit	Total rate £
400 mm	142.30	180.09	3.58	134.48	m	**314.57**
450 mm	430.12	544.35	4.55	170.91	m	**715.27**
500 mm	499.21	631.80	5.01	188.19	m	**820.00**
600 mm	854.54	1081.50	5.78	217.12	m	**1298.62**
45° bend						
110 mm	20.16	25.51	0.71	26.67	m	**52.18**
160 mm	26.70	33.79	0.93	34.93	m	**68.73**
200 mm	29.86	37.79	1.14	42.82	m	**80.62**
225 mm	36.53	46.23	1.41	52.96	m	**99.20**
250 mm	41.69	52.76	1.62	60.85	m	**113.61**
315 mm	66.34	83.97	1.93	72.50	m	**156.46**
355 mm	85.72	108.48	2.22	83.39	m	**191.88**
400 mm	111.93	141.66	2.85	107.06	m	**248.72**
450 mm	251.92	318.83	3.77	141.61	m	**460.44**
500 mm	290.88	368.14	4.16	156.26	m	**524.41**
600 mm	436.48	552.41	4.81	180.68	m	**733.10**
Tee						
110 mm	66.20	83.79	1.04	39.07	m	**122.85**
160 mm	94.84	120.03	1.38	51.84	m	**171.87**
200 mm	129.91	164.41	1.79	67.24	m	**231.65**
225 mm	155.35	196.61	2.25	84.52	m	**281.13**
250 mm	187.10	236.79	2.62	98.42	m	**335.21**
315 mm	282.49	357.52	3.17	119.08	m	**476.59**
355 mm	387.94	490.98	3.73	140.11	m	**631.09**
400 mm	403.73	510.96	4.71	176.92	m	**687.88**
450 mm	447.62	566.51	5.72	214.86	m	**781.37**
500 mm	561.17	710.22	6.33	237.78	m	**947.99**
Coupler						
110 mm	13.08	16.56	0.70	26.29	m	**42.85**
160 mm	18.93	23.96	0.91	34.18	m	**58.14**
200 mm	24.80	31.39	1.12	42.07	m	**73.46**
225 mm	32.10	40.62	1.41	52.96	m	**93.58**
250 mm	37.11	46.97	1.60	60.10	m	**107.07**
315 mm	45.12	57.10	1.86	69.87	m	**126.97**
355 mm	49.57	62.73	2.43	91.28	m	**154.01**
400 mm	56.45	71.45	2.66	99.92	m	**171.37**
450 mm	123.14	155.85	3.20	120.20	m	**276.05**
500 mm	139.31	176.31	3.52	132.22	m	**308.53**
Damper						
110 mm	82.31	104.17	0.96	36.06	m	**140.23**
160 mm	86.74	109.78	1.26	47.33	m	**157.11**
200 mm	106.52	134.81	1.52	57.10	m	**191.91**
225 mm	114.24	144.59	1.88	70.62	m	**215.21**
250 mm	115.47	146.14	2.16	81.14	m	**227.28**
315 mm	136.38	172.61	2.94	110.44	m	**283.05**
355 mm	151.72	192.01	3.42	128.47	m	**320.48**
400 mm	159.63	202.02	3.79	142.37	m	**344.39**
Reducer						
160 × 110 mm	15.03	19.03	0.85	31.93	m	**50.96**
200 × 110 mm	19.28	24.40	0.98	36.81	m	**61.21**
200 × 160 mm	37.11	46.97	1.07	40.19	m	**87.16**

38 VENTILATION/AIR CONDITIONING SYSTEMS

Item	Net Price £	Material £	Labour hours	Labour £	Unit	Total rate £
DUCTWORK: PLASTIC – cont						
Extra over fittings – cont						
Reducer – cont						
225 × 200 mm	47.90	60.62	1.38	51.84	m	**112.46**
250 × 160 mm	48.89	61.87	1.38	51.84	m	**113.71**
250 × 200 mm	51.74	65.48	1.48	55.59	m	**121.07**
250 × 225 mm	56.31	71.27	1.58	59.35	m	**130.62**
315 × 200 mm	56.52	71.53	1.63	61.23	m	**132.76**
315 × 250 mm	61.86	78.29	1.84	69.12	m	**147.41**
355 × 200 mm	92.86	117.52	1.97	74.00	m	**191.52**
355 × 250 mm	100.01	126.58	1.99	74.75	m	**201.33**
355 × 315 mm	100.01	126.58	2.16	81.14	m	**207.71**
400 × 315 mm	109.12	138.10	2.73	102.55	m	**240.65**
400 × 355 mm	131.13	165.96	2.75	103.30	m	**269.26**
Flange						
110 mm	18.93	23.96	0.70	26.29	m	**50.26**
160 mm	23.02	29.14	0.91	34.18	m	**63.32**
200 mm	25.57	32.36	1.12	42.07	m	**74.43**
225 mm	26.29	33.28	1.35	50.71	m	**83.99**
250 mm	27.93	35.35	1.55	58.22	m	**93.57**
315 mm	40.40	51.13	1.85	69.49	m	**120.62**
355 mm	43.67	55.27	2.06	77.38	m	**132.65**
400 mm	47.36	59.94	2.62	98.42	m	**158.36**
Polypropylene (PPS) DW154 circular section ductwork; filler rod welded joints; excludes couplers and supports (these are detailed separately); ductwork to conform to current HSE regulations						
Straight duct (standard length 5 m)						
110 mm	15.58	19.72	0.65	24.42	m	**44.14**
160 mm	23.40	29.61	0.75	28.17	m	**57.79**
200 mm	30.32	38.37	0.85	31.93	m	**70.30**
225 mm	39.66	50.19	1.07	40.19	m	**90.39**
250 mm	44.00	55.69	1.18	44.33	m	**100.01**
315 mm	84.38	106.79	1.41	52.96	m	**159.76**
355 mm	95.80	121.24	1.53	57.47	m	**178.71**
400 mm	128.93	163.17	1.73	64.99	m	**228.16**
Extra for supports (BZP finish); Horizontal – Maximum 2.4 m centre; Vertical – Maximum 4.0 m centre						
Duct size						
110 mm	62.30	78.85	0.42	15.78	m	**94.63**
160 mm	62.30	78.85	0.42	15.78	m	**94.63**
200 mm	62.30	78.85	0.42	15.78	m	**94.63**
225 mm	62.30	78.85	0.42	15.78	m	**94.63**
250 mm	62.30	78.85	0.42	15.78	m	**94.63**
315 mm	62.30	78.85	0.42	15.78	m	**94.63**
355 mm	62.30	78.85	0.55	20.66	m	**99.51**
400 mm	88.19	111.61	0.55	20.66	m	**132.27**

38 VENTILATION/AIR CONDITIONING SYSTEMS

Item	Net Price £	Material £	Labour hours	Labour £	Unit	Total rate £
Note: These are maximum figures and may be reduced subject to local conditions (i.e. a high number of changes of direction)						
Extra over fittings; Polypropylene (DW 154)						
90° bend						
110 mm	27.55	34.87	1.01	37.94	m	**72.81**
160 mm	38.83	49.14	1.30	48.83	m	**97.97**
200 mm	45.54	57.64	1.58	59.35	m	**116.99**
225 mm	61.56	77.92	2.10	78.88	m	**156.80**
250 mm	66.94	84.72	2.36	88.65	m	**173.37**
315 mm	164.22	207.83	3.14	117.95	m	**325.78**
355 mm	198.18	250.82	3.52	132.22	m	**383.05**
400 mm	212.18	268.54	4.39	164.90	m	**433.44**
45° bend						
110 mm	20.29	25.67	0.84	31.55	m	**57.23**
160 mm	32.71	41.40	1.13	42.45	m	**83.84**
200 mm	37.49	47.45	1.39	52.21	m	**99.66**
225 mm	46.20	58.47	1.73	64.99	m	**123.46**
250 mm	52.71	66.71	1.99	74.75	m	**141.46**
315 mm	133.62	169.11	2.57	96.54	m	**265.64**
355 mm	148.96	188.53	2.93	110.06	m	**298.59**
400 mm	157.49	199.32	3.60	135.23	m	**334.55**
Tee						
110 mm	103.00	130.36	1.55	58.22	m	**188.58**
160 mm	143.49	181.60	2.17	81.51	m	**263.12**
200 mm	181.44	229.63	2.68	100.67	m	**330.30**
225 mm	205.31	259.84	3.38	126.96	m	**386.80**
250 mm	268.90	340.32	3.95	148.38	m	**488.70**
315 mm	354.24	448.33	4.69	176.17	m	**624.50**
355 mm	432.13	546.90	5.44	204.35	m	**751.25**
400 mm	487.80	617.35	6.09	228.76	m	**846.12**
Coupler						
110 mm	21.56	27.29	0.84	31.55	m	**58.84**
160 mm	26.21	33.18	1.10	41.32	m	**74.50**
200 mm	31.56	39.94	1.36	51.09	m	**91.03**
225 mm	33.94	42.95	1.67	62.73	m	**105.68**
250 mm	36.22	45.83	1.91	71.75	m	**117.58**
315 mm	48.62	61.54	2.24	84.14	m	**145.68**
355 mm	62.40	78.97	2.58	96.91	m	**175.88**
400 mm	71.11	90.00	3.30	123.96	m	**213.96**
Damper						
110 mm	106.02	134.18	0.79	29.68	m	**163.85**
160 mm	117.03	148.11	1.13	42.45	m	**190.56**
200 mm	128.51	162.64	1.44	54.09	m	**216.73**
225 mm	137.50	174.02	1.81	67.99	m	**242.01**
250 mm	143.60	181.74	2.08	78.13	m	**259.88**
315 mm	163.10	206.42	2.40	90.15	m	**296.57**
355 mm	179.58	227.28	2.75	103.30	m	**330.58**
400 mm	193.99	245.51	3.57	134.10	m	**379.61**

38 VENTILATION/AIR CONDITIONING SYSTEMS

Item	Net Price £	Material £	Labour hours	Labour £	Unit	Total rate £
DUCTWORK: PLASTIC – cont						
Extra over fittings – cont						
Reducer						
160 × 110 mm	19.56	24.76	0.87	32.68	m	**57.44**
200 × 160 mm	18.82	23.82	1.16	43.57	m	**67.40**
225 × 200 mm	24.09	30.49	1.46	54.84	m	**85.33**
250 × 200 mm	24.55	31.07	1.61	60.48	m	**91.54**
250 × 225 mm	27.55	34.87	1.66	62.36	m	**97.22**
315 × 200 mm	38.50	48.72	1.70	63.86	m	**112.58**
315 × 250 mm	44.31	56.08	2.03	76.25	m	**132.34**
355 × 250 mm	61.06	77.28	2.05	77.01	m	**154.28**
355 × 315 mm	72.43	91.67	2.22	83.39	m	**175.06**
400 × 315 mm	77.27	97.80	2.88	108.18	m	**205.98**
400 × 355 mm	79.11	100.13	2.92	109.69	m	**209.81**
Flange						
110 mm	20.87	26.41	0.76	28.55	m	**54.96**
160 mm	27.25	34.49	1.02	38.31	m	**72.80**
200 mm	31.09	39.35	1.27	47.71	m	**87.05**
225 mm	32.88	41.61	1.55	58.22	m	**99.84**
250 mm	36.95	46.77	1.79	67.24	m	**114.01**
315 mm	44.16	55.88	2.08	78.13	m	**134.02**
355 mm	50.46	63.86	2.36	88.65	m	**152.52**
400 mm	55.67	70.45	3.07	115.32	m	**185.77**

38 VENTILATION/AIR CONDITIONING SYSTEMS

Item	Net Price £	Material £	Labour hours	Labour £	Unit	Total rate £
DUCTWORK: RECTANGULAR – CLASS B						
AIR DUCTLINES						
Galvanized sheet metal DW144 class B rectangular section ductwork; including all necessary stiffeners, joints, couplers in the running length and duct supports						
Ductwork up to 400 mm longest side						
Sum of two sides 200 mm	31.29	39.60	1.16	43.57	m	**83.17**
Sum of two sides 300 mm	33.21	42.03	1.16	43.57	m	**85.60**
Sum of two sides 400 mm	33.83	42.82	1.19	44.70	m	**87.52**
Sum of two sides 500 mm	35.11	44.43	1.19	44.70	m	**89.13**
Sum of two sides 600 mm	35.88	45.41	1.27	47.71	m	**93.11**
Sum of two sides 700 mm	37.95	48.03	1.27	47.71	m	**95.74**
Sum of two sides 800 mm	40.16	50.83	1.27	47.71	m	**98.53**
Extra over fittings; Rectangular ductwork class B; up to 400 mm longest side						
End cap						
Sum of two sides 200 mm	17.80	22.53	0.38	14.27	nr	**36.80**
Sum of two sides 300 mm	19.84	25.11	0.38	14.27	nr	**39.38**
Sum of two sides 400 mm	21.86	27.67	0.38	14.27	nr	**41.94**
Sum of two sides 500 mm	23.90	30.25	0.38	14.27	nr	**44.52**
Sum of two sides 600 mm	25.94	32.83	0.38	14.27	nr	**47.10**
Sum of two sides 700 mm	27.96	35.39	0.38	14.27	nr	**49.67**
Sum of two sides 800 mm	30.01	37.99	0.38	14.27	nr	**52.26**
Reducer						
Sum of two sides 200 mm	27.92	35.33	1.40	52.59	nr	**87.92**
Sum of two sides 300 mm	31.52	39.89	1.40	52.59	nr	**92.48**
Sum of two sides 400 mm	58.43	73.94	1.42	53.34	nr	**127.28**
Sum of two sides 500 mm	63.24	80.04	1.42	53.34	nr	**133.38**
Sum of two sides 600 mm	68.06	86.14	1.69	63.48	nr	**149.62**
Sum of two sides 700 mm	72.88	92.24	1.69	63.48	nr	**155.72**
Sum of two sides 800 mm	77.65	98.27	1.92	72.12	nr	**170.39**
Offset						
Sum of two sides 200 mm	40.97	51.85	1.63	61.23	nr	**113.08**
Sum of two sides 300 mm	46.30	58.60	1.63	61.23	nr	**119.82**
Sum of two sides 400 mm	75.33	95.34	1.65	61.98	nr	**157.32**
Sum of two sides 500 mm	81.89	103.65	1.65	61.98	nr	**165.63**
Sum of two sides 600 mm	87.26	110.44	1.92	72.12	nr	**182.56**
Sum of two sides 700 mm	93.25	118.02	1.92	72.12	nr	**190.14**
Sum of two sides 800 mm	98.45	124.60	1.92	72.12	nr	**196.72**
Square to round						
Sum of two sides 200 mm	36.58	46.29	1.63	61.23	nr	**107.52**
Sum of two sides 300 mm	41.11	52.03	1.63	61.23	nr	**113.26**
Sum of two sides 400 mm	58.15	73.59	1.65	61.98	nr	**135.57**
Sum of two sides 500 mm	63.27	80.08	1.65	61.98	nr	**142.06**
Sum of two sides 600 mm	68.41	86.58	1.92	72.12	nr	**158.70**
Sum of two sides 700 mm	73.54	93.08	1.92	72.12	nr	**165.20**
Sum of two sides 800 mm	78.62	99.50	1.92	72.12	nr	**171.62**

38 VENTILATION/AIR CONDITIONING SYSTEMS

Item	Net Price £	Material £	Labour hours	Labour £	Unit	Total rate £
DUCTWORK: RECTANGULAR – CLASS B – cont						
Extra over fittings – cont						
90° radius bend						
Sum of two sides 200 mm	26.92	34.07	1.22	45.83	nr	**79.90**
Sum of two sides 300 mm	29.04	36.75	1.22	45.83	nr	**82.58**
Sum of two sides 400 mm	55.37	70.07	1.25	46.95	nr	**117.03**
Sum of two sides 500 mm	58.91	74.56	1.25	46.95	nr	**121.51**
Sum of two sides 600 mm	64.11	81.14	1.33	49.96	nr	**131.10**
Sum of two sides 700 mm	68.24	86.37	1.33	49.96	nr	**136.33**
Sum of two sides 800 mm	72.90	92.26	1.40	52.59	nr	**144.85**
45° radius bend						
Sum of two sides 200 mm	29.72	37.62	0.89	33.43	nr	**71.05**
Sum of two sides 300 mm	32.62	41.29	1.10	41.32	nr	**82.61**
Sum of two sides 400 mm	58.48	74.02	1.10	41.32	nr	**115.34**
Sum of two sides 500 mm	62.68	79.33	1.12	42.07	nr	**121.40**
Sum of two sides 600 mm	67.93	85.97	1.16	43.57	nr	**129.54**
Sum of two sides 700 mm	72.43	91.66	1.16	43.57	nr	**135.24**
Sum of two sides 800 mm	77.18	97.67	1.22	45.83	nr	**143.50**
90° mitre bend						
Sum of two sides 200 mm	46.94	59.41	1.29	48.46	nr	**107.87**
Sum of two sides 300 mm	51.58	65.28	1.29	48.46	nr	**113.74**
Sum of two sides 400 mm	82.12	103.94	1.29	48.46	nr	**152.39**
Sum of two sides 500 mm	88.68	112.24	1.29	48.46	nr	**160.70**
Sum of two sides 600 mm	99.30	125.67	1.39	52.21	nr	**177.88**
Sum of two sides 700 mm	108.68	137.55	1.39	52.21	nr	**189.76**
Sum of two sides 800 mm	119.85	151.68	1.46	54.84	nr	**206.52**
Branch						
Sum of two sides 200 mm	47.00	59.48	0.92	34.56	nr	**94.04**
Sum of two sides 300 mm	52.22	66.09	0.92	34.56	nr	**100.65**
Sum of two sides 400 mm	68.14	86.24	0.95	35.69	nr	**121.93**
Sum of two sides 500 mm	74.22	93.94	0.95	35.69	nr	**129.62**
Sum of two sides 600 mm	80.13	101.41	1.03	38.69	nr	**140.10**
Sum of two sides 700 mm	86.05	108.90	1.03	38.69	nr	**147.59**
Sum of two sides 800 mm	91.96	116.39	1.03	38.69	nr	**155.08**
Grille neck						
Sum of two sides 200 mm	53.82	68.11	1.10	41.32	nr	**109.43**
Sum of two sides 300 mm	60.27	76.27	1.10	41.32	nr	**117.59**
Sum of two sides 400 mm	66.70	84.42	1.16	43.57	nr	**127.99**
Sum of two sides 500 mm	73.14	92.57	1.16	43.57	nr	**136.14**
Sum of two sides 600 mm	79.60	100.74	1.18	44.33	nr	**145.07**
Sum of two sides 700 mm	86.03	108.88	1.18	44.33	nr	**153.20**
Sum of two sides 800 mm	92.47	117.03	1.18	44.33	nr	**161.35**
Ductwork 401 to 600 mm longest side						
Sum of two sides 600 mm	40.95	51.83	1.27	47.71	m	**99.53**
Sum of two sides 700 mm	44.10	55.81	1.27	47.71	m	**103.52**
Sum of two sides 800 mm	47.04	59.54	1.27	47.71	m	**107.24**
Sum of two sides 900 mm	49.85	63.09	1.27	47.71	m	**110.79**
Sum of two sides 1000 mm	52.65	66.64	1.37	51.46	m	**118.10**
Sum of two sides 1100 mm	55.90	70.74	1.37	51.46	m	**122.20**
Sum of two sides 1200 mm	58.70	74.29	1.37	51.46	m	**125.75**

38 VENTILATION/AIR CONDITIONING SYSTEMS

Item	Net Price £	Material £	Labour hours	Labour £	Unit	Total rate £
Extra over fittings; Ductwork 401 to 600 mm longest side						
End cap						
Sum of two sides 600 mm	26.40	33.41	0.38	14.27	nr	**47.68**
Sum of two sides 700 mm	28.50	36.07	0.38	14.27	nr	**50.35**
Sum of two sides 800 mm	30.59	38.71	0.38	14.27	nr	**52.99**
Sum of two sides 900 mm	32.70	41.38	0.58	21.79	nr	**63.17**
Sum of two sides 1000 mm	34.80	44.04	0.58	21.79	nr	**65.82**
Sum of two sides 1100 mm	36.89	46.69	0.58	21.79	nr	**68.48**
Sum of two sides 1200 mm	38.99	49.35	0.58	21.79	nr	**71.13**
Reducer						
Sum of two sides 600 mm	65.86	83.35	1.69	63.48	nr	**146.83**
Sum of two sides 700 mm	70.61	89.36	1.69	63.48	nr	**152.84**
Sum of two sides 800 mm	75.58	95.65	1.92	72.12	nr	**167.78**
Sum of two sides 900 mm	80.35	101.69	1.92	72.12	nr	**173.81**
Sum of two sides 1000 mm	85.63	108.38	2.18	81.89	nr	**190.27**
Sum of two sides 1100 mm	90.83	114.95	2.18	81.89	nr	**196.84**
Sum of two sides 1200 mm	95.63	121.02	2.18	81.89	nr	**202.91**
Offset						
Sum of two sides 600 mm	89.28	113.00	1.92	72.12	nr	**185.12**
Sum of two sides 700 mm	95.86	121.32	1.92	72.12	nr	**193.44**
Sum of two sides 800 mm	101.07	127.91	1.92	72.12	nr	**200.03**
Sum of two sides 900 mm	106.35	134.59	1.92	72.12	nr	**206.71**
Sum of two sides 1000 mm	112.22	142.03	2.18	81.89	nr	**223.92**
Sum of two sides 1100 mm	117.70	148.96	2.18	81.89	nr	**230.84**
Sum of two sides 1200 mm	122.83	155.46	2.18	81.89	nr	**237.34**
Square to round						
Sum of two sides 600 mm	65.97	83.49	1.33	49.96	nr	**133.45**
Sum of two sides 700 mm	71.04	89.91	1.33	49.96	nr	**139.87**
Sum of two sides 800 mm	76.03	96.22	1.40	52.59	nr	**148.81**
Sum of two sides 900 mm	81.52	103.17	1.40	52.59	nr	**155.76**
Sum of two sides 1000 mm	86.60	109.61	1.82	68.37	nr	**177.97**
Sum of two sides 1100 mm	91.79	116.17	1.82	68.37	nr	**184.54**
Sum of two sides 1200 mm	96.90	122.63	1.82	68.37	nr	**191.00**
90° radius bend						
Sum of two sides 600 mm	63.75	80.68	1.16	43.57	nr	**124.25**
Sum of two sides 700 mm	67.27	85.13	1.16	43.57	nr	**128.71**
Sum of two sides 800 mm	72.50	91.76	1.22	45.83	nr	**137.58**
Sum of two sides 900 mm	78.31	99.11	1.22	45.83	nr	**144.94**
Sum of two sides 1000 mm	82.80	104.80	1.40	52.59	nr	**157.38**
Sum of two sides 1100 mm	88.47	111.97	1.40	52.59	nr	**164.56**
Sum of two sides 1200 mm	93.89	118.82	1.40	52.59	nr	**171.41**
45° bend						
Sum of two sides 600 mm	68.04	86.11	1.16	43.57	nr	**129.69**
Sum of two sides 700 mm	72.24	91.43	1.39	52.21	nr	**143.65**
Sum of two sides 800 mm	77.72	98.36	1.46	54.84	nr	**153.20**
Sum of two sides 900 mm	82.93	104.96	1.46	54.84	nr	**159.80**
Sum of two sides 1000 mm	87.67	110.95	1.88	70.62	nr	**181.57**
Sum of two sides 1100 mm	93.73	118.62	1.88	70.62	nr	**189.24**
Sum of two sides 1200 mm	98.98	125.27	1.88	70.62	nr	**195.89**

38 VENTILATION/AIR CONDITIONING SYSTEMS

Item	Net Price £	Material £	Labour hours	Labour £	Unit	Total rate £
DUCTWORK: RECTANGULAR – CLASS B – cont						
Extra over fittings – cont						
90° mitre bend						
Sum of two sides 600 mm	110.47	139.81	1.39	52.21	nr	**192.02**
Sum of two sides 700 mm	117.05	148.13	2.16	81.14	nr	**229.27**
Sum of two sides 800 mm	129.70	164.15	2.26	84.89	nr	**249.05**
Sum of two sides 900 mm	143.00	180.98	2.26	84.89	nr	**265.87**
Sum of two sides 1000 mm	152.08	192.47	3.01	113.07	nr	**305.54**
Sum of two sides 1100 mm	167.82	212.39	3.01	113.07	nr	**325.45**
Sum of two sides 1200 mm	178.50	225.90	3.01	113.07	nr	**338.97**
Branch						
Sum of two sides 600 mm	81.95	103.72	1.03	38.69	nr	**142.41**
Sum of two sides 700 mm	88.07	111.46	1.03	38.69	nr	**150.15**
Sum of two sides 800 mm	94.20	119.22	1.03	38.69	nr	**157.91**
Sum of two sides 900 mm	100.35	127.00	1.03	38.69	nr	**165.69**
Sum of two sides 1000 mm	106.49	134.77	1.29	48.46	nr	**183.23**
Sum of two sides 1100 mm	112.88	142.86	1.29	48.46	nr	**191.32**
Sum of two sides 1200 mm	119.01	150.62	1.29	48.46	nr	**199.08**
Grille neck						
Sum of two sides 600 mm	81.76	103.47	1.18	44.33	nr	**147.80**
Sum of two sides 700 mm	88.52	112.03	1.18	44.33	nr	**156.35**
Sum of two sides 800 mm	95.29	120.60	1.18	44.33	nr	**164.92**
Sum of two sides 900 mm	102.05	129.15	1.18	44.33	nr	**173.47**
Sum of two sides 1000 mm	108.82	137.72	1.44	54.09	nr	**191.81**
Sum of two sides 1100 mm	115.57	146.27	1.44	54.09	nr	**200.36**
Sum of two sides 1200 mm	122.34	154.84	1.44	54.09	nr	**208.93**
Ductwork 601 to 800 mm longest side						
Sum of two sides 900 mm	56.30	71.25	1.27	47.71	m	**118.96**
Sum of two sides 1000 mm	59.13	74.84	1.37	51.46	m	**126.30**
Sum of two sides 1100 mm	61.95	78.41	1.37	51.46	m	**129.87**
Sum of two sides 1200 mm	65.10	82.39	1.37	51.46	m	**133.85**
Sum of two sides 1300 mm	67.93	85.97	1.40	52.59	m	**138.56**
Sum of two sides 1400 mm	70.54	89.28	1.40	52.59	m	**141.87**
Sum of two sides 1500 mm	73.36	92.85	1.48	55.59	m	**148.44**
Sum of two sides 1600 mm	76.18	96.42	1.55	58.22	m	**154.64**
Extra over fittings: Ductwork 601 to 800 mm longest side						
End cap						
Sum of two sides 900 mm	32.70	41.38	0.58	21.79	nr	**63.17**
Sum of two sides 1000 mm	34.80	44.04	0.58	21.79	nr	**65.82**
Sum of two sides 1100 mm	36.89	46.69	0.58	21.79	nr	**68.48**
Sum of two sides 1200 mm	38.99	49.35	0.58	21.79	nr	**71.13**
Sum of two sides 1300 mm	41.10	52.01	0.58	21.79	nr	**73.80**
Sum of two sides 1400 mm	42.90	54.29	0.58	21.79	nr	**76.08**
Sum of two sides 1500 mm	49.83	63.06	0.58	21.79	nr	**84.85**
Sum of two sides 1600 mm	56.78	71.86	0.58	21.79	nr	**93.65**

38 VENTILATION/AIR CONDITIONING SYSTEMS

Item	Net Price £	Material £	Labour hours	Labour £	Unit	Total rate £
Reducer						
Sum of two sides 900 mm	81.66	103.35	1.92	72.12	nr	**175.47**
Sum of two sides 1000 mm	86.46	109.42	2.18	81.89	nr	**191.31**
Sum of two sides 1100 mm	91.29	115.53	2.18	81.89	nr	**197.42**
Sum of two sides 1200 mm	96.46	122.08	2.18	81.89	nr	**203.97**
Sum of two sides 1300 mm	101.26	128.16	2.30	86.40	nr	**214.55**
Sum of two sides 1400 mm	105.42	133.41	2.30	86.40	nr	**219.81**
Sum of two sides 1500 mm	120.81	152.90	2.47	92.78	nr	**245.69**
Sum of two sides 1600 mm	136.19	172.36	2.47	92.78	nr	**265.14**
Offset						
Sum of two sides 900 mm	109.65	138.77	1.92	72.12	nr	**210.89**
Sum of two sides 1000 mm	114.56	144.99	2.18	81.89	nr	**226.87**
Sum of two sides 1100 mm	119.46	151.18	2.18	81.89	nr	**233.07**
Sum of two sides 1200 mm	124.55	157.63	2.18	81.89	nr	**239.51**
Sum of two sides 1300 mm	129.34	163.69	2.30	86.40	nr	**250.08**
Sum of two sides 1400 mm	134.04	169.64	2.30	86.40	nr	**256.04**
Sum of two sides 1500 mm	152.64	193.18	2.47	92.78	nr	**285.97**
Sum of two sides 1600 mm	172.47	218.28	2.47	92.78	nr	**311.07**
Square to round						
Sum of two sides 900 mm	80.84	102.31	1.40	52.59	nr	**154.90**
Sum of two sides 1000 mm	85.95	108.78	1.82	68.37	nr	**177.14**
Sum of two sides 1100 mm	91.04	115.22	1.82	68.37	nr	**183.59**
Sum of two sides 1200 mm	96.23	121.79	1.82	68.37	nr	**190.15**
Sum of two sides 1300 mm	101.33	128.25	2.15	80.76	nr	**209.01**
Sum of two sides 1400 mm	105.67	133.73	2.15	80.76	nr	**214.49**
Sum of two sides 1500 mm	123.20	155.92	2.38	89.40	nr	**245.32**
Sum of two sides 1600 mm	140.74	178.12	2.38	89.40	nr	**267.52**
90° radius bend						
Sum of two sides 900 mm	72.83	92.17	1.22	45.83	nr	**138.00**
Sum of two sides 1000 mm	78.28	99.07	1.40	52.59	nr	**151.66**
Sum of two sides 1100 mm	84.04	106.36	1.40	52.59	nr	**158.95**
Sum of two sides 1200 mm	90.08	114.00	1.40	52.59	nr	**166.59**
Sum of two sides 1300 mm	95.57	120.95	1.91	71.75	nr	**192.70**
Sum of two sides 1400 mm	99.28	125.64	1.91	71.75	nr	**197.39**
Sum of two sides 1500 mm	114.46	144.85	2.11	79.26	nr	**224.11**
Sum of two sides 1600 mm	130.31	164.93	2.11	79.26	nr	**244.19**
45° bend						
Sum of two sides 900 mm	81.93	103.69	1.22	45.83	nr	**149.51**
Sum of two sides 1000 mm	87.20	110.37	1.40	52.59	nr	**162.95**
Sum of two sides 1100 mm	92.47	117.03	1.88	70.62	nr	**187.65**
Sum of two sides 1200 mm	98.15	124.21	1.88	70.62	nr	**194.83**
Sum of two sides 1300 mm	102.80	130.10	2.26	84.89	nr	**214.99**
Sum of two sides 1400 mm	107.45	135.98	2.26	84.89	nr	**220.88**
Sum of two sides 1500 mm	123.33	156.09	2.49	93.53	nr	**249.62**
Sum of two sides 1600 mm	137.82	174.42	2.49	93.53	nr	**267.96**
90° mitre bend						
Sum of two sides 900 mm	128.96	163.21	1.22	45.83	nr	**209.04**
Sum of two sides 1000 mm	143.62	181.77	1.40	52.59	nr	**234.36**
Sum of two sides 1100 mm	159.79	202.23	3.01	113.07	nr	**315.30**
Sum of two sides 1200 mm	171.27	216.76	3.01	113.07	nr	**329.83**
Sum of two sides 1300 mm	189.44	239.76	3.67	137.86	nr	**377.61**

38 VENTILATION/AIR CONDITIONING SYSTEMS

Item	Net Price £	Material £	Labour hours	Labour £	Unit	Total rate £
DUCTWORK: RECTANGULAR – CLASS B – cont						
Extra over fittings – cont						
90° mitre bend – cont						
Sum of two sides 1400 mm	198.61	251.36	3.67	137.86	nr	**389.22**
Sum of two sides 1500 mm	228.40	289.07	4.07	152.88	nr	**441.95**
Sum of two sides 1600 mm	258.21	326.79	4.07	152.88	nr	**479.68**
Branch						
Sum of two sides 900 mm	103.82	131.39	1.22	45.83	nr	**177.22**
Sum of two sides 1000 mm	110.14	139.40	1.29	48.46	nr	**187.86**
Sum of two sides 1100 mm	116.47	147.41	1.29	48.46	nr	**195.86**
Sum of two sides 1200 mm	123.09	155.78	1.39	52.21	nr	**207.99**
Sum of two sides 1300 mm	129.43	163.80	1.39	52.21	nr	**216.02**
Sum of two sides 1400 mm	134.93	170.77	1.40	52.59	nr	**223.36**
Sum of two sides 1500 mm	154.75	195.85	1.64	61.60	nr	**257.46**
Sum of two sides 1600 mm	174.60	220.97	1.64	61.60	nr	**282.57**
Grille neck						
Sum of two sides 900 mm	102.05	129.15	1.22	45.83	nr	**174.98**
Sum of two sides 1000 mm	108.82	137.72	1.40	52.59	nr	**190.31**
Sum of two sides 1100 mm	115.57	146.27	1.44	54.09	nr	**200.36**
Sum of two sides 1200 mm	122.34	154.84	1.44	54.09	nr	**208.93**
Sum of two sides 1300 mm	128.83	163.05	1.69	63.48	nr	**226.53**
Sum of two sides 1400 mm	135.57	171.58	1.69	63.48	nr	**235.06**
Sum of two sides 1500 mm	156.40	197.94	1.79	67.24	nr	**265.18**
Sum of two sides 1600 mm	177.75	224.96	1.79	67.24	nr	**292.20**
Ductwork 801 to 1000 mm longest side						
Sum of two sides 1100 mm	91.45	115.74	1.37	51.46	m	**167.20**
Sum of two sides 1200 mm	96.87	122.59	1.37	51.46	m	**174.06**
Sum of two sides 1300 mm	102.27	129.43	1.40	52.59	m	**182.02**
Sum of two sides 1400 mm	108.13	136.85	1.40	52.59	m	**189.44**
Sum of two sides 1500 mm	113.54	143.70	1.48	55.59	m	**199.30**
Sum of two sides 1600 mm	118.96	150.56	1.55	58.22	m	**208.78**
Sum of two sides 1700 mm	124.38	157.41	1.55	58.22	m	**215.64**
Sum of two sides 1800 mm	130.24	164.83	1.61	60.48	m	**225.31**
Sum of two sides 1900 mm	135.65	171.68	1.61	60.48	m	**232.16**
Sum of two sides 2000 mm	141.06	178.52	1.61	60.48	m	**239.00**
Extra over fittings; Ductwork 801 to 1000 mm longest side						
End cap						
Sum of two sides 1100 mm	36.89	46.69	1.44	54.09	nr	**100.78**
Sum of two sides 1200 mm	38.99	49.35	1.44	54.09	nr	**103.44**
Sum of two sides 1300 mm	41.10	52.01	1.44	54.09	nr	**106.11**
Sum of two sides 1400 mm	42.90	54.29	1.44	54.09	nr	**108.38**
Sum of two sides 1500 mm	49.83	63.06	1.44	54.09	nr	**117.15**
Sum of two sides 1600 mm	56.78	71.86	1.44	54.09	nr	**125.95**
Sum of two sides 1700 mm	63.72	80.65	1.44	54.09	nr	**134.74**
Sum of two sides 1800 mm	70.67	89.44	1.44	54.09	nr	**143.54**
Sum of two sides 1900 mm	77.61	98.23	1.44	54.09	nr	**152.32**
Sum of two sides 2000 mm	84.56	107.01	1.44	54.09	nr	**161.10**

38 VENTILATION/AIR CONDITIONING SYSTEMS

Item	Net Price £	Material £	Labour hours	Labour £	Unit	Total rate £
Reducer						
Sum of two sides 1100 mm	70.42	89.12	1.44	54.09	nr	**143.21**
Sum of two sides 1200 mm	73.80	93.40	1.44	54.09	nr	**147.49**
Sum of two sides 1300 mm	77.19	97.69	1.69	63.48	nr	**161.17**
Sum of two sides 1400 mm	80.28	101.60	1.69	63.48	nr	**165.08**
Sum of two sides 1500 mm	93.46	118.29	2.47	92.78	nr	**211.07**
Sum of two sides 1600 mm	106.66	134.99	2.47	92.78	nr	**227.77**
Sum of two sides 1700 mm	119.86	151.69	2.47	92.78	nr	**244.48**
Sum of two sides 1800 mm	133.36	168.77	2.59	97.29	nr	**266.06**
Sum of two sides 1900 mm	146.54	185.46	2.71	101.80	nr	**287.26**
Sum of two sides 2000 mm	159.75	202.18	2.71	101.80	nr	**303.98**
Offset						
Sum of two sides 1100 mm	107.69	136.30	1.44	54.09	nr	**190.39**
Sum of two sides 1200 mm	108.70	137.57	1.44	54.09	nr	**191.66**
Sum of two sides 1300 mm	109.81	138.97	1.69	63.48	nr	**202.45**
Sum of two sides 1400 mm	111.61	141.25	1.69	63.48	nr	**204.74**
Sum of two sides 1500 mm	124.01	156.94	2.47	92.78	nr	**249.73**
Sum of two sides 1600 mm	138.98	175.89	2.47	92.78	nr	**268.68**
Sum of two sides 1700 mm	148.90	188.44	2.59	97.29	nr	**285.73**
Sum of two sides 1800 mm	162.97	206.26	2.61	98.04	nr	**304.30**
Sum of two sides 1900 mm	178.69	226.15	2.71	101.80	nr	**327.95**
Sum of two sides 2000 mm	192.12	243.15	2.71	101.80	nr	**344.95**
Square to round						
Sum of two sides 1100 mm	70.14	88.77	1.44	54.09	nr	**142.86**
Sum of two sides 1200 mm	73.82	93.42	1.44	54.09	nr	**147.52**
Sum of two sides 1300 mm	77.51	98.10	1.69	63.48	nr	**161.58**
Sum of two sides 1400 mm	80.52	101.91	1.69	63.48	nr	**165.39**
Sum of two sides 1500 mm	95.65	121.06	2.38	89.40	nr	**210.46**
Sum of two sides 1600 mm	111.96	141.70	2.38	89.40	nr	**231.10**
Sum of two sides 1700 mm	127.25	161.05	2.55	95.79	nr	**256.84**
Sum of two sides 1800 mm	142.56	180.42	2.55	95.79	nr	**276.21**
Sum of two sides 1900 mm	157.84	199.76	2.83	106.31	nr	**306.07**
Sum of two sides 2000 mm	173.12	219.10	2.83	106.31	nr	**325.41**
90° radius bend						
Sum of two sides 1100 mm	52.06	65.88	1.44	54.09	nr	**119.97**
Sum of two sides 1200 mm	56.18	71.11	1.44	54.09	nr	**125.20**
Sum of two sides 1300 mm	60.31	76.33	1.69	63.48	nr	**139.81**
Sum of two sides 1400 mm	64.03	81.03	1.69	63.48	nr	**144.52**
Sum of two sides 1500 mm	77.10	97.57	2.11	79.26	nr	**176.83**
Sum of two sides 1600 mm	92.15	116.63	2.11	79.26	nr	**195.89**
Sum of two sides 1700 mm	105.52	133.55	2.26	84.89	nr	**218.44**
Sum of two sides 1800 mm	119.02	150.64	2.26	84.89	nr	**235.53**
Sum of two sides 1900 mm	130.00	164.53	2.48	93.16	nr	**257.69**
Sum of two sides 2000 mm	143.37	181.44	2.48	93.16	nr	**274.60**
45° bend						
Sum of two sides 1100 mm	74.93	94.84	1.44	54.09	nr	**148.93**
Sum of two sides 1200 mm	79.55	100.68	1.44	54.09	nr	**154.77**
Sum of two sides 1300 mm	84.15	106.50	1.69	63.48	nr	**169.99**
Sum of two sides 1400 mm	88.49	111.99	1.69	63.48	nr	**175.47**
Sum of two sides 1500 mm	103.03	130.39	2.49	93.53	nr	**223.92**
Sum of two sides 1600 mm	117.59	148.82	2.49	93.53	nr	**242.35**
Sum of two sides 1700 mm	132.13	167.22	2.67	100.29	nr	**267.51**

38 VENTILATION/AIR CONDITIONING SYSTEMS

Item	Net Price £	Material £	Labour hours	Labour £	Unit	Total rate £
DUCTWORK: RECTANGULAR – CLASS B – cont						
Extra over fittings – cont						
45° bend – cont						
Sum of two sides 1800 mm	146.99	186.03	2.67	100.29	nr	**286.33**
Sum of two sides 1900 mm	158.60	200.72	3.06	114.94	nr	**315.67**
Sum of two sides 2000 mm	173.02	218.97	3.06	114.94	nr	**333.91**
90° mitre bend						
Sum of two sides 1100 mm	125.01	158.21	1.44	54.09	nr	**212.30**
Sum of two sides 1200 mm	134.09	169.70	1.44	54.09	nr	**223.79**
Sum of two sides 1300 mm	150.61	190.61	1.69	63.48	nr	**254.09**
Sum of two sides 1400 mm	164.78	208.54	1.69	63.48	nr	**272.03**
Sum of two sides 1500 mm	191.61	242.50	2.07	77.76	nr	**320.25**
Sum of two sides 1600 mm	218.43	276.45	2.07	77.76	nr	**354.21**
Sum of two sides 1700 mm	248.79	314.87	2.80	105.18	nr	**420.05**
Sum of two sides 1800 mm	276.13	349.47	2.67	100.29	nr	**449.76**
Sum of two sides 1900 mm	300.17	379.90	2.95	110.81	nr	**490.71**
Sum of two sides 2000 mm	327.53	414.52	2.95	110.81	nr	**525.34**
Branch						
Sum of two sides 1100 mm	116.73	147.74	1.44	54.09	nr	**201.83**
Sum of two sides 1200 mm	123.09	155.78	1.44	54.09	nr	**209.87**
Sum of two sides 1300 mm	129.43	163.80	1.64	61.60	nr	**225.41**
Sum of two sides 1400 mm	135.20	171.10	1.64	61.60	nr	**232.71**
Sum of two sides 1500 mm	155.04	196.22	1.64	61.60	nr	**257.82**
Sum of two sides 1600 mm	174.12	220.36	1.64	61.60	nr	**281.97**
Sum of two sides 1700 mm	192.81	244.02	1.69	63.48	nr	**307.50**
Sum of two sides 1800 mm	212.72	269.22	1.69	63.48	nr	**332.70**
Sum of two sides 1900 mm	232.37	294.09	1.85	69.49	nr	**363.58**
Sum of two sides 2000 mm	252.01	318.94	1.85	69.49	nr	**388.44**
Grille neck						
Sum of two sides 1100 mm	115.57	146.27	1.44	54.09	nr	**200.36**
Sum of two sides 1200 mm	122.34	154.84	1.44	54.09	nr	**208.93**
Sum of two sides 1300 mm	128.83	163.05	1.69	63.48	nr	**226.53**
Sum of two sides 1400 mm	135.12	171.01	1.69	63.48	nr	**234.49**
Sum of two sides 1500 mm	156.44	197.99	1.79	67.24	nr	**265.23**
Sum of two sides 1600 mm	177.77	224.98	1.79	67.24	nr	**292.22**
Sum of two sides 1700 mm	199.08	251.95	1.86	69.87	nr	**321.82**
Sum of two sides 1800 mm	220.40	278.94	2.02	75.88	nr	**354.82**
Sum of two sides 1900 mm	241.71	305.91	2.02	75.88	nr	**381.79**
Sum of two sides 2000 mm	263.03	332.89	2.02	75.88	nr	**408.76**
Ductwork 1001 to 1250 mm longest side						
Sum of two sides 1300 mm	121.60	153.90	1.40	52.59	m	**206.48**
Sum of two sides 1400 mm	128.16	162.19	1.40	52.59	m	**214.78**
Sum of two sides 1500 mm	134.33	170.01	1.48	55.59	m	**225.61**
Sum of two sides 1600 mm	140.74	178.12	1.55	58.22	m	**236.35**
Sum of two sides 1700 mm	147.14	186.23	1.55	58.22	m	**244.45**
Sum of two sides 1800 mm	153.32	194.04	1.61	60.48	m	**254.52**
Sum of two sides 1900 mm	159.50	201.86	1.61	60.48	m	**262.34**

38 VENTILATION/AIR CONDITIONING SYSTEMS

Item	Net Price £	Material £	Labour hours	Labour £	Unit	Total rate £
Sum of two sides 2000 mm	166.12	210.24	1.61	60.48	m	**270.72**
Sum of two sides 2100 mm	172.31	218.07	2.17	81.51	m	**299.59**
Sum of two sides 2200 mm	178.49	225.89	2.19	82.26	m	**308.16**
Sum of two sides 2300 mm	185.49	234.75	2.19	82.26	m	**317.02**
Sum of two sides 2400 mm	191.66	242.57	2.38	89.40	m	**331.97**
Sum of two sides 2500 mm	198.29	250.95	2.38	89.40	m	**340.35**
Extra over fittings; Ductwork 1001 to 1250 mm longest side						
End cap						
Sum of two sides 1300 mm	41.93	53.07	1.69	63.48	nr	**116.55**
Sum of two sides 1400 mm	43.80	55.43	1.69	63.48	nr	**118.91**
Sum of two sides 1500 mm	50.81	64.30	1.69	63.48	nr	**127.79**
Sum of two sides 1600 mm	57.83	73.19	1.69	63.48	nr	**136.67**
Sum of two sides 1700 mm	64.83	82.04	1.69	63.48	nr	**145.53**
Sum of two sides 1800 mm	71.84	90.92	1.69	63.48	nr	**154.40**
Sum of two sides 1900 mm	78.86	99.80	1.69	63.48	nr	**163.28**
Sum of two sides 2000 mm	85.84	108.64	1.69	63.48	nr	**172.13**
Sum of two sides 2100 mm	92.85	117.51	1.69	63.48	nr	**181.00**
Sum of two sides 2200 mm	99.87	126.40	1.69	63.48	nr	**189.88**
Sum of two sides 2300 mm	106.88	135.27	1.69	63.48	nr	**198.75**
Sum of two sides 2400 mm	113.88	144.13	1.69	63.48	nr	**207.61**
Sum of two sides 2500 mm	120.90	153.01	1.69	63.48	nr	**216.49**
Reducer						
Sum of two sides 1300 mm	74.67	94.51	1.69	63.48	nr	**157.99**
Sum of two sides 1400 mm	77.45	98.02	1.69	63.48	nr	**161.50**
Sum of two sides 1500 mm	90.58	114.64	2.47	92.78	nr	**207.42**
Sum of two sides 1600 mm	103.50	130.98	2.47	92.78	nr	**223.77**
Sum of two sides 1700 mm	116.57	147.53	2.47	92.78	nr	**240.31**
Sum of two sides 1800 mm	129.64	164.07	2.59	97.29	nr	**261.36**
Sum of two sides 1900 mm	142.71	180.61	2.71	101.80	nr	**282.41**
Sum of two sides 2000 mm	156.09	197.54	2.59	97.29	nr	**294.83**
Sum of two sides 2100 mm	169.16	214.08	2.92	109.69	nr	**323.77**
Sum of two sides 2200 mm	182.24	230.64	2.92	109.69	nr	**340.32**
Sum of two sides 2300 mm	195.38	247.27	2.92	109.69	nr	**356.95**
Sum of two sides 2400 mm	208.69	264.11	3.12	117.20	nr	**381.31**
Sum of two sides 2500 mm	222.05	281.03	3.12	117.20	nr	**398.23**
Offset						
Sum of two sides 1300 mm	161.49	204.38	1.69	63.48	nr	**267.86**
Sum of two sides 1400 mm	167.03	211.39	1.69	63.48	nr	**274.87**
Sum of two sides 1500 mm	183.62	232.39	2.47	92.78	nr	**325.17**
Sum of two sides 1600 mm	203.14	257.09	2.47	92.78	nr	**349.88**
Sum of two sides 1700 mm	216.92	274.53	2.59	97.29	nr	**371.82**
Sum of two sides 1800 mm	235.08	297.52	2.61	98.04	nr	**395.56**
Sum of two sides 1900 mm	252.87	320.03	2.71	101.80	nr	**421.83**
Sum of two sides 2000 mm	262.82	332.62	2.71	101.80	nr	**434.42**
Sum of two sides 2100 mm	279.46	353.68	2.92	109.69	nr	**463.36**
Sum of two sides 2200 mm	298.54	377.83	3.26	122.46	nr	**500.29**
Sum of two sides 2300 mm	320.42	405.52	3.26	122.46	nr	**527.98**
Sum of two sides 2400 mm	342.22	433.11	3.48	130.72	nr	**563.83**
Sum of two sides 2500 mm	364.10	460.80	3.48	130.72	nr	**591.53**

38 VENTILATION/AIR CONDITIONING SYSTEMS

Item	Net Price £	Material £	Labour hours	Labour £	Unit	Total rate £
DUCTWORK: RECTANGULAR – CLASS B – cont						
Extra over fittings – cont						
Square to round						
Sum of two sides 1300 mm	74.27	93.99	1.69	63.48	nr	**157.47**
Sum of two sides 1400 mm	77.21	97.72	1.69	63.48	nr	**161.20**
Sum of two sides 1500 mm	92.23	116.72	2.38	89.40	nr	**206.12**
Sum of two sides 1600 mm	107.29	135.78	2.38	89.40	nr	**225.18**
Sum of two sides 1700 mm	122.30	154.78	2.55	95.79	nr	**250.57**
Sum of two sides 1800 mm	140.26	177.52	2.55	95.79	nr	**273.31**
Sum of two sides 1900 mm	155.59	196.92	2.83	106.31	nr	**303.23**
Sum of two sides 2000 mm	170.98	216.39	2.83	106.31	nr	**322.70**
Sum of two sides 2100 mm	186.31	235.79	3.85	144.62	nr	**380.41**
Sum of two sides 2200 mm	201.65	255.21	4.18	157.02	nr	**412.22**
Sum of two sides 2300 mm	217.02	274.66	4.22	158.52	nr	**433.18**
Sum of two sides 2400 mm	232.36	294.08	4.68	175.80	nr	**469.88**
Sum of two sides 2500 mm	247.73	313.52	4.70	176.55	nr	**490.07**
90° radius bend						
Sum of two sides 1300 mm	45.24	57.25	1.69	63.48	nr	**120.73**
Sum of two sides 1400 mm	46.30	58.59	1.69	63.48	nr	**122.08**
Sum of two sides 1500 mm	58.74	74.34	2.11	79.26	nr	**153.60**
Sum of two sides 1600 mm	74.23	93.95	2.11	79.26	nr	**173.21**
Sum of two sides 1700 mm	89.20	112.89	2.19	82.26	nr	**195.16**
Sum of two sides 1800 mm	102.44	129.65	2.19	82.26	nr	**211.92**
Sum of two sides 1900 mm	115.71	146.44	2.26	84.89	nr	**231.33**
Sum of two sides 2000 mm	131.96	167.01	2.48	93.16	nr	**260.17**
Sum of two sides 2100 mm	145.49	184.14	2.48	93.16	nr	**277.29**
Sum of two sides 2200 mm	159.07	201.31	2.48	93.16	nr	**294.47**
Sum of two sides 2300 mm	166.72	211.00	2.48	93.16	nr	**304.16**
Sum of two sides 2400 mm	180.31	228.21	3.90	146.50	nr	**374.70**
Sum of two sides 2500 mm	193.93	245.44	3.90	146.50	nr	**391.93**
45° bend						
Sum of two sides 1300 mm	76.76	97.15	1.69	63.48	nr	**160.63**
Sum of two sides 1400 mm	79.34	100.41	1.69	63.48	nr	**163.89**
Sum of two sides 1500 mm	93.87	118.81	2.49	93.53	nr	**212.34**
Sum of two sides 1600 mm	108.72	137.59	2.49	93.53	nr	**231.13**
Sum of two sides 1700 mm	123.27	156.01	2.67	100.29	nr	**256.30**
Sum of two sides 1800 mm	137.80	174.40	2.67	100.29	nr	**274.70**
Sum of two sides 1900 mm	152.36	192.83	3.06	114.94	nr	**307.77**
Sum of two sides 2000 mm	167.20	211.60	3.06	114.94	nr	**326.55**
Sum of two sides 2100 mm	181.76	230.03	4.05	152.13	nr	**382.16**
Sum of two sides 2200 mm	196.30	248.44	4.05	152.13	nr	**400.57**
Sum of two sides 2300 mm	207.75	262.92	4.39	164.90	nr	**427.83**
Sum of two sides 2400 mm	222.31	281.35	4.85	182.18	nr	**463.53**
Sum of two sides 2500 mm	237.15	300.14	4.85	182.18	nr	**482.32**
90° mitre bend						
Sum of two sides 1300 mm	187.13	236.83	1.69	63.48	nr	**300.31**
Sum of two sides 1400 mm	195.27	247.14	1.69	63.48	nr	**310.62**
Sum of two sides 1500 mm	232.17	293.83	2.80	105.18	nr	**399.01**

38 VENTILATION/AIR CONDITIONING SYSTEMS

Item	Net Price £	Material £	Labour hours	Labour £	Unit	Total rate £
Sum of two sides 1600 mm	264.49	334.74	2.80	105.18	nr	**439.92**
Sum of two sides 1700 mm	302.24	382.51	2.95	110.81	nr	**493.33**
Sum of two sides 1800 mm	335.15	424.16	2.95	110.81	nr	**534.98**
Sum of two sides 1900 mm	368.03	465.78	4.05	152.13	nr	**617.91**
Sum of two sides 2000 mm	406.40	514.34	4.05	152.13	nr	**666.47**
Sum of two sides 2100 mm	439.73	556.52	4.07	152.88	nr	**709.40**
Sum of two sides 2200 mm	473.07	598.71	4.07	152.88	nr	**751.60**
Sum of two sides 2300 mm	497.62	629.78	4.39	164.90	nr	**794.69**
Sum of two sides 2400 mm	531.23	672.32	4.85	182.18	nr	**854.51**
Sum of two sides 2500 mm	564.84	714.86	4.85	182.18	nr	**897.05**
Branch						
Sum of two sides 1300 mm	132.75	168.01	1.44	54.09	nr	**222.10**
Sum of two sides 1400 mm	138.45	175.22	1.44	54.09	nr	**229.31**
Sum of two sides 1500 mm	158.49	200.59	1.64	61.60	nr	**262.19**
Sum of two sides 1600 mm	178.81	226.31	1.64	61.60	nr	**287.91**
Sum of two sides 1700 mm	196.92	249.22	1.64	61.60	nr	**310.82**
Sum of two sides 1800 mm	216.77	274.34	1.64	61.60	nr	**335.95**
Sum of two sides 1900 mm	236.61	299.45	1.69	63.48	nr	**362.93**
Sum of two sides 2000 mm	256.72	324.90	1.69	63.48	nr	**388.39**
Sum of two sides 2100 mm	275.21	348.31	1.85	69.49	nr	**417.80**
Sum of two sides 2200 mm	294.94	373.28	1.85	69.49	nr	**442.77**
Sum of two sides 2300 mm	314.97	398.62	2.61	98.04	nr	**496.66**
Sum of two sides 2400 mm	334.71	423.61	2.61	98.04	nr	**521.65**
Sum of two sides 2500 mm	354.73	448.95	2.61	98.04	nr	**546.99**
Grille neck						
Sum of two sides 1300 mm	133.74	169.27	1.79	67.24	nr	**236.51**
Sum of two sides 1400 mm	139.94	177.11	1.79	67.24	nr	**244.35**
Sum of two sides 1500 mm	161.61	204.53	1.79	67.24	nr	**271.77**
Sum of two sides 1600 mm	183.32	232.01	1.79	67.24	nr	**299.25**
Sum of two sides 1700 mm	205.00	259.45	1.86	69.87	nr	**329.32**
Sum of two sides 1800 mm	226.69	286.90	2.02	75.88	nr	**362.78**
Sum of two sides 1900 mm	248.41	314.38	2.02	75.88	nr	**390.26**
Sum of two sides 2000 mm	270.07	341.81	2.02	75.88	nr	**417.68**
Sum of two sides 2100 mm	291.76	369.26	2.61	98.04	nr	**467.30**
Sum of two sides 2200 mm	313.44	396.69	2.61	98.04	nr	**494.74**
Sum of two sides 2300 mm	335.15	424.16	2.61	98.04	nr	**522.20**
Sum of two sides 2400 mm	356.82	451.60	2.88	108.18	nr	**559.78**
Sum of two sides 2500 mm	378.52	479.05	2.88	108.18	nr	**587.23**
Ductwork 1251 to 1600 mm longest side						
Sum of two sides 1700 mm	180.11	227.95	1.55	58.22	m	**286.17**
Sum of two sides 1800 mm	187.00	236.66	1.61	60.48	m	**297.14**
Sum of two sides 1900 mm	194.12	245.68	1.61	60.48	m	**306.16**
Sum of two sides 2000 mm	200.81	254.14	1.61	60.48	m	**314.62**
Sum of two sides 2100 mm	207.49	262.60	2.17	81.51	m	**344.12**
Sum of two sides 2200 mm	214.18	271.06	2.19	82.26	m	**353.33**
Sum of two sides 2300 mm	221.31	280.09	2.19	82.26	m	**362.35**
Sum of two sides 2400 mm	228.19	288.80	2.38	89.40	m	**378.20**
Sum of two sides 2500 mm	234.88	297.26	2.38	89.40	m	**386.66**
Sum of two sides 2600 mm	241.56	305.72	2.64	99.17	m	**404.89**
Sum of two sides 2700 mm	248.25	314.18	2.66	99.92	m	**414.10**

38 VENTILATION/AIR CONDITIONING SYSTEMS

Item	Net Price £	Material £	Labour hours	Labour £	Unit	Total rate £
DUCTWORK: RECTANGULAR – CLASS B – cont						
Ductwork 1251 to 1600 mm longest side – cont						
Sum of two sides 2800 mm	255.13	322.90	2.95	110.81	m	**433.71**
Sum of two sides 2900 mm	284.92	360.59	2.96	111.19	m	**471.78**
Sum of two sides 3000 mm	291.60	369.05	3.15	118.33	m	**487.37**
Sum of two sides 3100 mm	298.55	377.85	3.15	118.33	m	**496.17**
Sum of two sides 3200 mm	305.23	386.30	3.18	119.45	m	**505.76**
Extra over fittings; Ductwork 1251 to 1600 mm longest side						
End cap						
Sum of two sides 1700 mm	64.83	82.04	0.58	21.79	nr	**103.83**
Sum of two sides 1800 mm	71.84	90.92	0.58	21.79	nr	**112.70**
Sum of two sides 1900 mm	78.86	99.80	0.58	21.79	nr	**121.59**
Sum of two sides 2000 mm	85.84	108.64	0.58	21.79	nr	**130.43**
Sum of two sides 2100 mm	92.85	117.51	0.87	32.68	nr	**150.19**
Sum of two sides 2200 mm	99.87	126.40	0.87	32.68	nr	**159.08**
Sum of two sides 2300 mm	106.88	135.27	0.87	32.68	nr	**167.95**
Sum of two sides 2400 mm	113.88	144.13	0.87	32.68	nr	**176.81**
Sum of two sides 2500 mm	120.90	153.01	0.87	32.68	nr	**185.69**
Sum of two sides 2600 mm	127.91	161.88	0.87	32.68	nr	**194.56**
Sum of two sides 2700 mm	134.91	170.74	0.87	32.68	nr	**203.42**
Sum of two sides 2800 mm	141.93	179.62	1.16	43.57	nr	**223.20**
Sum of two sides 2900 mm	148.94	188.50	1.16	43.57	nr	**232.07**
Sum of two sides 3000 mm	155.95	197.37	1.30	48.83	nr	**246.20**
Sum of two sides 3100 mm	247.10	312.73	1.80	67.61	nr	**380.34**
Sum of two sides 3200 mm	134.72	170.51	1.80	67.61	nr	**238.12**
Reducer						
Sum of two sides 1700 mm	75.55	95.61	2.47	92.78	nr	**188.40**
Sum of two sides 1800 mm	87.15	110.30	2.59	97.29	nr	**207.59**
Sum of two sides 1900 mm	98.89	125.15	2.71	101.80	nr	**226.95**
Sum of two sides 2000 mm	112.17	141.96	2.71	101.80	nr	**243.76**
Sum of two sides 2100 mm	123.88	156.78	2.92	109.69	nr	**266.47**
Sum of two sides 2200 mm	135.59	171.61	2.92	109.69	nr	**281.29**
Sum of two sides 2300 mm	147.51	186.68	2.92	109.69	nr	**296.37**
Sum of two sides 2400 mm	159.29	201.59	3.12	117.20	nr	**318.79**
Sum of two sides 2500 mm	171.00	216.42	3.12	117.20	nr	**333.61**
Sum of two sides 2600 mm	182.71	231.24	3.12	117.20	nr	**348.44**
Sum of two sides 2700 mm	194.40	246.04	3.12	117.20	nr	**363.23**
Sum of two sides 2800 mm	201.11	254.52	3.95	148.38	nr	**402.90**
Sum of two sides 2900 mm	206.18	260.94	3.97	149.13	nr	**410.07**
Sum of two sides 3000 mm	212.54	268.99	4.52	169.79	nr	**438.77**
Sum of two sides 3100 mm	221.06	279.78	4.52	169.79	nr	**449.57**
Sum of two sides 3200 mm	228.87	289.66	4.52	169.79	nr	**459.45**
Offset						
Sum of two sides 1700 mm	211.49	267.66	2.59	97.29	nr	**364.95**
Sum of two sides 1800 mm	234.98	297.39	2.61	98.04	nr	**395.43**
Sum of two sides 1900 mm	250.79	317.40	2.71	101.80	nr	**419.20**

38 VENTILATION/AIR CONDITIONING SYSTEMS

Item	Net Price £	Material £	Labour hours	Labour £	Unit	Total rate £
Sum of two sides 2000 mm	260.80	330.07	2.71	101.80	nr	**431.87**
Sum of two sides 2100 mm	275.19	348.28	2.92	109.69	nr	**457.96**
Sum of two sides 2200 mm	284.90	360.57	3.26	122.46	nr	**483.02**
Sum of two sides 2300 mm	298.10	377.27	3.26	122.46	nr	**499.73**
Sum of two sides 2400 mm	320.04	405.04	3.47	130.35	nr	**535.39**
Sum of two sides 2500 mm	330.41	418.17	3.48	130.72	nr	**548.89**
Sum of two sides 2600 mm	351.63	445.03	3.49	131.10	nr	**576.12**
Sum of two sides 2700 mm	372.85	471.87	3.50	131.47	nr	**603.35**
Sum of two sides 2800 mm	388.72	491.96	4.34	163.03	nr	**654.99**
Sum of two sides 2900 mm	394.19	498.88	4.76	178.80	nr	**677.68**
Sum of two sides 3000 mm	409.67	518.48	5.32	199.84	nr	**718.32**
Sum of two sides 3100 mm	425.91	539.03	5.35	200.97	nr	**740.00**
Sum of two sides 3200 mm	441.00	558.13	5.35	200.97	nr	**759.09**
Square to round						
Sum of two sides 1700 mm	82.99	105.03	2.55	95.79	nr	**200.82**
Sum of two sides 1800 mm	96.42	122.03	2.55	95.79	nr	**217.82**
Sum of two sides 1900 mm	109.76	138.91	2.83	106.31	nr	**245.21**
Sum of two sides 2000 mm	126.65	160.29	2.83	106.31	nr	**266.59**
Sum of two sides 2100 mm	140.42	177.71	3.85	144.62	nr	**322.33**
Sum of two sides 2200 mm	154.16	195.11	4.18	157.02	nr	**352.12**
Sum of two sides 2300 mm	170.68	216.01	4.22	158.52	nr	**374.53**
Sum of two sides 2400 mm	184.72	233.79	4.68	175.80	nr	**409.59**
Sum of two sides 2500 mm	198.70	251.47	4.70	176.55	nr	**428.02**
Sum of two sides 2600 mm	212.69	269.17	4.70	176.55	nr	**445.72**
Sum of two sides 2700 mm	226.68	286.89	4.71	176.92	nr	**463.81**
Sum of two sides 2800 mm	237.41	300.46	8.19	307.65	nr	**608.11**
Sum of two sides 2900 mm	240.73	304.67	8.62	323.80	nr	**628.47**
Sum of two sides 3000 mm	251.16	317.86	8.75	328.68	nr	**646.54**
Sum of two sides 3100 mm	261.39	330.82	8.75	328.68	nr	**659.50**
Sum of two sides 3200 mm	275.19	348.28	8.75	328.68	nr	**676.96**
90° radius bend						
Sum of two sides 1700 mm	198.22	250.87	2.19	82.26	nr	**333.13**
Sum of two sides 1800 mm	216.73	274.29	2.19	82.26	nr	**356.55**
Sum of two sides 1900 mm	245.10	310.20	2.26	84.89	nr	**395.09**
Sum of two sides 2000 mm	272.94	345.43	2.26	84.89	nr	**430.32**
Sum of two sides 2100 mm	300.78	380.67	2.48	93.16	nr	**473.83**
Sum of two sides 2200 mm	328.63	415.92	2.48	93.16	nr	**509.08**
Sum of two sides 2300 mm	356.99	451.81	2.48	93.16	nr	**544.97**
Sum of two sides 2400 mm	374.88	474.44	3.90	146.50	nr	**620.94**
Sum of two sides 2500 mm	402.69	509.64	3.90	146.50	nr	**656.14**
Sum of two sides 2600 mm	430.52	544.86	4.26	160.02	nr	**704.88**
Sum of two sides 2700 mm	458.32	580.05	4.55	170.91	nr	**750.97**
Sum of two sides 2800 mm	467.14	591.21	4.55	170.91	nr	**762.12**
Sum of two sides 2900 mm	475.73	602.08	6.87	258.06	nr	**860.14**
Sum of two sides 3000 mm	494.46	625.79	7.00	262.95	nr	**888.74**
Sum of two sides 3100 mm	515.59	652.53	7.00	262.95	nr	**915.48**
Sum of two sides 3200 mm	535.19	677.33	7.00	262.95	nr	**940.28**
45° bend						
Sum of two sides 1700 mm	95.25	120.55	2.67	100.29	nr	**220.84**
Sum of two sides 1800 mm	104.38	132.10	2.67	100.29	nr	**232.40**
Sum of two sides 1900 mm	118.55	150.04	3.06	114.94	nr	**264.98**

38 VENTILATION/AIR CONDITIONING SYSTEMS

Item	Net Price £	Material £	Labour hours	Labour £	Unit	Total rate £
DUCTWORK: RECTANGULAR – CLASS B – cont						
Extra over fittings – cont						
45° bend – cont						
Sum of two sides 2000 mm	132.49	167.68	3.06	114.94	nr	282.62
Sum of two sides 2100 mm	146.43	185.33	4.05	152.13	nr	337.46
Sum of two sides 2200 mm	160.37	202.97	4.05	152.13	nr	355.10
Sum of two sides 2300 mm	174.56	220.92	4.39	164.90	nr	385.82
Sum of two sides 2400 mm	183.37	232.07	4.85	182.18	nr	414.25
Sum of two sides 2500 mm	197.29	249.69	4.85	182.18	nr	431.88
Sum of two sides 2600 mm	211.22	267.32	4.87	182.93	nr	450.25
Sum of two sides 2700 mm	225.14	284.94	4.87	182.93	nr	467.87
Sum of two sides 2800 mm	225.68	285.63	8.81	330.94	nr	616.56
Sum of two sides 2900 mm	233.70	295.77	8.81	330.94	nr	626.71
Sum of two sides 3000 mm	240.29	304.11	9.31	349.72	nr	653.82
Sum of two sides 3100 mm	250.76	317.37	9.31	349.72	nr	667.08
Sum of two sides 3200 mm	260.50	329.69	9.39	352.72	nr	682.41
90° mitre bend						
Sum of two sides 1700 mm	254.29	321.82	2.67	100.29	nr	422.12
Sum of two sides 1800 mm	271.47	343.58	2.80	105.18	nr	448.76
Sum of two sides 1900 mm	303.75	384.42	2.95	110.81	nr	495.23
Sum of two sides 2000 mm	336.08	425.35	2.95	110.81	nr	536.16
Sum of two sides 2100 mm	380.60	481.69	4.05	152.13	nr	633.82
Sum of two sides 2200 mm	414.02	523.99	4.05	152.13	nr	676.12
Sum of two sides 2300 mm	449.78	569.24	4.39	164.90	nr	734.15
Sum of two sides 2400 mm	465.46	589.09	4.85	182.18	nr	771.27
Sum of two sides 2500 mm	499.09	631.65	4.85	182.18	nr	813.84
Sum of two sides 2600 mm	544.87	689.59	4.87	182.93	nr	872.52
Sum of two sides 2700 mm	579.28	733.14	4.87	182.93	nr	916.07
Sum of two sides 2800 mm	597.79	756.56	8.81	330.94	nr	1087.49
Sum of two sides 2900 mm	603.17	763.38	14.81	556.32	nr	1319.69
Sum of two sides 3000 mm	637.80	807.20	15.20	570.97	nr	1378.17
Sum of two sides 3100 mm	666.28	843.25	15.60	585.99	nr	1429.24
Sum of two sides 3200 mm	693.57	877.78	15.60	585.99	nr	1463.77
Branch						
Sum of two sides 1700 mm	196.92	249.22	1.69	63.48	nr	312.70
Sum of two sides 1800 mm	216.77	274.34	1.69	63.48	nr	337.82
Sum of two sides 1900 mm	236.41	299.21	1.85	69.49	nr	368.70
Sum of two sides 2000 mm	256.72	324.90	1.85	69.49	nr	394.40
Sum of two sides 2100 mm	275.21	348.31	2.61	98.04	nr	446.35
Sum of two sides 2200 mm	294.94	373.28	2.61	98.04	nr	471.32
Sum of two sides 2300 mm	314.97	398.62	2.61	98.04	nr	496.66
Sum of two sides 2400 mm	334.71	423.61	2.88	108.18	nr	531.79
Sum of two sides 2500 mm	352.72	446.40	2.88	108.18	nr	554.58
Sum of two sides 2600 mm	372.36	471.26	2.88	108.18	nr	579.45
Sum of two sides 2700 mm	392.01	496.12	2.88	108.18	nr	604.31
Sum of two sides 2800 mm	411.66	521.00	3.94	148.00	nr	669.00
Sum of two sides 2900 mm	431.58	546.20	3.94	148.00	nr	694.20
Sum of two sides 3000 mm	451.22	571.07	4.83	181.43	nr	752.50
Sum of two sides 3100 mm	466.93	590.94	4.83	181.43	nr	772.38
Sum of two sides 3200 mm	481.23	609.05	4.83	181.43	nr	790.48

38 VENTILATION/AIR CONDITIONING SYSTEMS

Item	Net Price £	Material £	Labour hours	Labour £	Unit	Total rate £
Grille neck						
Sum of two sides 1700 mm	205.00	259.45	1.86	69.87	nr	**329.32**
Sum of two sides 1800 mm	226.69	286.90	2.02	75.88	nr	**362.78**
Sum of two sides 1900 mm	248.41	314.38	2.02	75.88	nr	**390.26**
Sum of two sides 2000 mm	270.07	341.81	2.02	75.88	nr	**417.68**
Sum of two sides 2100 mm	291.76	369.26	2.61	98.04	nr	**467.30**
Sum of two sides 2200 mm	313.44	396.69	2.61	98.04	nr	**494.74**
Sum of two sides 2300 mm	335.15	424.16	2.61	98.04	nr	**522.20**
Sum of two sides 2400 mm	355.65	450.12	2.88	108.18	nr	**558.30**
Sum of two sides 2500 mm	378.52	479.05	2.88	108.18	nr	**587.23**
Sum of two sides 2600 mm	398.89	504.84	2.88	108.18	nr	**613.02**
Sum of two sides 2700 mm	420.51	532.20	2.88	108.18	nr	**640.38**
Sum of two sides 2800 mm	442.13	559.56	3.94	148.00	nr	**707.56**
Sum of two sides 2900 mm	463.74	586.91	4.12	154.76	nr	**741.67**
Sum of two sides 3000 mm	485.36	614.27	5.00	187.82	nr	**802.09**
Sum of two sides 3100 mm	502.65	636.15	5.00	187.82	nr	**823.97**
Sum of two sides 3200 mm	518.40	656.09	5.00	187.82	nr	**843.90**
Ductwork 1601 to 2000 mm longest side						
Sum of two sides 2100 mm	233.00	294.89	2.17	81.51	m	**376.40**
Sum of two sides 2200 mm	240.26	304.07	2.17	81.51	m	**385.58**
Sum of two sides 2300 mm	247.76	313.56	2.19	82.26	m	**395.83**
Sum of two sides 2400 mm	254.78	322.46	2.38	89.40	m	**411.86**
Sum of two sides 2500 mm	262.28	331.95	2.38	89.40	m	**421.35**
Sum of two sides 2600 mm	269.34	340.87	2.64	99.17	m	**440.04**
Sum of two sides 2700 mm	276.39	349.80	2.66	99.92	m	**449.72**
Sum of two sides 2800 mm	283.44	358.73	2.95	110.81	m	**469.54**
Sum of two sides 2900 mm	290.94	368.22	2.96	111.19	m	**479.40**
Sum of two sides 3000 mm	297.99	377.14	2.96	111.19	m	**488.33**
Sum of two sides 3100 mm	321.63	407.06	2.96	111.19	m	**518.25**
Sum of two sides 3200 mm	328.69	415.99	3.15	118.33	m	**534.31**
Sum of two sides 3300 mm	360.24	455.92	3.15	118.33	m	**574.24**
Sum of two sides 3400 mm	367.30	464.86	3.15	118.33	m	**583.19**
Sum of two sides 3500 mm	374.37	473.80	3.15	118.33	m	**592.13**
Sum of two sides 3600 mm	381.44	482.75	3.18	119.45	m	**602.20**
Sum of two sides 3700 mm	388.95	492.26	3.18	119.45	m	**611.71**
Sum of two sides 3800 mm	396.02	501.20	3.18	119.45	m	**620.65**
Sum of two sides 3900 mm	403.09	510.15	3.18	119.45	m	**629.60**
Sum of two sides 4000 mm	410.15	519.09	3.18	119.45	m	**638.54**
Extra over fittings; Ductwork 1601 to 2000 mm longest side						
End cap						
Sum of two sides 2100 mm	92.85	117.51	0.87	32.68	nr	**150.19**
Sum of two sides 2200 mm	99.87	126.40	0.87	32.68	nr	**159.08**
Sum of two sides 2300 mm	106.88	135.27	0.87	32.68	nr	**167.95**
Sum of two sides 2400 mm	113.88	144.13	0.87	32.68	nr	**176.81**
Sum of two sides 2500 mm	120.90	153.01	0.87	32.68	nr	**185.69**
Sum of two sides 2600 mm	127.91	161.88	0.87	32.68	nr	**194.56**
Sum of two sides 2700 mm	134.91	170.74	0.87	32.68	nr	**203.42**

38 VENTILATION/AIR CONDITIONING SYSTEMS

Item	Net Price £	Material £	Labour hours	Labour £	Unit	Total rate £
DUCTWORK: RECTANGULAR – CLASS B – cont						
Extra over fittings – cont						
End cap – cont						
Sum of two sides 2800 mm	141.93	179.62	1.16	43.57	nr	**223.20**
Sum of two sides 2900 mm	148.94	188.50	1.16	43.57	nr	**232.07**
Sum of two sides 3000 mm	155.95	197.37	1.73	64.99	nr	**262.35**
Sum of two sides 3100 mm	161.06	203.82	1.80	67.61	nr	**271.43**
Sum of two sides 3200 mm	166.17	210.29	1.80	67.61	nr	**277.90**
Sum of two sides 3300 mm	171.28	216.77	1.80	67.61	nr	**284.39**
Sum of two sides 3400 mm	176.31	223.13	1.80	67.61	nr	**290.75**
Sum of two sides 3500 mm	181.36	229.53	1.80	67.61	nr	**297.14**
Sum of two sides 3600 mm	186.40	235.90	1.80	67.61	nr	**303.52**
Sum of two sides 3700 mm	191.44	242.28	1.80	67.61	nr	**309.89**
Sum of two sides 3800 mm	196.49	248.67	1.80	67.61	nr	**316.29**
Sum of two sides 3900 mm	201.52	255.05	1.80	67.61	nr	**322.66**
Sum of two sides 4000 mm	206.59	261.45	1.80	67.61	nr	**329.07**
Reducer						
Sum of two sides 2100 mm	123.16	155.88	2.61	98.04	nr	**253.92**
Sum of two sides 2200 mm	134.74	170.53	2.61	98.04	nr	**268.57**
Sum of two sides 2300 mm	146.53	185.45	2.61	98.04	nr	**283.49**
Sum of two sides 2400 mm	157.17	198.92	2.88	108.18	nr	**307.10**
Sum of two sides 2500 mm	169.69	214.77	3.12	117.20	nr	**331.96**
Sum of two sides 2600 mm	181.27	229.42	3.12	117.20	nr	**346.62**
Sum of two sides 2700 mm	192.84	244.06	3.12	117.20	nr	**361.26**
Sum of two sides 2800 mm	204.41	258.71	3.95	148.38	nr	**407.08**
Sum of two sides 2900 mm	216.20	273.63	3.97	149.13	nr	**422.76**
Sum of two sides 3000 mm	221.12	279.84	4.52	169.79	nr	**449.63**
Sum of two sides 3100 mm	224.77	284.46	4.52	169.79	nr	**454.25**
Sum of two sides 3200 mm	227.78	288.28	4.52	169.79	nr	**458.07**
Sum of two sides 3300 mm	229.27	290.16	4.52	169.79	nr	**459.95**
Sum of two sides 3400 mm	233.83	295.93	4.52	169.79	nr	**465.72**
Sum of two sides 3500 mm	236.83	299.74	4.52	169.79	nr	**469.52**
Sum of two sides 3600 mm	244.42	309.34	4.52	169.79	nr	**479.12**
Sum of two sides 3700 mm	271.40	343.48	4.52	169.79	nr	**513.27**
Sum of two sides 3800 mm	279.19	353.34	4.52	169.79	nr	**523.13**
Sum of two sides 3900 mm	286.96	363.17	4.52	169.79	nr	**532.96**
Sum of two sides 4000 mm	294.76	373.05	4.52	169.79	nr	**542.83**
Offset						
Sum of two sides 2100 mm	315.98	399.91	2.61	98.04	nr	**497.95**
Sum of two sides 2200 mm	339.78	430.02	2.61	98.04	nr	**528.06**
Sum of two sides 2300 mm	348.34	440.87	2.61	98.04	nr	**538.91**
Sum of two sides 2400 mm	387.42	490.32	2.88	108.18	nr	**598.50**
Sum of two sides 2500 mm	394.84	499.70	3.48	130.72	nr	**630.43**
Sum of two sides 2600 mm	407.08	515.19	3.49	131.10	nr	**646.29**
Sum of two sides 2700 mm	418.86	530.11	3.50	131.47	nr	**661.58**
Sum of two sides 2800 mm	421.78	533.81	4.34	163.03	nr	**696.84**
Sum of two sides 2900 mm	432.51	547.39	4.76	178.80	nr	**726.19**
Sum of two sides 3000 mm	437.51	553.72	5.32	199.84	nr	**753.55**
Sum of two sides 3100 mm	440.31	557.26	5.35	200.97	nr	**758.23**

38 VENTILATION/AIR CONDITIONING SYSTEMS

Item	Net Price £	Material £	Labour hours	Labour £	Unit	Total rate £
Sum of two sides 3200 mm	442.61	560.17	5.35	200.97	nr	**761.13**
Sum of two sides 3300 mm	452.89	573.18	5.35	200.97	nr	**774.14**
Sum of two sides 3400 mm	456.12	577.27	5.35	200.97	nr	**778.23**
Sum of two sides 3500 mm	471.34	596.53	5.35	200.97	nr	**797.49**
Sum of two sides 3600 mm	486.55	615.78	5.35	200.97	nr	**816.75**
Sum of two sides 3700 mm	519.72	657.75	5.35	200.97	nr	**858.72**
Sum of two sides 3800 mm	536.33	678.78	5.35	200.97	nr	**879.74**
Sum of two sides 3900 mm	551.72	698.25	5.35	200.97	nr	**899.22**
Sum of two sides 4000 mm	567.13	717.76	5.35	200.97	nr	**918.73**
Square to round						
Sum of two sides 2100 mm	173.68	219.81	2.61	98.04	nr	**317.85**
Sum of two sides 2200 mm	190.47	241.05	2.61	98.04	nr	**339.09**
Sum of two sides 2300 mm	207.20	262.23	2.61	98.04	nr	**360.27**
Sum of two sides 2400 mm	227.77	288.27	2.88	108.18	nr	**396.45**
Sum of two sides 2500 mm	244.78	309.80	4.70	176.55	nr	**486.34**
Sum of two sides 2600 mm	261.80	331.34	4.70	176.55	nr	**507.89**
Sum of two sides 2700 mm	278.85	352.91	4.71	176.92	nr	**529.83**
Sum of two sides 2800 mm	295.88	374.46	8.19	307.65	nr	**682.11**
Sum of two sides 2900 mm	315.26	398.99	8.19	307.65	nr	**706.64**
Sum of two sides 3000 mm	332.42	420.71	8.19	307.65	nr	**728.36**
Sum of two sides 3100 mm	332.77	421.16	8.19	307.65	nr	**728.80**
Sum of two sides 3200 mm	342.29	433.20	8.19	307.65	nr	**740.84**
Sum of two sides 3300 mm	344.32	435.78	8.19	307.65	nr	**743.42**
Sum of two sides 3400 mm	347.97	440.39	8.62	323.80	nr	**764.18**
Sum of two sides 3500 mm	359.32	454.75	8.62	323.80	nr	**778.55**
Sum of two sides 3600 mm	370.69	469.14	8.62	323.80	nr	**792.94**
Sum of two sides 3700 mm	394.34	499.08	8.62	323.80	nr	**822.88**
Sum of two sides 3800 mm	405.90	513.71	8.75	328.68	nr	**842.39**
Sum of two sides 3900 mm	417.45	528.33	8.75	328.68	nr	**857.01**
Sum of two sides 4000 mm	429.01	542.96	8.75	328.68	nr	**871.64**
90° radius bend						
Sum of two sides 2100 mm	266.86	337.74	2.61	98.04	nr	**435.78**
Sum of two sides 2200 mm	535.52	677.75	2.61	98.04	nr	**775.79**
Sum of two sides 2300 mm	580.29	734.42	2.61	98.04	nr	**832.46**
Sum of two sides 2400 mm	588.67	745.02	2.88	108.18	nr	**853.21**
Sum of two sides 2500 mm	632.99	801.11	3.90	146.50	nr	**947.61**
Sum of two sides 2600 mm	676.40	856.05	4.26	160.02	nr	**1016.07**
Sum of two sides 2700 mm	719.80	910.98	4.55	170.91	nr	**1081.90**
Sum of two sides 2800 mm	763.21	965.92	4.55	170.91	nr	**1136.84**
Sum of two sides 2900 mm	807.49	1021.96	6.87	258.06	nr	**1280.03**

38 VENTILATION/AIR CONDITIONING SYSTEMS

Item	Net Price £	Material £	Labour hours	Labour £	Unit	Total rate £
DUCTWORK: RECTANGULAR – CLASS B – cont						
Extra over fittings – cont						
90° radius bend – cont						
Sum of two sides 3000 mm	850.90	1076.90	6.87	258.06	nr	1334.96
Sum of two sides 3100 mm	860.61	1089.19	6.87	258.06	nr	1347.25
Sum of two sides 3200 mm	893.15	1130.37	6.87	258.06	nr	1388.43
Sum of two sides 3300 mm	874.92	1107.30	6.87	258.06	nr	1365.37
Sum of two sides 3400 mm	910.43	1152.24	7.00	262.95	nr	1415.18
Sum of two sides 3500 mm	941.62	1191.72	7.00	262.95	nr	1454.66
Sum of two sides 3600 mm	972.78	1231.15	7.00	262.95	nr	1494.10
Sum of two sides 3700 mm	1056.91	1337.63	7.00	262.95	nr	1600.57
Sum of two sides 3800 mm	1088.48	1377.59	7.00	262.95	nr	1640.53
Sum of two sides 3900 mm	1120.08	1417.57	7.00	262.95	nr	1680.52
Sum of two sides 4000 mm	1151.65	1457.53	7.00	262.95	nr	1720.48
45° bend						
Sum of two sides 2100 mm	126.89	160.59	2.61	98.04	nr	258.64
Sum of two sides 2200 mm	385.85	488.33	2.61	98.04	nr	586.37
Sum of two sides 2300 mm	416.03	526.53	2.61	98.04	nr	624.57
Sum of two sides 2400 mm	429.59	543.69	2.88	108.18	nr	651.87
Sum of two sides 2500 mm	459.73	581.83	4.85	182.18	nr	764.02
Sum of two sides 2600 mm	489.20	619.14	4.87	182.93	nr	802.07
Sum of two sides 2700 mm	518.66	656.42	4.87	182.93	nr	839.36
Sum of two sides 2800 mm	548.13	693.71	8.81	330.94	nr	1024.64
Sum of two sides 2900 mm	578.29	731.88	8.81	330.94	nr	1062.82
Sum of two sides 3000 mm	607.75	769.17	9.31	349.72	nr	1118.88
Sum of two sides 3100 mm	617.89	782.01	9.31	349.72	nr	1131.72
Sum of two sides 3200 mm	635.73	804.57	9.31	349.72	nr	1154.29
Sum of two sides 3300 mm	639.31	809.12	9.31	349.72	nr	1158.83
Sum of two sides 3400 mm	656.93	831.41	9.31	349.72	nr	1181.12
Sum of two sides 3500 mm	678.12	858.22	9.39	352.72	nr	1210.95
Sum of two sides 3600 mm	699.31	885.04	9.39	352.72	nr	1237.76
Sum of two sides 3700 mm	756.15	956.99	9.39	352.72	nr	1309.71
Sum of two sides 3800 mm	777.59	984.11	9.39	352.72	nr	1336.84
Sum of two sides 3900 mm	799.03	1011.25	9.39	352.72	nr	1363.98
Sum of two sides 4000 mm	820.49	1038.41	9.39	352.72	nr	1391.13
90° mitre bend						
Sum of two sides 2100 mm	673.62	852.54	2.61	98.04	nr	950.58
Sum of two sides 2200 mm	711.70	900.72	2.61	98.04	nr	998.76
Sum of two sides 2300 mm	768.92	973.15	2.61	98.04	nr	1071.19
Sum of two sides 2400 mm	785.96	994.72	2.88	108.18	nr	1102.90
Sum of two sides 2500 mm	843.44	1067.46	4.85	182.18	nr	1249.64
Sum of two sides 2600 mm	901.26	1140.63	4.87	182.93	nr	1323.57
Sum of two sides 2700 mm	980.47	1240.89	4.87	182.93	nr	1423.82
Sum of two sides 2800 mm	1039.58	1315.69	8.81	330.94	nr	1646.63
Sum of two sides 2900 mm	1098.32	1390.04	14.81	556.32	nr	1946.36
Sum of two sides 3000 mm	1176.95	1489.55	15.20	570.97	nr	2060.52
Sum of two sides 3100 mm	1192.98	1509.84	15.20	570.97	nr	2080.81
Sum of two sides 3200 mm	1239.66	1568.92	15.20	570.97	nr	2139.88
Sum of two sides 3300 mm	1252.76	1585.49	15.20	570.97	nr	2156.46
Sum of two sides 3400 mm	1315.91	1665.41	15.20	570.97	nr	2236.38
Sum of two sides 3500 mm	1363.16	1725.22	15.60	585.99	nr	2311.21

38 VENTILATION/AIR CONDITIONING SYSTEMS

Item	Net Price £	Material £	Labour hours	Labour £	Unit	Total rate £
Sum of two sides 3600 mm	1410.42	1785.03	15.60	585.99	nr	**2371.02**
Sum of two sides 3700 mm	1496.32	1893.74	15.60	585.99	nr	**2479.74**
Sum of two sides 3800 mm	1544.35	1954.53	15.60	585.99	nr	**2540.52**
Sum of two sides 3900 mm	1592.40	2015.34	15.60	585.99	nr	**2601.33**
Sum of two sides 4000 mm	1640.43	2076.13	15.60	585.99	nr	**2662.12**
Branch						
Sum of two sides 2100 mm	282.61	357.67	2.61	98.04	nr	**455.72**
Sum of two sides 2200 mm	302.37	382.67	2.61	98.04	nr	**480.72**
Sum of two sides 2300 mm	322.41	408.05	2.61	98.04	nr	**506.09**
Sum of two sides 2400 mm	341.85	432.65	2.88	108.18	nr	**540.83**
Sum of two sides 2500 mm	361.91	458.03	2.88	108.18	nr	**566.22**
Sum of two sides 2600 mm	381.70	483.08	2.88	108.18	nr	**591.26**
Sum of two sides 2700 mm	401.48	508.11	2.88	108.18	nr	**616.29**
Sum of two sides 2800 mm	421.25	533.14	3.94	148.00	nr	**681.14**
Sum of two sides 2900 mm	441.30	558.51	3.94	148.00	nr	**706.51**
Sum of two sides 3000 mm	461.07	583.53	3.94	148.00	nr	**731.53**
Sum of two sides 3100 mm	476.91	603.58	3.94	148.00	nr	**751.58**
Sum of two sides 3200 mm	491.31	621.80	3.94	148.00	nr	**769.80**
Sum of two sides 3300 mm	505.99	640.38	3.94	148.00	nr	**788.38**
Sum of two sides 3400 mm	520.41	658.63	4.83	181.43	nr	**840.06**
Sum of two sides 3500 mm	534.81	676.85	4.83	181.43	nr	**858.29**
Sum of two sides 3600 mm	549.23	695.11	4.83	181.43	nr	**876.54**
Sum of two sides 3700 mm	570.51	722.04	4.83	181.43	nr	**903.47**
Sum of two sides 3800 mm	584.93	740.29	4.83	181.43	nr	**921.72**
Sum of two sides 3900 mm	599.33	758.51	4.83	181.43	nr	**939.95**
Sum of two sides 4000 mm	613.75	776.77	4.83	181.43	nr	**958.20**
Grille neck						
Sum of two sides 2100 mm	291.76	369.26	2.61	98.04	nr	**467.30**
Sum of two sides 2200 mm	313.44	396.69	2.61	98.04	nr	**494.74**
Sum of two sides 2300 mm	335.15	424.16	2.61	98.04	nr	**522.20**
Sum of two sides 2400 mm	356.82	451.60	2.88	108.18	nr	**559.78**
Sum of two sides 2500 mm	378.52	479.05	2.88	108.18	nr	**587.23**
Sum of two sides 2600 mm	398.89	504.84	2.88	108.18	nr	**613.02**
Sum of two sides 2700 mm	420.51	532.20	2.88	108.18	nr	**640.38**
Sum of two sides 2800 mm	442.13	559.56	3.94	148.00	nr	**707.56**
Sum of two sides 2900 mm	463.74	586.91	4.12	154.76	nr	**741.67**
Sum of two sides 3000 mm	485.36	614.27	4.12	154.76	nr	**769.03**
Sum of two sides 3100 mm	502.65	636.15	4.12	154.76	nr	**790.91**
Sum of two sides 3200 mm	518.40	656.09	4.12	154.76	nr	**810.85**
Sum of two sides 3300 mm	534.13	675.99	4.12	154.76	nr	**830.75**
Sum of two sides 3400 mm	549.91	695.97	5.00	187.82	nr	**883.78**
Sum of two sides 3500 mm	565.67	715.91	5.00	187.82	nr	**903.73**
Sum of two sides 3600 mm	581.41	735.83	5.00	187.82	nr	**923.65**
Sum of two sides 3700 mm	597.17	755.78	5.00	187.82	nr	**943.60**
Sum of two sides 3800 mm	612.91	775.70	5.00	187.82	nr	**963.52**
Sum of two sides 3900 mm	628.67	795.65	5.00	187.82	nr	**983.47**
Sum of two sides 4000 mm	644.43	815.60	5.00	187.82	nr	**1003.41**
Ductwork 2001 to 2500 mm longest side						
Sum of two sides 2500 mm	343.84	435.17	2.38	89.40	m	**524.57**
Sum of two sides 2600 mm	352.70	446.37	2.64	99.17	m	**545.54**
Sum of two sides 2700 mm	360.33	456.03	2.66	99.92	m	**555.95**

38 VENTILATION/AIR CONDITIONING SYSTEMS

Item	Net Price £	Material £	Labour hours	Labour £	Unit	Total rate £
DUCTWORK: RECTANGULAR – CLASS B – cont						
Ductwork 2001 to 2500 mm longest side – cont						
Sum of two sides 2800 mm	369.75	467.96	2.95	110.81	m	578.77
Sum of two sides 2900 mm	377.38	477.62	2.96	111.19	m	588.80
Sum of two sides 3000 mm	384.59	486.74	3.15	118.33	m	605.06
Sum of two sides 3100 mm	392.07	496.20	3.15	118.33	m	614.53
Sum of two sides 3200 mm	399.28	505.33	3.15	118.33	m	623.65
Sum of two sides 3300 mm	406.90	514.97	3.15	118.33	m	633.29
Sum of two sides 3400 mm	414.11	524.09	2.66	99.92	m	624.01
Sum of two sides 3500 mm	441.70	559.02	3.15	118.33	m	677.34
Sum of two sides 3600 mm	448.91	568.14	3.18	119.45	m	687.59
Sum of two sides 3700 mm	480.84	608.55	3.18	119.45	m	728.01
Sum of two sides 3800 mm	488.05	617.68	3.18	119.45	m	737.13
Sum of two sides 3900 mm	495.26	626.80	3.18	119.45	m	746.25
Sum of two sides 4000 mm	504.71	638.76	3.18	119.45	m	758.21
Extra over fittings; Ductwork 2001 to 2500 mm longest side						
End cap						
Sum of two sides 2500 mm	120.90	153.01	0.87	32.68	nr	185.69
Sum of two sides 2600 mm	127.91	161.88	0.87	32.68	nr	194.56
Sum of two sides 2700 mm	134.91	170.74	0.87	32.68	nr	203.42
Sum of two sides 2800 mm	141.93	179.62	1.16	43.57	nr	223.20
Sum of two sides 2900 mm	148.94	188.50	1.16	43.57	nr	232.07
Sum of two sides 3000 mm	155.95	197.37	1.73	64.99	nr	262.35
Sum of two sides 3100 mm	161.06	203.82	1.73	64.99	nr	268.81
Sum of two sides 3200 mm	166.17	210.29	1.73	64.99	nr	275.28
Sum of two sides 3300 mm	171.28	216.77	1.73	64.99	nr	281.76
Sum of two sides 3400 mm	176.31	223.13	1.80	67.61	nr	290.75
Sum of two sides 3500 mm	181.36	229.53	1.80	67.61	nr	297.14
Sum of two sides 3600 mm	186.40	235.90	1.80	67.61	nr	303.52
Sum of two sides 3700 mm	191.44	242.28	1.80	67.61	nr	309.89
Sum of two sides 3800 mm	196.49	248.67	1.80	67.61	nr	316.29
Sum of two sides 3900 mm	201.52	255.05	1.80	67.61	nr	322.66
Sum of two sides 4000 mm	206.59	261.45	1.80	67.61	nr	329.07
Reducer						
Sum of two sides 2500 mm	125.99	159.46	3.12	117.20	nr	276.65
Sum of two sides 2600 mm	136.48	172.73	3.12	117.20	nr	289.93
Sum of two sides 2700 mm	147.21	186.31	3.12	117.20	nr	303.51
Sum of two sides 2800 mm	162.01	205.03	3.95	148.38	nr	353.41
Sum of two sides 2900 mm	173.03	218.99	3.97	149.13	nr	368.12
Sum of two sides 3000 mm	183.86	232.69	3.97	149.13	nr	381.82
Sum of two sides 3100 mm	191.85	242.81	3.97	149.13	nr	391.93
Sum of two sides 3200 mm	199.10	251.98	3.97	149.13	nr	401.11
Sum of two sides 3300 mm	204.16	258.39	3.97	149.13	nr	407.51
Sum of two sides 3400 mm	206.61	261.48	4.52	169.79	nr	431.27
Sum of two sides 3500 mm	209.78	265.49	4.52	169.79	nr	435.28
Sum of two sides 3600 mm	213.86	270.66	4.52	169.79	nr	440.44
Sum of two sides 3700 mm	217.75	275.58	4.52	169.79	nr	445.37
Sum of two sides 3800 mm	219.88	278.28	4.52	169.79	nr	448.07
Sum of two sides 3900 mm	219.88	278.28	4.52	169.79	nr	448.07
Sum of two sides 4000 mm	227.06	287.37	4.52	169.79	nr	457.15

38 VENTILATION/AIR CONDITIONING SYSTEMS

Item	Net Price £	Material £	Labour hours	Labour £	Unit	Total rate £
Offset						
Sum of two sides 2500 mm	368.62	466.52	3.48	130.72	nr	**597.24**
Sum of two sides 2600 mm	391.62	495.64	3.49	131.10	nr	**626.74**
Sum of two sides 2700 mm	400.46	506.83	3.50	131.47	nr	**638.30**
Sum of two sides 2800 mm	424.74	537.55	4.34	163.03	nr	**700.57**
Sum of two sides 2900 mm	430.38	544.69	4.76	178.80	nr	**723.49**
Sum of two sides 3000 mm	434.38	549.75	5.32	199.84	nr	**749.59**
Sum of two sides 3100 mm	437.60	553.83	5.32	199.84	nr	**753.67**
Sum of two sides 3200 mm	438.18	554.56	5.32	199.84	nr	**754.40**
Sum of two sides 3300 mm	441.55	558.82	5.32	199.84	nr	**758.66**
Sum of two sides 3400 mm	443.89	561.79	5.34	200.59	nr	**762.38**
Sum of two sides 3500 mm	444.53	562.60	5.35	200.97	nr	**763.57**
Sum of two sides 3600 mm	444.85	563.00	5.35	200.97	nr	**763.97**
Sum of two sides 3700 mm	445.99	564.44	5.35	200.97	nr	**765.41**
Sum of two sides 3800 mm	445.99	564.45	5.35	200.97	nr	**765.41**
Sum of two sides 3900 mm	460.57	582.90	5.35	200.97	nr	**783.87**
Sum of two sides 4000 mm	475.10	601.29	5.35	200.97	nr	**802.26**
Square to round						
Sum of two sides 2500 mm	201.14	254.57	4.70	176.55	nr	**431.12**
Sum of two sides 2600 mm	217.14	274.81	4.70	176.55	nr	**451.36**
Sum of two sides 2700 mm	233.11	295.02	4.71	176.92	nr	**471.95**
Sum of two sides 2800 mm	254.32	321.87	8.19	307.65	nr	**629.51**
Sum of two sides 2900 mm	270.61	342.49	8.19	307.65	nr	**650.14**
Sum of two sides 3000 mm	286.95	363.17	8.19	307.65	nr	**670.81**
Sum of two sides 3100 mm	299.16	378.62	8.19	307.65	nr	**686.26**
Sum of two sides 3200 mm	310.23	392.62	8.62	323.80	nr	**716.42**
Sum of two sides 3300 mm	321.30	406.64	8.62	323.80	nr	**730.44**
Sum of two sides 3400 mm	330.97	418.88	8.62	323.80	nr	**742.67**
Sum of two sides 3500 mm	332.35	420.62	8.62	323.80	nr	**744.41**
Sum of two sides 3600 mm	340.65	431.12	8.75	328.68	nr	**759.80**
Sum of two sides 3700 mm	343.38	434.58	8.75	328.68	nr	**763.26**
Sum of two sides 3800 mm	344.91	436.51	8.75	328.68	nr	**765.20**
Sum of two sides 3900 mm	355.76	450.25	8.75	328.68	nr	**778.93**
Sum of two sides 4000 mm	366.52	463.86	8.75	328.68	nr	**792.54**
90° radius bend						
Sum of two sides 2500 mm	507.63	642.45	3.90	146.50	nr	**788.95**
Sum of two sides 2600 mm	530.48	671.37	4.26	160.02	nr	**831.40**
Sum of two sides 2700 mm	564.40	714.31	4.55	170.91	nr	**885.22**
Sum of two sides 2800 mm	571.98	723.89	4.55	170.91	nr	**894.81**
Sum of two sides 2900 mm	616.31	780.00	6.87	258.06	nr	**1038.06**
Sum of two sides 3000 mm	656.96	831.45	6.87	258.06	nr	**1089.51**
Sum of two sides 3100 mm	688.80	871.74	6.87	258.06	nr	**1129.80**
Sum of two sides 3200 mm	718.36	909.15	6.87	258.06	nr	**1167.21**
Sum of two sides 3300 mm	748.69	947.54	6.87	258.06	nr	**1205.60**
Sum of two sides 3400 mm	778.27	984.98	6.87	258.06	nr	**1243.04**
Sum of two sides 3500 mm	778.50	985.27	7.00	262.95	nr	**1248.21**
Sum of two sides 3600 mm	808.08	1022.70	7.00	262.95	nr	**1285.65**
Sum of two sides 3700 mm	783.92	992.12	7.00	262.95	nr	**1255.07**
Sum of two sides 3800 mm	814.91	1031.36	7.00	262.95	nr	**1294.30**
Sum of two sides 3900 mm	843.94	1068.10	7.00	262.95	nr	**1331.04**
Sum of two sides 4000 mm	852.01	1078.31	7.00	262.95	nr	**1341.25**

38 VENTILATION/AIR CONDITIONING SYSTEMS

Item	Net Price £	Material £	Labour hours	Labour £	Unit	Total rate £
DUCTWORK: RECTANGULAR – CLASS B – cont						
Extra over fittings – cont						
45° bend						
Sum of two sides 2500 mm	394.05	498.71	4.85	182.18	nr	680.90
Sum of two sides 2600 mm	412.75	522.37	4.87	182.93	nr	705.31
Sum of two sides 2700 mm	441.62	558.91	4.87	182.93	nr	741.85
Sum of two sides 2800 mm	449.46	568.84	8.81	330.94	nr	899.77
Sum of two sides 2900 mm	478.29	605.32	8.81	330.94	nr	936.26
Sum of two sides 3000 mm	506.48	641.00	9.31	349.72	nr	990.72
Sum of two sides 3100 mm	528.67	669.08	9.31	349.72	nr	1018.80
Sum of two sides 3200 mm	549.16	695.01	9.31	349.72	nr	1044.73
Sum of two sides 3300 mm	570.31	721.79	9.31	349.72	nr	1071.51
Sum of two sides 3400 mm	590.82	747.74	9.31	349.72	nr	1097.46
Sum of two sides 3500 mm	595.97	754.26	9.31	349.72	nr	1103.98
Sum of two sides 3600 mm	616.49	780.23	9.39	352.72	nr	1132.95
Sum of two sides 3700 mm	608.75	770.44	9.39	352.72	nr	1123.16
Sum of two sides 3800 mm	628.98	796.03	9.39	352.72	nr	1148.75
Sum of two sides 3900 mm	649.20	821.62	9.39	352.72	nr	1174.35
Sum of two sides 4000 mm	658.48	833.37	9.39	352.72	nr	1186.09
90° mitre bend						
Sum of two sides 2500 mm	679.44	859.90	4.85	182.18	nr	1042.09
Sum of two sides 2600 mm	720.75	912.18	4.87	182.93	nr	1095.11
Sum of two sides 2700 mm	775.55	981.54	4.87	182.93	nr	1164.47
Sum of two sides 2800 mm	777.16	983.57	8.81	330.94	nr	1314.51
Sum of two sides 2900 mm	832.18	1053.21	14.81	556.32	nr	1609.53
Sum of two sides 3000 mm	908.29	1149.54	14.81	556.32	nr	1705.85
Sum of two sides 3100 mm	956.10	1210.04	15.20	570.97	nr	1781.01
Sum of two sides 3200 mm	1023.31	1295.10	15.20	570.97	nr	1866.07
Sum of two sides 3300 mm	1068.14	1351.84	15.20	570.97	nr	1922.80
Sum of two sides 3400 mm	1131.60	1432.15	15.20	570.97	nr	2003.11
Sum of two sides 3500 mm	1131.81	1432.42	15.20	570.97	nr	2003.39
Sum of two sides 3600 mm	1194.36	1511.58	15.60	585.99	nr	2097.57
Sum of two sides 3700 mm	1200.94	1519.91	15.60	585.99	nr	2105.90
Sum of two sides 3800 mm	1247.59	1578.95	15.60	585.99	nr	2164.94
Sum of two sides 3900 mm	1294.23	1637.98	15.60	585.99	nr	2223.97
Sum of two sides 4000 mm	1308.05	1655.47	15.60	585.99	nr	2241.46
Branch						
Sum of two sides 2500 mm	361.85	457.96	2.88	108.18	nr	566.14
Sum of two sides 2600 mm	381.56	482.90	2.88	108.18	nr	591.08
Sum of two sides 2700 mm	401.56	508.22	2.88	108.18	nr	616.40
Sum of two sides 2800 mm	420.97	532.78	3.94	148.00	nr	680.79
Sum of two sides 2900 mm	440.98	558.11	3.94	148.00	nr	706.11
Sum of two sides 3000 mm	460.72	583.09	3.94	148.00	nr	731.09
Sum of two sides 3100 mm	476.50	603.05	3.94	148.00	nr	751.06
Sum of two sides 3200 mm	490.89	621.27	3.94	148.00	nr	769.27
Sum of two sides 3300 mm	505.54	639.81	3.94	148.00	nr	787.81
Sum of two sides 3400 mm	519.92	658.01	3.94	148.00	nr	806.02
Sum of two sides 3500 mm	535.01	677.11	4.83	181.43	nr	858.54
Sum of two sides 3600 mm	549.39	695.31	4.83	181.43	nr	876.74
Sum of two sides 3700 mm	564.05	713.87	4.83	181.43	nr	895.30
Sum of two sides 3800 mm	578.42	732.05	4.83	181.43	nr	913.48
Sum of two sides 3900 mm	592.82	750.27	4.83	181.43	nr	931.70

38 VENTILATION/AIR CONDITIONING SYSTEMS

Item	Net Price £	Material £	Labour hours	Labour £	Unit	Total rate £
Sum of two sides 4000 mm	607.42	768.75	4.83	181.43	nr	**950.18**
Grille neck						
Sum of two sides 2500 mm	378.52	479.05	2.88	108.18	nr	**587.23**
Sum of two sides 2600 mm	398.89	504.84	2.88	108.18	nr	**613.02**
Sum of two sides 2700 mm	420.51	532.20	2.88	108.18	nr	**640.38**
Sum of two sides 2800 mm	442.13	559.56	3.94	148.00	nr	**707.56**
Sum of two sides 2900 mm	463.62	586.75	3.94	148.00	nr	**734.75**
Sum of two sides 3000 mm	485.36	614.27	3.94	148.00	nr	**762.27**
Sum of two sides 3100 mm	502.65	636.15	4.12	154.76	nr	**790.91**
Sum of two sides 3200 mm	518.40	656.09	4.12	154.76	nr	**810.85**
Sum of two sides 3300 mm	534.14	676.00	4.12	154.76	nr	**830.77**
Sum of two sides 3400 mm	549.91	695.97	4.12	154.76	nr	**850.73**
Sum of two sides 3500 mm	565.67	715.91	5.00	187.82	nr	**903.73**
Sum of two sides 3600 mm	581.41	735.83	5.00	187.82	nr	**923.65**
Sum of two sides 3700 mm	597.17	755.78	5.00	187.82	nr	**943.60**
Sum of two sides 3800 mm	612.91	775.70	5.00	187.82	nr	**963.52**
Sum of two sides 3900 mm	628.67	795.65	5.00	187.82	nr	**983.47**
Sum of two sides 4000 mm	644.43	815.60	5.00	187.82	nr	**1003.41**
Ductwork 2501 to 4000 mm longest side						
Sum of two sides 3000 mm	528.05	668.30	2.38	89.40	m	**757.70**
Sum of two sides 3100 mm	540.69	684.30	2.38	89.40	m	**773.70**
Sum of two sides 3200 mm	552.67	699.46	2.38	89.40	m	**788.86**
Sum of two sides 3300 mm	564.64	714.61	2.38	89.40	m	**804.02**
Sum of two sides 3400 mm	579.46	733.37	2.64	99.17	m	**832.54**
Sum of two sides 3500 mm	591.44	748.53	2.66	99.92	m	**848.45**
Sum of two sides 3600 mm	603.43	763.70	2.95	110.81	m	**874.51**
Sum of two sides 3700 mm	618.64	782.96	2.96	111.19	m	**894.14**
Sum of two sides 3800 mm	630.62	798.11	3.15	118.33	m	**916.44**
Sum of two sides 3900 mm	664.62	841.14	3.15	118.33	m	**959.46**
Sum of two sides 4000 mm	676.59	856.30	3.15	118.33	m	**974.62**
Sum of two sides 4100 mm	688.98	871.97	3.35	125.84	m	**997.81**
Sum of two sides 4200 mm	700.95	887.13	3.35	125.84	m	**1012.96**
Sum of two sides 4300 mm	737.45	933.31	3.60	135.23	m	**1068.54**
Sum of two sides 4400 mm	740.47	937.13	3.60	135.23	m	**1072.36**
Sum of two sides 4500 mm	743.06	940.42	3.60	135.23	m	**1075.65**
Extra over fittings; Ductwork 2501 to 4000 mm longest side						
End cap						
Sum of two sides 3000 mm	157.66	199.53	1.73	64.99	nr	**264.51**
Sum of two sides 3100 mm	163.27	206.64	1.73	64.99	nr	**271.63**
Sum of two sides 3200 mm	168.38	213.10	1.73	64.99	nr	**278.09**
Sum of two sides 3300 mm	173.40	219.45	1.73	64.99	nr	**284.43**
Sum of two sides 3400 mm	178.60	226.04	1.73	64.99	nr	**291.03**
Sum of two sides 3500 mm	183.73	232.53	1.73	64.99	nr	**297.52**
Sum of two sides 3600 mm	188.84	238.99	1.73	64.99	nr	**303.98**
Sum of two sides 3700 mm	193.96	245.47	1.73	64.99	nr	**310.46**
Sum of two sides 3800 mm	199.06	251.93	1.73	64.99	nr	**316.92**
Sum of two sides 3900 mm	204.18	258.41	1.80	67.61	nr	**326.02**

38 VENTILATION/AIR CONDITIONING SYSTEMS

Item	Net Price £	Material £	Labour hours	Labour £	Unit	Total rate £
DUCTWORK: RECTANGULAR – CLASS B – cont						
Extra over fittings – cont						
End cap – cont						
Sum of two sides 4000 mm	209.28	264.87	1.80	67.61	nr	**332.49**
Sum of two sides 4100 mm	214.40	271.35	1.80	67.61	nr	**338.96**
Sum of two sides 4200 mm	219.51	277.81	1.80	67.61	nr	**345.42**
Sum of two sides 4300 mm	224.61	284.27	1.88	70.62	nr	**354.89**
Sum of two sides 4400 mm	229.73	290.75	1.88	70.62	nr	**361.37**
Sum of two sides 4500 mm	234.84	297.21	1.88	70.62	nr	**367.83**
Reducer						
Sum of two sides 3000 mm	105.83	133.94	3.12	117.20	nr	**251.14**
Sum of two sides 3100 mm	110.82	140.25	3.12	117.20	nr	**257.45**
Sum of two sides 3200 mm	118.67	150.18	3.12	117.20	nr	**267.38**
Sum of two sides 3300 mm	123.12	155.82	3.12	117.20	nr	**273.02**
Sum of two sides 3400 mm	127.51	161.38	3.12	117.20	nr	**278.58**
Sum of two sides 3500 mm	131.96	167.01	3.12	117.20	nr	**284.21**
Sum of two sides 3600 mm	136.40	172.63	3.95	148.38	nr	**321.01**
Sum of two sides 3700 mm	138.31	175.04	3.97	149.13	nr	**324.17**
Sum of two sides 3800 mm	140.98	178.42	4.52	169.79	nr	**348.21**
Sum of two sides 3900 mm	141.49	179.07	4.52	169.79	nr	**348.86**
Sum of two sides 4000 mm	145.41	184.03	4.52	169.79	nr	**353.82**
Sum of two sides 4100 mm	146.28	185.13	4.52	169.79	nr	**354.92**
Sum of two sides 4200 mm	150.88	190.96	4.52	169.79	nr	**360.75**
Sum of two sides 4300 mm	155.33	196.59	4.92	184.81	nr	**381.40**
Sum of two sides 4400 mm	164.31	207.95	4.92	184.81	nr	**392.76**
Sum of two sides 4500 mm	168.75	213.57	5.12	192.33	nr	**405.90**
Offset						
Sum of two sides 3000 mm	365.39	462.43	3.48	130.72	nr	**593.16**
Sum of two sides 3100 mm	365.90	463.08	3.48	130.72	nr	**593.80**
Sum of two sides 3200 mm	387.33	490.20	3.48	130.72	nr	**620.92**
Sum of two sides 3300 mm	397.46	503.02	3.48	130.72	nr	**633.74**
Sum of two sides 3400 mm	398.42	504.24	3.49	131.10	nr	**635.34**
Sum of two sides 3500 mm	402.89	509.90	3.50	131.47	nr	**641.37**
Sum of two sides 3600 mm	409.75	518.58	4.25	159.65	nr	**678.22**
Sum of two sides 3700 mm	415.67	526.07	4.76	178.80	nr	**704.87**
Sum of two sides 3800 mm	415.90	526.37	5.32	199.84	nr	**726.21**
Sum of two sides 3900 mm	420.65	532.37	5.35	200.97	nr	**733.34**
Sum of two sides 4000 mm	422.86	535.17	5.35	200.97	nr	**736.13**
Sum of two sides 4100 mm	428.50	542.31	5.85	219.75	nr	**762.06**
Sum of two sides 4200 mm	432.10	546.86	5.85	219.75	nr	**766.61**
Sum of two sides 4300 mm	433.50	548.64	6.15	231.02	nr	**779.66**
Sum of two sides 4400 mm	440.61	557.63	6.15	231.02	nr	**788.65**
Sum of two sides 4500 mm	446.08	564.56	6.30	236.65	nr	**801.21**
Square to round						
Sum of two sides 3000 mm	275.83	349.09	4.70	176.55	nr	**525.64**
Sum of two sides 3100 mm	287.67	364.07	4.70	176.55	nr	**540.62**
Sum of two sides 3200 mm	298.22	377.43	4.70	176.55	nr	**553.97**
Sum of two sides 3300 mm	308.78	390.79	4.70	176.55	nr	**567.34**
Sum of two sides 3400 mm	319.34	404.16	4.71	176.92	nr	**581.08**
Sum of two sides 3500 mm	329.89	417.51	4.71	176.92	nr	**594.43**
Sum of two sides 3600 mm	340.44	430.86	8.19	307.65	nr	**738.51**

38 VENTILATION/AIR CONDITIONING SYSTEMS

Item	Net Price £	Material £	Labour hours	Labour £	Unit	Total rate £
Sum of two sides 3700 mm	350.99	444.22	8.62	323.80	nr	**768.01**
Sum of two sides 3800 mm	361.55	457.58	8.62	323.80	nr	**781.38**
Sum of two sides 3900 mm	456.89	578.24	8.62	323.80	nr	**902.04**
Sum of two sides 4000 mm	469.88	594.69	8.75	328.68	nr	**923.37**
Sum of two sides 4100 mm	482.43	610.56	11.23	421.84	nr	**1032.40**
Sum of two sides 4200 mm	495.85	627.54	11.23	421.84	nr	**1049.38**
Sum of two sides 4300 mm	508.82	643.97	11.25	422.59	nr	**1066.56**
Sum of two sides 4400 mm	521.81	660.40	11.25	422.59	nr	**1082.99**
Sum of two sides 4500 mm	534.80	676.85	11.26	422.97	nr	**1099.81**
90° radius bend						
Sum of two sides 3000 mm	783.91	992.12	3.90	146.50	nr	**1138.62**
Sum of two sides 3100 mm	821.18	1039.29	3.90	146.50	nr	**1185.78**
Sum of two sides 3200 mm	833.49	1054.87	4.26	160.02	nr	**1214.89**
Sum of two sides 3300 mm	854.62	1081.61	4.26	160.02	nr	**1241.63**
Sum of two sides 3400 mm	863.87	1093.32	4.26	160.02	nr	**1253.34**
Sum of two sides 3500 mm	864.01	1093.48	4.55	170.91	nr	**1264.40**
Sum of two sides 3600 mm	867.25	1097.59	4.55	170.91	nr	**1268.50**
Sum of two sides 3700 mm	888.06	1123.93	6.87	258.06	nr	**1381.99**
Sum of two sides 3800 mm	890.07	1126.47	6.87	258.06	nr	**1384.53**
Sum of two sides 3900 mm	892.82	1129.96	6.87	258.06	nr	**1388.02**
Sum of two sides 4000 mm	895.12	1132.86	7.00	262.95	nr	**1395.81**
Sum of two sides 4100 mm	900.24	1139.35	7.20	270.46	nr	**1409.81**
Sum of two sides 4200 mm	909.97	1151.65	7.20	270.46	nr	**1422.11**
Sum of two sides 4300 mm	925.49	1171.30	7.41	278.35	nr	**1449.65**
Sum of two sides 4400 mm	933.25	1181.12	7.41	278.35	nr	**1459.47**
Sum of two sides 4500 mm	942.54	1192.87	7.55	283.61	nr	**1476.48**
45° bend						
Sum of two sides 3000 mm	380.31	481.32	4.85	182.18	nr	**663.50**
Sum of two sides 3100 mm	396.01	501.19	4.85	182.18	nr	**683.38**
Sum of two sides 3200 mm	398.85	504.79	4.85	182.18	nr	**686.97**
Sum of two sides 3300 mm	409.73	518.55	4.87	182.93	nr	**701.48**
Sum of two sides 3400 mm	410.87	519.99	4.87	182.93	nr	**702.93**
Sum of two sides 3500 mm	410.93	520.08	4.87	182.93	nr	**703.01**
Sum of two sides 3600 mm	415.51	525.87	8.81	330.94	nr	**856.80**
Sum of two sides 3700 mm	422.10	534.20	8.81	330.94	nr	**865.14**
Sum of two sides 3800 mm	422.10	534.20	9.31	349.72	nr	**883.92**
Sum of two sides 3900 mm	423.80	536.36	9.31	349.72	nr	**886.08**
Sum of two sides 4000 mm	426.22	539.43	9.31	349.72	nr	**889.14**
Sum of two sides 4100 mm	426.98	540.39	9.31	349.72	nr	**890.11**
Sum of two sides 4200 mm	431.17	545.69	10.01	376.01	nr	**921.70**
Sum of two sides 4300 mm	441.00	558.13	10.01	376.01	nr	**934.14**
Sum of two sides 4400 mm	443.03	560.69	10.52	395.17	nr	**955.86**
Sum of two sides 4500 mm	447.00	565.72	10.52	395.17	nr	**960.89**
90° mitre bend						
Sum of two sides 3000 mm	1140.50	1443.42	4.85	182.18	nr	**1625.60**
Sum of two sides 3100 mm	1196.72	1514.56	4.85	182.18	nr	**1696.75**
Sum of two sides 3200 mm	1275.62	1614.42	4.87	182.93	nr	**1797.36**
Sum of two sides 3300 mm	1296.70	1641.11	4.87	182.93	nr	**1824.04**
Sum of two sides 3400 mm	1302.80	1648.83	4.87	182.93	nr	**1831.76**
Sum of two sides 3500 mm	1302.80	1648.83	8.81	330.94	nr	**1979.76**
Sum of two sides 3600 mm	1329.60	1682.74	8.81	330.94	nr	**2013.68**
Sum of two sides 3700 mm	1375.05	1740.26	14.81	556.32	nr	**2296.58**

38 VENTILATION/AIR CONDITIONING SYSTEMS

Item	Net Price £	Material £	Labour hours	Labour £	Unit	Total rate £
DUCTWORK: RECTANGULAR – CLASS B – cont						
Extra over fittings – cont						
90° mitre bend – cont						
Sum of two sides 3800 mm	1377.13	1742.90	14.81	556.32	nr	**2299.21**
Sum of two sides 3900 mm	1382.57	1749.78	14.81	556.32	nr	**2306.10**
Sum of two sides 4000 mm	1430.22	1810.09	15.20	570.97	nr	**2381.05**
Sum of two sides 4100 mm	1434.48	1815.48	16.30	612.29	nr	**2427.77**
Sum of two sides 4200 mm	1443.47	1826.86	16.50	619.80	nr	**2446.66**
Sum of two sides 4300 mm	1508.30	1908.90	17.01	638.96	nr	**2547.86**
Sum of two sides 4400 mm	1509.86	1910.88	17.01	638.96	nr	**2549.83**
Sum of two sides 4500 mm	1518.32	1921.59	17.01	638.96	nr	**2560.54**
Branch						
Sum of two sides 3000 mm	475.79	602.17	2.88	108.18	nr	**710.35**
Sum of two sides 3100 mm	492.25	622.99	2.88	108.18	nr	**731.17**
Sum of two sides 3200 mm	507.10	641.79	2.88	108.18	nr	**749.97**
Sum of two sides 3300 mm	521.93	660.56	2.88	108.18	nr	**768.74**
Sum of two sides 3400 mm	536.72	679.27	2.88	108.18	nr	**787.46**
Sum of two sides 3500 mm	551.56	698.06	3.94	148.00	nr	**846.06**
Sum of two sides 3600 mm	566.42	716.86	3.94	148.00	nr	**864.86**
Sum of two sides 3700 mm	581.43	735.86	3.94	148.00	nr	**883.86**
Sum of two sides 3800 mm	596.29	754.66	4.83	181.43	nr	**936.09**
Sum of two sides 3900 mm	611.72	774.19	4.83	181.43	nr	**955.62**
Sum of two sides 4000 mm	626.57	792.99	4.83	181.43	nr	**974.42**
Sum of two sides 4100 mm	641.67	812.09	5.44	204.35	nr	**1016.44**
Sum of two sides 4200 mm	656.51	830.88	5.44	204.35	nr	**1035.22**
Sum of two sides 4300 mm	671.34	849.65	5.85	219.75	nr	**1069.39**
Sum of two sides 4400 mm	675.63	855.07	5.85	219.75	nr	**1074.82**
Sum of two sides 4500 mm	701.20	887.43	5.85	219.75	nr	**1107.18**
Grille neck						
Sum of two sides 3000 mm	502.07	635.41	2.88	108.18	nr	**743.60**
Sum of two sides 3100 mm	519.93	658.02	2.88	108.18	nr	**766.20**
Sum of two sides 3200 mm	536.23	678.65	2.88	108.18	nr	**786.83**
Sum of two sides 3300 mm	552.54	699.29	2.88	108.18	nr	**807.47**
Sum of two sides 3400 mm	568.85	719.93	3.94	148.00	nr	**867.93**
Sum of two sides 3500 mm	585.16	740.58	3.94	148.00	nr	**888.58**
Sum of two sides 3600 mm	611.43	773.83	3.94	148.00	nr	**921.83**
Sum of two sides 3700 mm	617.74	781.82	4.12	154.76	nr	**936.58**
Sum of two sides 3800 mm	634.05	802.46	4.12	154.76	nr	**957.22**
Sum of two sides 3900 mm	650.36	823.10	4.12	154.76	nr	**977.86**
Sum of two sides 4000 mm	666.66	843.73	5.00	187.82	nr	**1031.55**
Sum of two sides 4100 mm	682.96	864.36	5.00	187.82	nr	**1052.18**
Sum of two sides 4200 mm	699.27	885.00	5.00	187.82	nr	**1072.82**
Sum of two sides 4300 mm	715.58	905.64	5.23	196.46	nr	**1102.10**
Sum of two sides 4400 mm	731.88	926.27	5.23	196.46	nr	**1122.73**
Sum of two sides 4500 mm	748.19	946.91	5.39	202.47	nr	**1149.38**

38 VENTILATION/AIR CONDITIONING SYSTEMS

Item	Net Price £	Material £	Labour hours	Labour £	Unit	Total rate £
DUCTWORK: RECTANGULAR – CLASS C						
AIR DUCTLINES						
Galvanized sheet metal DW144 class C rectangular section ductwork; including all necessary stiffeners, joints, couplers in the running length and duct supports						
Ductwork up to 400 mm longest side						
Sum of two sides 200 mm	34.19	43.27	1.16	43.57	m	**86.85**
Sum of two sides 300 mm	34.44	43.59	1.17	43.95	m	**87.54**
Sum of two sides 400 mm	37.24	47.13	1.17	43.95	m	**91.08**
Sum of two sides 500 mm	37.25	47.15	1.18	44.33	m	**91.47**
Sum of two sides 600 mm	40.02	50.66	1.19	44.70	m	**95.36**
Sum of two sides 700 mm	42.81	54.18	1.19	44.70	m	**98.88**
Sum of two sides 800 mm	45.73	57.88	1.19	44.70	m	**102.58**
Extra over fittings; Ductwork up to 400 mm longest side						
End cap						
Sum of two sides 200 mm	17.95	22.71	0.38	14.27	nr	**36.99**
Sum of two sides 300 mm	20.05	25.38	0.38	14.27	nr	**39.65**
Sum of two sides 400 mm	22.15	28.03	0.38	14.27	nr	**42.30**
Sum of two sides 500 mm	24.24	30.68	0.38	14.27	nr	**44.95**
Sum of two sides 600 mm	26.34	33.34	0.38	14.27	nr	**47.61**
Sum of two sides 700 mm	28.45	36.00	0.38	14.27	nr	**50.28**
Sum of two sides 800 mm	30.53	38.64	0.38	14.27	nr	**52.91**
Reducer						
Sum of two sides 200 mm	28.02	35.47	1.40	52.59	nr	**88.06**
Sum of two sides 300 mm	31.62	40.02	1.40	52.59	nr	**92.61**
Sum of two sides 400 mm	58.38	73.88	1.42	53.34	nr	**127.22**
Sum of two sides 500 mm	63.17	79.95	1.42	53.34	nr	**133.29**
Sum of two sides 600 mm	67.98	86.04	1.69	63.48	nr	**149.52**
Sum of two sides 700 mm	72.77	92.09	1.69	63.48	nr	**155.58**
Sum of two sides 800 mm	77.52	98.10	1.92	72.12	nr	**170.23**
Offset						
Sum of two sides 200 mm	41.11	52.03	1.63	61.23	nr	**113.26**
Sum of two sides 300 mm	46.47	58.82	1.63	61.23	nr	**120.05**
Sum of two sides 400 mm	70.85	89.67	1.65	61.98	nr	**151.65**
Sum of two sides 500 mm	77.03	97.49	1.65	61.98	nr	**159.47**
Sum of two sides 600 mm	82.04	103.84	1.92	72.12	nr	**175.96**
Sum of two sides 700 mm	87.62	110.89	1.92	72.12	nr	**183.02**
Sum of two sides 800 mm	92.49	117.06	1.92	72.12	nr	**189.18**
Square to round						
Sum of two sides 200 mm	36.55	46.26	1.22	45.83	nr	**92.08**
Sum of two sides 300 mm	41.07	51.97	1.22	45.83	nr	**97.80**
Sum of two sides 400 mm	56.94	72.07	1.25	46.95	nr	**119.02**
Sum of two sides 500 mm	61.96	78.42	1.25	46.95	nr	**125.37**
Sum of two sides 600 mm	66.95	84.74	1.33	49.96	nr	**134.70**
Sum of two sides 700 mm	71.99	91.12	1.33	49.96	nr	**141.08**
Sum of two sides 800 mm	76.95	97.39	1.40	52.59	nr	**149.98**

38 VENTILATION/AIR CONDITIONING SYSTEMS

Item	Net Price £	Material £	Labour hours	Labour £	Unit	Total rate £
DUCTWORK: RECTANGULAR – CLASS C – cont						
Extra over fittings – cont						
90° radius bend						
Sum of two sides 200 mm	27.00	34.18	1.10	41.32	nr	**75.50**
Sum of two sides 300 mm	29.13	36.87	1.10	41.32	nr	**78.19**
Sum of two sides 400 mm	55.54	70.29	1.12	42.07	nr	**112.36**
Sum of two sides 500 mm	59.06	74.75	1.12	42.07	nr	**116.82**
Sum of two sides 600 mm	63.77	80.71	1.16	43.57	nr	**124.28**
Sum of two sides 700 mm	68.19	86.31	1.16	43.57	nr	**129.88**
Sum of two sides 800 mm	72.70	92.01	1.22	45.83	nr	**137.84**
45° radius bend						
Sum of two sides 200 mm	29.81	37.73	1.29	48.46	nr	**86.19**
Sum of two sides 300 mm	32.73	41.42	1.29	48.46	nr	**89.88**
Sum of two sides 400 mm	58.31	73.79	1.29	48.46	nr	**122.25**
Sum of two sides 500 mm	62.91	79.62	1.29	48.46	nr	**128.08**
Sum of two sides 600 mm	67.71	85.69	1.39	52.21	nr	**137.90**
Sum of two sides 700 mm	72.18	91.35	1.39	52.21	nr	**143.57**
Sum of two sides 800 mm	76.92	97.34	1.46	54.84	nr	**152.19**
90° mitre bend						
Sum of two sides 200 mm	46.84	59.28	2.04	76.63	nr	**135.91**
Sum of two sides 300 mm	51.45	65.12	2.04	76.63	nr	**141.75**
Sum of two sides 400 mm	82.40	104.28	2.09	78.51	nr	**182.79**
Sum of two sides 500 mm	88.95	112.58	2.09	78.51	nr	**191.09**
Sum of two sides 600 mm	97.38	123.25	2.15	80.76	nr	**204.01**
Sum of two sides 700 mm	109.54	138.63	2.15	80.76	nr	**219.39**
Sum of two sides 800 mm	118.45	149.91	2.26	84.89	nr	**234.80**
Branch						
Sum of two sides 200 mm	47.13	59.64	0.92	34.56	nr	**94.20**
Sum of two sides 300 mm	52.46	66.39	0.92	34.56	nr	**100.95**
Sum of two sides 400 mm	68.56	86.77	0.95	35.69	nr	**122.46**
Sum of two sides 500 mm	75.17	95.13	0.95	35.69	nr	**130.82**
Sum of two sides 600 mm	81.22	102.79	1.03	38.69	nr	**141.48**
Sum of two sides 700 mm	87.29	110.47	1.03	38.69	nr	**149.16**
Sum of two sides 800 mm	93.35	118.14	1.03	38.69	nr	**156.83**
Grille neck						
Sum of two sides 200 mm	54.57	69.07	1.10	41.32	nr	**110.39**
Sum of two sides 300 mm	61.33	77.62	1.10	41.32	nr	**118.94**
Sum of two sides 400 mm	68.07	86.15	1.16	43.57	nr	**129.72**
Sum of two sides 500 mm	74.83	94.70	1.16	43.57	nr	**138.27**
Sum of two sides 600 mm	81.76	103.47	1.18	44.33	nr	**147.80**
Sum of two sides 700 mm	88.52	112.03	1.18	44.33	nr	**156.35**
Sum of two sides 800 mm	95.29	120.60	1.18	44.33	nr	**164.92**
Ductwork 401 to 600 mm longest side						
Sum of two sides 600 mm	56.72	71.78	1.17	43.95	m	**115.73**
Sum of two sides 700 mm	62.39	78.96	1.17	43.95	m	**122.91**
Sum of two sides 800 mm	67.90	85.94	1.17	43.95	m	**129.89**
Sum of two sides 900 mm	73.29	92.76	1.27	47.71	m	**140.46**
Sum of two sides 1000 mm	78.68	99.58	1.48	55.59	m	**155.17**
Sum of two sides 1100 mm	84.50	106.94	1.49	55.97	m	**162.91**
Sum of two sides 1200 mm	89.89	113.76	1.49	55.97	m	**169.73**

38 VENTILATION/AIR CONDITIONING SYSTEMS

Item	Net Price £	Material £	Labour hours	Labour £	Unit	Total rate £
Extra over fittings: Ductwork 401 to 600 mm longest side						
End cap						
Sum of two sides 600 mm	26.40	33.41	0.38	14.27	nr	**47.68**
Sum of two sides 700 mm	28.50	36.07	0.38	14.27	nr	**50.35**
Sum of two sides 800 mm	30.59	38.71	0.38	14.27	nr	**52.99**
Sum of two sides 900 mm	32.70	41.38	0.38	14.27	nr	**55.66**
Sum of two sides 1000 mm	34.80	44.04	0.38	14.27	nr	**58.31**
Sum of two sides 1100 mm	36.89	46.69	0.38	14.27	nr	**60.97**
Sum of two sides 1200 mm	38.99	49.35	0.38	14.27	nr	**63.62**
Reducer						
Sum of two sides 600 mm	58.71	74.30	1.69	63.48	nr	**137.78**
Sum of two sides 700 mm	62.48	79.08	1.69	63.48	nr	**142.56**
Sum of two sides 800 mm	66.19	83.76	1.92	72.12	nr	**155.89**
Sum of two sides 900 mm	70.00	88.59	1.92	72.12	nr	**160.71**
Sum of two sides 1000 mm	73.77	93.36	2.18	81.89	nr	**175.25**
Sum of two sides 1100 mm	77.92	98.62	2.18	81.89	nr	**180.51**
Sum of two sides 1200 mm	81.70	103.40	2.18	81.89	nr	**185.29**
Offset						
Sum of two sides 600 mm	80.95	102.45	1.92	72.12	nr	**174.57**
Sum of two sides 700 mm	86.42	109.37	1.92	72.12	nr	**181.49**
Sum of two sides 800 mm	87.77	111.08	1.92	72.12	nr	**183.20**
Sum of two sides 900 mm	90.52	114.56	1.92	72.12	nr	**186.68**
Sum of two sides 1000 mm	94.29	119.34	2.18	81.89	nr	**201.23**
Sum of two sides 1100 mm	94.30	119.35	2.18	81.89	nr	**201.23**
Sum of two sides 1200 mm	96.15	121.69	2.18	81.89	nr	**203.58**
Square to round						
Sum of two sides 600 mm	58.70	74.29	1.33	49.96	nr	**124.25**
Sum of two sides 700 mm	62.80	79.47	1.33	49.96	nr	**129.43**
Sum of two sides 800 mm	66.84	84.59	1.40	52.59	nr	**137.18**
Sum of two sides 900 mm	70.94	89.78	1.40	52.59	nr	**142.37**
Sum of two sides 1000 mm	75.06	94.99	1.82	68.37	nr	**163.36**
Sum of two sides 1100 mm	79.24	100.29	1.82	68.37	nr	**168.66**
Sum of two sides 1200 mm	83.35	105.49	1.82	68.37	nr	**173.86**
90° radius bend						
Sum of two sides 600 mm	54.07	68.43	1.16	43.57	nr	**112.00**
Sum of two sides 700 mm	54.49	68.96	1.16	43.57	nr	**112.53**
Sum of two sides 800 mm	58.20	73.65	1.22	45.83	nr	**119.48**
Sum of two sides 900 mm	62.01	78.47	1.22	45.83	nr	**124.30**
Sum of two sides 1000 mm	64.32	81.40	1.40	52.59	nr	**133.99**
Sum of two sides 1100 mm	68.35	86.50	1.40	52.59	nr	**139.09**
Sum of two sides 1200 mm	72.10	91.25	1.40	52.59	nr	**143.84**
45° bend						
Sum of two sides 600 mm	62.18	78.70	1.39	52.21	nr	**130.91**
Sum of two sides 700 mm	65.14	82.44	1.39	52.21	nr	**134.65**
Sum of two sides 800 mm	69.37	87.79	1.46	54.84	nr	**142.64**
Sum of two sides 900 mm	73.67	93.24	1.46	54.84	nr	**148.09**
Sum of two sides 1000 mm	77.19	97.69	1.88	70.62	nr	**168.31**
Sum of two sides 1100 mm	81.87	103.61	1.88	70.62	nr	**174.23**
Sum of two sides 1200 mm	86.17	109.06	1.88	70.62	nr	**179.68**

38 VENTILATION/AIR CONDITIONING SYSTEMS

Item	Net Price £	Material £	Labour hours	Labour £	Unit	Total rate £
DUCTWORK: RECTANGULAR – CLASS C – cont						
Extra over fittings – cont						
90° mitre bend						
Sum of two sides 600 mm	99.97	126.53	2.15	80.76	nr	**207.29**
Sum of two sides 700 mm	103.88	131.47	2.15	80.76	nr	**212.24**
Sum of two sides 800 mm	114.00	144.28	2.26	84.89	nr	**229.18**
Sum of two sides 900 mm	125.24	158.50	2.26	84.89	nr	**243.40**
Sum of two sides 1000 mm	131.61	166.57	3.03	113.82	nr	**280.38**
Sum of two sides 1100 mm	145.18	183.74	3.03	113.82	nr	**297.56**
Sum of two sides 1200 mm	153.94	194.83	3.03	113.82	nr	**308.65**
Branch						
Sum of two sides 600 mm	81.95	103.72	1.03	38.69	nr	**142.41**
Sum of two sides 700 mm	88.07	111.46	1.03	38.69	nr	**150.15**
Sum of two sides 800 mm	94.20	119.22	1.03	38.69	nr	**157.91**
Sum of two sides 900 mm	100.35	127.00	1.03	38.69	nr	**165.69**
Sum of two sides 1000 mm	106.49	134.77	1.29	48.46	nr	**183.23**
Sum of two sides 1100 mm	112.88	142.86	1.29	48.46	nr	**191.32**
Sum of two sides 1200 mm	119.01	150.62	1.29	48.46	nr	**199.08**
Grille neck						
Sum of two sides 600 mm	81.76	103.47	1.18	44.33	nr	**147.80**
Sum of two sides 700 mm	88.52	112.03	1.18	44.33	nr	**156.35**
Sum of two sides 800 mm	95.29	120.60	1.18	44.33	nr	**164.92**
Sum of two sides 900 mm	102.05	129.15	1.18	44.33	nr	**173.47**
Sum of two sides 1000 mm	108.82	137.72	1.44	54.09	nr	**191.81**
Sum of two sides 1100 mm	115.32	145.95	1.44	54.09	nr	**200.04**
Sum of two sides 1200 mm	122.34	154.84	1.44	54.09	nr	**208.93**
Ductwork 601 to 800 mm longest side						
Sum of two sides 900 mm	79.62	100.76	1.27	47.71	m	**148.47**
Sum of two sides 1000 mm	85.03	107.62	1.48	55.59	m	**163.21**
Sum of two sides 1100 mm	90.45	114.47	1.49	55.97	m	**170.44**
Sum of two sides 1200 mm	96.31	121.89	1.49	55.97	m	**177.86**
Sum of two sides 1300 mm	101.72	128.74	1.51	56.72	m	**185.46**
Sum of two sides 1400 mm	107.14	135.60	1.55	58.22	m	**193.82**
Sum of two sides 1500 mm	112.56	142.45	1.61	60.48	m	**202.93**
Sum of two sides 1600 mm	117.96	149.29	1.62	60.85	m	**210.14**
Extra over fittings; Ductwork 601 to 800 mm longest side						
End cap						
Sum of two sides 900 mm	32.70	41.38	0.38	14.27	nr	**55.66**
Sum of two sides 1000 mm	34.80	44.04	0.38	14.27	nr	**58.31**
Sum of two sides 1100 mm	36.89	46.69	0.38	14.27	nr	**60.97**
Sum of two sides 1200 mm	38.99	49.35	0.38	14.27	nr	**63.62**
Sum of two sides 1300 mm	41.10	52.01	0.38	14.27	nr	**66.29**
Sum of two sides 1400 mm	42.90	54.29	0.38	14.27	nr	**68.57**
Sum of two sides 1500 mm	49.81	63.03	0.38	14.27	nr	**77.31**
Sum of two sides 1600 mm	56.78	71.86	0.38	14.27	nr	**86.14**

38 VENTILATION/AIR CONDITIONING SYSTEMS

Item	Net Price £	Material £	Labour hours	Labour £	Unit	Total rate £
Reducer						
Sum of two sides 900 mm	70.95	89.79	1.92	72.12	nr	**161.92**
Sum of two sides 1000 mm	74.75	94.60	2.18	81.89	nr	**176.49**
Sum of two sides 1100 mm	78.53	99.39	2.18	81.89	nr	**181.28**
Sum of two sides 1200 mm	82.69	104.65	2.18	81.89	nr	**186.54**
Sum of two sides 1300 mm	86.15	109.03	2.30	86.40	nr	**195.43**
Sum of two sides 1400 mm	89.32	113.04	2.30	86.40	nr	**199.44**
Sum of two sides 1500 mm	103.19	130.60	2.47	92.78	nr	**223.38**
Sum of two sides 1600 mm	117.06	148.15	2.47	92.78	nr	**240.93**
Offset						
Sum of two sides 900 mm	97.73	123.68	1.92	72.12	nr	**195.80**
Sum of two sides 1000 mm	99.19	125.54	2.18	81.89	nr	**207.43**
Sum of two sides 1100 mm	99.31	125.69	2.18	81.89	nr	**207.58**
Sum of two sides 1200 mm	100.69	127.43	2.18	81.89	nr	**209.32**
Sum of two sides 1300 mm	101.33	128.24	2.47	92.78	nr	**221.02**
Sum of two sides 1400 mm	102.72	130.00	2.47	92.78	nr	**222.79**
Sum of two sides 1500 mm	115.27	145.89	2.47	92.78	nr	**238.67**
Sum of two sides 1600 mm	129.68	164.13	2.47	92.78	nr	**256.91**
Square to round						
Sum of two sides 900 mm	69.95	88.52	1.40	52.59	nr	**141.11**
Sum of two sides 1000 mm	74.03	93.70	1.82	68.37	nr	**162.06**
Sum of two sides 1100 mm	78.12	98.87	1.82	68.37	nr	**167.24**
Sum of two sides 1200 mm	82.30	104.16	1.82	68.37	nr	**172.52**
Sum of two sides 1300 mm	86.39	109.33	2.32	87.15	nr	**196.48**
Sum of two sides 1400 mm	89.74	113.58	2.32	87.15	nr	**200.72**
Sum of two sides 1500 mm	105.62	133.68	2.56	96.16	nr	**229.84**
Sum of two sides 1600 mm	121.52	153.80	2.58	96.91	nr	**250.71**
90° radius bend						
Sum of two sides 900 mm	53.21	67.34	1.22	45.83	nr	**113.16**
Sum of two sides 1000 mm	56.73	71.79	1.40	52.59	nr	**124.38**
Sum of two sides 1100 mm	60.24	76.24	1.40	52.59	nr	**128.82**
Sum of two sides 1200 mm	63.96	80.94	1.40	52.59	nr	**133.53**
Sum of two sides 1300 mm	66.83	84.58	1.91	71.75	nr	**156.33**
Sum of two sides 1400 mm	67.85	85.87	1.91	71.75	nr	**157.61**
Sum of two sides 1500 mm	80.57	101.98	2.11	79.26	nr	**181.23**
Sum of two sides 1600 mm	93.30	118.08	2.11	79.26	nr	**197.34**
45° bend						
Sum of two sides 900 mm	70.34	89.02	1.46	54.84	nr	**143.86**
Sum of two sides 1000 mm	74.52	94.31	1.88	70.62	nr	**164.93**
Sum of two sides 1100 mm	78.71	99.61	1.88	70.62	nr	**170.23**
Sum of two sides 1200 mm	83.24	105.35	1.88	70.62	nr	**175.97**
Sum of two sides 1300 mm	87.17	110.32	2.26	84.89	nr	**195.22**
Sum of two sides 1400 mm	89.67	113.49	2.44	91.66	nr	**205.14**
Sum of two sides 1500 mm	103.82	131.40	2.44	91.66	nr	**223.05**
Sum of two sides 1600 mm	118.01	149.35	2.68	100.67	nr	**250.02**
90° mitre bend						
Sum of two sides 900 mm	106.94	135.34	2.26	84.89	nr	**220.23**
Sum of two sides 1000 mm	118.92	150.51	3.03	113.82	nr	**264.32**
Sum of two sides 1100 mm	131.08	165.89	3.03	113.82	nr	**279.71**
Sum of two sides 1200 mm	143.31	181.37	3.03	113.82	nr	**295.19**
Sum of two sides 1300 mm	156.81	198.46	3.85	144.62	nr	**343.08**

38 VENTILATION/AIR CONDITIONING SYSTEMS

Item	Net Price £	Material £	Labour hours	Labour £	Unit	Total rate £
DUCTWORK: RECTANGULAR – CLASS C – cont						
Extra over fittings – cont						
90° mitre bend – cont						
Sum of two sides 1400 mm	162.29	205.39	3.85	144.62	nr	**350.01**
Sum of two sides 1500 mm	188.87	239.04	4.25	159.65	nr	**398.69**
Sum of two sides 1600 mm	215.43	272.65	4.26	160.02	nr	**432.68**
Branch						
Sum of two sides 900 mm	103.82	131.39	1.03	38.69	nr	**170.08**
Sum of two sides 1000 mm	110.14	139.40	1.29	48.46	nr	**187.86**
Sum of two sides 1100 mm	116.47	147.41	1.29	48.46	nr	**195.86**
Sum of two sides 1200 mm	123.09	155.78	1.29	48.46	nr	**204.24**
Sum of two sides 1300 mm	129.43	163.80	1.39	52.21	nr	**216.02**
Sum of two sides 1400 mm	134.93	170.77	1.39	52.21	nr	**222.98**
Sum of two sides 1500 mm	154.75	195.85	1.64	61.60	nr	**257.46**
Sum of two sides 1600 mm	174.60	220.97	1.64	61.60	nr	**282.57**
Grille neck						
Sum of two sides 900 mm	102.05	129.15	1.18	44.33	nr	**173.47**
Sum of two sides 1000 mm	108.82	137.72	1.44	54.09	nr	**191.81**
Sum of two sides 1100 mm	115.32	145.95	1.44	54.09	nr	**200.04**
Sum of two sides 1200 mm	122.34	154.84	1.44	54.09	nr	**208.93**
Sum of two sides 1300 mm	128.83	163.05	1.69	63.48	nr	**226.53**
Sum of two sides 1400 mm	135.12	171.01	1.69	63.48	nr	**234.49**
Sum of two sides 1500 mm	156.44	197.99	1.79	67.24	nr	**265.23**
Sum of two sides 1600 mm	177.04	224.06	1.79	67.24	nr	**291.30**
Ductwork 801 to 1000 mm longest side						
Sum of two sides 1100 mm	91.45	115.74	1.49	55.97	m	**171.71**
Sum of two sides 1200 mm	96.87	122.59	1.49	55.97	m	**178.56**
Sum of two sides 1300 mm	102.27	129.43	1.51	56.72	m	**186.15**
Sum of two sides 1400 mm	108.13	136.85	1.55	58.22	m	**195.07**
Sum of two sides 1500 mm	113.54	143.70	1.61	60.48	m	**204.18**
Sum of two sides 1600 mm	118.96	150.56	1.62	60.85	m	**211.41**
Sum of two sides 1700 mm	124.38	157.41	1.74	65.36	m	**222.77**
Sum of two sides 1800 mm	130.24	164.83	1.76	66.11	m	**230.94**
Sum of two sides 1900 mm	135.65	171.68	1.81	67.99	m	**239.67**
Sum of two sides 2000 mm	141.06	178.52	1.82	68.37	m	**246.89**
Extra over fittings; Ductwork 801 to 1000 mm longest side						
End cap						
Sum of two sides 1100 mm	36.88	46.68	0.38	14.27	nr	**60.95**
Sum of two sides 1200 mm	38.99	49.35	0.38	14.27	nr	**63.62**
Sum of two sides 1300 mm	41.10	52.01	0.38	14.27	nr	**66.29**
Sum of two sides 1400 mm	42.91	54.31	0.38	14.27	nr	**68.58**
Sum of two sides 1500 mm	49.83	63.06	0.38	14.27	nr	**77.34**
Sum of two sides 1600 mm	56.78	71.86	0.38	14.27	nr	**86.14**
Sum of two sides 1700 mm	63.72	80.65	0.58	21.79	nr	**102.43**
Sum of two sides 1800 mm	70.67	89.44	0.58	21.79	nr	**111.23**
Sum of two sides 1900 mm	77.61	98.23	0.58	21.79	nr	**120.02**
Sum of two sides 2000 mm	84.56	107.01	0.58	21.79	nr	**128.80**

38 VENTILATION/AIR CONDITIONING SYSTEMS

Item	Net Price £	Material £	Labour hours	Labour £	Unit	Total rate £
Reducer						
Sum of two sides 1100 mm	70.40	89.10	2.18	81.89	nr	**170.99**
Sum of two sides 1200 mm	73.80	93.40	2.18	81.89	nr	**175.28**
Sum of two sides 1300 mm	77.19	97.69	2.30	86.40	nr	**184.08**
Sum of two sides 1400 mm	80.28	101.60	2.30	86.40	nr	**188.00**
Sum of two sides 1500 mm	93.48	118.30	2.47	92.78	nr	**211.09**
Sum of two sides 1600 mm	106.65	134.98	2.47	92.78	nr	**227.76**
Sum of two sides 1700 mm	119.86	151.69	2.59	97.29	nr	**248.98**
Sum of two sides 1800 mm	133.36	168.77	2.59	97.29	nr	**266.06**
Sum of two sides 1900 mm	146.54	185.46	2.71	101.80	nr	**287.26**
Sum of two sides 2000 mm	159.75	202.18	2.71	101.80	nr	**303.98**
Offset						
Sum of two sides 1100 mm	107.69	136.30	2.18	81.89	nr	**218.18**
Sum of two sides 1200 mm	108.70	137.57	2.18	81.89	nr	**219.46**
Sum of two sides 1300 mm	109.81	138.97	2.47	92.78	nr	**231.75**
Sum of two sides 1400 mm	111.61	141.25	2.47	92.78	nr	**234.04**
Sum of two sides 1500 mm	124.01	156.94	2.47	92.78	nr	**249.73**
Sum of two sides 1600 mm	138.98	175.89	2.47	92.78	nr	**268.68**
Sum of two sides 1700 mm	148.90	188.44	2.61	98.04	nr	**286.48**
Sum of two sides 1800 mm	162.96	206.24	2.61	98.04	nr	**304.28**
Sum of two sides 1900 mm	178.69	226.15	2.71	101.80	nr	**327.95**
Sum of two sides 2000 mm	192.12	243.15	2.71	101.80	nr	**344.95**
Square to round						
Sum of two sides 1100 mm	70.15	88.78	1.82	68.37	nr	**157.15**
Sum of two sides 1200 mm	73.82	93.42	1.82	68.37	nr	**161.79**
Sum of two sides 1300 mm	77.51	98.10	2.32	87.15	nr	**185.24**
Sum of two sides 1400 mm	80.52	101.91	2.32	87.15	nr	**189.06**
Sum of two sides 1500 mm	95.64	121.04	2.56	96.16	nr	**217.21**
Sum of two sides 1600 mm	111.96	141.70	2.58	96.91	nr	**238.61**
Sum of two sides 1700 mm	127.24	161.04	2.84	106.68	nr	**267.72**
Sum of two sides 1800 mm	142.56	180.42	2.84	106.68	nr	**287.10**
Sum of two sides 1900 mm	157.84	199.76	3.13	117.57	nr	**317.33**
Sum of two sides 2000 mm	173.13	219.11	3.13	117.57	nr	**336.69**
90° radius bend						
Sum of two sides 1100 mm	52.06	65.88	1.40	52.59	nr	**118.47**
Sum of two sides 1200 mm	56.18	71.11	1.40	52.59	nr	**123.70**
Sum of two sides 1300 mm	60.32	76.34	1.91	71.75	nr	**148.09**
Sum of two sides 1400 mm	64.03	81.03	1.91	71.75	nr	**152.78**
Sum of two sides 1500 mm	77.10	97.57	2.11	79.26	nr	**176.83**
Sum of two sides 1600 mm	92.15	116.63	2.11	79.26	nr	**195.89**
Sum of two sides 1700 mm	105.52	133.55	2.55	95.79	nr	**229.33**
Sum of two sides 1800 mm	119.02	150.64	2.55	95.79	nr	**246.42**
Sum of two sides 1900 mm	129.99	164.51	2.80	105.18	nr	**269.69**
Sum of two sides 2000 mm	143.34	181.42	2.80	105.18	nr	**286.59**
45° bend						
Sum of two sides 1100 mm	74.93	94.84	1.88	70.62	nr	**165.45**
Sum of two sides 1200 mm	79.54	100.66	1.88	70.62	nr	**171.28**
Sum of two sides 1300 mm	84.17	106.52	2.26	84.89	nr	**191.41**
Sum of two sides 1400 mm	88.49	111.99	2.44	91.66	nr	**203.65**
Sum of two sides 1500 mm	103.04	130.41	2.68	100.67	nr	**231.08**
Sum of two sides 1600 mm	117.58	148.80	2.69	101.05	nr	**249.85**
Sum of two sides 1700 mm	132.13	167.22	2.96	111.19	nr	**278.41**

38 VENTILATION/AIR CONDITIONING SYSTEMS

Item	Net Price £	Material £	Labour hours	Labour £	Unit	Total rate £
DUCTWORK: RECTANGULAR – CLASS C – cont						
Extra over fittings – cont						
45° bend – cont						
Sum of two sides 1800 mm	146.98	186.02	2.96	111.19	nr	**297.21**
Sum of two sides 1900 mm	158.60	200.72	3.26	122.46	nr	**323.18**
Sum of two sides 2000 mm	173.02	218.97	3.26	122.46	nr	**341.43**
90° mitre bend						
Sum of two sides 1100 mm	125.02	158.23	3.03	113.82	nr	**272.04**
Sum of two sides 1200 mm	134.09	169.70	3.03	113.82	nr	**283.52**
Sum of two sides 1300 mm	150.61	190.61	3.85	144.62	nr	**335.23**
Sum of two sides 1400 mm	164.78	208.54	3.85	144.62	nr	**353.16**
Sum of two sides 1500 mm	191.61	242.50	4.25	159.65	nr	**402.14**
Sum of two sides 1600 mm	218.43	276.45	4.26	160.02	nr	**436.47**
Sum of two sides 1700 mm	248.79	314.87	4.68	175.80	nr	**490.67**
Sum of two sides 1800 mm	276.13	349.47	4.68	175.80	nr	**525.27**
Sum of two sides 1900 mm	300.16	379.88	4.87	182.93	nr	**562.81**
Sum of two sides 2000 mm	327.53	414.52	4.87	182.93	nr	**597.46**
Branch						
Sum of two sides 1100 mm	116.75	147.75	1.29	48.46	nr	**196.21**
Sum of two sides 1200 mm	123.09	155.78	1.29	48.46	nr	**204.24**
Sum of two sides 1300 mm	129.43	163.80	1.39	52.21	nr	**216.02**
Sum of two sides 1400 mm	135.18	171.09	1.39	52.21	nr	**223.30**
Sum of two sides 1500 mm	155.04	196.22	1.64	61.60	nr	**257.82**
Sum of two sides 1600 mm	174.86	221.30	1.64	61.60	nr	**282.90**
Sum of two sides 1700 mm	192.81	244.02	1.69	63.48	nr	**307.50**
Sum of two sides 1800 mm	212.73	269.23	1.69	63.48	nr	**332.72**
Sum of two sides 1900 mm	232.37	294.09	1.85	69.49	nr	**363.58**
Sum of two sides 2000 mm	252.01	318.94	1.85	69.49	nr	**388.44**
Grille neck						
Sum of two sides 1100 mm	115.57	146.27	1.44	54.09	nr	**200.36**
Sum of two sides 1200 mm	122.34	154.84	1.44	54.09	nr	**208.93**
Sum of two sides 1300 mm	128.83	163.05	1.69	63.48	nr	**226.53**
Sum of two sides 1400 mm	135.12	171.01	1.69	63.48	nr	**234.49**
Sum of two sides 1500 mm	156.44	197.99	1.79	67.24	nr	**265.23**
Sum of two sides 1600 mm	177.17	224.23	1.79	67.24	nr	**291.47**
Sum of two sides 1700 mm	199.08	251.95	1.86	69.87	nr	**321.82**
Sum of two sides 1800 mm	219.67	278.01	1.86	69.87	nr	**347.88**
Sum of two sides 1900 mm	241.71	305.91	2.02	75.88	nr	**381.79**
Sum of two sides 2000 mm	263.03	332.89	2.02	75.88	nr	**408.76**
Ductwork 1001 to 1250 mm longest side						
Sum of two sides 1300 mm	146.41	185.29	1.51	56.72	m	**242.02**
Sum of two sides 1400 mm	153.38	194.12	1.55	58.22	m	**252.35**
Sum of two sides 1500 mm	160.07	202.58	1.61	60.48	m	**263.06**
Sum of two sides 1600 mm	167.20	211.60	1.62	60.85	m	**272.46**
Sum of two sides 1700 mm	173.88	220.06	1.74	65.36	m	**285.43**
Sum of two sides 1800 mm	180.57	228.52	1.76	66.11	m	**294.64**
Sum of two sides 1900 mm	187.25	236.98	1.81	67.99	m	**304.97**

38 VENTILATION/AIR CONDITIONING SYSTEMS

Item	Net Price £	Material £	Labour hours	Labour £	Unit	Total rate £
Sum of two sides 2000 mm	194.38	246.01	1.82	68.37	m	**314.37**
Sum of two sides 2100 mm	201.06	254.47	2.53	95.04	m	**349.50**
Sum of two sides 2200 mm	229.66	290.66	2.55	95.79	m	**386.45**
Sum of two sides 2300 mm	237.16	300.15	2.56	96.16	m	**396.31**
Sum of two sides 2400 mm	243.85	308.61	2.76	103.68	m	**412.29**
Sum of two sides 2500 mm	250.97	317.63	2.77	104.05	m	**421.68**
Extra over fittings; Ductwork 1001 to 1250 mm longest side						
End cap						
Sum of two sides 1300 mm	41.93	53.07	0.38	14.27	nr	**67.34**
Sum of two sides 1400 mm	43.81	55.45	0.38	14.27	nr	**69.72**
Sum of two sides 1500 mm	50.81	64.30	0.38	14.27	nr	**78.58**
Sum of two sides 1600 mm	57.83	73.19	0.38	14.27	nr	**87.46**
Sum of two sides 1700 mm	64.83	82.04	0.58	21.79	nr	**103.83**
Sum of two sides 1800 mm	71.84	90.92	0.58	21.79	nr	**112.70**
Sum of two sides 1900 mm	78.83	99.77	0.58	21.79	nr	**121.56**
Sum of two sides 2000 mm	108.08	136.79	0.58	21.79	nr	**158.57**
Sum of two sides 2100 mm	92.85	117.51	0.87	32.68	nr	**150.19**
Sum of two sides 2200 mm	99.87	126.40	0.87	32.68	nr	**159.08**
Sum of two sides 2300 mm	106.88	135.27	0.87	32.68	nr	**167.95**
Sum of two sides 2400 mm	113.88	144.13	0.87	32.68	nr	**176.81**
Sum of two sides 2500 mm	120.89	153.00	0.87	32.68	nr	**185.68**
Reducer						
Sum of two sides 1300 mm	60.10	76.06	2.30	86.40	nr	**162.45**
Sum of two sides 1400 mm	62.49	79.09	2.30	86.40	nr	**165.49**
Sum of two sides 1500 mm	75.20	95.18	2.47	92.78	nr	**187.96**
Sum of two sides 1600 mm	88.23	111.67	2.47	92.78	nr	**204.45**
Sum of two sides 1700 mm	100.97	127.78	2.59	97.29	nr	**225.07**
Sum of two sides 1800 mm	113.71	143.91	2.59	97.29	nr	**241.20**
Sum of two sides 1900 mm	126.43	160.01	2.71	101.80	nr	**261.81**
Sum of two sides 2000 mm	139.45	176.49	2.71	101.80	nr	**278.29**
Sum of two sides 2100 mm	151.23	191.39	2.92	109.69	nr	**301.08**
Sum of two sides 2200 mm	152.20	192.62	2.92	109.69	nr	**302.31**
Sum of two sides 2300 mm	164.00	207.56	2.92	109.69	nr	**317.24**
Sum of two sides 2400 mm	176.49	223.36	3.12	117.20	nr	**340.56**
Sum of two sides 2500 mm	189.23	239.49	3.12	117.20	nr	**356.69**
Offset						
Sum of two sides 1300 mm	153.23	193.93	2.47	92.78	nr	**286.71**
Sum of two sides 1400 mm	157.79	199.70	2.47	92.78	nr	**292.48**
Sum of two sides 1500 mm	175.41	222.00	2.47	92.78	nr	**314.79**
Sum of two sides 1600 mm	188.32	238.34	2.47	92.78	nr	**331.12**
Sum of two sides 1700 mm	204.61	258.95	2.61	98.04	nr	**356.99**
Sum of two sides 1800 mm	216.24	273.67	2.61	98.04	nr	**371.71**
Sum of two sides 1900 mm	231.37	292.82	2.71	101.80	nr	**394.62**
Sum of two sides 2000 mm	242.58	307.01	2.71	101.80	nr	**408.81**
Sum of two sides 2100 mm	246.61	312.10	2.92	109.69	nr	**421.79**
Sum of two sides 2200 mm	256.61	324.76	3.26	122.46	nr	**447.22**
Sum of two sides 2300 mm	267.53	338.58	3.26	122.46	nr	**461.04**
Sum of two sides 2400 mm	288.37	364.96	3.47	130.35	nr	**495.30**
Sum of two sides 2500 mm	309.28	391.42	3.48	130.72	nr	**522.14**

38 VENTILATION/AIR CONDITIONING SYSTEMS

Item	Net Price £	Material £	Labour hours	Labour £	Unit	Total rate £
DUCTWORK: RECTANGULAR – CLASS C – cont						
Extra over fittings – cont						
Square to round						
Sum of two sides 1300 mm	59.74	75.61	2.32	87.15	nr	**162.76**
Sum of two sides 1400 mm	62.30	78.85	2.32	87.15	nr	**166.00**
Sum of two sides 1500 mm	76.68	97.05	2.56	96.16	nr	**193.22**
Sum of two sides 1600 mm	91.11	115.31	2.58	96.91	nr	**212.22**
Sum of two sides 1700 mm	107.83	136.47	2.84	106.68	nr	**243.15**
Sum of two sides 1800 mm	122.53	155.08	2.84	106.68	nr	**261.76**
Sum of two sides 1900 mm	137.24	173.69	3.13	117.57	nr	**291.27**
Sum of two sides 2000 mm	151.98	192.35	3.13	117.57	nr	**309.93**
Sum of two sides 2100 mm	166.68	210.95	4.26	160.02	nr	**370.98**
Sum of two sides 2200 mm	171.49	217.04	4.27	160.40	nr	**377.44**
Sum of two sides 2300 mm	186.25	235.71	4.30	161.52	nr	**397.24**
Sum of two sides 2400 mm	200.97	254.34	4.77	179.18	nr	**433.52**
Sum of two sides 2500 mm	215.71	273.00	4.79	179.93	nr	**452.93**
90° radius bend						
Sum of two sides 1300 mm	17.13	21.68	1.91	71.75	nr	**93.43**
Sum of two sides 1400 mm	17.44	22.07	1.91	71.75	nr	**93.82**
Sum of two sides 1500 mm	30.36	38.43	2.11	79.26	nr	**117.69**
Sum of two sides 1600 mm	44.28	56.05	2.11	79.26	nr	**135.30**
Sum of two sides 1700 mm	57.76	73.10	2.55	95.79	nr	**168.89**
Sum of two sides 1800 mm	72.91	92.28	2.55	95.79	nr	**188.07**
Sum of two sides 1900 mm	85.31	107.96	2.80	105.18	nr	**213.14**
Sum of two sides 2000 mm	97.82	123.80	2.80	105.18	nr	**228.98**
Sum of two sides 2100 mm	99.77	126.27	2.80	105.18	nr	**231.45**
Sum of two sides 2200 mm	102.94	130.28	2.80	105.18	nr	**235.46**
Sum of two sides 2300 mm	110.20	139.47	4.35	163.40	nr	**302.87**
Sum of two sides 2400 mm	114.76	145.24	4.35	163.40	nr	**308.64**
Sum of two sides 2500 mm	126.60	160.23	4.35	163.40	nr	**323.63**
45° bend						
Sum of two sides 1300 mm	62.33	78.89	2.26	84.89	nr	**163.78**
Sum of two sides 1400 mm	64.02	81.03	2.44	91.66	nr	**172.68**
Sum of two sides 1500 mm	78.21	98.98	2.68	100.67	nr	**199.65**
Sum of two sides 1600 mm	92.70	117.32	2.69	101.05	nr	**218.37**
Sum of two sides 1700 mm	106.88	135.27	2.96	111.19	nr	**246.46**
Sum of two sides 1800 mm	121.07	153.22	2.96	111.19	nr	**264.41**
Sum of two sides 1900 mm	135.28	171.20	3.26	122.46	nr	**293.66**
Sum of two sides 2000 mm	149.76	189.53	3.26	122.46	nr	**311.99**
Sum of two sides 2100 mm	162.83	206.08	7.50	281.73	nr	**487.81**
Sum of two sides 2200 mm	163.94	207.49	7.50	281.73	nr	**489.21**
Sum of two sides 2300 mm	171.74	217.36	7.55	283.61	nr	**500.96**
Sum of two sides 2400 mm	185.58	234.87	8.13	305.39	nr	**540.26**
Sum of two sides 2500 mm	199.66	252.68	8.30	311.78	nr	**564.46**
90° mitre bend						
Sum of two sides 1300 mm	161.31	204.16	3.85	144.62	nr	**348.78**
Sum of two sides 1400 mm	171.35	216.86	3.85	144.62	nr	**361.48**
Sum of two sides 1500 mm	207.52	262.64	4.25	159.65	nr	**422.28**
Sum of two sides 1600 mm	239.58	303.21	4.26	160.02	nr	**463.23**
Sum of two sides 1700 mm	271.64	343.78	4.68	175.80	nr	**519.58**
Sum of two sides 1800 mm	312.21	395.13	4.68	175.80	nr	**570.93**

38 VENTILATION/AIR CONDITIONING SYSTEMS

Item	Net Price £	Material £	Labour hours	Labour £	Unit	Total rate £
Sum of two sides 1900 mm	345.15	436.83	4.87	182.93	nr	**619.76**
Sum of two sides 2000 mm	378.14	478.57	4.87	182.93	nr	**661.51**
Sum of two sides 2100 mm	411.08	520.27	7.50	281.73	nr	**801.99**
Sum of two sides 2200 mm	415.06	525.30	7.50	281.73	nr	**807.03**
Sum of two sides 2300 mm	433.46	548.58	7.55	283.61	nr	**832.19**
Sum of two sides 2400 mm	466.62	590.56	8.13	305.39	nr	**895.95**
Sum of two sides 2500 mm	499.79	632.53	8.30	311.78	nr	**944.31**
Branch						
Sum of two sides 1300 mm	132.75	168.01	1.39	52.21	nr	**220.22**
Sum of two sides 1400 mm	138.45	175.22	1.39	52.21	nr	**227.43**
Sum of two sides 1500 mm	158.49	200.59	1.64	61.60	nr	**262.19**
Sum of two sides 1600 mm	178.81	226.31	1.64	61.60	nr	**287.91**
Sum of two sides 1700 mm	196.92	249.22	1.69	63.48	nr	**312.70**
Sum of two sides 1800 mm	215.69	272.98	1.69	63.48	nr	**336.47**
Sum of two sides 1900 mm	236.61	299.45	1.85	69.49	nr	**368.94**
Sum of two sides 2000 mm	256.72	324.90	1.85	69.49	nr	**394.40**
Sum of two sides 2100 mm	275.21	348.31	2.61	98.04	nr	**446.35**
Sum of two sides 2200 mm	294.93	373.27	2.61	98.04	nr	**471.31**
Sum of two sides 2300 mm	314.97	398.62	2.61	98.04	nr	**496.66**
Sum of two sides 2400 mm	334.71	423.61	2.88	108.18	nr	**531.79**
Sum of two sides 2500 mm	354.73	448.95	2.88	108.18	nr	**557.13**
Grille neck						
Sum of two sides 1300 mm	133.74	169.27	1.69	63.48	nr	**232.75**
Sum of two sides 1400 mm	139.94	177.11	1.69	63.48	nr	**240.59**
Sum of two sides 1500 mm	161.64	204.58	1.79	67.24	nr	**271.82**
Sum of two sides 1600 mm	183.32	232.01	1.79	67.24	nr	**299.25**
Sum of two sides 1700 mm	205.00	259.45	1.86	69.87	nr	**329.32**
Sum of two sides 1800 mm	226.69	286.90	1.86	69.87	nr	**356.77**
Sum of two sides 1900 mm	248.39	314.37	2.02	75.88	nr	**390.25**
Sum of two sides 2000 mm	270.07	341.81	2.02	75.88	nr	**417.68**
Sum of two sides 2100 mm	291.76	369.26	2.61	98.04	nr	**467.30**
Sum of two sides 2200 mm	313.46	396.71	2.80	105.18	nr	**501.89**
Sum of two sides 2300 mm	335.15	424.16	2.80	105.18	nr	**529.34**
Sum of two sides 2400 mm	356.87	451.66	3.06	114.94	nr	**566.60**
Sum of two sides 2500 mm	378.52	479.05	3.06	114.94	nr	**593.99**
Ductwork 1251 to 1600 mm longest side						
Sum of two sides 1700 mm	186.53	236.07	1.74	65.36	m	**301.43**
Sum of two sides 1800 mm	193.78	245.25	1.76	66.11	m	**311.36**
Sum of two sides 1900 mm	201.28	254.74	1.81	67.99	m	**322.73**
Sum of two sides 2000 mm	208.33	263.67	1.82	68.37	m	**332.03**
Sum of two sides 2100 mm	215.39	272.59	2.53	95.04	m	**367.63**
Sum of two sides 2200 mm	222.44	281.52	2.55	95.79	m	**377.31**
Sum of two sides 2300 mm	229.94	291.01	2.56	96.16	m	**387.17**
Sum of two sides 2400 mm	237.21	300.21	2.76	103.68	m	**403.89**
Sum of two sides 2500 mm	244.26	309.14	2.77	104.05	m	**413.19**
Sum of two sides 2600 mm	273.85	346.58	2.97	111.56	m	**458.15**
Sum of two sides 2700 mm	280.90	355.51	2.99	112.32	m	**467.82**

38 VENTILATION/AIR CONDITIONING SYSTEMS

Item	Net Price £	Material £	Labour hours	Labour £	Unit	Total rate £
DUCTWORK: RECTANGULAR – CLASS C – cont						
Ductwork 1251 to 1600 mm longest side – cont						
Sum of two sides 2800 mm	288.32	364.90	3.30	123.96	m	**488.86**
Sum of two sides 2900 mm	295.82	374.39	3.31	124.34	m	**498.73**
Sum of two sides 3000 mm	302.87	383.32	3.53	132.60	m	**515.92**
Sum of two sides 3100 mm	310.20	392.58	3.55	133.35	m	**525.93**
Sum of two sides 3200 mm	317.25	401.51	3.56	133.73	m	**535.24**
Extra over fittings; Ductwork 1251 to 1600 mm longest side						
End cap						
Sum of two sides 1700 mm	64.83	82.04	0.58	21.79	nr	**103.83**
Sum of two sides 1800 mm	71.89	90.99	0.58	21.79	nr	**112.78**
Sum of two sides 1900 mm	78.83	99.77	0.58	21.79	nr	**121.56**
Sum of two sides 2000 mm	86.25	109.16	0.58	21.79	nr	**130.95**
Sum of two sides 2100 mm	92.78	117.42	0.87	32.68	nr	**150.10**
Sum of two sides 2200 mm	99.29	125.66	0.87	32.68	nr	**158.34**
Sum of two sides 2300 mm	100.53	127.24	0.87	32.68	nr	**159.92**
Sum of two sides 2400 mm	108.10	136.81	0.87	32.68	nr	**169.49**
Sum of two sides 2500 mm	114.75	145.23	0.87	32.68	nr	**177.91**
Sum of two sides 2600 mm	121.43	153.68	0.87	32.68	nr	**186.36**
Sum of two sides 2700 mm	128.06	162.07	0.87	32.68	nr	**194.75**
Sum of two sides 2800 mm	141.38	178.93	1.16	43.57	nr	**222.50**
Sum of two sides 2900 mm	161.52	204.42	1.16	43.57	nr	**247.99**
Sum of two sides 3000 mm	166.57	210.81	1.73	64.99	nr	**275.79**
Sum of two sides 3100 mm	247.10	312.73	1.73	64.99	nr	**377.71**
Sum of two sides 3200 mm	134.72	170.51	1.73	64.99	nr	**235.49**
Reducer						
Sum of two sides 1700 mm	84.72	107.22	2.59	97.29	nr	**204.51**
Sum of two sides 1800 mm	85.85	108.66	2.59	97.29	nr	**205.95**
Sum of two sides 1900 mm	96.30	121.88	2.71	101.80	nr	**223.68**
Sum of two sides 2000 mm	108.08	136.79	2.71	101.80	nr	**238.58**
Sum of two sides 2100 mm	143.60	181.74	2.92	109.69	nr	**291.42**
Sum of two sides 2200 mm	156.26	197.76	2.92	109.69	nr	**307.44**
Sum of two sides 2300 mm	169.15	214.07	2.92	109.69	nr	**323.76**
Sum of two sides 2400 mm	170.90	216.29	3.12	117.20	nr	**333.49**
Sum of two sides 2500 mm	172.88	218.80	3.12	117.20	nr	**336.00**
Sum of two sides 2600 mm	181.84	230.14	3.16	118.70	nr	**348.84**
Sum of two sides 2700 mm	194.50	246.16	3.16	118.70	nr	**364.86**
Sum of two sides 2800 mm	197.46	249.90	4.00	150.25	nr	**400.16**
Sum of two sides 2900 mm	209.61	265.28	4.01	150.63	nr	**415.91**
Sum of two sides 3000 mm	245.18	310.29	4.56	171.29	nr	**481.59**
Sum of two sides 3100 mm	251.65	318.49	4.56	171.29	nr	**489.78**
Sum of two sides 3200 mm	253.05	320.26	4.56	171.29	nr	**491.55**
Offset						
Sum of two sides 1700 mm	229.28	290.18	2.61	98.04	nr	**388.22**
Sum of two sides 1800 mm	253.20	320.45	2.61	98.04	nr	**418.49**
Sum of two sides 1900 mm	269.03	340.49	2.71	101.80	nr	**442.28**

38 VENTILATION/AIR CONDITIONING SYSTEMS

Item	Net Price £	Material £	Labour hours	Labour £	Unit	Total rate £
Sum of two sides 2000 mm	278.67	352.69	2.71	101.80	nr	**454.49**
Sum of two sides 2100 mm	292.99	370.80	2.92	109.69	nr	**480.49**
Sum of two sides 2200 mm	306.84	388.33	3.26	122.46	nr	**510.79**
Sum of two sides 2300 mm	315.64	399.47	3.26	122.46	nr	**521.93**
Sum of two sides 2400 mm	337.61	427.28	3.47	130.35	nr	**557.63**
Sum of two sides 2500 mm	341.98	432.81	3.48	130.72	nr	**563.53**
Sum of two sides 2600 mm	350.05	443.02	3.49	131.10	nr	**574.12**
Sum of two sides 2700 mm	363.20	459.67	3.50	131.47	nr	**591.14**
Sum of two sides 2800 mm	384.34	486.43	4.33	162.65	nr	**649.08**
Sum of two sides 2900 mm	422.91	535.24	4.74	178.05	nr	**713.29**
Sum of two sides 3000 mm	444.38	562.41	5.31	199.46	nr	**761.87**
Sum of two sides 3100 mm	461.05	583.50	5.34	200.59	nr	**784.09**
Sum of two sides 3200 mm	476.55	603.12	5.35	200.97	nr	**804.09**
Square to round						
Sum of two sides 1700 mm	84.60	107.07	2.84	106.68	nr	**213.75**
Sum of two sides 1800 mm	97.78	123.74	2.84	106.68	nr	**230.43**
Sum of two sides 1900 mm	114.07	144.36	3.13	117.57	nr	**261.94**
Sum of two sides 2000 mm	127.59	161.48	3.13	117.57	nr	**279.06**
Sum of two sides 2100 mm	141.10	178.57	4.26	160.02	nr	**338.59**
Sum of two sides 2200 mm	157.21	198.97	4.27	160.40	nr	**359.37**
Sum of two sides 2300 mm	170.92	216.32	4.30	161.52	nr	**377.84**
Sum of two sides 2400 mm	184.70	233.76	4.77	179.18	nr	**412.94**
Sum of two sides 2500 mm	198.46	251.17	4.79	179.93	nr	**431.10**
Sum of two sides 2600 mm	195.76	247.76	4.95	185.94	nr	**433.70**
Sum of two sides 2700 mm	209.30	264.89	4.95	185.94	nr	**450.83**
Sum of two sides 2800 mm	222.78	281.95	8.49	318.92	nr	**600.87**
Sum of two sides 2900 mm	246.00	311.33	8.88	333.56	nr	**644.90**
Sum of two sides 3000 mm	259.77	328.76	9.02	338.82	nr	**667.59**
Sum of two sides 3100 mm	269.94	341.64	9.02	338.82	nr	**680.46**
Sum of two sides 3200 mm	279.17	353.32	9.09	341.45	nr	**694.78**
90° radius bend						
Sum of two sides 1700 mm	218.58	276.63	2.55	95.79	nr	**372.42**
Sum of two sides 1800 mm	236.60	299.44	2.55	95.79	nr	**395.23**
Sum of two sides 1900 mm	265.13	335.55	2.61	98.04	nr	**433.59**
Sum of two sides 2000 mm	293.09	370.94	2.62	98.42	nr	**469.36**
Sum of two sides 2100 mm	321.07	406.34	2.80	105.18	nr	**511.52**
Sum of two sides 2200 mm	349.04	441.74	2.80	105.18	nr	**546.92**
Sum of two sides 2300 mm	377.54	477.81	2.63	98.79	nr	**576.60**
Sum of two sides 2400 mm	390.83	494.64	4.34	163.03	nr	**657.66**
Sum of two sides 2500 mm	413.34	523.13	4.35	163.40	nr	**686.53**
Sum of two sides 2600 mm	418.47	529.61	4.53	170.16	nr	**699.77**
Sum of two sides 2700 mm	440.50	557.49	4.53	170.16	nr	**727.66**
Sum of two sides 2800 mm	455.18	576.08	7.13	267.83	nr	**843.90**
Sum of two sides 2900 mm	516.25	653.37	7.17	269.33	nr	**922.70**
Sum of two sides 3000 mm	543.84	688.28	7.26	272.71	nr	**960.99**
Sum of two sides 3100 mm	565.20	715.32	7.26	272.71	nr	**988.03**
Sum of two sides 3200 mm	585.00	740.38	7.31	274.59	nr	**1014.97**
45° bend						
Sum of two sides 1700 mm	103.88	131.47	2.96	111.19	nr	**242.66**
Sum of two sides 1800 mm	112.64	142.56	2.96	111.19	nr	**253.75**
Sum of two sides 1900 mm	126.72	160.37	3.26	122.46	nr	**282.83**

38 VENTILATION/AIR CONDITIONING SYSTEMS

Item	Net Price £	Material £	Labour hours	Labour £	Unit	Total rate £
DUCTWORK: RECTANGULAR – CLASS C – cont						
Extra over fittings – cont						
45° bend – cont						
Sum of two sides 2000 mm	140.55	177.88	3.26	122.46	nr	**300.34**
Sum of two sides 2100 mm	154.35	195.34	7.50	281.73	nr	**477.07**
Sum of two sides 2200 mm	168.17	212.84	7.50	281.73	nr	**494.57**
Sum of two sides 2300 mm	182.26	230.66	7.55	283.61	nr	**514.27**
Sum of two sides 2400 mm	190.64	241.27	8.13	305.39	nr	**546.66**
Sum of two sides 2500 mm	199.94	253.04	8.30	311.78	nr	**564.82**
Sum of two sides 2600 mm	204.42	258.72	8.56	321.54	nr	**580.26**
Sum of two sides 2700 mm	213.97	270.80	8.62	323.80	nr	**594.60**
Sum of two sides 2800 mm	221.12	279.85	9.09	341.45	nr	**621.31**
Sum of two sides 2900 mm	252.18	319.17	9.09	341.45	nr	**660.62**
Sum of two sides 3000 mm	265.97	336.62	9.62	361.36	nr	**697.98**
Sum of two sides 3100 mm	276.60	350.06	9.62	361.36	nr	**711.42**
Sum of two sides 3200 mm	286.49	362.58	9.62	361.36	nr	**723.94**
90° mitre bend						
Sum of two sides 1700 mm	270.35	342.16	4.68	175.80	nr	**517.96**
Sum of two sides 1800 mm	287.11	363.37	4.68	175.80	nr	**539.16**
Sum of two sides 1900 mm	319.56	404.43	4.87	182.93	nr	**587.37**
Sum of two sides 2000 mm	363.71	460.31	4.87	182.93	nr	**643.25**
Sum of two sides 2100 mm	397.33	502.86	7.50	281.73	nr	**784.58**
Sum of two sides 2200 mm	430.94	545.40	7.50	281.73	nr	**827.13**
Sum of two sides 2300 mm	464.45	587.80	7.55	283.61	nr	**871.41**
Sum of two sides 2400 mm	481.89	609.89	8.13	305.39	nr	**915.28**
Sum of two sides 2500 mm	521.82	660.41	8.30	311.78	nr	**972.19**
Sum of two sides 2600 mm	525.11	664.58	8.56	321.54	nr	**986.12**
Sum of two sides 2700 mm	556.22	703.95	8.62	323.80	nr	**1027.75**
Sum of two sides 2800 mm	571.39	723.15	15.20	570.97	nr	**1294.12**
Sum of two sides 2900 mm	632.61	800.63	15.20	570.97	nr	**1371.60**
Sum of two sides 3000 mm	676.17	855.76	15.60	585.99	nr	**1441.75**
Sum of two sides 3100 mm	705.05	892.31	16.04	602.52	nr	**1494.83**
Sum of two sides 3200 mm	732.71	927.31	16.04	602.52	nr	**1529.84**
Branch						
Sum of two sides 1700 mm	202.96	256.87	1.69	63.48	nr	**320.35**
Sum of two sides 1800 mm	222.81	281.99	1.69	63.48	nr	**345.47**
Sum of two sides 1900 mm	242.97	307.50	1.85	69.49	nr	**376.99**
Sum of two sides 2000 mm	262.83	332.64	1.85	69.49	nr	**402.13**
Sum of two sides 2100 mm	282.72	357.81	2.61	98.04	nr	**455.85**
Sum of two sides 2200 mm	302.62	382.99	2.61	98.04	nr	**481.03**
Sum of two sides 2300 mm	322.76	408.49	2.61	98.04	nr	**506.53**
Sum of two sides 2400 mm	342.59	433.58	2.88	108.18	nr	**541.77**
Sum of two sides 2500 mm	362.49	458.77	2.88	108.18	nr	**566.95**
Sum of two sides 2600 mm	382.36	483.92	2.88	108.18	nr	**592.10**
Sum of two sides 2700 mm	402.25	509.09	2.88	108.18	nr	**617.27**
Sum of two sides 2800 mm	422.08	534.18	3.94	148.00	nr	**682.18**
Sum of two sides 2900 mm	448.71	567.89	3.94	148.00	nr	**715.89**
Sum of two sides 3000 mm	468.60	593.06	4.83	181.43	nr	**774.49**
Sum of two sides 3100 mm	484.50	613.18	4.83	181.43	nr	**794.61**
Sum of two sides 3200 mm	498.98	631.51	4.83	181.43	nr	**812.94**

38 VENTILATION/AIR CONDITIONING SYSTEMS

Item	Net Price £	Material £	Labour hours	Labour £	Unit	Total rate £
Grille neck						
Sum of two sides 1700 mm	205.00	259.45	1.86	69.87	nr	**329.32**
Sum of two sides 1800 mm	226.69	286.90	1.86	69.87	nr	**356.77**
Sum of two sides 1900 mm	248.39	314.37	2.02	75.88	nr	**390.25**
Sum of two sides 2000 mm	270.07	341.81	2.02	75.88	nr	**417.68**
Sum of two sides 2100 mm	291.76	369.26	2.80	105.18	nr	**474.44**
Sum of two sides 2200 mm	313.46	396.71	2.80	105.18	nr	**501.89**
Sum of two sides 2300 mm	335.15	424.16	2.80	105.18	nr	**529.34**
Sum of two sides 2400 mm	356.82	451.60	3.06	114.94	nr	**566.54**
Sum of two sides 2500 mm	378.52	479.05	3.06	114.94	nr	**593.99**
Sum of two sides 2600 mm	398.01	503.72	3.08	115.70	nr	**619.41**
Sum of two sides 2700 mm	420.51	532.20	3.08	115.70	nr	**647.90**
Sum of two sides 2800 mm	441.14	558.31	4.13	155.14	nr	**713.44**
Sum of two sides 2900 mm	463.74	586.91	4.13	155.14	nr	**742.05**
Sum of two sides 3000 mm	485.36	614.27	5.02	188.57	nr	**802.84**
Sum of two sides 3100 mm	501.56	634.77	5.02	188.57	nr	**823.34**
Sum of two sides 3200 mm	518.40	656.09	5.02	188.57	nr	**844.65**
Ductwork 1601 to 2000 mm longest side						
Sum of two sides 2100 mm	249.31	315.53	2.53	95.04	m	**410.57**
Sum of two sides 2200 mm	257.37	325.72	2.55	95.79	m	**421.51**
Sum of two sides 2300 mm	265.65	336.20	2.55	95.79	m	**431.99**
Sum of two sides 2400 mm	273.45	346.08	2.56	96.16	m	**442.24**
Sum of two sides 2500 mm	281.73	356.56	2.76	103.68	m	**460.24**
Sum of two sides 2600 mm	289.57	366.48	2.77	104.05	m	**470.53**
Sum of two sides 2700 mm	297.41	376.40	2.97	111.56	m	**487.96**
Sum of two sides 2800 mm	305.25	386.32	2.99	112.32	m	**498.64**
Sum of two sides 2900 mm	313.53	396.80	3.30	123.96	m	**520.76**
Sum of two sides 3000 mm	321.37	406.72	3.31	124.34	m	**531.06**
Sum of two sides 3100 mm	369.06	467.08	3.53	132.60	m	**599.68**
Sum of two sides 3200 mm	376.90	477.00	3.53	132.60	m	**609.60**
Sum of two sides 3300 mm	385.18	487.48	3.53	132.60	m	**620.08**
Sum of two sides 3400 mm	393.02	497.40	3.55	133.35	m	**630.76**
Sum of two sides 3500 mm	400.86	507.32	3.55	133.35	m	**640.68**
Sum of two sides 3600 mm	408.70	517.25	3.55	133.35	m	**650.60**
Sum of two sides 3700 mm	416.98	527.73	3.56	133.73	m	**661.45**
Sum of two sides 3800 mm	424.82	537.65	3.56	133.73	m	**671.37**
Sum of two sides 3900 mm	432.66	547.57	3.56	133.73	m	**681.30**
Sum of two sides 4000 mm	440.49	557.49	3.56	133.73	m	**691.22**
Extra over fittings; Ductwork 1601 to 2000 mm longest side						
End cap						
Sum of two sides 2100 mm	94.06	119.04	0.87	32.68	nr	**151.72**
Sum of two sides 2200 mm	101.13	127.99	0.87	32.68	nr	**160.67**
Sum of two sides 2300 mm	108.20	136.94	0.87	32.68	nr	**169.62**
Sum of two sides 2400 mm	115.27	145.89	0.87	32.68	nr	**178.57**
Sum of two sides 2500 mm	122.34	154.83	0.87	32.68	nr	**187.51**
Sum of two sides 2600 mm	129.40	163.77	0.87	32.68	nr	**196.45**
Sum of two sides 2700 mm	136.46	172.70	0.87	32.68	nr	**205.38**

38 VENTILATION/AIR CONDITIONING SYSTEMS

Item	Net Price £	Material £	Labour hours	Labour £	Unit	Total rate £
DUCTWORK: RECTANGULAR – CLASS C – cont						
Extra over fittings – cont						
End cap – cont						
Sum of two sides 2800 mm	143.52	181.64	1.16	43.57	nr	**225.22**
Sum of two sides 2900 mm	150.60	190.60	1.16	43.57	nr	**234.17**
Sum of two sides 3000 mm	157.66	199.53	1.73	64.99	nr	**264.51**
Sum of two sides 3100 mm	163.27	206.64	1.73	64.99	nr	**271.63**
Sum of two sides 3200 mm	168.38	213.10	1.73	64.99	nr	**278.09**
Sum of two sides 3300 mm	173.50	219.58	1.73	64.99	nr	**284.56**
Sum of two sides 3400 mm	178.62	226.06	1.73	64.99	nr	**291.04**
Sum of two sides 3500 mm	183.73	232.53	1.73	64.99	nr	**297.52**
Sum of two sides 3600 mm	188.84	238.99	1.73	64.99	nr	**303.98**
Sum of two sides 3700 mm	193.96	245.47	1.73	64.99	nr	**310.46**
Sum of two sides 3800 mm	199.05	251.92	1.73	64.99	nr	**316.90**
Sum of two sides 3900 mm	204.18	258.41	1.73	64.99	nr	**323.39**
Sum of two sides 4000 mm	209.28	264.87	1.73	64.99	nr	**329.86**
Reducer						
Sum of two sides 2100 mm	121.86	154.22	2.92	109.69	nr	**263.91**
Sum of two sides 2200 mm	133.35	168.76	2.92	109.69	nr	**278.45**
Sum of two sides 2300 mm	145.03	183.55	2.92	109.69	nr	**293.24**
Sum of two sides 2400 mm	156.34	197.87	3.12	117.20	nr	**315.06**
Sum of two sides 2500 mm	168.03	212.65	3.12	117.20	nr	**329.85**
Sum of two sides 2600 mm	179.51	227.19	3.16	118.70	nr	**345.89**
Sum of two sides 2700 mm	190.99	241.71	3.16	118.70	nr	**360.42**
Sum of two sides 2800 mm	202.46	256.24	4.00	150.25	nr	**406.49**
Sum of two sides 2900 mm	204.25	258.50	4.01	150.63	nr	**409.13**
Sum of two sides 3000 mm	211.75	267.98	4.01	150.63	nr	**418.62**
Sum of two sides 3100 mm	214.15	271.03	4.01	150.63	nr	**421.66**
Sum of two sides 3200 mm	225.66	285.59	4.56	171.29	nr	**456.88**
Sum of two sides 3300 mm	237.76	300.91	4.56	171.29	nr	**472.20**
Sum of two sides 3400 mm	245.47	310.66	4.56	171.29	nr	**481.95**
Sum of two sides 3500 mm	253.17	320.41	4.56	171.29	nr	**491.70**
Sum of two sides 3600 mm	260.89	330.18	4.56	171.29	nr	**501.47**
Sum of two sides 3700 mm	268.81	340.21	4.56	171.29	nr	**511.50**
Sum of two sides 3800 mm	276.53	349.97	4.56	171.29	nr	**521.27**
Sum of two sides 3900 mm	284.23	359.73	4.56	171.29	nr	**531.02**
Sum of two sides 4000 mm	291.95	369.49	4.56	171.29	nr	**540.78**
Offset						
Sum of two sides 2100 mm	313.82	397.17	2.92	109.69	nr	**506.86**
Sum of two sides 2200 mm	337.48	427.11	3.26	122.46	nr	**549.57**
Sum of two sides 2300 mm	350.94	444.16	3.26	122.46	nr	**566.61**
Sum of two sides 2400 mm	384.82	487.03	3.47	130.35	nr	**617.37**
Sum of two sides 2500 mm	397.80	503.46	3.48	130.72	nr	**634.18**
Sum of two sides 2600 mm	402.16	508.97	3.49	131.10	nr	**640.07**
Sum of two sides 2700 mm	413.69	523.56	3.50	131.47	nr	**655.03**
Sum of two sides 2800 mm	418.91	530.18	4.33	162.65	nr	**692.83**
Sum of two sides 2900 mm	418.91	530.18	4.74	178.05	nr	**708.23**
Sum of two sides 3000 mm	421.91	533.97	5.31	199.46	nr	**733.43**
Sum of two sides 3100 mm	423.42	535.88	5.34	200.59	nr	**736.47**

38 VENTILATION/AIR CONDITIONING SYSTEMS

Item	Net Price £	Material £	Labour hours	Labour £	Unit	Total rate £
Sum of two sides 3200 mm	429.45	543.51	5.34	200.59	nr	**744.10**
Sum of two sides 3300 mm	455.78	576.84	5.35	200.97	nr	**777.80**
Sum of two sides 3400 mm	471.08	596.19	5.35	200.97	nr	**797.16**
Sum of two sides 3500 mm	486.38	615.56	5.35	200.97	nr	**816.53**
Sum of two sides 3600 mm	501.65	634.89	5.35	200.97	nr	**835.85**
Sum of two sides 3700 mm	517.02	654.34	5.35	200.97	nr	**855.31**
Sum of two sides 3800 mm	532.30	673.68	5.35	200.97	nr	**874.65**
Sum of two sides 3900 mm	547.59	693.02	5.35	200.97	nr	**893.99**
Sum of two sides 4000 mm	562.88	712.38	5.35	200.97	nr	**913.34**
Square to round						
Sum of two sides 2100 mm	174.00	220.21	4.26	160.02	nr	**380.23**
Sum of two sides 2200 mm	190.86	241.56	4.27	160.40	nr	**401.95**
Sum of two sides 2300 mm	207.64	262.79	4.30	161.52	nr	**424.31**
Sum of two sides 2400 mm	224.58	284.23	4.77	179.18	nr	**463.41**
Sum of two sides 2500 mm	244.02	308.83	4.79	179.93	nr	**488.76**
Sum of two sides 2600 mm	261.02	330.34	4.95	185.94	nr	**516.28**
Sum of two sides 2700 mm	278.02	351.86	4.95	185.94	nr	**537.80**
Sum of two sides 2800 mm	295.03	373.39	8.49	318.92	nr	**692.30**
Sum of two sides 2900 mm	311.99	394.86	8.88	333.56	nr	**728.43**
Sum of two sides 3000 mm	314.67	398.24	9.02	338.82	nr	**737.07**
Sum of two sides 3100 mm	325.41	411.84	9.02	338.82	nr	**750.66**
Sum of two sides 3200 mm	326.08	412.69	9.09	341.45	nr	**754.14**
Sum of two sides 3300 mm	343.37	434.57	9.09	341.45	nr	**776.02**
Sum of two sides 3400 mm	354.75	448.98	9.09	341.45	nr	**790.43**
Sum of two sides 3500 mm	366.17	463.42	9.09	341.45	nr	**804.87**
Sum of two sides 3600 mm	377.57	477.85	9.09	341.45	nr	**819.31**
Sum of two sides 3700 mm	393.21	497.64	9.09	341.45	nr	**839.09**
Sum of two sides 3800 mm	404.73	512.23	9.09	341.45	nr	**853.68**
Sum of two sides 3900 mm	416.26	526.82	9.09	341.45	nr	**868.27**
Sum of two sides 4000 mm	427.78	541.40	9.09	341.45	nr	**882.85**
90° radius bend						
Sum of two sides 2100 mm	270.08	341.81	2.61	98.04	nr	**439.85**
Sum of two sides 2200 mm	539.12	682.31	2.62	98.42	nr	**780.73**
Sum of two sides 2300 mm	584.94	740.29	2.63	98.79	nr	**839.09**
Sum of two sides 2400 mm	593.03	750.54	4.34	163.03	nr	**913.57**
Sum of two sides 2500 mm	638.48	808.07	4.35	163.40	nr	**971.47**
Sum of two sides 2600 mm	683.04	864.45	4.53	170.16	nr	**1034.62**
Sum of two sides 2700 mm	727.59	920.84	4.53	170.16	nr	**1091.00**
Sum of two sides 2800 mm	772.14	977.23	7.13	267.83	nr	**1245.05**
Sum of two sides 2900 mm	817.60	1034.75	7.17	269.33	nr	**1304.08**

38 VENTILATION/AIR CONDITIONING SYSTEMS

Item	Net Price £	Material £	Labour hours	Labour £	Unit	Total rate £
DUCTWORK: RECTANGULAR – CLASS C – cont						
Extra over fittings – cont						
90° radius bend – cont						
Sum of two sides 3000 mm	839.07	1061.92	7.26	272.71	nr	**1334.63**
Sum of two sides 3100 mm	848.31	1073.62	7.26	272.71	nr	**1346.33**
Sum of two sides 3200 mm	857.76	1085.58	7.31	274.59	nr	**1360.17**
Sum of two sides 3300 mm	944.21	1195.00	7.31	274.59	nr	**1469.59**
Sum of two sides 3400 mm	976.92	1236.39	7.31	274.59	nr	**1510.98**
Sum of two sides 3500 mm	1009.63	1277.78	7.31	274.59	nr	**1552.37**
Sum of two sides 3600 mm	1042.33	1319.18	7.31	274.59	nr	**1593.77**
Sum of two sides 3700 mm	1075.92	1361.68	7.31	274.59	nr	**1636.27**
Sum of two sides 3800 mm	1108.62	1403.07	7.31	274.59	nr	**1677.66**
Sum of two sides 3900 mm	1141.33	1444.46	7.31	274.59	nr	**1719.05**
Sum of two sides 4000 mm	1174.01	1485.83	7.31	274.59	nr	**1760.42**
45° bend						
Sum of two sides 2100 mm	128.85	163.07	7.50	281.73	nr	**444.80**
Sum of two sides 2200 mm	391.07	494.94	7.50	281.73	nr	**776.67**
Sum of two sides 2300 mm	422.10	534.21	7.55	283.61	nr	**817.82**
Sum of two sides 2400 mm	431.00	545.48	8.13	305.39	nr	**850.87**
Sum of two sides 2500 mm	461.70	584.33	8.30	311.78	nr	**896.11**
Sum of two sides 2600 mm	491.74	622.34	8.56	321.54	nr	**943.89**
Sum of two sides 2700 mm	521.75	660.33	8.62	323.80	nr	**984.12**
Sum of two sides 2800 mm	551.79	698.35	9.09	341.45	nr	**1039.80**
Sum of two sides 2900 mm	582.51	737.22	9.09	341.45	nr	**1078.67**
Sum of two sides 3000 mm	605.13	765.85	9.62	361.36	nr	**1127.22**
Sum of two sides 3100 mm	605.26	766.02	9.62	361.36	nr	**1127.38**
Sum of two sides 3200 mm	619.70	784.29	9.62	361.36	nr	**1145.65**
Sum of two sides 3300 mm	676.10	855.68	9.62	361.36	nr	**1217.04**
Sum of two sides 3400 mm	698.12	883.54	9.62	361.36	nr	**1244.90**
Sum of two sides 3500 mm	720.14	911.41	9.62	361.36	nr	**1272.77**
Sum of two sides 3600 mm	742.13	939.24	9.62	361.36	nr	**1300.60**
Sum of two sides 3700 mm	764.84	967.98	9.62	361.36	nr	**1329.34**
Sum of two sides 3800 mm	786.84	995.83	9.62	361.36	nr	**1357.19**
Sum of two sides 3900 mm	808.85	1023.68	9.62	361.36	nr	**1385.04**
Sum of two sides 4000 mm	830.84	1051.51	9.62	361.36	nr	**1412.88**
90° mitre bend						
Sum of two sides 2100 mm	671.13	849.39	7.50	281.73	nr	**1131.11**
Sum of two sides 2200 mm	708.98	897.29	7.50	281.73	nr	**1179.01**
Sum of two sides 2300 mm	766.07	969.54	7.55	283.61	nr	**1253.15**
Sum of two sides 2400 mm	782.81	990.73	8.13	305.39	nr	**1296.12**
Sum of two sides 2500 mm	840.13	1063.27	8.30	311.78	nr	**1375.05**
Sum of two sides 2600 mm	897.79	1136.24	8.56	321.54	nr	**1457.78**
Sum of two sides 2700 mm	973.86	1232.52	8.62	323.80	nr	**1556.32**
Sum of two sides 2800 mm	1032.60	1306.86	15.20	570.97	nr	**1877.83**
Sum of two sides 2900 mm	1106.25	1400.07	15.20	570.97	nr	**1971.04**
Sum of two sides 3000 mm	1135.89	1437.58	16.04	602.52	nr	**2040.10**
Sum of two sides 3100 mm	1165.84	1475.49	15.60	585.99	nr	**2061.48**
Sum of two sides 3200 mm	1182.14	1496.12	16.04	602.52	nr	**2098.64**
Sum of two sides 3300 mm	1276.35	1615.35	16.04	602.52	nr	**2217.87**
Sum of two sides 3400 mm	1323.38	1674.87	16.04	602.52	nr	**2277.39**
Sum of two sides 3500 mm	1370.39	1734.37	16.04	602.52	nr	**2336.89**

38 VENTILATION/AIR CONDITIONING SYSTEMS

Item	Net Price £	Material £	Labour hours	Labour £	Unit	Total rate £
Sum of two sides 3600 mm	1417.42	1793.89	16.04	602.52	nr	**2396.41**
Sum of two sides 3700 mm	1492.40	1888.78	16.04	602.52	nr	**2491.30**
Sum of two sides 3800 mm	1540.34	1949.45	16.04	602.52	nr	**2551.97**
Sum of two sides 3900 mm	1588.26	2010.10	16.04	602.52	nr	**2612.62**
Sum of two sides 4000 mm	1636.18	2070.75	16.04	602.52	nr	**2673.28**
Branch						
Sum of two sides 2100 mm	287.75	364.17	2.61	98.04	nr	**462.21**
Sum of two sides 2200 mm	307.77	389.52	2.61	98.04	nr	**487.56**
Sum of two sides 2300 mm	328.09	415.23	2.61	98.04	nr	**513.27**
Sum of two sides 2400 mm	347.81	440.18	2.88	108.18	nr	**548.37**
Sum of two sides 2500 mm	368.15	465.94	2.88	108.18	nr	**574.12**
Sum of two sides 2600 mm	388.19	491.30	2.88	108.18	nr	**599.48**
Sum of two sides 2700 mm	408.23	516.66	2.88	108.18	nr	**624.84**
Sum of two sides 2800 mm	428.32	542.08	3.94	148.00	nr	**690.08**
Sum of two sides 2900 mm	448.63	567.79	3.94	148.00	nr	**715.79**
Sum of two sides 3000 mm	468.71	593.20	4.83	181.43	nr	**774.63**
Sum of two sides 3100 mm	484.77	613.52	4.83	181.43	nr	**794.96**
Sum of two sides 3200 mm	499.43	632.08	4.83	181.43	nr	**813.51**
Sum of two sides 3300 mm	520.99	659.37	4.83	181.43	nr	**840.80**
Sum of two sides 3400 mm	535.66	677.93	4.83	181.43	nr	**859.36**
Sum of two sides 3500 mm	550.32	696.48	4.83	181.43	nr	**877.92**
Sum of two sides 3600 mm	564.98	715.03	4.83	181.43	nr	**896.47**
Sum of two sides 3700 mm	579.91	733.93	4.83	181.43	nr	**915.36**
Sum of two sides 3800 mm	594.55	752.47	4.83	181.43	nr	**933.90**
Sum of two sides 3900 mm	609.25	771.06	4.83	181.43	nr	**952.49**
Sum of two sides 4000 mm	623.89	789.60	4.83	181.43	nr	**971.03**
Grille neck						
Sum of two sides 2100 mm	297.80	376.90	2.80	105.18	nr	**482.08**
Sum of two sides 2200 mm	319.77	404.70	2.80	105.18	nr	**509.87**
Sum of two sides 2300 mm	341.75	432.52	2.80	105.18	nr	**537.70**
Sum of two sides 2400 mm	363.74	460.35	3.06	114.94	nr	**575.30**
Sum of two sides 2500 mm	385.68	488.12	3.06	114.94	nr	**603.07**
Sum of two sides 2600 mm	407.67	515.95	3.08	115.70	nr	**631.65**
Sum of two sides 2700 mm	429.64	543.75	3.08	115.70	nr	**659.44**
Sum of two sides 2800 mm	451.61	571.56	4.13	155.14	nr	**726.70**
Sum of two sides 2900 mm	473.59	599.38	4.13	155.14	nr	**754.51**
Sum of two sides 3000 mm	502.07	635.41	5.02	188.57	nr	**823.98**
Sum of two sides 3100 mm	513.19	649.49	5.02	188.57	nr	**838.06**
Sum of two sides 3200 mm	529.30	669.88	5.02	188.57	nr	**858.45**
Sum of two sides 3300 mm	552.54	699.29	5.02	188.57	nr	**887.86**
Sum of two sides 3400 mm	568.85	719.93	5.02	188.57	nr	**908.50**
Sum of two sides 3500 mm	577.57	730.97	5.02	188.57	nr	**919.54**
Sum of two sides 3600 mm	593.66	751.34	5.02	188.57	nr	**939.90**
Sum of two sides 3700 mm	617.74	781.82	5.02	188.57	nr	**970.39**
Sum of two sides 3800 mm	634.05	802.46	5.02	188.57	nr	**991.03**
Sum of two sides 3900 mm	641.93	812.43	5.02	188.57	nr	**1001.00**
Sum of two sides 4000 mm	666.66	843.73	5.02	188.57	nr	**1032.30**
Ductwork 2001 to 2500 mm longest side						
Sum of two sides 2500 mm	460.81	583.20	2.77	104.05	m	**687.25**
Sum of two sides 2600 mm	474.37	600.36	2.97	111.56	m	**711.93**
Sum of two sides 2700 mm	486.75	616.04	2.99	112.32	m	**728.35**

38 VENTILATION/AIR CONDITIONING SYSTEMS

Item	Net Price £	Material £	Labour hours	Labour £	Unit	Total rate £
DUCTWORK: RECTANGULAR – CLASS C – cont						
Ductwork 2001 to 2500 mm longest side – cont						
Sum of two sides 2800 mm	500.85	633.87	3.30	123.96	m	757.83
Sum of two sides 2900 mm	513.23	649.54	3.31	124.34	m	773.88
Sum of two sides 3000 mm	525.21	664.70	3.53	132.60	m	797.30
Sum of two sides 3100 mm	537.44	680.19	3.55	133.35	m	813.54
Sum of two sides 3200 mm	549.43	695.36	3.56	133.73	m	829.09
Sum of two sides 3300 mm	561.80	711.02	3.56	133.73	m	844.74
Sum of two sides 3400 mm	573.78	726.17	3.56	133.73	m	859.90
Sum of two sides 3500 mm	628.50	795.43	3.56	133.73	m	929.16
Sum of two sides 3600 mm	640.48	810.59	3.56	133.73	m	944.32
Sum of two sides 3700 mm	652.86	826.26	3.56	133.73	m	959.99
Sum of two sides 3800 mm	664.84	841.42	3.56	133.73	m	975.15
Sum of two sides 3900 mm	676.82	856.58	3.56	133.73	m	990.30
Sum of two sides 4000 mm	690.93	874.45	3.56	133.73	m	1008.17
Extra over fittings; Ductwork 2001 to 2500 mm longest side						
End cap						
Sum of two sides 2500 mm	122.34	154.83	0.87	32.68	nr	187.51
Sum of two sides 2600 mm	129.40	163.77	0.87	32.68	nr	196.45
Sum of two sides 2700 mm	136.46	172.70	0.87	32.68	nr	205.38
Sum of two sides 2800 mm	143.52	181.64	1.16	43.57	nr	225.22
Sum of two sides 2900 mm	150.60	190.60	1.16	43.57	nr	234.17
Sum of two sides 3000 mm	157.66	199.53	1.73	64.99	nr	264.51
Sum of two sides 3100 mm	163.27	206.64	1.73	64.99	nr	271.63
Sum of two sides 3200 mm	168.38	213.10	1.73	64.99	nr	278.09
Sum of two sides 3300 mm	173.50	219.58	1.73	64.99	nr	284.56
Sum of two sides 3400 mm	178.62	226.06	1.73	64.99	nr	291.04
Sum of two sides 3500 mm	183.73	232.53	1.73	64.99	nr	297.52
Sum of two sides 3600 mm	188.84	238.99	1.73	64.99	nr	303.98
Sum of two sides 3700 mm	193.96	245.47	1.73	64.99	nr	310.46
Sum of two sides 3800 mm	199.05	251.92	1.73	64.99	nr	316.90
Sum of two sides 3900 mm	204.18	258.41	1.73	64.99	nr	323.39
Sum of two sides 4000 mm	209.28	264.87	1.73	64.99	nr	329.86
Reducer						
Sum of two sides 2500 mm	64.20	81.25	3.12	117.20	nr	198.45
Sum of two sides 2600 mm	70.39	89.08	3.16	118.70	nr	207.79
Sum of two sides 2700 mm	76.73	97.11	3.16	118.70	nr	215.81
Sum of two sides 2800 mm	91.24	115.47	4.00	150.25	nr	265.72
Sum of two sides 2900 mm	98.23	124.32	4.01	150.63	nr	274.95
Sum of two sides 3000 mm	105.07	132.97	4.56	171.29	nr	304.26
Sum of two sides 3100 mm	107.54	136.10	4.56	171.29	nr	307.39
Sum of two sides 3200 mm	111.42	141.01	4.56	171.29	nr	312.30
Sum of two sides 3300 mm	113.05	143.07	4.56	171.29	nr	314.36
Sum of two sides 3400 mm	113.26	143.34	4.56	171.29	nr	314.63
Sum of two sides 3500 mm	117.46	148.65	4.56	171.29	nr	319.94
Sum of two sides 3600 mm	117.67	148.93	4.56	171.29	nr	320.22
Sum of two sides 3700 mm	126.92	160.64	4.56	171.29	nr	331.93
Sum of two sides 3800 mm	131.22	166.07	4.56	171.29	nr	337.36
Sum of two sides 3900 mm	135.49	171.48	4.56	171.29	nr	342.77

38 VENTILATION/AIR CONDITIONING SYSTEMS

Item	Net Price £	Material £	Labour hours	Labour £	Unit	Total rate £
Sum of two sides 4000 mm	139.90	177.05	4.56	171.29	nr	**348.34**
Offset						
Sum of two sides 2500 mm	309.65	391.89	3.48	130.72	nr	**522.61**
Sum of two sides 2600 mm	318.40	402.96	3.49	131.10	nr	**534.06**
Sum of two sides 2700 mm	320.95	406.20	3.50	131.47	nr	**537.67**
Sum of two sides 2800 mm	321.51	406.90	4.33	162.65	nr	**569.55**
Sum of two sides 2900 mm	325.38	411.80	4.74	178.05	nr	**589.85**
Sum of two sides 3000 mm	329.05	416.45	5.31	199.46	nr	**615.91**
Sum of two sides 3100 mm	330.75	418.60	5.34	200.59	nr	**619.19**
Sum of two sides 3200 mm	332.92	421.35	5.35	200.97	nr	**622.31**
Sum of two sides 3300 mm	335.41	424.50	5.35	200.97	nr	**625.46**
Sum of two sides 3400 mm	336.19	425.48	5.35	200.97	nr	**626.44**
Sum of two sides 3500 mm	339.91	430.19	5.35	200.97	nr	**631.16**
Sum of two sides 3600 mm	340.68	431.17	5.35	200.97	nr	**632.13**
Sum of two sides 3700 mm	340.87	431.41	5.35	200.97	nr	**632.37**
Sum of two sides 3800 mm	341.20	431.82	5.35	200.97	nr	**632.78**
Sum of two sides 3900 mm	341.33	431.98	5.35	200.97	nr	**632.95**
Sum of two sides 4000 mm	342.02	432.87	5.35	200.97	nr	**633.83**
Square to round						
Sum of two sides 2500 mm	127.42	161.27	4.79	179.93	nr	**341.20**
Sum of two sides 2600 mm	143.99	182.24	4.95	185.94	nr	**368.18**
Sum of two sides 2700 mm	156.17	197.65	4.95	185.94	nr	**383.59**
Sum of two sides 2800 mm	172.83	218.74	8.49	318.92	nr	**537.65**
Sum of two sides 2900 mm	185.31	234.52	8.88	333.56	nr	**568.09**
Sum of two sides 3000 mm	202.95	256.86	9.02	338.82	nr	**595.68**
Sum of two sides 3100 mm	212.09	268.42	9.02	338.82	nr	**607.24**
Sum of two sides 3200 mm	219.02	277.19	9.09	341.45	nr	**618.64**
Sum of two sides 3300 mm	220.21	278.70	9.09	341.45	nr	**620.15**
Sum of two sides 3400 mm	222.58	281.70	9.09	341.45	nr	**623.15**
Sum of two sides 3500 mm	224.45	284.06	9.09	341.45	nr	**625.51**
Sum of two sides 3600 mm	228.29	288.93	9.09	341.45	nr	**630.38**
Sum of two sides 3700 mm	237.87	301.05	9.09	341.45	nr	**642.50**
Sum of two sides 3800 mm	245.77	311.05	9.09	341.45	nr	**652.51**
Sum of two sides 3900 mm	253.69	321.07	9.09	341.45	nr	**662.52**
Sum of two sides 4000 mm	261.55	331.02	9.09	341.45	nr	**672.48**
90° radius bend						
Sum of two sides 2500 mm	371.75	470.48	4.35	163.40	nr	**633.88**
Sum of two sides 2600 mm	382.38	483.94	4.53	170.16	nr	**654.10**
Sum of two sides 2700 mm	398.58	504.44	4.53	170.16	nr	**674.61**
Sum of two sides 2800 mm	400.97	507.46	7.13	267.83	nr	**775.29**
Sum of two sides 2900 mm	433.13	548.17	7.17	269.33	nr	**817.50**
Sum of two sides 3000 mm	464.39	587.73	7.26	272.71	nr	**860.44**
Sum of two sides 3100 mm	488.49	618.23	7.26	272.71	nr	**890.95**
Sum of two sides 3200 mm	520.92	659.27	7.31	274.59	nr	**933.86**
Sum of two sides 3300 mm	527.94	668.16	7.31	274.59	nr	**942.75**
Sum of two sides 3400 mm	534.15	676.01	7.31	274.59	nr	**950.60**
Sum of two sides 3500 mm	537.53	680.30	7.31	274.59	nr	**954.89**

38 VENTILATION/AIR CONDITIONING SYSTEMS

Item	Net Price £	Material £	Labour hours	Labour £	Unit	Total rate £
DUCTWORK: RECTANGULAR – CLASS C – cont						
Extra over fittings – cont						
90° radius bend – cont						
Sum of two sides 3600 mm	544.27	688.82	7.31	274.59	nr	963.41
Sum of two sides 3700 mm	610.91	773.17	7.31	274.59	nr	1047.76
Sum of two sides 3800 mm	633.63	801.92	7.31	274.59	nr	1076.51
Sum of two sides 3900 mm	651.74	824.84	7.31	274.59	nr	1099.43
Sum of two sides 4000 mm	656.34	830.66	7.31	274.59	nr	1105.25
45° bend						
Sum of two sides 2500 mm	330.26	417.98	8.30	311.78	nr	729.75
Sum of two sides 2600 mm	342.44	433.39	8.56	321.54	nr	754.93
Sum of two sides 2700 mm	365.47	462.53	8.62	323.80	nr	786.33
Sum of two sides 2800 mm	366.98	464.45	9.09	341.45	nr	805.91
Sum of two sides 2900 mm	389.79	493.32	9.09	341.45	nr	834.77
Sum of two sides 3000 mm	413.57	523.42	9.62	361.36	nr	884.78
Sum of two sides 3100 mm	432.17	546.96	9.62	361.36	nr	908.32
Sum of two sides 3200 mm	449.37	568.73	9.62	361.36	nr	930.09
Sum of two sides 3300 mm	467.10	591.16	9.62	361.36	nr	952.53
Sum of two sides 3400 mm	470.61	595.60	9.62	361.36	nr	956.96
Sum of two sides 3500 mm	484.28	612.91	9.62	361.36	nr	974.27
Sum of two sides 3600 mm	487.78	617.33	9.62	361.36	nr	978.69
Sum of two sides 3700 mm	527.59	667.71	9.62	361.36	nr	1029.07
Sum of two sides 3800 mm	544.78	689.47	9.62	361.36	nr	1050.83
Sum of two sides 3900 mm	561.95	711.20	9.62	361.36	nr	1072.56
Sum of two sides 4000 mm	564.39	714.29	9.62	361.36	nr	1075.65
90° mitre bend						
Sum of two sides 2500 mm	486.42	615.61	8.30	311.78	nr	927.39
Sum of two sides 2600 mm	509.47	644.78	8.56	321.54	nr	966.32
Sum of two sides 2700 mm	543.03	687.26	8.62	323.80	nr	1011.06
Sum of two sides 2800 mm	555.10	702.53	15.20	570.97	nr	1273.50
Sum of two sides 2900 mm	573.36	725.64	15.20	570.97	nr	1296.61
Sum of two sides 3000 mm	636.11	805.06	15.20	570.97	nr	1376.02
Sum of two sides 3100 mm	706.06	893.59	16.04	602.52	nr	1496.11
Sum of two sides 3200 mm	742.45	939.64	16.04	602.52	nr	1542.17
Sum of two sides 3300 mm	766.19	969.69	16.04	602.52	nr	1572.21
Sum of two sides 3400 mm	790.54	1000.50	16.04	602.52	nr	1603.02
Sum of two sides 3500 mm	803.15	1016.46	16.04	602.52	nr	1618.98
Sum of two sides 3600 mm	827.49	1047.28	16.04	602.52	nr	1649.80
Sum of two sides 3700 mm	884.74	1119.72	16.04	602.52	nr	1722.25
Sum of two sides 3800 mm	933.47	1181.39	16.04	602.52	nr	1783.91
Sum of two sides 3900 mm	959.66	1214.55	16.04	602.52	nr	1817.07
Sum of two sides 4000 mm	971.56	1229.60	16.04	602.52	nr	1832.12
Branch						
Sum of two sides 2500 mm	375.97	475.83	2.88	108.18	nr	584.01
Sum of two sides 2600 mm	395.97	501.14	2.88	108.18	nr	609.33
Sum of two sides 2700 mm	416.28	526.84	2.88	108.18	nr	635.03
Sum of two sides 2800 mm	436.00	551.80	3.94	148.00	nr	699.80
Sum of two sides 2900 mm	456.30	577.50	3.94	148.00	nr	725.50
Sum of two sides 3000 mm	476.35	602.87	4.83	181.43	nr	784.30
Sum of two sides 3100 mm	492.57	623.39	4.83	181.43	nr	804.83

38 VENTILATION/AIR CONDITIONING SYSTEMS

Item	Net Price £	Material £	Labour hours	Labour £	Unit	Total rate £
Sum of two sides 3200 mm	507.43	642.20	4.83	181.43	nr	**823.63**
Sum of two sides 3300 mm	522.55	661.33	4.83	181.43	nr	**842.77**
Sum of two sides 3400 mm	537.39	680.12	4.83	181.43	nr	**861.56**
Sum of two sides 3500 mm	552.93	699.79	4.83	181.43	nr	**881.22**
Sum of two sides 3600 mm	567.78	718.58	4.83	181.43	nr	**900.01**
Sum of two sides 3700 mm	589.43	745.98	4.83	181.43	nr	**927.41**
Sum of two sides 3800 mm	604.29	764.80	4.83	181.43	nr	**946.23**
Sum of two sides 3900 mm	619.14	783.59	4.83	181.43	nr	**965.02**
Sum of two sides 4000 mm	634.28	802.75	4.83	181.43	nr	**984.18**
Grille neck						
Sum of two sides 2500 mm	385.68	488.12	3.06	114.94	nr	**603.07**
Sum of two sides 2600 mm	407.67	515.95	3.08	115.70	nr	**631.65**
Sum of two sides 2700 mm	429.64	543.75	3.08	115.70	nr	**659.44**
Sum of two sides 2800 mm	451.61	571.56	4.13	155.14	nr	**726.70**
Sum of two sides 2900 mm	473.59	599.38	4.13	155.14	nr	**754.51**
Sum of two sides 3000 mm	502.07	635.41	5.02	188.57	nr	**823.98**
Sum of two sides 3100 mm	513.19	649.49	5.02	188.57	nr	**838.06**
Sum of two sides 3200 mm	529.30	669.88	5.02	188.57	nr	**858.45**
Sum of two sides 3300 mm	552.54	699.29	5.02	188.57	nr	**887.86**
Sum of two sides 3400 mm	568.85	719.93	5.02	188.57	nr	**908.50**
Sum of two sides 3500 mm	577.57	730.97	5.02	188.57	nr	**919.54**
Sum of two sides 3600 mm	593.66	751.34	5.02	188.57	nr	**939.90**
Sum of two sides 3700 mm	617.74	781.82	5.02	188.57	nr	**970.39**
Sum of two sides 3800 mm	634.05	802.46	5.02	188.57	nr	**991.03**
Sum of two sides 3900 mm	641.93	812.43	5.02	188.57	nr	**1001.00**
Sum of two sides 4000 mm	666.66	843.73	5.02	188.57	nr	**1032.30**
Ductwork 2501 to 4000 mm longest side						
Sum of two sides 3000 mm	528.05	668.30	2.38	89.40	m	**757.70**
Sum of two sides 3100 mm	540.69	684.30	2.38	89.40	m	**773.70**
Sum of two sides 3200 mm	552.67	699.46	2.38	89.40	m	**788.86**
Sum of two sides 3300 mm	564.64	714.61	2.38	89.40	m	**804.02**
Sum of two sides 3400 mm	579.46	733.37	2.64	99.17	m	**832.54**
Sum of two sides 3500 mm	591.44	748.53	2.66	99.92	m	**848.45**
Sum of two sides 3600 mm	603.43	763.70	2.95	110.81	m	**874.51**
Sum of two sides 3700 mm	618.64	782.96	2.96	111.19	m	**894.14**
Sum of two sides 3800 mm	630.62	798.11	3.15	118.33	m	**916.44**
Sum of two sides 3900 mm	664.62	841.14	3.15	118.33	m	**959.46**
Sum of two sides 4000 mm	676.59	856.30	3.15	118.33	m	**974.62**
Sum of two sides 4100 mm	688.98	871.97	3.35	125.84	m	**997.81**
Sum of two sides 4200 mm	700.95	887.13	3.35	125.84	m	**1012.96**
Sum of two sides 4300 mm	737.45	933.31	3.60	135.23	m	**1068.54**
Sum of two sides 4400 mm	740.47	937.13	3.60	135.23	m	**1072.36**
Sum of two sides 4500 mm	743.06	940.42	3.60	135.23	m	**1075.65**
Extra over fittings; Ductwork 2501 to 4000 mm longest side						
End cap						
Sum of two sides 3000 mm	157.66	199.53	1.73	64.99	nr	**264.51**
Sum of two sides 3100 mm	163.27	206.64	1.73	64.99	nr	**271.63**
Sum of two sides 3200 mm	168.38	213.10	1.73	64.99	nr	**278.09**
Sum of two sides 3300 mm	173.40	219.45	1.73	64.99	nr	**284.43**

38 VENTILATION/AIR CONDITIONING SYSTEMS

Item	Net Price £	Material £	Labour hours	Labour £	Unit	Total rate £
DUCTWORK: RECTANGULAR – CLASS C – cont						
Extra over fittings – cont						
End cap – cont						
Sum of two sides 3400 mm	178.60	226.04	1.73	64.99	nr	**291.03**
Sum of two sides 3500 mm	183.73	232.53	1.73	64.99	nr	**297.52**
Sum of two sides 3600 mm	188.84	238.99	1.73	64.99	nr	**303.98**
Sum of two sides 3700 mm	193.96	245.47	1.73	64.99	nr	**310.46**
Sum of two sides 3800 mm	199.06	251.93	1.73	64.99	nr	**316.92**
Sum of two sides 3900 mm	204.18	258.41	1.80	67.61	nr	**326.02**
Sum of two sides 4000 mm	209.28	264.87	1.80	67.61	nr	**332.49**
Sum of two sides 4100 mm	214.40	271.35	1.80	67.61	nr	**338.96**
Sum of two sides 4200 mm	219.51	277.81	1.80	67.61	nr	**345.42**
Sum of two sides 4300 mm	224.61	284.27	1.88	70.62	nr	**354.89**
Sum of two sides 4400 mm	229.73	290.75	1.88	70.62	nr	**361.37**
Sum of two sides 4500 mm	234.84	297.21	1.88	70.62	nr	**367.83**
Reducer						
Sum of two sides 3000 mm	105.83	133.94	3.12	117.20	nr	**251.14**
Sum of two sides 3100 mm	110.82	140.25	3.12	117.20	nr	**257.45**
Sum of two sides 3200 mm	118.67	150.18	3.12	117.20	nr	**267.38**
Sum of two sides 3300 mm	123.12	155.82	3.12	117.20	nr	**273.02**
Sum of two sides 3400 mm	127.51	161.38	3.12	117.20	nr	**278.58**
Sum of two sides 3500 mm	131.96	167.01	3.12	117.20	nr	**284.21**
Sum of two sides 3600 mm	136.40	172.63	3.95	148.38	nr	**321.01**
Sum of two sides 3700 mm	138.31	175.04	3.97	149.13	nr	**324.17**
Sum of two sides 3800 mm	140.98	178.42	4.52	169.79	nr	**348.21**
Sum of two sides 3900 mm	141.49	179.07	4.52	169.79	nr	**348.86**
Sum of two sides 4000 mm	145.41	184.03	4.52	169.79	nr	**353.82**
Sum of two sides 4100 mm	146.28	185.13	4.52	169.79	nr	**354.92**
Sum of two sides 4200 mm	150.88	190.96	4.52	169.79	nr	**360.75**
Sum of two sides 4300 mm	155.33	196.59	4.92	184.81	nr	**381.40**
Sum of two sides 4400 mm	164.31	207.95	4.92	184.81	nr	**392.76**
Sum of two sides 4500 mm	168.75	213.57	5.12	192.33	nr	**405.90**
Offset						
Sum of two sides 3000 mm	365.39	462.43	3.48	130.72	nr	**593.16**
Sum of two sides 3100 mm	365.90	463.08	3.48	130.72	nr	**593.80**
Sum of two sides 3200 mm	387.33	490.20	3.48	130.72	nr	**620.92**
Sum of two sides 3300 mm	397.46	503.02	3.48	130.72	nr	**633.74**
Sum of two sides 3400 mm	398.42	504.24	3.50	131.47	nr	**635.72**
Sum of two sides 3500 mm	402.89	509.90	3.50	131.47	nr	**641.37**
Sum of two sides 3600 mm	409.75	518.58	4.25	159.65	nr	**678.22**
Sum of two sides 3700 mm	415.67	526.07	4.76	178.80	nr	**704.87**
Sum of two sides 3800 mm	415.90	526.37	5.32	199.84	nr	**726.21**
Sum of two sides 3900 mm	420.65	532.37	5.35	200.97	nr	**733.34**
Sum of two sides 4000 mm	422.86	535.17	5.35	200.97	nr	**736.13**
Sum of two sides 4100 mm	428.50	542.31	5.85	219.75	nr	**762.06**
Sum of two sides 4200 mm	432.10	546.86	5.85	219.75	nr	**766.61**
Sum of two sides 4300 mm	433.50	548.64	6.15	231.02	nr	**779.66**
Sum of two sides 4400 mm	440.61	557.63	6.30	236.65	nr	**794.28**
Sum of two sides 4500 mm	446.08	564.56	6.15	231.02	nr	**795.58**

38 VENTILATION/AIR CONDITIONING SYSTEMS

Item	Net Price £	Material £	Labour hours	Labour £	Unit	Total rate £
Square to round						
Sum of two sides 3000 mm	275.83	349.09	4.70	176.55	nr	**525.64**
Sum of two sides 3100 mm	287.67	364.07	4.70	176.55	nr	**540.62**
Sum of two sides 3200 mm	298.22	377.43	4.70	176.55	nr	**553.97**
Sum of two sides 3300 mm	308.78	390.79	4.70	176.55	nr	**567.34**
Sum of two sides 3400 mm	319.34	404.16	4.71	176.92	nr	**581.08**
Sum of two sides 3500 mm	329.89	417.51	4.71	176.92	nr	**594.43**
Sum of two sides 3600 mm	340.44	430.86	8.19	307.65	nr	**738.51**
Sum of two sides 3700 mm	350.99	444.22	8.62	323.80	nr	**768.01**
Sum of two sides 3800 mm	361.55	457.58	8.62	323.80	nr	**781.38**
Sum of two sides 3900 mm	456.89	578.24	8.62	323.80	nr	**902.04**
Sum of two sides 4000 mm	469.88	594.69	8.75	328.68	nr	**923.37**
Sum of two sides 4100 mm	482.43	610.56	11.23	421.84	nr	**1032.40**
Sum of two sides 4200 mm	495.85	627.54	11.23	421.84	nr	**1049.38**
Sum of two sides 4300 mm	508.82	643.97	11.25	422.59	nr	**1066.56**
Sum of two sides 4400 mm	521.81	660.40	11.25	422.59	nr	**1082.99**
Sum of two sides 4500 mm	534.80	676.85	11.26	422.97	nr	**1099.81**
90° radius bend						
Sum of two sides 3000 mm	783.91	992.12	3.90	146.50	nr	**1138.62**
Sum of two sides 3100 mm	821.18	1039.29	3.90	146.50	nr	**1185.78**
Sum of two sides 3200 mm	833.49	1054.87	4.26	160.02	nr	**1214.89**
Sum of two sides 3300 mm	854.62	1081.61	4.26	160.02	nr	**1241.63**
Sum of two sides 3400 mm	863.87	1093.32	4.26	160.02	nr	**1253.34**
Sum of two sides 3500 mm	864.01	1093.48	4.55	170.91	nr	**1264.40**
Sum of two sides 3600 mm	867.25	1097.59	4.55	170.91	nr	**1268.50**
Sum of two sides 3700 mm	888.06	1123.93	6.87	258.06	nr	**1381.99**
Sum of two sides 3800 mm	890.07	1126.47	6.87	258.06	nr	**1384.53**
Sum of two sides 3900 mm	892.82	1129.96	6.87	258.06	nr	**1388.02**
Sum of two sides 4000 mm	895.12	1132.86	7.00	262.95	nr	**1395.81**
Sum of two sides 4100 mm	900.24	1139.35	7.20	270.46	nr	**1409.81**
Sum of two sides 4200 mm	909.97	1151.65	7.20	270.46	nr	**1422.11**
Sum of two sides 4300 mm	925.49	1171.30	7.41	278.35	nr	**1449.65**
Sum of two sides 4400 mm	933.25	1181.12	7.41	278.35	nr	**1459.47**
Sum of two sides 4500 mm	942.54	1192.87	7.55	283.61	nr	**1476.48**
45° bend						
Sum of two sides 3000 mm	380.31	481.32	4.85	182.18	nr	**663.50**
Sum of two sides 3100 mm	396.01	501.19	4.85	182.18	nr	**683.38**
Sum of two sides 3200 mm	398.85	504.79	4.85	182.18	nr	**686.97**
Sum of two sides 3300 mm	409.73	518.55	4.87	182.93	nr	**701.48**
Sum of two sides 3400 mm	410.87	519.99	4.87	182.93	nr	**702.93**
Sum of two sides 3500 mm	410.93	520.08	4.87	182.93	nr	**703.01**
Sum of two sides 3600 mm	415.51	525.87	8.81	330.94	nr	**856.80**
Sum of two sides 3700 mm	422.10	534.20	8.81	330.94	nr	**865.14**
Sum of two sides 3800 mm	422.10	534.20	9.31	349.72	nr	**883.92**
Sum of two sides 3900 mm	423.80	536.36	9.31	349.72	nr	**886.08**
Sum of two sides 4000 mm	426.22	539.43	9.31	349.72	nr	**889.14**
Sum of two sides 4100 mm	426.98	540.39	10.01	376.01	nr	**916.40**
Sum of two sides 4200 mm	431.17	545.69	10.01	376.01	nr	**921.70**
Sum of two sides 4300 mm	441.00	558.13	10.01	376.01	nr	**934.14**
Sum of two sides 4400 mm	443.03	560.69	10.52	395.17	nr	**955.86**
Sum of two sides 4500 mm	447.00	565.72	10.52	395.17	nr	**960.89**

38 VENTILATION/AIR CONDITIONING SYSTEMS

Item	Net Price £	Material £	Labour hours	Labour £	Unit	Total rate £
DUCTWORK: RECTANGULAR – CLASS C – cont						
Extra over fittings – cont						
90° mitre bend						
Sum of two sides 3000 mm	1140.50	1443.42	4.85	182.18	nr	**1625.60**
Sum of two sides 3100 mm	1196.72	1514.56	4.85	182.18	nr	**1696.75**
Sum of two sides 3200 mm	1275.62	1614.42	4.87	182.93	nr	**1797.36**
Sum of two sides 3300 mm	1296.70	1641.11	4.87	182.93	nr	**1824.04**
Sum of two sides 3400 mm	1302.80	1648.83	4.87	182.93	nr	**1831.76**
Sum of two sides 3500 mm	1302.80	1648.83	8.81	330.94	nr	**1979.76**
Sum of two sides 3600 mm	1329.60	1682.74	8.81	330.94	nr	**2013.68**
Sum of two sides 3700 mm	1375.05	1740.26	14.81	556.32	nr	**2296.58**
Sum of two sides 3800 mm	1377.13	1742.90	14.81	556.32	nr	**2299.21**
Sum of two sides 3900 mm	1382.57	1749.78	14.81	556.32	nr	**2306.10**
Sum of two sides 4000 mm	1430.22	1810.09	15.20	570.97	nr	**2381.05**
Sum of two sides 4100 mm	1434.48	1815.48	16.30	612.29	nr	**2427.77**
Sum of two sides 4200 mm	1443.47	1826.86	16.50	619.80	nr	**2446.66**
Sum of two sides 4300 mm	1508.30	1908.90	17.01	638.96	nr	**2547.86**
Sum of two sides 4400 mm	1509.86	1910.88	17.01	638.96	nr	**2549.83**
Sum of two sides 4500 mm	1518.32	1921.59	17.01	638.96	nr	**2560.54**
Branch						
Sum of two sides 3000 mm	475.79	602.17	2.88	108.18	nr	**710.35**
Sum of two sides 3100 mm	492.25	622.99	2.88	108.18	nr	**731.17**
Sum of two sides 3200 mm	507.10	641.79	2.88	108.18	nr	**749.97**
Sum of two sides 3300 mm	521.93	660.56	2.88	108.18	nr	**768.74**
Sum of two sides 3400 mm	536.72	679.27	2.88	108.18	nr	**787.46**
Sum of two sides 3500 mm	551.56	698.06	3.94	148.00	nr	**846.06**
Sum of two sides 3600 mm	566.42	716.86	3.94	148.00	nr	**864.86**
Sum of two sides 3700 mm	581.43	735.86	3.94	148.00	nr	**883.86**
Sum of two sides 3800 mm	596.29	754.66	4.83	181.43	nr	**936.09**
Sum of two sides 3900 mm	611.72	774.19	4.83	181.43	nr	**955.62**
Sum of two sides 4000 mm	626.57	792.99	4.83	181.43	nr	**974.42**
Sum of two sides 4100 mm	641.67	812.09	5.44	204.35	nr	**1016.44**
Sum of two sides 4200 mm	656.51	830.88	5.44	204.35	nr	**1035.22**
Sum of two sides 4300 mm	671.34	849.65	5.85	219.75	nr	**1069.39**
Sum of two sides 4400 mm	675.63	855.07	5.85	219.75	nr	**1074.82**
Sum of two sides 4500 mm	701.20	887.43	5.85	219.75	nr	**1107.18**
Grille neck						
Sum of two sides 3000 mm	502.07	635.41	2.88	108.18	nr	**743.60**
Sum of two sides 3100 mm	519.93	658.02	2.88	108.18	nr	**766.20**
Sum of two sides 3200 mm	536.23	678.65	2.88	108.18	nr	**786.83**
Sum of two sides 3300 mm	552.54	699.29	2.88	108.18	nr	**807.47**
Sum of two sides 3400 mm	568.85	719.93	3.94	148.00	nr	**867.93**
Sum of two sides 3500 mm	585.16	740.58	3.94	148.00	nr	**888.58**
Sum of two sides 3600 mm	611.43	773.83	3.94	148.00	nr	**921.83**
Sum of two sides 3700 mm	617.74	781.82	4.12	154.76	nr	**936.58**
Sum of two sides 3800 mm	634.05	802.46	4.12	154.76	nr	**957.22**
Sum of two sides 3900 mm	650.36	823.10	4.12	154.76	nr	**977.86**
Sum of two sides 4000 mm	666.66	843.73	5.00	187.82	nr	**1031.55**

38 VENTILATION/AIR CONDITIONING SYSTEMS

Item	Net Price £	Material £	Labour hours	Labour £	Unit	Total rate £
Sum of two sides 4100 mm	682.96	864.36	5.00	187.82	nr	**1052.18**
Sum of two sides 4200 mm	699.27	885.00	5.00	187.82	nr	**1072.82**
Sum of two sides 4300 mm	715.58	905.64	5.23	196.46	nr	**1102.10**
Sum of two sides 4400 mm	731.88	926.27	5.23	196.46	nr	**1122.73**
Sum of two sides 4500 mm	748.19	946.91	5.39	202.47	nr	**1149.38**

DUCTWORK ANCILLARIES: ACCESS DOORS
Refer to ancillaries in Ductwork Ancillaries: Access Hatches

38 VENTILATION/AIR CONDITIONING SYSTEMS

Item	Net Price £	Material £	Labour hours	Labour £	Unit	Total rate £
DUCTWORK ANCILLARIES: VOLUME CONTROL AND FIRE DAMPERS						
Volume control damper; opposed blade; galvanized steel casing; aluminium aerofoil blades; manually operated						
Rectangular						
Sum of two sides 200 mm	32.23	40.79	1.60	60.10	nr	**100.89**
Sum of two sides 300 mm	34.47	43.63	1.60	60.10	nr	**103.73**
Sum of two sides 400 mm	37.69	47.71	1.60	60.10	nr	**107.81**
Sum of two sides 500 mm	41.26	52.22	1.60	60.10	nr	**112.32**
Sum of two sides 600 mm	45.62	57.73	1.70	63.88	nr	**121.62**
Sum of two sides 700 mm	50.14	63.45	2.10	78.91	nr	**142.37**
Sum of two sides 800 mm	54.77	69.31	2.15	80.76	nr	**150.07**
Sum of two sides 900 mm	60.11	76.07	2.30	86.40	nr	**162.47**
Sum of two sides 1000 mm	65.27	82.61	2.40	90.15	nr	**172.76**
Sum of two sides 1100 mm	71.10	89.99	2.60	97.67	nr	**187.65**
Sum of two sides 1200 mm	80.42	101.77	2.80	105.18	nr	**206.95**
Sum of two sides 1300 mm	86.71	109.74	3.10	116.45	nr	**226.19**
Sum of two sides 1400 mm	93.49	118.32	3.25	122.08	nr	**240.40**
Sum of two sides 1500 mm	101.20	128.08	3.40	127.72	nr	**255.80**
Sum of two sides 1600 mm	108.80	137.69	3.45	129.59	nr	**267.29**
Sum of two sides 1700 mm	116.05	146.87	3.60	135.23	nr	**282.10**
Sum of two sides 1800 mm	124.73	157.86	3.90	146.50	nr	**304.36**
Sum of two sides 1900 mm	132.81	168.08	4.20	157.83	nr	**325.91**
Sum of two sides 2000 mm	142.46	180.30	4.33	162.65	nr	**342.95**
Circular						
100 mm dia.	45.82	58.00	0.80	30.05	nr	**88.05**
150 mm dia.	52.34	66.24	0.80	30.05	nr	**96.29**
160 mm dia.	54.56	69.06	0.90	33.81	nr	**102.86**
200 mm dia.	59.20	74.93	1.05	39.46	nr	**114.39**
250 mm dia.	65.91	83.41	1.20	45.10	nr	**128.51**
300 mm dia.	72.96	92.34	1.35	50.71	nr	**143.05**
315 mm dia.	76.20	96.44	1.35	50.76	nr	**147.21**
350 mm dia.	80.16	101.45	1.65	61.99	nr	**163.44**
400 mm dia.	87.52	110.76	1.90	71.41	nr	**182.18**
450 mm dia.	94.54	119.65	2.10	78.91	nr	**198.57**
500 mm dia.	102.99	130.35	2.94	110.44	nr	**240.78**
550 mm dia.	111.22	140.76	2.94	110.44	nr	**251.20**
600 mm dia.	120.47	152.47	2.95	110.81	nr	**263.28**
650 mm dia.	130.09	164.64	4.55	170.91	nr	**335.55**
700 mm dia.	140.56	177.89	5.20	195.33	nr	**373.22**
750 mm dia.	151.70	191.99	5.20	195.33	nr	**387.32**
800 mm dia.	163.23	206.58	5.80	217.87	nr	**424.45**
850 mm dia.	175.23	221.77	5.80	217.87	nr	**439.64**
900 mm dia.	187.75	237.61	6.40	240.41	nr	**478.02**
950 mm dia.	200.63	253.92	6.40	240.41	nr	**494.33**
1000 mm dia.	213.66	270.41	7.00	262.95	nr	**533.36**

38 VENTILATION/AIR CONDITIONING SYSTEMS

Item	Net Price £	Material £	Labour hours	Labour £	Unit	Total rate £
Flat oval						
345 × 102 mm	67.31	85.19	1.20	45.08	nr	**130.27**
508 × 102 mm	75.82	95.96	1.60	60.10	nr	**156.07**
559 × 152 mm	85.68	108.43	1.90	71.41	nr	**179.84**
531 × 203 mm	93.59	118.45	1.90	71.41	nr	**189.86**
851 × 203 mm	98.94	125.22	4.55	170.91	nr	**296.13**
582 × 254 mm	103.89	131.48	2.10	78.91	nr	**210.39**
823 × 254 mm	113.74	143.95	4.10	154.01	nr	**297.96**
632 × 305 mm	114.47	144.87	2.95	110.81	nr	**255.68**
765 × 356 mm	128.04	162.05	4.55	170.91	nr	**332.96**
737 × 406 mm	136.40	172.63	4.55	170.91	nr	**343.54**
818 × 406 mm	139.26	176.24	5.20	195.33	nr	**371.57**
978 × 406 mm	140.79	178.19	5.50	206.60	nr	**384.79**
709 × 457 mm	141.94	179.64	4.50	169.21	nr	**348.84**
678 × 508 mm	150.14	190.02	4.55	170.91	nr	**360.94**
919 × 508 mm	168.49	213.25	6.00	225.38	nr	**438.63**
Fire damper; galvanized steel casing; stainless steel folding shutter; fusible link with manual reset; BS 476 4 hour fire-rated						
Rectangular						
Sum of two sides 200 mm	59.06	74.74	1.60	60.10	nr	**134.84**
Sum of two sides 300 mm	60.94	77.13	1.60	60.10	nr	**137.23**
Sum of two sides 400 mm	63.55	80.43	1.60	60.10	nr	**140.53**
Sum of two sides 500 mm	69.52	87.98	1.60	60.10	nr	**148.08**
Sum of two sides 600 mm	76.15	96.38	1.70	63.88	nr	**160.27**
Sum of two sides 700 mm	82.98	105.02	2.10	78.91	nr	**183.94**
Sum of two sides 800 mm	90.07	113.99	2.15	80.78	nr	**194.78**
Sum of two sides 900 mm	97.19	123.00	2.30	86.55	nr	**209.55**
Sum of two sides 1000 mm	104.45	132.20	2.40	90.30	nr	**222.49**
Sum of two sides 1100 mm	110.93	140.39	2.60	97.82	nr	**238.21**
Sum of two sides 1200 mm	118.66	150.17	2.80	105.22	nr	**255.39**
Sum of two sides 1300 mm	126.56	160.18	3.10	116.45	nr	**276.63**
Sum of two sides 1400 mm	135.09	170.97	3.25	122.08	nr	**293.05**
Sum of two sides 1500 mm	142.50	180.35	3.40	127.77	nr	**308.12**
Sum of two sides 1600 mm	150.68	190.71	3.45	129.59	nr	**320.30**
Sum of two sides 1700 mm	158.92	201.13	3.60	135.23	nr	**336.36**
Sum of two sides 1800 mm	167.58	212.09	3.90	146.50	nr	**358.59**
Sum of two sides 1900 mm	175.79	222.48	4.20	157.83	nr	**380.31**
Sum of two sides 2000 mm	183.52	232.26	4.33	162.65	nr	**394.91**
Sum of two sides 2100 mm	195.29	247.16	4.43	166.41	nr	**413.57**
Sum of two sides 2200 mm	206.61	261.49	4.55	170.91	nr	**432.40**
Circular						
100 mm dia.	71.95	91.06	0.80	30.05	nr	**121.11**
150 mm dia.	75.70	95.81	0.80	30.05	nr	**125.86**
160 mm dia.	76.15	96.38	0.90	33.81	nr	**130.19**
200 mm dia.	79.92	101.15	1.05	39.44	nr	**140.59**
250 mm dia.	89.07	112.73	1.20	45.10	nr	**157.83**
300 mm dia.	99.00	125.30	1.35	50.71	nr	**176.01**
315 mm dia.	103.31	130.74	1.35	50.76	nr	**181.51**
350 mm dia.	109.57	138.68	1.65	61.99	nr	**200.67**
400 mm dia.	121.06	153.21	1.90	71.41	nr	**224.63**

38 VENTILATION/AIR CONDITIONING SYSTEMS

Item	Net Price £	Material £	Labour hours	Labour £	Unit	Total rate £
DUCTWORK ANCILLARIES: VOLUME CONTROL AND FIRE DAMPERS – cont						
Fire damper – cont						
Circular – cont						
450 mm dia.	132.38	167.54	2.10	78.91	nr	**246.46**
500 mm dia.	144.81	183.27	2.94	110.44	nr	**293.71**
550 mm dia.	157.27	199.04	2.95	110.81	nr	**309.85**
600 mm dia.	170.46	215.73	4.55	170.91	nr	**386.65**
650 mm dia.	184.45	233.43	4.55	170.91	nr	**404.35**
700 mm dia.	198.79	251.58	5.20	195.33	nr	**446.91**
750 mm dia.	213.85	270.65	5.20	195.33	nr	**465.98**
800 mm dia.	229.26	290.15	5.80	217.87	nr	**508.02**
850 mm dia.	245.45	310.65	5.80	217.87	nr	**528.52**
900 mm dia.	262.12	331.73	6.40	240.41	nr	**572.14**
950 mm dia.	279.22	353.38	6.40	240.41	nr	**593.78**
1000 mm dia.	297.16	376.09	7.00	262.95	nr	**639.03**
Flat oval						
345 × 102 mm	91.62	115.95	1.20	45.10	nr	**161.05**
427 × 102 mm	99.49	125.91	1.35	50.76	nr	**176.67**
508 × 102 mm	102.85	130.17	1.60	60.10	nr	**190.27**
559 × 152 mm	112.12	141.90	1.90	71.41	nr	**213.31**
531 × 203 mm	117.94	149.26	1.90	71.41	nr	**220.67**
851 × 203 mm	148.20	187.56	4.55	170.91	nr	**358.47**
582 × 254 mm	156.38	197.91	2.10	78.91	nr	**276.83**
632 × 305 mm	164.09	207.68	2.95	110.81	nr	**318.49**
765 × 356 mm	188.55	238.63	4.55	170.91	nr	**409.54**
737 × 406 mm	196.69	248.94	4.55	170.91	nr	**419.85**
818 × 406 mm	208.78	264.24	5.20	195.33	nr	**459.57**
978 × 406 mm	226.98	287.27	5.50	206.60	nr	**493.87**
709 × 457 mm	198.19	250.82	4.50	169.21	nr	**420.03**
678 × 508 mm	206.18	260.94	4.55	170.91	nr	**431.85**
Smoke/fire damper; galvanized steel casing; stainless steel folding shutter; fusible link and 24 V DC electromagnetic shutter release mechanism; spring operated; BS 476 4 hour fire-rating						
Rectangular						
Sum of two sides 200 mm	452.35	572.50	1.60	60.10	nr	**632.60**
Sum of two sides 300 mm	453.65	574.14	1.60	60.10	nr	**634.24**
Sum of two sides 400 mm	454.96	575.80	1.60	60.10	nr	**635.90**
Sum of two sides 500 mm	465.86	589.59	1.60	60.10	nr	**649.69**
Sum of two sides 600 mm	477.26	604.03	1.70	63.88	nr	**667.91**
Sum of two sides 700 mm	489.17	619.10	2.10	78.91	nr	**698.01**
Sum of two sides 800 mm	501.37	634.53	2.15	80.78	nr	**715.31**
Sum of two sides 900 mm	514.10	650.64	2.30	86.55	nr	**737.19**
Sum of two sides 1000 mm	527.26	667.31	2.40	90.30	nr	**757.60**
Sum of two sides 1100 mm	540.79	684.42	2.60	97.82	nr	**782.24**
Sum of two sides 1200 mm	554.78	702.13	2.80	105.22	nr	**807.35**

38 VENTILATION/AIR CONDITIONING SYSTEMS

Item	Net Price £	Material £	Labour hours	Labour £	Unit	Total rate £
Sum of two sides 1300 mm	569.29	720.49	3.10	116.45	nr	**836.94**
Sum of two sides 1400 mm	584.09	739.23	3.25	122.08	nr	**861.31**
Sum of two sides 1500 mm	599.38	758.57	3.40	127.77	nr	**886.34**
Sum of two sides 1600 mm	615.18	778.57	3.45	129.59	nr	**908.16**
Sum of two sides 1700 mm	631.31	798.98	3.60	135.23	nr	**934.21**
Sum of two sides 1800 mm	647.92	820.00	3.90	146.50	nr	**966.50**
Sum of two sides 1900 mm	664.82	841.40	4.20	157.83	nr	**999.23**
Sum of two sides 2000 mm	682.24	863.44	4.33	162.65	nr	**1026.09**
Circular						
100 mm dia.	230.85	292.17	0.80	30.05	nr	**322.22**
150 mm dia.	230.85	292.17	0.90	33.81	nr	**325.97**
200 mm dia.	230.85	292.17	1.05	39.44	nr	**331.61**
250 mm dia.	243.40	308.05	1.20	45.10	nr	**353.14**
300 mm dia.	256.45	324.56	1.35	50.76	nr	**375.32**
350 mm dia.	291.86	369.37	1.65	61.99	nr	**431.36**
400 mm dia.	320.96	406.21	1.90	71.41	nr	**477.63**
450 mm dia.	334.70	423.59	2.10	78.91	nr	**502.51**
500 mm dia.	364.51	461.32	2.95	110.81	nr	**572.13**
550 mm dia.	395.30	500.29	2.95	110.81	nr	**611.10**
600 mm dia.	476.19	602.67	4.55	170.91	nr	**773.58**
650 mm dia.	515.25	652.10	4.55	170.91	nr	**823.02**
700 mm dia.	557.75	705.89	5.20	195.33	nr	**901.22**
750 mm dia.	584.15	739.30	5.20	195.33	nr	**934.63**
800 mm dia.	634.97	803.61	5.80	217.87	nr	**1021.48**
850 mm dia.	680.73	861.54	5.80	217.87	nr	**1079.41**
900 mm dia.	706.36	893.97	6.40	240.41	nr	**1134.38**
950 mm dia.	753.82	954.04	6.40	240.41	nr	**1194.45**
1000 mm dia.	787.82	997.07	7.00	262.95	nr	**1260.01**
Flat oval						
531 × 203 mm	362.46	458.73	1.90	71.41	nr	**530.14**
851 × 203 mm	404.63	512.10	4.55	170.91	nr	**683.02**
582 × 254 mm	394.56	499.36	2.10	78.91	nr	**578.27**
632 × 305 mm	420.62	532.33	2.95	110.81	nr	**643.14**
765 × 356 mm	453.77	574.29	4.55	170.91	nr	**745.21**
737 × 406 mm	474.25	600.21	4.55	170.91	nr	**771.12**
818 × 406 mm	480.90	608.62	5.20	195.33	nr	**803.96**
978 × 406 mm	508.00	642.92	5.50	206.60	nr	**849.52**
709 × 457 mm	487.87	617.44	4.50	169.21	nr	**786.65**
678 × 508 mm	507.87	642.76	4.55	170.91	nr	**813.68**

38 VENTILATION/AIR CONDITIONING SYSTEMS

Item	Net Price £	Material £	Labour hours	Labour £	Unit	Total rate £
DUCTWORK ANCILLARIES: ACCESS HATCHES						
Access doors, hollow steel construction; 25 mm mineral wool insulation; removeable or hinged; fixed with cams; including subframe and integral sealing gaskets						
Rectangular duct						
150 × 150 mm	23.31	29.51	1.25	46.95	nr	**76.46**
200 × 200 mm	25.65	32.47	1.25	46.95	nr	**79.42**
300 × 150 mm	26.37	33.38	1.25	46.95	nr	**80.33**
300 × 300 mm	29.67	37.55	1.25	46.95	nr	**84.50**
400 × 400 mm	33.85	42.84	1.35	50.76	nr	**93.60**
450 × 300 mm	33.85	42.84	1.50	56.40	nr	**99.24**
450 × 450 mm	37.29	47.20	1.50	56.40	nr	**103.60**
Access doors, hollow steel construction; 25 mm mineral wool insulation; removeable or hinged; fixed with cams; including subframe and integral sealing gaskets						
Flat oval duct						
235 × 90 mm	44.61	56.46	1.25	46.95	nr	**103.42**
235 × 140 mm	47.54	60.17	1.35	50.76	nr	**110.93**
335 × 235 mm	54.32	68.75	1.50	56.40	nr	**125.15**
535 × 235 mm	61.11	77.34	1.50	56.40	nr	**133.74**

38 VENTILATION/AIR CONDITIONING SYSTEMS

Item	Net Price £	Material £	Labour hours	Labour £	Unit	Total rate £
GRILLES/DIFFUSERS/LOUVRES						
Supply grilles; single deflection; extruded aluminium alloy frame and adjustable horizontal vanes; silver grey polyester powder coated; screw fixed						
Rectangular; for duct, ceiling and sidewall applications						
150 × 100 mm	18.59	23.53	0.60	22.54	nr	**46.07**
150 × 150 mm	22.73	28.77	0.60	22.54	nr	**51.30**
200 × 150 mm	23.27	29.45	0.65	24.42	nr	**53.86**
200 × 200 mm	26.45	33.48	0.72	27.05	nr	**60.53**
300 × 100 mm	20.05	25.38	0.72	27.05	nr	**52.42**
300 × 150 mm	23.38	29.59	0.80	30.05	nr	**59.64**
300 × 200 mm	27.02	34.20	0.88	33.06	nr	**67.25**
300 × 300 mm	32.81	41.53	1.04	39.07	nr	**80.59**
400 × 100 mm	22.47	28.44	0.88	33.06	nr	**61.50**
400 × 150 mm	25.97	32.87	0.94	35.31	nr	**68.18**
400 × 200 mm	29.51	37.35	1.04	39.07	nr	**76.42**
400 × 300 mm	36.59	46.31	1.12	42.07	nr	**88.38**
600 × 200 mm	45.02	56.98	1.26	47.33	nr	**104.31**
600 × 300 mm	72.86	92.21	1.40	52.59	nr	**144.80**
600 × 400 mm	86.77	109.81	1.61	60.48	nr	**170.29**
600 × 500 mm	100.73	127.48	1.76	66.11	nr	**193.60**
600 × 600 mm	114.67	145.12	2.17	81.51	nr	**226.63**
800 × 300 mm	85.27	107.91	1.76	66.11	nr	**174.02**
800 × 400 mm	101.51	128.47	2.17	81.51	nr	**209.98**
800 × 600 mm	133.97	169.55	3.00	112.69	nr	**282.24**
1000 × 300 mm	97.71	123.67	2.60	97.67	nr	**221.33**
1000 × 400 mm	116.24	147.11	3.00	112.69	nr	**259.80**
1000 × 600 mm	153.30	194.01	3.80	142.74	nr	**336.76**
1000 × 800 mm	190.34	240.90	3.80	142.74	nr	**383.64**
1200 × 600 mm	172.59	218.42	4.61	173.17	nr	**391.59**
1200 × 800 mm	214.23	271.14	4.61	173.17	nr	**444.30**
1200 × 1000 mm	255.87	323.83	4.61	173.17	nr	**497.00**
Rectangular; for duct, ceiling and sidewall applications; including opposed blade damper volume regulator						
150 × 100 mm	34.67	43.87	0.72	27.06	nr	**70.94**
150 × 150 mm	40.15	50.81	0.72	27.06	nr	**77.88**
200 × 150 mm	43.09	54.54	0.83	31.20	nr	**85.74**
200 × 200 mm	46.59	58.97	0.90	33.81	nr	**92.78**
300 × 100 mm	49.48	62.62	0.90	33.81	nr	**96.43**
300 × 150 mm	52.74	66.74	0.98	36.81	nr	**103.56**
300 × 200 mm	56.83	71.92	1.06	39.83	nr	**111.76**
300 × 300 mm	64.96	82.21	1.20	45.10	nr	**127.31**
400 × 100 mm	64.31	81.39	1.06	39.83	nr	**121.22**
400 × 150 mm	67.90	85.94	1.13	42.44	nr	**128.38**
400 × 200 mm	68.23	86.35	1.20	45.10	nr	**131.45**
400 × 300 mm	76.84	97.25	1.34	50.35	nr	**147.60**
600 × 200 mm	93.64	118.51	1.50	56.40	nr	**174.91**
600 × 300 mm	124.39	157.43	1.66	62.40	nr	**219.83**
600 × 400 mm	141.40	178.96	1.80	67.68	nr	**246.64**
600 × 500 mm	160.15	202.69	2.00	75.13	nr	**277.82**
600 × 600 mm	177.16	224.21	2.60	97.82	nr	**322.04**

38 VENTILATION/AIR CONDITIONING SYSTEMS

Item	Net Price £	Material £	Labour hours	Labour £	Unit	Total rate £
GRILLES/DIFFUSERS/LOUVRES – cont						
Supply grilles – cont						
Rectangular – cont						
800 × 300 mm	170.59	215.90	2.00	75.13	nr	**291.02**
800 × 400 mm	191.28	242.08	2.60	97.82	nr	**339.91**
800 × 600 mm	234.98	297.39	3.61	135.61	nr	**433.00**
1000 × 300 mm	198.49	251.21	3.00	112.80	nr	**364.01**
1000 × 400 mm	222.41	281.48	3.61	135.61	nr	**417.09**
1000 × 600 mm	273.09	345.62	4.61	173.10	nr	**518.73**
1000 × 800 mm	403.87	511.14	4.61	173.10	nr	**684.24**
1200 × 600 mm	303.37	383.94	5.62	211.03	nr	**594.97**
1200 × 800 mm	445.15	563.38	6.10	229.14	nr	**792.52**
1200 × 1000 mm	506.16	640.59	6.50	244.16	nr	**884.76**
Supply grilles; double deflection; extruded aluminium alloy frame and adjustable horizontal and vertical vanes; white polyester powder coated; screw fixed						
Rectangular; for duct, ceiling and sidewall applications						
150 × 100 mm	18.59	23.53	0.88	33.06	nr	**56.59**
150 × 150 mm	24.00	30.38	0.88	33.06	nr	**63.44**
200 × 150 mm	27.02	34.20	1.08	40.57	nr	**74.77**
200 × 200 mm	30.11	38.10	1.25	46.95	nr	**85.06**
300 × 100 mm	29.33	37.12	1.25	46.95	nr	**84.07**
300 × 150 mm	33.11	41.90	1.50	56.35	nr	**98.25**
300 × 200 mm	36.90	46.71	1.75	65.74	nr	**112.44**
300 × 300 mm	44.54	56.37	2.15	80.76	nr	**137.13**
400 × 100 mm	34.70	43.91	1.75	65.74	nr	**109.65**
400 × 150 mm	39.21	49.63	1.95	73.25	nr	**122.88**
400 × 200 mm	43.74	55.36	2.15	80.76	nr	**136.13**
400 × 300 mm	52.78	66.80	2.55	95.79	nr	**162.59**
600 × 200 mm	97.52	123.42	3.01	113.07	nr	**236.48**
600 × 300 mm	117.84	149.14	3.36	126.21	nr	**275.35**
600 × 400 mm	138.19	174.89	3.80	142.74	nr	**317.63**
600 × 500 mm	158.48	200.57	4.20	157.77	nr	**358.34**
600 × 600 mm	178.80	226.29	4.51	169.41	nr	**395.70**
800 × 300 mm	145.95	184.71	4.20	157.77	nr	**342.48**
800 × 400 mm	171.20	216.67	4.51	169.41	nr	**386.08**
800 × 600 mm	221.71	280.60	5.10	191.57	nr	**472.17**
1000 × 300 mm	174.03	220.25	4.80	180.31	nr	**400.56**
1000 × 400 mm	204.22	258.46	5.10	191.57	nr	**450.04**
1000 × 600 mm	264.58	334.85	5.72	214.86	nr	**549.71**
1000 × 800 mm	324.98	411.29	6.33	237.78	nr	**649.07**
1200 × 600 mm	307.46	389.12	5.72	214.86	nr	**603.98**
1200 × 800 mm	377.71	478.04	6.33	237.78	nr	**715.81**
1200 × 1000 mm	447.94	566.91	6.33	237.78	nr	**804.69**
Rectangular; for duct, ceiling and sidewall applications; including opposed blade damper volume regulator						
150 × 100 mm	37.90	47.96	1.00	37.56	nr	**85.52**
150 × 150 mm	44.96	56.90	1.00	37.56	nr	**94.47**
200 × 150 mm	49.59	62.77	1.26	47.33	nr	**110.10**

38 VENTILATION/AIR CONDITIONING SYSTEMS

Item	Net Price £	Material £	Labour hours	Labour £	Unit	Total rate £
200 × 200 mm	53.29	67.44	1.43	53.72	nr	**121.16**
300 × 100 mm	58.76	74.36	1.43	53.72	nr	**128.08**
300 × 150 mm	62.62	79.25	1.68	63.11	nr	**142.36**
300 × 200 mm	67.28	85.15	1.93	72.50	nr	**157.65**
300 × 300 mm	76.68	97.05	2.31	86.77	nr	**183.82**
400 × 100 mm	76.54	96.87	1.93	72.50	nr	**169.37**
400 × 150 mm	81.13	102.68	2.14	80.39	nr	**183.07**
400 × 200 mm	82.42	104.31	2.31	86.77	nr	**191.08**
400 × 300 mm	93.75	118.65	2.77	104.05	nr	**222.70**
600 × 200 mm	146.98	186.02	3.25	122.08	nr	**308.10**
600 × 300 mm	170.52	215.81	3.62	135.98	nr	**351.79**
600 × 400 mm	194.10	245.65	3.99	149.88	nr	**395.53**
600 × 500 mm	219.38	277.64	4.44	166.78	nr	**444.42**
600 × 600 mm	242.93	307.45	4.94	185.56	nr	**493.01**
800 × 300 mm	232.83	294.67	4.44	166.78	nr	**461.45**
800 × 400 mm	262.70	332.47	4.94	185.56	nr	**518.03**
800 × 600 mm	324.84	411.11	5.71	214.49	nr	**625.60**
1000 × 300 mm	276.63	350.10	5.20	195.33	nr	**545.43**
1000 × 400 mm	312.40	395.38	5.71	214.49	nr	**609.86**
1000 × 600 mm	386.86	489.61	6.53	245.29	nr	**734.90**
1000 × 800 mm	542.19	686.19	6.53	245.29	nr	**931.48**
1200 × 600 mm	440.99	558.11	7.34	275.72	nr	**833.83**
1200 × 800 mm	612.67	775.39	8.80	330.56	nr	**1105.95**
1200 × 1000 mm	702.83	889.51	8.80	330.56	nr	**1220.07**
Floor grille suitable for mounting in raised access floors; heavy duty; extruded alumiinium; standard mill finish; complete with opposed blade volume control damper						
Diffuser						
600 mm × 600 mm	213.60	270.33	0.70	26.29	nr	**296.63**
Extra for nylon coated black finish	32.69	41.37		–	nr	**41.37**
Exhaust grilles; aluminium						
0° fixed blade core						
150 × 150 mm	22.63	28.64	0.60	22.62	nr	**51.27**
200 × 200 mm	26.62	33.69	0.72	27.06	nr	**60.76**
250 × 250 mm	30.99	39.21	0.80	30.05	nr	**69.27**
300 × 300 mm	35.73	45.22	1.00	37.56	nr	**82.78**
350 × 350 mm	40.84	51.69	1.20	45.08	nr	**96.77**
0° fixed blade core; including opposed blade damper volume regulator						
150 × 150 mm	42.61	53.93	0.62	23.30	nr	**77.23**
200 × 200 mm	48.70	61.64	0.72	27.06	nr	**88.70**
250 × 250 mm	55.44	70.17	0.80	30.05	nr	**100.22**
300 × 300 mm	66.33	83.95	1.00	37.56	nr	**121.51**
350 × 350 mm	74.26	93.99	1.20	45.08	nr	**139.06**
45° fixed blade core						
150 × 150 mm	22.63	28.64	0.62	23.30	nr	**51.94**
200 × 200 mm	26.62	33.69	0.72	27.06	nr	**60.76**
250 × 250 mm	30.99	39.21	0.80	30.05	nr	**69.27**

38 VENTILATION/AIR CONDITIONING SYSTEMS

Item	Net Price £	Material £	Labour hours	Labour £	Unit	Total rate £
GRILLES/DIFFUSERS/LOUVRES – cont						
Exhaust grilles – cont						
45° fixed blade core – cont						
300 × 300 mm	35.73	45.22	1.00	37.56	nr	**82.78**
350 × 350 mm	40.84	51.69	1.20	45.08	nr	**96.77**
45° fixed blade core; including opposed blade damper volume regulator						
150 × 150 mm	42.61	53.93	0.62	23.30	nr	**77.23**
200 × 200 mm	48.70	61.64	0.72	27.06	nr	**88.70**
250 × 250 mm	55.44	70.17	0.80	30.05	nr	**100.22**
300 × 300 mm	66.33	83.95	1.00	37.56	nr	**121.51**
350 × 350 mm	74.26	93.99	1.20	45.08	nr	**139.06**
Eggcrate core						
150 × 150 mm	15.18	19.21	0.62	23.30	nr	**42.51**
200 × 200 mm	18.76	23.75	0.80	30.05	nr	**53.80**
250 × 250 mm	22.86	28.93	1.00	37.56	nr	**66.49**
300 × 300 mm	29.09	36.81	1.00	37.56	nr	**74.38**
350 × 350 mm	32.50	41.13	1.20	45.08	nr	**86.21**
Eggcrate core; including opposed blade damper volume regulator						
150 × 150 mm	35.15	44.48	0.62	23.30	nr	**67.78**
200 × 200 mm	40.84	51.69	0.72	27.06	nr	**78.75**
250 × 250 mm	47.27	59.83	0.80	30.05	nr	**89.88**
300 × 300 mm	59.70	75.56	1.00	37.56	nr	**113.13**
350 × 350 mm	65.95	83.47	1.20	45.08	nr	**128.54**
Mesh/perforated plate core						
150 × 150 mm	16.70	21.13	0.62	23.29	nr	**44.42**
200 × 200 mm	20.63	26.11	0.72	27.05	nr	**53.16**
250 × 250 mm	25.15	31.83	0.80	30.05	nr	**61.88**
300 × 300 mm	32.02	40.52	1.00	37.56	nr	**78.09**
350 × 350 mm	35.79	45.29	1.20	45.08	nr	**90.37**
Mesh/perforated plate core; including opposed blade damper volume regulator						
150 × 150 mm	36.68	46.42	0.62	23.29	nr	**69.71**
200 × 200 mm	42.73	54.07	0.72	27.05	nr	**81.12**
250 × 250 mm	49.58	62.75	0.80	30.05	nr	**92.80**
300 × 300 mm	62.64	79.27	0.80	30.05	nr	**109.32**
350 × 350 mm	69.18	87.55	1.20	45.08	nr	**132.63**
Plastic air diffusion system						
Eggcrate grilles						
150 × 150 mm	6.81	8.62	0.62	24.39	nr	**33.01**
200 × 200 mm	11.92	15.09	0.72	28.33	nr	**43.43**
250 × 250 mm	13.61	17.22	0.80	31.46	nr	**48.68**
300 × 300 mm	13.61	17.22	1.00	39.33	nr	**56.55**
Single deflection grilles						
150 × 150 mm	10.72	13.57	0.62	24.39	nr	**37.96**
200 × 200 mm	13.28	16.81	0.72	28.33	nr	**45.15**
250 × 250 mm	14.61	18.50	0.80	31.46	nr	**49.96**
300 × 300 mm	18.38	23.26	1.00	39.33	nr	**62.59**

38 VENTILATION/AIR CONDITIONING SYSTEMS

Item	Net Price £	Material £	Labour hours	Labour £	Unit	Total rate £
Double deflection grilles						
150 × 150 mm	13.35	16.90	0.62	24.39	nr	**41.29**
200 × 200 mm	17.72	22.42	0.72	28.33	nr	**50.76**
250 × 250 mm	20.25	25.63	0.80	31.46	nr	**57.09**
300 × 300 mm	28.59	36.19	1.00	39.33	nr	**75.51**
Door transfer grilles						
150 × 150 mm	20.07	25.40	0.62	24.39	nr	**49.79**
200 × 200 mm	23.31	29.50	0.72	28.33	nr	**57.84**
250 × 250 mm	24.15	30.56	0.80	31.46	nr	**62.02**
300 × 300 mm	26.20	33.16	1.00	39.33	nr	**72.48**
Opposed blade dampers						
150 × 150 mm	11.87	15.03	0.62	24.39	nr	**39.42**
200 × 200 mm	13.96	17.67	0.72	28.33	nr	**46.01**
250 × 250 mm	20.94	26.50	0.80	31.46	nr	**57.96**
300 × 300 mm	22.78	28.83	1.00	39.33	nr	**68.16**
Ceiling mounted diffusers; circular aluminium multi-cone diffuser						
Circular; for ceiling mounting						
141 mm dia. neck	83.34	105.47	0.80	30.05	nr	**135.52**
197 mm dia. neck	122.00	154.41	1.10	41.32	nr	**195.73**
309 mm dia. neck	165.25	209.14	1.40	52.61	nr	**261.75**
365 mm dia. neck	194.11	245.66	1.50	56.40	nr	**302.07**
457 mm dia. neck	222.89	282.09	2.00	75.13	nr	**357.22**
Circular; for ceiling mounting; including louvre damper volume control						
141 mm dia. neck	124.56	157.64	1.00	37.56	nr	**195.20**
197 mm dia. neck	163.22	206.58	1.20	45.10	nr	**251.67**
309 mm dia. neck	215.35	272.55	1.60	60.10	nr	**332.65**
365 mm dia. neck	250.33	316.82	1.90	71.41	nr	**388.23**
457 mm dia. neck	286.26	362.29	2.40	90.30	nr	**452.59**
Ceiling mounted diffusers; rectangular aluminium multi-cone diffuser; four way flow						
Rectangular; for ceiling mounting						
150 × 150 mm neck	31.93	40.42	1.80	67.68	nr	**108.10**
300 × 150 mm neck	48.64	61.56	2.30	86.40	nr	**147.96**
300 × 300 mm neck	54.47	68.93	2.80	105.18	nr	**174.11**
450 × 150 mm neck	67.83	85.85	2.80	105.18	nr	**191.03**
450 × 300 mm neck	67.93	85.98	3.20	120.20	nr	**206.18**
450 × 450 mm neck	69.41	87.84	3.40	127.77	nr	**215.61**
600 × 150 mm neck	84.54	107.00	3.20	120.20	nr	**227.20**
600 × 300 mm neck	85.01	107.59	3.50	131.47	nr	**239.06**
600 × 600 mm neck	94.33	119.38	4.00	150.25	nr	**269.64**
Rectangular; for ceiling mounting; including opposed blade damper volume regulator						
150 × 150 mm neck	40.98	51.87	1.80	67.68	nr	**119.55**
300 × 150 mm neck	82.56	104.49	2.30	86.40	nr	**190.89**
300 × 300 mm neck	85.97	108.81	2.80	105.22	nr	**214.03**

38 VENTILATION/AIR CONDITIONING SYSTEMS

Item	Net Price £	Material £	Labour hours	Labour £	Unit	Total rate £
GRILLES/DIFFUSERS/LOUVRES – cont						
Ceiling mounted diffusers – cont						
Rectangular; for ceiling mounting; including opposed blade damper volume regulator – cont						
450 × 150 mm neck	96.57	122.22	2.80	105.18	nr	**227.40**
450 × 300 mm neck	108.66	137.52	3.30	123.96	nr	**261.48**
450 × 450 mm neck	110.69	140.08	3.51	131.80	nr	**271.89**
600 × 150 mm neck	134.72	170.50	3.30	123.96	nr	**294.46**
600 × 300 mm neck	135.72	171.77	4.00	150.25	nr	**322.03**
600 × 600 mm neck	138.87	175.75	5.62	211.03	nr	**386.78**
Slot diffusers; continuous aluminium slot diffuser with flanged frame (1500 mm sections)						
Diffuser						
1 slot	49.81	63.04	3.76	141.22	m	**204.26**
2 slot	60.63	76.73	3.76	141.18	m	**217.91**
3 slot	74.62	94.44	3.76	141.18	m	**235.62**
4 slot	83.99	106.30	4.50	169.21	m	**275.50**
6 slot	110.65	140.04	4.50	169.21	m	**309.25**
Diffuser; including equalizing deflector						
1 slot	103.10	130.48	5.26	197.70	m	**328.19**
2 slot	127.91	161.88	5.26	197.70	m	**359.59**
3 slot	156.44	198.00	5.26	197.70	m	**395.70**
4 slot	175.68	222.34	6.33	237.74	m	**460.08**
6 slot	231.09	292.47	6.33	237.74	m	**530.21**
Extra over for ends						
1 slot	5.56	7.04	1.00	37.56	nr	**44.61**
2 slot	5.86	7.42	1.00	37.56	nr	**44.98**
3 slot	6.21	7.86	1.00	37.56	nr	**45.42**
4 slot	6.52	8.25	1.30	48.83	nr	**57.08**
6 slot	7.45	9.42	1.40	52.59	nr	**62.01**
Plenum boxes; 1.0 m long; circular spigot; including cord operated flap damper						
1 slot	70.98	89.83	2.75	103.48	nr	**193.31**
2 slot	72.83	92.18	2.75	103.48	nr	**195.66**
3 slot	73.77	93.37	2.75	103.48	nr	**196.85**
4 slot	80.48	101.86	3.51	131.80	nr	**233.66**
6 slot	84.45	106.87	3.51	131.80	nr	**238.68**
Plenum boxes; 2.0 m long; circular spigot; including cord operated flap damper						
1 slot	78.59	99.46	3.26	122.36	nr	**221.81**
2 slot	82.40	104.29	3.26	122.36	nr	**226.64**
3 slot	83.57	105.77	3.26	122.36	nr	**228.12**
4 slot	93.07	117.79	3.76	141.22	nr	**259.01**
6 slot	99.24	125.60	3.76	141.22	nr	**266.82**

38 VENTILATION/AIR CONDITIONING SYSTEMS

Item	Net Price £	Material £	Labour hours	Labour £	Unit	Total rate £
Perforated diffusers; rectangular face aluminium perforated diffuser; quick release face plate; for integration with rectangular ceiling tiles						
Circular spigot; rectangular diffuser						
150 mm dia. spigot; 300 × 300 mm diffuser	72.55	91.82	1.00	37.56	nr	**129.38**
300 mm dia. spigot; 600 × 600 mm diffuser	108.26	137.02	1.40	52.61	nr	**189.63**
Circular spigot; rectangular diffuser; including louvre damper volume regulator						
150 mm dia. spigot; 300 × 300 mm diffuser	108.97	137.92	1.00	37.56	nr	**175.48**
300 mm dia. spigot; 600 × 600 mm diffuser	152.52	193.03	1.60	60.10	nr	**253.13**
Rectangular spigot; rectangular diffuser						
150 × 150 mm dia. spigot; 300 × 300 mm diffuser	72.55	91.82	1.00	39.33	nr	**131.15**
300 × 150 mm dia. spigot; 600 × 300 mm diffuser	97.66	123.59	1.20	47.19	nr	**170.79**
300 × 300 mm dia. spigot; 600 × 600 mm diffuser	108.26	137.02	1.40	55.06	nr	**192.07**
600 × 300 mm dia. spigot; 1200 × 600 mm diffuser	113.07	143.11	1.60	62.92	nr	**206.03**
Rectangular spigot; rectangular diffuser; including opposed blade damper volume regulator						
150 × 150 mm dia. spigot; 300 × 300 mm diffuser	108.97	137.92	1.20	47.19	nr	**185.11**
300 × 150 mm dia. spigot; 600 × 300 mm diffuser	145.28	183.86	1.40	55.08	nr	**238.94**
300 × 300 mm dia. spigot; 600 × 600 mm diffuser	152.52	193.03	1.60	62.92	nr	**255.95**
600 × 300 mm dia. spigot; 1200 × 600 mm diffuser	178.14	225.45	1.80	70.79	nr	**296.24**
Floor swirl diffuser; manual adjustment of air discharge direction; complete with damper and dirt trap						
Plastic diffuser						
150 mm dia.	42.36	53.61	0.50	18.78	nr	**72.39**
200 mm dia.	42.36	53.61	0.50	18.78	nr	**72.39**
Aluminium diffuser						
150 mm dia.	45.34	57.39	0.50	18.78	nr	**76.17**
200 mm dia.	66.28	83.88	0.50	18.78	nr	**102.66**
Plastic air diffusion system						
Cellular diffusers						
300 × 300 mm	18.48	23.39	2.80	110.16	nr	**133.54**
600 × 600 mm	42.78	54.14	4.00	157.30	nr	**211.45**
Multi-cone diffusers						
300 × 300 mm	14.92	18.88	2.80	110.16	nr	**129.04**
450 × 450 mm	24.54	31.06	3.40	133.76	nr	**164.82**
500 × 500 mm	25.94	32.83	3.80	149.53	nr	**182.36**
600 × 600 mm	35.90	45.43	4.00	157.30	nr	**202.74**
625 × 625 mm	61.76	78.16	4.26	167.34	nr	**245.50**
Opposed blade dampers						
300 × 300 mm	19.44	24.60	1.20	47.21	nr	**71.81**
450 × 450 mm	21.37	27.05	1.50	59.05	nr	**86.09**
600 × 600 mm	45.27	57.29	2.60	102.41	nr	**159.70**
Plenum boxes						
300 mm	12.61	15.96	2.80	110.16	nr	**126.11**
450 mm	19.57	24.77	3.40	133.76	nr	**158.53**
600 mm	31.34	39.67	4.00	157.30	nr	**196.97**

38 VENTILATION/AIR CONDITIONING SYSTEMS

Item	Net Price £	Material £	Labour hours	Labour £	Unit	Total rate £
GRILLES/DIFFUSERS/LOUVRES – cont						
Plastic air diffusion system – cont						
Plenum spigot reducer						
600 mm	24.86	31.47	1.00	39.33	nr	**70.79**
Blanking kits for cellular diffusers						
300 mm	8.80	11.14	0.88	34.62	nr	**45.76**
600 mm	10.46	13.23	1.10	43.26	nr	**56.50**
Blanking kits for multi-cone diffusers						
300 mm	4.65	5.89	0.88	34.62	nr	**40.50**
450 mm	13.92	17.62	0.90	35.40	nr	**53.02**
600 mm	16.25	20.57	1.10	43.26	nr	**63.83**
Acoustic louvres; opening mounted; 300 mm deep steel louvres with blades packed with acoustic infill; 12 mm galvanized mesh birdscreen; screw fixing in opening						
Louvre units; self-finished galvanized steel						
900 mm high × 600 mm wide	191.60	242.49	3.00	117.98	nr	**360.47**
900 mm high × 900 mm wide	237.71	300.85	3.00	117.98	nr	**418.83**
900 mm high × 1200 mm wide	281.45	356.20	3.34	131.35	nr	**487.55**
900 mm high × 1500 mm wide	367.28	464.83	3.34	131.35	nr	**596.17**
900 mm high × 1800 mm wide	411.82	521.19	3.34	131.35	nr	**652.54**
900 mm high × 2100 mm wide	456.35	577.56	3.34	131.35	nr	**708.91**
900 mm high × 2400 mm wide	500.06	632.87	3.68	144.72	nr	**777.59**
900 mm high × 2700 mm wide	565.25	715.39	3.68	144.72	nr	**860.11**
900 mm high × 3000 mm wide	604.99	765.68	3.68	144.72	nr	**910.40**
1200 mm high × 600 mm wide	251.22	317.95	3.00	117.98	nr	**435.93**
1200 mm high × 900 mm wide	309.26	391.40	3.34	131.35	nr	**522.75**
1200 mm high × 1200 mm wide	366.50	463.84	3.34	131.35	nr	**595.19**
1200 mm high × 1500 mm wide	485.75	614.77	3.34	131.35	nr	**746.12**
1200 mm high × 1800 mm wide	542.98	687.19	3.68	144.72	nr	**831.91**
1200 mm high × 2100 mm wide	601.82	761.66	3.68	144.72	nr	**906.38**
1200 mm high × 2400 mm wide	659.07	834.12	3.68	144.72	nr	**978.84**
1500 mm high × 600 mm wide	311.64	394.41	3.00	117.98	nr	**512.39**
1500 mm high × 900 mm wide	382.43	484.00	3.34	131.35	nr	**615.35**
1500 mm high × 1200 mm wide	453.15	573.51	3.34	131.35	nr	**704.86**
1500 mm high × 1500 mm wide	604.23	764.71	3.68	144.72	nr	**909.43**
1500 mm high × 1800 mm wide	674.98	854.25	3.68	144.72	nr	**998.97**
1500 mm high × 2100 mm wide	746.54	944.82	4.00	157.30	nr	**1102.12**
1800 mm high × 600 mm wide	370.46	468.86	3.34	131.35	nr	**600.21**
1800 mm high × 900 mm wide	454.74	575.52	3.34	131.35	nr	**706.87**
1800 mm high × 1200 mm wide	539.79	683.16	3.68	144.72	nr	**827.88**
1800 mm high × 1500 mm wide	723.47	915.62	3.68	144.72	nr	**1060.34**
Louvre units; polyester powder coated steel						
900 mm high × 600 mm wide	278.27	352.18	3.00	117.98	nr	**470.16**
900 mm high × 900 mm wide	366.50	463.84	3.00	117.98	nr	**581.82**
900 mm high × 1200 mm wide	453.15	573.51	3.34	131.35	nr	**704.86**
900 mm high × 1500 mm wide	581.95	736.52	3.34	131.35	nr	**867.87**
900 mm high × 1800 mm wide	669.38	847.17	3.34	131.35	nr	**978.52**
900 mm high × 2100 mm wide	757.65	958.89	3.34	131.35	nr	**1090.23**
900 mm high × 2400 mm wide	844.28	1068.52	3.68	144.72	nr	**1213.24**

38 VENTILATION/AIR CONDITIONING SYSTEMS

Item	Net Price £	Material £	Labour hours	Labour £	Unit	Total rate £
900 mm high × 2700 mm wide	951.62	1204.37	3.68	144.72	nr	**1349.09**
900 mm high × 3000 mm wide	1034.30	1309.01	3.68	144.72	nr	**1453.73**
1200 mm high × 600 mm wide	365.69	462.82	3.00	117.98	nr	**580.80**
1200 mm high × 900 mm wide	480.97	608.71	3.34	131.35	nr	**740.06**
1200 mm high × 1200 mm wide	596.25	754.62	3.34	131.35	nr	**885.97**
1200 mm high × 1500 mm wide	771.94	976.97	3.34	131.35	nr	**1108.32**
1200 mm high × 1800 mm wide	886.42	1121.86	3.68	144.72	nr	**1266.58**
1200 mm high × 2100 mm wide	1002.52	1268.79	3.68	144.72	nr	**1413.51**
1200 mm high × 2400 mm wide	1116.97	1413.64	3.68	144.72	nr	**1558.36**
1500 mm high × 600 mm wide	454.74	575.52	3.00	117.98	nr	**693.49**
1500 mm high × 900 mm wide	597.03	755.61	3.34	131.35	nr	**886.95**
1500 mm high × 1200 mm wide	739.36	935.73	3.34	131.35	nr	**1067.08**
1500 mm high × 1500 mm wide	961.96	1217.46	3.68	144.72	nr	**1362.18**
1500 mm high × 1800 mm wide	1105.06	1398.57	3.68	144.72	nr	**1543.28**
1500 mm high × 2100 mm wide	1247.38	1578.69	4.00	157.30	nr	**1735.99**
1800 mm high × 600 mm wide	542.16	686.15	3.34	131.35	nr	**817.50**
1800 mm high × 900 mm wide	712.32	901.51	3.34	131.35	nr	**1032.86**
1800 mm high × 1200 mm wide	883.26	1117.86	3.68	144.72	nr	**1262.58**
1800 mm high × 1500 mm wide	1152.77	1458.95	3.68	144.72	nr	**1603.67**
Weather louvres; opening mounted; 300 mm deep galvanized steel louvres; screw fixing in position						
Louvre units; including 12 mm galvanized mesh birdscreen						
900 × 600 mm	175.16	221.69	2.25	88.57	nr	**310.26**
900 × 900 mm	253.09	320.32	2.25	88.57	nr	**408.89**
900 × 1200 mm	304.53	385.41	2.50	98.31	nr	**483.73**
900 × 1500 mm	370.61	469.04	2.50	98.31	nr	**567.35**
900 × 1800 mm	436.70	552.68	2.50	98.31	nr	**651.00**
900 × 2100 mm	547.40	692.79	2.50	98.31	nr	**791.10**
900 × 2400 mm	588.03	744.21	2.76	108.63	nr	**852.84**
900 × 2700 mm	663.88	840.20	2.76	108.63	nr	**948.83**
900 × 3000 mm	725.01	917.57	2.76	108.63	nr	**1026.21**
1200 × 600 mm	237.25	300.26	2.25	88.57	nr	**388.83**
1200 × 900 mm	331.49	419.54	2.50	98.31	nr	**517.85**
1200 × 1200 mm	426.49	539.76	2.50	98.31	nr	**638.08**
1200 × 1500 mm	497.82	630.04	2.50	98.31	nr	**728.35**
1200 × 1800 mm	651.05	823.96	2.76	108.63	nr	**932.60**
1200 × 2100 mm	740.42	937.07	2.76	108.63	nr	**1045.71**
1200 × 2400 mm	773.23	978.60	2.76	108.63	nr	**1087.23**
1500 × 600 mm	280.40	354.87	2.25	88.57	nr	**443.44**
1500 × 900 mm	386.40	489.02	2.50	98.31	nr	**587.34**
1500 × 1200 mm	508.95	644.13	2.50	98.31	nr	**742.44**
1500 × 1500 mm	596.71	755.19	2.76	108.63	nr	**863.83**
1500 × 1800 mm	703.60	890.47	2.76	108.63	nr	**999.11**
1500 × 2100 mm	879.98	1113.70	3.00	118.10	nr	**1231.80**
1800 × 600 mm	313.65	396.95	2.50	98.31	nr	**495.27**
1800 × 900 mm	433.60	548.76	2.50	98.31	nr	**647.07**
1800 × 1200 mm	568.55	719.56	2.76	108.63	nr	**828.20**
1800 × 1500 mm	672.54	851.17	3.00	118.10	nr	**969.26**

38 VENTILATION/AIR CONDITIONING SYSTEMS

Item	Net Price £	Material £	Labour hours	Labour £	Unit	Total rate £
PLANT/EQUIPMENT: FANS						
Axial flow fan; including ancillaries, anti-vibration mountings, mounting feet, matching flanges, flexible connectors and clips; 415 V, 3 phase, 50 Hz motor; includes fixing in position; electrical work elsewhere						
Aerofoil blade fan unit; short duct case						
315 mm dia.; 0.47 m³/s duty; 147 Pa	1128.22	1427.87	4.50	176.97	nr	1604.84
500 mm dia.; 1.89 m³/s duty; 500 Pa	1871.16	2368.13	5.00	196.63	nr	2564.76
560 mm dia.; 2.36 m³/s duty; 147 Pa	1790.36	2265.88	5.50	216.29	nr	2482.18
710 mm dia.; 5.67 m³/s duty; 245 Pa	2403.88	3042.35	6.00	235.95	nr	3278.31
Aerofoil blade fan unit; long duct case						
315 mm dia.; 0.47 m³/s duty; 147 Pa	1128.22	1427.87	4.50	176.97	nr	1604.84
500 mm dia.; 1.89 m³/s duty; 500 Pa	1871.16	2368.13	5.00	196.63	nr	2564.76
560 mm dia.; 2.36 m³/s duty; 147 Pa	1790.36	2265.88	5.50	216.29	nr	2482.18
710 mm dia.; 5.67 m³/s duty; 245 Pa	2403.88	3042.35	6.00	235.95	nr	3278.31
Aerofoil blade fan unit; two stage parallel fan arrangement; long duct case						
315 mm; 0.47 m³/s @ 500 Pa	1128.22	1427.87	4.50	176.97	nr	1604.84
355 mm; 0.83 m³/s @ 147 Pa	1871.16	2368.13	4.75	186.80	nr	2554.93
710 mm; 3.77 m³/s @ 431 Pa	1790.36	2265.88	6.00	235.95	nr	2501.84
710 mm; 6.61 m³/s @ 500 Pa	2403.88	3042.35	6.00	235.95	nr	3278.31
Axial flow fan; suitable for operation at 300°C for 90 minutes; including ancillaries, anti-vibration mountings, mounting feet, matching flanges, flexible connectors and clips; 415 V, 3 phase, 50 Hz motor; includes fixing in position; price includes air operated damper; electrical work elsewhere						
450 mm; 2.0 m³/s @ 300 Pa	2410.16	3050.30	5.00	196.63	nr	3246.93
630 mm; 4.6 m³/s @ 200 Pa	3389.80	4290.13	5.50	216.29	nr	4506.42
900 mm; 9.0 m³/s @ 300 Pa	5395.42	6828.45	6.50	255.62	nr	7084.07
1000 mm; 15.0 m³/s @ 400 Pa	8315.85	10524.54	7.50	294.94	nr	10819.48
Bifurcated fan; suitable for temperature up to 200°C with motor protection to IP55; including ancillaries, anti-vibration mountings, mounting feet, matching flanges, flexible connectors and clips; 415 V, 3 phase, 50 Hz motor; includes fixing in position; electrical work elsewhere						
300 mm; 0.50 m³/s @ 100 Pa	2198.53	2782.46	4.50	176.97	nr	2959.43
400 mm; 1.97 m³/s @ 200 Pa	2735.62	3462.21	5.00	196.63	nr	3658.83
800 mm; 3.86 m³/s @ 200 Pa	4382.02	5545.89	5.50	216.29	nr	5762.18
1000 mm; 6.10 m³/s @ 400 Pa	6168.74	7807.15	6.50	255.62	nr	8062.77

38 VENTILATION/AIR CONDITIONING SYSTEMS

Item	Net Price £	Material £	Labour hours	Labour £	Unit	Total rate £
Duct mounted in line fan with backward curved centrifugal impellor; including ancillaries, matching flanges, flexible connectors and clips; 415 V, 3 phase, 50 Hz motor; includes fixing in position; electrical work elsewhere (NB: inverter included)						
0.5 m³/s @ 200 Pa	4526.32	5728.51	4.50	176.97	nr	**5905.47**
1.0 m³/s @ 300 Pa	4526.32	5728.51	5.00	196.63	nr	**5925.14**
3.0 m³/s @ 500 Pa	4364.02	5523.10	5.50	216.29	nr	**5739.39**
5.0 m³/s @ 750 Pa	8376.40	10601.17	6.50	255.62	nr	**10856.78**
7.0 m³/s @ 1000 Pa	10530.43	13327.31	7.00	275.28	nr	**13602.59**
Twin fan extract unit; belt driven; located internally; complete with anti-vibration mounts and non-return shutter; including ancillaries, matching flanges, flexible connectors and clips; 3 phase, 50 Hz motor; includes fixing in position; electrical work elsewhere (NB: inverter and auto change over included)						
0.25 m³/s @ 150 Pa	6837.85	8653.98	4.50	176.97	nr	**8830.95**
1.00 m³/s @ 200 Pa	6837.85	8653.98	5.00	196.63	nr	**8850.61**
2.00 m³/s @ 250 Pa	7149.16	9047.98	6.50	255.62	nr	**9303.60**
Roof mounted extract fan; including ancillaries, fibreglass cowling, fitted shutters and bird guard; 415 V, 3 phase, 50 Hz motor; includes fixing in position; electrical work elsewhere (NB: inverter included)						
Flat roof installation, fixed to curb						
355 mm	2681.31	3393.47	4.50	176.97	nr	**3570.44**
400 mm	2937.68	3717.93	5.50	216.29	nr	**3934.23**
450 mm	2937.68	3717.93	7.00	275.28	nr	**3993.21**
500 mm	3464.67	4384.89	8.00	314.61	nr	**4699.50**
Centrifugal fan; single speed for internal domestic kitchens/utility rooms; fitted with standard overload protection; complete with housing; includes placing in position; electrical work elsewhere						
Window mounted						
245 m³/hr	417.79	528.76	0.50	19.66	nr	**548.42**
500 m³/hr	808.15	1022.80	0.50	19.66	nr	**1042.46**
Wall mounted						
245 m³/hr	528.32	668.64	0.83	32.77	nr	**701.41**
500 m³/hr	1023.29	1295.08	0.83	32.77	nr	**1327.85**

38 VENTILATION/AIR CONDITIONING SYSTEMS

Item	Net Price £	Material £	Labour hours	Labour £	Unit	Total rate £
PLANT/EQUIPMENT: FANS – cont						
Centrifugal fan; various speeds, simultaneous ventilation from separate areas fitted with standard overload protection; complete with housing; includes placing in position; ducting and electrical work elsewhere						
Fan unit						
147–300 m³/hr	833.57	1054.96	1.00	39.33	nr	**1094.29**
175–411 m³/hr	833.57	1054.96	1.00	39.33	nr	**1094.29**
Toilet extract units; centrifugal fan; various speeds for internal domestic bathrooms/WCs, complete with housing; includes placing in position; electrical work elsewhere						
Fan unit; fixed to wall; including shutter						
Single speed 85 m³/hr	147.71	186.95	0.75	29.49	nr	**216.44**
Two speed 60–85 m³/hr	194.15	245.72	0.83	32.77	nr	**278.49**
Humidity controlled; autospeed; fixed to wall; including shutter						
30–60–85 m³/hr	343.27	434.45	1.00	39.33	nr	**473.77**

38 VENTILATION/AIR CONDITIONING SYSTEMS

Item	Net Price £	Material £	Labour hours	Labour £	Unit	Total rate £
PLANT/EQUIPMENT: AIR FILTRATION						
High efficiency duct mounted filters; 99.997% H13 (EU13); tested to BS 3928						
Standard; 1700 m³/hr air volume; continuous rating up to 80°C; sealed wood case, aluminium spacers, neoprene gaskets; water repellant filter media; includes placing in position						
610 × 610 × 292 mm	404.58	512.04	1.00	37.56	nr	**549.60**
Side withdrawal frame	153.27	193.98	2.50	93.91	nr	**287.89**
High capacity; 3400 m³/hr air volume; continuous rating up to 80°C; anti-corrosion coated mild steel frame, polyurethane sealant and neoprene gaskets; water repellant filter media; includes placing in position						
610 × 610 × 292 mm	439.90	556.74	1.00	37.56	nr	**594.30**
Side withdrawal frame	153.27	193.98	2.50	93.91	nr	**287.89**
Bag filters; 40/60% F5 (EU5); tested to BSEN 779						
Duct mounted bag filter; continuous rating up to 60°C; rigid filter assembly; sealed into one piece coated mild steel header with sealed pocket separators; includes placing in position						
6 pocket, 592 × 592 × 25 mm header; pockets 380 mm long; 1690 m³/hr	28.58	36.18	1.00	37.56	nr	**73.74**
Side withdrawal frame	26.53	33.58	2.00	75.13	nr	**108.70**
6 pocket, 592 × 592 × 25 mm header; pockets 500 mm long; 2550 m³/hr	29.04	36.76	1.50	56.35	nr	**93.10**
Side withdrawal frame	26.53	33.58	2.50	93.91	nr	**127.49**
6 pocket, 592 × 592 × 25 mm header; pockets 600 mm long; 3380 m³/hr	34.27	43.37	1.50	56.35	nr	**99.72**
Side withdrawal frame	26.53	33.58	3.00	112.69	nr	**146.27**
Bag filters; 80/90% F7, (EU7); tested to BSEN 779						
Duct mounted bag filter; continuous rating up to 60°C; rigid filter assembly; sealed into one piece coated mild steel header with sealed pocket separators; includes placing in position						
6 pocket, 592 × 592 × 25 mm header; pockets 500 mm long; 1688 m³/hr	29.04	36.76	1.00	37.56	nr	**74.32**
Side withdrawal frame	26.53	33.58	2.00	75.13	nr	**108.70**
6 pocket, 592 × 592 × 25 mm header; pockets 635 mm long; 2047 m³/hr	34.27	43.37	1.50	56.35	nr	**99.72**
Side withdrawal frame	26.53	33.58	2.50	93.91	nr	**127.49**
6 pocket, 592 × 592 × 25 mm header; pockets 762 mm long; 2729 m³/hr	39.51	50.00	1.50	56.35	nr	**106.35**
Side withdrawal frame	26.53	33.58	3.00	112.69	nr	**146.27**

38 VENTILATION/AIR CONDITIONING SYSTEMS

Item	Net Price £	Material £	Labour hours	Labour £	Unit	Total rate £
PLANT/EQUIPMENT: AIR FILTRATION – cont						
Grease filters, washable; minimum 65%						
Double sided extract unit; lightweight stainless steel construction; demountable composite filter media of woven metal mat and expanded metal mesh supports; for mounting on hood and extract systems (hood not included); includes placing in position						
500 × 686 × 565 mm, 4080 m³/hr	596.30	754.68	2.00	75.13	nr	**829.80**
1000 × 686 × 565 mm, 8160 m³/hr	909.33	1150.85	3.00	112.69	nr	**1263.54**
1500 × 686 × 565 mm, 12240 m³/hr	1247.28	1578.55	3.50	131.47	nr	**1710.03**
Panel filters; 82% G3 (EU3); tested to BS EN779						
Modular duct mounted filter panels; continuous rating up to 100°C; graduated density media; rigid cardboard frame; includes placing in position						
596 × 596 × 47 mm, 2360 m³/hr	7.22	9.14	1.00	37.56	nr	**46.70**
Side withdrawal frame	51.80	65.56	2.50	93.91	nr	**159.47**
596 × 287 × 47 mm, 1140 m³/hr	4.99	6.31	1.00	37.56	nr	**43.88**
Side withdrawal frame	51.80	65.56	2.50	93.91	nr	**159.47**
Panel filters; 90% G4 (EU4); tested to BS EN779						
Modular duct mounted filter panels; continuous rating up to 100°C; pleated media with wire support; rigid cardboard frame; includes placing in position						
596 × 596 × 47 mm, 2560 m³/hr	10.92	13.83	1.00	37.56	nr	**51.39**
side withdrawal frame	65.38	82.75	3.00	112.69	nr	**195.44**
596 × 287 × 47 mm, 1230 m³/hr	7.54	9.55	1.00	37.56	nr	**47.11**
Side withdrawal frame	51.80	65.56	3.00	112.69	nr	**178.25**
Carbon filters; standard duty disposable carbon filters; steel frame with bonded carbon panels; for fixing to ductwork; including placing in position						
12 panels						
597 × 597 × 298 mm, 1460 m³/hr	508.20	643.18	0.33	12.40	nr	**655.58**
597 × 597 × 451 mm, 2200 m³/hr	572.90	725.06	0.33	12.40	nr	**737.46**
597 × 597 × 597 mm, 2930 m³/hr	638.89	808.58	0.33	12.40	nr	**820.98**
8 panels						
451 × 451 × 298 mm, 740 m³/hr	383.12	484.88	0.29	10.89	nr	**495.78**
451 × 451 × 451 mm, 1105 m³/hr	424.04	536.67	0.29	10.89	nr	**547.56**
451 × 451 × 597 mm, 1460 m³/hr	463.08	586.07	0.29	10.89	nr	**596.96**
6 panels						
298 × 298 × 298 mm, 365 m³/hr	271.49	343.59	0.25	9.39	nr	**352.98**
298 × 298 × 451 mm, 550 m³/hr	290.67	367.87	0.25	9.39	nr	**377.26**
298 × 298 × 597 mm, 780 m³/hr	309.03	391.10	0.25	9.39	nr	**400.49**

38 VENTILATION/AIR CONDITIONING SYSTEMS

Item	Net Price £	Material £	Labour hours	Labour £	Unit	Total rate £
PLANT/EQUIPMENT: AIR CURTAINS						
The selection of air curtains requires consideration of the particular conditions involved; climatic conditions, wind influence, construction and position all influence selection; consultation with a specialist manufacturer is therefore advisable						
Commercial grade air curtains; recessed or exposed units with rigid sheet steel casing; aluminium grilles; high quality motor/centrifugal fan assembly; includes fixing in position; electrical work elsewhere						
Ambient temperature; 240 V single phase supply; mounting height 2.40 m						
1000 × 590 × 270 mm	3931.10	4975.20	12.05	473.81	nr	**5449.01**
1500 × 590 × 270 mm	5031.54	6367.92	12.05	473.81	nr	**6841.73**
2000 × 590 × 270 mm	6087.02	7703.73	12.05	473.88	nr	**8177.61**
2500 × 590 × 270 mm	6766.57	8563.77	13.00	511.24	nr	**9075.01**
Ambient temperature; 240 V single phase supply; mounting height 2.80 m						
1000 × 590 × 270 mm	4541.56	5747.80	16.13	634.29	nr	**6382.08**
1500 × 590 × 270 mm	5797.84	7337.74	16.13	634.29	nr	**7972.03**
2000 × 590 × 270 mm	7227.63	9147.29	16.13	634.29	nr	**9781.58**
2500 × 590 × 270 mm	8357.00	10576.61	17.10	672.47	nr	**11249.09**
Ambient temperature 240 V single phase supply; mounting height 3.30 m						
1000 × 774 × 370 mm	5932.79	7508.54	17.24	678.03	nr	**8186.57**
1500 × 774 × 370 mm	8022.85	10153.72	17.24	678.03	nr	**10831.75**
2000 × 774 × 370 mm	9923.33	12558.97	17.24	678.03	nr	**13237.00**
2500 × 774 × 370 mm	12543.54	15875.10	18.30	719.66	nr	**16594.76**
Ambient temperature; 240 V single phase supply; mounting height 4.00 m						
1000 × 774 × 370 mm	6552.91	8293.36	19.10	751.12	nr	**9044.49**
1500 × 774 × 370 mm	8596.36	10879.56	19.10	751.12	nr	**11630.68**
2000 × 774 × 370 mm	10702.49	13545.07	19.10	751.12	nr	**14296.20**
2500 × 774 × 370 mm	12543.54	15875.10	19.90	782.58	nr	**16657.68**
Water heated; 240 V single phase supply; mounting height 2.40 m						
1000 × 590 × 270 mm; 2.30–9.40 kW output	4625.11	5853.54	12.05	473.81	nr	**6327.35**
1500 × 590 × 270 mm; 3.50–14.20 kW output	5919.95	7492.29	12.05	473.81	nr	**7966.10**
2000 × 590 × 270 mm; 4.70–19.00 kW output	7161.76	9063.93	12.05	473.81	nr	**9537.73**
2500 × 590 × 270 mm; 5.90–23.70 kW output	7961.80	10076.46	13.00	511.24	nr	**10587.69**
Water heated; 240 V single phase supply; mounting height 2.80 m						
1000 × 590 × 270 mm; 3.30–11.90 kW output	5344.81	6764.39	16.13	634.29	nr	**7398.68**
1500 × 590 × 270 mm; 5.00–17.90 kW output	6822.80	8634.94	16.13	634.29	nr	**9269.23**
2000 × 590 × 270 mm; 6.70–23.90 kW output	8504.80	10763.68	16.13	634.29	nr	**11397.96**
2500 × 590 × 270 mm; 8.30–29.80 kW output	9833.37	12445.11	17.10	672.47	nr	**13117.58**

38 VENTILATION/AIR CONDITIONING SYSTEMS

Item	Net Price £	Material £	Labour hours	Labour £	Unit	Total rate £
PLANT/EQUIPMENT: AIR CURTAINS – cont						
Commercial grade air curtains – cont						
Water heated; 240 V single phase supply; mounting height 3.30 m						
1000 × 774 × 370 mm; 6.10–21.80 kW output	6980.23	8834.18	17.24	678.03	nr	**9512.21**
1500 × 774 × 370 mm; 9.20–32.80 kW output	9439.78	11946.98	17.24	678.03	nr	**12625.02**
2000 × 774 × 370 mm; 12.30–43.70 kW output	11676.03	14777.18	17.24	678.03	nr	**15455.21**
2500 × 774 × 370 mm; 15.30–54.60 kW output	13820.69	17491.47	18.30	719.66	nr	**18211.13**
Water heated; 240 V single phase supply; mounting height 4.00 m						
1000 × 774 × 370 mm; 7.20–24.20 kW output	7709.58	9757.25	19.10	751.12	nr	**10508.37**
1500 × 774 × 370 mm; 10.90–36.30 kW output	10114.50	12800.91	19.10	751.12	nr	**13552.03**
2000 × 774 × 370 mm; 14.50–48.40 kW output	12591.73	15936.10	19.10	751.12	nr	**16687.22**
2500 × 774 × 370 mm; 18.10–60.60 kW output	14757.30	18676.83	19.90	782.58	nr	**19459.42**
Electrically heated; 415 V three phase supply; mounting height 2.40 m						
1000 × 590 × 270 mm; 2.30–9.40 kW output	5625.96	7120.21	12.05	473.81	nr	**7594.02**
1500 × 590 × 270 mm; 3.50–14.20 kW output	6999.50	8858.57	12.05	473.81	nr	**9332.38**
2000 × 590 × 270 mm; 4.70–19.00 kW output	8342.54	10558.32	12.05	550.65	nr	**11108.97**
2500 × 590 × 270 mm; 5.90–23.70 kW output	9484.77	12003.93	13.00	594.15	nr	**12598.07**
Electrically heated; 415 V three phase supply; mounting height 2.80 m						
1000 × 590 × 270 mm; 3.30–11.90 kW output	6623.60	8382.82	16.13	634.29	nr	**9017.11**
1500 × 590 × 270 mm; 5.00–17.90 kW output	8144.94	10308.23	16.13	634.29	nr	**10942.52**
2000 × 590 × 270 mm; 6.70–23.90 kW output	10284.80	13016.44	16.13	634.29	nr	**13650.72**
2500 × 590 × 270 mm; 8.30–29.80 kW output	11494.50	14547.43	17.10	672.47	nr	**15219.90**
Electrically heated; 415 V three phase supply; mounting height 3.30 m						
1000 × 774 × 370 mm; 6.10–21.80 kW output	10064.70	12737.89	17.24	677.98	nr	**13415.86**
1500 × 774 × 370 mm; 9.20–32.80 kW output	13484.95	17066.55	17.24	677.98	nr	**17744.52**
2000 × 774 × 370 mm; 12.30–43.70 kW output	16275.43	20598.19	17.24	677.98	nr	**21276.17**
2500 × 774 × 370 mm; 15.30–54.60 kW output	19266.73	24383.97	18.30	719.66	nr	**25103.63**
Electrically heated; 415 V three phase supply; mounting height 4.00 m						
1000 × 774 × 370 mm; 7.20–24.20 kW output	11107.32	14057.43	19.10	751.12	nr	**14808.55**
1500 × 774 × 370 mm; 10.90–36.30 kW output	14410.29	18237.66	19.10	751.12	nr	**18988.78**
2000 × 774 × 370 mm; 14.50–48.40 kW output	17428.90	22058.02	19.10	751.12	nr	**22809.14**
2500 × 774 × 370 mm; 18.10–60.60 kW output	20645.09	26128.43	19.90	782.58	nr	**26911.02**
Industrial grade air curtains; recessed or exposed units with rigid sheet steel casing; aluminium grilles; high quality motor/centrifugal fan assembly; includes fixing in position; electrical work elsewhere						
Ambient temperature; 230 V single phase supply; including wiring between multiple units; horizontally or vertically mounted; opening maximum 6.00 m						
1500 × 585 × 853 mm; 3.0 A supply	7795.21	9865.61	17.24	678.03	nr	**10543.65**
2000 × 585 × 853 mm; 4.0 A supply	9446.48	11955.47	17.24	678.03	nr	**12633.50**

38 VENTILATION/AIR CONDITIONING SYSTEMS

Item	Net Price £	Material £	Labour hours	Labour £	Unit	Total rate £
Water heated; 230 V single phase supply; including wiring between multiple units; horizontally or vertically mounted; opening maximum 6.00 m						
1500 × 585 × 956 mm; 3.0 A supply; 34.80 kW output	8935.06	11308.21	17.24	678.03	nr	**11986.24**
2000 × 585 × 956 mm; 4.0 A supply; 50.70 kW output	10917.27	13816.89	17.24	678.03	nr	**14494.92**
Water heated; 230 V single phase supply; including wiring between multiple units; vertically mounted in single bank for openings maximum 6.00 m wide or opposing twin banks for openings maximum 10.00 m wide						
1500 × 585 mm; 3.0 A supply; 41.1 kW output	8935.06	11308.21	17.24	678.03	nr	**11986.24**
2000 × 585 mm; 4.0 A supply; 57.7 kW output	10917.27	13816.89	17.24	678.03	nr	**14494.92**
Remote mounted electronic controller unit; excluding wiring to units						
0–10 V control	137.07	173.47	5.00	196.63	nr	**370.10**

38 VENTILATION/AIR CONDITIONING SYSTEMS

Item	Net Price £	Material £	Labour hours	Labour £	Unit	Total rate £
SILENCERS/ACOUSTIC TREATMENT						
Attenuators; DW144 galvanized construction c/w splitters; self-securing; fitted to ductwork						
To suit rectangular ducts; unit length 600 mm						
100 × 100 mm	188.15	238.13	0.75	28.17	nr	266.30
150 × 150 mm	196.39	248.55	0.75	28.17	nr	276.73
200 × 200 mm	204.63	258.98	0.75	28.17	nr	287.15
300 × 300 mm	228.93	289.74	0.75	28.17	nr	317.91
400 × 400 mm	257.00	325.26	1.00	37.56	nr	362.82
500 × 500 mm	299.94	379.61	1.25	46.95	nr	426.56
600 × 300 mm	288.92	365.66	1.25	46.95	nr	412.61
600 × 600 mm	404.36	511.76	1.25	46.95	nr	558.71
700 × 300 mm	306.34	387.71	1.50	56.35	nr	444.05
700 × 700 mm	456.14	577.30	1.50	56.35	nr	633.64
800 × 300 mm	317.33	401.62	2.00	75.13	nr	476.74
800 × 800 mm	526.26	666.04	2.00	75.13	nr	741.16
1000 × 1000 mm	673.64	852.56	3.00	112.69	nr	965.25
To suit rectangular ducts; unit length 1200 mm						
200 × 200 mm	225.66	285.60	1.00	37.56	nr	323.16
300 × 300 mm	278.98	353.08	1.00	37.56	nr	390.64
400 × 400 mm	335.43	424.51	1.33	49.96	nr	474.47
500 × 500 mm	404.86	512.39	1.66	62.36	nr	574.74
600 × 300 mm	386.02	488.55	1.66	62.36	nr	550.91
600 × 600 mm	564.80	714.82	1.66	62.36	nr	777.17
700 × 300 mm	413.80	523.71	2.00	75.13	nr	598.83
700 × 700 mm	646.33	817.99	2.00	75.13	nr	893.12
800 × 300 mm	432.62	547.52	2.66	99.92	nr	647.44
800 × 800 mm	745.79	943.87	2.66	99.92	nr	1043.79
1000 × 1000 mm	1009.58	1277.72	4.00	150.25	nr	1427.98
1300 × 1300 mm	1732.97	2193.25	8.00	300.51	nr	2493.76
1500 × 1500 mm	1983.59	2510.43	8.00	300.51	nr	2810.94
1800 × 1800 mm	2714.02	3434.86	10.66	400.43	nr	3835.29
2000 × 2000 mm	3045.44	3854.31	13.33	500.72	nr	4355.03
To suit rectangular ducts; unit length 1800 mm						
200 × 200 mm	284.79	360.43	1.00	37.56	nr	398.00
300 × 300 mm	378.42	478.93	1.00	37.56	nr	516.50
400 × 400 mm	452.36	572.51	1.33	49.96	nr	622.47
500 × 500 mm	527.17	667.19	1.66	62.36	nr	729.55
600 × 300 mm	551.36	697.80	1.66	62.36	nr	760.15
600 × 600 mm	770.45	975.08	1.66	62.36	nr	1037.44
700 × 300 mm	584.51	739.76	2.00	75.13	nr	814.89
700 × 700 mm	920.51	1164.99	2.00	75.13	nr	1240.12
800 × 300 mm	618.09	782.25	2.66	99.92	nr	882.17
800 × 800 mm	1071.03	1355.50	2.66	99.92	nr	1455.42
1000 × 1000 mm	1439.67	1822.04	4.00	150.25	nr	1972.30
1300 × 1300 mm	2404.99	3043.76	8.00	300.51	nr	3344.27
1500 × 1500 mm	2725.51	3449.40	8.00	300.51	nr	3749.91
1800 × 1800 mm	3783.86	4788.86	10.66	400.43	nr	5189.29
2000 × 2000 mm	4284.69	5422.70	13.33	500.72	nr	5923.43
2300 × 2300 mm	5355.77	6778.27	16.00	601.02	nr	7379.28
2500 × 2500 mm	7470.66	9454.87	18.66	700.94	nr	10155.80

38 VENTILATION/AIR CONDITIONING SYSTEMS

Item	Net Price £	Material £	Labour hours	Labour £	Unit	Total rate £
To suit rectangular ducts; unit length 2400 mm						
500 × 500 mm	681.72	862.79	2.08	78.25	nr	**941.03**
600 × 300 mm	650.83	823.69	2.08	78.25	nr	**901.94**
600 × 600 mm	969.35	1226.81	2.08	78.25	nr	**1305.06**
700 × 300 mm	705.49	892.87	2.50	93.91	nr	**986.78**
700 × 700 mm	1145.00	1449.11	2.50	93.91	nr	**1543.02**
800 × 300 mm	745.79	943.87	3.33	125.09	nr	**1068.95**
800 × 800 mm	1325.06	1676.99	3.33	125.09	nr	**1802.08**
1000 × 1000 mm	1787.75	2262.58	5.00	187.82	nr	**2450.40**
1300 × 1300 mm	2895.59	3664.65	10.00	375.64	nr	**4040.29**
1500 × 1500 mm	3304.80	4182.55	10.00	375.64	nr	**4558.19**
1800 × 1800 mm	4715.31	5967.70	13.33	500.65	nr	**6468.35**
2000 × 2000 mm	5291.36	6696.75	16.66	625.81	nr	**7322.56**
2300 × 2300 mm	6546.59	8285.37	20.00	751.27	nr	**9036.64**
2500 × 2500 mm	9192.39	11633.88	23.32	876.13	nr	**12510.02**
To suit circular ducts; unit length 600 mm						
100 mm dia.	163.39	206.78	0.75	28.17	nr	**234.96**
200 mm dia.	192.25	243.31	0.75	28.17	nr	**271.48**
250 mm dia.	217.92	275.81	0.75	28.17	nr	**303.98**
315 mm dia.	253.66	321.04	1.00	37.56	nr	**358.60**
355 mm dia.	293.53	371.49	1.00	37.56	nr	**409.06**
400 mm dia.	315.53	399.33	1.00	37.56	nr	**436.89**
450 mm dia.	381.02	482.22	1.00	37.56	nr	**519.78**
500 mm dia.	419.53	530.95	1.26	47.26	nr	**578.21**
630 mm dia.	494.62	625.99	1.50	56.35	nr	**682.33**
710 mm dia.	538.63	681.69	1.50	56.35	nr	**738.03**
800 mm dia.	599.10	758.22	2.00	75.13	nr	**833.35**
1000 mm dia.	804.19	1017.79	3.00	112.69	nr	**1130.48**
To suit circular ducts; unit length 1200 mm						
100 mm dia.	240.39	304.23	1.00	37.56	nr	**341.80**
200 mm dia.	263.72	333.77	1.00	37.56	nr	**371.33**
250 mm dia.	296.28	374.97	1.00	37.56	nr	**412.54**
315 mm dia.	338.88	428.89	1.33	49.96	nr	**478.85**
355 mm dia.	406.23	514.13	1.33	49.96	nr	**564.09**
400 mm dia.	430.98	545.45	1.33	49.96	nr	**595.41**
450 mm dia.	499.21	631.80	1.33	49.96	nr	**681.76**
500 mm dia.	511.63	647.51	1.66	62.36	nr	**709.87**
630 mm dia.	677.46	857.39	2.00	75.13	nr	**932.52**
710 mm dia.	747.54	946.09	2.00	75.13	nr	**1021.21**
800 mm dia.	748.92	947.83	2.66	99.92	nr	**1047.75**
1000 mm dia.	1003.53	1270.06	4.00	150.25	nr	**1420.32**
1250 mm dia.	1890.41	2392.50	8.00	300.51	nr	**2693.01**
1400 mm dia.	2055.31	2601.20	8.00	300.51	nr	**2901.71**
1600 mm dia.	2278.60	2883.80	10.66	400.43	nr	**3284.23**
To suit circular ducts; unit length 1800 mm						
200 mm dia.	369.57	467.72	1.25	46.95	nr	**514.68**
250 mm dia.	413.56	523.40	1.25	46.95	nr	**570.36**
315 mm dia.	491.47	622.00	1.66	62.36	nr	**684.36**
355 mm dia.	579.42	733.32	1.66	62.36	nr	**795.67**
400 mm dia.	599.56	758.81	1.66	62.36	nr	**821.16**
450 mm dia.	612.43	775.09	1.66	62.36	nr	**837.44**
500 mm dia.	673.26	852.08	2.08	78.13	nr	**930.21**

38 VENTILATION/AIR CONDITIONING SYSTEMS

Item	Net Price £	Material £	Labour hours	Labour £	Unit	Total rate £
SILENCERS/ACOUSTIC TREATMENT – cont						
Attenuators – cont						
To suit circular ducts – cont						
630 mm dia.	744.79	942.60	2.50	93.91	nr	**1036.51**
710 mm dia.	871.23	1102.63	2.50	93.91	nr	**1196.54**
800 mm dia.	901.47	1140.90	3.33	125.09	nr	**1265.99**
1000 mm dia.	2353.59	2978.70	5.00	187.82	nr	**3166.52**
1250 mm dia.	2392.10	3027.44	6.66	250.17	nr	**3277.61**
1400 mm dia.	2642.84	3344.77	6.66	250.17	nr	**3594.95**
1600 mm dia.	2644.43	3346.80	9.90	371.88	nr	**3718.68**
THERMAL INSULATION						
Concealed ductwork						
Flexible wrap; 20 kg–45 kg Bright Class O aluminium foil faced; Bright Class O foil taped joints; 62 mm metal pins and washers; aluminium bands						
40 mm thick insulation	21.77	27.55	0.40	15.73	m^2	**43.28**
Semi-rigid slab; 45 kg Bright Class O aluminium foil faced mineral fibre; Bright Class O foil taped joints; 62 mm metal pins and washers; aluminium bands						
40 mm thick insulation	28.47	36.03	0.65	25.56	m^2	**61.59**
Plantroom ductwork						
Semi-rigid slab; 45 kg Bright Class O aluminium foil faced mineral fibre; Bright Class O foil taped joints; 62 mm metal pins and washers; 22 SWG plain/embossed aluminium cladding; pop rivited						
50 mm thick insulation	71.82	90.90	1.50	58.99	m^2	**149.89**
External ductwork						
Semi-rigid slab; 45 kg Bright Class O aluminium foil faced mineral fibre; Bright Class O foil taped joints; 62 mm metal pins and washers; 0.8 mm polyisobutylene sheeting; welded joints						
50 mm thick insulation	52.39	66.30	1.25	49.16	m^2	**115.46**

38 VENTILATION/AIR CONDITIONING SYSTEMS

Item	Net Price £	Material £	Labour hours	Labour £	Unit	Total rate £
DUCTWORK: FIRE RATED						
DUCTLINES						
The relevant BS requires that the fire-rating of ductwork meets 3 criteria; stability (hours), integrity (hours) and insulation (hours). The least of the 3 periods defines the fire-rating. The BS does however allow stability and integrity to be considered in isolation. Rates are therefore provided for both types of system.						
Care should to be taken when using the rates within this section to ensure that the requirements for stability, integrity and insulation are known and the appropriate rates are used.						
High density single layer mineral wool fire-rated ductwork slab, in accordance with BSEN1366, including ducts 'Type A' and 'Type B'; 165 kg class O foil faced mineral fibre; 100 mm wide bright class O foil taped joints; welded pins; includes protection to all supports						
½ hour stability, integrity and insulation						
25 mm thick, vertical and horizontal ductwork	126.41	159.99	1.25	49.16	m²	**209.15**
1 hour stability, integrity and insulation						
30 mm thick, vertical ductwork	148.40	187.81	1.50	58.99	m²	**246.80**
40 mm thick, horizontal ductwork	174.54	220.90	1.50	58.99	m²	**279.89**
1½ hour stability, integrity and insulation						
50 mm thick, vertical ductwork	208.50	263.88	1.75	68.82	m²	**332.70**
70 mm thick, horizontal ductwork	260.81	330.08	1.75	68.82	m²	**398.90**
2 hour stability, integrity and insulation						
70 mm, vertical ductwork	268.72	340.09	2.00	78.65	m²	**418.74**
90 mm horizontal ductwork	321.02	406.28	2.00	78.65	m²	**484.93**
Kitchen extract, 1 hour stability, integrity and insulation						
90 mm, vertical and horizontal	321.02	406.28	2.00	78.65	m²	**484.93**
Galvanized sheet metal rectangular section ductwork to BSEN1366, ducts 'Type A' and 'Type B'; provides 2 hours stability and 2 hours integrity at 1100°C (no insulation); including all necessary stiffeners, joints and supports in the running length						
Ductwork up to 600 mm longest side						
Sum of two sides 200 mm	165.80	209.84	2.91	109.31	m	**319.15**
Sum of two sides 300 mm	180.01	227.82	2.99	112.32	m	**340.13**
Sum of two sides 400 mm	194.17	245.75	3.17	119.08	m	**364.82**
Sum of two sides 500 mm	208.38	263.73	3.37	126.59	m	**390.32**
Sum of two sides 600 mm	222.55	281.65	3.54	132.98	m	**414.63**
Sum of two sides 700 mm	236.77	299.65	3.72	139.74	m	**439.39**
Sum of two sides 800 mm	250.96	317.61	3.90	146.50	m	**464.11**
Sum of two sides 900 mm	265.13	335.54	5.04	189.32	m	**524.86**
Sum of two sides 1000 mm	279.35	353.54	5.58	209.61	m	**563.14**
Sum of two sides 1100 mm	293.51	371.47	5.84	219.37	m	**590.84**
Sum of two sides 1200 mm	307.68	389.40	6.11	229.51	m	**618.91**

38 VENTILATION/AIR CONDITIONING SYSTEMS

Item	Net Price £	Material £	Labour hours	Labour £	Unit	Total rate £
DUCTWORK: FIRE RATED – cont						
Extra over fittings; Ductwork up to 600 mm longest side						
End cap						
Sum of two sides 200 mm	47.09	59.60	0.81	30.43	nr	**90.03**
Sum of two sides 300 mm	49.95	63.22	0.84	31.55	nr	**94.78**
Sum of two sides 400 mm	52.80	66.83	0.87	32.68	nr	**99.51**
Sum of two sides 500 mm	55.65	70.44	0.90	33.81	nr	**104.24**
Sum of two sides 600 mm	58.50	74.04	0.93	34.93	nr	**108.98**
Sum of two sides 700 mm	61.39	77.70	0.96	36.06	nr	**113.76**
Sum of two sides 800 mm	64.31	81.39	0.98	36.81	nr	**118.20**
Sum of two sides 900 mm	67.19	85.03	1.17	43.95	nr	**128.98**
Sum of two sides 1000 mm	70.00	88.59	1.22	45.83	nr	**134.42**
Sum of two sides 1100 mm	72.90	92.26	1.25	46.95	nr	**139.22**
Sum of two sides 1200 mm	75.74	95.85	1.28	48.08	nr	**143.93**
Reducer						
Sum of two sides 200 mm	186.72	236.31	2.23	83.77	nr	**320.08**
Sum of two sides 300 mm	195.31	247.18	2.37	89.03	nr	**336.21**
Sum of two sides 400 mm	203.84	257.98	2.51	94.28	nr	**352.27**
Sum of two sides 500 mm	212.44	268.87	2.65	99.54	nr	**368.41**
Sum of two sides 600 mm	221.01	279.71	2.79	104.80	nr	**384.51**
Sum of two sides 700 mm	229.58	290.56	2.87	107.81	nr	**398.37**
Sum of two sides 800 mm	238.18	301.45	3.01	113.07	nr	**414.51**
Sum of two sides 900 mm	246.76	312.30	3.06	114.94	nr	**427.24**
Sum of two sides 1000 mm	255.31	323.12	3.17	119.08	nr	**442.20**
Sum of two sides 1100 mm	263.86	333.94	3.22	120.95	nr	**454.89**
Sum of two sides 1200 mm	272.49	344.86	3.28	123.21	nr	**468.07**
Offset						
Sum of two sides 200 mm	450.64	570.33	2.95	110.81	nr	**681.15**
Sum of two sides 300 mm	461.81	584.47	3.11	116.82	nr	**701.29**
Sum of two sides 400 mm	472.98	598.60	3.23	121.33	nr	**719.93**
Sum of two sides 500 mm	484.18	612.78	3.26	122.46	nr	**735.24**
Sum of two sides 600 mm	495.36	626.93	3.42	128.47	nr	**755.40**
Sum of two sides 700 mm	506.56	641.10	3.45	129.59	nr	**770.69**
Sum of two sides 800 mm	517.74	655.25	3.57	134.10	nr	**789.35**
Sum of two sides 900 mm	528.96	669.45	3.67	137.86	nr	**807.31**
Sum of two sides 1000 mm	540.11	683.56	3.73	140.11	nr	**823.67**
Sum of two sides 1100 mm	551.26	697.68	3.78	141.99	nr	**839.67**
Sum of two sides 1200 mm	562.51	711.91	3.89	146.12	nr	**858.03**
90° radius bend						
Sum of two sides 200 mm	182.71	231.24	2.06	77.38	nr	**308.62**
Sum of two sides 300 mm	196.56	248.77	2.21	83.02	nr	**331.78**
Sum of two sides 400 mm	210.37	266.25	2.36	88.65	nr	**354.90**
Sum of two sides 500 mm	224.26	283.83	2.51	94.28	nr	**378.11**
Sum of two sides 600 mm	238.12	301.36	2.66	99.92	nr	**401.28**
Sum of two sides 700 mm	251.89	318.79	2.81	105.55	nr	**424.34**
Sum of two sides 800 mm	265.78	336.37	2.97	111.56	nr	**447.94**
Sum of two sides 900 mm	279.55	353.80	2.99	112.32	nr	**466.11**
Sum of two sides 1000 mm	293.48	371.43	3.04	114.19	nr	**485.63**
Sum of two sides 1100 mm	307.25	388.86	3.07	115.32	nr	**504.18**
Sum of two sides 1200 mm	321.07	406.34	3.09	116.07	nr	**522.41**

38 VENTILATION/AIR CONDITIONING SYSTEMS

Item	Net Price £	Material £	Labour hours	Labour £	Unit	Total rate £
45° radius bend						
Sum of two sides 200 mm	225.29	285.12	1.52	57.10	nr	**342.22**
Sum of two sides 300 mm	230.89	292.22	1.59	59.73	nr	**351.94**
Sum of two sides 400 mm	236.54	299.36	1.66	62.36	nr	**361.71**
Sum of two sides 500 mm	242.11	306.42	1.73	64.99	nr	**371.40**
Sum of two sides 600 mm	247.67	313.46	1.80	67.61	nr	**381.07**
Sum of two sides 700 mm	253.32	320.60	1.87	70.24	nr	**390.84**
Sum of two sides 800 mm	258.85	327.61	1.93	72.50	nr	**400.10**
Sum of two sides 900 mm	264.48	334.73	1.99	74.75	nr	**409.48**
Sum of two sides 1000 mm	270.06	341.79	2.05	77.01	nr	**418.79**
Sum of two sides 1100 mm	275.66	348.88	2.11	79.26	nr	**428.14**
Sum of two sides 1200 mm	281.24	355.94	2.17	81.51	nr	**437.45**
90° mitre bend						
Sum of two sides 200 mm	189.43	239.74	2.47	92.78	nr	**332.52**
Sum of two sides 300 mm	219.55	277.86	2.65	99.54	nr	**377.40**
Sum of two sides 400 mm	249.64	315.94	2.84	106.68	nr	**422.62**
Sum of two sides 500 mm	279.81	354.13	3.02	113.44	nr	**467.57**
Sum of two sides 600 mm	309.90	392.21	3.20	120.20	nr	**512.41**
Sum of two sides 700 mm	340.04	430.36	3.38	126.96	nr	**557.32**
Sum of two sides 800 mm	370.19	468.51	3.56	133.73	nr	**602.24**
Sum of two sides 900 mm	400.29	506.61	3.59	134.85	nr	**641.46**
Sum of two sides 1000 mm	430.48	544.81	3.66	137.48	nr	**682.30**
Sum of two sides 1100 mm	460.53	582.84	3.69	138.61	nr	**721.45**
Sum of two sides 1200 mm	490.74	621.08	3.72	139.74	nr	**760.82**
Branch (Side-on Shoe)						
Sum of two sides 200 mm	70.32	89.00	0.98	36.81	nr	**125.81**
Sum of two sides 300 mm	70.71	89.49	1.06	39.82	nr	**129.30**
Sum of two sides 400 mm	71.10	89.99	1.14	42.82	nr	**132.81**
Sum of two sides 500 mm	71.50	90.49	1.22	45.83	nr	**136.31**
Sum of two sides 600 mm	71.86	90.95	1.30	48.83	nr	**139.78**
Sum of two sides 700 mm	72.23	91.42	1.37	51.46	nr	**142.88**
Sum of two sides 800 mm	72.60	91.88	1.38	51.84	nr	**143.72**
Sum of two sides 900 mm	73.16	92.59	1.46	54.84	nr	**147.43**
Sum of two sides 1000 mm	73.35	92.83	1.46	54.84	nr	**147.68**
Sum of two sides 1100 mm	73.75	93.33	1.50	56.35	nr	**149.68**
Sum of two sides 1200 mm	74.13	93.82	1.55	58.22	nr	**152.04**
Ductwork 601 to 800 mm longest side						
Sum of two sides 900 mm	265.08	335.49	5.04	189.32	m	**524.81**
Sum of two sides 1000 mm	279.32	353.51	5.58	209.61	m	**563.11**
Sum of two sides 1100 mm	293.51	371.47	5.84	219.37	m	**590.84**
Sum of two sides 1200 mm	307.68	389.40	6.11	229.51	m	**618.91**
Sum of two sides 1300 mm	321.97	407.48	6.38	239.66	m	**647.14**
Sum of two sides 1400 mm	336.13	425.41	6.65	249.80	m	**675.21**
Sum of two sides 1500 mm	350.39	443.46	6.91	259.56	m	**703.02**
Sum of two sides 1600 mm	364.59	461.42	7.18	269.71	m	**731.13**

38 VENTILATION/AIR CONDITIONING SYSTEMS

Item	Net Price £	Material £	Labour hours	Labour £	Unit	Total rate £
DUCTWORK: FIRE RATED – cont						
Extra over fittings; Ductwork 601 to 800 mm longest side						
End cap						
Sum of two sides 900 mm	62.39	78.96	1.17	43.95	nr	**122.91**
Sum of two sides 1000 mm	66.86	84.62	1.22	45.83	nr	**130.45**
Sum of two sides 1100 mm	71.35	90.30	1.25	46.95	nr	**137.25**
Sum of two sides 1200 mm	75.79	95.92	1.28	48.08	nr	**144.00**
Sum of two sides 1300 mm	80.24	101.55	1.31	49.21	nr	**150.75**
Sum of two sides 1400 mm	84.71	107.21	1.34	50.34	nr	**157.54**
Sum of two sides 1500 mm	89.15	112.83	1.36	51.09	nr	**163.92**
Sum of two sides 1600 mm	93.62	118.49	1.39	52.21	nr	**170.70**
Reducer						
Sum of two sides 900 mm	233.55	295.58	3.06	114.94	nr	**410.52**
Sum of two sides 1000 mm	246.53	312.01	3.17	119.08	nr	**431.08**
Sum of two sides 1100 mm	259.48	328.40	3.22	120.95	nr	**449.35**
Sum of two sides 1200 mm	272.52	344.90	3.28	123.21	nr	**468.10**
Sum of two sides 1300 mm	285.52	361.36	3.33	125.09	nr	**486.44**
Sum of two sides 1400 mm	298.52	377.80	3.38	126.96	nr	**504.77**
Sum of two sides 1500 mm	311.48	394.21	3.44	129.22	nr	**523.43**
Sum of two sides 1600 mm	324.50	410.69	3.49	131.10	nr	**541.79**
Offset						
Sum of two sides 900 mm	496.67	628.59	3.45	129.59	nr	**758.18**
Sum of two sides 1000 mm	520.31	658.51	3.67	137.86	nr	**796.37**
Sum of two sides 1100 mm	543.98	688.46	3.78	141.99	nr	**830.45**
Sum of two sides 1200 mm	567.62	718.38	3.89	146.12	nr	**864.50**
Sum of two sides 1300 mm	591.22	748.25	4.00	150.25	nr	**898.51**
Sum of two sides 1400 mm	614.89	778.21	4.11	154.39	nr	**932.59**
Sum of two sides 1500 mm	638.53	808.13	4.23	158.89	nr	**967.02**
Sum of two sides 1600 mm	662.91	838.98	4.34	163.03	nr	**1002.00**
90° radius bend						
Sum of two sides 900 mm	248.31	314.27	2.99	112.32	nr	**426.58**
Sum of two sides 1000 mm	272.73	345.17	3.04	114.19	nr	**459.36**
Sum of two sides 1100 mm	297.06	375.95	3.07	115.32	nr	**491.27**
Sum of two sides 1200 mm	321.46	406.84	3.09	116.07	nr	**522.91**
Sum of two sides 1300 mm	345.81	437.66	3.12	117.20	nr	**554.86**
Sum of two sides 1400 mm	370.19	468.51	3.15	118.33	nr	**586.84**
Sum of two sides 1500 mm	394.57	499.36	3.17	119.08	nr	**618.44**
Sum of two sides 1600 mm	418.96	530.23	3.20	120.20	nr	**650.44**
45° bend						
Sum of two sides 900 mm	246.57	312.06	1.74	65.36	nr	**377.42**
Sum of two sides 1000 mm	258.13	326.69	1.85	69.49	nr	**396.18**
Sum of two sides 1100 mm	269.73	341.38	1.91	71.75	nr	**413.12**
Sum of two sides 1200 mm	281.27	355.97	1.96	73.62	nr	**429.60**
Sum of two sides 1300 mm	292.83	370.61	2.02	75.88	nr	**446.48**
Sum of two sides 1400 mm	304.40	385.25	2.07	77.76	nr	**463.01**
Sum of two sides 1500 mm	315.95	399.87	2.13	80.01	nr	**479.88**
Sum of two sides 1600 mm	327.54	414.54	2.18	81.89	nr	**496.43**

38 VENTILATION/AIR CONDITIONING SYSTEMS

Item	Net Price £	Material £	Labour hours	Labour £	Unit	Total rate £
90° mitre bend						
Sum of two sides 900 mm	371.78	470.53	3.59	134.85	nr	**605.38**
Sum of two sides 1000 mm	411.45	520.72	3.66	137.48	nr	**658.21**
Sum of two sides 1100 mm	451.07	570.87	3.69	138.61	nr	**709.48**
Sum of two sides 1200 mm	490.74	621.08	3.72	139.74	nr	**760.82**
Sum of two sides 1300 mm	530.31	671.15	3.75	140.86	nr	**812.02**
Sum of two sides 1400 mm	569.97	721.35	3.78	141.99	nr	**863.34**
Sum of two sides 1500 mm	609.57	771.48	3.81	143.12	nr	**914.59**
Sum of two sides 1600 mm	649.23	821.67	3.84	144.24	nr	**965.92**
Branch (Side-on Shoe)						
Sum of two sides 900 mm	69.00	87.33	1.37	51.46	nr	**138.79**
Sum of two sides 1000 mm	70.71	89.49	1.46	54.84	nr	**144.33**
Sum of two sides 1100 mm	72.45	91.69	1.50	56.35	nr	**148.04**
Sum of two sides 1200 mm	74.16	93.85	1.55	58.22	nr	**152.07**
Sum of two sides 1300 mm	74.55	94.35	1.59	59.73	nr	**154.08**
Sum of two sides 1400 mm	75.89	96.04	1.63	61.23	nr	**157.27**
Sum of two sides 1500 mm	77.69	98.32	1.68	63.11	nr	**161.43**
Sum of two sides 1600 mm	81.09	102.63	1.72	64.61	nr	**167.24**
Ductwork 801 to 1000 mm longest side						
Sum of two sides 1100 mm	293.51	371.47	5.84	219.37	m	**590.84**
Sum of two sides 1200 mm	307.68	389.40	6.11	229.51	m	**618.91**
Sum of two sides 1300 mm	321.90	407.39	6.38	239.66	m	**647.05**
Sum of two sides 1400 mm	336.10	425.37	6.65	249.80	m	**675.17**
Sum of two sides 1500 mm	350.30	443.34	6.91	259.56	m	**702.90**
Sum of two sides 1600 mm	364.50	461.32	7.18	269.71	m	**731.02**
Sum of two sides 1700 mm	378.74	479.33	7.45	279.85	m	**759.18**
Sum of two sides 1800 mm	392.94	497.31	7.71	289.62	m	**786.93**
Sum of two sides 1900 mm	407.15	515.29	7.98	299.76	m	**815.05**
Sum of two sides 2000 mm	421.37	533.29	8.25	309.90	m	**843.19**
Extra over fittings; Ductwork 801 to 1000 mm longest side						
End cap						
Sum of two sides 1100 mm	71.35	90.30	1.25	46.95	nr	**137.25**
Sum of two sides 1200 mm	75.79	95.92	1.28	48.08	nr	**144.00**
Sum of two sides 1300 mm	80.24	101.55	1.31	49.21	nr	**150.75**
Sum of two sides 1400 mm	84.71	107.21	1.34	50.34	nr	**157.54**
Sum of two sides 1500 mm	89.15	112.83	1.36	51.09	nr	**163.92**
Sum of two sides 1600 mm	93.58	118.44	1.39	52.21	nr	**170.65**
Sum of two sides 1700 mm	98.04	124.08	1.42	53.34	nr	**177.42**
Sum of two sides 1800 mm	102.51	129.74	1.45	54.47	nr	**184.21**
Sum of two sides 1900 mm	106.94	135.35	1.48	55.59	nr	**190.94**
Sum of two sides 2000 mm	111.39	140.97	1.51	56.72	nr	**197.70**
Reducer						
Sum of two sides 1100 mm	259.48	328.40	3.22	120.95	nr	**449.35**
Sum of two sides 1200 mm	272.57	344.96	3.28	123.21	nr	**468.17**
Sum of two sides 1300 mm	285.62	361.48	3.33	125.09	nr	**486.56**
Sum of two sides 1400 mm	298.72	378.06	3.38	126.96	nr	**505.02**
Sum of two sides 1500 mm	311.73	394.52	3.44	129.22	nr	**523.74**
Sum of two sides 1600 mm	324.84	411.12	3.49	131.10	nr	**542.22**
Sum of two sides 1700 mm	337.88	427.62	3.54	132.98	nr	**560.59**

38 VENTILATION/AIR CONDITIONING SYSTEMS

Item	Net Price £	Material £	Labour hours	Labour £	Unit	Total rate £
DUCTWORK: FIRE RATED – cont						
Extra over fittings – cont						
Reducer – cont						
Sum of two sides 1800 mm	350.94	444.15	3.60	135.23	nr	**579.38**
Sum of two sides 1900 mm	364.05	460.75	3.65	137.11	nr	**597.85**
Sum of two sides 2000 mm	383.59	485.47	3.70	138.99	nr	**624.46**
Offset						
Sum of two sides 1100 mm	539.69	683.03	3.78	141.99	nr	**825.02**
Sum of two sides 1200 mm	562.51	711.91	3.89	146.12	nr	**858.03**
Sum of two sides 1300 mm	585.29	740.74	4.00	150.25	nr	**891.00**
Sum of two sides 1400 mm	608.11	769.63	4.11	154.39	nr	**924.02**
Sum of two sides 1500 mm	630.94	798.51	4.23	158.89	nr	**957.41**
Sum of two sides 1600 mm	653.79	827.43	4.34	163.03	nr	**990.46**
Sum of two sides 1700 mm	676.57	856.27	4.45	167.16	nr	**1023.43**
Sum of two sides 1800 mm	699.39	885.15	4.56	171.29	nr	**1056.44**
Sum of two sides 1900 mm	722.16	913.97	4.67	175.42	nr	**1089.39**
Sum of two sides 2000 mm	756.40	957.30	5.53	207.73	nr	**1165.02**
90° radius bend						
Sum of two sides 1100 mm	296.61	375.38	3.07	115.32	nr	**490.71**
Sum of two sides 1200 mm	321.02	406.29	3.09	116.07	nr	**522.36**
Sum of two sides 1300 mm	345.48	437.24	3.12	117.20	nr	**554.44**
Sum of two sides 1400 mm	369.92	468.17	3.15	118.33	nr	**586.49**
Sum of two sides 1500 mm	394.35	499.09	3.17	119.08	nr	**618.16**
Sum of two sides 1600 mm	418.77	529.99	3.20	120.20	nr	**650.19**
Sum of two sides 1700 mm	443.18	560.89	3.22	120.95	nr	**681.85**
Sum of two sides 1800 mm	467.62	591.82	3.25	122.08	nr	**713.90**
Sum of two sides 1900 mm	492.05	622.74	3.28	123.21	nr	**745.95**
Sum of two sides 2000 mm	515.74	652.73	3.30	123.96	nr	**776.69**
45° bend						
Sum of two sides 1100 mm	271.49	343.60	1.91	71.75	nr	**415.35**
Sum of two sides 1200 mm	282.84	357.96	1.96	73.62	nr	**431.58**
Sum of two sides 1300 mm	294.19	372.33	2.02	75.88	nr	**448.21**
Sum of two sides 1400 mm	305.58	386.74	2.07	77.76	nr	**464.50**
Sum of two sides 1500 mm	316.93	401.11	2.13	80.01	nr	**481.12**
Sum of two sides 1600 mm	328.28	415.47	2.18	81.89	nr	**497.36**
Sum of two sides 1700 mm	339.62	429.82	2.24	84.14	nr	**513.97**
Sum of two sides 1800 mm	350.98	444.20	2.30	86.40	nr	**530.59**
Sum of two sides 1900 mm	362.29	458.52	2.35	88.27	nr	**546.79**
Sum of two sides 2000 mm	373.72	472.98	2.76	103.68	nr	**576.66**
90° mitre bend						
Sum of two sides 1100 mm	451.07	570.87	3.69	138.61	nr	**709.48**
Sum of two sides 1200 mm	490.74	621.08	3.72	139.74	nr	**760.82**
Sum of two sides 1300 mm	530.31	671.15	3.75	140.86	nr	**812.02**
Sum of two sides 1400 mm	569.97	721.35	3.78	141.99	nr	**863.34**
Sum of two sides 1500 mm	609.66	771.58	3.81	143.12	nr	**914.70**
Sum of two sides 1600 mm	649.23	821.67	3.84	144.24	nr	**965.92**
Sum of two sides 1700 mm	688.90	871.87	3.87	145.37	nr	**1017.24**
Sum of two sides 1800 mm	728.54	922.04	3.90	146.50	nr	**1068.54**
Sum of two sides 1900 mm	768.16	972.19	3.93	147.63	nr	**1119.81**
Sum of two sides 2000 mm	807.81	1022.37	3.96	148.75	nr	**1171.12**

38 VENTILATION/AIR CONDITIONING SYSTEMS

Item	Net Price £	Material £	Labour hours	Labour £	Unit	Total rate £
Branch (Side-on Shoe)						
Sum of two sides 1100 mm	72.40	91.62	1.50	56.35	nr	**147.97**
Sum of two sides 1200 mm	74.06	93.73	1.55	58.22	nr	**151.95**
Sum of two sides 1300 mm	75.86	96.01	1.59	59.73	nr	**155.73**
Sum of two sides 1400 mm	77.63	98.25	1.63	61.23	nr	**159.48**
Sum of two sides 1500 mm	79.38	100.46	1.68	63.11	nr	**163.57**
Sum of two sides 1600 mm	81.09	102.63	1.72	64.61	nr	**167.24**
Sum of two sides 1700 mm	82.88	104.89	1.77	66.49	nr	**171.38**
Sum of two sides 1800 mm	84.64	107.12	1.81	67.99	nr	**175.11**
Sum of two sides 1900 mm	86.38	109.33	1.86	69.87	nr	**179.20**
Sum of two sides 2000 mm	88.99	112.62	1.90	71.37	nr	**184.00**
Ductwork 1001 to 1250 mm longest side						
Sum of two sides 1300 mm	438.55	555.03	6.38	239.66	m	**794.68**
Sum of two sides 1400 mm	457.45	578.94	6.65	249.80	m	**828.74**
Sum of two sides 1500 mm	476.38	602.91	6.91	259.56	m	**862.48**
Sum of two sides 1600 mm	490.54	620.82	6.91	259.56	m	**880.39**
Sum of two sides 1700 mm	504.69	638.73	7.45	279.85	m	**918.58**
Sum of two sides 1800 mm	523.52	662.56	7.71	289.62	m	**952.18**
Sum of two sides 1900 mm	542.40	686.46	7.98	299.76	m	**986.22**
Sum of two sides 2000 mm	561.31	710.39	8.25	309.90	m	**1020.29**
Sum of two sides 2100 mm	580.18	734.27	9.62	361.36	m	**1095.64**
Sum of two sides 2200 mm	594.36	752.22	9.62	361.36	m	**1113.58**
Sum of two sides 2300 mm	608.50	770.11	10.07	378.27	m	**1148.38**
Sum of two sides 2400 mm	627.38	794.01	10.47	393.29	m	**1187.30**
Sum of two sides 2500 mm	646.26	817.91	10.87	408.32	m	**1226.23**
Extra over fittings; Ductwork 1001 to 1250 mm longest side						
End cap						
Sum of two sides 1300 mm	80.26	101.58	1.31	49.21	nr	**150.79**
Sum of two sides 1400 mm	84.71	107.21	1.34	50.34	nr	**157.54**
Sum of two sides 1500 mm	89.10	112.76	1.36	51.09	nr	**163.85**
Sum of two sides 1600 mm	93.58	118.44	1.39	52.21	nr	**170.65**
Sum of two sides 1700 mm	98.04	124.08	1.42	53.34	nr	**177.42**
Sum of two sides 1800 mm	102.45	129.65	1.45	54.47	nr	**184.12**
Sum of two sides 1900 mm	106.94	135.35	1.48	55.59	nr	**190.94**
Sum of two sides 2000 mm	111.39	140.97	1.51	56.72	nr	**197.70**
Sum of two sides 2100 mm	115.85	146.62	2.66	99.92	nr	**246.54**
Sum of two sides 2200 mm	120.29	152.24	2.80	105.18	nr	**257.42**
Sum of two sides 2300 mm	124.76	157.90	2.95	110.81	nr	**268.71**
Sum of two sides 2400 mm	129.18	163.49	3.10	116.45	nr	**279.94**
Sum of two sides 2500 mm	133.63	169.12	3.24	121.71	nr	**290.82**
Reducer						
Sum of two sides 1300 mm	285.52	361.36	3.33	125.09	nr	**486.44**
Sum of two sides 1400 mm	298.37	377.61	3.38	126.96	nr	**504.58**
Sum of two sides 1500 mm	311.22	393.88	3.44	129.22	nr	**523.10**
Sum of two sides 1600 mm	324.13	410.22	3.49	131.10	nr	**541.32**
Sum of two sides 1700 mm	337.09	426.62	3.54	132.98	nr	**559.59**
Sum of two sides 1800 mm	350.02	442.99	3.60	135.23	nr	**578.22**
Sum of two sides 1900 mm	362.89	459.28	3.65	137.11	nr	**596.39**

38 VENTILATION/AIR CONDITIONING SYSTEMS

Item	Net Price £	Material £	Labour hours	Labour £	Unit	Total rate £
DUCTWORK: FIRE RATED – cont						
Extra over fittings – cont						
Reducer – cont						
Sum of two sides 2000 mm	375.82	475.64	3.70	138.99	nr	**614.62**
Sum of two sides 2100 mm	388.72	491.96	3.75	140.86	nr	**632.82**
Sum of two sides 2200 mm	401.60	508.27	3.80	142.74	nr	**651.01**
Sum of two sides 2300 mm	414.53	524.62	3.85	144.62	nr	**669.24**
Sum of two sides 2400 mm	427.44	540.96	3.90	146.50	nr	**687.46**
Sum of two sides 2500 mm	440.39	557.36	3.95	148.38	nr	**705.73**
Offset						
Sum of two sides 1300 mm	583.34	738.28	4.00	150.25	nr	**888.53**
Sum of two sides 1400 mm	606.51	767.59	4.11	154.39	nr	**921.98**
Sum of two sides 1500 mm	629.63	796.86	4.23	158.89	nr	**955.75**
Sum of two sides 1600 mm	652.78	826.16	4.34	163.03	nr	**989.18**
Sum of two sides 1700 mm	675.90	855.42	4.45	167.16	nr	**1022.58**
Sum of two sides 1800 mm	699.03	884.69	4.56	171.29	nr	**1055.98**
Sum of two sides 1900 mm	722.14	913.93	4.67	175.42	nr	**1089.36**
Sum of two sides 2000 mm	745.26	943.20	5.53	207.73	nr	**1150.93**
Sum of two sides 2100 mm	768.37	972.45	5.76	216.37	nr	**1188.81**
Sum of two sides 2200 mm	791.11	1001.23	5.99	225.01	nr	**1226.23**
Sum of two sides 2300 mm	814.63	1030.99	6.22	233.65	nr	**1264.64**
Sum of two sides 2400 mm	837.72	1060.22	6.45	242.29	nr	**1302.51**
Sum of two sides 2500 mm	860.90	1089.56	6.68	250.93	nr	**1340.48**
90° radius bend						
Sum of two sides 1300 mm	345.02	436.66	3.12	117.20	nr	**553.86**
Sum of two sides 1400 mm	383.93	485.90	3.15	118.33	nr	**604.23**
Sum of two sides 1500 mm	394.06	498.72	3.17	119.08	nr	**617.80**
Sum of two sides 1600 mm	418.51	529.66	3.20	120.20	nr	**649.87**
Sum of two sides 1700 mm	443.02	560.69	3.22	120.95	nr	**681.64**
Sum of two sides 1800 mm	467.51	591.68	3.25	122.08	nr	**713.76**
Sum of two sides 1900 mm	492.01	622.69	3.28	123.21	nr	**745.89**
Sum of two sides 2000 mm	516.52	653.71	3.30	123.96	nr	**777.67**
Sum of two sides 2100 mm	541.02	684.72	3.32	124.71	nr	**809.43**
Sum of two sides 2200 mm	565.49	715.69	3.34	125.46	nr	**841.15**
Sum of two sides 2300 mm	590.02	746.73	3.36	126.21	nr	**872.95**
Sum of two sides 2400 mm	614.54	777.76	3.38	126.96	nr	**904.72**
Sum of two sides 2500 mm	639.02	808.75	3.40	127.72	nr	**936.46**
45° bend						
Sum of two sides 1300 mm	279.47	353.70	2.02	75.88	nr	**429.57**
Sum of two sides 1400 mm	292.05	369.62	2.07	77.76	nr	**447.38**
Sum of two sides 1500 mm	304.61	385.51	2.13	80.01	nr	**465.52**
Sum of two sides 1600 mm	317.21	401.46	2.18	81.89	nr	**483.35**
Sum of two sides 1700 mm	329.76	417.35	2.24	84.14	nr	**501.49**
Sum of two sides 1800 mm	342.38	433.31	2.30	86.40	nr	**519.71**
Sum of two sides 1900 mm	354.93	449.20	2.35	88.27	nr	**537.48**
Sum of two sides 2000 mm	367.54	465.16	2.76	103.68	nr	**568.84**
Sum of two sides 2100 mm	380.14	481.11	2.89	108.56	nr	**589.67**
Sum of two sides 2200 mm	392.74	497.05	3.01	113.07	nr	**610.12**
Sum of two sides 2300 mm	405.30	512.94	3.13	117.57	nr	**630.52**
Sum of two sides 2400 mm	417.92	528.92	3.26	122.46	nr	**651.38**
Sum of two sides 2500 mm	430.48	544.81	3.37	126.59	nr	**671.40**

38 VENTILATION/AIR CONDITIONING SYSTEMS

Item	Net Price £	Material £	Labour hours	Labour £	Unit	Total rate £
90° mitre bend						
Sum of two sides 1300 mm	529.11	669.64	3.75	140.86	nr	**810.50**
Sum of two sides 1400 mm	568.39	719.35	3.78	141.99	nr	**861.34**
Sum of two sides 1500 mm	609.22	771.03	3.81	143.12	nr	**914.14**
Sum of two sides 1600 mm	649.21	821.64	3.84	144.24	nr	**965.88**
Sum of two sides 1700 mm	689.18	872.23	3.87	145.37	nr	**1017.60**
Sum of two sides 1800 mm	729.16	922.82	3.90	146.50	nr	**1069.32**
Sum of two sides 1900 mm	769.17	973.46	3.93	147.63	nr	**1121.09**
Sum of two sides 2000 mm	809.22	1024.14	3.96	148.75	nr	**1172.89**
Sum of two sides 2100 mm	849.19	1074.73	3.99	149.88	nr	**1224.61**
Sum of two sides 2200 mm	889.21	1125.38	4.02	151.01	nr	**1276.38**
Sum of two sides 2300 mm	929.22	1176.02	4.05	152.13	nr	**1328.15**
Sum of two sides 2400 mm	969.14	1226.54	4.08	153.26	nr	**1379.80**
Sum of two sides 2500 mm	1009.14	1277.17	4.11	154.39	nr	**1431.56**
Branch (Side-on Shoe)						
Sum of two sides 1300 mm	75.98	96.16	1.59	59.73	nr	**155.89**
Sum of two sides 1400 mm	77.74	98.39	1.63	61.23	nr	**159.62**
Sum of two sides 1500 mm	79.46	100.56	1.68	63.11	nr	**163.67**
Sum of two sides 1600 mm	81.16	102.72	1.72	64.61	nr	**167.33**
Sum of two sides 1700 mm	82.94	104.96	1.77	66.49	nr	**171.45**
Sum of two sides 1800 mm	84.71	107.21	1.81	67.99	nr	**175.20**
Sum of two sides 1900 mm	86.43	109.38	1.86	69.87	nr	**179.25**
Sum of two sides 2000 mm	88.20	111.62	1.90	71.37	nr	**182.99**
Sum of two sides 2100 mm	89.92	113.80	2.58	96.91	nr	**210.71**
Sum of two sides 2200 mm	91.62	115.95	2.61	98.04	nr	**214.00**
Sum of two sides 2300 mm	93.42	118.23	2.64	99.17	nr	**217.40**
Sum of two sides 2400 mm	95.17	120.44	2.88	108.18	nr	**228.62**
Sum of two sides 2500 mm	96.87	122.60	2.91	109.31	nr	**231.91**
Ductwork 1251 to 2000 mm longest side						
Sum of two sides 1700 mm	536.92	679.52	7.45	279.85	m	**959.37**
Sum of two sides 1800 mm	571.34	723.09	7.71	289.62	m	**1012.71**
Sum of two sides 1900 mm	605.74	766.63	7.98	299.76	m	**1066.38**
Sum of two sides 2000 mm	640.11	810.13	8.25	309.90	m	**1120.03**
Sum of two sides 2100 mm	674.58	853.75	9.62	361.36	m	**1215.11**
Sum of two sides 2200 mm	708.94	897.23	9.66	362.86	m	**1260.10**
Sum of two sides 2300 mm	743.29	940.71	10.07	378.27	m	**1318.98**
Sum of two sides 2400 mm	777.69	984.25	10.47	393.29	m	**1377.54**
Sum of two sides 2500 mm	812.09	1027.78	10.87	408.32	m	**1436.10**
Sum of two sides 2600 mm	846.57	1071.42	11.27	423.34	m	**1494.76**
Sum of two sides 2700 mm	880.93	1114.90	11.67	438.37	m	**1553.27**
Sum of two sides 2800 mm	915.34	1158.46	12.08	453.77	m	**1612.22**
Sum of two sides 2900 mm	949.70	1201.94	12.48	468.79	m	**1670.73**
Sum of two sides 3000 mm	984.10	1245.47	12.88	483.82	m	**1729.29**
Sum of two sides 3100 mm	1018.52	1289.04	13.26	498.09	m	**1787.14**
Sum of two sides 3200 mm	1052.92	1332.58	13.69	514.25	m	**1846.82**
Sum of two sides 3300 mm	1087.34	1376.13	14.09	529.27	m	**1905.40**
Sum of two sides 3400 mm	1121.75	1419.69	14.49	544.30	m	**1963.99**
Sum of two sides 3500 mm	1156.17	1463.25	14.89	559.32	m	**2022.57**

38 VENTILATION/AIR CONDITIONING SYSTEMS

Item	Net Price £	Material £	Labour hours	Labour £	Unit	Total rate £
DUCTWORK: FIRE RATED – cont						
Ductwork 1251 to 2000 mm longest side – cont						
Sum of two sides 3600 mm	1190.57	1506.79	15.29	574.35	m	**2081.13**
Sum of two sides 3700 mm	1224.99	1550.34	15.69	589.37	m	**2139.72**
Sum of two sides 3800 mm	1259.40	1593.90	16.09	604.40	m	**2198.30**
Sum of two sides 3900 mm	1293.81	1637.44	16.49	619.42	m	**2256.86**
Sum of two sides 4000 mm	1332.00	1685.78	16.89	634.45	m	**2320.23**
Extra over fittings; Ductwork 1251 to 2000 mm longest side						
End cap						
Sum of two sides 1700 mm	145.64	184.32	1.42	53.34	nr	**237.66**
Sum of two sides 1800 mm	155.02	196.19	1.45	54.47	nr	**250.66**
Sum of two sides 1900 mm	164.37	208.03	1.48	55.59	nr	**263.62**
Sum of two sides 2000 mm	173.76	219.92	1.51	56.72	nr	**276.64**
Sum of two sides 2100 mm	183.09	231.72	2.66	99.92	nr	**331.64**
Sum of two sides 2200 mm	192.50	243.62	2.80	105.18	nr	**348.80**
Sum of two sides 2300 mm	201.81	255.41	2.95	110.81	nr	**366.22**
Sum of two sides 2400 mm	211.19	267.28	3.10	116.45	nr	**383.73**
Sum of two sides 2500 mm	220.56	279.14	3.24	121.71	nr	**400.84**
Sum of two sides 2600 mm	229.95	291.02	3.39	127.34	nr	**418.37**
Sum of two sides 2700 mm	239.25	302.79	3.54	132.98	nr	**435.77**
Sum of two sides 2800 mm	248.64	314.68	3.68	138.23	nr	**452.92**
Sum of two sides 2900 mm	258.08	326.62	3.83	143.87	nr	**470.49**
Sum of two sides 3000 mm	267.40	338.42	3.98	149.50	nr	**487.93**
Sum of two sides 3100 mm	276.77	350.28	4.12	154.76	nr	**505.04**
Sum of two sides 3200 mm	286.16	362.17	4.27	160.40	nr	**522.56**
Sum of two sides 3300 mm	295.55	374.04	4.42	166.03	nr	**540.08**
Sum of two sides 3400 mm	304.94	385.93	4.57	171.67	nr	**557.60**
Sum of two sides 3500 mm	314.32	397.80	4.72	177.30	nr	**575.10**
Sum of two sides 3600 mm	323.71	409.69	4.87	182.93	nr	**592.62**
Sum of two sides 3700 mm	333.09	421.56	5.02	188.57	nr	**610.13**
Sum of two sides 3800 mm	342.48	433.44	5.17	194.20	nr	**627.65**
Sum of two sides 3900 mm	351.86	445.31	5.32	199.84	nr	**645.15**
Sum of two sides 4000 mm	361.25	457.20	5.47	205.47	nr	**662.67**
Reducer						
Sum of two sides 1700 mm	314.52	398.06	2.92	109.69	nr	**507.74**
Sum of two sides 1800 mm	352.83	446.54	3.10	116.45	nr	**562.99**
Sum of two sides 1900 mm	391.10	494.98	3.29	123.58	nr	**618.56**
Sum of two sides 2000 mm	429.43	543.48	3.48	130.72	nr	**674.21**
Sum of two sides 2100 mm	467.70	591.92	3.54	132.98	nr	**724.89**
Sum of two sides 2200 mm	506.01	640.41	3.60	135.23	nr	**775.64**
Sum of two sides 2300 mm	544.29	688.86	3.65	137.11	nr	**825.97**
Sum of two sides 2400 mm	582.61	737.35	3.66	137.48	nr	**874.83**
Sum of two sides 2500 mm	620.85	785.75	3.70	138.99	nr	**924.73**
Sum of two sides 2600 mm	659.19	834.27	3.85	144.62	nr	**978.89**
Sum of two sides 2700 mm	697.48	882.74	4.03	151.38	nr	**1034.12**
Sum of two sides 2800 mm	735.80	931.22	4.22	158.52	nr	**1089.74**
Sum of two sides 2900 mm	774.09	979.69	4.40	165.28	nr	**1144.97**
Sum of two sides 3000 mm	831.45	1052.29	4.59	172.42	nr	**1224.70**
Sum of two sides 3100 mm	831.45	1052.29	4.78	179.55	nr	**1231.84**

38 VENTILATION/AIR CONDITIONING SYSTEMS

Item	Net Price £	Material £	Labour hours	Labour £	Unit	Total rate £
Sum of two sides 3200 mm	869.79	1100.81	4.96	186.32	nr	**1287.12**
Sum of two sides 3300 mm	882.61	1117.03	5.15	193.45	nr	**1310.48**
Sum of two sides 3400 mm	901.83	1141.36	5.34	200.59	nr	**1341.95**
Sum of two sides 3500 mm	921.06	1165.69	5.53	207.73	nr	**1373.42**
Sum of two sides 3600 mm	940.28	1190.02	5.72	214.86	nr	**1404.88**
Sum of two sides 3700 mm	959.51	1214.35	5.91	222.00	nr	**1436.35**
Sum of two sides 3800 mm	978.73	1238.68	6.10	229.14	nr	**1467.82**
Sum of two sides 3900 mm	997.96	1263.01	6.29	236.28	nr	**1499.29**
Sum of two sides 4000 mm	1017.18	1287.34	6.48	243.41	nr	**1530.76**
Offset						
Sum of two sides 1700 mm	314.52	398.06	4.45	167.16	nr	**565.22**
Sum of two sides 1800 mm	340.44	430.86	4.56	171.29	nr	**602.15**
Sum of two sides 1900 mm	366.41	463.73	4.67	175.42	nr	**639.15**
Sum of two sides 2000 mm	392.21	496.38	5.53	207.73	nr	**704.10**
Sum of two sides 2100 mm	418.14	529.20	5.76	216.37	nr	**745.56**
Sum of two sides 2200 mm	444.06	562.00	5.99	225.01	nr	**787.01**
Sum of two sides 2300 mm	470.02	594.85	6.22	233.65	nr	**828.50**
Sum of two sides 2400 mm	495.89	627.60	6.45	242.29	nr	**869.89**
Sum of two sides 2500 mm	521.81	660.40	6.68	250.93	nr	**911.33**
Sum of two sides 2600 mm	547.76	693.24	6.91	259.56	nr	**952.81**
Sum of two sides 2700 mm	573.68	726.04	7.14	268.20	nr	**994.25**
Sum of two sides 2800 mm	599.62	758.88	7.37	276.84	nr	**1035.72**
Sum of two sides 2900 mm	625.46	791.58	7.60	285.48	nr	**1077.06**
Sum of two sides 3000 mm	651.43	824.45	7.83	294.12	nr	**1118.57**
Sum of two sides 3100 mm	677.36	857.27	8.06	302.76	nr	**1160.03**
Sum of two sides 3200 mm	703.27	890.05	8.29	311.40	nr	**1201.46**
Sum of two sides 3300 mm	729.19	922.86	8.52	320.04	nr	**1242.90**
Sum of two sides 3400 mm	755.10	955.66	8.75	328.68	nr	**1284.34**
Sum of two sides 3500 mm	781.02	988.46	8.98	337.32	nr	**1325.78**
Sum of two sides 3600 mm	806.94	1021.27	9.21	345.96	nr	**1367.23**
Sum of two sides 3700 mm	832.86	1054.07	9.44	354.60	nr	**1408.67**
Sum of two sides 3800 mm	858.78	1086.87	9.67	363.24	nr	**1450.11**
Sum of two sides 3900 mm	884.70	1119.68	9.90	371.88	nr	**1491.56**
Sum of two sides 4000 mm	910.62	1152.48	10.13	380.52	nr	**1533.00**
90° radius bend						
Sum of two sides 1700 mm	437.72	553.98	3.22	120.95	nr	**674.93**
Sum of two sides 1800 mm	477.27	604.03	3.25	122.08	nr	**726.11**
Sum of two sides 1900 mm	503.72	637.51	3.28	123.21	nr	**760.72**
Sum of two sides 2000 mm	516.85	654.12	3.30	123.96	nr	**778.08**
Sum of two sides 2100 mm	539.62	682.94	3.31	124.34	nr	**807.28**
Sum of two sides 2200 mm	636.60	805.68	3.32	124.71	nr	**930.39**
Sum of two sides 2300 mm	675.08	854.39	3.36	126.21	nr	**980.60**
Sum of two sides 2400 mm	714.57	904.36	3.38	126.96	nr	**1031.32**
Sum of two sides 2500 mm	754.17	954.48	3.40	127.72	nr	**1082.20**
Sum of two sides 2600 mm	793.75	1004.58	3.42	128.47	nr	**1133.04**
Sum of two sides 2700 mm	833.33	1054.67	3.44	129.22	nr	**1183.89**
Sum of two sides 2800 mm	872.90	1104.74	3.46	129.97	nr	**1234.71**
Sum of two sides 2900 mm	912.37	1154.69	3.48	130.72	nr	**1285.42**
Sum of two sides 3000 mm	951.96	1204.80	3.50	131.47	nr	**1336.28**
Sum of two sides 3100 mm	991.50	1254.84	3.52	132.22	nr	**1387.07**
Sum of two sides 3200 mm	997.01	1261.81	3.54	132.98	nr	**1394.79**
Sum of two sides 3300 mm	1018.78	1289.37	3.56	133.73	nr	**1423.10**

38 VENTILATION/AIR CONDITIONING SYSTEMS

Item	Net Price £	Material £	Labour hours	Labour £	Unit	Total rate £
DUCTWORK: FIRE RATED – cont						
Extra over fittings – cont						
90° radius bend – cont						
Sum of two sides 3400 mm	1040.56	1316.93	3.58	134.48	nr	**1451.40**
Sum of two sides 3500 mm	1062.33	1344.48	3.60	135.23	nr	**1479.71**
Sum of two sides 3600 mm	1084.10	1372.03	3.62	135.98	nr	**1508.01**
Sum of two sides 3700 mm	1105.87	1399.59	3.64	136.73	nr	**1536.32**
Sum of two sides 3800 mm	1127.64	1427.14	3.66	137.48	nr	**1564.62**
Sum of two sides 3900 mm	1149.41	1454.69	3.68	138.23	nr	**1592.93**
Sum of two sides 4000 mm	1171.18	1482.25	3.70	138.99	nr	**1621.23**
45° bend						
Sum of two sides 1700 mm	324.50	410.69	2.24	84.14	nr	**494.83**
Sum of two sides 1800 mm	360.37	456.09	2.30	86.40	nr	**542.48**
Sum of two sides 1900 mm	396.12	501.33	2.35	88.27	nr	**589.60**
Sum of two sides 2000 mm	432.03	546.78	2.76	103.68	nr	**650.46**
Sum of two sides 2100 mm	539.62	682.94	2.89	108.56	nr	**791.50**
Sum of two sides 2200 mm	556.40	704.18	3.10	116.45	nr	**820.63**
Sum of two sides 2300 mm	595.94	754.22	3.13	117.57	nr	**871.80**
Sum of two sides 2400 mm	603.15	763.34	3.26	122.46	nr	**885.80**
Sum of two sides 2500 mm	611.26	773.62	3.37	126.59	nr	**900.21**
Sum of two sides 2600 mm	634.43	802.93	3.49	131.10	nr	**934.03**
Sum of two sides 2700 mm	647.05	818.91	3.62	135.98	nr	**954.89**
Sum of two sides 2800 mm	659.75	834.97	3.74	140.49	nr	**975.46**
Sum of two sides 2900 mm	682.98	864.38	3.86	145.00	nr	**1009.37**
Sum of two sides 3000 mm	685.08	867.03	3.98	149.50	nr	**1016.54**
Sum of two sides 3100 mm	710.42	899.11	4.10	154.01	nr	**1053.12**
Sum of two sides 3200 mm	718.81	909.72	4.22	158.52	nr	**1068.24**
Sum of two sides 3300 mm	735.69	931.09	4.34	163.03	nr	**1094.11**
Sum of two sides 3400 mm	754.64	955.07	4.46	167.53	nr	**1122.60**
Sum of two sides 3500 mm	761.03	963.16	4.58	172.04	nr	**1135.20**
Sum of two sides 3600 mm	786.32	995.17	4.70	176.55	nr	**1171.72**
Sum of two sides 3700 mm	790.48	1000.43	4.82	181.06	nr	**1181.49**
Sum of two sides 3800 mm	811.70	1027.28	4.94	185.56	nr	**1212.85**
Sum of two sides 3900 mm	826.34	1045.81	5.06	190.07	nr	**1235.89**
Sum of two sides 4000 mm	862.14	1091.13	5.18	194.58	nr	**1285.71**
90° mitre bend						
Sum of two sides 1700 mm	655.59	829.71	3.82	143.49	nr	**973.21**
Sum of two sides 1800 mm	768.89	973.10	3.83	143.87	nr	**1116.97**
Sum of two sides 1900 mm	882.21	1116.53	3.83	143.87	nr	**1260.40**
Sum of two sides 2000 mm	995.43	1259.81	3.84	144.24	nr	**1404.06**
Sum of two sides 2100 mm	1108.77	1403.25	3.85	144.62	nr	**1547.87**
Sum of two sides 2200 mm	1222.05	1546.63	3.85	144.62	nr	**1691.25**
Sum of two sides 2300 mm	1335.38	1690.05	3.86	145.00	nr	**1835.05**
Sum of two sides 2400 mm	1448.67	1833.44	3.86	145.00	nr	**1978.44**
Sum of two sides 2500 mm	1561.96	1976.81	3.87	145.37	nr	**2122.19**
Sum of two sides 2600 mm	1675.19	2120.12	3.87	145.37	nr	**2265.49**
Sum of two sides 2700 mm	1788.50	2263.53	3.87	145.37	nr	**2408.90**
Sum of two sides 2800 mm	1901.80	2406.92	3.88	145.75	nr	**2552.66**
Sum of two sides 2900 mm	2015.10	2550.31	3.88	145.75	nr	**2696.05**
Sum of two sides 3000 mm	2128.38	2693.68	3.89	146.12	nr	**2839.80**
Sum of two sides 3100 mm	2241.72	2837.12	3.90	146.50	nr	**2983.62**

38 VENTILATION/AIR CONDITIONING SYSTEMS

Item	Net Price £	Material £	Labour hours	Labour £	Unit	Total rate £
Sum of two sides 3200 mm	2354.98	2980.46	3.90	146.50	nr	**3126.96**
Sum of two sides 3300 mm	2468.28	3123.85	3.91	146.87	nr	**3270.73**
Sum of two sides 3400 mm	2581.57	3267.24	3.92	147.25	nr	**3414.49**
Sum of two sides 3500 mm	2694.87	3410.62	3.93	147.63	nr	**3558.25**
Sum of two sides 3600 mm	2808.16	3554.01	3.93	147.63	nr	**3701.64**
Sum of two sides 3700 mm	2921.46	3697.40	3.94	148.00	nr	**3845.40**
Sum of two sides 3800 mm	3034.75	3840.78	3.95	148.38	nr	**3989.16**
Sum of two sides 3900 mm	3148.05	3984.17	3.95	148.38	nr	**4132.54**
Sum of two sides 4000 mm	3261.34	4127.55	3.96	148.75	nr	**4276.30**
Branch (Side-on Shoe)						
Sum of two sides 1700 mm	74.33	94.07	1.77	66.49	nr	**160.56**
Sum of two sides 1800 mm	81.19	102.75	1.81	67.99	nr	**170.74**
Sum of two sides 1900 mm	88.16	111.57	1.86	69.87	nr	**181.44**
Sum of two sides 2000 mm	95.07	120.32	1.90	71.37	nr	**191.69**
Sum of two sides 2100 mm	101.98	129.07	2.58	96.91	nr	**225.98**
Sum of two sides 2200 mm	108.87	137.78	2.61	98.04	nr	**235.82**
Sum of two sides 2300 mm	115.82	146.58	2.65	99.54	nr	**246.13**
Sum of two sides 2400 mm	122.71	155.30	2.68	100.67	nr	**255.97**
Sum of two sides 2500 mm	129.63	164.06	2.71	101.80	nr	**265.86**
Sum of two sides 2600 mm	136.57	172.84	2.75	103.30	nr	**276.14**
Sum of two sides 2700 mm	143.42	181.51	2.78	104.43	nr	**285.93**
Sum of two sides 2800 mm	150.38	190.32	2.81	105.55	nr	**295.88**
Sum of two sides 2900 mm	157.29	199.07	2.84	106.68	nr	**305.75**
Sum of two sides 3000 mm	164.19	207.80	2.87	107.81	nr	**315.61**
Sum of two sides 3100 mm	168.15	212.81	2.90	108.93	nr	**321.74**
Sum of two sides 3200 mm	171.11	216.55	2.93	110.06	nr	**326.61**
Sum of two sides 3300 mm	174.44	220.77	2.93	110.06	nr	**330.83**
Sum of two sides 3400 mm	178.05	225.33	3.00	112.69	nr	**338.03**
Sum of two sides 3500 mm	180.73	228.73	3.03	113.82	nr	**342.55**
Sum of two sides 3600 mm	187.02	236.70	3.06	114.94	nr	**351.64**
Sum of two sides 3700 mm	193.32	244.66	3.09	116.07	nr	**360.73**
Sum of two sides 3800 mm	199.61	252.62	3.12	117.20	nr	**369.82**
Sum of two sides 3900 mm	205.90	260.59	3.15	118.33	nr	**378.91**
Sum of two sides 4000 mm	212.19	268.55	3.18	119.45	nr	**388.00**

Rectangular section ductwork to BSEN1366 (Duraduct LT), ducts 'Type A' and 'Type B'; manufactured from 6 mm thick laminate fire board consisting of steel circular hole punched facings pressed to a fibre cement core; provides upto 4 hours stability, 4 hours integrity and 32 minutes insulation; including all necessary stiffeners, joints and supports in the running length

Item	Net Price £	Material £	Labour hours	Labour £	Unit	Total rate £
Ductwork up to 600 mm longest side						
Sum of two sides 200 mm	667.75	845.10	4.00	150.25	m	**995.35**
Sum of two sides 400 mm	684.04	865.72	4.00	150.25	m	**1015.98**
Sum of two sides 600 mm	714.82	904.67	4.00	150.25	m	**1054.92**
Sum of two sides 800 mm	943.27	1193.80	5.50	206.60	m	**1400.40**
Sum of two sides 1000 mm	1270.76	1608.28	5.50	206.60	m	**1814.88**
Sum of two sides 1200 mm	1313.50	1662.37	6.00	225.38	m	**1887.75**

38 VENTILATION/AIR CONDITIONING SYSTEMS

Item	Net Price £	Material £	Labour hours	Labour £	Unit	Total rate £
DUCTWORK: FIRE RATED – cont						
Extra over fittings; Ductwork up to 600 mm longest side						
End cap						
Sum of two sides 200 mm	158.47	200.56	0.81	30.43	m	**230.99**
Sum of two sides 400 mm	158.47	200.56	0.87	32.68	m	**233.24**
Sum of two sides 600 mm	158.47	200.56	0.93	34.93	m	**235.50**
Sum of two sides 800 mm	255.97	323.96	0.98	36.81	m	**360.77**
Sum of two sides 1000 mm	255.97	323.96	1.22	45.83	m	**369.79**
Sum of two sides 1200 mm	390.01	493.60	1.28	48.08	m	**541.68**
Reducer						
Sum of two sides 200 mm	140.14	177.36	2.23	83.77	m	**261.13**
Sum of two sides 400 mm	140.14	177.36	2.51	94.28	m	**271.64**
Sum of two sides 600 mm	140.14	177.36	2.79	104.80	m	**282.16**
Sum of two sides 800 mm	274.24	347.08	3.01	113.07	m	**460.15**
Sum of two sides 1000 mm	274.24	347.08	3.17	119.08	m	**466.16**
Sum of two sides 1200 mm	347.39	439.65	3.28	123.21	m	**562.86**
Offset						
Sum of two sides 200 mm	140.14	177.36	2.95	110.81	m	**288.17**
Sum of two sides 400 mm	140.14	177.36	3.23	121.33	m	**298.69**
Sum of two sides 600 mm	140.14	177.36	3.26	122.46	m	**299.82**
Sum of two sides 800 mm	274.24	347.08	3.57	134.10	m	**481.18**
Sum of two sides 1000 mm	274.24	347.08	3.67	137.86	m	**484.94**
Sum of two sides 1200 mm	347.39	439.65	3.89	146.12	m	**585.78**
90° radius bend						
Sum of two sides 200 mm	469.23	593.86	2.06	77.38	m	**671.24**
Sum of two sides 400 mm	469.23	593.86	2.36	88.65	m	**682.51**
Sum of two sides 600 mm	469.23	593.86	2.66	99.92	m	**693.78**
Sum of two sides 800 mm	548.48	694.16	2.97	111.56	m	**805.72**
Sum of two sides 1000 mm	548.48	694.16	3.04	114.19	m	**808.35**
Sum of two sides 1200 mm	621.63	786.73	3.09	116.07	m	**902.80**
45° radius bend						
Sum of two sides 200 mm	469.23	593.86	1.52	57.10	m	**650.96**
Sum of two sides 400 mm	469.23	593.86	1.66	62.36	m	**656.22**
Sum of two sides 600 mm	469.23	593.86	1.80	67.61	m	**661.48**
Sum of two sides 800 mm	548.48	694.16	1.93	72.50	m	**766.66**
Sum of two sides 1000 mm	548.48	694.16	2.05	77.01	m	**771.16**
Sum of two sides 1200 mm	621.63	786.73	2.17	81.51	m	**868.25**
90° mitre bend						
Sum of two sides 200 mm	469.23	593.86	2.47	92.78	m	**686.64**
Sum of two sides 400 mm	469.23	593.86	2.84	106.68	m	**700.54**
Sum of two sides 600 mm	469.23	593.86	3.20	120.20	m	**714.06**
Sum of two sides 800 mm	548.48	694.16	3.56	133.73	m	**827.88**
Sum of two sides 1000 mm	548.48	694.16	3.66	137.48	m	**831.64**
Sum of two sides 1200 mm	621.63	786.73	3.72	139.74	m	**926.47**
Branch						
Sum of two sides 200 mm	170.65	215.97	0.98	36.81	m	**252.79**
Sum of two sides 400 mm	170.65	215.97	1.14	42.82	m	**258.80**
Sum of two sides 600 mm	170.65	215.97	1.30	48.83	m	**264.81**

38 VENTILATION/AIR CONDITIONING SYSTEMS

Item	Net Price £	Material £	Labour hours	Labour £	Unit	Total rate £
Sum of two sides 800 mm	213.29	269.93	1.46	54.84	m	**324.78**
Sum of two sides 1000 mm	232.56	294.33	1.46	54.84	m	**349.17**
Sum of two sides 1200 mm	286.38	362.44	1.55	58.22	m	**420.66**
Ductwork 601 to 1000 mm longest side						
Sum of two sides 1000 mm	943.27	1193.80	6.00	225.38	m	**1419.18**
Sum of two sides 1100 mm	1313.50	1662.37	6.00	225.38	m	**1887.75**
Sum of two sides 1300 mm	1335.42	1690.11	6.00	225.38	m	**1915.49**
Sum of two sides 1500 mm	1366.56	1729.51	6.00	225.38	m	**1954.90**
Sum of two sides 1700 mm	1651.59	2090.25	6.00	225.38	m	**2315.63**
Sum of two sides 1900 mm	1892.68	2395.38	6.00	225.38	m	**2620.76**
Extra over fittings; Ductwork 601 to 1000 mm longest side						
End cap						
Sum of two sides 1000 mm	255.97	323.96	1.25	46.95	m	**370.92**
Sum of two sides 1100 mm	390.01	493.60	1.25	46.95	m	**540.55**
Sum of two sides 1300 mm	390.01	493.60	1.31	49.21	m	**542.81**
Sum of two sides 1500 mm	390.29	493.95	1.36	51.09	m	**545.03**
Sum of two sides 1700 mm	688.65	871.55	1.42	53.34	m	**924.89**
Sum of two sides 1900 mm	688.65	871.55	1.48	55.59	m	**927.14**
Reducer						
Sum of two sides 1000 mm	274.24	347.08	3.22	120.95	m	**468.03**
Sum of two sides 1100 mm	347.39	439.65	3.22	120.95	m	**560.61**
Sum of two sides 1300 mm	347.39	439.65	3.33	125.09	m	**564.74**
Sum of two sides 1500 mm	347.39	439.65	3.44	129.22	m	**568.87**
Sum of two sides 1700 mm	426.62	539.93	3.54	132.98	m	**672.91**
Sum of two sides 1900 mm	426.62	539.93	3.65	137.11	m	**677.04**
Offset						
Sum of two sides 1000 mm	274.24	347.08	3.78	141.99	m	**489.07**
Sum of two sides 1100 mm	347.39	439.65	3.78	141.99	m	**581.64**
Sum of two sides 1300 mm	347.39	439.65	4.00	150.25	m	**589.91**
Sum of two sides 1500 mm	347.39	439.65	4.23	158.89	m	**598.55**
Sum of two sides 1700 mm	426.62	539.93	4.45	167.16	m	**707.09**
Sum of two sides 1900 mm	426.62	539.93	4.67	175.42	m	**715.36**
90° radius bend						
Sum of two sides 1000 mm	548.48	694.16	3.07	115.32	m	**809.48**
Sum of two sides 1100 mm	621.63	786.73	3.12	117.20	m	**903.93**
Sum of two sides 1300 mm	621.63	786.73	3.17	119.08	m	**905.81**
Sum of two sides 1500 mm	621.63	786.73	3.22	120.95	m	**907.69**
Sum of two sides 1700 mm	816.63	1033.53	3.28	123.21	m	**1156.74**
Sum of two sides 1900 mm	816.63	1033.53	3.78	141.99	m	**1175.52**
45° bend						
Sum of two sides 1000 mm	621.63	786.73	1.91	71.75	m	**858.48**
Sum of two sides 1100 mm	621.63	786.73	2.00	75.13	m	**861.86**
Sum of two sides 1300 mm	621.63	786.73	2.02	75.88	m	**862.61**
Sum of two sides 1500 mm	621.63	786.73	2.13	80.01	m	**866.74**
Sum of two sides 1700 mm	816.63	1033.53	2.24	84.14	m	**1117.67**
Sum of two sides 1900 mm	816.63	1033.53	2.35	88.27	m	**1121.81**

38 VENTILATION/AIR CONDITIONING SYSTEMS

Item	Net Price £	Material £	Labour hours	Labour £	Unit	Total rate £
DUCTWORK: FIRE RATED – cont						
Extra over fittings – cont						
90° mitre bend						
Sum of two sides 1000 mm	548.48	694.16	3.69	138.61	m	832.77
Sum of two sides 1100 mm	621.63	786.73	3.75	140.86	m	927.60
Sum of two sides 1300 mm	621.63	786.73	3.78	141.99	m	928.72
Sum of two sides 1500 mm	621.63	786.73	3.81	143.12	m	929.85
Sum of two sides 1700 mm	816.63	1033.53	3.93	147.63	m	1181.16
Sum of two sides 1900 mm	816.63	1033.53	3.87	145.37	m	1178.90
Branch						
Sum of two sides 1000 mm	221.76	280.66	1.50	56.35	m	337.00
Sum of two sides 1100 mm	286.38	362.44	1.59	59.73	m	422.17
Sum of two sides 1300 mm	286.38	362.44	1.68	63.11	m	425.55
Sum of two sides 1500 mm	286.38	362.44	1.77	66.49	m	428.93
Sum of two sides 1700 mm	365.65	462.77	1.86	69.87	m	532.64
Sum of two sides 1900 mm	365.65	462.77	2.38	89.40	m	552.17
Ductwork 1001 to 1250 mm longest side						
Sum of two sides 1300 mm	1313.50	1662.37	6.00	225.38	m	1887.75
Sum of two sides 1500 mm	1396.10	1766.91	6.00	225.38	m	1992.29
Sum of two sides 1700 mm	1651.59	2090.25	6.00	225.38	m	2315.63
Sum of two sides 1900 mm	1679.11	2125.08	6.00	225.38	m	2350.46
Sum of two sides 2100 mm	2200.38	2784.81	6.50	244.16	m	3028.97
Sum of two sides 2300 mm	2319.93	2936.10	6.50	244.16	m	3180.26
Sum of two sides 2500 mm	2761.77	3495.30	6.50	244.16	m	3739.46
Extra over fittings; Ductwork 1001 to 1250 mm longest side						
End cap						
Sum of two sides 1300 mm	390.01	493.60	1.31	49.21	m	542.81
Sum of two sides 1500 mm	390.01	493.60	1.36	51.09	m	544.68
Sum of two sides 1700 mm	688.65	871.55	1.42	53.34	m	924.89
Sum of two sides 1900 mm	688.65	871.55	1.48	55.59	m	927.14
Sum of two sides 2100 mm	938.49	1187.75	2.66	99.92	m	1287.67
Sum of two sides 2300 mm	938.49	1187.75	2.95	110.81	m	1298.57
Sum of two sides 2500 mm	938.49	1187.75	3.24	121.71	m	1309.46
Reducer						
Sum of two sides 1300 mm	347.39	439.65	3.33	125.09	m	564.74
Sum of two sides 1500 mm	347.39	439.65	3.44	129.22	m	568.87
Sum of two sides 1700 mm	426.62	539.93	3.54	132.98	m	672.91
Sum of two sides 1900 mm	426.62	539.93	3.65	137.11	m	677.04
Sum of two sides 2100 mm	505.79	640.13	3.75	140.86	m	780.99
Sum of two sides 2300 mm	505.79	640.13	3.85	144.62	m	784.75
Sum of two sides 2500 mm	505.79	640.13	3.95	148.38	m	788.51
Offset						
Sum of two sides 1300 mm	347.39	439.65	4.00	150.25	m	589.91
Sum of two sides 1500 mm	347.39	439.65	4.23	158.89	m	598.55
Sum of two sides 1700 mm	426.62	539.93	4.45	167.16	m	707.09

38 VENTILATION/AIR CONDITIONING SYSTEMS

Item	Net Price £	Material £	Labour hours	Labour £	Unit	Total rate £
Sum of two sides 1900 mm	426.62	539.93	4.67	175.42	m	**715.36**
Sum of two sides 2100 mm	505.79	640.13	5.76	216.37	m	**856.50**
Sum of two sides 2300 mm	505.79	640.13	6.22	233.65	m	**873.78**
Sum of two sides 2500 mm	505.79	640.13	6.68	250.93	m	**891.06**
90° radius bend						
Sum of two sides 1300 mm	621.63	786.73	3.12	117.20	m	**903.93**
Sum of two sides 1500 mm	621.63	786.73	3.17	119.08	m	**905.81**
Sum of two sides 1700 mm	816.63	1033.53	3.22	120.95	m	**1154.49**
Sum of two sides 1900 mm	816.63	1033.53	3.28	123.21	m	**1156.74**
Sum of two sides 2100 mm	834.90	1056.65	3.32	124.71	m	**1181.36**
Sum of two sides 2300 mm	834.90	1056.65	3.36	126.21	m	**1182.86**
Sum of two sides 2500 mm	834.90	1056.65	3.40	127.72	m	**1184.37**
45° bend						
Sum of two sides 1300 mm	621.63	786.73	2.02	75.88	m	**862.61**
Sum of two sides 1500 mm	621.63	786.73	2.13	80.01	m	**866.74**
Sum of two sides 1700 mm	816.63	1033.53	2.24	84.14	m	**1117.67**
Sum of two sides 1900 mm	816.63	1033.53	2.35	88.27	m	**1121.81**
Sum of two sides 2100 mm	834.90	1056.65	2.89	108.56	m	**1165.21**
Sum of two sides 2300 mm	834.90	1056.65	3.13	117.57	m	**1174.22**
Sum of two sides 2500 mm	834.90	1056.65	3.37	126.59	m	**1183.24**
90° mitre bend						
Sum of two sides 1300 mm	621.63	786.73	3.75	140.86	m	**927.60**
Sum of two sides 1500 mm	621.63	786.73	3.81	143.12	m	**929.85**
Sum of two sides 1700 mm	816.63	1033.53	3.87	145.37	m	**1178.90**
Sum of two sides 1900 mm	816.63	1033.53	3.93	147.63	m	**1181.16**
Sum of two sides 2100 mm	834.90	1056.65	3.99	149.88	m	**1206.53**
Sum of two sides 2300 mm	834.90	1056.65	4.05	152.13	m	**1208.78**
Sum of two sides 2500 mm	834.90	1056.65	4.11	154.39	m	**1211.04**
Branch						
Sum of two sides 1300 mm	286.38	362.44	1.59	59.73	m	**422.17**
Sum of two sides 1500 mm	286.38	362.44	1.68	63.11	m	**425.55**
Sum of two sides 1700 mm	365.65	462.77	1.77	66.49	m	**529.26**
Sum of two sides 1900 mm	365.65	462.77	1.86	69.87	m	**532.64**
Sum of two sides 2100 mm	383.88	485.84	2.58	96.91	m	**582.75**
Sum of two sides 2300 mm	383.88	485.84	2.64	99.17	m	**585.01**
Sum of two sides 2500 mm	383.88	485.84	2.91	109.31	m	**595.15**
Ductwork 1251 to 2000 mm longest side						
Sum of two sides 1800 mm	1651.59	2090.25	6.00	225.38	m	**2315.63**
Sum of two sides 2000 mm	1926.81	2438.57	6.00	225.38	m	**2663.95**
Sum of two sides 2200 mm	2200.38	2784.81	6.50	244.16	m	**3028.97**
Sum of two sides 2400 mm	2474.44	3131.65	6.50	244.16	m	**3375.82**
Sum of two sides 2600 mm	2606.87	3299.26	6.66	250.17	m	**3549.43**
Sum of two sides 2800 mm	2859.47	3618.95	6.66	250.17	m	**3869.12**
Sum of two sides 3000 mm	3110.52	3936.68	6.66	250.17	m	**4186.85**
Sum of two sides 3200 mm	3133.78	3966.11	9.00	338.07	m	**4304.18**
Sum of two sides 3400 mm	3447.15	4362.71	9.00	338.07	m	**4700.78**
Sum of two sides 3600 mm	3509.39	4441.48	11.70	439.49	m	**4880.98**
Sum of two sides 3800 mm	3821.65	4836.67	11.70	439.49	m	**5276.17**
Sum of two sides 4000 mm	4421.06	5595.30	11.70	439.49	m	**6034.79**

38 VENTILATION/AIR CONDITIONING SYSTEMS

Item	Net Price £	Material £	Labour hours	Labour £	Unit	Total rate £
DUCTWORK: FIRE RATED – cont						
Extra over fittings; Ductwork 1251 to 2000 mm longest sides						
End cap						
Sum of two sides 1800 mm	688.65	871.55	1.45	54.47	m	**926.02**
Sum of two sides 2000 mm	688.65	871.55	1.51	56.72	m	**928.27**
Sum of two sides 2200 mm	938.49	1187.75	2.80	105.18	m	**1292.93**
Sum of two sides 2400 mm	938.49	1187.75	3.10	116.45	m	**1304.20**
Sum of two sides 2600 mm	1285.80	1627.31	3.39	127.34	m	**1754.65**
Sum of two sides 2800 mm	1285.80	1627.31	3.68	138.23	m	**1765.54**
Sum of two sides 3000 mm	1285.80	1627.31	3.98	149.50	m	**1776.81**
Sum of two sides 3200 mm	1736.76	2198.05	4.27	160.40	m	**2358.45**
Sum of two sides 3400 mm	1736.76	2198.05	4.57	171.67	m	**2369.72**
Sum of two sides 3600 mm	2138.94	2707.04	4.87	182.93	m	**2889.98**
Sum of two sides 3800 mm	2138.94	2707.04	5.17	194.20	m	**2901.25**
Sum of two sides 4000 mm	2138.94	2707.04	5.47	205.47	m	**2912.51**
Reducer						
Sum of two sides 1800 mm	426.62	539.93	3.10	116.45	m	**656.38**
Sum of two sides 2000 mm	426.62	539.93	3.48	130.72	m	**670.66**
Sum of two sides 2200 mm	505.79	640.13	3.60	135.23	m	**775.36**
Sum of two sides 2400 mm	505.79	640.13	3.70	138.99	m	**779.12**
Sum of two sides 2600 mm	700.78	886.91	3.85	144.62	m	**1031.53**
Sum of two sides 2800 mm	700.78	886.91	4.22	158.52	m	**1045.43**
Sum of two sides 3000 mm	700.78	886.91	4.59	172.42	m	**1059.33**
Sum of two sides 3200 mm	780.01	987.18	4.96	186.32	m	**1173.49**
Sum of two sides 3400 mm	780.01	987.18	5.34	200.59	m	**1187.77**
Sum of two sides 3600 mm	853.17	1079.77	5.72	214.86	m	**1294.63**
Sum of two sides 3800 mm	853.17	1079.77	6.10	229.14	m	**1308.91**
Sum of two sides 4000 mm	853.17	1079.77	6.48	243.41	m	**1323.18**
Offset						
Sum of two sides 1800 mm	426.62	539.93	4.56	171.29	m	**711.22**
Sum of two sides 2000 mm	426.62	539.93	5.53	207.73	m	**747.66**
Sum of two sides 2200 mm	505.79	640.13	5.99	225.01	m	**865.14**
Sum of two sides 2400 mm	505.79	640.13	6.45	242.29	m	**882.42**
Sum of two sides 2600 mm	700.78	886.91	6.91	259.56	m	**1146.48**
Sum of two sides 2800 mm	700.78	886.91	7.37	276.84	m	**1163.76**
Sum of two sides 3000 mm	700.78	886.91	7.83	294.12	m	**1181.04**
Sum of two sides 3200 mm	780.01	987.18	8.29	311.40	m	**1298.58**
Sum of two sides 3400 mm	780.01	987.18	8.75	328.68	m	**1315.86**
Sum of two sides 3600 mm	853.17	1079.77	9.21	345.96	m	**1425.73**
Sum of two sides 3800 mm	853.17	1079.77	9.67	363.24	m	**1443.01**
Sum of two sides 4000 mm	853.17	1079.77	10.13	380.52	m	**1460.29**
90° radius bend						
Sum of two sides 1800 mm	816.63	1033.53	3.25	122.08	m	**1155.61**
Sum of two sides 2000 mm	816.63	1033.53	3.30	123.96	m	**1157.49**
Sum of two sides 2200 mm	834.90	1056.65	3.34	125.46	m	**1182.11**
Sum of two sides 2400 mm	834.90	1056.65	3.38	126.96	m	**1183.61**
Sum of two sides 2600 mm	1279.74	1619.63	3.42	128.47	m	**1748.10**
Sum of two sides 2800 mm	1279.74	1619.63	3.46	129.97	m	**1749.61**
Sum of two sides 3000 mm	1279.74	1619.63	3.50	131.47	m	**1751.11**

38 VENTILATION/AIR CONDITIONING SYSTEMS

Item	Net Price £	Material £	Labour hours	Labour £	Unit	Total rate £
Sum of two sides 3200 mm	1535.68	1943.56	3.54	132.98	m	**2076.54**
Sum of two sides 3400 mm	1535.68	1943.56	3.58	134.48	m	**2078.04**
Sum of two sides 3600 mm	1797.68	2275.14	3.62	135.98	m	**2411.13**
Sum of two sides 3800 mm	1797.68	2275.14	3.66	137.48	m	**2412.63**
Sum of two sides 4000 mm	1797.68	2275.14	3.70	138.99	m	**2414.13**
45° bend						
Sum of two sides 1800 mm	816.63	1033.53	2.30	86.40	m	**1119.93**
Sum of two sides 2000 mm	816.63	1033.53	2.76	103.68	m	**1137.21**
Sum of two sides 2200 mm	834.90	1056.65	3.01	113.07	m	**1169.72**
Sum of two sides 2400 mm	834.90	1056.65	3.26	122.46	m	**1179.11**
Sum of two sides 2600 mm	1279.74	1619.63	3.49	131.10	m	**1750.73**
Sum of two sides 2800 mm	1279.74	1619.63	3.74	140.49	m	**1760.12**
Sum of two sides 3000 mm	1279.74	1619.63	3.98	149.50	m	**1769.14**
Sum of two sides 3200 mm	1535.68	1943.56	4.22	158.52	m	**2102.08**
Sum of two sides 3400 mm	1535.68	1943.56	4.46	167.53	m	**2111.10**
Sum of two sides 3600 mm	1797.68	2275.14	4.70	176.55	m	**2451.69**
Sum of two sides 3800 mm	1797.68	2275.14	4.94	185.56	m	**2460.71**
Sum of two sides 4000 mm	1797.68	2275.14	5.18	194.58	m	**2469.72**
90° mitre bend						
Sum of two sides 1800 mm	816.63	1033.53	3.83	143.87	m	**1177.40**
Sum of two sides 2000 mm	816.63	1033.53	3.84	144.24	m	**1177.78**
Sum of two sides 2200 mm	834.90	1056.65	3.85	144.62	m	**1201.27**
Sum of two sides 2400 mm	1110.12	1404.97	3.86	145.00	m	**1549.96**
Sum of two sides 2600 mm	1279.74	1619.63	3.87	145.37	m	**1765.01**
Sum of two sides 2800 mm	1279.74	1619.63	3.88	145.75	m	**1765.38**
Sum of two sides 3000 mm	1279.74	1619.63	3.90	146.50	m	**1766.13**
Sum of two sides 3200 mm	1535.68	1943.56	3.90	146.50	m	**2090.06**
Sum of two sides 3400 mm	1535.68	1943.56	3.92	147.25	m	**2090.81**
Sum of two sides 3600 mm	1797.68	2275.14	3.94	148.00	m	**2423.15**
Sum of two sides 3800 mm	1797.68	2275.14	3.95	148.38	m	**2423.52**
Sum of two sides 4000 mm	1797.68	2275.14	3.96	148.75	m	**2423.90**
Branch						
Sum of two sides 1800 mm	365.65	462.77	1.81	67.99	m	**530.76**
Sum of two sides 2000 mm	365.65	462.77	1.90	71.37	m	**534.14**
Sum of two sides 2200 mm	383.88	485.84	2.61	98.04	m	**583.88**
Sum of two sides 2400 mm	383.88	485.84	2.68	100.67	m	**586.51**
Sum of two sides 2600 mm	457.03	578.41	2.75	103.30	m	**681.71**
Sum of two sides 2800 mm	457.03	578.41	2.81	105.55	m	**683.97**
Sum of two sides 3000 mm	457.03	578.41	2.87	107.81	m	**686.22**
Sum of two sides 3200 mm	475.36	601.62	2.93	110.06	m	**711.68**
Sum of two sides 3400 mm	475.36	601.62	3.00	112.69	m	**714.31**
Sum of two sides 3600 mm	609.42	771.29	3.06	114.94	m	**886.23**
Sum of two sides 3800 mm	609.42	771.29	3.12	117.20	m	**888.49**
Sum of two sides 4000 mm	609.42	771.29	3.18	119.45	m	**890.74**

38 VENTILATION/AIR CONDITIONING SYSTEMS

Item	Net Price £	Material £	Labour hours	Labour £	Unit	Total rate £
LOW VELOCITY AIR CONDITIONING: AIR HANDLING UNITS						
Supply air handling unit; inlet with motorized damper, LTHW frost coil (at −5°C to +5°C), panel filter (EU4), bag filter (EU6), cooling coil (at 28°C db/20°C wb to 12°C db/11.5°C wb), LTHW heating coil (at 5°C to 21°C), supply fan, outlet plenum; includes access sections; all units located internally; Includes placing in position and fitting of sections together; electrical work elsewhere						
Volume, external pressure						
2 m³/s at 350 Pa	11777.95	14906.18	40.00	1573.03	nr	**16479.21**
2 m³/s at 700 Pa	12583.07	15925.13	40.00	1573.03	nr	**17498.17**
5 m³/s at 350 Pa	19859.99	25134.80	65.00	2556.18	nr	**27690.98**
5 m³/s at 700 Pa	20554.72	26014.06	65.00	2556.18	nr	**28570.24**
8 m³/s at 350 Pa	29746.04	37646.59	77.00	3028.09	nr	**40674.68**
8 m/³s at 700 Pa	30209.76	38233.47	77.00	3028.09	nr	**41261.55**
10 m/³ at 350 Pa	32756.12	41456.15	100.00	3932.58	nr	**45388.73**
10 m³/s at 700 Pa	34319.98	43435.37	100.00	3932.58	nr	**47367.95**
13 m³/s at 350 Pa	42025.20	53187.09	108.00	4247.19	nr	**57434.28**
13 m³/s at 700 Pa	43323.15	54829.78	108.00	4247.19	nr	**59076.97**
15 m³/s at 350 Pa	48128.12	60910.94	120.00	4719.10	nr	**65630.04**
15 m³/s at 700 Pa	49003.83	62019.25	120.00	4719.10	nr	**66738.34**
18 m³/s at 350 Pa	54613.05	69118.28	133.00	5230.33	nr	**74348.61**
18 m³/s at 700 Pa	55617.01	70388.89	133.00	5230.33	nr	**75619.22**
20 m³/s at 350 Pa	63287.25	80096.35	142.00	5584.27	nr	**85680.61**
20 m³/s at 700 Pa	64197.39	81248.22	142.00	5584.27	nr	**86832.48**
Extra for inlet and discharge attenuators at 900 mm long						
2 m³/s at 350 Pa	3638.94	4605.44	5.00	196.63	nr	**4802.07**
5 m³/s at 350 Pa	7637.28	9665.74	10.00	393.26	nr	**10058.99**
10 m³/s at 700 Pa	10866.23	13752.30	13.00	511.24	nr	**14263.53**
15 m³/s at 700 Pa	16477.36	20853.75	16.00	629.21	nr	**21482.96**
20 m³/s at 700 Pa	21264.14	26911.89	20.00	786.52	nr	**27698.41**
Extra for locating units externally						
2 m³/s at 350 Pa	2696.27	3412.40	–	–	nr	**3412.40**
5 m³/s at 350 Pa	3421.05	4329.68	–	–	nr	**4329.68**
10 m³/s at 700 Pa	5473.02	6926.66	–	–	nr	**6926.66**
15 m³/s at 700 Pa	7909.32	10010.04	–	–	nr	**10010.04**
20 m³/s at 700 Pa	13709.41	17350.63	–	–	nr	**17350.63**

38 VENTILATION/AIR CONDITIONING SYSTEMS

Item	Net Price £	Material £	Labour hours	Labour £	Unit	Total rate £
Modular air handling unit with supply and extract sections. Supply side; inlet with motorized damper, LTHW frost coil (at −5°C to 5°C), panel filter (EU4), bag filter (EU6), cooling coil at 28°C db/20°C wb to 12°C db/ 11.5°C wb), LTHW heating coil (at 5°C to 21°C), supply fan, outlet plenum. Extract side; inlet with motorized damper, extract fan; includes access sections; placing in position and fitting of sections together; electrical work elsewhere						
Volume, external pressure						
2 m³/s at 350 Pa	16453.36	20823.37	50.00	1966.29	nr	**22789.66**
2 m³/s at 700 Pa	17249.48	21830.94	50.00	1966.29	nr	**23797.23**
5 m³/s at 350 Pa	28356.55	35888.05	86.00	3382.02	nr	**39270.07**
5 m³/s at 700 Pa	28888.80	36561.66	86.00	3382.02	nr	**39943.68**
8 m³/s at 350 Pa	39377.15	49835.72	105.00	4129.21	nr	**53964.93**
8 m³/s at 700 Pa	40322.37	51031.99	105.00	4129.21	nr	**55161.20**
10 m³/s at 350 Pa	45468.86	57545.39	120.00	4719.10	nr	**62264.49**
10 m³/s at 700 Pa	46448.40	58785.10	120.00	4719.10	nr	**63504.19**
13 m³/s at 350 Pa	58511.70	74052.41	130.00	5112.36	nr	**79164.77**
13 m³/s at 700 Pa	60346.95	76375.10	130.00	5112.36	nr	**81487.45**
15 m³/s at 350 Pa	65382.71	82748.36	145.00	5702.24	nr	**88450.60**
15 m³/s at 700 Pa	68475.86	86663.04	145.00	5702.24	nr	**92365.29**
18 m³/s at 350 Pa	74740.31	94591.33	160.00	6292.13	nr	**100883.46**
18 m³/s at 700 Pa	76490.16	96805.95	160.00	6292.13	nr	**103098.08**
20 m³/s at 350 Pa	81688.12	103384.48	175.00	6882.02	nr	**110266.50**
20 m³/s at 700 Pa	84527.95	106978.58	175.00	6882.02	nr	**113860.59**
Extra for inlet and discharge attenuators at 900 mm long						
2 m³/s at 350 Pa	6633.98	8395.96	8.00	314.61	nr	**8710.57**
5 m³/s at 350 Pa	13100.21	16579.63	10.00	393.26	nr	**16972.89**
10 m³/s at 700 Pa	18745.31	23724.07	13.00	511.24	nr	**24235.30**
15 m³/s at 700 Pa	32064.70	40581.08	16.00	629.21	nr	**41210.29**
20 m³/s at 700 Pa	37430.37	47371.88	20.00	786.52	nr	**48158.39**
Extra for locating units externally						
2 m³/s at 350 Pa	5310.21	6720.60	–	–	nr	**6720.60**
5 m³/s at 350 Pa	7540.45	9543.20	–	–	nr	**9543.20**
10 m³/s at 700 Pa	10763.48	13622.26	–	–	nr	**13622.26**
15 m³/s at 700 Pa	20437.04	25865.11	–	–	nr	**25865.11**
20 m³/s at 700 Pa	31417.99	39762.61	–	–	nr	**39762.61**
Extra for humidifier, self-generating type						
2 m³/s at 350 Pa (10 kg/hr)	4627.48	5856.54	5.00	196.63	nr	**6053.16**
5 m³/s at 350 Pa (18 kg/hr)	5726.59	7247.57	5.00	196.63	nr	**7444.20**
10 m³/s at 700 Pa (30 kg/hr)	6516.57	8247.37	6.00	235.95	nr	**8483.33**
15 m³/s at 700 Pa (60 kg/hr)	12068.28	15273.62	8.00	314.61	nr	**15588.23**
20 m³/s at 700 Pa (90 kg/hr)	18144.60	22963.80	10.00	393.26	nr	**23357.06**
Extra for mixing box						
2 m³/s at 350 Pa	2222.02	2812.19	4.00	157.30	nr	**2969.49**
5 m³/s at 350 Pa	3132.05	3963.92	4.00	157.30	nr	**4121.23**
10 m³/s at 700 Pa	4387.98	5553.43	5.00	196.63	nr	**5750.06**
15 m³/s at 700 Pa	5856.32	7411.76	6.00	235.95	nr	**7647.72**
20 m³/s at 700 Pa	8927.81	11299.03	6.00	235.95	nr	**11534.99**

38 VENTILATION/AIR CONDITIONING SYSTEMS

Item	Net Price £	Material £	Labour hours	Labour £	Unit	Total rate £
LOW VELOCITY AIR CONDITIONING: AIR HANDLING UNITS – cont						
Modular air handling unit with supply and extract sections – cont						
Extra for runaround coil, including pump and associated pipework; typical outputs in brackets (based on minimal distance between the supply and extract units)						
2 m³/s at 350 Pa (26 kW)	6652.02	8418.79	30.00	1179.77	nr	**9598.57**
5 m³/s at 350 Pa (37 kW)	11794.16	14926.69	30.00	1179.77	nr	**16106.47**
10 m³/s at 700 Pa (85 kW)	20419.79	25843.29	40.00	1573.03	nr	**27416.32**
15 m³/s at 700 Pa (151 kW)	28731.38	36362.44	50.00	1966.29	nr	**38328.73**
20 m³/s at 700 Pa (158 kW)	39164.94	49567.15	60.00	2359.55	nr	**51926.70**
Extra for thermal wheel (typical outputs in brackets)						
2 m³/s at 350 Pa (37 kW)	13758.81	17413.15	12.00	471.91	nr	**17885.06**
5 m³/s at 350 Pa (65 kW)	19760.98	25009.50	12.00	471.91	nr	**25481.41**
10 m³/s at 700 Pa (127 kW)	26352.72	33352.00	15.00	589.89	nr	**33941.88**
15 m³/s at 700 Pa (160 kW)	49373.80	62487.48	17.00	668.54	nr	**63156.02**
20 m³/s at 700 Pa (262 kW)	57674.34	72992.65	19.00	747.19	nr	**73739.84**
Extra for plate heat exchanger, including additional filtration in extract leg (typical outputs in brackets)						
2 m³/s at 350 Pa (25 kW)	8720.07	11036.12	12.00	471.91	nr	**11508.03**
5 m³/s at 350 Pa (51 kW)	17046.64	21574.23	12.00	471.91	nr	**22046.14**
10 m³/s at 700 Pa (98 kW)	25012.16	31655.39	15.00	589.89	nr	**32245.28**
15 m³/s at 700 Pa (160 kW)	43193.92	54666.23	17.00	668.54	nr	**55334.77**
20 m³/s at 700 Pa (190 kW)	55781.34	70596.87	19.00	747.19	nr	**71344.06**
Extra for electric heating in lieu of LTHW						
2 m³/s at 350 Pa	3022.28	3824.99	–	–	nr	**3824.99**
5 m³/s at 350 Pa	5662.56	7166.54	–	–	nr	**7166.54**
10 m³/s at 700 Pa	7645.35	9675.96	–	–	nr	**9675.96**
15 m³/s at 700 Pa	8016.87	10146.16	–	–	nr	**10146.16**
20 m³/s at 700 Pa	9461.05	11973.91	–	–	nr	**11973.91**

38 VENTILATION/AIR CONDITIONING SYSTEMS

Item	Net Price £	Material £	Labour hours	Labour £	Unit	Total rate £
VAV AIR CONDITIONING						
VAV terminal boxes						
VAV terminal box; integral acoustic silencer; factory installed and prewired control components (excluding electronic controller); selected at 200 Pa at entry to unit; includes fixing in position; electrical work elsewhere						
80 l/s–110 l/s	757.74	959.00	2.00	75.13	nr	**1034.13**
Extra for secondary silencer	184.28	233.22	0.50	18.78	nr	**252.00**
Extra for 2 row LTHW heating coil	455.25	576.17	–	–	nr	**576.17**
150 l/s–190 l/s	797.36	1009.13	2.00	75.13	nr	**1084.26**
Extra for secondary silencer	202.34	256.09	0.50	18.78	nr	**274.87**
Extra for 2 row LTHW heating coil	487.75	617.30	–	–	nr	**617.30**
250 l/s–310 l/s	889.49	1125.74	2.00	75.13	nr	**1200.87**
Extra for secondary silencer	267.81	338.94	0.50	18.78	nr	**357.72**
Extra for 2 row LTHW heating coil	552.80	699.62	–	–	nr	**699.62**
420 l/s–520 l/s	963.55	1219.47	2.00	75.13	nr	**1294.59**
Extra for secondary silencer	304.81	385.77	0.50	18.78	nr	**404.55**
Extra for 2 row LTHW heating coil	673.11	851.89	–	–	nr	**851.89**
650 l/s–790 l/s	1166.76	1476.65	2.00	75.13	nr	**1551.77**
Extra for secondary silencer	390.07	493.67	0.50	18.78	nr	**512.45**
Extra for 2 row LTHW heating coil	705.64	893.05	–	–	nr	**893.05**
1130 l/s–1370 l/s	1355.34	1715.32	2.00	75.13	nr	**1790.45**
Extra for secondary silencer	545.06	689.83	0.50	18.78	nr	**708.61**
Extra for 2 row LTHW heating coil	738.13	934.18	–	–	nr	**934.18**
Extra for electric heater & thyristor controls, 3 kW/lph (per box)	601.56	761.33	–	–	nr	**761.33**
Fan assisted VAV terminal box; factory installed and prewired control components (excluding electronic controller); selected at 40 Pa external static pressure; includes fixing in position, electrical work elsewhere						
100 l/s–175 l/s	1677.40	2122.92	3.00	112.69	nr	**2235.61**
Extra for secondary silencer	199.76	252.82	0.50	18.78	nr	**271.60**
Extra for 1 row LTHW heating coil	536.54	679.04	–	–	nr	**679.04**
170 l/s–360 l/s	1829.79	2315.78	3.00	112.69	nr	**2428.47**
Extra for secondary silencer	253.16	320.40	0.50	18.78	nr	**339.18**
Extra for 1 row LTHW heating coil	569.06	720.20	–	–	nr	**720.20**
300 l/s–640 l/s	2097.59	2654.70	3.00	112.69	nr	**2767.40**
Extra for secondary silencer	387.50	490.42	0.50	18.78	nr	**509.20**
Extra for 1 row LTHW heating coil	617.83	781.93	–	–	nr	**781.93**
620 l/s–850 l/s	2097.59	2654.70	3.00	112.69	nr	**2767.40**
Extra for secondary silencer	387.50	490.42	0.50	18.78	nr	**509.20**
Extra for 1 row LTHW heating coil	666.61	843.66	–	–	nr	**843.66**
Extra for electric heater plus thyristor controls 3 kW/lph (per box)	617.83	781.93	–	–	nr	**781.93**
Extra for fitting free issue controller	–	–	2.13	75.57	nr	**75.57**

38 VENTILATION/AIR CONDITIONING SYSTEMS

Item	Net Price £	Material £	Labour hours	Labour £	Unit	Total rate £
FAN COIL AIR CONDITIONING						
All selections based on summer return air condition of 23°C @ 50% RH, CHW @ 6°/12°C, LTHW @ 82°/71°C (where applicable), medium speed, external resistance of 30 Pa						
All selections are based on heating and cooling units. For waterside control units there is no significant reduction in cost between 4 pipe heating and cooling and 2 pipe cooling only units (excluding controls). For airside control units, there is a marginal reduction (less than 5%) between 4 pipe heating and cooling units and 2 pipe cooling only units (excluding controls)						
Ceiling void mounted horizontal waterside control fan coil unit; cooling coil; LTHW heating coil; multi-tapped speed transformer; fine wire mesh filter; includes fixing in position; electrical work elsewhere						
Total cooling load, heating load						
2800 W, 1000 W	639.89	809.84	4.00	157.30	nr	**967.14**
4000 W, 1700 W	672.33	850.90	4.00	157.30	nr	**1008.20**
4500 W, 1900 W	992.48	1256.08	4.00	157.30	nr	**1413.38**
6000 W, 2600 W	1298.09	1642.86	4.00	157.30	nr	**1800.16**
Ceiling void mounted horizontal waterside control fan coil unit; cooling coil; electric heating coil; multi-tapped speed transformer; fine wire mesh filter; includes fixing in position; electrical work elsewhere + thyristor and 2 No. HTCOs						
Total cooling load, heating load						
2800 W, 1500 W	932.55	1180.23	4.00	157.30	nr	**1337.53**
4000 W, 2000 W	964.86	1221.13	4.00	157.30	nr	**1378.43**
4500 W, 2000 W	1228.55	1554.85	4.00	157.30	nr	**1712.15**
6000 W, 3000 W	1517.79	1920.92	4.00	157.30	nr	**2078.22**
Ceiling void mounted horizontal airside control fan coil unit; cooling coil; LHTW heating coil; multi-tapped speed transformer; fine wire mesh filter, damper actuator & fixing kit; includes fixing in position; electrical work elsewhere						
Total cooling load, heating load						
2600 W, 2200 W	1056.51	1337.12	4.00	157.30	nr	**1494.42**
3600 W, 3200 W	1087.07	1375.80	4.00	157.30	nr	**1533.10**
4000 W, 3600 W	1216.59	1539.72	4.00	157.30	nr	**1697.02**
5400 W, 5000 W	1452.34	1838.08	4.00	157.30	nr	**1995.39**
Ceiling void mounted horizontal airside control fan coil unit; cooling coil; electric heating coil; multi-tapped speed transformer; fine wire mesh filter, damper actuator & fixing kit; includes fixing in position; electrical work elsewhere + thyristor and 2 No. HTCOs						
Total cooling load, heating load						
2600 W, 1500 W	1124.91	1423.68	4.00	157.30	nr	**1580.98**
3600 W, 2000 W	1154.01	1460.51	4.00	157.30	nr	**1617.82**
4000 W, 2000 W	1276.26	1615.23	4.00	157.30	nr	**1772.53**
5400 W, 3000 W	1516.36	1919.11	4.00	157.30	nr	**2076.41**

38 VENTILATION/AIR CONDITIONING SYSTEMS

Item	Net Price £	Material £	Labour hours	Labour £	Unit	Total rate £
Ceiling void mounted slimline horizontal waterside control fan coil unit, 170 mm deep; cooilng coil; LTHW heating coil; multi-tapped speed transformer; fine wire mesh filter; includes fixing in position; electrical work elsewhere						
Total cooling load, heating load						
1100 W, 1500 W	829.49	1049.80	3.50	137.64	nr	**1187.44**
3200 W, 3700 W	1584.76	2005.67	4.00	157.30	nr	**2162.97**
Ceiling void mounted slimline horizontal waterside control fan coil unit, 170 mm deep; cooilng coil; electric heating coil; multi-tapped speed transformer; fine wire mesh filter; includes fixing in position; electrical work elsewhere						
Total cooling load, heating load						
1100 W, 1000 W	612.66	775.38	3.50	137.64	nr	**913.02**
3400 W, 2000 W	1725.93	2184.33	4.00	157.30	nr	**2341.64**
Ceiling void mounted slimline horizontal airside control fan coil unit, 170 mm deep; cooling coil; multi-tapped speed transformer; fine wire mesh filter,damper actuator & fixing kit; includes fixing in position; electrical work elsewhere						
Total cooling load						
1000 W	1076.88	1362.90	3.50	137.64	nr	**1500.54**
3000 W	1436.34	1817.83	4.00	157.30	nr	**1975.13**
Ceiling void mounted slimline horizontal airside control fan coil unit, 170 mm deep; cooilng coil; electric heating coil; multi-tapped speed transformer; fine wire mesh filter, damper actuator & fixing kit; includes fixing in position; electrical work elsewhere						
Total cooling load, heating load						
1000 W, 1000 W	797.48	1009.29	3.50	137.64	nr	**1146.93**
3000 W, 2000 W	1577.48	1996.46	4.00	157.30	nr	**2153.77**
Low level perimeter waterside control fan coil unit; cooling coil; LTHW heating coil; multi-tapped speed transformer; fine wire mesh filter; includes fixing in position; electrical work elsewhere						
Total cooling load, heating load						
1700 W, 1400 W	730.54	924.58	3.50	137.64	nr	**1062.22**
Extra over for standard cabinet	289.17	365.98	1.00	39.33	nr	**405.30**
2200 W, 1900 W	944.45	1195.30	3.50	137.64	nr	**1332.94**
Extra over for standard cabinet	352.24	445.79	1.00	39.33	nr	**485.12**
2600 W, 2200 W	944.45	1195.30	3.50	137.64	nr	**1332.94**
Extra over for standard cabinet	360.93	456.79	1.00	39.33	nr	**496.12**
3900 W, 3200 W	1203.49	1523.14	3.50	137.64	nr	**1660.78**
Extra over for standard cabinet	393.55	498.08	1.00	39.33	nr	**537.40**

38 VENTILATION/AIR CONDITIONING SYSTEMS

Item	Net Price £	Material £	Labour hours	Labour £	Unit	Total rate £
FAN COIL AIR CONDITIONING – cont						
Low level perimeter waterside control fan coil unit; cooling coil; electric heating coil; multi-tapped speed transformer; fine wire mesh filter; includes fixing in position; electrical work elsewhere						
Total cooling load, heating load						
1700 W, 1500 W	868.78	1099.53	3.50	137.64	nr	**1237.17**
Extra over for standard cabinet	289.17	365.98	1.00	39.33	nr	**405.30**
2200 W, 2000 W	1085.61	1373.95	3.50	137.64	nr	**1511.59**
Extra over for standard cabinet	352.24	445.79	1.00	39.33	nr	**485.12**
2600 W, 2000 W	1085.61	1373.95	3.50	137.64	nr	**1511.59**
Extra over for standard cabinet	360.93	456.79	1.00	39.33	nr	**496.12**
3800 W, 3000 W	1347.56	1705.47	3.50	137.64	nr	**1843.11**
Extra over for standard cabinet	393.55	498.08	1.00	39.33	nr	**537.40**

38 VENTILATION/AIR CONDITIONING SYSTEMS

Item	Net Price £	Material £	Labour hours	Labour £	Unit	Total rate £
MECHANICAL VENTILATION WITH HEAT RECOVERY (MVHR) UNITS						
A ventilation unit with high efficiency (up to 130%) rotary heat exchanger (rotary heat exchangers do not require a condense drain or heater battery installed). The unit recovers hot, cold and humid air. Includes main unit, built-in controller and delivery						
Air flow rate						
40–200 m³/h, 55 l/s	1778.34	2250.67	3.00	117.98	nr	**2368.65**
50–250 m³/h, 69 l/s	1959.67	2480.16	3.00	117.98	nr	**2598.14**
60–300 m³/h, 83 l/s	2144.07	2713.53	3.00	117.98	nr	**2831.51**
60–350 m³/h, 97 l/s	2249.22	2846.61	3.00	117.98	nr	**2964.58**
70–450 m³/h, 125 l/s	2451.88	3103.10	3.00	117.98	nr	**3221.08**
High performance ventilation unit with enthalpy exchanger, class leading efficiency, automatic modulating true summer bypass, integrated humidity sensor, four variable speed flow rate set points, dial-a-duty motor						
Air flow rate						
70–600 m³/h	5342.35	6761.28	3.00	117.98	nr	**6879.26**
Extra over for additional cooling module with 1.5 kW cooling	5320.83	6734.05	–	–	nr	**6734.05**
Extra over for post treatment air to water exchanger						
300 m³/h	841.36	1064.83	3.00	117.98	nr	**1182.81**
400 m³/h	967.56	1224.54	3.00	117.98	nr	**1342.52**
500 m³/h	1088.76	1377.94	3.00	117.98	nr	**1495.91**
600 m³/h	1277.59	1616.92	3.00	117.98	nr	**1734.90**
Extra over for subsoil heat exchanger	5582.20	7064.83	3.00	117.98	nr	**7182.81**

Construction Project Manager's Pocket Book, 3rd edition

By Duncan Cartlidge

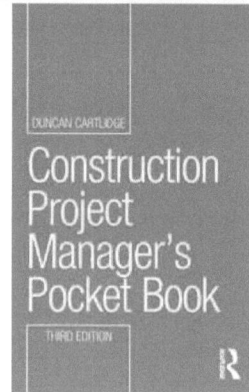

The third edition of the Construction Project Manager's Pocket Book continues to guide and educate readers on the broad range of essential skills required to be a successful construction project manager. The book introduces the generic skills required by any project manager, before tackling the core skills and activities of a construction project manager with direct reference to the RIBA Plan of Work and the OGC Gateway. Key features and coverage in the new edition include:

- a step-by-step explanation of construction project management from pre-construction to occupancy,
- hard and soft skills, including ethics, leadership, team building,
- procurement strategies,
- supply chain and contract management,
- feasibility studies / development appraisals,
- environmental issues,
- digital tools and
- occupancy activities.

The updates in this new edition take account of all regulatory and legislative changes, and also changing market conditions and working trends. This is the ideal concise reference that no project manager, construction manager, architect or quantity surveyor should be without.

July 2024: 270 pp
ISBN: 9781032761350

To Order
Tel:+44 (0) 1235 400524
Email: tandf@hachette.co.uk

For a complete listing of all our titles visit:
www.routledge.com

Taylor & Francis
Taylor & Francis Group

PART 4

Material Costs/Measured Work Prices –
Electrical Installations

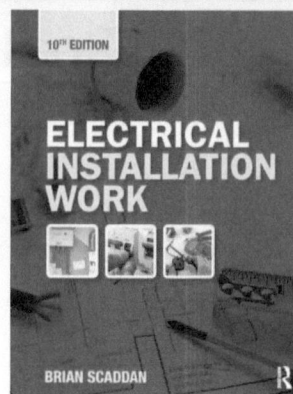

Material Costs/Measured Work Prices

DIRECTIONS

The following explanations are given for each of the column headings and letter codes.

Unit	Prices for each unit are given as singular (i.e. 1 metre, 1 nr) unless stated otherwise.
Net price	Industry tender prices, plus nominal allowance for fixings, waste and applicable trade discounts.
Material cost	Net price plus percentage allowance for overheads (7%), profit (5%) and preliminaries (13%).
Labour norms	In man-hours for each operation.
Labour cost	Labour constant multiplied by the appropriate all-in man-hour cost based on gang rate (See also relevant Rates of Wages Section) plus percentage allowance for overheads, profit and preliminaries.
Measured work	Material cost plus Labour cost.
Price (total rate)	

MATERIAL COSTS

The Material Costs given are based at Third Quarter 2025 but exclude any charges in respect of VAT. The average rate of copper during this quarter is US $9,093 / UK £6,864 per tonne.

MEASURED WORK PRICES

These prices are intended to apply to new work in the London area. The prices are for reasonable quantities of work and the user should make suitable adjustments if the quantities are especially small or especially large. Adjustments may also be required for locality (e.g. outside London – refer to cost indices in approximate estimating section for details of adjustment factors) and for the market conditions (e.g. volume of work secured or being tendered) at the time of use.

ELECTRICAL INSTALLATIONS

The labour rate has been based on average gang rates per man hour effective from 6 January 2020 including allowances for all other emoluments and expenses. To this rate, has been added 13% and 7% to cover site and head office overheads and preliminary items together with a further 5% for profit, resulting in an inclusive rate of £44.93 per man hour. The rate has been calculated on a working year of 2,007 hours; a detailed build-up of the rate is given at the end of these directions.

DIRECTIONS

In calculating the 'Measured Work Prices' the following assumptions have been made:

(a)   That the work is carried out as a subcontract under the Standard Form of Building Contract.

(b)   That, unless otherwise stated, the work is being carried out in open areas at a height which would not require more than simple scaffolding.

(c)   That the building in which the work is being carried out is no more than six storey's high.

Where these assumptions are not valid, as for example where work is carried out in ducts and similar confined spaces or in multi-storey structures when additional time is needed to get to and from upper floors, then an appropriate adjustment must be made to the prices. Such adjustment will normally be to the labour element only.

DIRECTIONS

LABOUR RATE – ELECTRICAL

The annual cost of a notional 11 man gang

	TECHNICIAN 1 NR	APPROVED ELECTRICIANS 4 NR	ELECTRICIANS 4 NR	LABOURERS 2 NR	SUB-TOTALS
Hourly Rate from 6 January 2025	**25.69**	**22.83**	**21.06**	**16.91**	
Working hours per annum per man	1,672.50	1,672.50	1,672.50	1,672.50	
× Hourly rate × nr of men = £ per annum	**42,966.53**	**152,732.70**	**140,891.40**	**56,563.95**	**393,154.58**
Overtime Rate	38.54	34.25	31.59	25.37	
Overtime hours per annum per man	334.50	334.50	334.50	334.50	
× Hourly rate × nr of men = £ per annum	**12,889.96**	**45,819.81**	**42,267.42**	**16,969.19**	**117,946.37**
Total	**55,856.48**	**198,552.51**	**183,158.82**	**73,533.14**	**511,100.95**
Incentive schemes (insert percentage) 0.00%	**0.00**	**0.00**	**0.00**	**0.00**	**0.00**
Daily Travel Allowance (15–20 miles each way)	4.80	4.80	4.80	4.80	
Days per annum per man	223.00	223.00	223.00	223.00	
× nr of men = £ per annum	**1,070.40**	**4,281.60**	**4,281.60**	**2,140.80**	**11,774.40**
Daily Travel Allowance (20 miles each way)	8.80	8.80	8.80	8.80	
Days per annum per man	223.00	223.00	223.00	223.00	
× nr of men = £ per annum	**1,962.40**	**7,849.60**	**7,849.60**	**3,924.80**	**21,586.40**
JIB Pension Scheme @ 3.0%	**2,030.54**	**7,268.87**	**6,740.79**	**2,751.32**	**18,791.51**
JIB combined benefits scheme (nr of weeks per man)	52.00	52.00	52.00	52.00	
Benefit Credit effective from 6 January 2020	88.24	80.19	75.05	63.52	
× nr of men = £ per annum	**4,588.23**	**16,679.99**	**15,609.68**	**6,605.62**	**43,483.52**
Holiday Top-up Funding	**80.90**	**71.79**	**66.31**	**52.94**	
× nr of men @ 7.5 hrs per day = £ per annum	**4,207.05**	**14,931.85**	**13,793.20**	**5,506.22**	**38,438.32**
National Insurance Contributions					
Annual gross pay (subject to NI) each	67,684.56	242,295.55	224,692.90	91,710.58	
% of NI Contributions	15	15	15	15	
£ Contributions/annum	**8,004.17**	**27,750.29**	**25,109.90**	**9,459.57**	**70,323.93**
				SUBTOTAL	**715,499.02**

TRAINING (INCLUDING ANY TRADE REGISTRATIONS) – SAY	1.00%	7,154.99
SEVERANCE PAY AND SUNDRY COSTS – SAY	1.50%	10,839.81
EMPLOYER'S LIABILITY AND THIRD PARTY INSURANCE – SAY	2.00%	14,669.88
ANNUAL COST OF NOTIONAL GANG		748,163.70
THEREFORE ANNUAL COST PER PRODUCTIVE MAN		71,253.69
MEN ACTUALLY WORKING = 10.5 AVERAGE NR OF HOURS WORKED 2007 PER MAN =		
THEREFORE ALL-IN MAN-HOUR		35.50
PRELIMINARY ITEMS – SAY	13%	4.62
SITE AND HEAD OFFICE OVERHEADS AND PROFIT (7% & 5% RESPECTIVELY) – SAY	12%	4.81
THEREFORE INCLUSIVE MAN-HOUR		44.93

DIRECTIONS

Notes:

(1)   Hourly wage rates are those effective from 6 January 2025.

(2)   The following assumptions have been made in the above calculations:

      (a)   Hourly rates are based on London rate and job reporting own transport.

      (b)   The working week of 37.5 hours is made up of 7.5 hours Monday to Friday.

      (c)   Five days in the year are lost through sickness or similar reason.

      (d)   A working year of 2,007 hours (including overtime).

      (e)   Annual holiday entitlement calculated at 24 days.

(3)   The incentive scheme addition of 0% is intended to reflect bonus schemes typically in use.

(4)   National insurance contributions are those effective from 6 April 2025.

(5)   Allowance only for weekly JIB Combined Benefit Credit Scheme.

(6)   Overtime is paid after 37.5 hours.

ELECTRICAL SUPPLY/POWER/LIGHTING SYSTEMS

Item	Net Price £	Material £	Labour hours	Labour £	Unit	Total rate £
ELECTRICAL GENERATION PLANT						
STANDBY GENERATORS						
Standby diesel generating sets; supply and installation; fixing to base; all supports and fixings; all necessary connections to equipment						
Three phase, 400 volt, four wire 50 Hz packaged standby diesel generating set, complete with radio and television suppressors, daily service fuel tank and associated piping, 4 metres of exhaust pipe and primary exhaust silencer, control panel, mains failure relay, starting battery with charger, all internal wiring, interconnections, earthing and labels. Rated for standby duty; including UK delivery, installation and commissioning.						
60 kVA	25907.24	32788.21	100.00	4570.36	nr	**37358.56**
100 kVA	26169.90	33120.62	100.00	4570.36	nr	**37690.98**
150 kVA	36319.33	45965.75	100.00	4570.36	nr	**50536.10**
300 kVA	40099.75	50750.24	120.00	5484.43	nr	**56234.67**
500 kVA	74179.10	93881.07	120.00	5484.43	nr	**99365.50**
750 kVA	150214.48	190111.45	140.00	6398.50	nr	**196509.95**
1000 kVA	214621.50	271624.96	140.00	6398.50	nr	**278023.47**
1500 kVA	252588.76	319676.34	170.00	7769.61	nr	**327445.95**
2000 kVA	389081.16	492421.12	170.00	7769.61	nr	**500190.73**
2500 kVA	532704.65	674191.01	210.00	9597.75	nr	**683788.76**
Extra for residential silencer; peformance 75 dBA at 1 m; including connection to exhaust pipe						
60 kVA	1530.07	1936.45	10.00	457.04	nr	**2393.49**
100 kVA	1828.28	2313.87	10.00	457.04	nr	**2770.91**
150 kVA	2096.66	2653.54	10.00	457.04	nr	**3110.57**
315 kVA	4078.63	5161.91	15.00	685.55	nr	**5847.47**
500 kVA	5216.44	6601.92	15.00	685.55	nr	**7287.48**
750 kVA	7771.85	9836.05	20.00	914.07	nr	**10750.12**
1000 kVA	10467.26	13247.37	20.00	914.07	nr	**14161.44**
1500 kVA	16674.64	21103.43	20.00	914.07	nr	**22017.50**
2000 kVA	18060.20	22856.98	30.00	1371.11	nr	**24228.09**
2500 kVA	20650.05	26134.70	30.00	1371.11	nr	**27505.81**
Synchronization panel for paralleling generators – not generators to mains; including interconnecting cables; commissioning and testing; fixing to backgrounds						
2 × 60 kVA	12860.99	16276.87	80.00	3656.29	nr	**19933.16**
2 × 100 kVA	13826.27	17498.52	80.00	3656.29	nr	**21154.81**
2 × 150 kVA	17393.35	22013.02	80.00	3656.29	nr	**25669.30**
2 × 315 kVA	29569.19	37422.76	80.00	3656.29	nr	**41079.05**
2 × 500 kVA	42689.45	54027.77	80.00	3656.29	nr	**57684.05**
2 × 750 kVA	48173.18	60967.97	80.00	3656.29	nr	**64624.26**
2 × 1000 kVA	51918.52	65708.08	120.00	5484.43	nr	**71192.51**
2 × 1500 kVA	61462.46	77786.89	120.00	5484.43	nr	**83271.32**
2 × 2000 kVA	66594.34	84281.79	120.00	5484.43	nr	**89766.22**
2 × 2500 kVA	77638.24	98258.95	120.00	5484.43	nr	**103743.38**

ELECTRICAL SUPPLY/POWER/LIGHTING SYSTEMS

Item	Net Price £	Material £	Labour hours	Labour £	Unit	Total rate £
ELECTRICAL GENERATION PLANT – cont						
Standby diesel generating sets – cont						
Prefabricated drop-over acoustic housing; performance 85 dBA at 1 m over the range from 60 kVA to 315 kVA, 75 dBA from 500 kVA to 2500 kVA						
100 kVA	5566.38	7044.82	7.00	319.93	nr	**7364.74**
150 kVA	9038.96	11439.71	15.00	685.55	nr	**12125.27**
315 kVA	19286.23	24408.65	25.00	1142.59	nr	**25551.24**
500 kVA	51855.04	65627.74	40.00	1828.14	nr	**67455.88**
750 kVA	81718.53	103422.97	40.00	1828.14	nr	**105251.11**
1000 kVA	103385.31	130844.45	40.00	1828.14	nr	**132672.60**
1500 kVA	154229.46	195192.80	40.00	1828.14	nr	**197020.94**
2000 kVA	205073.60	259541.15	60.00	2742.21	nr	**262283.36**
2500 kVA	246494.86	311963.90	70.00	3199.25	nr	**315163.15**
COMBINED HEAT AND POWER (CHP) UNITS						
Gas fired engine; skid-mounted with acoustic enclosure complete with exhaust fan and attenuators; exhaust gas attenuation; includes 6 m long pipe connections; dry air cooler for secondary water circuit to reject excess heat; controls and panel; commissioning						
Electrical output; Heat output						
18 kW; 30 kW	52089.93	65925.02	–	–	nr	**65925.02**
18 kW; 34 kW	53994.93	68335.98	–	–	nr	**68335.98**
18 kW; 43 kW	56025.37	70905.71	–	–	nr	**70905.71**
43 kW; 63 kW	102017.31	129113.11	–	–	nr	**129113.11**
50 kW; 79 kW	104643.90	132437.32	–	–	nr	**132437.32**
70 kW; 113 kW	115341.19	145975.82	–	–	nr	**145975.82**
104 kW, 136 kW	151626.08	191897.97	–	–	nr	**191897.97**
142 kW, 193 kW	152501.98	193006.50	–	–	nr	**193006.50**
166 kW, 248 kW	209304.45	264895.71	–	–	nr	**264895.71**
183 kW, 263 kW	210303.57	266160.20	–	–	nr	**266160.20**
264 kW, 375 kW	240019.77	303769.02	–	–	nr	**303769.02**
281 kW, 426 kW	285187.98	360933.91	–	–	nr	**360933.91**
324 kW, 497 kW	288815.91	365525.42	–	–	nr	**365525.42**
349 kW; 516 kW	324289.29	410420.52	–	–	nr	**410420.52**
385 kW; 581 kW	324972.02	411284.59	–	–	nr	**411284.59**
418 kW, 665 kW	335359.59	424431.10	–	–	nr	**424431.10**
439 kW, 687 kW	338237.07	428072.84	–	–	nr	**428072.84**
Heat dump; emergency cooling						
18 kW; 30 kW	4208.54	5326.32		–	nr	**5326.32**
18 kW; 34 kW	4208.54	5326.32		–	nr	**5326.32**
18 kW; 43 kW	4208.54	5326.32	–	–	nr	**5326.32**
43 kW; 63 kW	4379.50	5542.69	–	–	nr	**5542.69**
50 kW; 79 kW	4712.54	5964.19	–	–	nr	**5964.19**
70 kW; 113 kW	5953.67	7534.97	–	–	nr	**7534.97**
104 kW, 136 kW	6577.57	8324.58	–	–	nr	**8324.58**
142 kW, 193 kW	7816.49	9892.55	–	–	nr	**9892.55**
166 kW, 248 kW	11203.52	14179.18	–	–	nr	**14179.18**
183 kW, 263 kW	11203.52	14179.18	–	–	nr	**14179.18**
264 kW, 375 kW	16599.90	21008.84	–	–	nr	**21008.84**

ELECTRICAL SUPPLY/POWER/LIGHTING SYSTEMS

Item	Net Price £	Material £	Labour hours	Labour £	Unit	Total rate £
281 kW, 426 kW	18456.06	23357.99	–	–	nr	**23357.99**
324 kW, 497 kW	18456.06	23357.99	–	–	nr	**23357.99**
349 kW; 516 kW	20553.11	26012.01		–	nr	**26012.01**
385 kW; 581 kW	20553.11	26012.01		–	nr	**26012.01**
418 kW, 665 kW	23302.92	29492.18	–	–	nr	**29492.18**
439 kW, 687 kW	23302.92	29492.18	–	–	nr	**29492.18**
Note: The costs detailed are based on a specialist subcontract package, as part of the M&E contract works, and include installation						
Upgraded catalytic converter						
43 kW; 63 kW	1352.15	1711.28	–	–	nr	**1711.28**
50 kW; 79 kW	1352.15	1711.28	–	–	nr	**1711.28**
70 kW; 113 kW	1352.15	1711.28	–	–	nr	**1711.28**
142 kW, 193 kW	4013.15	5079.04	–	–	nr	**5079.04**
264 kW, 375 kW	7149.29	9048.15	–	–	nr	**9048.15**
Flexible connections						
43 kW; 63 kW	767.11	970.85	–	–	nr	**970.85**
50 kW; 79 kW	767.11	970.85	–	–	nr	**970.85**
70 kW; 113 kW	982.47	1243.42	–	–	nr	**1243.42**
104 kW, 136 kW	1323.29	1674.75	–	–	nr	**1674.75**
142 kW, 193 kW	1392.11	1761.86	–	–	nr	**1761.86**
166 kW, 248 kW	1873.91	2371.63	–	–	nr	**2371.63**
183 kW, 263 kW	1931.64	2444.69	–	–	nr	**2444.69**
264 kW, 375 kW	1936.08	2450.31	–	–	nr	**2450.31**
281 kW, 426 kW	2113.70	2675.10	–	–	nr	**2675.10**
324 kW, 497 kW	2113.70	2675.10	–	–	nr	**2675.10**
349 kW; 516 kW	2375.70	3006.68		–	nr	**3006.68**
385 kW; 581 kW	2375.70	3006.68		–	nr	**3006.68**
418 kW, 665 kW	2725.39	3449.25	–	–	nr	**3449.25**
439 kW, 687 kW	2725.39	3449.25	–	–	nr	**3449.25**

ELECTRICAL SUPPLY/POWER/LIGHTING SYSTEMS

Item	Net Price £	Material £	Labour hours	Labour £	Unit	Total rate £
HV SUPPLY						
Cable; 6350/11000 volts, 3 core, XLPE; stranded copper conductors; steel wire armoured; LSOH to BS 7835						
Laid in trench/duct including marker tape (cable tiles measured elsewhere)						
95 mm²	46.53	58.88	0.23	10.51	m	69.40
120 mm²	57.76	73.10	0.23	10.51	m	83.61
150 mm²	66.33	83.94	0.25	11.43	m	95.37
185 mm²	89.51	113.29	0.25	11.43	m	124.71
240 mm²	108.10	136.81	0.27	12.34	m	149.15
300 mm²	135.23	171.15	0.29	13.25	m	184.40
Pile tape; ES1–12–23; 1 m	12.47	15.78	0.01	0.46	m	16.24
Clipped direct to backgrounds including cleats						
95 mm²	46.53	58.88	0.47	21.48	m	80.36
120 mm²	57.76	73.10	0.50	22.85	m	95.95
150 mm²	66.33	83.94	0.53	24.22	m	108.16
185 mm²	89.51	113.29	0.55	25.14	m	138.42
240 mm²	108.10	136.81	0.60	27.42	m	164.23
300 mm²	135.23	171.15	0.68	31.08	m	202.23
Terminations for above cables, including heat-shrink kit and glanding off						
95 × 3 Core XLPE SWA LSF 11 kV Termination	655.03	829.01	4.75	217.09	nr	1046.10
120 × 3 Core XLPE SWA LSF 11 kV Termination	670.56	848.65	5.00	228.52	nr	1077.17
150 × 3 Core XLPE SWA LSF 11 kV Termination	708.15	896.24	6.00	274.22	nr	1170.46
185 × 3 Core XLPE SWA LSF 11 kV Termination	716.56	906.88	6.90	315.35	nr	1222.23
240 × 3 Core XLPE SWA LSF 11 kV Termination	752.88	952.84	7.50	342.78	nr	1295.62
300 × 3 Core XLPE SWA LSF 11 kV Termination	824.16	1043.06	8.75	399.91	nr	1442.96
Cable tiles; single width; laid in trench above cables on prepared sand bed (cost of excavation excluded); reinforced concrete covers; concave/convex ends						
914 × 152 × 63/38 mm	12.47	15.78	0.11	5.03	m	20.81
914 × 229 × 63/38 mm	16.83	21.30	0.11	5.03	m	26.33
914 × 305 × 63/38 mm	22.20	28.10	0.11	5.03	m	33.12

HV SWITCHGEAR AND TRANSFORMERS

HV circuit breakers; installed on prepared foundations including all supports, fixings and inter panel connections where relevant. Excludes main and multicore cabling and heat shrink cable termination kits
Three phase 11 kV, 630 amp, Air or SF6 insulated, with fixed pattern vacuum or SF6 circuit breaker panels; hand charged spring closing operation; prospective fault level up to 25 kA for 3 seconds. Feeders include ammeter with selector switch VIP relays, overcurrent and earth fault relays with necessary current relays with necessary current transformers; incomers include 3 phase VT, voltmeter and phase selector switch; Includes IDMT overcurrent and earth fault relays/CTs

ELECTRICAL SUPPLY/POWER/LIGHTING SYSTEMS

Item	Net Price £	Material £	Labour hours	Labour £	Unit	Total rate £
Single panel with cable chamber	52525.26	66475.96	31.70	1448.80	nr	**67924.77**
Three panel with one incomer and two feeders; with cable chambers	97050.30	122826.86	67.83	3100.07	nr	**125926.93**
Five panel with two incoming, two feeders and a bus section; with cable chambers	173860.16	220037.42	99.17	4532.42	nr	**224569.85**
Tripping batteries						
Battery chargers; switchgear tripping and closing; double wound transfomer and earth screen; including fixing to background, commissioning and testing						
Valve regulated lead acid battery NGTS 3.12.2; BS 6290 Part 2; TPS 9/3; IEEE485						
30 volt; 19 Ah; 3 A	4114.10	5206.80	6.50	297.07	nr	**5503.88**
40 volt; 29 Ah; 3 A	6591.10	8341.70	8.50	388.48	nr	**8730.18**
110 volt; 19 Ah; 3 A	6723.58	8509.36	6.50	297.07	nr	**8806.44**
110 volt; 29 Ah; 3 A	7336.21	9284.71	8.50	388.48	nr	**9673.19**
100 volt; 38 Ah; 3 A	7839.42	9921.57	10.00	457.04	nr	**10378.60**
Step down transformers; 11/0.415 kV, Dyn 11, 50 Hz. Complete with lifting lugs, mounting skids, provisions for wheels, undrilled gland plates to air-filled cable boxes, off load tapping facility, including UK delivery						
Oil-filled in free breathing ventilated steel tank						
500 kVA	15208.86	19248.33	30.00	1371.11	nr	**20619.44**
800 kVA	17320.50	21920.82	30.00	1371.11	nr	**23291.93**
1000 kVA	19630.53	24844.40	30.00	1371.11	nr	**26215.51**
1250 kVA	23853.72	30189.27	35.00	1599.63	nr	**31788.90**
1500 kVA	28266.84	35774.51	35.00	1599.63	nr	**37374.14**
2000 kVA	36713.23	46464.26	35.00	1599.63	nr	**48063.89**
MIDEL – filled ingasket-sealed steel tank						
500 kVA	20144.35	25494.69	30.00	1371.11	nr	**26865.80**
800 kVA	22959.83	29057.96	30.00	1371.11	nr	**30429.07**
1000 kVA	26141.26	33084.37	30.00	1371.11	nr	**34455.48**
1250 kVA	31596.18	39988.13	35.00	1599.63	nr	**41587.76**
1500 kVA	37593.05	47577.77	35.00	1599.63	nr	**49177.39**
2000 kVA	48678.91	61608.03	40.00	1828.14	nr	**63436.17**
Extra for						
Fluid temperature indicator with 2 N/O contacts	539.42	682.69	2.00	91.41	nr	**774.09**
Winding temperature indicator with 2 N/O contacts	1321.59	1672.60	2.00	91.41	nr	**1764.01**
Dehydrating breather	134.85	170.67	2.00	91.41	nr	**262.08**
Plain rollers	404.56	512.01	2.00	91.41	nr	**603.42**
Pressure relief device with 1 N/O contact	809.11	1024.01	2.00	91.41	nr	**1115.42**

ELECTRICAL SUPPLY/POWER/LIGHTING SYSTEMS

Item	Net Price £	Material £	Labour hours	Labour £	Unit	Total rate £
HV SUPPLY – cont						
Step down transformers; 11/0.415 kV, Dyn 11, 50 Hz. Complete with lifting lugs, mounting skids, provisions for wheels, undrilled gland plates to air-filled cable boxes, off load tapping facility, including UK delivery						
Cast Resin type in ventilated steel encloure, AN – Air Natural including winding temperture indicator with 2 N/O contacts						
500 kVA	21552.07	27276.30	40.00	1828.14	nr	**29104.44**
800 kVA	25071.39	31730.35	40.00	1828.14	nr	**33558.50**
1000 kVA	30329.24	38384.69	40.00	1828.14	nr	**40212.83**
1250 kVA	32827.94	41547.05	45.00	2056.66	nr	**43603.71**
1600 kVA	37241.14	47132.39	45.00	2056.66	nr	**49189.05**
2000 kVA	42168.22	53368.10	50.00	2285.18	nr	**55653.28**
Cast Resin type in ventilated steel enclosure with temperature controlled fans to achieve 40% increase to AN/AF rating. Includes winding temperature indicator with 2 N/O contacts						
500/700 kVA	23663.64	29948.71	42.00	1919.55	nr	**31868.26**
800/1120 kVA	27534.93	34848.20	42.00	1919.55	nr	**36767.75**
1000/1400 kVA	34068.14	43116.64	42.00	1919.55	nr	**45036.19**
1250/1750 kVA	35475.85	44898.24	47.00	2148.07	nr	**47046.30**
1600/2240 kVA	40079.05	50724.04	47.00	2148.07	nr	**52872.11**
2000/2800 kVA	45709.95	57850.52	52.00	2376.59	nr	**60227.10**

ELECTRICAL SUPPLY/POWER/LIGHTING SYSTEMS

Item	Net Price £	Material £	Labour hours	Labour £	Unit	Total rate £
LV DISTRIBUTION: CONDUIT AND CABLE TRUNKING						
Heavy gauge, screwed drawn steel; surface fixed on saddles to backgrounds, with standard pattern boxes and fittings including all fixings and supports (forming holes, conduit entry, draw wires etc. and components for earth continuity are included)						
Black enamelled						
20 mm dia.	4.92	6.23	0.49	22.39	m	**28.63**
25 mm dia.	6.69	8.47	0.56	25.59	m	**34.06**
32 mm dia.	14.36	18.18	0.64	29.25	m	**47.43**
38 mm dia.	18.61	23.56	0.73	33.36	m	**56.92**
50 mm dia.	35.29	44.66	1.04	47.53	m	**92.19**
Galvanized						
20 mm dia.	5.07	6.42	0.49	22.39	m	**28.81**
25 mm dia.	6.93	8.77	0.56	25.59	m	**34.37**
32 mm dia.	13.28	16.81	0.64	29.25	m	**46.06**
38 mm dia.	17.21	21.79	0.73	33.36	m	**55.15**
50 mm dia.	33.09	41.88	1.04	47.53	m	**89.42**
High impact PVC; surface fixed on saddles to backgrounds; with standard pattern boxes and fittings; including all fixings and supports						
Light gauge						
16 mm dia.	4.49	5.69	0.27	12.34	m	**18.03**
20 mm dia.	6.65	8.41	0.28	12.80	m	**21.21**
25 mm dia.	9.33	11.81	0.33	15.08	m	**26.89**
32 mm dia.	13.11	16.60	0.38	17.37	m	**33.96**
38 mm dia.	16.36	20.70	0.44	20.11	m	**40.81**
50 mm dia.	22.83	28.90	0.48	21.94	m	**50.83**
Heavy gauge						
16 mm dia.	6.07	7.68	0.27	12.34	m	**20.02**
20 mm dia.	7.58	9.60	0.28	12.80	m	**22.39**
25 mm dia.	9.46	11.98	0.33	15.08	m	**27.06**
32 mm dia.	12.12	15.34	0.38	17.37	m	**32.71**
38 mm dia.	14.41	18.24	0.44	20.11	m	**38.35**
50 mm dia.	18.97	24.01	0.48	21.94	m	**45.95**
Flexible conduits; including adaptors and locknuts (for connections to equipment)						
Metallic, PVC covered conduit; not exceeding 1 m long; including zinc plated mild steel adaptors, lock nuts and earth conductor						
16 mm dia.	31.30	39.61	0.42	19.20	nr	**58.81**
20 mm dia.	33.24	42.07	0.43	19.65	nr	**61.72**
25 mm dia.	49.33	62.43	0.46	21.02	nr	**83.46**
32 mm dia.	76.32	96.59	0.51	23.31	nr	**119.90**
38 mm dia.	130.09	164.64	0.56	25.59	nr	**190.24**
50 mm dia.	363.18	459.64	0.82	37.48	nr	**497.12**

ELECTRICAL SUPPLY/POWER/LIGHTING SYSTEMS

Item	Net Price £	Material £	Labour hours	Labour £	Unit	Total rate £
LV DISTRIBUTION: CONDUIT AND CABLE TRUNKING – cont						
Flexible conduits – cont						
PVC conduit; not exceeding 1 m long; including nylon adaptors, lock nuts						
16 mm dia.	8.50	10.76	0.46	21.02	nr	**31.78**
20 mm dia.	8.50	10.76	0.48	21.94	nr	**32.70**
25 mm dia.	10.66	13.49	0.50	22.85	nr	**36.35**
32 mm.dia.	15.49	19.60	0.58	26.51	nr	**46.11**
PVC adaptable boxes; fixed to backgrounds; including all supports and fixings (cutting and connecting conduit to boxes is included)						
Square pattern						
75 × 75 × 53 mm	6.62	8.38	0.69	31.54	nr	**39.92**
100 × 100 × 75 mm	11.23	14.21	0.71	32.45	nr	**46.66**
150 × 150 × 75 mm	13.72	17.36	0.80	36.56	nr	**53.92**
Terminal strips to be fixed in metal or polythene adaptable boxes)						
20 amp high density polythene						
2 way	1.79	2.27	0.23	10.51	nr	**12.78**
3 way	2.24	2.84	0.23	10.51	nr	**13.35**
4 way	2.99	3.78	0.23	10.51	nr	**14.29**
5 way	3.73	4.72	0.23	10.51	nr	**15.23**
6 way	4.53	5.73	0.25	11.43	nr	**17.16**
7 way	5.23	6.62	0.25	11.43	nr	**18.04**
8 way	6.02	7.62	0.29	13.25	nr	**20.87**
9 way	6.71	8.50	0.30	13.71	nr	**22.21**
10 way	7.53	9.54	0.34	15.54	nr	**25.07**
11 way	8.20	10.38	0.34	15.54	nr	**25.92**
13 way	9.72	12.30	0.37	16.91	nr	**29.21**
14 way	10.45	13.22	0.37	16.91	nr	**30.13**
15 way	11.19	14.16	0.39	17.82	nr	**31.98**
16 way	11.94	15.12	0.45	20.57	nr	**35.68**
18 way	13.42	16.98	0.45	20.57	nr	**37.55**

ELECTRICAL SUPPLY/POWER/LIGHTING SYSTEMS

Item	Net Price £	Material £	Labour hours	Labour £	Unit	Total rate £
LV DISTRIBUTION: TRUNKING						
Galvanized steel trunking; fixed to backgrounds; jointed with standard connectors (including plates for air gap between trunking and background); earth continuity straps included						
Single compartment						
50 × 50 mm	10.33	13.07	0.39	17.82	m	**30.89**
75 × 50 mm	15.58	19.72	0.44	20.11	m	**39.83**
75 × 75 mm	18.29	23.14	0.47	21.48	m	**44.62**
100 × 50 mm	18.95	23.99	0.50	22.85	m	**46.84**
100 × 75 mm	20.29	25.68	0.57	26.05	m	**51.73**
100 × 100 mm	20.39	25.81	0.62	28.34	m	**54.15**
150 × 50 mm	27.10	34.30	0.78	35.65	m	**69.95**
150 × 100 mm	32.31	40.89	0.78	35.65	m	**76.54**
150 × 150 mm	36.62	46.34	0.84	38.39	m	**84.73**
225 × 75 mm	53.78	68.07	0.86	39.31	m	**107.37**
225 × 150 mm	67.23	85.08	0.88	40.22	m	**125.30**
225 × 225 mm	68.01	86.07	0.99	45.25	m	**131.32**
300 × 75 mm	55.18	69.83	0.96	43.88	m	**113.71**
300 × 100 mm	63.52	80.39	0.99	45.25	m	**125.64**
300 × 150 mm	71.84	90.93	0.99	45.25	m	**136.17**
300 × 225 mm	84.33	106.73	1.09	49.82	m	**156.54**
300 × 300 mm	89.13	112.81	1.16	53.02	m	**165.83**
Double compartment						
100 × 50 mm	15.29	19.35	0.54	24.68	m	**44.03**
100 × 75 mm	16.49	20.88	0.62	28.34	m	**49.21**
100 × 100 mm	28.80	36.45	0.66	30.16	m	**66.62**
150 × 50 mm	24.24	30.68	0.70	31.99	m	**62.67**
150 × 100 mm	30.03	38.00	0.83	37.93	m	**75.93**
150 × 150 mm	46.50	58.85	0.92	42.05	m	**100.90**
Triple compartment						
150 × 50 mm	28.44	35.99	0.78	35.65	m	**71.64**
150 × 100 mm	29.19	36.94	0.79	36.11	m	**73.05**
150 × 150 mm	55.55	70.31	1.01	46.16	m	**116.47**
Galvanized steel trunking fittings; cutting and jointing trunking to fittings is included						
Stop end						
50 × 50 mm	4.25	5.38	0.19	8.68	nr	**14.07**
75 × 50 mm	4.52	5.72	0.20	9.14	nr	**14.86**
75 × 75 mm	4.51	5.71	0.21	9.60	nr	**15.30**
100 × 50 mm	4.70	5.95	0.27	12.34	nr	**18.29**
100 × 75 mm	5.05	6.39	0.27	12.34	nr	**18.73**
100 × 100 mm	4.98	6.30	0.28	12.80	nr	**19.09**
150 × 50 mm	5.54	7.01	0.30	13.71	nr	**20.73**
150 × 100 mm	6.03	7.63	0.31	14.17	nr	**21.80**
150 × 150 mm	6.19	7.84	0.32	14.63	nr	**22.46**
225 × 75 mm	6.84	8.66	0.35	16.00	nr	**24.65**
225 × 150 mm	7.60	9.62	0.37	16.91	nr	**26.53**
225 × 225 mm	11.02	13.95	0.38	17.37	nr	**31.31**

ELECTRICAL SUPPLY/POWER/LIGHTING SYSTEMS

Item	Net Price £	Material £	Labour hours	Labour £	Unit	Total rate £
LV DISTRIBUTION: TRUNKING – cont						
Galvanized steel trunking fittings – cont						
Stop end – cont						
300 × 75 mm	11.07	14.01	0.42	19.20	nr	**33.21**
300 × 100 mm	11.79	14.93	0.42	19.20	nr	**34.12**
300 × 150 mm	12.53	15.85	0.43	19.65	nr	**35.51**
300 × 225 mm	13.28	16.81	0.45	20.57	nr	**37.38**
300 × 300 mm	13.97	17.68	0.48	21.94	nr	**39.61**
Flanged connector						
50 × 50 mm	4.54	5.74	0.19	8.68	nr	**14.43**
75 × 50 mm	6.91	8.75	0.20	9.14	nr	**17.89**
75 × 75 mm	7.04	8.91	0.21	9.60	nr	**18.51**
100 × 50 mm	7.20	9.12	0.26	11.88	nr	**21.00**
100 × 75 mm	7.42	9.39	0.27	12.34	nr	**21.73**
100 × 100 mm	7.68	9.72	0.27	12.34	nr	**22.06**
150 × 50 mm	7.76	9.82	0.28	12.80	nr	**22.62**
150 × 100 mm	7.83	9.92	0.30	13.71	nr	**23.63**
150 × 150 mm	7.90	10.00	0.32	14.63	nr	**24.62**
225 × 75 mm	8.05	10.18	0.35	16.00	nr	**26.18**
225 × 150 mm	8.24	10.43	0.37	16.91	nr	**27.34**
225 × 225 mm	11.14	14.09	0.38	17.37	nr	**31.46**
300 × 75 mm	11.17	14.14	0.42	19.20	nr	**33.33**
300 × 100 mm	11.34	14.35	0.42	19.20	nr	**33.54**
300 × 150 mm	12.19	15.43	0.43	19.65	nr	**35.09**
300 × 225 mm	12.83	16.24	0.45	20.57	nr	**36.81**
300 × 300 mm	13.99	17.70	0.48	21.94	nr	**39.64**
Bends 90°; single compartment						
50 × 50 mm	19.44	24.60	0.42	19.20	nr	**43.80**
75 × 50 mm	23.21	29.38	0.45	20.57	nr	**49.94**
75 × 75 mm	23.69	29.98	0.48	21.94	nr	**51.92**
100 × 50 mm	24.16	30.58	0.53	24.22	nr	**54.80**
100 × 75 mm	24.49	30.99	0.56	25.59	nr	**56.58**
100 × 100 mm	24.62	31.17	0.58	26.51	nr	**57.67**
150 × 50 mm	29.70	37.59	0.64	29.25	nr	**66.84**
150 × 100 mm	36.43	46.11	0.76	34.73	nr	**80.84**
150 × 150 mm	35.24	44.59	0.82	37.48	nr	**82.07**
225 × 75 mm	49.72	62.93	0.84	38.39	nr	**101.32**
225 × 150 mm	61.79	78.20	0.85	38.85	nr	**117.05**
225 × 225 mm	72.75	92.07	0.89	40.68	nr	**132.75**
300 × 75 mm	73.36	92.84	0.90	41.13	nr	**133.98**
300 × 100 mm	74.23	93.95	0.91	41.59	nr	**135.54**
300 × 150 mm	75.00	94.92	0.96	43.88	nr	**138.79**
300 × 225 mm	82.49	104.40	0.98	44.79	nr	**149.19**
300 × 300 mm	84.33	106.73	1.06	48.45	nr	**155.18**
Bends 90°; double compartment						
100 × 50 mm	20.53	25.98	0.53	24.22	nr	**50.20**
100 × 75 mm	21.63	27.37	0.56	25.59	nr	**52.96**
100 × 100 mm	35.30	44.68	0.58	26.51	nr	**71.19**
150 × 50 mm	43.19	54.66	0.65	29.71	nr	**84.37**
150 × 100 mm	45.51	57.60	0.69	31.54	nr	**89.14**
150 × 150 mm	57.89	73.26	0.73	33.36	nr	**106.62**

ELECTRICAL SUPPLY/POWER/LIGHTING SYSTEMS

Item	Net Price £	Material £	Labour hours	Labour £	Unit	Total rate £
Bends 90°; triple compartment						
150 × 50 mm	30.00	37.97	0.68	31.08	nr	**69.05**
150 × 100 mm	41.95	53.09	0.73	33.36	nr	**86.45**
150 × 150 mm	68.29	86.43	0.77	35.19	nr	**121.62**
Tees; single compartment						
50 × 50 mm	26.44	33.46	0.56	25.59	nr	**59.05**
75 × 50 mm	29.88	37.82	0.57	26.05	nr	**63.87**
75 × 75 mm	31.13	39.40	0.60	27.42	nr	**66.82**
100 × 50 mm	32.29	40.87	0.65	29.71	nr	**70.57**
100 × 75 mm	33.62	42.54	0.71	32.45	nr	**74.99**
100 × 100 mm	34.13	43.20	0.72	32.91	nr	**76.10**
150 × 50 mm	38.83	49.14	0.82	37.48	nr	**86.62**
150 × 100 mm	45.83	58.00	0.84	38.39	nr	**96.39**
150 × 150 mm	44.59	56.43	0.91	41.59	nr	**98.02**
225 × 75 mm	60.33	76.35	0.94	42.96	nr	**119.32**
225 × 150 mm	81.49	103.14	1.01	46.16	nr	**149.30**
225 × 225 mm	95.36	120.68	1.02	46.62	nr	**167.30**
300 × 75 mm	97.24	123.07	1.07	48.90	nr	**171.97**
300 × 100 mm	100.74	127.50	1.07	48.90	nr	**176.40**
300 × 150 mm	108.53	137.36	1.14	52.10	nr	**189.46**
300 × 225 mm	117.53	148.75	1.19	54.39	nr	**203.13**
300 × 300 mm	127.13	160.89	1.26	57.59	nr	**218.48**
Tees; double compartment						
100 × 50 mm	21.45	27.14	0.65	29.71	nr	**56.85**
100 × 75 mm	50.50	63.92	0.71	32.45	nr	**96.37**
100 × 100 mm	51.50	65.18	0.72	32.91	nr	**98.09**
150 × 50 mm	26.49	33.53	0.82	37.48	nr	**71.00**
150 × 100 mm	54.25	68.65	0.85	38.85	nr	**107.50**
150 × 150 mm	82.01	103.79	0.91	41.59	nr	**145.38**
Tees; triple compartment						
150 × 50 mm	29.78	37.69	0.87	39.76	nr	**77.45**
150 × 100 mm	32.72	41.41	0.89	40.68	nr	**82.09**
150 × 150 mm	95.44	120.79	0.96	43.88	nr	**164.67**
Crossovers; single compartment						
50 × 50 mm	16.27	20.59	0.65	29.71	nr	**50.30**
75 × 50 mm	22.05	27.91	0.66	30.16	nr	**58.08**
75 × 75 mm	22.56	28.55	0.69	31.54	nr	**60.09**
100 × 50 mm	27.87	35.28	0.74	33.82	nr	**69.10**
100 × 75 mm	27.99	35.42	0.80	36.56	nr	**71.98**
100 × 100 mm	28.10	35.56	0.81	37.02	nr	**72.58**
150 × 50 mm	30.68	38.83	0.91	41.59	nr	**80.42**
150 × 100 mm	33.76	42.73	0.94	42.96	nr	**85.69**
150 × 150 mm	42.44	53.71	0.99	45.25	nr	**98.95**
225 × 75 mm	52.58	66.55	1.01	46.16	nr	**112.71**
225 × 150 mm	57.39	72.63	1.08	49.36	nr	**121.99**
225 × 225 mm	78.08	98.82	1.09	49.82	nr	**148.63**
300 × 75 mm	79.88	101.09	1.14	52.10	nr	**153.20**
300 × 100 mm	82.28	104.13	1.16	53.02	nr	**157.15**
300 × 150 mm	84.76	107.28	1.19	54.39	nr	**161.66**
300 × 225 mm	86.32	109.25	1.21	55.30	nr	**164.55**
300 × 300 mm	92.03	116.47	1.29	58.96	nr	**175.43**

ELECTRICAL SUPPLY/POWER/LIGHTING SYSTEMS

Item	Net Price £	Material £	Labour hours	Labour £	Unit	Total rate £
LV DISTRIBUTION: TRUNKING – cont						
Galvanized steel trunking fittings – cont						
Crossovers; double compartment						
100 × 50 mm	15.36	19.44	0.74	33.82	nr	**53.26**
100 × 75 mm	16.11	20.39	0.80	36.56	nr	**56.95**
100 × 100 mm	52.66	66.64	0.81	37.02	nr	**103.66**
150 × 50 mm	18.19	23.02	0.86	39.31	nr	**62.33**
150 × 100 mm	18.25	23.10	0.94	42.96	nr	**66.06**
150 × 150 mm	46.63	59.02	1.00	45.70	nr	**104.72**
Crossovers; triple compartment						
150 × 50 mm	19.91	25.19	0.97	44.33	nr	**69.53**
150 × 100 mm	21.90	27.72	0.99	45.25	nr	**72.97**
150 × 150 mm	102.53	129.76	1.06	48.45	nr	**178.20**
Galvanized steel flush floor trunking; fixed to backgrounds; supports and fixings; standard coupling joints; earth continuity straps included						
Triple compartment						
350 × 60 mm	71.86	90.94	1.32	60.33	m	**151.27**
Four compartment						
350 × 60 mm	74.81	94.68	1.32	60.33	m	**155.01**
Galvanized steel flush floor trunking; fittings (cutting and jointing trunking to fittings is included)						
Stop end; triple compartment						
350 × 60 mm	7.60	9.62	0.53	24.22	nr	**33.84**
Stop end; four compartment						
350 × 60 mm	17.62	22.30	0.53	24.22	nr	**46.52**
Rising bend; standard; triple compartment						
350 × 60 mm	59.80	75.68	1.30	59.41	nr	**135.10**
Rising bend; standard; four compartment						
350 × 60 mm	62.66	79.30	1.30	59.41	nr	**138.71**
Rising bend; skirting; triple compartment						
350 × 60 mm	117.32	148.48	1.33	60.79	nr	**209.26**
Rising bend; skirting; four compartment						
350 × 60 mm	133.15	168.52	1.33	60.79	nr	**229.30**
Junction box; triple compartment						
350 × 60 mm	77.97	98.68	1.16	53.02	nr	**151.69**
Junction box; four compartment						
350 × 60 mm	81.10	102.64	1.16	53.02	nr	**155.65**
Body coupler (pair)						
3 and 4 compartment	4.02	5.09	0.16	7.31	nr	**12.40**
Service outlet module comprising flat lid with flanged carpet trim; twin 13 A outlet and drilled plate for mounting 2 telephone outlets; one blank plate; triple compartment						
3 compartment	96.72	122.41	0.47	21.48	nr	**143.89**
Service outlet module comprising flat lid with flanged carpet trim; twin 13 A outlet and drilled plate for mounting 2 telephone outlets; two blank plates; four compartment						
4 compartment	107.73	136.34	0.47	21.48	nr	**157.82**

ELECTRICAL SUPPLY/POWER/LIGHTING SYSTEMS

Item	Net Price £	Material £	Labour hours	Labour £	Unit	Total rate £
Single compartment PVC trunking; grey finish; clip on lid; fixed to backgrounds; including supports and fixings (standard coupling joints)						
Dimensions						
50 × 50 mm	23.24	29.42	0.27	12.34	m	**41.76**
75 × 50 mm	25.20	31.89	0.28	12.80	m	**44.69**
75 × 75 mm	28.54	36.13	0.29	13.25	m	**49.38**
100 × 50 mm	39.07	49.45	0.34	15.54	m	**64.99**
100 × 75 mm	42.87	54.25	0.37	16.91	m	**71.16**
100 × 100 mm	49.03	62.05	0.37	16.91	m	**78.96**
150 × 50 mm	42.55	53.85	0.41	18.74	m	**72.59**
150 × 75 mm	75.84	95.98	0.44	20.11	m	**116.09**
150 × 100 mm	91.27	115.52	0.44	20.11	m	**135.62**
150 × 150 mm	93.37	118.17	0.48	21.94	m	**140.11**
Single compartment PVC trunking; fittings (cutting and jointing trunking to fittings is included)						
Crossover						
50 × 50 mm	36.99	46.82	0.29	13.25	nr	**60.07**
75 × 50 mm	41.01	51.90	0.30	13.71	nr	**65.61**
75 × 75 mm	44.50	56.32	0.31	14.17	nr	**70.49**
100 × 50 mm	59.32	75.08	0.35	16.00	nr	**91.08**
100 × 75 mm	74.83	94.71	0.36	16.45	nr	**111.16**
100 × 100 mm	81.47	103.11	0.40	18.28	nr	**121.39**
150 × 75 mm	84.78	107.30	0.45	20.57	nr	**127.87**
150 × 100 mm	101.71	128.72	0.46	21.02	nr	**149.74**
150 × 150 mm	157.88	199.81	0.47	21.48	nr	**221.29**
Stop end						
50 × 50 mm	3.35	4.24	0.12	5.48	nr	**9.73**
75 × 50 mm	4.81	6.09	0.12	5.48	nr	**11.57**
75 × 75 mm	6.31	7.99	0.13	5.94	nr	**13.93**
100 × 50 mm	7.29	9.23	0.16	7.31	nr	**16.54**
100 × 75 mm	7.44	9.41	0.16	7.31	nr	**16.72**
100 × 100 mm	7.64	9.67	0.18	8.23	nr	**17.90**
150 × 75 mm	17.88	22.63	0.20	9.14	nr	**31.77**
150 × 100 mm	22.14	28.02	0.21	9.60	nr	**37.62**
150 × 150 mm	22.59	28.59	0.22	10.05	nr	**38.65**
Flanged coupling						
50 × 50 mm	9.27	11.73	0.32	14.63	nr	**26.35**
75 × 50 mm	10.60	13.42	0.33	15.08	nr	**28.50**
75 × 75 mm	12.72	16.10	0.34	15.54	nr	**31.64**
100 × 50 mm	14.32	18.12	0.44	20.11	nr	**38.23**
100 × 75 mm	16.21	20.52	0.45	20.57	nr	**41.09**
100 × 100 mm	17.26	21.85	0.46	21.02	nr	**42.87**
150 × 75 mm	18.29	23.15	0.57	26.05	nr	**49.20**
150 × 100 mm	19.28	24.40	0.57	26.05	nr	**50.46**
150 × 150 mm	20.23	25.61	0.59	26.97	nr	**52.57**
Internal coupling						
50 × 50 mm	3.27	4.14	0.07	3.20	nr	**7.34**
75 × 50 mm	3.88	4.91	0.07	3.20	nr	**8.11**
75 × 75 mm	3.90	4.93	0.07	3.20	nr	**8.13**

ELECTRICAL SUPPLY/POWER/LIGHTING SYSTEMS

Item	Net Price £	Material £	Labour hours	Labour £	Unit	Total rate £
LV DISTRIBUTION: TRUNKING – cont						
Single compartment PVC trunking – cont						
Internal coupling – cont						
100 × 50 mm	5.21	6.59	0.08	3.66	nr	**10.25**
100 × 75 mm	5.84	7.39	0.08	3.66	nr	**11.04**
100 × 100 mm	6.44	8.15	0.08	3.66	nr	**11.81**
External coupling						
50 × 50 mm	3.57	4.52	0.09	4.11	nr	**8.64**
75 × 50 mm	4.28	5.42	0.09	4.11	nr	**9.53**
75 × 75 mm	5.38	6.81	0.09	4.11	nr	**10.92**
100 × 50 mm	5.69	7.21	0.10	4.57	nr	**11.78**
100 × 75 mm	6.48	8.20	0.10	4.57	nr	**12.77**
100 × 100 mm	7.07	8.94	0.10	4.57	nr	**13.51**
150 × 75 mm	8.10	10.25	0.11	5.03	nr	**15.27**
150 × 100 mm	8.44	10.68	0.11	5.03	nr	**15.71**
150 × 150 mm	8.72	11.04	0.11	5.03	nr	**16.07**
Angle; flat cover						
50 × 50 mm	9.43	11.93	0.18	8.23	nr	**20.16**
75 × 50 mm	12.56	15.90	0.19	8.68	nr	**24.58**
75 × 75 mm	14.45	18.29	0.20	9.14	nr	**27.43**
100 × 50 mm	29.60	37.46	0.23	10.51	nr	**47.97**
100 × 75 mm	45.08	57.05	0.26	11.88	nr	**68.93**
100 × 100 mm	68.54	86.74	0.26	11.88	nr	**98.63**
150 × 75 mm	56.88	71.99	0.30	13.71	nr	**85.70**
150 × 100 mm	67.66	85.63	0.33	15.08	nr	**100.72**
150 × 150 mm	102.53	129.77	0.34	15.54	nr	**145.30**
Angle; internal or external cover						
50 × 50 mm	15.97	20.21	0.18	8.23	nr	**28.44**
75 × 50 mm	22.01	27.85	0.19	8.68	nr	**36.54**
75 × 75 mm	28.27	35.78	0.20	9.14	nr	**44.92**
100 × 50 mm	30.45	38.54	0.23	10.51	nr	**49.05**
100 × 75 mm	48.96	61.97	0.26	11.88	nr	**73.85**
100 × 100 mm	49.19	62.25	0.26	11.88	nr	**74.13**
150 × 75 mm	60.13	76.10	0.30	13.71	nr	**89.81**
150 × 100 mm	70.87	89.70	0.33	15.08	nr	**104.78**
150 × 150 mm	99.54	125.98	0.34	15.54	nr	**141.52**
Tee; flat cover						
50 × 50 mm	10.98	13.90	0.24	10.97	nr	**24.87**
75 × 50 mm	16.76	21.21	0.25	11.43	nr	**32.64**
75 × 75 mm	17.22	21.80	0.26	11.88	nr	**33.68**
100 × 50 mm	37.21	47.10	0.32	14.63	nr	**61.72**
100 × 75 mm	39.54	50.04	0.33	15.08	nr	**65.12**
100 × 100 mm	51.59	65.29	0.34	15.54	nr	**80.83**
150 × 75 mm	68.23	86.35	0.41	18.74	nr	**105.09**
150 × 100 mm	87.51	110.75	0.42	19.20	nr	**129.95**
150 × 150 mm	119.05	150.67	0.44	20.11	nr	**170.78**
Tee; internal or external cover						
50 × 50 mm	31.14	39.41	0.24	10.97	nr	**50.37**
75 × 50 mm	34.20	43.29	0.25	11.43	nr	**54.72**
75 × 75 mm	38.04	48.14	0.26	11.88	nr	**60.03**

ELECTRICAL SUPPLY/POWER/LIGHTING SYSTEMS

Item	Net Price £	Material £	Labour hours	Labour £	Unit	Total rate £
100 × 50 mm	48.82	61.79	0.32	14.63	nr	**76.42**
100 × 75 mm	55.77	70.58	0.33	15.08	nr	**85.66**
100 × 100 mm	62.57	79.19	0.34	15.54	nr	**94.73**
150 × 75 mm	81.07	102.60	0.41	18.74	nr	**121.34**
150 × 100 mm	97.71	123.66	0.42	19.20	nr	**142.85**
150 × 150 mm	130.46	165.11	0.44	20.11	nr	**185.22**
Division strip (1.8 m long)						
50 mm	16.88	21.36	0.07	3.20	nr	**24.56**
75 mm	21.63	27.37	0.07	3.20	nr	**30.57**
100 mm	27.48	34.78	0.08	3.66	nr	**38.44**
PVC miniature trunking; white finish; fixed to backgrounds; including supports and fixing; standard coupling joints						
Single compartment						
16 × 16 mm	2.70	3.41	0.20	9.14	m	**12.56**
25 × 16 mm	2.70	3.41	0.21	9.60	m	**13.01**
38 × 16 mm	2.70	3.41	0.24	10.97	m	**14.38**
38 × 25 mm	3.16	3.99	0.25	11.43	m	**15.42**
Compartmented						
38 × 16 mm	6.04	7.64	0.24	10.97	m	**18.61**
38 × 25 mm	7.20	9.11	0.25	11.43	m	**20.54**
PVC miniature trunking fittings; single compartment; white finish; cutting and jointing trunking to fittings is included						
Coupling						
16 × 16 mm	2.66	3.36	0.10	4.57	nr	**7.93**
25 × 16 mm	2.66	3.36	0.12	5.26	nr	**8.62**
38 × 16 mm	2.66	3.36	0.12	5.48	nr	**8.85**
38 × 25 mm	3.14	3.97	0.14	6.40	nr	**10.37**
Stop end						
16 × 16 mm	2.66	3.36	0.12	5.48	nr	**8.85**
25 × 16 mm	2.66	3.36	0.13	5.80	nr	**9.17**
38 × 16 mm	2.66	3.36	0.15	6.86	nr	**10.22**
38 × 25 mm	3.14	3.97	0.17	7.77	nr	**11.74**
Bend; flat, internal or external						
16 × 16 mm	4.50	5.70	0.18	8.23	nr	**13.92**
25 × 16 mm	4.50	5.70	0.20	9.05	nr	**14.74**
38 × 16 mm	4.50	5.70	0.21	9.60	nr	**15.29**
38 × 25 mm	5.74	7.26	0.23	10.51	nr	**17.77**
Tee						
16 × 16 mm	1.75	2.21	0.19	8.68	nr	**10.90**
25 × 16 mm	1.75	2.21	0.23	10.51	nr	**12.73**
38 × 16 mm	1.75	2.21	0.26	11.88	nr	**14.10**
38 × 25 mm	2.45	3.10	0.29	13.25	nr	**16.35**
PVC bench trunking; white or grey finish; fixed to backgrounds; including supports and fixings; standard coupling joints						
Trunking						
90 × 90 mm	38.23	48.38	0.33	15.08	m	**63.46**

ELECTRICAL SUPPLY/POWER/LIGHTING SYSTEMS

Item	Net Price £	Material £	Labour hours	Labour £	Unit	Total rate £
LV DISTRIBUTION: TRUNKING – cont						
PVC bench trunking fittings; white or grey finish; cutting and jointing trunking to fittings is included						
Stop end						
90 × 90 mm	8.38	10.60	0.09	4.11	nr	**14.72**
Coupling						
90 × 90 mm	5.19	6.57	0.09	4.11	nr	**10.68**
Internal or external bend						
90 × 90 mm	28.46	36.02	0.28	12.80	nr	**48.82**
Socket plate						
90 × 90 mm – 1 gang	1.36	1.72	0.10	4.57	nr	**6.29**
90 × 90 mm – 2 gang	1.63	2.06	0.10	4.57	nr	**6.63**
PVC underfloor trunking; single compartment; fitted in floor screed; standard coupling joints						
Trunking						
60 × 25 mm	27.86	35.27	0.22	10.05	m	**45.32**
90 × 35 mm	40.14	50.80	0.27	12.34	m	**63.14**
PVC underfloor trunking fittings; single compartment; fitted in floor screed (cutting and jointing trunking to fittings is included)						
Jointing sleeve						
60 × 25 mm	1.86	2.36	0.08	3.66	nr	**6.01**
90 × 35 mm	3.10	3.92	0.10	4.57	nr	**8.49**
Duct connector						
90 × 35 mm	1.44	1.82	0.17	7.77	nr	**9.59**
Socket reducer						
90 × 35 mm	2.56	3.24	0.12	5.48	nr	**8.72**
Vertical access box; 2 compartment						
Shallow	144.15	182.43	0.37	16.91	nr	**199.34**
Duct bend; vertical						
60 × 25 mm	28.11	35.58	0.27	12.34	nr	**47.92**
90 × 35 mm	31.72	40.15	0.35	16.00	nr	**56.15**
Duct bend; horizontal						
60 × 25 mm	33.16	41.97	0.30	13.71	nr	**55.68**
90 × 35 mm	33.65	42.59	0.37	16.91	nr	**59.50**
Zinc coated steel underfloor ducting; fixed to backgrounds; standard coupling joints; earth continuity straps (Including supports and fixing, packing shims where required)						
Double compartment						
150 × 25 mm	14.34	18.15	0.57	26.05	m	**44.20**
Triple compartment						
225 × 25 mm	25.40	32.15	0.93	42.50	m	**74.65**

ELECTRICAL SUPPLY/POWER/LIGHTING SYSTEMS

Item	Net Price £	Material £	Labour hours	Labour £	Unit	Total rate £
Zinc coated steel underfloor ducting fittings (cutting and jointing to fittings is included)						
Stop end; double compartment						
150 × 25 mm	8.77	11.10	0.31	14.17	nr	**25.26**
Stop end; triple compartment						
225 × 25 mm	10.04	12.70	0.37	16.91	nr	**29.61**
Rising bend; double compartment; standard trunking						
150 × 25 mm	58.03	73.44	0.71	32.45	nr	**105.89**
Rising bend; triple compartment; standard trunking						
225 × 25 mm	63.22	80.01	0.85	38.85	nr	**118.86**
Rising bend; double compartment; to skirting						
150 × 25 mm	37.33	47.24	0.90	41.14	nr	**88.38**
Rising bend; triple compartment; to skirting						
225 × 25 mm	47.67	60.33	0.95	43.42	nr	**103.75**
Horizontal bend; double compartment						
150 × 25 mm	46.64	59.03	0.64	29.25	nr	**88.28**
Horizontal bend; triple compartment						
225 × 25 mm	66.74	84.46	0.77	35.19	nr	**119.66**
Junction or service outlet boxes; terminal; double compartment						
150 mm	45.67	57.80	0.91	41.59	nr	**99.39**
Junction or service outlet boxes; terminal; triple compartment						
225 mm	52.19	66.05	1.11	50.73	nr	**116.78**
Junction or service outlet boxes; through or angle; double compartment						
150 mm	60.64	76.75	0.97	44.33	nr	**121.08**
Junction or service outlet boxes; through or angle; triple compartment						
225 mm	67.37	85.26	1.17	53.47	nr	**138.73**
Junction or service outlet boxes; tee; double compartment						
150 mm	60.64	76.75	1.02	46.62	nr	**123.37**
Junction or service outlet boxes; tee; triple compartment						
225 mm	67.37	85.26	1.22	55.76	nr	**141.02**
Junction or service outlet boxes; cross; double compartment						
up to 150 mm	60.64	76.75	1.03	47.07	nr	**123.83**
Junction or service outlet boxes; cross; triple compartment						
225 mm	67.37	85.26	1.23	56.22	nr	**141.47**
Plates for junction/inspection boxes; double and triple compartment						
Blank plate	2.18	2.76	0.92	42.05	nr	**44.81**
Conduit entry plate	9.39	11.88	0.86	39.31	nr	**51.19**
Trunking entry plate	9.39	11.88	0.86	39.31	nr	**51.19**
Service outlet box comprising flat lid with flanged carpet trim; twin 13 A outlet and drilled plate for mounting 2 telephone outlets and terminal blocks; terminal outlet box; double compartment						
150 × 25 mm trunking	197.32	249.73	1.68	76.78	nr	**326.51**

ELECTRICAL SUPPLY/POWER/LIGHTING SYSTEMS

Item	Net Price £	Material £	Labour hours	Labour £	Unit	Total rate £
LV DISTRIBUTION: TRUNKING – cont						
Zinc coated steel underfloor ducting fittings – cont						
Service outlet box comprising flat lid with flanged carpet trim; twin 13 A outlet and drilled plate for mounting 2 telephone outlets and terminal blocks; terminal outlet box; triple compartment						
225 × 25 mm trunking	274.99	348.03	1.93	88.21	nr	**436.24**
PVC skirting/dado modular trunking; white (cutting and jointing trunking to fittings and backplates for fixing to walls is included)						
Main carrier/backplate						
50 × 170 mm	34.11	43.17	0.22	10.05	m	**53.22**
62 × 190 mm	37.87	47.93	0.22	10.05	m	**57.98**
Extension carrier/backplate						
50 × 42 mm	20.72	26.23	0.58	26.51	m	**52.73**
Carrier/backplate						
Including cover seal	12.10	15.31	0.53	24.22	m	**39.53**
Chamfered covers for fixing to backplates						
50 × 42 mm	7.03	8.89	0.33	15.08	m	**23.98**
Square covers for fixing to backplates						
50 × 42 mm	14.09	17.83	0.33	15.08	m	**32.91**
Plain covers for fixing to backplates						
85 mm	7.03	8.89	0.34	15.54	m	**24.43**
Retainers-clip to backplates to hold cables						
For chamfered covers	1.94	2.46	0.07	3.20	m	**5.66**
For square-recessed covers	1.68	2.13	0.07	3.20	m	**5.33**
For plain covers	6.48	8.20	0.07	3.20	m	**11.40**
Prepackaged corner assemblies						
Internal; for 170 × 50 Assy	13.63	17.25	0.51	23.31	nr	**40.55**
Internal; for 190 × 62 Assy	15.14	19.16	0.51	23.31	nr	**42.47**
Internal; for 215 × 50 Assy	16.94	21.44	0.53	24.22	nr	**45.67**
Internal; for 254 × 50 Assy	20.17	25.53	0.53	24.22	nr	**49.75**
External; for 170 × 50 Assy	13.63	17.25	0.56	25.59	nr	**42.84**
External; for 190 × 62 Assy	15.14	19.16	0.56	25.59	nr	**44.75**
External; for 215 × 50 Assy	16.94	21.44	0.58	26.51	nr	**47.95**
External; for 254 × 50 Assy	20.17	25.53	0.58	26.51	nr	**52.04**
Clip on end caps						
170 × 50 Assy	8.16	10.33	0.11	5.03	nr	**15.36**
215 × 50 Assy	9.66	12.22	0.11	5.03	nr	**17.25**
254 × 50 Assy	11.32	14.33	0.11	5.03	nr	**19.36**
190 × 62 Assy	11.41	14.44	0.11	5.03	nr	**19.47**
Outlet box						
1 gang; in horizontal trunking; clip in	7.29	9.22	0.34	15.54	nr	**24.76**
2 gang; in horizontal trunking; clip in	9.09	11.50	0.34	15.54	nr	**27.04**
1 gang; in vertical trunking; clip in	7.29	9.22	0.34	15.54	nr	**24.76**

ELECTRICAL SUPPLY/POWER/LIGHTING SYSTEMS

Item	Net Price £	Material £	Labour hours	Labour £	Unit	Total rate £
Sheet steel adaptable boxes; with plain or knockout sides; fixed to backgrounds; including supports and fixings (cutting and connecting conduit to boxes is included)						
Square pattern – black						
75 × 75 × 37 mm	3.03	3.83	0.69	31.54	nr	**35.37**
75 × 75 × 50 mm	3.66	4.63	0.69	31.54	nr	**36.16**
75 × 75 × 75 mm	5.29	6.69	0.69	31.54	nr	**38.23**
100 × 100 × 50 mm	6.19	7.83	0.71	32.45	nr	**40.28**
150 × 150 × 50 mm	7.11	8.99	0.79	36.11	nr	**45.10**
150 × 150 × 75 mm	7.93	10.03	0.80	36.56	nr	**46.60**
150 × 150 × 100 mm	8.97	11.35	0.80	36.56	nr	**47.91**
200 × 200 × 50 mm	8.77	11.09	0.80	36.56	nr	**47.66**
225 × 225 × 50 mm	9.19	11.63	0.93	42.50	nr	**54.13**
225 × 225 × 100 mm	12.19	15.42	0.94	42.96	nr	**58.39**
300 × 300 × 100 mm	13.08	16.55	0.99	45.25	nr	**61.80**
Square pattern – galvanized						
75 × 75 × 37 mm	2.78	3.52	0.69	31.54	nr	**35.05**
75 × 75 × 50 mm	3.25	4.11	0.69	31.54	nr	**35.64**
75 × 75 × 75 mm	3.78	4.78	0.70	31.99	nr	**36.78**
100 × 100 × 50 mm	3.80	4.81	0.71	32.45	nr	**37.26**
150 × 150 × 50 mm	4.01	5.08	0.80	36.56	nr	**41.64**
150 × 150 × 75 mm	4.83	6.12	0.80	36.56	nr	**42.68**
150 × 150 × 100 mm	5.85	7.40	0.84	38.39	nr	**45.79**
225 × 225 × 50 mm	7.56	9.57	0.93	42.50	nr	**52.07**
225 × 225 × 100 mm	9.19	11.63	0.94	42.96	nr	**54.59**
300 × 300 × 100 mm	15.62	19.77	0.96	43.88	nr	**63.65**
Rectangular pattern – black						
100 × 75 × 50 mm	5.12	6.48	0.69	31.54	nr	**38.02**
150 × 75 × 50 mm	5.37	6.79	0.70	31.99	nr	**38.79**
150 × 75 × 75 mm	5.82	7.37	0.71	32.45	nr	**39.81**
150 × 100 × 75 mm	12.97	16.41	0.71	32.45	nr	**48.86**
225 × 75 × 50 mm	11.20	14.18	0.78	35.65	nr	**49.82**
225 × 150 × 75 mm	17.90	22.65	0.81	37.02	nr	**59.67**
225 × 150 × 100 mm	33.59	42.51	0.81	37.02	nr	**79.53**
300 × 150 × 50 mm	33.59	42.51	0.93	42.50	nr	**85.02**
300 × 150 × 75 mm	33.59	42.51	0.94	42.96	nr	**85.47**
300 × 150 × 100 mm	33.59	42.51	0.96	43.88	nr	**86.39**
Rectangular pattern – galvanized						
100 × 75 × 50 mm	4.86	6.15	0.69	31.54	nr	**37.69**
150 × 75 × 50 mm	5.09	6.45	0.70	31.99	nr	**38.44**
150 × 75 × 75 mm	5.53	7.00	0.71	32.45	nr	**39.45**
150 × 100 × 75 mm	12.30	15.56	0.71	32.45	nr	**48.01**
225 × 75 × 50 mm	10.65	13.48	0.81	37.02	nr	**50.50**
225 × 150 × 75 mm	16.99	21.51	0.81	37.02	nr	**58.53**
225 × 150 × 100 mm	31.92	40.40	0.89	40.68	nr	**81.07**
300 × 150 × 50 mm	31.92	40.40	0.93	42.50	nr	**82.90**
300 × 150 × 75 mm	31.92	40.40	0.94	42.96	nr	**83.36**
300 × 150 × 100 mm	31.92	40.40	0.96	43.88	nr	**84.27**

ELECTRICAL SUPPLY/POWER/LIGHTING SYSTEMS

Item	Net Price £	Material £	Labour hours	Labour £	Unit	Total rate £
LV DISTRIBUTION: CABLES AND WIRING						
ARMOURED CABLE						
Cable; XLPE insulated; PVC sheathed; copper stranded conductors to BS 5467; laid in trench/duct including marker tape (cable tiles measured elsewhere)						
600/1000 volt grade; single core (aluminium wire armour)						
70 mm²	10.17	12.87	0.18	8.23	m	**21.09**
95 mm²	13.64	17.26	0.20	9.14	m	**26.40**
120 mm²	16.76	21.21	0.22	10.05	m	**31.26**
150 mm²	20.87	26.41	0.24	10.97	m	**37.38**
185 mm²	25.20	31.89	0.26	11.88	m	**43.77**
240 mm²	33.46	42.35	0.30	13.71	m	**56.06**
300 mm²	40.74	51.56	0.31	14.17	m	**65.73**
400 mm²	52.89	66.93	0.38	17.37	m	**84.30**
500 mm²	67.40	85.30	0.44	20.11	m	**105.41**
630 mm²	87.32	110.52	0.52	23.77	m	**134.28**
600/1000 volt grade; two core (galvanized steel wire armour)						
1.5 mm²	1.76	2.23	0.06	2.74	m	**4.97**
2.5 mm²	2.16	2.74	0.06	2.74	m	**5.48**
4 mm²	2.67	3.38	0.08	3.66	m	**7.03**
6 mm²	3.31	4.20	0.08	3.66	m	**7.85**
10 mm²	4.54	5.74	0.10	4.57	m	**10.31**
16 mm²	6.56	8.30	0.10	4.57	m	**12.87**
25 mm²	7.88	9.98	0.15	6.86	m	**16.83**
35 mm²	10.94	13.85	0.15	6.86	m	**20.70**
50 mm²	13.86	17.54	0.17	7.77	m	**25.31**
70 mm²	18.33	23.20	0.18	8.23	m	**31.42**
95 mm²	25.78	32.62	0.20	9.14	m	**41.76**
120 mm²	32.03	40.53	0.22	10.05	m	**50.59**
150 mm²	39.93	50.54	0.24	10.97	m	**61.51**
185 mm²	51.37	65.02	0.26	11.88	m	**76.90**
240 mm²	66.37	84.00	0.30	13.71	m	**97.71**
300 mm²	87.71	111.01	0.31	14.17	m	**125.18**
600/1000 volt grade; three core (galvanized steel wire armour)						
1.5 mm²	1.98	2.51	0.07	3.20	m	**5.71**
2.5 mm²	2.45	3.09	0.07	3.20	m	**6.29**
4 mm²	3.24	4.10	0.09	4.11	m	**8.21**
6 mm²	3.79	4.80	0.10	4.57	m	**9.37**
10 mm²	5.96	7.55	0.11	5.03	m	**12.58**
16 mm²	8.66	10.95	0.11	5.03	m	**15.98**
25 mm²	11.73	14.84	0.16	7.31	m	**22.15**
35 mm²	15.75	19.93	0.16	7.31	m	**27.24**
50 mm²	21.11	26.72	0.19	8.68	m	**35.40**
70 mm²	29.87	37.80	0.21	9.60	m	**47.40**
95 mm²	39.18	49.58	0.23	10.51	m	**60.09**

ELECTRICAL SUPPLY/POWER/LIGHTING SYSTEMS

Item	Net Price £	Material £	Labour hours	Labour £	Unit	Total rate £
120 mm²	48.45	61.32	0.24	10.97	m	**72.29**
150 mm²	66.94	84.71	0.27	12.34	m	**97.05**
185 mm²	76.08	96.29	0.30	13.71	m	**110.00**
240 mm²	98.00	124.03	0.33	15.08	m	**139.11**
300 mm²	123.89	156.80	0.35	16.00	m	**172.80**
600/1000 volt grade; four core (galvanized steel wire armour)						
1.5 mm²	1.86	2.36	0.08	3.66	m	**6.01**
2.5 mm²	2.57	3.26	0.09	4.11	m	**7.37**
4 mm²	3.58	4.53	0.10	4.57	m	**9.10**
6 mm²	5.12	6.48	0.10	4.57	m	**11.05**
10 mm²	7.58	9.60	0.12	5.48	m	**15.08**
16 mm²	11.20	14.18	0.12	5.48	m	**19.66**
25 mm²	15.07	19.08	0.18	8.23	m	**27.30**
35 mm²	20.28	25.67	0.19	8.68	m	**34.35**
50 mm²	26.96	34.12	0.21	9.60	m	**43.72**
70 mm²	38.88	49.21	0.23	10.51	m	**59.72**
95 mm²	52.80	66.83	0.26	11.88	m	**78.71**
120 mm²	64.31	81.39	0.28	12.80	m	**94.19**
150 mm²	78.61	99.49	0.32	14.63	m	**114.11**
185 mm²	98.50	124.66	0.35	16.00	m	**140.66**
240 mm²	131.75	166.74	0.36	16.45	m	**183.20**
300 mm²	165.90	209.97	0.40	18.28	m	**228.25**
600/1000 volt grade; seven core (galvanized steel wire armour)						
1.5 mm²	3.03	3.83	0.10	4.57	m	**8.41**
2.5 mm²	4.28	5.42	0.10	4.57	m	**9.99**
600/1000 volt grade; twelve core (galvanized steel wire armour)						
1.5 mm²	4.94	6.25	0.11	5.03	m	**11.27**
2.5 mm²	6.69	8.47	0.11	5.03	m	**13.49**
600/1000 volt grade; nineteen core (galvanized steel wire armour)						
1.5 mm²	7.11	8.99	0.13	5.94	m	**14.94**
2.5 mm²	9.87	12.50	0.14	6.40	m	**18.89**
600/1000 volt grade; twenty-seven core (galvanized steel wire armour)						
1.5 mm²	9.88	12.50	0.14	6.40	m	**18.90**
2.5 mm²	16.07	20.33	0.16	7.31	m	**27.65**
600/1000 volt grade; thirty-seven core (galvanized steel wire armour)						
1.5 mm²	13.02	16.48	0.15	6.86	m	**23.33**
2.5 mm²	18.98	24.03	0.17	7.77	m	**31.80**

Cable; XLPE insulated; PVC sheathed copper stranded conductors to BS 5467; clipped direct to backgrounds including cleat

ELECTRICAL SUPPLY/POWER/LIGHTING SYSTEMS

Item	Net Price £	Material £	Labour hours	Labour £	Unit	Total rate £
LV DISTRIBUTION: CABLES AND WIRING – cont						
600/1000 volt grade; single core (aluminium wire armour)						
70 mm²	13.22	16.73	0.39	17.82	m	**34.55**
95 mm²	19.68	24.91	0.42	19.20	m	**44.10**
120 mm²	21.95	27.78	0.47	21.48	m	**49.27**
150 mm²	26.43	33.45	0.51	23.31	m	**56.75**
185 mm²	30.33	38.39	0.59	26.97	m	**65.35**
240 mm²	38.78	49.08	0.68	31.08	m	**80.16**
300 mm²	45.99	58.21	0.74	33.82	m	**92.03**
400 mm²	57.65	72.96	0.88	40.22	m	**113.18**
500 mm²	74.34	94.08	0.88	40.22	m	**134.30**
630 mm²	94.95	120.16	1.05	47.99	m	**168.15**
600/1000 volt grade; two core (galvanized steel wire armour)						
1.5 mm²	2.27	2.87	0.20	9.14	m	**12.01**
2.5 mm²	2.54	3.22	0.20	9.14	m	**12.36**
4.0 mm²	3.14	3.97	0.21	9.60	m	**13.57**
6.0 mm²	3.84	4.86	0.22	10.05	m	**14.91**
10.0 mm²	5.36	6.78	0.24	10.97	m	**17.75**
16.0 mm²	7.39	9.36	0.25	11.43	m	**20.78**
25 mm²	9.32	11.80	0.35	16.00	m	**27.79**
35 mm²	12.10	15.31	0.36	16.45	m	**31.76**
50 mm²	15.03	19.02	0.37	16.91	m	**35.93**
70 mm²	18.94	23.97	0.39	17.82	m	**41.79**
95 mm²	26.44	33.46	0.42	19.20	m	**52.65**
120 mm²	32.68	41.37	0.47	21.48	m	**62.85**
150 mm²	40.96	51.84	0.51	23.31	m	**75.15**
185 mm²	52.58	66.55	0.59	26.97	m	**93.51**
240 mm²	66.05	83.59	0.68	31.08	m	**114.67**
300 mm²	84.77	107.28	0.74	33.82	m	**141.10**
600/1000 volt grade; three core (galvanized steel wire armour)						
1.5 mm²	2.52	3.19	0.20	9.14	m	**12.33**
2.5 mm²	2.98	3.77	0.21	9.60	m	**13.36**
4.0 mm²	3.75	4.75	0.22	10.05	m	**14.80**
6.0 mm²	4.62	5.85	0.22	10.05	m	**15.91**
10.0 mm²	7.06	8.93	0.25	11.43	m	**20.36**
16.0 mm²	9.70	12.28	0.26	11.88	m	**24.16**
25 mm²	12.97	16.42	0.37	16.91	m	**33.33**
35 mm²	17.05	21.57	0.39	17.82	m	**39.40**
50 mm²	23.22	29.39	0.40	18.28	m	**47.67**
70 mm²	29.94	37.89	0.42	19.20	m	**57.09**
95 mm²	39.53	50.03	0.45	20.57	m	**70.60**
120 mm²	52.25	66.13	0.52	23.77	m	**89.90**
150 mm²	67.04	84.84	0.55	25.14	m	**109.98**
185 mm²	77.34	97.88	0.63	28.79	m	**126.67**
240 mm²	99.22	125.57	0.71	32.45	m	**158.02**
300 mm²	126.24	159.76	0.78	35.65	m	**195.41**

ELECTRICAL SUPPLY/POWER/LIGHTING SYSTEMS

Item	Net Price £	Material £	Labour hours	Labour £	Unit	Total rate £
600/1000 volt grade; four core (galvanized steel wire armour)						
1.5 mm²	2.70	3.42	0.21	9.60	m	**13.02**
2.5 mm²	3.39	4.29	0.22	10.05	m	**14.35**
4.0 mm²	4.44	5.62	0.22	10.05	m	**15.68**
6.0 mm²	6.21	7.86	0.23	10.51	m	**18.37**
10.0 mm²	8.98	11.36	0.26	11.88	m	**23.25**
16.0 mm²	12.56	15.90	0.26	11.88	m	**27.78**
25 mm²	15.97	20.21	0.39	17.82	m	**38.04**
35 mm²	20.97	26.54	0.40	18.28	m	**44.82**
50 mm²	27.44	34.73	0.41	18.74	m	**53.47**
70 mm²	38.17	48.31	0.45	20.57	m	**68.88**
95 mm²	51.41	65.07	0.50	22.85	m	**87.92**
120 mm²	64.99	82.26	0.54	24.68	m	**106.94**
150 mm²	80.51	101.89	0.60	27.42	m	**129.32**
185 mm²	100.55	127.26	0.67	30.62	m	**157.88**
240 mm²	128.79	162.99	0.75	34.28	m	**197.27**
300 mm²	161.92	204.92	0.83	37.93	m	**242.86**
600/1000 volt grade; seven core (galvanized steel wire armour)						
1.5 mm²	3.84	4.86	0.20	9.14	m	**14.00**
2.5 mm²	5.06	6.40	0.20	9.14	m	**15.54**
600/1000 volt grade; twelve core (galvanized steel wire armour)						
1.5 mm²	6.92	8.76	0.23	10.51	m	**19.27**
2.5 mm²	8.36	10.57	0.24	10.97	m	**21.54**
600/1000 volt grade; nineteen core (galvanized steel wire armour)						
1.5 mm²	7.77	9.83	0.26	11.88	m	**21.71**
2.5 mm²	11.35	14.36	0.28	12.80	m	**27.16**
600/1000 volt grade; twenty-seven core (galvanized steel wire armour)						
1.5 mm²	11.48	14.53	0.29	13.25	m	**27.78**
2.5 mm²	17.61	22.29	0.30	13.71	m	**36.00**
600/1000 volt grade; thirty-seven core (galvanized steel wire armour)						
1.5 mm²	14.35	18.16	0.32	14.63	m	**32.79**
2.5 mm²	21.18	26.81	0.33	15.08	m	**41.89**
Cable termination; brass weatherproof gland with inner and outer seal, shroud, brass locknut and earth ring (including drilling and cutting mild steel gland plate)						
600/1000 volt grade; single core (aluminium wire armour)						
70 mm²	10.94	13.84	2.12	96.89	nr	**110.73**
95 mm²	11.00	13.92	2.39	109.23	nr	**123.15**
120 mm²	11.03	13.97	2.47	112.89	nr	**126.85**
150 mm²	22.27	28.18	2.73	124.77	nr	**152.95**
185 mm²	22.36	28.30	3.05	139.40	nr	**167.69**
240 mm²	22.61	28.62	3.45	157.68	nr	**186.30**
300 mm²	23.00	29.10	3.84	175.50	nr	**204.61**

ELECTRICAL SUPPLY/POWER/LIGHTING SYSTEMS

Item	Net Price £	Material £	Labour hours	Labour £	Unit	Total rate £
LV DISTRIBUTION: CABLES AND WIRING – cont						
Cable termination – cont						
600/1000 volt grade; single core – cont						
400 mm²	32.84	41.57	4.21	192.41	nr	**233.98**
500 mm²	35.04	44.35	5.70	260.51	m	**304.86**
630 mm²	42.73	54.08	6.20	283.36	m	**337.44**
600/1000 volt grade; two core (galvanized steel wire armour)						
1.5 mm²	8.76	11.09	0.58	26.51	nr	**37.60**
2.5 mm²	8.76	11.09	0.58	26.51	nr	**37.60**
4 mm²	8.76	11.09	0.58	26.55	nr	**37.65**
6 mm²	8.79	11.12	0.67	30.62	nr	**41.75**
10 mm²	9.76	12.35	1.00	45.70	nr	**58.06**
16 mm²	9.99	12.65	1.11	50.73	nr	**63.38**
25 mm²	11.39	14.41	1.70	77.70	nr	**92.11**
35 mm²	11.48	14.52	1.79	81.81	nr	**96.33**
50 mm²	20.98	26.56	2.06	94.15	nr	**120.70**
70 mm²	21.08	26.68	2.12	96.89	nr	**123.58**
95 mm²	21.87	27.67	2.39	109.23	nr	**136.90**
120 mm²	22.04	27.90	2.47	112.89	nr	**140.79**
150 mm²	32.91	41.65	2.73	124.77	nr	**166.42**
185 mm²	33.65	42.59	3.05	139.40	nr	**181.98**
240 mm²	35.77	45.27	3.45	157.68	nr	**202.95**
300 mm²	52.08	65.92	3.84	175.50	nr	**241.42**
600/1000 volt grade; three core (galvanized steel wire armour)						
1.5 mm²	8.79	11.12	0.61	28.09	nr	**39.22**
2.5 mm²	8.79	11.12	0.62	28.34	nr	**39.46**
4 mm²	8.79	11.12	0.62	28.34	nr	**39.46**
6 mm²	9.78	12.38	0.71	32.45	nr	**44.83**
10 mm²	11.32	14.32	1.06	48.45	nr	**62.77**
16 mm²	11.66	14.75	1.19	54.39	nr	**69.14**
25 mm²	20.88	26.42	1.81	82.72	nr	**109.14**
35 mm²	21.06	26.66	1.99	90.95	nr	**117.61**
50 mm²	21.18	26.80	2.23	101.92	nr	**128.72**
70 mm²	21.71	27.47	2.40	109.69	nr	**137.16**
95 mm²	21.74	27.52	2.63	120.20	nr	**147.72**
120 mm²	21.78	27.57	2.83	129.34	nr	**156.91**
150 mm²	34.09	43.15	3.22	147.17	nr	**190.31**
185 mm²	35.15	44.48	3.44	157.22	nr	**201.70**
240 mm²	53.34	67.51	3.83	175.04	nr	**242.56**
300 mm²	54.85	69.42	4.28	195.61	nr	**265.03**
600/1000 volt grade; four core (galvanized steel wire armour)						
1.5 mm²	9.06	11.47	0.67	30.62	nr	**42.09**
2.5 mm²	9.06	11.47	0.67	30.62	nr	**42.09**
4 mm²	9.07	11.47	0.71	32.45	nr	**43.92**
6 mm²	11.63	14.72	0.76	34.73	nr	**49.46**
10 mm²	11.65	14.74	1.14	52.10	nr	**66.84**
16 mm²	12.10	15.32	1.29	58.96	nr	**74.28**
25 mm²	21.17	26.80	1.99	90.95	nr	**117.75**

ELECTRICAL SUPPLY/POWER/LIGHTING SYSTEMS

Item	Net Price £	Material £	Labour hours	Labour £	Unit	Total rate £
35 mm²	21.51	27.22	2.16	98.72	nr	**125.94**
50 mm²	22.24	28.14	2.49	113.80	nr	**141.94**
70 mm²	22.31	28.24	2.65	121.11	nr	**149.35**
95 mm²	30.94	39.16	2.98	136.20	nr	**175.36**
120 mm²	30.95	39.17	3.15	143.97	nr	**183.14**
150 mm²	36.27	45.91	3.50	159.96	nr	**205.87**
185 mm²	52.81	66.83	3.72	170.02	nr	**236.85**
240 mm²	57.13	72.30	4.33	197.90	nr	**270.20**
300 mm²	81.79	103.52	4.86	222.12	nr	**325.64**
600/1000 volt grade; seven core (galvanized steel wire armour)						
1.5 mm²	10.71	13.56	0.81	37.02	nr	**50.58**
2.5 mm²	10.71	13.56	0.85	38.85	nr	**52.41**
600/1000 volt grade; twelve core (galvanized steel wire armour)						
1.5 mm²	12.27	15.53	1.13	51.65	nr	**67.18**
2.5 mm²	13.94	17.64	1.14	52.10	nr	**69.75**
600/1000 volt grade; nineteen core (galvanized steel wire armour)						
1.5 mm²	15.04	19.03	1.54	70.38	nr	**89.42**
2.5 mm²	15.07	19.07	1.54	70.38	nr	**89.45**
600/1000 volt grade; twenty-seven core (galvanized steel wire armour)						
1.5 mm²	16.53	20.92	1.94	88.66	nr	**109.58**
2.5 mm²	27.36	34.62	2.31	105.58	nr	**140.20**
600/1000 volt grade; thirty-seven core (galvanized steel wire armour)						
1.5 mm²	27.02	34.20	2.53	115.63	nr	**149.83**
2.5 mm²	27.02	34.20	2.87	131.17	nr	**165.37**

ELECTRICAL SUPPLY/POWER/LIGHTING SYSTEMS

Item	Net Price £	Material £	Labour hours	Labour £	Unit	Total rate £
LV DISTRIBUTION: CABLES AND WIRING – cont						
Cable; XLPE Insulated; LSOH sheathed (LSF); copper stranded conductors to BS 6724; laid in trench/duct including marker tape (cable tiles measured elsewhere)						
600/1000 volt grade; single core (aluminium wire armour)						
50 mm²	6.70	8.48	0.17	7.77	m	**16.25**
70 mm²	9.91	12.54	0.18	8.23	m	**20.77**
95 mm²	12.93	16.37	0.20	9.14	m	**25.51**
120 mm²	15.81	20.01	0.22	10.05	m	**30.07**
150 mm²	20.17	25.53	0.24	10.97	m	**36.50**
185 mm²	23.94	30.30	0.26	11.88	m	**42.19**
240 mm²	30.97	39.19	0.30	13.71	m	**52.90**
300 mm²	38.50	48.73	0.31	14.17	m	**62.89**
400 mm²	50.24	63.58	0.35	16.00	m	**79.58**
500 mm²	63.64	80.54	0.44	20.11	m	**100.65**
630 mm²	79.83	101.04	0.52	23.77	m	**124.80**
800 mm²	93.77	118.68	0.62	28.34	m	**147.02**
1000 mm²	119.69	151.47	0.65	29.71	m	**181.18**
600/1000 volt grade; two core (galvanized steel wire armour)						
1.5 mm²	1.58	1.99	0.06	2.74	m	**4.74**
2.5 mm²	1.93	2.45	0.06	2.74	m	**5.19**
4 mm²	2.40	3.03	0.08	3.66	m	**6.69**
6 mm²	3.09	3.92	0.08	3.66	m	**7.57**
10 mm²	4.14	5.24	0.10	4.57	m	**9.81**
16 mm²	6.00	7.59	0.10	4.57	m	**12.16**
25 mm²	7.95	10.06	0.15	6.86	m	**16.92**
35 mm²	10.63	13.46	0.15	6.86	m	**20.31**
50 mm²	13.50	17.09	0.17	7.77	m	**24.86**
70 mm²	17.77	22.50	0.18	8.23	m	**30.72**
95 mm²	25.18	31.87	0.20	9.14	m	**41.01**
120 mm²	31.17	39.44	0.22	10.05	m	**49.50**
150 mm²	37.35	47.27	0.24	10.97	m	**58.24**
185 mm²	49.98	63.25	0.26	11.88	m	**75.14**
240 mm²	51.14	64.73	0.30	13.71	m	**78.44**
300 mm²	54.75	69.30	0.31	14.17	m	**83.47**
400 mm²	79.85	101.06	0.35	16.00	m	**117.06**

ELECTRICAL SUPPLY/POWER/LIGHTING SYSTEMS

Item	Net Price £	Material £	Labour hours	Labour £	Unit	Total rate £
600/1000 volt grade; three core (galvanized steel wire armour)						
1.5 mm²	1.69	2.14	0.07	3.20	m	**5.34**
2.5 mm²	2.17	2.75	0.07	3.20	m	**5.95**
4 mm²	2.83	3.59	0.09	4.11	m	**7.70**
6 mm²	3.49	4.42	0.10	4.57	m	**8.99**
10 mm²	5.47	6.92	0.11	5.03	m	**11.95**
16 mm²	8.04	10.18	0.11	5.03	m	**15.21**
25 mm²	10.93	13.83	0.16	7.31	m	**21.15**
35 mm²	14.75	18.66	0.16	7.31	m	**25.98**
50 mm²	19.88	25.16	0.19	8.68	m	**33.84**
70 mm²	27.38	34.66	0.21	9.60	m	**44.25**
95 mm²	37.47	47.43	0.23	10.51	m	**57.94**
120 mm²	46.00	58.22	0.24	10.97	m	**69.19**
150 mm²	63.14	79.91	0.27	12.34	m	**92.25**
185 mm²	71.53	90.53	0.30	13.71	m	**104.24**
240 mm²	92.21	116.70	0.33	15.08	m	**131.79**
300 mm²	116.59	147.55	0.35	16.00	m	**163.55**
400 mm²	143.46	181.57	0.41	18.74	m	**200.31**
600/1000 volt grade; four core (galvanized steel wire armour)						
1.5 mm²	1.70	2.16	0.08	3.66	m	**5.81**
2.5 mm²	2.31	2.92	0.09	4.11	m	**7.03**
4 mm²	3.18	4.03	0.10	4.57	m	**8.60**
6 mm²	4.45	5.63	0.10	4.57	m	**10.20**
10 mm²	6.99	8.85	0.12	5.48	m	**14.33**
16 mm²	10.39	13.14	0.12	5.48	m	**18.63**
25 mm²	13.79	17.46	0.18	8.23	m	**25.68**
35 mm²	18.59	23.52	0.19	8.68	m	**32.21**
50 mm²	24.57	31.10	0.21	9.60	m	**40.70**
70 mm²	35.39	44.79	0.23	10.51	m	**55.31**
95 mm²	48.26	61.08	0.26	11.88	m	**72.96**
120 mm²	62.15	78.66	0.28	12.80	m	**91.46**
150 mm²	74.69	94.53	0.32	14.63	m	**109.16**
185 mm²	93.13	117.86	0.35	16.00	m	**133.86**
240 mm²	120.60	152.63	0.36	16.45	m	**169.08**
300 mm²	151.33	191.52	0.40	18.28	m	**209.80**
400 mm²	194.83	246.57	0.45	20.57	m	**267.14**
600/1000 volt grade; seven core (galvanized steel wire armour)						
1.5 mm²	2.69	3.41	0.10	4.57	m	**7.98**
2.5 mm²	3.83	4.84	0.10	4.57	m	**9.41**
4 mm²	6.74	8.52	0.11	5.03	m	**13.55**
600/1000 volt grade; twelve core (galvanized steel wire armour)						
1.5 mm²	4.19	5.31	0.11	5.03	m	**10.34**
2.5 mm²	6.15	7.78	0.11	5.03	m	**12.81**
600/1000 volt grade; nineteen core (galvanized steel wire armour)						
1.5 mm²	6.04	7.64	0.13	5.94	m	**13.58**
2.5 mm²	9.33	11.81	0.14	6.40	m	**18.21**

ELECTRICAL SUPPLY/POWER/LIGHTING SYSTEMS

Item	Net Price £	Material £	Labour hours	Labour £	Unit	Total rate £
LV DISTRIBUTION: CABLES AND WIRING – cont						
Cable – cont						
600/1000 volt grade; twenty-seven core (galvanized steel wire armour)						
1.5 mm²	8.96	11.35	0.14	6.40	m	**17.74**
2.5 mm²	12.73	16.11	0.16	7.31	m	**23.42**
600/1000 volt grade; thirty-seven core (galvanized steel wire armour)						
1.5 mm²	11.13	14.09	0.15	6.86	m	**20.94**
2.5 mm²	17.16	21.71	0.17	7.77	m	**29.48**
Cable; XLPE insulated; LSOH sheathed (LSF) copper stranded conductors to BS 6724; clipped direct to backgrounds including cleat						
600/1000 volt grade; single core (aluminium wire armour)						
50 mm²	6.48	8.20	0.37	16.91	m	**25.11**
70 mm²	9.17	11.60	0.39	17.82	m	**29.43**
95 mm²	12.82	16.23	0.42	19.20	m	**35.43**
120 mm²	15.43	19.52	0.47	21.48	m	**41.00**
150 mm²	19.42	24.58	0.51	23.31	m	**47.88**
185 mm²	23.67	29.96	0.59	26.97	m	**56.92**
240 mm²	31.81	40.26	0.68	31.08	m	**71.34**
300 mm²	37.98	48.07	0.74	33.82	m	**81.89**
400 mm²	50.03	63.32	0.81	37.02	m	**100.34**
500 mm²	64.24	81.31	0.88	40.22	m	**121.53**
630 mm²	79.02	100.01	1.05	47.99	m	**148.00**
800 mm²	97.96	123.98	1.33	60.79	m	**184.77**
1000 mm²	121.35	153.58	1.40	63.99	m	**217.57**

ELECTRICAL SUPPLY/POWER/LIGHTING SYSTEMS

Item	Net Price £	Material £	Labour hours	Labour £	Unit	Total rate £
600/1000 volt grade; two core (galvanized steel wire armour)						
1.5 mm²	1.69	2.14	0.20	9.14	m	**11.28**
2.5 mm²	2.05	2.59	0.20	9.14	m	**11.73**
4.0 mm²	2.52	3.19	0.21	9.60	m	**12.79**
6.0 mm²	3.49	4.42	0.22	10.05	m	**14.48**
10.0 mm²	4.69	5.93	0.24	10.97	m	**16.90**
16.0 mm²	6.54	8.28	0.25	11.43	m	**19.71**
25 mm²	7.31	9.25	0.35	16.00	m	**25.25**
35 mm²	9.94	12.58	0.36	16.45	m	**29.03**
50 mm²	12.88	16.31	0.37	16.91	m	**33.22**
70 mm²	17.43	22.07	0.39	17.82	m	**39.89**
95 mm²	24.79	31.37	0.42	19.20	m	**50.57**
120 mm²	30.80	38.98	0.47	21.48	m	**60.46**
150 mm²	37.09	46.94	0.51	23.31	m	**70.25**
185 mm²	49.49	62.63	0.59	26.97	m	**89.60**
240 mm²	49.68	62.87	0.68	31.08	m	**93.95**
300 mm²	65.30	82.65	0.74	33.82	m	**116.47**
400 mm²	91.38	115.65	0.81	37.02	m	**152.67**
600/1000 volt grade; three core (galvanized steel wire armour)						
1.5 mm²	1.90	2.40	0.20	9.14	m	**11.54**
2.5 mm²	2.49	3.15	0.21	9.60	m	**12.74**
4.0 mm²	3.07	3.89	0.22	10.05	m	**13.94**
6.0 mm²	4.35	5.50	0.22	10.05	m	**15.56**
10.0 mm²	6.28	7.95	0.25	11.43	m	**19.37**
16.0 mm²	8.83	11.18	0.26	11.88	m	**23.06**
25 mm²	11.17	14.13	0.37	16.91	m	**31.04**
35 mm²	14.42	18.24	0.39	17.82	m	**36.07**
50 mm²	19.25	24.36	0.40	18.28	m	**42.64**
70 mm²	27.14	34.35	0.42	19.20	m	**53.54**
95 mm²	38.12	48.24	0.45	20.57	m	**68.81**
120 mm²	47.45	60.05	0.52	23.77	m	**83.82**
150 mm²	58.57	74.12	0.55	25.14	m	**99.26**
185 mm²	70.98	89.84	0.63	28.79	m	**118.63**
240 mm²	92.27	116.78	0.71	32.45	m	**149.23**
300 mm²	111.33	140.90	0.78	35.65	m	**176.54**
400 mm²	147.26	186.37	0.87	39.76	m	**226.14**
600/1000 volt grade; four core (galvanized steel wire armour)						
1.5 mm²	2.20	2.79	0.21	9.60	m	**12.38**
2.5 mm²	2.83	3.58	0.22	10.05	m	**13.64**
4.0 mm²	4.00	5.07	0.22	10.05	m	**15.12**
6.0 mm²	5.34	6.76	0.23	10.51	m	**17.27**
10.0 mm²	7.88	9.98	0.26	11.88	m	**21.86**
16.0 mm²	11.31	14.31	0.26	11.88	m	**26.19**
25 mm²	13.37	16.92	0.39	17.82	m	**34.74**
35 mm²	18.08	22.88	0.40	18.28	m	**41.16**
50 mm²	24.09	30.49	0.41	18.74	m	**49.23**
70 mm²	34.86	44.11	0.45	20.57	m	**64.68**
95 mm²	47.39	59.98	0.50	22.85	m	**82.83**

ELECTRICAL SUPPLY/POWER/LIGHTING SYSTEMS

Item	Net Price £	Material £	Labour hours	Labour £	Unit	Total rate £
LV DISTRIBUTION: CABLES AND WIRING – cont						
Cable – cont						
600/1000 volt grade; four core – cont						
120 mm²	60.02	75.97	0.54	24.68	m	**100.65**
150 mm²	73.82	93.43	0.60	27.42	m	**120.85**
185 mm²	92.66	117.27	0.67	30.62	m	**147.89**
240 mm²	118.77	150.31	0.75	34.28	m	**184.59**
300 mm²	150.21	190.11	0.83	37.93	m	**228.04**
400 mm²	198.41	251.11	0.91	41.59	m	**292.70**
600/1000 volt grade; seven core (galvanized steel wire armour)						
1.5 mm²	3.11	3.93	0.20	9.14	m	**13.08**
2.5 mm²	4.58	5.80	0.20	9.14	m	**14.94**
4.0 mm²	5.58	7.07	0.23	10.51	m	**17.58**
600/1000 volt grade; twelve core (galvanized steel wire armour)						
1.5 mm²	4.91	6.22	0.23	10.51	m	**16.73**
2.5 mm²	7.00	8.87	0.24	10.97	m	**19.83**
600/1000 volt grade; nineteen core (galvanized steel wire armour)						
1.5 mm²	6.87	8.70	0.26	11.88	m	**20.58**
2.5 mm²	10.25	12.97	0.28	12.80	m	**25.77**
600/1000 volt grade; twenty-seven core (galvanized steel wire armour)						
1.5 mm²	9.89	12.51	0.29	13.25	m	**25.77**
2.5 mm²	13.72	17.37	0.30	13.71	m	**31.08**
600/1000 volt grade; thirty-seven core (galvanized steel wire armour)						
1.5 mm²	11.84	14.99	0.32	14.63	m	**29.62**
2.5 mm²	17.76	22.47	0.33	15.08	m	**37.56**
Cable termination; brass weatherproof gland with inner and outer seal, shroud, brass locknut and earth ring (including drilling and cutting mild steel gland plate)						
600/1000 volt grade; single core (aluminium wire armour)						
25 mm²	4.45	5.64	1.70	77.70	nr	**83.33**
35 mm²	4.85	6.14	1.79	81.81	nr	**87.95**
50 mm²	5.38	6.81	2.06	94.15	nr	**100.96**
70 mm²	7.84	9.92	2.12	96.89	nr	**106.81**
95 mm²	10.15	12.85	2.39	109.23	nr	**122.08**
120 mm²	12.03	15.22	2.47	112.89	nr	**128.11**
150 mm²	15.09	19.10	2.73	124.77	nr	**143.87**
185 mm²	18.87	23.88	3.05	139.40	nr	**163.28**
240 mm²	23.87	30.21	3.45	157.68	nr	**187.89**
300 mm²	29.60	37.47	3.84	175.50	nr	**212.97**
400 mm²	41.14	52.07	4.21	192.41	nr	**244.48**
500 mm²	51.84	65.61	5.70	260.51	m	**326.12**
630 mm²	66.07	83.62	6.20	283.36	m	**366.98**
800 mm²	84.44	106.87	7.50	342.78	m	**449.64**
1000 mm²	94.43	119.51	10.00	457.04	m	**576.55**

ELECTRICAL SUPPLY/POWER/LIGHTING SYSTEMS

Item	Net Price £	Material £	Labour hours	Labour £	Unit	Total rate £
600/1000 volt grade; two core (galvanized steel wire armour)						
1.5 mm²	2.08	2.63	0.52	23.63	nr	**26.26**
2.5 mm²	2.37	3.00	0.58	26.51	nr	**29.51**
4 mm²	2.93	3.71	0.58	26.51	nr	**30.21**
6 mm²	3.34	4.22	0.67	30.62	nr	**34.85**
10 mm²	4.30	5.44	1.00	45.70	nr	**51.15**
16 mm²	5.09	6.44	1.11	50.73	nr	**57.17**
25 mm²	6.19	7.84	1.70	77.70	nr	**85.53**
35 mm²	11.61	14.69	1.79	81.81	nr	**96.50**
50 mm²	13.14	16.63	2.06	94.15	nr	**110.77**
70 mm²	19.53	24.72	2.12	96.89	nr	**121.61**
95 mm²	24.43	30.91	2.39	109.23	nr	**140.15**
120 mm²	29.80	37.72	2.47	112.89	nr	**150.61**
150 mm²	37.21	47.09	2.73	124.77	nr	**171.86**
185 mm²	47.37	59.95	3.05	139.40	nr	**199.34**
240 mm²	57.98	73.38	3.45	157.68	nr	**231.05**
300 mm²	81.18	102.74	3.84	175.50	nr	**278.24**
400 mm²	101.92	128.99	4.21	192.41	nr	**321.40**
600/1000 volt grade; three core (galvanized steel wire armour)						
1.5 mm²	2.03	2.57	0.62	28.34	nr	**30.91**
2.5 mm²	2.40	3.03	0.62	28.34	nr	**31.37**
4 mm²	2.98	3.77	0.62	28.34	nr	**32.11**
6 mm²	3.71	4.70	0.71	32.45	nr	**37.15**
10 mm²	5.35	6.78	1.06	48.45	nr	**55.22**
16 mm²	7.04	8.91	1.19	54.39	nr	**63.30**
25 mm²	10.86	13.75	1.81	82.72	nr	**96.47**
35 mm²	13.72	17.36	1.99	90.95	nr	**108.31**
50 mm²	19.11	24.18	2.23	101.92	nr	**126.10**
70 mm²	27.32	34.58	2.40	109.69	nr	**144.27**
95 mm²	36.40	46.07	2.63	120.20	nr	**166.27**
120 mm²	46.24	58.52	2.83	129.34	nr	**187.86**
150 mm²	54.94	69.53	3.22	147.17	nr	**216.69**
185 mm²	65.95	83.46	3.44	157.22	nr	**240.68**
240 mm²	88.97	112.60	3.83	175.04	nr	**287.65**
300 mm²	106.63	134.95	4.28	195.61	nr	**330.56**
400 mm²	132.96	168.27	5.00	228.52	nr	**396.79**
600/1000 volt grade; four core (galvanized steel wire armour)						
1.5 mm²	2.20	2.79	0.67	30.62	nr	**33.41**
2.5 mm²	2.70	3.42	0.67	30.62	nr	**34.04**
4 mm²	3.25	4.11	0.71	32.45	nr	**36.56**
6 mm²	4.58	5.79	0.76	34.73	nr	**40.53**
10 mm²	5.90	7.46	1.14	52.10	nr	**59.57**
16 mm²	7.86	9.94	1.29	58.96	nr	**68.90**
25 mm²	13.73	17.37	1.99	90.95	nr	**108.32**
35 mm²	17.67	22.37	2.16	98.72	nr	**121.09**

ELECTRICAL SUPPLY/POWER/LIGHTING SYSTEMS

Item	Net Price £	Material £	Labour hours	Labour £	Unit	Total rate £
LV DISTRIBUTION: CABLES AND WIRING – cont						
Cable termination – cont						
600/1000 volt grade; four core – cont						
50 mm²	24.74	31.31	2.49	113.80	nr	**145.12**
70 mm²	36.33	45.98	2.65	121.11	nr	**167.10**
95 mm²	49.03	62.05	2.98	136.20	nr	**198.25**
120 mm²	58.82	74.45	3.15	143.97	nr	**218.41**
150 mm²	72.93	92.30	3.50	159.96	nr	**252.26**
185 mm²	87.52	110.76	3.72	170.02	nr	**280.78**
240 mm²	109.73	138.87	4.33	197.90	nr	**336.77**
300 mm²	133.05	168.39	4.86	222.12	nr	**390.51**
400 mm²	146.39	185.28	5.46	249.54	nr	**434.82**
600/1000 volt grade; seven core (galvanized steel wire armour)						
1.5 mm²	3.58	4.53	0.81	37.02	nr	**41.55**
2.5 mm²	4.77	6.04	0.85	38.85	nr	**44.88**
4 mm²	8.78	11.11	0.93	42.50	nr	**53.62**
600/1000 volt grade; twelve core (galvanized steel wire armour)						
1.5 mm²	4.73	5.99	1.13	51.65	nr	**57.63**
2.5 mm²	7.11	9.00	1.14	52.10	nr	**61.10**
600/1000 volt grade; nineteen core (galvanized steel wire armour)						
1.5 mm²	6.74	8.53	1.54	70.38	nr	**78.91**
2.5 mm²	9.63	12.19	1.54	70.38	nr	**82.58**
600/1000 volt grade; twenty-seven core (galvanized steel wire armour)						
1.5 mm²	8.76	11.09	1.94	88.66	nr	**99.76**
2.5 mm²	12.00	15.18	2.31	105.58	nr	**120.76**
600/1000 volt grade; thirty-seven core (galvanized steel wire armour)						
1.5 mm²	10.04	12.70	2.53	115.63	nr	**128.33**
2.5 mm²	13.68	17.31	2.87	131.17	nr	**148.48**
UN-ARMOURED CABLE						
Cable: XLPE insulated; PVC sheathed 90c copper to CMA Code 6181e; for internal wiring; clipped to backgrounds (Supports and fixings included)						
300/500 volt grade; single core						
6.0 mm²	1.22	1.54	0.09	4.11	m	**5.65**
10 mm²	1.99	2.52	0.10	4.57	m	**7.09**
16 mm²	3.02	3.82	0.12	5.48	m	**9.31**

ELECTRICAL SUPPLY/POWER/LIGHTING SYSTEMS

Item	Net Price £	Material £	Labour hours	Labour £	Unit	Total rate £
Cable; LSF insulated to CMA Code 6491B; non-sheathed copper; laid/drawn in trunking/conduit						
450/750 volt grade; single core						
1.5 mm²	0.19	0.23	0.03	1.37	m	**1.61**
2.5 mm²	0.29	0.37	0.03	1.37	m	**1.74**
4.0 mm²	0.46	0.58	0.03	1.37	m	**1.95**
6.0 mm²	0.68	0.86	0.04	1.83	m	**2.69**
10.0 mm²	1.12	1.42	0.04	1.83	m	**3.25**
16.0 mm²	1.83	2.32	0.05	2.29	m	**4.60**
25.0 mm²	2.76	3.50	0.06	2.74	m	**6.24**
35.0 mm²	3.80	4.80	0.06	2.74	m	**7.55**
50.0 mm²	5.25	6.64	0.07	3.20	m	**9.84**
70.0 mm²	7.49	9.47	0.08	3.66	m	**13.13**
95.0 mm²	10.28	13.01	0.08	3.66	m	**16.67**
120.0 mm²	13.00	16.45	0.10	4.57	m	**21.02**
150.0 mm²	16.06	20.33	0.13	5.94	m	**26.27**
185.0 mm²	20.17	25.53	0.13	5.94	m	**31.47**
240.0 mm²	26.53	33.58	0.13	5.94	m	**39.52**
300.0 mm²	32.82	41.54	0.13	5.94	m	**47.48**
Cable; twin & earth to CMA code 6242Y; clipped to backgrounds						
300/500 volt grade; PVC/PVC						
1.5 mm² 2C+E	0.65	0.82	0.01	0.46	m	**1.28**
1.5 mm² 3C+E	0.79	0.99	0.02	0.91	m	**1.91**
2.5 mm² 2C+E	1.00	1.26	0.02	0.91	m	**2.18**
4.0 mm² 2C+E	1.42	1.80	0.02	0.91	m	**2.71**
6.0 mm² 2C+E	2.09	2.65	0.02	0.91	m	**3.56**
10.0 mm² 2C+E	3.49	4.41	0.03	1.37	m	**5.79**
16.0 mm² 2C+E	5.45	6.90	0.03	1.37	m	**8.27**
300/500 volt grade; LSF/LSF						
1.5 mm² 2C+E	0.78	0.99	0.01	0.46	m	**1.45**
1.5 mm² 3C+E	1.12	1.41	0.02	0.91	m	**2.33**
2.5 mm² 2C+E	1.31	1.65	0.02	0.91	m	**2.57**
4.0 mm² 2C+E	1.51	1.91	0.02	0.91	m	**2.82**
6.0 mm² 2C+E	2.16	2.74	0.02	0.91	m	**3.65**
10.0 mm² 2C+E	4.88	6.18	0.03	1.37	m	**7.55**
16.0 mm² 2C+E	8.09	10.23	0.03	1.37	m	**11.61**
EARTH CABLE						
Cable; LSF insulated to CMA Code 6491B; non-sheathed copper; laid/drawn in trunking/conduit						
450/750 volt grade; single core						
1.5 mm²	0.16	0.20	0.03	1.37	m	**1.57**
2.5 mm²	0.25	0.32	0.03	1.37	m	**1.69**
4.0 mm²	0.38	0.49	0.03	1.37	m	**1.86**
6.0 mm²	0.56	0.70	0.04	1.83	m	**2.53**
10.0 mm²	0.99	1.25	0.04	1.83	m	**3.08**
16.0 mm²	1.52	1.93	0.05	2.29	m	**4.21**
25.0 mm²	2.37	3.00	0.06	2.74	m	**5.74**

ELECTRICAL SUPPLY/POWER/LIGHTING SYSTEMS

Item	Net Price £	Material £	Labour hours	Labour £	Unit	Total rate £
LV DISTRIBUTION: CABLES AND WIRING – cont						
Cable – cont						
450/750 volt grade; single core – cont						
35.0 mm²	3.22	4.08	0.06	2.74	m	6.82
50.0 mm²	4.44	5.62	0.07	3.20	m	8.82
70.0 mm²	6.32	8.00	0.08	3.66	m	11.65
95.0 mm²	8.94	11.32	0.08	3.66	m	14.97
120.0 mm²	11.04	13.97	0.10	4.57	m	18.54
150.0 mm²	13.72	17.37	0.13	5.94	m	23.31
185.0 mm²	16.84	21.32	0.16	7.31	m	28.63
240.0 mm²	22.01	27.85	0.20	9.14	m	36.99
300.0 mm²	27.66	35.01	0.13	5.94	m	40.95
FLEXIBLE CABLE						
Flexible cord; PVC insulated; PVC sheathed; copper stranded to CMA Code 218Y (laid loose)						
300 volt grade; two core						
0.50 mm²	0.24	0.30	0.07	3.20	m	3.50
0.75 mm²	0.31	0.39	0.07	3.20	m	3.59
300 volt grade; three core						
0.50 mm²	0.35	0.44	0.07	3.20	m	3.64
0.75 mm²	0.48	0.61	0.07	3.20	m	3.80
1.0 mm²	0.51	0.64	0.07	3.20	m	3.84
1.5 mm²	0.80	1.01	0.07	3.20	m	4.21
2.5 mm²	1.08	1.37	0.08	3.66	m	5.03
Flexible cord; PVC insulated; PVC sheathed; copper stranded to CMA Code 318Y (laid loose)						
300/500 volt grade; two core						
1.0 mm²	2.11	2.67	0.07	3.20	m	5.87
1.5 mm²	2.79	3.53	0.07	3.20	m	6.73
2.5 mm²	5.04	6.38	0.07	3.20	m	9.58
300/500 volt grade; three core						
0.75 mm²	1.37	1.73	0.07	3.20	m	4.93
1.5 mm²	2.38	3.01	0.07	3.20	m	6.21
2.5 mm²	6.49	8.21	0.08	3.66	m	11.87
300/500 volt grade; four core						
0.75 mm²	2.91	3.68	0.08	3.66	m	7.34
1.0 mm²	3.78	4.78	0.08	3.66	m	8.44
1.5 mm²	5.46	6.91	0.08	3.66	m	10.56
2.5 mm²	8.77	11.10	0.09	4.11	m	15.21
Flexible cord; PVC insulated; PVC sheathed for use in high temperature zones; copper stranded to CMA Code 309Y (laid loose)						
300/500 volt grade; two core						
0.50 mm²	1.49	1.89	0.07	3.20	m	5.09
0.75 mm²	1.99	2.52	0.07	3.20	m	5.72

ELECTRICAL SUPPLY/POWER/LIGHTING SYSTEMS

Item	Net Price £	Material £	Labour hours	Labour £	Unit	Total rate £
300/500 volt grade; three core						
1.0 mm²	1.96	2.48	0.07	3.20	m	**5.68**
1.5 mm²	2.72	3.44	0.07	3.20	m	**6.64**
2.5 mm²	4.09	5.18	0.07	3.20	m	**8.38**
Flexible cord; rubber insulated; rubber sheathed; copper stranded to CMA code 318 (laid loose)						
300/500 volt grade; three core						
0.75 mm²	0.98	1.24	0.07	3.20	m	**4.43**
1.0 mm²	1.18	1.49	0.07	3.20	m	**4.69**
1.5 mm²	1.71	2.17	0.07	3.20	m	**5.37**
2.5 mm²	2.63	3.33	0.07	3.20	m	**6.53**
300/500 volt grade; four core						
0.75 mm²	1.21	1.53	0.08	3.66	m	**5.19**
1.0 mm²	1.54	1.95	0.08	3.66	m	**5.60**
1.5 mm²	2.17	2.74	0.08	3.66	m	**6.40**
2.5 mm²	3.49	4.41	0.08	3.66	m	**8.07**
Flexible cord; rubber insulated; rubber sheathed; for 90C operation; copper stranded to CMA Code 318 (laid loose)						
450/750 volt grade; two core						
1.0 mm²	0.81	1.03	0.07	3.20	m	**4.23**
1.5 mm²	1.00	1.26	0.07	3.20	m	**4.46**
2.5 mm²	1.46	1.85	0.07	3.20	m	**5.05**
450/750 volt grade; three core						
1.0 mm²	1.15	1.45	0.07	3.20	m	**4.65**
1.5 mm²	1.65	2.08	0.07	3.20	m	**5.28**
2.5 mm²	2.43	3.07	0.07	3.20	m	**6.27**
450/750 volt grade; four core						
0.75 mm²	0.88	1.11	0.07	3.20	m	**4.31**
1.0 mm²	1.14	1.45	0.08	3.66	m	**5.10**
1.5 mm²	1.71	2.16	0.08	3.66	m	**5.82**
2.5 mm²	2.50	3.17	0.08	3.66	m	**6.82**
Heavy flexible cable; rubber insulated; rubber sheathed; copper stranded to CMA Code 638P (laid loose)						
450/750 volt grade; two core						
1.0 mm²	0.70	0.89	0.08	3.66	m	**4.55**
1.5 mm²	0.84	1.06	0.08	3.66	m	**4.71**
2.5 mm²	1.15	1.45	0.08	3.66	m	**5.11**
450/750 volt grade; three core						
1.0 mm²	1.02	1.29	0.08	3.66	m	**4.95**
1.5 mm²	1.26	1.60	0.08	3.66	m	**5.25**
2.5 mm²	1.46	1.85	0.08	3.66	m	**5.51**
450/750 volt grade; four core						
1.0 mm²	1.12	1.42	0.08	3.66	m	**5.07**
1.5 mm²	1.23	1.56	0.08	3.66	m	**5.22**
2.5 mm²	1.69	2.14	0.08	3.66	m	**5.80**

ELECTRICAL SUPPLY/POWER/LIGHTING SYSTEMS

Item	Net Price £	Material £	Labour hours	Labour £	Unit	Total rate £
LV DISTRIBUTION: CABLES AND WIRING – cont						
FIRE RATED CABLE						
Cable, mineral insulated; copper sheathed with						
copper conductors; fixed with clips to backgrounds.						
BASEC approval to BS 6207 Part 1 1995; complies						
with BS 6387 Category CWZ						
Light duty 500 volt grade; bare						
2L 1.0	6.62	8.38	0.23	10.51	m	**18.89**
2L 1.5	7.55	9.55	0.23	10.51	m	**20.06**
2L 2.5	9.46	11.97	0.25	11.43	m	**23.40**
2L 4.0	13.89	17.58	0.25	11.43	m	**29.01**
3L 1.0	7.92	10.03	0.25	11.43	m	**21.45**
3L 1.5	9.47	11.99	0.25	11.43	m	**23.42**
3L 2.5	13.66	17.29	0.25	11.43	m	**28.71**
4L 1.0	9.21	11.65	0.25	11.43	m	**23.08**
4L 1.5	10.29	13.02	0.25	11.43	m	**24.44**
4L 2.5	16.35	20.70	0.26	11.88	m	**32.58**
7L 1.5	16.18	20.48	0.28	12.80	m	**33.28**
7L 2.5	20.59	26.06	0.28	12.80	m	**38.85**
Light duty 500 volt grade; LSF sheathed						
2L 1.0	4.91	6.22	0.23	10.51	m	**16.73**
2L 1.5	5.60	7.09	0.23	10.51	m	**17.60**
2L 4.0	5.66	7.16	0.25	11.43	m	**18.59**
3L 1.0	6.71	8.49	0.25	11.43	m	**19.92**
3L 1.5	7.05	8.92	0.25	11.43	m	**20.35**
4L 1.0	6.73	8.52	0.25	11.43	m	**19.94**
4L 1.5	8.14	10.30	0.25	11.43	m	**21.73**
4L 2.5	11.55	14.62	0.26	11.88	m	**26.50**
7L 1.5	9.93	12.57	0.28	12.80	m	**25.37**
7L 2.5	12.42	15.72	0.28	12.80	m	**28.52**
7L 1.0	8.15	10.31	0.27	12.34	m	**22.65**
Heavy duty 750 volt grade; bare						
1H 10	14.82	18.76	0.25	11.43	m	**30.19**
1H 16	18.65	23.60	0.26	11.88	m	**35.48**
1H 25	25.93	32.81	0.27	12.34	m	**45.15**
1H 35	37.22	47.11	0.32	14.63	m	**61.73**
1H 50	42.31	53.55	0.35	16.00	m	**69.55**
1H 70	56.45	71.44	0.38	17.37	m	**88.81**
1H 95	80.56	101.96	0.41	18.74	m	**120.69**
1H 120	87.08	110.20	0.46	21.02	m	**131.23**
1H 150	106.80	135.17	0.50	22.85	m	**158.02**
1H 185	130.25	164.84	0.56	25.59	m	**190.43**
1H 240	169.45	214.46	0.69	31.54	m	**246.00**
2H 1.5	12.59	15.93	0.25	11.43	m	**27.36**
2H 2.5	15.08	19.08	0.26	11.88	m	**30.97**
2H 4.0	18.65	23.60	0.26	11.88	m	**35.48**
2H 6.0	24.96	31.59	0.29	13.25	m	**44.84**

ELECTRICAL SUPPLY/POWER/LIGHTING SYSTEMS

Item	Net Price £	Material £	Labour hours	Labour £	Unit	Total rate £
2H 10.0	32.01	40.51	0.34	15.54	m	**56.05**
2H 16.0	45.81	57.98	0.40	18.28	m	**76.26**
2H 25.0	63.03	79.77	0.44	20.11	m	**99.88**
3H 1.5	13.90	17.59	0.25	11.43	m	**29.02**
3H 2.5	17.17	21.73	0.25	11.43	m	**33.16**
3H 4.0	22.05	27.91	0.27	12.34	m	**40.25**
3H 6.0	27.33	34.59	0.30	13.71	m	**48.30**
3H 10.0	40.20	50.88	0.35	16.00	m	**66.87**
3H 16.0	53.68	67.94	0.41	18.74	m	**86.68**
3H 25.0	77.65	98.28	0.47	21.48	m	**119.76**
4H 1.5	16.98	21.49	0.25	11.43	m	**32.92**
4H 2.5	21.51	27.22	0.26	11.88	m	**39.10**
4H 4.0	26.68	33.76	0.29	13.25	m	**47.02**
4H 6.0	33.69	42.63	0.31	14.17	m	**56.80**
4H 10.0	47.06	59.55	0.37	16.91	m	**76.46**
4H 16.0	67.31	85.19	0.44	20.11	m	**105.30**
4H 25.0	94.94	120.16	0.52	23.77	m	**143.92**
7H 1.5	23.53	29.77	0.30	13.71	m	**43.49**
7H 2.5	31.75	40.19	0.32	14.63	m	**54.81**
12H 2.5	54.22	68.62	0.39	17.82	m	**86.45**
19H 1.5	77.74	98.39	0.42	19.20	m	**117.59**
Heavy duty 750 volt grade; LSF sheathed						
1H 10	10.95	13.86	0.25	11.43	m	**25.29**
1H 16	13.86	17.54	0.26	11.88	m	**29.42**
1H 25	19.04	24.10	0.27	12.34	m	**36.44**
1H 35	25.70	32.52	0.32	14.63	m	**47.14**
1H 50	32.40	41.00	0.35	16.00	m	**57.00**
1H 70	41.13	52.06	0.38	17.37	m	**69.43**
1H 95	55.50	70.24	0.41	18.74	m	**88.98**
2H 1.5	9.41	11.91	0.25	11.43	m	**23.34**
2H 2.5	11.21	14.18	0.26	11.88	m	**26.06**
2H 4.0	13.88	17.57	0.26	11.88	m	**29.45**
2H 6.0	18.49	23.40	0.29	13.25	m	**36.65**
2H 10.0	24.52	31.03	0.34	15.54	m	**46.57**
2H 16.0	33.66	42.60	0.40	18.28	m	**60.88**
2H 25.0	46.74	59.15	0.44	20.11	m	**79.26**
3H 1.5	10.31	13.05	0.25	11.43	m	**24.47**
3H 2.5	12.82	16.22	0.25	11.43	m	**27.65**
3H 4.0	16.45	20.82	0.27	12.34	m	**33.16**
3H 6.0	21.21	26.84	0.30	13.71	m	**40.55**
3H 10.0	23.81	30.13	0.35	16.00	m	**46.13**
3H 16.0	33.42	42.30	0.41	18.74	m	**61.04**
3H 25.0	47.36	59.94	0.47	21.48	m	**81.42**

ELECTRICAL SUPPLY/POWER/LIGHTING SYSTEMS

Item	Net Price £	Material £	Labour hours	Labour £	Unit	Total rate £
LV DISTRIBUTION: CABLES AND WIRING – cont						
Cable, mineral insulated – cont						
Heavy duty 750 volt grade – cont						
4H 1.5	12.58	15.93	0.24	10.97	m	**26.90**
4H 2.5	15.99	20.24	0.26	11.88	m	**32.12**
4H 4.0	20.43	25.86	0.29	13.25	m	**39.11**
4H 6.0	25.87	32.75	0.31	14.17	m	**46.91**
4H 10.0	28.97	36.66	0.37	16.91	m	**53.58**
4H 16.0	40.80	51.63	0.44	20.11	m	**71.74**
4H 25.0	57.00	72.14	0.52	23.77	m	**95.91**
7H 1.5	14.42	18.25	0.30	13.71	m	**31.96**
7H 2.5	19.73	24.96	0.32	14.63	m	**39.59**
12H 1.5	24.76	31.34	0.39	17.82	m	**49.17**
12H 2.5	33.00	41.77	0.39	17.82	m	**59.59**
19H 1.5	46.82	59.26	0.42	19.20	m	**78.45**
Cable terminations for MI Cable; Polymeric one piece moulding; containing grey sealing compound; testing; phase marking and connection						
Light duty 500 volt grade; brass gland; polymeric one moulding containing grey sealing compound; coloured conductor sleeving; Earth tag; plastic gland shroud						
2L 1.5	15.60	19.74	0.27	12.34	m	**32.08**
2L 2.5	15.60	19.74	0.27	12.34	m	**32.08**
3L 1.5	15.22	19.26	0.27	12.34	m	**31.60**
4L 1.5	15.22	19.26	0.27	12.34	m	**31.60**
Cable terminations; for MI copper sheathed cable. Certified for installation in potentially explosive atmospheres; testing; phase marking and connection; BS 6207 Part 2 1995						
Light duty 500 volt grade; brass gland; brass pot with earth tail; pot closure; sealing compound; conductor sleving; plastic gland shroud; identification markers						
2L 1.5	13.43	17.00	0.41	18.74	nr	**35.73**
2L 2.5	13.36	16.91	0.44	19.93	nr	**36.83**
2L 4.0	13.36	16.91	0.46	21.02	nr	**37.93**
3L 1.5	13.40	16.96	0.44	20.00	nr	**36.95**
3L 2.5	13.40	16.96	0.44	20.11	nr	**37.07**
4L 1.5	13.40	16.96	0.47	21.69	nr	**38.64**
4L 2.5	13.40	16.96	0.50	22.85	nr	**39.81**
7L 1.5	33.25	42.08	0.70	31.99	nr	**74.08**
7L 2.5	33.25	42.08	0.74	33.82	nr	**75.90**
Heavy duty 750 volt grade; brass gland; brass pot with earth tail; pot closure; sealing compound; conductor sleeving; plastic gland shroud; identification markers						
1H 10	14.64	18.52	0.37	16.91	nr	**35.43**
1H 16	14.64	18.52	0.39	17.82	nr	**36.35**
1H 25	14.64	18.52	0.56	25.59	nr	**44.12**

ELECTRICAL SUPPLY/POWER/LIGHTING SYSTEMS

Item	Net Price £	Material £	Labour hours	Labour £	Unit	Total rate £
1H 35	14.64	18.52	0.57	26.05	nr	**44.57**
1H 50	35.65	45.12	0.60	27.42	nr	**72.54**
1H 70	37.54	47.51	0.67	30.62	nr	**78.13**
1H 95	37.48	47.43	0.75	34.28	nr	**81.71**
2H 1.5	14.64	18.52	0.42	19.20	nr	**37.72**
2H 2.5	14.64	18.52	0.44	20.25	nr	**38.77**
2H 4	14.64	18.52	0.47	21.48	nr	**40.00**
2H 6	14.64	18.52	0.54	24.68	nr	**43.20**
2H 10	35.65	45.12	0.58	26.51	nr	**71.63**
2H 16	35.65	45.12	0.69	31.54	nr	**76.66**
2H 25	57.53	72.81	0.77	35.19	nr	**108.01**
3H 1.5	14.64	18.53	0.44	20.11	nr	**38.64**
3H 2.5	14.64	18.53	0.47	21.62	nr	**40.15**
3H 4	14.64	18.53	0.57	26.05	nr	**44.58**
3H 6	35.76	45.26	0.61	27.88	nr	**73.14**
3H 10	35.76	45.26	0.65	29.71	nr	**74.97**
3H 16	35.76	45.26	0.78	35.65	nr	**80.91**
3H 25	86.17	109.06	0.85	38.85	nr	**147.91**
4H 1.5	13.72	17.37	0.52	23.77	nr	**41.13**
4H 2.5	–	–	0.53	24.22	nr	**24.22**
4H 4	35.48	44.91	0.60	27.42	nr	**72.33**
4H 6	35.76	45.26	0.65	29.71	nr	**74.97**
4H 10	35.76	45.26	0.69	31.54	nr	**76.79**
4H 16	57.53	72.81	0.88	40.22	nr	**113.03**
4H 25	86.90	109.97	0.93	42.50	nr	**152.48**
7H 1.5	35.65	45.12	0.71	32.45	nr	**77.57**
7H 2.5	35.34	44.73	0.74	33.82	nr	**78.55**
12H 1.5	57.44	72.70	0.85	38.85	nr	**111.55**
12H 2.5	57.11	72.28	1.00	45.70	nr	**117.98**
19H 2.5	86.90	109.97	1.11	50.73	nr	**160.71**
Cable; FP100; LOSH insulated; non-sheathed fire-resistant to LPCB Approved to BS 6387 Catergory CWZ; in conduit or trunking including terminations						
450/750 volt grade; single core						
1.5 mm²	1.67	2.11	0.13	5.94	m	**8.05**
2.5 mm²	2.00	2.53	0.13	5.94	m	**8.47**
4.0 mm²	2.41	3.05	0.13	5.94	m	**9.00**
6.0 mm²	2.57	3.25	0.14	6.26	m	**9.51**
10 mm²	3.82	4.84	0.16	7.31	m	**12.15**
Cable; FP200; Insudite insulated; LSOH sheathed screened fire-resistant BASEC Approved to BS 7629; fixed with clips to backgrounds						
300/500 volt grade; two core						
1.0 mm²	3.81	4.83	0.18	8.23	m	**13.05**
1.5 mm²	4.38	5.55	0.18	8.23	m	**13.77**
2.5 mm²	6.07	7.68	0.18	8.23	m	**15.90**
4.0 mm²	9.38	11.87	0.19	8.68	m	**20.55**

ELECTRICAL SUPPLY/POWER/LIGHTING SYSTEMS

Item	Net Price £	Material £	Labour hours	Labour £	Unit	Total rate £
LV DISTRIBUTION: CABLES AND WIRING – cont						
Cable – cont						
300/500 volt grade; three core						
1.0 mm²	5.12	6.47	0.19	8.68	m	**15.16**
1.5 mm²	6.18	7.82	0.19	8.68	m	**16.50**
2.5 mm²	7.46	9.44	0.19	8.68	m	**18.12**
4.0 mm²	12.09	15.31	0.20	9.14	m	**24.45**
300/500 volt grade; four core						
1.0 mm²	5.81	7.35	0.20	9.14	m	**16.50**
1.5 mm²	7.36	9.32	0.20	9.14	m	**18.46**
2.5 mm²	10.04	12.70	0.20	9.14	m	**21.84**
4.0 mm	15.42	19.51	0.21	9.60	m	**29.11**
Terminations; including glanding-off, connection to equipment						
Two core						
1.0 mm²	0.75	0.95	0.28	12.80	nr	**13.74**
1.5 mm²	0.75	0.95	0.28	12.80	nr	**13.74**
2.5 mm²	0.75	0.95	0.28	12.80	nr	**13.74**
4.0 mm²	0.75	0.95	0.28	12.80	nr	**13.74**
Three core						
1.0 mm²	0.75	0.95	0.35	16.00	nr	**16.94**
1.5 mm²	0.75	0.95	0.35	16.00	nr	**16.94**
2.5 mm²	0.75	0.95	0.35	16.00	nr	**16.94**
4.0 mm²	0.75	0.95	0.35	16.00	nr	**16.94**
Four core						
1.0 mm²	0.75	0.95	0.42	19.20	nr	**20.14**
1.5 mm²	0.75	0.95	0.42	19.20	nr	**20.14**
2.5 mm²	0.75	0.95	0.42	19.20	nr	**20.14**
4.0 mm²	0.75	0.95	0.42	19.20	nr	**20.14**
Cable; FP400; polymeric insulated; LSOH sheathed fire-resistant; armoured; with copper stranded copper conductors; BASEC Approved to BS 7846; fixed with clips to backgrounds						
600/1000 volt grade; two core						
1.5 mm²	5.60	7.09	0.20	9.14	m	**16.23**
2.5 mm²	6.49	8.21	0.20	9.14	m	**17.35**
4.0 mm²	7.18	9.09	0.21	9.60	m	**18.69**
6.0 mm²	8.85	11.21	0.22	10.05	m	**21.26**
10 mm²	9.55	12.09	0.24	10.97	m	**23.05**
16 mm²	14.52	18.38	0.25	11.43	m	**29.81**
25 mm²	20.07	25.40	0.35	16.00	m	**41.40**
600/1000 volt grade; three core						
1.5 mm²	6.21	7.86	0.20	9.14	m	**17.00**
2.5 mm²	7.24	9.16	0.21	9.60	m	**18.76**
4.0 mm²	8.54	10.81	0.22	10.05	m	**20.86**
6.0 mm²	9.05	11.45	0.22	10.05	m	**21.50**
10 mm²	10.98	13.89	0.25	11.43	m	**25.32**
16 mm²	17.99	22.76	0.26	11.88	m	**34.65**
25 mm²	22.46	28.42	0.37	16.91	m	**45.33**

ELECTRICAL SUPPLY/POWER/LIGHTING SYSTEMS

Item	Net Price £	Material £	Labour hours	Labour £	Unit	Total rate £
600/1000 volt grade; four core						
1.5 mm²	7.62	9.65	0.21	9.60	m	**19.25**
2.5 mm²	9.63	12.19	0.22	10.05	m	**22.25**
4.0 mm²	11.79	14.92	0.22	10.05	m	**24.98**
6.0 mm²	14.62	18.51	0.23	10.51	m	**29.02**
10 mm²	21.59	27.32	0.25	11.43	m	**38.75**
16 mm²	28.16	35.64	0.26	11.88	m	**47.52**
25 mm²	32.21	40.77	0.39	17.82	m	**58.59**
Terminations; including glanding-off, connection to equipment						
Two core						
1.5 mm²	9.24	11.70	0.58	26.51	nr	**38.20**
2.5 mm²	9.80	12.41	0.58	26.51	nr	**38.92**
4.0 mm²	10.72	13.56	0.61	28.02	nr	**41.58**
6.0 mm²	11.84	14.99	0.67	30.62	nr	**45.61**
10 mm²	13.02	16.47	1.00	45.70	nr	**62.18**
16 mm²	14.32	18.12	1.11	50.73	nr	**68.85**
25 mm²	15.70	19.88	1.70	77.70	nr	**97.57**
Three core						
1.5 mm²	13.93	17.63	0.62	28.34	nr	**45.96**
2.5 mm²	14.62	18.50	0.62	28.34	nr	**46.84**
4.0 mm²	16.14	20.43	0.66	30.35	nr	**50.77**
6.0 mm²	17.70	22.40	0.71	32.45	nr	**54.85**
10 mm²	19.44	24.60	1.06	48.45	nr	**73.04**
16 mm²	21.48	27.18	1.19	54.39	nr	**81.57**
25 mm²	23.56	29.81	1.81	82.72	nr	**112.54**
Four core						
1.5 mm²	18.53	23.45	0.67	30.62	nr	**54.07**
2.5 mm²	24.26	30.70	0.95	43.42	nr	**74.12**
4.0 mm²	25.39	32.13	1.00	45.70	nr	**77.83**
6.0 mm²	27.01	34.19	1.15	52.56	nr	**86.75**
10 mm²	27.91	35.33	1.34	61.24	nr	**96.57**
16 mm²	29.06	36.78	1.86	85.01	nr	**121.79**
25 mm²	43.61	55.20	1.96	89.58	nr	**144.78**
Cable; FP600; polymeric insulated; LSOH sheathed fire-resistant; armoured; with copper stranded copper conductors; BASEC Approved to BS 7846; fixed with clips to backgrounds						
600/1000 volt grade; two core						
4.0 mm²	8.82	11.16	0.22	10.05	m	**21.22**
10 mm²	10.79	13.66	0.25	11.43	m	**25.09**
16 mm²	11.83	14.97	0.37	16.91	m	**31.88**
600/1000 volt grade; three core						
4.0 mm²	8.86	11.21	0.23	10.51	m	**21.72**
6.0 mm²	9.52	12.05	0.23	10.51	m	**22.56**
10 mm²	11.74	14.86	0.26	11.88	m	**26.74**
16 mm²	14.97	18.95	0.35	16.00	m	**34.94**
25 mm²	18.96	23.99	0.40	18.28	m	**42.27**
35 mm²	24.59	31.12	0.40	18.28	m	**49.40**
50 mm²	31.16	39.43	0.41	18.74	m	**58.17**

ELECTRICAL SUPPLY/POWER/LIGHTING SYSTEMS

Item	Net Price £	Material £	Labour hours	Labour £	Unit	Total rate £
LV DISTRIBUTION: CABLES AND WIRING – cont						
Cable – cont						
600/1000 volt grade; three core – cont						
70 mm²	41.54	52.57	0.61	27.88	m	**80.45**
95 mm²	63.77	80.71	0.65	29.71	m	**110.41**
600/1000 volt grade; four core						
4.0 mm²	9.69	12.27	0.23	10.51	nr	**22.78**
6.0 mm²	10.65	13.48	0.24	10.97	nr	**24.45**
10 mm²	13.93	17.64	0.26	11.88	nr	**29.52**
16 mm²	18.93	23.96	0.36	16.45	nr	**40.41**
25 mm²	22.29	28.21	0.44	20.11	nr	**48.32**
35 mm²	29.40	37.21	0.45	20.57	m	**57.78**
50 mm²	37.26	47.16	0.54	24.68	m	**71.84**
70 mm²	50.81	64.31	0.68	31.08	m	**95.39**
95 mm²	66.65	84.35	0.73	33.36	m	**117.72**
120 mm²	83.63	105.85	1.05	47.99	m	**153.84**
150 mm²	100.76	127.52	1.19	54.39	m	**181.91**
185 mm²	124.59	157.68	1.58	72.21	m	**229.89**
240 mm²	160.44	203.06	1.84	84.09	m	**287.15**
300 mm²	199.88	252.97	1.98	90.49	m	**343.46**
400 mm²	255.39	323.22	2.15	98.26	m	**421.48**
Terminations; including glanding-off, connection to equipment						
Two core						
4.0 mm²	4.44	5.62	0.92	42.05	nr	**47.67**
10 mm²	5.16	6.53	1.01	46.16	nr	**52.69**
16 mm²	4.82	6.09	1.50	68.56	nr	**74.65**
Three core						
4.0 mm²	4.76	6.02	0.99	45.25	nr	**51.27**
6.0 mm²	4.83	6.11	1.07	48.90	nr	**55.01**
10 mm²	6.05	7.65	1.59	72.67	nr	**80.32**
16 mm²	7.27	9.20	1.79	81.81	nr	**91.01**
25 mm²	9.67	12.24	1.87	85.47	nr	**97.70**
35 mm²	11.66	14.75	1.97	90.04	m	**104.79**
50 mm²	15.50	19.62	2.07	94.61	m	**114.23**
70 mm²	18.95	23.98	2.17	99.18	m	**123.16**
95 mm²	28.00	35.44	2.28	104.20	m	**139.65**
Four core						
4.0 mm²	9.03	11.43	1.07	48.90	nr	**60.33**
6.0 mm²	9.99	12.64	1.14	52.10	nr	**64.74**
10 mm²	11.13	14.09	1.71	78.15	nr	**92.24**
16 mm²	12.77	16.16	1.94	88.66	nr	**104.83**
25 mm²	15.59	19.73	2.03	92.78	nr	**112.51**
35 mm²	19.95	25.24	2.13	97.35	m	**122.59**
50 mm²	24.78	31.36	2.23	101.92	m	**133.28**
70 mm²	32.03	40.54	2.34	106.95	m	**147.49**
95 mm²	42.40	53.66	2.46	112.43	m	**166.09**
120 mm²	52.63	66.61	2.58	117.92	m	**184.53**
150 mm²	63.25	80.05	2.71	123.86	m	**203.91**

ELECTRICAL SUPPLY/POWER/LIGHTING SYSTEMS

Item	Net Price £	Material £	Labour hours	Labour £	Unit	Total rate £
185 mm²	77.80	98.47	2.85	130.26	m	**228.72**
240 mm²	99.99	126.55	2.99	136.65	m	**263.20**
300 mm²	122.49	155.03	3.14	143.51	m	**298.54**
400 mm²	153.46	194.22	3.30	150.82	m	**345.04**
Cable; Firetuff fire-resistant to BS 6387; fixed with clips to backgrounds						
Two core						
1.5 mm²	3.71	4.69	0.20	9.14	m	**13.83**
2.5 mm²	4.48	5.68	0.20	9.14	m	**14.82**
4.0 mm²	5.82	7.37	0.21	9.60	m	**16.97**
Three core						
1.5 mm²	4.53	5.73	0.20	9.14	m	**14.87**
2.5 mm²	5.12	6.48	0.21	9.60	m	**16.08**
4.0 mm²	7.40	9.37	0.22	10.05	m	**19.42**
Four core						
1.5 mm²	5.06	6.41	0.21	9.60	m	**16.00**
2.5 mm²	6.20	7.85	0.22	10.05	m	**17.90**
4.0 mm²	8.45	10.70	0.22	10.05	m	**20.75**

ELECTRICAL SUPPLY/POWER/LIGHTING SYSTEMS

Item	Net Price £	Material £	Labour hours	Labour £	Unit	Total rate £
LV DISTRIBUTION: MODULAR WIRING						
Modular wiring systems; including commissioning						
Master distribution box; steel; fixed to backgrounds; 6 Port						
4.0 mm 18 core armoured home run cable	242.03	306.31	0.90	41.13	nr	**347.44**
4.0 mm 24 core armoured home run cable	242.03	306.31	0.95	43.42	nr	**349.73**
4.0 mm 18 core armoured home run cable & data cable	257.12	325.42	0.95	43.42	nr	**368.83**
6.0 mm 18 core armoured home run cable	242.03	306.31	1.00	45.70	nr	**352.02**
6.0 mm 24 core armoured home run cable	242.03	306.31	1.10	50.27	nr	**356.59**
6.0 mm 18 core armoured home run cable & data cable	257.12	325.42	1.10	50.27	nr	**375.69**
Master distribution box; steel; fixed to backgrounds; 9 Port						
4.0 mm 27 core armoured home run cable	302.51	382.85	1.30	59.41	nr	**442.27**
4.0 mm 27 core armoured home run cable & data cable	302.51	382.85	1.45	66.27	nr	**449.12**
6.0 mm 27 core armoured home run cable	317.64	402.01	1.45	66.27	nr	**468.28**
6.0 mm 27 core armoured home run cable & data cable	317.64	402.01	1.55	70.84	nr	**472.85**
Metal clad cable; BSEN 60439 Part 2 1993; BASEC approved						
4.0 mm 18 core	24.38	30.86	0.30	13.71	m	**44.57**
4.0 mm 24 core	37.53	47.49	0.32	14.63	m	**62.12**
4.0 mm 27 core	37.83	47.87	0.35	16.00	m	**63.87**
6.0 mm 18 core	32.49	41.11	0.32	14.63	m	**55.74**
6.0 mm 27 core	45.25	57.27	0.35	16.00	m	**73.27**
Metal clad data cable						
Single twisted pair	4.76	6.02	0.18	8.23	m	**14.25**
Twin twisted pair	7.66	9.70	0.18	8.23	m	**17.93**
Distribution cables; armoured; BSEN 60439 Part 2 1993; BASEC approved						
3 wire; 6.1 metre long	106.75	135.10	0.92	42.05	nr	**177.15**
4 wire; 6.1 metre long	120.40	152.38	0.96	43.88	nr	**196.26**
Extender cables; armoured; BSEN 60439 Part 2 1993; BASEC approved						
3 wire						
0.9 metre long	63.83	80.78	0.13	5.94	nr	**86.72**
1.5 metre long	70.28	88.94	0.23	10.51	nr	**99.46**
2.1 metre long	76.72	97.10	0.31	14.17	nr	**111.26**
2.7 metre long	83.18	105.27	0.40	18.28	nr	**123.55**
3.4 metre long	90.69	114.78	0.51	23.31	nr	**138.09**
4.6 metre long	106.16	134.36	0.69	31.54	nr	**165.89**
6.1 metre long	125.63	158.99	0.92	42.05	nr	**201.04**
7.6 metre long	144.80	183.26	1.14	52.10	nr	**235.36**
9.1 metre long	164.48	208.17	1.37	62.61	nr	**270.78**
10.7 metre long	183.10	231.73	1.61	73.58	nr	**305.31**

ELECTRICAL SUPPLY/POWER/LIGHTING SYSTEMS

Item	Net Price £	Material £	Labour hours	Labour £	Unit	Total rate £
4 wire						
0.9 metre long	68.32	86.46	0.14	6.40	nr	**92.86**
1.5 metre long	76.04	96.24	0.24	10.97	nr	**107.20**
2.1 metre long	83.74	105.98	0.32	14.63	nr	**120.60**
2.7 metre long	91.44	115.72	0.43	19.65	nr	**135.37**
3.4 metre long	100.44	127.11	0.51	23.31	nr	**150.42**
4.6 metre long	118.72	150.25	0.67	30.62	nr	**180.87**
6.1 metre long	142.26	180.04	0.92	42.05	nr	**222.09**
7.6 metre long	164.58	208.30	1.22	55.76	nr	**264.05**
9.1 metre long	187.98	237.91	1.46	66.73	nr	**304.64**
10.7 metre long	210.25	266.09	1.71	78.15	nr	**344.24**
3 wire; including twisted pair						
0.9 metre long	87.50	110.73	0.13	5.94	nr	**116.68**
1.5 metre long	98.31	124.43	0.23	10.51	nr	**134.94**
2.1 metre long	109.10	138.08	0.31	14.17	nr	**152.25**
2.7 metre long	119.89	151.74	0.40	18.28	nr	**170.02**
3.4 metre long	132.50	167.70	0.51	23.31	nr	**191.01**
4.6 metre long	157.90	199.83	0.69	31.54	nr	**231.37**
6.1 metre long	188.39	238.42	0.92	42.05	nr	**280.47**
7.6 metre long	221.83	280.74	1.14	52.10	nr	**332.85**
9.1 metre long	254.50	322.10	1.37	62.61	nr	**384.71**
10.7 metre long	285.65	361.52	1.61	73.58	nr	**435.10**
Extender whip ended cables; armoured; BSEN 60439 Part 2 1993; BASEC approved						
3 wire; 3.0 metre long	61.11	77.35	0.30	13.71	nr	**91.06**
4 wire; 3.0 metre long	67.75	85.74	0.30	13.71	nr	**99.45**
T connectors						
3 wire						
6 pole 20a tee	95.62	121.01	0.10	4.57	nr	**125.58**
Snap fix	48.73	61.68	0.10	4.57	nr	**66.25**
1.5 metre flexible cable	56.51	71.52	0.10	4.57	nr	**76.09**
1.5 metre armoured cable	63.80	80.74	0.15	6.86	nr	**87.60**
1.5 metre armoured cable with twisted pair	81.31	102.91	0.15	6.86	nr	**109.77**
4 wire						
Snap fix	50.11	63.42	0.10	4.57	nr	**67.99**
1.5 metre flexible cable	58.84	74.47	0.10	4.57	nr	**79.04**
1.5 metre armoured cable	66.13	83.70	0.18	8.23	nr	**91.92**
Splitters						
5 wire	45.38	57.44	0.20	9.14	nr	**66.58**
5 wire converter	58.50	74.04	0.20	9.14	nr	**83.18**
Switch modules						
3 wire; 6.1 metre long armoured cable	84.61	107.08	0.75	34.28	nr	**141.36**
4 wire; 6.1 metre long armoured cable	90.73	114.82	0.80	36.56	nr	**151.39**
Distribution cables; unarmoured; IEC 998 DIN/VDE 0628						
3 wire; 6.1 metre long	41.67	52.73	0.70	31.99	nr	**84.73**
4 wire; 6.1 metre long	52.12	65.97	0.75	34.28	nr	**100.24**

ELECTRICAL SUPPLY/POWER/LIGHTING SYSTEMS

Item	Net Price £	Material £	Labour hours	Labour £	Unit	Total rate £
LV DISTRIBUTION: MODULAR WIRING – cont						
Extender cables; unarmoured; IEC 998 DIN/VDE 0628						
3 wire						
0.9 metre long	24.26	30.70	0.07	3.20	nr	**33.90**
1.5 metre long	26.53	33.58	0.12	5.48	nr	**39.06**
2.1 metre long	28.85	36.51	0.17	7.77	nr	**44.28**
2.7 metre long	38.76	49.05	0.22	10.05	nr	**59.11**
3.4 metre long	37.06	46.91	0.27	12.34	nr	**59.25**
4.6 metre long	39.43	49.90	0.37	16.91	nr	**66.81**
6.1 metre long	46.43	58.76	0.49	22.39	nr	**81.15**
7.6 metre long	53.72	67.98	0.61	27.88	nr	**95.86**
9.1 metre long	60.36	76.39	0.73	33.36	nr	**109.76**
10.7 metre long	67.55	85.49	0.86	39.31	nr	**124.80**
4 wire						
0.9 metre long	29.19	36.95	0.08	3.66	nr	**40.60**
1.5 metre long	32.20	40.75	0.14	6.40	nr	**47.14**
2.1 metre long	35.20	44.55	0.19	8.68	nr	**53.24**
2.7 metre long	38.76	49.05	0.24	10.97	nr	**60.02**
3.4 metre long	41.56	52.60	0.31	14.17	nr	**66.77**
4.6 metre long	48.90	61.89	0.41	18.74	nr	**80.62**
6.1 metre long	57.93	73.32	0.55	25.14	nr	**98.46**
7.6 metre long	67.34	85.23	0.68	31.08	nr	**116.31**
9.1 metre long	75.97	96.15	0.82	37.48	nr	**133.63**
10.7 metre long	85.30	107.95	0.96	43.88	nr	**151.83**
5 wire						
0.9 metre long	44.28	56.04	0.09	4.11	nr	**60.16**
1.5 metre long	48.16	60.95	0.15	6.86	nr	**67.81**
2.1 metre long	52.03	65.84	0.21	9.60	nr	**75.44**
2.7 metre long	56.83	71.92	0.27	12.34	nr	**84.26**
3.4 metre long	61.44	77.76	0.34	15.54	nr	**93.30**
4.6 metre long	69.88	88.44	0.46	21.02	nr	**109.46**
6.1 metre long	81.74	103.45	0.61	27.88	nr	**131.33**
7.6 metre long	94.09	119.08	0.76	34.73	nr	**153.82**
9.1 metre long	105.30	133.26	0.91	41.59	nr	**174.85**
10.7 metre long	117.37	148.54	1.07	48.90	nr	**197.45**
Extender whip ended cables; armoured; IEC 998 DIN/ VDE 0628						
3 wire; 2.5 mm; 3.0 metre long	24.31	30.77	0.30	13.71	nr	**44.48**
4 wire; 2.5 mm; 3.0 metre long	28.61	36.21	0.30	13.71	nr	**49.92**
T connectors						
3 wire						
5 pin; direct fix	25.67	32.49	0.10	4.57	nr	**37.06**
5 pin; 1.5 mm flexible cable; 0.3 metre long	35.96	45.51	0.15	6.86	nr	**52.36**
4 wire						
5 pin; direct fix	28.10	35.56	0.20	9.14	nr	**44.70**
5 pin; 1.5 mm flexible cable; 0.3 metre long	34.96	44.25	0.20	9.14	nr	**53.39**

ELECTRICAL SUPPLY/POWER/LIGHTING SYSTEMS

Item	Net Price £	Material £	Labour hours	Labour £	Unit	Total rate £
5 wire						
5 pin; direct fix	37.51	47.48	0.20	9.14	nr	**56.62**
Splitters						
3 way; 5 pin	19.04	24.10	0.25	11.43	nr	**35.53**
Switch modules						
3 wire	54.68	69.20	0.20	9.14	nr	**78.34**
4 wire	56.36	71.33	0.22	10.05	nr	**81.38**

ELECTRICAL SUPPLY/POWER/LIGHTING SYSTEMS

Item	Net Price £	Material £	Labour hours	Labour £	Unit	Total rate £
LV DISTRIBUTION: BUSBAR TRUNKING						
MAINS BUSBAR						
Low impedance busbar trunking; fixed to backgrounds including supports, fixings and connections/jointing to equipment						
Straight copper busbar						
1000 amp TP&N	603.36	763.61	3.41	155.85	m	**919.46**
1350 amp TP&N	741.61	938.58	3.58	163.62	m	**1102.20**
2000 amp TP&N	930.16	1177.21	5.00	228.52	m	**1405.73**
2500 amp TP&N	1405.33	1778.58	5.90	269.65	m	**2048.24**
Extra for fittings mains bus bar						
IP54 protection						
1000 amp TP&N	41.15	52.07	2.16	98.72	m	**150.79**
1350 amp TP&N	44.21	55.96	2.61	119.29	m	**175.24**
2000 amp TP&N	61.33	77.62	3.51	160.42	m	**238.04**
2500 amp TP&N	72.83	92.17	3.96	180.99	m	**273.15**
End cover						
1000 amp TP&N	47.32	59.89	0.56	25.59	nr	**85.49**
1350 amp TP&N	49.35	62.46	0.56	25.59	nr	**88.05**
2000 amp TP&N	79.19	100.23	0.66	30.16	nr	**130.39**
2500 amp TP&N	80.22	101.53	0.66	30.16	nr	**131.69**
Edge elbow						
1000 amp TP&N	677.80	857.82	2.01	91.86	nr	**949.69**
1350 amp TP&N	762.81	965.42	2.01	91.86	nr	**1057.28**
2000 amp TP&N	1183.13	1497.36	2.40	109.69	nr	**1607.05**
2500 amp TP&N	1553.95	1966.68	2.40	109.69	nr	**2076.37**
Flat elbow						
1000 amp TP&N	588.06	744.24	2.01	91.86	nr	**836.11**
1350 amp TP&N	637.64	807.00	2.01	91.86	nr	**898.86**
2000 amp TP&N	904.49	1144.73	2.40	109.69	nr	**1254.41**
2500 amp TP&N	1126.49	1425.69	2.40	109.69	nr	**1535.37**
Offset						
1000 amp TP&N	1180.81	1494.44	3.00	137.11	nr	**1631.55**
1350 amp TP&N	1452.38	1838.13	3.00	137.11	nr	**1975.24**
2000 amp TP&N	2288.41	2896.22	3.50	159.96	nr	**3056.18**
2500 amp TP&N	2604.82	3296.66	3.50	159.96	nr	**3456.63**
Edge Z unit						
1000 amp TP&N	1768.84	2238.64	3.00	137.11	nr	**2375.76**
1350 amp TP&N	2264.79	2866.31	3.00	137.11	nr	**3003.43**
2000 amp TP&N	3393.65	4295.00	3.50	159.96	nr	**4454.96**
2500 amp TP&N	3875.40	4904.71	3.50	159.96	nr	**5064.67**
Flat Z unit						
1000 amp TP&N	1912.48	2420.43	3.00	137.11	nr	**2557.54**
1350 amp TP&N	1912.92	2420.99	3.00	137.11	nr	**2558.10**
2000 amp TP&N	2848.10	3604.56	3.50	159.96	nr	**3764.52**
2500 amp TP&N	3424.33	4333.83	3.50	159.96	nr	**4493.79**

ELECTRICAL SUPPLY/POWER/LIGHTING SYSTEMS

Item	Net Price £	Material £	Labour hours	Labour £	Unit	Total rate £
Edge tee						
1000 amp TP&N	1768.84	2238.64	2.20	100.55	nr	**2339.19**
1350 amp TP&N	2264.79	2866.31	2.20	100.55	nr	**2966.86**
2000 amp TP&N	3393.65	4295.00	2.60	118.83	nr	**4413.83**
2500 amp TP&N	3876.55	4906.17	2.60	118.83	nr	**5025.00**
Tap off; TP&N integral contactor/breaker						
18 amp	287.71	364.13	0.82	37.48	nr	**401.61**
Tap off; TP&N fusable withon-load switch; excludes fuses						
32 amp	814.05	1030.27	0.82	37.48	nr	**1067.74**
63 amp	831.76	1052.68	0.88	40.22	nr	**1092.90**
100 amp	1017.20	1287.37	1.18	53.93	nr	**1341.30**
160 amp	1156.29	1463.41	1.41	64.44	nr	**1527.85**
250 amp	1493.08	1889.65	1.76	80.44	nr	**1970.09**
315 amp	1763.06	2231.33	2.06	94.15	nr	**2325.48**
Tap off; TP&N MCCB						
63 amp	1029.48	1302.90	0.88	40.22	nr	**1343.12**
125 amp	1234.01	1561.76	1.18	53.93	nr	**1615.69**
160 amp	1347.19	1705.00	1.41	64.44	nr	**1769.44**
250 amp	1733.06	2193.37	1.76	80.44	nr	**2273.81**
400 amp	2208.95	2795.64	2.06	94.15	nr	**2889.79**
RISING MAINS BUSBAR						
Rising mains busbar; insulated supports, earth continuity bar; including couplers; fixed to backgrounds						
Straight aluminium bar						
200 amp TP&N	262.91	332.74	2.13	97.35	m	**430.09**
315 amp TP&N	294.55	372.78	2.15	98.26	m	**471.04**
400 amp TP&N	341.12	431.72	2.15	98.26	m	**529.98**
630 amp TP&N	421.01	532.83	2.47	112.89	m	**645.72**
800 amp TP&N	632.32	800.27	2.88	131.63	m	**931.90**
Extra for fittings rising busbar						
End feed unit						
200 amp TP&N	525.85	665.51	2.57	117.46	nr	**782.97**
315 amp TP&N	527.49	667.59	2.76	126.14	nr	**793.73**
400 amp TP&N	589.07	745.53	2.76	126.14	nr	**871.67**
630 amp TP&N	590.73	747.62	3.64	166.36	nr	**913.99**
800 amp TP&N	655.64	829.77	4.54	207.49	nr	**1037.27**
Top feeder unit						
200 amp TP&N	525.85	665.51	2.57	117.46	nr	**782.97**
315 amp TP&N	527.49	667.59	2.76	126.14	nr	**793.73**
400 amp TP&N	589.07	745.53	2.76	126.14	nr	**871.67**
630 amp TP&N	590.73	747.62	3.64	166.36	nr	**913.99**
800 amp TP&N	655.64	829.77	4.54	207.49	nr	**1037.27**
End cap						
200 amp TP&N	44.93	56.86	0.18	8.23	nr	**65.08**
315 amp TP&N	44.93	56.86	0.27	12.34	nr	**69.20**
400 amp TP&N	49.91	63.17	0.27	12.34	nr	**75.51**
630 amp TP&N	49.91	63.17	0.41	18.74	nr	**81.91**
800 amp TP&N	144.76	183.21	0.41	18.74	nr	**201.95**

ELECTRICAL SUPPLY/POWER/LIGHTING SYSTEMS

Item	Net Price £	Material £	Labour hours	Labour £	Unit	Total rate £
LV DISTRIBUTION: BUSBAR TRUNKING – cont						
Extra for fittings rising busbar – cont						
Edge elbow						
200 amp TP&N	61.57	77.92	0.55	25.14	nr	103.06
315 amp TP&N	63.24	80.03	0.94	42.96	nr	123.00
400 amp TP&N	447.62	566.51	0.94	42.96	nr	609.47
630 amp TP&N	470.92	596.00	1.45	66.27	nr	662.27
800 amp TP&N	472.59	598.11	1.45	66.27	nr	664.38
Flat elbow						
200 amp TP&N	203.01	256.93	0.55	25.14	nr	282.07
315 amp TP&N	203.01	256.93	0.94	42.96	nr	299.90
400 amp TP&N	274.57	347.49	0.94	42.96	nr	390.45
630 amp TP&N	276.22	349.59	1.45	66.27	nr	415.86
800 amp TP&N	381.07	482.28	1.45	66.27	nr	548.55
Edge tee						
200 amp TP&N	284.54	360.11	0.61	27.88	nr	387.99
315 amp TP&N	366.09	463.32	1.02	46.62	nr	509.94
400 amp TP&N	401.04	507.56	1.02	46.62	nr	554.18
630 amp TP&N	401.04	507.56	1.57	71.75	nr	579.31
800 amp TP&N	655.64	829.77	1.57	71.75	nr	901.53
Flat tee						
200 amp TP&N	286.22	362.24	0.61	27.88	nr	390.12
315 amp TP&N	366.09	463.32	1.02	46.62	nr	509.94
400 amp TP&N	575.76	728.68	1.02	46.62	nr	775.30
630 amp TP&N	577.43	730.79	1.57	71.75	nr	802.55
800 amp TP&N	807.05	1021.41	1.57	71.75	nr	1093.16
Tap off units						
TP&N fusable with on-load switch; excludes fuses						
32 amp	266.64	337.46	0.82	37.48	nr	374.93
63 amp	352.73	446.42	0.88	40.22	nr	486.64
100 amp	473.63	599.42	1.18	53.93	nr	653.35
250 amp	710.45	899.15	1.41	64.44	nr	963.59
400 amp	1035.04	1309.94	2.06	94.15	nr	1404.09
TP&N MCCB						
32 amp	271.60	343.74	0.82	37.48	nr	381.21
63 amp	375.93	475.78	0.88	40.22	nr	516.00
100 amp	602.79	762.90	1.18	53.93	nr	816.83
250 amp	1021.79	1293.17	1.41	64.44	nr	1357.62
400 amp	1795.18	2271.98	2.06	94.15	nr	2366.12
LIGHTING BUSBAR						
Prewired busbar, plug-in trunking for lighting; galvanized sheet steel housing (PE); tin-plated copper conductors with tap-off units at 1 m intervals						
Straight lengths – 25 amp						
2 pole & PE	46.94	59.41	0.16	7.31	m	66.73
4 pole & PE	52.15	66.00	0.16	7.31	m	73.32

ELECTRICAL SUPPLY/POWER/LIGHTING SYSTEMS

Item	Net Price £	Material £	Labour hours	Labour £	Unit	Total rate £
Straight lengths – 40 amp						
2 pole & PE	46.94	59.41	0.16	7.31	m	**66.73**
4 pole & PE	62.57	79.19	0.16	7.31	m	**86.50**
Components for prewired busbars, plug-in trunking for lighting						
Plug-in tap off units						
10 amp 4 pole & PE; 3 m of cable	37.21	47.10	0.10	4.57	nr	**51.67**
16 amp 4 pole & PE; 3 m of cable	40.46	51.21	0.10	4.57	nr	**55.78**
16 amp with phase selection, 2 pole & PE; no cable	30.75	38.92	0.10	4.57	nr	**43.49**
Trunking components						
End feed unit & cover; 4 pole & PE	46.93	59.39	0.23	10.51	nr	**69.90**
Centre feed unit	234.68	297.01	0.29	13.25	nr	**310.27**
Right hand, intermediate terminal box feed unit	50.17	63.50	0.23	10.51	nr	**74.01**
End cover (for R/hand feed)	12.96	16.40	0.06	2.74	nr	**19.14**
Flexible elbow unit	111.68	141.35	0.12	5.48	nr	**146.83**
Fixing bracket – universal	8.11	10.26	0.10	4.57	nr	**14.83**
Suspension bracket – flat	6.46	8.18	0.10	4.57	nr	**12.75**
UNDERFLOOR BUSBAR						
Prewired busbar, plug-in trunking for underfloor power distribution; galvanized sheet steel housing (PE); copper conductors with tap-off units at 300 mm intervals						
Straight lengths – 63 amp						
2 pole & PE	29.40	37.21	0.28	12.80	m	**50.00**
3 pole & PE; Clean Earth System	38.07	48.18	0.28	12.80	m	**60.97**
Components for prewired busbars, plug-in trunking for underfloor power distribution						
Plug-in tap off units						
32 amp 2 pole & PE; 3 m metal flexible prewired conduit	48.45	61.32	0.25	11.43	nr	**72.74**
32 amp 3 pole & PE; clean earth; 3 m metal flexible prewired conduit	58.83	74.45	0.28	12.80	nr	**87.25**
Trunking components						
End feed unit & cover; 2 pole & PE	51.90	65.69	0.35	16.00	nr	**81.68**
End feed unit & cover; 3 pole & PE; clean earth	57.09	72.25	0.38	17.37	nr	**89.61**
End cover; 2 pole & PE	15.58	19.72	0.11	5.03	nr	**24.74**
End cover; 3 pole & PE	17.29	21.88	0.11	5.03	nr	**26.91**
Flexible interlink/corner; 2 pole & PE; 1 m long	89.95	113.85	0.34	15.54	nr	**129.39**
Flexible interlink/corner; 3 pole & PE; 1 m long	102.08	129.19	0.35	16.00	nr	**145.19**
Flexible interlink/corner; 2 pole & PE; 2 m long	110.72	140.12	0.37	16.91	nr	**157.03**
Flexible interlink/corner; 3 pole & PE; 2 m long	121.10	153.26	0.37	16.91	nr	**170.17**

ELECTRICAL SUPPLY/POWER/LIGHTING SYSTEMS

Item	Net Price £	Material £	Labour hours	Labour £	Unit	Total rate £
LV DISTRIBUTION: CABLE SUPPORTS						
LADDER RACK						
Light duty galvanized steel ladder rack; fixed to backgrounds; including supports, fixings and brackets; earth continuity straps						
Straight lengths						
150 mm wide ladder	41.95	53.09	0.69	31.54	m	**84.62**
300 mm wide ladder	45.26	57.29	0.88	40.22	m	**97.51**
450 mm wide ladder	46.41	58.73	1.26	57.59	m	**116.32**
600 mm wide ladder	51.37	65.02	1.51	69.01	m	**134.03**
750 mm wide ladder	62.19	78.71	1.69	77.24	m	**155.95**
900 mm wide ladder	66.22	83.81	1.75	79.98	m	**163.79**
Extra over (cutting and jointing racking to fittings is included)						
Inside riser bend						
150 mm wide ladder	157.64	199.51	0.33	15.08	nr	**214.59**
300 mm wide ladder	160.83	203.54	0.56	25.59	nr	**229.13**
450 mm wide ladder	214.33	271.26	0.85	38.85	nr	**310.11**
600 mm wide ladder	218.63	276.70	0.99	45.25	nr	**321.95**
750 mm wide ladder	252.53	319.60	1.07	48.90	nr	**368.51**
900 mm wide ladder	264.58	334.85	1.12	51.19	nr	**386.04**
Outside riser bend						
150 mm wide ladder	157.64	199.51	0.43	19.65	nr	**219.16**
300 mm wide ladder	160.83	203.54	0.43	19.65	nr	**223.19**
450 mm wide ladder	214.66	271.68	0.73	33.36	nr	**305.04**
600 mm wide ladder	218.07	275.98	0.86	39.31	nr	**315.29**
750 mm wide ladder	251.00	317.67	0.97	44.33	nr	**362.00**
900 mm wide ladder	264.16	334.32	1.15	52.56	nr	**386.87**
Equal tee						
150 mm wide ladder	222.27	281.30	0.62	28.34	nr	**309.64**
300 mm wide ladder	239.60	303.23	0.62	28.34	nr	**331.57**
450 mm wide ladder	255.99	323.97	1.09	49.82	nr	**373.79**
600 mm wide ladder	278.41	352.35	1.12	51.19	nr	**403.54**
750 mm wide ladder	353.72	447.67	1.16	53.02	nr	**500.69**
900 mm wide ladder	368.17	465.95	1.21	55.30	nr	**521.25**
Unequal tee						
150 mm wide ladder	195.89	247.92	0.57	26.05	nr	**273.97**
300 mm wide ladder	200.94	254.32	0.57	26.05	nr	**280.37**
450 mm wide ladder	211.35	267.49	1.17	53.47	nr	**320.96**
600 mm wide ladder	232.53	294.29	1.17	53.47	nr	**347.76**
750 mm wide ladder	277.15	350.76	1.37	62.61	nr	**413.37**
900 mm wide ladder	294.05	372.15	1.37	62.61	nr	**434.76**
4 way crossovers						
150 mm wide ladder	335.63	424.78	0.72	32.91	nr	**457.68**
300 mm wide ladder	344.91	436.52	0.72	32.91	nr	**469.43**
450 mm wide ladder	383.43	485.26	1.13	51.65	nr	**536.91**
600 mm wide ladder	833.21	1054.51	1.29	58.96	nr	**1113.47**
750 mm wide ladder	505.97	640.36	1.41	64.44	nr	**704.80**
900 mm wide ladder	525.96	665.66	1.64	74.95	nr	**740.61**

ELECTRICAL SUPPLY/POWER/LIGHTING SYSTEMS

Item	Net Price £	Material £	Labour hours	Labour £	Unit	Total rate £
Flat bend (light duty)						
150 mm wide ladder	158.21	200.23	0.36	16.45	nr	**216.69**
300 mm wide ladder	161.41	204.29	0.40	18.28	nr	**222.57**
450 mm wide ladder	164.92	208.72	0.42	19.20	nr	**227.92**
600 mm wide ladder	165.70	209.71	0.59	26.97	nr	**236.67**
750 mm wide ladder	180.86	228.90	0.78	35.65	nr	**264.55**
900 mm wide ladder	208.44	263.80	0.86	39.31	nr	**303.11**
Heavy duty galvanized steel ladder rack; fixed to backgrounds; including supports, fixings and brackets; earth continuity straps						
Straight lengths						
150 mm wide ladder	58.24	73.71	0.68	31.08	m	**104.79**
300 mm wide ladder	62.80	79.48	0.79	36.11	m	**115.59**
450 mm wide ladder	67.16	85.00	1.07	48.90	m	**133.90**
600 mm wide ladder	72.01	91.13	1.24	56.67	m	**147.81**
750 mm wide ladder	83.45	105.61	1.49	68.10	m	**173.71**
900 mm wide ladder	88.04	111.43	1.67	76.32	m	**187.75**
Extra over (cutting and jointing racking to fittings is included)						
Flat bend						
150 mm wide ladder	165.76	209.79	0.34	15.54	nr	**225.32**
300 mm wide ladder	174.96	221.43	0.39	17.82	nr	**239.26**
450 mm wide ladder	192.93	244.17	0.43	19.65	nr	**263.83**
600 mm wide ladder	214.97	272.07	0.61	27.88	nr	**299.95**
750 mm wide ladder	237.95	301.15	0.82	37.48	nr	**338.63**
900 mm wide ladder	258.31	326.91	0.97	44.33	nr	**371.25**
Inside riser bend						
150 mm wide ladder	217.75	275.58	0.27	12.34	nr	**287.92**
300 mm wide ladder	220.76	279.39	0.45	20.57	nr	**299.96**
450 mm wide ladder	236.45	299.25	0.65	29.71	nr	**328.96**
600 mm wide ladder	250.36	316.86	0.81	37.02	nr	**353.88**
750 mm wide ladder	260.14	329.24	0.92	42.05	nr	**371.28**
900 mm wide ladder	284.45	360.01	1.06	48.45	nr	**408.45**
Outside riser bend						
150 mm wide ladder	217.75	275.58	0.27	12.34	nr	**287.92**
300 mm wide ladder	220.76	279.39	0.33	15.08	nr	**294.48**
450 mm wide ladder	236.45	299.25	0.61	27.88	nr	**327.13**
600 mm wide ladder	250.36	316.86	0.76	34.73	nr	**351.59**
750 mm wide ladder	260.14	329.24	0.94	42.96	nr	**372.20**
900 mm wide ladder	284.45	360.01	1.05	47.99	nr	**407.99**
Equal tee						
150 mm wide ladder	252.79	319.93	0.37	16.91	nr	**336.84**
300 mm wide ladder	287.16	363.43	0.57	26.05	nr	**389.48**
450 mm wide ladder	307.23	388.83	0.83	37.93	nr	**426.77**
600 mm wide ladder	336.20	425.49	0.92	42.05	nr	**467.54**
750 mm wide ladder	413.98	523.93	1.13	51.65	nr	**575.57**
900 mm wide ladder	435.75	551.49	1.20	54.84	nr	**606.33**

ELECTRICAL SUPPLY/POWER/LIGHTING SYSTEMS

Item	Net Price £	Material £	Labour hours	Labour £	Unit	Total rate £
LV DISTRIBUTION: CABLE SUPPORTS – cont						
Extra over (cutting and jointing racking to fittings is included) – cont						
Unequal tee						
300 mm wide ladder	210.36	266.23	0.57	26.05	nr	**292.29**
450 mm wide ladder	231.60	293.11	1.17	53.47	nr	**346.59**
600 mm wide ladder	257.47	325.85	1.17	53.47	nr	**379.32**
750 mm wide ladder	321.92	407.43	1.25	57.13	nr	**464.56**
900 mm wide ladder	336.58	425.97	1.33	60.79	nr	**486.76**
4 way crossovers						
150 mm wide ladder	390.77	494.56	0.50	22.85	nr	**517.41**
300 mm wide ladder	406.99	515.08	0.67	30.62	nr	**545.71**
450 mm wide ladder	494.85	626.28	0.92	42.05	nr	**668.33**
600 mm wide ladder	570.93	722.57	1.07	48.90	nr	**771.47**
750 mm wide ladder	573.22	725.47	1.25	57.13	nr	**782.60**
900 mm wide ladder	938.82	1188.17	1.36	62.16	nr	**1250.33**
Double set						
150 mm wide ladder	332.13	420.34	0.51	23.31	nr	**443.65**
300 mm wide ladder	430.66	545.05	0.63	28.79	nr	**573.84**
450 mm wide ladder	461.73	584.37	0.86	39.31	nr	**623.67**
600 mm wide ladder	518.37	656.05	0.99	45.25	nr	**701.30**
750 mm wide ladder	570.54	722.08	1.16	53.02	nr	**775.09**
900 mm wide ladder	622.42	787.73	1.31	59.87	nr	**847.60**
Extra heavy duty galvanized steel ladder rack; fixed to backgrounds; including supports, fixings and brackets; earth continuity straps						
Straight lengths						
150 mm wide ladder	67.05	84.86	0.63	28.79	m	**113.65**
300 mm wide ladder	70.37	89.06	0.70	31.99	m	**121.05**
450 mm wide ladder	73.86	93.47	0.83	37.93	m	**131.41**
600 mm wide ladder	74.85	94.73	0.89	40.68	m	**135.40**
750 mm wide ladder	75.05	94.99	1.22	55.76	m	**150.74**
900 mm wide ladder	78.32	99.12	1.44	65.81	m	**164.94**
Extra over (cutting and jointing racking to fittings is included)						
Flat bend						
150 mm wide ladder	161.76	204.72	0.36	16.45	nr	**221.17**
300 mm wide ladder	168.32	213.03	0.39	17.82	nr	**230.85**
450 mm wide ladder	181.76	230.03	0.43	19.65	nr	**249.69**
600 mm wide ladder	205.61	260.22	0.61	27.88	nr	**288.10**
750 mm wide ladder	222.90	282.11	0.82	37.48	nr	**319.58**
900 mm wide ladder	241.41	305.53	0.97	44.33	nr	**349.87**
Inside riser bend						
150 mm wide ladder	168.89	213.74	0.36	16.45	nr	**230.20**
300 mm wide ladder	171.62	217.21	0.39	17.82	nr	**235.03**
450 mm wide ladder	181.41	229.60	0.43	19.65	nr	**249.25**
600 mm wide ladder	200.09	253.23	0.61	27.88	nr	**281.11**
750 mm wide ladder	205.27	259.79	0.82	37.48	nr	**297.27**
900 mm wide ladder	221.34	280.13	0.97	44.33	nr	**324.46**

ELECTRICAL SUPPLY/POWER/LIGHTING SYSTEMS

Item	Net Price £	Material £	Labour hours	Labour £	Unit	Total rate £
Outside riser bend						
150 mm wide ladder	169.33	214.31	0.36	16.45	nr	**230.76**
300 mm wide ladder	171.62	217.21	0.39	17.82	nr	**235.03**
450 mm wide ladder	181.41	229.60	0.41	18.74	nr	**248.34**
600 mm wide ladder	200.09	253.23	0.57	26.05	nr	**279.28**
750 mm wide ladder	205.27	259.79	0.82	37.48	nr	**297.27**
900 mm wide ladder	221.34	280.13	0.93	42.50	nr	**322.63**
Equal tee						
150 mm wide ladder	218.69	276.77	0.37	16.91	nr	**293.68**
300 mm wide ladder	241.44	305.57	0.57	26.05	nr	**331.62**
450 mm wide ladder	255.73	323.65	0.83	37.93	nr	**361.58**
600 mm wide ladder	288.23	364.78	0.92	42.05	nr	**406.83**
750 mm wide ladder	304.32	385.14	1.13	51.65	nr	**436.79**
900 mm wide ladder	348.97	441.65	1.20	54.84	nr	**496.50**
Unequal tee						
150 mm wide ladder	167.83	212.41	0.37	16.91	nr	**229.32**
300 mm wide ladder	184.83	233.92	0.57	26.05	nr	**259.97**
450 mm wide ladder	195.51	247.43	1.17	53.47	nr	**300.91**
600 mm wide ladder	219.95	278.37	1.17	53.47	nr	**331.84**
750 mm wide ladder	348.39	440.92	1.25	57.13	nr	**498.05**
900 mm wide ladder	270.37	342.18	1.33	60.79	nr	**402.97**
4 way crossovers						
150 mm wide ladder	283.82	359.20	0.50	22.85	nr	**382.05**
300 mm wide ladder	294.95	373.29	0.67	30.62	nr	**403.91**
450 mm wide ladder	345.04	436.69	0.92	42.05	nr	**478.73**
600 mm wide ladder	383.59	485.47	1.07	48.90	nr	**534.37**
750 mm wide ladder	398.74	504.65	1.25	57.13	nr	**561.78**
900 mm wide ladder	684.40	866.17	1.36	62.16	nr	**928.33**

ELECTRICAL SUPPLY/POWER/LIGHTING SYSTEMS

Item	Net Price £	Material £	Labour hours	Labour £	Unit	Total rate £
LV DISTRIBUTION: CABLE TRAYS						
Galvanized steel cable tray to BS 729; including standard coupling joints, fixings and earth continuity straps (supports and hangers are excluded)						
Light duty tray						
Straight lengths						
50 mm wide	5.60	7.09	0.19	8.68	m	**15.78**
75 mm wide	7.06	8.94	0.23	10.51	m	**19.45**
100 mm wide	8.70	11.01	0.31	14.17	m	**25.18**
150 mm wide	11.39	14.42	0.33	15.08	m	**29.50**
225 mm wide	21.76	27.54	0.39	17.82	m	**45.36**
300 mm wide	30.60	38.73	0.49	22.39	m	**61.12**
450 mm wide	37.07	46.91	0.60	27.42	m	**74.34**
600 mm wide	43.55	55.11	0.79	36.11	m	**91.22**
750 mm wide	55.32	70.01	1.04	47.53	m	**117.54**
900 mm wide	68.79	87.07	1.26	57.59	m	**144.65**
Extra over (cutting and jointing tray to fittings is included)						
Straight reducer						
75 mm wide	27.69	35.04	0.22	10.05	nr	**45.10**
100 mm wide	28.23	35.73	0.25	11.43	nr	**47.16**
150 mm wide	36.91	46.71	0.27	12.34	nr	**59.05**
225 mm wide	48.45	61.32	0.34	15.54	nr	**76.86**
300 mm wide	61.69	78.07	0.39	17.82	nr	**95.89**
450 mm wide	69.84	88.39	0.49	22.39	nr	**110.78**
600 mm wide	78.00	98.71	0.54	24.68	nr	**123.39**
750 mm wide	102.47	129.69	0.61	27.88	nr	**157.56**
900 mm wide	119.05	150.67	0.69	31.54	nr	**182.21**
Flat bend; 90°						
50 mm wide	13.85	17.53	0.19	8.68	nr	**26.21**
75 mm wide	14.42	18.25	0.24	10.97	nr	**29.22**
100 mm wide	16.22	20.53	0.28	12.80	nr	**33.32**
150 mm wide	17.54	22.20	0.30	13.71	nr	**35.91**
225 mm wide	22.77	28.82	0.36	16.45	nr	**45.27**
300 mm wide	33.30	42.15	0.44	20.11	nr	**62.26**
450 mm wide	43.49	55.04	0.57	26.05	nr	**81.09**
600 mm wide	53.68	67.93	0.69	31.54	nr	**99.47**
750 mm wide	76.02	96.21	0.81	37.02	nr	**133.23**
900 mm wide	111.81	141.50	0.94	42.96	nr	**184.46**
Adjustable riser						
50 mm wide	33.39	42.25	0.26	11.88	nr	**54.14**
75 mm wide	34.97	44.25	0.29	13.25	nr	**57.51**
100 mm wide	36.12	45.72	0.32	14.63	nr	**60.34**
150 mm wide	40.48	51.23	0.36	16.45	nr	**67.68**
225 mm wide	53.30	67.46	0.44	20.11	nr	**87.57**
300 mm wide	57.11	72.28	0.52	23.77	nr	**96.05**
450 mm wide	66.69	84.40	0.66	30.16	nr	**114.56**

ELECTRICAL SUPPLY/POWER/LIGHTING SYSTEMS

Item	Net Price £	Material £	Labour hours	Labour £	Unit	Total rate £
600 mm wide	76.28	96.54	0.79	36.11	nr	**132.64**
750 mm wide	98.17	124.24	1.03	47.07	nr	**171.31**
900 mm wide	115.53	146.22	1.10	50.27	nr	**196.49**
Inside riser; 90°						
50 mm wide	10.49	13.28	0.28	12.80	nr	**26.08**
75 mm wide	10.97	13.88	0.31	14.17	nr	**28.05**
100 mm wide	12.06	15.26	0.33	15.08	nr	**30.34**
150 mm wide	14.32	18.12	0.37	16.91	nr	**35.03**
225 mm wide	17.60	22.28	0.44	20.11	nr	**42.39**
300 mm wide	18.62	23.57	0.53	24.22	nr	**47.79**
450 mm wide	29.80	37.71	0.67	30.62	nr	**68.33**
600 mm wide	76.09	96.30	0.79	36.11	nr	**132.41**
750 mm wide	93.86	118.79	0.95	43.42	nr	**162.21**
900 mm wide	111.52	141.14	1.11	50.73	nr	**191.87**
Outside riser; 90°						
50 mm wide	10.49	13.28	0.28	12.80	nr	**26.08**
75 mm wide	10.97	13.88	0.31	14.17	nr	**28.05**
100 mm wide	12.06	15.26	0.33	15.08	nr	**30.34**
150 mm wide	14.32	18.12	0.37	16.91	nr	**35.03**
225 mm wide	17.60	22.28	0.44	20.11	nr	**42.39**
300 mm wide	18.62	23.57	0.53	24.22	nr	**47.79**
450 mm wide	29.80	37.71	0.67	30.62	nr	**68.33**
600 mm wide	42.31	53.55	0.79	36.11	nr	**89.66**
750 mm wide	57.03	72.17	0.95	43.42	nr	**115.59**
900 mm wide	57.45	72.71	1.11	50.73	nr	**123.44**
Equal tee						
50 mm wide	25.10	31.77	0.30	13.71	nr	**45.48**
75 mm wide	26.09	33.02	0.31	14.17	nr	**47.19**
100 mm wide	26.87	34.01	0.35	16.00	nr	**50.01**
150 mm wide	32.62	41.28	0.36	16.45	nr	**57.74**
225 mm wide	43.18	54.65	0.44	20.11	nr	**74.76**
300 mm wide	59.99	75.92	0.54	24.68	nr	**100.60**
450 mm wide	72.47	91.71	0.71	32.45	nr	**124.16**
600 mm wide	84.95	107.52	0.92	42.05	nr	**149.56**
750 mm wide	127.35	161.18	1.19	54.39	nr	**215.56**
900 mm wide	181.13	229.24	1.44	65.81	nr	**295.05**
Unequal tee						
75 mm wide	16.17	20.47	0.38	17.37	nr	**37.84**
100 mm wide	16.17	20.47	0.39	17.82	nr	**38.29**
150 mm wide	16.65	21.07	0.43	19.65	nr	**40.72**
225 mm wide	19.32	24.45	0.50	22.85	nr	**47.30**
300 mm wide	24.34	30.80	0.63	28.79	nr	**59.59**
450 mm wide	32.28	40.85	0.80	36.56	nr	**77.42**
600 mm wide	50.63	64.08	1.02	46.62	nr	**110.70**
750 mm wide	144.63	183.04	1.12	51.19	nr	**234.23**
900 mm wide	190.65	241.29	1.35	61.70	nr	**302.99**

ELECTRICAL SUPPLY/POWER/LIGHTING SYSTEMS

Item	Net Price £	Material £	Labour hours	Labour £	Unit	Total rate £
LV DISTRIBUTION: CABLE TRAYS – cont						
Extra over (cutting and jointing tray to fittings is included) – cont						
4 way crossovers						
50 mm wide	36.84	46.62	0.38	17.37	nr	**63.99**
75 mm wide	37.54	47.50	0.40	18.28	nr	**65.79**
100 mm wide	39.49	49.98	0.40	18.28	nr	**68.26**
150 mm wide	46.79	59.21	0.44	20.11	nr	**79.32**
225 mm wide	63.03	79.77	0.53	24.22	nr	**103.99**
300 mm wide	86.75	109.80	0.64	29.25	nr	**139.05**
450 mm wide	100.29	126.92	0.84	38.39	nr	**165.32**
600 mm wide	113.82	144.05	1.03	47.07	nr	**191.12**
750 mm wide	127.35	161.18	1.13	51.65	nr	**212.82**
900 mm wide	181.13	229.24	1.36	62.16	nr	**291.39**
Medium duty tray with return flange						
Straight lengths						
50 mm wide	12.53	15.85	0.33	15.08	m	**30.94**
75 mm wide	12.92	16.35	0.33	15.08	m	**31.43**
100 mm wide	13.31	16.85	0.35	16.00	m	**32.85**
150 mm wide	16.14	20.43	0.39	17.82	m	**38.26**
225 mm wide	20.87	26.42	0.45	20.57	m	**46.98**
300 mm wide	32.23	40.79	0.57	26.05	m	**66.84**
450 mm wide	45.04	57.00	0.69	31.54	m	**88.54**
600 mm wide	60.31	76.33	0.91	41.59	m	**117.92**
Extra over (cutting and jointing tray to fittings is included)						
Straight reducer						
100 mm wide	53.53	67.74	0.25	11.43	nr	**79.17**
150 mm wide	58.48	74.01	0.27	12.34	nr	**86.35**
225 mm wide	67.37	85.26	0.34	15.54	nr	**100.80**
300 mm wide	78.01	98.73	0.39	17.82	nr	**116.55**
450 mm wide	100.82	127.60	0.49	22.39	nr	**149.99**
600 mm wide	124.91	158.08	0.54	24.68	nr	**182.76**
Flat bend; 90°						
75 mm wide	71.01	89.87	0.24	10.97	nr	**100.84**
100 mm wide	74.82	94.70	0.28	12.80	nr	**107.49**
150 mm wide	83.87	106.15	0.30	13.71	nr	**119.86**
225 mm wide	96.50	122.13	0.36	16.45	nr	**138.58**
300 mm wide	115.77	146.52	0.44	20.11	nr	**166.63**
450 mm wide	124.17	157.16	0.57	26.05	nr	**183.21**
600 mm wide	198.41	251.11	0.69	31.54	nr	**282.64**
Adjustable bend						
75 mm wide	81.62	103.29	0.29	13.25	nr	**116.55**
100 mm wide	88.83	112.42	0.32	14.63	nr	**127.04**
150 mm wide	101.59	128.57	0.36	16.45	nr	**145.02**
225 mm wide	113.84	144.07	0.44	20.11	nr	**164.18**
300 mm wide	128.65	162.82	0.52	23.77	nr	**186.59**

ELECTRICAL SUPPLY/POWER/LIGHTING SYSTEMS

Item	Net Price £	Material £	Labour hours	Labour £	Unit	Total rate £
Adjustable riser						
75 mm wide	72.80	92.14	0.29	13.25	nr	**105.39**
100 mm wide	74.00	93.65	0.32	14.63	nr	**108.27**
150 mm wide	81.56	103.23	0.36	16.45	nr	**119.68**
225 mm wide	86.52	109.50	0.44	20.11	nr	**129.61**
300 mm wide	92.24	116.74	0.52	23.77	nr	**140.51**
450 mm wide	121.00	153.13	0.66	30.16	nr	**183.30**
600 mm wide	148.40	187.82	0.79	36.11	nr	**223.92**
Inside riser; 90°						
75 mm wide	43.74	55.35	0.31	14.17	nr	**69.52**
100 mm wide	44.16	55.90	0.33	15.08	nr	**70.98**
150 mm wide	50.30	63.66	0.37	16.91	nr	**80.57**
225 mm wide	61.45	77.78	0.44	20.11	nr	**97.89**
300 mm wide	74.64	94.47	0.53	24.22	nr	**118.69**
450 mm wide	107.74	136.35	0.67	30.62	nr	**166.98**
600 mm wide	172.90	218.82	0.79	36.11	nr	**254.93**
Outside riser; 90°						
75 mm wide	43.74	55.35	0.31	14.17	nr	**69.52**
100 mm wide	44.16	55.90	0.33	15.08	nr	**70.98**
150 mm wide	50.30	63.66	0.37	16.91	nr	**80.57**
225 mm wide	61.45	77.78	0.44	20.11	nr	**97.89**
300 mm wide	74.64	94.47	0.53	24.22	nr	**118.69**
450 mm wide	107.74	136.35	0.67	30.62	nr	**166.98**
600 mm wide	172.90	218.82	0.79	36.11	nr	**254.93**
Equal tee						
75 mm wide	98.14	124.21	0.31	14.17	nr	**138.38**
100 mm wide	107.13	135.58	0.35	16.00	nr	**151.58**
150 mm wide	116.61	147.58	0.36	16.45	nr	**164.03**
225 mm wide	125.90	159.34	0.54	24.68	nr	**184.02**
300 mm wide	154.60	195.67	0.71	32.45	nr	**228.11**
450 mm wide	200.01	253.14	0.74	33.82	nr	**286.96**
600 mm wide	285.87	361.79	0.92	42.05	nr	**403.84**
Unequal tee						
100 mm wide	103.66	131.19	0.39	17.82	nr	**149.01**
150 mm wide	103.66	131.19	0.43	19.65	nr	**150.84**
225 mm wide	121.24	153.44	0.50	22.85	nr	**176.29**
300 mm wide	147.17	186.26	0.63	28.79	nr	**215.05**
450 mm wide	193.86	245.35	0.80	36.56	nr	**281.92**
600 mm wide	278.50	352.47	1.02	46.62	nr	**399.09**
4 way crossovers						
75 mm wide	137.84	174.45	0.40	18.28	nr	**192.73**
100 mm wide	148.55	188.01	0.40	18.28	nr	**206.29**
150 mm wide	163.80	207.31	0.44	20.11	nr	**227.42**
225 mm wide	186.67	236.25	0.53	24.22	nr	**260.47**
300 mm wide	217.20	274.89	0.64	29.25	nr	**304.14**
450 mm wide	276.44	349.86	0.84	38.39	nr	**388.25**
600 mm wide	402.97	510.00	1.03	47.07	nr	**557.07**

ELECTRICAL SUPPLY/POWER/LIGHTING SYSTEMS

Item	Net Price £	Material £	Labour hours	Labour £	Unit	Total rate £
LV DISTRIBUTION: CABLE TRAYS – cont						
Heavy duty tray with return flange						
Straight lengths						
75 mm	22.14	28.02	0.34	15.54	m	**43.56**
100 mm	22.46	28.43	0.36	16.45	m	**44.88**
150 mm	24.83	31.42	0.40	18.28	m	**49.70**
225 mm	30.12	38.12	0.46	21.02	m	**59.14**
300 mm	34.19	43.27	0.58	26.51	m	**69.77**
450 mm	60.14	76.12	0.70	31.99	m	**108.11**
600 mm	83.81	106.07	0.92	42.05	m	**148.12**
750 mm	115.91	146.70	1.01	46.16	m	**192.86**
900 mm	121.33	153.56	1.14	52.10	m	**205.66**
Extra over (cutting and jointing tray to fittings is included)						
Straight reducer						
100 mm wide	78.79	99.72	0.25	11.43	nr	**111.15**
150 mm wide	81.81	103.54	0.27	12.34	nr	**115.88**
225 mm wide	92.92	117.60	0.34	15.54	nr	**133.14**
300 mm wide	103.58	131.09	0.39	17.82	nr	**148.91**
450 mm wide	148.15	187.50	0.49	22.39	nr	**209.89**
600 mm wide	164.71	208.46	0.54	24.68	nr	**233.14**
750 mm wide	208.89	264.37	0.60	27.42	nr	**291.79**
900 mm wide	229.14	290.00	0.66	30.16	nr	**320.16**
Flat bend; 90°						
75 mm wide	99.01	125.31	0.24	10.97	nr	**136.28**
100 mm wide	107.27	135.76	0.28	12.80	nr	**148.55**
150 mm wide	116.83	147.86	0.30	13.71	nr	**161.57**
225 mm wide	118.59	150.09	0.36	16.45	nr	**166.55**
300 mm wide	134.66	170.42	0.44	20.11	nr	**190.53**
450 mm wide	216.27	273.71	0.57	26.05	nr	**299.76**
600 mm wide	294.38	372.56	0.69	31.54	nr	**404.10**
750 mm wide	375.66	475.43	0.83	37.93	nr	**513.37**
900 mm wide	392.68	496.98	1.01	46.16	nr	**543.14**
Adjustable bend						
75 mm wide	91.67	116.01	0.29	13.25	nr	**129.27**
100 mm wide	101.35	128.26	0.32	14.63	nr	**142.89**
150 mm wide	107.08	135.52	0.36	16.45	nr	**151.97**
225 mm wide	117.86	149.16	0.44	20.11	nr	**169.27**
300 mm wide	139.82	176.96	0.52	23.77	nr	**200.73**
Adjustable riser						
75 mm wide	80.78	102.24	0.29	13.25	nr	**115.50**
100 mm wide	84.86	107.40	0.32	14.63	nr	**122.02**
150 mm wide	93.00	117.71	0.36	16.45	nr	**134.16**
225 mm wide	99.60	126.05	0.44	20.11	nr	**146.16**
300 mm wide	107.87	136.52	0.52	23.77	nr	**160.28**
450 mm wide	131.29	166.17	0.66	30.16	nr	**196.33**
600 mm wide	156.42	197.97	0.79	36.11	nr	**234.07**
750 mm wide	189.42	239.74	1.03	47.07	nr	**286.81**
900 mm wide	219.97	278.40	1.10	50.27	nr	**328.67**

ELECTRICAL SUPPLY/POWER/LIGHTING SYSTEMS

Item	Net Price £	Material £	Labour hours	Labour £	Unit	Total rate £
Inside riser; 90°						
75 mm wide	72.05	91.19	0.31	14.17	nr	**105.35**
100 mm wide	72.99	92.38	0.33	15.08	nr	**107.46**
150 mm wide	79.07	100.07	0.37	16.91	nr	**116.98**
225 mm wide	82.66	104.61	0.44	20.11	nr	**124.72**
300 mm wide	85.26	107.91	0.53	24.22	nr	**132.13**
450 mm wide	146.30	185.15	0.67	30.62	nr	**215.77**
600 mm wide	179.49	227.16	0.79	36.11	nr	**263.27**
750 mm wide	222.15	281.16	0.95	43.42	nr	**324.58**
900 mm wide	261.90	331.46	1.11	50.73	nr	**382.20**
Outside riser; 90°						
75 mm wide	72.05	91.19	0.31	14.17	nr	**105.35**
100 mm wide	72.99	92.38	0.33	15.08	nr	**107.46**
150 mm wide	79.07	100.07	0.37	16.91	nr	**116.98**
225 mm wide	82.66	104.61	0.44	20.11	nr	**124.72**
300 mm wide	85.26	107.91	0.53	24.22	nr	**132.13**
450 mm wide	146.30	185.15	0.67	30.62	nr	**215.77**
600 mm wide	179.49	227.16	0.79	36.11	nr	**263.27**
750 mm wide	222.15	281.16	0.95	43.42	nr	**324.58**
900 mm wide	261.90	331.46	1.11	50.73	nr	**382.20**
Equal tee						
75 mm wide	131.19	166.04	0.31	14.17	nr	**180.20**
100 mm wide	139.88	177.03	0.35	16.00	nr	**193.02**
150 mm wide	151.87	192.21	0.36	16.45	nr	**208.66**
225 mm wide	168.56	213.32	0.44	20.11	nr	**233.43**
300 mm wide	181.69	229.95	0.54	24.68	nr	**254.63**
450 mm wide	263.81	333.87	0.71	32.45	nr	**366.32**
600 mm wide	360.81	456.64	0.92	42.05	nr	**498.69**
750 mm wide	445.10	563.32	1.19	54.39	nr	**617.70**
900 mm wide	543.18	687.45	1.45	66.27	nr	**753.72**
Unequal tee						
75 mm wide	125.51	158.85	0.38	17.37	nr	**176.21**
100 mm wide	140.19	177.42	0.39	17.82	nr	**195.24**
150 mm wide	153.68	194.49	0.43	19.65	nr	**214.14**
225 mm wide	176.84	223.81	0.50	22.85	nr	**246.67**
300 mm wide	191.21	242.00	0.63	28.79	nr	**270.79**
450 mm wide	277.62	351.36	0.80	36.56	nr	**387.92**
600 mm wide	366.54	463.90	1.02	46.62	nr	**510.52**
750 mm wide	483.44	611.84	1.12	51.19	nr	**663.03**
900 mm wide	584.80	740.13	1.35	61.70	nr	**801.83**
4 way crossovers						
75 mm wide	127.38	161.21	0.40	18.28	nr	**179.50**
100 mm wide	182.59	231.09	0.40	18.28	nr	**249.37**
150 mm wide	182.59	231.09	0.44	20.11	nr	**251.20**
225 mm wide	253.87	321.30	0.53	24.22	nr	**345.52**
300 mm wide	279.94	354.29	0.64	29.25	nr	**383.54**
450 mm wide	402.69	509.64	0.84	38.39	nr	**548.03**
600 mm wide	547.30	692.67	1.03	47.07	nr	**739.74**
750 mm wide	667.53	844.82	1.13	51.65	nr	**896.47**
900 mm wide	805.22	1019.09	1.36	62.16	nr	**1081.25**

ELECTRICAL SUPPLY/POWER/LIGHTING SYSTEMS

Item	Net Price £	Material £	Labour hours	Labour £	Unit	Total rate £
LV DISTRIBUTION: CABLE TRAYS – cont						
GRP cable tray including standard coupling joints and fixings (supports and hangers excluded)						
Tray						
100 mm wide	56.78	71.86	0.34	15.54	m	**87.40**
200 mm wide	72.72	92.03	0.39	17.82	m	**109.85**
400 mm wide	114.02	144.30	0.53	24.22	m	**168.53**
Cover						
100 mm wide	31.99	40.49	0.10	4.57	m	**45.06**
200 mm wide	42.38	53.64	0.11	5.03	m	**58.67**
400 mm wide	72.37	91.59	0.14	6.40	m	**97.99**
Extra for (cutting and jointing to fittings included)						
Reducer						
200 mm wide	150.88	190.96	0.23	10.51	nr	**201.47**
400 mm wide	197.12	249.48	0.30	13.71	nr	**263.19**
Reducer cover						
200 mm wide	94.45	119.53	0.25	11.43	nr	**130.96**
400 mm wide	138.02	174.68	0.28	12.80	nr	**187.48**
Bend						
100 mm wide	125.87	159.30	0.32	14.63	nr	**173.93**
200 mm wide	146.11	184.92	0.34	15.54	nr	**200.45**
400 mm wide	187.78	237.66	0.40	18.28	nr	**255.94**
Bend cover						
100 mm wide	62.69	79.34	0.10	4.57	nr	**83.91**
200 mm wide	83.74	105.98	0.10	4.57	nr	**110.55**
400 mm wide	115.68	146.41	0.13	5.94	nr	**152.35**
Tee						
100 mm wide	160.59	203.24	0.37	16.91	nr	**220.15**
200 mm wide	177.07	224.10	0.43	19.65	nr	**243.75**
400 mm wide	219.63	277.96	0.56	25.59	nr	**303.56**
Tee cover						
100 mm wide	80.91	102.40	0.27	12.34	nr	**114.74**
200 mm wide	97.39	123.26	0.31	14.17	nr	**137.42**
400 mm wide	137.37	173.86	0.37	16.91	nr	**190.77**

ELECTRICAL SUPPLY/POWER/LIGHTING SYSTEMS

Item	Net Price £	Material £	Labour hours	Labour £	Unit	Total rate £
LV DISTRIBUTION: BASKET TRAY						
Mild steel cable basket; zinc plated including standard coupling joints, fixings and earth continuity straps (supports and hangers are excluded)						
Basket 54 mm deep						
100 mm wide	6.33	8.02	0.22	10.05	m	**18.07**
150 mm wide	7.20	9.12	0.25	11.43	m	**20.54**
200 mm wide	7.75	9.81	0.28	12.80	m	**22.60**
300 mm wide	10.28	13.01	0.34	15.54	m	**28.55**
450 mm wide	14.97	18.95	0.44	20.11	m	**39.06**
600 mm wide	20.42	25.84	0.70	31.99	m	**57.83**
Extra for (cutting and jointing to fittings is included)						
Reducer						
150 mm wide	7.41	9.38	0.25	11.43	nr	**20.80**
200 mm wide	9.63	12.18	0.28	12.80	nr	**24.98**
300 mm wide	9.78	12.38	0.38	17.37	nr	**29.75**
450 mm wide	11.43	14.46	0.48	21.94	nr	**36.40**
600 mm wide	15.53	19.66	0.48	21.94	nr	**41.59**
Bend						
100 mm wide	10.45	13.22	0.23	10.51	nr	**23.74**
150 mm wide	12.04	15.24	0.26	11.88	nr	**27.12**
200 mm wide	13.56	17.16	0.30	13.71	nr	**30.87**
300 mm wide	14.37	18.18	0.35	16.00	nr	**34.18**
450 mm wide	19.04	24.09	0.50	22.85	nr	**46.95**
600 mm wide	24.60	31.13	0.58	26.51	nr	**57.64**
Tee						
100 mm wide	13.11	16.59	0.28	12.80	nr	**29.38**
150 mm wide	14.12	17.87	0.30	13.71	nr	**31.58**
200 mm wide	16.16	20.45	0.33	15.08	nr	**35.54**
300 mm wide	17.03	21.55	0.39	17.82	nr	**39.37**
450 mm wide	21.75	27.52	0.56	25.59	nr	**53.11**
600 mm wide	27.18	34.40	0.65	29.71	nr	**64.10**
Crossovers						
100 mm wide	14.30	18.10	0.40	18.28	nr	**36.38**
150 mm wide	14.54	18.40	0.42	19.20	nr	**37.60**
200 mm wide	17.71	22.41	0.46	21.02	nr	**43.44**
300 mm wide	18.49	23.39	0.51	23.31	nr	**46.70**
450 mm wide	19.13	24.21	0.74	33.82	nr	**58.03**
600 mm wide	23.69	29.99	0.82	37.48	nr	**67.46**
Mild steel cable basket; epoxy coated including standard coupling joints, fixings and earth continuity straps (supports and hangers are excluded)						
Basket 54 mm deep						
100 mm wide	18.49	23.40	0.22	10.05	m	**33.45**
150 mm wide	20.90	26.45	0.25	11.43	m	**37.88**
200 mm wide	23.28	29.46	0.28	12.80	m	**42.26**
300 mm wide	26.43	33.46	0.34	15.54	m	**48.99**
450 mm wide	32.63	41.29	0.44	20.11	m	**61.40**
600 mm wide	37.16	47.03	0.70	31.99	m	**79.02**

ELECTRICAL SUPPLY/POWER/LIGHTING SYSTEMS

Item	Net Price £	Material £	Labour hours	Labour £	Unit	Total rate £
LV DISTRIBUTION: BASKET TRAY – cont						
Extra for (cutting and jointing to fittings is included)						
Reducer						
150 mm wide	38.66	48.93	0.28	12.80	nr	**61.73**
200 mm wide	43.32	54.83	0.28	12.80	nr	**67.63**
300 mm wide	46.21	58.49	0.38	17.37	nr	**75.86**
450 mm wide	54.02	68.37	0.48	21.94	nr	**90.30**
600 mm wide	67.33	85.21	0.48	21.94	nr	**107.15**
Bend						
100 mm wide	37.49	47.44	0.23	10.51	nr	**57.95**
150 mm wide	41.88	53.01	0.26	11.88	nr	**64.89**
200 mm wide	42.41	53.68	0.30	13.71	nr	**67.39**
300 mm wide	47.72	60.39	0.35	16.00	nr	**76.39**
450 mm wide	60.55	76.64	0.50	22.85	nr	**99.49**
600 mm wide	72.72	92.03	0.58	26.51	nr	**118.54**
Tee						
100 mm wide	44.06	55.76	0.28	12.80	nr	**68.56**
150 mm wide	46.10	58.34	0.30	13.71	nr	**72.05**
200 mm wide	46.76	59.17	0.33	15.08	nr	**74.26**
300 mm wide	58.15	73.60	0.39	17.82	nr	**91.42**
450 mm wide	74.10	93.78	0.56	25.59	nr	**119.38**
600 mm wide	83.12	105.20	0.65	29.71	nr	**134.90**
Crossovers						
100 mm wide	55.89	70.74	0.40	18.28	nr	**89.02**
150 mm wide	56.53	71.54	0.42	19.20	nr	**90.74**
200 mm wide	60.55	76.64	0.46	21.02	nr	**97.66**
300 mm wide	69.51	87.97	0.51	23.31	nr	**111.28**
450 mm wide	83.71	105.95	0.74	33.82	nr	**139.77**
600 mm wide	90.45	114.47	0.82	37.48	nr	**151.95**

ELECTRICAL SUPPLY/POWER/LIGHTING SYSTEMS

Item	Net Price £	Material £	Labour hours	Labour £	Unit	Total rate £
LV DISTRIBUTION: SWITCHGEAR AND DISTRIBUTION BOARDS						
LV switchboard components, factory-assembled modular construction to IP41; form 4, type 5; 2400 mm high, with front and rear access; top cable entry/exit; includes delivery, offloading, positioning and commissioning (hence separate labour costs are not detailed below); excludes cabling and cable terminations						
Air circuit breakers (ACBs) to BSEN60947–2, withdrawable type, fitted with adjustable instantaneous and overload protection. Includes enclosure and copper links, assembled into LV switchboard						
ACB–100 kA fault rated						
4 pole, 6300 A (1600 mm wide)	57257.88	72465.58	–	–	nr	**72465.58**
4 pole, 5000 A (1600 mm wide)	45442.77	57512.37	–	–	nr	**57512.37**
4 pole, 4000 A (1600 mm wide)	26039.16	32955.16	–	–	nr	**32955.16**
4 pole, 3200 A (1600 mm wide)	20831.32	26364.12	–	–	nr	**26364.12**
4 pole, 2500 A (1600 mm wide)	15136.69	19156.99	–	–	nr	**19156.99**
4 pole, 2000 A (1600 mm wide) 85ka	12109.35	15325.59	–	–	nr	**15325.59**
4 pole, 1600 A (1600 mm wide) 85ka	11004.85	13927.74	–	–	nr	**13927.74**
4 pole, 1250 A (1600 mm wide)	8597.55	10881.06	–	–	nr	**10881.06**
4 pole, 1000 A (1600 mm wide)	8435.88	10676.44	–	–	nr	**10676.44**
4 pole, 800 A (1600 mm wide) 85ka	8048.26	10185.88	–	–	nr	**10185.88**
3 pole, 6300 A (1600 mm wide)	47392.35	59979.76	–	–	nr	**59979.76**
3 pole, 5000 A (1600 mm wide)	37612.98	47602.98	–	–	nr	**47602.98**
3 pole, 4000 A (1600 mm wide)	20013.36	25328.91	–	–	nr	**25328.91**
3 pole, 3200 A (1600 mm wide)	16010.69	20263.13	–	–	nr	**20263.13**
3 pole, 2500 A (1600 mm wide)	12508.33	15830.55	–	–	nr	**15830.55**
3 pole, 2000 A (1600 mm wide)	11084.08	14028.01	–	–	nr	**14028.01**
3 pole, 1600 A (1600 mm wide)	8867.26	11222.40	–	–	nr	**11222.40**
3 pole, 1250 A (1600 mm wide)	6927.56	8767.52	–	–	nr	**8767.52**
3 pole, 1000 A (1600 mm wide)	6825.45	8638.29	–	–	nr	**8638.29**
3 pole, 800 A (1600 mm wide)	5460.36	6910.64	–	–	nr	**6910.64**
ACB–65 kA fault rated						
4 pole, 4000 A (1600 mm wide)	57257.88	72465.58	–	–	nr	**72465.58**
4 pole, 3200 A (1600 mm wide)	19217.09	24321.15	–	–	nr	**24321.15**
4 pole, 2500 A (1600 mm wide)	14393.27	18216.12	–	–	nr	**18216.12**
4 pole, 2000 A (1600 mm wide)	11514.62	14572.90	–	–	nr	**14572.90**
4 pole, 1600 A (1600 mm wide)	10410.14	13175.07	–	–	nr	**13175.07**
4 pole, 1250 A (1600 mm wide)	9818.02	12425.68	–	–	nr	**12425.68**
4 pole, 1000 A (1600 mm wide)	9539.96	12073.77	–	–	nr	**12073.77**
4 pole, 800 A (1600 mm wide)	7631.96	9659.01	–	–	nr	**9659.01**
3 pole, 6300 A (1600 mm wide)	47392.35	59979.76	–	–	nr	**59979.76**
3 pole, 4000 A (1600 mm wide)	30090.37	38082.38	–	–	nr	**38082.38**
3 pole, 3200 A (1600 mm wide)	14821.24	18757.76	–	–	nr	**18757.76**
3 pole, 2500 A (1600 mm wide)	11579.09	14654.50	–	–	nr	**14654.50**
3 pole, 2000 A (1600 mm wide)	10468.15	13248.49	–	–	nr	**13248.49**
3 pole, 1600 A (1600 mm wide)	8374.50	10598.77	–	–	nr	**10598.77**

ELECTRICAL SUPPLY/POWER/LIGHTING SYSTEMS

Item	Net Price £	Material £	Labour hours	Labour £	Unit	Total rate £
LV DISTRIBUTION: SWITCHGEAR AND DISTRIBUTION BOARDS – cont						
Air circuit breakers (ACBs) to BSEN60947–2, withdrawable type, fitted with adjustable instantaneous and overload protection – cont						
ACB–65 kA fault rated – cont						
3 pole, 1250 A (1600 mm wide)	6542.57	8280.28	–	–	nr	**8280.28**
3 pole, 1000 A (1600 mm wide)	6517.49	8248.54	–	–	nr	**8248.54**
3 pole, 800 A (1600 mm wide)	5213.99	6598.82	–	–	nr	**6598.82**
Extra for						
Cable box (one per ACB for form 4, types 6 & 7)	502.44	635.89	–	–	nr	**635.89**
Opening coil (shunt trip)	228.54	289.25	–	–	nr	**289.25**
Closing coil	151.40	191.62	–	–	nr	**191.62**
Undervoltage release	271.02	343.00	–	–	nr	**343.00**
Motor operator	911.93	1154.14	–	–	nr	**1154.14**
Mechnical interlock (per ACB)	630.41	797.84	–	–	nr	**797.84**
ACB Fortress/Castell adaptor kit (one per ACB)	135.94	172.05	–	–	nr	**172.05**
Fortress/Castell ACB lock (one per ACB)	428.43	542.23	–	–	nr	**542.23**
Fortress/Castell key	106.68	135.02	–	–	nr	**135.02**
Moulded case circuit breakers (MCCBs) to BS EN 60947–2; plug-in type, fitted with electronic trip unit. Includes metalwork section and copper links, assembled into LV switchboard						
MCCB–150 kA fault rated						
4 pole, 630 A (800 mm wide, 600 mm high)	3860.30	4885.59	–	–	nr	**4885.59**
4 pole, 400 A (800 mm wide, 400 mm high)	2584.91	3271.46	–	–	nr	**3271.46**
4 pole, 250 A (800 mm wide, 400 mm high)	2577.29	3261.82	–	–	nr	**3261.82**
4 pole, 160 A (800 mm wide, 300 mm high)	2386.85	3020.80	–	–	nr	**3020.80**
4 pole, 100 A (800 mm wide, 200 mm high)	1634.98	2069.23	–	–	nr	**2069.23**
3 pole, 630 A (800 mm wide, 600 mm high)	3050.32	3860.49	–	–	nr	**3860.49**
3 pole, 400 A (800 mm wide, 400 mm high)	2076.21	2627.65	–	–	nr	**2627.65**
3 pole, 250 A (800 mm wide, 400 mm high)	1357.79	1718.42	–	–	nr	**1718.42**
3 pole, 160 A (800 mm wide, 300 mm high)	975.59	1234.71	–	–	nr	**1234.71**
3 pole, 100 A (800 mm wide, 200 mm high)	793.88	1004.73	–	–	nr	**1004.73**
MCCB–70kA fault rated						
4 pole, 630 A (800 mm wide, 600 mm high)	3223.94	4080.22	–	–	nr	**4080.22**
4 pole, 400 A (800 mm wide, 400 mm high)	2216.12	2804.72	–	–	nr	**2804.72**
4 pole, 250 A (800 mm wide, 400 mm high)	1613.86	2042.50	–	–	nr	**2042.50**
4 pole, 160 A (800 mm wide, 300 mm high)	1216.37	1539.44	–	–	nr	**1539.44**
4 pole, 100 A (800 mm wide, 200 mm high)	1002.34	1268.56	–	–	nr	**1268.56**
3 pole, 630 A (800 mm wide, 600 mm high)	2362.80	2990.36	–	–	nr	**2990.36**
3 pole, 400 A (800 mm wide, 400 mm high)	1594.73	2018.29	–	–	nr	**2018.29**
3 pole, 250 A (800 mm wide, 400 mm high)	1112.14	1407.52	–	–	nr	**1407.52**
3 pole, 160 A (800 mm wide, 300 mm high)	825.56	1044.83	–	–	nr	**1044.83**
3 pole, 100 A (800 mm wide, 200 mm high)	653.83	827.49	–	–	nr	**827.49**
MCCB–45kA fault rated						
4 pole, 630 A (800 mm wide, 600 mm, LI, SSKA high)	2969.18	3757.80	–	–	nr	**3757.80**
4 pole, 400 A (800 mm wide, 400 mm, LI, SSKA high)	2061.23	2608.69	–	–	nr	**2608.69**
3 pole, 630 A (800 mm wide, 600 mm, LI, SSKA high)	2213.48	2801.38	–	–	nr	**2801.38**
3 pole, 400 A (800 mm wide, 400 mm, LI, SSKA high)	1519.90	1923.59	–	–	nr	**1923.59**

ELECTRICAL SUPPLY/POWER/LIGHTING SYSTEMS

Item	Net Price £	Material £	Labour hours	Labour £	Unit	Total rate £
MCCB–36kA fault rated						
4 pole, 250 A (800 mm wide, 400 mm high)	2471.87	3128.40	–	–	nr	**3128.40**
4 pole, 160 A (800 mm wide, 300 mm high)	1872.50	2369.83	–	–	nr	**2369.83**
3 pole, 250 A (800 mm wide, 400 mm high)	1404.37	1777.37	–	–	nr	**1777.37**
3 pole, 160 A (800 mm wide, 300 mm high)	1236.79	1565.28	–	–	nr	**1565.28**
Extra for						
Cable box (one per MCCB for form 4, types 6 & 7)	212.49	268.93	–	–	nr	**268.93**
Shunt trip (for ratings 100 A to 630 A)	65.50	82.89	–	–	nr	**82.89**
Undervoltage release (for ratings 100 A to 630 A)	84.96	107.53	–	–	nr	**107.53**
Motor operator for 630 A MCCB	156.33	197.86	–	–	nr	**197.86**
Motor operator for 400 A MCCB	156.33	197.86	–	–	nr	**197.86**
Motor operator for 250 A MCCB	75.88	96.03	–	–	nr	**96.03**
Motor operator for 160 A/100 A MCCB	67.46	85.37	–	–	nr	**85.37**
Door handle for 630/400 A MCCB	133.00	168.32	–	–	nr	**168.32**
Door handle for 250/160/100 A MCCB	105.38	133.37	–	–	nr	**133.37**
MCCB earth fault protection	801.51	1014.39	–	–	nr	**1014.39**
LV switchboard busbar						
Copper busbar assembled into LV switchboard, ASTA type tested to appropriate fault level. Busbar length may be estimated by adding the widths of the ACB sections to the width of the MCCB sections. ACBs up to 2000 A rating may be stacked two high; larger ratings are one per section. To determine the number of MCCB sections, add together all the MCCB heights and divide by 1800 mm, rounding up as necessary						
6000 A (6 × 10 mm × 100 mm)	4170.62	5278.34	–	–	nr	**5278.34**
5000 A (4 × 10 mm × 100 mm)	3428.00	4338.47	–	–	nr	**4338.47**
4000 A (4 × 10 mm × 100 mm)	3428.00	4338.47	–	–	nr	**4338.47**
3200 A (3 × 10 mm × 100 mm)	2380.98	3013.36	–	–	nr	**3013.36**
2500 A (2 × 10 mm × 100 mm)	2008.81	2542.35	–	–	nr	**2542.35**
2000 A (2 × 10 mm × 80 mm)	1476.58	1868.76	–	–	nr	**1868.76**
1600 A (2 × 10 mm × 50 mm)	1071.35	1355.90	–	–	nr	**1355.90**
1250 A (2 × 10 mm × 40 mm)	840.04	1063.15	–	–	nr	**1063.15**
1000 A (2 × 10 mm × 30 mm)	840.04	1063.15	–	–	nr	**1063.15**
800 A (2 × 10 mm × 20 mm)	653.92	827.61	–	–	nr	**827.61**
630 A (2 × 10 mm × 20 mm)	653.92	827.61	–	–	nr	**827.61**
400 A (2 × 10 mm × 10 mm)	560.03	708.78	–	–	nr	**708.78**

ELECTRICAL SUPPLY/POWER/LIGHTING SYSTEMS

Item	Net Price £	Material £	Labour hours	Labour £	Unit	Total rate £
LV DISTRIBUTION: SWITCHGEAR AND DISTRIBUTION BOARDS – cont						
LV switchboard busbar – cont						
Automatic power factor correction (PFC); floor standing steel enclosure to IP 42, complete with microprocessor based relay and status indication; includes delivery, offloading, positioning and commissioning; excludes cabling and cable terminations						
Standard PFC (no detuning)						
100 kVAr	8177.79	10349.81	–		nr	**10349.81**
200 kVAr	11437.06	14474.74	–		nr	**14474.74**
400 kVAr	19524.38	24710.05	–		nr	**24710.05**
600 kVAr	26743.84	33847.01	–		nr	**33847.01**
PFC with detuning reactors						
100 kVAr	13304.97	16838.76	–		nr	**16838.76**
200 kVAr	18593.90	23532.44	–		nr	**23532.44**
400 kVAr	32046.69	40558.30	–		nr	**40558.30**
600 kVAr	47129.13	59646.63	–		nr	**59646.63**

ELECTRICAL SUPPLY/POWER/LIGHTING SYSTEMS

Item	Net Price £	Material £	Labour hours	Labour £	Unit	Total rate £
LV DISTRIBUTION: AUTOMATIC TRANSFER SWITCHES						
Automatic transfer switches; steel enclosure; solenoid operating; programmable controller, keypad and LCD display; fixed to backgrounds; including commissioning and testing						
Panel mounting type 4 pole M6 s; non BMS connection						
40 amp	2792.05	3533.62	2.40	109.69	nr	**3643.31**
63 amp	2940.55	3721.56	2.50	114.26	nr	**3835.82**
80 amp	3002.37	3799.80	2.60	118.83	nr	**3918.63**
100 amp	3043.45	3851.79	2.60	118.83	nr	**3970.62**
125 amp	3113.50	3940.44	2.70	123.40	nr	**4063.84**
160 amp	3255.62	4120.31	2.80	127.97	nr	**4248.28**
Panel mounting type 4 pole M6e; BMS connection						
40 amp	3852.83	4876.15	2.50	114.26	nr	**4990.41**
63 amp	4049.89	5125.55	2.60	118.83	nr	**5244.38**
80 amp	4132.05	5229.52	2.70	123.40	nr	**5352.92**
100 amp	4186.69	5298.68	2.80	127.97	nr	**5426.65**
125 amp	4279.60	5416.27	2.90	132.54	nr	**5548.81**
160 amp	4468.32	5655.10	3.00	137.11	nr	**5792.21**
Enclosed type 3 pole or 4 pole						
125 amp	5062.00	6406.46	2.60	118.83	nr	**6525.29**
160 amp	5349.49	6770.32	2.90	132.54	nr	**6902.86**
200 amp	5665.03	7169.66	3.30	150.82	nr	**7320.48**
250 amp	5986.40	7576.39	4.30	196.53	nr	**7772.91**
300 amp	6665.50	8435.86	4.84	221.21	nr	**8657.07**
400 amp	7217.22	9134.12	5.12	234.00	nr	**9368.12**
500 amp	7764.97	9827.34	5.50	251.37	nr	**10078.71**
630 amp	8640.64	10935.59	6.00	274.22	nr	**11209.81**
800 amp	9841.98	12456.01	6.20	283.36	nr	**12739.37**
1000 amp	11203.09	14178.63	6.90	315.35	nr	**14493.99**
1250 amp	12231.66	15480.39	7.70	351.92	nr	**15832.30**
1600 amp	13536.54	17131.85	8.50	388.48	nr	**17520.33**
2000 amp	20846.93	26383.87		411.33	nr	**26795.20**
Enclosed type 3 pole or 4 pole; with single bypass						
40 amp	8417.44	10653.11	2.60	118.83	nr	**10771.94**
63 amp	8580.84	10859.92	2.60	118.83	nr	**10978.75**
80 amp	8689.51	10997.44	2.60	118.83	nr	**11116.27**
100 amp	8808.63	11148.20	2.60	118.83	nr	**11267.03**
125 amp	9264.26	11724.85	2.90	132.54	nr	**11857.39**
160 amp	10181.40	12885.58	2.90	132.54	nr	**13018.12**
250 amp	13139.18	16628.95	3.30	150.82	nr	**16779.77**
400 amp	13859.13	17540.12	4.30	196.53	nr	**17736.64**
630 amp	18968.41	24006.42	4.84	221.21	nr	**24227.63**
800 amp	20260.74	25641.99	5.12	234.00	nr	**25875.99**
1000 amp	28728.65	36358.98	5.50	251.37	nr	**36610.34**
1250 amp	34841.06	44094.84	6.00	274.22	nr	**44369.06**
1600 amp	42300.63	53535.68	6.20	283.36	nr	**53819.04**
2000 amp	51562.07	65256.95	6.90	315.35	nr	**65572.31**
2500 amp	60196.15	76184.25	7.70	351.92	nr	**76536.17**
3200 amp	70982.87	89835.91	8.50	388.48	nr	**90224.39**

ELECTRICAL SUPPLY/POWER/LIGHTING SYSTEMS

Item	Net Price £	Material £	Labour hours	Labour £	Unit	Total rate £
LV DISTRIBUTION: AUTOMATIC TRANSFER SWITCHES – cont						
Automatic transfer switches – cont						
Enclosed type 3 pole or 4 pole; with dual bypass						
40 amp	11021.33	13948.60	2.60	118.83	nr	**14067.43**
63 amp	11441.12	14479.88	2.60	118.83	nr	**14598.71**
80 amp	11582.63	14658.97	2.60	118.83	nr	**14777.80**
100 amp	11744.87	14864.31	2.60	118.83	nr	**14983.14**
125 amp	12352.35	15633.14	2.90	132.54	nr	**15765.68**
160 amp	13460.45	17035.54	2.90	132.54	nr	**17168.08**
250 amp	17471.11	22111.44	3.30	150.82	nr	**22262.26**
400 amp	18438.55	23335.83	4.30	196.53	nr	**23532.35**
630 amp	25111.41	31781.00	4.84	221.21	nr	**32002.20**
800 amp	27081.25	34274.03	5.12	234.00	nr	**34508.03**
1000 amp	38252.26	48412.06	5.50	251.37	nr	**48663.43**
1250 amp	44584.45	56426.08	6.00	274.22	nr	**56700.30**
1600 amp	52418.15	66340.41	6.20	283.36	nr	**66623.78**
2000 amp	62488.45	79085.38	6.90	315.35	nr	**79400.73**
2500 amp	70878.87	89704.29	7.70	351.92	nr	**90056.21**
3200 amp	83765.93	106014.17	8.50	388.48	nr	**106402.65**

ELECTRICAL SUPPLY/POWER/LIGHTING SYSTEMS

Item	Net Price £	Material £	Labour hours	Labour £	Unit	Total rate £
LV DISTRIBUTION: BREAKERS AND FUSES						
MCCB panelboards; IP4X construction, 50 kA busbars and fully-rated neutral; fitted with doorlock, removable glandplate; form 3b Type2; BSEN60439–1; including fixing to backgrounds						
Panelboards cubicle with MCCB incomer						
Up to 250 A						
4 way TP&N	1636.22	2070.79	1.00	45.70	nr	**2116.50**
Extra over for integral incomer metering	1675.81	2120.91	1.50	68.56	nr	**2189.46**
Up to 630 A						
6 way TP&N	2943.31	3725.05	2.00	91.41	nr	**3816.46**
12 way TP&N	3324.18	4207.08	2.50	114.26	nr	**4321.34**
18 way TP&N	3869.57	4897.33	3.00	137.11	nr	**5034.44**
Extra over for integral incomer metering	1980.48	2506.50	1.50	68.56	nr	**2575.05**
Up to 800 A						
6 way TP&N	5036.55	6374.25	2.00	91.41	nr	**6465.66**
12 way TP&N	6042.07	7646.84	2.50	114.26	nr	**7761.10**
18 way TP&N	6432.05	8140.41	3.00	137.11	nr	**8277.52**
Extra over for integral incomer metering	1980.48	2506.50	1.50	68.56	nr	**2575.05**
Up to 1200 A						
20 way TP&N	13476.54	17055.91	3.50	159.96	nr	**17215.88**
Up to 1600 A						
28 way TP&N	17452.73	22088.17	3.50	159.96	nr	**22248.13**
Up to 2000 A						
28 way TP&N	18997.52	24043.26	4.00	182.81	nr	**24226.07**
Feeder MCCBs						
Single pole						
32 A	138.20	174.91	0.75	34.28	nr	**209.18**
63 A	141.35	178.89	0.75	34.28	nr	**213.17**
100 A	144.47	182.85	0.75	34.28	nr	**217.12**
160 A	153.91	194.79	1.00	45.70	nr	**240.49**
Double pole						
32 A	207.31	262.38	0.75	34.28	nr	**296.65**
63 A	210.44	266.33	0.75	34.28	nr	**300.61**
100 A	307.82	389.58	0.75	34.28	nr	**423.86**
160 A	383.19	484.97	1.00	45.70	nr	**530.67**
Triple pole						
32 A	276.43	349.85	0.75	34.28	nr	**384.13**
63 A	282.67	357.75	0.75	34.28	nr	**392.03**
100 A	367.48	465.09	0.75	34.28	nr	**499.37**
160 A	474.32	600.30	1.00	45.70	nr	**646.00**
250 A	713.03	902.41	1.00	45.70	nr	**948.11**
400 A	964.32	1220.44	1.25	57.13	nr	**1277.57**
630 A	1583.18	2003.67	1.50	68.56	nr	**2072.23**

ELECTRICAL SUPPLY/POWER/LIGHTING SYSTEMS

Item	Net Price £	Material £	Labour hours	Labour £	Unit	Total rate £
LV DISTRIBUTION: BREAKERS AND FUSES – cont						
MCB distribution boards; IP3X external protection enclosure; removable earth and neutral bars and DIN rail; 125/250 amp incomers; including fixing to backgrounds						
SP&N						
6 way	277.35	351.01	2.00	91.41	nr	442.42
8 way	333.14	421.62	2.50	114.26	nr	535.88
12 way	448.31	567.38	3.00	137.11	nr	704.49
16 way	549.31	695.21	4.00	182.81	nr	878.02
24 way	829.82	1050.21	5.00	228.52	nr	1278.73
TP&N						
4 way	1059.54	1340.96	3.00	137.11	nr	1478.07
6 way	1100.01	1392.17	3.50	159.96	nr	1552.13
8 way	1239.36	1568.53	4.00	182.81	nr	1751.35
12 way	1513.66	1915.69	4.00	182.81	nr	2098.51
16 way	1952.01	2470.46	5.00	228.52	nr	2698.98
24 way	2787.14	3527.40	6.40	292.50	nr	3819.90
Miniature circuit breakers for distribution boards; BS EN 60 898; DIN rail mounting; including connecting to circuit						
SP&N; including connecting of wiring						
6 amp	34.12	43.18	0.10	4.57	nr	47.75
10–40 amp	35.48	44.91	0.10	4.57	nr	49.48
50–63 amp	34.13	43.19	0.14	6.40	nr	49.59
TP&N; including connecting of wiring						
6 amp	140.65	178.01	0.30	13.71	nr	191.72
10–40 amp	138.82	175.69	0.45	20.57	nr	196.25
50–63 amp	157.10	198.83	0.45	20.57	nr	219.39
Residual current circuit breakers for distribution boards; DIN rail mounting; including connecting to circuit						
SP&N						
10 mA						
6 amp	110.43	139.77	0.21	9.60	nr	149.36
10–32 amp	112.24	142.06	0.26	11.88	nr	153.94
45 amp	112.58	142.48	0.26	11.88	nr	154.36
30 mA						
6 amp	110.43	139.77	0.21	9.60	nr	149.36
10–40 amp	112.58	142.48	0.21	9.60	nr	152.08
50–63 amp	114.05	144.35	0.26	11.88	nr	156.23
100 mA						
6 amp	208.16	263.44	0.21	9.60	nr	273.04
10–40 amp	208.16	263.44	0.23	10.65	nr	274.09
50–63 amp	208.16	263.44	0.26	11.88	nr	275.33

ELECTRICAL SUPPLY/POWER/LIGHTING SYSTEMS

Item	Net Price £	Material £	Labour hours	Labour £	Unit	Total rate £
HRC fused distribution boards; IP4X external protection enclosure; including earth and neutral bars; fixing to backgrounds						
SP&N						
20 amp incomer						
4 way	276.21	349.57	1.00	45.70	nr	**395.28**
6 way	333.47	422.04	1.20	54.84	nr	**476.88**
8 way	390.92	494.75	1.40	63.99	nr	**558.74**
12 way	505.87	640.23	1.80	82.27	nr	**722.50**
32 amp incomer						
4 way	332.56	420.89	1.00	45.70	nr	**466.59**
6 way	437.34	553.50	1.20	54.84	nr	**608.35**
8 way	514.55	651.21	1.40	63.99	nr	**715.20**
12 way	662.45	838.40	1.80	82.27	nr	**920.67**
TP&N						
20 amp incomer						
4 way	550.67	696.92	1.50	68.56	nr	**765.48**
6 way	696.45	881.43	2.10	95.98	nr	**977.41**
8 way	823.44	1042.14	2.70	123.40	nr	**1165.54**
12 way	1155.26	1462.10	3.90	178.24	nr	**1640.34**
32 amp incomer						
4 way	658.76	833.73	1.50	68.56	nr	**902.29**
6 way	884.43	1119.33	2.10	95.98	nr	**1215.31**
8 way	1080.37	1367.31	2.70	123.40	nr	**1490.71**
12 way	1498.46	1896.45	3.90	178.24	nr	**2074.69**
63 amp incomer						
4 way	1400.31	1772.24	2.17	99.18	nr	**1871.41**
6 way	1795.90	2272.89	2.57	117.46	nr	**2390.34**
8 way	2162.67	2737.07	2.83	129.34	nr	**2866.42**
100 amp incomer						
4 way	2214.58	2802.78	2.40	109.69	nr	**2912.47**
6 way	2894.66	3663.48	2.73	124.77	nr	**3788.25**
8 way	3539.47	4479.55	3.87	176.87	nr	**4656.42**
200 amp incomer						
4 way	5485.31	6942.21	5.36	244.97	nr	**7187.18**
6 way	7250.02	9175.62	6.17	281.99	nr	**9457.61**
HRC fuse; includes fixing to fuse holder						
2–30 amp	4.93	6.24	0.10	4.57	nr	**10.81**
35–63 amp	10.63	13.45	0.12	5.48	nr	**18.94**
80 amp	15.71	19.88	0.15	6.86	nr	**26.73**
100 amp	18.90	23.92	0.15	6.86	nr	**30.78**
125 amp	28.54	36.12	0.15	6.86	nr	**42.98**
160 amp	29.94	37.89	0.15	6.86	nr	**44.74**
200 amp	31.01	39.24	0.15	6.86	nr	**46.10**

ELECTRICAL SUPPLY/POWER/LIGHTING SYSTEMS

Item	Net Price £	Material £	Labour hours	Labour £	Unit	Total rate £
LV DISTRIBUTION: BREAKERS AND FUSES – cont						
Consumer units; fixed to backgrounds; including supports, fixings, connections/jointing to equipment						
Switched and insulated; moulded plastic case, 63 amp 230 volt SP&N; earth and neutral bars; 30 mA RCCB protection; fitted MCBs						
2 way	241.45	305.58	1.59	72.67	nr	378.25
4 way	270.91	342.87	1.67	76.32	nr	419.19
6 way	294.50	372.71	2.50	114.26	nr	486.97
8 way	317.79	402.20	3.00	137.11	nr	539.31
12 way	369.16	467.21	4.00	182.81	nr	650.02
16 way	446.05	564.52	5.50	251.37	nr	815.89
Switched and insulated; moulded plastic case, 100 amp 230 volt SP&N; earth and neutral bars; 30 mA RCCB protection; fitted MCBs						
2 way	241.45	305.58	1.59	72.67	nr	378.25
4 way	270.91	342.87	1.67	76.32	nr	419.19
6 way	294.50	372.71	2.50	114.26	nr	486.97
8 way	317.79	402.20	3.00	137.11	nr	539.31
12 way	369.16	467.21	4.00	182.81	nr	650.02
16 way	446.05	564.52	5.50	251.37	nr	815.89
Extra for						
Residual current device; double pole; 230 volt/30 mA tripping current						
16 amp	117.60	148.84	0.22	10.05	nr	158.89
30 amp	119.47	151.20	0.22	10.05	nr	161.26
40 amp	121.33	153.55	0.22	10.05	nr	163.61
63 amp	150.22	190.12	0.22	10.05	nr	200.17
80 amp	167.10	211.48	0.22	10.05	nr	221.53
100 amp	205.66	260.28	0.25	11.43	nr	271.71
Residual current device; double pole; 230 volt/100 mA tripping current						
63 amp	137.33	173.80	0.22	10.05	nr	183.86
80 amp	158.87	201.06	0.22	10.05	nr	211.12
100 amp	205.69	260.32	0.25	11.43	nr	271.75
Heavy duty fuse switches; with HRC fuses BS 5419; short circuit rating 65 kA, 500 volt; including retractable operating switches						
SP&N						
63 amp	495.87	627.57	1.30	59.41	nr	686.99
100 amp	724.86	917.39	1.95	89.12	nr	1006.51
TP&N						
63 amp	624.73	790.65	1.83	83.64	nr	874.29
100 amp	877.86	1111.02	2.48	113.34	nr	1224.37
200 amp	1354.15	1713.82	3.13	143.05	nr	1856.87
300 amp	2352.55	2977.38	4.45	203.38	nr	3180.76
400 amp	2582.66	3268.61	4.45	203.38	nr	3471.99
600 amp	3898.40	4933.82	5.72	261.42	nr	5195.24
800 amp	6089.76	7707.21	7.88	360.14	nr	8067.35

ELECTRICAL SUPPLY/POWER/LIGHTING SYSTEMS

Item	Net Price £	Material £	Labour hours	Labour £	Unit	Total rate £
Switch disconnectors to BSEN60947–3; in sheet steel case; IP41 with door interlock fixed to backgrounds						
Double pole						
20 amp	94.66	119.81	1.02	46.62	nr	**166.42**
32 amp	114.10	144.41	1.02	46.62	nr	**191.02**
63 amp	382.06	483.54	1.21	55.30	nr	**538.84**
100 amp	416.36	526.94	1.86	85.01	nr	**611.95**
TP&N						
20 amp	118.84	150.40	1.29	58.96	nr	**209.36**
32 amp	138.29	175.01	1.83	83.64	nr	**258.65**
63 amp	469.25	593.88	2.48	113.34	nr	**707.23**
100 amp	479.94	607.41	2.48	113.34	nr	**720.76**
125 amp	493.49	624.56	2.48	113.34	nr	**737.90**
160 amp	1135.44	1437.01	2.48	113.34	nr	**1550.35**
Enclosed switch disconnector to BSEN60947–3; enclosure minimum IP55 rating; complete with earth connection bar; fixed to backgrounds						
TP						
20 amp	94.66	119.81	1.02	46.62	nr	**166.42**
32 amp	114.10	144.41	1.02	46.62	nr	**191.02**
63 amp	382.06	483.54	1.21	55.30	nr	**538.84**
TP&N						
20 amp	118.84	150.40	1.29	58.96	nr	**209.36**
32 amp	138.29	175.01	1.83	83.64	nr	**258.65**
63 amp	469.25	593.88	2.48	113.34	nr	**707.23**
Busbar chambers; fixed to background including all supports, fixings, connections/jointing to equipment						
Sheet steel case enclosing 4 pole 550 volt copper bars, detachable metal end plates						
600 mm long						
200 amp	862.36	1091.40	2.62	119.74	nr	**1211.15**
300 amp	1108.16	1402.49	3.03	138.48	nr	**1540.97**
500 amp	1898.80	2403.12	4.48	204.75	nr	**2607.88**
900 mm long						
200 amp	1242.17	1572.08	3.04	138.94	nr	**1711.02**
300 amp	1463.94	1852.76	3.59	164.08	nr	**2016.83**
500 amp	2165.42	2740.55	4.42	202.01	nr	**2942.56**
1350 mm long						
200 amp	1696.70	2147.35	3.38	154.48	nr	**2301.82**
300 amp	1996.99	2527.39	3.94	180.07	nr	**2707.46**
500 amp	3196.60	4045.62	4.82	220.29	nr	**4265.91**

ELECTRICAL SUPPLY/POWER/LIGHTING SYSTEMS

Item	Net Price £	Material £	Labour hours	Labour £	Unit	Total rate £
LV DISTRIBUTION: BREAKERS AND FUSES – cont						
Contactor relays; pressed steel enclosure; fixed to backgrounds including supports, fixings, connections/jointing to equipment						
Relays						
6 amp, 415/240 volt, 4 pole N/O	99.87	126.39	0.52	23.77	nr	**150.16**
6 amp, 415/240 volt, 8 pole N/O	122.15	154.59	0.85	38.85	nr	**193.44**
Push button stations; heavy gauge pressed steel enclosure; polycarbonate cover; IP65; fixed to backgrounds including supports, fixings, connections/joining to equipment						
Standard units						
One button (start or stop)	126.41	159.99	0.39	17.82	nr	**177.81**
Two button (start or stop)	134.16	169.80	0.47	21.48	nr	**191.28**
Three button (forward-reverse-stop)	190.15	240.65	0.57	26.05	nr	**266.71**
Weatherproof junction boxes; enclosures with rail mounted terminal blocks; side hung door to receive padlock; fixed to backgrounds, including all supports and fixings (Suitable for cable up to 2.5 mm²; including glandplates and gaskets.)						
Sheet steel with zinc spray finish enclosure						
Overall size 229 × 152; suitable to receive 3 × 20(A) glands per gland plate	128.01	162.00	1.43	65.36	nr	**227.36**
Overall size 306 × 306; suitable to receive 14 × 20(A) glands per gland plate	171.73	217.34	2.17	99.18	nr	**316.52**
Overall size 458 × 382; suitable to receive 18 × 20(A) glands per gland plate	250.06	316.48	3.51	160.42	nr	**476.90**
Overall size 762 x508; suitable to receive 26 × 20(A) glands per gland plate	264.48	334.72	4.85	221.66	nr	**556.39**
Overall size 914 × 610; suitable to receive 45 × 20(A) glands per gland plate	292.33	369.97	7.01	320.38	nr	**690.35**
Weatherproof junction boxes; enclosures with rail mounted terminal blocks; screw fixed lid; fixed to backgrounds, including all supports and fixings (suitable for cable up to 2.5 mm²; including glandplates and gaskets)						
Glassfibre reinforced polycarbonate enclosure						
Overall size 190 × 190 × 130 mm	168.34	213.05	1.43	65.36	nr	**278.41**
Overall size 190 × 190 × 180 mm	246.49	311.96	1.53	69.93	nr	**381.88**
Overall size 280 × 190 × 130 mm	278.41	352.36	2.17	99.18	nr	**451.53**
Overall size 280 × 190 × 180 mm	311.89	394.73	2.37	108.32	nr	**503.05**
Overall size 380 × 190 × 130 mm	348.00	440.43	3.30	150.82	nr	**591.25**
Overall size 380 × 190 × 180 mm	373.74	473.00	3.33	152.19	nr	**625.20**
Overall size 380 × 280 × 130 mm	399.54	505.66	4.66	212.98	nr	**718.64**
Overall size 380 × 280 × 180 mm	430.46	544.79	5.36	244.97	nr	**789.76**
Overall size 560 × 280 × 130 mm	518.12	655.73	7.01	320.38	nr	**976.11**
Overall size 560 × 380 × 180 mm	533.57	675.29	7.67	350.55	nr	**1025.84**

ELECTRICAL SUPPLY/POWER/LIGHTING SYSTEMS

Item	Net Price £	Material £	Labour hours	Labour £	Unit	Total rate £
GENERAL LIGHTING						
LED tube luminaires; surface fixed to backgrounds						
Batten type; surface mounted						
600 mm single – 10 W	46.08	58.32	0.58	26.51	nr	84.83
600 mm twin – 20 W	62.61	79.24	0.59	26.97	nr	106.21
1200 mm single – 20 W	62.11	78.60	0.76	34.73	nr	113.34
1200 mm twin – 40 W	98.94	125.22	0.84	38.39	nr	163.61
1500 mm single – 26 W	75.72	95.84	0.77	35.19	nr	131.03
1500 mm twin – 50 W	120.19	152.11	0.85	38.85	nr	190.96
1800 mm single – 30 W	86.30	109.22	1.05	47.99	nr	157.21
1800 mm twin – 60 W	155.51	196.81	1.06	48.45	nr	245.25
Extra over cost for emergency (remote)	126.69	160.34	–	–	nr	160.34
Extra over cost for DALI dimming (remote)	45.79	57.95	–	–	nr	57.95
Surface mounted LED batten, non-dimmable, opal diffuser						
600 mm single – 10 W	91.58	115.91	0.62	28.34	nr	144.24
600 mm twin – 20 W	133.15	168.51	0.62	28.34	nr	196.84
1200 mm single – 20 W	105.50	133.52	0.79	36.11	nr	169.63
1200 mm twin – 40 W	137.56	174.09	0.80	36.56	nr	210.65
1500 mm single – 26 W	120.58	152.61	0.88	40.22	nr	192.83
1500 mm twin – 50 W	146.27	185.12	0.90	41.13	nr	226.26
1800 mm single – 30 W	125.71	159.10	1.09	49.82	nr	208.92
1800 mm twin – 60 W	168.39	213.11	1.10	50.27	nr	263.39
Extra over cost for emergency (remote)	–	160.34	–	–	nr	160.34
Extra over cost for DALI dimming (remote)	–	81.13	–	–	nr	81.13
Surface mounted linear LED batten, non-dimmable, MPO diffuser						
600 mm single – 10 W	96.16	121.70	0.79	36.11	nr	157.81
600 mm twin – 20 W	140.96	178.40	0.79	36.11	nr	214.51
1200 mm single – 20 W	111.68	141.35	0.90	41.13	nr	182.48
1200 mm twin – 40 W	145.19	183.75	1.09	49.82	nr	233.57
1500 mm single – 26 W	125.16	158.41	1.25	57.13	nr	215.54
1500 mm twin – 50 W	–	193.12	1.25	57.13	nr	250.25
1800 mm single – 30 W	133.34	168.76	1.25	57.13	nr	225.89
1800 mm twin – 60 W	179.27	226.89	1.25	57.13	nr	284.02
Extra over cost for emergency (remote)	126.69	160.34	–	–	nr	160.34
Extra over cost for DALI dimming (remote)	64.11	81.13		–	nr	81.13
Modular recessed LED; non-dimmable, low brightness; MPO diffuser fitted to exposed T grid ceiling						
300 × 300 mm, 2 W	48.84	61.82	0.84	38.39	nr	100.21
Over cost for emergency	122.11	154.54		–	nr	154.54
Over cost for DALI dimming	45.79	57.95		–	nr	57.95
600 × 600 mm, 40 W	53.42	67.61	0.87	39.76	nr	107.37
Over cost for emergency	122.11	154.54		–	nr	154.54
Over cost for DALI dimming	64.11	81.13		–	nr	81.13
300 × 1200 mm, 70 W	122.87	155.51	0.87	39.76	nr	195.27
Over cost for emergency (remote)	167.90	212.50		–	nr	212.50
Over cost for DALI dimming (remote)	68.69	86.93		–	nr	86.93

ELECTRICAL SUPPLY/POWER/LIGHTING SYSTEMS

Item	Net Price £	Material £	Labour hours	Labour £	Unit	Total rate £
GENERAL LIGHTING – cont						
LED tube luminaires – cont						
Ceiling recessed round asymetric LED downlighter; for wall washing with non dimmable gear						
1 × 12 W	109.90	139.09	0.75	34.28	nr	**173.37**
1 × 18 W	149.58	189.31	0.75	34.28	nr	**223.59**
1 × 26 W	187.74	237.61	0.75	34.28	nr	**271.89**
1 × 38 W	176.30	223.12	0.75	34.28	nr	**257.40**
1 × 50 W	222.09	281.07	0.75	34.28	nr	**315.35**
Extra over cost for emergency (remote)	221.32	280.11	–		nr	**280.11**
Extra over cost for DALI dimming (remote)	64.11	81.13	–		nr	**81.13**
Ceiling recessed asymetric LED wall washer; non-dimmable control gear; linear 80 mm × 600 mm						
1 × 35 watt	335.80	424.99	0.75	34.28	nr	**459.27**
Extra over cost for emergency (integral)	221.32	280.11	–		nr	**280.11**
Extra over cost for DALI dimming (integral)	64.11	81.13	–		nr	**81.13**
Wall mounted LED uplighter; Integral driver; non-dimmable						
1 × 15 W	167.90	212.50	0.84	38.39	nr	**250.89**
1 × 30 W	221.32	280.11	0.84	38.39	nr	**318.50**
1 × 45 W	267.11	338.06	0.84	38.39	nr	**376.45**
Extra over cost for emergency (remote)	221.32	280.11	–		nr	**280.11**
Extra over cost for DALI dimming (remote)	64.11	81.13	–		nr	**81.13**
Suspended linear LED up/downlight 1.5 m Integral gear; MPO diffuser cut-off; 30% uplight, 70% downlight						
1 × 45 W (downlight), 1 × 30 W (uplight)	366.33	463.63	0.75	34.28	nr	**497.90**
Extra over cost for emergency (remote)	221.32	280.11	–		nr	**280.11**
Extra over cost for DALI dimming (remote)	64.11	81.13	–		nr	**81.13**
Recessed 'architectural' LED, Integral gear, Bat wing, low brightness, delivers direct, ceiling and graduated wall washing illumination						
600 × 600 mm, 40 W	335.80	424.99	0.87	39.76	nr	**464.75**
600 × 600 mm, 50 W	427.38	540.90	0.87	39.76	nr	**580.66**
Extra over cost for emergency (remote)	221.32	280.11	–		nr	**280.11**
Extra over cost for DALI dimming (remote)	61.05	77.27	–		nr	**77.27**
500 × 500 mm, 70 W	305.27	386.35	0.87	39.76	nr	**426.12**
500 × 500 mm, 85 W	335.80	424.99	0.87	39.76	nr	**464.75**
Extra over cost for emergency (remote)	274.75	347.72	–		nr	**347.72**
Extra over cost for DALI dimming (remote)	76.32	96.59	–		nr	**96.59**
Ceiling recessed round fixed LED downlighter with non-dimmable gear						
1 × 12 W	94.64	119.77	0.75	34.28	nr	**154.05**
1 × 18 W	134.32	170.00	0.75	34.28	nr	**204.27**
1 × 26 W	160.27	202.84	0.75	34.28	nr	**237.11**
1 × 38 W	170.95	216.36	0.75	34.28	nr	**250.64**
1 × 50 W	206.06	260.79	0.75	34.28	nr	**295.07**
Extra over cost for emergency (remote)	221.32	280.11	–		nr	**280.11**
Extra over cost for DALI dimming (remote)	64.11	81.13	–		nr	**81.13**

ELECTRICAL SUPPLY/POWER/LIGHTING SYSTEMS

Item	Net Price £	Material £	Labour hours	Labour £	Unit	Total rate £
Ceiling recessed round adjustable LED downlighter with non-dimmable gear						
1 × 12 W	114.48	144.88	0.75	34.28	nr	**179.16**
1 × 18 W	140.43	177.72	0.75	34.28	nr	**212.00**
1 × 26 W	176.60	223.51	0.75	34.28	nr	**257.78**
1 × 38 W	199.95	253.06	0.75	34.28	nr	**287.34**
1 × 50 W	251.85	318.74	0.75	34.28	nr	**353.02**
Extra over cost for emergency (remote)	221.32	280.11	–		nr	**280.11**
Extra over cost for DALI dimming (remote)	64.11	81.13	–		nr	**81.13**
LUMINAIRES						
High/low bay, LED non-dimmable; aluminium reflector						
35 W	320.54	405.67	1.50	68.56	nr	**474.23**
45 W	373.96	473.28	1.50	68.56	nr	**541.84**
55 W	404.49	511.92	1.50	68.56	nr	**580.48**
LED batten fitting						
Corrosion resistant GRP body; gasket sealed; acrylic diffuser						
600 mm – 35 W	116.97	148.04	0.49	22.39	nr	**170.44**
1200 mm – 40 W	156.92	198.59	0.64	29.25	nr	**227.84**
1500 mm – 45 W	196.86	249.14	0.72	32.91	nr	**282.05**
1800 mm – 50 W	241.08	305.11	0.94	42.96	nr	**348.07**
Extra over cost for emergency (Integral)	221.32	280.11	–		nr	**280.11**
Flameproof to IIA/IIB, I.P. 64; Aluminium Body; BS 229 and 899						
600 mm – 35 W	557.77	705.91	1.04	47.53	nr	**753.44**
1200 mm – 40 W	667.61	844.93	1.31	59.87	nr	**904.80**
1500 mm – 45 W	748.92	947.83	1.64	74.95	nr	**1022.79**
1800 mm – 50 W	830.23	1050.74	1.97	90.04	nr	**1140.78**
Extra over cost for emergency (Integral)	221.32	280.11	–		nr	**280.11**
LED downlight/spotlight; track mounted (3ct); adjustable fixing; including driver, polycarbonate optical lenses/diffuser, heat sink and 3000 K colour temperature 100 mm dia.						
20 W	303.75	384.42	0.75	34.28	nr	**418.70**
30 W	335.80	424.99	0.75	34.28	nr	**459.27**
Extra for emergency pack (track mounted remote)	228.96	289.77	–	–	nr	**289.77**
Extra for DALI lighting control (with DALI track)	68.69	86.93	–	–	nr	**86.93**
LED square modular spotlight; recessed into ceiling; including driver, polycarbonate optical lenses/diffuser, heat sink and 3000 K colour temperature						
1 × 15 W	236.59	299.43	0.84	38.39	nr	**337.82**
Extra for emergency pack 1 lamp (remote)	228.96	289.77	–	–	nr	**289.77**
Extra for DALI lighting control	45.79	57.95	–	–	nr	**57.95**
2 × 15 W	320.54	405.67	0.89	40.68	nr	**446.35**
Extra for emergency pack 1 lamp (remote)	228.96	289.77	–	–	nr	**289.77**
Extra for DALI lighting control	83.95	106.25	–	–	nr	**106.25**
3 × 15 W	435.02	550.56	0.89	40.68	nr	**591.23**
Extra for emergency pack 1 lamp (remote)	228.96	289.77	–	–	nr	**289.77**
Extra for DALI lighting control	145.01	183.52	–	–	nr	**183.52**

ELECTRICAL SUPPLY/POWER/LIGHTING SYSTEMS

Item	Net Price £	Material £	Labour hours	Labour £	Unit	Total rate £
GENERAL LIGHTING – cont						
LUMINAIRES – cont						
LED downlight/spotlight; track mounted; adjustable fixing; including driver, polycarbonate optical lenses/diffuser, heat sink, RGB colour						
100 mm dia.	430.44	544.76	0.75	34.28	nr	**579.04**
Internal flexible linear LED, IP20, including driver, extrusion and opal cover white 3000 K colour temperature, per/m						
7.2 W	84.87	107.41	0.25	11.43	nr	**118.83**
14.4 W	110.81	140.25	0.25	11.43	nr	**151.67**
17.3 W	130.66	165.36	0.25	11.43	nr	**176.79**
Extra for DALI lighting control (remote)	76.32	96.59	–	–	nr	**96.59**
Internal flexible linear LED, IP20, including DMX driver and opal cover RGB, per/m						
6 W	114.48	144.88	0.25	11.43	nr	**156.31**
9 W	126.69	160.34	0.25	11.43	nr	**171.76**
12 W	145.01	183.52	0.25	11.43	nr	**194.94**
15 W	167.90	212.50	0.25	11.43	nr	**223.92**
LED in-ground adjustable performance uplighters, IP68, including driver, polycarbonate optical lenses/diffuser, heat sink, 4000 K colour temperature						
15 W	488.44	618.17	0.25	11.43	nr	**629.59**
30 W	541.86	685.78	0.25	11.43	nr	**697.21**
45 W	602.92	763.05	0.25	11.43	nr	**774.48**
Extra for DALI lighting control	76.32	96.59	–	–	nr	**96.59**
LED in-ground adjustable performance uplighters, IP68, including driver, polycarbonate optical lenses/diffuser, heat sink, RGB						
15 W	511.33	647.14	0.25	11.43	nr	**658.57**
30 W	564.76	714.76	0.25	11.43	nr	**726.18**
45 W	641.08	811.35	0.25	11.43	nr	**822.77**
LED in-ground fixed decorative uplighters, IP68, including driver, polycarbonate optical lenses/diffuser, heat sink, 4000 K colour, 50 mm dia.						
3 W	282.38	357.38	0.25	11.43	nr	**368.80**
6 W	305.27	386.35	0.25	11.43	nr	**397.78**
9 W	335.80	424.99	0.25	11.43	nr	**436.42**
Extra for DALI lighting control	83.95	106.25	–	–	nr	**106.25**
LED in-ground fixed decorative uplighters, IP68, including driver, polycarbonate optical lenses/diffuser, heat sink, RGB						
50 mm dia.						
3 W	236.59	299.43	0.25	11.43	nr	**310.85**
6 W	251.85	318.74	0.25	11.43	nr	**330.17**
9 W	267.11	338.06	0.25	11.43	nr	**349.49**

ELECTRICAL SUPPLY/POWER/LIGHTING SYSTEMS

Item	Net Price £	Material £	Labour hours	Labour £	Unit	Total rate £
External flexible linear LED, IP66, including driver, extrusion and opal cover, 3000 K colour temperature, per/m						
6 W	129.74	164.20	0.25	11.43	nr	**175.63**
9 W	148.06	187.38	0.25	11.43	nr	**198.81**
12 W	167.90	212.50	0.25	11.43	nr	**223.92**
15 W	190.80	241.47	0.25	11.43	nr	**252.90**
Extra for DALI lighting control	76.32	96.59	–	–	nr	**96.59**
External flexible linear LED, IP66, including driver, extrusion and opal cover, RGB, per/m						
6 W	145.01	183.52	0.25	11.43	nr	**194.94**
9 W	166.37	210.56	0.25	11.43	nr	**221.99**
12 W	184.69	233.74	0.25	11.43	nr	**245.17**
15 W	212.17	268.52	0.25	11.43	nr	**279.94**
Extra for DALI lighting control	76.32	96.59	–	–	nr	**96.59**
Handrail puck, LED, IP66, including driver and controller, optical lenses/diffuser 3000 K colour temperature						
9 W asymmetric	221.32	280.11	0.25	11.43	nr	**291.53**
9 W down	251.85	318.74	0.25	11.43	nr	**330.17**
Extra for emergency pack (per unit)	228.96	289.77	–	–	nr	**289.77**
Extra for DALI lighting control	76.32	96.59	–	–	nr	**96.59**
Handrail puck, LED, IP66, including driver and controller, optical lenses/diffuser, RGB						
9 W asymmetric	244.22	309.08	0.25	11.43	nr	**320.51**
9 W down	271.69	343.86	0.25	11.43	nr	**355.28**
Extra for emergency pack (per unit)	228.96	289.77	–	–	nr	**289.77**
Extra for DALI lighting control	76.32	96.59	–	–	nr	**96.59**
External lighting						
Bulkhead; aluminium body and polycarbonate bowl; vandal-resistant; IP65						
15 W	290.01	367.04	0.75	34.28	nr	**401.32**
25 W	320.54	405.67	0.75	34.28	nr	**439.95**
40 W	358.70	453.97	0.75	34.28	nr	**488.24**
Extra for emergency pack (Integral)	228.96	289.77	–	–	nr	**289.77**
Photocell	38.16	48.29	0.75	34.28	nr	**82.57**
1500 mm high LED circular bollard; polycarbonate visor; vandal-resistant; IP54						
15 W	473.17	598.85	1.75	79.98	nr	**678.83**
25 W	511.33	647.14	1.75	79.98	nr	**727.13**
40 W	541.86	685.78	1.75	79.98	nr	**765.76**
Floodlight; enclosed high performance LED; integeral control gear; reflector; toughened glass; IP65						
20 W	282.38	357.38	1.25	57.13	nr	**414.51**
35 W	320.54	405.67	1.25	57.13	nr	**462.80**
50 W	373.96	473.28	1.25	57.13	nr	**530.41**
65 W	419.75	531.24	1.25	57.13	nr	**588.37**
80 W	465.54	589.19	1.25	57.13	nr	**646.32**
Extra for Photocell	38.16	48.29	–	–	nr	**48.29**

ELECTRICAL SUPPLY/POWER/LIGHTING SYSTEMS

Item	Net Price £	Material £	Labour hours	Labour £	Unit	Total rate £
GENERAL LIGHTING – cont						
Lighting track						
Single circuit; 25 A 2 P&E steel trunking; low voltage with copper conductors; including couplers and supports; suspended						
Straight track/m	33.58	42.50	0.50	22.85	m	**65.35**
Live end feed unit complete with end stop	9.92	12.56	0.33	15.08	nr	**27.64**
Flexible couplers 0.5 m	13.00	16.46	0.33	15.08	nr	**31.54**
Tap off complete with 0.8 m of cable	14.27	18.06	0.25	11.43	nr	**29.49**
Three circuit; 25 A 2 P&E steel trunking; low voltage with copper conductors; including couplers and supports incorporating integral twisted pair comms bus bracket; fixed to backgrounds						
Straight track	38.69	48.97	0.75	34.28	nr	**83.25**
Live end feed unit complete with end stop	11.45	14.49	0.50	22.85	nr	**37.34**
Flexible couplers 0.5 m	15.10	19.11	0.45	20.57	nr	**39.67**
Tap off complete with 0.8 m of cable	15.80	19.99	0.25	11.43	nr	**31.42**
LIGHTING ACCESSORIES						
Switches						
6 amp metal clad surface mounted switch, gridswitch; one way						
1 gang	16.95	21.46	0.43	19.65	nr	**41.11**
2 gang	23.13	29.28	0.55	25.14	nr	**54.42**
3 gang	40.61	51.40	0.77	35.19	nr	**86.59**
4 gang	43.05	54.49	0.88	40.22	nr	**94.71**
6 gang	74.17	93.88	1.00	45.70	nr	**139.58**
8 gang	102.13	129.26	1.28	58.50	nr	**187.76**
10 gang	161.11	203.90	1.67	76.32	nr	**280.23**
Extra for						
10 amp – two way switch	6.05	7.66	0.03	1.37	nr	**9.03**
20 amp – two way switch	7.88	9.97	0.04	1.83	nr	**11.80**
20 amp – intermediate	14.80	18.73	0.08	3.66	nr	**22.39**
20 amp – one way SP switch	5.87	7.43	0.08	3.66	nr	**11.09**
Steel blank plate; 1 gang	3.58	4.53	0.07	3.20	nr	**7.73**
Steel blank plate; 2 gang	6.07	7.69	0.08	3.66	nr	**11.34**
6 amp modular type switch; galvanized steel box, bronze or satin chrome coverplate; metalclad switches; flush mounting; one way						
1 gang	46.70	59.11	0.43	19.65	nr	**78.76**
2 gang	52.90	66.95	0.55	25.14	nr	**92.09**
3 gang	76.42	96.71	0.77	35.19	nr	**131.90**
4 gang	110.07	139.30	0.88	40.22	nr	**179.52**
6 gang	185.56	234.85	1.18	53.93	nr	**288.78**
8 gang	197.39	249.81	1.63	74.50	nr	**324.31**
9 gang	277.65	351.39	1.83	83.64	nr	**435.03**
12 gang	281.09	355.75	2.29	104.66	nr	**460.41**

ELECTRICAL SUPPLY/POWER/LIGHTING SYSTEMS

Item	Net Price £	Material £	Labour hours	Labour £	Unit	Total rate £
6 amp modular type swtich; galvanized steel box; bronze or satin chrome coverplate; flush mounting; two way						
1 gang	48.56	61.46	0.43	19.65	nr	**81.12**
2 gang	66.55	84.22	0.55	25.14	nr	**109.36**
3 gang	96.13	121.67	0.77	35.19	nr	**156.86**
4 gang	102.06	129.17	0.88	40.22	nr	**169.39**
6 gang	192.97	244.23	1.18	53.93	nr	**298.16**
8 gang	199.51	252.50	1.63	74.50	nr	**327.00**
9 gang	247.87	313.71	1.83	83.64	nr	**397.34**
12 gang	292.34	369.98	2.22	101.46	nr	**471.44**
Plate switches; 10 amp flush mounted, white plastic fronted; 16 mm metal box; fitted brass earth terminal						
1 gang, 1 way, single pole	19.17	24.26	0.28	12.80	nr	**37.06**
1 gang, 2 way, single pole	21.47	27.18	0.33	15.08	nr	**42.26**
2 gang, 2 way, single pole	26.83	33.95	0.44	20.11	nr	**54.06**
3 gang, 2 way, single pole	31.49	39.85	0.56	25.59	nr	**65.45**
1 gang intermediate	42.46	53.74	0.43	19.65	nr	**73.39**
1 gang, 1 way, double pole	36.71	46.46	0.33	15.08	nr	**61.55**
1 gang single pole with bell symbol	29.37	37.18	0.23	10.51	nr	**47.69**
1 gang single pole marked PRESS	23.50	29.74	0.23	10.51	nr	**40.25**
Time delay switch, suppressed	130.06	164.60	0.49	22.39	nr	**187.00**
Plate switches; 6 amp flush mounted white plastic fronted; 25 mm metal box; fitted brass earth terminal						
4 gang, 2 way, single pole	57.33	72.55	0.42	19.20	nr	**91.75**
6 gang, 2 way, single way	91.04	115.22	0.47	21.48	nr	**136.70**
Architrave plate switches; 6 amp flush mounted, white plastic fronted; 27 mm metal box; brass earth terminal						
1 gang, 2 way, single pole	7.68	9.71	0.30	13.71	nr	**23.43**
2 gang, 2 way, single pole	15.56	19.69	0.36	16.45	nr	**36.14**
Ceiling switches, white moulded plastic, pull cord; standard unit						
6 amp, 1 way, single pole	8.44	10.68	0.32	14.63	nr	**25.31**
6 amp, 2 way, single pole	9.85	12.46	0.34	15.54	nr	**28.00**
16 amp, 1 way, double pole	14.67	18.56	0.37	16.91	nr	**35.47**
45 amp, 1 way, double pole with neon indicator	27.01	34.18	0.47	21.48	nr	**55.66**
10 amp splash proof moulded switch with plain, threaded or PVC entry						
1 gang, 2 way; single pole	26.87	34.01	0.34	15.54	nr	**49.55**
2 gang, 1 way; single pole	30.37	38.43	0.36	16.45	nr	**54.88**
2 gang, 2 way; single pole	38.47	48.68	0.40	18.28	nr	**66.96**
6 amp watertight switch; metalclad; BS 3676; ingress protected to IP65 surface mounted						
1 gang, 2 way; terminal entry	36.01	45.57	0.41	18.74	nr	**64.31**
1 gang, 2 way; through entry	36.01	45.57	0.42	19.20	nr	**64.77**
2 gang, 2 way; terminal entry	43.90	55.56	0.54	24.68	nr	**80.24**
2 gang, 2 way; through entry	43.90	55.56	0.53	24.22	nr	**79.78**
2 way replacement switch	32.37	40.97	0.10	4.57	nr	**45.54**
15 amp watertight switch; metalclad; BS 3676; ingress protected to IP65; surface mounted						
1 gang, 2 way; terminal entry	54.59	69.09	0.42	19.20	nr	**88.28**
1 gang, 2 way; through entry	54.59	69.09	0.43	19.65	nr	**88.74**
2 gang, 2 way; terminal entry	128.13	162.16	0.54	24.68	nr	**186.84**

ELECTRICAL SUPPLY/POWER/LIGHTING SYSTEMS

Item	Net Price £	Material £	Labour hours	Labour £	Unit	Total rate £
GENERAL LIGHTING – cont						
Switches – cont						
15 amp watertight switch – cont						
2 gang, 2 way; through entry	128.13	162.16	0.55	25.14	nr	**187.30**
Intermediate interior only	32.37	40.97	0.11	5.03	nr	**46.00**
2 way interior only	32.37	40.97	0.11	5.03	nr	**46.00**
Double pole interior only	32.37	40.97	0.11	5.03	nr	**46.00**
Electrical accessories; fixed to backgrounds						
(Including fixings)						
Dimmer switches; rotary action; for individual lights;						
moulded plastic case; metal backbox; flush mounted						
1 gang, 1 way; 250 watt	27.10	34.29	0.28	12.80	nr	**47.09**
1 gang, 1 way; 400 watt	35.70	45.18	0.28	12.80	nr	**57.97**
Dimmer switches; push on/off action; for individual lights;						
moulded plastic case; metal backbox; flush mounted						
1 gang, 2 way; 250 watt	41.42	52.42	0.34	15.54	nr	**67.96**
3 gang, 2 way; 250 watt	61.05	77.27	0.48	21.94	nr	**99.20**
4 gang, 2 way; 250 watt	80.33	101.67	0.57	26.05	nr	**127.72**
Dimmer switches; rotary action; metal cald; metal						
backbox; BS 5518 and BS 800; flush mounted						
1 gang, 1 way; 400 watt	75.24	95.22	0.33	15.08	nr	**110.31**
Ceiling roses						
Ceiling rose: white moulded plastic; flush fixed to conduit						
box						
Plug in type; ceiling socket with 2 terminals,loop-in						
and ceiling plug with 3 terminals and cover	18.21	23.05	0.34	15.54	nr	**38.59**
BC lampholder; white moulded plastic; heat resistent						
PVC insulated and sheathed cable; flush fixed						
2 core; 0.75 mm²	5.76	7.29	0.33	15.08	nr	**22.38**
Batten holder: white moulded plastic; 3 terminals; BS						
5042; fixed to conduit						
Straight pattern; 2 terminals withloop-in and Earth	8.71	11.02	0.29	13.25	nr	**24.28**
Angled pattern; looped in terminal	–	11.02	0.29	13.25	nr	**24.28**
LIGHTING CONTROL MODULES						
SPV.92 Lighting control module; plug in; 9 output, 9						
channel – switching						
Base and lid assembly	160.63	203.29	2.05	93.69	nr	**296.99**
SPV.92 Lighting control module; plug in; 9 output, 9						
channel – dimming (DALI, DSI, 1–10 V)						
Base and lid assembly	160.81	203.52	2.05	93.69	nr	**297.21**
SPH.12+ Lighting control module; hard wired; 4 circuit						
switching						
Base and lid assembly	256.04	324.04	1.85	84.55	nr	**408.60**
SPH.27r Lighting control module; hard wired; 2 circuit 40						
ballast drive DALI (with relays)						
Base and lid assembly	650.79	823.65	1.85	84.55	nr	**908.20**

ELECTRICAL SUPPLY/POWER/LIGHTING SYSTEMS

Item	Net Price £	Material £	Labour hours	Labour £	Unit	Total rate £
SPH.27 Lighting control module; hard wired; 2 circuit 40 ballast drive DALI (without relays)						
Base and lid assembly	238.06	301.29	1.85	84.55	nr	**385.84**
SPV.56 Compact lighting control module; 3 output 18 ballast drive; dimmable (DALI)						
Base and lid assembly	154.97	196.14	1.85	84.55	nr	**280.69**
SPB.82 Blind control module; 8 outputs						
Base and lid assembly	220.92	279.60	1.85	84.55	nr	**364.15**
Interfaces						
SPO.20 DALI to DMX converter	956.57	1210.63	1.85	84.55	nr	**1295.18**
Dual bus presence detectors (DALI &E-Bus)						
SPU.7-S Dual-bus presence detector and infra-red sensor	47.21	59.75	0.60	27.42	nr	**87.17**
SPU.6-S Dual-bus universal sensor	47.36	59.94	0.60	27.42	nr	**87.36**
SPU.6-C Compact Dual-bus Universal Sensor	–	122.72	0.60	27.42	nr	**150.14**
SPU.6-C1 Compact Dual-bus Universal Sensor conduit/surface mount	53.01	67.08	0.60	27.42	nr	**94.51**
Scene switch plate; anodized aluminium finish						
SPK.9–4+4 Eight button intelligent switch plate	208.27	263.59	2.00	91.41	nr	**355.00**
SPK.9–4-S Four button intelligent switch plate	206.77	261.68	1.60	73.13	nr	**334.81**
SPK.9–2-S Two button intelligent switch plate	112.87	142.85	1.20	54.84	nr	**197.69**
Input Devices/Interfaces						
SPF.4d Compact DALI interface for 4 gang retractive switch plate	175.18	221.70	1.20	54.84	nr	**276.55**
SPL.20 External Photocell 20,000 Lux range	80.43	101.79	1.50	68.56	nr	**170.35**
SPF.6-T AV interface; RS232 interface	220.43	278.98	1.85	84.55	nr	**363.53**
Sub-addressing for DALI strings						
Sub-addressing charge per SPH.27R (40 Sub-addresses)	628.30	795.18	3.00	137.11	nr	**932.29**
Sub-addressing charge per SPV.56 (20 Sub-addresses)	626.71	793.16	3.00	137.11	nr	**930.27**
Commissioning						
Not included – worked out based on complexity of project						

ELECTRICAL SUPPLY/POWER/LIGHTING SYSTEMS

Item	Net Price £	Material £	Labour hours	Labour £	Unit	Total rate £
GENERAL LV POWER						
ACCESSORIES						
OUTLETS						
Socket outlet: unswitched; 13 amp metal clad; BS 1363; galvanized steel box and coverplate with white plastic inserts; fixed surface mounted						
1 gang	20.11	25.45	0.41	18.74	nr	**44.18**
2 gang	33.50	42.39	0.41	18.74	nr	**61.13**
Socket outlet: switched; 13 amp metal clad; BS 1363; galvanized steel box and coverplate with white plastic inserts; fixed surface mounted						
1 gang	23.49	29.73	0.43	19.65	nr	**49.38**
2 gang	42.76	54.12	0.45	20.57	nr	**74.68**
Socket outlet: switched with neon indicator; 13 amp metal clad; BS 1363; galvanized steel box and coverplate with white plastic inserts; fixed surface mounted						
1 gang	49.72	62.92	0.43	19.65	nr	**82.57**
2 gang	90.39	114.40	0.45	20.57	nr	**134.97**
Socket outlet: unswitched; 13 amp; BS 1363; white moulded plastic box and coverplate; fixed surface mounted						
1 gang	11.69	14.79	0.41	18.74	nr	**33.53**
2 gang	23.06	29.18	0.41	18.74	nr	**47.92**
Socket outlet; switched; 13 amp; BS 1363; white moulded plastic box and coverplate; fixed surface mounted						
1 gang	8.53	10.79	0.43	19.65	nr	**30.45**
2 gang	13.69	17.32	0.45	20.57	nr	**37.89**
Socket outlet: switched with neon indicator; 13 amp; BS 1363; white moulded plastic box and coverplate; fixed surface mounted						
1 gang	38.94	49.28	0.43	19.65	nr	**68.93**
2 gang	52.50	66.44	0.45	20.57	nr	**87.01**
Socket outlet: switched; 13 amp; BS 1363; galvanized steel box, white moulded coverplate; flush fitted						
1 gang	10.62	13.44	0.43	19.65	nr	**33.09**
2 gang	29.30	37.09	0.45	20.57	nr	**57.65**
Socket outlet: switched with neon indicator; 13 amp; BS 1363; galvanized steel box, white moulded coverplate; flush fixed						
1 gang	38.94	49.28	0.43	19.65	nr	**68.93**
2 gang	67.33	85.21	0.45	20.57	nr	**105.77**
Socket outlet: switched; 13 amp; BS 1363; galvanized steel box, satin chrome coverplate; BS 4662; flush fixed						
1 gang	44.73	56.61	0.43	19.65	nr	**76.26**
2 gang	56.58	71.61	0.45	20.57	nr	**92.17**

ELECTRICAL SUPPLY/POWER/LIGHTING SYSTEMS

Item	Net Price £	Material £	Labour hours	Labour £	Unit	Total rate £
Socket outlet: switched with neon indicator; 13 amp; BS 1363; steel backbox, satin chrome coverplate; BS 4662; flush fixed						
1 gang	56.34	71.30	0.43	19.65	nr	**90.95**
2 gang	101.59	128.57	0.45	20.57	nr	**149.14**
RCD protected socket outlets, 13 amp, to BS 1363; galvanized steel box, white moulded cover plate; flush fitted						
2 gang, 10 mA tripping (active control)	252.99	320.18	0.45	20.57	nr	**340.75**
2 gang, 30 mA tripping (active control)	225.37	285.23	0.45	20.57	nr	**305.79**
2 gang, 30 mA tripping (passive control)	225.37	285.23	0.45	20.57	nr	**305.79**
Filtered socket outlets, 13 amp, to BS 1363, with separate 'clean earth' terminal; galvanized steel box, white moulded cover plate; flush fitted						
2 gang (spike protected)	190.49	241.09	0.50	22.85	nr	**263.94**
2 gang (spike and RFI protected)	241.12	305.16	0.55	25.14	nr	**330.30**
Replacement filter cassette	66.83	84.58	0.15	6.86	nr	**91.43**
Non-standard socket outlets, 13 amp, to BS 1363, with separate 'clean earth' terminal; for plugs with T-shaped earth pin; galvanized steel box, white moulded cover plate; flush fitted						
1 gang	35.36	44.75	0.43	19.65	nr	**64.40**
2 gang	63.63	80.53	0.43	19.65	nr	**100.18**
2 gang coloured RED	88.37	111.84	0.43	19.65	nr	**131.49**
Weatherproof socket outlet: 40 amp; switched; single gang; RCD protected; water and dust protected to I.P.66; surface mounted						
40 A 30 mA tripping current protecting 1 socket	114.10	144.41	0.52	23.77	nr	**168.18**
40 A 30 mA tripping current protecting 2 sockets	256.72	324.90	0.64	29.25	nr	**354.15**
Plug for weatherproof socket outlet: protected to I.P.66						
13 amp plug	11.24	14.22	0.21	9.60	nr	**23.82**
Floor service outlet box; comprising flat lid with flanged carpet trim; twin 13 A switched socket outlets; punched plate for mounting 2 telephone outlets; one blank plate; triple compartment						
3 compartment	118.91	150.50	0.88	40.22	nr	**190.72**
Floor service outlet box; comprising flat lid with flanged carpet trim; 2 twin 13 A switched socket outlets; punched plate for mounting 1 telephone outlet; one blank plate; triple compartment						
3 compartment	106.12	134.31	0.88	40.22	nr	**174.53**
Floor service outlet box; comprising flat lid with flanged carpet trim; twin 13 A switched socket outlets; punched plate for mounting 2 telephone outlets; two blank plates; four compartment						
4 compartment	139.57	176.63	0.88	40.22	nr	**216.85**
Floor service outlet box; comprising flat lid with flanged carpet trim; single 13 A unswitched socket outlet; single compartment; circular						
1 compartment	148.88	188.43	0.79	36.11	nr	**224.53**

ELECTRICAL SUPPLY/POWER/LIGHTING SYSTEMS

Item	Net Price £	Material £	Labour hours	Labour £	Unit	Total rate £
GENERAL LV POWER – cont						
OUTLETS – cont						
Floor service grommet, comprising flat lid with flanged carpet trim; circular						
Floor grommet	67.34	85.23	0.49	22.39	nr	**107.62**
POWER POSTS/POLES/PILLARS						
Power post						
Power post; aluminium painted body; PVC-u cover; 5 nr outlets	717.96	908.65	4.00	182.81	nr	**1091.47**
Power pole						
Power pole; 3.6 metres high; aluminium painted body; PVC-u cover; 6 nr outlets	920.64	1165.16	4.00	182.81	nr	**1347.98**
Extra for						
Power pole extension bar; 900 mm long	100.60	127.32	1.50	68.56	nr	**195.88**
Vertical multi-compartment pillar; PVC-u; BS 4678 Part 4 EN60529; excludes accessories						
Single						
630 mm long	320.67	405.84	2.00	91.41	nr	**497.24**
3000 mm long	924.10	1169.54	2.00	91.41	nr	**1260.94**
Double						
630 mm long	320.67	405.84	3.00	137.11	nr	**542.95**
3000 mm long	980.78	1241.27	3.00	137.11	nr	**1378.38**
CONNECTION UNITS						
Connection units: moulded pattern; BS 5733; moulded plastic box; white coverplate; knockout for flex outlet; surface mounted – standard fused						
DP switched	20.46	25.89	0.49	22.39	nr	**48.29**
DP unswitched	18.80	23.80	0.49	22.39	nr	**46.19**
DP switched with neon indicator	25.96	32.86	0.49	22.39	nr	**55.26**
Connection units: moulded pattern; BS 5733; galvanized steel box; white coverplate; knockout for flex outlet; surface mounted						
DP switched	25.78	32.63	0.49	22.39	nr	**55.02**
DP unswitched	24.12	30.53	0.49	22.39	nr	**52.93**
DP switched with neon indicator	31.30	39.62	0.49	22.39	nr	**62.01**
Connection units: galvanized pressed steel pattern; galvanized steel box; satin chrome or satin brass finish; white moulded plastic inserts; flush mounted – standard fused						
DP switched	35.65	45.12	0.49	22.39	nr	**67.52**
DP unswitched	33.51	42.42	0.49	22.39	nr	**64.81**
DP switched with neon indicator	48.11	60.89	0.49	22.39	nr	**83.29**
Connection units: galvanized steel box; satin chrome or satin brass finish; white moulded plastic inserts; flex outlet; flush mounted – standard fused						
DP switched	34.20	43.28	0.49	22.39	nr	**65.68**
DP unswitched	32.44	41.06	0.49	22.39	nr	**63.46**
DP switched with neon indicator	44.01	55.69	0.49	22.39	nr	**78.09**

ELECTRICAL SUPPLY/POWER/LIGHTING SYSTEMS

Item	Net Price £	Material £	Labour hours	Labour £	Unit	Total rate £
SHAVER SOCKETS						
Shaver unit: self-setting overload device; 200/250 voltage supply; white moulded plastic faceplate; unswitched						
Surface type with moulded plastic box	61.05	77.26	0.55	25.14	nr	**102.40**
Flush type with galvanized steel box	64.09	81.11	0.57	26.05	nr	**107.16**
Shaver unit: dual voltage supply unit; white moulded plastic faceplate; unswitched						
Surface type with moulded plastic box	73.61	93.17	0.62	28.34	nr	**121.50**
Flush type with galvanized steel box	76.52	96.85	0.64	29.25	nr	**126.10**
COOKER CONTROL UNITS						
Cooker control unit: BS 4177; 45 amp DP main switch; 13 amp switched socket outlet; metal coverplate; plastic inserts; neon indicators						
Surface mounted with mounting box	94.87	120.07	0.61	27.88	nr	**147.95**
Flush mounted with galvanized steel box	90.68	114.77	0.61	27.88	nr	**142.64**
Cooker control unit: BS 4177; 45 amp DP main switch; 13 amp switched socket outlet; moulded plastic box and coverplate; surface mounted						
Standard	65.83	83.32	0.61	27.88	nr	**111.20**
With neon indicators	77.24	97.75	0.61	27.88	nr	**125.63**
CONTROL COMPONENTS						
Connector unit: moulded white plastic cover and block; galvanized steel back box; to immersion heaters						
3 kW up to 915 mm long; fitted to thermostat	58.17	73.62	0.75	34.29	nr	**107.91**
Water heater switch: 20 amp; switched with neon indicator						
DP switched with neon indicator	31.94	40.43	0.45	20.57	nr	**60.99**
SWITCH DISCONNECTORS						
Switch disconnectors; moulded plastic enclosure; fixed to backgrounds						
3 pole; IP54; Grey						
16 amp	60.30	76.31	0.80	36.56	nr	**112.87**
25 amp	71.39	90.35	0.80	36.56	nr	**126.91**
40 amp	116.25	147.13	0.80	36.56	nr	**183.69**
63 amp	180.95	229.02	1.00	45.70	nr	**274.72**
80 amp	313.71	397.03	1.25	57.13	nr	**454.16**
6 pole; IP54; Grey						
25 amp	100.30	126.94	1.00	45.70	nr	**172.64**
63 amp	169.37	214.36	1.25	57.13	nr	**271.49**
80 amp	321.06	406.34	1.80	82.27	nr	**488.60**
3 pole; IP54; Yellow						
16 amp	66.27	83.87	0.80	36.56	nr	**120.44**
25 amp	78.35	99.17	0.80	36.56	nr	**135.73**
40 amp	127.12	160.88	0.80	36.56	nr	**197.45**
63 amp	197.77	250.29	1.00	45.70	nr	**296.00**

ELECTRICAL SUPPLY/POWER/LIGHTING SYSTEMS

Item	Net Price £	Material £	Labour hours	Labour £	Unit	Total rate £
GENERAL LV POWER – cont						
SWITCH DISCONNECTORS – cont						
6 pole; IP54; Yellow						
25 amp	98.39	124.52	1.00	45.70	nr	**170.22**
INDUSTRIAL SOCKETS/PLUGS						
Plugs; Splashproof; 100–130 volts, 50–60 Hz; IP 44 (Yellow)						
2 pole and earth						
16 amp	9.00	11.40	0.55	25.14	nr	**36.53**
32 amp	29.91	37.86	0.60	27.42	nr	**65.28**
3 pole and earth						
16 amp	16.35	20.70	0.65	29.71	nr	**50.40**
32 amp	17.23	21.81	0.72	32.91	nr	**54.71**
3 pole; neutral and earth						
16 amp	12.91	16.34	0.72	32.91	nr	**49.25**
32 amp	19.66	24.88	0.78	35.65	nr	**60.53**
Connectors; Splashproof; 100–130 volts, 50–60 Hz; IP 44 (Yellow)						
2 pole and earth						
16 amp	16.22	20.52	0.42	19.20	nr	**39.72**
32 amp	28.17	35.65	0.50	22.85	nr	**58.50**
3 pole and earth						
16 amp	30.52	38.62	0.48	21.94	nr	**60.56**
32 amp	32.92	41.67	0.58	26.51	nr	**68.18**
3 pole; neutral and earth						
16 amp	42.76	54.12	0.52	23.77	nr	**77.89**
32 amp	43.35	54.86	0.73	33.36	nr	**88.22**
Angled sockets; surface mounted; Splashproof; 100–130 volts, 50–60 Hz; IP 44 (Yellow)						
2 pole and earth						
16 amp	25.83	32.69	0.55	25.14	nr	**57.83**
32 amp	53.22	67.36	0.60	27.42	nr	**94.78**
3 pole and earth						
16 amp	28.73	36.36	0.65	29.71	nr	**66.06**
32 amp	47.59	60.23	0.72	32.91	nr	**93.14**
3 pole; neutral and earth						
16 amp	37.06	46.90	0.72	32.91	nr	**79.81**
32 amp	46.00	58.22	0.78	35.65	nr	**93.87**
Plugs; Watertight; 100–130 volts, 50–60 Hz; IP67 (Yellow)						
2 pole and earth						
16 amp	25.66	32.48	0.55	25.14	nr	**57.61**
32 amp	44.71	56.59	0.60	27.42	nr	**84.01**
63 amp	122.61	155.17	0.75	34.28	nr	**189.45**

ELECTRICAL SUPPLY/POWER/LIGHTING SYSTEMS

Item	Net Price £	Material £	Labour hours	Labour £	Unit	Total rate £
Connectors; Watertight; 100–130 volts, 50–60 Hz; IP 67 (Yellow)						
2 pole and earth						
16 amp	52.42	66.34	0.42	19.20	nr	**85.53**
32 amp	88.82	112.42	0.50	22.85	nr	**135.27**
63 amp	214.99	272.09	0.67	30.62	nr	**302.71**
Angled sockets; surface mounted; Watertight; 100–130 volts, 50–60 Hz; IP 67 (Yellow)						
2 pole and earth						
16 amp	43.31	54.81	0.55	25.14	nr	**79.95**
32 amp	84.70	107.20	0.60	27.42	nr	**134.62**
Plugs; Splashproof; 200–250 volts, 50–60 Hz; IP 44 (Blue)						
2 pole and earth						
16 amp	6.18	7.82	0.55	25.14	nr	**32.95**
32 amp	22.16	28.04	0.60	27.42	nr	**55.46**
63 amp	106.75	135.10	0.75	34.28	nr	**169.38**
3 pole and earth						
16 amp	24.52	31.04	0.65	29.71	nr	**60.75**
32 amp	32.94	41.69	0.72	32.91	nr	**74.60**
63 amp	107.13	135.59	0.83	37.93	nr	**173.52**
3 pole; neutral and earth						
16 amp	26.15	33.09	0.72	32.91	nr	**66.00**
32 amp	39.67	50.20	0.78	35.65	nr	**85.85**
Connectors; Splashproof; 200–250 volts, 50–60 Hz; IP 44 (Blue)						
2 pole and earth						
16 amp	9.00	11.39	0.42	19.20	nr	**30.58**
32 amp	21.24	26.88	0.50	22.85	nr	**49.73**
63 amp	110.51	139.87	0.67	30.62	nr	**170.49**
3 pole and earth						
16 amp	28.19	35.68	0.48	21.94	nr	**57.62**
32 amp	38.83	49.14	0.58	26.51	nr	**75.65**
63 amp	54.12	68.49	0.75	34.28	nr	**102.77**
3 pole; neutral and earth						
16 amp	30.92	39.13	0.52	23.77	nr	**62.89**
32 amp	32.10	40.62	0.73	33.36	nr	**73.98**
Angled sockets; surface mounted; Splashproof; 200–250 volts, 50–60 Hz; IP 44 (Blue)						
2 pole and earth						
16 amp	23.65	29.94	0.55	25.14	nr	**55.07**
32 amp	33.10	41.89	0.60	27.42	nr	**69.31**
63 amp	158.98	201.21	0.75	34.28	nr	**235.49**
3 pole and earth						
16 amp	45.63	57.75	0.65	29.71	nr	**87.46**
32 amp	74.98	94.90	0.72	32.91	nr	**127.80**
63 amp	92.03	116.47	0.83	37.93	nr	**154.41**

ELECTRICAL SUPPLY/POWER/LIGHTING SYSTEMS

Item	Net Price £	Material £	Labour hours	Labour £	Unit	Total rate £
GENERAL LV POWER – cont						
Angled sockets – cont						
3 pole; neutral and earth						
16 amp	52.38	66.30	0.72	32.91	nr	**99.20**
32 amp	77.65	98.27	0.78	35.65	nr	**133.92**
Plugs; Watertight; 200–250 volts, 50–60 Hz; IP67 (Blue)						
2 pole and earth						
16 amp	26.19	33.15	0.41	18.74	nr	**51.88**
32 amp	48.09	60.86	0.50	22.85	nr	**83.71**
63 amp	131.97	167.02	0.66	30.16	nr	**197.18**
125 amp	338.14	427.95	0.86	39.31	nr	**467.26**
Connectors; Watertight; 200–250 volts, 50–60 Hz; IP 67 (Blue)						
2 pole and earth						
16 amp	26.19	33.15	0.41	18.74	nr	**51.88**
32 amp	44.64	56.50	0.50	22.85	nr	**79.35**
63 amp	114.61	145.05	0.67	30.62	nr	**175.67**
125 amp	455.62	576.63	0.87	39.76	nr	**616.39**
Angled sockets; surface mounted; Watertight; 200–250 volts, 50–60 Hz; IP 67 (Blue)						
2 pole and earth						
16 amp	43.31	54.81	0.55	25.14	nr	**79.95**
32 amp	84.68	107.17	0.60	27.42	nr	**134.60**
125 amp	616.79	780.61	1.00	45.70	nr	**826.31**

ELECTRICAL SUPPLY/POWER/LIGHTING SYSTEMS

Item	Net Price £	Material £	Labour hours	Labour £	Unit	Total rate £
UNINTERRUPTIBLE POWER SUPPLY						
Uninterruptible power supply; sheet steel enclosure; self-contained battery pack; including installation, testing and commissioning						
Single phase input and output; 5 year battery life; standard 13 A socket outlet connection						
1.0 kVA (10 minute supply)	1492.56	1888.98	0.30	13.71	nr	**1902.70**
1.0 kVA (30 minute supply)	2127.13	2692.09	0.50	22.85	nr	**2714.94**
2.0 kVA (10 minute supply)	2662.86	3370.11	0.50	22.85	nr	**3392.96**
2.0 kVA (60 minute supply)	4469.04	5656.02	0.50	22.85	nr	**5678.87**
3.0 kVA (10 minute supply)	3518.72	4453.29	0.50	22.85	nr	**4476.15**
3.0 kVA (40 minute supply)	4543.52	5750.28	1.00	45.70	nr	**5795.98**
5.0 kVA (30 minute supply)	5759.50	7289.23	1.00	45.70	nr	**7334.93**
8.0 kVA (10 minute supply)	6928.02	8768.11	2.00	91.41	nr	**8859.51**
8.0 kVA (30 minute supply)	9368.29	11856.51	2.00	91.41	nr	**11947.91**
Uninterruptible power supply; including final connections and testing and commissioning						
Medium-sized static; single phase input and output; 10 year battery life; in cubicle						
10.0 kVA (10 minutes supply)	7096.61	8981.47	10.00	457.04	nr	**9438.50**
10.0 kVA (30 minutes supply)	9596.35	12145.14	15.00	685.55	nr	**12830.69**
15.0 kVA (10 minutes supply)	11583.40	14659.96	10.00	457.04	nr	**15116.99**
15.0 kVA (30 minutes supply)	16010.82	20263.29	15.00	685.55	nr	**20948.84**
20.0 kVA (10 minutes supply)	11583.40	14659.96	10.00	457.04	nr	**15116.99**
20.0 kVA (30 minutes supply)	20609.13	26082.91	15.00	685.55	nr	**26768.47**
Medium-sized static; three phase input and output; 10 year battery life; in cubicle						
10.0 kVA (10 minutes supply)	9578.70	12122.80	10.00	457.04	nr	**12579.84**
10.0 kVA (30 minutes supply)	11999.40	15186.44	15.00	685.55	nr	**15871.99**
15.0 kVA (10 minutes supply)	10924.39	13825.91	15.00	685.55	nr	**14511.46**
15.0 kVA (30 minutes supply)	13499.87	17085.43	20.00	914.07	nr	**17999.50**
20.0 kVA (10 minutes supply)	11029.38	13958.79	20.00	914.07	nr	**14872.86**
20.0 kVA (30 minutes supply)	16687.95	21120.27	25.00	1142.59	nr	**22262.85**
30.0 kVA (10 minutes supply)	16505.32	20889.13	25.00	1142.59	nr	**22031.72**
30.0 kVA (30 minutes supply)	20748.92	26259.84	30.00	1371.11	nr	**27630.94**
Large-sized static; three phase input and output; 10 year battery life; in cubicle						
40 kVA (10 minutes supply)	16505.32	20889.13	30.00	1371.11	nr	**22260.24**
40 kVA (30 minutes supply)	24930.89	31552.54	30.00	1371.11	nr	**32923.64**
60 kVA (10 minutes supply)	26549.22	33600.69	35.00	1599.63	nr	**35200.32**
60 kVA (30 minutes supply)	30922.66	39135.71	35.00	1599.63	nr	**40735.34**
100 kVA (10 minutes supply)	39017.37	49380.38	40.00	1828.14	nr	**51208.53**
200 kVA (10 minutes supply)	64391.81	81494.28	40.00	1828.14	nr	**83322.42**
300 kVA (10 minutes supply)	104792.70	132625.64	40.00	1828.14	nr	**134453.79**
400 kVA (10 minutes supply)	122956.01	155613.13	50.00	2285.18	nr	**157898.30**
500 kVA (10 minutes supply)	144298.19	182623.78	60.00	2742.21	nr	**185366.00**
600 kVA (10 minutes supply)	194131.62	245692.98	70.00	3199.25	nr	**248892.23**
800 kVA (10 minutes supply)	242446.78	306840.64	80.00	3656.29	nr	**310496.93**

ELECTRICAL SUPPLY/POWER/LIGHTING SYSTEMS

Item	Net Price £	Material £	Labour hours	Labour £	Unit	Total rate £
UNINTERRUPTIBLE POWER SUPPLY – cont						
Uninterruptible power supply – cont						
Integral diesel rotary; 400 V three phase input and output; no break supply; including ventilation and acoustic attenuation oil day tank and interconnecting pipework						
300 kVA	722257.90	914089.60	120.00	5484.43	nr	**919574.03**
500 kVA	743837.22	941400.39	120.00	5484.43	nr	**946884.82**
800 kVA	812172.10	1027885.02	140.00	6398.50	nr	**1034283.52**
1000 kVA	901052.25	1140371.73	140.00	6398.50	nr	**1146770.23**
1670 kVA	1073294.78	1358361.87	170.00	7769.61	nr	**1366131.48**
2000 kVA	1307532.93	1654813.67	200.00	9140.72	nr	**1663954.39**
2500 kVA	1511076.22	1912418.06	230.00	10511.82	nr	**1922929.88**
Uninterruptible power supply (UPS) consisting of a single cabinet, sheet steel enclosure with single 3 phase input and output. Internally consists of multiple power modules to provide the rated output, with an additional power module for redundancy. 10 or 30 minute back-up, 10 year design life VRLA battery either internal or external (sizes dependent), including delivery, testing and commissioning						
20 kVA N+1 (10 mins)	17148.87	21703.60	10.00	457.04	nr	**22160.64**
20 kVA N+1 (30 mins)	18911.02	23933.79	10.00	457.04	nr	**24390.82**
40 kVA N+1 (10 mins)	23135.64	29280.46	30.00	1371.11	nr	**30651.57**
40 kVA N+1 (30 mins)	29257.65	37028.48	30.00	1371.11	nr	**38399.59**
60 kVA N+1 (10 mins)	31611.65	40007.70	35.00	1599.63	nr	**41607.33**
60 kVA N+1 (30 mins)	38874.51	49199.58	35.00	1599.63	nr	**50799.20**
100 kVA N+1 (10 mins)	45711.56	57852.55	40.00	1828.14	nr	**59680.69**
160 kVA N+1 (10 mins)	66521.31	84189.37	40.00	1828.14	nr	**86017.52**
200 kVA N+1 (10 mins)	85005.18	107582.56	40.00	1828.14	nr	**109410.70**

ELECTRICAL SUPPLY/POWER/LIGHTING SYSTEMS

Item	Net Price £	Material £	Labour hours	Labour £	Unit	Total rate £
COMMERCIAL BATTERY STORAGE SYSTEM						
Battery storage system; self-contained compact battery pack; flexible layout; rated for outdoor use; includes labour, materials, civils, design and project management						
100 kW nominal/169 kWh batteries						
(1.6 hr capacity)	127745.97	161675.29	48.00	2193.77	nr	**163869.07**
500 kW nominal/845 kWh batteries						
(1.6 hr capacity)	465536.25	589182.68	58.00	2650.81	nr	**591833.49**
1000 kW nominal/1690 kWh batteries						
(1.6 hr capacity)	880990.56	1114981.65	76.00	3473.47	nr	**1118455.12**

ELECTRICAL SUPPLY/POWER/LIGHTING SYSTEMS

Item	Net Price £	Material £	Labour hours	Labour £	Unit	Total rate £
EMERGENCY LIGHTING						
24 volt/50 volt/110 volt fluorescent slave luminaires						
For use with DC central battery systems						
Indoor, 8 watt	63.76	80.69	0.80	36.56	nr	**117.26**
Indoor, exit sign box	79.42	100.52	0.80	36.56	nr	**137.08**
Outdoor, 8 watt weatherproof	71.06	89.93	0.80	36.56	nr	**126.49**
Conversion module AC/DC	79.42	100.52	0.25	11.43	nr	**111.94**
Self-contained; polycarbonate base and diffuser; LED charging light to European sign directive; 3 hour standby						
Non-maintained						
Indoor, 8 watt	70.39	89.09	1.00	45.70	nr	**134.79**
Outdoor, 8 watt weatherproof, vandal-resistant IP65	101.08	127.93	1.00	45.70	nr	**173.63**
Maintained						
Indoor, 8 watt	50.39	63.78	1.00	45.70	nr	**109.48**
Outdoor, 8 watt weatherproof, vandal-resistant IP65	149.09	188.69	1.00	45.70	nr	**234.40**
Exit signage						
Exit sign; gold effect, pendular including brackets						
Non-maintained, 8 watt	191.44	242.29	1.00	45.70	nr	**287.99**
Maintained, 8 watt	206.05	260.78	1.00	45.70	nr	**306.48**
Modification kit						
Module & battery for 58 W fluorescent modification from mains fitting to emergency; 3 hour standby	65.99	83.51	0.50	22.85	nr	**106.37**
Extra for remote box (when fitting is too small for modification)	32.77	41.47	0.50	22.85	nr	**64.32**
12 volt low voltage lighting; non-maintained; 3 hour standby						
2 × 20 watt lamp load	227.36	287.75	1.20	54.84	nr	**342.60**
1 × 50 watt lamp load	248.21	314.13	1.00	45.70	nr	**359.83**
Maintained 3 hour standby						
2 × 20 watt lamp load	215.23	272.40	1.20	54.84	nr	**327.24**
1 × 50 watt lamp load	236.06	298.76	1.00	45.70	nr	**344.46**
DC central battery systems BS5266 compliant 24/50/ 110 volt						
DC supply to luminaires on mains failure; metal cubicle with battery charger, changeover device and battery as integral unit; ICEL 1001 compliant; 10 year design life valve regulated lead acid battery; 24 hour recharge; LCD display & LED indication; ICEL alarm pack; Includes on-site commissioning on 110 volt systems only						
24 volt, wall mounted						
300 W maintained, 1 hour	3625.60	4588.57	4.00	182.81	nr	**4771.38**
635 W maintained, 3 hour	4398.53	5566.78	6.00	274.22	nr	**5841.00**
470 W non-maintained, 1 hour	3232.36	4090.88	4.00	182.81	nr	**4273.69**
780 W non-maintained, 3 hour	3818.83	4833.11	6.00	274.22	nr	**5107.34**

ELECTRICAL SUPPLY/POWER/LIGHTING SYSTEMS

Item	Net Price £	Material £	Labour hours	Labour £	Unit	Total rate £
50 volt						
935 W maintained, 3 hour	3473.08	4395.53	8.00	365.63	nr	**4761.16**
1965 W maintained, 3 hour	4834.14	6118.08	8.00	365.63	nr	**6483.71**
1311 W non-maintained, 3 hour	4732.45	5989.38	8.00	365.63	nr	**6355.01**
2510 W non-maintained 3, hour	7227.49	9147.11	8.00	365.63	nr	**9512.74**
110 volt						
1603 W maintained, 3 hour	6812.21	8621.54	8.00	365.63	nr	**8987.16**
4446 W maintained, 3 hour	7780.07	9846.46	10.00	457.04	nr	**10303.49**
2492 W non-maintained, 3 hour	8642.82	10938.35	10.00	457.04	nr	**11395.39**
5429 W non-maintained, 3 hour	11797.21	14930.55	12.00	548.44	nr	**15479.00**
DC central battery systems; BS EN 50171 compliant; 24/50/110 volt						
Central power systems						
DC supply to luminaires on mains failure; metal cubicle with battery charger, changeover device and battery as integral unit; 10 year design life valve regulated lead acid battery; 12 hour recharge to 80% of specified duty; low volts discount; LCD display & LED indication; includes on-site commissioning for CPS systems only; battery sized for 'end of life' @ 20°C test push button						
24 volt, floor standing						
400 W non-maintained, 1 hour	3405.26	4309.70	4.00	182.81	nr	**4492.51**
600 W maintained, 3 hour	4378.20	5541.05	6.00	274.22	nr	**5815.27**
50 volt						
2133 W non-maintained, 3 hour	7654.63	9687.70	8.00	365.63	nr	**10053.33**
1900 W maintained, 3 hour	7315.62	9258.65	8.00	365.63	nr	**9624.28**
110 volt						
2200 W non-maintained, 3 hour	7080.01	8960.47	8.00	365.63	nr	**9326.10**
4000 W maintained, 3 hour	7902.10	10000.90	12.00	548.44	nr	**10549.34**
Low power systems						
DC supply to luminaires on mains failure; metal cubicle with battery charger, changeover device and battery as integral unit; 5 year design life valve regulated lead acid battery; low volts discount; LED display & LED indication; battery sized for 'end of life' @ 20°C test push button						
24 volt, floor standing						
300 W non-maintained, 1 hour	3734.09	4725.87	4.00	182.81	nr	**4908.68**
600 W maintained, 3 hour	4134.11	5232.13	6.00	274.22	nr	**5506.35**
AC static inverter system; BS5266 compliant; one hour standby						
Central system supplying AC power on mains failure to mains luminaires; ICEL 1001 compliant metal cubicle(s) with changeover device, battery charger, battery & static inverter; 10 year design life valve regulated lead acid battery; 24 hour recharge; LED indication and LCD display; pure sinewave output						

ELECTRICAL SUPPLY/POWER/LIGHTING SYSTEMS

Item	Net Price £	Material £	Labour hours	Labour £	Unit	Total rate £
EMERGENCY LIGHTING – cont						
AC static inverter system – cont						
One hour						
750 VA, 600 W single phase I/P & O/P	3821.34	4836.28	6.00	274.22	nr	**5110.51**
3 kVA, 2.55 kW single phase I/P & O/P	6268.57	7933.50	8.00	365.63	nr	**8299.13**
5 kVA, 4.25 kW single phase I/P & O/P	8278.33	10477.06	10.00	457.04	nr	**10934.09**
8 kVA, 6.80 kW single phase I/P & O/P	9752.23	12342.42	12.00	548.44	nr	**12890.87**
10 kVA, 8.5 kW single phase I/P & O/P	12589.98	15933.88	14.00	639.85	nr	**16573.73**
13 kVA, 11.05 kW single phase I/P & O/P	16842.91	21316.38	16.00	731.26	nr	**22047.64**
15 kVA, 12.75 kW single phase I/P & O/P	17456.55	22093.01	30.00	1371.11	nr	**23464.12**
20 kVA, 17.0 kW 3 phase I/P & single phase O/P	24109.75	30513.30	40.00	1828.14	nr	**32341.44**
30 kVA, 25.5 kW 3 phase I/P & O/P	32581.85	41235.59	60.00	2742.21	nr	**43977.80**
40 kVA, 34.0 kW 3 phase I/P & O/P	40619.41	51407.92	80.00	3656.29	nr	**55064.21**
50 kVA, 42.5 kW 3 phase I/P & O/P	52977.46	67048.27	90.00	4113.32	nr	**71161.59**
65 kVA, 55.25 kW 3 phase I/P & O/P	64989.03	82250.11	100.00	4570.36	nr	**86820.47**
90 kVA, 68.85 kW 3 phase I/P & O/P	85880.82	108690.77	120.00	5484.43	nr	**114175.20**
120 kVA, 102 kW 3 phase I/P & O/P	95863.52	121324.87	150.00	6855.54	nr	**128180.41**
Three hour						
750 VA, 600 W single phase I/P & O/P	4471.66	5659.34	6.00	274.22	nr	**5933.56**
3 kVA, 2.55 kW single phase I/P & O/P	7525.20	9523.90	8.00	365.63	nr	**9889.52**
5 kVA, 4.25 kW single phase I/P & O/P	11929.36	15097.80	10.00	457.04	nr	**15554.84**
8 kVA, 6.80 kW single phase I/P & O/P	14841.98	18784.01	12.00	548.44	nr	**19332.45**
10 kVA, 8.5 kW single phase I/P & O/P	19577.90	24777.79	14.00	639.85	nr	**25417.64**
13 kVA, 11.05 kW single phase I/P & O/P	24140.59	30552.33	16.00	731.26	nr	**31283.59**
15 kVA, 12.75 kW single phase I/P & O/P	24536.96	31053.98	30.00	1371.11	nr	**32425.09**
20 kVA, 17.0 kW 3 phase I/P & single phase O/P	35532.63	44970.10	40.00	1828.14	nr	**46798.24**
30 kVA, 25.5 kW 3 phase I/P & O/P	53394.36	67575.91	60.00	2742.21	nr	**70318.12**
40 kVA, 34.0 kW 3 phase I/P & O/P	63836.61	80791.61	80.00	3656.29	nr	**84447.90**
50 kVA, 42.5 kW 3 phase I/P & O/P	86101.01	108969.44	90.00	4113.32	nr	**113082.76**
65 kVA, 55.25 kW 3 phase I/P & O/P	104105.15	131755.48	100.00	4570.36	nr	**136325.84**
90 kVA, 68.85 kW 3 phase I/P & O/P	123208.86	155933.13	120.00	5484.43	nr	**161417.56**
120 kVA, 102 kW 3 phase I/P & O/P	151857.83	192191.27	150.00	6855.54	nr	**199046.80**
AC static inverter system; BS EN 50171 compliant; one hour standby; Low power system (typically wall mounted)						
Central system supplying AC power on mains failure to mains luminaires; metal cubicle(s) with changeover device, battery charger, battery & static inverter; 5 year design life valve regulated lead acid battery; LED indication and LCD display; 12 hour recharge to 80% duty; inverter rated for 120% of load for 100% of duty; battery sized for 'end of life' @ 20°C test push button						
One hour						
300 VA, 240 W single phase I/P & O/P	1708.82	2162.68	3.00	137.11	nr	**2299.79**
600 VA, 480 W single phase I/P & O/P	2059.67	2606.72	4.00	182.81	nr	**2789.54**
750 VA, 600 W single phase I/P & O/P	3800.78	4810.27	6.00	274.22	nr	**5084.49**

ELECTRICAL SUPPLY/POWER/LIGHTING SYSTEMS

Item	Net Price £	Material £	Labour hours	Labour £	Unit	Total rate £
Three hour						
150 VA, 120 W single phase I/P & O/P	1883.50	2383.76	3.00	137.11	nr	**2520.87**
450 VA, 360 W single phase I/P & O/P	2234.37	2827.82	4.00	182.81	nr	**3010.63**
750 VA, 600 W single phase I/P & O/P	4448.17	5629.60	6.00	274.22	nr	**5903.82**
AC static inverter system central power system; CPS BS EN 50171 compliant; one hour standby						
Central system supplying AC power on mains failure to mains luminaires; metal cubicle(s) with changeover device, battery charger, battery & static inverter; LED indication and LCD display; pure sinewave output; 10 year design life valve regulated lead acid battery; 12 hour recharge to 80% duty specified; inverter rated for 120% of load for 100% of duty; battery sized for 'end of life' @ 20°C test push button; includes on-site commissioning						
One hour						
750 VA, 600 W single phase I/P & O/P	4252.93	5382.50	6.00	274.22	nr	**5656.72**
3 kVA, 2.55 kW single phase I/P & O/P	6776.50	8576.34	8.00	365.63	nr	**8941.97**
5 kVA, 4.25 kW single phase I/P & O/P	8774.52	11105.03	10.00	457.04	nr	**11562.07**
8 kVA, 6.80 kW single phase I/P & O/P	10241.10	12961.14	12.00	548.44	nr	**13509.58**
10 kVA, 8.5 kW single phase I/P & O/P	13150.77	16643.61	14.00	639.85	nr	**17283.46**
13 kVA, 11.05 kW single phase I/P & O/P	17381.69	21998.26	16.00	731.26	nr	**22729.52**
15 kVA, 12.75 kW single phase I/P & O/P	17990.93	22769.32	30.00	1371.11	nr	**24140.43**
20 kVA, 17.0 kW 3 phase I/P & single phase O/P	24783.58	31366.09	40.00	1828.14	nr	**33194.24**
30 kVA, 25.5 kW 3 phase I/P & O/P	33384.88	42251.90	60.00	2742.21	nr	**44994.11**
40 kVA, 34.0 kW 3 phase I/P & O/P	41378.40	52368.51	80.00	3656.29	nr	**56024.79**
50 kVA, 42.5 kW 3 phase I/P & O/P	53889.10	68202.05	90.00	4113.32	nr	**72315.37**
65 kVA, 55.25 kW 3 phase I/P & O/P	66559.82	84238.10	100.00	4570.36	nr	**88808.46**
90 kVA, 68.85 kW 3 phase I/P & O/P	87340.04	110537.56	120.00	5484.43	nr	**116021.99**
120 kVA, 102 kW 3 phase I/P & O/P	97269.92	123104.81	150.00	6855.54	nr	**129960.34**
Three hour						
750 VA, 600 W single phase I/P & O/P	4920.91	6227.90	6.00	274.22	nr	**6502.12**
3 kVA, 2.55 kW single phase I/P & O/P	8025.82	10157.47	8.00	365.63	nr	**10523.10**
5 kVA, 4.25 kW single phase I/P & O/P	12435.83	15738.79	10.00	457.04	nr	**16195.82**
8 kVA, 6.80 kW single phase I/P & O/P	15304.39	19369.24	12.00	548.44	nr	**19917.68**
10 kVA, 8.5 kW single phase I/P & O/P	20103.45	25442.93	14.00	639.85	nr	**26082.78**
13 kVA, 11.05 kW single phase I/P & O/P	24641.18	31185.88	16.00	731.26	nr	**31917.13**
15 kVA, 12.75 kW single phase I/P & O/P	25033.14	31681.94	30.00	1371.11	nr	**33053.05**
20 kVA, 17.0 kW 3 phase I/P & single phase O/P	36479.53	46168.49	40.00	1828.14	nr	**47996.64**
30 kVA, 25.5 kW 3 phase I/P & O/P	54084.34	68449.15	60.00	2742.21	nr	**71191.36**
40 kVA, 34.0 kW 3 phase I/P & O/P	64472.26	81596.09	80.00	3656.29	nr	**85252.37**
50 kVA, 42.5 kW 3 phase I/P & O/P	86835.04	109898.42	90.00	4113.32	nr	**114011.74**
65 kVA, 55.25 kW 3 phase I/P & O/P	105467.50	133479.67	100.00	4570.36	nr	**138050.03**
90 kVA, 68.85 kW 3 phase I/P & O/P	131120.16	165945.67	120.00	5484.43	nr	**171430.10**
120 kVA, 102 kW 3 phase I/P & O/P	152961.79	193588.44	150.00	6855.54	nr	**200443.97**

COMMUNICATIONS/SECURITY/CONTROL

Item	Net Price £	Material £	Labour hours	Labour £	Unit	Total rate £
TELECOMMUNICATIONS						
CABLES						
Multipair internal telephone cable; BS 6746; loose laid on tray/basket						
0.5 millimetre dia. conductor LSZH insulated and sheathed multipair cables; BT specification CW 1308						
3 pair	0.28	0.35	0.03	1.37	m	**1.72**
4 pair	0.32	0.40	0.03	1.37	m	**1.77**
6 pair	0.49	0.62	0.03	1.37	m	**1.99**
10 pair	0.83	1.05	0.05	2.29	m	**3.34**
15 pair	1.25	1.58	0.06	2.74	m	**4.33**
20 pair + 1 wire	1.78	2.25	0.06	2.74	m	**4.99**
25 pair	2.25	2.85	0.08	3.66	m	**6.51**
40 pair + earth	3.19	4.03	0.08	3.66	m	**7.69**
50 pair + earth	3.83	4.85	0.10	4.57	m	**9.42**
80 pair + earth	5.12	6.48	0.10	4.57	m	**11.05**
100 pair + earth	8.16	10.33	0.12	5.48	m	**15.82**
Multipair internal telephone cable; BS 6746; installed in conduit/trunking						
0.5 millimetre dia. conductor LSZH insulated and sheathed multipair cables; BT specification CW 1308						
3 pair	0.28	0.35	0.05	2.29	m	**2.64**
4 pair	0.32	0.40	0.06	2.74	m	**3.14**
6 pair	0.49	0.62	0.06	2.74	m	**3.36**
10 pair	0.83	1.05	0.07	3.20	m	**4.25**
15 pair	1.25	1.58	0.07	3.20	m	**4.78**
20 pair + 1 wire	1.78	2.25	0.09	4.11	m	**6.36**
25 pair	2.25	2.85	0.10	4.57	m	**7.42**
40 pair + earth	3.19	4.03	0.12	5.48	m	**9.52**
50 pair + earth	3.83	4.85	0.14	6.40	m	**11.25**
80 pair + earth	5.12	6.48	0.14	6.40	m	**12.88**
100 pair + earth	8.16	10.33	0.15	6.86	m	**17.19**
Low speed data; unshielded twisted pair; solid copper conductors; LSOH sheath; nominal impedance 100 ohm; Category 3 to ISO IS 1801/EIA/TIA 568B and EN50173/ 50174 standards to current revisions						
Installed in riser						
25 pair 24 AWG	2.48	3.13	0.03	1.37	m	**4.50**
50 pair 24 AWG	4.16	5.27	0.06	2.74	m	**8.01**
100 pair 24 AWG	9.28	11.75	0.10	4.57	m	**16.32**
Installed below floor						
25 pair 24 AWG	2.48	3.13	0.02	0.91	m	**4.05**
50 pair 24 AWG	4.16	5.27	0.05	2.29	m	**7.55**
100 pair 24 AWG	9.28	11.75	0.08	3.66	m	**15.41**

COMMUNICATIONS/SECURITY/CONTROL

Item	Net Price £	Material £	Labour hours	Labour £	Unit	Total rate £
ACCESSORIES						
Telephone outlet: moulded plastic plate with box; fitted and connected; flush or surface mounted						
Single master outlet	10.84	13.72	0.35	16.00	nr	**29.71**
Single secondary outlet	8.01	10.14	0.35	16.00	nr	**26.13**
Telephone outlet: bronze or satin chromeplate; with box; fitted and connected; flush or surface mounted						
Single master outlet	17.63	22.31	0.35	16.00	nr	**38.31**
Single secondary outlet	19.26	24.37	0.35	16.00	nr	**40.37**
Frames and Box Connections						
Provision and Installation of a Dual Vertical Krone 108 A Voice Distribution Frame that can accommodate a total of 138 × Krone 237 A Strips	366.41	463.73	1.50	68.56	nr	**532.28**
Label Frame (Traffolyte style)	2.31	2.93	0.27	12.34	nr	**15.27**
Provision and Installation of a Box Connection 301 A Voice Termination Unit that can accommodate a total of 10 × Krone 237 A Strips	31.77	40.21	0.25	11.43	nr	**51.64**
Label Frame (Traffolyte style)	0.77	0.98	0.08	3.79	nr	**4.77**
Provision and Installation of a Box Connection 201 Voice Termination Unit that can accommodate 20 pairs	16.90	21.39	0.17	7.59	nr	**28.98**
Label Frame (Traffolyte style)	0.77	0.98	0.08	3.79	nr	**4.77**
Terminate, Test and Label Voice Multicore System						
Patch panels						
Voice; 19" wide fully loaded, finished in black including termination and forming of cables (assuming 2 pairs per port)						
25 port – RJ45 UTP – Krone	79.50	100.61	2.60	118.83	nr	**219.44**
50 port – RJ45 UTP – Krone	108.37	137.16	4.65	212.52	nr	**349.68**
900 pair fully loaded Systimax style of frame including forming and termination of 9 × 100 pair cables	895.52	1133.37	25.00	1142.59	nr	**2275.96**
Installation and termination of Krone Strip (10 pair block – 237 A) including designation label strip	5.35	6.76	0.50	22.85	nr	**29.62**
Patch panel and Outlet labelling per port (Traffolyte style)	0.27	0.34	0.02	0.91	nr	**1.26**
Provision and installation of a voice jumper, for cross termination on Krone Termination Strips	0.08	0.10	0.04	1.83	nr	**1.93**
CW1308/Cat 3 cable circuit test per pair	–	–	0.05	2.29	nr	**2.29**

COMMUNICATIONS/SECURITY/CONTROL

Item	Net Price £	Material £	Labour hours	Labour £	Unit	Total rate £
RADIO AND TELEVISION						
RADIO						
Cables						
Radio frequency cable; BS 2316; PVC sheathed; laid loose						
7/0.41 mm tinned copper inner conductor; solid polyethylene dielectric insulation; bare copper wire braid; PVC sheath; 75 ohm impedance						
Cable	1.65	2.09	0.05	2.29	m	**4.37**
Twin 1/0.58 mm copper covered steel solid core wire conductor; solid polyethylene dielectric insulation; barecopper wire braid; PVC sheath; 75 ohm impedance						
Cable	0.38	0.49	0.05	2.29	m	**2.77**
TELEVISION						
Cables						
Television aerial cable; coaxial; PVC sheathed; fixed to backgrounds						
General purpose TV aerial downlead; copper stranded inner conductor; cellular polythene insulation; copper braid outer conductor; 75 ohm impedance						
7/0.25 mm	0.38	0.48	0.06	2.74	m	**3.22**
Low loss TV aerial downlead; solid copper inner conductor; cellular polythene insulation; copper braid outer; conductor; 75 ohm impedance						
1/1.12 mm	0.39	0.50	0.06	2.74	m	**3.24**
Low loss air spaced; solid copper inner conductor; air spaced polythene insulation; copper braid outer conductor; 75 ohm impedance						
1/1.00 mm	0.52	0.66	0.06	2.74	m	**3.40**
Satellite aerial downlead; solid copper inner conductor; air spaced polythene insulation; copper tape and braid outer conductor; 75 ohm impedance						
1/1.00 mm	0.45	0.57	0.06	2.74	m	**3.31**
Satellite TV coaxial; solid copper inner conductor; semi air spaced polyethylene dielectric insulation; plain annealed copper foil and copper braid screen in outer conductor; PVC sheath; 75 ohm impedance						
1/1.25 mm	0.74	0.94	0.08	3.66	m	**4.59**
Satelite TV coaxial; solid copper inner conductor; air spaced polyethylene dielectric insulation; plain annealed copper foil and copper braid screen in outer conductor; PVC sheath; 75 ohm impedance						
1/1.67 mm	1.22	1.54	0.09	4.11	m	**5.65**

COMMUNICATIONS/SECURITY/CONTROL

Item	Net Price £	Material £	Labour hours	Labour £	Unit	Total rate £
Video cable; PVC flame retardant sheath; laid loose						
7/0.1 mm silver coated copper covered annealed steel wire conductor; polyethylene dielectric insulation with tin coated copper wire braid; 75 ohm impedance						
Cable	0.28	0.36	0.05	2.29	m	**2.65**
ACCESSORIES						
TV coaxial socket outlet: moulded plastic box; flush or surface mounted						
One way direct connection	12.12	15.34	0.35	16.00	nr	**31.34**
Two way direct connection	16.89	21.37	0.35	16.00	nr	**37.37**
One way isolated UHF/VHF	21.34	27.01	0.35	16.00	nr	**43.00**
Two way isolated UHF/VHF	29.05	36.77	0.35	16.00	nr	**52.76**

COMMUNICATIONS/SECURITY/CONTROL

Item	Net Price £	Material £	Labour hours	Labour £	Unit	Total rate £
CLOCKS						
Master clock						
Master clock module 230 V with programming circuits, NTP client and server for synchronization over Ethernet network, sounders Harmonys or Melodys, control of DHF and NTP clock systems						
Rack mounted	1283.59	1624.51	3.00	137.11	nr	**1761.62**
Wall mounted	1154.61	1461.28	5.00	228.52	nr	**1689.80**
Master clock module 230 V with programming circuits for control of wired and DHF clock systems, relays or sounders, NTP client and server for synchronization over Ethernet network						
Rack mounted	1733.15	2193.47	3.00	137.11	nr	**2330.58**
Wall mounted	1407.64	1781.52	5.00	228.52	nr	**2010.03**
Clocks and bells, master clock module 230 V with 3 programming circuits for control of wired and DHF clock systems, relays and sounders						
Rack mounted	1118.99	1416.20	3.00	137.11	nr	**1553.31**
Wall mounted	884.38	1119.28	5.00	228.52	nr	**1347.79**
Clocks only, master clock module 230 V for control of wired and DHF clock systems						
Rack mounted	846.31	1071.09	3.00	137.11	nr	**1208.20**
Wall mounted	657.15	831.68	5.00	228.52	nr	**1060.20**
Extra over for						
GPS antenna with 20 m cable	257.95	326.46	2.50	114.26	nr	**440.71**
DHF transmitter	379.55	480.36	1.50	68.56	nr	**548.91**
DHF repeater, signal booster up to 200 m	434.82	550.31	3.00	137.11	nr	**687.42**
UPS	1123.90	1422.41	5.00	228.52	nr	**1650.93**
Rack mounted external AC/DC power supply	646.09	817.69	3.00	137.11	nr	**954.80**
Indoor clock, 30 cm dia.						
Stand-alone						
HMS display, quartz movement, battery operated	61.42	77.73	0.80	36.56	nr	**114.29**
DHF wireless						
Hour and minute display, battery operated	140.03	177.22	0.80	36.56	nr	**213.78**
Hour, minute and second display, battery operated	153.54	194.32	0.80	36.56	nr	**230.88**
HM display, TBT	144.94	183.44	0.80	36.56	nr	**220.00**
HMS display, TBT	156.00	197.43	0.80	36.56	nr	**233.99**
Slave wired						
Hour and minute display, 24 V	83.53	105.71	1.00	45.70	nr	**151.41**
Hour, minute and second display, 24 V	100.72	127.47	1.00	45.70	nr	**173.18**
Hour, minute display, TBT, AFNOR	138.80	175.66	1.00	45.70	nr	**221.37**
Hour, minute and second display, TBT AFNOR	153.54	194.32	1.00	45.70	nr	**240.02**
Hour, minute display, POE, NTP	194.07	245.62	1.00	45.70	nr	**291.32**
Hour, minute and second display, POE, NTP	206.36	261.16	1.00	45.70	nr	**306.87**

COMMUNICATIONS/SECURITY/CONTROL

Item	Net Price £	Material £	Labour hours	Labour £	Unit	Total rate £
Indoor clock, 40 cm dia.						
Stand-alone						
HMS display, quartz movement, battery operated	116.69	147.68	0.80	36.56	nr	**184.25**
DHF wireless						
Hour and minute display, battery operated	205.13	259.61	0.80	36.56	nr	**296.17**
Hour, minute and second display, battery operated	211.27	267.38	0.80	36.56	nr	**303.95**
HM display, TBT	205.13	259.61	0.80	36.56	nr	**296.17**
HMS display, TBT	219.87	278.26	0.80	36.56	nr	**314.83**
Slave wired						
Hour and minute display, 24 V	154.77	195.87	1.00	45.70	nr	**241.58**
Hour, minute and second display, 24 V	142.48	180.33	1.00	45.70	nr	**226.03**
Hour, minute display, TBT, AFNOR	217.41	275.16	1.00	45.70	nr	**320.86**
Hour, minute and second display, TBT, AFNOR	224.78	284.48	1.00	45.70	nr	**330.19**
Hour, minute display, POE, NTP	256.72	324.90	1.00	45.70	nr	**370.60**
Hour, minute and second display, POE, NTP	264.09	334.23	1.00	45.70	nr	**379.93**
Digital clock, 7 cm digit height, 230 V						
Independent quartz movement						
Red display	257.95	326.46	0.80	36.56	nr	**363.02**
Green display	257.95	326.46	0.80	36.56	nr	**363.02**
Yellow display	257.95	326.46	0.80	36.56	nr	**363.02**
Blue display	350.07	443.05	0.80	36.56	nr	**479.61**
White display	350.07	443.05	0.80	36.56	nr	**479.61**
AFNOR, 230 V						
Red display	264.09	334.23	0.80	36.56	nr	**370.79**
Green display	264.09	334.23	0.80	36.56	nr	**370.79**
Yellow display	264.09	334.23	0.80	36.56	nr	**370.79**
Blue display	356.21	450.82	0.80	36.56	nr	**487.38**
White display	356.21	450.82	0.80	36.56	nr	**487.38**
DHF wireless, 230 V						
Red display	282.51	357.55	0.80	36.56	nr	**394.11**
Green display	282.51	357.55	0.80	36.56	nr	**394.11**
Yellow display	282.51	357.55	0.80	36.56	nr	**394.11**
Blue display	374.63	474.14	0.80	36.56	nr	**510.70**
White display	374.63	474.14	0.80	36.56	nr	**510.70**
NTP POE						
Red display	307.08	388.64	0.80	36.56	nr	**425.20**
Green display	307.08	388.64	0.80	36.56	nr	**425.20**
Yellow display	307.08	388.64	0.80	36.56	nr	**425.20**
Blue display	399.20	505.23	0.80	36.56	nr	**541.79**
White display	399.20	505.23	0.80	36.56	nr	**541.79**
Digital clock, 5 cm digit height, 230 V						
Independent quartz movement						
Red display	214.95	272.05	0.80	36.56	nr	**308.61**
Green display	214.95	272.05	0.80	36.56	nr	**308.61**
Yellow display	214.95	272.05	0.80	36.56	nr	**308.61**
Blue display	282.51	357.55	0.80	36.56	nr	**394.11**
White display	282.51	357.55	0.80	36.56	nr	**394.11**

COMMUNICATIONS/SECURITY/CONTROL

Item	Net Price £	Material £	Labour hours	Labour £	Unit	Total rate £
CLOCKS – cont						
Digital clock, 5 cm digit height, 230 V – cont						
AFNOR, 230 V						
Red display	214.95	272.05	0.80	36.56	nr	**308.61**
Green display	214.95	272.05	0.80	36.56	nr	**308.61**
Yellow display	214.95	272.05	0.80	36.56	nr	**308.61**
Blue display	282.51	357.55	0.80	36.56	nr	**394.11**
White display	282.51	357.55	0.80	36.56	nr	**394.11**
DHF wireless, 230 V						
Red display	251.80	318.68	0.80	36.56	nr	**355.25**
Green display	251.80	318.68	0.80	36.56	nr	**355.25**
Yellow display	251.80	318.68	0.80	36.56	nr	**355.25**
Blue display	319.36	404.18	0.80	36.56	nr	**440.75**
White display	319.36	404.18	0.80	36.56	nr	**440.75**
NTP POE						
Red display	270.23	342.00	0.80	36.56	nr	**378.56**
Green display	270.23	342.00	0.80	36.56	nr	**378.56**
Yellow display	270.23	342.00	0.80	36.56	nr	**378.56**
Blue display	337.79	427.50	0.80	36.56	nr	**464.06**
White display	337.79	427.50	0.80	36.56	nr	**464.06**
Digital clock, 10 cm digit height, 230 V						
Independent quartz movement						
Red display	460.62	582.96	0.80	36.56	nr	**619.52**
Green display	460.62	582.96	0.80	36.56	nr	**619.52**
Yellow display	460.62	582.96	0.80	36.56	nr	**619.52**
Blue display	571.16	722.87	0.80	36.56	nr	**759.43**
White display	571.16	722.87	0.80	36.56	nr	**759.43**
AFNOR, 230 V						
Red display	460.62	582.96	0.80	36.56	nr	**619.52**
Green display	460.62	582.96	0.80	36.56	nr	**619.52**
Yellow display	460.62	582.96	0.80	36.56	nr	**619.52**
Blue display	571.16	722.87	0.80	36.56	nr	**759.43**
White display	571.16	722.87	0.80	36.56	nr	**759.43**
DHF wireless, 230 V						
Red display	479.04	606.27	0.80	36.56	nr	**642.84**
Green display	479.04	606.27	0.80	36.56	nr	**642.84**
Yellow display	479.04	606.27	0.80	36.56	nr	**642.84**
Blue display	589.59	746.18	0.80	36.56	nr	**782.75**
White display	589.59	746.18	0.80	36.56	nr	**782.75**
NTP POE						
Red display	497.47	629.59	0.80	36.56	nr	**666.16**
Green display	497.47	629.59	0.80	36.56	nr	**666.16**
Yellow display	497.47	629.59	0.80	36.56	nr	**666.16**
Blue display	608.01	769.50	0.80	36.56	nr	**806.07**
White display	608.01	769.50	0.80	36.56	nr	**806.07**

COMMUNICATIONS/SECURITY/CONTROL

Item	Net Price £	Material £	Labour hours	Labour £	Unit	Total rate £
DATA TRANSMISSION						
Cabinets						
Floor standing; suitable for 19" patch panels with glass lockable doors, metal rear doors, side panels, vertical cable management, 2 × 4 way PDUs, 4 way fan, earth bonding kit; installed on raised floor						
600 mm wide × 800 mm deep – 18U	1084.33	1372.32	3.00	137.11	nr	**1509.44**
600 mm wide × 800 mm deep – 24U	1129.49	1429.48	3.00	137.11	nr	**1566.59**
600 mm wide × 800 mm deep – 33U	1213.07	1535.27	4.00	182.81	nr	**1718.08**
600 mm wide × 800 mm deep – 42U	1301.18	1646.78	4.00	182.81	nr	**1829.59**
600 mm wide × 800 mm deep – 47U	1373.61	1738.44	4.00	182.81	nr	**1921.25**
800 mm wide × 800 mm deep – 42U	1472.90	1864.10	4.00	182.81	nr	**2046.91**
800 mm wide × 800 mm deep – 47U	1536.20	1944.21	4.00	182.81	nr	**2127.03**
Label cabinet	3.34	4.23	0.25	11.43	nr	**15.65**
Wall mounted; suitable for 19" patch panels with glass lockable doors, side panels, vertical cable management, 2 × 4 way PDUs, 4 way fan, earth bonding kit; fixed to wall						
19 mm wide × 500 mm deep – 9U	723.21	915.30	3.00	137.11	nr	**1052.41**
19 mm wide × 500 mm deep – 12U	741.00	937.81	3.00	137.11	nr	**1074.92**
19 mm wide × 500 mm deep – 15U	767.90	971.85	3.00	137.11	nr	**1108.96**
19 mm wide × 500 mm deep – 18U	903.61	1143.60	3.00	137.11	nr	**1280.72**
19 mm wide × 500 mm deep – 21U	940.00	1189.67	3.00	137.11	nr	**1326.78**
Label cabinet	3.48	4.40	0.25	11.43	nr	**15.83**
Frames						
Floor standing; suitable for 19" patch panels with supports, vertical cable management, earth bonding kit; installed on raised floor						
19 mm wide × 500 mm deep – 25U	813.41	1029.45	2.50	114.26	nr	**1143.71**
19 mm wide × 500 mm deep – 39U	993.79	1257.74	2.50	114.26	nr	**1372.00**
19 mm wide × 500 mm deep – 42U	1084.40	1372.41	2.50	114.26	nr	**1486.67**
19 mm wide × 500 mm deep – 47U	1175.00	1487.08	2.50	114.26	nr	**1601.34**
Label frame	3.48	4.40	0.25	11.43	nr	**15.83**
Copper data cabling						
Unshielded twisted pair; solid copper conductors; PVC insulation; nominal impedance 100 ohm; Cat 5e to ISO 11801, EIA/TIA 568B and EN 50173/50174 standards to the current revisions						
4 pair 24 AWG; nominal outside dia. 5.6 mm; installed above ceiling	0.32	0.40	0.02	0.91	m	**1.32**
4 pair 24 AWG; nominal outside dia. 5.6 mm; installed in riser	0.32	0.40	0.02	0.91	m	**1.32**
4 pair 24 AWG; nominal outside dia. 5.6 mm; installed below floor	0.32	0.40	0.01	0.46	m	**0.86**
4 pair 24 AWG; nominal outside dia. 5.6 mm; installed in trunking	0.32	0.40	0.02	0.91	m	**1.32**

COMMUNICATIONS/SECURITY/CONTROL

Item	Net Price £	Material £	Labour hours	Labour £	Unit	Total rate £
DATA TRANSMISSION – cont						
Copper data cabling – cont						
Unshielded twisted pair; solid copper conductors; LSOH sheathed; nominal impedance 100 ohm; Cat 5e to ISO 11801, EIA/TIA 568B and EN 50173/50174 standards to the current revisions						
4 pair 24 AWG; nominal outside dia. 5.6 mm; installed above ceiling	0.34	0.43	0.03	1.14	m	**1.58**
4 pair 24 AWG; nominal outside dia. 5.6 mm; installed in riser	0.34	0.43	0.03	1.14	m	**1.58**
4 pair 24 AWG; nominal outside dia. 5.6 mm; installed below floor	0.34	0.43	0.03	1.14	m	**1.58**
4 pair 24 AWG; nominal outside dia. 5.6 mm; installed in trunking	0.34	0.43	0.03	1.14	m	**1.58**
Unshielded twisted pair; solid copper conductors; PVC insulation; nominal impedance 100 ohm; Cat 6 to ISO 11801, EIA/TIA 568B and EN 50173/50174 standards to the current revisions						
4 pair 24 AWG; nominal outside dia. 5.6 mm; installed above ceiling	0.51	0.64	0.02	0.69	m	**1.33**
4 pair 24 AWG; nominal outside dia. 5.6 mm; installed in riser	0.51	0.64	0.02	0.69	m	**1.33**
4 pair 24 AWG; nominal outside dia. 5.6 mm; installed below floor	0.51	0.64	0.01	0.37	m	**1.01**
4 pair 24 AWG; nominal outside dia. 5.6 mm; installed in trunking	0.51	0.64	0.02	0.91	m	**1.56**
Unshielded twisted pair; solid copper conductors; LSOH sheathed; nominal impedance 100 ohm; Cat 6 to ISO 11801, EIA/TIA 568B and EN 50173/50174 standards to the current revisions						
4 pair 24 AWG; nominal outside dia. 5.6 mm; installed above ceiling	0.47	0.60	0.02	1.01	m	**1.61**
4 pair 24 AWG; nominal outside dia. 5.6 mm; installed in riser	0.47	0.60	0.02	1.01	m	**1.61**
4 pair 24 AWG; nominal outside dia. 5.6 mm; installed below floor	0.47	0.60	0.01	0.50	m	**1.10**
4 pair 24 AWG; nominal outside dia. 5.6 mm; installed in trunking	0.47	0.60	0.02	1.01	m	**1.61**
Patch panels						
Category 5e; 19″ wide fully loaded, finished in black including termination and forming of cables						
24 port – RJ45 UTP – Krone/110	77.32	97.85	4.75	217.09	nr	**314.94**
48 port – RJ45 UTP – Krone/110	145.02	183.53	9.35	427.33	nr	**610.86**
Patch panel labelling per port	0.35	0.45	0.02	0.91	nr	**1.36**
Category 6; 19″ wide fully loaded, finished in black including termination and forming of cables						
24 port – RJ45 UTP – Krone/110	147.67	186.88	5.00	228.52	nr	**415.40**
48 port – RJ45 UTP – Krone/110	269.33	340.86	9.80	447.90	nr	**788.76**
Patch panel labelling per port	0.35	0.45	0.02	0.91	nr	**1.36**

COMMUNICATIONS/SECURITY/CONTROL

Item	Net Price £	Material £	Labour hours	Labour £	Unit	Total rate £
Workstation						
Category 5e RJ45 data outlet plate and multiway outlet boxes for wall, ceiling and below floor installations including label to ISO 11801 standards						
Wall mounted; fully loaded						
One gang LSOH PVC plate	7.33	9.28	0.15	6.86	nr	**16.14**
Two gang LSOH PVC plate	10.50	13.29	0.20	9.14	nr	**22.43**
Four gang LSOH PVC plate	17.98	22.76	0.40	18.28	nr	**41.04**
One gang satin brass plate	23.80	30.12	0.25	11.43	nr	**41.55**
Two gang satin brass plate	26.86	33.99	0.33	15.08	nr	**49.08**
Ceiling mounted; fully loaded						
One gang metal clad plate	12.64	16.00	0.33	15.08	nr	**31.08**
Two gang metal clad plate	17.17	21.73	0.45	20.57	nr	**42.30**
Below floor; fully loaded						
Four way outlet box, 5 m length 20 mm flexible conduit with glands and starin relief bracket	42.75	54.10	0.80	36.56	nr	**90.66**
Six way outlet box, 5 m length 25 mm flexible conduit with glands and strain relief bracket	55.26	69.94	1.20	54.84	nr	**124.78**
Eight way outlet box, 5 m length 25 mm flexible conduit with glands and strain relief bracket	69.58	88.06	1.40	63.99	nr	**152.04**
Installation of outlet boxes to desks	–	–	0.40	18.28	nr	**18.28**
Category 6 RJ45 data outlet plate and multiway outlet boxes for wall, ceiling and below floor installations including label to ISO 11801 standards						
Wall mounted; fully loaded						
One gang LSOH PVC plate	9.36	11.85	0.17	7.54	nr	**19.39**
Two gang LSOH PVC plate	13.75	17.40	0.22	10.05	nr	**27.45**
Four gang LSOH PVC plate	24.82	31.41	0.44	20.11	nr	**51.52**
One gang satin brass plate	22.88	28.96	0.28	12.57	nr	**41.53**
Two gang satin brass plate	28.63	36.24	0.36	16.59	nr	**52.83**
Ceiling mounted; fully loaded						
One gang metal clad plate	15.17	19.20	0.36	16.59	nr	**35.79**
Two gang metal clad plate	21.42	27.11	0.50	22.62	nr	**49.73**
Below floor; fully loaded						
Four way outlet box, 5 m length 20 mm flexible conduit with glands and strain relief bracket	52.39	66.31	0.88	40.22	nr	**106.53**
Six way outlet box, 5 m length 25 mm flexible conduit with glands and strain relief bracket	69.69	88.20	1.32	60.33	nr	**148.53**
Eight way outlet box, 5 m length 25 mm flexible conduit with glands and strain relief bracket	88.75	112.33	1.54	70.38	nr	**182.71**
Installation of outlet boxes to desks	–	–	0.40	18.28	nr	**18.28**
Category 5e cable test	–	–	0.10	4.57	nr	**4.57**
Category 6 cable test	–	–	0.11	5.03	nr	**5.03**
Copper patch leads						
Category 5e; straight through booted RJ45 UTP – RJ45 UTP						
Patch lead 1 m length	2.58	3.26	0.09	4.11	nr	**7.38**
Patch lead 3 m length	3.07	3.89	0.09	4.11	nr	**8.00**
Patch lead 5 m length	4.35	5.50	0.10	4.57	nr	**10.07**
Patch lead 7 m length	6.38	8.07	0.10	4.57	nr	**12.64**

COMMUNICATIONS/SECURITY/CONTROL

Item	Net Price £	Material £	Labour hours	Labour £	Unit	Total rate £
DATA TRANSMISSION – cont						
Copper patch leads – cont						
Category 6; straight through booted RJ45 UTP – RJ45 UTP						
Patch lead 1 m length	6.84	8.66	0.09	4.11	nr	**12.78**
Patch lead 3 m length	10.33	13.07	0.09	4.11	nr	**17.18**
Patch lead 5 m length	12.11	15.32	0.10	4.57	nr	**19.89**
Patch lead 7 m length	15.35	19.43	0.10	4.57	nr	**24.00**
Note 1 – With an intelligent Patching System, add a 25% uplift to the Cat 5e and Cat 6 System price. This would be dependent on the System and IMS solution chosen, i.e. iPatch, RIT or iTRACs.						
Note 2 – For a Cat 6 augmented solution (10 Gbps capable), add 25% to a Cat 6 System price.						
Fibre infrastructure						
Fibreoptic cable, tight buffered, internal/external application, single mode, LSOH sheathed						
4 core fibreoptic cable	2.00	2.53	0.10	4.57	m	**7.10**
8 core fibreoptic cable	3.34	4.22	0.10	4.57	m	**8.79**
12 core fibreoptic cable	4.13	5.23	0.10	4.57	m	**9.80**
16 core fibreoptic cable	5.03	6.36	0.10	4.57	m	**10.93**
24 core fibreoptic cable	6.07	7.68	0.10	4.57	m	**12.25**
Fibreoptic cable OM1 and OM2, tight buffered, internal/external application, 62.5/125 multimode fibre, LSOH sheathed						
4 core fibreoptic cable	2.13	2.70	0.10	4.57	m	**7.27**
8 core fibreoptic cable	3.53	4.47	0.10	4.57	m	**9.04**
12 core fibreoptic cable	4.52	5.72	0.10	4.57	m	**10.29**
16 core fibreoptic cable	5.60	7.08	0.10	4.57	m	**11.65**
24 core fibreoptic cable	6.49	8.22	0.10	4.57	m	**12.79**
Fibreoptic cable OM3, tight buffered, internal only application, 50/125 multimode fibre, LSOH sheathed						
4 core fibreoptic cable	2.65	3.35	0.10	4.57	nr	**7.92**
8 core fibreoptic cable	4.13	5.23	0.10	4.57	nr	**9.80**
12 core fibreoptic cable	5.99	7.58	0.10	4.57	nr	**12.15**
24 core fibreoptic cable	11.44	14.48	0.10	4.57	nr	**19.05**
Fibreoptic single and multimode connectors and couplers including termination						
ST singlemode booted connector	10.08	12.76	0.25	11.43	nr	**24.18**
ST multimode booted connector	4.87	6.16	0.25	11.43	nr	**17.59**
SC simplex singlemode booted connector	14.29	18.08	0.25	11.43	nr	**29.51**
SC simplex multimode booted connector	5.21	6.59	0.25	11.43	nr	**18.02**
SC duplex multimode booted connector	10.39	13.15	0.25	11.43	nr	**24.58**
ST – SC duplex adaptor	25.25	31.95	0.01	0.46	nr	**32.41**
ST inline bulkhead coupler	5.42	6.85	0.01	0.46	nr	**7.31**
SC duplex coupler	10.10	12.79	0.01	0.46	nr	**13.25**
MTRJ small form factor duplex connector	10.07	12.74	0.25	11.43	nr	**24.16**
LC simplex multimode booted connector	–	–	0.25	11.43	nr	**11.43**
Singlemode core test per core	–	–	0.20	9.14	nr	**9.14**
Multimode core test per core	–	–	0.20	9.14	nr	**9.14**

COMMUNICATIONS/SECURITY/CONTROL

Item	Net Price £	Material £	Labour hours	Labour £	Unit	Total rate £
Fibre; 19" wide fully loaded, labled, alluminium alloy c/w couplers, fibre management and glands (excludes termination of fibre cores)						
8 way ST; fixed drawer	108.46	137.27	0.50	22.85	nr	**160.12**
16 way ST; fixed drawer	162.66	205.86	0.50	22.85	nr	**228.71**
24 way ST; fixed drawer	225.99	286.01	0.50	22.85	nr	**308.86**
8 way ST; sliding drawer	137.98	174.63	0.50	22.85	nr	**197.48**
16 way ST; sliding drawer	189.60	239.95	0.50	22.85	nr	**262.81**
24 way ST; sliding drawer	249.52	315.80	0.50	22.85	nr	**338.65**
8 way (4 duplex) SC; fixed drawer	129.29	163.63	0.50	22.85	nr	**186.49**
16 way (8 duplex) SC; fixed drawer	189.22	239.48	0.50	22.85	nr	**262.33**
24 way (12 duplex) SC; fixed drawer	232.59	294.36	0.50	22.85	nr	**317.21**
8 way (4 duplex) SC; sliding drawer	163.60	207.05	0.50	22.85	nr	**229.90**
16 way (8 duplex) SC; sliding drawer	223.91	283.38	0.50	22.85	nr	**306.23**
24 way (12 duplex) SC; sliding drawer	266.87	337.76	0.50	22.85	nr	**360.61**
8 way (4 duplex) MTRJ; fixed drawer	146.65	185.60	0.50	22.85	nr	**208.45**
16 way (8 duplex) MTRJ; fixed drawer	223.91	283.38	0.50	22.85	nr	**306.23**
24 way (12 duplex) MTRJ; fixed drawer	266.87	337.76	0.50	22.85	nr	**360.61**
8 way (4 duplex) FC/PC; fixed drawer	153.74	194.57	0.50	22.85	nr	**217.42**
16 way (8 duplex) FC/PC; fixed drawer	234.95	297.35	0.50	22.85	nr	**320.21**
24 way (12 duplex) FC/PC; fixed drawer	280.29	354.74	0.50	22.85	nr	**377.59**
Patch panel label per way	0.35	0.45	0.02	0.91	nr	**1.36**
Fibre patch leads						
Single mode						
Duplex OS1 LC – LC						
Fibre patch lead 1 m length	33.26	42.10	0.08	3.66	nr	**45.75**
Fibre patch lead 3 m length	33.52	42.42	0.08	3.66	nr	**46.08**
Fibre patch lead 5 m length	33.09	41.88	0.10	4.57	nr	**46.45**
Multimode						
Duplex 50/125 OM3 MTRJ – MTRJ						
Fibre patch lead 1 m length	29.20	36.95	0.08	3.66	nr	**40.61**
Fibre patch lead 3 m length	31.73	40.16	0.08	3.66	nr	**43.81**
Fibre patch lead 5 m length	33.33	42.18	0.10	4.57	nr	**46.75**
Duplex 50/125 OM3 ST – ST						
Fibre patch lead 1 m length	36.41	46.07	0.08	3.66	nr	**49.73**
Fibre patch lead 3 m length	39.60	50.12	0.08	3.66	nr	**53.77**
Fibre patch lead 5 m length	41.59	52.63	0.10	4.57	nr	**57.20**
Duplex 50/125 OM3 SC – SC						
Fibre patch lead 1 m length	32.66	41.34	0.08	3.66	nr	**45.00**
Firbe patch lead 3 m length	35.55	44.99	0.08	3.66	nr	**48.65**
Fibre patch lead 5 m length	37.31	47.22	0.10	4.57	nr	**51.80**
Duplex 50/125 OM3 LC – LC						
Fibre patch lead 1 m length	35.00	44.30	0.08	3.66	nr	**47.96**
Fibre patch lead 3 m length	38.08	48.19	0.08	3.66	nr	**51.85**
Fibre patch lead 5 m length	39.99	50.61	0.10	4.57	nr	**55.18**

COMMUNICATIONS/SECURITY/CONTROL

Item	Net Price £	Material £	Labour hours	Labour £	Unit	Total rate £
ACCESS CONTROL						
Equipment to control the movement of personnel into defined spaces; includes fixing to backgrounds, termination of power and data cables; excludes cable containment and cable installation						
Access control						
Proximity reader	166.39	210.58	1.50	68.56	nr	279.13
Exit button	26.97	34.13	1.50	68.56	nr	102.69
Exit PIR	76.69	97.06	2.00	91.41	nr	188.46
Emergency break glass double pole	26.31	33.30	1.00	45.70	nr	79.01
Alarm contact flush	9.00	11.39	1.00	45.70	nr	57.10
Alarm contact surface	9.10	11.52	1.00	45.70	nr	57.22
Reader controller 16 door	5633.84	7130.19	4.00	182.81	nr	7313.00
Reader controller 8 door	3279.02	4149.93	4.00	182.81	nr	4332.74
Reader controller 2 door	1806.95	2286.87	4.00	182.81	nr	2469.69
Reader interface	703.43	890.26	1.50	68.56	nr	958.82
Lock power supply 12 volt 3 amp	214.18	271.07	2.00	91.41	nr	362.48
Lock power supply 24 volt 3 amp	213.95	270.78	2.00	91.41	nr	362.18
Rechargeable battery 12 volt 7 Ahr	14.90	18.86	0.25	11.43	nr	30.29
Lock equipment						
Single slimline magnetic lock, monitored	149.23	188.86	2.00	91.41	nr	280.27
Single slimline magnetic lock, unmonitored	140.55	177.88	1.75	79.98	nr	257.86
Double slimline magnetic lock, monitored	297.39	376.38	4.00	182.81	nr	559.20
Double slimline magnetic lock, unmonitored	277.73	351.49	4.50	205.67	nr	557.16
Standard single magnetic lock, monitored	161.30	204.14	1.25	57.13	nr	261.27
Standard single magnetic lock, unmonitored	149.97	189.80	1.00	45.70	nr	235.50
Standard single magnetic lock, double monitored	192.33	243.41	1.50	68.56	nr	311.97
Standard double magnetic lock, monitored	322.27	407.86	1.50	68.56	nr	476.42
Standard double magnetic lock, unmonitored	298.06	377.22	1.25	57.13	nr	434.35
12 V electric release fail safe, monitored	250.46	316.99	1.25	57.13	nr	374.12
12 V electric release fail secure, monitored	250.46	316.99	1.00	45.70	nr	362.69
Solenoid bolt	322.26	407.86	1.50	68.56	nr	476.41
Electric mortice lock	164.34	207.99	1.00	45.70	nr	253.69

COMMUNICATIONS/SECURITY/CONTROL

Item	Net Price £	Material £	Labour hours	Labour £	Unit	Total rate £
SECURITY DETECTION AND ALARM						
Detection and alarm systems for the protection of property and persons; includes fixing of equipment to backgrounds and termination of power and data cabling; excludes cable containment and cable installation						
Detection, alarm equipment						
Alarm contact flush	7.72	9.77	1.00	45.70	nr	55.48
Alarm contact surface	7.72	9.77	1.00	45.70	nr	55.48
Roller shutter contact	24.73	31.30	1.00	45.70	nr	77.00
Personal attack button	26.27	33.24	1.00	45.70	nr	78.95
Acoustic break glass detectors	35.54	44.98	2.00	91.41	nr	136.39
Vibration detectors	46.35	58.66	2.00	91.41	nr	150.07
12 metre PIR detector	54.09	68.45	1.50	68.56	nr	137.01
15 metre dual detector	89.63	113.43	1.50	68.56	nr	181.99
8 zone alarm panel	475.93	602.34	3.00	137.11	nr	739.45
8–24 zone end station	467.69	591.90	4.00	182.81	nr	774.72
Remote keypad	157.61	199.48	2.00	91.41	nr	290.88
8 zone expansion	120.54	152.55	1.50	68.56	nr	221.10
Final exit set button	12.85	16.27	1.00	45.70	nr	61.97
Self-contained external sounder	115.89	146.67	2.00	91.41	nr	238.08
Internal loudspeaker	23.18	29.33	2.00	91.41	nr	120.74
Rechargeable battery 12 volt 7 Ahr (ampere hours)	18.55	23.47	0.25	11.43	nr	34.90
Surveillance equipment						
Vandal-resistant camera, colour	248.34	314.30	3.00	137.11	nr	451.41
External camera, colour	496.68	628.60	5.00	228.52	nr	857.12
Auto dome external, colour	1241.71	1571.50	5.00	228.52	nr	1800.02
Auto dome external, colour/monochrome	1241.71	1571.50	5.00	228.52	nr	1800.02
Auto dome internal, colour	496.68	628.60	4.00	182.81	nr	811.42
Auto dome internal colour/monochrome	496.68	628.60	4.00	182.81	nr	811.42
Mini internal domes	824.76	1043.82	3.00	137.11	nr	1180.93
Camera switcher, 32 inputs	1789.93	2265.34	4.00	182.81	nr	2448.15
Full function keyboard	336.64	426.05	1.00	45.70	nr	471.76
16 CH multiplexors, duplex	992.01	1255.49	2.00	91.41	nr	1346.90
16 way DVR, 250 GB harddrive	1241.71	1571.50	3.00	137.11	nr	1708.61
Additional 250 GB harddrive	331.12	419.07	3.00	137.11	nr	556.18
16 way DVR, 500 GB harddrive	1655.61	2095.34	3.00	137.11	nr	2232.45
16 way DVR, 750 GB harddrive	4966.82	6286.01	1.00	45.70	nr	6331.71
10" colour monitor	215.23	272.39	2.00	91.41	nr	363.80
15" colour monitor	248.34	314.30	2.00	91.41	nr	405.71
17" colour monitor, high resolution	298.01	377.16	2.00	91.41	nr	468.57
21" colour monitor, high resolution	331.12	419.07	2.00	91.41	nr	510.47

COMMUNICATIONS/SECURITY/CONTROL

Item	Net Price £	Material £	Labour hours	Labour £	Unit	Total rate £
SECURITY DETECTION AND ALARM – cont						
Detection and alarm systems for the protection of property and persons – cont						
IP CCTV						
Internal fixed dome IP camera h.264	551.87	698.45	–	–	nr	**698.45**
Internal fixed dome IP dome camera 1MP	551.87	698.45	–	–	nr	**698.45**
External fixed dome IP	620.86	785.76	–	–	nr	**785.76**
External PTZ IP	2207.48	2793.78	–	–	nr	**2793.78**
24 port PoE switch	2414.43	3055.71	–	–	nr	**3055.71**
16 base channel network video recorder with 4TB storage	2759.35	3492.23	–	–	nr	**3492.23**
IP camera channel licence	110.37	139.69	–	–	nr	**139.69**

COMMUNICATIONS/SECURITY/CONTROL

Item	Net Price £	Material £	Labour hours	Labour £	Unit	Total rate £
FIRE DETECTION AND ALARM						
STANDARD FIRE DETECTION						
Control panel						
Zone control panel; 16 zone, mild steel case						
Single loop	622.73	788.13	3.00	137.11	nr	**925.24**
Single loop (flush mounted)	656.15	830.42	0.25	11.43	nr	**841.85**
Loop extension card (1 loop)	325.64	412.14	3.51	160.36	nr	**572.50**
Two loop	930.19	1177.25	4.00	182.81	nr	**1360.07**
Two loop (flush mounted)	963.61	1219.54	5.00	228.52	nr	**1448.06**
Repeater panels						
Standard network card	540.29	683.79	4.00	182.81	nr	**866.60**
FR – focus repeater	540.29	683.79	4.00	182.81	nr	**866.60**
Focus repeater with controls	595.98	754.28	0.25	11.43	nr	**765.70**
Equipment						
Manual call point units: plastic covered						
Surface mounted						
Call point	59.47	75.27	0.50	22.85	nr	**98.12**
Call point; weatherproof	215.39	272.60	0.80	36.56	nr	**309.16**
Flush mounted						
Call point	59.47	75.27	0.56	25.60	nr	**100.87**
Call point; weatherproof	215.39	272.60	0.86	39.33	nr	**311.93**
Detectors						
Smoke, ionization type with mounting base	24.28	30.73	0.75	34.29	nr	**65.02**
Smoke, optical type with mounting base	23.39	29.60	0.75	34.29	nr	**63.89**
Fixed temperature heat detector with mounting base (60°C)	27.86	35.26	0.75	34.29	nr	**69.55**
Rate of rise heat detector with mounting base (90°C)	31.72	40.15	0.75	34.29	nr	**74.43**
Duct detector including optical smoke detector and base	335.35	424.42	2.00	91.41	nr	**515.82**
Remote smoke detector LED indicator with base	7.98	10.10	0.50	22.85	nr	**32.95**
Sounders						
Intelligent wall sounder beacon	126.05	159.53	0.75	34.29	nr	**193.81**
Waterproof intelligent sounder beacon	213.69	270.44	0.75	34.28	nr	**304.72**
Siren; 230 V	104.71	132.52	1.25	57.13	nr	**189.65**
Magnetic door holder; 230 V; surface fixed	98.76	124.99	1.50	68.62	nr	**193.61**

COMMUNICATIONS/SECURITY/CONTROL

Item	Net Price £	Material £	Labour hours	Labour £	Unit	Total rate £
FIRE DETECTION AND ALARM – cont						
ADDRESSABLE FIRE DETECTION						
Control panel						
Analogue addressable panel; BS EN54 Part 2 and 4 1998; incorporating 120 addresses per loop (maximum 1–2 km length); sounders wired on loop; sealed lead acid integral battery standby providing 48 hour standby; 24 volt DC; mild steel case; surface fixed						
1 loop; 4 × 12 volt batteries	2084.36	2637.96	6.00	274.22	nr	**2912.18**
Extra for 1 loop panel						
Loop expander card	117.84	149.14	1.00	45.70	nr	**194.85**
Repeater panel	1447.05	1831.38	6.00	274.22	nr	**2105.60**
Network nodes	2476.22	3133.90	6.00	274.22	nr	**3408.12**
Interface unit; for other systems						
Mains powered	741.55	938.50	1.50	68.56	nr	**1007.06**
Loop powered	325.49	411.94	1.00	45.70	nr	**457.65**
Single channel I/O	169.15	214.08	1.00	45.70	nr	**259.78**
Zone module	210.38	266.26	1.50	68.56	nr	**334.81**
4 loop; 4 × 12 volt batteries; 24 hour standby; 30 minute alarm	3494.38	4422.49	8.00	365.63	nr	**4788.12**
Extra for 4 loop panel						
Loop card	732.76	927.38	1.00	45.70	nr	**973.09**
Repeater panel	2074.97	2626.09	6.00	274.22	nr	**2900.31**
Mimic panel	5094.30	6447.34	5.00	228.52	nr	**6675.86**
Network nodes	2474.30	3131.48	6.00	274.22	nr	**3405.70**
Interface unit; for other systems						
Mains powered	741.55	938.50	1.50	68.56	nr	**1007.06**
Loop powered	325.49	411.94	1.00	45.70	nr	**457.65**
Single channel I/O	169.15	214.08	1.00	45.70	nr	**259.78**
Zone module	210.38	266.26	1.50	68.56	nr	**334.81**
Line modules	38.79	49.09	1.00	45.70	nr	**94.79**
8 loop; 4 × 12 volt batteries; 24 hour standby; 30 minute alarm	7123.46	9015.45	12.00	548.44	nr	**9563.89**
Extra for 8 loop panel						
Loop card	732.76	927.38	1.00	45.70	nr	**973.09**
Repeater panel	2074.97	2626.09	6.00	274.22	nr	**2900.31**
Mimic panel	5094.30	6447.34	5.00	228.52	nr	**6675.86**
Network nodes	2474.30	3131.48	6.00	274.22	nr	**3405.70**

COMMUNICATIONS/SECURITY/CONTROL

Item	Net Price £	Material £	Labour hours	Labour £	Unit	Total rate £
Interface unit; for other systems						
Mains powered	741.55	938.50	1.50	68.56	nr	**1007.06**
Loop powered	325.53	411.99	1.00	45.70	nr	**457.70**
Single channel I/O	71.01	89.87	1.00	45.70	nr	**135.57**
Zone module	210.38	266.26	1.50	68.56	nr	**334.81**
Line modules	38.79	49.09	1.00	45.70	nr	**94.79**
Equipment						
Manual call point						
Surface mounted						
Call point	97.70	123.65	1.00	45.70	nr	**169.35**
Call point; weatherproof	358.02	453.11	1.25	57.13	nr	**510.24**
Flush mounted						
Call point	97.70	123.65	1.00	45.70	nr	**169.35**
Call point; weatherproof	358.02	453.11	1.25	57.13	nr	**510.24**
Detectors						
Smoke, ionization type with mounting base	99.49	125.92	0.75	34.29	nr	**160.20**
Smoke, optical type with mounting base	98.49	124.65	0.75	34.29	nr	**158.94**
Fixed temperature heat detector with mounting base (60°C)	98.99	125.28	0.75	34.29	nr	**159.57**-->
Rate of rise heat detector with mounting base (90°C)	98.99	125.28	0.75	34.29	nr	**159.57**
Duct detector including optical smoke detector and addressable base	630.85	798.41	2.00	91.41	nr	**889.82**
Beam smoke detector with transmitter and receiver unit	972.56	1230.87	2.00	91.41	nr	**1322.28**
Zone short circuit isolator	62.55	79.17	0.75	34.28	nr	**113.44**
Plant interface unit	46.09	58.33	0.50	22.85	nr	**81.18**
Sounders						
Xenon flasher, 24 volt, conduit box	105.05	132.95	0.50	22.85	nr	**155.80**
Xenon flasher, 24 volt, conduit box; weatherproof	156.37	197.90	0.50	22.85	nr	**220.75**
6" bell, conduit box, weatherproof	126.05	159.53	0.75	34.29	nr	**193.81**
6" bell, conduit box	85.22	107.86	0.75	34.28	nr	**142.13**
Siren; 24 V polarized	106.39	134.65	1.00	45.70	nr	**180.35**
Siren; 240 V	104.71	132.52	1.25	57.13	nr	**189.65**
Magnetic door holder; 240 V; surface fixed	98.76	124.99	1.50	68.62	nr	**193.61**

COMMUNICATIONS/SECURITY/CONTROL

Item	Net Price £	Material £	Labour hours	Labour £	Unit	Total rate £
FIRE DETECTION AND ALARM – cont						
Addressable wireless fire detection system, BS 5839 and EN54 Part 25 compliant, 868 mHz operating frequency						
Control panel						
Wireless fire alarm panel, complete with back-up battery						
8 zone	697.60	882.88	4.00	182.81	nr	**1065.70**
20 zone	940.95	1190.87	5.00	228.52	nr	**1419.39**
Equipment						
Smoke detector	186.57	236.13	0.50	22.85	nr	**258.98**
Combined smoke detector and sounder	356.91	451.70	0.50	22.85	nr	**474.55**
Combined smoke detector, sounder and beacon	421.81	533.84	0.50	22.85	nr	**556.69**
Heat detector	178.45	225.85	0.50	22.85	nr	**248.70**
Combined heat detector and sounder	348.81	441.46	0.50	22.85	nr	**464.31**
Combined heat detector, sounder and beacon	405.58	513.30	0.50	22.85	nr	**536.15**
Manual call point	186.57	236.13	0.50	22.85	nr	**258.98**
Manual call point (weatherproof)	438.03	554.37	0.75	34.28	nr	**588.65**
Sounder	251.47	318.26	0.50	22.85	nr	**341.12**
Beacon	308.24	390.10	0.50	22.85	nr	**412.96**
Combined sounder and beacon	316.34	400.36	0.50	22.85	nr	**423.22**
Remote silence button	178.45	225.85	0.50	22.85	nr	**248.70**
Output unit	332.59	420.92	0.50	22.85	nr	**443.77**
Signal booster panel	616.49	780.23	0.50	22.85	nr	**803.08**
Extra over for additional wired antenna	324.46	410.64	1.00	45.70	nr	**456.35**

COMMUNICATIONS/SECURITY/CONTROL

Item	Net Price £	Material £	Labour hours	Labour £	Unit	Total rate £
EARTHING AND BONDING						
Earth bar; polymer insulators and base mounting; including connections						
Non-disconnect link						
6 way	108.34	137.11	0.81	37.02	nr	**174.13**
8 way	119.96	151.82	0.81	37.02	nr	**188.84**
10 way	139.24	176.22	0.81	37.02	nr	**213.24**
Disconnect link						
6 way	122.05	154.46	1.01	46.16	nr	**200.62**
8 way	134.94	170.78	1.01	46.16	nr	**216.94**
10 way	149.64	189.38	1.01	46.16	nr	**235.54**
Soild earth bar; including connections						
150 × 50 × 6 mm	33.02	41.79	1.01	46.16	nr	**87.95**
Extra for earthing						
Disconnecting link						
300 × 50 × 6 mm	37.42	47.36	1.16	53.02	nr	**100.38**
500 × 50 × 6 mm	41.17	52.11	1.16	53.02	nr	**105.13**
Crimp lugs; including screws and connections to cable						
25 mm	0.14	0.18	0.31	14.17	nr	**14.35**
35 mm	0.19	0.24	0.31	14.17	nr	**14.41**
50 mm	0.22	0.28	0.32	14.63	nr	**14.90**
70 mm	0.39	0.50	0.32	14.63	nr	**15.12**
95 mm	0.66	0.83	0.46	21.02	nr	**21.86**
120 mm	0.92	1.17	1.25	57.13	nr	**58.30**
Earth clamps; connection to pipework						
15 mm to 32 mm dia.	2.57	3.26	0.15	6.86	nr	**10.11**
32 mm to 50 mm dia.	3.22	4.07	0.18	8.23	nr	**12.30**
50 mm to 75 mm dia.	4.59	5.81	0.20	9.14	nr	**14.95**

COMMUNICATIONS/SECURITY/CONTROL

Item	Net Price £	Material £	Labour hours	Labour £	Unit	Total rate £
LIGHTNING PROTECTION						
CONDUCTOR TAPE						
PVC sheathed copper tape						
25 × 3 mm	51.83	65.59	0.30	13.71	m	**79.30**
25 × 6 mm	89.90	113.78	0.30	13.71	m	**127.49**
50 × 6 mm	185.20	234.38	0.30	13.71	m	**248.10**
PVC sheathed copper solid circular conductor						
8 mm	28.15	35.62	0.50	22.85	m	**58.47**
Bare copper tape						
20 × 3 mm	14.76	18.68	0.30	13.71	m	**32.39**
25 × 3 mm	16.91	21.40	0.30	13.71	m	**35.11**
25 × 6 mm	31.02	39.25	0.40	18.28	m	**57.53**
50 × 6 mm	60.43	76.48	0.50	22.85	m	**99.33**
Bare copper solid circular conductor						
8 mm	26.42	33.43	0.50	22.85	m	**56.29**
Tape fixings; flat; metalic						
PVC sheathed copper						
25 × 3 mm	13.89	17.57	0.33	15.08	nr	**32.66**
25 × 6 mm	18.60	23.54	0.33	15.08	nr	**38.63**
50 × 6 mm	28.03	35.47	0.33	15.08	nr	**50.55**
8 mm	15.58	19.71	0.50	22.85	nr	**42.56**
Bare copper						
20 × 3 mm	9.52	12.05	0.30	13.71	nr	**25.76**
25 × 3 mm	10.26	12.99	0.30	13.71	nr	**26.70**
25 × 6 mm	8.70	11.01	0.40	18.28	nr	**29.29**
50 × 6 mm	11.39	14.41	0.50	22.85	nr	**37.26**
8 mm	27.74	35.11	0.50	22.85	nr	**57.96**
Tape fixings; flat; non-metallic; PVC sheathed copper						
25 × 3 mm	13.70	17.34	0.30	13.71	nr	**31.05**
Tape fixings; flat; non-metallic; bare copper						
20 × 3 mm	1.51	1.92	0.30	13.71	nr	**15.63**
25 × 3 mm	1.55	1.96	0.30	13.71	nr	**15.67**
50 × 6 mm	3.95	5.00	0.30	13.71	nr	**18.71**
Puddle flanges; copper						
600 mm long	215.98	273.34	0.93	42.50	nr	**315.85**
AIR RODS						
Pointed air rod fixed to structure; copper						
10 mm dia.						
500 mm long	36.01	45.57	1.00	45.70	nr	**91.27**
1000 mm long	60.62	76.72	1.50	68.56	nr	**145.28**
Extra for						
Air terminal base	114.20	144.53	0.35	16.00	nr	**160.52**
Strike pad	47.69	60.35	0.35	16.00	nr	**76.35**
16 mm dia.						
500 mm long	56.99	72.12	0.91	41.59	nr	**113.71**
1000 mm long	104.08	131.73	1.75	79.98	nr	**211.71**
2000 mm long	190.08	240.57	2.50	114.26	nr	**354.83**

COMMUNICATIONS/SECURITY/CONTROL

Item	Net Price £	Material £	Labour hours	Labour £	Unit	Total rate £
Extra for						
Multiple point	95.76	121.19	0.35	16.00	nr	**137.19**
Air terminal base	114.29	144.65	0.35	16.00	nr	**160.64**
Ridge saddle	98.81	125.05	0.35	16.00	nr	**141.05**
Side mounting bracket	98.33	124.44	0.50	22.85	nr	**147.29**
Rod to tape coupling	39.68	50.22	0.50	22.85	nr	**73.07**
Strike pad	47.69	60.35	0.35	16.00	nr	**76.35**
AIR TERMINALS						
16 mm dia.						
500 mm long	56.99	72.12	0.65	29.71	nr	**101.83**
1000 mm long	104.08	131.73	0.78	35.65	nr	**167.38**
2000 mm long	190.08	240.57	1.50	68.56	nr	**309.12**
Extra for						
Multiple point	95.76	121.19	0.35	16.00	nr	**137.19**
Flat saddle	100.46	127.14	0.35	16.00	nr	**143.14**
Side bracket	97.73	123.69	0.50	22.85	nr	**146.54**
Rod to cable coupling	39.69	50.23	0.50	22.85	nr	**73.09**
BONDS AND CLAMPS						
Bond to flat surface; copper						
26 mm	8.55	10.82	0.45	20.57	nr	**31.39**
8 mm dia.	35.23	44.59	0.33	15.08	nr	**59.67**
Pipe bond						
26 mm	8.55	10.82	0.45	20.57	nr	**31.39**
8 mm dia.	87.59	110.86	0.33	15.08	nr	**125.94**
Rod to tape clamp						
26 mm	38.86	49.18	0.45	20.57	nr	**69.75**
Square clamp; copper						
25 × 3 mm	9.30	11.76	0.33	15.08	nr	**26.85**
50 × 6 mm	64.42	81.53	0.50	22.85	nr	**104.39**
8 mm dia.	19.21	24.31	0.33	15.08	nr	**39.39**
Test clamp; copper						
26 × 8 mm; oblong	15.94	20.17	0.50	22.85	nr	**43.02**
26 × 8 mm; plate type	81.42	103.04	0.50	22.85	nr	**125.90**
26 × 8 mm; screw down	73.18	92.61	0.50	22.85	nr	**115.47**
Cast in earth points						
2 hole	64.12	81.15	0.75	34.28	nr	**115.43**
4 hole	125.01	158.21	1.00	45.70	nr	**203.91**
Extra for cast in earth points						
Cover plate; 25 × 3 mm	40.85	51.70	0.25	11.43	nr	**63.13**
Cover plate; 8 mm	40.85	51.70	0.25	11.43	nr	**63.13**
Rebar clamp; 8 mm	45.98	58.19	0.25	11.43	nr	**69.61**
Static earth receptacle	312.95	396.07	0.50	22.85	nr	**418.92**
Copper braided bonds						
25 × 3 mm						
200 mm hole centres	14.24	18.02	0.33	15.08	nr	**33.11**
400 mm holes centres	45.05	57.02	0.40	18.28	nr	**75.30**

COMMUNICATIONS/SECURITY/CONTROL

Item	Net Price £	Material £	Labour hours	Labour £	Unit	Total rate £
LIGHTNING PROTECTION – cont						
BONDS AND CLAMPS – cont						
U bolt clamps						
16 mm	28.22	35.72	0.33	15.08	nr	**50.80**
20 mm	29.04	36.75	0.33	15.08	nr	**51.83**
25 mm	35.00	44.30	0.33	15.08	nr	**59.38**
EARTH PITS/MATS						
Earth inspection pit; hand to others for fixing						
Concrete	36.45	46.13	1.00	45.70	nr	**91.83**
Polypropylene	40.31	51.01	1.00	45.70	nr	**96.71**
Extra for						
5 hole copper earth bar; concrete pit	119.78	151.59	0.35	16.00	nr	**167.59**
5 hole earth bar; polypropylene	29.60	37.47	0.35	16.00	nr	**53.46**
Waterproof electrode seal						
Single flange	422.07	534.17	0.93	42.50	nr	**576.67**
Double flange	710.20	898.83	0.93	42.50	nr	**941.33**
Earth electrode mat; laid in ground and connected						
Copper tape lattice						
600 × 600 × 3 mm	269.66	341.28	0.93	42.50	nr	**383.79**
900 × 900 × 3 mm	484.64	613.36	0.93	42.50	nr	**655.86**
Copper tape plate						
600 × 600 × 1.5 mm	235.76	298.37	0.93	42.50	nr	**340.88**
600 × 600 × 3 mm	508.72	643.84	0.93	42.50	nr	**686.34**
900 × 900 × 1.5 mm	552.29	698.98	0.93	42.50	nr	**741.49**
900 × 900 × 3 mm	987.42	1249.68	0.93	42.50	nr	**1292.19**
EARTH RODS						
Solid cored copper earth electrodes driven into ground and connected						
15 mm dia.						
1200 mm long	39.22	49.64	0.93	42.50	nr	**92.14**
Extra for						
Coupling	3.24	4.11	0.06	2.74	nr	**6.85**
Driving stud	6.29	7.97	0.06	2.74	nr	**10.71**
Spike	6.43	8.14	0.06	2.74	nr	**10.88**
Rod clamp; flat tape	7.29	9.22	0.25	11.43	nr	**20.65**
Rod clamp; solid conductor	23.37	29.58	0.25	11.43	nr	**41.00**
20 mm dia.						
1200 mm long	193.37	244.73	0.98	44.79	nr	**289.52**
Extra for						
Coupling	3.24	4.11	0.06	2.74	nr	**6.85**
Driving stud	6.32	8.00	0.06	2.74	nr	**10.74**
Spike	5.40	6.83	0.06	2.74	nr	**9.57**
Rod clamp; flat tape	7.29	9.22	0.25	11.43	nr	**20.65**
Rod clamp; solid conductor	23.38	29.60	0.25	11.43	nr	**41.02**

COMMUNICATIONS/SECURITY/CONTROL

Item	Net Price £	Material £	Labour hours	Labour £	Unit	Total rate £
Stainless steel earth electrodes driven into ground and connected						
16 mm dia.						
1200 mm long	192.65	243.82	0.93	42.50	nr	**286.32**
Extra for						
Coupling	3.26	4.12	0.06	2.74	nr	**6.87**
Driving head	6.30	7.98	0.06	2.74	nr	**10.72**
Spike	5.39	6.82	0.06	2.74	nr	**9.56**
Rod clamp; flat tape	7.29	9.22	0.25	11.43	nr	**20.65**
Rod clamp; solid conductor	39.65	50.18	0.25	11.43	nr	**61.61**
SURGE PROTECTION						
Single phase; including connection to equipment						
90–150 V	468.37	592.77	5.00	228.52	nr	**821.29**
200–280 V	443.21	560.93	5.00	228.52	nr	**789.45**
Three phase; including connection to equipment						
156–260 V	996.92	1261.70	10.00	457.04	nr	**1718.74**
346–484 V	928.94	1175.67	10.00	457.04	nr	**1632.70**
346–484 V; remote display	886.04	1121.37	10.00	457.04	nr	**1578.40**
346–484 V; 60 kA	1718.15	2174.49	10.00	457.04	nr	**2631.52**
346–484 V; 120 kA	3751.31	4747.66	10.00	457.04	nr	**5204.70**

COMMUNICATIONS/SECURITY/CONTROL

Item	Net Price £	Material £	Labour hours	Labour £	Unit	Total rate £
CENTRAL CONTROL/BUILDING MANAGEMENT						
EQUIPMENT						
Switches/sensors; includes fixing in position; electrical work elsewhere.						
Note: These are normally free issued to the mechanical contractor for fitting. The labour times applied assume the installation has been prepared for the fitting of the component.						
Pressure devices						
Liquid differential pressure sensor	306.65	388.10	0.50	22.85	nr	410.95
Liquid differential pressure switch	50.46	63.86	0.50	22.85	nr	86.72
Air differential pressure transmitter	118.53	150.01	0.50	22.85	nr	172.87
Air differential pressure switch	11.81	14.94	0.50	22.85	nr	37.80
Liquid level switch	107.61	136.19	0.50	22.85	nr	159.04
Static pressure sensor	116.74	147.75	0.50	22.85	nr	170.60
High pressure switch	65.43	82.80	0.50	22.85	nr	105.65
Low pressure switch	65.43	82.80	0.50	22.85	nr	105.65
Water pressure switch	168.41	213.14	0.50	22.85	nr	235.99
Duct averaging temperature sensor	72.97	92.35	1.00	45.70	nr	138.06
Temperature devices						
Return air sensor (fan coils)	4.28	5.41	1.00	45.70	nr	51.12
Frost thermostat	62.40	78.97	0.50	22.85	nr	101.83
Immersion thermostat	133.75	169.27	0.50	22.85	nr	192.12
Temperature high limit	133.75	169.27	0.50	22.85	nr	192.12
Temperature sensor with averaging element	76.06	96.26	0.50	22.85	nr	119.11
Immersion temperature sensor	27.33	34.59	0.50	22.85	nr	57.44
Space temperature sensor	5.70	7.22	1.00	45.70	nr	52.92
Combined space temperature & humidity sensor	52.28	66.16	1.00	45.70	nr	111.87
Outside air temperature sensor	21.32	26.99	2.00	91.41	nr	118.39
Outside air temperature & humidity sensor	94.32	119.37	2.00	91.41	nr	210.77
Duct humidity sensor	84.57	107.03	0.50	22.85	nr	129.89
Space humidity sensor	52.28	66.16	1.00	45.70	nr	111.87
Immersion water flow sensor	44.98	56.92	0.50	22.85	nr	79.77
Rain sensor	114.62	145.06	2.00	91.41	nr	236.47
Wind speed and direction sensor	706.92	894.67	2.00	91.41	nr	986.08
Controllers; includes fixing in position; electrical work elsewhere						
Zone						
Fan coil controller	234.21	296.41	2.00	91.41	nr	387.82
VAV controller	306.26	387.61	2.00	91.41	nr	479.01
Plant						
Controller, 96 I/O points (exact configuration is dependent upon the number of I/O boards added)	5376.54	6804.54	0.50	22.85	nr	6827.40
Controller, 48 I/O points (exact configuration is dependent upon the number of I/O boards added)	4040.34	5113.45	0.50	22.85	nr	5136.30
Controller, 32 I/O points (exact configuration is dependent upon the number of I/O boards added)	3199.90	4049.80	0.50	22.85	nr	4072.65

COMMUNICATIONS/SECURITY/CONTROL

Item	Net Price £	Material £	Labour hours	Labour £	Unit	Total rate £
Additional Digital Input Boards (12 DI)	520.14	658.29	0.20	9.14	nr	**667.43**
Additional Digital Output Boards (6 DO)	446.00	564.46	0.20	9.14	nr	**573.60**
Additional Analogue Input Boards (8 AI)	446.00	564.46	0.20	9.14	nr	**573.60**
Additional Analogue Output Boards (8 AO)	446.00	564.46	0.20	9.14	nr	**573.60**
Outstation Enclosure (fitted in riser with space allowance for controller and network device)	701.54	887.87	5.00	228.52	nr	**1116.39**
Damper actuator; electrical work elsewhere						
Damper actuator 0–10 V	129.39	163.75	1.00	45.70	nr	**209.45**
Damper actuator with auxiliary switches	85.84	108.64	1.00	45.70	nr	**154.34**
Frequency inverters: not mounted within MCC; includes fixing in position; electrical work elsewhere						
2.2 kW	596.89	755.43	2.00	91.41	nr	**846.83**
3 kW	645.08	816.41	2.00	91.41	nr	**907.82**
7.5 kW	881.90	1116.13	2.00	91.41	nr	**1207.54**
11 kW	1160.98	1469.34	2.00	91.41	nr	**1560.75**
15 kW	1472.85	1864.03	2.00	91.41	nr	**1955.44**
18.5 kW	1653.08	2092.14	2.50	114.26	nr	**2206.40**
20 kW	1964.47	2486.23	2.50	114.26	nr	**2600.49**
30 kW	2350.56	2974.87	2.50	114.26	nr	**3089.13**
55 kW	4594.83	5815.22	3.00	137.11	nr	**5952.33**
Miscellaneous; includes fixing in position; electrical work elsewhere						
1 kW Thyristor	81.31	102.91	2.00	91.41	nr	**194.32**
10 kW Thyristor	286.67	362.81	2.00	91.41	nr	**454.22**
Front end and networking; electrical work elsewhere						
PC/monitor	595.29	753.39	2.00	91.41	nr	**844.80**
Printer	64.11	81.14	2.00	91.41	nr	**172.55**
PC software	–	–	–	–	nr	**–**
Network server software	3129.65	3960.89	–	–	nr	**3960.89**
Router (allows connection to a network)	947.64	1199.34	–	–	nr	**1199.34**

COMMUNICATIONS/SECURITY/CONTROL

Item	Net Price £	Material £	Labour hours	Labour £	Unit	Total rate £
Wi-Fi						
Zone director controller, 1000 Mbps, RJ 45, 2 ports, auto MDX and suto sensing						
Up to 2,000 clients	1111.14	1406.26	4.00	182.81	nr	**1589.07**
Up to 5,000 clients, 500 access points	7690.40	9732.97	4.00	182.81	nr	**9915.79**
Up to 20,000 clients, 1,000 access points	9124.02	11547.37	4.00	182.81	nr	**11730.18**
Wi-Fi access points, suitable for large developments (commercial buildings, stadiums, hotels, education)						
Indoor access points, 802.11ac Wi-Fi, dual band, 2.4 GHz and 5 GHz						
Up to 100 connected devices	370.84	469.33	1.20	54.84	nr	**524.18**
Up to 100 connected devices, 867 Mbps	502.20	635.59	1.20	54.84	nr	**690.43**
Up to 300 connected devices, 867 Mbps	654.10	827.83	1.20	54.84	nr	**882.67**
Up to 400 connected devices, 1300 Mbps	805.98	1020.05	1.20	54.84	nr	**1074.90**
Up to 500 connected devices, 1300 Mbps	1008.51	1276.38	1.20	54.84	nr	**1331.22**
Up to 500 connected devices, 1733 Mbps	1327.35	1679.90	1.20	54.84	nr	**1734.74**
Outdoor access points, 802.11ac Wi-Fi, dual band, 2.4 GHz and 5 GHz, IP67, plastic enclosure, with flexible wall or pole mounting						
Standard	1312.29	1660.84	1.80	82.27	nr	**1743.10**
Up to 500 Mbps	2422.08	3065.38	1.80	82.27	nr	**3147.65**
Wi-Fi access points, suitable for small/medium sized and residential						
Indoor access points, 802.11ac Wi-Fi, dual band, 2.4 GHz and 5 GHz						
Up to 100 connected devices, 867 Mbps	502.20	635.59	1.20	54.84	nr	**690.43**
Up to 300 connected devices, 567 Mbps	654.10	827.83	1.20	54.84	nr	**882.67**
Up to 400 connected devices, 1300 Mbps	805.98	1020.05	1.20	54.84	nr	**1074.90**
Outdoor access points, 802.11ac Wi-Fi, dual band, 2.4 GHz and 5 GHz, IP67, plastic enclosure, with flexible wall or pole mounting						
Standard	1312.29	1660.84	1.80	82.27	nr	**1743.10**

Energy Efficiency Applications in Buildings

Bertug Ozarisoy, Hasim Altan,
Young Ki Kim, Wei Shi

Energy Efficiency Applications in Buildings presents an investigation into the energy use and measures to improve the energy efficiency of existing building stock in the UK. The aim of this research is to assess the domestic energy use of statistically representative residential buildings and their occupants' thermal comfort by considering the significant impact of overheating risks on energy consumption and occupants' well-being.

Divided into two volumes, these books present energy consumption and thermal comfort in the construction sector as a complex socio-technical problem that involves the analysis of an intrinsic interrelationship amongst dwellings, occupants, and the environment. Using case studies, the authors demonstrate the significance of improving energy efficiency and its impact on occupants during long-term heatwaves in the summer. Additionally, the volumes demonstrate how dynamic thermal energy simulation can be used as a learning laboratory for future trends in housing energy consumption reduction.

Volume 1: Theories, Methods, and Tools presents the background to the research and the assessment methods and tools adopted by the research team.

*Volume 2: Performance Evaluation and Retrofitting Strategie*s describes the case studies and building performance and post-occupancy evaluations, before making strategic policy and retrofitting recommendations.

The research and roadmap presented in these volumes can be used as a guidance tool for building energy modelling and performance simulation, enabling architects, building engineers, and other practitioners to close the gap between the current understanding and the actual performance of existing building stocks.

June 2025: 296pp
ISBN: 9781032762296

To Order
Tel: +44 (0) 1235 400524
Email: book.orders@tandf.co.uk

For a complete listing of all our titles visit:
www.tandf.co.uk

Taylor & Francis Group
an **informa** business

Mechanical Installations

Rates of Wages

HEATING, VENTILATING, AIR CONDITIONING, PIPING AND DOMESTIC ENGINEERING INDUSTRY

For full details of the wage agreement and the Heating Ventilating Air Conditioning Piping and Domestic Engineering Industry's National Working Rule Agreement, contact:

Building & Engineering Services Association
Lincoln House,
137 – 143 Hammersmith Road,
London W14 0QL
Tel: 020 7313 4900
Internet: www.b-es.org

WAGE RATES, ALLOWANCES AND OTHER PROVISIONS

Hourly rates of wages

All districts of the United Kingdom

Main Grades	From 7 October 2024 p/hr
Foreman	20.29
Senior Craftsman (+2nd welding skill)	18.10
Senior Craftsman	16.78
Craftsman (+2nd welding skill)	16.78
Craftsman	15.45
Operative	13.95
Adult Trainee	11.76
Mate (18 and over)	11.76
Modern Apprentices	
Junior Apprentice	7.62
Intermediate Apprentice	10.82
Senior Apprentice	13.95

Note: Ductwork Erection Operatives are entitled to the same rates and allowances as the parallel Fitter grades shown.

HEATING, VENTILATING, AIR CONDITIONING, PIPING AND DOMESTIC ENGINEERING INDUSTRY

Trainee Rates of Pay

Junior Ductwork Trainees (Probationary)

Age at entry	From 7 October 2024 p/hr
17	6.92
18	6.92
19	6.92
20	6.92

Junior Ductwork Erectors (Year of Training)

	From 7 October 2024		
	1 yr	2 yr	3 yr
Age at entry	p/h	p/hr	p/hr
17	8.62	10.72	12.15
18	8.62	10.72	12.15
19	8.62	10.72	12.15
20	8.62	10.72	12.15

Responsibility Allowance (Craftsmen)	*From 7 October 2024 p/hr*
Second welding skill or supervisory responsibility (one unit)	0.66
Second welding skill and supervisory responsibility (two units)	0.66

Responsibility Allowance (Senior Craftsmen)	*From 7 October 2024 p/hr*
Second welding skill	0.66
Supervising responsibility	1.32
Second welding skill and supervisory responsibility	0.66

Daily travelling allowance – Scale 2

C:	Craftsmen including Installers
M&A:	Mates, Apprentices and Adult Trainees

Direct distance from centre to job in miles

		From 7 October 2024	
Over	*Not exceeding*	*C p/hr*	*p/hr M&A*
15	20	3.24	2.77
20	30	8.29	7.19
30	40	11.94	10.32
40	50	15.73	13.43

HEATING, VENTILATING, AIR CONDITIONING, PIPING AND DOMESTIC ENGINEERING INDUSTRY

Daily travelling allowance – Scale 1

C:	Craftsmen including Installers
M&A:	Mates, Apprentices and Adult Trainees

Direct distance from centre to job in miles

		From 7 October 2024	
Over	Not exceeding	C p/hr	M&A p/hr
0	15	8.82	8.82
15	20	12.05	11.60
20	30	17.12	15.99
30	40	20.75	19.16
40	50	24.53	22.30

Weekly Holiday Credit and Welfare Contributions

	From 7 October 2019						
	£	£	£	£	£	£	£
	a	b	c	d	e	f	g
Weekly Holiday Credit	82.43	76.25	73.55	70.84	68.12	65.45	62.78
Combined Weekly/Welfare Holiday Credit and Contribution	92.08	85.90	83.20	80.49	77.77	75.10	72.43

	From 7 October 2019			
	£	£	£	£
	h	i	j	k
Weekly Holiday Credit	56.63	47.74	43.92	30.99
Combined Weekly/Welfare Holiday Credit and Contribution	66.28	57.39	53.57	40.64

HEATING, VENTILATING, AIR CONDITIONING, PIPING AND DOMESTIC ENGINEERING INDUSTRY

The grades of H&V Operatives entitled to the different rates of Weekly Holiday Credit and Welfare Contribution are as follows:

a	b	c	d
Foreman	Senior Craftsman (RAS & RAW)	Senior Craftsman (RAS)	Senior Craftsman (RAW)
e	f	g	h
Senior Craftsman Craftsman (+2 RA)	Craftsman (+ 1RA)	Craftsman	Installer Senior Modern Apprentice
i	j	k	l
Adult Trainee Mate (over 18)	Intermediate Modern Apprentice	Junior Modern Apprentice	No grade allocated to this Credit Value Category

Daily abnormal conditions money	*From 7 October 2024*
Per day	3.96

Lodging allowance	*From 1 October 2018*
Per night	40.40

Explanatory Notes

1.	Working Hours

	The normal working week (Monday to Friday) shall be 37.5 hours.

2.	Overtime

	Time worked in excess of 37.5 hours during the normal working week shall be paid at time and a half until 12 hours have been worked since the actual starting time. Thereafter double time shall be paid until normal starting time the following morning. Weekend overtime shall be paid at time and a half for the first 5 hours worked on a Saturday and at double time thereafter until normal starting time on Monday morning.

3.	Lodging Allowance

	At the time of review, the rate of nightly Lodging Allowance for the periods from 4 October 2021 and 3 October 2022 are still subject to HMRC approval.

PLUMBING MECHANICAL ENGINEERING SERVICES INDUSTRY

The Joint Industry Board for Plumbing Mechanical Engineering Services has agreed a two-year wage agreement for 2024 and 2025 with effect from 1 January 2024 and subsequently from 6 January 2025.

For full details of this wage agreement and the JIB PMES National Working Rules, contact:

The Joint Industry Board for Plumbing Mechanical Engineering Services in England and Wales
JIB-PMES
PO Box 267
PE19 9DN
Tel: 01480 476925
Email: info@jib-pmes.org.uk

WAGE RATES, ALLOWANCES AND OTHER PROVISIONS

EFFECTIVE FROM 6 January 2025.

Basic Rates of Hourly Pay

Applicable in England and Wales

	Hourly rate £
Operatives	
Technical plumber and gas service technician	20.90
Advanced plumber and gas service engineer	18.82
Trained plumber and gas service fitter	16.16
Apprentices	
4th year of training with NVQ level 2	14.15
4th year of training	12.45
3rd year of training with NVQ level 2	12.31
3rd year of training	10.04
2nd year of training	8.98
1st year of training	7.81
Adult Trainees	
3rd 6 months of employment	14.07
2nd 6 months of employment	13.51
1st 6 months of employment	12.61

PLUMBING MECHANICAL ENGINEERING SERVICES INDUSTRY

Major Projects Agreement

Where a job is designated as being a Major Project then the following Major Project Performance Payment hourly rate supplement shall be payable:

Employee Category

	National Payment £	London* Payment £
Technical Plumber and Gas Service Technician	2.20	3.57
Advanced Plumber and Gas Service Engineer	2.20	3.57
Trained Plumber and Gas Service Fitter	2.20	3.57
All 4 year apprentices	1.76	2.86
All 3 year apprentices	1.32	2.68
2 year apprentice	1.21	1.96
1 year apprentice	0.88	1.43
All adult trainees	1.76	2.86

* The London Payment Supplement applies only to designated Major Projects that are within the M25 London orbital motorway and are effective from 1 February 2007.

* National payment hourly rates are unchanged for 2007 and will continue to be at the rates shown in Promulgation 138A issued 14 October 2003.

PLUMBING MECHANICAL ENGINEERING SERVICES INDUSTRY

Allowances

Daily travel time allowance plus return fares

All daily travel allowances are to be paid at the daily rate as follows:

Over	Not exceeding	All Operatives	3rd & 4th Year Apprentices	1st & 2nd Year Apprentices
20	30	5.91	3.80	2.37
30	40	13.79	8.87	5.69
40	50	15.76	9.41	5.91

Responsibility/Incentive Pay Allowance

As from 3 September 2003, Employers may, in consultation with the employees concerned, enhance the basic graded rates of pay by the payment of an additional amount, as per the bands shown below, where it is agreed that their work involves extra responsibility, productivity or flexibility.

Band 1 –   an additional rate of up to £ 0.38 per hour
Band 2 –   an additional rate of up to £ 0.64 per hour
Band 3 –   an additional rate of up to £ 0.95 per hour
Band 4 –   an additional rate of up to £ 1.24 per hour

This allowance forms part of an operative's basic rate of pay and shall be used to calculate premium payments.

Mileage allowance	£0.45 per mile
Lodging allowance	£51.29 per night
Subsistence allowance (London only)	£6.79 per night

Plumbers welding supplement

Possession of Gas or Arc Certificate	£0.42 per hour
Possession of Gas and Arc Certificate	£0.66 per hour

PLUMBING MECHANICAL ENGINEERING SERVICES INDUSTRY

JIB-PMES Additional Holiday Pay (AHP) (From 6 January 2025)

All PHMES Operatives, Apprentices, etc. who are in current membership of 'Unite the Union' at the time a holiday is taken, shall also be entitled to receive an additional payment of AHP from the JIB-PMES – to be paid via their employer – for each Credit funding their holiday pay. 62 Credits is the maximum number payable in a year.

The amount of AHP payable per credit for all holidays is as follows:

From January 2019

> Operatives etc. £3.02 per credit (Max – £193.28)
>
> All Apprentices £1.53 per credit (Max – £97.92)

Explanatory Notes

1. Working Hours

   The normal working week (Monday to Friday) shall be 37½ hours, with 45 hours to be worked in the same period before overtime rates become applicable.

2. Overtime

   Overtime shall be paid at time and a half up to 8.00pm (Monday to Friday) and up to 1.00pm (Saturday). Overtime worked after these times shall be paid at double time.

3. Major Projects Agreement

   Under the Major Projects Agreement the normal working week shall be 38 hours (Monday to Friday) with overtime rates payable for all hours worked in excess of 38 hours in accordance with 2 above. However, it should be noted that the hourly rate supplement shall be paid for each hour worked but does not attract premium time enhancement.

4. Pension

   In addition to their hourly rates of pay, plumbing employees are entitled to inclusion within the Industry Pension Scheme (or one providing equivalent benefits). The current levels of industry scheme contributions are 6½% (employers) and 3¼% (employees).

5. Additional Holiday Pay (AHP) Contributions

   As last year, JIB-PMES Holiday Pay Schemes shall apply for the Holiday Credit Period 2024–2025, with 64 credits being the maximum number payable in a year.

   For full details please refer to the Additional Holiday Pay (AHP) section as published by the JIB for Plumbing Mechanical Engineering Services in England and Wales.

Electrical Installations

Rates of Wages

ELECTRICAL CONTRACTING INDUSTRY

For full details of this wage agreement and the Joint Industry Board for the Electrical Contracting Industry's National Working Rules, contact:

Joint Industry Board
PO Box 127
Swanley
Kent
BR8 9BH
Telephone: 0333 321 230
Internet: www.jib.org.uk

WAGES (Graded Operatives)

Rates

Since 7 January 2002 two different wage rates have applied to JIB Graded Operatives working on site, depending on whether the Employer transports them to site or whether they provide their own transport. The two categories are:

Job Employed (Transport Provided)

Payable to an Operative who is transported to and from the job by his Employer. The Operative shall also be entitled to payment for Travel Time, when travelling in his own time, as detailed in the appropriate scale.

Job Employed (Own Transport)

Payable to an Operative who travels by his own means to and from the job. The Operative shall be entitled to payment for Travel Allowance and also Travel Time, when travelling in his own time, as detailed in the appropriate scale.

ELECTRICAL CONTRACTING INDUSTRY

The JIB rates of wages are set out below:

From and including 6 January 2025, the JIB hourly rates of wages shall be as set out below:

(i)   National Standard Rate:

Grade	Transport Provided	Own Transport
Technician (or equivalent specialist grade)	£ 21.84	£ 22.96
Approved Electrician (or equivalent specialist grade)	£ 19.32	£ 20.38
Electrician (or equivalent specialist grade)	£ 17.68	£ 18.80
Senior Graded Electrical Trainee	£ 16.84	£ 17.86
Electrical Improver	£ 15.91	£ 16.95
Labourer	£ 14.05	£ 15.09
Adult Trainee	£ 14.05	£ 15.09

(ii)   London Rate:

Grade	Transport Provided	Own Transport
Technician (or equivalent specialist grade)	£ 24.50	£ 25.69
Approved Electrician (or equivalent specialist grade)	£ 21.63	£ 22.83
Electrician (or equivalent specialist grade)	£ 19.80	£ 21.06
Senior Graded Electrical Trainee	£ 18.81	£ 20.01
Electrical Improver	£ 17.82	£ 18.96
Labourer	£ 15.75	£ 16.91
Adult Trainee	£ 15.75	£ 16.91

From and including 6 January 2025, the JIB hourly rates for Job Employed apprentices shall be:

(i)   National Standard Rates

	At College	At Work
Stage 1	£ 6.40	£ 6.44
Stage 2	£ 8.60	£ 9.09
Stage 3	£ 11.92	£ 13.02
Stage 4	£ 12.69	£ 13.99

(ii)   London Rate

	At College	At Work
Stage 1	£ 6.40	£ 7.21
Stage 2	£ 9.24	£ 10.21
Stage 3	£ 13.35	£ 14.58
Stage 4	£ 14.23	£ 15.69

ELECTRICAL CONTRACTING INDUSTRY

Travelling Time and Travel Allowances

Lodging Allowances

£51.29 from and including 6 January 2025

Lodgings weekend retention fee, maximum reimbursement

£51.29 from and including 6 January 2025

Annual Holiday Lodging Allowance Retention

Maximum £16.87 per night (£ 118.09 per week) from and including 6 January 2025

Responsibility money

From and including 30 March 1998 the minimum payment increased to 10p per hour and the maximum to £1.00 per hour (no change)

From and including 4 January 1992 responsibility payments are enhanced by overtime and shift premiums where appropriate (no change)

Combined JIB Benefits Stamp Value (from week commencing 6 January 2020)

JIB grade	Weekly JIB combined credit value £	Holiday value £
Technician	£ 66.93	£ 59.61
Approved Electrician	£ 59.70	£ 52.71
Electrician	£ 55.43	£ 48.27
Labourer & Adult Trainee	£ 44.93	£ 38.31

Explanatory Notes

1. Working Hours

    The normal working week (Monday to Friday) shall be 37½ hours, with 38 hours to be worked in the same period before overtime rates become applicable.

2. Overtime

    Overtime shall be paid at time and a half for all weekday overtime. Saturday overtime shall be paid at time and a half for the first 6 hours, or up to 3.00pm (whichever comes first). Thereafter double time shall be paid until normal starting time on Monday.

Estimator's Pocket Book
3rd edition

By Duncan Cartlidge

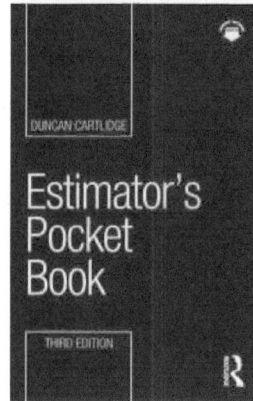

The Estimator's Pocket Book, Third Edition is a concise and practical reference covering the main approaches to pricing, as well as useful information such as how to process sub-contractor quotations, tender settlement and adjudication. It is fully up to date with the New Rules of Measurement (NRM2) (2nd Edition) throughout and based on up-to-date wage rates, legislative changes and guidance notes.

The book includes instructions on how to carry out:

- an NRM order of cost estimate,
- unit-rate pricing for a range different trades,
- pro rata pricing for variations, and
- the preparation and pricing of builders' quantities and approximate quantities.

This book is an essential source of reference for quantity surveyors, cost managers, project managers and anybody else with estimating responsibilities.

June 2024: 312 pp
ISBN: 9781032661520

To Order
Tel:+44 (0) 1235 400524
Email: tandf@hachette.co.uk

For a complete listing of all our titles visit:
www.routledge.com

Taylor & Francis
Taylor & Francis Group

PART 6

Daywork

When work is carried out in connection with a contract that cannot be valued in any other way, it is usual to assess the value on a cost basis with suitable allowances to cover overheads and profit. The basis of costing is a matter for agreement between the parties concerned but definitions of prime cost for the Heating and Ventilating and Electrical Industries have been published jointly by the Royal Institution of Chartered Surveyors and the appropriate bodies of the industries concerned, for those who wish to use them.

These, together with a schedule of basic plant hire charges are reproduced on the following pages, with the kind permission of the Royal Institution of Chartered Surveyors, who own the copyright.

Fire Safe Use of Wood in Buildings

By Andy Buchanan, *et al.*

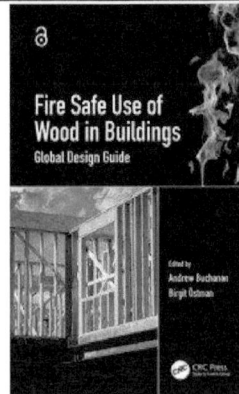

This Open Access book is the first set of global guidelines for fire safe timber structures and wood products in buildings, developed by the Fire Safe Use of Wood global network. It is based on best current scientific knowledge, current codes and standards such as the Fire Part of Eurocode 5, and practical guidance. It provides details on separating, and on the load-bearing functions of timber structures, with information on the reaction to fire of timber and the performance of wood products according to different classification systems, with practical examples stressing the importance proper detailing in building design, active fire protection, and quality workmanship and inspection.

August 2022: 432 pp
ISBN: 9781032040394

To Order
Tel:+44 (0) 1235 400524
Email: tandf@hachette.co.uk

For a complete listing of all our titles visit:
www.tandf.co.uk

Taylor & Francis
Taylor & Francis Group

HEATING AND VENTILATING INDUSTRY

DEFINITION OF PRIME COST OF DAYWORK CARRIED OUT UNDER A HEATING, VENTILATING, AIR CONDITIONING, REFRIGERATION, PIPEWORK AND/OR DOMESTIC ENGINEERING CONTRACT (JULY 1980 EDITION)

This Definition of Prime Cost is published by the Royal Institution of Chartered Surveyors and the Heating and Ventilating Contractors Association for convenience, and for use by people who choose to use it. Members of the Heating and Ventilating Contractors Association are not in any way debarred from defining Prime Cost and rendering accounts for work carried out on that basis in any way they choose. Building owners are advised to reach agreement with contractors on the Definition of Prime Cost to be used prior to entering into a contract or subcontract.

SECTION 1: APPLICATION

1.1 This Definition provides a basis for the valuation of daywork executed under such heating, ventilating, air conditioning, refrigeration, pipework and or domestic engineering contracts as provide for its use.
1.2 It is not applicable in any other circumstances, such as jobbing or other work carried out as a separate or main contract nor in the case of daywork executed after a date of practical completion.
1.3 The terms 'contract' and 'contractor' herein shall be read as 'subcontract' and 'subcontractor' as applicable.

SECTION 2: COMPOSITION OF TOTAL CHARGES

2.1 The Prime Cost of daywork comprises the sum of the following costs:
  (a) Labour as defined in Section 3.
  (b) Materials and goods as defined in Section 4.
  (c) Plant as defined in Section 5.
2.2 Incidental costs, overheads and profit as defined in Section 6, as provided in the contract and expressed therein as percentage adjustments, are applicable to each of 2.1 (a)–(c).

SECTION 3: LABOUR

3.1 The standard wage rates, emoluments and expenses referred to below and the standard working hours referred to in 3.2 are those laid down for the time being in the rules or decisions or agreements of the Joint Conciliation Committee of the Heating, Ventilating and Domestic Engineering Industry applicable to the works (or those of such other body as may be appropriate) and to the grade of operative concerned at the time when and the area where the daywork is executed.
3.2 Hourly base rates for labour are computed by dividing the annual prime cost of labour, based upon the standard working hours and as defined in 3.4, by the number of standard working hours per annum. See example.
3.3 The hourly rates computed in accordance with 3.2 shall be applied in respect of the time spent by operatives directly engaged on daywork, including those operating mechanical plant and transport and erecting and dismantling other plant (unless otherwise expressly provided in the contract) and handling and distributing the materials and goods used in the daywork.
3.4 The annual prime cost of labour comprises the following:
  (a) Standard weekly earnings (i.e. the standard working week as determined at the appropriate rate for the operative concerned).
  (b) Any supplemental payments.
  (c) Any guaranteed minimum payments (unless included in Section 6.1(a)–(p)).
  (d) Merit money.
  (e) Differentials or extra payments in respect of skill, responsibility, discomfort, inconvenience or risk (excluding those in respect of supervisory responsibility – see 3.5)
  (f) Payments in respect of public holidays.
  (g) Any amounts which may become payable by the contractor to or in respect of operatives arising from the rules etc. referred to in 3.1 which are not provided for in 3.4 (a)–(f) nor in Section 6.1 (a)–(p).
  (h) Employers contributions to the WELPLAN, the HVACR Welfare and Holiday Scheme or payments in lieu thereof.
  (i) Employers National Insurance contributions as applicable to 3.4 (a)–(h).
  (j) Any contribution, levy or tax imposed by Statute, payable by the contractor in his capacity as an employer.
3.5 Differentials or extra payments in respect of supervisory responsibility are excluded from the annual prime cost (see Section 6). The time of principals, staff, foremen, chargehands and the like when working manually is admissible under this Section at the rates for the appropriate grades.

HEATING AND VENTILATING INDUSTRY

SECTION 4: MATERIALS AND GOODS

4.1   The prime cost of materials and goods obtained specifically for the daywork is the invoice cost after deducting all trade discounts and any portion of cash discounts in excess of 5%.

4.2   The prime cost of all other materials and goods used in the daywork is based upon the current market prices plus any appropriate handling charges.

4.3   The prime cost referred to in 4.1 and 4.2 includes the cost of delivery to site.

4.4   Any Value Added Tax which is treated, or is capable of being treated, as input tax (as defined by the Finance Act 1972, or any re-enactment or amendment thereof or substitution therefore) by the contractor is excluded.

SECTION 5: PLANT

5.1   Unless otherwise stated in the contract, the prime cost of plant comprises the cost of the following:
   (a)   use or hire of mechanically-operated plant and transport for the time employed on and/or provided or retained for the daywork;
   (b)   use of non-mechanical plant (excluding non-mechanical hand tools) for the time employed on and/or provided or retained for the daywork;
   (c)   transport to and from the site and erection and dismantling where applicable.

5.2   The use of non-mechanical hand tools and of erected scaffolding, staging, trestles or the like is excluded (see Section 6), unless specifically retained for the daywork.

SECTION 6: INCIDENTAL COSTS, OVERHEADS AND PROFIT

6.1   The percentage adjustments provided in the contract which are applicable to each of the totals of Sections 3, 4 and 5 comprise the following:
   (a)   Head office charges.
   (b)   Site staff including site supervision.
   (c)   The additional cost of overtime (other than that referred to in 6.2).
   (d)   Time lost due to inclement weather.
   (e)   The additional cost of bonuses and all other incentive payments in excess of any included in 3.4.
   (f)   Apprentices' study time.
   (g)   Fares and travelling allowances.
   (h)   Country, lodging and periodic allowances.
   (i)   Sick pay or insurances in respect thereof, other than as included in 3.4.
   (j)   Third party and employers' liability insurance.
   (k)   Liability in respect of redundancy payments to employees.
   (l)   Employer's National Insurance contributions not included in 3.4.
   (m)   Use and maintenance of non-mechanical hand tools.
   (n)   Use of erected scaffolding, staging, trestles or the like (but see 5.2).
   (o)   Use of tarpaulins, protective clothing, artificial lighting, safety and welfare facilities, storage and the like that may be available on site.
   (p)   Any variation to basic rates required by the contractor in cases where the contract provides for the use of a specified schedule of basic plant charges (to the extent that no other provision is made for such variation – see 5.1).
   (q)   In the case of a subcontract which provides that the subcontractor shall allow a cash discount, such provision as is necessary for the allowance of the prescribed rate of discount.
   (r)   All other liabilities and obligations whatsoever not specifically referred to in this Section nor chargeable under any other Section.
   (s)   Profit.

6.2   The additional cost of overtime where specifically ordered by the Architect/Supervising Officer shall only be chargeable in the terms of a prior written agreement between the parties.

HEATING AND VENTILATING INDUSTRY

MECHANICAL INSTALLATIONS

Calculation of Hourly Base Rate of Labour for Typical Main Grades applicable from 7 October 2024, refer to notes within Section Four – Rates of Wages.

	Foreman	Senior Craftsman (+ 2nd Welding Skill)	Senior Craftsman	Craftsman	Installer	Mate over 18
Hourly Rate from 7 October 2024	20.29	18.10	16.78	15.45	13.95	11.76
Annual standard earnings excluding all holidays, 45.6 weeks × 38 hours	34,695.90	30,951.00	28,693.80	26,419.50	23,854.50	20,109.60
Employers national insurance contributions for this year	2,695.65	2,342.34	2,132.66	1,914.39	1,672.36	1,318.82
Weekly holiday credit and welfare contributions (52 weeks) for this year	5,503.64	5,335.38	4,657.76	4,335.90	3,974.60	3,448.65
Annual prime cost of labour	43,633.47	38,863.74	36,033.02	33,086.37	29,819.02	25,046.18
Hourly base rate	25.52	22.73	21.07	19.35	17.44	14.65

Notes:

(1) Annual industry holiday (4.8 weeks × 37.5 hours) and public holidays (1.6 weeks × 37.5 hours) are paid through weekly holiday credit and welfare stamp scheme.

(2) Where applicable, Merit money and other variables (e.g. daily abnormal conditions money), which attract Employer's National Insurance contribution, should be included.

(3) Contractors In Northern Ireland should add the appropriate amount of CITB Levy to the annual prime cost of labour prior to calculating the hourly base rate.

(4) Hourly rate based on 1,710.00 hours per annum and calculated as follows:

52 Weeks @ 37.5 hrs/wk	=		1,950.00
Less			
Public Holiday = 8/5 = 1.6 weeks @ 3.75 hrs/wk	=	60	
Annual holidays = 4.6 weeks @ 37.5 hrs/wk	=	180.00	240.00
Hours	=		1,710.00

(5) For calculation of Holiday Credits and ENI refer to detailed labour rate evaluation

(6) National Insurance contributions are those effective from 6 April 2025.

(7) Allowance only for Weekly Holiday Credit/Welfare Stamp

(8) Hourly rates of wages are those assumed and effective from 7 October 2024.

ELECTRICAL INDUSTRY

DEFINITION OF PRIME COST OF DAYWORK CARRIED OUT UNDER AN ELECTRICAL CONTRACT (MARCH 1981 EDITION)

This Definition of Prime Cost is published by The Royal Institution of Chartered Surveyors and The Electrical Contractors' Associations for convenience and for use by people who choose to use it. Members of The Electrical Contractors' Association are not in any way debarred from defining Prime Cost and rendering accounts for work carried out on that basis in any way they choose. Building owners are advised to reach agreement with contractors on the Definition of Prime Cost to be used prior to entering into a contract or subcontract.

SECTION 1: APPLICATION

1.1　This Definition provides a basis for the valuation of daywork executed under such electrical contracts as provide for its use.
1.2　It is not applicable in any other circumstances, such as jobbing, or other work carried out as a separate or main contract, nor in the case of daywork executed after the date of practical completion.
1.3　The terms 'contract' and 'contractor' herein shall be read as 'subcontract' and 'subcontractor' as the context may require.

SECTION 2: COMPOSITION OF TOTAL CHARGES

2.1　The Prime Cost of daywork comprises the sum of the following costs:
　　(a)　Labour as defined in Section 3.
　　(b)　Materials and goods as defined in Section 4.
　　(c)　Plant as defined in Section 5.
2.2　Incidental costs, overheads and profit as defined in Section 6, as provided in the contract and expressed therein as percentage adjustments, are applicable to each of 2.1 (a)–(c).

SECTION 3: LABOUR

3.1　The standard wage rates, emoluments and expenses referred to below and the standard working hours referred to in 3.2 are those laid down for the time being in the rules and determinations or decisions of the Joint Industry Board or the Scottish Joint Industry Board for the Electrical Contracting Industry (or those of such other body as may be appropriate) applicable to the works and relating to the grade of operative concerned at the time when and in the area where daywork is executed.
3.2　Hourly base rates for labour are computed by dividing the annual prime cost of labour, based upon the standard working hours and as defined in 3.4 by the number of standard working hours per annum. See examples.
3.3　The hourly rates computed in accordance with 3.2 shall be applied in respect of the time spent by operatives directly engaged on daywork, including those operating mechanical plant and transport and erecting and dismantling other plant (unless otherwise expressly provided in the contract) and handling and distributing the materials and goods used in the daywork.
3.4　The annual prime cost of labour comprises the following:
　　(a)　Standard weekly earnings (i.e. the standard working week as determined at the appropriate rate for the operative concerned).
　　(b)　Payments in respect of public holidays.
　　(c)　Any amounts which may become payable by the Contractor to or in respect of operatives arising from operation of the rules etc. referred to in 3.1 which are not provided for in 3.4(a) and (b) nor in Section 6.
　　(d)　Employer's National Insurance Contributions as applicable to 3.4 (a)–(c).
　　(e)　Employer's contributions to the Joint Industry Board Combined Benefits Scheme or Scottish Joint Industry Board Holiday and Welfare Stamp Scheme, and holiday payments made to apprentices in compliance with the Joint Industry Board National Working Rules and Industrial Determinations as an employer.
　　(f)　Any contribution, levy or tax imposed by Statute, payable by the Contractor in his capacity as an employer.
3.5　Differentials or extra payments in respect of supervisory responsibility are excluded from the annual prime cost (see Section 6). The time of principals and similar categories, when working manually, is admissible under this Section at the rates for the appropriate grades.

ELECTRICAL INDUSTRY

SECTION 4: MATERIALS AND GOODS

4.1 The prime cost of materials and goods obtained specifically for the daywork is the invoice cost after deducting all trade discounts and any portion of cash discounts in excess of 5%.

4.2 The prime cost of all other materials and goods used in the daywork is based upon the current market prices plus any appropriate handling charges.

4.3 The prime cost referred to in 4.1 and 4.2 includes the cost of delivery to site.

4.4 Any Value Added Tax which is treated, or is capable of being treated, as input tax (as defined by the Finance Act 1972, or any re-enactment or amendment thereof or substitution therefore) by the Contractor is excluded.

SECTION 5: PLANT

5.1 Unless otherwise stated in the contract, the prime cost of plant comprises the cost of the following:
   (a) Use or hire of mechanically-operated plant and transport for the time employed on and/or provided or retained for the daywork;
   (b) Use of non-mechanical plant (excluding non-mechanical hand tools) for the time employed on and/or provided or retained for the daywork;
   (c) Transport to and from the site and erection and dismantling where applicable.

5.2 The use of non-mechanical hand tools and of erected scaffolding, staging, trestles or the likes is excluded (see Section 6), unless specifically retained for daywork.

5.3 Note: Where hired or other plant is operated by the Electrical Contractor's operatives, such time is to be included under Section 3 unless otherwise provided in the contract.

SECTION 6: INCIDENTAL COSTS, OVERHEADS AND PROFIT

6.1 The percentage adjustments provided in the contract which are applicable to each of the totals of Sections 3, 4 and 5, compromise the following:
   (a) Head Office charges.
   (b) Site staff including site supervision.
   (c) The additional cost of overtime (other than that referred to in 6.2).
   (d) Time lost due to inclement weather.
   (e) The additional cost of bonuses and other incentive payments.
   (f) Apprentices' study time.
   (g) Travelling time and fares.
   (h) Country and lodging allowances.
   (i) Sick pay or insurance in lieu thereof, in respect of apprentices.
   (j) Third party and employers' liability insurance.
   (k) Liability in respect of redundancy payments to employees.
   (l) Employers' National Insurance Contributions not included in 3.4.
   (m) Use and maintenance of non-mechanical hand tools.
   (n) Use of erected scaffolding, staging, trestles or the like (but see 5.2).
   (o) Use of tarpaulins, protective clothing, artificial lighting, safety and welfare facilities, storage and the like that may be available on site.
   (p) Any variation to basic rates required by the Contractor in cases where the contract provides for the use of a specified schedule of basic plant charges (to the extent that no other provision is made for such variation – see 5.1).
   (q) All other liabilities and obligations whatsoever not specifically referred to in this Section nor chargeable under any other Section.
   (r) Profit.
   (s) In the case of a subcontract which provides that the subcontractor shall allow a cash discount, such provision as is necessary for the allowance of the prescribed rate of discount.

6.2 The additional cost of overtime where specifically ordered by the Architect/Supervising Officer shall only be chargeable in the terms of a prior written agreement between the parties.

ELECTRICAL INDUSTRY

ELECTRICAL INSTALLATIONS

Calculation of Hourly Base Rate of Labour for Typical Main Grades applicable from 1 January 2024.

	Technician	Approved Electrician	Electrician	Labourer
Hourly Rate from 1 January 2024 (London Rates)	24.47	21.74	20.06	16.10
Annual standard earnings excluding all holidays, 45.8 weeks × 37.5 hours	41,843.70	37,175.40	34,302.60	27,531.00
Employers national insurance contributions from 6 April 2024	5,484.26	4,749.58	4,297.50	3,231.86
JIB Combined benefits	4,370.34	3,970.90	3,717.12	3,144.60
Holiday top up funding	2,282.46	2,026.70	1,877.28	1,499.40
Annual prime cost of labour	53,980.73	47,922.58	44,194.50	35,406.86
Hourly base rate	31.57	28.02	25.84	20.71

Notes:

(1)   Annual industry holiday (4.8 weeks × 37.5 hours) and public holidays (1.6 weeks × 37.5 hours)
(2)   It should be noted that all labour costs incurred by the Contractor in his capacity as an Employer, other than those contained in the hourly rate above, must be taken into account under Section 6.
(3)   Public Holidays are paid through weekly holiday credit and welfare stamp scheme.
(4)   Contractors in Northern Ireland should add the appropriate amount of CITB Levy to the annual prime cost of labour prior to calculating the hourly base rate.
(5)   Hourly rate based on 1,710.00 hours per annum and calculated as follows:

52 Weeks @ 37.5 hrs/wk	=		1,950.00
Less			
Public Holiday = 8/5 = 1.6 weeks @ 37.5 hrs/wk	=	60	
Annual holidays = 4.8 weeks @ 37.5 hrs/wk	=	180	240
Hours	=		1,710.00

(6)   For calculation of holiday credits and ENI refer to detailed labour rate evaluation.
(7)   Hourly wage rates are those effective from 1 January 2024.
(8)   National Insurance contributions are those effective from 6 April 2024.

BUILDING INDUSTRY PLANT HIRE COSTS

SCHEDULE OF BASIC PLANT CHARGES (JULY 2010)

This Schedule is published by the Royal Institution of Chartered Surveyors and is for use in connection with Dayworks under a Building Contract.

EXPLANATORY NOTES

1   The rates in the Schedule are intended to apply solely to daywork carried out under and incidental to a Building Contract. They are NOT intended to apply to:
    (i)   jobbing or any other work carried out as a main or separate contract; or
    (ii)   work carried out after the date of commencement of the Defects Liability Period.

2   The rates apply to plant and machinery already on site, whether hired or owned by the Contractor.

3   The rates, unless otherwise stated, include the cost of fuel and power of every description, lubricating oils, grease, maintenance, sharpening of tools, replacement of spare parts, all consumable stores and for licences and insurances applicable to items of plant.

4   The rates, unless otherwise stated, do not include the costs of drivers and attendants (unless otherwise stated).

5   The rates in the Schedule are base costs and may be subject to an overall adjustment for price movement, overheads and profit, quoted by the Contractor prior to the placing of the Contract.

6   The rates should be applied to the time during which the plant is actually engaged in daywork.

7   Whether or not plant is chargeable on daywork depends on the daywork agreement in use and the inclusion of an item of plant in this schedule does not necessarily indicate that item is chargeable.

8   Rates for plant not included in the Schedule or which is not already on site and is specifically provided or hired for daywork shall be settled at prices which are reasonably related to the rates in the Schedule having regard to any overall adjustment quoted by the Contractor in the Conditions of Contract.

9   The data below is as per the RICS Schedule of Basic Plant Charges 2010 document published on BCIS. In order to adjust the data for current market (or forecast), please follow the steps below:

    −   Log into BCIS Online: https://service.bcis.co.uk/BCISOnline/Account/LogOn?ReturnUrl=%2fBCI-SOnline
    −   Search for 'Schedule of Basic Plant Charges'
    −   On the Basic parameters tab, click 'Edit' then select 'Adjustment selection'
    −   Then choose the quarter and year you would like to adjust the data to
    −   The results will then appear in the Resources section within the Results tab
    −   Now proceed to download where you can download the data that has been adjusted
    −   However, please note that only 1 section can be downloaded at a time

BUILDING INDUSTRY PLANT HIRE COSTS

MECHANICAL PLANT AND TOOLS

Item of plant	Size/Rating	Unit	Rate per Hour (£)
PUMPS			
Mobile Pumps			
Including pump hoses, values and strainers, etc.			
Diaphragm	50 mm dia.	Each	1.17
Diaphragm	76 mm dia.	Each	1.89
Diaphragm	102 mm dia.	Each	3.54
Submersible	50 mm dia.	Each	0.76
Submersible	76 mm dia.	Each	0.86
Submersible	102 mm dia.	Each	1.03
Induced Flow	50 mm dia.	Each	0.77
Induced Flow	76 mm dia.	Each	1.67
Centrifugal, self-priming	25 mm dia.	Each	1.30
Centrifugal, self-priming	50 mm dia.	Each	1.92
Centrifugal, self-priming	75 mm dia.	Each	2.74
Centrifugal, self-priming	102 mm dia.	Each	3.35
Centrifugal, self-priming	152 mm dia.	Each	4.27
SCAFFOLDING, SHORING, FENCING			
Complete Scaffolding			
Mobile working towers, single width	2.0 m × 0.72 m base × 7.45 m high	Each	3.36
Mobile working towers, single width	2.0 m × 0.72 m base × 8.84 m high	Each	3.79
Mobile working towers, double width	2.0 m × 1.35 m × 7.45 m high	Each	3.79
Mobile working towers, double width	2.0 m × 1.35 m × 15.8 m high	Each	7.13
Chimney scaffold, single unit		Each	1.92
Chimney scaffold, twin unit		Each	3.59
Push along access platform	1.63–3.1 m	Each	5.00
Push along access platform	1.80 m × 0.70 m	Each	1.79
Trestles			
Trestle, adjustable	Any height	Pair	0.41
Trestle, painters	1.8 m high	Pair	0.31
Trestle, painters	2.4 m high	Pair	0.36
Shoring, Planking and Strutting			
'Acrow' adjustable prop	Sizes up to 4.9 m (open)	Each	0.06
'Strong boy' support attachment		Each	0.22
Adjustable trench strut	Sizes up to 1.67 m (open)	Each	0.16

BUILDING INDUSTRY PLANT HIRE COSTS

Item of plant	Size/Rating	Unit	Rate per Hour (£)
Trench sheet		Metre	0.03
Backhoe trench box	Base unit	Each	1.23
Backhoe trench box	Top unit	Each	0.87
Temporary Fencing			
Including block and coupler			
Site fencing steel grid panel	3.5 m × 2.0 m	Each	0.05
Anti-climb site steel grid fence panel	3.5 m × 2.0 m	Each	0.08
Solid panel Heras	2.0 m × 2.0 m	Each	0.09
Pedestrian gate		Each	0.36
Roadway gate		Each	0.60

LIFTING APPLIANCES AND CONVEYORS

Cranes

Mobile Cranes

Rates are inclusive of drivers

Lorry mounted, telescopic jib

Item of plant	Size/Rating	Unit	Rate per Hour (£)
Two wheel drive	5 tonnes	Each	19.00
Two wheel drive	8 tonnes	Each	42.00
Two wheel drive	10 tonnes	Each	50.00
Two wheel drive	12 tonnes	Each	77.00
Two wheel drive	20 tonnes	Each	89.69
Four wheel drive	18 tonnes	Each	46.51
Four wheel drive	25 tonnes	Each	35.90
Four wheel drive	30 tonnes	Each	38.46
Four wheel drive	45 tonnes	Each	46.15
Four wheel drive	50 tonnes	Each	53.85
Four wheel drive	60 tonnes	Each	61.54
Four wheel drive	70 tonnes	Each	71.79

Static tower crane

Rates inclusive of driver

Note: Capacity equals maximum lift in tonnes times maximum radius at which it can be lifted

	Capacity (metre/tonnes)	Height under hook above ground (m)		
	Up to	Up to		
Tower crane	30	22	Each	22.23
Tower crane	40	22	Each	26.62
Tower crane	40	30	Each	33.33

BUILDING INDUSTRY PLANT HIRE COSTS

Item of plant	Size/Rating		Unit	Rate per Hour (£)
Tower crane	50	22	Each	29.16
Tower crane	60	22	Each	35.90
Tower crane	60	36	Each	35.90
Tower crane	70	22	Each	41.03
Tower crane	80	22	Each	39.12
Tower crane	90	42	Each	37.18
Tower crane	110	36	Each	47.62
Tower crane	140	36	Each	55.77
Tower crane	170	36	Each	64.11
Tower crane	200	36	Each	71.95
Tower crane	250	36	Each	84.77
Tower crane with luffing jig	30	25	Each	22.23
Tower crane with luffing jig	40	30	Each	26.62
Tower crane with luffing jig	50	30	Each	29.16
Tower crane with luffing jig	60	36	Each	41.03
Tower crane with luffing jig	65	30	Each	33.13
Tower crane with luffing jig	80	22	Each	48.72
Tower crane with luffing jig	100	45	Each	48.72
Tower crane with luffing jig	125	30	Each	53.85
Tower crane with luffing jig	160	50	Each	53.85
Tower crane with luffing jig	200	50	Each	74.36
Tower crane with luffing jig	300	60	Each	100.00
Crane Equipment				
Muck tipping skip	Up to 200 litres		Each	0.67
Muck tipping skip	500 litres		Each	0.82
Muck tipping skip	750 litres		Each	1.08
Muck tipping skip	1000 litres		Each	1.28
Muck tipping skip	1500 litres		Each	1.41
Muck tipping skip	2000 litres		Each	1.67
Mortar skip	250 litres, plastic		Each	0.41
Mortar skip	350 litres steel		Each	0.77
Boat skip	250 litres		Each	0.92
Boat skip	500 litres		Each	1.08
Boat skip	750 litres		Each	1.23
Boat skip	1000 litres		Each	1.38
Boat skip	1500 litres		Each	1.64
Boat skip	2000 litres		Each	1.90

BUILDING INDUSTRY PLANT HIRE COSTS

Item of plant	Size/Rating	Unit	Rate per Hour (£)
Boat skip	3000 litres	Each	2.82
Boat skip	4000 litres	Each	3.23
Master flow skip	250 litres	Each	0.77
Master flow skip	500 litres	Each	1.03
Master flow skip	750 litres	Each	1.28
Master flow skip	1000 litres	Each	1.44
Master flow skip	1500 litres	Each	1.69
Master flow skip	2000 litres	Each	1.85
Grand master flow skip	500 litres	Each	1.28
Grand master flow skip	750 litres	Each	1.64
Grand master flow skip	1000 litres	Each	1.69
Grand master flow skip	1500 litres	Each	1.95
Grand master flow skip	2000 litres	Each	2.21
Cone flow skip	500 litres	Each	1.33
Cone flow skip	1000 litres	Each	1.69
Geared rollover skip	500 litres	Each	1.28
Geared rollover skip	750 litres	Each	1.64
Geared rollover skip	1000 litres	Each	1.69
Geared rollover skip	1500 litres	Each	1.95
Geared rollover skip	2000 litres	Each	2.21
Multi skip, rope operated	200 mm outlet size, 500 litres	Each	1.49
Multi skip, rope operated	200 mm outlet size, 750 litres	Each	1.64
Multi skip, rope operated	200 mm outlet size, 1000 litres	Each	1.74
Multi skip, rope operated	200 mm outlet size, 1500 litres	Each	2.00
Multi skip, rope operated	200 mm outlet size, 2000 litres	Each	2.26
Multi skip, man riding	200 mm outlet size, 1000 litres	Each	2.00
Multi skip	4 point lifting frame	Each	0.90
Multi skip	Chain brothers	Set	0.87
Crane Accessories			
Multi-purpose crane forks	1.5 and 2 tonnes S.W.L.	Each	1.13
Self-levelling crane forks		Each	1.28
Man cage	1 man, 230 kg S.W.L.	Each	1.90
Man cage	2 man, 500 kg S.W.L.	Each	1.95
Man cage	4 man, 750 kg S.W.L.	Each	2.15
Man cage	8 man, 1000 kg S.W.L.	Each	3.33
Stretcher cage	500 kg, S.W.L.	Each	2.69
Goods carrying cage	1500 kg, S.W.L.	Each	1.33
Goods carrying cage	3000 kg, S.W.L.	Each	1.85

BUILDING INDUSTRY PLANT HIRE COSTS

Item of plant	Size/Rating		Unit	Rate per Hour (£)
Builders' skip lifting cradle	12 tonnes, S.W.L.		Each	2.31
Board/pallet fork	1600 kg, S.W.L.		Each	1.90
Gas bottle carrier	500 kg, S.W.L.		Each	0.92
Hoists				
Scaffold hoist	200 kg		Each	2.46
Rack and pinion (goods only)	500 kg		Each	4.56
Rack and pinion (goods only)	1100 kg		Each	5.90
Rack and pinion (goods and passenger)	8 person, 80 kg		Each	7.44
Rack and pinion (goods and passenger)	14 person, 1400 kg		Each	8.72
Wheelbarrow chain sling			Each	1.67
Conveyors				
Belt conveyors				
Conveyor	8 m long × 450 mm wide		Each	5.90
Miniveyor, control box and loading hopper	3 m unit		Each	4.49
Other Conveying Equipment				
Wheelbarrow			Each	0.62
Hydraulic superlift			Each	4.56
Pavac slab lifter (tile hoist)			Each	4.49
High lift pallet truck			Each	3.08
Lifting Trucks	Payload	Maximum Lift		
Fork lift, two wheel drive	1100 kg	up to 3.0 m	Each	5.64
Fork lift, two wheel drive	2540 kg	up to 3.7 m	Each	5.64
Fork lift, four wheel drive	1524 kg	up to 6.0 m	Each	5.64
Fork lift, four wheel drive	2600 kg	up to 5.4 m	Each	7.44
Fork life, four wheel drive	4000 kg	up to 17 m	Each	10.77
Lifting Platforms				
Hydraulic platform (Cherry picker)	9 m		Each	4.62
Hydraulic platform (Cherry picker)	12 m		Each	7.56
Hydraulic platform (Cherry picker)	15 m		Each	10.13
Hydraulic platform (Cherry picker)	17 m		Each	15.63
Hydraulic platform (Cherry picker)	20 m		Each	18.13
Hydraulic platform (Cherry picker)	25.6 m		Each	32.38
Scissor lift	7.6 m, electric		Each	3.85
Scissor lift	7.8 m, electric		Each	5.13
Scissor lift	9.7 m, electric		Each	4.23
Scissor lift	10 m, diesel		Each	6.41
Telescopic handler	7 m, 2 tonnes		Each	5.13

BUILDING INDUSTRY PLANT HIRE COSTS

Item of plant	Size/Rating	Unit	Rate per Hour (£)
Telescopic handler	13 m, 3 tonnes	Each	7.18
Lifting and Jacking Gear			
Pipe winch including gantry	1 tonne	Set	1.92
Pipe winch including gantry	3 tonnes	Set	3.21
Chain block	1 tonne	Each	0.35
Chain block	2 tonnes	Each	0.58
Chain block	5 tonnes	Each	1.14
Pull lift (Tirfor winch)	1 tonne	Each	0.64
Pull lift (Tirfor winch)	1.6 tonnes	Each	0.90
Pull lift (Tirfor winch)	3.2 tonnes	Each	1.15
Brother or chain slings, two legs	not exceeding 3.1 tonnes	Set	0.21
Brother or chain slings, two legs	not exceeding 4.25 tonnes	Set	0.31
Brother or chain slings, four legs	not exceeding 11.2 tonnes	Set	1.09
CONSTRUCTION VEHICLES			
Lorries			
Plated lorries (Rates are inclusive of driver)			
Platform lorry	7.5 tonnes	Each	16.21
Platform lorry	17 tonnes	Each	22.90
Platform lorry	24 tonnes	Each	30.68
Extra for lorry with crane attachment	up to 2.5 tonnes	Each	3.25
Extra for lorry with crane attachment	up to 5 tonnes	Each	6.00
Extra for lorry with crane attachment	up to 7.5 tonnes	Each	9.10
Tipper Lorries			
(Rates are inclusive of driver)			
Tipper lorry	up to 11 tonnes	Each	15.78
Tipper lorry	up to 17 tonnes	Each	23.95
Tipper lorry	up to 25 tonnes	Each	31.35
Tipper lorry	up to 31 tonnes	Each	37.79
Dumpers			
Site use only (excl. tax, insurance and extra cost of DERV etc. when operating on highway)	Makers Capacity		
Two wheel drive	1 tonne	Each	1.71
Four wheel drive	2 tonnes	Each	2.43
Four wheel drive	3 tonnes	Each	2.44
Four wheel drive	5 tonnes	Each	3.08

BUILDING INDUSTRY PLANT HIRE COSTS

Item of plant	Size/Rating	Unit	Rate per Hour (£)
Four wheel drive	6 tonnes	Each	3.85
Four wheel drive	9 tonnes	Each	5.65
Tracked	0.5 tonnes	Each	3.33
Tracked	1.5 tonnes	Each	4.23
Tracked	3.0 tonnes	Each	8.33
Tracked	6.0 tonnes	Each	16.03
Dumper Trucks (*Rates are inclusive of drivers*)			
Dumper truck	up to 15 tonnes	Each	28.56
Dumper truck	up to 17 tonnes	Each	32.82
Dumper truck	up to 23 tonnes	Each	54.64
Dumper truck	up to 30 tonnes	Each	63.50
Dumper truck	up to 35 tonnes	Each	73.02
Dumper truck	up to 40 tonnes	Each	87.84
Dumper truck	up to 50 tonnes	Each	133.44
Tractors			
Agricultural Type			
Wheeled, rubber-clad tyred	up to 40 kW	Each	8.63
Wheeled, rubber-clad tyred	up to 90 kW	Each	25.31
Wheeled, rubber-clad tyred	up to 140 kW	Each	36.49
Crawler Tractors			
With bull or angle dozer	up to 70 kW	Each	29.38
With bull or angle dozer	up to 85 kW	Each	38.63
With bull or angle dozer	up to 100 kW	Each	52.59
With bull or angle dozer	up to 115 kW	Each	55.85
With bull or angle dozer	up to 135 kW	Each	60.43
With bull or angle dozer	up to 185 kW	Each	76.44
With bull or angle dozer	up to 200 kW	Each	96.43
With bull or angle dozer	up to 250 kW	Each	117.68
With bull or angle dozer	up to 350 kW	Each	160.03
With bull or angle dozer	up to 450 kW	Each	219.86
With loading shovel	0.8 m³	Each	26.92
With loading shovel	1.0 m³	Each	32.59
With loading shovel	1.2 m³	Each	37.53
With loading shovel	1.4 m³	Each	42.89
With loading shovel	1.8 m³	Each	52.22
With loading shovel	2.0 m³	Each	57.22
With loading shovel	2.1 m³	Each	60.12

BUILDING INDUSTRY PLANT HIRE COSTS

Item of plant	Size/Rating	Unit	Rate per Hour (£)
With loading shovel	3.5 m³	Each	87.26
Light vans			
VW Caddivan or the like		Each	5.26
VW Transport transit or the like	1.0 tonne	Each	6.03
Luton Box Van or the like	1.8 tonnes	Each	9.87
Water/Fuel Storage			
Mobile water container	110 litres	Each	0.62
Water bowser	1100 litres	Each	0.72
Water bowser	3000 litres	Each	0.87
Mobile fuel container	110 litres	Each	0.62
Fuel bowser	1100 litres	Each	1.23
Fuel bowser	3000 litres	Each	1.87
EXCAVATIONS AND LOADERS			
Excavators			
Wheeled, hydraulic	up to 11 tonnes	Each	25.86
Wheeled, hydraulic	up to 14 tonnes	Each	30.82
Wheeled, hydraulic	up to 16 tonnes	Each	34.50
Wheeled, hydraulic	up to 21 tonnes	Each	39.10
Wheeled, hydraulic	up to 25 tonnes	Each	43.81
Wheeled, hydraulic	up to 30 tonnes	Each	55.30
Crawler, hydraulic	up to 11 tonnes	Each	25.86
Crawler, hydraulic	up to 14 tonnes	Each	30.82
Crawler, hydraulic	up to 17 tonnes	Each	34.50
Crawler, hydraulic	up to 23 tonnes	Each	39.10
Crawler, hydraulic	up to 30 tonnes	Each	43.81
Crawler, hydraulic	up to 35 tonnes	Each	55.30
Crawler, hydraulic	up to 38 tonnes	Each	71.73
Crawler, hydraulic	up to 55 tonnes	Each	95.63
Mini excavator	1000/1500 kg	Each	4.87
Mini excavator	2150/2400 kg	Each	6.67
Mini excavator	2700/3500 kg	Each	7.31
Mini excavator	3500/4500 kg	Each	8.21
Mini excavator	4500/6000 kg	Each	9.23
Mini excavator	7000 kg	Each	14.10
Micro excavator	725 mm wide	Each	5.13

BUILDING INDUSTRY PLANT HIRE COSTS

Item of plant	Size/Rating	Unit	Rate per Hour (£)
Loaders			
Shovel loader	0.4 m³	Each	7.69
Shovel loader	1.57 m³	Each	8.97
Shovel loader, four wheel drive	1.7 m³	Each	4.83
Shovel loader, four wheel drive	2.3 m³	Each	4.38
Shovel loader, four wheel drive	3.3 m³	Each	5.06
Skid steer loader wheeled	300/400 kg payload	Each	7.31
Skid steer loader wheeled	625 kg payload	Each	7.67
Tracked skip loader	650 kg	Each	4.42
Excavator Loaders			
Wheeled tractor type with black-hoe excavator			
Four wheel drive			
Four wheel drive, 2 wheel steer	6 tonnes	Each	6.41
Four wheel drive, 2 wheel steer	8 tonnes	Each	8.59
Attachments			
Breakers for excavator		Each	8.72
Breakers for mini excavator		Each	1.75
Breakers for back-hoe excavator/loader		Each	5.13
COMPACTION EQUIPMENT			
Rollers			
Vibrating roller	368–420 kg	Each	1.43
Single roller	533 kg	Each	1.94
Single roller	750 kg	Each	3.43
Twin roller	up to 650 kg	Each	6.03
Twin roller	up to 950 kg	Each	6.62
Twin roller with seat end steering wheel	up to 1400 kg	Each	7.68
Twin roller with seat end steering wheel	up to 2500 kg	Each	10.61
Pavement roller	3–4 tonnes dead weight	Each	6.00
Pavement roller	4–6 tonnes	Each	6.86
Pavement roller	6–10 tonnes	Each	7.17
Pavement roller	10–13 tonnes	Each	19.86
Rammers			
Tamper rammer 2 stroke-petrol	225 mm–275 mm	Each	1.52

BUILDING INDUSTRY PLANT HIRE COSTS

Item of plant	Size/Rating	Unit	Rate per Hour (£)
Soil Compactors			
Plate compactor	75 mm–400 mm	Each	1.53
Plate compactor rubber pad	375 mm–1400 mm	Each	1.53
Plate compactor reversible plate-petrol	400 mm	Each	2.44
CONCRETE EQUIPMENT			
Concrete/Mortar Mixers			
Open drum without hopper	0.09/0.06 m³	Each	0.61
Open drum without hopper	0.12/0.09 m³	Each	1.22
Open drum without hopper	0.15/0.10 m³	Each	0.72
Concrete/Mortar Transport Equipment			
Concrete pump incl. hose, valve and couplers			
Lorry mounted concrete pump	24 m max. distance	Each	50.00
Lorry mounted concrete pump	34 m max. distance	Each	66.00
Lorry mounted concrete pump	42 m max. distance	Each	91.50
Concrete Equipment			
Vibrator, poker, petrol type	up to 75 mm dia.	Each	0.69
Air vibrator (*excluding compressor and hose*)	up to 75 mm dia.	Each	0.64
Extra poker heads	25/36/60 mm dia.	Each	0.76
Vibrating screed unit with beam	5.00 m	Each	2.48
Vibrating screed unit with adjustable beam	3.00–5.00 m	Each	3.54
Power float	725 mm–900 mm	Each	2.56
Power float finishing pan		Each	0.62
Floor grinder	660 × 1016 mm, 110 V electric	Each	4.31
Floor plane	450 × 1100 mm	Each	4.31
TESTING EQUIPMENT			
Pipe Testing Equipment			
Pressure testing pump, electric		Set	2.19
Pressure test pump		Set	0.80
			Rate per Hour (£)
SITE ACCOMODATION AND TEMPORARY SERVICES			
Heating equipment			
Space heater – propane	80,000 Btu/hr	Each	1.03
Space heater – propane/electric	125,000 Btu/hr	Each	2.09

BUILDING INDUSTRY PLANT HIRE COSTS

Item of plant	Size/Rating	Unit	Rate per Hour (£)
Space heater – propane/electric	250,000 Btu/hr	Each	2.33
Space heater – propane	125,000 Btu/hr	Each	1.54
Space heater – propane	260,000 Btu/hr	Each	1.88
Cabinet heater		Each	0.82
Cabinet heater, catalytic		Each	0.57
Electric halogen heater		Each	1.27
Ceramic heater	3 kW	Each	0.99
Fan heater	3 kW	Each	0.66
Cooling fan		Each	1.92
Mobile cooling unit, small		Each	3.60
Mobile cooling unit, large		Each	4.98
Air conditioning unit		Each	2.81
Site Lighting and Equipment			
Tripod floodlight	500 W	Each	0.48
Tripod floodlight	1000 W	Each	0.62
Towable floodlight	4 × 100 W	Each	3.85
Hand held floodlight	500 W	Each	0.51
Rechargeable light		Each	0.41
Inspection light		Each	0.37
Plasterer's light		Each	0.65
Lighting mast		Each	2.87
Festoon light string	25 m	Each	0.55
Site Electrical Equipment			
Extension leads	240 V/14 m	Each	0.26
Extension leads	110 V/14 m	Each	0.36
Cable reel	25 m 110 V/240 V	Each	0.46
Cable reel	50 m 110 V240 V	Each	0.88
4 way junction box	110V	Each	0.56
Power Generating Units			
Generator – petrol	2 kVA	Each	1.23
Generator – silenced petrol	2 kVA	Each	2.87
Generator – petrol	3 kVA	Each	1.47
Generator – diesel	5 kVA	Each	2.44
Generator – silenced diesel	10 kVA	Each	1.90
Generator – silenced diesel	15 kVA	Each	2.26
Generator – silenced diesel	30 kVA	Each	3.33
Generator – silenced diesel	50 kVA	Each	4.10
Generator – silenced diesel	75 kVA	Each	4.62

BUILDING INDUSTRY PLANT HIRE COSTS

Item of plant	Size/Rating	Unit	Rate per Hour (£)
Generator – silenced diesel	100 kVA	Each	5.64
Generator – silenced diesel	150 kVA	Each	7.18
Generator – silenced diesel	200 kVA	Each	9.74
Generator – silenced diesel	250 kVA	Each	11.28
Generator – silenced diesel	350 kVA	Each	14.36
Generator – silenced diesel	500 kVA	Each	15.38
Tail adaptor	240 V	Each	0.10
Transformers			
Transformer	3 kVA	Each	0.32
Transformer	5 kVA	Each	1.23
Transformer	7.5 kVA	Each	0.59
Transformer	10 kVA	Each	2.00
Rubbish Collection and Disposal Equipment			
<u>Rubbish Chutes</u>			
Standard plastic module	1 m section	Each	0.15
Steel liner insert		Each	0.30
Steel top hopper		Each	0.22
Plastic side entry hopper		Each	0.22
Plastic side entry hopper liner		Each	0.22
Dust Extraction Plant			
Dust extraction unit, light duty		Each	2.97
Dust extraction unit, heavy duty		Each	2.97
SITE EQUIPMENT – Welding Equipment			
<u>Arc-(Electric) Complete With Leads</u>			
Welder generator – petrol	200 amp	Each	3.53
Welder generator – diesel	300/350 amp	Each	3.78
Welder generator – diesel	4000 amp	Each	7.92
Extra welding lead sets		Each	0.69
<u>Gas-Oxy Welder</u>			
Welding and cutting set (including oxygen and acetylene, excluding underwater equipment and thermic boring)			
Small		Each	2.24
Large		Each	3.75
Lead burning gun		Each	0.50

BUILDING INDUSTRY PLANT HIRE COSTS

Item of plant	Size/Rating	Unit	Rate per Hour (£)
Mig welder		Each	1.38
Fume extractor		Each	2.46
Road Works Equipment			
Traffic lights, mains/generator	2-way	Set	10.94
Traffic lights, mains/generator	3-way	Set	11.56
Traffic lights, mains/generator	4-way	Set	12.19
Flashing light		Each	0.10
Road safety cone	450 mm	Each	0.08
Safety cone	750 mm	Each	0.10
Safety barrier plank	1.25 m	Each	0.13
Safety barrier plank	2 m	Each	0.15
Safety barrier plank post		Each	0.13
Safety barrier plank post base		Each	0.10
Safety four gate barrier	1 m each gate	Set	0.77
Guard barrier	2 m	Each	0.19
Road sign	750 mm	Each	0.23
Road sign	900 mm	Each	0.31
Road sign	1200 mm	Each	0.42
Speed ramp/cable protection	500 mm section	Each	0.14
Hose ramp open top	3 m section	Each	0.07
DPC Equipment			
Damp-proofing injection machine		Each	2.56
Cleaning Equipment			
Vacuum cleaner (industrial wet) single motor		Each	1.08
Vacuum cleaner (industrial wet) twin motor	30 litre capacity	Each	1.79
Vacuum cleaner (industrial wet) twin motor	70 litre capacity	Each	2.21
Steam cleaner	Diesel/electric 1 phase	Each	3.33
Steam cleaner	Diesel/electric 3 phase	Each	3.85
Pressure washer, light duty electric	1450 PSI	Each	0.72
Pressure washer, heavy duty, diesel	2500 PSI	Each	1.33
Pressure washer, heavy duty, diesel	4000 PSI	Each	2.18
Cold pressure washer, electric		Each	2.39
Hot pressure washer, petrol		Each	4.19
Hot pressure washer, electric		Each	5.13
Cold pressure washer, petrol		Each	2.92

BUILDING INDUSTRY PLANT HIRE COSTS

Item of plant	Size/Rating	Unit	Rate per Hour (£)
Sandblast attachment to last washer		Each	1.23
Drain cleaning attachment to last washer		Each	1.03
Surface Preparation Equipment			
Rotavator	5 h.p.	Each	2.46
Rotavator	9 h.p.	Each	5.00
Scabbler, up to three heads		Each	1.53
Scabbler, pole		Each	2.68
Scrabbler, multi-headed floor		Each	3.89
Floor preparation machine		Each	1.05
Compressors and Equipment			
<u>Portable Compressors</u>			
Compressor – electric	4 cfm	Each	1.36
Compressor – electric	8 cfm lightweight	Each	1.31
Compressor – electric	8 cfm	Each	1.36
Compressor – electric	14 cfm	Each	1.56
Compressor – petrol	24 cfm	Each	2.15
Compressor – electric	25 cfm	Each	2.10
Compressor – electric	30 cfm	Each	2.36
Compressor – diesel	100 cfm	Each	2.56
Compressor – diesel	250 cfm	Each	5.54
Compressor – diesel	400 cfm	Each	8.72
<u>Mobile Compressors</u>			
Lorry mounted compressor			
(machine plus lorry only)	up to 3 m³	Each	41.47
(machine plus lorry only)	up to 5 m³	Each	48.94
Tractor mounted compressor			
(machine plus rubber tyred tractor)	up to 4 m³	Each	21.03
<u>Accessories (Pneumatic Tools) *(with</u>			
<u>and including up to 15 m of air hose)*</u>			
Demolition pick, medium duty		Each	0.90
Demolition pick, heavy duty		Each	1.03
Breakers (with six steels) light	up to 150 kg	Each	1.19
Breakers (with six steels) medium	295 kg	Each	1.24
Breakers (with six steels) heavy	386 kg	Each	1.44
Rock drill (for use with compressor) hand held		Each	1.18
Additional hoses	15 m	Each	0.09

BUILDING INDUSTRY PLANT HIRE COSTS

Item of plant	Size/Rating	Unit	Rate per Hour (£)
Breakers			
Demolition hammer drill, heavy duty, electric		Each	1.54
Road breaker, electric		Each	2.41
Road breaker, 2 stroke, petrol		Each	4.06
Hydraulic breaker unit, light duty, petrol		Each	3.06
Hydraulic breaker unit, heavy duty, petrol		Each	3.46
Hydraulic breaker unit, heavy duty, diesel		Each	4.62
Quarrying and Tooling Equipment			
Block and stone splitter, hydraulic	600 mm × 600 mm	Each	1.90
Block and stone splitter, manual		Each	1.64
Steel Reinforcement Equipment			
Bar bending machine – manual	up to 13 mm dia. rods	Each	1.03
Bar bending machine – manual	up to 20 mm dia. rods	Each	1.41
Bar shearing machine – electric	up to 38 mm dia. rods	Each	3.08
Bar shearing machine – electric	up to 40 mm dia. rods	Each	4.62
Bar cropper machine – electric	up to 13 mm dia. rods	Each	2.05
Bar cropper machine – electric	up to 20 mm dia. rods	Each	2.56
Bar cropper machine – electric	up to 40 mm dia. rods	Each	4.62
Bar cropper machine – 3 phase	up to 40 mm dia. rods	Each	4.62
Dehumidifiers			
110/240 V Water	68 litres extraction per 24 hours	Each	2.46
110/240 V Water	90 litres extraction per 24 hours	Each	3.38
SMALL TOOLS			
Saws			
Masonry bench saw	350 mm–500 mm dia.	Each	1.13
Floor saw	125 mm max. cut	Each	1.15
Floor saw	150 mm max. cut	Each	3.83
Floor saw, reversible	350 mm max. cut	Each	3.32
Wall saw, electric		Each	2.05
Chop/cut off saw, electric	350 mm dia.	Each	1.79
Circular saw, electric	230 mm dia.	Each	0.72
Tyrannosaw		Each	1.74
Reciprocating saw		Each	0.79
Door trimmer		Each	1.17
Stone saw	300 mm	Each	1.44

BUILDING INDUSTRY PLANT HIRE COSTS

Item of plant	Size/Rating	Unit	Rate per Hour (£)
Chainsaw, petrol	500 mm	Each	3.92
Full chainsaw safety kit		Each	0.41
Worktop jig		Each	1.08
Pipe Work Equipment			
Pipe bender	15 mm–22 mm	Each	0.92
Pipe bender, hydraulic	50 mm	Each	1.76
Pipe bender, electric	50 mm–150 mm dia.	Each	2.19
Pipe cutter, hydraulic		Each	0.46
Tripod pipe vice		Set	0.75
Ratchet threader	12 mm–32 mm	Each	0.93
Pipe threading machine, electric	12 mm–75 mm	Each	3.07
Pipe threading machine, electric	12 mm–100 mm	Each	4.93
Impact wrench, electric		Each	1.33
Hand-held Drills and Equipment			
Impact or hammer drill	up to 25 mm dia.	Each	1.03
Impact or hammer drill	35 mm dia.	Each	1.29
Dry diamond core cutter		Each	0.99
Angle head drill		Each	0.90
Stirrer, mixer drill		Each	1.13
Paint, Insulation Application Equipment			
Airless spray unit		Each	4.13
Portaspray unit		Each	1.16
HPVL turbine spray unit		Each	2.23
Compressor and spray gun		Each	1.91
Other Handtools			
Staple gun		Each	0.96
Air nail gun	110 V	Each	1.01
Cartridge hammer		Each	1.08
Tongue and groove nailer complete with mallet		Each	1.59
Diamond wall chasing machine		Each	2.63
Masonry chain saw	300 mm	Each	5.49
Floor grinder		Each	3.99
Floor plane		Each	1.79
Diamond concrete planer		Each	1.93
Autofeed screwdriver, electric		Each	1.38
Laminate trimmer		Each	0.91
Biscuit jointer		Each	1.49

BUILDING INDUSTRY PLANT HIRE COSTS

Item of plant	Size/Rating	Unit	Rate per Hour (£)
Random orbital sander		Each	0.97
Floor sander		Each	1.54
Palm, delta, flap or belt sander		Each	0.75
Disc cutter, electric	300 mm	Each	1.49
Disc cutter, 2 stroke petrol	300 mm	Each	1.24
Dust suppressor for petrol disc cutter		Each	0.51
Cutter cart for petrol disc cutter		Each	1.21
Grinder, angle or cutter	up to 225 mm	Each	0.50
Grinder, angle or cutter	300 mm	Each	1.41
Mortar raking tool attachment		Each	0.19
Floor/polisher scrubber	325 mm	Each	1.76
Floor tile stripper		Each	2.44
Wallpaper stripper, electric		Each	0.81
Hot air paint stripper		Each	0.50
Electric diamond tile cutter	all sizes	Each	2.42
Hand tile cutter		Each	0.82
Electric needle gun		Each	1.29
Needle chipping gun		Each	1.85
Pedestrian floor sweeper	250 mm dia.	Each	0.82
Pedestrian floor sweeper	Petrol	Each	2.20
Diamond tile saw		Each	1.84
Blow lamp equipment and glass		Set	0.50

PART 7

Tables and Memoranda

This part contains the following sections:

Metric Handbook: Planning and Design Data, 7th edition

Edited by Pamela Buxton

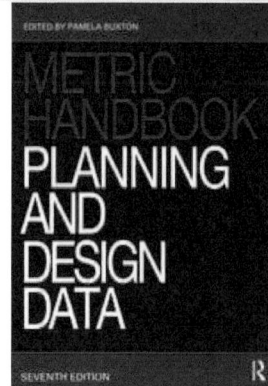

- Consistently updated since 2015 by expert authors in the field
- Significantly revised in reference to changing building types and construction standards
- New chapters added on data centres and logistics facilities
- Sustainable design integrated into chapters throughout
- Over 100,000 copies sold to successive generations of architects and designers
- This book belongs in every design office.

The *Metric Handbook* is the major handbook of planning and design data for architects and architecture students. Covering basic design data for all the major building types it is the ideal starting point for any project. For each building type, the book gives the basic design requirements and all the principal dimensional data, and succinct guidance on how to use the information and what regulations the designer needs to be aware of.

As well as buildings, the *Metric Handbook* deals with broader aspects of design such as materials, acoustics and lighting, and general design data on human dimensions and space requirements. The *Metric Handbook* is the unique reference for solving everyday planning problems.

November 2021: 880 pp
ISBN: 9780367511395

To Order
Tel:+44 (0) 1235 400524
Email: tandf@hachette.co.uk

For a complete listing of all our titles visit:
www.tandf.co.uk

Taylor & Francis
Taylor & Francis Group

CONVERSION TABLES

CONVERSION TABLES

Length	Unit	Conversion factors			
Millimetre	mm	1 in	= 25.4 mm	1 mm	= 0.0394 in
Centimetre	cm	1 in	= 2.54 cm	1 cm	= 0.3937 in
Metre	m	1 ft	= 0.3048 m	1 m	= 3.2808 ft
		1 yd	= 0.9144 m		= 1.0936 yd
Kilometre	km	1 mile	= 1.6093 km	1 km	= 0.6214 mile

Note:	1 cm	= 10 mm	1 ft	= 12 in
	1 m	= 1 000 mm	1 yd	= 3 ft
	1 km	= 1 000 m	1 mile	= 1 760 yd

Area	Unit	Conversion factors			
Square Millimetre	mm^2	1 in^2	= 645.2 mm^2	1 mm^2	= 0.0016 in^2
Square Centimetre	cm^2	1 in^2	= 6.4516 cm^2	1 cm^2	= 1.1550 in^2
Square Metre	m^2	1 ft^2	= 0.0929 m^2	1 m^2	= 10.764 ft^2
		1 yd^2	= 0.8361 m^2	1 m^2	= 1.1960 yd^2
Square Kilometre	km^2	1 $mile^2$	= 2.590 km^2	1 km^2	= 0.3861 $mile^2$

Note:	1 cm^2	= 100 mm^2	1 ft^2	= 144 in^2
	1 m^2	= 10 000 cm^2	1 yd^2	= 9 ft^2
	1 km^2	= 100 hectares	1 acre	= 4 840 yd^2
			1 $mile^2$	= 640 acres

Volume	Unit	Conversion factors			
Cubic Centimetre	cm^3	1 cm^3	= 0.0610 in^3	1 in^3	= 16.387 cm^3
Cubic Decimetre	dm^3	1 dm^3	= 0.0353 ft^3	1 ft^3	= 28.329 dm^3
Cubic Metre	m^3	1 m^3	= 35.3147 ft^3	1 ft^3	= 0.0283 m^3
		1 m^3	= 1.3080 yd^3	1 yd^3	= 0.7646 m^3
Litre	l	1 l	= 1.76 pint	1 pint	= 0.5683 l
			= 2.113 US pt		= 0.4733 US l

Note:	1 dm^3	= 1 000 cm^3	1 ft^3	= 1 728 in^3	1 pint	= 20 fl oz
	1 m^3	= 1 000 dm^3	1 yd^3	= 27 ft^3	1 gal	= 8 pints
	1 l	= 1 dm^3				

Neither the Centimetre nor Decimetre are SI units, and as such their use, particularly that of the Decimetre, is not widespread outside educational circles.

Mass	Unit	Conversion factors			
Milligram	mg	1 mg	= 0.0154 grain	1 grain	= 64.935 mg
Gram	g	1 g	= 0.0353 oz	1 oz	= 28.35 g
Kilogram	kg	1 kg	= 2.2046 lb	1 lb	= 0.4536 kg
Tonne	t	1 t	= 0.9842 ton	1 ton	= 1.016 t

Note:	1 g	= 1000 mg	1 oz	= 437.5 grains	1 cwt	= 112 lb
	1 kg	= 1000 g	1 lb	= 16 oz	1 ton	= 20 cwt
	1 t	= 1000 kg	1 stone	= 14 lb		

Force	Unit	Conversion factors			
Newton	N	1 lbf	= 4.448 N	1 kgf	= 9.807 N
Kilonewton	kN	1 lbf	= 0.004448 kN	1 ton f	= 9.964 kN
Meganewton	MN	100 tonf	= 0.9964 MN		

CONVERSION TABLES

Pressure and stress	Unit	Conversion factors	
Kilonewton per square metre	kN/m^2	1 lbf/in^2	= 6.895 kN/m^2
		1 bar	= 100 kN/m^2
Meganewton per square metre	MN/m^2	1 $tonf/ft^2$	= 107.3 kN/m^2 = 0.1073 MN/m^2
		1 kgf/cm^2	= 98.07 kN/m^2
		1 lbf/ft^2	= 0.04788 kN/m^2

Coefficient of consolidation (Cv) or swelling	Unit	Conversion factors	
Square metre per year	$m^2/year$	1 cm^2/s	= 3 154 $m^2/year$
		1 $ft^2/year$	= 0.0929 $m^2/year$

Coefficient of permeability	Unit	Conversion factors	
Metre per second	m/s	1 cm/s	= 0.01 m/s
Metre per year	m/year	1 ft/year	= 0.3048 m/year
			= 0.9651 × $(10)^8$ m/s

Temperature	Unit	Conversion factors	
Degree Celsius	°C	°C = 5/9 × (°F − 32)	°F = (9 × °C)/ 5 + 32

CONVERSION TABLES

SPEED CONVERSION

km/h	m/min	mph	fpm
1	16.7	0.6	54.7
2	33.3	1.2	109.4
3	50.0	1.9	164.0
4	66.7	2.5	218.7
5	83.3	3.1	273.4
6	100.0	3.7	328.1
7	116.7	4.3	382.8
8	133.3	5.0	437.4
9	150.0	5.6	492.1
10	166.7	6.2	546.8
11	183.3	6.8	601.5
12	200.0	7.5	656.2
13	216.7	8.1	710.8
14	233.3	8.7	765.5
15	250.0	9.3	820.2
16	266.7	9.9	874.9
17	283.3	10.6	929.6
18	300.0	11.2	984.3
19	316.7	11.8	1038.9
20	333.3	12.4	1093.6
21	350.0	13.0	1148.3
22	366.7	13.7	1203.0
23	383.3	14.3	1257.7
24	400.0	14.9	1312.3
25	416.7	15.5	1367.0
26	433.3	16.2	1421.7
27	450.0	16.8	1476.4
28	466.7	17.4	1531.1
29	483.3	18.0	1585.7
30	500.0	18.6	1640.4
31	516.7	19.3	1695.1
32	533.3	19.9	1749.8
33	550.0	20.5	1804.5
34	566.7	21.1	1859.1
35	583.3	21.7	1913.8
36	600.0	22.4	1968.5
37	616.7	23.0	2023.2
38	633.3	23.6	2077.9
39	650.0	24.2	2132.5
40	666.7	24.9	2187.2
41	683.3	25.5	2241.9
42	700.0	26.1	2296.6
43	716.7	26.7	2351.3
44	733.3	27.3	2405.9
45	750.0	28.0	2460.6

CONVERSION TABLES

km/h	m/min	mph	fpm
46	766.7	28.6	2515.3
47	783.3	29.2	2570.0
48	800.0	29.8	2624.7
49	816.7	30.4	2679.4
50	833.3	31.1	2734.0

GEOMETRY

Two dimensional figures

Figure	Diagram of figure	Surface area	Perimeter
Square		a^2	$4a$
Rectangle		ab	$2(a+b)$
Triangle		$\frac{1}{2}ch$	$a+b+c$
Circle		πr^2 $\frac{1}{4}\pi d^2$ where $2r = d$	$2\pi r$ πd
Parallelogram		ah	$2(a+b)$
Trapezium		$\frac{1}{2}h(a+b)$	$a+b+c+d$
Ellipse		Approximately πab	$\pi(a+b)$
Hexagon		$2.6 \times a^2$	

GEOMETRY

Figure	Diagram of figure	Surface area	Perimeter
Octagon		$4.83 \times a^2$	6a
Sector of a circle		$\frac{1}{2}rb$ or $\frac{q}{360}\pi r^2$ note b = angle $\frac{q}{360} \times \pi 2r$	
Segment of a circle		$S - T$ where S = area of sector, T = area of triangle	
Bellmouth		$\frac{3}{14} \times r^2$	

GEOMETRY

Three dimensional figures

Figure	Diagram of figure	Surface area	Volume
Cube		$6a^2$	a^3
Cuboid/ rectangular block		$2(ab + ac + bc)$	abc
Prism/ triangular block		$bd + hc + dc + ad$	$\frac{1}{2}hcd$
Cylinder		$2\pi r^2 + 2\pi h$	$\pi r^2 h$
Sphere		$4\pi r^2$	$\frac{4}{3}\pi r^3$
Segment of sphere		$2\pi Rh$	$\frac{1}{6}\pi h(3r^2 + h^2)$ $\frac{1}{3}\pi h^2(3R - H)$
Pyramid		$(a + b)l + ab$	$\frac{1}{3}abh$

GEOMETRY

Figure	Diagram of figure	Surface area	Volume
Frustum of a pyramid		$l(a + b + c + d) + \sqrt{(ab + cd)}$ [rectangular figure only]	$\frac{h}{3}(ab + cd + \sqrt{abcd})$
Cone		πrl (excluding base) $\pi rl + \pi r^2$ (including base)	$\frac{1}{3}\pi r^2 h$ $\frac{1}{12}\pi d^2 h$
Frustum of a cone		$\pi r^2 + \pi R^2 + \pi l(R + r)$	$\frac{1}{3}\pi(R^2 + Rr + r^2)$

FORMULAE

Formulae

Formula	Description
Pythagoras Theorem	$A^2 = B^2 + C^2$ where A is the hypotenuse of a right-angled triangle and B and C are the two adjacent sides
Simpsons Rule	The Area is divided into an even number of strips of equal width, and therefore has an odd number of ordinates at the division points $$\text{area} = \frac{S(A + 2B + 4C)}{3}$$ where S = common interval (strip width) A = sum of first and last ordinates B = sum of remaining odd ordinates C = sum of the even ordinates The Volume can be calculated by the same formula, but by substituting the area of each coordinate rather than its length
Trapezoidal Rule	A given trench is divided into two equal sections, giving three ordinates, the first, the middle and the last $$\text{volume} = \frac{S \times (A + B + 2C)}{2}$$ where S = width of the strips A = area of the first section B = area of the last section C = area of the rest of the sections
Prismoidal Rule	A given trench is divided into two equal sections, giving three ordinates, the first, the middle and the last $$\text{volume} = \frac{L \times (A + 4B + C)}{6}$$ where L = total length of trench A = area of the first section B = area of the middle section C = area of the last section

TYPICAL THERMAL CONDUCTIVITY OF BUILDING MATERIALS

(Always check manufacturer's details – variation will occur depending on product and nature of materials)

	Thermal conductivity (W/mK)		Thermal conductivity (W/mK)
Acoustic plasterboard	0.25	Oriented strand board	0.13
Aerated concrete slab (500 kg/m^3)	0.16	Outer leaf brick	0.77
Aluminium	237	Plasterboard	0.22
Asphalt (1700 kg/m^3)	0.5	Plaster dense (1300 kg/m^3)	0.5
Bitumen-impregnated fibreboard	0.05	Plaster lightweight (600 kg/m^3)	0.16
Blocks (standard grade 600 kg/m^3)	0.15	Plywood (950 kg/m^3)	0.16
Blocks (solar grade 460 kg/m^3)	0.11	Prefabricated timber wall panels (check manufacturer)	0.12
Brickwork (outer leaf 1700 kg/m^3)	0.84	Screed (1200 kg/m^3)	0.41
Brickwork (inner leaf 1700 kg/m^3)	0.62	Stone chippings (1800 kg/m^3)	0.96
Dense aggregate concrete block 1800 kg/m^3 (exposed)	1.21	Tile hanging (1900 kg/m^3)	0.84
Dense aggregate concrete block 1800 kg/m^3 (protected)	1.13	Timber (650 kg/m^3)	0.14
Calcium silicate board (600 kg/m^3)	0.17	Timber flooring (650 kg/m^3)	0.14
Concrete general	1.28	Timber rafters	0.13
Concrete (heavyweight 2300 kg/m^3)	1.63	Timber roof or floor joists	0.13
Concrete (dense 2100 kg/m^3 typical floor)	1.4	Roof tile (1900 kg/m^3)	0.84
Concrete (dense 2000 kg/m^3 typical floor)	1.13	Timber blocks (650 kg/m^3)	0.14
Concrete (medium 1400 kg/m^3)	0.51	Cellular glass	0.045
Concrete (lightweight 1200 kg/m^3)	0.38	Expanded polystyrene	0.034
Concrete (lightweight 600 kg/m^3)	0.19	Expanded polystyrene slab (25 kg/m^3)	0.035
Concrete slab (aerated 500 kg/m^3)	0.16	Extruded polystyrene	0.035
Copper	390	Glass mineral wool	0.04
External render sand/cement finish	1	Mineral quilt (12 kg/m^3)	0.04
External render (1300 kg/m^3)	0.5	Mineral wool slab (25 kg/m^3)	0.035
Felt – Bitumen layers (1700 kg/m^3)	0.5	Phenolic foam	0.022
Fibreboard (300 kg/m^3)	0.06	Polyisocyanurate	0.025
Glass	0.93	Polyurethane	0.025
Marble	3	Rigid polyurethane	0.025
Metal tray used in wriggly tin concrete floors (7800 kg/m^3)	50	Rock mineral wool	0.038
Mortar (1750 kg/m^3)	0.8		

EARTHWORK

Weights of Typical Materials Handled by Excavators

The weight of the material is that of the state in its natural bed and includes moisture.
Adjustments should be made to allow for loose or compacted states

Material	Mass (kg/m^3)	Mass (lb/cu yd)
Ashes, dry	610	1028
Ashes, wet	810	1365
Basalt, broken	1954	3293
Basalt, solid	2933	4943
Bauxite, crushed	1281	2159
Borax, fine	849	1431
Caliche	1440	2427
Cement, clinker	1415	2385
Chalk, fine	1221	2058
Chalk, solid	2406	4055
Cinders, coal, ash	641	1080
Cinders, furnace	913	1538
Clay, compacted	1746	2942
Clay, dry	1073	1808
Clay, wet	1602	2700
Coal, anthracite, solid	1506	2538
Coal, bituminous	1351	2277
Coke	610	1028
Dolomite, lumpy	1522	2565
Dolomite, solid	2886	4864
Earth, dense	2002	3374
Earth, dry, loam	1249	2105
Earth, Fullers, raw	673	1134
Earth, moist	1442	2430
Earth, wet	1602	2700
Felsite	2495	4205
Fieldspar, solid	2613	4404
Fluorite	3093	5213
Gabbro	3093	5213
Gneiss	2696	4544
Granite	2690	4534
Gravel, dry ¼ to 2 inch	1682	2835
Gravel, dry, loose	1522	2565
Gravel, wet ¼ to 2 inch	2002	3374
Gypsum, broken	1450	2444
Gypsum, solid	2787	4697
Hardcore (consolidated)	1928	3249
Lignite, dry	801	1350
Limestone, broken	1554	2619
Limestone, solid	2596	4375
Magnesite, magnesium ore	2993	5044
Marble	2679	4515
Marl, wet	2216	3735
Mica, broken	1602	2700
Mica, solid	2883	4859
Peat, dry	400	674
Peat, moist	700	1179

EARTHWORK

Material	Mass (kg/m^3)	Mass (lb/cu yd)
Peat, wet	1121	1889
Potash	1281	2159
Pumice, stone	640	1078
Quarry waste	1438	2423
Quartz sand	1201	2024
Quartz, solid	2584	4355
Rhyolite	2400	4045
Sand and gravel, dry	1650	2781
Sand and gravel, wet	2020	3404
Sand, dry	1602	2700
Sand, wet	1831	3086
Sandstone, solid	2412	4065
Shale, solid	2637	4444
Slag, broken	2114	3563
Slag, furnace granulated	961	1619
Slate, broken	1370	2309
Slate, solid	2667	4495
Snow, compacted	481	810
Snow, freshly fallen	160	269
Taconite	2803	4724
Trachyte	2400	4045
Trap rock, solid	2791	4704
Turf	400	674
Water	1000	1685

Transport Capacities

Type of vehicle	Capacity of vehicle	
	Payload	Heaped capacity
Wheelbarrow	150	0.10
1 tonne dumper	1250	1.00
2.5 tonne dumper	4000	2.50
Articulated dump truck (Volvo A20 6 × 4)	18500	11.00
Articulated dump truck (Volvo A35 6 × 6)	32000	19.00
Large capacity rear dumper (Euclid R35)	35000	22.00
Large capacity rear dumper (Euclid R85)	85000	50.00

EARTHWORK

Machine Volumes for Excavating and Filling

Machine type	Cycles per minute	Volume per minute (m³)
1.5 tonne excavator	1	0.04
	2	0.08
	3	0.12
3 tonne excavator	1	0.13
	2	0.26
	3	0.39
5 tonne excavator	1	0.28
	2	0.56
	3	0.84
7 tonne excavator	1	0.28
	2	0.56
	3	0.84
21 tonne excavator	1	1.21
	2	2.42
	3	3.63
Backhoe loader JCB3CX excavator	1	0.28
Rear bucket capacity 0.28 m³	2	0.56
	3	0.84
Backhoe loader JCB3CX loading	1	1.00
Front bucket capacity 1.00 m³	2	2.00

Machine Volumes for Excavating and Filling

Machine type	Loads per hour	Volume per hour (m³)
1 tonne high tip skip loader	5	2.43
Volume 0.485 m³	7	3.40
	10	4.85
3 tonne dumper	4	7.60
Max volume 2.40 m³	5	9.50
Available volume 1.9 m³	7	13.30
	10	19.00
6 tonne dumper	4	15.08
Max volume 3.40 m³	5	18.85
Available volume 3.77 m³	7	26.39
	10	37.70

EARTHWORK

Bulkage of Soils (after excavation)

Type of soil	Approximate bulking of 1 m^3 after excavation
Vegetable soil and loam	25–30%
Soft clay	30–40%
Stiff clay	10–20%
Gravel	20–25%
Sand	40–50%
Chalk	40–50%
Rock, weathered	30–40%
Rock, unweathered	50–60%

Shrinkage of Materials (on being deposited)

Type of soil	Approximate bulking of 1 m^3 after excavation
Clay	10%
Gravel	8%
Gravel and sand	9%
Loam and light sandy soils	12%
Loose vegetable soils	15%

Voids in Material Used as Subbases or Beddings

Material	m^3 of voids/m^3
Alluvium	0.37
River grit	0.29
Quarry sand	0.24
Shingle	0.37
Gravel	0.39
Broken stone	0.45
Broken bricks	0.42

Angles of Repose

Type of soil		Degrees
Clay	– dry	30
	– damp, well drained	45
	– wet	15–20
Earth	– dry	30
	– damp	45
Gravel	– moist	48
Sand	– dry or moist	35
	– wet	25
Loam		40

EARTHWORK

Slopes and Angles

Ratio of base to height	Angle in degrees
5:1	11
4:1	14
3:1	18
2:1	27
1½:1	34
1:1	45
1:1½	56
1:2	63
1:3	72
1:4	76
1:5	79

Grades (in Degrees and Percents)

Degrees	Percent	Degrees	Percent
1	1.8	24	44.5
2	3.5	25	46.6
3	5.2	26	48.8
4	7.0	27	51.0
5	8.8	28	53.2
6	10.5	29	55.4
7	12.3	30	57.7
8	14.0	31	60.0
9	15.8	32	62.5
10	17.6	33	64.9
11	19.4	34	67.4
12	21.3	35	70.0
13	23.1	36	72.7
14	24.9	37	75.4
15	26.8	38	78.1
16	28.7	39	81.0
17	30.6	40	83.9
18	32.5	41	86.9
19	34.4	42	90.0
20	36.4	43	93.3
21	38.4	44	96.6
22	40.4	45	100.0
23	42.4		

EARTHWORK

Bearing Powers

Ground conditions		Bearing power		
		kg/m²	lb/in²	Metric t/m²
Rock,	broken	483	70	50
	solid	2415	350	240
Clay,	dry or hard	380	55	40
	medium dry	190	27	20
	soft or wet	100	14	10
Gravel,	cemented	760	110	80
Sand,	compacted	380	55	40
	clean dry	190	27	20
Swamp and alluvial soils		48	7	5

Earthwork Support

Maximum depth of excavation in various soils without the use of earthwork support

Ground conditions	Feet (ft)	Metres (m)
Compact soil	12	3.66
Drained loam	6	1.83
Dry sand	1	0.3
Gravelly earth	2	0.61
Ordinary earth	3	0.91
Stiff clay	10	3.05

It is important to note that the above table should only be used as a guide. Each case must be taken on its merits and, as the limited distances given above are approached, careful watch must be kept for the slightest signs of caving in

CONCRETE WORK

Weights of Concrete and Concrete Elements

Type of material		kg/m³	lb/cu ft
Ordinary concrete (dense aggregates)			
Non-reinforced plain or mass concrete			
Nominal weight		2305	144
Aggregate	– limestone	2162 to 2407	135 to 150
	– gravel	2244 to 2407	140 to 150
	– broken brick	2000 (av)	125 (av)
	– other crushed stone	2326 to 2489	145 to 155
Reinforced concrete			
Nominal weight		2407	150
Reinforcement	– 1%	2305 to 2468	144 to 154
	– 2%	2356 to 2519	147 to 157
	– 4%	2448 to 2703	153 to 163
Special concretes			
Heavy concrete			
Aggregates	– barytes, magnetite	3210 (min)	200 (min)
	– steel shot, punchings	5280	330
Lean mixes			
Dry-lean (gravel aggregate)		2244	140
Soil-cement (normal mix)		1601	100

CONCRETE WORK

Type of material		kg/m² per mm thick	lb/sq ft per inch thick
Ordinary concrete (dense aggregates)			
Solid slabs (floors, walls etc.)			
Thickness:	75 mm or 3 in	184	37.5
	100 mm or 4 in	245	50
	150 mm or 6 in	378	75
	250 mm or 10 in	612	125
	300 mm or 12 in	734	150
Ribbed slabs			
Thickness:	125 mm or 5 in	204	42
	150 mm or 6 in	219	45
	225 mm or 9 in	281	57
	300 mm or 12 in	342	70
Special concretes			
Finishes etc.			
	Rendering, screed etc. Granolithic, terrazzo	1928 to 2401	10 to 12.5
	Glass-block (hollow) concrete	1734 (approx)	9 (approx)
Prestressed concrete		Weights as for reinforced concrete (upper limits)	
Air-entrained concrete		Weights as for plain or reinforced concrete	

CONCRETE WORK

Average Weight of Aggregates

Materials	Voids %	Weight kg/m³
Sand	39	1660
Gravel 10–20 mm	45	1440
Gravel 35–75 mm	42	1555
Crushed stone	50	1330
Crushed granite (over 15 mm)	50	1345
(n.e. 15 mm)	47	1440
'All-in' ballast	32	1800–2000

Material	kg/m³	lb/cu yd
Vermiculite (aggregate)	64–80	108–135
All-in aggregate	1999	125

Applications and Mix Design

Site mixed concrete

Recommended mix	Class of work suitable for	Cement (kg)	Sand (kg)	Coarse aggregate (kg)	Nr 25 kg bags cement per m³ of combined aggregate
1:3:6	Roughest type of mass concrete such as footings, road haunching over 300 mm thick	208	905	1509	8.30
1:2.5:5	Mass concrete of better class than 1:3:6 such as bases for machinery, walls below ground etc.	249	881	1474	10.00
1:2:4	Most ordinary uses of concrete, such as mass walls above ground, road slabs etc. and general reinforced concrete work	304	889	1431	12.20
1:1.5:3	Watertight floors, pavements and walls, tanks, pits, steps, paths, surface of 2 course roads, reinforced concrete where extra strength is required	371	801	1336	14.90
1:1:2	Works of thin section such as fence posts and small precast work	511	720	1206	20.40

CONCRETE WORK

Ready mixed concrete

Application	Designated concrete	Standardized prescribed concrete	Recommended consistence (nominal slump class)
Foundations			
Mass concrete fill or blinding	GEN 1	ST2	S3
Strip footings	GEN 1	ST2	S3
Mass concrete foundations			
Single storey buildings	GEN 1	ST2	S3
Double storey buildings	GEN 3	ST4	S3
Trench fill foundations			
Single storey buildings	GEN 1	ST2	S4
Double storey buildings	GEN 3	ST4	S4
General applications			
Kerb bedding and haunching	GEN 0	ST1	S1
Drainage works – immediate support	GEN 1	ST2	S1
Other drainage works	GEN 1	ST2	S3
Oversite below suspended slabs	GEN 1	ST2	S3
Floors			
Garage and house floors with no embedded steel	GEN 3	ST4	S2
Wearing surface: Light foot and trolley traffic	RC30	ST4	S2
Wearing surface: General industrial	RC40	N/A	S2
Wearing surface: Heavy industrial	RC50	N/A	S2
Paving			
House drives, domestic parking and external parking	PAV 1	N/A	S2
Heavy-duty external paving	PAV 2	N/A	S2

CONCRETE WORK

Prescribed Mixes for Ordinary Structural Concrete

Weights of cement and total dry aggregates in kg to produce approximately one cubic metre of fully compacted concrete together with the percentages by weight of fine aggregate in total dry aggregates

Conc. grade	Nominal max size of aggregate (mm)	40		20		14		10	
	Workability	Med.	High	Med.	High	Med.	High	Med.	High
	Limits to slump that may be expected (mm)	50–100	100–150	25–75	75–125	10–50	50–100	10–25	25–50
7	Cement (kg)	180	200	210	230	–	–	–	–
	Total aggregate (kg)	1950	1850	1900	1800	–	–	–	–
	Fine aggregate (%)	30–45	30–45	35–50	35–50	–	–	–	–
10	Cement (kg)	210	230	240	260	–	–	–	–
	Total aggregate (kg)	1900	1850	1850	1800	–	–	–	–
	Fine aggregate (%)	30–45	30–45	35–50	35–50	–	–	–	–
15	Cement (kg)	250	270	280	310	–	–	–	–
	Total aggregate (kg)	1850	1800	1800	1750	–	–	–	–
	Fine aggregate (%)	30–45	30–45	35–50	35–50	–	–	–	–
20	Cement (kg)	300	320	320	350	340	380	360	410
	Total aggregate (kg)	1850	1750	1800	1750	1750	1700	1750	1650
	Sand								
	Zone 1 (%)	35	40	40	45	45	50	50	55
	Zone 2 (%)	30	35	35	40	40	45	45	50
	Zone 3 (%)	30	30	30	35	35	40	40	45
25	Cement (kg)	340	360	360	390	380	420	400	450
	Total aggregate (kg)	1800	1750	1750	1700	1700	1650	1700	1600
	Sand								
	Zone 1 (%)	35	40	40	45	45	50	50	55
	Zone 2 (%)	30	35	35	40	40	45	45	50
	Zone 3 (%)	30	30	30	35	35	40	40	45
30	Cement (kg)	370	390	400	430	430	470	460	510
	Total aggregate (kg)	1750	1700	1700	1650	1700	1600	1650	1550
	Sand								
	Zone 1 (%)	35	40	40	45	45	50	50	55
	Zone 2 (%)	30	35	35	40	40	45	45	50
	Zone 3 (%)	30	30	30	35	35	40	40	45

REINFORCEMENT

Weights of Bar Reinforcement

Nominal sizes (mm)	Cross-sectional area (mm²)	Mass (kg/m)	Length of bar (m/tonne)
6	28.27	0.222	4505
8	50.27	0.395	2534
10	78.54	0.617	1622
12	113.10	0.888	1126
16	201.06	1.578	634
20	314.16	2.466	405
25	490.87	3.853	260
32	804.25	6.313	158
40	1265.64	9.865	101
50	1963.50	15.413	65

Weights of Bars (at specific spacings)

Weights of metric bars in kilogrammes per square metre

Size (mm)	Spacing of bars in millimetres									
	75	100	125	150	175	200	225	250	275	300
6	2.96	2.220	1.776	1.480	1.27	1.110	0.99	0.89	0.81	0.74
8	5.26	3.95	3.16	2.63	2.26	1.97	1.75	1.58	1.44	1.32
10	8.22	6.17	4.93	4.11	3.52	3.08	2.74	2.47	2.24	2.06
12	11.84	8.88	7.10	5.92	5.07	4.44	3.95	3.55	3.23	2.96
16	21.04	15.78	12.63	10.52	9.02	7.89	7.02	6.31	5.74	5.26
20	32.88	24.66	19.73	16.44	14.09	12.33	10.96	9.87	8.97	8.22
25	51.38	38.53	30.83	25.69	22.02	19.27	17.13	15.41	14.01	12.84
32	84.18	63.13	50.51	42.09	36.08	31.57	28.06	25.25	22.96	21.04
40	131.53	98.65	78.92	65.76	56.37	49.32	43.84	39.46	35.87	32.88
50	205.51	154.13	123.31	102.76	88.08	77.07	68.50	61.65	56.05	51.38

Basic weight of steelwork taken as 7850 kg/m³
Basic weight of bar reinforcement per metre run = 0.00785 kg/mm²
The value of π has been taken as 3.141592654

REINFORCEMENT

Fabric Reinforcement

Preferred range of designated fabric types and stock sheet sizes

Fabric reference	Longitudinal wires			Cross wires			
	Nominal wire size (mm)	Pitch (mm)	Area (mm²/m)	Nominal wire size (mm)	Pitch (mm)	Area (mm²/m)	Mass (kg/m²)
Square mesh							
A393	10	200	393	10	200	393	6.16
A252	8	200	252	8	200	252	3.95
A193	7	200	193	7	200	193	3.02
A142	6	200	142	6	200	142	2.22
A98	5	200	98	5	200	98	1.54
Structural mesh							
B1131	12	100	1131	8	200	252	10.90
B785	10	100	785	8	200	252	8.14
B503	8	100	503	8	200	252	5.93
B385	7	100	385	7	200	193	4.53
B283	6	100	283	7	200	193	3.73
B196	5	100	196	7	200	193	3.05
Long mesh							
C785	10	100	785	6	400	70.8	6.72
C636	9	100	636	6	400	70.8	5.55
C503	8	100	503	5	400	49.0	4.34
C385	7	100	385	5	400	49.0	3.41
C283	6	100	283	5	400	49.0	2.61
Wrapping mesh							
D98	5	200	98	5	200	98	1.54
D49	2.5	100	49	2.5	100	49	0.77

Stock sheet size 4.8 m × 2.4 m, Area 11.52 m²

Average weight kg/m³ of steelwork reinforcement in concrete for various building elements

Substructure	kg/m³ concrete	Substructure	kg/m³ concrete
Pile caps	110–150	Plate slab	150–220
Tie beams	130–170	Cant slab	145–210
Ground beams	230–330	Ribbed floors	130–200
Bases	125–180	Topping to block floor	30–40
Footings	100–150	Columns	210–310
Retaining walls	150–210	Beams	250–350
Raft	60–70	Stairs	130–170
Slabs – one way	120–200	Walls – normal	40–100
Slabs – two way	110–220	Walls – wind	70–125

Note: For exposed elements add the following %:
Walls 50%, Beams 100%, Columns 15%

FORMWORK

Formwork Stripping Times – Normal Curing Periods

Conditions under which concrete is maturing	Minimum periods of protection for different types of cement					
	Number of days (where the average surface temperature of the concrete exceeds 10°C during the whole period)			Equivalent maturity (degree hours) calculated as the age of the concrete in hours multiplied by the number of degrees Celsius by which the average surface temperature of the concrete exceeds 10°C		
	Other	SRPC	OPC or RHPC	Other	SRPC	OPC or RHPC
1. Hot weather or drying winds	7	4	3	3500	2000	1500
2. Conditions not covered by 1	4	3	2	2000	1500	1000

KEY
OPC – Ordinary Portland Cement
RHPC – Rapid-hardening Portland Cement
SRPC – Sulphate-resisting Portland Cement

Minimum Period before Striking Formwork

	Minimum period before striking		
	Surface temperature of concrete		
	16°C	17°C	t°C (0–25)
Vertical formwork to columns, walls and large beams	12 hours	18 hours	300 hours t+10
Soffit formwork to slabs	4 days	6 days	100 days t+10
Props to slabs	10 days	15 days	250 days t+10
Soffit formwork to beams	9 days	14 days	230 days t+10
Props to beams	14 days	21 days	360 days t+10

MASONRY

Number of Bricks Required for Various Types of Work per m² of Walling

Description	Brick size	
	215 × 102.5 × 50 mm	215 × 102.5 × 65 mm
Half brick thick		
Stretcher bond	74	59
English bond	108	86
English garden wall bond	90	72
Flemish bond	96	79
Flemish garden wall bond	83	66
One brick thick and cavity wall of two half brick skins		
Stretcher bond	148	119

Quantities of Bricks and Mortar Required per m² of Walling

	Unit	No of bricks required	Mortar required (cubic metres)		
Standard bricks			**No frogs**	**Single frogs**	**Double frogs**
Brick size 215 × 102.5 × 50 mm					
half brick wall (103 mm)	m²	72	0.022	0.027	0.032
2 × half brick cavity wall (270 mm)	m²	144	0.044	0.054	0.064
one brick wall (215 mm)	m²	144	0.052	0.064	0.076
one and a half brick wall (322 mm)	m²	216	0.073	0.091	0.108
Mass brickwork	m³	576	0.347	0.413	0.480
Brick size 215 × 102.5 × 65 mm					
half brick wall (103 mm)	m²	58	0.019	0.022	0.026
2 × half brick cavity wall (270 mm)	m²	116	0.038	0.045	0.055
one brick wall (215 mm)	m²	116	0.046	0.055	0.064
one and a half brick wall (322 mm)	m²	174	0.063	0.074	0.088
Mass brickwork	m³	464	0.307	0.360	0.413
Metric modular bricks			**Perforated**		
Brick size 200 × 100 × 75 mm					
90 mm thick	m²	67	0.016	0.019	
190 mm thick	m²	133	0.042	0.048	
290 mm thick	m²	200	0.068	0.078	
Brick size 200 × 100 × 100 mm					
90 mm thick	m²	50	0.013	0.016	
190 mm thick	m²	100	0.036	0.041	
290 mm thick	m²	150	0.059	0.067	
Brick size 300 × 100 × 75 mm					
90 mm thick	m²	33	–	0.015	
Brick size 300 × 100 × 100 mm					
90 mm thick	m²	44	0.015	0.018	

Note: Assuming 10 mm thick joints

MASONRY

Mortar Required per m² Blockwork (9.88 blocks/m²)

Wall thickness	75	90	100	125	140	190	215
Mortar m³/m²	0.005	0.006	0.007	0.008	0.009	0.013	0.014

Mortar Group	Cement: lime: sand	Masonry cement: sand	Cement: sand with plasticizer
1	1:0–0.25:3		
2	1:0.5:4–4.5	1:2.5-3.5	1:3–4
3	1:1:5–6	1:4–5	1:5–6
4	1:2:8–9	1:5.5–6.5	1:7–8
5	1:3:10–12	1:6.5–7	1:8

Group 1: strong inflexible mortar
Group 5: weak but flexible

All mixes within a group are of approximately similar strength
Frost resistance increases with the use of plasticizers
Cement: lime: sand mixes give the strongest bond and greatest resistance to rain penetration
Masonry cement equals ordinary Portland cement plus a fine neutral mineral filler and an air entraining agent

Calcium Silicate Bricks

Type	Strength	Location
Class 2 crushing strength	14.0 N/mm²	not suitable for walls
Class 3	20.5 N/mm²	walls above dpc
Class 4	27.5 N/mm²	cappings and copings
Class 5	34.5 N/mm²	retaining walls
Class 6	41.5 N/mm²	walls below ground
Class 7	48.5 N/mm²	walls below ground

The Class 7 calcium silicate bricks are therefore equal in strength to Class B bricks
Calcium silicate bricks are not suitable for DPCs

Durability of Bricks	
FL	Frost resistant with low salt content
FN	Frost resistant with normal salt content
ML	Moderately frost resistant with low salt content
MN	Moderately frost resistant with normal salt content

MASONRY

Brickwork Dimensions

No. of horizontal bricks	Dimensions (mm)	No. of vertical courses	Height of vertical courses (mm)
½	112.5	1	75
1	225.0	2	150
1½	337.5	3	225
2	450.0	4	300
2½	562.5	5	375
3	675.0	6	450
3½	787.5	7	525
4	900.0	8	600
4½	1012.5	9	675
5	1125.0	10	750
5½	1237.5	11	825
6	1350.0	12	900
6½	1462.5	13	975
7	1575.0	14	1050
7½	1687.5	15	1125
8	1800.0	16	1200
8½	1912.5	17	1275
9	2025.0	18	1350
9½	2137.5	19	1425
10	2250.0	20	1500
20	4500.0	24	1575
40	9000.0	28	2100
50	11250.0	32	2400
60	13500.0	36	2700
75	16875.0	40	3000

TIMBER

Weights of Timber

Material	kg/m³	lb/cu ft
General	806 (avg)	50 (avg)
Douglas fir	479	30
Yellow pine, spruce	479	30
Pitch pine	673	42
Larch, elm	561	35
Oak (English)	724 to 959	45 to 60
Teak	643 to 877	40 to 55
Jarrah	959	60
Greenheart	1040 to 1204	65 to 75
Quebracho	1285	80
Material	**kg/m² per mm thickness**	**lb/sq ft per inch thickness**
Wooden boarding and blocks		
Softwood	0.48	2.5
Hardwood	0.76	4
Hardboard	1.06	5.5
Chipboard	0.76	4
Plywood	0.62	3.25
Blockboard	0.48	2.5
Fibreboard	0.29	1.5
Wood-wool	0.58	3
Plasterboard	0.96	5
Weather boarding	0.35	1.8

TIMBER

Conversion Tables (for timber only)

Inches	Millimetres	Feet	Metres
1	25	1	0.300
2	50	2	0.600
3	75	3	0.900
4	100	4	1.200
5	125	5	1.500
6	150	6	1.800
7	175	7	2.100
8	200	8	2.400
9	225	9	2.700
10	250	10	3.000
11	275	11	3.300
12	300	12	3.600
13	325	13	3.900
14	350	14	4.200
15	375	15	4.500
16	400	16	4.800
17	425	17	5.100
18	450	18	5.400
19	475	19	5.700
20	500	20	6.000
21	525	21	6.300
22	550	22	6.600
23	575	23	6.900
24	600	24	7.200

Planed Softwood

The finished end section size of planed timber is usually 3/16" less than the original size from which it is produced. This however varies slightly depending upon availability of material and origin of the species used.

Standards (timber) to cubic metres and cubic metres to standards (timber)

Cubic metres	Cubic metres standards	Standards
4.672	1	0.214
9.344	2	0.428
14.017	3	0.642
18.689	4	0.856
23.361	5	1.070
28.033	6	1.284
32.706	7	1.498
37.378	8	1.712
42.050	9	1.926
46.722	10	2.140
93.445	20	4.281
140.167	30	6.421
186.890	40	8.561
233.612	50	10.702
280.335	60	12.842
327.057	70	14.982
373.779	80	17.122

1 cu metre = 35.3148 cu ft = 0.21403 std

TIMBER

1 cu ft = 0.028317 cu metres

1 std = 4.67227 cu metres

Basic sizes of sawn softwood available (cross-sectional areas)

Thickness (mm)	Width (mm)								
	75	100	125	150	175	200	225	250	300
16	X	X	X	X					
19	X	X	X	X					
22	X	X	X	X					
25	X	X	X	X	X	X	X	X	X
32	X	X	X	X	X	X	X	X	X
36	X	X	X	X					
38	X	X	X	X	X	X	X		
44	X	X	X	X	X	X	X	X	X
47*	X	X	X	X	X	X	X	X	X
50	X	X	X	X	X	X	X	X	X
63	X	X	X	X	X	X	X		
75	X	X	X	X	X	X	X	X	
100		X		X		X		X	X
150				X		X			X
200						X			
250								X	
300									X

* This range of widths for 47 mm thickness will usually be found to be available in construction quality only

Note: The smaller sizes below 100 mm thick and 250 mm width are normally but not exclusively of European origin. Sizes beyond this are usually of North and South American origin

Basic lengths of sawn softwood available (metres)

1.80	2.10	3.00	4.20	5.10	6.00	7.20
	2.40	3.30	4.50	5.40	6.30	
	2.70	3.60	4.80	5.70	6.60	
		3.90			6.90	

Note: Lengths of 6.00 m and over will generally only be available from North American species and may have to be recut from larger sizes

TIMBER

Reductions from basic size to finished size by planning of two opposed faces

	Reductions from basic sizes for timber			
Purpose	15–35 mm	36–100 mm	101–150 mm	over 150 mm
a) Constructional timber	3 mm	3 mm	5 mm	6 mm
b) Matching interlocking boards	4 mm	4 mm	6 mm	6 mm
c) Wood trim not specified in BS 584	5 mm	7 mm	7 mm	9 mm
d) Joinery and cabinet work	7 mm	9 mm	11 mm	13 mm

Note: The reduction of width or depth is overall the extreme size and is exclusive of any reduction of the face by the machining of a tongue or lap joints

Maximum Spans for Various Roof Trusses

Maximum permissible spans for rafters for Fink trussed rafters

Basic size (mm)	Actual size (mm)	Pitch (degrees)								
		15 (m)	17.5 (m)	20 (m)	22.5 (m)	25 (m)	27.5 (m)	30 (m)	32.5 (m)	35 (m)
38 × 75	35 × 72	6.03	6.16	6.29	6.41	6.51	6.60	6.70	6.80	6.90
38 × 100	35 × 97	7.48	7.67	7.83	7.97	8.10	8.22	8.34	8.47	8.61
38 × 125	35 × 120	8.80	9.00	9.20	9.37	9.54	9.68	9.82	9.98	10.16
44 × 75	41 × 72	6.45	6.59	6.71	6.83	6.93	7.03	7.14	7.24	7.35
44 × 100	41 × 97	8.05	8.23	8.40	8.55	8.68	8.81	8.93	9.09	9.22
44 × 125	41 × 120	9.38	9.60	9.81	9.99	10.15	10.31	10.45	10.64	10.81
50 × 75	47 × 72	6.87	7.01	7.13	7.25	7.35	7.45	7.53	7.67	7.78
50 × 100	47 × 97	8.62	8.80	8.97	9.12	9.25	9.38	9.50	9.66	9.80
50 × 125	47 × 120	10.01	10.24	10.44	10.62	10.77	10.94	11.00	11.00	11.00

TIMBER

Sizes of Internal and External Doorsets

Description	Internal size (mm)	Permissible deviation	External size (mm)	Permissible deviation
Coordinating dimension: height of door leaf height sets	2100		2100	
Coordinating dimension: height of ceiling height set	2300 2350 2400 2700 3000		2300 2350 2400 2700 3000	
Coordinating dimension: width of all doorsets S = Single leaf set D = Double leaf set	600 S 700 S 800 S&D 900 S&D 1000 S&D 1200 D 1500 D 1800 D 2100 D		900 S 1000 S 1200 D 1800 D 2100 D	
Work size: height of door leaf height set	2090	± 2.0	2095	± 2.0
Work size: height of ceiling height set	2285 2335 2385 2685 2985	± 2.0	2295 2345 2395 2695 2995	± 2.0
Work size: width of all doorsets S = Single leaf set D = Double leaf set	590 S 690 S 790 S&D 890 S&D 990 S&D 1190 D 1490 D 1790 D 2090 D	± 2.0	895 S 995 S 1195 D 1495 D 1795 D 2095 D	± 2.0
Width of door leaf in single leaf sets F = Flush leaf P = Panel leaf	526 F 626 F 726 F&P 826 F&P 926 F&P	± 1.5	806 F&P 906 F&P	± 1.5
Width of door leaf in double leaf sets F = Flush leaf P = Panel leaf	362 F 412 F 426 F 562 F&P 712 F&P 826 F&P 1012 F&P	± 1.5	552 F&P 702 F&P 852 F&P 1002 F&P	± 1.5
Door leaf height for all doorsets	2040	± 1.5	1994	± 1.5

ROOFING

Total Roof Loadings for Various Types of Tiles/Slates

	Roof load (slope) kg/m²		
	Slate/Tile	Roofing underlay and battens²	Total dead load kg/m
Asbestos cement slate (600 × 300)	21.50	3.14	24.64
Clay tile interlocking	67.00	5.50	72.50
plain	43.50	2.87	46.37
Concrete tile interlocking	47.20	2.69	49.89
plain	78.20	5.50	83.70
Natural slate (18" × 10")	35.40	3.40	38.80
	Roof load (plan) kg/m²		
Asbestos cement slate (600 × 300)	28.45	76.50	104.95
Clay tile interlocking	53.54	76.50	130.04
plain	83.71	76.50	60.21
Concrete tile interlocking	57.60	76.50	134.10
plain	96.64	76.50	173.14

ROOFING

Tiling Data

Product		Lap (mm)	Gauge of battens	No. slates per m²	Battens (m/m²)	Weight as laid (kg/m²)
CEMENT SLATES						
Eternit slates	600 × 300 mm	100	250	13.4	4.00	19.50
(Duracem)		90	255	13.1	3.92	19.20
		80	260	12.9	3.85	19.00
		70	265	12.7	3.77	18.60
	600 × 350 mm	100	250	11.5	4.00	19.50
		90	255	11.2	3.92	19.20
	500 × 250 mm	100	200	20.0	5.00	20.00
		90	205	19.5	4.88	19.50
		80	210	19.1	4.76	19.00
		70	215	18.6	4.65	18.60
	400 × 200 mm	90	155	32.3	6.45	20.80
		80	160	31.3	6.25	20.20
		70	165	30.3	6.06	19.60
CONCRETE TILES/SLATES						
Redland Roofing						
Stonewold slate	430 × 380 mm	75	355	8.2	2.82	51.20
Double Roman tile	418 × 330 mm	75	355	8.2	2.91	45.50
Grovebury pantile	418 × 332 mm	75	343	9.7	2.91	47.90
Norfolk pantile	381 × 227 mm	75	306	16.3	3.26	44.01
		100	281	17.8	3.56	48.06
Renown interlocking tile	418 × 330 mm	75	343	9.7	2.91	46.40
'49' tile	381 × 227 mm	75	306	16.3	3.26	44.80
		100	281	17.8	3.56	48.95
Plain, vertical tiling	265 × 165 mm	35	115	52.7	8.70	62.20
Marley Roofing						
Bold roll tile	420 × 330 mm	75	344	9.7	2.90	47.00
		100	–	10.5	3.20	51.00
Modern roof tile	420 × 330 mm	75	338	10.2	3.00	54.00
		100	–	11.0	3.20	58.00
Ludlow major	420 × 330 mm	75	338	10.2	3.00	45.00
		100	–	11.0	3.20	49.00
Ludlow plus	387 × 229 mm	75	305	16.1	3.30	47.00
		100	–	17.5	3.60	51.00
Mendip tile	420 × 330 mm	75	338	10.2	3.00	47.00
		100	–	11.0	3.20	51.00
Wessex	413 × 330 mm	75	338	10.2	3.00	54.00
		100	–	11.0	3.20	58.00
Plain tile	267 × 165 mm	65	100	60.0	10.00	76.00
		75	95	64.0	10.50	81.00
		85	90	68.0	11.30	86.00
Plain vertical	267 × 165 mm	35	110	53.0	8.70	67.00
tiles (feature)		34	115	56.0	9.10	71.00

ROOFING

Slate Nails, Quantity per Kilogram

Length	Type			
	Plain wire	Galvanized wire	Copper nail	Zinc nail
28.5 mm	325	305	325	415
34.4 mm	286	256	254	292
50.8 mm	242	224	194	200

Metal Sheet Coverings

Thicknesses and weights of sheet metal coverings								
Lead to BS 1178								
BS Code No	3	4	5	6	7	8		
Colour code	Green	Blue	Red	Black	White	Orange		
Thickness (mm)	1.25	1.80	2.24	2.50	3.15	3.55		
Density kg/m^2	14.18	20.41	25.40	30.05	35.72	40.26		
Copper to BS 2870								
Thickness (mm)		0.60	0.70					
Bay width								
Roll (mm)		500	650					
Seam (mm)		525	600					
Standard width to form bay	600	750						
Normal length of sheet	1.80	1.80						
Zinc to BS 849								
Zinc Gauge (Nr)	9	10	11	12	13	14	15	16
Thickness (mm)	0.43	0.48	0.56	0.64	0.71	0.79	0.91	1.04
Density (kg/m^2)	3.1	3.2	3.8	4.3	4.8	5.3	6.2	7.0
Aluminium to BS 4868								
Thickness (mm)	0.5	0.6	0.7	0.8	0.9	1.0	1.2	
Density (kg/m^2)	12.8	15.4	17.9	20.5	23.0	25.6	30.7	

ROOFING

Type of felt	Nominal mass per unit area (kg/10 m)	Nominal mass per unit area of fibre base (g/m²)	Nominal length of roll (m)
Class 1			
1B fine granule	14	220	10 or 20
surfaced bitumen	18	330	10 or 20
	25	470	10
1E mineral surfaced bitumen	38	470	10
1F reinforced bitumen	15	160 (fibre) 110 (hessian)	15
1F reinforced bitumen, aluminium faced	13	160 (fibre) 110 (hessian)	15
Class 2			
2B fine granule surfaced bitumen asbestos	18	500	10 or 20
2E mineral surfaced bitumen asbestos	38	600	10
Class 3			
3B fine granule surfaced bitumen glass fibre	18	60	20
3E mineral surfaced bitumen glass fibre	28	60	10
3E venting base layer bitumen glass fibre	32	60*	10
3H venting base layer bitumen glass fibre	17	60*	20

* Excluding effect of perforations

GLAZING

Nominal thickness (mm)	Tolerance on thickness (mm)	Approximate weight (kg/m²)	Normal maximum size (mm)
Float and polished plate glass			
3	+ 0.2	7.50	2140 × 1220
4	+ 0.2	10.00	2760 × 1220
5	+ 0.2	12.50	3180 × 2100
6	+ 0.2	15.00	4600 × 3180
10	+ 0.3	25.00)	6000 × 3300
12	+ 0.3	30.00)	
15	+ 0.5	37.50	3050 × 3000
19	+ 1.0	47.50)	3000 × 2900
25	+ 1.0	63.50)	
Clear sheet glass			
2 *	+ 0.2	5.00	1920 × 1220
3	+ 0.3	7.50	2130 × 1320
4	+ 0.3	10.00	2760 × 1220
5 *	+ 0.3	12.50)	2130 × 2400
6 *	+ 0.3	15.00)	
Cast glass			
3	+ 0.4		
	− 0.2	6.00)	2140 × 1280
4	+ 0.5	7.50)	
5	+ 0.5	9.50	2140 × 1320
6	+ 0.5	11.50)	3700 × 1280
10	+ 0.8	21.50)	
Wired glass			
(Cast wired glass)			
6	+ 0.3	−)	
	− 0.7	)	3700 × 1840
7	+ 0.7	−)	
(Polished wire glass)			
6	+ 1.0	−	330 × 1830

* The 5 mm and 6 mm thickness are known as *thick drawn sheet*. Although 2 mm sheet glass is available it is not recommended for general glazing purposes

METAL

Weights of Metals

Material	kg/m³	lb/cu ft
Metals, steel construction, etc.		
Iron		
– cast	7207	450
– wrought	7687	480
– ore – general	2407	150
– (crushed) Swedish	3682	230
Steel	7854	490
Copper		
– cast	8731	545
– wrought	8945	558
Brass	8497	530
Bronze	8945	558
Aluminium	2774	173
Lead	11322	707
Zinc (rolled)	7140	446
	g/mm² per metre	lb/sq ft per foot
Steel bars	7.85	3.4
Structural steelwork	Net weight of member @ 7854 kg/m³	
riveted	+ 10% for cleats, rivets, bolts, etc.	
welded	+ 1.25% to 2.5% for welds, etc.	
Rolled sections		
beams	+ 2.5%	
stanchions	+ 5% (extra for caps and bases)	
Plate		
web girders	+ 10% for rivets or welds, stiffeners, etc.	
	kg/m	lb/ft
Steel stairs: industrial type		
1 m or 3 ft wide	84	56
Steel tubes		
50 mm or 2 in bore	5 to 6	3 to 4
Gas piping		
20 mm or ¾ in	2	1¼

METAL

Universal Beams BS 4: Part 1: 2005

Designation	Mass (kg/m)	Depth of section (mm)	Width of section (mm)	Thickness		Surface area (m²/m)
				Web (mm)	Flange (mm)	
1016 × 305 × 487	487.0	1036.1	308.5	30.0	54.1	3.20
1016 × 305 × 438	438.0	1025.9	305.4	26.9	49.0	3.17
1016 × 305 × 393	393.0	1016.0	303.0	24.4	43.9	3.15
1016 × 305 × 349	349.0	1008.1	302.0	21.1	40.0	3.13
1016 × 305 × 314	314.0	1000.0	300.0	19.1	35.9	3.11
1016 × 305 × 272	272.0	990.1	300.0	16.5	31.0	3.10
1016 × 305 × 249	249.0	980.2	300.0	16.5	26.0	3.08
1016 × 305 × 222	222.0	970.3	300.0	16.0	21.1	3.06
914 × 419 × 388	388.0	921.0	420.5	21.4	36.6	3.44
914 × 419 × 343	343.3	911.8	418.5	19.4	32.0	3.42
914 × 305 × 289	289.1	926.6	307.7	19.5	32.0	3.01
914 × 305 × 253	253.4	918.4	305.5	17.3	27.9	2.99
914 × 305 × 224	224.2	910.4	304.1	15.9	23.9	2.97
914 × 305 × 201	200.9	903.0	303.3	15.1	20.2	2.96
838 × 292 × 226	226.5	850.9	293.8	16.1	26.8	2.81
838 × 292 × 194	193.8	840.7	292.4	14.7	21.7	2.79
838 × 292 × 176	175.9	834.9	291.7	14.0	18.8	2.78
762 × 267 × 197	196.8	769.8	268.0	15.6	25.4	2.55
762 × 267 × 173	173.0	762.2	266.7	14.3	21.6	2.53
762 × 267 × 147	146.9	754.0	265.2	12.8	17.5	2.51
762 × 267 × 134	133.9	750.0	264.4	12.0	15.5	2.51
686 × 254 × 170	170.2	692.9	255.8	14.5	23.7	2.35
686 × 254 × 152	152.4	687.5	254.5	13.2	21.0	2.34
686 × 254 × 140	140.1	383.5	253.7	12.4	19.0	2.33
686 × 254 × 125	125.2	677.9	253.0	11.7	16.2	2.32
610 × 305 × 238	238.1	635.8	311.4	18.4	31.4	2.45
610 × 305 × 179	179.0	620.2	307.1	14.1	23.6	2.41
610 × 305 × 149	149.1	612.4	304.8	11.8	19.7	2.39
610 × 229 × 140	139.9	617.2	230.2	13.1	22.1	2.11
610 × 229 × 125	125.1	612.2	229.0	11.9	19.6	2.09
610 × 229 × 113	113.0	607.6	228.2	11.1	17.3	2.08
610 × 229 × 101	101.2	602.6	227.6	10.5	14.8	2.07
533 × 210 × 122	122.0	544.5	211.9	12.7	21.3	1.89
533 × 210 × 109	109.0	539.5	210.8	11.6	18.8	1.88
533 × 210 × 101	101.0	536.7	210.0	10.8	17.4	1.87
533 × 210 × 92	92.1	533.1	209.3	10.1	15.6	1.86
533 × 210 × 82	82.2	528.3	208.8	9.6	13.2	1.85
457 × 191 × 98	98.3	467.2	192.8	11.4	19.6	1.67
457 × 191 × 89	89.3	463.4	191.9	10.5	17.7	1.66
457 × 191 × 82	82.0	460.0	191.3	9.9	16.0	1.65
457 × 191 × 74	74.3	457.0	190.4	9.0	14.5	1.64
457 × 191 × 67	67.1	453.4	189.9	8.5	12.7	1.63
457 × 152 × 82	82.1	465.8	155.3	10.5	18.9	1.51
457 × 152 × 74	74.2	462.0	154.4	9.6	17.0	1.50
457 × 152 × 67	67.2	458.0	153.8	9.0	15.0	1.50
457 × 152 × 60	59.8	454.6	152.9	8.1	13.3	1.50
457 × 152 × 52	52.3	449.8	152.4	7.6	10.9	1.48
406 × 178 × 74	74.2	412.8	179.5	9.5	16.0	1.51
406 × 178 × 67	67.1	409.4	178.8	8.8	14.3	1.50
406 × 178 × 60	60.1	406.4	177.9	7.9	12.8	1.49

METAL

Designation	Mass (kg/m)	Depth of section (mm)	Width of section (mm)	Thickness		Surface area (m²/m)
				Web (mm)	Flange (mm)	
406 × 178 × 50	54.1	402.6	177.7	7.7	10.9	1.48
406 × 140 × 46	46.0	403.2	142.2	6.8	11.2	1.34
406 × 140 × 39	39.0	398.0	141.8	6.4	8.6	1.33
356 × 171 × 67	67.1	363.4	173.2	9.1	15.7	1.38
356 × 171 × 57	57.0	358.0	172.2	8.1	13.0	1.37
356 × 171 × 51	51.0	355.0	171.5	7.4	11.5	1.36
356 × 171 × 45	45.0	351.4	171.1	7.0	9.7	1.36
356 × 127 × 39	39.1	353.4	126.0	6.6	10.7	1.18
356 × 127 × 33	33.1	349.0	125.4	6.0	8.5	1.17
305 × 165 × 54	54.0	310.4	166.9	7.9	13.7	1.26
305 × 165 × 46	46.1	306.6	165.7	6.7	11.8	1.25
305 × 165 × 40	40.3	303.4	165.0	6.0	10.2	1.24
305 × 127 × 48	48.1	311.0	125.3	9.0	14.0	1.09
305 × 127 × 42	41.9	307.2	124.3	8.0	12.1	1.08
305 × 127 × 37	37.0	304.4	123.3	7.1	10.7	1.07
305 × 102 × 33	32.8	312.7	102.4	6.6	10.8	1.01
305 × 102 × 28	28.2	308.7	101.8	6.0	8.8	1.00
305 × 102 × 25	24.8	305.1	101.6	5.8	7.0	0.992
254 × 146 × 43	43.0	259.6	147.3	7.2	12.7	1.08
254 × 146 × 37	37.0	256.0	146.4	6.3	10.9	1.07
254 × 146 × 31	31.1	251.4	146.1	6.0	8.6	1.06
254 × 102 × 28	28.3	260.4	102.2	6.3	10.0	0.904
254 × 102 × 25	25.2	257.2	101.9	6.0	8.4	0.897
254 × 102 × 22	22.0	254.0	101.6	5.7	6.8	0.890
203 × 133 × 30	30.0	206.8	133.9	6.4	9.6	0.923
203 × 133 × 25	25.1	203.2	133.2	5.7	7.8	0.915
203 × 102 × 23	23.1	203.2	101.8	5.4	9.3	0.790
178 × 102 × 19	19.0	177.8	101.2	4.8	7.9	0.738
152 × 89 × 16	16.0	152.4	88.7	4.5	7.7	0.638
127 × 76 × 13	13.0	127.0	76.0	4.0	7.6	0.537

METAL

Universal Columns BS 4: Part 1: 2005

Designation	Mass (kg/m)	Depth of section (mm)	Width of section (mm)	Thickness		Surface area (m²/m)
				Web (mm)	Flange (mm)	
356 × 406 × 634	633.9	474.7	424.0	47.6	77.0	2.52
356 × 406 × 551	551.0	455.6	418.5	42.1	67.5	2.47
356 × 406 × 467	467.0	436.6	412.2	35.8	58.0	2.42
356 × 406 × 393	393.0	419.0	407.0	30.6	49.2	2.38
356 × 406 × 340	339.9	406.4	403.0	26.6	42.9	2.35
356 × 406 × 287	287.1	393.6	399.0	22.6	36.5	2.31
356 × 406 × 235	235.1	381.0	384.8	18.4	30.2	2.28
356 × 368 × 202	201.9	374.6	374.7	16.5	27.0	2.19
356 × 368 × 177	177.0	368.2	372.6	14.4	23.8	2.17
356 × 368 × 153	152.9	362.0	370.5	12.3	20.7	2.16
356 × 368 × 129	129.0	355.6	368.6	10.4	17.5	2.14
305 × 305 × 283	282.9	365.3	322.2	26.8	44.1	1.94
305 × 305 × 240	240.0	352.5	318.4	23.0	37.7	1.91
305 × 305 × 198	198.1	339.9	314.5	19.1	31.4	1.87
305 × 305 × 158	158.1	327.1	311.2	15.8	25.0	1.84
305 × 305 × 137	136.9	320.5	309.2	13.8	21.7	1.82
305 × 305 × 118	117.9	314.5	307.4	12.0	18.7	1.81
305 × 305 × 97	96.9	307.9	305.3	9.9	15.4	1.79
254 × 254 × 167	167.1	289.1	265.2	19.2	31.7	1.58
254 × 254 × 132	132.0	276.3	261.3	15.3	25.3	1.55
254 × 254 × 107	107.1	266.7	258.8	12.8	20.5	1.52
254 × 254 × 89	88.9	260.3	256.3	10.3	17.3	1.50
254 × 254 × 73	73.1	254.1	254.6	8.6	14.2	1.49
203 × 203 × 86	86.1	222.2	209.1	12.7	20.5	1.24
203 × 203 × 71	71.0	215.8	206.4	10.0	17.3	1.22
203 × 203 × 60	60.0	209.6	205.8	9.4	14.2	1.21
203 × 203 × 52	52.0	206.2	204.3	7.9	12.5	1.20
203 × 203 × 46	46.1	203.2	203.6	7.2	11.0	1.19
152 × 152 × 37	37.0	161.8	154.4	8.0	11.5	0.912
152 × 152 × 30	30.0	157.6	152.9	6.5	9.4	0.901
152 × 152 × 23	23.0	152.4	152.2	5.8	6.8	0.889

METAL

Joists BS 4: Part 1: 2005 (retained for reference, Corus have ceased manufacture in UK)

Designation	Mass	Depth of section	Width of section	Thickness		Surface area
				Web	Flange	
	(kg/m)	(mm)	(mm)	(mm)	(mm)	(m²/m)
254 × 203 × 82	82.0	254.0	203.2	10.2	19.9	1.210
203 × 152 × 52	52.3	203.2	152.4	8.9	16.5	0.932
152 × 127 × 37	37.3	152.4	127.0	10.4	13.2	0.737
127 × 114 × 29	29.3	127.0	114.3	10.2	11.5	0.646
127 × 114 × 27	26.9	127.0	114.3	7.4	11.4	0.650
102 × 102 × 23	23.0	101.6	101.6	9.5	10.3	0.549
102 × 44 × 7	7.5	101.6	44.5	4.3	6.1	0.350
89 × 89 × 19	19.5	88.9	88.9	9.5	9.9	0.476
76 × 76 × 13	12.8	76.2	76.2	5.1	8.4	0.411

Parallel Flange Channels

Designation	Mass	Depth of section	Width of section	Thickness		Surface area
				Web	Flange	
	(kg/m)	(mm)	(mm)	(mm)	(mm)	(m²/m)
430 × 100 × 64	64.4	430	100	11.0	19.0	1.23
380 × 100 × 54	54.0	380	100	9.5	17.5	1.13
300 × 100 × 46	45.5	300	100	9.0	16.5	0.969
300 × 90 × 41	41.4	300	90	9.0	15.5	0.932
260 × 90 × 35	34.8	260	90	8.0	14.0	0.854
260 × 75 × 28	27.6	260	75	7.0	12.0	0.79
230 × 90 × 32	32.2	230	90	7.5	14.0	0.795
230 × 75 × 26	25.7	230	75	6.5	12.5	0.737
200 × 90 × 30	29.7	200	90	7.0	14.0	0.736
200 × 75 × 23	23.4	200	75	6.0	12.5	0.678
180 × 90 × 26	26.1	180	90	6.5	12.5	0.697
180 × 75 × 20	20.3	180	75	6.0	10.5	0.638
150 × 90 × 24	23.9	150	90	6.5	12.0	0.637
150 × 75 × 18	17.9	150	75	5.5	10.0	0.579
125 × 65 × 15	14.8	125	65	5.5	9.5	0.489
100 × 50 × 10	10.2	100	50	5.0	8.5	0.382

METAL

Equal Angles BS EN 10056-1

Designation	Mass (kg/m)	Surface area (m²/m)
200 × 200 × 24	71.1	0.790
200 × 200 × 20	59.9	0.790
200 × 200 × 18	54.2	0.790
200 × 200 × 16	48.5	0.790
150 × 150 × 18	40.1	0.59
150 × 150 × 15	33.8	0.59
150 × 150 × 12	27.3	0.59
150 × 150 × 10	23.0	0.59
120 × 120 × 15	26.6	0.47
120 × 120 × 12	21.6	0.47
120 × 120 × 10	18.2	0.47
120 × 120 × 8	14.7	0.47
100 × 100 × 15	21.9	0.39
100 × 100 × 12	17.8	0.39
100 × 100 × 10	15.0	0.39
100 × 100 × 8	12.2	0.39
90 × 90 × 12	15.9	0.35
90 × 90 × 10	13.4	0.35
90 × 90 × 8	10.9	0.35
90 × 90 × 7	9.61	0.35
90 × 90 × 6	8.30	0.35

Unequal Angles BS EN 10056-1

Designation	Mass (kg/m)	Surface area (m²/m)
200 × 150 × 18	47.1	0.69
200 × 150 × 15	39.6	0.69
200 × 150 × 12	32.0	0.69
200 × 100 × 15	33.7	0.59
200 × 100 × 12	27.3	0.59
200 × 100 × 10	23.0	0.59
150 × 90 × 15	26.6	0.47
150 × 90 × 12	21.6	0.47
150 × 90 × 10	18.2	0.47
150 × 75 × 15	24.8	0.44
150 × 75 × 12	20.2	0.44
150 × 75 × 10	17.0	0.44
125 × 75 × 12	17.8	0.40
125 × 75 × 10	15.0	0.40
125 × 75 × 8	12.2	0.40
100 × 75 × 12	15.4	0.34
100 × 75 × 10	13.0	0.34
100 × 75 × 8	10.6	0.34
100 × 65 × 10	12.3	0.32
100 × 65 × 8	9.94	0.32
100 × 65 × 7	8.77	0.32

METAL

Structural Tees Split from Universal Beams BS 4: Part 1: 2005

Designation	Mass (kg/m)	Surface area (m²/m)
305 × 305 × 90	89.5	1.22
305 × 305 × 75	74.6	1.22
254 × 343 × 63	62.6	1.19
229 × 305 × 70	69.9	1.07
229 × 305 × 63	62.5	1.07
229 × 305 × 57	56.5	1.07
229 × 305 × 51	50.6	1.07
210 × 267 × 61	61.0	0.95
210 × 267 × 55	54.5	0.95
210 × 267 × 51	50.5	0.95
210 × 267 × 46	46.1	0.95
210 × 267 × 41	41.1	0.95
191 × 229 × 49	49.2	0.84
191 × 229 × 45	44.6	0.84
191 × 229 × 41	41.0	0.84
191 × 229 × 37	37.1	0.84
191 × 229 × 34	33.6	0.84
152 × 229 × 41	41.0	0.76
152 × 229 × 37	37.1	0.76
152 × 229 × 34	33.6	0.76
152 × 229 × 30	29.9	0.76
152 × 229 × 26	26.2	0.76

Universal Bearing Piles BS 4: Part 1: 2005

Designation	Mass (kg/m)	Depth of Section (mm)	Width of section (mm)	Thickness	
				Web (mm)	Flange (mm)
356 × 368 × 174	173.9	361.4	378.5	20.3	20.4
356 × 368 × 152	152.0	356.4	376.0	17.8	17.9
356 × 368 × 133	133.0	352.0	373.8	15.6	15.7
356 × 368 × 109	108.9	346.4	371.0	12.8	12.9
305 × 305 × 223	222.9	337.9	325.7	30.3	30.4
305 × 305 × 186	186.0	328.3	320.9	25.5	25.6
305 × 305 × 149	149.1	318.5	316.0	20.6	20.7
305 × 305 × 126	126.1	312.3	312.9	17.5	17.6
305 × 305 × 110	110.0	307.9	310.7	15.3	15.4
305 × 305 × 95	94.9	303.7	308.7	13.3	13.3
305 × 305 × 88	88.0	301.7	307.8	12.4	12.3
305 × 305 × 79	78.9	299.3	306.4	11.0	11.1
254 × 254 × 85	85.1	254.3	260.4	14.4	14.3
254 × 254 × 71	71.0	249.7	258.0	12.0	12.0
254 × 254 × 63	63.0	247.1	256.6	10.6	10.7
203 × 203 × 54	53.9	204.0	207.7	11.3	11.4
203 × 203 × 45	44.9	200.2	205.9	9.5	9.5

METAL

Hot Formed Square Hollow Sections EN 10210 S275J2H & S355J2H

Size (mm)	Wall thickness (mm)	Mass (kg/m)	Superficial area (m²/m)
40 × 40	2.5	2.89	0.154
	3.0	3.41	0.152
	3.2	3.61	0.152
	3.6	4.01	0.151
	4.0	4.39	0.150
	5.0	5.28	0.147
50 × 50	2.5	3.68	0.194
	3.0	4.35	0.192
	3.2	4.62	0.192
	3.6	5.14	0.191
	4.0	5.64	0.190
	5.0	6.85	0.187
	6.0	7.99	0.185
	6.3	8.31	0.184
60 × 60	3.0	5.29	0.232
	3.2	5.62	0.232
	3.6	6.27	0.231
	4.0	6.90	0.230
	5.0	8.42	0.227
	6.0	9.87	0.225
	6.3	10.30	0.224
	8.0	12.50	0.219
70 × 70	3.0	6.24	0.272
	3.2	6.63	0.272
	3.6	7.40	0.271
	4.0	8.15	0.270
	5.0	9.99	0.267
	6.0	11.80	0.265
	6.3	12.30	0.264
	8.0	15.00	0.259
80 × 80	3.2	7.63	0.312
	3.6	8.53	0.311
	4.0	9.41	0.310
	5.0	11.60	0.307
	6.0	13.60	0.305
	6.3	14.20	0.304
	8.0	17.50	0.299
90 × 90	3.6	9.66	0.351
	4.0	10.70	0.350
	5.0	13.10	0.347
	6.0	15.50	0.345
	6.3	16.20	0.344
	8.0	20.10	0.339
100 × 100	3.6	10.80	0.391
	4.0	11.90	0.390
	5.0	14.70	0.387
	6.0	17.40	0.385
	6.3	18.20	0.384
	8.0	22.60	0.379
	10.0	27.40	0.374
120 × 120	4.0	14.40	0.470
	5.0	17.80	0.467
	6.0	21.20	0.465

METAL

Size (mm)	Wall thickness (mm)	Mass (kg/m)	Superficial area (m²/m)
	6.3	22.20	0.464
	8.0	27.60	0.459
	10.0	33.70	0.454
	12.0	39.50	0.449
	12.5	40.90	0.448
140 × 140	5.0	21.00	0.547
	6.0	24.90	0.545
	6.3	26.10	0.544
	8.0	32.60	0.539
	10.0	40.00	0.534
	12.0	47.00	0.529
	12.5	48.70	0.528
150 × 150	5.0	22.60	0.587
	6.0	26.80	0.585
	6.3	28.10	0.584
	8.0	35.10	0.579
	10.0	43.10	0.574
	12.0	50.80	0.569
	12.5	52.70	0.568
Hot formed from seamless hollow	16.0	65.2	0.559
160 × 160	5.0	24.10	0.627
	6.0	28.70	0.625
	6.3	30.10	0.624
	8.0	37.60	0.619
	10.0	46.30	0.614
	12.0	54.60	0.609
	12.5	56.60	0.608
	16.0	70.20	0.599
180 × 180	5.0	27.30	0.707
	6.0	32.50	0.705
	6.3	34.00	0.704
	8.0	42.70	0.699
	10.0	52.50	0.694
	12.0	62.10	0.689
	12.5	64.40	0.688
	16.0	80.20	0.679
200 × 200	5.0	30.40	0.787
	6.0	36.20	0.785
	6.3	38.00	0.784
	8.0	47.70	0.779
	10.0	58.80	0.774
	12.0	69.60	0.769
	12.5	72.30	0.768
	16.0	90.30	0.759
250 × 250	5.0	38.30	0.987
	6.0	45.70	0.985
	6.3	47.90	0.984
	8.0	60.30	0.979
	10.0	74.50	0.974
	12.0	88.50	0.969
	12.5	91.90	0.968
	16.0	115.00	0.959

METAL

Size (mm)	Wall thickness (mm)	Mass (kg/m)	Superficial area (m²/m)
300 × 300	6.0	55.10	1.18
	6.3	57.80	1.18
	8.0	72.80	1.18
	10.0	90.20	1.17
	12.0	107.00	1.17
	12.5	112.00	1.17
	16.0	141.00	1.16
350 × 350	8.0	85.40	1.38
	10.0	106.00	1.37
	12.0	126.00	1.37
	12.5	131.00	1.37
	16.0	166.00	1.36
400 × 400	8.0	97.90	1.58
	10.0	122.00	1.57
	12.0	145.00	1.57
	12.5	151.00	1.57
	16.0	191.00	1.56
(Grade S355J2H only)	20.00*	235.00	1.55

Note: * SAW process

METAL

Hot Formed Square Hollow Sections JUMBO RHS: JIS G3136

Size (mm)	Wall thickness (mm)	Mass (kg/m)	Superficial area (m²/m)
350 × 350	19.0	190.00	1.33
	22.0	217.00	1.32
	25.0	242.00	1.31
400 × 400	22.0	251.00	1.52
	25.0	282.00	1.51
450 × 450	12.0	162.00	1.76
	16.0	213.00	1.75
	19.0	250.00	1.73
	22.0	286.00	1.72
	25.0	321.00	1.71
	28.0 *	355.00	1.70
	32.0 *	399.00	1.69
500 × 500	12.0	181.00	1.96
	16.0	238.00	1.95
	19.0	280.00	1.93
	22.0	320.00	1.92
	25.0	360.00	1.91
	28.0 *	399.00	1.90
	32.0 *	450.00	1.89
	36.0 *	498.00	1.88
550 × 550	16.0	263.00	2.15
	19.0	309.00	2.13
	22.0	355.00	2.12
	25.0	399.00	2.11
	28.0 *	443.00	2.10
	32.0 *	500.00	2.09
	36.0 *	555.00	2.08
	40.0 *	608.00	2.06
600 × 600	25.0 *	439.00	2.31
	28.0 *	487.00	2.30
	32.0 *	550.00	2.29
	36.0 *	611.00	2.28
	40.0 *	671.00	2.26
700 × 700	25.0 *	517.00	2.71
	28.0 *	575.00	2.70
	32.0 *	651.00	2.69
	36.0 *	724.00	2.68
	40.0 *	797.00	2.68

Note: * SAW process

METAL

Hot Formed Rectangular Hollow Sections: EN10210 S275J2h & S355J2H

Size (mm)	Wall thickness (mm)	Mass (kg/m)	Superficial area (m²/m)
50 × 30	2.5	2.89	0.154
	3.0	3.41	0.152
	3.2	3.61	0.152
	3.6	4.01	0.151
	4.0	4.39	0.150
	5.0	5.28	0.147
60 × 40	2.5	3.68	0.194
	3.0	4.35	0.192
	3.2	4.62	0.192
	3.6	5.14	0.191
	4.0	5.64	0.190
	5.0	6.85	0.187
	6.0	7.99	0.185
	6.3	8.31	0.184
80 × 40	3.0	5.29	0.232
	3.2	5.62	0.232
	3.6	6.27	0.231
	4.0	6.90	0.230
	5.0	8.42	0.227
	6.0	9.87	0.225
	6.3	10.30	0.224
	8.0	12.50	0.219
76.2 × 50.8	3.0	5.62	0.246
	3.2	5.97	0.246
	3.6	6.66	0.245
	4.0	7.34	0.244
	5.0	8.97	0.241
	6.0	10.50	0.239
	6.3	11.00	0.238
	8.0	13.40	0.233
90 × 50	3.0	6.24	0.272
	3.2	6.63	0.272
	3.6	7.40	0.271
	4.0	8.15	0.270
	5.0	9.99	0.267
	6.0	11.80	0.265
	6.3	12.30	0.264
	8.0	15.00	0.259
100 × 50	3.0	6.71	0.292
	3.2	7.13	0.292
	3.6	7.96	0.291
	4.0	8.78	0.290
	5.0	10.80	0.287
	6.0	12.70	0.285
	6.3	13.30	0.284
	8.0	16.30	0.279

METAL

Size (mm)	Wall thickness (mm)	Mass (kg/m)	Superficial area (m²/m)
100 × 60	3.0	7.18	0.312
	3.2	7.63	0.312
	3.6	8.53	0.311
	4.0	9.41	0.310
	5.0	11.60	0.307
	6.0	13.60	0.305
	6.3	14.20	0.304
	8.0	17.50	0.299
120 × 60	3.6	9.70	0.351
	4.0	10.70	0.350
	5.0	13.10	0.347
	6.0	15.50	0.345
	6.3	16.20	0.344
	8.0	20.10	0.339
120 × 80	3.6	10.80	0.391
	4.0	11.90	0.390
	5.0	14.70	0.387
	6.0	17.40	0.385
	6.3	18.20	0.384
	8.0	22.60	0.379
	10.0	27.40	0.374
150 × 100	4.0	15.10	0.490
	5.0	18.60	0.487
	6.0	22.10	0.485
	6.3	23.10	0.484
	8.0	28.90	0.479
	10.0	35.30	0.474
	12.0	41.40	0.469
	12.5	42.80	0.468
160 × 80	4.0	14.40	0.470
	5.0	17.80	0.467
	6.0	21.20	0.465
	6.3	22.20	0.464
	8.0	27.60	0.459
	10.0	33.70	0.454
	12.0	39.50	0.449
	12.5	40.90	0.448
200 × 100	5.0	22.60	0.587
	6.0	26.80	0.585
	6.3	28.10	0.584
	8.0	35.10	0.579
	10.0	43.10	0.574
	12.0	50.80	0.569
	12.5	52.70	0.568
	16.0	65.20	0.559
250 × 150	5.0	30.40	0.787
	6.0	36.20	0.785
	6.3	38.00	0.784
	8.0	47.70	0.779
	10.0	58.80	0.774
	12.0	69.60	0.769
	12.5	72.30	0.768
	16.0	90.30	0.759

METAL

Size (mm)	Wall thickness (mm)	Mass (kg/m)	Superficial area (m²/m)
300 × 200	5.0	38.30	0.987
	6.0	45.70	0.985
	6.3	47.90	0.984
	8.0	60.30	0.979
	10.0	74.50	0.974
	12.0	88.50	0.969
	12.5	91.90	0.968
	16.0	115.00	0.959
400 × 200	6.0	55.10	1.18
	6.3	57.80	1.18
	8.0	72.80	1.18
	10.0	90.20	1.17
	12.0	107.00	1.17
	12.5	112.00	1.17
	16.0	141.00	1.16
450 × 250	8.0	85.40	1.38
	10.0	106.00	1.37
	12.0	126.00	1.37
	12.5	131.00	1.37
	16.0	166.00	1.36
500 × 300	8.0	98.00	1.58
	10.0	122.00	1.57
	12.0	145.00	1.57
	12.5	151.00	1.57
	16.0	191.00	1.56
	20.0	235.00	1.55

METAL

Hot Formed Circular Hollow Sections EN 10210 S275J2H & S355J2H

Outside diameter (mm)	Wall thickness (mm)	Mass (kg/m)	Superficial area (m²/m)
21.3	3.2	1.43	0.067
26.9	3.2	1.87	0.085
33.7	3.0	2.27	0.106
	3.2	2.41	0.106
	3.6	2.67	0.106
	4.0	2.93	0.106
42.4	3.0	2.91	0.133
	3.2	3.09	0.133
	3.6	3.44	0.133
	4.0	3.79	0.133
48.3	2.5	2.82	0.152
	3.0	3.35	0.152
	3.2	3.56	0.152
	3.6	3.97	0.152
	4.0	4.37	0.152
	5.0	5.34	0.152
60.3	2.5	3.56	0.189
	3.0	4.24	0.189
	3.2	4.51	0.189
	3.6	5.03	0.189
	4.0	5.55	0.189
	5.0	6.82	0.189
76.1	2.5	4.54	0.239
	3.0	5.41	0.239
	3.2	5.75	0.239
	3.6	6.44	0.239
	4.0	7.11	0.239
	5.0	8.77	0.239
	6.0	10.40	0.239
	6.3	10.80	0.239
88.9	2.5	5.33	0.279
	3.0	6.36	0.279
	3.2	6.76	0.27
	3.6	7.57	0.279
	4.0	8.38	0.279
	5.0	10.30	0.279
	6.0	12.30	0.279
	6.3	12.80	0.279
114.3	3.0	8.23	0.359
	3.2	8.77	0.359
	3.6	9.83	0.359
	4.0	10.09	0.359
	5.0	13.50	0.359
	6.0	16.00	0.359
	6.3	16.80	0.359

METAL

Outside diameter (mm)	Wall thickness (mm)	Mass (kg/m)	Superficial area (m²/m)
139.7	3.2	10.80	0.439
	3.6	12.10	0.439
	4.0	13.40	0.439
	5.0	16.60	0.439
	6.0	19.80	0.439
	6.3	20.70	0.439
	8.0	26.00	0.439
	10.0	32.00	0.439
168.3	3.2	13.00	0.529
	3.6	14.60	0.529
	4.0	16.20	0.529
	5.0	20.10	0.529
	6.0	24.00	0.529
	6.3	25.20	0.529
	8.0	31.60	0.529
	10.0	39.00	0.529
	12.0	46.30	0.529
	12.5	48.00	0.529
193.7	5.0	23.30	0.609
	6.0	27.80	0.609
	6.3	29.10	0.609
	8.0	36.60	0.609
	10.0	45.30	0.609
	12.0	53.80	0.609
	12.5	55.90	0.609
219.1	5.0	26.40	0.688
	6.0	31.50	0.688
	6.3	33.10	0.688
	8.0	41.60	0.688
	10.0	51.60	0.688
	12.0	61.30	0.688
	12.5	63.70	0.688
	16.0	80.10	0.688
244.5	5.0	29.50	0.768
	6.0	35.30	0.768
	6.3	37.00	0.768
	8.0	46.70	0.768
	10.0	57.80	0.768
	12.0	68.80	0.768
	12.5	71.50	0.768
	16.0	90.20	0.768
273.0	5.0	33.00	0.858
	6.0	39.50	0.858
	6.3	41.40	0.858
	8.0	52.30	0.858
	10.0	64.90	0.858
	12.0	77.20	0.858
	12.5	80.30	0.858
	16.0	101.00	0.858
323.9	5.0	39.30	1.02
	6.0	47.00	1.02
	6.3	49.30	1.02
	8.0	62.30	1.02
	10.0	77.40	1.02

METAL

Outside diameter (mm)	Wall thickness (mm)	Mass (kg/m)	Superficial area (m²/m)
	12.0	92.30	1.02
	12.5	96.00	1.02
	16.0	121.00	1.02
355.6	6.3	54.30	1.12
	8.0	68.60	1.12
	10.0	85.30	1.12
	12.0	102.00	1.12
	12.5	106.00	1.12
	16.0	134.00	1.12
406.4	6.3	62.20	1.28
	8.0	79.60	1.28
	10.0	97.80	1.28
	12.0	117.00	1.28
	12.5	121.00	1.28
	16.0	154.00	1.28
457.0	6.3	70.00	1.44
	8.0	88.60	1.44
	10.0	110.00	1.44
	12.0	132.00	1.44
	12.5	137.00	1.44
	16.0	174.00	1.44
508.0	6.3	77.90	1.60
	8.0	98.60	1.60
	10.0	123.00	1.60
	12.0	147.00	1.60
	12.5	153.00	1.60
	16.0	194.00	1.60

METAL

Spacing of Holes in Angles

Nominal leg length (mm)	Spacing of holes						Maximum diameter of bolt or rivet		
	A	B	C	D	E	F	A	B and C	D, E and F
200		75	75	55	55	55		30	20
150		55	55					20	
125		45	60					20	
120									
100	55						24		
90	50						24		
80	45						20		
75	45						20		
70	40						20		
65	35						20		
60	35						16		
50	28						12		
45	25								
40	23								
30	20								
25	15								

KERBS, PAVING, ETC.

KERBS/EDGINGS/CHANNELS

Precast Concrete Kerbs to BS 7263

Straight kerb units: length from 450 to 915 mm

150 mm high × 125 mm thick		
bullnosed	type BN	
half battered	type HB3	
255 mm high × 125 mm thick		
45° splayed	type SP	
half battered	type HB2	
305 mm high × 150 mm thick		
half battered	type HB1	
Quadrant kerb units		
150 mm high × 305 and 455 mm radius to match	type BN	type QBN
150 mm high × 305 and 455 mm radius to match	type HB2, HB3	type QHB
150 mm high × 305 and 455 mm radius to match	type SP	type QSP
255 mm high × 305 and 455 mm radius to match	type BN	type QBN
255 mm high × 305 and 455 mm radius to match	type HB2, HB3	type QHB
225 mm high × 305 and 455 mm radius to match	type SP	type QSP
Angle kerb units		
305 × 305 × 225 mm high × 125 mm thick		
bullnosed external angle	type XA	
splayed external angle to match type SP	type XA	
bullnosed internal angle	type IA	
splayed internal angle to match type SP	type IA	
Channels		
255 mm wide × 125 mm high flat	type CS1	
150 mm wide × 125 mm high flat type	CS2	
255 mm wide × 125 mm high dished	type CD	

KERBS, PAVING, ETC.

Transition kerb units		
from kerb type SP to HB	left handed	type TL
	right handed	type TR
from kerb type BN to HB	left handed	type DL1
	right handed	type DR1
from kerb type BN to SP	left handed	type DL2
	right handed	type DR2

Number of kerbs required per quarter circle (780 mm kerb lengths)

Radius (m)	Number in quarter circle
12	24
10	20
8	16
6	12
5	10
4	8
3	6
2	4
1	2

Precast Concrete Edgings

Round top type ER	Flat top type EF	Bullnosed top type EBN
150 × 50 mm	150 × 50 mm	150 × 50 mm
200 × 50 mm	200 × 50 mm	200 × 50 mm
250 × 50 mm	250 × 50 mm	250 × 50 mm

KERBS, PAVING, ETC.

BASES

Cement Bound Material for Bases and Subbases

CBM1:	very carefully graded aggregate from 37.5–75 mm, with a 7-day strength of 4.5 N/mm^2
CBM2:	same range of aggregate as CBM1 but with more tolerance in each size of aggregate with a 7-day strength of 7.0 N/mm^2
CBM3:	crushed natural aggregate or blast furnace slag, graded from 37.5–150 mm for 40 mm aggregate, and from 20–75 mm for 20 mm aggregate, with a 7-day strength of 10 N/mm^2
CBM4:	crushed natural aggregate or blast furnace slag, graded from 37.5–150 mm for 40 mm aggregate, and from 20–75 mm for 20 mm aggregate, with a 7-day strength of 15 N/mm^2

INTERLOCKING BRICK/BLOCK ROADS/PAVINGS

Sizes of Precast Concrete Paving Blocks

Type R blocks
200 × 100 × 60 mm
200 × 100 × 65 mm
200 × 100 × 80 mm
200 × 100 × 100 mm

Type S
Any shape within a 295 mm space

Sizes of clay brick pavers
200 × 100 × 50 mm
200 × 100 × 65 mm
210 × 105 × 50 mm
210 × 105 × 65 mm
215 × 102.5 × 50 mm
215 × 102.5 × 65 mm

Type PA: 3 kN
Footpaths and pedestrian areas, private driveways, car parks, light vehicle traffic and over-run

Type PB: 7 kN
Residential roads, lorry parks, factory yards, docks, petrol station forecourts, hardstandings, bus stations

KERBS, PAVING, ETC.

PAVING AND SURFACING

Weights and Sizes of Paving and Surfacing

Description of item	Size	Quantity per tonne
Paving 50 mm thick	900 × 600 mm	15
Paving 50 mm thick	750 × 600 mm	18
Paving 50 mm thick	600 × 600 mm	23
Paving 50 mm thick	450 × 600 mm	30
Paving 38 mm thick	600 × 600 mm	30
Path edging	914 × 50 × 150 mm	60
Kerb (including radius and tapers)	125 × 254 × 914 mm	15
Kerb (including radius and tapers)	125 × 150 × 914 mm	25
Square channel	125 × 254 × 914 mm	15
Dished channel	125 × 254 × 914 mm	15
Quadrants	300 × 300 × 254 mm	19
Quadrants	450 × 450 × 254 mm	12
Quadrants	300 × 300 × 150 mm	30
Internal angles	300 × 300 × 254 mm	30
Fluted pavement channel	255 × 75 × 914 mm	25
Corner stones	300 × 300 mm	80
Corner stones	360 × 360 mm	60
Cable covers	914 × 175 mm	55
Gulley kerbs	220 × 220 × 150 mm	60
Gulley kerbs	220 × 200 × 75 mm	120

KERBS, PAVING, ETC.

Weights and Sizes of Paving and Surfacing

Material	kg/m³	lb/cu yd
Tarmacadam	2306	3891
Macadam (waterbound)	2563	4325
Vermiculite (aggregate)	64–80	108–135
Terracotta	2114	3568
Cork – compressed	388	24
	kg/m²	lb/sq ft
Clay floor tiles, 12.7 mm	27.3	5.6
Pavement lights	122	25
Damp-proof course	5	1
	kg/m² per mm thickness	lb/sq ft per inch thickness
Paving slabs (stone)	2.3	12
Granite setts	2.88	15
Asphalt	2.30	12
Rubber flooring	1.68	9
Polyvinyl chloride	1.94 (avg)	10 (avg)

Coverage (m²) Per Cubic Metre of Materials Used as Subbases or Capping Layers

Consolidated thickness laid in (mm)	Square metre coverage		
	Gravel	Sand	Hardcore
50	15.80	16.50	–
75	10.50	11.00	–
100	7.92	8.20	7.42
125	6.34	6.60	5.90
150	5.28	5.50	4.95
175	–	–	4.23
200	–	–	3.71
225	–	–	3.30
300	–	–	2.47

KERBS, PAVING, ETC.

Approximate Rate of Spreads

Average thickness of course (mm)	Description	Approximate rate of spread			
		Open Textured		Dense, Medium & Fine Textured	
		(kg/m^2)	(m^2/t)	(kg/m^2)	(m^2/t)
35	14 mm open textured or dense wearing course	60–75	13–17	70–85	12–14
40	20 mm open textured or dense base course	70–85	12–14	80–100	10–12
45	20 mm open textured or dense base course	80–100	10–12	95–100	9–10
50	20 mm open textured or dense, or 28 mm dense base course	85–110	9–12	110–120	8–9
60	28 mm dense base course, 40 mm open textured of dense base course or 40 mm single course as base course		8–10	130–150	7–8
65	28 mm dense base course, 40 mm open textured or dense base course or 40 mm single course	100–135	7–10	140–160	6–7
75	40 mm single course, 40 mm open textured or dense base course, 40 mm dense roadbase	120–150	7–8	165–185	5–6
100	40 mm dense base course or roadbase	–	–	220–240	4–4.5

KERBS, PAVING, ETC.

Surface Dressing Roads: Coverage (m²) per Tonne of Material

Size in mm	Sand	Granite chips	Gravel	Limestone chips
Sand	168	–	–	–
3	–	148	152	165
6	–	130	133	144
9	–	111	114	123
13	–	85	87	95
19	–	68	71	78

Sizes of Flags

Reference	Nominal size (mm)	Thickness (mm)
A	600 × 450	50 and 63
B	600 × 600	50 and 63
C	600 × 750	50 and 63
D	600 × 900	50 and 63
E	450 × 450	50 and 70 chamfered top surface
F	400 × 400	50 and 65 chamfered top surface
G	300 × 300	50 and 60 chamfered top surface

Sizes of Natural Stone Setts

Width (mm)		Length (mm)		Depth (mm)
100	×	100	×	100
75	×	150 to 250	×	125
75	×	150 to 250	×	150
100	×	150 to 250	×	100
100	×	150 to 250	×	150

SEEDING/TURFING AND PLANTING

Topsoil Quality

Topsoil grade	Properties
Premium	Natural topsoil, high fertility, loamy texture, good soil structure, suitable for intensive cultivation.
General purpose	Natural or manufactured topsoil of lesser quality than Premium, suitable for agriculture or amenity landscape, may need fertilizer or soil structure improvement.
Economy	Selected subsoil, natural mineral deposit such as river silt or greensand. The grade comprises two subgrades; 'Low clay' and 'High clay' which is more liable to compaction in handling. This grade is suitable for low-production agricultural land and amenity woodland or conservation planting areas.

Forms of Trees

Standards:	Shall be clear with substantially straight stems. Grafted and budded trees shall have no more than a slight bend at the union. Standards shall be designated as Half, Extra light, Light, Standard, Selected standard, Heavy, and Extra heavy.
Sizes of Standards	
Heavy standard	12–14 cm girth × 3.50 to 5.00 m high
Extra Heavy standard	14–16 cm girth × 4.25 to 5.00 m high
Extra Heavy standard	16–18 cm girth × 4.25 to 6.00 m high
Extra Heavy standard	18–20 cm girth × 5.00 to 6.00 m high
Semi-mature trees:	Between 6.0 m and 12.0 m tall with a girth of 20 to 75 cm at 1.0 m above ground.
Feathered trees:	Shall have a defined upright central leader, with stem furnished with evenly spread and balanced lateral shoots down to or near the ground.
Whips:	Shall be without significant feather growth as determined by visual inspection.
Multi-stemmed trees:	Shall have two or more main stems at, near, above or below ground.

Seedlings grown from seed and not transplanted shall be specified when ordered for sale as:

1+0	one year old seedling
2+0	two year old seedling
1+1	one year seed bed, one year transplanted = two year old seedling
1+2	one year seed bed, two years transplanted = three year old seedling
2+1	two years seed bed, one year transplanted = three year old seedling
1u1	two years seed bed, undercut after 1 year = two year old seedling
2u2	four years seed bed, undercut after 2 years = four year old seedling

SEEDING/TURFING AND PLANTING

Cuttings

The age of cuttings (plants grown from shoots, stems, or roots of the mother plant) shall be specified when ordered for sale. The height of transplants and undercut seedlings/cuttings (which have been transplanted or undercut at least once) shall be stated in centimetres. The number of growing seasons before and after transplanting or undercutting shall be stated.

0 + 1	one year cutting
0 + 2	two year cutting
0 + 1 + 1	one year cutting bed, one year transplanted = two year old seedling
0 + 1 + 2	one year cutting bed, two years transplanted = three year old seedling

Grass Cutting Capacities in m² per hour

Speed mph	Width of cut in metres												
	0.5	0.7	1.0	1.2	1.5	1.7	2.0	2.0	2.1	2.5	2.8	3.0	3.4
1.0	724	1127	1529	1931	2334	2736	3138	3219	3380	4023	4506	4828	5472
1.5	1086	1690	2293	2897	3500	4104	4707	4828	5069	6035	6759	7242	8208
2.0	1448	2253	3058	3862	4667	5472	6276	6437	6759	8047	9012	9656	10944
2.5	1811	2816	3822	4828	5834	6840	7846	8047	8449	10058	11265	12070	13679
3.0	2173	3380	4587	5794	7001	8208	9415	9656	10139	12070	13518	14484	16415
3.5	2535	3943	5351	6759	8167	9576	10984	11265	11829	14082	15772	16898	19151
4.0	2897	4506	6115	7725	9334	10944	12553	12875	13518	16093	18025	19312	21887
4.5	3259	5069	6880	8690	10501	12311	14122	14484	15208	18105	20278	21726	24623
5.0	3621	5633	7644	9656	11668	13679	15691	16093	16898	20117	22531	24140	27359
5.5	3983	6196	8409	10622	12834	15047	17260	17703	18588	22128	24784	26554	30095
6.0	4345	6759	9173	11587	14001	16415	18829	19312	20278	24140	27037	28968	32831
6.5	4707	7322	9938	12553	15168	17783	20398	20921	21967	26152	29290	31382	35566
7.0	5069	7886	10702	13518	16335	19151	21967	22531	23657	28163	31543	33796	38302

Number of Plants per m² at the following offset spacings

(All plants equidistant horizontally and vertically)

Distance mm	Nr of Plants
100	114.43
200	28.22
250	18.17
300	12.35
400	6.72
500	4.29
600	3.04
700	2.16
750	1.88
900	1.26
1000	1.05
1200	0.68
1500	0.42
2000	0.23

SEEDING/TURFING AND PLANTING

Grass Clippings Wet: Based on 3.5 m³/tonne

Annual kg/100 m²		Average 20 cuts kg/100 m²			m²/tonne	m²/m³
32.0		1.6			61162.1	214067.3

Nr of cuts	22	20	18	16	12	4
kg/cut	1.45	1.60	1.78	2.00	2.67	8.00
Area capacity of 3 tonne vehicle per load						
m²	206250	187500	168750	150000	112500	37500
Load m³	**100 m² units/m³ of vehicle space**					
1	196.4	178.6	160.7	142.9	107.1	35.7
2	392.9	357.1	321.4	285.7	214.3	71.4
3	589.3	535.7	482.1	428.6	321.4	107.1
4	785.7	714.3	642.9	571.4	428.6	142.9
5	982.1	892.9	803.6	714.3	535.7	178.6

Transportation of Trees

To unload large trees a machine with the necessary lifting strength is required. The weight of the trees must therefore be known in advance. The following table gives a rough overview. The additional columns with root ball dimensions and the number of plants per trailer provide additional information, for example about preparing planting holes and calculating unloading times.

Girth in cm	Rootball diameter in cm	Ball height in cm	Weight in kg	Numbers of trees per trailer
16–18	50–60	40	150	100–120
18–20	60–70	40–50	200	80–100
20–25	60–70	40–50	270	50–70
25–30	80	50–60	350	50
30–35	90–100	60–70	500	12–18
35–40	100–110	60–70	650	10–15
40–45	110–120	60–70	850	8–12
45–50	110–120	60–70	1100	5–7
50–60	130–140	60–70	1600	1–3
60–70	150–160	60–70	2500	1
70–80	180–200	70	4000	1
80–90	200–220	70–80	5500	1
90–100	230–250	80–90	7500	1
100–120	250–270	80–90	9500	1

Data supplied by Lorenz von Ehren GmbH
The information in the table is approximate; deviations depend on soil type, genus and weather

FENCING AND GATES

FENCING AND GATES

Types of Preservative

Creosote (tar oil) can be 'factory' applied	by pressure to BS 144: pts 1&2
	by immersion to BS 144: pt 1
	by hot and cold open tank to BS 144: pts 1&2
Copper/chromium/arsenic (CCA)	by full cell process to BS 4072 pts 1&2
Organic solvent (OS)	by double vacuum (vacvac) to BS 5707 pts 1&3
	by immersion to BS 5057 pts 1&3
Pentachlorophenol (PCP)	by heavy oil double vacuum to BS 5705 pts 2&3

Boron diffusion process (treated with disodium octaborate to BWPA Manual 1986)

Note: Boron is used on green timber at source and the timber is supplied dry

Cleft Chestnut Pale Fences

Pales	Pale spacing	Wire lines	
900 mm	75 mm	2	temporary protection
1050 mm	75 or 100 mm	2	light protective fences
1200 mm	75 mm	3	perimeter fences
1350 mm	75 mm	3	perimeter fences
1500 mm	50 mm	3	narrow perimeter fences
1800 mm	50 mm	3	light security fences

Close-Boarded Fences

Close-boarded fences 1.05 to 1.8 m high
Type BCR (recessed) or BCM (morticed) with concrete posts 140 × 115 mm tapered and Type BW with timber posts

Palisade Fences

Wooden palisade fences
Type WPC with concrete posts 140 × 115 mm tapered and Type WPW with timber posts

For both types of fence:
Height of fence 1050 mm: two rails
Height of fence 1200 mm: two rails
Height of fence 1500 mm: three rails
Height of fence 1650 mm: three rails
Height of fence 1800 mm: three rails

FENCING AND GATES

Post and Rail Fences

Wooden post and rail fences
Type MPR 11/3 morticed rails and Type SPR 11/3 nailed rails
Height to top of rail 1100 mm
Rails: three rails 87 mm, 38 mm

Type MPR 11/4 morticed rails and Type SPR 11/4 nailed rails
Height to top of rail 1100 mm
Rails: four rails 87 mm, 38 mm

Type MPR 13/4 morticed rails and Type SPR 13/4 nailed rails
Height to top of rail 1300 mm
Rail spacing 250 mm, 250 mm, and 225 mm from top
Rails: four rails 87 mm, 38 mm

Steel Posts

Rolled steel angle iron posts for chain link fencing

Posts	Fence height	Strut	Straining post
1500 × 40 × 40 × 5 mm	900 mm	1500 × 40 × 40 × 5 mm	1500 × 50 × 50 × 6 mm
1800 × 40 × 40 × 5 mm	1200 mm	1800 × 40 × 40 × 5 mm	1800 × 50 × 50 × 6 mm
2000 × 45 × 45 × 5 mm	1400 mm	2000 × 45 × 45 × 5 mm	2000 × 60 × 60 × 6 mm
2600 × 45 × 45 × 5 mm	1800 mm	2600 × 45 × 45 × 5 mm	2600 × 60 × 60 × 6 mm
3000 × 50 × 50 × 6 mm	1800 mm	2600 × 45 × 45 × 5 mm	3000 × 60 × 60 × 6 mm
with arms			

Concrete Posts

Concrete posts for chain link fencing

Posts and straining posts	Fence height	Strut
1570 mm 100 × 100 mm	900 mm	1500 mm × 75 × 75 mm
1870 mm 125 × 125 mm	1200 mm	1830 mm × 100 × 75 mm
2070 mm 125 × 125 mm	1400 mm	1980 mm × 100 × 75 mm
2620 mm 125 × 125 mm	1800 mm	2590 mm × 100 × 85 mm
3040 mm 125 × 125 mm	1800 mm	2590 mm × 100 × 85 mm (with arms)

Rolled Steel Angle Posts

Rolled steel angle posts for rectangular wire mesh (field) fencing

Posts	Fence height	Strut	Straining post
1200 × 40 × 40 × 5 mm	600 mm	1200 × 75 × 75 mm	1350 × 100 × 100 mm
1400 × 40 × 40 × 5 mm	800 mm	1400 × 75 × 75 mm	1550 × 100 × 100 mm
1500 × 40 × 40 × 5 mm	900 mm	1500 × 75 × 75 mm	1650 × 100 × 100 mm
1600 × 40 × 40 × 5 mm	1000 mm	1600 × 75 × 75 mm	1750 × 100 × 100 mm
1750 × 40 × 40 × 5 mm	1150 mm	1750 × 75 × 100 mm	1900 × 125 × 125 mm

Concrete Posts

Concrete posts for rectangular wire mesh (field) fencing

Posts	Fence height	Strut	Straining post
1270 × 100 × 100 mm	600 mm	1200 × 75 × 75 mm	1420 × 100 × 100 mm
1470 × 100 × 100 mm	800 mm	1350 × 75 × 75 mm	1620 × 100 × 100 mm
1570 × 100 × 100 mm	900 mm	1500 × 75 × 75 mm	1720 × 100 × 100 mm
1670 × 100 × 100 mm	600 mm	1650 × 75 × 75 mm	1820 × 100 × 100 mm
1820 × 125 × 125 mm	1150 mm	1830 × 75 × 100 mm	1970 × 125 × 125 mm

Cleft Chestnut Pale Fences

Timber Posts

Timber posts for wire mesh and hexagonal wire netting fences

Round timber for general fences

Posts	Fence height	Strut	Straining post
1300 × 65 mm dia.	600 mm	1200 × 80 mm dia.	1450 × 100 mm dia.
1500 × 65 mm dia.	800 mm	1400 × 80 mm dia.	1650 × 100 mm dia.
1600 × 65 mm dia.	900 mm	1500 × 80 mm dia.	1750 × 100 mm dia.
1700 × 65 mm dia.	1050 mm	1600 × 80 mm dia.	1850 × 100 mm dia.
1800 × 65 mm dia.	1150 mm	1750 × 80 mm dia.	2000 × 120 mm dia.

Squared timber for general fences

Posts	Fence height	Strut	Straining post
1300 × 75 × 75 mm	600 mm	1200 × 75 × 75 mm	1450 × 100 × 100 mm
1500 × 75 × 75 mm	800 mm	1400 × 75 × 75 mm	1650 × 100 × 100 mm
1600 × 75 × 75 mm	900 mm	1500 × 75 × 75 mm	1750 × 100 × 100 mm
1700 × 75 × 75 mm	1050 mm	1600 × 75 × 75 mm	1850 × 100 × 100 mm
1800 × 75 × 75 mm	1150 mm	1750 × 75 × 75 mm	2000 × 125 × 100 mm

FENCING AND GATES

Steel Fences to BS 1722: Part 9: 1992

	Fence height	Top/bottom rails and flat posts	Vertical bars
Light	1000 mm	40 × 10 mm 450 mm in ground	12 mm dia. at 115 mm cs
	1200 mm	40 × 10 mm 550 mm in ground	12 mm dia. at 115 mm cs
	1400 mm	40 × 10 mm 550 mm in ground	12 mm dia. at 115 mm cs
Light	1000 mm	40 × 10 mm 450 mm in ground	16 mm dia. at 120 mm cs
	1200 mm	40 × 10 mm 550 mm in ground	16 mm dia. at 120 mm cs
	1400 mm	40 × 10 mm 550 mm in ground	16 mm dia. at 120 mm cs
Medium	1200 mm	50 × 10 mm 550 mm in ground	20 mm dia. at 125 mm cs
	1400 mm	50 × 10 mm 550 mm in ground	20 mm dia. at 125 mm cs
	1600 mm	50 × 10 mm 600 mm in ground	22 mm dia. at 145 mm cs
	1800 mm	50 × 10 mm 600 mm in ground	22 mm dia. at 145 mm cs
Heavy	1600 mm	50 × 10 mm 600 mm in ground	22 mm dia. at 145 mm cs
	1800 mm	50 × 10 mm 600 mm in ground	22 mm dia. at 145 mm cs
	2000 mm	50 × 10 mm 600 mm in ground	22 mm dia. at 145 mm cs
	2200 mm	50 × 10 mm 600 mm in ground	22 mm dia. at 145 mm cs

Notes: Mild steel fences: round or square verticals; flat standards and horizontals. Tops of vertical bars may be bow-top, blunt, or pointed. Round or square bar railings

Timber Field Gates to BS 3470: 1975

Gates made to this standard are designed to open one way only
All timber gates are 1100 mm high
Width over stiles 2400, 2700, 3000, 3300, 3600, and 4200 mm
Gates over 4200 mm should be made in two leaves

Steel Field Gates to BS 3470: 1975

All steel gates are 1100 mm high
Heavy duty: width over stiles 2400, 3000, 3600 and 4500 mm
Light duty: width over stiles 2400, 3000, and 3600 mm

FENCING AND GATES

Domestic Front Entrance Gates to BS 4092: Part 1: 1966

Metal gates:	Single gates are 900 mm high minimum, 900 mm, 1000 mm and 1100 mm wide

Domestic Front Entrance Gates to BS 4092: Part 2: 1966

Wooden gates:	All rails shall be tenoned into the stiles Single gates are 840 mm high minimum, 801 mm and 1020 mm wide Double gates are 840 mm high minimum, 2130, 2340 and 2640 mm wide

Timber Bridle Gates to BS 5709:1979 (Horse or Hunting Gates)

Gates open one way only Minimum width between posts Minimum height	1525 mm 1100 mm

Timber Kissing Gates to BS 5709:1979

Minimum width	700 mm
Minimum height	1000 mm
Minimum distance between shutting posts	600 mm
Minimum clearance at mid-point	600 mm

Metal Kissing Gates to BS 5709:1979

Sizes are the same as those for timber kissing gates Maximum gaps between rails 120 mm

Categories of Pedestrian Guard Rail to BS 3049:1976

Class A for normal use Class B where vandalism is expected Class C where crowd pressure is likely

DRAINAGE

Width Required for Trenches for Various Diameters of Pipes

Pipe diameter (mm)	Trench n.e. 1.50 m deep	Trench over 1.50 m deep
n.e. 100 mm	450 mm	600 mm
100–150 mm	500 mm	650 mm
150–225 mm	600 mm	750 mm
225–300 mm	650 mm	800 mm
300–400 mm	750 mm	900 mm
400–450 mm	900 mm	1050 mm
450–600 mm	1100 mm	1300 mm

Weights and Dimensions – Vitrified Clay Pipes

Product	Nominal diameter (mm)	Effective length (mm)	BS 65 limits of tolerance min (mm)	max (mm)	Crushing strength (kN/m)	Weight (kg/pipe)	(kg/m)
Supersleve	100	1600	96	105	35.00	14.71	9.19
	150	1750	146	158	35.00	29.24	16.71
Hepsleve	225	1850	221	236	28.00	84.03	45.42
	300	2500	295	313	34.00	193.05	77.22
	150	1500	146	158	22.00	37.04	24.69
Hepseal	225	1750	221	236	28.00	85.47	48.84
	300	2500	295	313	34.00	204.08	81.63
	400	2500	394	414	44.00	357.14	142.86
	450	2500	444	464	44.00	454.55	181.63
	500	2500	494	514	48.00	555.56	222.22
	600	2500	591	615	57.00	796.23	307.69
	700	3000	689	719	67.00	1111.11	370.45
	800	3000	788	822	72.00	1351.35	450.45
Hepline	100	1600	95	107	22.00	14.71	9.19
	150	1750	145	160	22.00	29.24	16.71
	225	1850	219	239	28.00	84.03	45.42
	300	1850	292	317	34.00	142.86	77.22
Hepduct (conduit)	90	1500	–	–	28.00	12.05	8.03
	100	1600	–	–	28.00	14.71	9.19
	125	1750	–	–	28.00	20.73	11.84
	150	1750	–	–	28.00	29.24	16.71
	225	1850	–	–	28.00	84.03	45.42
	300	1850	–	–	34.00	142.86	77.22

DRAINAGE

Weights and Dimensions – Vitrified Clay Pipes

Nominal internal diameter (mm)	Nominal wall thickness (mm)	Approximate weight (kg/m)
150	25	45
225	29	71
300	32	122
375	35	162
450	38	191
600	48	317
750	54	454
900	60	616
1200	76	912
1500	89	1458
1800	102	1884
2100	127	2619

Wall thickness, weights and pipe lengths vary, depending on type of pipe required

The particulars shown above represent a selection of available diameters and are applicable to strength class 1 pipes with flexible rubber ring joints

Tubes with Ogee joints are also available

DRAINAGE

Weights and Dimensions – PVC-u Pipes

	Nominal size	Mean outside diameter (mm)		Wall thickness	Weight
		min	max	(mm)	(kg/m)
Standard pipes	82.4	82.4	82.7	3.2	1.2
	110.0	110.0	110.4	3.2	1.6
	160.0	160.0	160.6	4.1	3.0
	200.0	200.0	200.6	4.9	4.6
	250.0	250.0	250.7	6.1	7.2
Perforated pipes heavy grade	As above	As above	As above	As above	As above
thin wall	82.4	82.4	82.7	1.7	–
	110.0	110.0	110.4	2.2	–
	160.0	160.0	160.6	3.2	–

Width of Trenches Required for Various Diameters of Pipes

Pipe diameter (mm)	Trench n.e. 1.5 m deep (mm)	Trench over 1.5 m deep (mm)
n.e. 100	450	600
100–150	500	650
150–225	600	750
225–300	650	800
300–400	750	900
400–450	900	1050
450–600	1100	1300

DRAINAGE

DRAINAGE BELOW GROUND AND LAND DRAINAGE

Flow of Water Which Can Be Carried by Various Sizes of Pipe

Clay or concrete pipes

Pipe size	Gradient of pipeline							
	1:10	1:20	1:30	1:40	1:50	1:60	1:80	1:100
	Flow in litres per second							
DN 100 15.0	8.5	6.8	5.8	5.2	4.7	4.0	3.5	
DN 150 28.0	19.0	16.0	14.0	12.0	11.0	9.1	8.0	
DN 225 140.0	95.0	76.0	66.0	58.0	53.0	46.0	40.0	

Plastic pipes

Pipe size	Gradient of pipeline							
	1:10	1:20	1:30	1:40	1:50	1:60	1:80	1:100
	Flow in litres per second							
82.4 mm i/dia.	12.0	8.5	6.8	5.8	5.2	4.7	4.0	3.5
110 mm i/dia.	28.0	19.0	16.0	14.0	12.0	11.0	9.1	8.0
160 mm i/dia.	76.0	53.0	43.0	37.0	33.0	29.0	25.0	22.0
200 mm i/dia.	140.0	95.0	76.0	66.0	58.0	53.0	46.0	40.0

Vitrified (Perforated) Clay Pipes and Fittings to BS En 295-5 1994

Length not specified		
75 mm bore	250 mm bore	600 mm bore
100	300	700
125	350	800
150	400	1000
200	450	1200
225	500	

Precast Concrete Pipes: Prestressed Non-pressure Pipes and Fittings: Flexible Joints to BS 5911: Pt. 103: 1994

Rationalized metric nominal sizes: 450, 500	
Length:	500–1000 by 100 increments
	1000–2200 by 200 increments
	2200–2800 by 300 increments
Angles: length:	450–600 angles 45, 22.5,11.25°
	600 or more angles 22.5, 11.25°

DRAINAGE

Precast Concrete Pipes: Unreinforced and Circular Manholes and Soakaways to BS 5911: Pt. 200: 1994

Nominal sizes:	
Shafts:	675, 900 mm
Chambers:	900, 1050, 1200, 1350, 1500, 1800, 2100, 2400, 2700, 3000 mm
Large chambers:	To have either tapered reducing rings or a flat reducing slab in order to accept the standard cover
Ring depths:	1. 300–1200 mm by 300 mm increments except for bottom slab and rings below cover slab, these are by 150 mm increments
	2. 250–1000 mm by 250 mm increments except for bottom slab and rings below cover slab, these are by 125 mm increments
Access hole:	750 × 750 mm for DN 1050 chamber 1200 × 675 mm for DN 1350 chamber

Calculation of Soakaway Depth

The following formula determines the depth of concrete ring soakaway that would be required for draining given amounts of water.

$$h = \frac{4ar}{3\pi D^2}$$

h = depth of the chamber below the invert pipe
a = the area to be drained
r = the hourly rate of rainfall (50 mm per hour)
π = pi
D = internal diameter of the soakaway

This table shows the depth of chambers in each ring size which would be required to contain the volume of water specified. These allow a recommended storage capacity of ⅓ (one third of the hourly rainfall figure).

Table Showing Required Depth of Concrete Ring Chambers in Metres

Area m²	50	100	150	200	300	400	500
Ring size							
0.9	1.31	2.62	3.93	5.24	7.86	10.48	13.10
1.1	0.96	1.92	2.89	3.85	5.77	7.70	9.62
1.2	0.74	1.47	2.21	2.95	4.42	5.89	7.37
1.4	0.58	1.16	1.75	2.33	3.49	4.66	5.82
1.5	0.47	0.94	1.41	1.89	2.83	3.77	4.72
1.8	0.33	0.65	0.98	1.31	1.96	2.62	3.27
2.1	0.24	0.48	0.72	0.96	1.44	1.92	2.41
2.4	0.18	0.37	0.55	0.74	1.11	1.47	1.84
2.7	0.15	0.29	0.44	0.58	0.87	1.16	1.46
3.0	0.12	0.24	0.35	0.47	0.71	0.94	1.18

DRAINAGE

Precast Concrete Inspection Chambers and Gullies to BS 5911: Part 230: 1994

Nominal sizes:	375 diameter, 750, 900 mm deep
	450 diameter, 750, 900, 1050, 1200 mm deep
Depths:	from the top for trapped or untrapped units:
	centre of outlet 300 mm
	invert (bottom) of the outlet pipe 400 mm
Depth of water seal for trapped gullies:	
	85 mm, rodding eye int. dia. 100 mm
Cover slab:	65 mm min

Bedding Flexible Pipes: PVC-u Or Ductile Iron

Type 1 =	100 mm fill below pipe, 300 mm above pipe: single size material
Type 2 =	100 mm fill below pipe, 300 mm above pipe: single size or graded material
Type 3 =	100 mm fill below pipe, 75 mm above pipe with concrete protective slab over
Type 4 =	100 mm fill below pipe, fill laid level with top of pipe
Type 5 =	200 mm fill below pipe, fill laid level with top of pipe
Concrete =	25 mm sand blinding to bottom of trench, pipe supported on chocks, 100 mm concrete under the pipe, 150 mm concrete over the pipe

DRAINAGE

Bedding Rigid Pipes: Clay or Concrete
(for vitrified clay pipes the manufacturer should be consulted)

Class D:	Pipe laid on natural ground with cut-outs for joints, soil screened to remove stones over 40 mm and returned over pipe to 150 m min depth. Suitable for firm ground with trenches trimmed by hand.
Class N:	Pipe laid on 50 mm granular material of graded aggregate to Table 4 of BS 882, or 10 mm aggregate to Table 6 of BS 882, or as dug light soil (not clay) screened to remove stones over 10 mm. Suitable for machine dug trenches.
Class B:	As Class N, but with granular bedding extending half way up the pipe diameter.
Class F:	Pipe laid on 100 mm granular fill to BS 882 below pipe, minimum 150 mm granular fill above pipe: single size material. Suitable for machine dug trenches.
Class A:	Concrete 100 mm thick under the pipe extending half way up the pipe, backfilled with the appropriate class of fill. Used where there is only a very shallow fall to the drain. Class A bedding allows the pipes to be laid to an exact gradient.
Concrete surround:	25 mm sand blinding to bottom of trench, pipe supported on chocks, 100 mm concrete under the pipe, 150 mm concrete over the pipe. It is preferable to bed pipes under slabs or wall in granular material.

PIPED SUPPLY SYSTEMS

Identification of Service Tubes From Utility to Dwellings

Utility	Colour	Size	Depth
British Telecom	grey	54 mm od	450 mm
Electricity	black	38 mm od	450 mm
Gas	yellow	42 mm od rigid 60 mm od convoluted	450 mm
Water	may be blue	(normally untubed)	750 mm

ELECTRICAL SUPPLY/POWER/LIGHTING SYSTEMS

Electrical Insulation Class En 60.598 BS 4533

Class 1:	luminaires comply with class 1 (I) earthed electrical requirements
Class 2:	luminaires comply with class 2 (II) double insulated electrical requirements
Class 3:	luminaires comply with class 3 (III) electrical requirements

Protection to Light Fittings

BS EN 60529:1992 Classification for degrees of protection provided by enclosures.
(IP Code – International or ingress Protection)

1st characteristic: against ingress of solid foreign objects

The figure	2	indicates that fingers cannot enter
	3	that a 2.5 mm diameter probe cannot enter
	4	that a 1.0 mm diameter probe cannot enter
	5	the fitting is dust proof (no dust around live parts)
	6	the fitting is dust tight (no dust entry)

2nd characteristic: ingress of water with harmful effects

The figure	0	indicates unprotected
	1	vertically dripping water cannot enter
	2	water dripping 15° (tilt) cannot enter
	3	spraying water cannot enter
	4	splashing water cannot enter
	5	jetting water cannot enter
	6	powerful jetting water cannot enter
	7	proof against temporary immersion
	8	proof against continuous immersion
Optional additional codes:		A–D protects against access to hazardous parts
	H	high voltage apparatus
	M	fitting was in motion during water test
	S	fitting was static during water test
	W	protects against weather
Marking code arrangement:		(example) IPX5S = IP (International or Ingress Protection)
		X (denotes omission of first characteristic)
		5 = jetting
		S = static during water test

RAIL TRACKS

RAIL TRACKS

	kg/m of track	lb/ft of track
Standard gauge Bull-head rails, chairs, transverse timber (softwood) sleepers etc.	245	165
Main lines Flat-bottom rails, transverse prestressed concrete sleepers, etc.	418	280
Add for electric third rail	51	35
Add for crushed stone ballast	2600	1750
	kg/m²	**lb/sq ft**
Overall average weight – rails connections, sleepers, ballast, etc.	733	150
	kg/m of track	**lb/ft of track**
Bridge rails, longitudinal timber sleepers, etc.	112	75

RAIL TRACKS

Heavy Rails

British Standard Section No.	Rail height (mm)	Foot width (mm)	Head width (mm)	Min web thickness (mm)	Section weight (kg/m)
Flat Bottom Rails					
60 A	114.30	109.54	57.15	11.11	30.62
70 A	123.82	111.12	60.32	12.30	34.81
75 A	128.59	114.30	61.91	12.70	37.45
80 A	133.35	117.47	63.50	13.10	39.76
90 A	142.88	127.00	66.67	13.89	45.10
95 A	147.64	130.17	69.85	14.68	47.31
100 A	152.40	133.35	69.85	15.08	50.18
110 A	158.75	139.70	69.85	15.87	54.52
113 A	158.75	139.70	69.85	20.00	56.22
50 'O'	100.01	100.01	52.39	10.32	24.82
80 'O'	127.00	127.00	63.50	13.89	39.74
60R	114.30	109.54	57.15	11.11	29.85
75R	128.59	122.24	61.91	13.10	37.09
80R	133.35	127.00	63.50	13.49	39.72
90R	142.88	136.53	66.67	13.89	44.58
95R	147.64	141.29	68.26	14.29	47.21
100R	152.40	146.05	69.85	14.29	49.60
95N	147.64	139.70	69.85	13.89	47.27
Bull Head Rails					
95R BH	145.26	69.85	69.85	19.05	47.07

Light Rails

British Standard Section No.	Rail height (mm)	Foot width (mm)	Head width (mm)	Min web thickness (mm)	Section weight (kg/m)
Flat Bottom Rails					
20M	65.09	55.56	30.96	6.75	9.88
30M	75.41	69.85	38.10	9.13	14.79
35M	80.96	76.20	42.86	9.13	17.39
35R	85.73	82.55	44.45	8.33	17.40
40	88.11	80.57	45.64	12.3	19.89
Bridge Rails					
13	48.00	92	36.00	18.0	13.31
16	54.00	108	44.50	16.0	16.06
20	55.50	127	50.00	20.5	19.86
28	67.00	152	50.00	31.0	28.62
35	76.00	160	58.00	34.5	35.38
50	76.00	165	58.50	–	50.18
Crane Rails					
A65	75.00	175.00	65.00	38.0	43.10
A75	85.00	200.00	75.00	45.0	56.20
A100	95.00	200.00	100.00	60.0	74.30
A120	105.00	220.00	120.00	72.0	100.00
175CR	152.40	152.40	107.95	38.1	86.92

RAIL TRACKS

Fish Plates

British Standard Section No.	Overall plate length		Hole diameter	Finished weight per pair	
	4 Hole (mm)	6 Hole (mm)	(mm)	4 Hole (kg/pair)	6 Hole (kg/pair)
For British Standard Heavy Rails: Flat Bottom Rails					
60 A	406.40	609.60	20.64	9.87	14.76
70 A	406.40	609.60	22.22	11.15	16.65
75 A	406.40	–	23.81	11.82	17.73
80 A	406.40	609.60	23.81	13.15	19.72
90 A	457.20	685.80	25.40	17.49	26.23
100 A	508.00	–	pear	25.02	–
110 A (shallow)	507.00	–	27.00	30.11	54.64
113 A (heavy)	507.00	–	27.00	30.11	54.64
50 'O' (shallow)	406.40	–	–	6.68	10.14
80 'O' (shallow)	495.30	–	23.81	14.72	22.69
60R (shallow)	406.40	609.60	20.64	8.76	13.13
60R (angled)	406.40	609.60	20.64	11.27	16.90
75R (shallow)	406.40	–	23.81	10.94	16.42
75R (angled)	406.40	–	23.81	13.67	–
80R (shallow)	406.40	609.60	23.81	11.93	17.89
80R (angled)	406.40	609.60	23.81	14.90	22.33
For British Standard Heavy Rails: Bull head rails					
95R BH (shallow)	–	457.20	27.00	14.59	14.61
For British Standard Light Rails: Flat Bottom Rails					
30M	355.6	–	–	–	2.72
35M	355.6	–	–	–	2.83
40	355.6	–	–	3.76	–

FRACTIONS, DECIMALS AND MILLIMETRE EQUIVALENTS

Fractions	Decimals	(mm)	Fractions	Decimals	(mm)
1/64	0.015625	0.396875	33/64	0.515625	13.096875
1/32	0.03125	0.79375	17/32	0.53125	13.49375
3/64	0.046875	1.190625	35/64	0.546875	13.890625
1/16	0.0625	1.5875	9/16	0.5625	14.2875
5/64	0.078125	1.984375	37/64	0.578125	14.684375
3/32	0.09375	2.38125	19/32	0.59375	15.08125
7/64	0.109375	2.778125	39/64	0.609375	15.478125
1/8	0.125	3.175	5/8	0.625	15.875
9/64	0.140625	3.571875	41/64	0.640625	16.271875
5/32	0.15625	3.96875	21/32	0.65625	16.66875
11/64	0.171875	4.365625	43/64	0.671875	17.065625
3/16	0.1875	4.7625	11/16	0.6875	17.4625
13/64	0.203125	5.159375	45/64	0.703125	17.859375
7/32	0.21875	5.55625	23/32	0.71875	18.25625
15/64	0.234375	5.953125	47/64	0.734375	18.653125
1/4	0.25	6.35	3/4	0.75	19.05
17/64	0.265625	6.746875	49/64	0.765625	19.446875
9/32	0.28125	7.14375	25/32	0.78125	19.84375
19/64	0.296875	7.540625	51/64	0.796875	20.240625
5/16	0.3125	7.9375	13/16	0.8125	20.6375
21/64	0.328125	8.334375	53/64	0.828125	21.034375
11/32	0.34375	8.73125	27/32	0.84375	21.43125
23/64	0.359375	9.128125	55/64	0.859375	21.828125
3/8	0.375	9.525	7/8	0.875	22.225
25/64	0.390625	9.921875	57/64	0.890625	22.621875
13/32	0.40625	10.31875	29/32	0.90625	23.01875
27/64	0.421875	10.71563	59/64	0.921875	23.415625
7/16	0.4375	11.1125	15/16	0.9375	23.8125
29/64	0.453125	11.50938	61/64	0.953125	24.209375
15/32	0.46875	11.90625	31/32	0.96875	24.60625
31/64	0.484375	12.30313	63/64	0.984375	25.003125
1/2	0.5	12.7	1.0	1	25.4

IMPERIAL STANDARD WIRE GAUGE (SWG)

SWG No.	Diameter (inches)	Diameter (mm)	SWG No.	Diameter (inches)	Diameter (mm)
7/0	0.5	12.7	23	0.024	0.61
6/0	0.464	11.79	24	0.022	0.559
5/0	0.432	10.97	25	0.02	0.508
4/0	0.4	10.16	26	0.018	0.457
3/0	0.372	9.45	27	0.0164	0.417
2/0	0.348	8.84	28	0.0148	0.376
1/0	0.324	8.23	29	0.0136	0.345
1	0.3	7.62	30	0.0124	0.315
2	0.276	7.01	31	0.0116	0.295
3	0.252	6.4	32	0.0108	0.274
4	0.232	5.89	33	0.01	0.254
5	0.212	5.38	34	0.009	0.234
6	0.192	4.88	35	0.008	0.213
7	0.176	4.47	36	0.008	0.193
8	0.16	4.06	37	0.007	0.173
9	0.144	3.66	38	0.006	0.152
10	0.128	3.25	39	0.005	0.132
11	0.116	2.95	40	0.005	0.122
12	0.104	2.64	41	0.004	0.112
13	0.092	2.34	42	0.004	0.102
14	0.08	2.03	43	0.004	0.091
15	0.072	1.83	44	0.003	0.081
16	0.064	1.63	45	0.003	0.071
17	0.056	1.42	46	0.002	0.061
18	0.048	1.22	47	0.002	0.051
19	0.04	1.016	48	0.002	0.041
20	0.036	0.914	49	0.001	0.031
21	0.032	0.813	50	0.001	0.025
22	0.028	0.711			

PIPES, WATER, STORAGE, INSULATION

WATER PRESSURE DUE TO HEIGHT

Imperial

Head (Feet)	Pressure (lb/in^2)		Head (Fcct)	Pressure (lb/in^2)
1	0.43		70	30.35
5	2.17		75	32.51
10	4.34		80	34.68
15	6.5		85	36.85
20	8.67		90	39.02
25	10.84		95	41.18
30	13.01		100	43.35
35	15.17		105	45.52
40	17.34		110	47.69
45	19.51		120	52.02
50	21.68		130	56.36
55	23.84		140	60.69
60	26.01		150	65.03
65	28.18			

Metric

Head (m)	Pressure (bar)		Head (m)	Pressure (bar)
0.5	0.049		18.0	1.766
1.0	0.098		19.0	1.864
1.5	0.147		20.0	1.962
2.0	0.196		21.0	2.06
3.0	0.294		22.0	2.158
4.0	0.392		23.0	2.256
5.0	0.491		24.0	2.354
6.0	0.589		25.0	2.453
7.0	0.687		26.0	2.551
8.0	0.785		27.0	2.649
9.0	0.883		28.0	2.747
10.0	0.981		29.0	2.845
11.0	1.079		30.0	2.943
12.0	1.177		32.5	3.188
13.0	1.275		35.0	3.434
14.0	1.373		37.5	3.679
15.0	1.472		40.0	3.924
16.0	1.57		42.5	4.169
17.0	1.668		45.0	4.415

1 bar	=	14.5038 lbf/in^2
1 lbf/in^2	=	0.06895 bar
1 metre	=	3.2808 ft or 39.3701 in
1 foot	=	0.3048 metres
1 in wg	=	2.5 mbar (249.1 N/m^2)

PIPES, WATER, STORAGE, INSULATION

Dimensions and Weights of Copper Pipes to BSEN 1057, BSEN 12499, BSEN 14251

Outside diameter (mm)	Internal diameter (mm)	Weight per metre (kg)	Internal diameter (mm)	Weight per metre (kg)	Internal siameter (mm)	Weight per metre (kg)
	Formerly Table X		Formerly Table Y		Formerly Table Z	
6	4.80	0.0911	4.40	0.1170	5.00	0.0774
8	6.80	0.1246	6.40	0.1617	7.00	0.1054
10	8.80	0.1580	8.40	0.2064	9.00	0.1334
12	10.80	0.1914	10.40	0.2511	11.00	0.1612
15	13.60	0.2796	13.00	0.3923	14.00	0.2031
18	16.40	0.3852	16.00	0.4760	16.80	0.2918
22	20.22	0.5308	19.62	0.6974	20.82	0.3589
28	26.22	0.6814	25.62	0.8985	26.82	0.4594
35	32.63	1.1334	32.03	1.4085	33.63	0.6701
42	39.63	1.3675	39.03	1.6996	40.43	0.9216
54	51.63	1.7691	50.03	2.9052	52.23	1.3343
76.1	73.22	3.1287	72.22	4.1437	73.82	2.5131
108	105.12	4.4666	103.12	7.3745	105.72	3.5834
133	130.38	5.5151	–	–	130.38	5.5151
159	155.38	8.7795	–	–	156.38	6.6056

Dimensions of Stainless Steel Pipes to BS 4127

Outside siameter (mm)	Maximum outside siameter (mm)	Minimum outside diameter (mm)	Wall thickness (mm)	Working pressure (bar)
6	6.045	5.940	0.6	330
8	8.045	7.940	0.6	260
10	10.045	9.940	0.6	210
12	12.045	11.940	0.6	170
15	15.045	14.940	0.6	140
18	18.045	17.940	0.7	135
22	22.055	21.950	0.7	110
28	28.055	27.950	0.8	121
35	35.070	34.965	1.0	100
42	42.070	41.965	1.1	91
54	54.090	53.940	1.2	77

PIPES, WATER, STORAGE, INSULATION

Dimensions of Steel Pipes to BS 1387

Nominal Size	Approx. Outside Diameter	Outside diameter				Thickness		
		Light		Medium & Heavy		Light	Medium	Heavy
		Max	Min	Max	Min			
(mm)	(mm)	(mm)	(mm)	(mm)	(mm)	(mm)	(mm)	(mm)
6	10.20	10.10	9.70	10.40	9.80	1.80	2.00	2.65
8	13.50	13.60	13.20	13.90	13.30	1.80	2.35	2.90
10	17.20	17.10	16.70	17.40	16.80	1.80	2.35	2.90
15	21.30	21.40	21.00	21.70	21.10	2.00	2.65	3.25
20	26.90	26.90	26.40	27.20	26.60	2.35	2.65	3.25
25	33.70	33.80	33.20	34.20	33.40	2.65	3.25	4.05
32	42.40	42.50	41.90	42.90	42.10	2.65	3.25	4.05
40	48.30	48.40	47.80	48.80	48.00	2.90	3.25	4.05
50	60.30	60.20	59.60	60.80	59.80	2.90	3.65	4.50
65	76.10	76.00	75.20	76.60	75.40	3.25	3.65	4.50
80	88.90	88.70	87.90	89.50	88.10	3.25	4.05	4.85
100	114.30	113.90	113.00	114.90	113.30	3.65	4.50	5.40
125	139.70	–	–	140.60	138.70	–	4.85	5.40
150	165.1*	–	–	166.10	164.10	–	4.85	5.40

* 165.1 mm (6.5in) outside diameter is not generally recommended except where screwing to BS 21 is necessary
All dimensions are in accordance with ISO R65 except approximate outside diameters which are in accordance with ISO R64
Light quality is equivalent to ISO R65 Light Series II

Approximate Metres Per Tonne of Tubes to BS 1387

Nom. size	BLACK						GALVANIZED					
	Plain/screwed ends			Screwed & socketed			Plain/screwed ends			Screwed & socketed		
	L	M	H	L	M	H	L	M	H	L	M	H
(mm)	(m)	(m)	(m)	(m)	(m)	(m)	(m)	(m)	(m)	(m)	(m)	(m)
6	2765	2461	2030	2743	2443	2018	2604	2333	1948	2584	2317	1937
8	1936	1538	1300	1920	1527	1292	1826	1467	1254	1811	1458	1247
10	1483	1173	979	1471	1165	974	1400	1120	944	1386	1113	939
15	1050	817	688	1040	811	684	996	785	665	987	779	661
20	712	634	529	704	628	525	679	609	512	673	603	508
25	498	410	336	494	407	334	478	396	327	474	394	325
32	388	319	260	384	316	259	373	308	254	369	305	252
40	307	277	226	303	273	223	296	268	220	292	264	217
50	244	196	162	239	194	160	235	191	158	231	188	157
65	172	153	127	169	151	125	167	149	124	163	146	122
80	147	118	99	143	116	98	142	115	97	139	113	96
100	101	82	69	98	81	68	98	81	68	95	79	67
125	–	62	56	–	60	55	–	60	55	–	59	54
150	–	52	47	–	50	46	–	51	46	–	49	45

The figures for 'plain or screwed ends' apply also to tubes to BS 1775 of equivalent size and thickness
Key:
L – Light
M – Medium
H – Heavy

PIPES, WATER, STORAGE, INSULATION

Flange Dimension Chart to BS 4504 & BS 10

Normal Pressure Rating (PN 6) 6 Bar

Nom. size	Flange outside dia.	Table 6/2 Forged Welding Neck	Table 6/3 Plate Slip on	Table 6/4 Forged Bossed Screwed	Table 6/5 Forged Bossed Slip on	Table 6/8 Plate Blank	Raised face Dia.	Raised face T'ness	Nr. bolt hole	Size of bolt
15	80	12	12	12	12	12	40	2	4	M10 × 40
20	90	14	14	14	14	14	50	2	4	M10 × 45
25	100	14	14	14	14	14	60	2	4	M10 × 45
32	120	14	16	14	14	14	70	2	4	M12 × 45
40	130	14	16	14	14	14	80	3	4	M12 × 45
50	140	14	16	14	14	14	90	3	4	M12 × 45
65	160	14	16	14	14	14	110	3	4	M12 × 45
80	190	16	18	16	16	16	128	3	4	M16 × 55
100	210	16	18	16	16	16	148	3	4	M16 × 55
125	240	18	20	18	18	18	178	3	8	M16 × 60
150	265	18	20	18	18	18	202	3	8	M16 × 60
200	320	20	22	–	20	20	258	3	8	M16 × 60
250	375	22	24	–	22	22	312	3	12	M16 × 65
300	440	22	24	–	22	22	365	4	12	M20 × 70

Normal Pressure Rating (PN 16) 16 Bar

Nom. size	Flange outside dia.	Table 6/2 Forged Welding Neck	Table 6/3 Plate Slip on	Table 6/4 Forged Bossed Screwed	Table 6/5 Forged Bossed Slip on	Table 6/8 Plate Blank	Raised face Dia.	Raised face T'ness	Nr. bolt hole	Size of bolt
15	95	14	14	14	14	14	45	2	4	M12 × 45
20	105	16	16	16	16	16	58	2	4	M12 × 50
25	115	16	16	16	16	16	68	2	4	M12 × 50
32	140	16	16	16	16	16	78	2	4	M16 × 55
40	150	16	16	16	16	16	88	3	4	M16 × 55
50	165	18	18	18	18	18	102	3	4	M16 × 60
65	185	18	18	18	18	18	122	3	4	M16 × 60
80	200	20	20	20	20	20	138	3	8	M16 × 60
100	220	20	20	20	20	20	158	3	8	M16 × 65
125	250	22	22	22	22	22	188	3	8	M16 × 70
150	285	22	22	22	22	22	212	3	8	M20 × 70
200	340	24	24	–	24	24	268	3	12	M20 × 75
250	405	26	26	–	26	26	320	3	12	M24 × 90
300	460	28	28	–	28	28	378	4	12	M24 × 90

PIPES, WATER, STORAGE, INSULATION

Minimum Distances Between Supports/Fixings

Material	BS Nominal pipe size		Pipes – Vertical	Pipes – Horizontal on to low gradients
	(inch)	(mm)	Support distance in metres	Support distance in metres
Copper	0.50	15.00	1.90	1.30
	0.75	22.00	2.50	1.90
	1.00	28.00	2.50	1.90
	1.25	35.00	2.80	2.50
	1.50	42.00	2.80	2.50
	2.00	54.00	3.90	2.50
	2.50	67.00	3.90	2.80
	3.00	76.10	3.90	2.80
	4.00	108.00	3.90	2.80
	5.00	133.00	3.90	2.80
	6.00	159.00	3.90	2.80
muPVC	1.25	32.00	1.20	0.50
	1.50	40.00	1.20	0.50
	2.00	50.00	1.20	0.60
Polypropylene	1.25	32.00	1.20	0.50
	1.50	40.00	1.20	0.50
uPVC	–	82.40	1.20	0.50
	–	110.00	1.80	0.90
	–	160.00	1.80	1.20
Steel	0.50	15.00	2.40	1.80
	0.75	20.00	3.00	2.40
	1.00	25.00	3.00	2.40
	1.25	32.00	3.00	2.40
	1.50	40.00	3.70	2.40
	2.00	50.00	3.70	2.40
	2.50	65.00	4.60	3.00
	3.00	80.40	4.60	3.00
	4.00	100.00	4.60	3.00
	5.00	125.00	5.50	3.70
	6.00	150.00	5.50	4.50
	8.00	200.00	8.50	6.00
	10.00	250.00	9.00	6.50
	12.00	300.00	10.00	7.00
	16.00	400.00	10.00	8.25

PIPES, WATER, STORAGE, INSULATION

Litres of Water Storage Required Per Person Per Building Type

Type of building	Storage (litres)
Houses and flats (up to 4 bedrooms)	120/bedroom
Houses and flats (more than 4 bedrooms)	100/bedroom
Hostels	90/bed
Hotels	200/bed
Nurses homes and medical quarters	120/bed
Offices with canteen	45/person
Offices without canteen	40/person
Restaurants	7/meal
Boarding schools	90/person
Day schools – Primary	15/person
Day schools – Secondary	20/person

Recommended Air Conditioning Design Loads

Building type	Design loading
Computer rooms	500 W/m² of floor area
Restaurants	150 W/m² of floor area
Banks (main area)	100 W/m² of floor area
Supermarkets	25 W/m² of floor area
Large office block (exterior zone)	100 W/m² of floor area
Large office block (interior zone)	80 W/m² of floor area
Small office block (interior zone)	80 W/m² of floor area

PIPES, WATER, STORAGE, INSULATION

Capacity and Dimensions of Galvanized Mild Steel Cisterns – BS 417

Capacity (litres)	BS type (SCM)	Dimensions		
		Length (mm)	Width (mm)	Depth (mm)
18	45	457	305	305
36	70	610	305	371
54	90	610	406	371
68	110	610	432	432
86	135	610	457	482
114	180	686	508	508
159	230	736	559	559
191	270	762	584	610
227	320	914	610	584
264	360	914	660	610
327	450/1	1220	610	610
336	450/2	965	686	686
423	570	965	762	787
491	680	1090	864	736
709	910	1070	889	889

Capacity of Cold Water Polypropylene Storage Cisterns – BS 4213

Capacity (litres)	BS type (PC)	Maximum height (mm)
18	4	310
36	8	380
68	15	430
91	20	510
114	25	530
182	40	610
227	50	660
273	60	660
318	70	660
455	100	760

PIPES, WATER, STORAGE, INSULATION

Minimum Insulation Thickness to Protect Against Freezing for Domestic Cold Water Systems (8 Hour Evaluation Period)

Pipe size (mm)	Insulation thickness (mm)					
	Condition 1			Condition 2		
	λ = 0.020	λ = 0.030	λ = 0.040	λ = 0.020	λ = 0.030	λ = 0.040
Copper pipes						
15	11	20	34	12	23	41
22	6	9	13	6	10	15
28	4	6	9	4	7	10
35	3	5	7	4	5	7
42	3	4	5	8	4	6
54	2	3	4	2	3	4
76	2	2	3	2	2	3
Steel pipes						
15	9	15	24	10	18	29
20	6	9	13	6	10	15
25	4	7	9	5	7	10
32	3	5	6	3	5	7
40	3	4	5	3	4	6
50	2	3	4	2	3	4
65	2	2	3	2	3	3

Condition 1: water temperature 7°C; ambient temperature –6°C; evaluation period 8 h; permitted ice formation 50%; normal installation, i.e. inside the building and inside the envelope of the structural insulation
Condition 2: water temperature 2°C; ambient temperature –6°C; evaluation period 8 h; permitted ice formation 50%; extreme installation, i.e. inside the building but outside the envelope of the structural insulation
λ = thermal conductivity [W/(mK)]

Insulation Thickness for Chilled And Cold Water Supplies to Prevent Condensation

On a Low Emissivity Outer Surface (0.05, i.e. Bright Reinforced Aluminium Foil) with an Ambient Temperature of +25°C and a Relative Humidity of 80%

Steel pipe size (mm)	*t* = +10			*t* = +5			*t* = 0		
	Insulation thickness (mm)			Insulation thickness (mm)			Insulation thickness (mm)		
	λ = 0.030	λ = 0.040	λ = 0.050	λ = 0.030	λ = 0.040	λ = 0.050	λ = 0.030	λ = 0.040	λ = 0.050
15	16	20	25	22	28	34	28	36	43
25	18	24	29	25	32	39	32	41	50
50	22	28	34	30	39	47	38	49	60
100	26	34	41	36	47	57	46	60	73
150	29	38	46	40	52	64	51	67	82
250	33	43	53	46	60	74	59	77	94
Flat surfaces	39	52	65	56	75	93	73	97	122

t = temperature of contents (°C)
λ = thermal conductivity at mean temperature of insulation [W/(mK)]

PIPES, WATER, STORAGE, INSULATION

Insulation Thickness for Non-domestic Heating Installations to Control Heat Loss

Steel pipe size (mm)	t = 75			t = 100			t = 150		
	Insulation thickness (mm)			Insulation thickness (mm)			Insulation thickness (mm)		
	λ = 0.030	λ = 0.040	λ = 0.050	λ = 0.030	λ = 0.040	λ = 0.050	λ = 0.030	λ = 0.040	λ = 0.050
10	18	32	55	20	36	62	23	44	77
15	19	34	56	21	38	64	26	47	80
20	21	36	57	23	40	65	28	50	83
25	23	38	58	26	43	68	31	53	85
32	24	39	59	28	45	69	33	55	87
40	25	40	60	29	47	70	35	57	88
50	27	42	61	31	49	72	37	59	90
65	29	43	62	33	51	74	40	63	92
80	30	44	62	35	52	75	42	65	94
100	31	46	63	37	54	76	45	68	96
150	33	48	64	40	57	77	50	73	100
200	35	49	65	42	59	79	53	76	103
250	36	50	66	43	61	80	55	78	105

t = hot face temperature (°C)
$λ$ = thermal conductivity at mean temperature of insulation [W/(mK)]

Index

Ebook Single-User Licence Agreement

We welcome you as a user of this Spon Price Book ebook and hope that you find it a useful and valuable tool. Please read this document carefully. **This is a legal agreement** between you (hereinafter referred to as the "Licensee") and Taylor and Francis Books Ltd. (the "Publisher"), which defines the terms under which you may use the Product. **By accessing and retrieving the access code on the label inside the front cover of this book you agree to these terms and conditions outlined herein. If you do not agree to these terms you must return the Product to your supplier intact, with the seal on the label unbroken and with the access code not accessed.**

1. **Definition of the Product**
   The product which is the subject of this Agreement, (the "Product") consists of online and offline access to the Vital-Source ebook edition of *Spon's Mechanical and Electrical Services Price Book 2026*.

2. **Commencement and licence**
   2.1 This Agreement commences upon the breaking open of the document containing the access code by the Licensee (the "Commencement Date").
   2.2 This is a licence agreement (the "Agreement") for the use of the Product by the Licensee, and not an agreement for sale.
   2.3 The Publisher licenses the Licensee on a non-exclusive and non-transferable basis to use the Product on condition that the Licensee complies with this Agreement. The Licensee acknowledges that it is only permitted to use the Product in accordance with this Agreement.

3. **Multiple use**
   Use of the Product is not provided or allowed for more than one user or for a wide area network or consortium.

4. **Installation and Use**
   4.1 The Licensee may provide access to the Product for individual study in the following manner: The Licensee may install the Product on a secure local area network on a single site for use by one user.
   4.2 The Licensee shall be responsible for installing the Product and for the effectiveness of such installation.
   4.3 Text from the Product may be incorporated in a coursepack. Such use is only permissible with the express permission of the Publisher in writing and requires the payment of the appropriate fee as specified by the Publisher and signature of a separate licence agreement.
   4.4 The Product is a free addition to the book and the Publisher is under no obligation to provide any technical support.

5. **Permitted Activities**
   5.1 The Licensee shall be entitled to use the Product for its own internal purposes;
   5.2 The Licensee acknowledges that its rights to use the Product are strictly set out in this Agreement, and all other uses (whether expressly mentioned in Clause 6 below or not) are prohibited.

6. **Prohibited Activities**
   The following are prohibited without the express permission of the Publisher:
   6.1 The commercial exploitation of any part of the Product.
   6.2 The rental, loan, (free or for money or money's worth) or hire purchase of this product, save with the express consent of the Publisher.
   6.3 Any activity which raises the reasonable prospect of impeding the Publisher's ability or opportunities to market the Product.
   6.4 Any networking, physical or electronic distribution or dissemination of the product save as expressly permitted by this Agreement.
   6.5 Any reverse engineering, decompilation, disassembly or other alteration of the Product save in accordance with applicable national laws.
   6.6 The right to create any derivative product or service from the Product save as expressly provided for in this Agreement.
   6.7 Any alteration, amendment, modification or deletion from the Product, whether for the purposes of error correction or otherwise.

7. **General Responsibilities of the License**
   7.1 The Licensee will take all reasonable steps to ensure that the Product is used in accordance with the terms and conditions of this Agreement.
   7.2 The Licensee acknowledges that damages may not be a sufficient remedy for the Publisher in the event of breach of this Agreement by the Licensee, and that an injunction may be appropriate.
   7.3 The Licensee undertakes to keep the Product safe and to use its best endeavours to ensure that the product does not fall into the hands of third parties, whether as a result of theft or otherwise.
   7.4 Where information of a confidential nature relating to the product of the business affairs of the Publisher comes into the possession of the Licensee pursuant to this Agreement (or otherwise), the Licensee agrees to use such information solely for the purposes of this Agreement, and under no circumstances to disclose any element of the information to any third party save strictly as permitted under this Agreement. For the avoidance of doubt, the Licensee's obligations under this sub-clause 7.4 shall survive the termination of this Agreement.

8. **Warrant and Liability**
   8.1 The Publisher warrants that it has the authority to enter into this agreement and that it has secured all rights and permissions necessary to enable the Licensee to use the Product in accordance with this Agreement.
   8.2 The Publisher warrants that the Product as supplied on the Commencement Date shall be free of defects in materials and workmanship, and undertakes to replace any defective Product within 28 days of notice of such defect being received provided such notice is received within 30 days of such supply. As an alternative to replacement, the Publisher agrees fully to refund the Licensee in such circumstances, if the Licensee so requests, provided that the Licensee returns this copy of *Spon's Mechanical and Electrical Services Price Book 2026* to the Publisher. The provisions of this sub-clause 8.2 do not apply where the defect results from an accident or from misuse of the product by the Licensee.
   8.3 Sub-clause 8.2 sets out the sole and exclusive remedy of the Licensee in relation to defects in the Product.
   8.4 The Publisher and the Licensee acknowledge that the Publisher supplies the Product on an "as is" basis. The Publisher gives no warranties:
   8.4.1 that the Product satisfies the individual requirements of the Licensee; or
   8.4.2 that the Product is otherwise fit for the Licensee's purpose; or
   8.4.3 that the Product is compatible with the Licensee's hardware equipment and software operating environment.
   8.5 The Publisher hereby disclaims all warranties and conditions, express or implied, which are not stated above.
   8.6 Nothing in this Clause 8 limits the Publisher's liability to the Licensee in the event of death or personal injury resulting from the Publisher's negligence.
   8.7 The Publisher hereby excludes liability for loss of revenue, reputation, business, profits, or for indirect or consequential losses, irrespective of whether the Publisher was advised by the Licensee of the potential of such losses.
   8.8 The Licensee acknowledges the merit of independently verifying the price book data prior to taking any decisions of material significance (commercial or otherwise) based on such data. It is agreed that the Publisher shall not be liable for any losses which result from the Licensee placing reliance on the data under any circumstances.
   8.9 Subject to sub-clause 8.6 above, the Publisher's liability under this Agreement shall be limited to the purchase price.

9. **Intellectual Property Rights**
   9.1 Nothing in this Agreement affects the ownership of copyright or other intellectual property rights in the Product.
   9.2 The Licensee agrees to display the Publishers' copyright notice in the manner described in the Product.
   9.3 The Licensee hereby agrees to abide by copyright and similar notice requirements required by the Publisher, details of which are as follows:
   "© 2026 Taylor & Francis. All rights reserved. All materials in *Spon's Mechanical and Electrical Services Price Book 2026* are copyright protected. All rights reserved. No such materials may be used, displayed, modified, adapted, distributed, transmitted, transferred, published or otherwise reproduced in any form or by any means now or hereafter developed other than strictly in accordance with the terms of the licence agreement enclosed with *Spon's Mechanical and Electrical Services Price Book 2026*. However, text and images may be printed and copied for research and private study within the preset program limitations. Please note the copyright notice above, and that any text or images printed or copied must credit the source."
   9.4 This Product contains material proprietary to and copyedited by the Publisher and others. Except for the licence granted herein, all rights, title and interest in the Product, in all languages, formats and media throughout the world, including copyrights therein, are and remain the property of the Publisher or other copyright holders identified in the Product.

10. Non-assignment

This Agreement and the licence contained within it may not be assigned to any other person or entity without the written consent of the Publisher.

11. Termination and Consequences of Termination.

11.1   The Publisher shall have the right to terminate this Agreement if:

11.1.1   the Licensee is in material breach of this Agreement and fails to remedy such breach (where capable of remedy) within 14 days of a written notice from the Publisher requiring it to do so; or

11.1.2   the Licensee becomes insolvent, becomes subject to receivership, liquidation or similar external administration; or

11.1.3   the Licensee ceases to operate in business.

11.2   The Licensee shall have the right to terminate this Agreement for any reason upon two month's written notice. The Licensee shall not be entitled to any refund for payments made under this Agreement prior to termination under this sub-clause 11.2.

11.3   Termination by either of the parties is without prejudice to any other rights or remedies under the general law to which they may be entitled, or which survive such termination (including rights of the Publisher under sub-clause 7.4 above).

11.4   Upon termination of this Agreement, or expiry of its terms, the Licensee must destroy all copies and any back up copies of the product or part thereof.

12. General

12.1   *Compliance with export provisions*
The Publisher hereby agrees to comply fully with all relevant export laws and regulations of the United Kingdom to ensure that the Product is not exported, directly or indirectly, in violation of English law.

12.2   *Force majeure*
The parties accept no responsibility for breaches of this Agreement occurring as a result of circumstances beyond their control.

12.3   *No waiver*
Any failure or delay by either party to exercise or enforce any right conferred by this Agreement shall not be deemed to be a waiver of such right.

12.4   *Entire agreement*
This Agreement represents the entire agreement between the Publisher and the Licensee concerning the Product. The terms of this Agreement supersede all prior purchase orders, written terms and conditions, written or verbal representations, advertising or statements relating in any way to the Product.

12.5   *Severability*
If any provision of this Agreement is found to be invalid or unenforceable by a court of law of competent jurisdiction, such a finding shall not affect the other provisions of this Agreement and all provisions of this Agreement unaffected by such a finding shall remain in full force and effect.

12.6   *Variations*
This agreement may only be varied in writing by means of variation signed in writing by both parties.

12.7   *Notices*
All notices to be delivered to: Spon's Price Books, Taylor & Francis Books Ltd., 4 Park Square, Milton Park, Abingdon, Oxfordshire, OX14 4RN, UK.

12.8   *Governing law*
This Agreement is governed by English law and the parties hereby agree that any dispute arising under this Agreement shall be subject to the jurisdiction of the English courts.

If you have any queries about the terms of this licence, please contact:

Spon's Price Books
Taylor & Francis Books Ltd.
4 Park Square, Milton Park, Abingdon, Oxfordshire, OX14 4RN
Tel: +44 (0) 20 7017 6000
https://www.routledge.com/spon-press

Spon Press
an imprint of Taylor & Francis

Ebook options

Your print copy comes with a free ebook on the VitalSource® Bookshelf platform. Further copies of the ebook are available from https://www.routledge.com/spon-press

To buy Spon's ebooks for five or more users of in your organisation please contact:

Spon's Price Books
eBooks & Online Sales
Taylor & Francis Books Ltd.
4 Park Square, Milton Park, Oxfordshire, OX14 4RN
onlinesales@informa.com